STATS 4e
Modeling the World
AP® EDITION

STATS 4e

Modeling the World
AP® EDITION

David E. Bock
Ithaca High School (Retired)

Paul F. Velleman
Cornell University

Richard D. De Veaux
Williams College

PEARSON

Boston   Columbus   Indianapolis   New York   San Francisco   Upper Saddle River
Amsterdam   Cape Town   Dubai   London   Madrid   Milan Munich   Paris   Montréal   Toronto
Delhi   Mexico City   São Paulo   Sydney   Hong Kong   Seoul   Singapore   Taipei   Tokyo

Editor-in-Chief: Deirdre Lynch
Executive Editor: Christopher Cummings
Senior Content Editor: Chere Bemelmans
Assistant Editor: Sonia Ashraf
Senior Managing Editor: Karen Wernholm
Associate Managing Editor: Tamela Ambush
Project Managers: Sherry Berg and Sheila Spinney
Digital Assets Manager: Marianne Groth
Supplements Production Coordinator: Katherine Roz
Manager, Multimedia Production: Christine Stavrou
Software Development: Bob Carroll and Mary Durnwald
Marketing Manager, AP and Electives: Jackie Flynn
Senior Marketing Manager: Erin K. Lane

Marketing Coordinator: Kathleen DeChavez
Senior Author Support/Technology Specialist: Joe Vetere
Rights and Permissions Advisor: Dana Weightman
Image Manager: Rachel Youdelman
Procurement Specialist: Debbie Rossi
Associate Director of Design, USHE EMSS/HSC/EDU:
  Andrea Nix
Senior Designer and Cover Design: Barbara Atkinson
Text Design: Studio Montage
Production Management, Composition, and Illustrations:
  PreMedia Global
Cover Image: iStockphoto/Thinkstock/Getty Images

The Pearson team would like to acknowledge Sheila Spinney and her many years of hard work and dedication to publishing. She was a joy to work with and a wonderful friend, and we will miss her greatly.

For permission to use copyrighted material, grateful acknowledgment is made to the copyright holders on pages A-63 to A-64, which is hereby made part of this copyright page.

Many of the designations used by manufacturers and sellers to distinguish their products are claimed as trademarks. Where those designations appear in this book, and Pearson Education was aware of a trademark claim, the designations have been printed in initial caps or all caps.

Library of Congress Cataloging-in-Publication Data

Bock, David E.
  Stats : modeling the world / David E. Bock, Paul F. Velleman, Richard D. De Veaux.— 4th ed.
    p. cm.
  Includes index.
  ISBN 978-0-321-85401-8
  1. Graphic calculators—Textbooks.   I. Velleman, Paul F., 1949-   II. De Veaux, Richard D.   III. Title.
  QA276.12.B628 2010
  519.5—dc22

                                        2012005942

2 3 4 5 6 7 8 9 10—V011—17 16 15 14

www.PearsonSchool.com/Advanced

ISBN 10:   0-13-315154-9
ISBN 13: 978-0-13-315154-1
(High School Binding)

David E. Bock taught mathematics at Ithaca High School for 35 years. He has taught Statistics at Ithaca High School, Tompkins-Cortland Community College, Ithaca College, and Cornell University. Dave has won numerous teaching awards, including the MAA's Edyth May Sliffe Award for Distinguished High School Mathematics Teaching (twice), Cornell University's Outstanding Educator Award (three times), and has been a finalist for New York State Teacher of the Year.

Dave holds degrees from the University at Albany in Mathematics (B.A.) and Statistics/Education (M.S.). Dave has been a reader and table leader for the AP Statistics exam, serves as a Statistics consultant to the College Board, and leads workshops and institutes for AP Statistics teachers. He also served as K–12 Education and Outreach Coordinator and senior lecturer for the Mathematics Department at Cornell University. His understanding of how students learn informs much of this book's approach.

Dave and his wife relax by biking or hiking, and when not at home near Ithaca can often be found in North Carolina's Blue Ridge Mountains. They have a son, a daughter, and four grandchildren.

Paul F. Velleman has an international reputation for innovative Statistics education. He is the author and designer of the multimedia Statistics program *ActivStats*, for which he was awarded the EDUCOM Medal for innovative uses of computers in teaching statistics, and the ICTCM Award for Innovation in Using Technology in College Mathematics. He also developed the award-winning statistics program, *Data Desk*, and the Internet site Data and Story Library (DASL) (lib.stat.cmu.edu/DASL/), which provides data sets for teaching Statistics. Paul's understanding of using and teaching with technology informs much of this book's approach.

Paul teaches Statistics at Cornell University in the Department of Statistical Sciences, for which he has been awarded the MacIntyre prize for Exemplary Teaching. He holds an A.B. from Dartmouth College in Mathematics and Social Science, and M.S. and Ph.D. degrees in Statistics from Princeton University, where he studied with John Tukey. His research often deals with statistical graphics and data analysis methods. Paul co-authored (with David Hoaglin) *ABCs of Exploratory Data Analysis*. Paul is a Fellow of the American Statistical Association and of the American Association for the Advancement of Science. Paul is the father of two boys.

Richard D. De Veaux is an internationally known educator and consultant. He has taught at the Wharton School and the Princeton University School of Engineering, where he won a "Lifetime Award for Dedication and Excellence in Teaching." Since 1994, he has taught at Williams College. He is currently the C. Carlisle and Margaret Tippit Professor of Statistics at Williams College. Dick has won both the Wilcoxon and Shewell awards from the American Society for Quality. He is an elected member of the International Statistics Institute (ISI) and a fellow of the American Statistical Association (ASA). In 2008, he was named Statistician of the Year by the Boston Chapter of the ASA. Dick is also well known in industry, where for more than 25 years he has consulted for such Fortune 500 companies as American Express, Hewlett-Packard, Alcoa, DuPont, Pillsbury, General Electric, and Chemical Bank. Because he consulted with Mickey Hart on his book *Planet Drum*, he has also sometimes been called the "Official Statistician for the Grateful Dead." His real-world experiences and anecdotes illustrate many of this book's chapters.

Dick holds degrees from Princeton University in Civil Engineering (B.S.E.) and Mathematics (A.B.) and from Stanford University in Dance Education (M.A.) and Statistics (Ph.D.), where he studied dance with Inga Weiss and Statistics with Persi Diaconis. His research focuses on the analysis of large data sets and data mining in science and industry.

In his spare time, he is an avid cyclist and swimmer. He also is the founder and bass for the doo-wop group, the Diminished Faculty, and is a frequent singer and soloist with various local choirs, including the Choeur Vittoria of Paris, France. Dick is the father of four children.

Table of Contents

---

*Optional chapter.

Preface

About the Book

Yes, a preface is supposed to be "about this book" – and we'll get there – but first we want to talk about the bigger picture: the ongoing growth of interest in Statistics. From the hit movie *Moneyball* to Nate Silver's success at predicting elections to *Wall Street Journal* and *New York Times* articles touting the explosion of job opportunities for graduates with degrees in Statistics, public awareness of the widespread applicability, power, and importance of statistical analysis has never been higher. Each year, more students sign up for Stats courses and discover what drew us to this field: it's interesting, stimulating, and even fun. Statistics helps students develop key tools and critical thinking skills needed to become well-informed consumers and citizens. We think Statistics isn't as much a math course as a civics course, and we're delighted that our books can play a role in preparing a generation for life in the Information Age.

New to the Fourth Edition

This new edition of *Stats: Modeling the World* extends the series of innovations pioneered in our books, teaching Statistics and statistical thinking as it is practiced today. We've made some important revisions and additions, each with the goal of making it even easier for students to put the concepts of Statistics together into a coherent whole.

- *Chapter 1 (and beyond).* Now Chapter 1 gets down to business immediately, looking at data rather than just presenting the book's features. And throughout the book we've rewritten many other sections to make them clearer and more interesting. Several chapters lead with new up-to-the-minute motivating examples and follow through with analyses of the data, and many other new examples provide a basis for sample problems and exercises.

- *What If.* We close most chapters by looking at a simulation that explores or extends an important concept. Starting with Chapter 2, students see the power of simulation as they gain additional insights or get a sneak preview of important ideas yet to come. These *What If* elements offer great fodder for class discussions while paving the way for better grasp of such critical concepts as independence, sampling variability, the Central Limit Theorem, and statistical significance.

- *Practice Exams.* At the end of each of the book's seven parts you'll find a practice exam, consisting of both multiple choice and free response questions. These cumulative exams encourage students to keep important concepts and skills in mind throughout the course while helping them synthesize their understanding as they build connections among the various topics.

- *What Have We Learned?* We've revised our chapter-ending study guides to better help students review the key concepts and terms.

- *Updated examples, exercises, and data.* We've updated our innovative *Think/Show/Tell Step-by-Step* examples with new contexts and data. We've added hundreds of new exercises and updated continuing exercises with the most recent data. Whenever possible, we've provided those data on the DVD and the book's website. Most of the examples and exercises are based on recent news stories, research articles, and other real-world sources. We've listed many of those sources so students can explore them further.

- *Updated TI Tips.* Each chapter's easy-to-read "TI Tips" now show students how to use TI-84 Plus statistics functions with the StatWizard operating system.

- *Streamlined design.* This edition sports a new design that clarifies the purpose of each text element. The major theme of each chapter is easier to follow without distraction.

To better help students know where to focus their study efforts, essential supporting material is shaded, while enriching—and often entertaining—side material is not.

Our Goal: Read This Book!

The best text in the world is of little value if students don't read it. Starting with the first edition, our goal has been to create a book that students would willingly read, easily learn from, and even like. We've been thrilled with the glowing feedback we've received from teachers and students using the first three editions of *Stats: Modeling the World*. Our conversational style, our interesting anecdotes and examples, and even our humor[1] engage students' interest as they learn statistical thinking. We hear from grateful teachers that their students actually do read this book (sometimes even voluntarily reading ahead of the assignments). And we hear from (often amazed) students that they actually enjoyed their textbook.

Here are some of the ways we have made *Stats: Modeling the World*, Fourth Edition engaging:

- *Readability.* You'll see immediately that this book doesn't read like other Statistics texts. The style is both colloquial and informative, enticing students to actually read the book to see what it says.

- *Informality.* Our informal style doesn't mean that the subject matter is covered superficially. Not only have we tried to be precise, but wherever possible we offer deeper explanations and justifications than those found in most introductory texts.

- *Focused lessons.* The chapters are shorter than in most other texts, making it easier for both teachers and students to focus on one topic at a time.

- *Consistency.* We've worked hard to demonstrate how to do Statistics well. From the very start and throughout the book we model the importance of plotting data, of checking assumptions and conditions, and of writing conclusions that are clear, concise, and in context.

- *The need to read.* Because the important concepts, definitions, and sample solutions aren't set in boxes, students won't find it easy to just to skim this book. We intend that it be read, so we've tried to make the experience enjoyable.

Continuing Features

Along with the improvements we've made, you'll still find the many engaging, innovative, and pedagogically effective features responsible for the success of our earlier editions.

- *Think, Show, Tell.* The worked examples repeat the mantra of *Think, Show,* and *Tell* in every chapter. They emphasize the importance of thinking about a Statistics question (What do we know? What do we hope to learn? Are the assumptions and conditions satisfied?) and reporting our findings (the *Tell* step). The *Show* step contains the mechanics of calculating results and conveys our belief that it is only one part of the process.

- *Step-by-Step* examples guide students through the process of analyzing a problem by showing the general explanation on the left and the worked-out solution on the right. The result: better understanding of the concept, not just number crunching.

---

[1] And, yes, those footnotes!

- *For Example.* In every chapter, an interconnected series of *For Example* elements present a continuing discussion, recapping a story and moving it forward to illustrate how to apply each new concept or skill.

- *Just Checking.* At key points in each chapter, we ask students to pause and think with questions designed to be a quick check that they understand the material they've just read. Answers are at the end of the exercise sets in each chapter so students can easily check themselves.

- *Updated TI Tips.* Each chapter's easy-to-read "TI Tips" now show students how to use TI-84 Plus statistics functions with the StatWizard operating system. (Help using a TI-Nspire appears in Appendix B, and help with a TI-89 is on the book's companion website www.pearsonhighered.com/bock.) As we strive for sound understanding of formulas and methods, we want students to use technology for actual calculations. We do emphasize that calculators are just for "Show"—they cannot Think about what to do nor Tell what it all means.

- *Math Boxes.* In many chapters we present the mathematical underpinnings of the statistical methods and concepts. By setting these proofs, derivations, and justifications apart from the narrative, we allow students to continue to follow the logical development of the topic at hand, yet also explore the underlying mathematics for greater depth.

- *ActivStats Pointers.* Margin pointers alert students to *ActivStats* videos, simulations, animations, and activities that enhance learning by paralleling the book's discussions.

- *TI-Nspire Activities.* Other margin pointers identify demonstrations and investigations for TI-Nspire handhelds to enhance each chapter. They're found at the book's website (www.pearsonhighered.com/bock).

- *What Can Go Wrong?* Each chapter still contains our innovative *What Can Go Wrong?* sections that highlight the most common errors people make and the misconceptions they have about Statistics. Our goals are to help students avoid these pitfalls and to arm them with the tools to detect statistical errors and to debunk misuses of statistics, whether intentional or not.

- *Exercises.* We've maintained the pairing of examples so that each odd-numbered exercise (with an answer in the back of the book) is followed by an even-numbered exercise illustrating the same concept. Exercises are ordered by approximate level of complexity.

- *Reality Check.* We regularly remind students that Statistics is about understanding the world with data. Results that make no sense are probably wrong, no matter how carefully we think we did the calculations. Mistakes are often easy to spot with a little thought, so we ask students to stop for a reality check before interpreting their result.

- *Notation Alerts.* Clear communication is essential in Statistics, and proper notation is part of the vocabulary students need to learn. We've found that it helps to call attention to the letters and symbols statisticians use to mean very specific things.

- *On the Computer.* Because real-world data analysis is done on computers, at the end of each chapter we summarize what students can find in most statistics software, usually with an annotated example.

Our Approach

We've been guided in the choice of topics and emphasis on clear communication by the requirements of the Advanced Placement Statistics course. In our order of presentation, we have tried to ensure that each new topic fits logically into the growing structure of understanding that we hope students will build.

GAISE Guidelines

We have worked to provide materials to help each class, in its own way, follow the guidelines of the GAISE (Guidelines for Assessment and Instruction in Statistics Education) project sponsored by the American Statistical Association. That report urges that Statistics education should

1. emphasize Statistical literacy and develop Statistical thinking,

2. use real data,

3. stress conceptual understanding rather than mere knowledge of procedures,

4. foster active learning,

5. use technology for developing concepts and analyzing data, and

6. make assessment a part of the learning process.

Mathematics

Mathematics traditionally appears in Statistics texts in several roles:

1. It can provide a concise, clear statement of important concepts.

2. It can embody proofs of fundamental results.

3. It can describe calculations to be performed with data.

Of these, we emphasize the first. Mathematics can make discussions of Statistics concepts, probability, and inference clear and concise. We have tried to be sensitive to those who are discouraged by equations by also providing verbal descriptions and numerical examples.

This book is not concerned with proving theorems about Statistics. Some of these theorems are quite interesting, and many are important. Often, though, their proofs are not enlightening to introductory Statistics students, and can distract the audience from the concepts we want them to understand. However, we have not shied away from the mathematics where we believed that it helped clarify without intimidating. You will find some important proofs, derivations, and justifications in the Math Boxes that accompany the development of many topics.

Nor do we concentrate on calculations. Although statistics calculations are generally straightforward, they are also usually tedious. And, more to the point, they are often unnecessary. Today, virtually all statistics are calculated with technology, so there is little need for students to work by hand. The equations we use have been selected for their focus on understanding concepts and methods.

Technology and Data

To experience the real world of Statistics, it's best to explore real data sets using modern technology. This fact permeates *Stats: Modeling the World*, Fourth Edition, where we use real data for the book's examples and exercises. Technology lets us focus on teaching statistical thinking rather than getting bogged down in calculations. The questions that motivate each of our hundreds of examples are not "How do you find the answer?" but "How do you think about the answer?"

Technology. We assume that students are using some form of technology in this Statistics course. That could include a graphing calculator along with a statistics package or spreadsheet. Rather than adopt any particular software, we discuss generic computer output. "TI-Tips"—included in most chapters—show students how to use statistics features of the TI-84 Plus series. The DVD includes *ActivStats* and the software package Data Desk. In Appendix B, we offer general guidance (by chapter) to help students get started on common software platforms (StatCrunch, Excel, MINITAB, Data Desk, JMP, and SPSS) and a TI-Nspire. The book's website includes additional guidance for students using a TI-89.

Data. Because we use technology for computing, we don't limit ourselves to small, artificial data sets. In addition to including some small data sets, we have built examples and exercises on real data with a moderate number of cases—usually more than you would want to enter by hand into a program or calculator. These data are included on the DVD as well as on the book's website, www.pearsonhighered.com/bock.

On the DVD

The DVD includes a wealth of supporting materials.

ActivStats. The award-winning ActivStats multimedia program complements the book with videos of real-word stories, worked examples, animated expositions of each of the major Statistics topics, and tools for performing simulations, visualizing inference, and learning to use statistics software. The new version of *ActivStats* includes

- Improved navigation and a cleaner design that makes it easier to find and use tools;
- More than 1000 homework exercises;
- Video clips, animated activities, teaching applets, and more than 300 data sets.

Data Desk. This full-featured statistics software package is both powerful and easy to use.

Additional tech guidance for the TI-89 calculators.

Additional chapters. Two additional chapters cover **Analysis of Variance** (Chapter 27) and **Multiple Regression** (Chapter 28). These topics point the way to further study in Statistics.

Supplements

For the Student

Some of the following student supplements are available for purchase:

Pearson Education AP® Test Prep Series: Statistics by Anne Carroll, Ruth Carver, Susan Peters, and Janice Ricks, is written specifically to complement *Stats: Modeling the World, Fourth Edition, AP® Edition*, and to help students prepare for the AP Statistics Exam. Students can review topics that are discussed in *Stats: Modeling the World, Fourth Edition AP® Edition*, and are likely to appear on the Advanced Placement Exam. The guide also contains test-taking strategies as well as practice tests.

Graphing Calculator Manual (Download Only) by John Diehl (Hinsdale Central High School) and Patricia Humphrey (Georgia Southern University) is organized to follow the sequence of topics in the text, and is an easy-to-follow, step-by-step guide on how to use the TI-84 Plus, TI-Nspire, and Casio graphing calculators. It provides worked-out examples to help students fully understand and use the graphing calculator. Available for download from within MyMathLab for School or from www .pearsonhighered.com/mathstatsresources.

Study card for the De Veaux/Velleman/Bock Statistics Series is a resource for students containing important formulas, definitions, and tables that correspond precisely to the De Veaux/Velleman/ Bock Statistics series. This card can work as a reference for completing homework assignments or as an aid in studying.

Videos for the Bock/Velleman/De Veaux Series, Fourth Edition, available to stream from within MyMathLab for School or on DVD.

For the Teacher

Many of the teacher supplements are available electronically, at no charge, to qualified adopters on the Instructor Resource Center (IRC). To obtain IRC access, you must first register and set up a user name and password. To register, visit PearsonSchool.com/ access_request and select We need IRC Access, then choose Option 1. You will be required to complete a brief, one-time registration subject to verification of educator status. Upon verification, access information and instructions will be sent via email. After you receive your confirmation, go to pearsonhighered.com and choose Support from the top right navigation, then select Educator. Enter the ISBN of your student edition in the search box. Once you locate your program select the Resources tab to preview a list of available online resources.

Teacher's Edition contains answers to all exercises.

Teacher's Solutions Manual (Download Only), by William Craine, contains detailed solutions to all of the exercises. The Teacher's Solutions Manual is available to download from within MyMathLab for School and in the Instructor Resource Center.

Test Bank and Resource Guide, by William Craine, with contributions from Corey Andreasen, Jared Derksen, John Diehl, and Jane Viau, is completely revised and expanded for the fourth edition. The Test Bank and Resource Guide contains chapter-by-chapter comments on major concepts; tips on presenting topics (and what to avoid); teaching examples; suggested assignments; Web links and lists of other resources; additional sets of chapter quizzes, unit tests, and investigative tasks; TI-Nspire activities; and suggestions for projects. We've added more worksheets on key topics, correspondence to AP exam questions in each chapter, and reading guides to the fourth edition. An indispensable guide to help teachers prepare for class, the previous editions were soundly praised by new teachers of Statistics and seasoned veterans alike. The Test Bank and Resource Guide is available on the Instructor's Resource CD, to download from within MyMathLab for School and in the Instructor Resource Center.

Teacher's Podcasts (10 points in 10 minutes). These audio podcasts focus on key points in each chapter to help you with class preparation. They can be easily downloaded from MyMathLab for School and the Instructor Resource Center.

Technology Resources

Instructor's Resource CD, packaged with every new Teacher's Edition, includes the Teacher's Solutions Manual, Test Bank and Resource Guide (which includes a correlation to the AP® Statistics course standards), Audio Podcasts, Guide to the TI-89 Graphing Calculator, StatCrunch Demos and Activities, and TI-Nspire Demos and Activities. A replacement CD is available for purchase.

Companion DVD

A multimedia program on DVD comes with the Teacher's Edition. It may be purchased separately. The DVD holds a number of supporting materials, including:

- *ActivStats® for Data Desk.* The award-winning *ActivStats* multimedia program supports learning chapter by chapter with the book. It complements the book with videos of real-word stories, worked examples, animated expositions of each of the major Statistics topics, and tools for performing simulations, visualizing inference, and learning to use statistics software. *ActivStats* includes 17 short video clips; 170 animated activities and teaching applets; 300 data sets; 1,000 homework exercises, many with links to Data Desk files; interactive graphs, simulations, activities for the TI-Nspire graphing calculator, visualization tools, and much more.

- *Data Desk* statistics package.

- *Data.* Data for exercises marked ⓣ are available on the DVD and website formatted for CSV, TXT, and TI calculators suitable for virtually any statistics software.

- *Additional Chapters.* Two additional chapters cover **Analysis of Variance** (Chapter 27) and **Multiple Regression** (Chapter 28). These topics point the way to further study in Statistics.

ActivStats® The award-winning ActivStats multimedia program supports learning chapter by chapter with the book and is available as a standalone DVD. It complements the book with videos of real-word stories, worked examples, animated expositions of each of the major Statistics topics, and tools for performing simulations, visualizing inference, and learning to use statistics software. ActivStats includes 17 short video clips; 170 animated activities and teaching applets; 300 data sets; 1,000 homework exercises; interactive graphs, simulations, visualization tools, and much more.

MathXL® for School (optional, for purchase only; access code required) www.mathxlforschool.com

MathXL for School is a powerful online homework, tutorial, and assessment supplement that aligns to Pearson Education's textbooks in mathematics or statistics. MathXL for School is the homework and assessment engine that runs MyMathLab for School. With MathXL for School, teachers can:

- Create, edit, and assign auto-graded online homework and tests correlated at the objective level to the textbook
- Utilize automatic grading to rapidly assess student understanding
- Track both student and group performance in an online gradebook
- Prepare students for high-stakes testing, including aligning assignments to state and Common Core State Standards, where available
- Deliver quality, effective instruction regardless of experience level

With MathXL for School, students can:

- Do their homework and receive immediate feedback
- Get self-paced assistance on problems in a variety of ways (guided solutions, step-by-step examples, video clips, animations)
- Have a large number of practice problems to choose from, helping them master a topic
- Receive personalized study plans and homework based on test results

For more information and to purchase student access codes after the first year, visit our Website at www.mathxlforschool.com, or contact your Pearson School representative.

MyMathLab for School Online Course (optional, for purchase only; access code required) www.mymathlabforschool.com

MyMathLab for School delivers proven results in helping individual students succeed. It provides engaging experiences that personalize, stimulate, and measure learning for each student. And it comes from a trusted partner with educational expertise and an eye on the future. To learn more about how MyMathLab for School combines proven learning applications with powerful assessment, visit www.

mymathlabforschool.com or contact your Pearson School representative. In this MyMathLab for School course, you have access to the most cutting-edge, innovative study solutions proven to increase student success. Noteworthy features include the following:

StatCrunch®

StatCrunch is powerful web-based statistical software that allows users to perform complex analyses, share data sets, and generate compelling reports of their data. The vibrant online community offers tens of thousands of data sets for students to analyze.

- **Collect.** Users can upload their own data to StatCrunch or search a large library of publicly shared data sets, spanning almost any topic of interest. Also, an online survey tool allows users to quickly collect data via web-based surveys.
- **Crunch.** A full range of numerical and graphical methods allow users to analyze and gain insights from any data set. Interactive graphics help users understand statistical concepts, and are available for export to enrich reports with visual representations of data.
- **Communicate.** Reporting options help users create a wide variety of visually-appealing representations of their data.

StatCrunch may be purchased in a Registration Packet of 10 redemptions. One redemption is for one student for 12 months beginning at the time of registration. Teacher access for StatCrunch adopters or those who wishing to preview the product may be obtained by filling out the form at www.PearsonSchool.com/Access_Request. For more information, visit our website at **www.StatCrunch.com**, or contact your Pearson School representative.

TestGen® (Download Only)

TestGen® (**www.pearsoned.com/testgen**) enables teachers to build, edit, print, and administer tests using a computerized bank of questions developed to cover all the objectives of the text. TestGen is algorithmically based, allowing teachers to create multiple but equivalent versions of the same question or test with the click of a button. Teachers can also modify test bank questions or add new questions. The software and testbank are available for download from Pearson Education's online catalog.

PowerPoint® Lecture Slides (Download Only)

PowerPoint® Lecture Slides provide an outline to use in a lecture setting, presenting definitions, key concepts, and figures from the text. These slides are available within MyMathLab for School and in the Instructor Resource Center.

Companion Website

Companion Website (www.pearsonhighered.com/bock) Provides additional resources for teachers and students.

Acknowledgments

Many people have contributed to this book in all four of its editions. This edition would have never seen the light of day without the assistance of the incredible team at Pearson. Our Editor in Chief, Deirdre Lynch, was central to the genesis, development, and realization of the book from day one. Chris Cummings, Executive Editor, provided much needed support. Chere Bemelmans, Senior Content Editor, kept us on task as much as humanly possible. Sheila Spinney, Senior Production Project Manager, and Sherry Berg, Project Manager, kept the cogs from getting into the wheels where they often wanted to wander. Marketing Managers Erin Lane and Jackie Flynn made sure the word got out. Sonia Ashraf, Assistant Editor, and Kathleen DeChavez, Marketing Assistant, were essential in managing all of the behind-the-scenes work that needed to be done. Christine Stavrou, Manager–Media Production, put together a top-notch media package for this book. Barbara T. Atkinson, Senior Designer, and Studio Montage are responsible for the wonderful way the book looks. Debbie Rossi, Manufacturing Buyer, worked miracles to get this book and DVD in your hands, and Greg Tobin, President, EMSS, was supportive and good-humored throughout all aspects of the project. Special thanks go out to PreMedia Global, the compositor, for the wonderful work they did on this book, and in particular to Nancy Kincade, Project Manager, for her close attention to detail.

We'd also like to thank our accuracy checkers whose monumental task was to make sure we said what we thought we were saying. They are Douglas Cashing, St. Bonaventure University; Mark Littlefield, Newburyport High School; Stanley Seltzer, Ithaca College; and Susan Blackwell, First Flight High School.

We extend our sincere thanks for the suggestions and contributions made by the following reviewers, focus group participants, and class-testers:

John Arko
Glenbrook South High School, IL

Kathleen Arthur
Shaker High School, NY

Allen Back
Cornell University, NY

Beverly Beemer
Ruben S. Ayala High School, CA

Judy Bevington
Santa Maria High School, CA

Susan Blackwell
First Flight High School, NC

Gail Brooks
McLennan Community College, TX

Walter Brown
Brackenridge High School, TX

Darin Clifft
Memphis University School, TN

Bill Craine
Lansing High School, NY

Sybil Coley
Woodward Academy, GA

Kevin Crowther
Lake Orion High School, MI

Caroline DiTullio
Summit High School, NJ

Jared Derksen
Rancho Cucamonga High School, CA

Sam Erickson
North High School, WI

Laura Estersohn
Scarsdale High School, NY

Laura Favata
Niskayuna High School, NY

David Ferris
Noblesville High School, IN

Linda Gann
Sandra Day O'Connor High School, TX

Randall Groth
Illinois State University, IL

Donnie Hallstone
Green River Community College, WA

Howard W. Hand
St. Marks School of Texas, TX

Bill Hayes
Foothill High School, CA

Miles Hercamp
New Palestine High School, IN

Michelle Hipke
Glen Burnie Senior High School, MD

Carol Huss
Independence High School, NC

Sam Jovell
Niskayuna High School, NY

Peter Kaczmar
Lower Merion High School, PA

John Kotmel
Lansing High School, NY

Beth Lazerick
St. Andrews School, FL

Michael Legacy
Greenhill School, TX

Guillermo Leon
Coral Reef High School, FL

John Lieb
The Roxbury Latin School, MA

Mark Littlefield
Newburyport High School, MA

Martha Lowther
*The Tatnall School,
DE*

John Maceli
Ithaca College, NY

Jim Miller
Alta High School, UT

Timothy E. Mitchell
*King Philip Regional High
School, MA*

Maxine Nesbitt
Carmel High School, IN

Elizabeth Ann Przybysz
Dr. Phillips High School, FL

Diana Podhrasky
Hillcrest High School, TX

Rochelle Robert
Nassau Community College, NY

Karl Ronning
Davis Senior High School, CA

Bruce Saathoff
Centennial High School, CA

Agatha Shaw
*Valencia Community College,
FL*

Murray Siegel
*Sam Houston State University,
TX*

Chris Sollars
*Alamo Heights High School,
TX*

Darren Starnes
The Webb Schools, CA

Jane Viau
*The Frederick Douglass
Academy, NY*

David Bock
Paul Velleman
Richard De Veaux

STATS 4e
Modeling the World
AP® EDITION

1

Stats Starts Here[1]

> "But where shall I begin?"
> asked Alice. "Begin at the
> beginning," the King said
> gravely, "and go on till you
> come to the end: then stop."
>
> —*Lewis Carroll,*
> *Alice's Adventures*
> *in Wonderland*

tatistics gets no respect. People say things like "You can prove anything with Statistics." People will write off a claim based on data as "just a statistical trick." And a Statistics course may not be your friends' first choice for a fun elective.

But Statistics *is* fun. That's probably not what you heard on the street, but it's true. Statistics is about how to think clearly with data. We'll talk about data in more detail soon, but for now, think of **data** as any collection of numbers, characters, images, or other items that provide information about something. Whenever there are data and a need for understanding the world, you'll find Statistics. A little practice thinking statistically is all it takes to start seeing the world more clearly and accurately.

So, What Is (Are?) Statistics?

Q: What is Statistics?
A: Statistics is a way of reasoning, along with a collection of tools and methods, designed to help us understand the world.
Q: What are statistics?
A: Statistics (plural) are particular calculations made from data.
Q: So what is data?
A: You mean, "what *are* data?" Data is the plural form. The singular is datum.
Q: OK, OK, so what are data?
A: Data are values along with their context.

It seems every time we turn around, someone is collecting data on us, from every purchase we make in the grocery store, to every click of our mouse as we surf the Web.

Consider the following:

- If you have a Facebook account, you have probably noticed that the ads you see online tend to match your interests and activities. Coincidence? Hardly. According to the *Wall Street Journal* (10/18/2010),[2] much of your personal information has probably been sold to marketing or tracking companies. Why would Facebook give you a free account and let you upload as much as you want to its site? Because your data are valuable! Using your Facebook profile, a company might build a profile of your

---

[1]We could have called this chapter "Introduction," but nobody reads the introduction, and we wanted you to read this. We feel safe admitting this here, in the footnote, because nobody reads footnotes either.

[2]blogs.wsj.com/digits/2010/10/18/referers-how-facebook-apps-leak-user-ids/

Frazz © 2013 Jef Mallett. Distributed by Universal Uclick. Reprinted with permission. All rights reserved.

interests and activities: what movies and sports you like; your age, sex, education level, and hobbies; where you live; and, of course, who your friends are and what *they* like. From Facebook's point of view, your data are a potential gold mine. Gold ore in the ground is neither very useful nor pretty. But with skill, it can be turned into something both beautiful and valuable. What we're going to talk about in this book is how you can mine your own data and learn valuable insights about the world.

- Like many other retailers, Target stores create customer profiles by collecting data about purchases using credit cards. Patterns the company discovers across similar customer profiles enable it to send you advertising and coupons that promote items you might be particularly interested in purchasing. As valuable to the company as these marketing insights can be, some may prove startling to individuals. Recently coupons Target sent to a Minneapolis girl's home revealed she was pregnant before her father knew![3]

- How dangerous is texting while driving? Researchers at the University of Utah tested drivers on simulators that could present emergency situations. They compared reaction times of sober drivers, drunk drivers, and texting drivers.[4] The results were striking. The texting drivers actually responded more slowly and were more dangerous than those who were above the legal limit for alcohol.

In this book, you'll learn how to design and analyze experiments like this. You'll learn how to interpret data and to communicate the message you see to others. You'll also learn how to spot deficiencies and weaknesses in conclusions drawn by others that you see in newspapers and on the Internet every day. Statistics can help you become a more informed citizen by giving you the tools to understand, question, and interpret data.

> **Are You a Statistic?**
> The ads say, "Don't drink and drive; you don't want to be a statistic." But you can't be a statistic.
>   We say: "Don't be a datum."

Statistics in a Word

> **Statistics Is about Variation**
> Data vary because we don't see everything and because even what we do see and measure, we measure imperfectly.
>   So, in a very basic way, Essential Statistics is about the real, imperfect world in which we live.

It can be fun, and sometimes useful, to summarize a discipline in only a few words. So,

Economics is about . . . *Money (and why it is good).*

Psychology: *Why we think what we think (we think).*

Biology: *Life.*

Anthropology: *Who?*

History: *What, where, and when?*

Philosophy: *Why?*

Engineering: *How?*

Accounting: *How much?*

In such a caricature, Statistics is about . . . ***Variation.***

---

[3]http://www.forbes.com/sites/kashmirhill/2012/02/16/how-target-figured-out-a-teen-girl-was-pregnant-before-her-father-did/

[4]"Text Messaging During Simulated Driving," Drews, F. A. et al. Human Factors: hfs.sagepub.com/content/51/5/762

Data vary. Ask different people the same question and you'll get a variety of answers. Statistics helps us to make sense of the world described by our data by seeing past the underlying variation to find patterns and relationships. This book will teach you skills to help with this task and ways of thinking about variation that are the foundation of sound reasoning about data.

But What *Are* Data?

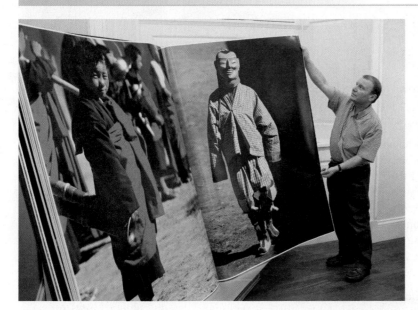

Amazon.com opened for business in July 1995, billing itself as "Earth's Biggest Bookstore." By 1997, Amazon had a catalog of more than 2.5 million book titles and had sold books to more than 1.5 million customers in 150 countries. In 2010, the company's sales reached $34.2 billion (a nearly 40% increase from the previous year). Amazon has sold a wide variety of merchandise, including a $400,000 necklace, yak cheese from Tibet, and the largest book in the world. How did Amazon become so successful and how can it keep track of so many customers and such a wide variety of products? The answer to both questions is *data.*

But what are data? Think about it for a minute. What exactly *do* we mean by "data"? Do data have to be numbers? The amount of your last purchase in dollars is numerical data. But your name and address in Amazon's database are also data even though they are not numerical. What about your ZIP code? That's a number, but would Amazon care about, say, the *average* ZIP code of its customers?

Let's look at some hypothetical values that Amazon might collect:

105-2686834-3759466	Ohio	Nashville	Kansas	10.99	440	N	B00000I5Y6	Katherine H.
105-9318443-4200264	Illinois	Orange County	Boston	16.99	312	Y	B000002BK9	Samuel P.
105-1872500-0198646	Massachusetts	Bad Blood	Chicago	15.98	413	N	B000068ZVQ	Chris G.
103-2628345-9238664	Canada	Let Go	Mammals	11.99	902	N	B0000010AA	Monique D.
002-1663369-6638649	Ohio	Best of Kansas	Kansas	10.99	440	N	B002MXA7Q0	Katherine H.

A S *Activity:* **What Is (Are) Data?** Do you really know what are data and what are just numbers?

Try to guess what they represent. Why is that hard? Because there is no *context*. If we don't know what values are measured and what is measured about them, the values are meaningless. We can make the meaning clear if we organize the values into a **data table** such as this one:

Order Number	Name	State/Country	Price	Area Code	Previous Album Download	Gift?	ASIN	New Purchase Artist
105-2686834-3759466	Katherine H.	Ohio	10.99	440	Nashville	N	B0000015Y6	Kansas
105-9318443-4200264	Samuel R	Illinois	16.99	312	Orange County	Y	B000002BK9	Boston
105-1372500-0198646	Chris G.	Massachusetts	15.98	413	Bad Blood	N	B000068ZVQ	Chicago
103-2628345-9238664	Monique D.	Canada	11.99	902	Let Go	N	B0000010AA	Mammals
002-1663369-6638649	Katherine H.	Ohio	10.99	440	Best of Kansas	N	B002MXA7Q0	Kansas

The W's:
  Who
  What
      and in what units
  When
  Where
  Why
  How

Now we can see that these are purchase records for album download orders from Amazon. The column titles tell what has been recorded. Each row is about a particular purchase.

What information would provide a **context**? Newspaper journalists know that the lead paragraph of a good story should establish the "Five W's": *who, what, when, where,* and (if possible) *why*. Often, we add *how* to the list as well. The answers to the first two questions are essential. If we don't know *what* values are measured and *who* those values are measured on, the values are meaningless.

Who and What

In general, the rows of a data table correspond to individual **cases** about *Who*m (or about which—if they're not people) we record some characteristics. Cases go by different names, depending on the situation.

- Individuals who answer a survey are called **respondents**.
- People on whom we experiment are **subjects** or (in an attempt to acknowledge the importance of their role in the experiment) **participants**.
- Animals, plants, websites, and other inanimate subjects are often called **experimental units**.
- Often we simply call cases what they are: for example, *customers, economic quarters,* or *companies.*
- In a database, rows are called **records**—in this example, purchase records. Perhaps the most generic term is *cases,* but in any event the rows represent the *who* of the data.

The characteristics recorded about each individual are called **variables.** These are usually shown as the columns of a data table, and they should have a name that identifies *What* has been measured. *Name, Price, Area Code,* and whether the purchase was a *Gift* are some of the variables Amazon collected data for. Variables may seem simple, but we'll need to take a closer look soon.

We must know *who* and *what* to analyze data. Without knowing these two, we don't have enough information to start. Of course, we'd always like to know more. The more we know about the data, the more we'll understand about the world. If possible, we'd like to know the *when* and *where* of data as well. Values recorded in 1803 may mean something different than similar values recorded last year. Values measured in Tanzania may differ in meaning from similar measurements made in Mexico. And knowing *why* the data were collected can tell us much about its reliability and quality.

Often, the cases are a **sample** of cases selected from some larger **population** that we'd like to understand. Amazon certainly cares about its customers, but also wants to know how to attract all those other Internet users who may never have made a purchase from Amazon's site. To be able to generalize from the sample of cases to the larger population, we'll want the sample to be *representative* of that population—a kind of snapshot image of the larger world.

A **S** *Activity*: **Consider the context**
. . .Can you tell who's *Who* and what's *What*? And *Why*? This activity offers real-world examples to help you practice identifying the context.

For Example IDENTIFYING THE "WHO"

In December 2011, *Consumer Reports* published an evaluation of 25 tablets from a variety of manufacturers.

QUESTION: Describe the population of interest, the sample, and the *Who* of the study.

ANSWER: The magazine is interested in the performance of tablets currently offered for sale. It tested a sample of 25 tablets, which are the "Who" for these data. Each tablet selected represents all tablets of that model offered by that manufacturer.

How the Data Are Collected

How the data are collected can make the difference between insight and nonsense. As we'll see later, data that come from a voluntary survey on the Internet are almost always worthless. One primary concern of Statistics is the design of sound methods for collecting data.[5] Throughout this book, whenever we introduce data, we'll provide a margin note listing the W's (and H) of the data. Identifying the W's is a habit we recommend.

The first step of any data analysis is to know what you are trying to accomplish and what you want to know. To help you use Statistics to understand the world and make decisions, we'll lead you through the entire process of *thinking* about the problem, *showing* what you've found, and *telling* others what you've learned. Every guided example in this book is broken into these three steps: *Think, Show,* and *Tell.* Identifying the problem and the *who* and *what* of the data is a key part of the *Think* step of any analysis. Make sure you know these before you proceed to *Show* or *Tell* anything about the data.

A S **Activity: Collect data in an experiment on yourself.** With the computer, you can experiment on yourself and then save the data. Go on to the subsequent related activities to check your understanding.

More About Variables (*What*?)

Privacy and the Internet You have many Identifiers: a social security number, a student ID number, possibly a passport number, a health insurance number, and probably a Facebook account name. Privacy experts are worried that Internet thieves may match your identity in these different areas of your life, allowing, for example, your health, education, and financial records to be merged. Even online companies such as Facebook and Google are able to link your online behavior to some of these identifiers, which carries with it both advantages and dangers. The National Strategy for Trusted Identities in Cyberspace (www.wired.com/images_blogs/threatlevel/2011/04/NSTIC strategy_041511.pdf) proposes ways that we may address this challenge in the near future.

The Amazon data table displays information about several variables: *Order Number, Name, State/Country, Price,* and so on. These identify *what* we know about each individual. Variables such as these can play different roles, depending on how we plan to use them. While some are merely identifiers, others may be categorical or quantitative. Making that distinction is an important step in our analysis.

Identifiers

For some variables, such as a *student ID*, each individual receives a unique value. We call a variable like this, an **identifier variable**. Identifiers are useful, but not typically for analysis.

Amazon wants to know who you are when you sign in again and doesn't want to confuse you with some other customer. So it assigns you a unique identifier. Amazon also wants to send you the right product, so it assigns a unique Amazon Standard Identification Number (ASIN) to each item it carries. Identifier variables themselves don't tell us anything useful about their categories because we know there is exactly one individual in each. You'll want to recognize when a categorical variable is playing the role of an identifier so you aren't tempted to analyze it.

Categorical Variables

Some variables just tell us what group or category each individual belongs to. Are you male or female? Pierced or not? What color are your eyes? We call variables like these **categorical variables**.[6] Some variables are clearly categorical, like the variable *State/Country.* Its values are text and those values tell us what category the particular case falls into. Descriptive responses to questions are often categories. For example, the responses to the questions "Who is your cell phone provider?" or "What is your marital status?" yield categorical values. But numerals are often used to label categories, so categorical variable values can also be numerals. For example, Amazon collects telephone area codes that *categorize* each phone number into a geographical region. So area code is considered a categorical variable even though it has numeric values.

---

[5]Coming attractions: to be discussed in Part III. We sense your excitement.

[6]You may also see them called *qualitative* variables.

Quantitative Variables

When a variable contains measured numerical values with measurement *units*, we call it a **quantitative variable**. Quantitative variables typically record an amount or degree of something. For a quantitative variable, its measurement **units** provide a meaning for the numbers. Even more important, units such as yen, cubits, carats, angstroms, nanoseconds, miles per hour, or degrees Celsius tell us the *scale* of measurement, so we know how far apart two values are. Without units, the values of a measured variable have no meaning. It does little good to be promised a raise of 5000 a year if you don't know whether it will be paid in Euros, dollars, pennies, yen, or Estonian krooni.

Either/Or?

Some variables with numeric values can be treated as either categorical or quantitative depending on what we want to know. Amazon could record your *Age* in years. That seems quantitative, and it would be if the company wanted to know the average age of those customers who visit their site after 3 A.M. But suppose Amazon wants to decide which album to feature on its site when you visit. Then thinking of your age in one of the categories Child, Teen, Adult, or Senior might be more useful. So, sometimes whether a variable is treated as categorical or quantitative is more about the question we want to ask rather than an intrinsic property of the variable itself.

Suppose a course evaluation survey asks, "How valuable do you think this course will be to you?" 1 = Worthless; 2 = Slightly; 3 = Middling; 4 = Reasonably; 5 = Invaluable. Is *Educational Value* categorical or quantitative? A teacher might just count the number of students who gave each response for her course, treating *Educational Value* as a categorical variable. Or if she wants to see whether the course is improving, she might treat the responses as the *amount* of perceived value—in effect, treating the variable as quantitative.

But what are the units? There is certainly an *order* of perceived worth: Higher numbers indicate higher perceived worth. A course that averages 4.5 seems more valuable than one that averages 2, but the teacher will have to imagine that it has "educational value units," whatever they are. Because there are no natural units, she should be cautious. Variables that report order without natural units are often called *ordinal variables*. But saying "that's an ordinal variable" doesn't get you off the hook. You must still look to the *why* of your study and understand what you want to learn from the variable to decide whether to treat it as categorical or quantitative.

A S *Activity*: **Recognize variables measured in a variety of ways.** This activity shows examples of the many ways to measure data.

A S *Activities*: **Variables.** Several activities show you how to begin working with data in your statistics package.

For Example   IDENTIFYING "WHAT" AND "WHY" OF TABLETS

RECAP: A *Consumer Reports* article about 25 tablet computers lists each tablet's manufacturer, cost, battery life (hrs.), operating system (iOS/Android/RIM), and overall performance score (0–100).

QUESTION: Are these variables categorical or quantitative? Include units where appropriate, and describe the "Why" of this investigation.

ANSWER: The variables are
- manufacturer (categorical)
- cost (quantitative, $)
- battery life (quantitative, hrs.)
- operating system (categorical)
- performance score (quantitative, no units)

The magazine hopes to provide consumers with the information to choose a good tablet.

Just Checking

In the 2004 Tour de France, Lance Armstrong made history by winning the race for an unprecedented sixth time. In 2005, he became the only 7-time winner and set a new record for the fastest average speed—41.65 kilometers per hour. A cancer survivor, Armstrong became an international celebrity. But it was all too good to be true. In 2012, following revelations of doping, the International Cycling Union stripped Armstrong of all of his titles and records and banned him from professional cycling for life.

You can find data on all the Tour de France races on the DVD. Keep in mind that the entire data set has over 100 entries.

1. List as many of the W's as you can for this data set.

2. Classify each variable as categorical or quantitative; if quantitative, identify the units.

Year	Winner	Country of Origin	Total Time (h/min/s)	Avg. Speed (km/h)	Stages	Total Distance Ridden (km)	Starting Riders	Finishing Riders
1903	Maurice Garin	France	94.33.00	25.3	6	2428	60	21
1904	Henri Cornet	France	96.05.00	24.3	6	2388	88	23
1905	Louis Trousseller	France	112.18.09	27.3	11	2975	60	24
⋮								
1999	Lance Armstrong (DQ)	USA	91.32.16	40.30	20	3687	180	141
2000	Lance Armstrong (DQ)	USA	92.33.08	39.56	21	3662	180	128
2001	Lance Armstrong (DQ)	USA	86.17.28	40.02	20	3453	189	144
2002	Lance Armstrong (DQ)	USA	82.05.12	39.93	20	3278	189	153
2003	Lance Armstrong (DQ)	USA	83.41.12	40.94	20	3427	189	147
2004	Lance Armstrong (DQ)	USA	83.36.02	40.53	20	3391	188	147
2005	Lance Armstrong (DQ)	USA	86.15.02	41.65	21	3608	189	155
⋮								
2011	Cadel Evans	Australia	86.12.22	39.788	21	3430	198	167
2012	Bradley Wiggins	Great Britain	87.34.47	39.928	20	3497	219	153
2013	Chris Froome	Great Britain	83.56.40	40.551	21	3404	219	170

A S *Self-Test*: Review concepts about data.
Like the Just Checking sections of this textbook, but interactive. (Usually, we won't reference the *ActivStats* self-tests here, but look for one whenever you'd like to check your understanding or review material.

There's a World of Data on the Internet These days, one of the richest sources of data is the Internet. With a bit of practice, you can learn to find data on almost any subject. Many of the data sets we use in this book were found in this way. The Internet has both advantages and disadvantages as a source of data. Among the advantages are the fact that often you'll be able to find even more current data than those we present. The disadvantage is that references to Internet addresses can "break" as sites evolve, move, and die.

Our solution to these challenges is to offer the best advice we can to help you search for the data, wherever they may be residing. We usually point you to a website. We'll sometimes suggest search terms and offer other guidance.

Some words of caution, though: Data found on Internet sites may not be formatted in the best way for use in statistics software. Although you may see a data table in standard form, an attempt to copy the data may leave you with a single column of values. You may have to work in your favorite statistics or spreadsheet program to reformat the data into variables. You will also probably want to remove commas from large numbers and extra symbols such as money indicators ($, ¥, £); few statistics packages can handle these.

WHAT CAN GO WRONG?

■ **Don't label a variable as categorical or quantitative without thinking about the question you want it to answer.** The same variable can sometimes take on different roles.

■ **Just because your variable's values are numbers, don't assume that it's quantitative.** Categories are often given numerical labels. Don't let that fool you into thinking they have quantitative meaning. Look at the context.

■ **Always be skeptical.** One reason to analyze data is to discover the truth. Even when you are told a context for the data, it may turn out that the truth is a bit (or even a lot) different. Think about *how* the data were collected. People who want to influence what you think may slant the context. A survey that seems to be about all students may in fact report just the opinions of those who visited a fan website. The question that respondents answered may be posed in a way that influences responses.

TI Tips   WORKING WITH DATA

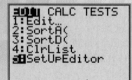

You'll need to be able to enter and edit data in your calculator. Here's how:

TO ENTER DATA: Hit the STAT button, and choose EDIT from the menu. You'll see a set of columns labeled L1, L2, and so on. Here is where you can enter, change, or delete a set of data.

Let's enter the heights (in inches) of the five starting players on a basketball team: 71, 75, 75, 76, and 80. Move the cursor to the space under L1, type in 71, and hit ENTER (or the down arrow). There's the first player. Now enter the data for the rest of the team.

TO CHANGE A DATUM: Suppose the 76" player grew since last season; his height should be listed as 78". Use the arrow keys to move the cursor onto the 76, then change the value and ENTER the correction.

TO ADD MORE DATA: We want to include the sixth man, 73" tall. It would be easy to simply add this new datum to the end of the list. However, sometimes the order of the data matters, so let's place this datum in numerical order. Move the cursor to the desired position (atop the first 75). Hit 2ND INS, then ENTER the 73 in the new space.

TO DELETE A DATUM: The 78" player just quit the team. Move the cursor there. Hit DEL. Bye.

TO CLEAR THE DATALIST: Finished playing basketball? Move the cursor atop the L1. Hit CLEAR, then ENTER (or down arrow). You should now have a blank datalist, ready for you to enter your next set of values.

LOST A DATALIST? Oops! Is L1 now missing entirely? Did you delete L1 by mistake, instead of just *clearing* it? Easy problem to fix: buy a new calculator. No? OK, then simply go to the STAT EDIT menu, and run SetUpEditor to recreate all the lists.

What Have We Learned?

We've learned that data are information in a context.

- The W's help nail down the context *Who*, *What*, *When*, *Why*, *Where*, and *hoW*.
- We must know at least the *Who*, *What*, and *hoW* to be able to say anything useful based on the data. The *Who* are the cases. The *What* are the *variables*. A variable gives information about each of the cases. The *hoW* helps us decide whether we can trust the data.

We treat variables in two basic ways: as *categorical* or *quantitative*.

- Categorical variables identify a category for each case. Usually, we think about the counts of cases that fall into each category. (An exception is an identifier variable that just names each case.)
- Quantitative variables record measurements or amounts of something; they must have *units*.
- Sometimes we treat a variable as categorical or quantitative depending on what we want to learn from it, which means that some variables can't be pigeonholed as one type or the other. That's an early hint that in Statistics we can't always pin things down precisely.

Terms

Data	Systematically recorded information, whether numbers or labels, together with its context. (p. 1)
Data table	An arrangement of data in which each row represents a case and each column represents a variable. (p. 3)
Context	The context ideally tells *Who* was measured, *What* was measured, *How* the data were collected, *Where* the data were collected, and *When* and *Why* the study was performed. (p. 4)
Case	A case is an individual about whom or which we have data. (*Who*). (p. 4)
Respondent	Someone who answers, or responds to, a survey. (p. 4)
Subject	A human experimental unit. Also called a participant. (p. 4)
Participant	A human experimental unit. Also called a subject. (p. 4)
Experimental unit	An individual in a study for which or for whom data values are recorded. Human experimental units are usually called subjects or participants. (p. 4)
Record	Information about an individual in a database. (p. 4)
Variable	A variable holds information about the same characteristic for many cases. (*What*). (p. 4)
Sample	The cases we actually examine in seeking to understand the much larger population. (p. 4)
Population	All the cases we wish we knew about. (p. 4)
Identifier variable	A categorical variable that records a unique value for each case, used to name or identify it. (p. 5)
Categorical variable	A variable that names categories (whether with words or numerals) is called categorical. (p. 5)
Quantitative variable	A variable in which the numbers act as numerical values is called quantitative. Quantitative variables always have units. (p. 6)
Units	A quantity or amount adopted as a standard of measurement, such as dollars, hours, or grams. (p. 6)

On the Computer DATA

A S   *Activity*: **Examine the Data.** Take a look at your own data from your experiment (p. 8) and get comfortable with your statistics package as you find out about the experiment test results.

"Computers are useless; they can only give you answers."

—*Pablo Picasso*

Most often we find statistics on a computer using a program, or *package*, designed for that purpose. There are many different statistics packages, but they all do essentially the same things. If you understand what the computer needs to know to do what you want and what it needs to show you in return, you can figure out the specific details of most packages pretty easily.

For example, to get your data into a computer statistics package, you need to tell the computer:

- Where to find the data. This usually means directing the computer to a file stored on your computer's disk or to data on a database. Or it might just mean that you have copied the data from a spreadsheet program or Internet site and it is currently on your computer's clipboard. Usually, the data should be in the form of a data table. Most computer statistics packages prefer the *delimiter* that marks the division between elements of a data table to be a *tab* character and the delimiter that marks the end of a case to be a *return* character.
- Where to put the data. (Usually this is handled automatically.)
- What to call the variables. Some data tables have variable names as the first row of the data, and often statistics packages can take the variable names from the first row automatically.

Exercises

1. **Voters** A February 2010 Gallup Poll question asked, "In politics, as of today, do you consider yourself a Republican, a Democrat, or an Independent?" The possible responses were "Democrat", "Republican", "Independent", "Other", and "No Response". What kind of variable is the response?

2. **Job hunting** A June 2011 Gallup Poll asked Americans, "Thinking about the job situation in America today, would you say that it is now a good time or a bad time to find a quality jobs?" The choices were "Good time" or "Bad time". What kind of variable is the response?

3. **Medicine** A pharmaceutical company conducts an experiment in which a subject takes 100 mg of a substance orally. The researchers measure how many minutes it takes for half of the substance to exit the bloodstream. What kind of variable is the company studying?

4. **Stress** A medical researcher measures the increase in heart rate of patients under a stress test. What kind of variable is the researcher studying?

(Exercises 5–12) For each description of data, identify Who and What were investigated and the population of interest.

5. **The news** Find a newspaper or magazine article in which some data are reported. For the data discussed in the article, answer the questions above. Include a copy of the article with your report.

6. **The Internet** Find an Internet source that reports on a study and describes the data. Print out the description and answer the questions above.

7. **Bicycle safety** Ian Walker, a psychologist at the University of Bath, wondered whether drivers treat bicycle riders differently when they wear helmets. He rigged his bicycle with an ultrasonic sensor that could measure how close each car was that passed him. He then rode on alternating days with and without a helmet. Out of 2500 cars passing him, he found that when he wore his helmet, motorists passed 3.35 inches closer to him, on average, than when his head was bare. [*NY Times*, Dec. 10, 2006]

8. **Investments** Some companies offer 401(k) retirement plans to employees, permitting them to shift part of their before-tax salaries into investments such as mutual funds. Employers typically match 50% of the employees'

contribution up to about 6% of salary. One company, concerned with what it believed was a low employee participation rate in its 401(k) plan, sampled 30 other companies with similar plans and asked for their 401(k) participation rates.

9. **Honesty** Coffee stations in offices often just ask users to leave money in a tray to pay for their coffee, but many people cheat. Researchers at Newcastle University alternately taped two posters over the coffee station. During one week, it was a picture of flowers; during the other, it was a pair of staring eyes. They found that the average contribution was significantly higher when the eyes poster was up than when the flowers were there. Apparently, the mere feeling of being watched—even by eyes that were not real—was enough to encourage people to behave more honestly. [*NY Times,* Dec. 10, 2006]

10. **Gaydar** A study conducted by a team of American and Canadian researchers found that during ovulation, a woman can tell whether a man is gay or straight by looking at his face. To explore the subject, the authors conducted three investigations, the first of which involved 40 undergraduate women who were asked to guess the sexual orientation of 80 men based on photos of their face. Half of the men were gay, and the other half were straight. All held similar expressions in the photos or were deemed to be equally attractive. None of the women were using any contraceptive drugs at the time of the test. The result: the closer a woman was to her peak ovulation the more accurate her guess.

    (*Source:* news.yahoo.com/does-ovulation-boost-womans-gaydar-210405621.html)

11. **Blindness:** A study begun in 2011 examines the use of stem cells in treating two forms of blindness, Stargardt's disease, and dry age-related macular degeneration. Each of the 24 patients entered one of two separate trials in which embryonic stem cells were to be used to treat the condition.

12. **Molten iron** The Cleveland Casting Plant is a large, highly automated producer of gray and nodular iron automotive castings for Ford Motor Company. The company is interested in keeping the pouring temperature of the molten iron (in degrees Fahrenheit) close to the specified value of 2550 degrees. Cleveland Casting measured the pouring temperature for 10 randomly selected crankshafts.

(*Exercises 13–26*) *For each description of data, identify the W's, name the variables, specify for each variable whether its use indicates that it should be treated as categorical or quantitative, and, for any quantitative variable, identify the units in which it was measured (or note that they were not provided).*

13. **Weighing bears** Because of the difficulty of weighing a bear in the woods, researchers caught and measured 54 bears, recording their weight, neck size, length, and sex.

They hoped to find a way to estimate weight from the other, more easily determined quantities.

14. **Schools** The State Education Department requires local school districts to keep these records on all students: age, race or ethnicity, days absent, current grade level, standardized test scores in reading and mathematics, and any disabilities or special educational needs.

15. **Arby's menu** A listing posted by the Arby's restaurant chain gives, for each of the sandwiches it sells, the type of meat in the sandwich, the number of calories, and the serving size in ounces. The data might be used to assess the nutritional value of the different sandwiches.

16. **Age and party** Gallup conducted a series of telephone polls involving 20,392 American adults during 2011. Among the reported results were the voters' gender, age, race, party affiliation, whether they were of Hispanic ethnicity, education, region, adults in the household, and phone status (cell phone only/landline only/both, cell phone mostly, and having an unlisted landline number).

17. **Babies** Medical researchers at a large city hospital investigating the impact of prenatal care on newborn health collected data from 882 births during 1998–2000. They kept track of the mother's age, the number of weeks the pregnancy lasted, the type of birth (cesarean, induced, natural), the level of prenatal care the mother had (none, minimal, adequate), the birth weight and sex of the baby, and whether the baby exhibited health problems (none, minor, major).

18. **Flowers** In a study appearing in the journal *Science,* a research team reports that plants in southern England are flowering earlier in the spring. Records of the first flowering dates for 385 species over a period of 47 years show that flowering has advanced an average of 15 days per decade, an indication of climate warming, according to the authors.

19. **Herbal medicine** Scientists at a major pharmaceutical firm conducted an experiment to study the effectiveness of an herbal compound to treat the common cold. They exposed each patient to a cold virus, then gave them either the herbal compound or a sugar solution known to have no effect on colds. Several days later they assessed each patient's condition, using a cold severity scale ranging from 0 to 5. They found no evidence of the benefits of the compound.

20. **Vineyards** Business analysts hoping to provide information helpful to American grape growers compiled these data about vineyards: size (acres), number of years in existence, state, varieties of grapes grown, average case price, gross sales, and percent profit.

21. **Streams** In performing research for an ecology class, students at a college in upstate New York collect data on local streams each year. They record a number of biological,

chemical, and physical variables, including the stream name, the substrate of the stream (limestone, shale, or mixed), the acidity of the water (pH), the temperature (°C), and the BCI (a numerical measure of biological diversity).

22. **Fuel economy** The Environmental Protection Agency (EPA) tracks fuel economy of automobiles based on information from the manufacturers (Ford, Toyota, etc.). Among the data the agency collects are the manufacturer, vehicle type (car, SUV, etc.), weight, horsepower, and gas mileage (mpg) for city and highway driving.

23. **Refrigerators** In 2012, *Consumer Reports* rated bottom-freezer refrigerators. It listed 102 models, giving the brand, cost, size (cu ft), temperature performance, noise (poor, fair, etc.), ease of use, energy efficiency, estimated annual energy cost, an overall rating (good, excellent, etc.), and the exterior dimensions.

24. **Walking in circles** People who get lost in the desert, mountains, or woods often seem to wander in circles rather than walk in straight lines. To see whether people naturally walk in circles in the absence of visual clues, researcher Andrea Axtell tested 32 people on a football field. One at a time, they stood at the center of one goal line, were blindfolded, and then tried to walk to the other goal line. She recorded each individual's sex, height, handedness, the number of yards each was able to walk before going out of bounds, and whether each wandered off course to the left or the right. No one made it all the way to the far end of the field without crossing one of the sidelines. [*STATS* No. 39, Winter 2004]

25. **Kentucky Derby 2012** The Kentucky Derby is a horse race that has been run every year since 1875 at Churchill Downs, Louisville, Kentucky. The race started as a 1.5-mile race, but in 1896, it was shortened to 1.25 miles because experts felt that 3-year-old horses shouldn't run such a long race that early in the season. (It has been run in May every year but one—1901—when it took place on April 29). Here are the data for the first four and several recent races.

Year	Winner	Jockey	Trainer	Owner	Time
2013	Orb	J. Rosario	C. McGaughey III	Phipps/Janney	2:02.89
2012	I'll Have Another	M. Gutierrez	D. O'Neill	Reddam Racing	2:01.83
2011	Animal Kingdom	J. Velazquez	H. G. Motion	Team Valor	2:02.04
2010	Super Saver	C. Borel	T. Pletcher	WinStar Farm	2:04.45
2009	Mine That Bird	C. Borel	B. Woolley	Double Eagle Ranch	2:02.66
. . .					
1878	Day Star	J. Carter	L. Paul	T.J. Nichols	2:37.25
1877	Baden Baden	W. Walker	E. Brown	Daniel Swigert	2:38
1876	Vagrant	R. Swim	J. Williams	William Astor	2:38.25
1875	Aristides	O. Lewis	A. Williams	H.P. McGrath	2:37.75

26. **Indy 2013** The 2.5-mile Indianapolis Motor Speedway has been the home to a race on Memorial Day weekend nearly every year since 1911. Even during the first race, there were controversies. Ralph Mulford was given the checkered flag first but took three extra laps just to make sure he'd completed 500 miles. When he finished, another driver, Ray Harroun, was being presented with the winner's trophy, and Mulford's protests were ignored. Harroun averaged 74.6 mph for the 500 miles. In 2013, the winner, Tony Kanaan, averaged 187.433 mph.

Here are the data for the first five races and five recent Indianapolis 500 races.

Year	Winner	Time	Average Speed (mph)
1911	Ray Harroun	6:42:08.039	74.602
1912	Joe Dawson	6:21:06.144	78.719
1913	Jules Goux	6:35:05.108	75.933
1914	René Thomas	6:03:45.060	82.474
1915	Ralph DePalma	5:33:55.619	89.840
. . .			
2009	Hélio Castroneves	3:19:34.6427	150.318
2010	Dario Franchitti	3:05:37.0131	161.623
2011	Dan Wheldon	2:56:11.7267	170.265
2012	Dario Franchitti	2:58:51	167.734
2013	Tony Kanaan	2:40:03.4181	187.433

Just Checking ANSWERS

1. Who—Tour de France races; What—year, winner, country of origin, total time, average speed, stages, total distance ridden, starting riders, finishing riders; How—official statistics at race; Where—France (for the most part); When—1903 to 2012; Why—not specified (To see progress in speeds of cycling racing?)

2.

Variable	Type	Units
Year	Quantitative or Categorical	Years
Winner	Categorical	
Country of Origin	Categorical	
Total Time	Quantitative	Hours/minutes/seconds
Average Speed	Quantitative	Kilometers per hour
Stages	Quantitative	Counts (stages)
Total Distance	Quantitative	Kilometers
Starting Riders	Quantitative	Counts (riders)
Finishing Riders	Quantitative	Counts (riders)

chapter 2

Displaying and Describing Categorical Data

Who	People on the *Titanic*
What	Survival status, age, sex, ticket class
When	April 14, 1912
Where	North Atlantic
How	A variety of sources and Internet sites
Why	Historical interest

What happened on the *Titanic* at 11:40 on the night of April 14, 1912, is well known. Frederick Fleet's cry of "Iceberg, right ahead" and the three accompanying pulls of the crow's nest bell signaled the beginning of a nightmare that has become legend. By 2:15 A.M., the *Titanic*, thought by many to be unsinkable, had sunk, leaving more than 1500 passengers and crew members on board to meet their icy fate.

Here are some data about the passengers and crew aboard the *Titanic*. Each case (row) of the data table represents a person on board the ship. The variables are the person's *Survival* status (Dead or Alive), *Age* (Adult or Child), *Sex* (Male or Female), and ticket *Class* (First, Second, Third, or Crew).

Table 2.1

Part of a data table showing four variables for nine people aboard the *Titanic*

Survival	Age	Sex	Class
Dead	Adult	Male	Third
Dead	Adult	Male	Crew
Dead	Adult	Male	Third
Dead	Adult	Male	Crew
Dead	Adult	Male	Crew
Dead	Adult	Male	Crew
Alive	Adult	Female	First
Dead	Adult	Male	Third
Dead	Adult	Male	Crew

A S *Video*: **The Incident** tells the story of the *Titanic*, and includes rare film footage.

The problem with a data table like this—and in fact with all data tables—is that you can't *see* what's going on. And seeing is just what we want to do. We need ways to show the data so that we can see patterns, relationships, trends, and exceptions.

The Three Rules of Data Analysis

So, what should we do with data like these? There are three things you should always do first with data:

1. **Make a picture.** A display of your data will reveal things you are not likely to see in a table of numbers and will help you to *Think* clearly about the patterns and relationships that may be hiding in your data.
2. **Make a picture.** A well-designed display will *Show* the important features and patterns in your data. A picture will also show you the things you did not expect to see: the extraordinary (possibly wrong) data values or unexpected patterns.
3. **Make a picture.** The best way to *Tell* others about your data is with a well-chosen picture.

These are the three rules of data analysis. There are pictures of data throughout the book, and new kinds keep showing up. These days, technology makes drawing pictures of data easy, so there is no reason not to follow the three rules.

Figure 2.1

A picture to tell a story In November 2012, Barack Obama was re-elected as president of the United States. News reports commonly showed the election results with maps like the one on top, coloring states won by Obama blue and those won by his opponent Mitt Romney red. Even though Romney lost, doesn't it look like there's more red than blue? That's because some of the larger states like Montana and Wyoming have far fewer voters than some of the smaller states like Maryland and Connecticut. The strange-looking map on the bottom cleverly distorts the states to resize them proportional to their populations. By sacrificing an accurate display of the land areas, we get a better impression of the votes cast, giving us a clear picture of Obama's victory.
(*Source:* www-personal.umich.edu/~mejn/election/2012/)

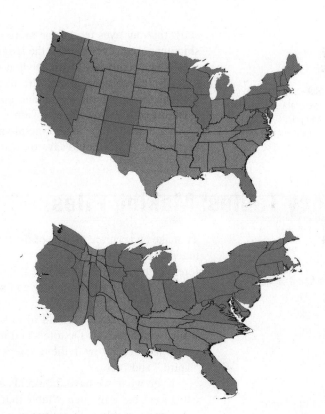

The Area Principle

The best data displays, like the distorted electoral map above, observe a fundamental principle of graphing data called the **area principle**. The area principle says that the area occupied by a part of the graph should correspond to the magnitude of the value it represents. But a bad picture can distort our understanding rather than help it. On the next page is a graph of the *Titanic* data. What impression do you get about who was aboard the ship?

It sure looks like most of the people on the *Titanic* were crew members, with a few passengers along for the ride. That doesn't seem right. What's wrong? The lengths of the ships *do* match the totals in the table. However, experience and psychological tests

Figure 2.2

How many people were in each class on the *Titanic*? From this display, it looks as though the service must have been great, since most aboard were crew members. Although the length of each ship here corresponds to the correct number, the impression is all wrong. in fact, only about 40% were crew.

A *S* *Activity*: **Make and Examine a Table of Counts.** Even data on something as simple as hair color can reveal surprises when you organize it in a data table.

show that our eyes tend to be more impressed by the *area* than by other aspects of each ship image. So, even though the *length* of each ship matches up with one of the totals, it's the associated *area* in the image that we notice. There were about 3 times as many crew as second-class passengers, and the ship depicting the number of crew is about 3 times longer than the ship depicting second-class passengers, but it occupies about 9 times the area. That just isn't a correct impression.

Violations of the area principle are a common way to lie (or, since most mistakes are unintentional, we should say err) with Statistics.

Frequency Tables: Making Piles

Class	Count
First	325
Second	285
Third	706
Crew	885

Table 2.2

A frequency table of the *Titanic* passengers

Class	%
First	14.77
Second	12.95
Third	32.08
Crew	40.21

Table 2.3

A relative frequency table for the same data

To make an accurate picture of data, the first thing we have to do is to make piles. We pile together things that seem to go together, so we can see how the cases distribute across different categories. For categorical data, piling is easy. We just count the number of cases corresponding to each category and put them in a table.

One way to put all 2201 people on the *Titanic* into piles is by ticket *Class*, counting up how many had each kind of ticket. We can organize these counts into a **frequency table,** which records the totals and the category names. We use the names of the categories to label each row in the frequency table. For ticket *Class*, these are "First," "Second," "Third," and "Crew."

Even when we have thousands of cases, a variable like ticket *Class,* with only a few categories, has a frequency table that's easy to read.

For a variable with dozens or hundreds of categories, a frequency table will be much harder to read. You might want to combine categories into larger headings. For example, instead of counting the number of students from each state, you might group the states into regions like "Northeast," "South," "Midwest," "Mountain States," and "West." If the number of cases in several categories is relatively small, you can put them together into one category labeled "Other."

Counts are useful, but sometimes we want to know the fraction or **proportion** of the data in each category, so we divide the counts by the total number of cases. Usually we multiply by 100 to express these proportions as **percentages.** A **relative frequency table** displays the *percentages,* rather than the counts, of the values in each category. Both types of tables show how the cases are distributed across the categories. In this way, they describe the **distribution** of a categorical variable because they name the possible categories and tell how frequently each occurs.

Bar Charts

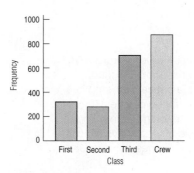

Figure 2.3

People on the *Titanic* by Ticket Class With the area principle satisfied, we can see the true distribution more clearly.

Here's a chart that obeys the area principle. It's not as visually entertaining as the ships, but it does give an *accurate* visual impression of the distribution. The height of each bar shows the count for its category. The bars are the same width, so their heights determine their areas, and the areas are proportional to the counts in each class. Now it's easy to see that the majority of people on board were *not* crew, as the ships picture led us to believe. We can also see that there were about 3 times as many crew as second-class passengers. And there were more than twice as many third-class passengers as either first- or second-class passengers, something you may have missed in the frequency table. Bar charts make these kinds of comparisons easy and natural.

A **bar chart** displays the distribution of a categorical variable, showing the counts for each category next to each other for easy comparison. Bar charts should have small spaces between the bars to indicate that these are freestanding bars that could be rearranged into any order. The bars are lined up along a common base.

Usually they stick up like this 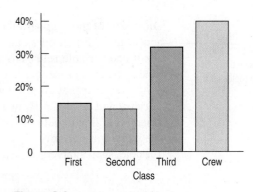 but sometimes they run

sideways like this

If we really want to draw attention to the relative *proportion* of passengers falling into each of these classes, we could replace the counts with percentages and use a **relative frequency bar chart**.

What a Bar Chart Is, and Isn't For some reason, some computer programs give the name "bar chart" to any graph that uses bars. And others use different names according to whether the bars are horizontal or vertical. Don't be misled. "Bar chart" is the term for a *display of counts of a categorical variable* with bars.

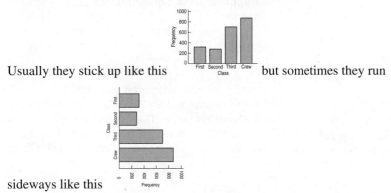

Figure 2.4

The relative frequency bar chart looks the same as the bar chart (Figure 2.3) but shows the proportion of people in each category rather than the counts.

Pie Charts

Another common display that shows how a whole group breaks into several categories is a pie chart. **Pie charts** show the whole group of cases as a circle. They slice the circle into pieces whose sizes are proportional to the fraction of the whole in each category.

Count

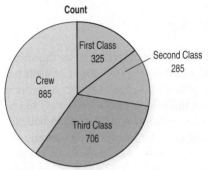

Figure 2.5

Number of *Titanic* passengers in each class

© 2013 Randall Munroe. Reprinted with permission. All rights reserved.

Pie charts give a quick impression of how a whole group is partitioned into smaller groups. Because we're used to cutting up pies into 2, 4, or 8 pieces, pie charts are good for seeing relative frequencies near 1/2, 1/4, or 1/8. For example, you may be able to tell that the pink slice, representing the second-class passengers, is very close to 1/8 of the total. It's harder to see that there were about twice as many third-class as first-class passengers. Which category had the most passengers? Were there more crew or more third-class passengers? Comparisons such as these are easier in a bar chart.

> **Think Before You Draw**
>
> Our first rule of data analysis is *Make a picture*. But what kind of picture? We don't have a lot of options—yet. There's more to Statistics than pie charts and bar charts, and knowing when to use each type of graph is a critical first step in data analysis. That decision depends in part on what type of data we have.
>
> It's important to check that the data are appropriate for whatever method of analysis you choose. Before you make a bar chart or a pie chart, always check the **Categorical Data Condition:** The data are counts or percentages of individuals in categories.
>
> If you want to make a relative frequency bar chart or a pie chart, you'll need to also make sure that the categories don't overlap so that no individual is counted twice. If the categories do overlap, you can still make a bar chart, but the percentages won't add up to 100%. For the *Titanic* data, either kind of display is appropriate because the categories don't overlap.
>
> Throughout this course, you'll see that doing Statistics right means selecting the proper methods. That means you have to *Think* about the situation at hand. An important first step, then, is to check that the type of analysis you plan is appropriate. The Categorical Data Condition is just the first of many such checks.

Contingency Tables: Children and First-Class Ticket Holders First?

A S   *Activity:* **Children at Risk.**
This activity looks at the fates of children aboard the *Titanic*; the subsequent activity shows how to make such tables on a computer.

Only 32% of those aboard the *Titanic* survived. Was that survival rate the same for men and women? For children and adults? For all ticket classes? It's often more interesting to ask if one variable relates to another. For example, was there a relationship between the kind of ticket a passenger held and the passenger's chances of making it into a lifeboat?

To answer that question we can arrange the counts for the two categorical variables, *Survival* and ticket *Class,* in a table. Table 2.4 shows each person aboard the *Titanic* classified according to both their ticket *Class* and their *Survival*. Because the table shows how the individuals are distributed along each variable, contingent on the value of the other variable, such a table is called a **contingency table**.

Table 2.4

Contingency table of ticket *Class* and *Survival* The bottom line of "Totals" is the same as the previous frequency table.

		Class				
		First	**Second**	**Third**	**Crew**	**Total**
Survival	**Alive**	203	118	178	212	**711**
	Dead	122	167	528	673	**1490**
	Total	325	285	706	885	**2201**

Each **cell** of the table gives the count for a combination of values of the two variables. The margins of the table, both on the right and at the bottom, give totals. The bottom line of the table is just the frequency distribution of ticket *Class*. The right column of the table is the frequency distribution of the variable *Survival*. When presented like this, in the

A bell-shaped artifact from the *Titanic*

margins of a contingency table, the frequency distribution of one of the variables is called its **marginal distribution**. The marginal distribution can be expressed either as counts or percentages.

If you look down the column for second-class passengers to the first row, you'll find the cell containing the 118 second-class passengers who survived. Looking at the cell to its right we see that more third-class passengers (178) survived. But, does that mean that third-class passengers were more *likely* to survive? It's true that *more* third-class passengers survived, but there were many more third-class passengers on board the ship. To compare the two numbers fairly, we need to express them as percentages—but as a percentage of what?

For any cell, there are three choices of percentage. We could express the 118 second-class survivors as 5.4% of all the passengers on the *Titanic* (the *overall percent*), as 16.6% of all the survivors (the *row percent*), or as 41.4% of all second-class passengers (the *column percent*). Each of these percentages is potentially interesting.

Statistics programs offer all three. Unfortunately, they often put them all together in each cell of the table. The resulting table holds lots of information, but it can be hard to understand:

Table 2.5

Another contingency table of ticket *Class* This time we see not only the counts for each combination of *Class* and *Survival* (in bold) but the percentages these counts represent. For each count, there are three choices for the percentage: by row, by column, and by table total. There's probably too much information here for this table to be useful.

			First	Second	Third	Crew	Total
Survival	Alive	Count	203	118	178	212	711
		% of Row	28.6%	16.6%	25.0%	29.8%	100%
		% of Column	62.5%	41.4%	25.2%	24.0%	32.3%
		% of Table	9.2%	5.4%	8.1%	9.6%	32.3%
	Dead	Count	122	167	528	673	1490
		% of Row	8.2%	11.2%	35.4%	45.2%	100%
		% of Column	37.5%	58.6%	74.8%	76.0%	67.7%
		% of Table	5.6%	7.6%	24.0%	30.6%	67.7%
	Total	Count	325	285	706	885	2201
		%of Row	14.8%	12.9%	32.1%	40.2%	100%
		% of Column	100%	100%	100%	100%	100%
		% of Table	14.8%	12.9%	32.1%	40.2%	100%

Column header above table: **Class**

To simplify the table, let's first pull out the percent of table values:

Table 2.6

A contingency table of *Class* by *Survival* with only the table percentages

		First	Second	Third	Crew	Total
Survival	Alive	9.2%	5.4%	8.1%	9.6%	32.3%
	Dead	5.6%	7.6%	24.0%	30.6%	67.7%
	Total	14.8%	12.9%	32.1%	40.2%	100%

Class

These percentages tell us what percent of *all* passengers belong to each combination of column and row category. For example, we see that although 8.1% of the people aboard the *Titanic* were surviving third-class ticket holders, only 5.4% were surviving second-class ticket holders. Is this fact useful? Comparing these percentages, you might think that the chances of surviving were better in third class than in second. But be careful. There were many more third-class than second-class passengers on the *Titanic,* so there were more third-class survivors. That group is a larger percentage of the passengers, but that's not really what we want to know. Overall percentages don't answer questions like this.

Percent of What? The English language can be tricky when we talk about percentages. If you're asked "What percent *of the survivors* were in second class?" it's pretty clear that we're interested only in survivors. It's as if we're restricting the *Who* in the question to the survivors, so we should look at the number of second-class passengers among all the survivors—in other words, the row percent.

But if you're asked "What percent were second-class passengers who survived?" you have a different question. Be careful; here, the *Who* is everyone on board, so 2201 should be the denominator, and the answer is the table percent.

And if you're asked "What percent of the second-class passengers survived?" you have a third question. Now the *Who* is the second-class passengers, so the denominator is the 285 second-class passengers, and the answer is the column percent.

Always be sure to ask "percent of what?" That will help you to know the *Who* and whether we want *row, column,* or *table* percentages.

For Example FINDING MARGINAL DISTRIBUTIONS

A recent Gallup poll asked 1008 Americans age 18 and over whether they planned to watch the upcoming Super Bowl. The pollster also asked those who planned to watch whether they were looking forward more to seeing the football game or the commercials. The results are summarized in the table:

QUESTION: What's the marginal distribution of the responses?

		Sex		
		Male	Female	Total
Response	Game	279	200	479
	Commercials	81	156	237
	Won't watch	132	160	292
	Total	492	516	1008

ANSWER: To determine the percentages for the three responses, divide the count for each response by the total number of people polled:

$$\frac{479}{1008} = 47.5\% \quad \frac{237}{1008} = 23.5\% \quad \frac{292}{1008} = 29.0\%$$

According to the poll, 47.5% of American adults were looking forward to watching the Super Bowl game, 23.5% were looking forward to watching the commercials, and 29% didn't plan to watch at all.

Conditional Distributions

Rather than look at the overall percentages, it's more interesting to ask whether the chance of surviving the *Titanic* sinking *depended* on ticket class. We can look at this question in two ways. First, we could ask how the distribution of ticket *Class* changes between survivors and nonsurvivors. To do that, we look at the *row percentages:*

Table 2.7

The conditional distribution of ticket *Class* conditioned on each value of *Survival: Alive* and *Dead*

		Class				
		First	Second	Third	Crew	Total
Survival	Alive	203	118	178	212	711
		28.6%	16.6%	25.0%	29.8%	100%
	Dead	122	167	528	673	1490
		8.2%	11.2%	35.4%	45.2%	100%

By focusing on each row separately, we see the distribution of class under the *condition* of surviving or not. The sum of the percentages in each row is 100%, and we divide that up by ticket class. In effect, we temporarily restrict the *Who* first to survivors and make a pie chart for them. Then we refocus the *Who* on the nonsurvivors and make

Figure 2.6

Pie charts of the conditional distributions of ticket *Class* for the survivors and nonsurvivors, separately Do the distributions appear to be the same? We're primarily concerned with percentages here, so pie charts are a reasonable choice.

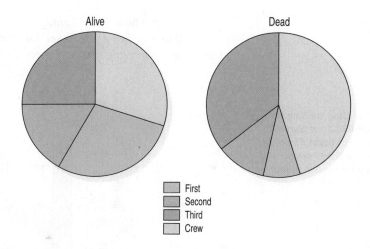

their pie chart. These pie charts show the distribution of ticket classes *for each row* of the table: survivors and nonsurvivors. The distributions we create this way are called **conditional distributions,** because they show the distribution of one variable for just those cases that satisfy a condition on another variable.

For Example FINDING CONDITIONAL DISTRIBUTIONS

RECAP: The table shows results of a poll asking adults whether they were looking forward to the Super Bowl game, looking forward to the commercials, or didn't plan to watch.

QUESTION: How do the conditional distributions of interest in the commercials differ for men and women?

		Sex		
		Male	**Female**	**Total**
Response	**Game**	279	200	**479**
	Commercials	81	156	**237**
	Won't watch	132	160	**292**
	Total	**492**	**516**	**1008**

ANSWER: Look at the group of people who responded "Commercials" and determine what percent of them were male and female:

$$\frac{81}{237} = 34.2\% \quad \frac{156}{237} = 65.8\%$$

Women make up a sizable majority of the adult Americans who look forward to seeing Super Bowl commercials more than the game itself. Nearly 66% of people who voiced a preference for the commercials were women, and only 34% were men.

But we can also turn the question around. We can look at the distribution of *Survival* for each category of ticket *Class*. To do this, we look at the *column percentages*. Those show us whether the chance of surviving was roughly the same *for each of the four classes*. Now the percentages in each column add to 100%, because we've restricted the *Who*, in turn, to each of the four ticket classes:

Table 2.8

A contingency table of *Class* by *Survival* with only counts and column percentages Each column represents the conditional distribution of *Survival* for a given category of ticket *Class*.

			Class				
			First	**Second**	**Third**	**Crew**	**Total**
Survival	**Alive**	**Count**	203	118	178	212	**711**
		% of Column	62.5%	41.4%	25.2%	24.0%	**32.3%**
	Dead	**Count**	122	167	528	673	**1490**
		% of Column	37.5%	58.6%	74.8%	76.0%	**67.7%**
	Total	**Count**	**325**	**285**	**706**	**885**	**2201**
			100%	**100%**	**100%**	**100%**	**100%**

Looking at how the percentages change across each row, it sure looks like ticket class mattered in whether a passenger survived. To make it more vivid, we could display the percentages surviving and not for each *Class* in a side-by-side bar chart:

Figure 2.7

Side-by-side bar chart showing the conditional distribution of *Survival* for each category of ticket *Class* The corresponding pie charts would have only two categories in each of four pies, so bar charts seem the better alternative.

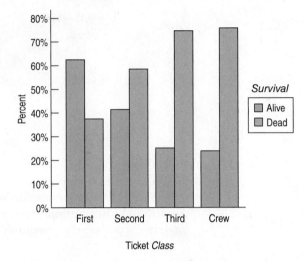

These bar charts are simple because, for the variable *Survival*, we have only two alternatives: Alive and Dead. When we have only two categories, we need to know only the percentage of one of them. We can simplify the display even more by dropping one category. Here are the percentages of dying *across the classes* displayed in one chart:

Figure 2.8

Bar chart showing just nonsurvivor percentages for each value of ticket *Class* Because we have only two values, the second bar doesn't add any information. Compare this chart to the side-by-side bar chart shown earlier.

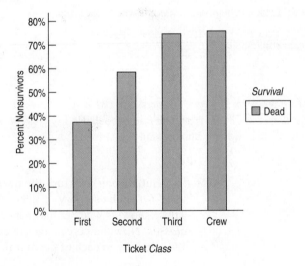

TI-*nspire*

Conditional distributions and association. Explore the *Titanic* data to see which passengers were most likely to survive.

Now it's easy to compare the risks. Among first-class passengers, 37.5% perished, compared to 58.6% for second-class ticket holders, 74.8% for those in third class, and 76.0% for crew members.

If the risk had been about the same across the ticket classes, we would have said that survival was *independent* of class. But it's not. The differences we see among these conditional distributions suggest that survival may have depended on ticket class. You may find it useful to consider conditioning on each variable in a contingency table in order to explore the dependence between them.

It is interesting to know that *Class* and *Survival* are associated. That's an important part of the *Titanic* story. And we know how important this is because the margins show us the actual numbers of people involved.

Variables can be associated in many ways and to different degrees. The best way to tell whether two variables are associated is to ask whether they are *not*.[1] In a contingency table, when the distribution of *one* variable is the same for all categories of another, we say that the variables are **independent**. That tells us there's no association between these variables. We'll see a way to check for independence formally later in the book. For now, we'll just compare the distributions.

For Example LOOKING FOR ASSOCIATIONS BETWEEN VARIABLES

RECAP: The table shows results of a poll asking adults whether they were looking forward to the Super Bowl game, looking forward to the commercials, or didn't plan to watch.

QUESTION: Does it seem that there's an association between interest in Super Bowl TV coverage and a person's sex?

	Sex		
Response	Male	Female	Total
Game	279	200	479
Commercials	81	156	237
Won't watch	132	160	292
Total	492	516	1008

ANSWER: First find the distribution of the three responses for the men (the column percentages):

$$\frac{279}{492} = 56.7\% \quad \frac{81}{492} = 16.5\% \quad \frac{132}{492} = 26.8\%$$

Then do the same for the women who were polled, and display the two distributions with a side-by-side bar chart:

Based on this poll it appears that women were only slightly less interested than men in watching the Super Bowl telecast: 31% of the women said they didn't plan to watch, compared to just under 27% of men. Among those who planned to watch, however, there appears to be an association between the viewer's sex and what the viewer is most looking forward to. While more women are interested in the game (39%) than the commercials (30%), the margin among men is much wider: 57% of men said they were looking forward to seeing the game, compared to only 16.5% who cited the commercials.

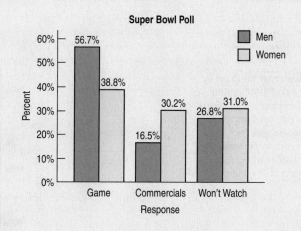

---

[1]This kind of "backwards" reasoning shows up surprisingly often in science—and in Statistics. We'll see it again.

Just Checking

A Statistics class reports the following data on *Sex* and *Eye Color* for students in the class:

		Eye Color			
		Blue	Brown	Green/Hazel /Other	Total
Sex	Males	6	20	6	32
	Females	4	16	12	32
	Total	10	36	18	64

1. What percent of females are brown-eyed?

2. What percent of brown-eyed students are female?

3. What percent of students are brown-eyed females?

4. What's the distribution of *Eye Color*?

5. What's the conditional distribution of *Eye Color* for the males?

6. Compare the percent who are female among the blue-eyed students to the percent of all students who are female.

7. Does it seem that *Eye Color* and *Sex* are independent? Explain.

Segmented Bar Charts

We could display the *Titanic* information by dividing up bars rather than circles. The resulting **segmented bar chart** treats each bar as the "whole" and divides it proportionally into segments corresponding to the percentage in each group. We can clearly see that the distributions of ticket *Class* are different, indicating again that *Survival* was not independent of ticket *Class*.

Figure 2.9

A segmented bar chart for *Class* by *Survival* Notice that although the totals for survivors and nonsurvivors are quite different, the bars are the same height because we have converted the numbers to *percentages*. Compare this display with the side-by-side pie charts of the same data in Figure 2.6.

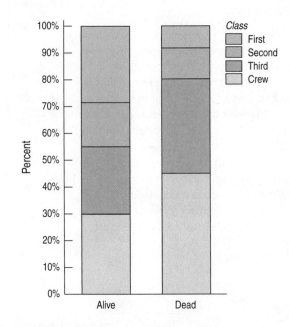

Step-by-Step Example EXAMINING CONTINGENCY TABLES

We asked for a picture of a man eating fish. This is what we got.

Medical researchers followed 6272 Swedish men for 30 years to see if there was any association between the amount of fish in their diet and prostate cancer ("Fatty Fish Consumption and Risk of Prostate Cancer," *Lancet*, June 2001). Their results are summarized in this table:

		Prostate Cancer	
		No	Yes
Fish Consumption	Never/seldom	110	14
	Small part of diet	2420	201
	Moderate part	2769	209
	Large part	507	42

Question: Is there an association between fish consumption and prostate cancer?

THINK ➡ **Plan** Be sure to state what the problem is about.

Variables Identify the variables and report the W's.

Be sure to check the appropriate condition.

I want to know if there is an association between fish consumption and prostate cancer.

The individuals are 6272 Swedish men followed by medical researchers for 30 years. The variables record their fish consumption and whether or not they were diagnosed with prostate cancer.

✓ **Categorical Data Condition:** I have counts for both fish consumption and cancer diagnosis. The categories of diet do not overlap, and the diagnoses do not overlap. It's okay to draw pie charts or bar charts.

SHOW ➡ **Mechanics** It's a good idea to check the marginal distributions first before looking at the two variables together.

		Prostate Cancer		
		No	Yes	Total
Fish Consumption	Never/seldom	110	14	124 (2.0%)
	Small part of diet	2420	201	2621 (41.8%)
	Moderate part	2769	209	2978 (47.5%)
	Large part	507	42	549 (8.8%)
	Total	5806 (92.6%)	466 (7.4%)	6272 (100%)

Two categories of the diet are quite small, with only 2.0% Never/Seldom eating fish and 8.8% in the "Large part" category. Overall, 7.4% of the men in this study had prostate cancer.

(continued)

Then, make appropriate displays to see whether there is a difference in the relative proportions. These pie charts compare fish consumption for men who have prostate cancer to fish consumption for men who don't.

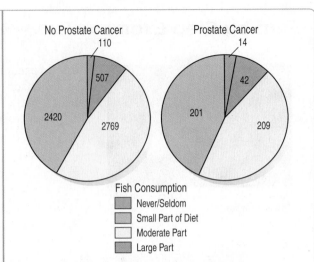

It's hard to see much difference in the pie charts. So, I made a display of the row percentages. Because there are only two alternatives, I chose to display the risk of prostate cancer for each group:

Both pie charts and bar charts can be used to compare conditional distributions. Here we compare prostate cancer rates based on differences in fish consumption.

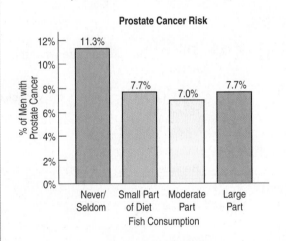

TELL ➡ Conclusion   Interpret the patterns in the table and displays in context. If you can, discuss possible real-world consequences. Be careful not to overstate what you see. The results may not generalize to other situations.

Overall, there is a 7.4% rate of prostate cancer among men in this study. Most of the men (89.3%) ate fish either as a moderate or small part of their diet. From the pie charts, it's hard to see a difference in cancer rates among the groups. But in the bar chart, it looks like the cancer rate for those who never/seldom ate fish may be somewhat higher.

However, only 124 of the 6272 men in the study fell into this category, and only 14 of them developed prostate cancer. More study would probably be needed before we would recommend that men change their diets.[2]

---

[2]The original study actually used pairs of twins, which enabled the researchers to discern that the risk of cancer for those who never ate fish actually *was* substantially greater. Using pairs is a special way of gathering data. We'll discuss such study design issues and how to analyze the data in the later chapters.

This study is an example of looking at a sample of data to learn something about a larger population, one of the main goals of this book. We care about more than these particular 6272 Swedish men. We hope that learning about their experiences will tell us something about the value of eating fish in general. That raises several questions. What population do we think this sample might represent? Do we hope to learn about all Swedish men? About all men? How do we know that other factors besides that amount of fish they ate weren't associated with prostate cancer? Perhaps men who eat fish often have other habits that distinguish them from the others and maybe those other habits are what actually kept their cancer rates lower.

Observational studies, like this one, often lead to contradictory results because we can't control all the other factors. In fact, a later paper, published in 2011, based on data from a cancer prevention trial on 3400 men from 1994 to 2003, showed that some fatty acids may actually increase the risk of prostate cancer.[3] We'll discuss the pros and cons of observational studies and experiments where we can control the factors in Chapter 11.

---

[3]"Serum phospholipid fatty acids and prostate cancer risk: Results from the Prostate Cancer Prevention Trial," *American Journal of Epidemiology,* 2011.

WHAT IF ●●● the variables really ARE independent?

We told you that Statistics is about variation, remember? That presents a problem when we must decide whether we think two variables are independent. If prostate cancer is independent of fish consumption, we expect to see *exactly* the same cancer rate regardless of the amount of fish in the Swedish men's diets. But the real world never cooperates "exactly"; there's always a bit of variation in such percentages. When the percentages are glaringly different (as for passengers' chances of surviving the *Titanic* disaster for various ticket classes) we conclude that there is an association. But when they're "close enough" (as for fish consumption and prostate cancer) we conclude that the variables seem independent. This raises an important question: How close is "close enough"?

Some people think a 5-year-old (or even a chimpanzee) could create modern art like this painting. To see whether there truly are discernable artistic qualities in modern art, researchers[4] paired paintings by real artists with similar works by children or animals. Then they showed those pairs to a group of art students and another group of students without any art expertise, asking which of each pair they preferred.

Here's a table that summarizes the results with counts and row percentages.

While we can see that both groups did prefer works by artists, did expertise play a role? True, the art students were more likely to choose the "real" art, but could this apparent difference be the result of random chance in this particular sample of people, or does it indicate that experts really can see artistic qualities the rest of us may miss? In other words, do these results provide evidence that the ability to distinguish modern art from that of children or animals is associated with the observer's art expertise?

STUDY		Preferred Painting Done by	
		Modern Artist	**Other**
Observer	**Art Student**	250 (62.5%)	150 (37.5%)
	Non-Art Student	180 (56.25%)	140 (43.75%)

We investigate by asking, "What if…" What if expertise and preference actually are independent, and these 720 observations just fell into the table randomly? How often would a difference in distributions

---

[4]Hawley-Dolan, Angelina, and Winner, Ellen, *"Seeing the Mind Behind the Art,"* © 2011 Psychological Science. http://pss.sagepub.com/content/22/4/435

this large (or larger) happen just by chance? To find out, we ran a computer simulation[5] that mimics this study. Our simulation replicates the same number of decisions by each group while maintaining the same overall level of preference for real art, but the simulation makes each decision randomly. The simulation's first trial produced the top table. Compare the results for the observers.

These percentages barely differ at all. If the study had come out like this, it certainly wouldn't have suggested an association.

But when we did it again, our simulation produced this table:

This time the association looks even stronger than what appeared in the real study, yet we know this is just random chance at work. Just how likely is something like this to happen?

We simulated again. And again. And… Well, we ran 10,000 trials! Tables like #2, where the apparent association was at least as strong as what actually showed up in the study, appeared 1,382 times. This suggests that we wouldn't be surprised to see an association this strong in our sample even if art and non-art students are equally likely to pick works by a modern artist. There's almost a 14% chance that the researchers' results could just have been random variability in their sample rather than meaningful evidence of any actual association.

TRIAL #1	Preferred Painting Done by	
	Modern Artist	**Other**
Art Student	241 (60.25%)	159 (39.75%)
Non-Art Student	189 (59.06%)	131 (40.94%)

TRIAL #2	Preferred Painting Done by	
	Modern Artist	**Other**
Art Student	266 (66.5%)	134 (33.5%)
Non-Art Student	164 (51.25%)	156 (48.75%)

So what does this mean? While 14% may seem somewhat low, it's not really all that unusual—about 1 chance in 7. We'll see later on that statisticians commonly consider observed results to be "statistically significant" only if there's less than a 5% chance (a 1-in-20 shot) they could have arisen by accident. They'd conclude that this study doesn't provide convincing evidence that there's an association between expertise and the ability to identify "real" modern art.

By the way, if you like the painting, you might be able to have one for peanuts. Check with the artist.[6]

---

[5]Simulations are really cool. You'll learn to do them in Chapter 10, so don't drop the course yet.

[6]Jojo, the elephant. http://www.elephantartgallery.com/paintings/1129.php

WHAT CAN GO WRONG?

■ **Don't violate the area principle.** This is probably the most common mistake in a graphical display. It is often made in the cause of artistic presentation. Here, for example, are two displays of the pie chart of the *Titanic* passengers by class:

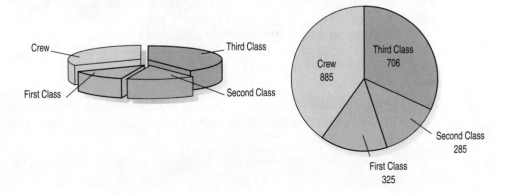

The one on the left looks pretty, doesn't it? But showing the pie on a slant violates the area principle and makes it much more difficult to compare fractions of the whole made up of each class—the principal feature that a pie chart ought to show.

■ **Keep it honest.** Here's a pie chart that displays data on the percentage of high school students who engage in specified dangerous behaviors as reported by the Centers for Disease Control. What's wrong with this plot?

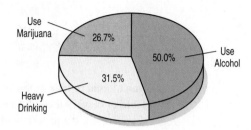

Try adding up the percentages. Or look at the 50% slice. Does it look right? Then think: What are these percentages of? Is there a "whole" that has been sliced up? In a pie chart, the proportions shown by each slice of the pie must add up to 100% and each individual must fall into only one category. Of course, showing the pie on a slant makes it even harder to detect the error.

The following chart shows the average number of texts in various time periods by American cell phone customers in the period 2006 to 2011.

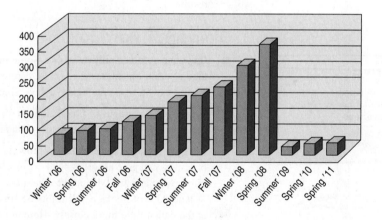

It may look as though text messaging decreased suddenly some time around 2010, which probably doesn't jibe with your experience. In fact, this chart has several problems. First, it's not a bar chart. Bar charts display counts of categories. This bar chart is a plot of a quantitative variable (average number of texts) against time—although to make it worse, some of the time periods are missing. Even though these flaws are already fatal, the worst mistake is one that can't be seen from the plot. In 2010, the company reporting the data switched from reporting the average number of texts per year (reported each quarter) to average number of texts per month. So, the numbers in the last three quarters should be multiplied by 12 to make them comparable to the rest.

■ **Don't confuse similar-sounding percentages.** These percentages sound similar but are different:

■ The percentage of the passengers who were both in first class and survived: This would be 203/2201, or 9.2%.
■ The percentage of the first-class passengers who survived: This is 203/325, or 62.5%.
■ The percentage of the survivors who were in first class: This is 203/711, or 28.6%.

In each instance, pay attention to the *Who* implicitly defined by the phrase. Often there is a restriction to a smaller group (all aboard the *Titanic,* those in first class, and those who survived, respectively) before a percentage is found. Your discussion of results must make these differences clear.

		Class				
		First	Second	Third	Crew	Total
Survival	Alive	203	118	178	212	**711**
	Dead	122	167	528	673	**1490**
	Total	**325**	**285**	**706**	**885**	**2201**

- **Don't forget to look at the variables separately, too.** When you make a contingency table or display a conditional distribution, be sure you also examine the marginal distributions. It's important to know how many cases are in each category.

- **Be sure to use enough individuals.** When you consider percentages, take care that they are based on a large enough number of individuals. Take care not to make a report such as this one:

  We found that 66.67% of the rats improved their performance with training. The other rat died.

- **Don't overstate your case.** Independence is an important concept, but it is rare for two variables to be *entirely* independent. We can't conclude that one variable has no effect whatsoever on another. Usually, all we know is that little effect was observed in our study. Other studies of other groups under other circumstances could find different results.

Simpson's Paradox

- **Don't use unfair or silly averages.** Sometimes averages can be misleading. Sometimes they just don't make sense at all. Be careful when averaging different variables that the quantities you're averaging are comparable. The Centerville sign says it all.

Entering Centerville

Established	1793
Population	7943
Elevation	710
Average	3482

When using averages of proportions across several different groups, it's important to make sure that the groups really are comparable.

It's easy to make up an example showing that averaging across very different values or groups can give absurd results. Here's how that might work: Suppose there are two pilots, Moe and Jill. Moe argues that he's the better pilot of the two, since he managed to land 83% of his last 120 flights on time compared with Jill's 78%. But let's look at the data a little more closely. Here are the results for each of their last 120 flights, broken down by the time of day they flew:

Table 2.9

On-time flights by *Time of Day* and *Pilot* Look at the percentages within each *Time of Day* category. Who has a better on-time record during the day? At night? Who is better overall?

		Time of Day		
		Day	Night	Overall
Pilot	Moe	90 out of 100 90%	10 out of 20 50%	**100 out of 120** **83%**
	Jill	19 out of 20 95%	75 out of 100 75%	**94 out of 120** **78%**

Look at the daytime and nighttime flights separately. For day flights, Jill had a 95% on-time rate and Moe only a 90% rate. At night, Jill was on time 75% of the time and Moe only 50%. So Moe is better "overall," but Jill is better both during the day and at night. How can this be?

What's going on here is a problem known as **Simpson's paradox,** named for the statistician who discovered it in the 1950s. It comes up rarely in real life, but there have been several well-publicized cases. As we can see from the pilot example, the problem is *unfair averaging* over different groups. Jill has mostly night flights, which are more difficult, so her *overall average* is heavily influenced by her nighttime average. Moe, on the other hand, benefits from flying mostly during the day, with its higher on-time percentage. With their very different patterns of flying conditions, taking an overall average is misleading. It's not a fair comparison.

The moral of Simpson's paradox is to be careful when you average across different levels of a second variable. It's always better to compare percentages or other averages *within* each level of the other variable. The overall average may be misleading.

> **Simpson's Paradox**   One famous example of Simpson's paradox arose during an investigation of admission rates for men and women at the University of California at Berkeley's graduate schools. As reported in an article in *Science*, about 45% of male applicants were admitted, but only about 30% of female applicants got in. It looked like a clear case of discrimination. However, when the data were broken down by school (Engineering, Law, Medicine, etc.), It turned out that, within each school, the women were admitted at nearly the same or, in some cases, much *higher* rates than the men. How could this be? Women applied in large numbers to schools with very low admission rates (Law and Medicine, for example, admitted fewer than 10%). Men tended to apply to Engineering and Science. Those schools have admission rates above 50%. When the *average* was taken, the women had a much lower *overall* rate, but the average didn't really make sense.

What Have We Learned?

We've learned to analyze categorical variables.

- The methods in this chapter apply to categorical variables only. We always check the Categorical Variable Condition before proceeding.
- We summarize categorical data by counting the number of cases in each category, sometimes expressing the resulting distribution as percents.
- We display the distributions in a pie chart or bar chart.

When we want to see how two categorical variables are related, we put the counts (and/or percentages) in a contingency table.

- We look at the marginal distribution of each variable.
- We also look at the conditional distribution of a variable within each category of the other variable.
- We compare these marginal and conditional distributions by using pie charts, bar charts, or segmented bar charts.
- We examine the association between categorical variables by comparing conditional and marginal distributions. If the conditional distributions of one variable are roughly the same for each category of the other, we say the variables are independent.

Terms

Area principle In a statistical display, each data value should be represented by the same amount of area. (p. 15)

Frequency table (Relative frequency table) A frequency table lists the categories in a categorical variable and gives the count (or percentage of observations for each category. (p. 16)

Distribution	The distribution of a variable gives ■ the possible values of the variable and ■ the relative frequency of each value. (p. 16)
Bar chart (Relative frequency bar chart)	Bar charts show a bar whose area represents the count (or percentage) of observations for each category of a categorical variable. (p. 17)
Pie chart	Pie charts show how a "whole" divides into categories by showing a wedge of a circle whose area corresponds to the proportion in each category. (p. 17)
Categorical Data Condition	The methods in this chapter are appropriate for displaying and describing categorical data. Be careful not to use them with quantitative data. (p. 18)
Contingency table	A contingency table displays counts and, sometimes, percentages of individuals falling into named categories on two or more variables. The table categorizes the individuals on all variables at once to reveal possible patterns in one variable that may be contingent on the category of the other. (p. 19)
Marginal distribution	In a contingency table, the distribution of either variable alone is called the marginal distribution. The counts or percentages are the totals found in the margins (last row or column) of the table. (p. 19)
Conditional distribution	The distribution of a variable restricting the *Who* to consider only a smaller group of individuals is called a conditional distribution. (p. 21)
Independence	Variables are said to be independent if the conditional distribution of one variable is the same for each category of the other. We'll show how to check for independence in a later chapter. (p. 23)
Segmented bar chart	A segmented bar chart displays the conditional distribution of a categorical variable within each category of another variable. (p. 24)
Simpson's paradox	When averages are taken across different groups, they can appear to contradict the overall averages. This is known as "Simpson's paradox". (p. 31)

On the Computer DISPLAYING CATEGORICAL DATA

Although every package makes a slightly different bar chart, they all have similar features:

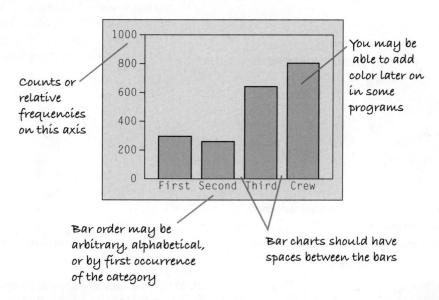

Counts or relative frequencies on this axis

You may be able to add color later on in some programs

Bar order may be arbitrary, alphabetical, or by first occurrence of the category

Bar charts should have spaces between the bars

Sometimes the count or a percentage is printed above or on top of each bar to give some additional information. You may find that your statistics package sorts category names in annoying orders by default. For example, many packages sort categories alphabetically or by the order the categories are seen in the data set. Often, neither of these is the best choice.

Exercises

1. Graphs in the news Find a bar graph of categorical data from a newspaper, a magazine, or the Internet.
  a) Is the graph clearly labeled?
  b) Does it violate the area principle?
  c) Does the accompanying article tell the W's of the variable?
  d) Do you think the article correctly interprets the data? Explain.

2. Graphs in the news II Find a pie chart of categorical data from a newspaper, a magazine, or the Internet.
  a) Is the graph clearly labeled?
  b) Does it violate the area principle?
  c) Does the accompanying article tell the W's of the variable?
  d) Do you think the article correctly interprets the data? Explain.

3. Tables in the news Find a frequency table of categorical data from a newspaper, a magazine, or the Internet.
  a) Is it clearly labeled?
  b) Does it display percentages or counts?
  c) Does the accompanying article tell the W's of the variable?
  d) Do you think the article correctly interprets the data? Explain.

4. Tables in the news II Find a contingency table of categorical data from a newspaper, a magazine, or the Internet.
  a) Is it clearly labeled?
  b) Does it display percentages or counts?
  c) Does the accompanying article tell the W's of the variables?
  d) Do you think the article correctly interprets the data? Explain.

5. **Movie genres** The pie chart summarizes the genres of 728 first-run movies released in 2011.

   a) Is this an appropriate display for the genres? Why/why not?

   b) Which genre was least common?

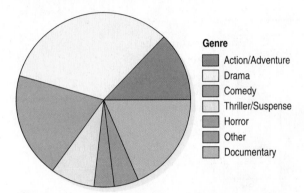

6. **Movie ratings** The pie chart shows the ratings assigned to 728 first-run movies released in 2011.

   a) Is this an appropriate display for these data? Explain.

   b) Which was the most common rating?

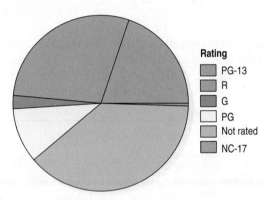

7. **Genres again** Here is a bar chart summarizing the 2011 movie genres, as seen in the pie chart in Exercise 5.

   a) Which genre was second most common?

   b) Is it easier to see that in the pie chart or the bar chart? Explain.

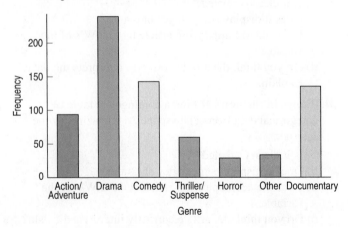

8. **Ratings again** Here is a bar chart summarizing the 2011 movie ratings, as seen in the pie chart in Exercise 6.

   a) Which was the least common rating?

   b) An editorial claimed that there's been a growth in PG-13 rated films that, according to the writer, "have too much sex and violence," at the expense of G-rated films that offer "good, clean fun." The writer offered the bar chart below as evidence to support his claim. Does the bar chart support his claim? Explain.

9. **Yearly ratings** The Motion Picture Association of America (MPAA) rates each film to designate the appropriate audience. The ratings are G (General Audiences. All Ages Admitted), PG (Parental Guidance Suggested. Some Material May Not Be Suitable For Children), PG-13 (Parents Strongly Cautioned. Some Material May Be Inappropriate For Children Under 13.), and R (Restricted. Children Under 17 Require Accompanying Parent or Adult Guardian.). The ratings of the 20 top grossing movies in the years 1992, 1996, 2006, and 2008 are shown in the bar charts below. The pie charts show the same data but are unlabeled. Match each pie chart with the correct year.

   i)

ii)

Top 20 Movies of 1996

iii)

Top 20 Movies of 2006

iv)

Top 20 Movies of 2008

10. Marriage in decline Changing attitudes about marriage and families prompted Pew Research to ask how people felt about particular recent trends (pewresearch.org/pubs/1802/decline-marriage-rise-new-families). For each trend, participants were asked whether the trend "is a good thing", "is a bad thing", or "makes no difference". Some participants said they didn't know or did chose not to respond. The bar charts below show the distribution of responses for each trend. The pie charts on the next page show the same data without the trends identified. Match each pie chart with the correct trend and bar chart.

i)

People Living Together without Being Married

ii)

Unmarried Couples Raising Children

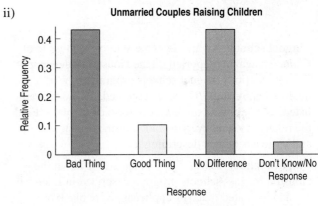

iii)

Single Women Having Children

iv)

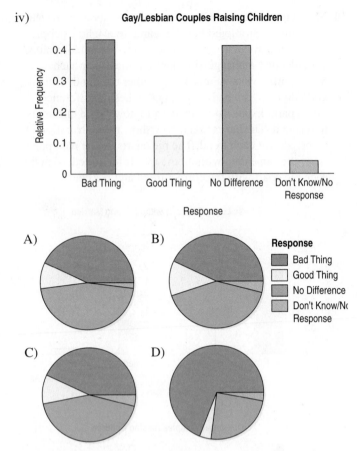

11. **Magnet schools** An article in the Winter 2003 issue of *Chance* magazine reported on the Houston Independent School District's magnet schools programs. Of the 1755 qualified applicants, 931 were accepted, 298 were wait-listed, and 526 were turned away for lack of space. Find the relative frequency distribution of the decisions made, and write a sentence describing it.

12. **Magnet schools again** The *Chance* article about the Houston magnet schools program described in Exercise 11 also indicated that 517 applicants were black or Hispanic, 292 Asian, and 946 white. Summarize the relative frequency distribution of ethnicity with a sentence or two (in the proper context, of course).

13. **Causes of death 2007** The Centers for Disease Control and Prevention (www.cdc.gov) lists causes of death in the United States during 2007:

Cause of Death	Percent
Heart disease	25.4
Cancer	23.2
Circulatory diseases and stroke	5.6
Chronic lower respiratory diseases	5.3
Accidents	5.1

a) Is it reasonable to conclude that heart or respiratory diseases were the cause of approximately 31% of U.S. deaths in 2007?
b) What percent of deaths were from causes not listed here?
c) Create an appropriate display for these data.

14. **Plane crashes** An investigation compiled information about recent nonmilitary plane crashes (www.plane crashinfo.com). The causes, to the extent that they could be determined, are summarized in the table.

Cause	Percent
Pilot error	40
Other human error	5
Weather	6
Mechanical failure	14
Sabotage	6

a) Is it reasonable to conclude that the weather or mechanical failures caused only about 20% of recent plane crashes?
b) In what percent of crashes were the causes not determined?
c) Create an appropriate display for these data.

15. **Oil spills 2010** Data from the International Tanker Owners Pollution Federation Limited (www.itopf.com) give the cause of spillage for 460 large oil tanker accidents from 1970–2010. Here are displays.

a) Write a brief report interpreting what the displays show.

b) Is a pie chart an appropriate display for these data? Why or why not?

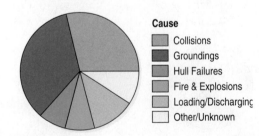

16. Winter Olympics 2010 Twenty-six countries won medals in the 2010 Winter Olympics. The table lists them, along with the total number of medals each won:

Country	Medals	Country	Medals
United States	37	Poland	6
Germany	30	Italy	5
Canada	26	Japan	5
Norway	23	Finland	5
Austria	16	Australia	3
Russia	15	Belarus	3
South Korea	14	Slovakia	3
China	11	Croatia	3
Sweden	11	Slovenia	3
France	11	Latvia	2
Switzerland	9	Great Britain	1
Netherlands	8	Estonia	1
Czech Republic	6	Kazakhstan	1

a) Try to make a display of these data. What problems do you encounter?
b) Can you find a way to organize the data so that the graph is more successful?

17. Global Warming The Pew Research Center for the People and the Press (http://people-press.org) has asked a representative sample of U.S. adults about global warming, repeating the question over time. In October 2010, the responses reflected a decreased belief that global warming is real and due to human activity. Here's a display of the percentages of respondents choosing each of the major alternatives offered:

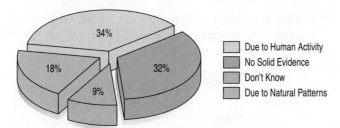

Due to Human Activity
No Solid Evidence
Don't Know
Due to Natural Patterns

List the errors in this display.

18. Death 2010 The January 2012 National Vital Statistics Report by the Centers for Disease Control and Prevention included information on deaths in the United States for the year 2010. The 4 most common causes-heart disease, cancer, respiratory diseases, and stroke-accounted for nearly 60% of all deaths. What problems do you see with the graph at the top of the next column?

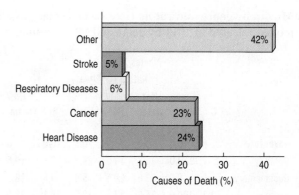

19. Teen smokers The organization Monitoring the Future (www.monitoringthefuture.org) asked 2048 eighth graders who said they smoked cigarettes what brands they preferred. The table below shows brand preferences for two regions of the country. Write a few sentences describing the similarities and differences in brand preferences among eighth graders in the two regions listed.

Brand Preference	South	West
Marlboro	58.4%	58.0%
Newport	22.5%	10.1%
Camel	3.3%	9.5%
Other (over 20 brands)	9.1%	9.5%
No usual brand	6.7%	12.9%

20. Handguns In an effort to reduce the number of gun-related homicides, some cities have run buyback programs in which the police offer cash (often $50) to anyone who turns in an operating handgun. *Chance* magazine looked at results from a four-year period in Milwaukee. The table below shows what types of guns were turned in and what types were used in homicides during a four-year period. Write a few sentences comparing the two distributions.

Caliber of Gun	Buyback	Homicide
Small (.22, .25, .32)	76.4%	20.3%
Medium (.357, .38, 9 mm)	19.3%	54.7%
Large (.40, .44, .45)	2.1%	10.8%
Other	2.2%	14.2%

21. Movies by genre and rating Here's a table that classifies movies released in 2011 by genre and MPAA rating:

Movies 2011							
Genre	Rating						Row Summary
	G	NC-17	Not Rated	PG	PG-13	R	
Action/ Adventure	22.2	0.0	4.6	34.2	20.3	10.0	12.6
Comedy	11.1	0.0	16.3	19.2	21.7	23.3	19.5
Documentary	50.0	0.0	33.0	20.5	5.6	4.8	18.5
Drama	11.1	100.0	34.0	17.8	32.2	38.1	32.8
Horror	0.0	0.0	1.8	1.4	3.5	8.1	3.8
Other	5.6	0.0	6.4	5.5	2.8	2.9	4.5
Thriller/ Suspense	0.0	0.0	3.9	1.4	14.0	12.9	8.1
Column Summary*	100.0	100.0	100.0	100.0	100.0	100.0	100.00

*S1 = columnProportion · 100

a) The table gives column percents. How could you tell that from the table itself?

b) What percentage of these movies were comedies?

c) What percentage of the PG-rated movies were comedies?

d) Which of the following can you learn from this table? Give the answer if you can find it from the table.
  i) The percentage of PG-13 movies that were comedies
  ii) The percentage of dramas that were R-rated
  iii) The percentage of horror movies that were G-rated
  iv) The percentage of 2011 movies that were PG-rated comedies

22. The last picture show Here's another table showing information about 728 movies released in 2011. This table gives percentages of the table total:

Movies 2011							
Genre	Rating						Row Summary
	G	NC-17	Not Rated	PG	PG-13	R	
Action/ Adventure	0.55	0.00	1.79	3.43	3.98	2.88	12.64
Comedy	0.27	0.00	6.32	1.92	4.26	6.73	19.51
Documentary	1.24	0.00	12.77	2.06	1.10	1.37	18.54
Drama	0.27	0.27	13.19	1.79	6.32	10.99	32.83
Horror	0.00	0.00	0.69	0.14	0.69	2.34	3.85
Other	0.14	0.00	2.47	0.55	0.55	0.82	4.53
Thriller/ Suspense	0.00	0.00	1.51	0.14	2.75	3.71	8.10
Column Summary*	2.47	0.27	38.74	10.03	19.64	28.85	100.00

$$*S1 = \frac{count()}{728} \cdot 100$$

a) How can you tell that this table holds table percentages (rather than row or column percentages)?

b) What was the most common genre/rating combination in 2011 movies?

c) How many of these movies were PG-rated comedies?

d) How many were G-rated?

e) An editorial about the movies noted, "More than three-quarters of the movies made today can be seen only by patrons 13 years old or older." Does this table support that assertion? Explain.

23. Seniors Prior to graduation, a high school class was surveyed about its plans. The following table displays the results for white and minority students (the "Minority" group included African-American, Asian, Hispanic, and Native American students):

Seniors		
Plans	White	Minority
4-year college	198	44
2-year college	36	6
Military	4	1
Employment	14	3
Other	16	3

a) What percent of the seniors are white?

b) What percent of the seniors are planning to attend a 2-year college?

c) What percent of the seniors are white and planning to attend a 2-year college?

d) What percent of the white seniors are planning to attend a 2-year college?

e) What percent of the seniors planning to attend a 2-year college are white?

24. Politics Students in an Intro Stats course were asked to describe their politics as "Liberal," "Moderate," or "Conservative." Here are the results:

Politics				
Sex	L	M	C	Total
Female	35	36	6	77
Male	50	44	21	115
Total	85	80	27	192

a) What percent of the class is male?

b) What percent of the class considers themselves to be "Conservative"?

c) What percent of the males in the class consider themselves to be "Conservative"?

d) What percent of all students in the class are males who consider themselves to be "Conservative"?

25. More about seniors Look again at the table of post-graduation plans for the senior class in Exercise 23.

a) Find the conditional distributions (percentages) of plans for the white students.
b) Find the conditional distributions (percentages) of plans for the minority students.
c) Create a graph comparing the plans of white and minority students.
d) Do you see any important differences in the post-graduation plans of white and minority students? Write a brief summary of what these data show, including comparisons of conditional distributions.

26. Politics revisited Look again at the table of political views for the Intro Stats students in Exercise 24.

a) Find the conditional distributions (percentages) of political views for the females.
b) Find the conditional distributions (percentages) of political views for the males.
c) Make a graphical display that compares the two distributions.
d) Do the variables *Politics* and *Sex* appear to be independent? Explain.

27. Magnet schools revisited The *Chance* magazine article described in Exercise 11 further examined the impact of an applicant's ethnicity on the likelihood of admission to the Houston Independent School District's magnet schools programs. Those data are summarized in the table below:

		Admission Decision			
		Accepted	Wait-listed	Turned away	Total
Ethnicity	Black/Hispanic	485	0	32	517
	Asian	110	49	133	292
	White	336	251	359	946
	Total	931	300	524	1755

a) What percent of all applicants were Asian?
b) What percent of the students accepted were Asian?
c) What percent of Asians were accepted?
d) What percent of all students were accepted?

28. More politics Look once more at the table summarizing the political views of Intro Stats students in Exercise 24.

a) Produce a graphical display comparing the conditional distributions of males and females among the three categories of politics.
b) Comment briefly on what you see from the display in a.

29. Back to school Examine the table about ethnicity and acceptance for the Houston Independent School District's magnet schools program, shown in Exercise 27. Does it appear that the admissions decisions are made independent of the applicant's ethnicity? Explain.

30. Cars A survey of autos parked in student and staff lots at a large university classified the brands by country of origin, as seen in the table.

		Driver	
		Student	Staff
Origin	American	107	105
	European	33	12
	Asian	55	47

a) What percent of all the cars surveyed were foreign?
b) What percent of the American cars were owned by students?
c) What percent of the students owned American cars?
d) What is the marginal distribution of origin?
e) What are the conditional distributions of origin by driver classification?
f) Do you think that the origin of the car is independent of the type of driver? Explain.

31. Weather forecasts Just how accurate are the weather forecasts we hear every day? The following table compares the daily forecast with a city's actual weather for a year:

		Actual Weather	
		Rain	No rain
Forecast	Rain	27	63
	No rain	7	268

a) On what percent of days did it actually rain?
b) On what percent of days was rain predicted?
c) What percent of the time was the forecast correct?
d) Do you see evidence of an association between the type of weather and the ability of forecasters to make an accurate prediction? Write a brief explanation, including an appropriate graph.

32. Twins In 2000, the *Journal of the American Medical Association (JAMA)* published a study that examined pregnancies that resulted in the birth of twins. Births were classified as preterm with intervention (induced labor or cesarean), preterm without procedures, or term/post-term. Researchers also classified the pregnancies by the level of prenatal medical care the mother received (inadequate, adequate, or intensive). The data, from the years 1995–1997, are summarized in the table on the next page. Figures are in thousands of births. (*JAMA* 284 [2000]:335–341)

	Twin Births 1995–1997 (In Thousands)			
Level of Prenatal Care		Preterm (induced or cesarean)	Preterm (without procedures)	Term or Post-Term	Total
	Intensive	18	15	28	61
	Adequate	46	43	65	154
	Inadequate	12	13	38	63
	Total	76	71	131	278

a) What percent of these mothers received inadequate medical care during their pregnancies?

b) What percent of all twin births were preterm?

c) Among the mothers who received inadequate medical care, what percent of the twin births were preterm?

d) Create an appropriate graph comparing the outcomes of these pregnancies by the level of medical care the mother received.

e) Write a few sentences describing the association between these two variables.

33. Blood pressure A company held a blood pressure screening clinic for its employees. The results are summarized in the table below by age group and blood pressure level:

		Age		
		Under 30	30–49	Over 50
Blood Pressure	Low	27	37	31
	Normal	48	91	93
	High	23	51	73

(handwritten: 95, 232, 147, 474; 98, 179, 197, 474)

a) Find the marginal distribution of blood pressure level.

b) Find the conditional distribution of blood pressure level within each age group.

c) Compare these distributions with a segmented bar graph.

d) Write a brief description of the association between age and blood pressure among these employees.

e) Does this prove that people's blood pressure increases as they age? Explain.

34. Obesity and exercise The Centers for Disease Control and Prevention (CDC) has estimated that 19.8% of Americans over 15 years old are obese. The CDC conducts a survey on obesity and various behaviors. Here is a table on self-reported exercise classified by body mass index (BMI):

		Body Mass Index		
		Normal (%)	Overweight (%)	Obese (%)
Physical Activity	Inactive	23.8	26.0	35.6
	Irregularly active	27.8	28.7	28.1
	Regular, not intense	31.6	31.1	27.2
	Regular, intense	16.8	14.2	9.1

a) Are these percentages column percentages, row percentages, or table percentages?

b) Use graphical displays to show different percentages of physical activities for the three BMI groups.

c) Do these data prove that lack of exercise causes obesity? Explain.

35. Anorexia Hearing anecdotal reports that some patients undergoing treatment for the eating disorder anorexia seemed to be responding positively to the antidepressant Prozac, medical researchers conducted an experiment to investigate. They found 93 women being treated for anorexia who volunteered to participate. For one year, 49 randomly selected patients were treated with Prozac and the other 44 were given an inert substance called a placebo. At the end of the year, patients were diagnosed as healthy or relapsed, as summarized in the table:

	Prozac	Placebo	Total
Healthy	35	32	67
Relapse	14	12	26
Total	49	44	93

Do these results provide evidence that Prozac might be helpful in treating anorexia? Explain.

36. Antidepressants and bone fractures For a period of five years, physicians at McGill University Health Center followed more than 5000 adults over the age of 50. The researchers were investigating whether people taking a certain class of antidepressants (SSRIs) might be at greater risk of bone fractures. Their observations are summarized in the table:

	Taking SSRI	No SSRI	Total
Experienced fractures	14	244	258
No fractures	123	4627	4750
Total	137	4871	5008

Do these results suggest there's an association between taking SSRI antidepressants and experiencing bone fractures? Explain.

CHAPTER 2 Displaying and Describing Categorical Data **41**

37. Drivers' licenses 2011 The following table shows the number of licensed U.S. drivers by age and by sex (www.dot.gov):

Age	Male Drivers (millions)	Female Drivers (millions)	Total
19 and Under	5.1	4.9	10.0
20–24	8.7	8.6	17.3
25–29	9.2	9.2	18.4
30–34	8.9	8.9	17.8
35–39	9.7	9.6	19.3
40–44	9.9	9.8	19.7
45–49	10.6	10.7	21.3
50–54	10.1	10.2	20.3
55–59	8.7	8.9	17.6
60–64	7.2	7.3	14.5
65–69	5.3	5.4	10.7
70–74	3.8	4.0	7.8
75–79	2.9	3.2	6.1
80–84	2.0	2.4	4.4
85 and Over	1.4	1.7	3.1
Total	103.5	104.8	208.3

a) What percent of total drivers are under 20?
b) What percent of total drivers are male?
c) Write a few sentences comparing the number of male and female licensed drivers in each age group.
d) Do a driver's age and sex appear to be independent? Explain?

38. Tattoos A study by the University of Texas Southwestern Medical Center examined 626 people to see if an increased risk of contracting hepatitis C was associated with having a tattoo. If the subject had a tattoo, researchers asked whether it had been done in a commercial tattoo parlor or elsewhere. Write a brief description of the association between tattooing and hepatitis C, including an appropriate graphical display.

	Tattoo Done in Commercial Parlor	Tattoo Done Elsewhere	No Tattoo
Has Hepatitis C	17	8	18
No Hepatitis C	35	53	495

39. Hospitals Most patients who undergo surgery make routine recoveries and are discharged as planned. Others suffer excessive bleeding, infection, or other postsurgical complications and have their discharges from the hospital delayed. Suppose your city has a large hospital and a small hospital, each performing major and minor surgeries. You collect data to see how many surgical patients have their discharges delayed by postsurgical

complications, and you find the results shown in the following table.

	Discharge Delayed	
	Large Hospital	Small Hospital
Major Surgery	120 of 800	10 of 50
Minor Surgery	10 of 200	20 of 250

a) Overall, for what percent of patients was discharge delayed?
b) Were the percentages different for major and minor surgery?
c) Overall, what were the discharge delay rates at each hospital?
d) What were the delay rates at each hospital for each kind of surgery?
e) The small hospital advertises that it has a lower rate of postsurgical complications. Do you agree?
f) Explain, in your own words, why this confusion occurs.

40. Delivery service A company must decide which of two delivery services it will contract with. During a recent trial period, the company shipped numerous packages with each service and kept track of how often deliveries did not arrive on time. Here are the results:

Delivery Service	Type of Service	Number of Deliveries	Number of Late Packages
Pack Rats	Regular	400	12
	Overnight	100	16
Boxes R Us	Regular	100	2
	Overnight	400	28

a) Compare the two services' overall percentage of late deliveries.
b) On the basis of the results in part a, the company has decided to hire Pack Rats. Do you agree that Pack Rats delivers on time more often? Explain.
c) The results here are an instance of what phenomenon?

41. Graduate admissions A 1975 article in the magazine *Science* examined the graduate admissions process at Berkeley for evidence of sex discrimination. The table below shows the number of applicants accepted to each of four graduate programs:

		Males Accepted (of applicants)	Females Accepted (of applicants)
Program	1	511 of 825	89 of 108
	2	352 of 560	17 of 25
	3	137 of 407	132 of 375
	4	22 of 373	24 of 341
	Total	1022 of 2165	262 of 849

a) What percent of total applicants were admitted?
b) Overall, was a higher percentage of males or females admitted?
c) Compare the percentage of males and females admitted in each program.
d) Which of the comparisons you made do you consider to be the most valid? Why?

42. Be a Simpson Can you design a Simpson's paradox? Two companies are vying for a city's "Best Local Employer" award, to be given to the company most committed to hiring local residents. Although both employers hired 300 new people in the past year, Company A brags that it deserves the award because 70% of its new jobs went to local residents, compared to only 60% for Company B. Company B concedes that those percentages are correct, but points out that most of its new jobs were full-time, while most of Company A's were part-time. Not only that, says Company B, but a higher percentage of its full-time jobs went to local residents than did Company A's, and the same was true for part-time jobs. Thus, Company B argues, it's a better local employer than Company A.

Show how it's possible for Company B to fill a higher percentage of both full-time and part-time jobs with local residents, even though Company A hired more local residents overall.

Just Checking ANSWERS

1. 50.0%

2. 44.4%

3. 25.0%

4. 15.6% Blue, 56.3% Brown, 28.1% Green/Hazel/Other

5. 18.8% Blue, 62.5% Brown, 18.8% Green/Hazel/Other

6. 40% of the blue-eyed students are female, while 50% of all students are female.

7. Since blue-eyed students appear less likely to be female, it seems that *Sex* and *Eye Color* may not be independent. (But the numbers are small.)

chapter 3 Displaying and Summarizing Quantitative Data

On March 11, 2011, the most powerful earthquake ever recorded in Japan created a wall of water that devastated the northeast coast of Japan and left 20,000 people dead or missing. Tsunamis like this are most often caused by earthquakes beneath the sea that shift the earth's crust, displacing a large mass of water. The 2011 tsunami in Japan was caused by a 9.0 magnitude earthquake and brought the Fukushima Daiichi nuclear power plant perilously close to a complete meltdown.

As disastrous as it was, the Japan tsunami was not nearly as deadly as the tsunami of December 26, 2004, off the west coast of Sumatra that killed an estimated 297,248 people, making it the most lethal tsunami on record. The earthquake that caused it was a magnitude 9.1 earthquake, more than 25% more powerful than the Japanese earthquake.[1] Were these earthquakes truly extraordinary, or did they just happen at unlucky times and places? The U.S. National Geophysical Data Center[2] has data on more than 5000 earthquakes dating back to 2150 B.C.E., and we have estimates of the magnitude of the underlying earthquake for the 1318 that were known to cause tsunamis. What can we learn from these data?

Histograms

Let's start with a picture. For categorical variables, it is easy to draw the distribution because each category is a natural "pile." But for quantitative variables, there's no obvious way to choose piles. So, usually, we slice up all the possible values into equal-width bins. We then count the number of cases that fall into each bin. The bins, together with these counts, give the **distribution** of the quantitative variable and provide the building blocks

---

[1]Earthquake magnitudes are measured on a log scale.
[2]www.ngdc.noaa.gov.

for the histogram. By representing the counts as heights of bars and plotting them against the bin values, the **histogram** displays the distribution at a glance.

For example, here are the *Magnitudes* (on the Richter scale) of the 1318 earthquakes in the NGDC data:

Who	1318 earthquakes known to have caused tsunamis for which we have data or good estimates
What	Magnitude (Richter scale), depth (m), date, location, and other variables
When	From 2150 B.C.E. to the present
Where	All over the earth

Figure 3.1

A histogram of earthquake magnitudes shows the number of earthquakes. with magnitudes (in Richter scale units) in each bin.

Like a bar chart, a histogram plots the bin counts as the heights of bars. In this histogram of earthquake magnitudes, each bin has a width of 0.2, so, for example, the height of the tallest bar says that there were about 200 earthquakes with magnitudes between 7.0 and 7.2. Does the distribution look as you expected? It is often a good idea to *imagine* what the distribution might look like before you make the display. That way you'll be less likely to be fooled by errors.

From the histogram, we can see that these earthquakes typically have magnitudes around 7. Most are between 5.5 and 8.5, and some are as small as 4 and as big as 9. Now we can answer the question about the Japan and Sumatra tsunamis. With values of 9.1 and 9.0 it's clear that these earthquakes were extraordinarily powerful—among the largest on record.

The bar charts of categorical variables we saw in Chapter 2 had spaces between the bars to separate the counts of different categories. But in a histogram, the bins slice up *all the values* of the quantitative variable, so any spaces in a histogram are actual **gaps** in the data, indicating a region where there are no values.

Sometimes it is useful to make a **relative frequency histogram,** replacing the counts on the vertical axis with the *percentage* of the total number of cases falling in each bin. Of course, the shape of the histogram is exactly the same; only the vertical scale is different.

Designing Your Histogram

Different features of the distribution may appear more obvious at different bin width choices. When you use technology, it's usually easy to vary the bin width interactively so you can make sure that a feature you think you see isn't just a consequence of a certain bin width choice.

Why so Many 7's?

One surprising feature of the earthquake magnitudes is the spike around magnitude 7.0. Only one other bin holds even half that many earthquakes. These values include historical data for which the magnitudes were estimated by experts and not measured by modern seismographs. Perhaps the experts thought 7 was a typical and reasonable value for a tsunami-causing earthquake when they lacked detailed information. That would explain the overabundance of magnitudes right at 7.0 rather than spread out near that value.

Figure 3.2

A relative frequency histogram looks just like a frequency histogram except for the labels on the y-axis, which now show the percentage of earthquakes in each bin.

TI Tips MAKING A HISTOGRAM

Your calculator can create histograms. First you need some data. For an agility test, fourth-grade children jump from side to side across a set of parallel lines, counting the number of lines they clear in 30 seconds. Here are their scores:

22, 17, 18, 29, 22, 22, 23, 24, 23, 17, 21, 25, 20
12, 19, 28, 24, 22, 21, 25, 26, 25, 16, 27, 22

Enter these data into L1.

Now set up the calculator's plot:

- Go to 2nd STATPLOT, choose Plot1, then ENTER.

- In the Plot1 screen choose On, select the little histogram icon, then specify Xlist:L1 and Freq:1.
- Be sure to turn off any other graphs the calculator may be set up for. Just hit the Y= button, and deactivate any functions seen there.

All set? To create your preliminary plot go to ZOOM, select 9:ZoomStat, and then ENTER.

You now see the calculator's initial attempt to create a histogram of these data. Not bad. We can see that the distribution is roughly symmetric. But it's hard to tell exactly what this histogram shows, right? Let's fix it up a bit.

- Under WINDOW, let's reset the bins to convenient, sensible values. Try Xmin = 12, Xmax = 30, and Xscl = 2. That specifies the interval of values along the x-axis and makes each bar span two lines.
- Hit GRAPH (*not* ZoomStat—this time we want control of the scale!).

There. We still see rough symmetry, but also see that one of the scores was much lower than the others. Note that you can now find out exactly what the bars indicate by activating TRACE and then moving across the histogram using the arrow keys. For each bar the calculator will indicate the interval of values and the number of data values in that bin. We see that 3 kids had agility scores of 20 or 21.

Play around with the WINDOW settings. A different Ymax will make the bars appear shorter or taller. What happens if you set the bar width (Xscl) smaller? Or larger? You don't want to lump lots of values into just a few bins or make so many bins that the overall shape of the histogram is not clear. Choosing the best bar width takes practice.

Finally, suppose the data are given as a frequency table. Consider a set of test scores, with two grades in the 60s, four in the 70s, seven in the 80s, five in the 90s, and one 100. Enter the group cutoffs 60, 70, 80, 90, 100 in L2 and the corresponding frequencies 2, 4, 7, 5, 1 in L3. When you set up the histogram STATPLOT, specify Xlist:L2 and Freq:L3. Can you specify the WINDOW settings to make this histogram look the way you want it? (By the way, if you get a DIM MISMATCH error, it means you can't count. Look at L2 and L3; you'll see the two lists don't have the same number of entries. Fix the problem by correcting the data you entered.)

Stem-and-Leaf Displays

Histograms provide an easy-to-understand summary of the distribution of a quantitative variable, but they don't show the data values themselves. Here's a histogram of the pulse rates of 24 women, taken by a researcher at a health clinic:

Figure 3.3

The pulse rates of 24 women at a health clinic.

The story seems pretty clear. We can see the entire span of the data and can easily see what a typical pulse rate might be. But is that all there is to these data?

A **stem-and-leaf display** is like a histogram, but it shows the individual values. It's also easier to make by hand. Here's a stem-and-leaf display of the same data:

```
8 | 8
8 | 000044
7 | 6666
7 | 2222
6 | 8888
6 | 0444
5 | 6
```
Pulse Rate
(8 | 8 means 88 beats/min)

> ### What's in a Name?
> The stem-and-leaf display was devised by John W. Tukey, one of the greatest statisticians of the 20th century. It is called a "Stemplot" in some texts and computer programs, but we prefer Tukey's original name for it.

A S *Activity*: **Stem-and-Leaf Displays.** As you might expect of something called "stem-and-leaf," these displays grow as you consider each data value.

Turn the stem-and-leaf on its side (or turn your head to the right) and squint at it. It should look roughly like the histogram of the same data. Does it? Well, it's backwards because now the higher values are on the left, but other than that, it has the same shape.[3]

What does the line at the top of the display that says 8 | 8 mean? It stands for a pulse of 88 beats per minute (bpm). We've taken the tens place of the number and made that the "stem." Then we sliced off the ones place and made it a "leaf." The next line down is 8 | 000044. That shows that there were four pulse rates of 80 and two of 84 bpm.

Stem-and-leaf displays are especially useful when you make them by hand for batches of fewer than a few hundred data values. They are a quick way to display—and even to record—numbers. Because the leaves show the individual values, we can sometimes see even more in the data than the distribution's shape. Take another look

---

[3]You could make the stem-and-leaf with the higher values on the bottom. Usually, though, higher on the top makes sense.

at all the leaves of the pulse data. See anything unusual? At a glance you can see that they are all even. With a bit more thought you can see that they are all multiples of 4—something you couldn't possibly see from a histogram. How do you think the nurse took these pulses? Counting beats for a full minute or counting for only 15 seconds and multiplying by 4?

How Do Stem-and-Leaf Displays Work?

Stem-and-leaf displays work like histograms, but they show more information. They use part of the number itself (called the stem) to name the bins. To make the "bars," they use the next digit of the number. For example, if we had a test score of 83, we could write it 8│3, where 8 serves as the stem and 3 as the leaf. Then, to display the scores 83, 76, and 88 together, we would write

```
8 | 38
7 | 6
```

For the pulse data, we have

```
8 | 0000448
7 | 22226666
6 | 04448888
5 | 6
```
Pulse Rate
(5│6 means 56 beats/min)

This display is OK, but a little crowded. A histogram might split each line into two bars. With a stem-and-leaf, we can do the same by putting the leaves 0–4 on one line and 5–9 on another, as we saw above:

```
8 | 8
8 | 000044
7 | 6666
7 | 2222
6 | 8888
6 | 0444
5 | 6
```
Pulse Rate
(8│8 means 88 beats/min)

For numbers with three or more digits, you'll often decide to truncate (or perhaps round) the number to two places, using the first digit as the stem and the second as the leaf. So, if you had 432, 540, 571, and 638, you might display them as shown below with an indication that 6│3 means 630–639. (We truncate; it's easier.)

```
6 | 3
5 | 47
4 | 3
```

When you make a stem-and-leaf by hand, make sure to give each digit the same width, in order to preserve the area principle. (That can lead to some fat 1's and thin 8's—but it makes the display honest.)

Dotplots

A **dotplot** is a simple display. It just places a dot along an axis for each case in the data. It's like a stem-and-leaf display, but with dots instead of digits for all the leaves. Dotplots are a great way to display a small data set (especially if you forget how to write the digits from 0 to 9). Here's a dotplot of the time (in seconds) that the winning horse took to win the Kentucky Derby in each race between the first Derby in 1875 and the 2011 Derby.

Dotplots show basic facts about the distribution. We can find the slowest and quickest races by finding times for the topmost and bottommost dots. It's also clear that there are two clusters of points, one just below 160 seconds and the other at about 122 seconds. Something strange happened to the Derby times. Once we know to look for it, we can find out that in 1896 the distance of the Derby race was changed from 1.5 miles to the current 1.25 miles. That explains the two clusters of winning times.

Some dotplots stretch out horizontally, with the counts on the vertical axis, like a histogram. Others, such as the one shown here, run vertically, like a stem-and-leaf display. Some dotplots place points next to each other when they would otherwise overlap. Others just place them on top of one another. Newspapers sometimes offer dotplots with the dots made up of little pictures. (There should be a key to tell you how many observations each picture represents.)

Figure 3.4

A dotplot of Kentucky Derby winning times plots each race as its own dot, showing the bimodal distribution.

Think Before You Draw, Again

Suddenly, we face a lot more options when it's time to invoke our first rule of data analysis and make a picture. You'll need to *Think* carefully to decide which type of graph to make. In the previous chapter you learned to check the Categorical Data Condition before making a pie chart or a bar chart. Now, before making a stem-and-leaf display, a histogram, or a dotplot, you need to check the

> **Quantitative Data Condition:** The data are values of a quantitative variable whose units are known.

Although a bar chart and a histogram may look somewhat similar, they're not the same display. You can't display categorical data in a histogram or quantitative data in a bar chart. Always check the condition that confirms what type of data you have before proceeding with your display.

The Shape of a Distribution

Step back from a histogram, dotplot, or stem-and-leaf display. What can you say about the distribution? When you describe a distribution, you should always tell about three things: its **shape**, **center**, and **spread**. Let's start with shape:

1. *Does the histogram have a single, central hump or several separated humps?* These humps are called **modes**.[4] The earthquake magnitudes have a single mode at just

---

[4]Well, technically, it's the value on the horizontal axis of the histogram that is the mode, but anyone asked to point to the mode would point to the hump.

What is the Mode?

The *mode* is sometimes defined as the single value that appears most often. That definition is fine for categorical variables because all we need to do is count the number of cases for each category. For quantitative variables, the mode is more ambiguous. What is the mode of the Kentucky Derby times? Well, seven races were timed at 122.2 seconds—more than any other race time. Should that be the mode? Probably not. For quantitative data, it makes more sense to use the term "mode" in the more general sense of the peak of the histogram rather than as a single summary value. In this sense, the important feature of the Kentucky Derby races is that there are two distinct modes, representing the two different versions of the race and warning us to consider those two versions separately.

about 7. A histogram with one peak, such as the earthquake magnitudes, is dubbed **unimodal**; histograms with two peaks are **bimodal**, and those with three or more are called **multimodal**.[5] For example, here's a bimodal histogram.

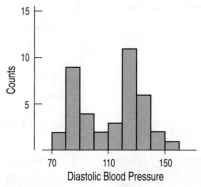

Figure 3.5

A bimodal histogram has two apparent peaks.

A histogram that doesn't appear to have any obvious mode and in which all the bars are approximately the same height is called **uniform.**

Figure 3.6

In a uniform histogram, the bars are all about the same height. The histogram doesn't appear to have a mode.

Pie à la Mode?

You've heard of pie à la mode. Is there a connection between pie and the mode of a distribution? Actually, there is! The mode of a distribution is a *popular* value near which a lot of the data values gather. And "à la mode" means "in style"—*not* "with ice cream." That just happened to be a *popular* way to have pie in Paris around 1900.

2. *Is the histogram* **symmetric?** Can you fold it along a vertical line through the middle and have the edges match pretty closely, or are more of the values on one side?

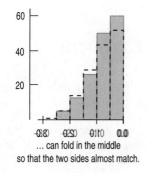

Figure 3.7

The (usually) thinner ends of a distribution are called the **tails.** If one tail stretches out farther than the other, the histogram is said to be **skewed** to the side of the longer tail.

---

[5]Apparently, statisticians don't like to count past two.

A S *Activity:* **Attributes of Distribution Shape.** This activity and the others on this page show off aspects of distribution shape through animation and example, then let you make and interpret histograms with your statistics package.

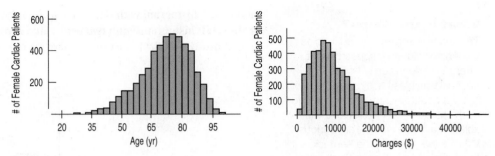

Figure 3.8

Two skewed histograms showing data on two variables for all female heart attack patients in New York state in one year. The blue one (age in years) is skewed to the left. The purple one (charges in $) is skewed to the right.

3. *Do any unusual features stick out?* Often such features tell us something interesting or exciting about the data. You should always mention any stragglers, or **outliers,** that stand off away from the body of the distribution. If you're collecting data on nose lengths and Pinocchio is in the group, you'd probably notice him, and you'd certainly want to mention it.

   Outliers can affect almost every method we discuss in this course. So we'll always be on the lookout for them. An outlier can be the most informative part of your data. Or it might just be an error. But don't throw it away without comment. Treat it specially and discuss it when you tell about your data. Or find the error and fix it if you can. Be sure to look for outliers. Always.

   Soon you'll learn a handy rule of thumb for deciding when a data value might be considered an outlier.

Figure 3.9

A histogram with outliers. There are three cities in the leftmost bar.

For Example DESCRIBING HISTOGRAMS

A credit card company wants to see how much customers in a particular segment of their market use their credit card. They have provided you with data[6] on the amount spent by 500 selected customers during a 3-month period and have asked you to summarize the expenditures. Of course, you begin by making a histogram.

QUESTION: Describe the shape of this distribution.

ANSWER: The distribution of expenditures is unimodal and skewed to the high end. There is an extraordinarily large value at about $7000, and some of the expenditures are negative.

---

[6]These data are real, but cannot be further identified for obvious privacy reasons.

Are there any gaps in the distribution? The Kentucky Derby data that we saw in the dotplot on page 48 has a large gap between two groups of times, one near 120 seconds and one near 160. Gaps help us see multiple modes and encourage us to notice when the data may come from different sources or contain more than one group.

Toto, I've a Feeling We're Not in Math Class Anymore . . .

When Dorothy and her dog Toto land in Oz, everything is more vivid and colorful, but also more dangerous and exciting. Dorothy has new choices to make. She can't always rely on the old definitions, and the yellow brick road has many branches. You may be coming to a similar realization about Statistics.

When we summarize data, our goal is usually more than just developing a detailed knowledge of the data we have at hand. We want to know what the data say about the world, so we'd like to know whether the patterns we see in histograms and summary statistics generalize to other individuals and situations. Scientists generally don't care about the particular guinea pigs in their experiment, but rather about what how they react to different treatments says about how other animals (and, perhaps, humans) might respond.

Because we want to see broad patterns, rather than focus on the details of the data set we're looking at, many of the most important concepts in Statistics are not precisely defined. Whether a histogram is symmetric or skewed, whether it has one or more modes, whether a case is far enough from the rest of the data to be considered an outlier—these are all somewhat vague concepts. They all require judgment.

You may be used to finding a single correct and precise answer, but in Statistics, there may be more than one interpretation. That may make you a little uncomfortable at first, but soon you'll see that leaving room for judgment brings you both power and responsibility. It means that your own knowledge about the world and your judgment matter. You'll use them, along with the statistical evidence, to draw conclusions and make decisions about the world.

 ## Just Checking

It's often a good idea to think about what the distribution of a data set might look like before we collect the data. What do you think the distribution of each of the following data sets will look like? Be sure to discuss its shape. Where do you think the center might be? How spread out do you think the values will be?

1. Number of miles run by Saturday morning joggers at a park.

2. Hours spent by U.S. adults watching football on Thanksgiving Day.

3. Amount of winnings of all people playing a particular state's lottery last week.

4. Ages of the faculty members at your school.

5. Last digit of phone numbers on your campus.

The Center of the Distribution: The Median

Let's return to the tsunami earthquakes. But this time, let's look at just 25 years of data: 207 earthquakes that occurred from 1987 through 2011. (See Figure 3.10). These should be more accurately measured than prehistoric quakes because seismographs were in wide use. Try to put your finger on the histogram at the value you think is typical. (Read the value from the horizontal axis and remember it.)

When we think of a typical value, we usually look for the center of the distribution. Where do you think the center of this distribution is? For a unimodal, symmetric distribution such as these earthquake data, it's easy. We'd all agree on the center of symmetry, where we would fold the histogram to match the two sides. But when the distribution is skewed or possibly multimodal, it's not immediately clear what we even mean by the center.

One natural choice for typical value is the value that is literally in the middle, with half the values below it and half above it.

Figure 3.10

Tsunami-causing earthquakes (1987–2011) The median splits the histogram into two halves of equal area.

Histograms follow the area principle, and each half of the data has about 88 earthquakes, so each colored region has the same area in the display. The middle value that divides the histogram into two equal areas is called the **median.**

The median has the same units as the data. Be sure to include the units whenever you discuss the median.

For the recent tsunamis, there are 207 earthquakes, so the median is found at the $(207 + 1)/2 = 104$th place in the sorted data. The median earthquake magnitude is 7.2.

How Do Medians Work?

Finding the median of a batch of *n* numbers is easy as long as you remember to order the values first. If *n* is odd, the median is the middle value.

Counting in from the ends, we find this value in the $\frac{n + 1}{2}$ position.

When *n* is even, there are two middle values. Then, the median is the average of the two values in positions $\frac{n}{2}$ and $\frac{n}{2} + 1$.

Here are two examples:
Suppose the batch has these values: 14.1, 3.2, 25.3, 2.8, −17.5, 13.9, 45.8.
First we order the values: −17.5, 2.8, 3.2, 13.9, 14.1, 25.3, 45.8.
Since there are 7 values, the median is the $(7 + 1)/2 = $ 4th value, counting from the top or bottom: 13.9. Notice that 3 values are lower, 3 higher.
Suppose we had the same batch with another value at 35.7. Then the ordered values are −17.5, 2.8, 3.2, 13.9, 14.1, 25.3, 35.7, 45.8.
The median is the average of the 8/2 or 4th, and the $(8/2) + 1$, or 5th, values. So the median is $(13.9 + 14.1)/2 = 14.0$. Four data values are lower, and four higher.

The median is one way to find the center of the data. But there are many others. We'll look at an even more important measure later in this chapter.

Knowing the median, we could say that a typical tsunami-causing earthquake, worldwide, was about 7.2 on the Richter scale. How much does that really say? How well does the median describe the data? After all, not every earthquake has a Richter scale value of 7.2. Whenever we find the center of data, the next step is always to ask how well it actually summarizes the data.

Spread: Home on the Range

What We Don't Know

Statistics pays close attention to what we *don't* know as well as what we do know. Understanding how spread out the data are is a first step in understanding what a summary *cannot* tell us about the data. It's the beginning of telling us what we don't know.

If every earthquake that caused a tsunami registered 7.2 on the Richter scale, then knowing the median would tell us everything about the distribution of earthquake magnitudes. The more the data vary, however, the less the median alone can tell us. So we need to measure how much the data values vary around the center. In other words, how spread out are they? When we describe a distribution numerically, we always report a measure of its spread along with its center. After all, Statistics is about variation—remember?

How should we measure the spread? We could simply look at the extent of the data. How far apart are the two extremes? The **range** of the data is defined as the *difference* between the maximum and minimum values:

$$Range = max - min.$$

Notice that the range is a *single number, not* an interval of values, as you might think from everyday speech. The maximum magnitude of these earthquakes is 9.1 and the minimum is 3.7, so the *range* is $9.1 - 3.7 = 5.4$.

The range has the disadvantage that a single extreme value can make it very large, giving a value that doesn't really represent the data overall.

Spread: The Interquartile Range

Percentiles

The lower and upper quartiles are also known as the 25th and 75th **percentiles** of the data, respectively, since the lower quartile falls above 25% of the data and the upper quartile falls above 75% of the data. If we count this way, the median is the 50th percentile. We could, of course, define and calculate any percentile that we want. For example, the 10th percentile would be the number that falls above the lowest 10% of the data values.

A better way to describe the spread of a variable might be to ignore the extremes and concentrate on the middle of the data. We could, for example, find the range of just the middle half of the data. What do we mean by the middle half? Divide the data in half at the median. Now divide both halves in half again, cutting the data into four quarters. We call these new dividing points **quartiles**. One quarter of the data lies below the **lower quartile,** and one quarter of the data lies above the **upper quartile,** so half the data lies between them. The quartiles border the middle half of the data.

How Do Quartiles Work?

A simple way to find the quartiles is to start by splitting the batch into two halves at the median. (When *n* is odd, some statisticians include the median in both halves; others omit it.) The lower quartile is the median of the lower half, and the upper quartile is the median of the upper half.

Here are our two examples again.

The ordered values of the first batch were −17.5, 2.8, 3.2, 13.9, 14.1, 25.3, and 45.8, with a median of 13.9. Excluding the median, the two halves of the list are −17.5, 2.8, 3.2 and 14.1, 25.3, 45.8.

Each half has 3 values, so the median of each is the middle one. The lower quartile is 2.8, and the upper quartile is 25.3.

The second batch of data had the ordered values −17.5, 2.8, 3.2, 13.9, 14.1, 25.3, 35.7, and 45.8.

Here *n* is even, so the two halves of 4 values are −17.5, 2.8, 3.2, 13.9 and 14.1, 25.3, 35.7, 45.8.

Now the lower quartile is $(2.8 + 3.2)/2 = 3.0$, and the upper quartile is $(25.3 + 35.7)/2 = 30.5$.

The difference between the quartiles tells us how much territory the middle half of the data covers and is called the **interquartile range.** It's commonly abbreviated IQR (and pronounced "eye-cue-are," not "ikker"):

$$IQR = upper\ quartile - lower\ quartile.$$

For the earthquakes, there are 103 values below the median and 103 values above the median. The midpoint of the lower half is the 52nd value in the ordered data; that turns out to be 6.7. In the upper half the third quartile is the 156th value, a magnitude of 7.6 as the third quartile. The *difference* between the quartiles gives the IQR:

$$IQR = 7.6 - 6.7 = 0.9.$$

Now we know that the middle half of the earthquake magnitudes extends across a (interquartile) range of 0.9 Richter scale units. This seems like a reasonable summary of the spread of the distribution, as we can see from this histogram:

Figure 3.11

The quartiles bound the middle 50% of the values of the distribution. This gives a visual indication of the spread of the data. Here we see that the IQR is 0.9 Richter scale units.

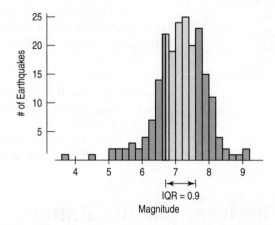

The IQR is almost always a reasonable summary of the spread of a distribution. Even if the distribution itself is skewed or has some outliers, the IQR should provide useful information. The one exception is when the data are strongly bimodal. For example, remember the dotplot of winning times in the Kentucky Derby (page 48)? Because the race distance changed, we really have data on two different races, and they shouldn't be summarized together.

So, What Is a Quartile Anyway?

Finding the quartiles sounds easy, but surprisingly, the quartiles are not well-defined. It's not always clear how to find a value such that exactly one quarter of the data lies above or below that value. We offered a simple rule for Finding Quartiles in the box on page 53: Find the median of each half of the data split by the median. When *n* is odd, we (and your TI calculator) omit the median from each of the halves. Some other texts include the median in both halves before finding the quartiles. Both methods are commonly used. If you are willing to do a bit more calculating, there are several other methods that locate a quartile somewhere between adjacent data values. We know of at least six different rules for finding quartiles. Remarkably, each one is in use in some software package or calculator.

So don't worry too much about getting the "exact" value for a quartile. All of the methods agree pretty closely when the data set is large. When the data set is small, different rules will disagree more, but in that case there's little need to summarize the data anyway.

Remember, Statistics is about understanding the world, not about calculating the right number. The "answer" to a statistical question is a sentence about the issue raised in the question.

5-Number Summary

■ **NOTATION ALERT**

We always use Q1 to label the lower (25%) quartile and Q3 to label the upper (75%) quartile. We skip the number 2 because the median would, by this system, naturally be labeled Q2—but we don't usually call it that.

The **5-number summary** of a distribution reports its median, quartiles, and extremes (maximum and minimum). The 5-number summary for the recent tsunami earthquake *Magnitudes* looks like this:

Max	9.1
Q3	7.6
Median	7.2
Q1	6.7
Min	3.7

It's good idea to report the number of data values and the identity of the cases (the *Who*). Here there are 207 earthquakes.

The 5-number summary provides a good overview of the distribution of magnitudes of these tsunami-causing earthquakes. For a start, we can see that the median magnitude is 7.2. Because the IQR is only 7.6 − 6.7 = 0.9, we see that many quakes are close to the median magnitude. Indeed, the quartiles show us that the middle half of these earthquakes had magnitudes between 6.7 and 7.6. One quarter of the earthquakes had magnitudes above 7.6, although one tsunami was caused by a quake measuring only 3.7 on the Richter scale.

Boxplots

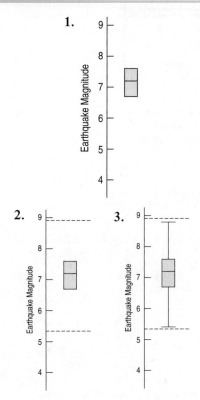

Once we have a 5-number summary of a (quantitative) variable, we can display that information in a **boxplot**. To make a boxplot of the earthquake magnitudes, follow these steps:

1. Draw a single vertical axis spanning the extent of the data.[7] Draw short horizontal lines at the lower and upper quartiles and at the median. Then connect them with vertical lines to form a box. The box can have any width that looks OK.[8]
2. To construct the boxplot, erect "fences" around the main part of the data. Place the upper fence 1.5 IQRs above the upper quartile and the lower fence 1.5 IQRs below the lower quartile. For the earthquake magnitude data, we compute

$$Upper\ fence = Q3 + 1.5\ IQR = 7.6 + 1.5 \times 0.9 = 8.95$$

and

$$Lower\ fence = Q1 + 1.5\ IQR = 6.7 - 1.5 \times 0.9 = 5.35.$$

The fences are just for construction and are not part of the display. We show them here with dotted lines for illustration. You should never include them in your boxplot.

3. We use the fences to grow "whiskers." Draw lines from the ends of the box up and down to *the most extreme data values found within the fences*. If a data value falls outside one of the fences, we do *not* connect it with a whisker.

---

[7]The axis could also run horizontally.

[8]Some computer programs draw wider boxes for larger data sets. That can be useful when comparing groups.

4.

 Boxplots.
Watch a boxplot under construction.

TI-nspire

Boxplots and dotplots. Drag data points around to explore what a boxplot shows (and doesn't).

Why 1.5 IQRs?
One of the authors asked the prominent statistician, John W. Tukey, the originator of the box-plot, why the outlier nomination rule cut at 1.5 IQRs beyond each quartile. He answered that the reason was that 1 IQR would be too small and 2 IQRs would be too large. That works for us.

4. Finally, we add the outliers by displaying any data values beyond the fences with special symbols. (We often use a different symbol for "**far outliers**"—data values farther than 3 IQRs from the quartiles.)

A boxplot highlights several features of the distribution. The central box shows the middle half of the data, between the quartiles. The height of the box is equal to the IQR. If the median is roughly centered between the quartiles, then the middle half of the data is roughly symmetric. If the median is not centered, the distribution is skewed. The whiskers show skewness as well if they are not roughly the same length. Any outliers are displayed individually, both to keep them out of the way for judging skewness and to encourage you to give them special attention. They may be mistakes, or they may be the most interesting cases in your data.

For the recent tsunami earthquake data, the central box contains all the earthquakes whose magnitudes are between 6.7 and 7.6 on the Richter scale. From the shape of the box, it looks like the central part of the distribution of earthquakes is roughly symmetric, but at the low end the longer whisker and outliers indicate that the distribution stretches out slightly to the left. We also see a few very small magnitude earthquakes and the two large earthquakes we've discussed. Boxplots are particularly good at pointing out outliers.

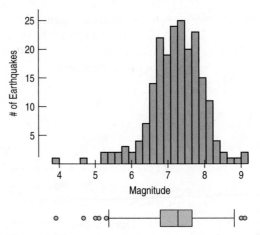

Figure 3.12
By turning the boxplot and putting it on the same scale as the histogram, we can compare both displays of the earthquake magnitudes and see how each represents the distribution.

Step-by-Step Example SHAPE, CENTER, AND SPREAD: FLIGHT CANCELLATIONS

The U.S. Bureau of Transportation Statistics (www.bts.gov) reports data on airline flights. Let's look at data giving the percentage of flights cancelled each month between 1995 and 2011.

Question: How often are flights cancelled?

Who	Months
What	Percentage of flights cancelled at U.S. airports
When	1995–2011
Where	United States

THINK ➡ **Variable** Identify the *variable*, and decide how you wish to display it.

To identify a variable, report the W's.

Select an appropriate display based on the nature of the data and what you want to know.

I want to learn about the monthly percentage of flight cancellations at U.S airports.

I have data from the U.S. Bureau of Transportation Statistics giving the percentage of flights cancelled at U.S. airports each month between 1995 and 2011.

✓ **Quantitative Data Condition:** Percentages are quantitative. A histogram and numerical summaries would be appropriate.

SHOW ➡ **Mechanics** We usually make histograms with a computer or graphing calculator.

The histogram shows a distribution skewed to the high end and one extreme outlier, a month in which more than 20% of flights were cancelled.

REALITY CHECK ➡ It's always a good idea to think about what you expect to see so that you can check whether the histogram looks like what you expected.

With 201 cases, we probably have more data than you'd choose to work with by hand. The results given here are from technology.

In most months, fewer than 5% of flights are cancelled and usually only about 2% or 3%. That seems reasonable.

Count	201
Max	20.24
Q3	2.54
Median	1.730
Q1	1.312
Min	0.540
IQR	1.227

(continued)

Tell ➡ **Interpretation**   Describe the shape, center, and spread of the distribution. Report on the symmetry, number of modes, and any gaps or outliers. You should also mention any concerns you may have about the data.

The distribution of cancellations is skewed to the right, and this makes sense: The values can't fall below 0%, but can increase almost arbitrarily due to bad weather or other events.

The median is 1.73% and the IQR is 1.23%. The low IQR indicates that in most months the cancellation rate is close to the median. In fact, it's between 1.31% and 2.54% in the middle 50% of all months, and in only 1/4 of the months were more than 2.54% of flights cancelled.

There is one extraordinary value: 20.2%. Looking it up, I find that the extraordinary month was September 2001. The attacks of September 11 shut down air travel for several days, accounting for this outlier.

TI Tips   MAKING A BOXPLOT (WITH A BONUS)

Last time we made a histogram to display the 4th-grade agility test data. It's just as easy to create a boxplot. You should still have those data in L1. (If not, enter them again—see page 45.)

Ready? First set up the plot:

- Go to 2nd STATPLOT, choose Plot 1, and hit ENTER.
- Turn the plot On;
- Select the first boxplot icon (you always want your plot to indicate outliers);
- Specify Xlist:L1 and Freq: 1;
- And select the Mark you want the calculator to use for displaying any outliers.

All set. ZoomStat will display the boxplot.

You can now TRACE the plot to see the statistics in the 5-number summary. Try it!

Summarizing Symmetric Distributions: The Mean

■ **NOTATION ALERT**

In Algebra you used letters to represent values in a problem, but it didn't matter what letter you picked. You could call the width of a rectangle *X* or you could call it *w* (or *Fred*, for that matter). But in Statistics, the notation is part of the vocabulary. For example, in Statistics *n* is always the number of data values. Always.

We have already begun to point out such special notation conventions: *n*, Q1, and Q3. Think of them as part of the terminology you need to learn in this course.

Here's another one: Whenever we put a bar over a symbol, it means "find the mean."

Medians do a good job of summarizing the center of a distribution, even when the shape is skewed or when there is an outlier, as with the flight cancellations. But when we have symmetric data, there's another alternative. You probably already know how to average values. In fact, to find the median when *n* is even, we said you should average the two middle values, and you didn't even flinch.

The earthquake magnitudes are pretty close to symmetric, so we can also summarize their center with a mean. The mean tsunami earthquake magnitude is 6.96—about what we might expect from the histogram. You already know how to average values, but this is a good place to introduce notation that we'll use throughout the book. We use the Greek capital letter sigma, Σ, to mean "sum" (sigma is "S" in Greek), and we'll write:

$$\bar{y} = \frac{Total}{n} = \frac{\sum y}{n}$$

The formula says to add up all the values of the variable and divide that sum by the number of data values, n—just as you've always done.[9]

Once we've averaged the data, you'd expect the result to be called the *average*, but that would be too easy. Informally, we speak of the "average person" but we don't add up people and divide by the number of people. A median is also a kind of average. To make this distinction, the value we calculated is called the mean, $\bar{y}$, and pronounced "y-bar."

The **mean** feels like the center because it is the point where the histogram balances:

Figure 3.13

The mean is located at the balancing point of the histogram.

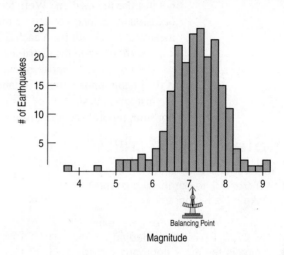

Average, or Mean?
In everyday language, sometimes "average" *does* mean what we want it to mean. We don't talk about your grade point mean or a baseball player's batting mean or the Dow Jones Industrial mean. So we'll continue to say "average" when that seems most natural. When we do, though, you may assume that what we mean is the mean.

Mean or Median?

Using the center of balance makes sense when the data are symmetric. But data are not always this well behaved. If the distribution is skewed or has outliers, the center is not so well defined and the mean may not be what we want. For example, the mean of the flight cancellations doesn't give a very good idea of the typical percentage of cancellations.

Figure 3.14

The median splits the area of the histogram in half at 1.755%. Because the distribution is skewed to the right, the mean (2.28%) is higher than the median. The points at the right have pulled the mean toward them away from the median.

TI-*nspire*

Mean, median, and outliers. Drag data points around to explore how outliers affect the mean and median.

A **S** *Activity*: **The Center of a Distribution.** Compare measures of center by dragging points up and down and seeing the consequences. Another activity shows how to find summaries with your statistics package.

The mean is 2.13%, but two-thirds of months had cancellation rates below that, so the mean doesn't feel like a good overall summary. Why is the balancing point so high? The large outlying value pulls it to the right. For data like these, the median is a better summary of the center.

---

[9]You may also see the variable called x and the equation written $\bar{x} = \dfrac{Total}{n} = \dfrac{\sum x}{n}$. Don't let that throw you.

You are free to name the variable anything you want, but we'll generally use y for variables like this that we want to summarize, model, or predict. (Later we'll talk about variables that are used to explain, model, or predict y. We'll call them x.)

Because the median considers only the order of the values, it is **resistant** to values that are extraordinarily large or small; it simply notes that they are one of the "big ones" or the "small ones" and ignores their distance from the center.

For the 1318 tsunami earthquake magnitudes, it doesn't seem to make much difference—the mean is 7.05; the median is 7.0. When the data are symmetric, the mean and median will be close, but when the data are skewed, the median is likely to be a better choice. So, why not just use the median? Well, for one, the median can go overboard. It's not just resistant to occasional outliers, but can be unaffected by changes in many data values. By contrast, the mean includes input from each data value and gives each one equal weight. It's also easier to work with, so when the distribution is unimodal and symmetric, we'll use the mean.

Of course, to choose between mean and median, we'll start by looking at the data. If the histogram is symmetric and there are no outliers, we'll prefer the mean. However, if the histogram is skewed or has outliers, we're usually better off with the median. If you're not sure, report both and discuss why they might differ.

For Example DESCRIBING CENTER

RECAP: You want to summarize the expenditures of 500 credit card company customers, and have looked at a histogram.

QUESTION: You have found the mean expenditure to be $478.19 and the median to be $216.28. Which is the more appropriate measure of center, and why?

ANSWER: Because the distribution of expenditures is skewed, the median is the more appropriate measure of center. Unlike the mean, it's not affected by the large outlying value or by the skewness. Half of these credit card customers had average monthly expenditures less than $216.28 and half more.

When To Expect Skewness

Even without making a histogram, we can expect some variables to be skewed. When values of a quantitative variable are bounded on one side but not the other, the distribution may be skewed. For example, incomes and waiting times can't be less than zero, so they are often skewed to the right. Amounts of things (dollars, employees) are often skewed to the right for the same reason. If a test is too easy, the distribution will be skewed to the left because many scores will bump against 100%. And combinations of things are often skewed. In the case of the cancelled flights, flights are more likely to be cancelled in January (due to snowstorms) and in August (thunderstorms). Combining values across months leads to a skewed distribution.

What About Spread? The Standard Deviation

A S *Activity: The Spread of a Distribution.* What happens to measures of spread when data values change may not be quite what you expect.

The IQR is always a reasonable summary of spread, but because it uses only the two quartiles of the data, it ignores much of the information about how individual values vary. A more powerful approach uses the **standard deviation,** which takes into account how far *each* value is from the mean. Like the mean, the standard deviation is most appropriate for symmetric data.

One way to think about spread is to examine how far each data value is from the mean. This difference is called a *deviation*. We could just average the deviations, but the positive and negative differences always cancel each other out. So the average deviation is always zero—not very helpful.

To keep them from canceling out, we *square* each deviation. After squaring there are no negative values, so the sum won't be zero. That's great. Squaring also emphasizes larger differences—a feature that turns out to be both good and bad.

When we add up these squared deviations and find their average (almost), we call the result the **variance:**

$$s^2 = \frac{\sum (y - \bar{y})^2}{n - 1}$$

Why almost? It *would* be a mean if we divided the sum by n. Instead, we divide by $n - 1$. Why? The simplest explanation is "to drive you crazy." But there are good technical reasons, some of which we'll see in this chapter's What If?

The variance will play an important role later in this book, but it has a problem as a measure of spread. Whatever the units of the original data are, the variance is in *squared* units. We want measures of spread to have the same units as the data. And we probably don't want to talk about squared dollars or mpg^2. So, to get back to the original units, we take the square root of s^2. The result, s, is the **standard deviation.**

Putting it all together, the standard deviation of the data is found by the following formula:

$$s = \sqrt{\frac{\sum (y - \bar{y})^2}{n - 1}}$$

Who	52 adults
What	Resting heart rates
Units	Beats per minute

You will almost always rely on a calculator or computer to do the calculating.

Understanding what the standard deviation really means will take some time, and we'll revisit the concept in later chapters. For now, have a look at this histogram of resting pulse rates. The distribution is roughly symmetric, so it's okay to choose the mean and standard deviation as our summaries of center and spread. The mean pulse rate is 72.7 beats per minute, and we can see that's a typical heart rate. We also see that some heart rates are higher and some lower—but how much? Well, the standard deviation of 6.5 beats per minute indicates that, on average, we might expect people's heart rates to differ from the mean rate by about 6.5 beats per minute. Looking at the histogram, we can see that 6.5 beats above or below the mean appears to be a typical deviation.

Measures of spread tell how well other summaries describe the data. That's why we always (always!) report a spread along with any summary of the center.

A S *Activity*: **Displaying Spread.**
What does the standard deviation look like on a histogram? How about the IQR?

Waiting in Line Why do banks favor a single line that feeds several teller windows rather than separate lines for each teller? The average waiting time is the same. But the time you can expect to wait is less variable when there is a single line, and people prefer consistency.

How Does Standard Deviation Work?

To find the standard deviation, start with the mean, $\bar{y}$. Then find the *deviations* by taking $\bar{y}$ from each value: $(y - \bar{y})$ Square each deviation: $(y - \bar{y})^2$.

Now you're nearly home. Just add these up and divide by $n - 1$. That gives you the variance, s^2. To find the standard deviation, s, take the square root. Here we go:

Suppose the batch of values is 14, 13, 20, 22, 18, 19, and 13.

The mean is $\bar{y} = 17$. So the deviations are found by subtracting 17 from each value:

Original Values	Deviations	Squared Deviations
14	$14 - 17 = -3$	$(-3)^2 = 9$
13	$13 - 17 = -4$	$(-4)^2 = 16$
20	$20 - 17 = 3$	9
22	$22 - 17 = 5$	25
18	$18 - 17 = 1$	1
19	$19 - 17 = 2$	4
13	$13 - 17 = -4$	16

Add up the squared deviations: $9 + 16 + 9 + 25 + 1 + 4 + 16 = 80$.

Now divide by $n - 1$: $80/6 = 13.33$.

Finally, take the square root: $s = \sqrt{13.33} = 3.65$

For Example DESCRIBING SPREAD

RECAP: The histogram has shown that the distribution of credit card expenditures is skewed, and you have used the median to describe the center. The quartiles are $73.84 and $624.80.

QUESTION: What is the IQR and why is it a suitable measure of spread?

ANSWER: *For these data, the interquartile range (IQR) is $624.80 − $73.84 = $550.96. Like the median, the IQR is not affected by the outlying value or by the skewness of the distribution, so it is an appropriate measure of spread for the given expenditures.*

✔ Just Checking

6. The U.S. Census Bureau reports the median family income in its summary of census data. Why do you suppose they use the median instead of the mean? What might be the disadvantages of reporting the mean?

7. You've just bought a new car that claims to get a highway fuel efficiency of 31 miles per gallon. Of course, your mileage will "vary." If you had to guess, would you expect the IQR of gas mileage attained by all cars like yours to be 30 mpg, 3 mpg, or 0.3 mpg? Why?

8. A company selling a new MP3 player advertises that the player has a mean lifetime of 5 years. If you were in charge of quality control at the factory, would you prefer that the standard deviation of lifespans of the players you produce be 2 years or 2 months? Why?

What to *Tell* about a Quantitative Variable

A S *Activity*: **Playing with Summaries.** Here's a Statistics game about summaries that even some experienced statisticians find . . . well, challenging. Your intuition may be better. Give it a try!

TI-*nspire*

Standard deviation, IQR, and outliers. Drag data points around to explore how outliers affect measures of spread.

What should you *Tell* about a quantitative variable?

- Start by making a histogram, dotplot, or stem-and-leaf display, and discuss the shape of the distribution.
- Next, discuss the center *and* spread.
  - Always pair the median with the IQR and the mean with the standard deviation. It's not useful to report one without the other. Reporting a center without a spread is dangerous. You may think you know more than you do about the distribution. Reporting only the spread leaves us wondering where we are.
  - If the shape is skewed, report the median and IQR. You may want to include the mean and standard deviation as well, but you should point out why the mean and median differ.
  - If the shape is symmetric, report the mean and standard deviation and possibly the median and IQR as well. For unimodal symmetric data, the IQR is usually a bit larger than the standard deviation. If that's not true of your data set, look again to make sure that the distribution isn't skewed and there are no outliers.

- Also, discuss any unusual features.
  - If there are multiple modes, try to understand why. If you can identify a reason for separate modes (for example, women and men typically have heart attacks at different ages), it may be a good idea to split the data into separate groups.
  - If there are any clear outliers, you should point them out. If you are reporting the mean and standard deviation, report them with the outliers present and with the outliers removed. The differences may be revealing. (Of course, the median and IQR won't be affected very much by the outliers.)

For Example CHOOSING SUMMARY STATISTICS

RECAP: You have provided the credit card company's board of directors with a histogram of customer expenditures, and you have summarized the center and spread with the median and IQR. Knowing a little Statistics, the directors now insist on having the mean and standard deviation as summaries of the spending data.

QUESTION: Although you know that the mean is $478.19 and the standard deviation is $741.87, you need to explain to them why these are not suitable summary statistics for these expenditures data. What would you give as reasons?

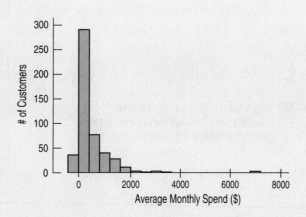

ANSWER: *The high outlier at $7000 pulls the mean up substantially and inflates the standard deviation. Locating the mean value on the histogram shows that it is not a typical value at all, and the standard deviation suggests that expenditures vary much more than they do. The median and IQR are more resistant to the presence of skewness and outliers, giving more realistic descriptions of center and spread.*

Step-by-Step Example SUMMARIZING A DISTRIBUTION

One of the authors owned a 1989 Nissan Maxima for 8 years. Being a statistician, he recorded the car's fuel efficiency (in mpg) each time he filled the tank. He wanted to know what fuel efficiency to expect as "ordinary" for his car. (Hey, he's a statistician. What would you expect?[10]) Knowing this, he was able to predict when he'd need to fill the tank again and to notice if the fuel efficiency suddenly got worse, which could be a sign of trouble.

Question: How would you describe the distribution of *Fuel efficiency* for this car?

(continued)

---

[10]He also recorded the time of day, temperature, price of gas, and phase of the moon. (OK, maybe not phase of the moon.) His data are on the DVD.

THINK ➡ **Plan** State what you want to find out.

I want to summarize the distribution of Nissan Maxima fuel efficiency.

Variable Identify the variable and report the W's.

The data are the fuel efficiency values in miles per gallon for the first 100 fill-ups of a 1989 Nissan Maxima between 1989 and 1992.

Be sure to check the appropriate condition.

✓ **Quantitative Data Condition:** The fuel efficiencies are quantitative with units of miles per gallon. Histograms and boxplots are appropriate displays for displaying the distribution. Numerical summaries are appropriate as well.

SHOW ➡ **Mechanics** Make a histogram. Based on the shape, choose appropriate numerical summaries.

REALITY CHECK ➡ A value of 22 mpg seems reasonable for such a car. The spread is reasonable, although the range looks a bit large.

A histogram of the data shows a fairly symmetric distribution with a low outlier.

Count	100
Mean	22.4 mpg
StdDev	2.45
Q1	20.8
Median	22.0
Q3	24.0
IQR	3.2

The mean and median are close, so the outlier doesn't seem to be a problem. I can use the mean and standard deviation.

TELL ➡ **Conclusion** Summarize and interpret your findings in context. Be sure to discuss the distribution's shape, center, spread, and unusual features (if any).

The distribution of mileage is unimodal and roughly symmetric with a mean of 22.4 mpg. There is a low outlier that should be investigated, but it does not influence the mean very much. The standard deviation suggests that from tankful to tankful, I can expect the car's fuel economy to differ from the mean by an average of about 2.45 mpg.

I Got a Different Answer: Did I Mess Up?

When you calculate a mean, the computation is clear: You sum all the values and divide by the sample size. You may round your answer less or more than someone else (we recommend one more decimal place than the data), but all books and technologies agree on how to find the mean. Some statistics, however, are more problematic. For example, we've already pointed out that methods of finding quartiles differ.

Differences in numeric results can also arise from decisions in the middle of calculations. For example, if you round off your value for the mean before you calculate the sum of squared deviations, your standard deviation probably won't agree with a computer program that calculates using many decimal places. (We do recommend that you do calculations using as many digits as you can to minimize this effect.)

Don't be overly concerned with these discrepancies, especially if the differences are small. They don't mean that your answer is "wrong," and they usually won't change any conclusion you might draw about the data. Sometimes (in footnotes and in the answers in the back of the book) we'll note alternative results, but we could never list all the possible values, so we'll rely on your common sense to focus on the meaning rather than on the digits. Remember: Answers are sentences—not single numbers!

TI Tips CALCULATING THE STATISTICS

Your calculator can easily find all the numerical summaries of data. To try it out, you simply need a set of values in one of your datalists. We'll illustrate using the boys' agility test results from this chapter's earlier TI Tips (still in L1), but you can use any data currently stored in your calculator.

- Under the STAT CALC menu, select 1-Var Stats and hit ENTER.
- Specify List:L1, leave Freqlist: blank, then go to Calculate and hit ENTER (OR on an older calculator, specify the location of your data by creating a command like 1-VarStats L1 and hit ENTER)."

Voilà! Everything you wanted to know, and more. Among all of the information shown, you are primarily interested in these statistics: $\bar{x}$ (the mean), Sx (the standard deviation), n (the count), and—scrolling down—minX (the smallest datum), Q_1 (the first quartile), Med (the median), Q_3 (the third quartile), and maxX (the largest datum).

Sorry, but the TI doesn't explicitly tell you the range or the IQR. Just subtract: IQR = Q_3 − Q_1 = 25 − 19.5 = 5.5. What's the range?

By the way, if the data come as a frequency table with the values stored in, say, L4 and the corresponding frequencies in L5, all you have to do is select 1-VarStats and specify List:L4, FreqList:L5, then go to Calculate and hit ENTER (OR on an older calculator ask for 1-VarStats L4,L5).

WHAT IF ●●● we divided by *n* instead of *n* − 1?

OK, we'll admit that at first it does seem strange that to find the variance (and therefore standard deviation) we must divide the sum of squared deviations from the mean by $n − 1$ instead of (more logically) n. What's wrong with n?

The key to understanding this issue rests in one of the fundamental aspects of Statistics. We collect data, make pictures, calculate summaries, and state conclusions. After all of that, we know a lot. *But we almost always wish we knew something else*. We rarely confine our interest to the specific data we have; we almost always wonder what those data say about a much bigger picture. What does the response of these patients tell us about how effective the treatment would be for everyone? What does the quality of a few items we inspect tell us about the entire production of the factory? What do these water specimens tell us about the safety of widespread hydrofracking? Our data come from samples; we wonder about populations.

We'll explore why this matters with another simulation.[11] In the Real World, we Must resort to looking at a sample because we don't have—and can't get—all the data from the entire population. But a simulation allows us to create a large "pretend" population; for ours, we set up a computer-generated list of 10,000 two-digit numbers. Let's see what happens when we try to estimate the mean and standard deviation of the whole population without looking at all of it.

We draw a random sample of size 5: 17 88 36 89 33. The mean is 52.6. If we calculate the standard deviation by dividing by $n = 5$, it comes out 30.02, but if we divide by $n - 1 = 4$ we get 33.56. Which method should we use? Well, we should go with the approach that tends to give the better estimate of the true standard deviation of the whole population. Which one is that? Let's try more samples to see what happens. Not just one or two more. Think big. We had the computer simulate 1000 samples! Here are a few:

Sample #	Sample data	Mean	Standard deviation dividing by	
			n	$n - 1$
1	17 88 36 89 33	52.6	30.02	33.56
2	54 46 70 67 86	64.6	13.79	15.42
...	...	...	...	...
1000	26 31 86 21 54	43.6	24.02	26.86
Average for all 1000 samples		**54.46**	**22.43**	**25.18**

So, what do we learn from this simulation? For one thing, we clearly see that information we get from samples is imperfect. All of the statistics vary depending on the particular sample data we have. That's why statisticians must be cautious in their conclusions; for example, you probably have heard pollsters talk about "sampling error." Here the values differ by a lot from one sample to the next; that's a big problem with small samples like this. The best universal advice a statistician can give anyone is, "Use a larger sample."

How well do samples like ours portray this population? Unlike the Real World, we can actually take a look at the population we created for this simulation. What we find is interesting.

1. The true mean of all 10,000 numbers was 54.49. It's no surprise that none of the samples got that exactly right, but there is some Good News: overall the sample estimates averaged 54.46. Almost dead on. Because of this, statisticians say that sample means are an *unbiased estimator* of the population mean.
2. The true standard deviation of our made-up population was 25.69. Look at the two options for finding the sample standard deviation. Overall, which one appears to be on target? Because dividing by n tends to underestimate the true value, statisticians say that method is *biased*. Our simulation confirms the wisdom of dividing by $n - 1$ when calculating the standard deviation.[12]

---

[11]You'll learn to play the "What If?" game with random numbers in Chapter 10. It's fun!

[12]Technically, it's the variance—the square of the standard deviation—that's unbiased when we use $n - 1$. But you don't need to worry about that in this course, so there's really no reason to read this footnote.

WHAT CAN GO WRONG?

A data display should tell a story about the data. To do that, it must speak in a clear language, making plain what variable is displayed, what any axis shows, and what the values of the data are. And it must be consistent in those decisions.

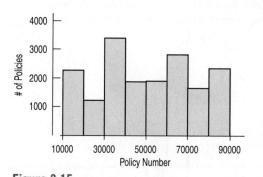

Figure 3.15

It's not appropriate to display these data with a histogram.

A display of quantitative data can go wrong in many ways. The most common failures arise from only a few basic errors:

- **Don't make a histogram of a categorical variable.** Just because the variable contains numbers doesn't mean that it's quantitative. Here's a histogram of the insurance policy numbers of some workers. It's not very informative because the policy numbers are just labels. A histogram or stem-and-leaf display of a categorical variable makes no sense. A bar chart or pie chart would be more appropriate.

- **Don't look for shape, center, and spread of a bar chart.** A bar chart showing the sizes of the piles displays the distribution of a categorical variable, but the bars could be arranged in any order left to right. Concepts like symmetry, center, and spread make sense only for quantitative variables.

- **Don't use bars in every display—save them for histograms and bar charts.** In a bar chart, the bars indicate how many cases of a categorical variable are piled in each category. Bars in a histogram indicate the number of cases piled in each interval of a quantitative variable. In both bar charts and histograms, the bars represent counts of data values. Some people create other displays that use bars to represent individual data values. Beware: Such graphs are neither bar charts nor histograms. For example, a student was asked to make a histogram from data showing the number of juvenile bald eagles seen during each of the 13 weeks in the winter of 2003–2004 at a site in Rock Island, IL. Instead, he made this plot:

Figure 3.16

This isn't a histogram or a bar chart. It's an ill-conceived graph that uses bars to represent individual data values (number of eagles sighted) week by week.

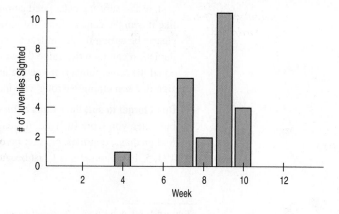

Look carefully. That's not a histogram. A histogram shows *What* we've measured along the horizontal axis and counts of the associated *Who*'s represented as bar heights. This student has it backwards: He used bars to show counts of birds for each week.[13] We need counts of weeks. A correct histogram should have a tall bar at "0" to show there were many weeks when no eagles were seen, like this:

Figure 3.17

A histogram of the eagle-sighting data shows the number of weeks in which different counts of eagles occurred. This display shows the distribution of juvenile-eagle sightings.

■ **Choose a bin width appropriate to the data.** Computer programs usually do a pretty good job of choosing histogram bin widths. Often there's an easy way to adjust the width, sometimes interactively. Here are the tsunami earthquakes with two (rather extreme) choices for the bin size:

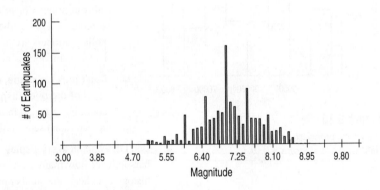

The task of summarizing a quantitative variable is relatively simple, and there is a simple path to follow. However, you need to watch out for certain features of the data that make summarizing them with a number dangerous. Here's some advice:

■ **Don't forget to do a reality check.** Don't let the computer or calculator do your thinking for you. Make sure the calculated summaries make sense. For example, does the mean look like it is in the center of the histogram? Think about the spread: An IQR of 50 mpg would clearly be wrong for gas mileage. And no measure of spread can be negative. The standard deviation can take the value 0, but only in the very unusual case that all the data values equal the same number. If you see an IQR or standard deviation equal to 0, it's probably a sign that something's wrong with the data.

■ **Don't forget to sort the values before finding the median or percentiles.** It seems obvious, but when you work by hand, it's easy to forget to sort the data first before counting in to find medians, quartiles, or other percentiles. Don't report that the median of the five values 194, 5, 1, 17, and 893 is 1 just because 1 is the middle number.

---

[13]Edward Tufte, in his book *The Visual Display of Quantitative Information*, proposes that graphs should have a high data-to-ink ratio. That is, we shouldn't waste a lot of ink to display a single number when a dot would do the job.

- **Don't worry about small differences when using different methods.** Finding the 10th percentile or the lower quartile in a data set sounds easy enough. But it turns out that the definitions are not exactly clear. If you compare different statistics packages or calculators, you may find that they give slightly different answers for the same data. These differences, though, are unlikely to be important in interpreting the data, the quartiles, or the IQR, so don't let them worry you.

Gold Card Customers—Regions National Banks		
Month	April 2007	May 2007
Average Zip Code	45,034.34	38,743.34

- **Don't compute numerical summaries of a categorical variable.** Neither the mean zip code nor the standard deviation of social security numbers is meaningful. If the variable is categorical, you should instead report summaries such as percentages of individuals in each category. It is easy to make this mistake when using technology to do the summaries for you. After all, the computer doesn't care what the numbers mean.

- **Don't report too many decimal places.** Statistical programs and calculators often report a ridiculous number of digits. A general rule for numerical summaries is to report one or two more digits than the number of digits in the data. For example, earlier we saw a dotplot of the Kentucky Derby race times. The mean and standard deviation of those times could be reported as:

$$\bar{y} = 130.63401639344262 \text{ sec} \qquad s = 13.66448201942662 \text{ sec}$$

But we knew the race times only to the nearest quarter second, so the extra digits are meaningless.

- **Don't round in the middle of a calculation.** Don't *report* too many decimal places, but it's best not to do any rounding until the end of your calculations. Even though you might report the mean of the earthquakes as 7.08, it's really 7.08339. Use the more precise number in your calculations if you're finding the standard deviation by hand—or be prepared to see small differences in your final result.

- **Watch out for multiple modes.** The summaries of the Kentucky Derby times are meaningless for another reason. As we saw in the dotplot, the Derby was initially a longer race. It would make much more sense to report that the old 1.5 mile Derby had a mean time of 159.6 seconds, while the current Derby has a mean time of 124.6 seconds. If the distribution has multiple modes, consider separating the data into different groups and summarizing each group separately.

- **Beware of outliers.** The median and IQR are resistant to outliers, but the mean and standard deviation are not. To help spot outliers . . .

- **Don't forget to: Make a picture (make a picture, make a picture).** The sensitivity of the mean and standard deviation to outliers is one reason you should always make a picture of the data. Summarizing a variable with its mean and standard deviation when you have not looked at a histogram or dotplot to check for outliers or skewness invites disaster. You may find yourself drawing absurd or dangerously wrong conclusions about the data. And, of course, you should demand no less of others. Don't accept a mean and standard deviation blindly without some evidence that the variable they summarize is unimodal, symmetric, and free of outliers.

What Have We Learned?

We've learned to *Think* about summarizing quantitative variables.

- All the methods of this chapter assume that the data are quantitative. The *Quantitative Data Condition* serves as a check that the data are, in fact, quantitative. One good way to be sure is to know the measurement units. You'll want those as part of the *Think* step of your answers.
- The median divides the data so that half of the data values are below the median and half are above.

- The mean is the point at which the histogram balances.
- The standard deviation summarizes how spread out all the data are around the mean.
- The median and IQR resist the effects of outliers, while the mean and standard deviation do not.
- In a skewed distribution, the mean is pulled in the direction of the skewness (toward the longer tail) relative to the median.
- We'll report the median and IQR when the distribution is skewed. If it's symmetric, we'll summarize the distribution with the mean and standard deviation (and possibly the median and IQR as well).

We've learned how to make a picture of quantitative data to help us see the story the data have to *Tell*.

- We can display the distribution of quantitative data with a *histogram*, a *stem-and-leaf* display, a *dotplot*, or a *boxplot*.

We've learned how to summarize distributions of quantitative variables numerically.

- Measures of center for a distribution include the median and the mean.
- Measures of spread include the range, IQR, and standard deviation.
- A 5-number summary includes the minimum and maximum values, the quartiles, and the median.
- We always pair the median with the IQR and the mean with the standard deviation.

We *Tell* what we see about the distribution by talking about *shape, center, spread*, and any *unusual features*.

Terms

Distribution	The distribution of a quantitative variable slices up all the possible values of the variable into equal-width bins and gives the number of values (or counts) falling into each bin. (p. 43)
Histogram (relative frequency histogram)	A histogram uses adjacent bars to show the distribution of a quantitative variable. Each bar represents the frequency (or relative frequency) of values falling in each bin. (p. 44)
Gap	A region of the distribution where there are no values. (p. 44)
Stem-and-leaf display	A stem-and-leaf display shows quantitative data values in a way that sketches the distribution of the data. It's best described in detail by example. (p. 46)
Dotplot	A dotplot graphs a dot for each case against a single axis. (p. 48)

Shape	To describe the shape of a distribution, look for
	
	  - single vs. multiple modes.
	  - symmetry vs. skewness.
	  - outliers and gaps. (p. 48)

Center	The place in the distribution of a variable that you'd point to if you wanted to attempt the impossible by summarizing the entire distribution with a single number. Measures of center include the mean and median. (p. 48)
Spread	A numerical summary of how tightly the values are clustered around the center. Measures of spread include the IQR and standard deviation. (p. 48)
Mode	A hump or local high point in the shape of the distribution of a variable. The apparent location of modes can change as the scale of a histogram is changed. (p. 48)
Unimodal (Bimodal)	Having one mode. This is a useful term for describing the shape of a histogram when it's generally mound-shaped. Distributions with two modes are called **bimodal**. Those with more than two are **multimodal.** (p. 49)

Uniform	A distribution that's roughly flat is said to be uniform. (p. 49)
Symmetric	A distribution is symmetric if the two halves on either side of the center look approximately like mirror images of each other. (p. 49)
Tails	The tails of a distribution are the parts that typically trail off on either side. Distributions can be characterized as having long tails (if they straggle off for some distance) or short tails (if they don't). (p. 49)
Skewed	A distribution is skewed if it's not symmetric and one tail stretches out farther than the other. Distributions are said to be **skewed left** when the longer tail stretches to the left, and **skewed right** when it goes to the right. (p. 49)
Outliers	Outliers are extreme values that don't appear to belong with the rest of the data. They may be unusual values that deserve further investigation, or they may be just mistakes; there's no obvious way to tell. Don't delete outliers automatically—you have to think about them. Outliers can affect many statistical analyses, so you should always be alert for them. (p. 50)
Median	The median is the middle value, with half of the data above and half below it. If n is even, it is the average of the two middle values. It is usually paired with the IQR. (p. 52)
Range	The difference between the lowest and highest values in a data set. $Range = max - min$. (p. 53)
Quartile	The **lower quartile** (Q1) is the value with a quarter of the data below it. The **upper quartile** (Q3) has three quarters of the data below it. The median and quartiles divide data into four parts with equal numbers of data values. (p. 53)
Interquartile range (IQR)	The IQR is the difference between the first and third quartiles. $IQR = Q3 - Q1$. It is usually reported along with the median. (p. 54)
Percentile	The ith percentile is the number that falls above $i\%$ of the data. (p. 53)
5-Number Summary	The 5-number summary of a distribution reports the minimum value, Q1, the median, Q3, and the maximum value. (p. 55)
Boxplot	A boxplot displays the 5-number summary as a central box, whiskers that extend to the nonoutlying data values, and any outliers shown. (p. 55)
Mean	The mean is found by summing all the data values and dividing by the count: $$\bar{y} = \frac{Total}{n} = \frac{\sum y}{n}.$$ It is usually paired with the standard deviation. (p. 59)
Resistant	A calculated summary is said to be resistant if outliers have only a small effect on it. (p. 60)
Standard deviation	The standard deviation is the square root of the variance: $$s = \sqrt{\frac{\sum (y - \bar{y})^2}{n - 1}}$$ It is usually reported along with the mean. (p. 60)
Variance	The variance is the sum of squared deviations from the mean, divided by the count minus 1: $$s^2 = \frac{\sum (y - \bar{y})^2}{n - 1}.$$ It is useful in calculations later in the book. (p. 61)

On the Computer DISPLAYING AND SUMMARIZING QUANTITATIVE VARIABLES

Almost any program that displays data can make a histogram, but some will do a better job of determining where the bars should start and how they should partition the span of the data.

The vertical scale may be counts or proportions. Sometimes it isn't clear which. But the shape of the histogram is the same either way.

Most packages choose the number of bars for you automatically. Often you can adjust that choice.

The axis should be clearly labeled so you can tell what "pile" each bar represents. You should be able to tell the lower and upper bounds of each bar.

Many statistics packages offer a prepackaged collection of summary measures. The result might look like this:

Variable: Weight

N = 234

Mean = 143.3   Median = 139

St. Dev = 11.1   IQR = 14

Alternatively, a package might make a table for several variables and summary measures:

A S *Case Study*: **Describing Distribution Shapes.** Who's safer in a crash—passengers or the driver? Investigate with your statistics package.

Variable	N	mean	median	stdev	IQR
Weight	234	143.3	139	11.1	14
Height	234	68.3	68.1	4.3	5
Score	234	86	88	9	5

It is usually easy to read the results and identify each computed summary. You should be able to read the summary statistics produced by any computer package.

Packages often provide many more summary statistics than you need. Of course, some of these may not be appropriate when the data are skewed or have outliers. It is your responsibility to check a histogram or stem-and-leaf display and decide which summary statistics to use.

It is common for packages to report summary statistics to many decimal places of "accuracy." Of course, it is rare data that have such accuracy in the original measurements. The ability to calculate to six or seven digits beyond the decimal point doesn't mean that those digits have any meaning. Generally it's a good idea to round these values, allowing perhaps one more digit of precision than was given in the original data.

Displays and summaries of quantitative variables are among the simplest things you can do in most statistics packages.

Exercises

1. Histogram Find a histogram that shows the distribution of a variable in a newspaper, a magazine, or the Internet.

a) Does the article identify the W's?

b) Discuss whether the display is appropriate.

c) Discuss what the display reveals about the variable and its distribution.

d) Does the article accurately describe and interpret the data? Explain.

2. Not a histogram Find a graph other than a histogram that shows the distribution of a quantitative variable in a newspaper, a magazine, or the Internet.

a) Does the article identify the W's?

b) Discuss whether the display is appropriate for the data.

c) Discuss what the display reveals about the variable and its distribution.

d) Does the article accurately describe and interpret the data? Explain.

3. In the news Find an article in a newspaper, a magazine, or the Internet that discusses an "average."

a) Does the article discuss the W's for the data?

b) What are the units of the variable?

c) Is the average used the median or the mean? How can you tell?

d) Is the choice of median or mean appropriate for the situation? Explain.

4. In the news II Find an article in a newspaper, a magazine, or the Internet that discusses a measure of spread.

a) Does the article discuss the W's for the data?

b) What are the units of the variable?

c) Does the article use the range, IQR, or standard deviation?

d) Is the choice of measure of spread appropriate for the situation? Explain.

5. Thinking about shape Would you expect distributions of these variables to be uniform, unimodal, or bimodal? Symmetric or skewed? Explain why.

a) The number of speeding tickets each student in the senior class of a college has ever had.

b) Players' scores (number of strokes) at the U.S. Open golf tournament in a given year.

c) Weights of female babies born in a particular hospital over the course of a year.

d) The length of the average hair on the heads of students in a large class.

6. More shapes Would you expect distributions of these variables to be uniform, unimodal, or bimodal? Symmetric or skewed? Explain why.

a) Ages of people at a Little League game.

b) Number of siblings of people in your class.

c) Pulse rates of college-age males.

d) Number of times each face of a die shows in 100 tosses.

7. Sugar in cereals The histogram displays the sugar content (as a percent of weight) of 49 brands of breakfast cereals.

a) Describe this distribution.

b) What do you think might account for this shape?

8. Singers The display shows the heights of some of the singers in a chorus, collected so that the singers could be positioned on stage with shorter ones in front and taller ones in back.

a) Describe the distribution.

b) Can you account for the features you see here?

9. Vineyards The histogram shows the sizes (in acres) of 36 vineyards in the Finger Lakes region of New York.

a) Approximately what percentage of these vineyards are under 60 acres?

b) Write a brief description of this distribution (shape, center, spread, unusual features).

10. Run times One of the authors collected the times (in minutes) it took him to run 4 miles on various courses during a 10-year period. Here is a histogram of the times.

Describe the distribution and summarize the important features. What is it about running that might account for the shape you see?

11. Heart attack stays The histogram shows the lengths of hospital stays (in days) for all the female patients admitted to hospitals in New York during one year with a primary diagnosis of acute myocardial infarction (heart attack).

a) From the histogram, would you expect the mean or median to be larger? Explain.
b) Write a few sentences describing this distribution (shape, center, spread, unusual features).
c) Which summary statistics would you choose to summarize the center and spread in these data? Why?

12. E-mails A university teacher saved every e-mail received from students in a large Introductory Statistics class during an entire term. He then counted, for each student who had sent him at least one e-mail, how many e-mails each student had sent.

a) From the histogram, would you expect the mean or the median to be larger? Explain.
b) Write a few sentences describing this distribution (shape, center, spread, unusual features).
c) Which summary statistics would you choose to summarize the center and spread in these data? Why?

13. Super Bowl points How many points do football teams score in the Super Bowl? Here are the total numbers of points scored by both teams in each of the first 46 Super Bowl games:

45, 47, 23, 30, 29, 27, 21, 31, 22, 38, 46, 37, 66, 50, 37, 47, 44, 47, 54, 56, 59, 52, 36, 65, 39, 61, 69, 43, 75, 44, 56, 55, 53, 39, 41, 37, 69, 61, 45, 31, 46, 31, 50, 48, 56, 38

a) Find the median.
b) Find the quartiles.
c) Are there any outliers?
d) Construct a boxplot of the data.
e) Write a description of the distribution based on the boxplot and the 5-number summary.

14. Super Bowl wins In the Super Bowl, by how many points does the winning team outscore the losers? Here are the winning margins for the first 46 Super Bowl games:

25, 19, 9, 16, 3, 21, 7, 17, 10, 4, 18, 17, 4, 12, 17, 5, 10, 29, 22, 36, 19, 32, 4, 45, 1, 13, 35, 17, 23, 10, 14, 7, 15, 7, 27, 3, 27, 3, 3, 11, 12, 3, 4, 14, 6, 4

a) Find the median.
b) Find the quartiles.
c) Are there any outliers?
d) Construct a boxplot of the data.
e) Write a description of the distribution based on the boxplot and the 5-number summary.

15. Details Here are histograms for four manufactured sets of numbers. The histograms look rather different, but all four sets have the same 5-number summary, so the boxplots for all four sets are identical to the one shown.

a) Using these plots as examples, identify some features of a distribution that a boxplot may not show.
b) What does this tell you about the limitations of using a boxplot to assess the shape of a distribution?

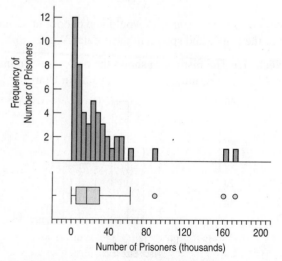

16. Opposites In a way, boxplots are the opposite of histograms. A histogram divides the number line into equal intervals and displays the number of data values in each interval. A boxplot divides the data into equal parts and displays the portion of the number line each part covers. These two plots display the number of incarcerated prisoners in each state as of June 2006.

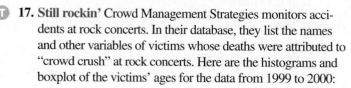

a) Explain how you could tell, by looking at a boxplot, where the tallest bars on the histogram would be located.
b) Explain how both the boxplot and the histogram can indicate a skewed distribution.
c) Identify one feature of the distribution that the histogram shows but a boxplot does not.
d) Identify one feature of the distribution that the boxplot shows but the histogram does not.

T 17. Still rockin' Crowd Management Strategies monitors accidents at rock concerts. In their database, they list the names and other variables of victims whose deaths were attributed to "crowd crush" at rock concerts. Here are the histograms and boxplot of the victims' ages for the data from 1999 to 2000:

a) What features of the distribution can you see in both the histogram and the boxplot?
b) What features of the distribution can you see in the histogram that you could not see in the boxplot?
c) What summary statistic would you choose to summarize the center of this distribution? Why?
d) What summary statistic would you choose to summarize the spread of this distribution? Why?

T 18. Slalom times The Men's Combined skiing event consists of a downhill and a slalom. Here are two displays of the slalom times in the Men's Combined at the 2006 Winter Olympics:

a) What features of the distribution can you see in both the histogram and the boxplot?
b) What features of the distribution can you see in the histogram that you could not see in the boxplot?

c) What summary statistic would you choose to summarize the center of this distribution? Why?

d) What summary statistic would you choose to summarize the spread of this distribution? Why?

19. Camp sites Shown below are the histogram and summary statistics for the number of camp sites at public parks in Vermont.

Count	46
Mean	62.8 sites
Median	43.5
StdDev	56.2
Min	0
Max	275
Q1	28
Q3	78

a) Which statistics would you use to identify the center and spread of this distribution? Why?

b) How many parks would you classify as outliers? Explain.

c) Create a boxplot for these data.

d) Write a few sentences describing the distribution.

20. Outliers The 5-number summary for the run times in minutes of the 150 highest grossing movies of 2010 looks like this:

Min	QI	Med	Q3	Max
43	98	104.5	116	160

a) Are there any outliers in these data? How can you tell?

b) Construct a boxplot. (You will be unable to mark outliers if they exist.) Based on your plot, say what you can about the shape of the distribution.

21. Standard deviation I For each lettered part, a through c, examine the two given sets of numbers. Without doing any calculations, decide which set has the larger standard deviation and explain why. Then check by finding the standard deviations *by hand.*

	Set 1	Set 2
a)	3, 5, 6, 7, 9	2, 4, 6, 8, 10
b)	10, 14, 15, 16, 20	10, 11, 15, 19, 20
c)	2, 6, 6, 9, 11, 14	82, 86, 86, 89, 91, 94

22. Standard deviation II For each lettered part, a through c, examine the two given sets of numbers. Without doing any calculations, decide which set has the larger standard deviation and explain why. Then check by finding the standard deviations *by hand.*

	Set 1	Set 2
a) 4, 7, 7, 7, 10	4, 6, 7, 8, 10
b) 100, 140, 150, 160, 200	10, 50, 60, 70, 110
c) 10, 16, 18, 20, 22, 28	48, 56, 58, 60, 62, 70

T **23. Pizza prices** The histogram shows the distribution of the prices of plain pizza slices (in $) for 156 weeks in Dallas, TX.

Which summary statistics would you choose to summarize the center and spread in these data? Why?

T **24. Neck size** The histogram shows the neck sizes (in inches) of 250 men recruited for a health study in Utah.

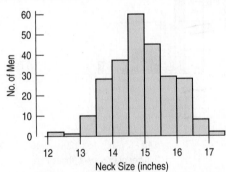

Which summary statistics would you choose to summarize the center and spread in these data? Why?

T **25. Pizza prices again** Look again at the histogram of the pizza prices in Exercise 23.

a) Is the mean closer to $2.40, $2.60, or $2.80? Why?

b) Is the standard deviation closer to $0.15, $0.50, or $1.00? Explain.

T **26. Neck sizes again** Look again at the histogram of men's neck sizes in Exercise 24.

a) Is the mean closer to 14, 15, or 16 inches? Why?

b) Is the standard deviation closer to 1 inch, 3 inches, or 5 inches? Explain.

27. Movie lengths The histogram shows the running times in minutes of 150 top grossing films of 2011.

a) You plan to see a movie this weekend. Based on these movies, how long do you expect a typical movie to run?
b) Would you be surprised to find that your movie ran for $2\frac{1}{2}$ hours (150 minutes)?
c) Which would you expect to be higher: the mean or the median run time for all movies? Why?

28. Golf drives 2011 The display shows the average drive distance (in yards) for 186 professional golfers on the men's PGA tour in 2011.

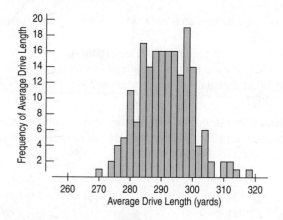

a) Describe this distribution.
b) Approximately what proportion of professional male golfers drive, on average, less than 280 yards?
c) Estimate the mean by examining the histogram.
d) Do you expect the mean to be smaller than, approximately equal to, or larger than the median? Why?

29. Movie lengths II Exercise 27 looked at the running times of movies released in 2011. The standard deviation of these running times is $s = 17.3$ minutes, and the quartiles are $Q_1 = 97$ minutes and $Q_3 = 115$ minutes.

a) Write a sentence or two describing the spread in running times based on
  i) the quartiles.
  ii) the standard deviation.
b) Do you have any concerns about using either of these descriptions of spread? Explain.

30. Golf drives II 2011 Exercise 28 looked at distances PGA golfers can hit the ball. The standard deviation of these average drive distances is 8.4 yards, and the quartiles are $Q_1 = 285.2$ yards and $Q_3 = 297.5$ yards.

a) Write a sentence or two describing the spread in distances based on
  i) the quartiles.
  ii) the standard deviation.
b) Do you have any concerns about using either of these descriptions of spread? Explain.

31. Fuel Economy The boxplot shows the fuel economy ratings for 67 model year 2012 subcompact cars. Some summary statistics are also provided. The extreme outlier is the **Mitsubishi i-MiEV**, an electric car whose electricity usage is *equivalent* to 112 miles per gallon.

Mean	StDev	Min	Q1	Med	Q3	Max	n
23.76	11.87	14	20	22	25	112	67

If that electric car is removed from the data set, how will the standard deviation be affected? The IQR?

32. Test scores correction After entering the test scores from her Statistics class of 25 students, the instructor calculated some statistics of the scores. Upon checking, she discovered that she had entered the top score as 46, but it should have been 56.

a) When she corrects this score, how will the mean and median be affected?
b) What effect will correcting the error have on the IQR and the standard deviation?

33. Mistake A clerk entering salary data into a company spreadsheet accidentally put an extra "0" in the boss's salary, listing it as $2,000,000 instead of $200,000. Explain how this error will affect these summary statistics for the company payroll:

a) measures of center: median and mean.
b) measures of spread: range, IQR, and standard deviation.

34. Cold weather A meteorologist preparing a talk about global warming compiled a list of weekly low temperatures (in degrees Fahrenheit) he observed at his southern Florida home last year. The coldest temperature for any week was 36°F, but he inadvertently recorded the Celsius value of 2°. Assuming that he correctly listed all the other temperatures, explain how this error will affect these summary statistics:

a) measures of center: mean and median.
b) measures of spread: range, IQR, and standard deviation.

35. Movie earnings The histogram shows total gross earnings (in millions of dollars) of major release movies in 2011.

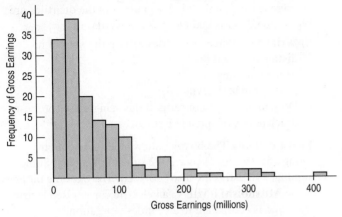

Gross Earnings (millions)

An industry publication reports that the average movie makes $41.7 million, but a watchdog group concerned with rising ticket prices says that the average earnings is $66.9 million. What statistic do you think each group is using? Explain.

36. Sick days During contract negotiations, a company seeks to change the number of sick days employees may take, saying that the annual "average" is 7 days of absence per employee. The union negotiators counter that the "average" employee misses only 3 days of work each year. Explain how both sides might be correct, identifying the measure of center you think each side is using and why the difference might exist.

37. Payroll A small warehouse employs a supervisor at $1200 a week, an inventory manager at $700 a week, six stock boys at $400 a week, and four drivers at $500 a week.

a) Find the mean and median wage.
b) How many employees earn more than the mean wage?
c) Which measure of center best describes a typical wage at this company: the mean or the median?
d) Which measure of spread would best describe the payroll: the range, the IQR, or the standard deviation? Why?

38. Singers The frequency table shows the heights (in inches) of 130 members of a choir.

Height	Count	Height	Count
60	2	69	5
61	6	70	11
62	9	71	8
63	7	72	9
64	5	73	4
65	20	74	2
66	18	75	4
67	7	76	1
68	12		

a) Find the median and IQR.
b) Find the mean and standard deviation.

c) Display these data with a histogram.
d) Write a few sentences describing the distribution.

39. Gasoline 2011 In October 2011, 16 gas stations in eastern Wisconsin, posted these prices for a gallon of regular gasoline:

3.43	3.46	3.43	3.59
3.65	3.63	3.62	3.65
3.66	3.31	3.35	3.42
3.41	3.46	3.47	3.48

a) Make a stem-and-leaf display of these gas prices. Use split stems; for example, use two 3.3 stems—one for prices between $3.30 and $3.34 and the other for prices from $3.35 to $3.39.
b) Describe the shape, center, and spread of this distribution.
c) What unusual feature do you see?

40. The Great One During his 20 seasons in the NHL, Wayne Gretzky scored 50% more points than anyone who ever played professional hockey. He accomplished this amazing feat while playing in 280 fewer games than Gordie Howe, the previous record holder. Here are the number of games Gretzky played during each season:

79, 80, 80, 80, 74, 80, 80, 79, 64, 78, 73, 78, 74, 45, 81, 48, 80, 82, 82, 70

a) Create a stem-and-leaf display for these data, using split stems.
b) Describe the shape of the distribution.
c) Describe the center and spread of this distribution.
d) What unusual feature do you see? What might explain this?

41. States The stem-and-leaf display shows populations of the 50 states and Washington, DC, in millions of people, according to the 2000 census.

```
3 | 4
2 |
2 | 1
1 | 69
1 | 0122
0 | 5555666667888
0 | 111111111111122222333333344444
```
State Populations (1 | 2 means 12 million)

a) What measures of center and spread are most appropriate?
b) Without doing any calculations, which must be larger: the median or the mean? Explain how you know.
c) From the stem-and-leaf display, find the median and the interquartile range.
d) Write a few sentences describing this distribution.

42. Wayne Gretzky In Exercise 40, you examined the number of games played by hockey great Wayne Gretzky during his 20-year career in the NHL.

a) Would you use the median or the mean to describe the center of this distribution? Why?

b) Find the median.

c) Without actually finding the mean, would you expect it to be higher or lower than the median? Explain.

43. A-Rod 2010 Alex Rodriguez (known to fans as A-Rod) was the youngest player ever to hit 500 home runs. Here is a stem-and-leaf display of the number of home runs hit by "A-Rod" during the 1994–2010 seasons (www .baseballreference.com/players/r/rodrial01.shtml.). Describe the distribution, mentioning its shape and any unusual features.

```
0 | 05
1 |
2 | 3
3 | 005566
4 | 12278
5 | 247
```
(5|2 means 52)

44. Bird species 2010 The Cornell Lab of Ornithology holds an annual Christmas Bird Count (www.birdsource .org), in which bird watchers at various locations around the country see how many different species of birds they can spot. Here are some of the counts reported from sites in Texas during the 2010 event:

150 216 177 150 166 156 159 160 164 169 175
150 178 183 199 154 164 203 158 231

a) Create a stem-and-leaf display of these data.

b) Write a brief description of the distribution. Be sure to discuss the overall shape as well as any unusual features.

T 45. Hurricanes 2010 The data below give the number of hurricanes classified as major hurricanes in the Atlantic Ocean each year from 1944 through 2006, as reported by *NOAA* (www.nhc.noaa.gov):

3, 3, 1, 2, 4, 3, 8, 5, 3, 4, 2, 6, 2, 2, 5, 2, 2, 7, 1, 2, 6, 1, 3, 1, 0, 5, 2, 1, 0, 1, 2, 3, 2, 1, 2, 2, 2, 3, 1, 1, 1, 3, 0, 1, 3, 2, 1, 2, 1, 1, 0, 5, 6, 1, 3, 5, 3, 4, 2, 3, 6, 7, 2, 2, 5, 2, 5

a) Create a dotplot of these data.

b) Describe the distribution.

T 46. Horsepower Create a stem-and-leaf display for these horsepowers of autos reviewed by *Consumer Reports* one year, and describe the distribution:

155	103	130	80	65
142	125	129	71	69
125	115	138	68	78
150	133	135	90	97
68	105	88	115	110
95	85	109	115	71
97	110	65	90	
75	120	80	70	

47. A-Rod 2010 again Students were asked to make a histogram of the number of home runs hit by Alex Rodriguez from 1995 to 2010 (see Exercise 43). One student submitted the following display:

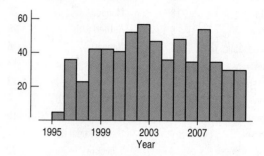

a) Comment on this graph.

b) Create your own histogram of the data.

48. Return of the birds Students were given the assignment to make a histogram of the data on bird counts reported in Exercise 44. One student submitted the following display:

a) Comment on this graph.

b) Create your own histogram of the data.

49. Acid rain Two researchers measured the pH (a scale on which a value of 7 is neutral and values below 7 are acidic) of water collected from rain and snow over a 6-month period in Allegheny County, PA. Describe their data with a graph and a few sentences:

4.57	5.62	4.12	5.29	4.64	4.31	4.30	4.39	4.45
5.67	4.39	4.52	4.26	4.26	4.40	5.78	4.73	4.56
5.08	4.41	4.12	5.51	4.82	4.63	4.29	4.60	

T 50. Marijuana 2007 In 2007 the Council of Europe published a report entitled *The European School Survey Project on Alcohol and Other Drugs* (www.espad.org). Among other issues, the survey investigated the percentages of 16-year-olds who had used marijuana. Shown on the next page are the results for 34 European countries. Create an appropriate graph of these data, and describe the distribution.

Country	Cannabis	Country	Cannabis
Armenia	3	Italy	23
Austria	17	Latvia	18
Belgium	24	Lithuania	18
Bulgaria	22	Malta	13
Croatia	18	Monaco	28
Cyprus	5	Netherlands	28
Czech Republic	45	Norway	6
Estonia	26	Poland	16
Faroe Islands	6	Portugal	13
Finland	8	Romania	4
France	31	Russia	19
Germany	20	Slovak Republic	32
Greece	6	Slovenia	22
Hungary	13	Sweden	7
Iceland	9	Switzerland	33
Ireland	20	Ukraine	14
Isle of Man	34	United Kingdom	29

51. Final grades A professor (of something other than Statistics!) distributed the following histogram to show the distribution of grades on his 200-point final exam. Comment on the display.

52. Final grades revisited After receiving many complaints about his final-grade histogram from students currently taking a Statistics course, the professor from Exercise 51 distributed the following revised histogram:

a) Comment on this display.
b) Describe the distribution of grades.

53. Zip codes Holes-R-Us, an Internet company that sells piercing jewelry, keeps transaction records on its sales. At a recent sales meeting, one of the staff presented a histogram of the zip codes of the last 500 customers, so that the staff might understand where sales are coming from. Comment on the usefulness and appropriateness of the display.

54. Zip codes revisited Here are some summary statistics to go with the histogram of the zip codes of 500 customers from the Holes-R-Us Internet Jewelry Salon that we saw in Exercise 53:

Count	500
Mean	64,970.0
StdDev	23,523.0
Median	64,871
IQR	44,183
Q1	46,050
Q3	90,233

What can these statistics tell you about the company's sales?

T 55. Math scores 2009 The National Center for Education Statistics (http://nces.ed.gov/nationsreportcard/) reported 2009 average mathematics achievement scores for eighth graders in all 50 states:

State	Score	State	Score
Alabama	269	Montana	292
Alaska	283	Nebraska	284
Arizona	277	Nevada	274
Arkansas	276	New Hampshire	292
California	270	New Jersey	293
Colorado	287	New Mexico	270
Connecticut	289	New York	283
Delaware	284	North Carolina	284
Florida	279	North Dakota	293
Georgia	278	Ohio	286
Hawaii	274	Oklahoma	276
Idaho	287	Oregon	285
Illinois	282	Pennsylvania	288
Indiana	287	Rhode Island	278
Iowa	284	South Carolina	280
Kansas	289	South Dakota	291
Kentucky	279	Tennessee	275
Louisiana	272	Texas	287
Maine	286	Utah	284
Maryland	288	Vermont	293
Massachusetts	299	Virginia	286
Michigan	278	Washington	289
Minnesota	294	West Virginia	270
Mississippi	265	Wisconsin	288
Missouri	286	Wyoming	286

a) Find the median, the IQR, the mean, and the standard deviation of these state averages.

b) Which summary statistics would you report for these data? Why?

c) Write a brief summary of the performance of eighth graders nationwide.

56. Boomtowns 2011 In 2011, the website NewGeography.com listed its ranking of the best cities for job growth in the United States. Here are the magazine's top 20 larger cites, along with their weighted job rating indices.

Metro Area	Job Rating Index
1 Austin-Round Rock-San Marcos, TX	89.1
2 New Orleans-Metairie-Kenner, LA	88.9
3 Houston-Sugar Land-Baytown, TX	82.1
4 San Antonio-New Braunfels, TX	80.7
5 Dallas-Plano-Irving, TX Metropolitan Division	80.6
6 Washington-Arlington-Alexandria, DC-VA-MD-WV Metropolitan Division	74.1
7 Northern Virginia, VA	73.8
8 Nashville-Davidson-Murfreesboro-Franklin, TN	73.1
9 New York City, NY	72.6
10 Philadelphia City, PA	71.9
11 Pittsburgh, PA	71.7
12 Bethesda-Rockville-Frederick, MD Metropolitan Division	70.7
13 Boston-Cambridge-Quincy, MA NECTA Division	70.4
14 Raleigh-Cary, NC	67.8
15 Fort Worth-Arlington, TX Metropolitan Division	67.8
16 Rochester, NY	66.7
17 Nassau-Suffolk, NY Metropolitan Division	65.5
18 Buffalo-Niagara Falls, NY	64.7
19 Columbus, OH	64.3
20 Salt Lake City, UT	64.2

a) Make a suitable display of the job rating indices.

b) Summarize the typical job rating index among these cities with a median and mean. Why do they differ?

c) Given what you know about the distribution, which of the measures in b) does the better job of summarizing the job rating indices? Why?

d) Summarize the spread of the job rating indices distribution with a standard deviation and with an IQR.

e) Given what you know about the distribution, which of the measures in d) does the better job of summarizing the job rating indices? Why?

f) Suppose we subtract from each of the preceding job rating indices the predicted U.S. average job rating index, so that we can look at how much these indices exceed the U.S. rate. How would this change the values of the summary statistics you calculated above?

g) If we were to omit the Austin and New Orleans metro areas from the data, how would you expect the mean, median, standard deviation, and IQR to change? Explain your expectations for each.

h) Write a brief report about all of these job rating indices.

57. Population growth The following data show the percentage change in population for the 50 states and the District of Columbia from 2000 to 2009. Using appropriate graphical displays and summary statistics, write a report on the percentage change in population by state.

State	% Inc	State	% Inc
Alabama	5.9	Montana	8.1
Alaska	11.4	Nebraska	5.0
Arizona	28.6	Nevada	32.3
Arkansas	8.1	New Hampshire	7.2
California	9.1	New Jersey	3.5
Colorado	16.8	New Mexico	10.5
Connecticut	3.3	New York	3.0
Delaware	13.0	North Carolina	16.6
District of Columbia	4.8	North Dakota	0.7
Florida	16.0	Ohio	1.7
Georgia	20.1	Oklahoma	6.9
Hawaii	6.9	Oregon	11.8
Idaho	19.5	Pennsylvania	2.6
Illinois	4.0	Rhode Island	0.5
Indiana	5.6	South Carolina	13.7
Iowa	2.8	South Dakota	7.6
Kansas	4.8	Tennessee	10.7
Kentucky	6.7	Texas	18.8
Louisiana	0.5	Utah	24.7
Maine	3.4	Vermont	2.1
Maryland	7.6	Virginia	11.4
Massachusetts	3.9	Washington	13.1
Michigan	0.3	West Virginia	0.6
Minnesota	7.0	Wisconsin	5.4
Mississippi	3.8	Wyoming	10.2
Missouri	7.0		

(*Source:* www.census.gov/compendia/statab/rankings.html)

58. Prisons 2006 A report from the U.S. Department of Justice (www.ojp.usdoj.gov/bjs/) reported the percent changes in federal prison populations in 21 northeastern and midwestern states during 2006. Using appropriate graphical displays and summary statistics, write a report on the changes in prison populations.

State	Percent Change	State	Percent Change
Connecticut	1.6	Iowa	0.9
Maine	−7.7	Kansas	−1.2
Massachusetts	4.8	Michigan	3.4
New Hampshire	2.5	Minnesota	6.4
New Jersey	1.1	Missouri	−2.8
New York	0.0	Nebraska	4.5
Pennsylvania	3.7	North Dakota	6.1
Rhode Island	7.8	Ohio	5.6
Vermont	8.3	South Dakota	5.3
Illinois	1.7	Wisconsin	−2.0
Indiana	3.8		

Just Checking ANSWERS

(Thoughts will vary.)

1. Slightly skewed to the right. Center around 3 miles? Few over 10 miles.

2. Bimodal. Center between 1 and 2 hours? Many people watch no football; others watch most of one or more games. Probably only a few values over 5 hours.

3. Strongly skewed to the right, with almost everyone at $0; a few small prizes, with the winner an outlier.

4. Fairly symmetric, somewhat uniform, perhaps slightly skewed to the right. Center in the 40s? Few ages below 25 or above 70.

5. Uniform, symmetric. Center near 5. Roughly equal counts for each digit 0–9.

6. Incomes are probably skewed to the right and not symmetric, making the median the more appropriate measure of center. The mean will be influenced by the high end of family incomes and not reflect the "typical" family income as well as the median would. It will give the impression that the typical income is higher than it is.

7. An IQR of 30 mpg would mean that only 50% of the cars get gas mileages in an interval 30 mpg wide. Fuel economy doesn't vary that much. 3 mpg is reasonable. It seems plausible that 50% of the cars will be within about 3 mpg of each other. An IQR of 0.3 mpg would mean that the gas mileage of half the cars varies little from the estimate. It's unlikely that cars, drivers, and driving conditions are that consistent.

8. We'd prefer a standard deviation of 2 months. Making a consistent product is important for quality. Customers want to be able to count on the MP3 player lasting somewhere close to 5 years, and a standard deviation of 2 years would mean that life-spans were highly variable.

4 Understanding and Comparing Distributions

Who	Days during 2011
What	Average daily wind speed (mph), Average barometric pressure (mb), Average daily temperature (deg Celsius)
When	2011
Where	Hopkins Forest, in Western Massachusetts
Why	Long-term observations to study ecology and climate

The Hopkins Memorial Forest is a 2500-acre reserve in Massachusetts, New York, and Vermont managed by the Williams College Center for Environmental Studies (CES). As part of their mission, CES monitors forest resources and conditions over the long term.[1]

One of the variables measured in the forest is wind speed. Three remote sensors record the minimum, maximum, and average wind speed (in mph) for each day.

Wind is caused as air flows from areas of high pressure to areas of low pressure. Centers of low pressure often accompany storms, so both high winds and low pressure are associated with some of the fiercest storms. Wind speeds can vary greatly during a day and from day to day, but if we step back a bit farther, we can see patterns. By modeling these patterns, we can understand things about *Average Wind Speed* that we may not have known.

In Chapter 2 we looked at the association between two categorical variables using contingency tables and displays. Here we'll explore different ways of examining the relationship between two variables when one is quantitative, and the other indicates groups to compare. We are given wind speed averages for each day of 2011. But we can collect the days together into different size groups and compare the wind speeds among them. If we partition *Time* in different ways, we'll gain enormous flexibility for our analysis. We'll discover new insights as we change from viewing the whole year's data at once, to comparing seasons, to looking for patterns across months, and, finally, to looking at the data day by day.

The Big Picture

Let's start with the "big picture." Here's a histogram, 5-number summary, and boxplot of the *Average Wind Speed* for every day in 2011. Because of the skewness, we'll report the median and IQR. We can see that the distribution of *Average Wind Speed* is unimodal and skewed to the right. Median daily wind speed is about 1.12 mph, and on half of the days,

[1]www.williams.edu/CES/hopkins.htm

the average wind speed is between 0.46 and 2.28 mph. We also see some outliers, including a rather windy 6.73-mph day. Were those unusual weather events, or just the windiest days of the year? To answer that, we'll need to work with the summaries a bit more.

Figure 4.1

A histogram of daily *Average Wind Speed* for 2011 It is unimodal and skewed to the right. The boxplot below the histogram suggests several possible outliers that may deserve our attention.

Max	6.73
Q3	2.28
Median	1.12
Q1	0.46
Min	0.00

Average Wind Speed (mph)

Comparing Groups with Histograms

It is almost always more interesting to compare groups. Is it windier in the winter or the summer? Are any months particularly windy? Are weekends a special problem? Let's split the year into two groups: April through September (Summer) and October through March (Winter). To compare the groups, we create two histograms, being careful to use the same scale. Here are displays of the average daily wind speed for Summer (on the left) and Winter (on the right):

Figure 4.2

Histograms of *Average Wind Speed* for days in Summer (left) and Winter (right) show very different patterns.

Average Wind Speed (mph)

The shapes, centers, and spreads of these two distributions are strikingly different. The distribution of summer wind speeds is strongly skewed to the right, while the winter wind speeds, although somewhat right-skewed, are more nearly uniform in distribution. Typical summer days have mean wind speeds less than 1 mph, lower than speeds for typical winter days. Summer wind speeds rarely average over 3 mph, but that's not unusual in the winter, and there's at least one very high value. And in the winter wind speeds are more variable (IQR = 1.91 mph) than in the summer (IQR = 0.79 mph).

| Summaries for *Average Wind Speed* by Season | | | | |
Group	Mean	StdDev	Median	IQR
Winter	2.17	1.33	2.07	1.91
Summer	0.85	0.74	0.62	0.79

For Example COMPARING GROUPS WITH STEM-AND-LEAF DISPLAYS

The Nest Egg Index, devised by the investment firm of A.G. Edwards, is a measure of saving and investment performance for each of the 50 states, based on 12 economic factors, including participation in retirement savings plans, personal debt levels, and home ownership. The average index is 100 and the numbers indicate the percentage above or below the average. There are only 50 values, so a back-to-back stem-and-leaf plot is an effective display. Here's one comparing the Nest Egg Index in the Northeast and Midwest states to those in the South and West. In this display, the stems run down the middle of the plot, with the leaves for the two regions to the left or right. Be careful when you read the values on the left: 5|8 means a Nest Egg Index of 85% for a southern or western state.

South and West		Northeast and Midwest
5778	8	8
12344	9	03
6667778899	9	67
02334	10	012233334
56	10	6779
	11	122444

(4|9|3 means 94% for a Northeast/Midwest state and 93% for a South/West state)

QUESTION: How do nest egg indices compare for these regions?

ANSWER: The distribution of Nest Egg Indices is nearly symmetric for the South and West, but skewed to the left for the Northeast and Midwest. Indices were generally higher and more variable in the Northeast/Midwest region than in the South/West. Nine northeastern and midwestern states had higher indices than any states in the South or West.

Comparing Groups with Boxplots

Video: **Can Diet Prolong Life?** Here's a subject that's been in the news: Can you live longer by eating less? (Or would it just seem longer?) Look at the data in subsequent activities, and you'll find that you can learn a lot by comparing two groups with boxplots.

Are some months windier than others? Even residents may not have a good idea of which parts of the year are the most windy. (Do you know for your hometown?) We're not interested just in the centers, but also in the spreads. Are wind speeds equally variable from month to month, or do some months show more variation?

Histograms or stem-and-leaf displays are a fine way to look at one distribution or two. But it would be hard to see patterns by comparing 12 histograms. Boxplots offer an ideal balance of information and simplicity, hiding the details while displaying the overall summary information. By placing boxplots side by side, we can easily see which groups have higher medians, which have the greater IQRs, where the central 50% of the data is located in each group, and which have the greater overall range. And, when the boxes are in an order, we can get a general idea of patterns in both the centers and the spreads. Equally important, we can see past any outliers in making these comparisons because they've been displayed separately.

Here are boxplots of the *Average Daily Wind Speed* by month:

Figure 4.3

Boxplots of the *Average Daily Wind Speed* plotted for each *Month* show seasonal patterns in both the centers and spreads New outliers appear because they are now judged relative to the *Month* in which they occurred.

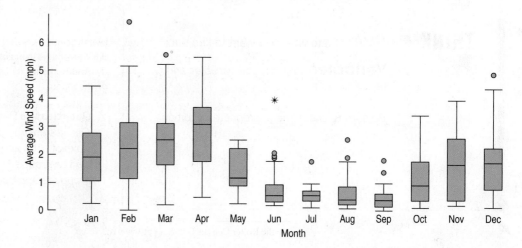

Here we see that wind speeds tend to decrease in the summer. The months in which the winds are both strongest and most variable are November through April.

When we looked at a boxplot of wind speeds for the entire year, there were only 5 outliers. But the monthly boxplots show different outliers than before because some days that seemed ordinary when placed against the entire year's data looked like outliers for the month that they're in. That windy day in August certainly wouldn't stand out in November or December, but for August, it was remarkable—as we'll soon see.

For Example COMPARING DISTRIBUTIONS

Roller coasters[2] are a thrill ride in many amusement parks worldwide. And thrill seekers want a coaster that goes fast. There are two main types of roller coasters: those with wooden tracks and those with steel tracks. Do they typically run at different speeds? Here are boxplots:

QUESTION: Compare the speeds of wood and steel roller coasters.

ANSWER: Overall, wooden-track roller coasters are slower than steel-track coasters. In fact, the fastest half of the steel coasters are faster than three quarters of the wooden coasters. Although the IQRs of the two groups are similar, the range of speeds among steel coasters is larger than the range for wooden coasters. The distribution of speeds of wooden coasters appears to be roughly symmetric, but the speeds of the steel coasters are skewed to the right, and there is a high outlier at 120 mph. We should look into why that steel coaster is so fast.

Step-by-Step Example COMPARING GROUPS

Most scientific studies compare two or more groups. It is almost always a good idea to start an analysis of data from such studies by comparing boxplots for the groups. Here's an example:

For her class project, a student compared the efficiency of various coffee containers. For her study, she decided to try 4 different containers and to test each of them 8 different times. Each time, she heated water to 180°F, poured it into a container, and sealed it. (We'll learn the details of how to set up experiments in Chapter 12.) After 30 minutes, she measured the temperature again and recorded the difference in temperature. Because these are temperature differences, smaller differences mean that the liquid stayed hot—just what we would want in a coffee mug.

Question: What can we say about the effectiveness of these four mugs?

THINK ➡ **Plan** State what you want to find out.

Variables Identify the *variables* and report the W's.

Be sure to check the appropriate condition.

I want to compare the effectiveness of the different mugs in maintaining temperature. I have 8 measurements of *Temperature Change* for each of the mugs.

✓ **Quantitative Data Condition:** The *Temperature Changes* are quantitative, with units of °F. Boxplots are appropriate displays for comparing the groups. Numerical summaries of each group are appropriate as well.

---

[2]See the Roller Coaster DataBase at www.rcdb.com.

Show ➡ **Mechanics** Report the 5-number summaries of the four groups. Including the IQR is a good idea as well.

	Min	Q1	Median	Q3	Max	IQR
CUPPS	6°F	6	8.25	14.25	18.50	8.25
Nissan	0	1	2	4.50	7	3.50
SIGG	9	11.50	14.25	21.75	24.50	10.25
Starbucks	6	6.50	8.50	14.25	17.50	7.75

Make a picture. Because we want to compare the distributions for four groups, boxplots are an appropriate choice.

Tell ➡ **Conclusion** Interpret what the boxplots and summaries say about the ability of these mugs to retain heat. Compare the shapes, centers, and spreads, and note any outliers.

The individual distributions of temperature changes are all slightly skewed to the high end. The Nissan cup does the best job of keeping liquids hot, with a median loss of only 2°F, and the SIGG cup does the worst, typically losing 14°F. The difference is large enough to be important: A coffee drinker would be likely to notice a 14° drop in temperature. And the mugs are clearly different: 75% of the Nissan tests showed less heat loss than any of the other mugs in the study. The IQR of results for the Nissan cup is also the smallest of these test cups, indicating that it is a consistent performer.

 Just Checking

The Bureau of Transportation Statistics of the U.S. Department of Transportation collects and publishes statistics on airline travel (www.transtats.bts.gov). Here are three displays of the % of flights arriving late each month from 1995 through 2005:

1. Describe what the histogram says about late arrivals.

2. What does the boxplot of late arrivals suggest that you can't see in the histogram?

3. Describe the patterns shown in the boxplots by month. At what time of year are flights least likely to be late? Can you suggest reasons for this pattern?

TI Tips   COMPARING GROUPS WITH BOXPLOTS

In the last chapter we looked at the performances of fourth-grade students on an agility test. Now let's make comparative boxplots for the boys' scores and the girls' scores:

> Boys: 22, 17, 18, 29, 22, 22, 23, 24, 23, 17, 21
> Girls: 25, 20, 12, 19, 28, 24, 22, 21, 25, 26, 25, 16, 27, 22

Enter these data in L1 (*Boys*) and L2 (*Girls*).

Set up STATPLOT's Plot1 to make a boxplot of the boys' data:
- Turn the plot On;
- Always choose the boxplot icon that shows outliers;
- Specify Xlist:L1 and Freq:1, and select an outlier Mark.

Use ZoomStat to display the boxplot for *Boys*.

As you did for the boys, set up Plot2 to display the girls' data. This time when you use ZoomStat with both plots turned on, the display shows the parallel boxplots. See the outlier?

This is a great opportunity to practice your "Tell" skills. How do these fourth graders compare in terms of agility?

Outliers

In the boxplots for the *Average Wind Speed* by *Month*, several days are nominated by the boxplots as possible outliers. Cases that stand out almost always deserve attention. An outlier is a value that doesn't fit with the rest of the data, but exactly how different it should be to receive special treatment is a judgment call. Boxplots provide a rule of thumb to highlight these unusual cases, but that rule is only a guide, and it doesn't tell you what to do with them. So, what *should* we do with outliers?

Outliers arise for many reasons. They *may* be the most important values in the data set, pointing out an exceptional case or illuminating a pattern by being the exception to the rule. They may be values that just happen to lie above the limits suggested by the box plot rule. Or they may be errors. A decimal point may have been misplaced, digits transposed, or digits repeated or omitted. Some outliers are obviously wrong. For example, if a class survey includes a student who claims to be 170 inches tall (about 14 feet, or 4.3 meters), you can be sure that's an error. Maybe the units were wrong. (Did the student mean 170 centimeters—about 65 inches?). Errors creep into data sets remarkably often. One important reason to look into outliers is to correct errors in your data.

The boxplots of *Average Wind Speed* by *Month* show possible outliers in several of the months. The windiest day in February was an outlier not only in that windy month, but for the entire year as well. The windiest day in June was a "far" outlier—lying more than 3 IQRs from the upper quartile for that month, but it wouldn't have been unusual for a winter day. And, the windiest day in August seemed much windier than the days around it even though it was relatively calm.

Many outliers are not wrong; they're just different. And most repay the effort to understand them. You can sometimes learn more from the extraordinary cases than from summaries of the entire data set. That blustery day in February turned out to be a blast that brought four days of subzero temperatures ($-17°$F) to the region.

What about that June day? A search for weather events on the Internet for June 2, 2011, finds a rare tornado in Western Massachusetts.

The extreme outlier in August wouldn't be remarkable in most other months. It turns out that that was Hurricane Irene, whose eye passed right over the Hopkins Forest.

In the aftermath of Hurricane Irene, the Hoosic River in Western Massachusetts rose more than 10 feet over its banks, swallowing portions of Williams College, including the soccer and baseball fields.

According to *The New York Times*, "it was probably the greatest number of people ever threatened by a single storm in the United States."[3]

Not all outliers are as dramatic as Hurricane Irene, but all deserve attention. If you can correct an error, you should do so (and note the correction). If you can't correct it, or if you confirm that it is correct, you can simply note its existence and leave it in the data set. But the safest path is to report summaries and analyses with *and* without the outlier so that a reader can judge the influence of the outlier for him- or herself.

There are two things you should *never* do with outliers. You should not leave an outlier in place and proceed as if nothing were unusual. Analyses of data with outliers are very likely to be wrong. Nor should you omit an outlier from the analysis without comment. If you want to exclude an outlier, you must announce your decision and, to the extent you can, justify it. Finally, keep in mind that a case lying just over the fence suggested by the boxplot may just be the largest (or smallest) value at the end of a stretched-out tail. A histogram is often a better way to examine how the outlier fits in (or doesn't) by seeing how large the gap is between it and the rest of the data.

For Example CHECKING OUT THE OUTLIERS

RECAP: We've looked at the speeds of roller coasters and found a difference between steel- and wooden-track coasters. We also noticed an extraordinary value.

QUESTION: The fastest coaster in this collection turns out to be the "Top Thrill Dragster" at Cedar Point amusement park. What might make this roller coaster unusual? You'll have to do some research, but that's often what happens with outliers.

ANSWER: *The Top Thrill Dragster is easy to find in an Internet search. We learn that it is a "hydraulic launch" coaster. That is, it doesn't get its remarkable speed just from gravity, but rather from a kick-start by a hydraulic piston. That could make it different from the other roller coasters.*

(You might also discover that it is no longer the fastest roller coaster in the world.)

Timeplots: Order, Please!

A S *Case Study*: **Are Passengers or Drivers Safer in a Crash?**
Practice the skills of this chapter by comparing these two groups.

Figure 4.4

A timeplot of Average Wind Speed shows the overall pattern and changes in variation.

The Hopkins Forest wind speeds are reported as daily averages. Previously, we grouped the days into months or seasons, but we could look at the wind speed values day by day. Whenever we have data measured over time, it is a good idea to look for patterns by plotting the data in time order. Here are the daily average wind speeds plotted over time:

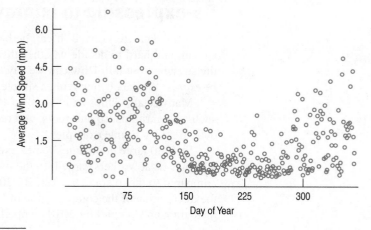

---

[3]Well, before Hurricane Sandy scored a direct hit on Connecticut, New York City, and New Jersey in October 2012, anyway.

A display of values against time is sometimes called a **timeplot**. This timeplot reflects the pattern that we saw when we plotted the wind speeds by month. But without the arbitrary divisions between months, we can see a calm period during the summer, starting around day 150 (the beginning of June), when the wind is relatively mild and doesn't vary greatly from day to day. We can also see that the wind becomes both more variable and stronger during the early and late parts of the year.

Looking into the Future

It is always tempting to try to extend what we see in a timeplot into the future. Sometimes that makes sense. Most likely, the Hopkins Forest climate follows regular seasonal patterns. It's probably safe to predict a less windy June next year and a windier November. But we certainly wouldn't predict another storm on November 21.

Other patterns are riskier to extend into the future. If a stock has been rising, will it continue to go up? No stock has ever increased in value indefinitely, and no stock analyst has consistently been able to forecast when a stock's value will turn around. Stock prices, unemployment rates, and other economic, social, or psychological concepts are much harder to predict than physical quantities. The path a ball will follow when thrown from a certain height at a given speed and direction is well understood. The path interest rates will take is much less clear. Unless we have strong (nonstatistical) reasons for doing otherwise, we should resist the temptation to think that any trend we see will continue, even into the near future.

Statistical models often tempt those who use them to think beyond the data. We'll pay close attention later in this book to understanding when, how, and how much we can justify doing that.

*Re-expressing Data: A First Look

Re-expressing to Improve Symmetry

When the data are skewed, it can be hard to summarize them simply with a center and spread, and hard to decide whether the most extreme values are outliers or just part of the stretched-out tail. How can we say anything useful about such data? The secret is to *re-express* the data by applying a simple function to each value.

Many relationships and "laws" in the sciences and social sciences include functions such as logarithms, square roots, and reciprocals. Similar relationships often show up in data. Here's a simple example:

In 1980 large companies' chief executive officers (CEOs) made, on average, about 42 times what workers earned. In the next two decades, CEO compensation soared when compared to the average worker. By 2008 that multiple had jumped[4] to 344. What does the distribution of the compensation of Fortune 500 companies' CEOs look like? Here's a histogram and boxplot for 2010 compensation:

---

[4]www.faireeconomy.org/files/executive-excess-2008.pdf

Figure 4.5

Compensation paid to CEOs of the Fortune 500 companies in 2010.

We have 500 CEOs and about 57 possible histogram bins, most of which are empty—but don't miss the tiny bars straggling out to the right. The boxplot indicates that some CEOs received extraordinarily high compensations, while the majority received relatively "little." But look at the values of the bins. The first bin, with about a quarter of the CEOs, covers incomes from $0 to $2,5000,000. Imagine receiving a salary survey with these categories:

> What is your income?
> **a.** $0 to $2,5000,000
> **b.** $2,5000,001 to $5,000,000
> **c.** $5,000,001 to $7,5000,000
> **d.** More than $7,5000,000

The reason that the histogram seems to leave so much of the area blank is that the salaries are spread all along the axis from about $35,000,000 to $150,000,000. After $50,000,000 there are so few for each bin that it's very hard to see the tiny bars. What we *can* see from this histogram and boxplot is that this distribution is highly skewed to the right.

It can be hard to decide what we mean by the "center" of a skewed distribution, so it's hard to pick a typical value to summarize the distribution. What would you say was a typical CEO total compensation? The mean value is $8,035,770, while the median is "only" $4,800,000. Each tells us something different about the data.

One approach is to **re-express,** or **transform,** the data by applying a simple function to make the skewed distribution more symmetric. For example, we could take the square root or logarithm of each compensation value. Taking logs works pretty well for the CEO compensations, as you can see:

The histogram of the logs of the total CEO compensations is much more nearly symmetric, so we can see that a typical log compensation is between 5, which corresponds to $100,000, and 7.5, corresponding to $31,6000,000. And it's easier to talk about a typical value for the logs. The mean log compensation is 6.68, while the median is 6.67. (That's $4,786,301 and $4,677,351, respectively but who's counting?)

Against the background of a generally symmetric main body of data, it's easier to decide whether the largest compensations are outliers. In fact, the three most highly compensated CEOs are identified as outliers by the boxplot rule of thumb even after this re-expression. It's perhaps impressive to be an outlier CEO in annual compensation. It's even more impressive to be an outlier in the log scale!

Figure 4.6

The logarithms of 2010 CEO compensations are much more nearly symmetric.

Dealing with Logarithms You have probably learned about logs in math courses and seen them in psychology or science classes. In this book, we use them only for making data behave better. Base 10 logs are the easiest to understand, but natural logs are often used as well. (Either one is fine.) You can think of base 10 logs as roughly one less than the number of digits you need to write the number. So 100, which is the smallest number to require 3 digits, has a $\log_{10}$ of 2. And 1000 has a $\log_{10}$ of 3. The $\log_{10}$ of 500 is between 2 and 3, but you'd need a calculator to find that it's approximately 2.7. All salaries of "six figures" have $\log_{10}$ between 5 and 6. Logs are incredibly useful for making skewed data more symmetric. But don't worry—nobody does logs without technology and neither should you. Often, remaking a histogram or other display of the data is as easy as pushing another button.

Re-expressing to Equalize Spread Across Groups

Researchers measured the concentration (nanograms per milliliter) of cotinine in the blood of three groups of people: nonsmokers who have not been exposed to smoke, nonsmokers who have been exposed to smoke (ETS), and smokers. Cotinine is left in the blood when the body metabolizes nicotine, so this measure gives a direct measurement of the effect of passive smoke exposure. The boxplots of the cotinine levels of the three groups tell us that the smokers have higher cotinine levels, but if we want to compare the levels of the passive smokers to those of the nonsmokers, we're in trouble, because on this scale, the cotinine levels for both nonsmoking groups are too low to be seen.

Figure 4.7

Cotinine levels (nanograms per milliliter) for three groups with different exposures to tobacco smoke Can you compare the ETS (exposed to smoke) and No-ETS groups?

Re-expressing can help alleviate the problem of comparing groups that have very different spreads. For measurements like the cotinine data, whose values can't be negative and whose distributions are skewed to the high end, a good first guess at a re-expression is the logarithm.

After taking logs, we can compare the groups and see that the nonsmokers exposed to environmental smoke (the ETS group) do show increased levels of (log) cotinine, although not the high levels found in the blood of smokers.

Notice that the same re-expression has also improved the symmetry of the cotinine distribution for smokers and pulled in most of the apparent outliers in all of the groups. It is not unusual for a re-expression that improves one aspect of data to improve others as well. We'll talk about other ways to re-express data as the need arises throughout the book. We'll explore some common re-expressions more thoroughly in Chapter 9.

Figure 4.8

Blood cotinine levels after taking logs What a difference a log makes!

WHAT IF ●●● steel and wooden roller coasters actually are equally fast?

In the boxplots it certainly appears that steel coasters are generally faster than the wooden ones. In fact, the median speed for the steel coasters is 68 mph, but it's only 56 mph for the wooden ones. A difference of 12 mph seems pretty large, but is it big enough to be *statistically significant*[5]? In other words, is the gap we see here so large that it's probably because there's a genuine difference in speeds between all coasters of these two types, or could it have arisen just by chance in this particular sample? To find out, we'll do (by now you probably know what's coming. . .) a simulation.

Here's how this simulation works. If the type of coaster really doesn't matter, then dividing them into these two categories was arbitrary and the resulting speed difference was just a fluke. To find out, we combine all the speeds into a single set of data, and then we randomly split them into two other groups to see how much difference in median speeds might arise just by chance.

Our computer simulation did exactly that—1,000 times. In the table you can compare some of the random results to the actual difference observed in the data.

| Simulation | Medians | | Difference in |
Trial #	Group A	Group B	medians (A − B)
1	66	65	+1.0 mph
2	65	66.3	−1.3 mph
3	66	60.3	+5.7 mph
. . .			
1,000	65	65.6	−0.6 mph
Actual Data	**Steel 68**	**Wooden 56**	+ 12.0 *mph*

The actual difference between steel and wooden coasters is far bigger than any of the random differences displayed here, but what about the rest of the 1000 trials? Instead of boring you with the whole table, here's a histogram showing how the actual coasters compared to all of the simulated differences.

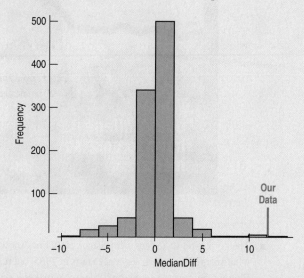

We can see that splitting the coasters into random groups rarely produced much of a difference in median speeds. Comparatively, the 12 mph difference in the actual data looks surprisingly large. In fact, only *once* in 1000 trials was the simulated difference that big. Statisticians would call a mere 1-in-1000 chance that the observed outcome could occur simply by natural sample-to-sample variation "statistically significant"[6]. This simulation provides strong evidence that steel roller coasters really *are* faster than wooden ones.

---

[5]In Statistics, when you see the word "significant" think "highly unusual". Keep that in mind as you read the rest of this story.

[6]In fact, in the simulation differences of 6 mph or more arose by chance in only 29, or 2.9%, of the 1000 trials, Statisticians would have considered even that much of a gap unusual enough to suggest a real difference between the two types of coasters. A common rule of thumb sets the bar for significance to include outcomes that could happen randomly less than 5% of the time.

WHAT CAN GO WRONG?

- **Avoid inconsistent scales.** Parts of displays should be mutually consistent—no fair changing scales in the middle or plotting two variables on different scales but on the same display. When comparing two groups, be sure to compare them on the same scale.

- **Label clearly.** Variables should be identified clearly and axes labeled so a reader knows what the plot displays.

Here's a remarkable example of a plot gone wrong. It illustrated a news story about rising college costs. It uses timeplots, but it gives a misleading impression. First think about the story you're being told by this display. Then try to figure out what has gone wrong.

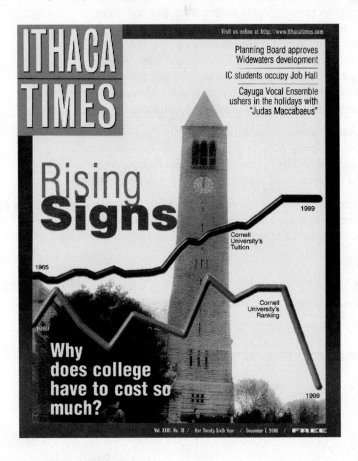

What's wrong? Just about everything.

- The horizontal scales are inconsistent. Both lines show trends over time, but exactly for what years? The tuition sequence starts in 1965, but rankings are graphed from 1989. Plotting them on the same (invisible) scale makes it seem that they're for the same years.
- The vertical axis isn't labeled. That hides the fact that it's inconsistent. Does it graph dollars (of tuition) or ranking (of Cornell University)?

This display violates three of the rules. And it's even worse than that: It violates a rule that we didn't even bother to mention.

- The two inconsistent scales for the vertical axis don't point in the same direction! The line for Cornell's rank shows that it has "plummeted" from 15th place to 6th place in academic rank. Most of us think that's an *improvement,* but that's not the message of this graph.

- **Beware of outliers.** If the data have outliers and you can correct them, you should do so. If they are clearly wrong or impossible, you should remove them and report on them. Otherwise, consider summarizing the data both with and without the outliers.

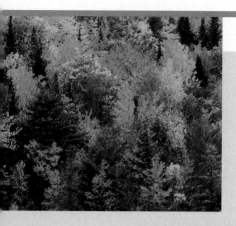

What Have We Learned?

- We've learned the value of comparing groups and looking for patterns among groups and over time.
- We've seen that histograms or stem-and-leaf plots can compare two distributions well, if drawn on the same scale. Boxplots are more effective for comparing several groups.
- When we compare groups, we've learned to compare their shape, center, spreads, and any unusual features.
- We've experienced the value of identifying and investigating outliers. And we've seen that when we group data in different ways, it can allow different cases to emerge as possible outliers.
- We've graphed data that have been measured over time against a time axis and looked for long-term trends.

Terms

Comparing distributions   When comparing the distributions of several groups using histograms or stem-and-leaf displays, compare their:

- Shape
- Center
- Spread (p. 86)

Comparing boxplots   When comparing groups with boxplots:

- Compare the shapes. Do the boxes look symmetric or skewed? Are there differences between groups?
- Compare the medians. Which group has the higher center? Is there any pattern to the medians?
- Compare the IQRs. Which group is more spread out? Is there any pattern to how the IQRs change?
- Using the IQRs as a background measure of variation, do the medians seem to be different, or do they just vary much as you'd expect from the overall variation?
- Check for possible outliers. Identify them if you can and discuss why they might be unusual. Of course, correct them if you find that they are errors. (p. 86)

Timeplot   A timeplot displays data that change over time to show long-term patterns and trends. (p. 89)

Re-express (Transform)   Applying a simple function (such as a logarithm or square root) to the data can make a skewed distribution more symmetric or equalize spread across groups. (p. 91)

On the Computer COMPARING DISTRIBUTIONS

Most programs for displaying and analyzing data can display plots to compare the distributions of different groups. Typically these are boxplots displayed side-by-side.

Side-by-side boxplots should be on the same y-axis scale so they can be compared.

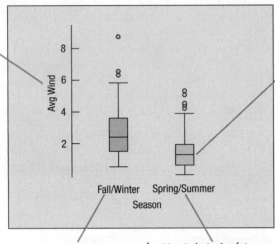

Some programs offer a graphical way to assess how much the medians differ by drawing a band around the median or by "notching" the boxes.

Boxes are typically labeled with a group name. Often they are placed in alphabetical order by group name—not the most useful order.

Exercises

1. **In the news** Find an article in a newspaper, magazine, or the Internet that compares two or more groups of data.
   a) Does the article discuss the W's?
   b) Is the chosen display appropriate? Explain.
   c) Discuss what the display reveals about the groups.
   d) Does the article accurately describe and interpret the data? Explain.

2. **In the news** Find an article in a newspaper, magazine, or the Internet that shows a time plot.
   a) Does the article discuss the W's?
   b) Is the timeplot appropriate for the data? Explain.
   c) Discuss what the timeplot reveals about the variable.
   d) Does the article accurately describe and interpret the data? Explain.

3. **Time on the Internet** Find data on the Internet (or elsewhere) that give results recorded over time. Make an appropriate display and discuss what it shows.

4. **Groups on the Internet** Find data on the Internet (or elsewhere) for two or more groups. Make appropriate displays to compare the groups, and interpret what you find.

5. **Pizza prices** A company that sells frozen pizza to stores in four markets in the United States (Denver, Baltimore, Dallas, and Chicago) wants to examine the prices that the stores charge for pizza slices. Here are boxplots comparing data from a sample of stores in each market:

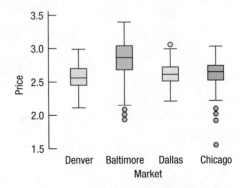

   a) Do prices appear to be the same in the four markets? Explain.
   b) Does the presence of any outliers affect your overall conclusions about prices in the four markets?

6. Costs To help travelers know what to expect, researchers collected the prices of commodities in 16 cities throughout the world. Here are boxplots comparing the prices of a ride on public transportation, a newspaper, and a cup of coffee in the 16 cities (prices are all in $US).

a) On average, which commodity is the most expensive?
b) Is a newspaper always more expensive than a ride on public transportation? Explain.
c) Does the presence of outliers affect your conclusions in a) or b)?

7. Hopkins 2007 Below are histograms and the five-number summaries for the average windspeeds in the Hopkins forest for the year 2007 and the year 2011, which was discussed in the chapter. Compare these distributions, and be sure to address shape (including outliers if there are any), center, and spread.

	Year	
	2007	**2011**
Min	0.12	0
Q1	1.28	0.46
Median	1.80	1.12
Q3	2.88	2.28
Max	6.82	6.73

8. Camping Here are summary statistics and histograms for the number of campsites at public parks in Wisconsin and Vermont. Write a few sentences comparing the numbers of campsites in these two states. Be sure to talk about shape (including outliers), center, and spread.

	Wisconsin	Vermont
Count	45	46
Mean	81.9	62.8
Median	60	43.5
StdDev	96.9	56.2
Min	0	0
Max	472	275
Q1	14	28
Q3	108	78

9. Cereals Sugar is a major ingredient in many breakfast cereals. The histogram displays the sugar content as a percentage of weight for 49 brands of cereal. The boxplot compares sugar content for adult and children's cereals.

a) What is the range of the sugar contents of these cereals?
b) Describe the shape of the distribution.
c) What aspect of breakfast cereals might account for this shape?

d) Are all children's cereals higher in sugar than adult cereals?

e) Which group of cereals varies more in sugar content? Explain.

10. Tendon transfers People with spinal cord injuries may lose function in some, but not all, of their muscles. The ability to push oneself up is particularly important for shifting position when seated and for transferring into and out of wheelchairs. Surgeons compared two operations to restore the ability to push up in children. The histogram shows scores rating pushing strength two years after surgery and boxplots compare results for the two surgical methods. (Mulcahey, Lutz, Kozen, Betz, "Prospective Evaluation of Biceps to Triceps and Deltoid to Triceps for Elbow Extension in Tetraplegia," *Journal of Hand Surgery*, 28, 6, 2003)

a) Describe the shape of the strength distribution.

b) What is the range of the strength scores?

c) What fact about results of the two procedures is hidden in the histogram?

d) Which method had the higher (better) median score?

e) Was that method always best?

f) Which method produced the most consistent results? Explain.

11. Population growth 2010 This "back-to-back" stem-and-leaf plot displays two data sets at once—one going to the left, one to the right. The plot compares the percent change in population for two regions of the United States (based on census figures for 2000 and 2010). The fastest growing state was Nevada at 35%. To show the distributions better, this display breaks each stem into two lines, putting leaves 0–4 on one stem and leaves 5–9 on the other.

NE/MW States		S/W States
4433332200	0	134
987777666555	0	7799
2	1	0023344
5	1	57889
	2	1
	2	5
	3	4
	3	5

Population Growth rate
(2|1|0 means 12% for a NE/NW state
and 10% for a S/W state)

a) Use the data displayed in the stem-and-leaf display to construct comparative boxplots.

b) Write a few sentences describing the difference in growth rates for the two regions of the United States.

12. Home Runs 2012 Here is a "back-to-back" stemplot that shows two data sets at once—one going to the left, one to the right. The display compares the number of home runs for Major League Baseball teams in the National League and the American League during the 2012 season.

National League		American League
Team Home Runs		
3	10	
6	11	
1	12	
977	13	116
96	14	9
98	15	
65	16	35
20	17	5
	18	7
4	19	58
2	20	0
	21	14
	22	
	23	
	24	5

Key: 2 | 20 | 0 means 202 HR
for a team in the NL and 200
for a team in the AL

a) Use the data in the stemplot to construct comparative boxplots.

b) Write a few sentences comparing the distributions in home runs for teams in the two leagues.

13. Hospital stays The U.S. National Center for Health Statistics compiles data on the length of stay by patients in short-term hospitals and publishes its findings in *Vital and Health Statistics*. Data from a sample of 39 male patients and 35 female patients on length of stay (in days) are displayed in the histograms on the next page.

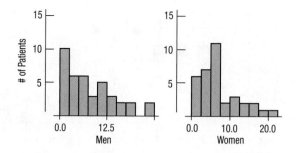

a) What would you suggest be changed about these histograms to make them easier to compare?

b) Describe these distributions by writing a few sentences comparing the duration of hospitalization for men and women.

c) Can you suggest a reason for the peak in women's length of stay?

14. Deaths 2009 A National Vital Statistics Report (www.cdc.gov/nchs/) indicated that nearly 290,000 black Americans died in 2009, compared with just over 2 million white Americans. Here are histograms displaying the distributions of their ages at death:

a) Describe the overall shapes of these distributions.

b) How do the distributions differ?

c) Look carefully at the bar definitions. Where do these plots violate the rules for statistical graphs?

T 15. Women's basketball Here are boxplots of the points scored during the first 10 games of the season for both Scyrine and Alexandra:

a) Summarize the similarities and differences in their performance so far.

b) The coach can take only one player to the state championship. Which one should she take? Why?

16. Gas prices 2011 Here are boxplots of weekly gas prices for regular gas in the United States as reported by the U.S. Energy Information Administration for 2009, 2010, and 2011.

a) Compare the distribution of prices over the three years.

b) In which year were the prices least stable? Explain.

T 17. Marriage age In 1975, did men and women marry at the same age? Here are boxplots of the age at first marriage for a sample of U.S. citizens then. Write a brief report discussing what these data show.

18. Fuel economy Describe what these boxplots tell you about the relationship between the number of cylinders a car's engine has and the car's fuel economy (mpg):

19. Fuel economy 2012 The Environmental Protection Agency provides fuel economy and pollution information on over 2000 car models. Here are boxplots of *Combined Fuel Economy* (using an average of driving conditions) in *miles per gallon* by vehicle *Type* (midsize car, standard pickup truck, or SUV) for 2012 model vehicles. Summarize what you see about the fuel economies of the three vehicle types.

20. Ozone Ozone levels (in parts per billion, ppb) were recorded at sites in New Jersey monthly between 1926 and 1971. Here are boxplots of the data for each month (over the 46 years), lined up in order (January = 1):

a) In what month was the highest ozone level ever recorded?
b) Which month has the largest IQR?
c) Which month has the smallest range?
d) Write a brief comparison of the ozone levels in January and June.
e) Write a report on the annual patterns you see in the ozone levels.

21. Load factors by month The Research and Innovative Technology Administration of the Bureau of Transportation Statistics (www.TranStats.bts.gov) reports load factors (passenger-miles as a percentage of available seat-miles) for commercial airlines for every month from 2000 through 2011. Here is a display of the load factors for international flights by month for the period from 2000 to 2011. Describe the patterns you see.

22. Load factors by year Here is a display of the load factors (passenger-miles as a percentage of available seat-miles) for domestic airline flights by year. Describe the patterns you see.

23. Test scores Three Statistics classes all took the same test. Histograms and boxplots of the scores for each class are shown below. Match each class with the corresponding boxplot.

24. Eye and hair color A survey of 1021 school-age children was conducted by randomly selecting children from several large urban elementary schools. Two of the questions concerned eye and hair color. In the survey, the following codes were used:

Hair Color	Eye Color
1 = Blond	1 = Blue
2 = Brown	2 = Green
3 = Black	3 = Brown
4 = Red	4 = Grey
5 = Other	5 = Other

The Statistics students analyzing the data were asked to study the relationship between eye and hair color. They produced this plot:

Is their graph appropriate? If so, summarize the findings. If not, explain why not.

25. Graduation? A survey of major universities asked what percentage of incoming freshmen usually graduate "on time" in 4 years. Use the summary statistics given to answer the questions that follow.

	% on Time
Count	48
Mean	68.35
Median	69.90
StdDev	10.20
Min	43.20
Max	87.40
Range	44.20
25th %tile	59.15
75th %tile	74.75

a) Would you describe this distribution as symmetric or skewed? Explain.
b) Are there any outliers? Explain.
c) Create a boxplot of these data.
d) Write a few sentences about the graduation rates.

T 26. Vineyards Here are summary statistics for the sizes (in acres) of Finger Lakes vineyards:

Count	36
Mean	46.50 acres
StdDev	47.76
Median	33.50
IQR	36.50
Min	6
Q1	18.50
Q3	55
Max	250

a) Would you describe this distribution as symmetric or skewed? Explain.
b) Are there any outliers? Explain.
c) Create a boxplot of these data.
d) Write a few sentences about the sizes of the vineyards.

27. Caffeine A student study of the effects of caffeine asked volunteers to take a memory test 2 hours after drinking soda. Some drank caffeine-free cola, some drank regular cola (with caffeine), and others drank a mixture of the two (getting a half-dose of caffeine). Here are the

5-number summaries for each group's scores (number of items recalled correctly) on the memory test:

	n	Min	Q1	Median	Q3	Max
No Caffeine	15	16	20	21	24	26
Low Caffeine	15	16	18	21	24	27
High Caffeine	15	12	17	19	22	24

a) Describe the W's for these data.
b) Name the variables and classify each as categorical or quantitative.
c) Create parallel boxplots to display these results as best you can with this information.
d) Write a few sentences comparing the performances of the three groups.

28. SAT scores Here are the summary statistics for Verbal SAT scores for a high school graduating class:

	n	Mean	Median	SD	Min	Max	Q1	Q3
Male	80	590	600	97.2	310	800	515	650
Female	82	602	625	102.0	360	770	530	680

a) Create parallel boxplots comparing the scores of boys and girls as best you can from the information given.
b) Write a brief report on these results. Be sure to discuss the shape, center, and spread of the scores.

29. Derby speeds 2011 How fast do horses run? Kentucky Derby winners top 30 miles per hour, as shown in this graph. The graph shows the percentage of Derby winners that have run *slower* than each given speed. Note that few have won running less than 33 miles per hour, but about 86% of the winning horses have run slower than 37 miles per hour. (A cumulative frequency graph like this is called an "ogive.")

a) Estimate the median winning speed.
b) Estimate the quartiles.
c) Estimate the range and the IQR.
d) Create a boxplot of these speeds.
e) Write a few sentences about the speeds of the Kentucky Derby winners.

30. Cholesterol The Framingham Heart Study recorded the cholesterol levels of more than 1400 men. Here is an ogive of the distribution of these cholesterol measures. (An ogive shows the percentage of cases at or below a certain value.) Construct a boxplot for these data, and write a few sentences describing the distribution.

31. Reading scores A class of fourth graders takes a diagnostic reading test, and the scores are reported by reading grade level. The 5-number summaries for the 14 boys and 11 girls are shown:

5-Number Summaries					
Boys	2.0	3.9	4.3	4.9	6.0
Girls	2.8	3.8	4.5	5.2	5.9

a) Which group had the highest score?
b) Which group had the greater range?
c) Which group had the greater interquartile range?
d) Which group's scores appear to be more skewed? Explain.
e) Which group generally did better on the test? Explain.
f) If the mean reading level for boys was 4.2 and for girls was 4.6, what is the overall mean for the class?

32. Rainmakers? In an experiment to determine whether seeding clouds with silver iodide increases rainfall, 52 clouds were randomly assigned to be seeded or not. The amount of rain they generated was then measured (in acre-feet). Here are the summary statistics:

	n	Mean	Median	SD	IQR	Q1	Q3
Unseeded	26	164.59	44.20	278.43	138.60	24.40	163
Seeded	26	441.98	221.60	650.79	337.60	92.40	430

a) Which of the summary statistics are most appropriate for describing these distributions? Why?

b) Do you see any evidence that seeding clouds may be effective? Explain.

T **33. Industrial experiment** Engineers at a computer production plant tested two methods for accuracy in drilling holes into a PC board. They tested how fast they could set the drilling machine by running 10 boards at each of two different speeds. To assess the results, they measured the distance (in inches) from the center of a target on the board to the center of the hole. The data and summary statistics are shown in the table:

	Distance (in.)	Speed		Distance (in.)	Speed
	0.000101	Fast		0.000098	Slow
	0.000102	Fast		0.000096	Slow
	0.000100	Fast		0.000097	Slow
	0.000102	Fast		0.000095	Slow
	0.000101	Fast		0.000094	Slow
	0.000103	Fast		0.000098	Slow
	0.000104	Fast		0.000096	Slow
	0.000102	Fast		0.975600	Slow
	0.000102	Fast		0.000097	Slow
	0.000100	Fast		0.000096	Slow
Mean	0.000102		**Mean**	0.097647	
StdDev	0.000001		**StdDev**	0.308481	

Write a report summarizing the findings of the experiment. Include appropriate visual and verbal displays of the distributions, and make a recommendation to the engineers if they are most interested in the accuracy of the method.

T **34. Cholesterol** A study examining the health risks of smoking measured the cholesterol levels of people who had smoked for at least 25 years and people of similar ages who had smoked for no more than 5 years and then stopped. Create appropriate graphical displays for both groups, and write a brief report comparing their cholesterol levels. Here are the data:

Smokers				Ex-Smokers		
225	211	209	284	250	134	300
258	216	196	288	249	213	310
250	200	209	280	175	174	328
225	256	243	200	160	188	321
213	246	225	237	213	257	292
232	267	232	216	200	271	227
216	243	200	155	238	163	263
216	271	230	309	192	242	249
183	280	217	305	242	267	243
287	217	246	351	217	267	218
200	280	209		217	183	228

T **35. MPG** A consumer organization wants to compare gas mileage figures for several models of cars made in the United States with autos manufactured in other countries. The data for a random sample of cars classified as "midsize" are shown in the table.

Gas Mileage (mpg)	Country	Gas Mileage (mpg)	Country
22	U.S.	17	Other
39	U.S.	26	Other
39	U.S.	18	Other
22	U.S.	20	Other
22	U.S.	24	Other
21	U.S.	22	Other
29	U.S.	28	Other
21	U.S.	23	Other
21	U.S.	30	Other
24	U.S.	19	Other
23	U.S.	27	Other
17	U.S.	21	Other
30	U.S.	22	Other
19	U.S.	29	Other
23	U.S.	29	Other
21	U.S.	28	Other
24	U.S.	26	Other
50	Other	21	Other
24	Other	20	Other
35	Other	21	Other

a) Create graphical displays for these two groups.

b) Write a few sentences comparing the distributions.

T **36. Baseball 2011** American League baseball teams play their games with the designated hitter rule, meaning that pitchers do not bat. The League believes that replacing the pitcher, typically a weak hitter, with another player in the batting order produces more runs and generates more interest among fans. Following are the average number of runs scored in American League and National League stadiums for the 2011 season:

American	Avg Runs/Game	National	Avg Runs/Game
TEX	11.06	COL	10.69
BOS	10.74	CIN	9.33
NYY	9.99	ARI	9.19
TOR	9.94	HOU	9.12
BAL	9.65	NYM	8.58
DET	9.52	MIL	8.56
KCR	9.17	STL	8.48
CLE	8.85	CHC	8.41
MIN	8.53	FLA	8.15
CHW	8.36	PIT	7.99
OAK	7.95	WSN	7.69
TBR	7.33	PHI	7.65
LAA	7.31	LAD	7.54
SEA	7.021	ATL	7.51
		SDP	6.69
		SFG	6.01

Source: www.baseball-reference.com

a) Create an appropriate graphical display of these data.

b) Write a few sentences comparing the average number of runs scored per game in the two leagues. (Remember: shape, center, spread, unusual features!)

c) The Texas Rangers ballpark in Arlington, Texas, was built in a "retro" style recalling older style parks. It is relatively small and has a reputation for home runs. Do you see evidence that the number of runs scored there is unusually high? Explain.

T 37. **Fruit flies** Researchers tracked a population of 1,203,646 fruit flies, counting how many died each day for 171 days. Here are three timeplots offering different views of these data. One shows the number of flies alive on each day, one the number who died that day, and the third the mortality rate—the fraction of the number alive who died. On the last day studied, the last 2 flies died, for a mortality rate of 1.0.

a) On approximately what day did the most flies die?

b) On what day during the first 100 days did the largest *proportion* of flies die?

c) When did the number of fruit flies alive stop changing very much from day to day?

T 38. **Drunk driving 2008** Accidents involving drunk drivers account for about 40% of all deaths on the nation's highways. The table below tracks the number of alcohol-related fatalities for 26 years. (www.alcoholalert.com)

Year	Deaths (thousands)	Year	Deaths (thousands)
1982	26.2	1996	17.7
1983	24.6	1997	16.7
1984	24.8	1998	16.7
1985	23.2	1999	16.6
1986	25.0	2000	17.4
1987	24.1	2001	17.4
1988	23.8	2002	17.5
1989	22.4	2003	17.1
1990	22.6	2004	16.9
1991	20.2	2005	16.9
1992	18.3	2006	15.8
1993	17.9	2007	15.4
1994	17.3	2008	13.8
1995	17.7		

a) Create a stem-and-leaf display or a histogram of these data.

b) Create a timeplot.

c) Using features apparent in the stem-and-leaf display (or histogram) and the timeplot, write a few sentences about deaths caused by drunk driving.

39. Assets Here is a histogram of the assets (in millions of dollars) of 79 companies chosen from the *Forbes* list of the nation's top corporations:

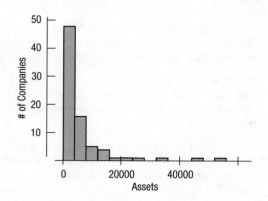

a) What aspect of this distribution makes it difficult to summarize, or to discuss, center and spread?

b) What would you suggest doing with these data if we want to understand them better?

40. Music library Students were asked how many songs they had in their digital music libraries. Here's a display of the responses:

a) What aspect of this distribution makes it difficult to summarize, or to discuss, center and spread?

b) What would you suggest doing with these data if we want to understand them better?

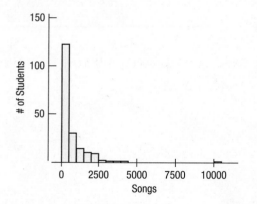

41. Assets again Here are the same data you saw in Exercise 37 after re-expressions as the square root of assets and the logarithm of assets:

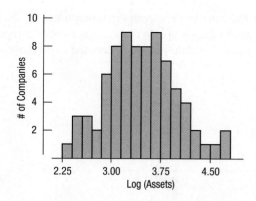

a) Which re-expression do you prefer? Why?

b) In the square root re-expression, what does the value 50 actually indicate about the company's assets?

c) In the logarithm re-expression, what does the value 3 actually indicate about the company's assets?

42. Rainmakers again The table lists the amount of rainfall (in acre-feet) from the 26 clouds seeded with silver iodide discussed in Exercise 32:

2745	703	302	242	119	40	7
1697	489	274	200	118	32	4
1656	430	274	198	115	31	
978	334	255	129	92	17	

a) Why is acre-feet a good way to measure the amount of precipitation produced by cloud seeding?

b) Plot these data, and describe the distribution.

c) Create a re-expression of these data that produces a more advantageous distribution.

d) Explain what your re-expressed scale means.

43. Stereograms Stereograms appear to be composed entirely of random dots. However, they contain separate images that a viewer can "fuse" into a three-dimensional (3D) image by staring at the dots while defocusing the eyes. An experiment was performed to determine whether knowledge of the embedded image affected the time required for subjects to fuse the images. One group of subjects (group NV) received no information or just verbal information about the shape of the embedded object. A second group (group VV) received both verbal information and visual information (specifically, a drawing of the object). The experimenters measured how many seconds it took for the subject to report that he or she saw the 3D image.

a) What two variables are discussed in this description?

b) For each variable, is it quantitative or categorical? If quantitative, what are the units?

c) The boxplots compare the fusion times for the two treatment groups. Write a few sentences comparing these distributions. What does the experiment show?

Just Checking ANSWERS

1. The % late arrivals have a unimodal, symmetric distribution centered at about 20%. In most months between 16% and 23% of the flights arrived late.

2. The boxplot of % late arrivals makes it easier to see that the median is just below 20%, with quartiles at about 17% and 22%. It nominates two months as high outliers.

3. The boxplots by month show a strong seasonal pattern. Flights are more likely to be late in the winter and summer and less likely to be late in the spring and fall. One likely reason for the pattern is snowstorms in the winter and thunderstorms in the summer.

T **44. Stereograms, revisited** Because of the skewness of the distributions of fusion times described in Exercise 43, we might consider a re-expression. Here are the boxplots of the *log* of fusion times. Is it better to analyze the original fusion times or the log fusion times? Explain.

5 The Standard Deviation as a Ruler and the Normal Model

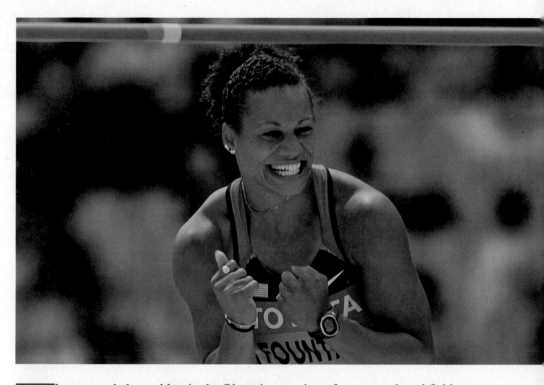

The women's heptathlon in the Olympics consists of seven track and field events: the 200-m and 800-m runs, 100-m high hurdles, shot put, javelin, high jump, and long jump. To determine who should get the gold medal, somehow the performances in all seven events have to be combined into one score. How can performances in such different events be compared? In the 2008 Olympics, Nataliya Dobrynska of the Ukraine posted a long jump of 6.63 meters—about half a meter farther than the mean distance for all contestants. Hyleas Fountain of the United States won the 200-m run with a time of 23.21 seconds—about a second and a half faster than the average. Which performance deserves more points? It's not clear how to compare them. They aren't measured in the same units, or even in the same *direction* (longer jumps are better but shorter times are better).

We want to compare individual values, not whole groups as we did earlier. To see which value is more extraordinary, we need a way to judge them against the background of data they come from. The tool we'll use is one you've already seen; the standard deviation.

The standard deviation tells us how the whole collection of values varies, so it's a natural ruler. Over and over during this course, we will ask questions such as "How far is this value from the mean?" or "How different are these two statistics?" The answer in every case will be to measure the distance or difference in standard deviations. This approach is one of the basic tools of statistical thinking.

Grading on a Curve
If you score 79% on an exam, what grade should you get? One teaching philosophy looks only at the raw percentage, 79, and bases the grade on that alone. Another looks at your *relative* performance and bases the grade on how you did compared with the rest of the class. Teachers and students still debate which method is better.

The Standard Deviation as a Ruler

In order to compare the two Olympic events, let's start with a picture. We'll use stem-and-leaf displays because they show individual values, and because it is easy to orient them either high-to-low or low-to-high so the best performances can be at the top of both displays.

Figure 5.1

Stem-and-leaf displays for the 200-m race and the long jump in the 2008 Olympic heptathlon. Hyleas Fountain (blue scores) won the 200-m, and Nataliya Dobrynska (yellow scores) won the long jump. The stems for the 200-m race run from faster to slower and the stems for the long jump from longer to shorter so that the best scores are at the top of each display.

200-m Race		Long Jump	
Stem	Leaf	Stem	Leaf
23	233	66	3
23	699	65	3
24	22333334	64	0578
24	5566677999	63	368
25	000234444	62	1
25	599	61	11235668
26	1	60	24689
		59	267778
		58	8

23|3=23.3 seconds   66|3=6.63 meters

	Long Jump	200 m
Mean (all contestants)	6.11 m	24.71 s
SD	0.238	0.700
n	36	38
Dobrynska	6.63	24.39
Fountain	6.38	23.21

Which of the two winning scores is the better one? Dobrynska's 6.63 m long jump is 0.52 m longer than the mean jump of 6.11 m. How many standard deviations better than the mean is that? The standard deviation for this event was 0.238 m, so her jump was $(6.63 - 6.11)/0.238 = 2.18$ standard deviations better than the mean. Fountain's winning 200-m run was 1.50 seconds faster than the mean, and that's $(23.21 - 24.71)/0.700 = -2.14$, or 2.14 standard deviations faster than the mean. That's a winning performance, but just a bit less impressive than Dobrynska's long jump.

Standardizing with z-Scores

Expressing a distance from the mean in standard deviations *standardizes* the performances. To **standardize** a value, we subtract the mean and then divide this difference by the standard deviation:

$$z = \frac{y - \bar{y}}{s}$$

■ NOTATION ALERT

We always use the letter *z* to denote values that have been standardized with the mean and standard deviation.

The values are called **standardized values**, and are commonly denoted with the letter z. Usually we just call them **z-scores**.

z-scores measure the distance of a value from the mean in standard deviations. A z-score of 2 says that a data value is 2 standard deviations above the mean. It doesn't matter whether the original variable was measured in fathoms, dollars, or carats; those units don't apply to z-scores. Data values below the mean have negative z-scores, so a z-score of −1.6 means that the data value was 1.6 standard deviations below the mean. Of course, regardless of the direction, the farther a data value is from the mean, the more unusual it is, making Dobrynska's long jump with a z-score of 2.18 better than Fountain's 200 m race with a z-score of −2.14.

For Example STANDARDIZING SKIING TIMES

The men's super combined skiing event debuted in the 2010 Winter Olympics in Vancouver. It consists of two races: a downhill and a slalom. Times for the two events are added together, and the skier with the lowest total time wins. At Vancouver, the mean slalom time was 52.67 seconds with a standard deviation of 1.614 seconds. The mean downhill time was 116.26 seconds with a standard deviation of 1.914 seconds. Bode Miller of the United States, who won the gold medal with a combined time of 164.92 seconds, skied the slalom in 51.01 seconds and the downhill in 113.91 seconds.

(continued)

QUESTION: On which race did he do better compared to the competition?

ANSWER: $z_{slalom} = \dfrac{y - \bar{y}}{s} = \dfrac{51.01 - 52.67}{1.614} = -1.03$

$z_{downhill} = \dfrac{113.91 - 116.26}{1.914} = -1.23$

Keeping in mind that faster times are *below* the mean, Miller's downhill time of 1.23 SDs below the mean is even more remarkable than his slalom time, which was 1.03 SDs below the mean.

Once the outcomes of each event have been standardized (removing units), we can evaluate the overall performances by simply adding the z-scores. To combine all seven Heptathlon events—each with its own scale—into a single score, Olympic judges use tables based on similar calculations. In the final standings (including all seven events), Dobrynska took the gold medal and Fountain the silver.

		Event	
		Long Jump	**200 m Run**
	Mean	6.11	24.71
	SD	0.238	0.700
Dobrynska	**Performance**	6.63 m	24.39 s
	z-score	$(6.63 - 6.11)/0.238 = 2.18$	$(24.39 - 24.71)/0.700 = -0.457$
	Total z-score	$2.18 + 0.457 = \mathbf{2.637}$	
Fountain	**Performance**	6.38 m	23.21 s
	z-score	$(6.38 - 6.11)/0.238 = 1.13$	$(23.21 - 24.71)/0.700 = -2.14$
	Total z-score	$1.13 + 2.14 = \mathbf{3.27}$	

For Example   COMBINING z-SCORES

At the 2010 Vancouver Winter Olympics, Bode Miller had earlier earned a bronze medal in the downhill and a silver medal in the super-G slalom before winning the gold in the men's super combined. Ted Ligety, the winner of the gold medal at the 2006 games, had super combined times in 2010 of 50.76 seconds in the slalom (fastest of the 35 finalists) and 115.06 seconds in the downhill for a total of 165.82 seconds, almost a second behind Miller. But he finished in fifth place in 2010. He beat Miller in the slalom, but Miller beat him in the downhill. The downhill is longer than the slalom and so, counts for more in the total.

QUESTION: Would the placement have changed if each event had been treated equally by standardizing each and adding the standardized scores?

ANSWER: We've seen that Miller's z-scores for slalom and downhill were -1.03 and -1.23, respectively. That's a total of -2.26. Ligety's z-scores were

$$z_{slalom} = \frac{y - \bar{y}}{s} = \frac{50.76 - 52.67}{1.614} = -1.18$$

$$z_{downhill} = \frac{115.06 - 116.26}{1.914} = -0.63$$

So his total z-score was -1.81. That's not as good as Miller's total of -2.26. So, although Ligety beat Miller in the slalom, Miller beat him by *more* in the downhill. Using the standardized scores would not have put Ligety ahead of Miller.

Just Checking

1. Your Statistics teacher has announced that the lower of your two tests will be dropped. You got a 90 on test 1 and an 80 on test 2. You're all set to drop the 80 until she announces that she grades "on a curve." She standardized the scores in order to decide which is the lower one. If the mean on the first test was 88 with a standard deviation of 4 and the mean on the second was 75 with a standard deviation of 5,
   a) Which one will be dropped?
   b) Does this seem "fair"?

When we standardize data to get a z-score, we do two things. First, we shift the data by subtracting the mean. Then, we rescale the values by dividing by their standard deviation. We often shift and rescale data. What happens to a grade distribution if *everyone* gets a five-point bonus? Everyone's grade goes up, but does the shape change? (*Hint:* Has anyone's distance from the mean changed?) If we switch from feet to meters, what happens to the distribution of heights of students in your class? Even though your intuition probably tells you the answers to these questions, we need to look at exactly how shifting and rescaling work.

Shifting Data: Move the Center

Who	80 male participants of the NHANES survey between the ages of 19 and 24 who measured between 68 and 70 inches tall
What	Their weights
Unit	Kilograms
When	2001–2002
Where	United States
Why	To study nutrition, and health issues and trends
How	National survey

Since the 1960s, the Centers for Disease Control's National Center for Health Statistics has been collecting health and nutritional information on people of all ages and backgrounds. The National Health and Nutrition Examination Survey (NHANES) 2001–2002,[1] measured a wide variety of variables, including body measurements, cardiovascular fitness, blood chemistry, and demographic information on more than 11,000 individuals.

Included in this group were 80 men between 19 and 24 years old of average height (between 5′8″ and 5′10″ tall). Here are a histogram and boxplot of their weights:

Figure 5.2

Histogram and boxplot for the men's weights. The shape is skewed to the right with several high outliers.

A S *Activity:* **Changing the Baseline.** What happens when we shift data? Do measures of center and spread change?

Their mean weight is 82.36 kg. For this age and height group, the National Institutes of Health recommends a maximum healthy weight of 74 kg, but we can see that some of the men are heavier than the recommended weight. To compare their weights to the

---

[1]www.cdc.gov/nchs/nhanes.htm

recommended maximum, we could subtract 74 kg from each of their weights. What would that do to the center, shape, and spread of the histogram? Here's the picture:

Figure 5.3

Subtracting 74 kilograms shifts the entire histogram down but leaves the spread and the shape exactly the same.

On average, they weigh 82.36 kg, so on average they're 8.36 kg overweight. And, after subtracting 74 from each weight, the mean of the new distribution is 82.36 − 74 = 8.36 kg. In fact, when we **shift** the data by adding (or subtracting) a constant to each value, all measures of position (center, percentiles, min, max) will increase (or decrease) by the same constant.

What about the spread? What does adding or subtracting a constant value do to the spread of the distribution? Look at the two histograms again. Adding or subtracting a constant changes each data value equally, so the entire distribution just shifts. Its shape doesn't change and neither does the spread. None of the measures of spread we've discussed—not the range, not the IQR, not the standard deviation—changes.

> *Adding (or subtracting) a constant to every data value adds (or subtracts) the same constant to measures of position, but leaves measures of spread unchanged.*

Shifting Heights
Doctors' height and weight charts sometimes give ideal weights for various heights that include 2-inch heels. If the mean height of adult women is 66 inches including 2-inch heels, what is the mean height of women without shoes? Each woman is shorter by 2 inches when barefoot, so the mean is decreased by 2 inches, to 64 inches.

Rescaling Data: Adjust the Scale

Not everyone thinks naturally in metric units. Suppose we want to look at the weights in pounds instead. We'd have to **rescale** the data. Because there are about 2.2 pounds in every kilogram, we'd convert the weights by multiplying each value by 2.2. Multiplying or dividing each value by a constant changes the measurement units. Here are histograms of the two weight distributions, plotted on the same scale, so you can see the effect of multiplying:

Figure 5.4

Men's weights in both kilograms and pounds. How do the distributions and numerical summaries change?

A S *Simulation:* **Changing the Units.**
Change the center and spread values for a distribution and watch the summaries change (or not, as the case may be).

What happens to the shape of the distribution? Although the histograms don't look exactly alike, we see that the shape really hasn't changed: Both are unimodal and skewed to the right.

What happens to the mean? Not too surprisingly, it gets multiplied by 2.2 as well. The men weigh 82.36 kg on average, which is 181.19 pounds. As the boxplots and 5-number summaries show, all measures of position act the same way. They all get multiplied by this same constant.

What happens to the spread? Take a look at the boxplots. The spread in pounds (on the right) is larger. How much larger? If you guessed 2.2 times, you've figured out how measures of spread get rescaled.

Figure 5.5

The boxplots (drawn on the same scale) show the weights measured in kilograms (on the left) and pounds (on the right). Because 1 kg is 2.2 lb, all the points in the right box are 2.2 times larger than the corresponding points in the left box. So each measure of position and spread is 2.2 times as large when measured in pounds rather than kilograms.

	Weight (kg)	Weight (lb)
Min	54.3	119.46
Q1	67.3	148.06
Median	76.85	169.07
Q3	92.3	203.06
Max	161.5	355.30
IQR	25	55
SD	22.27	48.99

When we multiply (or divide) all the data values by any constant, all measures of position (such as the mean, median, and percentiles) and measures of spread (such as the range, the IQR, and the standard deviation) are multiplied (or divided) by that same constant.

For Example RESCALING THE MEN'S COMBINED TIMES

RECAP: The times in the men's combined event at the winter Olympics are reported in minutes and seconds. The mean and standard deviation of the 34 final super combined times at the 2010 Olympics were 168.93 seconds and 2.90 seconds, respectively.

QUESTION: Suppose instead that we had reported the times in minutes—that is, that each individual time was divided by 60. What would the resulting mean and standard deviation be?

ANSWER: Dividing all the times by 60 would divide both the mean and the standard deviation by 60:

Mean = 168.93/60 = 2.816 minutes; SD = 2.90/60 = 0.048 minute.

Just Checking

2. In 1995 the Educational Testing Service (ETS) adjusted the scores of SAT tests. Before ETS recentered the SAT Verbal test, the mean of all test scores was 450.
   a) How would adding 50 points to each score affect the mean?
   b) The standard deviation was 100 points. What would the standard deviation be after adding 50 points?
   c) Suppose we drew boxplots of test takers' scores a year before and a year after the recentering. How would the boxplots of the two years differ?

3. A company manufactures wheels for in-line skates. The diameters of the wheels have a mean of 3 inches and a standard deviation of 0.1 inches. Because so many of their customers use the metric system, the company decided to report their production statistics in millimeters (1 inch = 25.4 mm). They report that the standard deviation is now 2.54 mm. A corporate executive is worried about this increase in variation. Should he be concerned? Explain.

Back to z-scores

A S **Activity: Standardizing.** What if we both shift and rescale? The result is so nice that we give it a name.

Standardizing data into z-scores is just shifting them by the mean and rescaling them by the standard deviation. Now we can see how standardizing affects the distribution. When we subtract the mean of the data from every data value, we shift the mean to zero. As we have seen, such a shift doesn't change the standard deviation.

When we *divide* each of these shifted values by s, however, the standard deviation should be divided by s as well. Since the standard deviation was s to start with, the new standard deviation becomes 1.

How, then, does standardizing affect the distribution of a variable? Let's consider the three aspects of a distribution: the shape, center, and spread.

> **Z-Scores**
>
> z-scores have mean 0 and standard deviation 1.

*Standardizing into z-scores does not change the **shape** of the distribution of a variable.*
*Standardizing into z-scores changes the **center** by making the mean 0.*
*Standardizing into z-scores changes the **spread** by making the standard deviation 1.*

Step-by-Step Example WORKING WITH STANDARDIZED VARIABLES

Many colleges and universities require applicants to submit scores on standardized tests such as the SAT Writing, Math, and Critical Reading (Verbal) tests. The college your little sister wants to apply to says that while there is no minimum score required, the middle 50% of their students have combined SAT scores between 1530 and 1850. You'd feel confident if you knew her score was in their top 25%, but unfortunately she took the ACT test, an alternative standardized test.

Question: How high does her ACT need to be to make it into the top quarter of equivalent SAT scores?

To answer that question you'll have to standardize all the scores, so you'll need to know the mean and standard deviations of scores for some group on both tests. The college doesn't report the mean or standard deviation for their applicants on either test, so we'll use the group of all test takers nationally. For college-bound seniors, the average combined SAT score is about 1500 and the standard deviation is about 250 points. For the same group, the ACT average is 20.8 with a standard deviation of 4.8.

THINK ➡ **Plan** State what you want to find out.	I want to know what ACT score corresponds to the upper-quartile SAT score. I know the mean and standard deviation for both the SAT and ACT scores based on all test takers, but I have no individual data values.
Variables Identify the variables and report the W's (if known).	
Check the appropriate conditions.	✓ **Quantitative Data Condition:** Scores for both tests are quantitative but have no meaningful units other than points.
SHOW ➡ **Mechanics** Standardize the variables.	The middle 50% of SAT scores at this college fall between 1530 and 1850 points. To be in the top quarter, my sister would have to have a score of at least 1850. That's a z-score of $$z = \frac{(1850 - 1500)}{250} = 1.40$$ So an SAT score of 1850 is 1.40 standard deviations above the mean of all test takers.
The y-value we seek is z standard deviations above the mean.	For the ACT, 1.40 standard deviations above the mean is $20.8 + 1.40(4.8) = 27.52$.

(continued)

<table>
<tr><td>

TELL ➡ **Conclusion**   Interpret your results in
context.

</td><td>

To be in the top quarter of applicants in terms of combined SAT score, she'd need to have an ACT score of at least 27.52.

</td></tr>
</table>

When Is a *z*-score BIG?

<table>
<tr><td>

Is Normal Normal?
Don't be misled. The name "Normal" doesn't mean that these are the *usual* shapes for histograms. The name follows a tradition of positive thinking in Mathematics and Statistics in which functions, equations, and relationships that are easy to work with or have other nice properties are called "normal," "common," "regular," "natural," or similar terms. It's as if by calling them ordinary, we could make them actually occur more often and simplify our lives.

</td><td>

Champion runners turn in exceptionally low times, and fishermen brag about "the big one."[2] Not only do things that are unusually large or unusually small catch our interest, they can provide especially important information about whatever we are investigating. Being far from typical, they have *z*-scores far from 0. How far from 0 does a *z*-score have to be to be interesting or unusual? There is no universal standard, but the larger the score is (negative or positive), the more unusual it is. We know that 50% of the data lie between the quartiles. For symmetric data, the standard deviation is usually a bit smaller than the IQR, and it's not uncommon for at least half of the data to have *z*-scores between −1 and 1. But no matter what the shape of the distribution, a *z*-score of 3 (plus or minus) or more is rare, and a *z*-score of 6 or 7 shouts out for attention.

To say more about how big we expect a *z*-score to be, we need to *model* the data's distribution. A model will let us say much more precisely how often we'd be likely to see *z*-scores of different sizes. Of course, like all models of the real world, the model will be wrong—wrong in the sense that it can't match reality exactly. But it can still be useful. Like a physical model, it's something we can look at and manipulate in order to learn more about the real world.

Models help our understanding in many ways. Just as a model of an airplane in a wind tunnel can give insights even though it doesn't show every rivet,[3] models of data give us summaries that we can learn from and use, even though they don't fit each data value exactly. It's important to remember that they're only *models* of reality and not reality itself. But without models, what we can learn about the world at large is limited to only what we can say about the data we have at hand.

There is no universal standard for *z*-scores, but there is a model that shows up over and over in Statistics. You may have heard of "bell-shaped curves." Statisticians call them Normal models. **Normal models** are appropriate for distributions whose shapes are unimodal and roughly symmetric. For these distributions, they provide a measure of how extreme a *z*-score is. Fortunately, there is a Normal model for every possible combination of mean and standard deviation. We write $N(\mu, \sigma)$ to represent a Normal model with a mean of μ and a standard deviation of σ. Why the Greek? Well, *this* mean and standard deviation are not numerical summaries of data. They are part of the model. They don't come from the data. Rather, they are numbers that we choose to help specify the model. Such numbers are called **parameters** of the model.

We don't want to confuse the parameters with summaries of the data such as $\bar{y}$ and s, so we use special symbols. In Statistics, we almost always use Greek letters for parameters. By contrast, summaries of data are called **statistics** and are usually written with Latin letters.

If we model data with a Normal model and standardize them using the corresponding μ and σ, we still call the standardized value a **z-score,** and we write

$$z = \frac{y - \mu}{\sigma}.$$

</td></tr>
</table>

■ **NOTATION ALERT**

$N(\mu, \sigma)$ always denotes a Normal model. The μ, pronounced "mew," is the Greek letter for "m" and always represents the mean in a model. The σ, sigma, is the lowercase Greek letter for "s" and always represents the standard deviation in a model.

Is the Standard Normal a standard?
Yes. We call it the "Standard Normal" because it models standardized values. It is also a "standard" because this is the particular Normal model that we almost always use.

Usually it's easier to standardize data first (using its mean and standard deviation). Then we need only the model $N(0,1)$. The Normal model with mean 0 and standard deviation 1 is called the **standard Normal model** (or the **standard Normal distribution**).

_____

[2]Especially if it got away.

[3]In fact, the model is useful *because* it doesn't have every rivet. It is because models offer a simpler view of reality that they are so useful as we try to understand reality.

> "All models are wrong—but some are useful."
>
> —*George Box, famous statistician*

A **S** *Activity*: **Working with Normal Models.** Learn more about the Normal model and see what data drawn at random from a Normal model might look like.

But be careful. You shouldn't use a Normal model for just any data set. Remember that standardizing won't change the shape of the distribution. If the distribution is not unimodal and symmetric to begin with, standardizing won't make it Normal.

When we use the Normal model, we assume that the distribution of the data is, well, Normal. Practically speaking, there's no way to check whether this **Normality Assumption** is true. In fact, it almost certainly is not true. Real data don't behave like mathematical models. Models are idealized; real data are real. The good news, however, is that to use a Normal model, it's sufficient to check the following condition:

> **Nearly Normal Condition.** The shape of the data's distribution is unimodal and symmetric. Check this by making a histogram (or a Normal probability plot, which we'll explain later).

Don't model data with a Normal model without checking whether the condition is satisfied.

All models make **assumptions.** Whenever we model—and we'll do that often—we'll be careful to point out the assumptions that we're making. And, what's even more important, we'll check the associated **conditions** in the data to make sure that those assumptions are reasonable.

The 68–95–99.7 Rule

One in a Million
These magic 68, 95, 99.7 values come from the Normal model. As a model, it can give us corresponding values for any *z*-score. For example, it tells us that fewer than 1 out of a million values have *z*-scores smaller than −5.0 or larger than +5.0. so if someone tells you you're "one in a million," they must really admire your *z*-score.

TI-*nspire*
The 68–95–99.7 Rule. See it work for yourself.

Normal models give us an idea of how extreme a value is by telling us how likely it is to find one that far from the mean. We'll soon show how to find these numbers precisely—but one simple rule is usually all we need.

It turns out that in a Normal model, about 68% of the values fall within 1 standard deviation of the mean, about 95% of the values fall within 2 standard deviations of the mean, and about 99.7%—almost all—of the values fall within 3 standard deviations of the mean. These facts are summarized in a rule that we call (let's see . . .) the **68–95–99.7 Rule.**[4]

Figure 5.6

Reaching out one, two, and three standard deviations on a Normal model gives the 68–95–99.7 Rule, seen as proportions of the area under the curve.

For Example USING THE 68–95–99.7 RULE

QUESTION: In the 2010 Winter Olympics men's slalom, Li Lei of China skied in a total time of 120.86 sec for two runs—about 1 standard deviation slower than the mean. If a Normal model is useful in describing slalom times, about how many of the 48 skiers finishing the event would you expect skied the slalom *slower* than Li Lei?

ANSWER: *From the 68–95–99.7 Rule, we expect 68% of the skiers to be within one standard deviation of the mean. Of the remaining 32%, we expect half on the high end and half on the low end. 16% of 48 is 7.7, so, conservatively, we'd expect about 7 skiers to do worse than Li Lei.*

---

[4]This rule is also called the "Empirical Rule" because it originally came from observation. The rule was first published by Abraham de Moivre in 1733, 75 years before the Normal model was discovered. Maybe it should be called "de Moivre's Rule," but that wouldn't help us remember the important numbers, 68, 95, and 99.7.

Just Checking

4. As a group, the Dutch are among the tallest people in the world. The average Dutch man is 184 cm tall—just over 6 feet (and the average Dutch woman is 170.8 cm tall—just over 5′7″). If a Normal model is appropriate and the standard deviation for men is about 8 cm, what percentage of all Dutch men will be over 2 meters (6′6″) tall?

5. Suppose it takes you 20 minutes, on average, to drive to school, with a standard deviation of 2 minutes.

Suppose a Normal model is appropriate for the distributions of driving times.

a) How often will you arrive at school in less than 22 minutes?
b) How often will it take you more than 24 minutes?
c) Do you think the distribution of your driving times is unimodal and symmetric?
d) What does this say about the accuracy of your predictions? Explain.

The First Three Rules for Working with Normal Models

A S *Activity*: **Working with Normal Models.** Well, actually playing with them. This interactive tool lets you do what this chapter's figures can't do, move them!

A S *Activity*: **Normal Models.** Normal models have several interesting properties—see them here.

1. Make a picture.
2. Make a picture.
3. Make a picture.

Although we're thinking about models, not histograms of data, the three rules don't change. To help you think clearly, a simple hand-drawn sketch is all you need. Even experienced statisticians sketch pictures to help them think about Normal models. You should too.

Of course, when we have data, we'll also need to make a histogram to check the **Nearly Normal Condition** to be sure we can use the Normal model to model the data's distribution. Other times, we may be told that a Normal model is appropriate based on prior knowledge of the situation or on theoretical considerations.

TI-*nspire*

Normal models. Watch the Normal model react as you change the mean and standard deviation.

How to Sketch a Normal Curve That Looks Normal

To sketch a good Normal curve, you need to remember only three things:

■ The Normal curve is bell-shaped and symmetric around its mean. Start at the middle, and sketch to the right and left from there.

■ Even though the Normal model extends forever on either side, you need to draw it only for 3 standard deviations. After that, there's so little left that it isn't worth sketching.

■ The place where the bell shape changes from curving downward to curving back up—the *inflection point*—is exactly one standard deviation away from the mean.

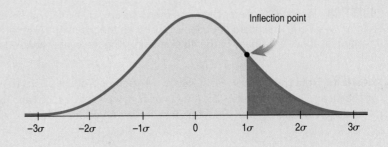

Step-by-Step Example WORKING WITH THE 68–95–99.7 RULE

The SAT Reasoning Test has three parts: Writing, Math, and Critical Reading (Verbal). Each part has a distribution that is roughly unimodal and symmetric and is designed to have an overall mean of about 500 and a standard deviation of 100 for all test takers. In any one year, the mean and standard deviation may differ from these target values by a small amount, but they are a good overall approximation.

Question: Suppose you earned a 600 on one part of your SAT. Where do you stand among all students who took that test?

You could calculate your z-score and find out that it's $z = (600 - 500)/100 = 1.0$, but what does that tell you about your percentile? You'll need the Normal model and the 68–95–99.7 Rule to answer that question.

THINK ➡ **Plan** State what you want to know.

I want to see how my SAT score compares with the scores of all other students. To do that, I'll need to model the distribution.

Variables Identify the variable and report the W's.

Let y = my SAT score. Scores are quantitative but have no meaningful units other than points.

Be sure to check the appropriate conditions.

✓ **Nearly Normal Condition:** If I had data, I would check the histogram. I have no data, but I am told that the SAT scores are roughly unimodal and symmetric.

Specify the parameters of your model.

I will model SAT score with a N(500, 100) model.

SHOW ➡ **Mechanics** Make a picture of this Normal model. (A simple sketch is all you need.)

Locate your score.

My score of 600 is 1 standard deviation above the mean. That corresponds to one of the points of the 68–95–99.7 Rule.

TELL ➡ **Conclusion** Interpret your result in context.

About 68% of those who took the test had scores that fell no more than 1 standard deviation from the mean, so 100% − 68% = 32% of all students had scores more than 1 standard deviation away. Only half of those were on the high side, so about 16% (half of 32%) of the test scores were better than mine. My score of 600 is higher than about 84% of all scores on this test, placing me at the 84th percentile.

The bounds of SAT scoring at 200 and 800 can also be explained by the 68−95−99.7 Rule. Since 200 and 800 are three standard deviations from 500, it hardly pays to extend the scoring any farther on either side. We'd get more information only on $100 − 99.7 = 0.3\%$ of students.

---

*The Worst-Case Scenario   Suppose we encounter an observation that's 5 standard deviations above the mean. Should we be surprised? We've just seen that when a Normal model is appropriate, such a value is exceptionally rare. After all, 99.7% of all the values should be within 3 standard deviations of the mean, so anything farther away would be unusual indeed.

But our handy 68−95−99.7 Rule applies only to Normal models, and the Normal is such a *nice* shape. What if we're dealing with a distribution that's strongly skewed (like the CEO salaries, in Chapter 4), or one that is uniform or bimodal or something really strange? A Normal model has 68% of its observations within one standard deviation of the mean, but a bimodal distribution could even be entirely empty in the middle. In that case could we still say anything at all about an observation 5 standard deviations above the mean?

Remarkably, even with really weird distributions, the worst case can't get all that bad. A Russian mathematician named Pafnuty Tchebycheff[5] answered the question by proving this theorem:

In any distribution, at least $1 − \dfrac{1}{k^2}$ of the values must lie within $\pm k$ standard deviations of the mean.

What does that mean?

For $k = 1$, $1 − \dfrac{1}{1^2} = 0$; if the distribution is far from Normal, it's possible that none of the values is within 1 standard deviation of the mean. We should be really cautious about saying anything about 68% unless we think a Normal model is justified. (Tchebycheff's theorem really is about the worst case; it tells us nothing about the middle; only about the extremes.)

For $k = 2$, $1 − \dfrac{1}{2^2} = \dfrac{3}{4}$; no matter how strange the shape of the distribution, at least 75% of the values must be within 2 standard deviations of the mean. Normal models may expect 95% in that 2-standard-deviation interval, but even in a worst-case scenario it can never go lower than 75%.

For $k = 3$, $1 − \dfrac{1}{3^2} = \dfrac{8}{9}$; in any distribution, at least 89% of the values lie within 3 standard deviations of the mean.

What we see is that values beyond 3 standard deviations from the mean are uncommon, Normal model or not. Tchebycheff tells us that at least 96% of all values must be within 5 standard deviations of the mean. While we can't always apply the 68−95−99.7 Rule, we can be sure that the observation we encountered 5 standard deviations above the mean is unusual.

---

[5]He may have made the worst case for deviations clear, but the English spelling of his name is not. You'll find his first name spelled Pavnutii or Pavnuty and his last name spelled Chebsheff, Cebysev, and other creative versions.

Finding Normal Percentiles

A S **Activity: Your Pulse z-Score.**
Is your pulse rate high or low? Find its
z-score with the Normal Model Tool.

A S **Activity: The Normal Table.**
Table Z just sits there, but this version
of the Normal table changes so it always
Makes a Picture that fits.

An SAT score of 600 is easy to assess, because we can think of it as one standard deviation above the mean. If your score was 680, though, where do you stand among the rest of the people tested? Your z-score is 1.80, so you're somewhere between 1 and 2 standard deviations above the mean. We figured out that no more than 16% of people score better than 600. By the same logic, no more than 2.5% of people score better than 700. Can we be more specific than "between 16% and 2.5%"?

When the value doesn't fall exactly 1, 2, or 3 standard deviations from the mean, we can look it up in a table of **Normal percentiles** or use technology.[6] Either way, we first convert our data to z-scores before using the table. Your SAT score of 680 has a z-score of $(680 - 500)/100 = 1.80$.

Figure 5.7
A table of Normal percentiles
(Table Z in Appendix G) lets us find
the percentage of individuals in a
Standard Normal distribution falling
below any specified z-score value.

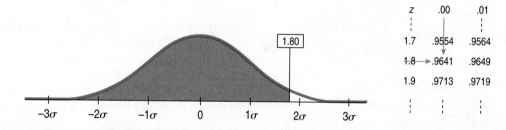

z	.00	.01
⋮	⋮	⋮
1.7	.9554	.9564
1.8 →	.9641	.9649
1.9	.9713	.9719
⋮	⋮	⋮

TI-nspire
Normal percentiles. Explore the
relationship between z-scores and
areas in a Normal model.

In the piece of the table shown, we find your z-score by looking down the left column for the first two digits, 1.8, and across the top row for the third digit, 0. The table gives the percentile as 0.9641. That means that 96.4% of the z-scores are less than 1.80. Only 3.6% of people, then, scored better than 680 on the SAT.

Most of the time, though, you'll do this with your calculator.

TI Tips FINDING NORMAL PERCENTAGES

Your calculator knows the Normal model. Have a look under 2nd `DISTR`. There you will see three "norm" functions, `normalpdf(`, `normalcdf(`, and `invNorm(`. Let's play with the first two.

- `normalpdf(` calculates y-values for graphing a Normal curve. You probably won't use this very often, if at all. If you want to try it, graph `Y1 = normalpdf(X)` in a graphing `WINDOW` with `Xmin = -4`, `Xmax = 4`, `Ymin = -0.1`, and `Ymax = 0.5`.
- `normalcdf(` finds the proportion of area under the curve between two z-score cut points, by specifying `normalcdf(zLeft, zRight)`. Do make friends with this function; you will use it often!

EXAMPLE 1: The Normal model shown shades the region between $z = -0.5$ and $z = 1.0$.

(continued)

---
[6]See Table Z in Appendix G, if you're curious. But your calculator (and any statistics computer package) does this, too—and more easily!

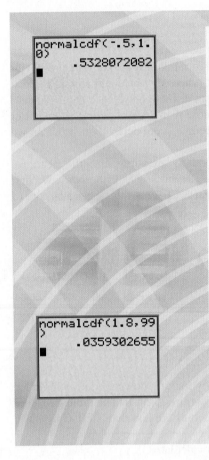

To find the shaded area:

- Under 2nd DISTR select normalcdf(and hit ENTER.
- Specify lower: −.5, uppr:1, μ:0, sigma:1, and then go to Paste and hit ENTER twice (OR on an older calculator, just enter normalcdf(-.5,1)).

There's the area. Approximately 53% of a Normal model lies between half a standard deviation below and one standard deviation above the mean.

EXAMPLE 2: In the example in the text we used Table Z to determine the fraction of SAT scores above your score of 680. Now let's do it again, this time using your TI.

First we need z-scores for the cut points:

- Since 680 is 1.8 standard deviations above the mean, your z-score is 1.8; that's the left cut point.
- Theoretically the standard Normal model extends rightward forever, but you can't tell the calculator to use infinity as the right cut point. Recall that for a Normal model almost all the area lies within ±3 standard deviations of the mean, so any upper cut point beyond, say, $z = 5$ does not cut off anything very important. We suggest you always use 99 (or −99) when you really want infinity as your cut point—it's easy to remember and way beyond any meaningful area.

Now you're ready. Use the command normalcdf(as above with lower:1.8, upper:99.

There you are! The Normal model estimates that approximately 3.6% of SAT scores are higher than 680.

Step-by-Step Example WORKING WITH NORMAL MODELS PART I

The Normal model is our first model for data. It's the first in a series of modeling situations where we step away from the data at hand to make more general statements about the world. We'll become more practiced in thinking about and learning the details of models as we progress through the book. To give you some practice in thinking about the Normal model, here are several problems that ask you to find percentiles in detail.

Question: What proportion of SAT scores fall between 450 and 600?

THINK ➡ **Plan** State the problem.

I want to know the proportion of SAT scores between 450 and 600.

Variables Name the variable.

Let y = SAT score.

Check the appropriate conditions and specify which Normal model to use.

✓ **Nearly Normal Condition:** We are told that SAT scores are nearly Normal.

I'll model SAT scores with a N(500, 100) model, using the mean and standard deviation specified for them.

(continued)

SHOW ➡ **Mechanics** Make a picture of this Normal model. Locate the desired values and shade the region of interest.

Find z-scores for the cut points 450 and 600. Use technology to find the desired proportions, represented by the area under the curve. (This was Example 1 in the TI Tips—take another look.)

Standardizing the two scores, I find that

$$z = \frac{(y - \mu)}{\sigma} = \frac{(600 - 500)}{100} = 1.00$$

and

$$z = \frac{(450 - 500)}{100} = -0.50$$

So,

$$\text{Area}(450 < y < 600) = \text{Area}(-0.5 < z < 1.0)$$
$$= 0.5328$$

(If you use a table, then you need to subtract the two areas to find the area *between* the cut points. If you think using the calculator is far easier...bingo!)

(**OR**: From Table Z, the area $(z < 1.0) = 0.8413$ and area $(z < -0.5) = 0.3085$, so the proportion of z-scores *between* them is $0.8413 - 0.3085 = 0.5328$, or 53.28%.)

TELL ➡ **Conclusion** Interpret your result in context.

The Normal model estimates that about 53.3% of SAT scores fall between 450 and 600.

From Percentiles to Scores: *z* in Reverse

Finding areas from z-scores is the simplest way to work with the Normal model. But sometimes we start with areas and are asked to work backward to find the corresponding z-score or even the original data value. For instance, what z-score cuts off the top 10% in a Normal model?

Make a picture like the one shown, shading the rightmost 10% of the area. Notice that this is the 90th percentile. If you're determined to do it the hard way, look in Table Z for an area of 0.900. The exact area is not there, but 0.8997 is pretty close. That shows up in the table with 1.2 in the left margin and .08 in the top margin. The z-score for the 90th percentile, then, is approximately $z = 1.28$.

Computers and calculators will determine the cut point more precisely (and more easily).

TI Tips FINDING NORMAL CUTPOINTS

To find the *z*-score at the 25th percentile, go to 2nd DISTR again. This time we'll use the third of the "norm" functions, invNorm(.

- Just specify the desired percentile using the invNorm(command with area: 0.25, mu:0, sigma:1, then go to Paste and hit ENTER. (Twice.)

The calculator says that the cut point for the leftmost 25% of a Normal model is approximately $z = -0.674$.

One more example: What *z*-score cuts off the highest 10% of a Normal model? That's easily done—just remember to specify the *percentile*. Since we want the cut point for the *highest* 10%, we know that the other 90% must be *below* that *z*-score. The cut point, then, must stand at the 90th percentile, so specify invNorm(.90,0,1).

Only 10% of the area in a Normal model is more than about 1.28 standard deviations above the mean.

Step-by-Step Example WORKING WITH NORMAL MODELS PART II

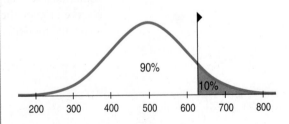

Question: Suppose a college says it admits only people with SAT Verbal test scores among the top 10%. How high a score does it take to be eligible?

THINK ➡ **Plan** State the problem.

How high an SAT Verbal score do I need to be in the top 10% of all test takers?

Variable Define the variable.

Let y = my SAT score.

Check to see if a Normal model is appropriate, and specify which Normal model to use.

✓ **Nearly Normal Condition:** I am told that SAT scores are nearly Normal. I'll model them with N(500, 100).

SHOW ➡ **Mechanics** Make a picture of this Normal model. Locate the desired percentile approximately by shading the rightmost 10% of the area.

The college takes the top 10%, so its cutoff score is the 90th percentile. Find the corresponding *z*-score using your calculator as shown in the TI Tips. (*OR*: Use Table Z as shown on p. 121.)

The cut point is $z = 1.28$.

Convert the *z*-score back to the original units.

A z-score of 1.28 is 1.28 standard deviations above the mean. Since the SD is 100, that's 128 SAT points. The cutoff is 128 points above the mean of 500, or 628.

(continued)

TELL ➡ **Conclusion** Interpret your results in the proper context.	Because the school wants SAT Verbal scores in the top 10%, the cutoff is 628. (Actually, since SAT scores are reported only in multiples of 10, I'd have to score at least a 630.)

Step-by-Step Example MORE WORKING WITH NORMAL MODELS

Working with Normal percentiles can be a little tricky, depending on how the problem is stated. Here are a few more worked examples of the kind you're likely to see.

A cereal manufacturer has a machine that fills the boxes. Boxes are labeled "16 ounces," so the company wants to have that much cereal in each box, but since no packaging process is perfect, there will be minor variations. If the machine is set at exactly 16 ounces and the Normal model applies (or at least the distribution is roughly symmetric), then about half of the boxes will be underweight, making consumers unhappy and exposing the company to bad publicity and possible lawsuits. To prevent underweight boxes, the manufacturer has to set the mean a little higher than 16.0 ounces.

Based on their experience with the packaging machine, the company believes that the amount of cereal in the boxes fits a Normal model with a standard deviation of 0.2 ounces. The manufacturer decides to set the machine to put an average of 16.3 ounces in each box. Let's use that model to answer a series of questions about these cereal boxes.

Question 1: What fraction of the boxes will be underweight?

THINK ➡ **Plan** State the problem.	What proportion of boxes weigh less than 16 ounces?
Variable Name the variable.	Let y = weight of cereal in a box.
Check to see if a Normal model is appropriate.	✓ **Nearly Normal Condition:** I have no data, so I cannot make a histogram, but I am told that the company believes the distribution of weights from the machine is Normal.
Specify which Normal model to use.	I'll use a $N(16.3, 0.2)$ model.

SHOW ➡ **Mechanics** Make a picture of this Normal model. Locate the value you're interested in on the picture, label it, and shade the appropriate region.	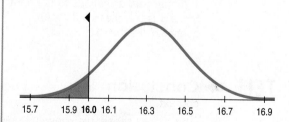
REALITY CHECK Estimate from the picture the percentage of boxes that are underweight. (This will be useful later to check that your answer makes sense.) It looks like a low percentage. Less than 20% for sure.	I want to know what fraction of the boxes will weigh less than 16 ounces.
Convert your cutoff value into a *z*-score.	$z = \dfrac{y - \mu}{\sigma} = \dfrac{16 - 16.3}{0.2} = -1.50$
Find the area with your calculator (or use the Normal table).	Area $(y < 16)$ = Area$(z < -1.50)$ = 0.0668

(continued)

TELL ➡ **Conclusion** State your conclusion, and check that it's consistent with your earlier guess. It's below 20%—seems okay.

I estimate that approximately 6.7% of the boxes will contain less than 16 ounces of cereal.

Question 2: The company's lawyers say that 6.7% is too high. They insist that no more than 4% of the boxes can be underweight. So the company needs to set the machine to put a little more cereal in each box. What mean setting do they need?

THINK ➡ **Plan** State the problem.

What mean weight will reduce the proportion of underweight boxes to 4%?

Variable Name the variable.

Let y = weight of cereal in a box.

Check to see if a Normal model is appropriate.

✓ **Nearly Normal Condition:** I am told that a Normal model applies.

Specify which Normal model to use. This time you are not given a value for the mean!

I don't know μ, the mean amount of cereal. The standard deviation for this machine is 0.2 ounces. The model is $N(\mu, 0.2)$.

REALITY CHECK We found out earlier that setting the machine to μ = 16.3 ounces made 6.7% of the boxes too light. We'll need to raise the mean a bit to reduce this fraction.

No more than 4% of the boxes can be below 16 ounces.

SHOW ➡ **Mechanics** Make a picture of this Normal model. Center it at μ (since you don't know the mean), and shade the region below 16 ounces.

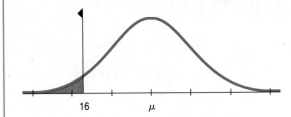

Using your calculator (or the Normal table), find the z-score that cuts off the lowest 4%.

The z-score that has 0.04 area to the left of it is $z = -1.75$.

Use this information to find μ. It's located 1.75 standard deviations to the right of 16. Since σ is 0.2, that's 1.75 × 0.2, or 0.35 ounces more than 16.

For 16 to be 1.75 standard deviations below the mean, the mean must be

$$16 + 1.75(0.2) = 16.35 \text{ ounces.}$$

TELL ➡ **Conclusion** Interpret your result in context.

(This makes sense; we knew it would have to be just a bit higher than 16.3.)

The company must set the machine to average 16.35 ounces of cereal per box.

Question 3 : The company president vetoes that plan, saying the company should give away less free cereal, not more. Her goal is to set the machine no higher than 16.2 ounces and still have only 4% underweight boxes. The only way to accomplish this is to reduce the standard deviation. What standard deviation must the company achieve, and what does that mean about the machine?

(continued)

THINK ➡ **Plan** State the problem.

What standard deviation will allow the mean to be 16.2 ounces and still have only 4% of boxes underweight?

Variable Name the variable.

Let y = weight of cereal in a box.

Check conditions to be sure that a Normal model is appropriate.

✓ **Nearly Normal Condition:** The company believes that the weights are described by a Normal model.

Specify which Normal model to use. This time you don't know σ.

I know the mean, but not the standard deviation, so my model is $N(16.2, \sigma)$.

REALITY CHECK We know the new standard deviation must be less than 0.2 ounces.

SHOW ➡ **Mechanics** Make a picture of this Normal model. Center it at 16.2, and shade the area you're interested in. We want 4% of the area to the left of 16 ounces.

Find the z-score that cuts off the lowest 4%.

I know that the z-score with 4% below it is $z = -1.75$.

Solve for σ. (We need 16 to be 1.75σ's below 16.2, so 1.75σ must be 0.2 ounces. You could just start with that equation.)

$$z = \frac{y - \mu}{\sigma}$$

$$-1.75 = \frac{16 - 16.2}{\sigma}$$

$$1.75\,\sigma = 0.2$$

$$\sigma = 0.114$$

TELL ➡ **Conclusion** Interpret your result in context.

As we expected, the standard deviation is lower than before—actually, quite a bit lower.

The company must get the machine to box cereal with a standard deviation of only 0.114 ounces. This means the machine must be more consistent (by nearly a factor of 2) in filling the boxes.

*Are You Normal? Find Out with a Normal Probability Plot

In the examples we've worked through, we've assumed that the underlying data distribution was roughly unimodal and symmetric, so that using a Normal model makes sense. When you have data, you must *check* to see whether a Normal model is reasonable. How? Make a picture, of course! Drawing a histogram of the data and looking at the shape is one good way to see if a Normal model might work.

There's a more specialized graphical display that can help you to decide whether the Normal model is appropriate: the **Normal probability plot.** If the distribution of the data is roughly Normal, the plot is roughly a diagonal straight line. Deviations from a straight line indicate that the distribution is not Normal. This plot is usually able to show deviations from Normality more clearly than the corresponding histogram, but it's usually easier to understand *how* a distribution fails to be Normal by looking at its histogram.

TI-*nspire*
Normal probability plots and histograms. See how a normal probability plot responds as you change the shape of a distribution.

Some data on a car's fuel efficiency provide an example of data that are nearly Normal. The overall pattern of the Normal probability plot is straight. The two trailing low values correspond to the values in the histogram that trail off the low end. They're not quite in line with the rest of the data set. The Normal probability plot shows us that they're a bit lower than we'd expect of the lowest two values in a Normal model.

Figure 5.8

Histogram and Normal probability plot for gas mileage (mpg) recorded by one of the authors over the 8 years he owned a 1989 Nissan Maxima. The vertical axes are the same, so each dot on the probability plot would fall into the bar on the histogram immediately to its left.

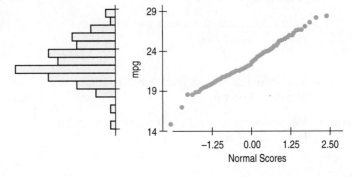

A S *Activity*: **Assessing Normality.**
This activity guides you through the process of checking the Nearly Normal condition using your statistics package.

Figure 5.9

Histogram and Normal probability plot for men's weights. Note how a skewed distribution corresponds to a bent probability plot.

By contrast, the Normal probability plot of the men's *Weight*s from the NHANES Study is far from straight. The weights are skewed to the high end, and the plot is curved. We'd conclude from these pictures that approximations using the 68–95–99.7 Rule for these data would not be very accurate.

A Normal probability plot works by comparing the data values' actual z-scores with those we'd expect to find in a data set of this size. When they match up well, the line is straight. We don't worry much if one or two points don't line up. But a plot that bends is a warning that the distribution is skewed (or strange in some other way) and we should not use a Normal model.

TI Tips *CREATING A NORMAL PROBABILITY PLOT

Let's make a Normal probability plot with the calculator. Here are the boys' agility test scores we looked at in Chapter 4; enter them in L1:

22, 17, 18, 29, 22, 23, 24, 23, 17, 21

Now you can create the plot:

- Turn a STATPLOT On.
- Tell it to make a Normal probability plot by choosing the last of the icons.
- Specify your datalist and which axis you want the data on. (We'll use Y so the plot looks like the others we showed you.)
- Specify the Mark you want the plot to use.
- Now ZoomStat does the rest.

The plot doesn't look very straight. Normality is certainly questionable here.

(Not that it matters in making this decision, but that vertical line is the *y*-axis. Points to the left have negative *z*-scores and points to the right have positive *z*-scores.)

WHAT IF ●●● samples don't behave, er, normally?

Samples can provide insights about population characteristics we wish we knew, but no sample is perfect. Different samples from the sample population will disagree somewhat—often a lot! One of the fundamental tasks we face involves peering through the fog of sample-to-sample variation to try to discern the "truth." To do that, we have to know how thick the fog is; in other words, we need to understand how sample statistics might vary from one sample to the next. We can use a simulation to investigate this sampling variability.

As we did in Chapter 3, we start by creating an artificial population. We had the computer generate 10,000 3-digit numbers. This allows us to actually look at the whole population, something we can't really do in the Real World. Then we can draw samples to see what the variability looks like.

Suppose we want to know what the largest value in the population is. (As in: What's the biggest fish in the lake?) The maximum value in a sample is almost certainly smaller (unless you get really lucky and hook The Big One), but how much? If we pick a random sample, could we get close to seeing the actual population max of 999?[7]

Our simulation randomly selected 25 values from the population. In the first trial, the sample max was 995. Not bad—pretty close to the truth. But the second trial chose a random sample with a maximum of only 884. Had this been our only sample (and in the Real World we'd just have one), we'd get a false impression.

What about other samples? The beauty of simulations is that we get to run lots of trials. This histogram shows the distribution of the maximum values that showed up in 500 different random samples.

We see that most samples would suggest the population max to be above 970, perhaps a decent estimate for many purposes. But we also see a warning signal: the distribution is skewed to the left, and that long tail suggests that a sample like these could greatly underestimate the true population maximum.

Now, suppose we want to know the mean value in the population. We looked; it's really 552. But in the Real World we can't look; we'd need to find the average for a sample. To see how successful that strategy might be, we again put our simulation to work picking random samples of 25.

The first sample had a mean of 562—close, just a little too high. We ran another trial. The mean of that second random sample was 524—too low, and not as close either. Once again we simulated 500 trials. On the next page there's a histogram showing the distribution of all those sample means.

---

[7]We admit it: we peeked. If we could actually do that with real populations, we wouldn't need a lot of Statistics!

We see that samples like these usually have means between 500 and 600. (Whether such an estimate would be close enough depends on our purposes; if not, we should use a bigger sample size.) But what's really striking about this histogram is its shape: roughly unimodal and symmetric. That's potentially great news, because if a Normal model is useful, then we'd be able to say a lot more about the behavior of sample means.[8]

MORAL: *Don't assume you can always use a Normal model.* While it appears that may be okay for means, the distribution of sample maxima is decidedly NOT Normal. All too often people jump to conclusions based on thinking about a Normal model without first checking to be sure that's appropriate. Don't go there!

[8]Hint: Exciting times lie ahead!

WHAT CAN GO WRONG?

- **Don't use a Normal model when the distribution is not unimodal and symmetric.** Normal models are so easy and useful that it is tempting to use them even when they don't describe the data very well. That can lead to wrong conclusions. Don't use a Normal model without first checking the **Nearly Normal Condition.** Look at a picture of the data to check that it is unimodal and symmetric. A histogram, or a Normal probability plot, can help you tell whether a Normal model is appropriate.

  The CEOs (p. 90) had a mean total compensation of $8.0372M and a standard deviation of $13.1376M. Using the Normal model rule, we should expect about 68% of the CEOs to have compensations between −$5.100M and $21.1748M. In fact, more than 93% of the CEOs have annual compensations in this interval. What went wrong? The distribution is skewed, not symmetric. Using the 68–95–99.7 Rule for data like these will lead to silly results.

Annual Compensation ($)

- **Don't use the mean and standard deviation when outliers are present.** Both means and standard deviations can be distorted by outliers, and no model based on distorted values will do a good job. A z-score calculated from a distribution with outliers may be misleading. It's always a good idea to check for outliers. How? Make a picture.

- **Don't round your results in the middle of a calculation.** We reported the mean of the heptathletes' 200 m run as 24.71 seconds. More precisely, it was 24.70894736842105 seconds.

  You should use all the precision available in the data for all the intermediate steps of a calculation. Using the more precise value for the mean (and also carrying 15 digits for the SD), the z-score calculation for Fountain's run comes out to

  $$z = \frac{23.21 - 24.70894736842105}{0.7002346690571983} = -2.140635753495674.$$

We'd likely report that as −2.141, as opposed to the rounded-off value of −2.14 we got earlier from the table.

■ **Don't worry about minor differences in results.** Because various calculators and programs may carry different precision in calculations, your answers may differ slightly from those we show in the text and in the Step-By-Steps, or even from the values given in the answers in the back of the book. Those differences aren't anything to worry about. They're not the main story Statistics tries to tell.

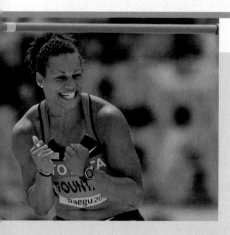

What Have We Learned?

We've learned that the story data can tell may be easier to understand after shifting or rescaling the data.

■ Shifting data by adding or subtracting the same amount from each value affects measures of center and position but not measures of spread.

■ Rescaling data by multiplying or dividing every value by a constant, changes all the summary statistics—center, position, and spread.

We've learned the power of standardizing data.

■ Standardizing uses the standard deviation as a ruler to measure distance from the mean, creating z-scores.

■ Using these z-scores, we can compare apples and oranges. Because standardizing eliminates units, standardized values can be compared and combined even if the original variables had different units and magnitudes.

■ And a z-score can identify unusual or surprising values among data.

We've learned that the 68–95–99.7 Rule can be a useful rule of thumb for understanding distributions.

■ For data that are unimodal and symmetric, about 68% fall within 1 SD of the mean, 95% fall within 2 SDs of the mean, and 99.7% fall within 3 SDs of the mean.

Again we've seen the importance of *Thinking* about whether a method will work.

■ Data can't be exactly Normal, so we check the Nearly Normal Condition by making a histogram (is it unimodal, symmetric, and free of outliers?) or a Normal probability plot (is it straight enough?).

Terms

Standardized value
A value found by subtracting the mean and dividing by the standard deviation. (p. 108)

Shifting
Adding a constant to each data value adds the same constant to the measures of position (mean, median, and quartiles), but does not change the measures of spread (standard deviation or IQR). (p. 111)

Rescale
Multiplying each data value by a constant multiplies both the measures of position (mean, median, and quartiles) and the measures of spread (standard deviation and IQR) by that constant. (p. 111)

Normal model
A useful family of models for unimodal, symmetric distributions. (p. 114)

Parameter
A numerically valued attribute of a model. For example, the values of μ and σ in a $N(\mu, \sigma)$ model are parameters. (p. 114)

Statistic
A value calculated from data to summarize aspects of the data. For example, the mean, $\bar{y}$, and standard deviation, s, are statistics. (p. 114)

z-score
A z-score tells how many standard deviations a value is from the mean, and in which direction; z-scores have a mean of 0 and a standard deviation of 1. When working with data, use the statistics $\bar{y}$ and s:

$$z = \frac{y - \bar{y}}{s}.$$

When working with models, use the parameters μ and σ:

$$z = \frac{y - \mu}{\sigma}. \quad \text{(p. 114)}$$

Standard Normal model   A Normal model, $N(\mu, \sigma)$ with mean $\mu = 0$ and standard deviation $\sigma = 1$. Also called the **standard Normal distribution**. (p. 114)

Normality Assumption   We must have a reason to believe a variable's distribution is Normal before applying a Normal model. (p. 115)

Nearly Normal Condition   A distribution is nearly Normal if it is unimodal and symmetric. We can check by looking at a histogram (or a Normal probability plot). (p. 115)

68–95–99.7 Rule   In a Normal model, about 68% of values fall within 1 standard deviation of the mean, about 95% fall within 2 standard deviations of the mean, and about 99.7% fall within 3 standard deviations of the mean. (p. 115)

Normal percentile   The Normal percentile corresponding to a z-score gives the percentage of values in a standard Normal distribution found at that z-score or below. (p. 119)

***Normal probability plot**   A display to help assess whether a distribution of data is approximately Normal. If the plot is nearly straight, the data satisfy the **Nearly Normal Condition**. (p. 125)

On the Computer THE NORMAL MODEL

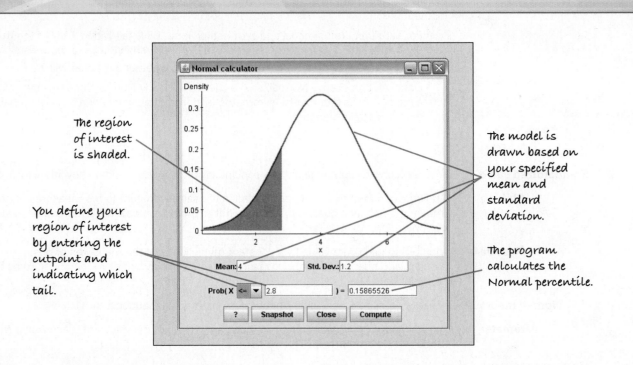

The region of interest is shaded.

You define your region of interest by entering the cutpoint and indicating which tail.

The model is drawn based on your specified mean and standard deviation.

The program calculates the Normal percentile.

The best way to tell whether your data can be modeled well by a Normal model is to make a picture or two. We've already talked about making histograms. Normal probability plots are almost never made by hand because the values of the Normal scores are tricky to find. But most statistics software make Normal plots, though various packages call the same plot by different names and array the information differently.

Exercises

1. **Stats test** Nicole's score on the Stats midterm was 80 points. The class average was 75 and the standard deviation was 5 points. What was her z-score?

2. **Horsepower** Cars currently sold in the United States have an average of 135 horsepower, with a standard deviation of 40 horsepower. What's the z-score for a car with 195 horsepower?

3. **Homeruns** In 1998, Mark McGwire hit 70 homeruns, to break the single season home run record of 61 home runs set by Roger Maris in 1961. McGwire later admitted to using performance enhancing steroids that season. Even so, how did McGwire's performance really compare to Maris's? The league was very different in 1998 than in 1961. The table below shows the mean and standard deviations for the number of home runs scored by all players with at least 502 plate appearances in their respective seasons. Use these to determine who's home run feat was more impressive.

	1961	1998
Mean	18.8	20.7
Standard deviation	13.37	12.74

source: mlb. mlb.com

4. **Teenage mothers** The rates of teenage births are reported as the number of births to mothers age 15–19 per 1,000 people age 15–19 in the state. Shown below are the mean and standard deviation of teenage birthrates for all 50 states and the District of Columbia, for Hispanic mothers, Non-Hispanic black mothers, and Non-Hispanic white mothers. The lowest rate for Hispanic mothers, 31.3 births per 1,000, occurs in Maine. The minimum for Non-Hispanic black mothers is Hawaii with a rate of 17.4 births per 1,000, and the minimum for Non-Hispanic white mothers is the District of Columbia with 4.3.

Which of these rates is the most extreme low rate compared to the other states?

	Hispanic	Non-Hispanic black	Non-Hispanic white
Mean	96.7	64.0	29.0
Standard deviation	31.60	16.12	11.55

5. **SAT or ACT?** Each year thousands of high school students take either the SAT or the ACT, standardized tests used in the college admissions process. Combined SAT Math and Verbal scores go as high as 1600, while the maximum ACT composite score is 36. Since the two exams use very different scales, comparisons of performance are difficult. A convenient rule of thumb is $SAT = 40 \times ACT + 150$; that is, multiply an ACT score by 40 and add 150 points to estimate the equivalent SAT score. An admissions officer reported the following statistics about the ACT scores of 2355 students who applied to her college one year. Find the summaries of equivalent SAT scores.

Lowest score = 19   Mean = 27   Standard deviation = 3

Q3 = 30   *Median* = 28   IQR = 6

6. **Cold U?** A high school senior uses the Internet to get information on February temperatures in the town where he'll be going to college. He finds a website with some statistics, but they are given in degrees Celsius. The conversion formula is $°F = 9/5 °C + 32$. Determine the Fahrenheit equivalents for the summary information below.

Maximum temperature = 11°C   Range = 33°

Mean = 1°   Standard deviation = 7°

Median = 2°   IQR = 16°

7. Stats test, Part II Suppose your Statistics professor reports test grades as z-scores, and you got a score of 2.20 on an exam. Write a sentence explaining what that means.

8. Checkup One of the authors has an adopted grandson whose birth family members are very short. After examining him at his 2-year checkup, the boy's pediatrician said that the z-score for his height relative to American 2-year-olds was -1.88. Write a sentence explaining what that means.

9. Stats test, part III The mean score on the Stats exam was 75 points with a standard deviation of 5 points, and Gregor's z-score was -2. How many points did he score?

10. Mensa People with z-scores above 2.5 on an IQ test are sometimes classified as geniuses. If IQ scores have a mean of 100 and a standard deviation of 16 points, what IQ score do you need to be considered a genius?

11. Temperatures A town's January high temperatures average 36°F with a standard deviation of 10°, while in July the mean high temperature is 74° and the standard deviation is 8°. In which month is it more unusual to have a day with a high temperature of 55°? Explain.

12. Placement exams An incoming freshman took her college's placement exams in French and mathematics. In French, she scored 82 and in math 86. The overall results on the French exam had a mean of 72 and a standard deviation of 8, while the mean math score was 68, with a standard deviation of 12. On which exam did she do better compared with the other freshmen?

13. Combining test scores The first Stats exam had a mean of 65 and a standard deviation of 10 points; the second had a mean of 80 and a standard deviation of 5 points. Derrick scored an 80 on both tests. Julie scored a 70 on the first test and a 90 on the second. They both totaled 160 points on the two exams, but Julie claims that her total is better. Explain.

14. Combining scores again The first Stat exam had a mean of 80 and a standard deviation of 4 points; the second had a mean of 70 and a standard deviation of 15 points. Reginald scored an 80 on the first test and an 85 on the second. Sara scored an 88 on the first but only a 65 on the second. Although Reginald's total score is higher, Sara feels she should get the higher grade. Explain her point of view.

15. Final exams Anna, a language major, took final exams in both French and Spanish and scored 83 on each. Her roommate Megan, also taking both courses, scored 77 on the French exam and 95 on the Spanish exam. Overall, student scores on the French exam had a mean of 81 and a standard deviation of 5, and the Spanish scores had a mean of 74 and a standard deviation of 15.

a) To qualify for language honors, a major must maintain at least an 85 average for all language courses taken. So far, which student qualifies?

b) Which student's overall performance was better?

16. MP3s Two companies market new batteries targeted at owners of personal music players. DuraTunes claims a mean battery life of 11 hours, while RockReady advertises 12 hours.

a) Explain why you would also like to know the standard deviations of the battery lifespans before deciding which brand to buy.

b) Suppose those standard deviations are 2 hours for DuraTunes and 1.5 hours for RockReady. You are headed for 8 hours at the beach. Which battery is most likely to last all day? Explain.

c) If your beach trip is all weekend, and you probably will have the music on for 16 hours, which battery is most likely to last? Explain.

17. Cattle The Virginia Cooperative Extension reports that the mean weight of yearling Angus steers is 1152 pounds. Suppose that weights of all such animals can be described by a Normal model with a standard deviation of 84 pounds.

a) How many standard deviations from the mean would a steer weighing 1000 pounds be?

b) Which would be more unusual, a steer weighing 1000 pounds or one weighing 1250 pounds?

T 18. Car speeds John Beale of Stanford, CA, recorded the speeds of cars driving past his house, where the speed limit was 20 mph. The mean of 100 readings was 23.84 mph, with a standard deviation of 3.56 mph. (He actually recorded every car for a two-month period. These are 100 representative readings.)

a) How many standard deviations from the mean would a car going under the speed limit be?

b) Which would be more unusual, a car traveling 34 mph or one going 10 mph?

19. More cattle Recall that the beef cattle described in Exercise 17 had a mean weight of 1152 pounds, with a standard deviation of 84 pounds.

a) Cattle buyers hope that yearling Angus steers will weigh at least 1000 pounds. To see how much over (or under) that goal the cattle are, we could subtract 1000 pounds from all the weights. What would the new mean and standard deviation be?

b) Suppose such cattle sell at auction for 40 cents a pound. Find the mean and standard deviation of the sale prices for all the steers.

T 20. Car speeds again For the car speed data of Exercise 18, recall that the mean speed recorded was 23.84 mph, with a standard deviation of 3.56 mph. To see how many cars are speeding, John subtracts 20 mph from all speeds.

a) What is the mean speed now? What is the new standard deviation?

b) His friend in Berlin wants to study the speeds, so John converts all the original miles-per-hour readings to kilometers per hour by multiplying all speeds by 1.609 (km per mile). What is the mean now? What is the new standard deviation?

21. **Cattle, part III** Suppose the auctioneer in Exercise 19 sold a herd of cattle whose minimum weight was 980 pounds, median was 1140 pounds, standard deviation 84 pounds, and IQR 102 pounds. They sold for 40 cents a pound, and the auctioneer took a $20 commission on each animal. Then, for example, a steer weighing 1100 pounds would net the owner $0.40(1100) - 20 = \$420$. Find the minimum, median, standard deviation, and IQR of the net sale prices.

22. **Caught speeding** Suppose police set up radar surveillance on the Stanford street described in Exercise 18. They handed out a large number of tickets to speeders going a mean of 28 mph, with a standard deviation of 2.4 mph, a maximum of 33 mph, and an IQR of 3.2 mph. Local law prescribes fines of $100, plus $10 per mile per hour over the 20 mph speed limit. For example, a driver convicted of going 25 mph would be fined $100 + 10(5) = \$150$. Find the mean, standard deviation, maximum, and IQR of all the potential fines.

23. **Professors** A friend tells you about a recent study dealing with the number of years of teaching experience among current college professors. He remembers the mean but can't recall whether the standard deviation was 6 months, 6 years, or 16 years. Tell him which one it must have been, and why.

24. **Rock concerts** A popular band on tour played a series of concerts in large venues. They always drew a large crowd, averaging 21,359 fans. While the band did not announce (and probably never calculated) the standard deviation, which of these values do you think is most likely to be correct: 20, 200, 2000, or 20,000 fans? Explain your choice.

25. **Guzzlers?** Environmental Protection Agency (EPA) fuel economy estimates for automobile models tested recently predicted a mean of 24.8 mpg and a standard deviation of 6.2 mpg for highway driving. Assume that a Normal model can be applied.

a) Draw the model for auto fuel economy. Clearly label it, showing what the 68–95–99.7 Rule predicts.

b) In what interval would you expect the central 68% of autos to be found?

c) About what percent of autos should get more than 31 mpg?

d) About what percent of cars should get between 31 and 37.2 mpg?

e) Describe the gas mileage of the worst 2.5% of all cars.

26. **IQ** Some IQ tests are standardized to a Normal model, with a mean of 100 and a standard deviation of 16.

a) Draw the model for these IQ scores. Clearly label it, showing what the 68–95–99.7 Rule predicts.

b) In what interval would you expect the central 95% of IQ scores to be found?

c) About what percent of people should have IQ scores above 116?

d) About what percent of people should have IQ scores between 68 and 84?

e) About what percent of people should have IQ scores above 132?

27. **Small steer** In Exercise 17 we suggested the model $N(1152, 84)$ for weights in pounds of yearling Angus steers. What weight would you consider to be unusually low for such an animal? Explain.

28. **High IQ** Exercise 26 proposes modeling IQ scores with $N(100, 16)$. What IQ would you consider to be unusually high? Explain.

29. **College hoops** The winning scores of all college men's basketball games for the 2011–12 season were approximately normally distributed with mean 77.5 points and standard deviation 12.5 points.

a) Draw the Normal model for winning scores.

b) What interval of winning scores would be the central 95% of all winning scores for the 2011–12 season?

c) About what percent of the winning scores should be less than 65 points?

d) About what percent of the winning scores should be between 65 and 102 points?

e) About what percent of the winning scores should be over 102 points?

30. **Rivets** A company that manufactures rivets believes the shear strength (in pounds) is modeled by $N(800, 50)$.

a) Draw and label the Normal model.

b) Would it be safe to use these rivets in a situation requiring a shear strength of 750 pounds? Explain.

c) About what percent of these rivets would you expect to fall below 900 pounds?

d) Rivets are used in a variety of applications with varying shear strength requirements. What is the maximum shear strength for which you would feel comfortable approving this company's rivets? Explain your reasoning.

31. **Trees** A forester measured the diameters of 27 trees in a woods, and from these made projections about the whole forest based on a Normal model. The histogram displays

his data. Do you think his analysis was justified? Explain, citing some specific concerns.

32. **Car speeds, the picture** For the car speed data of Exercise 18, here is the histogram, boxplot, and Normal probability plot of the 100 readings. Do you think it is appropriate to apply a Normal model here? Explain.

33. **Wisconsin ACT math** The histogram shows the distribution of mean ACT mathematics scores for all Wisconsin public schools in 2011. The vertical lines show the mean and one standard deviation above and below the mean. 78.8% of the data points are between the two outer lines.

   a) Give two reasons that a Normal model is not appropriate for these data.

   b) The lower cluster, that is, the scores less than about 17, is almost entirely made up of schools in the Milwaukee school district. If those scores are removed, what would be the shape of the new

distribution? What would happen to the mean and standard deviation?

34. **Wisconsin ACT math II** This plot shows the mean ACT scores for Wisconsin schools with the Milwaukee schools removed. The vertical lines show the mean, mean − 1s, and mean + 1s.

   a) Describe the shape of the distribution.
   b) Does a normal model seem appropriate for this distribution?
   c) 68.9% of the school average scores are between the two outer lines. Does that support or refute your answer to part b?
   d) Below, a normal model with mean 21.910 and standard deviation 1.539 is drawn over the histogram. Explain how this demonstrates George Box's statement, *"All models are wrong—but some are useful."*

35. **Winter Olympics 2010 downhill** Fifty-nine men qualified for the men's alpine downhill race in Vancouver. The gold medal winner finished in 114.3 seconds. All competitors' times (in seconds) are found in the following

list (espn.go.com/olympics/winter/2010/results/_/sport/1/
event/2):

114.3	115.0	115.7	116.7	118.6	119.8
114.4	115.2	115.8	116.7	118.7	120.0
114.4	115.2	116.0	117.0	118.8	120.1
114.5	115.3	116.1	117.2	118.9	120.6
114.6	115.3	116.2	117.2	119.2	121.7
114.7	115.4	116.2	117.4	119.5	121.7
114.8	115.4	116.3	117.7	119.6	122.6
114.8	115.5	116.4	117.9	119.7	123.4
114.9	115.6	116.6	118.1	119.8	124.4
114.9	115.7	116.7	118.4	119.8	

a) The mean time was 117.34 seconds, with a standard deviation of 2.465 seconds. If the Normal model is appropriate, what percent of times will be less than 114.875 seconds?
b) What is the actual percent of times less than 114.875 seconds?
c) Why do you think the two percentages don't agree?
d) Create a histogram of these times. What do you see?

T **36. Check the model** The mean of the 100 car speeds in Exercise 20 was 23.84 mph, with a standard deviation of 3.56 mph.

a) Using a Normal model, what values should border the middle 95% of all car speeds?
b) Here are some summary statistics.

Percentile		Speed
100%	**Max**	34.060
97.5%		30.976
90.0%		28.978
75.0%	**Q3**	25.785
50.0%	**Median**	23.525
25.0%	**Q1**	21.547
10.0%		19.163
2.5%		16.638
0.0%	**Min**	16.270

From your answer in part a, how well does the model do in predicting those percentiles? Are you surprised? Explain.

T **37. Receivers 2010** This histogram displays NFL data from the 2010 football season reporting the number of yards gained by each of the league's 191 wide receivers:

The mean is 397.15 yards, with a standard deviation of 362.4 yards.

a) According to the Normal model, what percent of receivers would you expect to gain more yards than 2 standard deviations above the mean number of yards?
b) For these data, what does that mean?
c) Explain the problem in using a Normal model here.

38. Customer database A large philanthropic organization keeps records on the people who have contributed to their cause. In addition to keeping records of past giving, the organization buys demographic data on neighborhoods from the U.S. Census Bureau. Eighteen of these variables concern the ethnicity of the neighborhood of the donor. Here are a histogram and summary statistics for the percentage of whites in the neighborhoods of 500 donors:

Count	500
Mean	83.59
Median	93
StdDev	22.26
IQR	17
Q1	80
Q3	97

a) Which is a better summary of the percentage of white residents in the neighborhoods, the mean or the median? Explain.
b) Which is a better summary of the spread, the IQR or the standard deviation? Explain.

c) From a Normal model, about what percentage of neighborhoods should have a percent white within one standard deviation of the mean?

d) What percentage of neighborhoods actually have a percent white within one standard deviation of the mean?

e) Explain the discrepancy between parts c and d.

39. **Normal cattle** Using $N(1152, 84)$, the Normal model for weights of Angus steers in Exercise 17, what percent of steers weigh

a) over 1250 pounds?

b) under 1200 pounds?

c) between 1000 and 1100 pounds?

40. **IQs revisited** Based on the Normal model $N(100, 16)$ describing IQ scores, what percent of people's IQs would you expect to be

a) over 80?

b) under 90?

c) between 112 and 132?

41. **More cattle** Based on the model $N(1152, 84)$ describing Angus steer weights, what are the cutoff values for

a) the highest 10% of the weights?

b) the lowest 20% of the weights?

c) the middle 40% of the weights?

42. **More IQs** In the Normal model $N(100, 16)$, what cutoff value bounds

a) the highest 5% of all IQs?

b) the lowest 30% of the IQs?

c) the middle 80% of the IQs?

43. **Cattle, finis** Consider the Angus weights model $N(1152, 84)$ one last time.

a) What weight represents the 40th percentile?

b) What weight represents the 99th percentile?

c) What's the IQR of the weights of these Angus steers?

44. **IQ, finis** Consider the IQ model $N(100, 16)$ one last time.

a) What IQ represents the 15th percentile?

b) What IQ represents the 98th percentile?

c) What's the IQR of the IQs?

45. **Cholesterol** Assume the cholesterol levels of adult American women can be described by a Normal model with a mean of 188 mg/dL and a standard deviation of 24.

a) Draw and label the Normal model.

b) What percent of adult women do you expect to have cholesterol levels over 200 mg/dL?

c) What percent of adult women do you expect to have cholesterol levels between 150 and 170 mg/dL?

d) Estimate the IQR of the cholesterol levels.

e) Above what value are the highest 15% of women's cholesterol levels?

46. **Tires** A tire manufacturer believes that the treadlife of its snow tires can be described by a Normal model with a mean of 32,000 miles and standard deviation of 2500 miles.

a) If you buy a set of these tires, would it be reasonable for you to hope they'll last 40,000 miles? Explain.

b) Approximately what fraction of these tires can be expected to last less than 30,000 miles?

c) Approximately what fraction of these tires can be expected to last between 30,000 and 35,000 miles?

d) Estimate the IQR of the treadlives.

e) In planning a marketing strategy, a local tire dealer wants to offer a refund to any customer whose tires fail to last a certain number of miles. However, the dealer does not want to take too big a risk. If the dealer is willing to give refunds to no more than 1 of every 25 customers, for what mileage can he guarantee these tires to last?

47. **Kindergarten** Companies that design furniture for elementary school classrooms produce a variety of sizes for kids of different ages. Suppose the heights of kindergarten children can be described by a Normal model with a mean of 38.2 inches and standard deviation of 1.8 inches.

a) What fraction of kindergarten kids should the company expect to be less than 3 feet tall?

b) In what height interval should the company expect to find the middle 80% of kindergarteners?

c) At least how tall are the biggest 10% of kindergarteners?

48. **Body temperatures** Most people think that the "normal" adult body temperature is 98.6°F. That figure, based on a 19th-century study, has recently been challenged. In a 1992 article in the *Journal of the American Medical Association,* researchers reported that a more accurate figure may be 98.2°F. Furthermore, the standard deviation appeared to be around 0.7°F. Assume that a Normal model is appropriate.

a) In what interval would you expect most people's body temperatures to be? Explain.

b) What fraction of people would be expected to have body temperatures above 98.6°F?

c) Below what body temperature are the coolest 20% of all people?

49. **Eggs** Hens usually begin laying eggs when they are about 6 months old. Young hens tend to lay smaller eggs, often weighing less than the desired minimum weight of 54 grams.

a) The average weight of the eggs produced by the young hens is 50.9 grams, and only 28% of their eggs exceed the desired minimum weight. If a Normal model is appropriate, what would the standard deviation of the egg weights be?

b) By the time these hens have reached the age of 1 year, the eggs they produce average 67.1 grams, and 98%

of them are above the minimum weight. What is the standard deviation for the appropriate Normal model for these older hens?

c) Are egg sizes more consistent for the younger hens or the older ones? Explain.

50. Tomatoes Agricultural scientists are working on developing an improved variety of Roma tomatoes. Marketing research indicates that customers are likely to bypass Romas that weigh less than 70 grams. The current variety of Roma plants produces fruit that averages 74 grams, but 11% of the tomatoes are too small. It is reasonable to assume that a Normal model applies.

a) What is the standard deviation of the weights of Romas now being grown?

b) Scientists hope to reduce the frequency of undersized tomatoes to no more than 4%. One way to accomplish this is to raise the average size of the fruit. If the standard deviation remains the same, what target mean should they have as a goal?

c) The researchers produce a new variety with a mean weight of 75 grams, which meets the 4% goal. What is the standard deviation of the weights of these new Romas?

d) Based on their standard deviations, compare the tomatoes produced by the two varieties.

Just Checking ANSWERS

1. a) On the first test, the mean is 88 and the SD is 4, so $z = (90 - 88)/4 = 0.5$. On the second test, the mean is 75 and the SD is 5, so $z = (80 - 75)/5 = 1.0$. The first test has the lower z-score, so it is the one that will be dropped.
   b) Yes. The second test is 1 standard deviation above the mean, farther away than the first test, so it's the better score relative to the class.

2. a) The mean would increase to 500.
   b) The standard deviation is still 100 points.
   c) The two boxplots would look nearly identical (the shape of the distribution would remain the same), but the later one would be shifted 50 points higher.

3. The standard deviation is now 2.54 millimeters, which is the same as 0.1 inches. Nothing has changed. The standard deviation has "increased" only because we're reporting it in millimeters now, not inches.

4. The mean is 184 centimeters, with a standard deviation of 8 centimeters. 2 meters is 200 centimeters, which is 2 standard deviations above the mean. We expect 5% of the men to be more than 2 standard deviations below or above the mean, so half of those, 2.5%, are likely to be above 2 meters.

5. a) We know that 68% of the time we'll be within 1 standard deviation (2 min) of 20. So 32% of the time we'll arrive in less than 18 or more than 22 minutes. Half of those times (16%) will be greater than 22 minutes, so 84% will be less than 22 minutes.
   b) 24 minutes is 2 standard deviations above the mean. Because of the 95% rule, we know 2.5% of the times will be more than 24 minutes.
   c) Traffic incidents may occasionally increase the time it takes to get to school, so the driving times may be skewed to the right, and there may be outliers.
   d) If so, the Normal model would not be appropriate and the percentages we predict would not be accurate.

Review of Part I

Exploring and Understanding Data

Quick Review

It's time to put it all together. Real data don't come tagged with instructions for use. So let's step back and look at how the key concepts and skills we've seen work together. This brief list and the review exercises that follow should help you check your understanding of Statistics so far.

- We treat data two ways: as categorical and as quantitative.

- To describe categorical data:

  - Make a picture. Bar graphs work well for comparing counts in categories.

  - Summarize the distribution with a table of counts or relative frequencies (percents) in each category.

  - Pie charts and segmented bar charts display divisions of a whole.

  - Compare distributions with plots side by side.

  - Look for associations between variables by comparing marginal and conditional distributions.

- To describe quantitative data:

  - Make a picture. Use histograms, boxplots, stem-and-leaf displays, or dotplots. Stem-and-leafs are great when working by hand and good for small data sets. Histograms are a good way to see the distribution. Boxplots are best for comparing several distributions.

  - Describe distributions in terms of their shape, center, and spread, and note any unusual features such as gaps or outliers.

  - The shape of most distributions you'll see will likely be uniform, unimodal, or bimodal. It may be multimodal. If it is unimodal, then it may be symmetric or skewed.

  - A 5-number summary makes a good numerical description of a distribution: min, Q1, median, Q3, and max.

- If the distribution is skewed, be sure to include the median and interquartile range (IQR) when you describe its center and spread.

- A distribution that is severely skewed may benefit from re-expressing the data. If it is skewed to the high end, taking logs often works well.

- If the distribution is unimodal and symmetric, describe its center and spread with the mean and standard deviation.

- Use the standard deviation as a ruler to tell how unusual an observed value may be, or to compare or combine measurements made on different scales.

- Shifting a distribution by adding or subtracting a constant affects measures of position but not measures of spread. Rescaling by multiplying or dividing by a constant affects both.

- When a distribution is roughly unimodal and symmetric, a Normal model may be useful. For Normal models, the 68–95–99.7 Rule is a good rule of thumb.

- If the Normal model fits well (check a histogram or Normal probability plot), then Normal percentile tables or functions found in most statistics technology can provide more detailed values.

Need more help with some of this? It never hurts to reread sections of the chapters! And in the following pages we offer you more opportunities[1] to review these concepts and skills.

The exercises that follow use the concepts and skills you've learned in the first five chapters. To be more realistic and more useful for your review, they don't tell you which of the concepts or methods you need. But neither will the exam.

---

[1]If you doubted that we are teachers, this should convince you. Only a teacher would call additional homework exercises "opportunities."

Review Exercises

1. Bananas Here are the prices (in cents per pound) of bananas reported from 15 markets surveyed by the U.S. Department of Agriculture.

51	52	45
48	53	52
50	49	52
48	43	46
45	42	50

a) Display these data with an appropriate graph.
b) Report appropriate summary statistics.
c) Write a few sentences about this distribution.

2. Prenatal care Results of a 1996 American Medical Association report about the infant mortality rate for twins carried for the full term of a normal pregnancy are shown on the next page, broken down by the level of prenatal care the mother had received.

Full-Term Pregnancies, Level of Prenatal Care	Infant Mortality Rate Among Twins (deaths per thousand live births)
Intensive	5.4
Adequate	3.9
Inadequate	6.1
Overall	5.1

a) Is the overall rate the average of the other three rates? Should it be? Explain.

b) Do these results indicate that adequate prenatal care is important for pregnant women? Explain.

c) Do these results suggest that a woman pregnant with twins should be wary of seeking too much medical care? Explain.

3. Singers The boxplots shown display the heights (in inches) of 130 members of a choir.

a) It appears that the median height for sopranos is missing, but actually the median and the upper quartile are equal. How could that happen?

b) Write a few sentences describing what you see.

4. Dialysis In a study of dialysis, researchers found that "of the three patients who were currently on dialysis, 67% had developed blindness and 33% had their toes amputated." What kind of display might be appropriate for these data? Explain.

5. Beanstalks Beanstalk Clubs are social clubs for very tall people. To join, a man must be over 6′2″ tall, and a woman over 5′10″. The National Health Survey suggests that heights of adults may be Normally distributed, with mean heights of 69.1″ for men and 64.0″ for women. The respective standard deviations are 2.8″ and 2.5″.

a) You are probably not surprised to learn that men are generally taller than women, but what does the greater standard deviation for men's heights indicate?

b) Who are more likely to qualify for Beanstalk membership, men or women? Explain.

6. Bread Clarksburg Bakery is trying to predict how many loaves to bake. In the last 100 days, they have sold between 95 and 140 loaves per day. Here is a histogram of the number of loaves they sold for the last 100 days.

a) Describe the distribution.

b) Which should be larger, the mean number of sales or the median? Explain.

c) Here are the summary statistics for Clarksburg Bakery's bread sales. Use these statistics and the histogram above to create a boxplot. You may approximate the values of any outliers.

Summary of Sales	
Median	100
Min	95
Max	140
25th %tile	97
75th %tile	105.5

d) For these data, the mean was 103 loaves sold per day, with a standard deviation of 9 loaves. Do these statistics suggest that Clarksburg Bakery should expect to sell between 94 and 112 loaves on about 68% of the days? Explain.

7. State University Public relations staff at State U. phoned 850 local residents. After identifying themselves, the callers asked the survey participants their ages, whether they had attended college, and whether they had a favorable opinion of the university. The official report to the university's directors claimed that, in general, people had very favorable opinions about their university.

a) Identify the W's of these data.

b) Identify the variables, classify each as categorical or quantitative, and specify units if relevant.

c) Are you confident about the report's conclusion? Explain.

8. Acid rain Based on long-term investigation, researchers have suggested that the acidity (pH) of rainfall in the Shenandoah Mountains can be described by the Normal model $N(4.9, 0.6)$.

a) Draw and carefully label the model.
b) What percent of storms produce rainfall with pH over 6?
c) What percent of storms produce rainfall with pH under 4?
d) The lower the pH, the more acidic the rain. What is the pH level for the most acidic 20% of all storms?
e) What is the pH level for the least acidic 5% of all storms?
f) What is the IQR for the pH of rainfall?

9. Fraud detection A credit card bank is investigating the incidence of fraudulent card use. The bank suspects that the type of product bought may provide clues to the fraud. To examine this situation, the bank looks at the Standard Industrial Code (SIC) of the business related to the transaction. This is a code that was used by the U.S. Census Bureau and Statistics Canada to identify the type of every registered business in North America.[2] For example, 1011 designates Meat and Meat Products (except Poultry), 1012 is Poultry Products, 1021 is Fish Products, 1031 is Canned and Preserved Fruits and Vegetables, and 1032 is Frozen Fruits and Vegetables.

A company intern produces the following histogram of the SIC codes for 1536 transactions:

He also reports that the mean SIC is 5823.13 with a standard deviation of 488.17.

a) Comment on any problems you see with the use of the mean and standard deviation as summary statistics.
b) How well do you think the Normal model will work on these data? Explain.

10. Streams As part of the course work, a class at an upstate NY college collects data on streams each year. Students record a number of biological, chemical, and physical

---
[2]Since 1997, the SIC has been replaced by the NAICS, a code of six letters.

variables, including the stream name, the substrate of the stream (*limestone, shale,* or *mixed*), the pH, the temperature (°C), and the BCI, a measure of biological diversity.

Group	Count	%
Limestone	77	44.8
Mixed	26	15.1
Shale	69	40.1

a) Name each variable, indicating whether it is categorical or quantitative, and giving the units if available.
b) These streams have been classified according to their substrate—the composition of soil and rock over which they flow—as summarized in the table. What kind of graph might be used to display these data?

11. Cramming One Thursday, researchers gave students enrolled in a section of basic Spanish a set of 50 new vocabulary words to memorize. On Friday the students took a vocabulary test. When they returned to class the following Monday, they were retested—without advance warning. Both sets of test scores for the 25 students are shown below.

Fri	Mon	Fri	Mon
42	36	50	47
44	44	34	34
45	46	38	31
48	38	43	40
44	40	39	41
43	38	46	32
41	37	37	36
35	31	40	31
43	32	41	32
48	37	48	39
43	41	37	31
45	32	36	41
47	44		

a) Create a graphical display to compare the two distributions of scores.
b) Write a few sentences about the scores reported on Friday and Monday.
c) Create a graphical display showing the distribution of the *changes* in student scores.
d) Describe the distribution of changes.

12. e-Books A study by the Pew Internet & American Life Project found that 78% of U.S. residents over 16 years old read a book in the past 12 months. They also found that 21% had read an e-book using a reader or computer during that period. A newspaper reporting on these findings concluded that 99% of U.S. adult residents had read

a book in some fashion in the past year. (http://libraries
.pewinternet.org/2012/04/04/the-rise-of-e-reading/) Do you
agree? Explain.

13. Let's play cards You pick a card from a deck (see
description in Chapter 10) and record its denomination
(7, say) and its suit (maybe spades).

a) Is the variable *suit* categorical or quantitative?
b) Name a game you might be playing for which you
would consider the variable *denomination* to be
categorical. Explain.
c) Name a game you might be playing for which you
would consider the variable *denomination* to be
quantitative. Explain.

14. Accidents Progressive Insurance asked customers who
had been involved in auto accidents how far they were
from home when the accident happened. The data are
summarized in the table.

Miles from Home	% of Accidents
Less than 1	23
1 to 5	29
6 to 10	17
11 to 15	8
16 to 20	6
Over 20	17

a) Create an appropriate graph of these data.
b) Do these data indicate that driving near home is
particularly dangerous? Explain.

15. Hard water In an investigation of environmental causes
of disease, data were collected on the annual mortality
rate (deaths per 100,000) for males in 61 large towns in
England and Wales. In addition, the water hardness was
recorded as the calcium concentration (parts per million,
ppm) in the drinking water.

a) What are the variables in this study? For each, indicate whether it is quantitative or categorical and what
the units are.
b) Here are histograms of calcium concentration and mortality. Describe the distributions of the two variables.

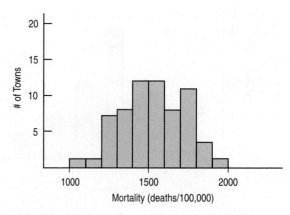

16. Hard water II The data set from England and Wales
also notes for each town whether it was south or north of
Derby. Here are some summary statistics and a comparative boxplot for the two regions.

Summary of Mortality				
Group	Count	Mean	Median	StdDev
North	34	1631.59	1631	138.470
South	27	1388.85	1369	151.114

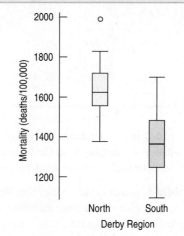

a) What is the overall mean mortality rate for the two
regions?
b) Do you see evidence of a difference in mortality
rates? Explain.

17. Seasons Average daily temperatures in January and July for
60 large U.S. cities are graphed in the following histograms.

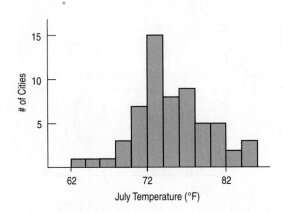

a) What aspect of these histograms makes it difficult to compare the distributions?

b) What differences do you see between the distributions of January and July average temperatures?

c) Differences in temperatures (July–January) for each of the cities are displayed in the boxplot above. Write a few sentences describing what you see.

T **18. Old Faithful** It is a common belief that Yellowstone's most famous geyser erupts once an hour at very predictable intervals. The histogram below shows the time gaps (in minutes) between 222 successive eruptions. Describe this distribution.

T **19. Old Faithful?** Does the duration of an eruption have an effect on the length of time that elapses before the next eruption?

a) The histogram below shows the duration (in minutes) of those 222 eruptions. Describe this distribution.

b) Explain why it is not appropriate to find summary statistics for this distribution.

c) Let's classify the eruptions as "long" or "short," depending upon whether or not they last at least 3 minutes. Describe what you see in the comparative boxplots.

20. Teen drivers 2008 In its *Teen Driver Crashes Report to Congress 2008*, the National Highway Traffic Safety Administration reported that 6.3% of licensed drivers were between the ages of 15 and 20, yet this age group was behind the wheel in 12.9% of all fatal crashes. Use these statistics to explain the concept of independence.

T **21. Liberty's nose** Is the Statue of Liberty's nose too long? Her nose measures, 4′6″, but she is a large statue, after all. Her arm is 42 feet long. That means her arm is $42/4.5 = 9.3$ times as long as her nose. Is that a reasonable ratio? Shown in the table are arm and nose lengths of 18 girls in a Statistics class, and the ratio of arm-to-nose length for each.

Arm (cm)	Nose (cm)	Arm/Nose Ratio
73.8	5.0	14.8
74.0	4.5	16.4
69.5	4.5	15.4
62.5	4.7	13.3
68.6	4.4	15.6
64.5	4.8	13.4
68.2	4.8	14.2
63.5	4.4	14.4
63.5	5.4	11.8
67.0	4.6	14.6
67.4	4.4	15.3
70.7	4.3	16.4
69.4	4.1	16.9
71.7	4.5	15.9
69.0	4.4	15.7
69.8	4.5	15.5
71.0	4.8	14.8
71.3	4.7	15.2

a) Make an appropriate plot and describe the distribution of the ratios.
b) Summarize the ratios numerically, choosing appropriate measures of center and spread.
c) Is the ratio of 9.3 for the Statue of Liberty unrealistically low? Explain.

T **22. Winter Olympics 2010 speed skating** The top 36 women's 500-m speed skating times are listed in the table.

a) The mean finishing time was 40.44 seconds, with a standard deviation of 10.03 seconds. If a Normal model were appropriate, what percent of the times should be within 2 seconds of the mean?
b) What percent of the times actually fall within this interval?
c) Explain the discrepancy between parts a and b.

Nation	Athlete	Result
Korea	Sang-Hwa Lee	38.249
Germany	Jenny Wolf	38.307
China	Beixing Wang	38.487
Netherlands	Margot Boer	38.511
China	Shuang Zhang	38.530
Japan	Sayuri Yoshii	38.566
Russian Federation	Yulia Nemaya	38.594
China	Peiyu Jin	38.686
United States	Heather Richardson	38.698
Germany	Monique Angermuller	38.761
China	Aihua Xing	38.792

Nation	Athlete	Result
Japan	Nao Kodaira	38.835
Canada	Christine Nesbitt	38.881
Netherlands	Thijsje Oenema	38.892
DPR Korea	Hyon-Suk Ko	38.893
Japan	Shihomi Shinya	38.964
Japan	Tomomi Okazaki	38.971
United States	Elli Ochowicz	39.002
Kazakhstan	Yekaterina Aydova	39.024
United States	Jennifer Rodriguez	39.182
Netherlands	Laurine van Riessen	39.302
Canada	Shannon Rempel	39.351
Germany	Judith Hesse	39.357
Russian Federation	Olga Fatkulina	39.359
Czech Republic	Karolina Erbanova	39.365
Korea	Bo-Ra Lee	39.396
Russian Federation	Svetlana Kaykan	39.422
Italy	Chiara Simionato	39.480
United States	Lauren Cholewinski	39.514
Korea	Jee-Min Ahn	39.595
Australia	Sophie Muir	39.649
Russian Federation	Yekaterina Malysheva	39.782
Korea	Min-Jee Oh	39.816
Canada	Anastasia Bucsis	39.879
Belarus	Svetlana Radkevich	39.899
Netherlands	Annette Gerritsen	97.952

23. Sample A study in South Africa focusing on the impact of health insurance identified 1590 children at birth and then sought to conduct follow-up health studies 5 years later. Only 416 of the original group participated in the 5-year follow-up study. This made researchers concerned that the follow-up group might not accurately resemble the total group in terms of health insurance. The table in the next column summarizes the two groups by race and by presence of medical insurance when the child was born. Carefully explain how this study demonstrates Simpson's paradox. (*Birth to Ten Study*, Medical Research Council, South Africa)

		Number (%) Insured	
		Follow-Up	Not Traced
Race	Black	36 of 404 (8.9%)	91 of 1048 (8.7%)
	White	10 of 12 (83.3%)	104 of 126 (82.5%)
	Overall	46 of 416 (11.1%)	195 of 1174 (16.6%)

(continued)

24. Sluggers Babe Ruth was the first great "slugger" in baseball. His record of 60 home runs in one season held for 34 years until Roger Maris hit 61 in 1961. Mark McGwire (with the aid of steroids) set a new standard of 70 in 1998. Listed below are the home run totals for each season McGwire played. Also listed are Babe Ruth's home run totals.

McGwire: 3*, 49, 32, 33, 39, 22, 42, 9*,9*, 39, 52, 58, 70, 65, 32*, 29*

Ruth: 54, 59, 35, 41, 46, 25, 47, 60, 54, 46, 49, 46, 41, 34, 22

a) Find the 5-number summary for McGwire's career.
b) Do any of his seasons appear to be outliers? Explain.
c) McGwire played in only 18 games at the end of his first big league season, and missed major portions of some other seasons because of injuries to his back and knees. Those seasons might not be representative of his abilities. They are marked with asterisks in the list above. Omit these values and make parallel boxplots comparing McGwire's career to Babe Ruth's.
d) Write a few sentences comparing the two sluggers.
e) Create a side-by-side stem-and-leaf display comparing the careers of the two players.
f) What aspects of the distributions are apparent in the stem-and-leaf displays that did not clearly show in the boxplots?

25. Be quick! Avoiding an accident when driving can depend on reaction time. That time, measured from the moment the driver first sees the danger until he or she steps on the brake pedal, is thought to follow a Normal model with a mean of 1.5 seconds and a standard deviation of 0.18 seconds.

a) Use the 68–95–99.7 Rule to draw the Normal model.
b) Write a few sentences describing driver reaction times.
c) What percent of drivers have a reaction time less than 1.25 seconds?
d) What percent of drivers have reaction times between 1.6 and 1.8 seconds?
e) What is the interquartile range of reaction times?
f) Describe the reaction times of the slowest 1/3 of all drivers.

26. Music and memory Is it a good idea to listen to music when studying for a big test? In a study conducted by some Statistics students, 62 people were randomly assigned to listen to rap music, Mozart, or no music while attempting to memorize objects pictured on a page. They were then asked to list all the objects they could remember. Here are the 5-number summaries for each group:

	n	Min	Q1	Median	Q3	Max
Rap	29	5	8	10	12	25
Mozart	20	4	7	10	12	27
None	13	8	9.5	13	17	24

a) Describe the W's for these data: *Who, What, Where, Why, When, How.*
b) Name the variables and classify each as categorical or quantitative.
c) Create parallel boxplots as best you can from these summary statistics to display these results.
d) Write a few sentences comparing the performances of the three groups.

27. Mail Here are the number of pieces of mail received at a school office for 36 days.

123	70	90	151	115	97
80	78	72	100	128	130
52	103	138	66	135	76
112	92	93	143	100	88
118	118	106	110	75	60
95	131	59	115	105	85

a) Plot these data.
b) Find appropriate summary statistics.
c) Write a brief description of the school's mail deliveries.
d) What percent of the days actually lie within one standard deviation of the mean? Comment.

28. Birth order Is your birth order related to your choice of major? A Statistics professor at a large university polled his students to find out what their majors were and what position they held in the family birth order. The results are summarized in the table.

a) What percent of these students are oldest or only children?
b) What percent of Humanities majors are oldest children?
c) What percent of oldest children are Humanities students?
d) What percent of the students are oldest children majoring in the Humanities?

		Birth Order*				
		1	2	3	4+	Total
Major	Math/Science	34	14	6	3	57
	Agriculture	52	27	5	9	93
	Humanities	15	17	8	3	43
	Other	12	11	1	6	30
	Total	113	69	20	21	223

*1 = oldest or only child

29. Herbal medicine Researchers for the Herbal Medicine Council collected information on people's experiences with a new herbal remedy for colds. They went to a store that sold natural health products. There they asked

100 customers whether they had taken the cold remedy and, if so, to rate its effectiveness (on a scale from 1 to 10) in curing their symptoms. The Council concluded that this product was highly effective in treating the common cold.

a) Identify the W's of these data.
b) Identify the variables, classify each as categorical or quantitative, and specify units if relevant.
c) Are you confident about the Council's conclusion? Explain.

T 30. **Birth order revisited** Consider again the data on birth order and college majors in Exercise 28.

a) What is the marginal distribution of majors?
b) What is the conditional distribution of majors for the oldest children?
c) What is the conditional distribution of majors for the children born second?
d) Do you think that college major appears to be independent of birth order? Explain.

31. **Engines** One measure of the size of an automobile engine is its "displacement," the total volume (in liters or cubic inches) of its cylinders. Summary statistics for several models of new cars are shown. These displacements were measured in cubic inches.

Summary of Displacement	
Count	38
Mean	177.29
Median	148.5
StdDev	88.88
Range	275
25th %tile	105
75th %tile	231

a) How many cars were measured?
b) Why might the mean be so much larger than the median?
c) Describe the center and spread of this distribution with appropriate statistics.
d) Your neighbor is bragging about the 227-cubic-inch engine he bought in his new car. Is that engine unusually large? Explain.
e) Are there any engines in this data set that you would consider to be outliers? Explain.
f) Is it reasonable to expect that about 68% of car engines measure between 88 and 266 cubic inches? (That's 177.289 $\pm$ 88.8767.) Explain.
g) We can convert all the data from cubic inches to cubic centimeters (cc) by multiplying by 16.4. For example, a 200-cubic-inch engine has a displacement of 3280 cc. How would such a conversion affect each of the summary statistics?

32. **Engines, again** Horsepower is another measure commonly used to describe auto engines. Here are the summary statistics and histogram displaying horsepowers of the same group of 38 cars discussed in Exercise 31.

Summary of Horsepower	
Count	38
Mean	101.7
Median	100
StdDev	26.4
Range	90
25th %tile	78
75th %tile	125

a) Describe the shape, center, and spread of this distribution.
b) What is the interquartile range?
c) Are any of these engines outliers in terms of horsepower? Explain.
d) Do you think the 68–95–99.7 Rule applies to the horsepower of auto engines? Explain.
e) From the histogram, make a rough estimate of the percentage of these engines whose horsepower is within one standard deviation of the mean.
f) A fuel additive boasts in its advertising that it can "add 10 horsepower to any car." Assuming that is true, what would happen to each of these summary statistics if this additive were used in all the cars?

33. **Age and party 2011** The Pew Research Center conducts surveys regularly asking respondents which political party they identify with or lean toward. Among their results is the following table relating preferred political party and age. (http://people-press.org/)

		Party			
		Republican/ Lean Rep.	Democrat/ Lean Dem.	Neither	Total
Age	18–29	318	424	73	815
	30–49	991	1058	203	2252
	50–64	1260	1407	264	2931
	65+	1136	1087	193	2416
	Total	3705	3976	733	8414

a) What percent of people surveyed were Republicans or leaned Republican?
b) Do you think this might be a reasonable estimate of the percentage of all voters who are Republicans or lean Republicans? Explain.
c) What percent of people surveyed were under 30 or over 65?
d) What percent of people were classified as "Neither" and under the age of 30?
e) What percent of the people classified as "Neither" were under 30?
f) What percent of people under 30 were classified as "Neither"?

34. **Pay** According to the Bureau of Labor Statistics, the mean hourly wage for Chief Executives in 2009 was $80.43 and the median hourly wage was $77.27. By contrast, for General and Operations Managers, the mean hourly wage was $53.15 and the median was $44.55. Are these wage distributions likely to be symmetric, skewed left, or skewed right? Explain.

35. **Age and party 2011 II** Consider again the Pew Research Center results on age and political party in Exercise 33.

a) What is the marginal distribution of party affiliation?
b) Create segmented bar graphs displaying the conditional distribution of party affiliation for each age group.
c) Summarize these poll results in a few sentences that might appear in a newspaper article about party affiliation in the United States.
d) Do you think party affiliation is independent of the voter's age? Explain.

36. **Bike safety 2005** The Bicycle Helmet Safety Institute website includes a report on the number of bicycle fatalities per year in the United States. The table below shows the counts for the years 1994–2005.

Year	Bicycle fatalities
1994	802
1995	833
1996	765
1997	814
1998	760
1999	754
2000	693
2001	732
2002	665
2003	629
2004	727
2005	784

a) What are the W's for these data?
b) Display the data in a stem-and-leaf display.

c) Display the data in a timeplot.
d) What is apparent in the stem-and-leaf display that is hard to see in the timeplot?
e) What is apparent in the timeplot that is hard to see in the stem-and-leaf display?
f) Write a few sentences about bicycle fatalities in the United States.

37. **Some assembly required** A company that markets build-it-yourself furniture sells a computer desk that is advertised with the claim "less than an hour to assemble." However, through postpurchase surveys the company has learned that only 25% of its customers succeeded in building the desk in under an hour. The mean time was 1.29 hours. The company assumes that consumer assembly time follows a Normal model.

a) Find the standard deviation of the assembly time model.
b) One way the company could solve this problem would be to change the advertising claim. What assembly time should the company quote in order that 60% of customers succeed in finishing the desk by then?
c) Wishing to maintain the "less than an hour" claim, the company hopes that revising the instructions and labeling the parts more clearly can improve the 1-hour success rate to 60%. If the standard deviation stays the same, what new lower mean time does the company need to achieve?
d) Months later, another postpurchase survey shows that new instructions and part labeling did lower the mean assembly time, but only to 55 minutes. Nonetheless, the company did achieve the 60%-in-an-hour goal, too. How was that possible?

38. **Profits** Here is a stem-and-leaf display showing profits as a percent of sales for 29 of the *Forbes* 500 largest U.S. corporations. The stems are split; each stem represents a span of 5%, from a loss of 9% to a profit of 25%.

```
-0 | 99
-0 | 1234
 0 | 11112 3444
 0 | 5555679
 1 | 00113
 1 |
 2 | 2
 2 | 5
```

Profits (% of sales)
(−0|3 means a loss of 3%)

a) Find the 5-number summary.
b) Draw a boxplot for these data.
c) Find the mean and standard deviation.
d) Describe the distribution of profits for these corporations.

Practice Exam

I. Multiple Choice

1. Below are summary statistics for infant mortality rates for Wisconsin counties in 2011. The numbers represent infant deaths per 1000 residents.

Mean	0.05958
Median	0.04492
Min	0
Q1	0
Q3	0.0877265
Max	0.324366

Which of the following statements is true?

A) About half the counties had more than 0.05958 infant deaths per 1000 residents.
B) Because the distribution is skewed right, more than half the counties had more than 0.04492 infant deaths per 1000 residents.
C) Because the distribution is skewed right, more than half the counties had less than 0.04492 infant deaths per 1000 residents.
D) At least one fourth of the counties had no infant deaths.
E) About half the counties had less than 0.08773 infant deaths per 1000 residents.

2. At a certain school 60 of the 100 boys and 60 of the 80 girls signed up for the senior trip. Is there an association between going on the trip and gender?

A) We can't tell, because the class doesn't have the same number of boys and girls.
B) Yes, because the same number of boys and girls signed up.
C) Yes, because a lower percentage of boys signed up than of girls.
D) No, because the people on the trip were 50% boys and 50% girls.
E) No, because the sign-up rate was higher among girls than among boys.

3. During the 2011–12 NBA season, LeBron James had an average of 30.3 points per game. The mean and standard deviation for the league were 7.65 points per game and 6.64 points per game, respectively. That same season the WNBA was led by two players; both Diana Taurasi and Angel McCoughtry averaged 21.6 points per game. The mean and standard deviation for the entire WNBA that year were 8.13 points per game and 4.77 points per game, respectively. Which is the more remarkable performance compared to the rest of their league?

A) LeBron James had the more remarkable performance because he scored more points than the women players did.
B) LeBron James had the more remarkable performance because the means show that it's harder to score points in the NBA.
C) Diana Taurasi and Angel McCoughtry had the more remarkable performances because the standard deviations show that there's less variability in scoring in the WNBA.
D) LeBron James had the more remarkable performance because his average was more standard deviations above the mean than the women's average.
E) You cannot compare these performances because the two leagues are so different.

4. Below is a histogram of the runlengths of the 150 top-grossing films from 2011. Also shown is an incorrectly drawn boxplot for the data.

Which of these is *not* a reason the boxplot is incorrect?

A) The median should be between 105 and 115, and the boxplot shows a different value.
B) The mean should be between 105 and 115, and the boxplot shows a different value.
C) There should be more outliers.
D) The box contains more than 50% of the data.
E) The IQR shown in the boxplot is too large.

5. The fuel economy in miles per gallon for 2012 model cars in the U.S. is summarized in the table below.

Mean	s	Q1	Median	Q3	IQR
22.0	6.69	18	21	25	7

In order to compare these U.S. models to European cars, these statistics will be converted to km per liter by multiplying the numerical mpg rating for each car by 0.425. Which of the following statements is true?

A) All of these summary statistics will be multiplied by 0.425.
B) All of these summary statistics will remain unchanged.
C) The mean and median will be multiplied by 0.425, but the other summary statistics will remain unchanged.
D) The standard deviation and *IQR* will be multiplied by 0.425, but the other statistics will remain unchanged.
E) The standard deviation and *IQR* will remain unchanged, but the other statistics will be multiplied by 0.425

6. Relatively minor repetitive impacts to the head such as those experienced by elite male soccer players when hitting a soccer ball with their heads may cause brain damage, according to the Journal of the American Medical Association. One indicator of mild brain injury is called radial diffusivity. The radial diffusivity measurements of twelve right-handed male soccer players were compared to those of eleven competitive swimmers (who are not likely to have repeated head impacts). Which type of plot would *not* be appropriate for assessing the difference between the two groups?

A) Stacked dotplots, one for each group, on the same axis
B) Side-by-side boxplots
C) Stacked bar graphs
D) A back-to-back stem-and-leaf plot
E) A pair of histograms, one above the other, on the same axis. www.medpagetoday.com/Neurology

7. American automobiles produced in 2012 and classified as "large" had a mean fuel economy of 19.6 miles per gallon with a standard deviation of 3.36 miles per gallon. A particular model on this list was rated at 23 miles per gallon, giving it a *z*-score of about 1.01. Which statement is true based on this information?

A) Because the standard deviation is small compared to the mean, a Normal model is appropriate and we can say that about 84.4% of "large" automobiles have a fuel economy of 23 miles per gallon or less.
B) Because a *z*-score was calculated, it is appropriate to use a Normal model to say that about 84.4% of "large" automobiles have a fuel economy of 23 miles per gallon or less.
C) Because 23 miles per gallon is greater than the mean of 19.6 miles per gallon, the distribution is skewed to the right. This means the *z*-score cannot be used to calculate a proportion.

D) Because no information was given about the shape of the distribution, it is not appropriate to use the *z*-score to calculate the proportion of automobiles with a fuel economy of 23 miles per gallon or less.
E) Because no information was given about the shape of the distribution, it is not appropriate to calculate a *z*-score, so the *z*-score has no meaning in this situation.

8. The losing teams in all college basketball games for 2011 had scores that are approximately normally distributed with mean 64 points and standard deviation about 11.7 points. Based on the Normal model, we'd expect that the middle 90% of losing teams' scores would be between about

A) 29 and 99 points
B) 41 and 87 points
C) 45 and 83 points
D) 49 and 79 points
E) 52 and 76 points

9. A survey of students at a Wisconsin high school asked the following question:

Whom do you most often text during class?

_____ *family members*   _____ *girlfriend/boyfriend*
_____ *friends inside school*   _____ *friends outside school*

The results, sorted by grade, are summarized in this table:

Grade	Family members	Girlfriend/ boyfriend	Friends inside school	Friends outside school	Total
9th	19	10	8	5	42
10th	17	12	5	1	35
11th	13	9	11	5	38
12th	8	21	10	6	45
Total	57	52	34	17	160

Which statement about these results is correct?

A) The proportion of 9th graders who said they text family members most is 19/57.
B) Among those who said they text girlfriend/boyfriend the most the proportion who are seniors is 21/45.
C) The proportion of 10th graders who said they text friends inside school the most is greater than the proportion of *all* students who said that.
D) A student who texts friends inside school is more likely to be a senior than is someone who texts friends outside school.
E) Of all the grade levels, 11th graders are least likely to text a girlfriend or boyfriend.

10. The pie charts below show the percentages of users of Facebook and users of Twitter that fall into various age groups.

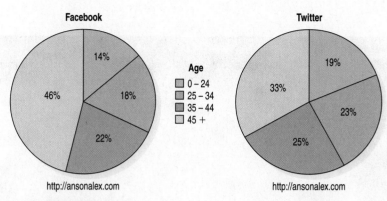

Facebook

Twitter

Age
- ☐ 0 – 24
- ☐ 25 – 34
- ☐ 35 – 44
- ☐ 45 +

http://ansonalex.com

http://ansonalex.com

Which of the following cannot be concluded from the pie charts?

A) Facebook has a larger proportion of users in the 45+ age range than Twitter.

B) The smallest age group for both Facebook and Twitter users is the 0–24 age group.

C) There are about the same number of Facebook users as Twitter users.

D) Both Twitter and Facebook have more people in the 45+ age group than any other age group.

E) A smaller proportion of Facebook users than Twitter users are in the 25 to 34 age group.

II. Free Response

1. The summary statistics below describe the daily high temperatures (°F) from June to November 2012 in Champaign, IL.

Mean	Median	s	Min	Q1	Q3	Max
75.8	79	16.33	31	63	89	101

a) Based on these statistics, what do you suspect to be true about the shape of the distribution? Explain.

b) Are there any outliers among these temperatures? Justify your conclusion.

c) Below is a boxplot for daily high temperatures in the same region from December 2011 to May 2012. Sketch a boxplot of the June to November temperatures above it on the same axes.

Maximum Temperature

d) Write a few sentences comparing these distributions.

2. Many electronic devices use disposable batteries, which eventually have to be replaced. Assume that for one particular brand and type of battery, the distribution of the hours of useful power can be approximated well by a Normal model. The mean is 140 hours, and the standard deviation is 4 hours.

a) Using the 68-95-99.7 Rule, sketch the appropriate model.

b) The marketing department wants to write a guarantee for battery life. What lifespan should they quote so that they can expect 98% of the batteries to meet the guarantee?

c) The company's research department has been given the task of improving the batteries, with a goal that at least 90% of the new batteries should last longer than the average of the current type. Initial research suggests that a mean of 143 hours is attainable. What other improvement must they make to achieve their goal?

d) Explain what the improvement you suggest in part c would mean about the new batteries.

6 Scatterplots, Association, and Correlation

Who	Years 1970–2010
What	Mean error in the position of Atlantic hurricanes as predicted 72 hours ahead by the NHC
Units	nautical miles
When	1970–2010
Where	Atlantic and Gulf of Mexico
Why	The NHC wants to improve prediction models

I n 2005 Hurricane Katrina made a direct hit on New Orleans, killing at least 1833 people and causing $108 billion in damage across the region. In 2012 Hurricane Sandy became the largest Atlantic storm on record, stretching 1100 miles in diameter at its peak. Sandy slammed into New York City, New Jersey, and Connecticut, where timely evacuations kept the death toll to 131, but over 8 million people lost power (many for several weeks) and damages totaled about $66 billion.

Where will a hurricane strike? People want to know if a hurricane is coming their way, and the National Hurricane Center (NHC) of the National Oceanic and Atmospheric Administration (NOAA) tries to predict the path a hurricane will take. But hurricanes tend to wander around aimlessly and are pushed by fronts and other weather phenomena in their area, so they are notoriously difficult to predict. Even relatively small changes in a hurricane's track can make big differences in the damage it causes.

To improve hurricane prediction, NOAA[1] relies on sophisticated computer models and has been working for decades to improve them. How well are they doing? Have predictions improved in recent years? Has the improvement been consistent? Here's a time-plot of the mean error, in nautical miles, of the NHC's 72-hour predictions of Atlantic hurricanes since 1970:

Figure 6.1

A scatterplot of the average error in nautical miles of the predicted position of Atlantic hurricanes for predictions made by the National Hurricane Center of NOAA, plotted against the *Year* in which the predictions were made.

[1]www.nhc.noaa.gov

Look, Ma, No Origin!
Scatterplots usually don't—and shouldn't—show the origin, because often neither variable has values near 0. The display should focus on the part of the coordinate plane that actually contains the data. In our example about hurricanes, none of the prediction errors or years were anywhere near 0, so the computer drew the scatterplot with axes that don't quite meet.

A S *Activity*: **Heights of Husbands and Wives.** Husbands are usually taller than their wives. Or are they?

Clearly, predictions have improved. The plot shows a fairly steady decline in the average error, from almost 500 nautical miles in the late 1970s to about 120 nautical miles in 2009. We can also see a few years when predictions were unusually good and that 1972 was a really bad year for predicting hurricane tracks.

This timeplot is an example of a more general kind of display called a **scatterplot**. Scatterplots may be the most common displays for data. By just looking at them, you can see patterns, trends, relationships, and even the occasional extraordinary value sitting apart from the others. As the great philosopher Yogi Berra[2] once said, "You can observe a lot by watching."[3] Scatterplots are the best way to start observing the relationship between two *quantitative* variables.

Relationships between variables are often at the heart of what we'd like to learn from data:

- Are grades actually higher now than they used to be?
- Do people tend to reach puberty at a younger age than in previous generations?
- Does applying magnets to parts of the body relieve pain? If so, are stronger magnets more effective?
- Do students learn better with more use of computer technology?

Questions such as these relate two quantitative variables and ask whether there is an **association** between them. Scatterplots are the ideal way to *picture* such associations.

Looking at Scatterplots

A S *Activity*: **Making and Understanding Scatterplots.** See the best way to make scatterplots—using a computer.

How would you describe the association of hurricane *Prediction Error* and *Year*? Everyone looks at scatterplots. But, if asked, many people would find it hard to say what to look for in a scatterplot. What do *you* see? Try to describe the scatterplot of *Prediction Error* against *Year*.

Look for Direction:
What's my sign—positive, negative, or neither?

You might say that the **direction** of the association is important. Over time, the NHC's prediction errors have decreased. A pattern like this that runs from the upper left to the lower right is said to be **negative**. A pattern running the other way is called **positive**.

Look for Form:
Is it straight, curved, something exotic, or no pattern?

The second thing to look for in a scatterplot is its **form**. If there is a straight line relationship, it will appear as a cloud or swarm of points stretched out in a generally consistent, straight form. For example, the scatterplot of *Prediction Error* vs. *Year* has such an underlying **linear** form, although some points stray away from it.

Scatterplots can reveal many kinds of patterns. Often they will not be straight, but straight line patterns are both the most common and the most useful for statistics.

If the relationship isn't straight, but curves gently, while still increasing or decreasing steadily, we can often find ways to make it more nearly straight. But if it curves sharply—up and then down, for example —there is much less we can say about it with the methods of this book.

Look for Strength:
How much scatter is there?

The third feature to look for in a scatterplot is how strong the relationship is. At one extreme, do the points appear tightly clustered in a single stream (whether straight, curved, or bending all over the place)? Or, at the other extreme, does the swarm of

---

[2]Hall of Fame catcher and manager of the New York Mets and Yankees.

[3]But then he also said, "I really didn't say everything I said." So we can't really be sure.

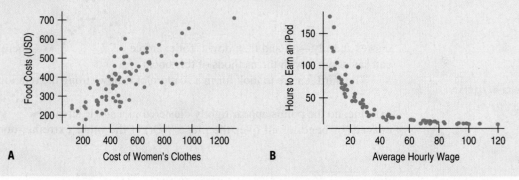

points seem to form a vague cloud　　　　through which we can barely discern any trend or pattern? The Predictio*n error* vs. *Year* plot shows moderate scatter around a generally straight form. This indicates that the linear trend of improving prediction is pretty consistent and moderately strong.

Finally, always look for the unexpected. Often the most interesting thing to see in a scatterplot is something you never thought to look for. One example of such a surprise is an **outlier** standing away from the overall pattern of the scatterplot. Such a point is almost always interesting and always deserves special attention. In the scatterplot of prediction errors, the year 1972 stands out as a year with very high prediction errors. An Internet search shows that it was a relatively quiet hurricane season. However, it included the very unusual—and deadly—Hurricane Agnes, which combined with another low-pressure center to ravage the northeastern United States, killing 122 and causing 1.3 billion 1972 dollars in damage. Possibly, Agnes was also unusually difficult to predict.

You should also look for clusters or subgroups that stand away from the rest of the plot or that show a trend in a different direction. Deviating groups should raise questions about why they are different. They may be a clue that you should split the data into subgroups instead of looking at them all together.

> **Look for Unusual Features:** Are there outliers or subgroups?

For Example　COMPARING PRICES WORLDWIDE

If you travel overseas, you know that what's really important is not the amount in your wallet but the amount it can buy. UBS (one of the largest banks in the world) prepared a report comparing prices, wages, and other economic conditions in cities around the world for their international clients. Some of the variables they measured in 73 cities are *Cost of Living*, *Food Costs*, *Average Hourly Wage*, average number of *Working Hours* per Year, average number of *Vacation Days*, hours of work (at the average wage) needed to buy an *iPod*, minutes of work needed to buy a *Big Mac*, and *Women's Clothing Costs*.[4] For your burger fix, you might want to live in Chicago, Toronto, or Tokyo where it takes only about 12 minutes of work to afford a Big Mac. In Mexico City, Jakarta, and Nairobi, you'd have to work more than two hours.

Of course, these variables are associated, but do they consistently reflect costs of living? Plotting pairs of variables can reveal how and even if they are associated. The variety of these associations illustrate different directions and kinds of association patterns you might see in other scatterplots.

QUESTION: Describe the patterns shown by each of these plots.

---

[4]Detail of the methodology can be found in the report *Prices and Earning: A comparison of purchasing power around the globe/2009 edition.* www.economist.com/node/14288808?story_id=14288808

11 Miles
27 mins

ANSWER: In Plot A, the association between *Food Costs* and *Cost of Women's Clothes* is positive, straight, and moderate in strength. In Plot B, the association between *Hours to Earn an iPod* and *Average Hourly Wage* is negative, curved, and strong. In Plot C, the association between the *Clothes Index* and *Working Hours* is negative, linear, and weak, with a high outlier. In Plot D, there does not appear to be any association between *Food Costs* and *Vacation Days*.

Roles for Variables

Which variable should go on the *x*-axis and which on the *y*-axis? What we want to know about the relationship can tell us how to make the plot. We often have questions such as:

- Do baseball teams that score more runs sell more tickets to their games?
- Do older houses sell for less than newer ones of comparable size and quality?
- Do students who score higher on their SAT tests have higher grade point averages in college?
- Can we estimate a person's percent body fat more simply by just measuring waist or wrist size?

In these examples, the two variables play different roles. We'll call the variable of interest the **response variable** and the other the **explanatory** or predictor variable.[5] We'll continue our practice of naming the variable of interest *y*. Naturally we'll plot it on the *y*-axis and place the explanatory variable on the *x*-axis. Sometimes, we'll call them the ***x*- and *y*-variables**. When you make a scatterplot, you can assume that those who view it will think this way, so choose which variables to assign to which axes carefully.

The roles that we choose for variables are more about how we *think* about them than about the variables themselves. Just placing a variable on the *x*-axis doesn't necessarily mean that it explains or predicts *anything*. And the variable on the *y*-axis may not respond to it in any way. We plotted prediction error on the *y*-axis against year on the *x*-axis because the National Hurricane Center is interested in how their predictions have changed over time. Could we have plotted them the other way? In this case, it's hard to imagine reversing the roles—knowing the prediction error and wanting to guess in what year it happened. But for some scatterplots, it can make sense to use either choice, so you have to think about how the choice of role helps to answer the question you have.

■ **NOTATION ALERT**

In Statistics, the assignment of variables to the *x*- and *y*-axes (and the choice of notation for them in formulas) often conveys information about their roles as predictor or response variable. So *x* and *y* are reserved letters as well, but not just for labeling the axes of a scatterplot.

A S *Self-Test*: **Scatterplot Check.**
Can you identify a scatterplot's direction, form, and strength?

---

[5]The *x*- and *y*-variables have sometimes been referred to as the *independent* and *dependent* variables, respectively. The idea was that the *y*-variable depended on the *x*-variable and the *x*-variable acted independently to make *y* respond. These names, however, conflict with other uses of the same terms in Statistics.

TI Tips CREATING A SCATTERPLOT

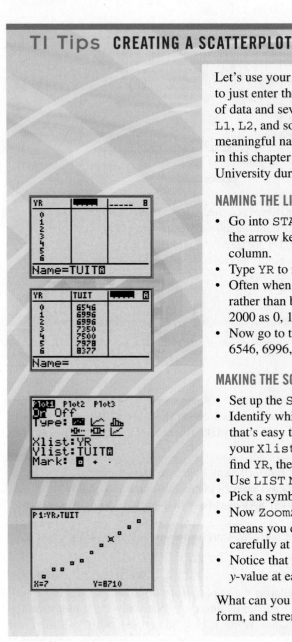

Let's use your calculator to make a scatterplot. First you need some data. It's okay to just enter the data in any two lists, but let's get fancy. When you are handling lots of data and several variables (as you will be soon), remembering what you stored in L1, L2, and so on can become confusing. You can—and should—give your variables meaningful names. To see how, let's store some data that you will use several times in this chapter and the next. They show the change in tuition costs at Arizona State University during the 1990s.

NAMING THE LISTS

- Go into STAT Edit, place the cursor on one of the list names (L1, say), and use the arrow key to move to the right across all the lists until you encounter a blank column.
- Type YR to name this first variable, then hit ENTER.
- Often when we work with years it makes sense to use values like "90" (or even "0") rather than big numbers like "1990." For these data enter the years 1990 through 2000 as 0, 1, 2, . . . , 10.
- Now go to the next blank column, name this variable TUIT, and enter these values: 6546, 6996, 6996, 7350, 7500, 7978, 8377, 8710, 9110, 9411, 9800.

MAKING THE SCATTERPLOT

- Set up the STATPLOT by choosing the scatterplot icon (the first option).
- Identify which lists you want as Xlist and Ylist. If the data are in L1 and L2, that's easy to do—but your data are stored in lists with special names. To specify your Xlist, go to 2nd LIST NAMES, scroll down the list of variables until you find YR, then hit ENTER.
- Use LIST NAMES again to specify Ylist:TUIT.
- Pick a symbol for displaying the points.
- Now ZoomStat to see your scatterplot. (Didn't work? ERR:DIM MISMATCH means you don't have the same number of x's and y's. Go to STAT Edit and look carefully at your two datalists. You can easily fix the problem once you find it.)
- Notice that if you TRACE the scatterplot the calculator will tell you the x- and y-value at each point.

What can you Tell about the trend in tuition costs at ASU? (Remember: direction, form, and strength!)

Correlation

Data collected from students in Statistics classes included their *Height* (in inches) and *Weight* (in pounds). It's no great surprise to discover that there is a positive association between the two. As you might suspect, taller students tend to weigh more. (If we had reversed the roles and chosen height as the explanatory variable, we might say that heavier students tend to be taller.[6]) And the form of the scatterplot is fairly straight as well, although there seems to be a high outlier, as the plot shows.

---

[6]The son of one of the authors, when told (as he often was) that he was tall for his age, used to point out that, actually, he was young for his height.

Figure 6.2

Weight *vs.* Height *of Statistics* students. Plotting *Weight* vs. *Height* in different units doesn't change the shape of the pattern.

Figure 6.3

In this scatterplot of *z*-scores, points are colored according to how they affect the association: green for positive, red for negative, and blue for neutral.

A S *Activity*: **Correlation.** Here's a good example of how correlation works to summarize the strength of a linear relationship and disregard scaling.

■ NOTATION ALERT

The letter *r* is always used for correlation, so you can't use it for anything else in Statistics. Whenever you see an *r*, it's safe to assume it's a correlation.

A S *Activity*: **Correlation and Relationship Strength.** What does a correlation of 0.8 look like? How about 0.3?

The pattern in the scatterplots looks straight and is clearly a positive association, but how strong is it? If you had to put a number (say, between 0 and 1) on the strength, what would it be? Whatever measure you use shouldn't depend on the choice of units for the variables. After all, if we measure heights and weights in centimeters and kilograms instead, it doesn't change the direction, form, or strength, so it shouldn't change the number.

To ensure the units don't matter when we measure the strength, we can remove them by standardizing each variable. Now, for each point, instead of the values (x, y) we'll have the standardized coordinates (z_x, z_y). Remember that to standardize values, we subtract the mean of each variable and then divide by its standard deviation:

$$(z_x, z_y) = \left(\frac{x - \bar{x}}{s_x}, \frac{y - \bar{y}}{s_y} \right).$$

Because standardizing makes the means of both variables 0, the center of the new scatterplot is at the origin. The scales on both axes are now standard deviation units, making the scaling consistent and providing a fairer impression of the strength of the association.

We've color-coded the points in this standardized plot. For the green points (in the first and third quadrants) the coordinates z_x and z_y are either both positive or both negative. Either way, the product $z_x z_y$ is positive, and these points are evidence of a positive association. Coordinates of the square red points (in the second and fourth quadrants) have opposite signs, so the product $z_x z_y$ is negative, evidence of a negative association. Points with larger *z*-scores will have larger products and offer greater evidence. The blue points on one of the axes offer no information on the direction of association, and the product $z_x z_y = 0$.

Now we add up the $z_x z_y$ products for every point in the scatterplot: $\Sigma z_x z_y$. This summarizes the direction *and* strength of the association for all the points. But, the sum gets bigger the more data we have. To adjust for this, the natural (for statisticians anyway) thing to do is to divide the sum by $n - 1$.[7] The result is the famous **correlation coefficient**:

$$r = \frac{\Sigma z_x z_y}{n - 1}.$$

For the students' heights and weights, the correlation is 0.644. Because it is based on *z*-scores, which have no units, correlation has no units either. It will stay the same regardless of how you measure height and weight.

---

[7]Yes, the same $n - 1$ as in the standard deviation calculation.

Correlation Conditions

A S *Simulation*: **Correlation and Linearity.** How much does straightness matter?

A S *Case Study*: **Mortality and Education.** Is the mortality rate lower in cities with higher education levels?

Correlation measures the strength of the *linear* association between two *quantitative* variables. Before you use correlation, you must check several *conditions*:

- **Quantitative Variables Condition** Don't make the common error of calling an association involving a categorical variable a correlation. Correlation is only about quantitative variables.
- **Straight Enough Condition** The best check for the assumption that the variables are truly linearly related is to look at the scatterplot to see whether it looks reasonably straight. That's a judgment call, but not a difficult one.
- **No Outliers Condition** Outliers can distort the correlation dramatically, making a weak association look strong or a strong one look weak. Outliers can even change the sign of the correlation. But it's easy to see outliers in the scatterplot, so to check this condition, just look.

Each of these conditions is easy to check with a scatterplot. Many correlations are reported without supporting data or plots. Nevertheless, you should still think about the conditions. And you should be cautious in interpreting (or accepting others' interpretations of) the correlation when you can't check the conditions for yourself.

For Example   CORRELATIONS FOR SCATTERPLOT PATTERNS

Look back at the scatterplots of the economic variables in cities around the world (p. 152). The correlations for those plots are 0.774, −0.783, −0.576, and −0.022, respectively.

QUESTION: Check the conditions for using correlation. If you feel they are satisfied, match the correlation to the plot, and interpret it.

ANSWER: In Plot A, the variables are quantitative, the relationship appears to be linear, and there are no outliers. The correlation of 0.774 indicates a moderately strong positive association between *Food Costs* and *Cost of Women's Clothes*.

In Plot B, the relationship is curved. It is not appropriate to compute a correlation.

In Plot C, the variables are quantitative, the relationship appears to be linear, and there may be an outlier. The correlation of −0.425 indicates a weak-to-moderate negative association between the *Clothes Index* and *Working Hours*.

In Plot D, the variables are quantitative and the relationship appears to be linear, but the small correlation of −0.022 suggests there may be no association between *Food Costs* and *Vacation Days*.

 ## Just Checking

Your Statistics teacher tells you that the correlation between the scores (points out of 50) on Exam 1 and Exam 2 was 0.75.

1. Before answering any questions about the correlation, what would you like to see? Why?

2. If she adds 10 points to each Exam 1 score, how will this change the correlation?

3. If she standardizes scores on each exam, how will this affect the correlation?

4. In general, if someone did poorly on Exam 1, are they likely to have done poorly or well on Exam 2? Explain.

5. If someone did poorly on Exam 1, can you be sure that they did poorly on Exam 2 as well? Explain.

Step-by-Step Example LOOKING AT ASSOCIATION

When your blood pressure is measured, it is reported as two values: systolic blood pressure and diastolic blood pressure.

Question: How are these variables related to each other? Do they tend to be both high or both low? How strongly associated are they?

THINK ➡ **Plan** State what you are trying to investigate.

I'll examine the relationship between two measures of blood pressure.

Variables Identify the two quantitative variables whose relationship we wish to examine. Report the W's, and be sure both variables are recorded for the same individuals.

The variables are systolic and diastolic blood pressure (*SBP* and *DBP*), recorded in millimeters of mercury (mm Hg) for each of 1406 participants in the Framingham Heart Study, a famous health study in Framingham, MA.[8]

Plot Make the scatterplot. Use a computer program or graphing calculator if you can.

Check the conditions.

✓ **Quantitative Variables Condition**: Both *SBP* and *DBP* are quantitative and measured in mm Hg.
✓ **Straight Enough Condition**: The scatterplot looks straight.
✓ **Outlier Condition**: There are a few straggling points, but none far enough from the body of the data to be called outliers.

REALITY CHECK Looks like a strong positive linear association. We shouldn't be surprised if the correlation coefficient is positive and fairly large.

I have two quantitative variables that satisfy the conditions, so correlation is a suitable measure of association.

SHOW ➡ **Mechanics** We usually calculate correlations with technology. Here we have 1406 cases, so we'd never try it by hand.

The correlation coefficient is $r = 0.792$.

TELL ➡ **Conclusion** Describe the direction, form, and strength you see in the plot, along with any unusual points or features. Be sure to state your interpretations in the proper context.

The scatterplot shows a positive direction, with higher *SBP* going with higher *DBP*. The plot is generally straight, with a moderate amount of scatter. The correlation of 0.792 is consistent with what I saw in the scatterplot.

---

[8]www.nhlbi.nih.gov/about/framingham

TI Tips   FINDING THE CORRELATION

Now let's use the calculator to find a correlation. Unfortunately, the statistics package on your TI calculator does not automatically do that. Correlations are one of the most important things we might want to do, so here's how to fix that, once and for all.

- Hit 2nd CATALOG (on the zero key). You now see a list of everything the calculator knows how to do. Impressive, huh?
- Scroll down until you find DiagnosticOn. Hit ENTER. Again. It should say Done.

Now and forevermore (or perhaps until you change batteries or clear memory) your calculator will find correlations.

FINDING THE CORRELATION

- *Always* check the conditions first. Look at the scatterplot for the Arizona State tuition data again. Does this association look linear? Are there outliers? This plot looks fine, but remember that correlation can be used to describe the strength of *linear* associations only, and outliers can distort the results. Eyeballing the scatterplot is an essential first step. (You should be getting used to checking on assumptions and conditions before jumping into a statistical procedure—it's always important.)
- Under the STAT CALC menu, select 8:LinReg(a+bx) and hit ENTER.

- Now specify Xlist:YR, Ylist:TUIT, leave both FreqList: and Store RegEQ: blank; then go to Calculate and hit ENTER.
  (OR on an older calculator, specify *x* and *y* by importing your variable names from the LIST NAMES menu, separated by a comma, to create the command LinReg(a + bx)LYR,LTUIT.)

Wow! A lot of stuff happened. If you suspect all those other numbers are important, too, you'll really enjoy the next chapter. But for now, it's the value of r you care about. What does this correlation, $r = 0.993$, say about the trend in tuition costs?

Correlation Properties

A S   *Activity*: **Construct Scatterplots with a Given Correlation.** Try to make a scatterplot that has a given correlation. How close can you get?

Here's a useful list of facts about the correlation coefficient:

- The sign of a correlation coefficient gives the direction of the association.
- Correlation is always between -1 and $+1$. Correlation *can* be exactly equal to -1.0 or $+1.0$, but these values are unusual in real data because they mean that all the data points fall *exactly* on a single straight line.
- Correlation treats *x* and *y* symmetrically. The correlation of *x* with *y* is the same as the correlation of *y* with *x*.
- Correlation has no units. This fact can be especially appropriate when the data's units are somewhat vague to begin with (IQ score, personality index, socialization, and so on). Correlation is sometimes given as a percentage, but you probably shouldn't do

that because it suggests a percentage of *something*—and correlation, lacking units, has no "something" of which to be a percentage.

- Correlation is not affected by changes in the center or scale of either variable. Changing the units or baseline of either variable has no effect on the correlation coefficient. Correlation depends only on the z-scores, and they are unaffected by changes in center or scale.
- Correlation measures the strength of the *linear* association between the two variable*s*. Variables can be strongly associated but still have a small correlation if the association isn't linear.
- Correlation is sensitive to outliers. A single outlying value can make a small correlation large or make a large one small.

Height and Weight, Again

We could have measured the students' weights in stones. In the now outdated UK system of measures, a stone is a measure equal to 14 pounds. And we could have measured heights in hands. Hands are still commonly used to measure the heights of horses. A hand is 4 inches. But no matter what *units* we use to measure the two variables, the *correlation* stays the same.

TI-*nspire*

Correlation and Scatterplots. See how the correlation changes as you drag data points around in a scatterplot.

How Strong Is Strong?

You'll often see correlations characterized as "weak," "moderate," or "strong," but be careful. There's no agreement on what those terms mean. The same numerical correlation might be strong in one context and weak in another. You might be thrilled to discover a correlation of 0.7 between the new summary of the economy you've come up with and stock market prices, but you'd consider it a design failure if you found a correlation of "only" 0.7 between two tests intended to measure the same skill. Deliberately vague terms like "weak," "moderate," or "strong" that describe a linear association can be useful additions to the numerical summary that correlation provides. But be sure to include the correlation and show a scatterplot, so others can judge for themselves.

For Example   CHANGING SCALES

RECAP: Several measures of prices and wages in cities around the world show a variety of relationships, some of which we can summarize with correlations.

QUESTION: Suppose that, instead of measuring prices in U.S. dollars and recording work time in hours, we had used Euros and minutes. How would those changes affect the conditions, the values of correlation, or our interpretation of the relationships involving those variables?

ANSWER: Not at all. Correlation is based on standardized values (z-scores), so the conditions, value of r, and interpretation are all unaffected by changes in units.

Warning: Correlation ≠ Causation

Whenever we have a strong correlation, it's tempting to try to explain it by imagining that the predictor variable has *caused* the response to change. Humans are like that; we tend to see causes and effects in everything.

Sometimes this tendency can be amusing. A scatterplot of the human population (y) of Oldenburg, Germany, in the beginning of the 1930s plotted against the number of storks nesting in the town (x) shows a tempting pattern.

Figure 6.4

The number of storks in Oldenburg, Germany, plotted against the population of the town for 7 years in the 1930s. The association is clear. How about the causation? (*Ornithologishe Monatsberichte*, 44, no. 2)

Anyone who has seen the beginning of the movie *Dumbo* remembers Mrs. Jumbo anxiously waiting for the stork to bring her new baby. Even though you know it's silly, you can't help but think for a minute that this plot shows that storks are the culprits. The two variables are obviously related to each other (the correlation is 0.97!), but that doesn't prove that storks bring babies.

It turns out that storks nest on house chimneys. More people means more houses, more nesting sites, and so more storks. The causation is actually in the *opposite* direction, but you can't tell from the scatterplot or correlation. You need additional information— not just the data—to determine the real mechanism.

A scatterplot of the damage (in dollars) caused to a house by fire would show a strong correlation with the number of firefighters at the scene. Surely the damage doesn't cause firefighters. And firefighters do seem to cause damage, spraying water all around and chopping holes. Does that mean we shouldn't call the fire department? Of course not. There is an underlying variable that leads to both more damage and more firefighters: the size of the blaze.

A hidden variable that stands behind a relationship and determines it by simultaneously affecting the other two variables is called a **lurking variable.** You can often debunk claims made about data by finding a lurking variable behind the scenes.

Scatterplots and correlation coefficients *never* prove causation. That's one reason it took so long for the U.S. Surgeon General to get warning labels on cigarettes. Although there was plenty of evidence that increased smoking was *associated* with increased levels of lung cancer, it took years to provide evidence that smoking actually *causes* lung cancer.

Does Cancer Cause Smoking?

Even if the correlation of two variables is due to a causal relationship, the correlation itself cannot tell us what causes what.

Sir Ronald Aylmer Fisher (1890–1962) was one of the greatest statisticians of the 20th century. Fisher testified in court (in testimony paid for by the tobacco companies) that a causal relationship might underlie the correlation of smoking and cancer:

> "Is it possible, then, that lung cancer . . . is one of the causes of smoking cigarettes? I don't think it can be excluded . . . the pre-cancerous condition is one involving a certain amount of slight chronic inflammation . . . .
>
> A slight cause of irritation . . . is commonly accompanied by pulling out a cigarette, and getting a little compensation for life's minor ills in that way. And . . . is not unlikely to be associated with smoking more frequently."

Ironically, the proof that smoking indeed is the cause of many cancers came from experiments conducted following the principles of experiment design and analysis that Fisher himself developed—and that we'll see in Chapter 12.

Correlation Tables

It is common in some fields to compute the correlations between every pair of variables in a collection of variables and arrange these correlations in a table. The rows and columns of the table name the variables, and the cells hold the correlations.

Correlation tables are compact and give a lot of summary information at a glance. They can be an efficient way to start to look at a large data set, but a dangerous one. By presenting all of these correlations without any checks for linearity and outliers, the correlation table risks showing truly small correlations that have been inflated by outliers, truly large correlations that are hidden by outliers, and correlations of any size that may be meaningless because the underlying form is not linear.

Table 6.1

A correlation table of data reported by _Forbes_ magazine for large companies. From this table, can you be sure that the variables are linearly associated and free from outliers?

	Assets	Sales	Market Value	Profits	Cash Flow	Employees
Assets	1.000					
Sales	0.746	1.000				
Market Value	0.682	0.879	1.000			
Profits	0.602	0.814	0.968	1.000		
Cash Flow	0.641	0.855	0.970	0.989	1.000	
Employees	0.594	0.924	0.818	0.762	0.787	1.000

The diagonal cells of a correlation table always show correlations of exactly 1. (Can you see why?) Correlation tables are commonly offered by statistics packages on computers. These same packages often offer simple ways to make all the scatterplots that go with these correlations.

Straightening Scatterplots

Correlation is a suitable measure of strength for straight relationships only. When a scatterplot shows a bent form that consistently increases or decreases, we can often straighten the form of the plot by re-expressing one or both variables.

Some camera lenses have an adjustable aperture, the hole that lets the light in. The size of the aperture is expressed in a mysterious number called the f/stop. Each increase of one f/stop number corresponds to a halving of the light that is allowed to come through. The f/stops of one digital camera are

f/stop: 2.8   4   5.6   8   11   16   22   32

When you halve the shutter speed, you cut down the light, so you have to open the aperture one notch. We could experiment to find the best f/stop value for each shutter speed. A table of recommended shutter speeds and f/stops for a camera lists the relationship like this:

Shutter speed: 1/1000 1/500 1/250 1/125 1/60 1/30 1/15 1/8
f/stop: 2.8   4   5.6   8   11   16   22   32

The correlation of these shutter speeds and f/stops is 0.979. That sounds pretty high. You might assume that there must be a strong linear relationship. But when we check the scatterplot (we _always_ check the scatterplot), it shows that something is not quite right:

Figure 6.5

A scatterplot of _f/stop_ vs. _Shutter Speed_ shows a bent relationship.

We can see that the f/stop is not *linearly* related to the shutter speed. Can we find a transformation of f/stop that straightens out the line? What if we look at the *square* of the f/stop against the shutter speed?

Figure 6.6

Re-expressing *f/stop* by squaring straightens the plot.

The second plot looks much more nearly straight. In fact, the correlation is now 0.998, but the increase in correlation is not important. (The original value of 0.979 should please almost anyone who sought a large correlation.) What is important is that the *form* of the plot is now straight, so the correlation is now an appropriate measure of association.[9]

We can often find transformations that straighten a scatterplot's form. Here, we found the square. Chapter 10 discusses simple ways to find a good re-expression.

TI Tips   STRAIGHTENING A CURVE

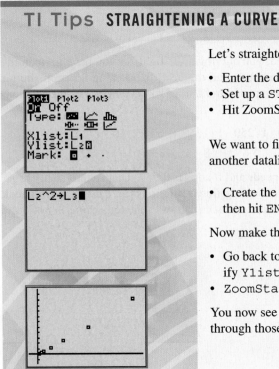

Let's straighten the f/stop scatterplot with your calculator.

- Enter the data in two lists, *shutterspeed* in L1 and *f/stop* in L2.
- Set up a STAT PLOT to create a scatterplot with Xlist:L1 and Ylist:L2.
- Hit ZoomStat. See the curve?

We want to find the squares of all the f/stops and save those re-expressed values in another datalist. That's easy to do.

- Create the command to square all the values in L2 and STOre those results in L3, then hit ENTER.

Now make the new scatterplot.

- Go back to STAT PLOT and change the setup. Xlist is still L1, but this time specify Ylist:L3.
- ZoomStat again.

You now see the straightened plot for these data. On deck: drawing the best line through those points!

---

[9]Sometimes we can do a "reality check" on our choice of re-expression. In this case, a bit of research reveals that f/stops are related to the diameter of the open shutter. Since the amount of light that enters is determined by the *area* of the open shutter, which is related to the diameter by squaring, the square re-expression seems reasonable. Not all re-expressions have such nice explanations, but it's a good idea to think about them.

WHAT IF ●●● correlations are variable?

The Finger Lakes region of central New York boasts 11 major lakes, several smaller ones, and numerous rivers and streams. That means there are a lot of highway bridges, over 700 in the counties bordering the large lakes. Bridge safety is an important concern, and highway engineers need to spot and repair problems before dangerous conditions arise. If they want to concentrate on bridges most likely to be unsafe, they need to identify factors that may signal risk. One of those may be the age of the bridge. Suppose that inspectors randomly select 20 bridges to investigate whether there's an association between age and condition. According to the New York State Department of Transportation's website:

> The NYSDOT condition rating scale ranges from 1 to 7, with 7 being in new condition and a rating of 5 or greater considered as good condition. . . NYSDOT defines a deficient bridge as one with a condition rating less than 5.0. A deficient condition rating indicates deterioration at a level that requires corrective maintenance or rehabilitation to restore the bridge.

Here's the scatterplot of *Condition* vs. *Age* for this random sample of 20 bridges:

This sample certainly confirms the suspicion that older bridges tend to be less safe. With a correlation of $r = -0.74$, engineers could reasonably conclude that there's a fairly strong negative relationship between a bridge's age and its condition.

But remember, this was just a sample. What about *all* bridges? We always wish we knew what's going on in the whole population but, unfortunately, entire populations are rarely accessible. We must settle for information from samples. In this sample, $r = -0.74$. What if the engineers had inspected a different sample? Let's have a look at another:

This time the correlation was only $r = -0.58$. Not as strong. Had this been the sample chosen, they might not have identified *Age* as being so important. (Wait, though. There's one old bridge that's in very good condition. See it? Does that outlier make an otherwise strong relationship just seem weaker?)

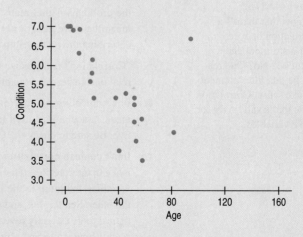

Just how much might correlations vary from sample to sample? To find out, we simulated 100 different samples of these bridges. In some the correlations were quite strong, in others rather weak. Here's the distribution:

By playing this "What If" game we get to see lots of samples. In the Real World, though, the engineers get just one. The histogram shows us that most samples would produce correlations of −0.5 or stronger. Any of those would make a pretty good case that age is an important factor in bridge safety. But we also see that sometimes a sample might produce a weaker correlation, one that could lead the engineers astray. That's the nature of Statistics: we can never be certain that our sample is telling us the truth.

What's to be done? For one thing, we can design sampling procedures that help get a better snapshot of the population; stick around for Chapter 11. If we can afford the time and money, we can pick a larger sample. And, importantly, we can develop a better understanding of how much our sample statistics may vary. That's a critical key to better insights about populations, coming your way in Chapter 17 and beyond. Don't go away.

WHAT CAN GO WRONG?

■ **Don't say "correlation" when you mean "association."** How often have you heard the word "correlation"? Chances are pretty good that when you've heard the term, it's been misused. When people want to sound scientific, they often say "correlation" when talking about the relationship between two variables. It's one of the most widely misused Statistics terms, and given how often statistics are misused, that's saying a lot. One of the problems is that many people use the specific term *correlation* when they really mean the more general term *association*. "Association" is a deliberately vague term describing the relationship between two variables.

"Correlation" is a precise term that measures the strength and direction of the linear relationship between quantitative variables.

■ **Don't correlate categorical variables.** People who misuse the term "correlation" to mean "association" often fail to notice whether the variables they discuss are quantitative. Be sure to check the Quantitative Variables Condition.

■ **Don't confuse correlation with causation.** One of the most common mistakes people make in interpreting statistics occurs when they observe a high correlation between two variables and jump to the perhaps tempting conclusion that one thing must be causing the other. Scatterplots and correlations *never* demonstrate causation. At best, these statistical tools can only reveal an association between variables, and that's a far cry from establishing cause and effect. While it's true that some associations may be causal, the nature and direction of the causation can be very hard to establish, and there's always the risk of overlooking lurking variables.

> **Not Correlation**
> Did you know that there's a strong correlation between playing an instrument and drinking coffee? No? One reason might be that the statement doesn't make sense. Correlation is a statistic that's valid only for *quantitative* variables.

■ **Make sure the association is linear.** Not all associations between quantitative variables are linear. Correlation can miss even a strong nonlinear association. A student project evaluating the quality of brownies baked at different temperatures reports a correlation of −0.05 between judges' scores and baking temperature. That seems to say there is no relationship—until we look at the scatterplot:

Figure 6.7

The relationship between brownie taste *Score* and *Baking Temperature* is strong, but not at all linear.

There is a strong association, but the relationship is not linear. Don't forget to check the Straight Enough Condition.

■ **Don't assume the relationship is linear just because the correlation coefficient is high.** Recall that the correlation of f/stops and shutter speeds is 0.979 and yet the relationship is clearly not straight. Although the relationship must be straight for the correlation to be an appropriate measure, a high correlation is no guarantee of straightness. Nor is it safe to use correlation to judge the best re-expression. It's always important to look at the scatterplot.

Figure 6.8

A scatterplot of *f/stop* vs. *Shutter Speed* shows a bent relationship even though the correlation is *r* = 0.979.

■ **Beware of outliers.** You can't interpret a correlation coefficient safely without a background check for outliers. Here's a silly example:

The relationship between IQ and shoe size among comedians shows a surprisingly strong positive correlation of 0.50. To check assumptions, we look at the scatterplot:

Figure 6.9

A scatterplot of *IQ* vs. *Shoe Size*. From this "study," what is the relationship between the two? The correlation is 0.50. Who does that point (the green x) in the upper right-hand corner belong to?

The outlier is Bozo the Clown, known for his large shoes, and widely acknowledged to be a comic "genius." Without Bozo, the correlation is near zero.

Even a single outlier can dominate the correlation value. That's why you need to check the Outlier Condition.

What Have We Learned?

In recent chapters we learned how to listen to the story told by data from a single variable. Now we've turned our attention to the more complicated (and more interesting) story we can discover in the association between two quantitative variables.

We've learned to begin our investigation by looking at a scatterplot. We're interested in the *direction* of the association, the *form* it takes, and its *strength*.

We've learned that, although not every relationship is linear, when the scatterplot is straight enough, the *correlation coefficient* is a useful numerical summary.

- The sign of the correlation tells us the direction of the association.
- The magnitude of the correlation tells us the *strength* of a linear association. Strong associations have correlations near −1 or +1 and very weak associations near 0.
- Correlation has no units, so shifting or scaling the data, standardizing, or even swapping the variables has no effect on the numerical value.

Once again we've learned that doing Statistics right means we have to *Think* about whether our choice of methods is appropriate.

- The correlation coefficient is appropriate only if the underlying relationship is linear.
- We'll check the Straight Enough Condition by looking at a scatterplot.
- And, as always, we'll watch out for outliers!

Finally, we've learned not to make the mistake of assuming that a high correlation or strong association is evidence of a cause-and-effect relationship. Beware of lurking variables!

A S *Simulation*: **Correlation, Center, and Scale.** If you have any lingering doubts that shifting and rescaling the data won't change the correlation, watch nothing happen right before your eyes!

Terms

Scatterplots	A scatterplot shows the relationship between two quantitative variables measured on the same cases. (p. 151)
Association	■ **Direction:** A positive direction or association means that, in general, as one variable increases, so does the other. When increases in one variable generally correspond to decreases in the other, the association is negative. ■ **Form:** The form we care about most is straight, but you should certainly describe other patterns you see in scatterplots. ■ **Strength:** A scatterplot is said to show a strong association if there is little scatter around the underlying relationship. (p. 151)
Outlier	A point that does not fit the overall pattern seen in the scatterplot. (p. 152)
Response variable, Explanatory variable, *x*-variable, *y*-variable	In a scatterplot, you must choose a role for each variable. Assign to the *y*-axis the response variable that you hope to predict or explain. Assign to the *x*-axis the explanatory or predictor variable that accounts for, explains, predicts, or is otherwise responsible for the *y*-variable. (p. 153)
Correlation coefficient	The correlation coefficient is a numerical measure of the direction and strength of a linear association. (p. 155)

$$r = \frac{\sum z_x z_y}{n - 1}$$

Lurking variable	A variable other than *x* and *y* that simultaneously affects both variables, accounting for the correlation between the two. (p. 160)

On the Computer SCATTERPLOTS AND CORRELATION

Statistics packages generally make it easy to look at a scatterplot to check whether the correlation is appropriate. Some packages make this easier than others.

Many packages allow you to modify or enhance a scatterplot, altering the axis labels, the axis numbering, the plot symbols, or the colors used. Some options, such as color and symbol choice, can be used to display additional information on the scatterplot.

Exercises

1. Association Suppose you were to collect data for each pair of variables. You want to make a scatterplot. Which variable would you use as the explanatory variable and which as the response variable? Why? What would you expect to see in the scatterplot? Discuss the likely direction, form, and strength.

a) Apples: weight in grams, weight in ounces
b) For each week: ice cream cone sales, air-conditioner sales
c) College freshmen: shoe size, grade point average
d) Gasoline: number of miles you drove since filling up, gallons remaining in your tank

2. Association Suppose you were to collect data for each pair of variables. You want to make a scatterplot. Which variable would you use as the explanatory variable and which as the response variable? Why? What would you expect to see in the scatterplot? Discuss the likely direction, form, and strength.

a) Cell phone data plans: file size, cost
b) Lightning strikes: distance from lightning, time delay of the thunder
c) A streetlight: its apparent brightness, your distance from it
d) Cars: weight of car, age of owner

3. Scatterplots Which of the four scatterplots show

a) little or no association?
b) a negative association?
c) a linear association?
d) a moderately strong association?
e) a very strong association?

(1)

(2)

(3)

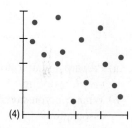
(4)

4. Scatterplots Which of the scatterplots below show

a) little or no association?
b) a negative association?
c) a linear association?
d) a moderately strong association?
e) a very strong association?

(1)

(2)

(3)

(4)

5. Bookstore sales Consider the following data from a small bookstore.

Number of Sales People Working	Sales (in $1000)
2	10
3	11
7	13
9	14
10	18
10	20
12	20
15	22
16	22
20	26
$\bar{x} = 10.4$	$\bar{y} = 17.6$
$SD(x) = 5.64$	$SD(y) = 5.34$

a) Prepare a scatterplot of *Sales* against *Number of sales people* working.
b) What can you say about the direction of the association?
c) What can you say about the form of the relationship?
d) What can you say about the strength of the relationship?
e) Does the scatterplot show any outliers?

6. Disk drives Disk drives have been getting larger. Their capacity is now often given in *terabytes* (TB) where 1 TB = 1000 gigabytes, or about a trillion bytes. A survey of prices for external disk drives found the following data:

Capacity (in TB)	Price (in $)
0.080	29.95
0.120	35.00
0.200	299.00
0.250	49.95
0.320	69.95
1.0	99.00
2.0	205.00
4.0	449.00

a) Prepare a scatterplot of *Price* against *Capacity*.
b) What can you say about the direction of the association?
c) What can you say about the form of the relationship?
d) What can you say about the strength of the relationship?
e) Does the scatterplot show any outliers?

7. Performance IQ scores vs. brain size A study examined brain size (measured as pixels counted in a digitized magnetic resonance image [MRI] of a cross section of the brain) and IQ (4 Performance scales of the Weschler IQ test) for college students. The scatterplot shows the Performance IQ scores vs. the brain size. Comment on the association between brain size and IQ.

8. Kentucky Derby 2011 The fastest horse in Kentucky Derby history was Secretariat in 1973. The scatterplot shows speed (in miles per hour) of the winning horses each year since 1875.

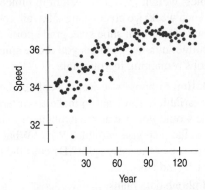

What do you see? In most sporting events, performances have improved and continue to improve, so surely we anticipate a positive direction. But what of the form? Has the performance increased at the same rate throughout the past 130 years?

9. Correlation facts If we assume that the conditions for correlation are met, which of the following are true? If false, explain briefly.
a) A correlation of −0.98 indicates a strong, negative association.
b) Multiplying every value of *x* by 2 will double the correlation.
c) The units of the correlation are the same as the units of *y*.

10. Correlation facts II If we assume that the conditions for correlation are met, which of the following are true? If false, explain briefly.

a) A correlation of 0.02 indicates a strong positive association.

b) Standardizing the variables will make the correlation 0.

c) Adding an outlier can dramatically change the correlation.

11. Bookstore sales again A larger firm is considering acquiring the bookstore of Exercise 5. An analyst for the firm, noting the relationship seen in Exercise 5, suggests that when they acquire the store they should hire more people because that will drive higher sales. Is his conclusion justified? What alternative explanations can you offer? Use appropriate statistics terminology.

12. Blizzards A study finds that during blizzards, online sales are highly associated with the number of snow plows on the road; the more plows, the more online purchases. The director of an association of online merchants suggests that the organization should encourage municipalities to send out more plows whenever it snows because, he says, that will increase business. Comment.

13. Firing pottery A ceramics factory can fire eight large batches of pottery a day. Sometimes a few of the pieces break in the process. In order to understand the problem better, the factory records the number of broken pieces in each batch for 3 days and then creates the scatterplot shown.

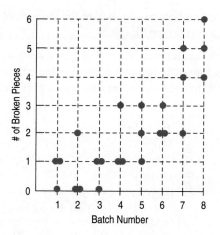

a) Make a histogram showing the distribution of the number of broken pieces in the 24 batches of pottery examined.

b) Describe the distribution as shown in the histogram. What feature of the problem is more apparent in the histogram than in the scatterplot?

c) What aspect of the company's problem is more apparent in the scatterplot?

14. Coffee sales Owners of a new coffee shop tracked sales for the first 20 days and displayed the data in a scatterplot (by day).

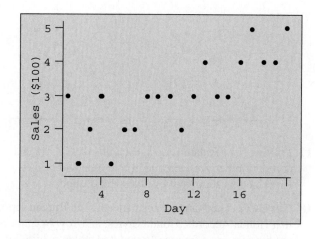

a) Make a histogram of the daily sales since the shop has been in business.

b) State one fact that is obvious from the scatterplot, but not from the histogram.

c) State one fact that is obvious from the histogram, but not from the scatterplot.

15. Matching Here are several scatterplots. The calculated correlations are −0.923, −0.487, 0.006, and 0.777. Which is which?

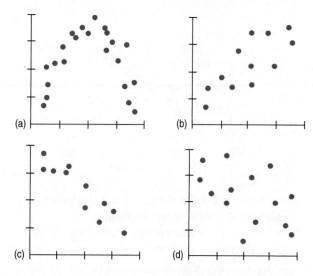

16. Matching Here and on the next page are several scatterplots. The calculated correlations are −0.977, −0.021, 0.736, and 0.951. Which is which?

(c)   (d)

17. Politics A candidate for office claims that "there is a correlation between television watching and crime." Criticize this statement on statistical grounds.

18. Car thefts The National Insurance Crime Bureau reports that Honda Accords, Honda Civics, and Toyota Camrys are the cars most frequently reported stolen, while Ford Tauruses, Pontiac Vibes, and Buick LeSabres are stolen least often. Is it reasonable to say that there's a correlation between the type of car you own and the risk that it will be stolen?

19. Roller coasters Roller coasters get all their speed by dropping down a steep initial incline, so it makes sense that the height of that drop might be related to the speed of the coaster. Here's a scatterplot of top *Speed* and largest *Drop* for 75 roller coasters around the world.

a) Does the scatterplot indicate that it is appropriate to calculate the correlation? Explain.
b) In fact, the correlation of *Speed* and *Drop* is 0.91. Describe the association.

20. Antidepressants A study compared the effectiveness of several antidepressants by examining the experiments in which they had passed the FDA requirements. Each of those experiments compared the active drug with a placebo, an inert pill given to some of the subjects. In each experiment some patients treated with the placebo had improved, a phenomenon called the *placebo effect*. Patients' depression levels were evaluated on the Hamilton Depression Rating Scale, where larger numbers indicate greater improvement. (The Hamilton scale is a widely accepted standard that was used in each of the independently run studies.) The scatterplot at the top of the next column compares mean improvement levels for the antidepressants and placebos for several experiments.

a) Is it appropriate to calculate the correlation? Explain.
b) The correlation is 0.898. Explain what we have learned about the results of these experiments.

21. Hard water In a study of streams in the Adirondack Mountains, the following relationship was found between the water's pH and its hardness (measured in grains):

Is it appropriate to summarize the strength of association with a correlation? Explain.

22. Traffic headaches A study of traffic delays in 68 U.S. cities found the following relationship between total delays (in total hours lost) and mean highway speed:

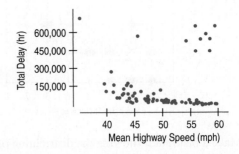

Is it appropriate to summarize the strength of association with a correlation? Explain.

23. Cold nights Is there an association between time of year and the nighttime temperature in North Dakota? A researcher assigned the numbers 1–365 to the days January 1–December 31 and recorded the temperature at 2:00 A.M. for each. What might you expect the correlation between *DayNumber* and *Temperature* to be? Explain.

24. Association A researcher investigating the association between two variables collected some data and was surprised when he calculated the correlation. He had expected to find a fairly strong association, yet the correlation was near 0. Discouraged, he didn't bother making a scatterplot. Explain to him how the scatterplot could still reveal the strong association he anticipated.

25. Prediction units The errors in predicting hurricane tracks (examined in this chapter) were given in nautical miles. An ordinary mile is 0.86898 nautical miles. Most people living on the Gulf Coast of the United States would prefer to know the prediction errors in miles rather than nautical miles. Explain why converting the errors to miles would not change the correlation between *Prediction Error* and *Year*.

26. More predictions Hurricane Katrina's hurricane force winds extended 120 miles from its center. Katrina was a big storm, and that affects how we think about the prediction errors. Suppose we add 120 miles to each error to get an idea of how far from the predicted track we might still find damaging winds. Explain what would happen to the correlation between *Prediction Error* and *Year*, and why.

27. Correlation errors Your Economics instructor assigns your class to investigate factors associated with the gross domestic product (*GDP*) of nations. Each student examines a different factor (such as *Life Expectancy, Literacy Rate*, etc.) for a few countries and reports to the class. Apparently, some of your classmates do not understand Statistics very well because you know several of their conclusions are incorrect. Explain the mistakes in their statements below.

a) "My very low correlation of −0.772 shows that there is almost no association between *GDP* and *Infant Mortality Rate*."

b) "There was a correlation of 0.44 between *GDP* and *Continent*."

28. More correlation errors Students in the Economics class discussed in Exercise 27 also wrote these conclusions. Explain the mistakes they made.

a) "There was a very strong correlation of 1.22 between *Life Expectancy* and *GDP*."

b) "The correlation between *Literacy Rate* and *GDP* was 0.83. This shows that countries wanting to increase their standard of living should invest heavily in education."

29. Height and reading A researcher studies children in elementary school and finds a strong positive linear association between height and reading scores.

a) Does this mean that taller children are generally better readers?

b) What might explain the strong correlation?

30. Cellular telephones and life expectancy A survey of the world's nations in 2010 shows a strong positive correlation between percentage of the country using cell phones and life expectancy in years at birth.

a) Does this mean that cell phones are good for your health?

b) What might explain the strong correlation?

31. Correlation conclusions I The correlation between *Age* and *Income* as measured on 100 people is $r = 0.75$. Explain whether or not each of these possible conclusions is justified:

a) When *Age* increases, *Income* increases as well.

b) The form of the relationship between *Age* and *Income* is straight.

c) There are no outliers in the scatterplot of *Income* vs. *Age*.

d) Whether we measure *Age* in years or months, the correlation will still be 0.75.

32. Correlation conclusions II The correlation between *Fuel Efficiency* (as measured by miles per gallon) and *Price* of 150 cars at a large dealership is $r = -0.34$. Explain whether or not each of these possible conclusions is justified:

a) The more you pay, the lower the fuel efficiency of your car will be.

b) The form of the relationship between *Fuel Efficiency* and *Price* is moderately straight.

c) There are several outliers that explain the low correlation.

d) If we measure *Fuel Efficiency* in kilometers per liter instead of miles per gallon, the correlation will increase.

33. Baldness and heart disease Medical researchers followed 1435 middle-aged men for a period of 5 years, measuring the amount of *Baldness* present (none = 1, little = 2, some = 3, much = 4, *extreme* = 5) and presence of *Heart Disease* (No = 0, Yes = 1). They found a correlation of 0.089 between the two variables. Comment on their conclusion that this shows that baldness is not a possible cause of heart disease.

34. Sample survey A polling organization is checking its database to see if the two data sources it used sampled the same zip codes. The variable *Datasource* = 1 if the data source is MetroMedia, 2 if the data source is DataQwest, and 3 if it's RollingPoll. The organization finds that the correlation between five-digit zip code and *Datasource* is −0.0229. It concludes that the correlation is low enough to state that there is no dependency between *Zip Code* and *Source of Data*. Comment.

T 35. Income and housing The Office of Federal Housing Enterprise Oversight (www.ofheo.gov) collects data on various aspects of housing costs around the United States. Here is a scatterplot of the *Housing Cost Index* versus the *Median Family Income* for each of the 50 states. The correlation is 0.65.

a) Describe the relationship between the *Housing Cost Index* and the *Median Family Income* by state.
b) If we standardized both variables, what would the correlation coefficient between the standardized variables be?
c) If we had measured *Median Family Income* in thousands of dollars instead of dollars, how would the correlation change?
d) Washington, DC, has a Housing Cost Index of 548 and a median income of about $45,000. If we were to include DC in the data set, how would that affect the correlation coefficient?
e) Do these data provide proof that by raising the median income in a state, the Housing Cost Index will rise as a result? Explain.

36. Interest rates and mortgages Since 1985, average mortgage interest rates have fluctuated from a low of under 6% to a high of over 14%. Is there a relationship between the amount of money people borrow and the interest rate that's offered? Here is a scatterplot of *Mortgage Loan Amount* in the United States (in thousands of dollars) versus *Interest Rate* at various times over the past 26 years. The correlation is −0.86.

a) Describe the relationship between *Mortgage Loan Amount* and *Interest Rate*.
b) If we standardized both variables, what would the correlation coefficient between the standardized variables be?
c) If we were to measure *Mortgage Loan Amount* in hundreds of dollars instead of thousands of dollars, how would the correlation coefficient change?

d) Suppose in another year, interest rates were 11% and mortgages totaled $250 thousand. How would including that year with these data affect the correlation coefficient?
e) Do these data provide proof that if mortgage rates are lowered, people will take out larger mortgages? Explain.

37. Fuel economy 2010 Here are advertised horsepower ratings and expected gas mileage for several 2010 vehicles. (www.kbb.com)

Car	hp	mpg
Audi A4	211	30
BMW 3 series	230	28
Buick LaCrosse	182	30
Chevy Cobalt	155	37
Chevy Suburban 1500	320	21
Ford Expedition	310	20
GMC Yukon	320	21
Honda Civic	140	34
Honda Accord	177	31
Hyundai Elantra	138	35
Lexus IS 350	306	25
Lincoln Navigator	310	20
Mazda Tribute	171	28
Toyota Camry	169	33
Volkswagen Beetle	150	28

a) Make a scatterplot for these data.
b) Describe the direction, form, and strength of the plot.
c) Find the correlation between horsepower and miles per gallon.
d) Write a few sentences telling what the plot says about fuel economy.

38. Drug abuse A survey was conducted in the United States and 10 countries of Western Europe to determine the percentage of teenagers who had used marijuana and other drugs. The results are summarized in the table.

Country	Percent Who Have Used	
	Marijuana	Other Drugs
Czech Rep.	22	4
Denmark	17	3
England	40	21
Finland	5	1
Ireland	37	16
Italy	19	8
No. Ireland	23	14
Norway	6	3
Portugal	7	3
Scotland	53	31
USA	34	24

a) Create a scatterplot.

b) What is the correlation between the percent of teens who have used marijuana and the percent who have used other drugs?

c) Write a brief description of the association.

d) Do these results confirm that marijuana is a "gateway drug," that is, that marijuana use leads to the use of other drugs? Explain.

39. Burgers Fast food is often considered unhealthy because much of it is high in both fat and sodium. But are the two related? Here are the fat and sodium contents of several brands of burgers. Analyze the association between fat content and sodium.

Fat (g)	19	31	34	35	39	39	43
Sodium (mg)	920	1500	1310	860	1180	940	1260

40. Burgers II In the previous exercise you analyzed the association between the amounts of fat and sodium in fast food hamburgers. What about fat and calories? Here are data for the same burgers:

Fat (g)	19	31	34	35	39	39	43
Calories	410	580	590	570	640	680	660

41. Attendance 2010 American League baseball games are played under the designated hitter rule, meaning that pitchers, often weak hitters, do not come to bat. Baseball owners believe that the designated hitter rule means more runs scored, which in turn means higher attendance. Is there evidence that more fans attend games if the teams score more runs? Data collected from American League games during the 2010 season indicate a correlation of 0.667 between runs scored and the number of people at the game. (mlb.mlb.com)

a) Does the scatterplot indicate that it's appropriate to calculate a correlation? Explain.

b) Describe the association between attendance and runs scored.

c) Does this association prove that the owners are right that more fans will come to games if the teams score more runs?

42. Second inning 2010 Perhaps fans are just more interested in teams that win. The displays below are based on American League teams for the 2010 season. (espn .go.com) Are the teams that win necessarily those which score the most runs?

	Wins	Runs	Attendance
Wins	1.00		
Runs	0.919	1.00	
Attendance	0.533	0.538	1.00

a) Do winning teams generally enjoy greater attendance at their home games? Describe the association.

b) Is attendance more strongly associated with winning or scoring runs? Explain.

c) How strongly is scoring more runs associated with winning more games?

43. Thrills 2011 Since 1994, the Best Roller Coaster Poll (www.ushsho.com/bestrollercoasterpoll.htm) has been ranking the world's best roller coasters. In 2011, Bizarro earned the top steel coaster rank for the sixth straight year. Here are data on the top 10 steel coasters from this poll:

Rank	Roller Coaster	Park	Location	Initial Drop (ft.)	Duration (sec)	Height (ft.)	Max Speed (mph)	Max Vert Angle (degrees)	Length (ft.)
1	Bizarro	Six Flags New England	MA	221	155	208	77	72	5400
2	Expedition GeForce	Holiday Park	DE	184	75	188	74.6	82	4003
3	Intimidator 305	Kings Dominion	VA	300	98	305	93	85	5100
4	Kawasemi	Tobu Zoo	JP		60	108	54	67.4	2454
5	Nemesis	Alton Towers	UK	104	80	43	50	40	2349
6	Piraten	Djurs Sommerland	DK	100	61	105	56	70	2477
7	Goliath	Walibi World	NL	152	92	155	66.9	70	3984
8	Millennium Force	Cedar Point	OH	300	120	310	93	80	6595
9	Katun	Mirabilandia	IT	148	142	169	65		3937
10	iSpeed	Mirabilandia	IT	181	60	180	74.6	90	3281

What do these data indicate about the *Length* of the track and the *Duration* of the ride you can expect?

44. Thrills II For the roller coaster data in Exercise 43:
a) Examine the relationship between *Initial Drop* and *Speed*.
b) Examine the relationship between *Initial Drop* and *Height*.
c) What conclusions can you safely draw about the initial drop of a roller coaster? Is *Initial Drop* strongly correlated with other variables as well?

45. Thrills III For the roller coaster data in Exercise 43:
a) Explain why in looking for a variable that explains rank, you will be hoping for a negative correlation.
b) Do any of the provided variables provide a strong predictor for roller coaster rank?
c) What other (unaccounted) for variables might help explain the rank?

46. Vehicle weights The Minnesota Department of Transportation hoped that they could measure the weights of big trucks without actually stopping the vehicles by using a newly developed "weight-in-motion" scale. To see if the new device was accurate, they conducted a calibration test. They weighed several stopped trucks (static weight) and assumed that this weight was correct. Then they weighed the trucks again while they were moving to see how well the new scale could estimate the actual weight. Their data are given in the following table.

Weights (1000s of lbs)	
Weight-in-Motion	Static Weight
26.0	27.9
29.9	29.1
39.5	38.0
25.1	27.0
31.6	30.3
36.2	34.5
25.1	27.8
31.0	29.6
35.6	33.1
40.2	35.5

a) Make a scatterplot for these data.
b) Describe the direction, form, and strength of the plot.
c) Write a few sentences telling what the plot says about the data. (*Note:* The sentences should be about weighing trucks, not about scatterplots.)
d) Find the correlation.
e) If the trucks were weighed in kilograms, how would this change the correlation? (1 kilogram = 2.2 pounds)

f) Do any points deviate from the overall pattern? What does the plot say about a possible recalibration of the weight-in-motion scale?

T **47. Planets (more or less)** On August 24, 2006, the International Astronomical Union voted that Pluto is not a planet. Some members of the public have been reluctant to accept that decision. Let's look at some of the data. (We'll see more in the next chapter.) Is there any pattern to the locations of the planets? The table shows the average distance of each of the traditional nine planets from the sun.

Planet	Position Number	Distance from Sun (million miles)
Mercury	1	36
Venus	2	67
Earth	3	93
Mars	4	142
Jupiter	5	484
Saturn	6	887
Uranus	7	1784
Neptune	8	2796
Pluto	9	3666

a) Make a scatterplot and describe the association. (Remember: direction, form, and strength!)
b) Why would you not want to talk about the correlation between a planet's *Position* and *Distance* from the sun?
c) Make a scatterplot showing the logarithm of *Distance* vs. *Position*. What is better about this scatterplot?

T **48. Flights 2010** Here are the number of domestic flights flown in each year from 2000 to 2010 (www.TranStats.bts.gov).

Year	Flights
2000	7,905,617
2001	7,626,312
2002	8,089,140
2003	9,458,818
2004	9,968,047
2005	10,038,373
2006	9,712,750
2007	9,839,578
2008	9,376,251
2009	8,753,295
2010	8,685,184

a) Find the correlation of *Flights* with *Year.*
b) Make a scatterplot and describe the trend.
c) Why is the correlation you found in part a not a suitable summary of the strength of the association?

 Just Checking ANSWERS

1. We know the scores are quantitative. We should check to see if the Straight Enough Condition and the Outlier Condition are satisfied by looking at a scatterplot of the two scores.

2. It won't change.

3. It won't change.

4. They are likely to have done poorly. The positive correlation means that low scores on Exam 1 are associated with low scores on Exam 2 (and similarly for high scores).

5. No. The general association is positive, but individual performances may vary.

7 Linear Regression

Who	Items on the Burger King (BK) menu
What	Protein content and total fat content
Units	Grams of protein and grams of fat
How	Supplied by BK on request or at their website

Figure 7.1

Total *Fat* versus *Protein* for 122 items on the BK menu. The Triple Whopper is in the upper right corner. It's extreme, but is it out of line?

A S *Video*: **Manatees and Motorboats.** Are motorboats killing more manatees in Florida? Here's the story on video.

he Whopper™ has been Burger King's signature sandwich since 1957. One Triple Whopper with cheese provides 71 grams of protein—all the protein you need in a day. It also supplies 1230 calories and 82 grams of fat. The Daily Value (based on a 2000-calorie diet) for fat is 65 grams. So after a Triple Whopper you'll want the rest of your calories that day to be fat-free.[1]

Of course, the Whopper™ isn't the only item Burger King (BK) sells. How are fat and protein related for the entire BK menu? The scatterplot of the *Fat* (in grams) versus the *Protein* (in grams) for foods sold at BK shows a positive, moderately strong, linear relationship.

---

[1]Sorry about the fries.

"Statisticians, like artists, have the bad habit of falling in love with their models."

—*George Box, famous statistician*

If you want 25 grams of protein in your lunch, how much fat should you expect to have to consume at Burger King? The correlation between *Fat* and *Protein* is 0.76, a sign that the linear association seen in the scatterplot is fairly strong. But *strength* of the relationship is only part of the picture. The correlation says, "The linear association between these two variables is fairly strong," but it doesn't tell us *what the line is.*

Obviously, there's no straight line that will pass through all of those points. But can't you imagine a line that would **model** the essence of the relationship, running across the scatter of points from the lower left to the upper right? We could use the equation of such a line to predict the fat content of any BK food from its protein. So that's our goal: a **linear model** is the equation of a straight line through the data.

Of course, no line can go through all the points, but a linear model can summarize the general pattern with only a couple of parameters. Like all models of the real world, the line will be wrong—wrong in the sense that it can't match reality *exactly.* But it can help us understand how the variables are associated.

Residuals

Not only can't we draw a line through all the points, the best line might not even hit *any* of the points. Then how can it be the "best" line? We want to find the line that somehow comes *closer* to all the points than any other line. Some of the points will be above the line and some below. For example, the line might suggest that a BK Tendercrisp chicken sand-wich without mayonnaise with 31 grams of protein[2] should have 36.6 grams of fat when, in fact, it actually has only 22. We call the estimate made from a model the **predicted value,** and write it as $\hat{y}$ (called *y-hat*) to distinguish it from the true value y (called, uh, y). The difference between the observed value and its associated predicted value is called the **residual.** The residual value tells us how far off the model's prediction is at that point. The BK Tendercrisp chicken residual would be $y - \hat{y} = 22 - 36.6 = -14.6$ g of fat.

Residual = Observed Value − Predicted Value

A *negative* residual means the predicted value is too big—an overestimate. And a *positive* residual shows that the model makes an underestimate. These may seem backwards until you think about them.

To find the residuals, we always subtract the predicted value from the observed one. The negative residual tells us that the actual fat content of the BK Tendercrisp is about 14.6 grams *less* than the model predicts for a typical BK menu item with 31 grams of protein. Our challenge now is how to find the right line.

"Best Fit" Means Least Squares

The size of the residuals tells us how well the line fits over data; a line that fits well will have very small residuals. But we can't assess how well the line fits by adding up the residuals—the positive and negative ones would just cancel each other out. We faced the same issue when we calculated a standard deviation to measure spread. And we deal with

---

[2]The sandwich comes with mayo unless you ask for it without. That adds an extra 24 grams of fat, which is more than the original sandwich contained.

it the same way here: by squaring the residuals. Squaring makes them all positive (or 0). Now we can add them up. Squaring also emphasizes the large residuals. After all, points near the line are consistent with the model, but we're more concerned about points far from the line. When we add all the squared residuals together, that sum indicates how well the line we drew fits the data—the smaller the sum, the better the fit. A different line will produce a different sum, maybe bigger, maybe smaller. The **line of best fit** is the line for which the sum of the squared residuals is smallest, the **least squares** line.

> **Who's on First**   In 1805, Legendre was the first to publish the "least squares" solution to the problem of fitting a line to data when the points don't all fall exactly on the line. The main challenge was how to distribute the errors "fairly." After considerable thought, he decided to minimize the sum of the squares of what we now call the residuals. When Legendre published his paper, though, Gauss claimed he had been using the method since 1795. Gauss later referred to the "least squares" solution as "*our* method" (*principium nostrum*), which certainly didn't help his relationship with Legendre.

You might think that finding this line would be pretty hard. Surprisingly, it's not, although it was an exciting mathematical discovery when Legendre published it in 1805.

TI-nspire

Least squares. Try to minimize the sum of areas of residual squares as you drag a line across a scatterplot.

Correlation and the Line

If you suspect that what we know about correlation can lead us to the equation of the linear model, you're headed in the right direction. It turns out that it's not a very big step. In Chapter 6 we learned a lot about how correlation worked by looking at a scatterplot of the standardized variables. Here's a scatterplot of z_y (standardized *Fat*) vs. z_x (standardized *Protein*).

Figure 7.2
The BK scatterplot in z-scores.

What line would you choose to model the relationship of the standardized values? Let's start by thinking about how much protein and fat a *typical* BK food item provides. If it has average protein content, $\bar{x}$, what about its fat content? If you guessed that its fat content should be about average, $\bar{y}$, as well, then you've discovered the first property of the line we're looking for. The line must go through the point $(\bar{x}, \bar{y})$. In the plot of z-scores, then, the line passes through the origin $(0, 0)$.

The equation for a line that passes through the origin can be written with just a slope and no intercept:

$$y = mx.$$

The coordinates of our standardized points aren't written (x, y); their coordinates are z-scores: (z_x, z_y). We'll need to change our equation to show that. And we'll need to indicate that the point on the line corresponding to a particular z_x is $\hat{z}_y$, the model's estimate of the actual value of z_y. So our equation becomes

$$\hat{z}_y = mz_x.$$

■ NOTATION ALERT

"Putting a hat on it" is standard Statistics notation to indicate that something has been predicted by a model. Whenever you see a hat over a variable name or symbol, you can assume it is the predicted version of that variable or symbol (and look around for the model).

Figure 7.3

Standardized fat vs. standardized protein with the regression line. Each one standard deviation change in *protein* results in a predicted change of r standard deviations in *fat*.

Many lines with different slopes pass through the origin. Which one fits our data the best? That is, which slope determines the line that minimizes the sum of the squared residuals? It turns out that the best choice for *m* is the correlation coefficient itself, *r*! (You must really wonder where that stunning assertion comes from. Check the Math Box.)

Wow! This line has an equation that's about as simple as we could possibly hope for:

$$\hat{z}_y = rz_x.$$

Great. It's simple, but what does it tell us? It says that in moving one standard deviation from the mean in *x*, we can expect to move about *r* standard deviations away from the mean in *y*. Let's be more specific. For the sandwiches, the correlation is 0.76. If we standardize both protein and fat, we can write

$$\hat{z}_{Fat} = 0.76z_{Protein}.$$

This model tells us that for every standard deviation above (or below) the mean a sandwich is in protein, we'll predict that its fat content is 0.76 standard deviations above (or below) the mean fat content. A double hamburger has 31 grams of protein, about 1 SD above the mean. Based on the model, you'd expect the fat content to be about 0.76 fat SDs above the mean fat level. Moving one standard deviation away from the mean in *x* moves our estimate *r* standard deviations away from the mean in *y*.

If *r* = 0, there's no linear relationship. The line is horizontal, and no matter how many standard deviations you move in *x*, the predicted value for *y* doesn't change. On the other hand, if *r* = 1.0 or −1.0, there's a perfect linear association. In that case, moving any number of standard deviations in *x* moves exactly the same number of standard deviations in *y*. In general, moving any number of standard deviations in *x* moves *r* times that number of standard deviations in *y*.

How Big Can Predicted Values Get?

Sir Francis Galton was the first to speak of "regression," although others had fit lines to data by the same method.

Suppose you were told that a new male student was about to join the class, and you were asked to guess his height in inches. What would be your guess? A safe guess would be the mean height of male students. Now suppose you are also told that this student has a grade point average (*GPA*) of 3.9—about 2 SDs above the mean *GPA*. Would that change your guess? Probably not. The correlation between *GPA* and *height* is near 0, so knowing the *GPA* value doesn't tell you anything and doesn't move your guess. (And the equation tells us that as well, since it says that we should move 0 × 2 SDs from the mean.)

On the other hand, suppose you were told that, measured in centimeters, the student's height was 2 SDs above the mean. There's a perfect correlation between *height in inches* and *height in centimeters,* so you'd know he's 2 SDs above mean height in inches as well. (The equation would tell us to move 1.0 × 2 SDs from the mean.)

What if you're told that the student is 2 SDs above the mean in *shoe size*? Would you still guess that he's of average *height*? You might guess that he's taller than average, since there's a positive correlation between *height* and *shoe size*. But would you guess that he's 2 SDs above the mean? When there was no correlation, we didn't move away from the mean at all. With a perfect correlation, we moved our guess the full 2 SDs. Any correlation between these extremes should lead us to move somewhere between 0 and 2 SDs above the mean. (To be exact, the equation tells us to move *r* × 2 standard deviations away from the mean.)

Notice that if *x* is 2 SDs above its mean, we won't ever guess more than 2 SDs away for *y*, since *r* can't be bigger than 1.0.[3] So, each predicted *y* tends to be closer to its mean (in standard deviations) than its corresponding *x* was. This property of the linear model is called **regression to the mean,** and the line is called the **regression line.**

---

[3]In the last chapter we asserted that correlations max out at 1, but we never actually *proved* that. Here's yet another reason to check out the Math Box on the next page.

The First Regression   Sir Francis Galton related the heights of sons to the heights of their fathers with a regression line. The slope of his line was less than 1. That is, sons of tall fathers were tall, but not as much above the average height as their fathers had been above their mean. Sons of short fathers were short, but generally not as far from their mean as their fathers. Galton interpreted the slope correctly as indicating a "regression" toward the mean height— and "regression" stuck as a description of the method he had used to find the line.

Just Checking

A scatterplot of house *Price* (in thousands of dollars) vs. house *Size* (in thousands of square feet) for houses sold recently in Saratoga, NY shows a relationship that is straight, with only moderate scatter and no outliers. The correlation between house *Price* and house *Size* is 0.77.

1. You go to an open house and find that the house is 1 standard deviation above the mean in size. What would you guess about its price?

2. You read an ad for a house priced 2 standard deviations below the mean. What would you guess about its size?

3. A friend tells you about a house whose size in square meters (he's European) is 1.5 standard deviations above the mean. What would you guess about its size in square feet?

MATH BOX

Where does the equation of the line of best fit come from? To write the equation of any line, we need to know a point on the line and the slope. The point is easy. Consider the BK menu example. Since it is logical to predict that a sandwich with average protein will contain average fat, the line passes through the point $(\bar{x}, \bar{y})$.[4]

To think about the slope, we look once again at the z-scores. We need to remember a few things:

1. The mean of any set of z-scores is 0. This tells us that the line that best fits the z-scores passes through the origin (0,0).

2. The standard deviation of a set of z-scores is 1, so the variance is also 1. This means that
$$\frac{\sum(z_y - \bar{z}_y)^2}{n-1} = \frac{\sum(z_y - 0)^2}{n-1} = \frac{\sum z_y^2}{n-1} = 1,$$ a fact that will be important soon.

3. The correlation is $r = \dfrac{\sum z_x z_y}{n-1}$, also important soon.

Ready? Remember that our objective is to find the slope of the best fit line. Because it passes through the origin, its equation will be of the form $\hat{z}_y = mz_x$. We want to find the value for m that will minimize the sum of the squared residuals. Actually we'll divide that sum by $n-1$ and minimize this "mean squared residual," or *MSR*. Here goes:

Minimize:
$$MSR = \frac{\sum(z_y - \hat{z}_y)^2}{n-1}$$

Since $\hat{z}_y = mz_x$:
$$MSR = \frac{\sum(z_y - mz_x)^2}{n-1}$$

Square the binomial:
$$= \frac{\sum(z_y^2 - 2mz_x z_y + m^2 z_x^2)}{n-1}$$

Rewrite the summation:
$$= \frac{\sum z_y^2}{n-1} - 2m\frac{\sum z_x z_y}{n-1} + m^2\frac{\sum z_x^2}{n-1}$$

4. Substitute from (2) and (3):
$$= 1 - 2mr + m^2$$

Wow! That simplified nicely! And as a bonus, the last expression is quadratic. Remember parabolas from algebra class? A parabola in the form $y = ax^2 + bx + c$ reaches its minimum at its turning point, which occurs when $x = \dfrac{-b}{2a}$. We can minimize the mean of squared residuals by choosing $m = \dfrac{-(-2r)}{2(1)} = r$.

(continued)

---

[4]It's actually not hard to prove this too.

Wow, again! The slope of the best fit line for z-scores is the correlation, r. This stunning fact immediately leads us to two important additional results, listed below. As you read on in the text, we explain them in the context of our continuing discussion of BK foods.

- A slope of r for z-scores means that for every increase of 1 standard deviation in z_x there is an increase of r standard deviations in $\hat{z}_y$. "Over one, up r," as you probably said in algebra class. Translate that back to the original x and y values: "Over one standard deviation in x, up r standard deviations in $\hat{y}$."

  That's it! In x- and y-values, the slope of the regression line is $b = \dfrac{rs_y}{s_x}$.

- We know choosing $m = r$ minimizes the sum of the squared residuals, but how small does that sum get? Equation (4) told us that the mean of the squared residuals is $1 - 2mr + m^2$. When $m = r$, $1 - 2mr + m^2 = 1 - 2r^2 + r^2 = 1 - r^2$. Think carefully about this. The variance in z_y was initially 1 (Equation 2). The line accounts for some of the variability in y, but $1 - r^2$ is left over, still unexplained. Therefore, the percentage of variability in y that is explained by x is r^2. This important fact will help us assess the strength of our models.

And there's still another bonus. Because r^2 is the percent of variability explained by our model, r^2 is at most 100%. If $r^2 \leq 1$, then $-1 \leq r \leq 1$, proving that correlations are always between -1 and $+1$. (Told you so!)

The Regression Line in Real Units

Why Is Correlation "r"? In his original paper on correlation, Galton used r for the "index of correlation" that we now call the correlation coefficient. He calculated it from the regression of y on x or of x on y after standardizing the variables, just as we have done. It's fairly clear from the text that he used r to stand for (standardized) regression.

Protein	Fat
$\bar{x} = 18.0\ g$	$\bar{y} = 24.8\ g$
$s_x = 13.5\ g$	$s_y = 16.2\ g$
$r = 0.76$	

Slope

$$b_1 = \frac{rs_y}{s_x}$$

Intercept

$$b_0 = \bar{y} - b_1\bar{x}$$

A S *Simulation*: Interpreting Equations. This demonstrates how to use and interpret linear equations.

When you read the BK menu, you probably don't think in z-scores. If you want to know the fat content in grams for a specific amount of protein in grams, you'd rather write the equation of the line for protein and fat—that is, the actual x and y values rather than their z-scores. In Algebra class you may have once seen lines written in the form $y = mx + b$. Statisticians do exactly the same thing, but with different notation:

$$\hat{y} = b_0 + b_1x.$$

In this equation, b_0 is the **y-intercept,** the value of y where the line crosses the y-axis, and b_1 is the **slope.**[5]

First we find the slope, using the formula we developed in the Math Box.[6] Remember? We know that our model predicts that for each increase of one standard deviation in protein we'll see an increase of about 0.76 standard deviations in fat.

In other words, the slope of the line in original units is

$$b_1 = \frac{rs_y}{s_x} = \frac{0.76 \times 16.2\ \text{g fat}}{13.5\ \text{g protein}} = 0.91 \text{ grams of fat per gram of protein.}$$

Next, how do we find the y-intercept, b_0? Remember that the line has to go through the mean-mean point $(\bar{x}, \bar{y})$. In other words, the model predicts $\bar{y}$ to be the value that corresponds to $\bar{x}$. We can put the means into the equation and write $\bar{y} = b_0 + b_1\bar{x}$. Solving for b_0, we see that the intercept is just

$$b_0 = \bar{y} - b_1\bar{x}.$$

For the BK foods, that comes out to

$$b_0 = 24.8\ \text{g fat} - 0.91\ \frac{\text{g fat}}{\text{g protein}} \times 18.0\ \text{g protein} = 8.4\ \text{g fat.}$$

Putting this back into the regression equation gives

$$\widehat{fat} = 8.4 + 0.91\ protein.$$

---

[5]We changed from $mx + b$ to $b_0 + b_1x$ for a reason—not just to be difficult. Eventually we'll want to add more x's to the model to make it more realistic and we don't want to use up the entire alphabet. What would we use after m? The next letter is n, and that one's already taken. o? See our point? Sometimes subscripts are the best approach.

[6]Several important results popped up in that Math Box. Check it out!

Units of *y* per unit of *x*

Get into the habit of identifying the units by writing down "*y*-units per *x*-unit," with the unit names put in place. You'll find it'll really help you to tell about the line in context.

Figure 7.4

Burger King menu items in their natural units with the regression line.

What does this mean? The slope, 0.91, says that an additional gram of protein is associated with an additional 0.91 grams of fat, on average. Less formally, we might say that the model predicts BK sandwiches pack about 0.91 additional grams of fat per additional gram of protein. Slopes are always expressed in *y*-units per *x*-unit. They tell how the *y*-variable changes (in its units) for a one-unit change in the *x*-variable. When you see a phrase like "students per teacher" or "kilobytes per second" think slope.

Changing the units of the variables doesn't change the *correlation*, but for the *slope*, units do matter. If children grow an average of 3 inches per year, that's the same as 0.21 millimeters per day. For the slope, it matters whether you express age in days or years and whether you measure height in inches or millimeters. How you choose to express *x* and *y*—what units you use—affects the slope directly. That's because changing units of *x* and *y* changes the standard deviations. The slope gets its units from the ratio of s_y to s_x. The units of the slope are always a ratio: the units of *y* per unit of *x*.

How about the **intercept** of the BK regression line, 8.4? Algebraically, that's the value the line takes when *x* is zero. Here, our model predicts that even a BK item with no protein would have, on average, about 8.4 grams of fat. But often 0 is not a plausible value for *x* (the year 0, a baby born weighing 0 grams, . . .). Then the intercept serves only as a starting value for our predictions and we don't interpret it as a meaningful predicted value.

For Example   A REGRESSION MODEL FOR HURRICANES

The barometric pressure at the center of a hurricane is often used to measure the strength of the hurricane because it can predict the maximum wind speed of the storm. A scatterplot shows that the relationship is straight, strong, and negative. It has a correlation of −0.879.

Using technology to fit the straight line, we find

$$\widehat{MaxWindSpeed} = 955.27 - 0.897 CentralPressure$$

QUESTION: Interpret this model. What does the slope mean in this context? Does the intercept have a meaningful interpretation?

ANSWER: *The negative slope says that as CentralPressure falls, MaxWindSpeed increases. That makes sense from our general understanding of how hurricanes work: Low central pressure pulls in moist air, driving the rotation and the resulting destructive winds. The slope's value says that, on average, the maximum wind speed increases by about 0.897 knots for every 1-millibar drop in central pressure.*

It's not meaningful, however, to interpret the intercept as the wind speed predicted for a central pressure of 0—that would be a vacuum. Instead, it is merely a starting value for the model.

With the estimated linear model, $\widehat{fat} = 8.4 + 0.91\ protein$, it's easy to predict fat content for any menu item we want. For example, for the BK Tendercrisp chicken sandwich with 31 grams of *protein,* we can plug in 31 grams for the amount of *protein* and see that the *predicted fat* content is $8.4 + 0.91(31) = 36.6$ grams of fat.[7] Because the BK Tendercrisp chicken sandwich actually has 22 grams of fat, its residual is

$$fat - \widehat{fat} = 22 - 36.6 = -14.6\ \text{g}.$$

---

[7]Actually, round off errors have caught up with us. Using full precision not shown here makes the predicted fat content 36.7 grams. *You* should always use full precision, and round off only at the final answer.

Just Checking

Let's look again at the relationship between house *Price* (in thousands of dollars) and house *Size* (in thousands of square feet) in Saratoga. The regression model is

$$\widehat{Price} = -3.117 + 94.454\ Size.$$

4. What does the slope of 94.454 mean?

5. What are the units of the slope?

6. Your house is 2000 sq ft bigger than your neighbor's house. How much more do you expect it to be worth?

7. Is the *y*-intercept of −3.117 meaningful? Explain.

To use a regression model, we should check the same conditions for regressions as we did for correlation: the **Quantitative Variables Condition,** the **Straight Enough Condition,** and the **Outlier Condition.**

Step-by-Step Example CALCULATING A REGRESSION EQUATION

During the evening rush hour of August 1, 2007, an eight-lane steel truss bridge spanning the Mississippi River in Minneapolis, Minnesota, collapsed without warning, sending cars plummeting into the river, killing 13 and injuring 145. Although similar events had brought attention to our aging infrastructure, this disaster put the spotlight on the problem and raised the awareness of the general public.

How can we tell which bridges are safe?

Most states conduct regular safety checks, giving a bridge a structural deficiency score on various scales. The New York State Department of Transportation uses a scale that runs from 1 to 7, with a score of 5 or less indicating "deficient." Many factors contribute to the deterioration of a bridge, including amount of traffic, material used in the bridge, weather, and bridge design.

New York has more than 17,000 bridges. We look at data on the 194 bridges of Tompkins County.[8]

One natural concern is the age of a bridge. A model that relates age to safety score might help the DOT to focus inspectors' efforts where they are most needed.

Question: Is there a relationship between the age of a bridge and its safety rating?

THINK ➡ Plan State the problem.	I want to know whether there is a relationship between the age of a bridge in Tompkins County, New York, and its safety rating.
Variables Identify the variables and report the W's.	I have data giving the Safety Score and Age for 194 bridges constructed since 1865.

(continued)

---

[8]Coincidentally, the home county of two of the authors.

Just as we did for correlation, check the conditions for a regression by making a picture. Never fit a regression without looking at the scatterplot first.

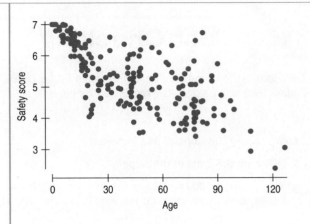

Conditions:

✓ **Quantitative Variables:** Yes, although Condition rating has no units. Age is in years.

✓ **Straight Enough:** Scatterplot looks straight.

✓ **No Outliers:** None are evident.

It is OK to use a linear regression to model this relationship.

SHOW ➡ **Mechanics** Find the equation of the regression line. Summary statistics give the building blocks of the calculation.

(We generally report summary statistics to one more digit of accuracy than the data. We do the same for intercept and predicted values, but for slopes we usually report an additional digit. Remember, though, not to round until you finish computing an answer.)[9]

Find the slope, b_1.

Find the intercept, b_0.

Write the equation of the model, using meaningful variable names.

Age

$\bar{x} = 44.9$

$s_x = 30.75$

Safety Score (points from 1 to 7)

$\bar{y} = 5.2779$

$s_y = 1.0297$

Correlation

$r = -0.681$

$b_1 = \dfrac{rs_y}{s_x}$

$\quad = \dfrac{(-0.681)(1.0297)}{30.75}$

$\quad = -0.0228$ points per year

$b_0 = \bar{y} - b_1\bar{x} = 5.2779 - (-0.0228)44.9$

$\quad = 6.30$

The least squares line is

$\hat{y} = 6.30 - 0.0228x$

or

$\widehat{SafetyScore} = 6.30 - 0.0228\ Age$

(continued)

---

[9]We warned you that we'll round in the intermediate steps of a calculation to show the steps more clearly, and we've done that here. If you repeat these calculations yourself on a calculator or statistics program, you may get somewhat different results. When calculated with more precision, the intercept is 6.40495 and the slope is -0.02565.

TELL ➡ **Conclusion**   Interpret what you have found in the context of the question. Discuss in terms of the variables and their units.

A/S   *Activity*: **Find a Regression Equation.** Now that we've done it by hand, try it with technology using the statistics package paired with your version of ActivStats.

The condition of the bridges in Tompkins County, New York, decreases with the age of the bridges at the time of inspection. Bridge condition declines by about 0.023 points on the scale from 1 to 7 for each year of age. The model uses a base of 6.3, which is quite reasonable because a new bridge (0 years of age) should have a safety score near the maximum of 7.

Because I have only data from one county, I can't tell from these data whether this model would apply to bridges in other counties of New York or in other locations.

Residuals Revisited

Why *e* for "Residual"?

The flip answer is that *r* is already taken, but the truth is that *e* stands for "error." No, that doesn't mean it's a mistake. Statisticians often refer to variability not explained by a model as error.

The residuals are the part of the data that *hasn't* been modeled. We see that in the definition:

$$Residual = Observed\ values - predicted\ value$$

Or, written in symbols,

$$e = y - \hat{y}.$$

When we want to know how well the model fits, we can ask instead what the model missed. No model is perfect, so it's important to know how and where it fails. To see that, we look at the residuals.

For Example KATRINA'S RESIDUAL

RECAP: The linear model relating hurricanes' wind speeds to their central pressures was

$$\widehat{MaxWindSpeed} = 955.27 - 0.897\,CentralPressure$$

Let's use this model to make predictions and see how those predictions do.

QUESTION: Hurricane Katrina had a central pressure measured at 920 millibars. What does our regression model predict for her maximum wind speed? How good is that prediction, given that Katrina's actual wind speed was measured at 110 knots?

ANSWER: Substituting 920 for the central pressure in the regression model equation gives

$$\widehat{MaxWindSpeed} = 955.27 - 0.897(920) = 130.03$$

(continued)

The regression model predicts a maximum wind speed of 130 knots for Hurricane Katrina.

The residual for this prediction is the observed value minus the predicted value:

$$110 - 130 = -20 \text{ kts.}$$

In the case of Hurricane Katrina, the model predicts a wind speed 20 knots higher than was actually observed.

Residuals help us to see whether the model makes sense. When a regression model is appropriate, it should capture the underlying relationship. Nothing interesting should be left behind. So after we fit a regression model, we usually plot the residuals in the hope of finding . . . nothing.

Figure 7.5

The residuals for the BK menu regression look appropriately boring. There are no obvious patterns, although there are a few points with large residuals. The negative ones turn out to be grilled chicken items, which are among the lowest fat (per protein) items on the menu. The two high outliers contain hash browns, which are high fat per protein items.

A scatterplot of the residuals should be the most boring scatterplot you've ever seen. It shouldn't have any interesting features, like a direction or shape. It should show about the same amount of scatter throughout with no bends or outliers. If you see any of these features, find out what the regression model missed.

Most computer statistics packages plot the residuals against the predicted values $\hat{y}$, but calculators plot them against x. When the slope is negative, the two versions are mirror images. When the slope is positive, they're identical except for the axis labels. Since all we care about is the patterns (or, better, lack of patterns) in the residuals, it really doesn't matter which way they are plotted.

 Just Checking

Our linear model for Saratoga homes uses the *Size* (in thousands of square feet) to estimate the *Price* (in thousands of dollars):

$$\widehat{Price} = -3.117 + 94.454 \; Size.$$

Suppose you're thinking of buying a home there.

8. Would you prefer to find a home with a negative or a positive residual? Explain.

9. You plan to look for a home of about 3000 square feet. How much should you expect to have to pay?

10. You find a nice home that size selling for $300,000. What's the residual?

The Residual Standard Deviation

Why n − 2?

Why $n - 2$ rather than $n - 1$? We used $n - 1$ for s when we estimated the mean. Now we're estimating both a slope and an intercept. Looks like a pattern—and it is. We subtract one more for each parameter we estimate.

If the residuals show no interesting pattern, we can look at how big they are. After all, we're trying to make them as small as possible. Since their mean is always zero, though, it's only sensible to look at how much they vary. The standard deviation of the residuals, s_e, gives us a measure of how for the points spread around the regression line. Of course, for this summary to make sense, the residuals should all share the same underlying spread, so we check to make sure that the residual plot has about the same amount of scatter throughout.

We estimate the standard deviation of the residuals in almost the way you'd expect:

$$s_e = \sqrt{\frac{\Sigma e^2}{n - 2}}$$

We don't need to subtract the mean because the mean of the residuals $\bar{e} = 0$.

For the Burger King foods, the standard deviation of the residuals is 10.6 grams of fat. That looks about right in the scatterplot of residuals. The residual for the BK Tendercrisp chicken was −14.7 grams, just under 1.5 standard deviations.

It's a good idea to make a histogram of the residuals. If we see a unimodal, symmetric histogram, then we can apply the 68–95–99.7 Rule to see how well the regression model describes the data. In particular, we know that 95% of the residuals should be within $2s_e$ of the mean (0). The BK residuals look like this:

Figure 7.6

The histogram of residuals is symmetric and unimodal, centered at 0, with a standard deviation of 10.6. Only a few values lie outside of 2 standard deviations. The low ones are the chicken items mentioned before. The high ones contain hash browns.

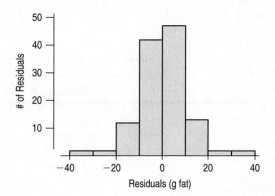

Sure enough, most are within 2(10.6), or 21.2, grams of fat from 0.

R^2—The Variation Accounted For

The variation in the residuals is the key to assessing how well the model fits. We compare the variation of the response variable with the variation of the residuals. The total *Fat* has a standard deviation of 16.2 grams. The standard deviation of the residuals is 10.6 grams. If the correlation were 1.0 and the model predicted the *Fat* values perfectly, the residuals would all be zero and have no variation. We couldn't possibly do any better than that.

On the other hand, if the correlation were zero, the model would simply predict 24.8 grams of *Fat* (the mean) for all menu items. The residuals from that prediction would just be the observed *Fat* values minus their mean. These residuals would have the same variability as the original data because, as we know, just subtracting the mean doesn't change the spread.

How well does the BK regression model do? Look at the boxplots. The variation in the residuals is smaller than in the data, but certainly bigger than zero. That's nice to know, but how much of the variation is still left in the residuals? If you had to put a number between 0% and 100% on the fraction of the variation left in the residuals, what would you say?

Figure 7.7

Compare the variability of total *Fat* with the *Residuals* from the regression. The means have been subtracted to make it easier to compare spreads. The variation left in the residuals is unaccounted for by the model, but it's less than the variation in the original data.

> **TI-*nspire***
>
> **Understanding R²**. Watch the unexplained variability decrease as you drag points closer to the regression line.

As we showed in the Math Box,[10] the squared correlation, r^2, gives the fraction of the data's variation accounted for by the model, and $1 - r^2$ is the fraction of the original variation left in the residuals. For the BK model, $r^2 = 0.76^2 = 0.58$, and $1 - r^2$ is 0.42, so 42% of the variability in total *Fat* has been left in the residuals. How close was that to your guess?

All regression analyses include this statistic, although by tradition, it is written with a capital letter, R^2, and pronounced "R-squared."

Because R^2 is a fraction of a whole, it is often given as a percentage.[11] For the BK data, R^2 is 58%. When interpreting a regression model, you need to *Tell* what R^2 means. According to our linear model, 58% of the variability in the fat content of BK sandwiches is accounted for by variation in the protein content.

> **Twice as Strong?**
>
> Is a correlation of 0.80 twice as strong as a correlation of 0.40? Not if you think in terms of R^2. A correlation of 0.80 means an R^2 of $0.80^2 = 64\%$. A correlation of 0.40 means an R^2 of $0.40^2 = 16\%$—only a quarter as much of the variability accounted for. A correlation of 0.80 gives an R^2 *four* times as strong as a correlation of 0.40 and accounts for four times as much of the variability.

> **How Can We See That R^2 is Really the Fraction of Variance Accounted for by the Model?**
>
> It's a simple calculation. The variance of the fat content of the BK foods is $16.2^2 = 262.44$. If we treat the residuals as data, the variance of the residuals is 112.36.[12] As a fraction, that's 112.36/262.44 or 42%. That's the fraction of the variance that is not accounted for by the model. The fraction that is accounted for is $100\% - 42\% = 58\%$, just the value we got for R^2.

For Example INTERPRETING R^2

RECAP: Our regression model that predicts maximum wind speed in hurricanes based on the storm's central pressure has $R^2 = 77.3\%$.

QUESTION: What does that say about our regression model?

ANSWER: An R^2 of 77.3% indicates that 77.3% of the variation in maximum wind speed can be accounted for by the hurricane's central pressure. Other factors, such as temperature and whether the storm is over water or land, may explain some of the remaining variation.

---

[10]Have you looked yet? Please do.

[11]By contrast, we usually give correlation coefficients as decimal values between -1.0 and 1.0.

[12]This isn't quite the same as squaring the s_e that we discussed on the previous page, but it's very close. We'll deal with the distinction in Chapter 26.

Just Checking

Back to our regression of house *Price* (in thousands of $) on house *Size* (in thousands of square feet). The R^2 value is reported as 59.5%, and the standard deviation of the residuals is 53.79.

11. What does the R^2 value mean about the relationship of *Price* and *Size*?

12. Is the correlation of *Price* and *Size* positive or negative? How do you know?

13. If we measure house *Size* in square meters instead, would R^2 change? Would the slope of the line change? Explain.

14. You find that your house in Saratoga is worth $50,000 more than the regression model predicts. Should you be very surprised (as well as pleased)?

How Big Should R^2 Be?

Some Extreme Tales

One major company developed a method to differentiate between proteins. To do so, they had to distinguish between regressions with R^2 of 99.99% and 99.98%. For this application, 99.98% was not high enough.

The president of a financial services company reports that although his regressions give R^2 below 2%, they are highly successful because those used by his competition are even lower.

R^2 is always between 0% and 100%. But what's a "good" R^2 value? The answer depends on the kind of data you are analyzing and on what you want to do with it. Just as with correlation, there is no value for R^2 that automatically determines that the regression is "good." As we've seen, an R^2 of 100% is a perfect fit, with no scatter around the line. The s_e would be zero. All of the variance is accounted for by the model and none is left in the residuals at all. This sounds great, but it's too good to be true for real data.[13] Data from scientific experiments often have R^2 in the 80% to 90% range and even higher. Data from observational studies and surveys, though, often show relatively weak associations because it's so difficult to measure responses reliably. An R^2 of 50% to 30% or even lower might be taken as evidence of a useful regression. R^2 is the first part of a regression that many people look at because, along with the scatterplot, it tells whether the regression model is even worth thinking about. The standard deviation of the residuals can give us more information about the usefulness of the regression by telling us how much scatter there is around the line.

More About Regression Assumptions and Conditions

Linear regression models may be the most widely used models in all of Statistics. They have everything we could want in a model: two easily estimated parameters, a meaningful measure of how well the model fits the data, and the ability to predict new values. They even provide a self-check in plots of the residuals to help us avoid silly mistakes.

Like all models, though, linear models are only appropriate if some assumptions are true. We can't confirm assumptions, but we often can check related conditions.

First, be sure that both variables are quantitative. It makes no sense to perform a regression on categorical variables. After all, what could the slope possibly mean? Always check the **Quantitative Variables Condition**.

Because a linear model only makes sense if the underlying relationship is linear you must consider whether a **Linearity Assumption** is justified. To see, check the associated **Straight Enough Condition**. Just look at the scatterplot of *y* vs. *x*. You don't need a *perfectly* straight plot, but it must be straight enough for the linear model to make sense. If you try to model a curved relationship with a straight line, you'll usually get just what you deserve. If the scatterplot is not straight enough, stop here. You can't use a linear model for *any* two variables, even if they are related.

---

[13]If you see an R^2 of 100%, it's a good idea to figure out what happened. You may have discovered a new law of Physics, but it's much more likely that you accidentally regressed two variables that measure the same thing.

For the standard deviation of the residuals to summarize the scatter of all the residuals, the residuals must share the same spread for each value of x. That's an assumption. But if the spread of the scatterplot from the line looks roughly the same everywhere and (often more vividly) if the *residual plot* of residuals *vs.* predicted values also has a consistent vertical spread, then that assumption is reasonable. The most common violation of this equal variance assumption is residuals that spread out more for *larger* values of x, so a good nickname for this check is the **Does the Plot Thicken? Condition**.

Outlying points can dramatically change a regression model. They can even change the sign of the slope, which would give a very different impression of the relationship between the variables if you only look at the regression model. So check the **Outlier Condition**. Check both the scatterplot of y against x and the residual plot to be sure there are no outliers. The residual plot often shows violations more clearly and may reveal other unexpected patterns or interesting quirks in the data. Of course, any outliers are likely to be interesting and informative, so be sure to look into why they are unusual.

To summarize:

Before starting, be sure to check the

- **Quantitative Variable Condition**   If either y or x is categorical, you can't make a scatterplot and you can't perform a regression. Stop.

From the scatterplot of y against x, check the

- **Straight Enough Condition**   Is the relationship between y and x straight enough to proceed with a linear regression model?
- **Outlier Condition**   Are there any outliers that might dramatically influence the fit of the least squares line?
- **Does the Plot Thicken? Condition**   Does the spread of the data around the generally straight relationship seem to be consistent for all values of x?

After fitting the regression model, make a plot of residuals and look for

- Any bends that would violate the **Straight Enough Condition**,
- Any outliers that weren't clear before, and
- Any change in the vertical spread of the residuals from one part of the plot to another.

> **Make a Picture (Or Two)**
>
> You can't check the conditions just by checking boxes. You need to examine both the original scatterplot of y against x before you fit the model, and the plot of residuals afterward. These plots can save you from making embarrassing errors.[14]

A Tale of Two Regressions

Regression equations may not behave exactly the way you'd expect. Our regression model for the BK sandwiches was $\widehat{fat} = 8.4 + 0.917\,protein$. That equation allowed us to estimate that a sandwich with 31 grams of protein would have 36.7 grams of fat. Suppose, though, that we knew the fat content and wanted to estimate the amount of protein. It might seem natural to think that by solving our equation for *protein* we'd get a model for predicting *protein* from *fat*. But that doesn't work.

Our original model is $\hat{y} = b_0 + b_1 x$, but the new one needs to evaluate an $\hat{x}$ based on a value of y. There's no y in our original model, only $\hat{y}$, and that makes all the difference. Our model doesn't fit the BK data values perfectly, and the least squares criterion focuses on the *vertical* errors the model makes in using x to model y—not on *horizontal* errors related to modeling x.

A quick look at the equations reveals why. Simply solving our equation for x would give a new line whose slope must be reciprocal. To model y in terms of x, our slope is $b_1 = \frac{rs_y}{s_x}$. To model x in terms of y, we'd need to use the slope $b_1 = \frac{rs_x}{s_y}$. Notice that is *not* the reciprocal of ours.

If we want to predict *protein* from *fat*, we need to create a different model. The slope is $b_1 = \frac{(0.76)(13.5)}{16.2} = 0.63$ grams of protein per gram of fat. The equation turns out to be $\widehat{protein} = 2.29 + 0.63\,fat$. Now we'd predict that a sandwich with 35.9 grams of fat should have 24.9 grams of protein—not the 30 grams that would arise from the first equation.

Protein	Fat
$\bar{x} = 18.0\ g$	$\bar{y} = 24.8\ g$
$s_x = 13.5\ g$	$s_y = 16.2\ g$
$r = 0.76$	

---

[14]And losing points on an exam!

Moral of the story: *Think.* (Where have you heard *that* before?) Decide which variable you want to use (*x*) to predict values for the other (*y*). Then find the model that does that. If, later, you want to make predictions in the other direction, you'll need to start over and create the other model from scratch.

Step-by-Step Example REGRESSION

Even if you hit the fast-food joints for lunch, you should have a good breakfast. Nutritionists, concerned about "empty calories" in breakfast cereals, recorded facts about 77 cereals, including their *Calories* per serving and *Sugar content (in grams).*

Question: How can we use sugar content to estimate calories in breakfast cereals?

THINK ➡ Plan State the problem and determine the role of the variables.

Variables Name the variables and report the W's.

Check the conditions for a regression by making a picture. Never fit a regression without looking at the scatterplot first.

I am interested in using cereals' sugar content (*x*) to estimate calories (*y*).

✓ **Quantitative Variables Condition:** I have two quantitative variables, *Calories* and *Sugar content per serving,* measured on 77 breakfast cereals. The units of measurement are calories and grams of sugar, respectively.

✓ **Outlier Condition:** There are no obvious outliers or groups.
✓ The **Straight Enough Condition** is satisfied; I will fit a regression model to these data.
✓ The **Does the Plot Thicken? Condition** is satisfied. The spread around the line looks about the same throughout.

SHOW ➡ Mechanics If there are no clear violations of the conditions, fit a straight line model of the form $\hat{y} = b_0 + b_1 x$ to the data. Summary statistics give the building blocks of the calculation.

Calories

$$\bar{y} = 107.0 \text{ calories}$$

$$s_y = 19.5 \text{ calories}$$

Sugar

$$\bar{x} = 7.0 \text{ grams}$$

$$s_x = 4.4 \text{ grams}$$

(continued)

Find the slope.

Correlation

$r = 0.564$

$$b_1 = \frac{rs_y}{s_x} = \frac{0.564(19.5)}{4.4}$$

$$= 2.50 \text{ calories per gram of sugar.}$$

Find the intercept.

$b_0 = \bar{y} - b_1\bar{x} = 107 - 2.50(7) = 89.5 \text{ calories.}$

So the least squares line is

Write the equation, using meaningful variable names.

$$\hat{y} = 89.5 + 2.50\, x \text{ or}$$

$$\widehat{Calories} = 89.5 + 2.50\, Sugar.$$

Squaring the correlation gives

State the value of R^2.

$$R^2 = 0.564^2 = 0.318 \text{ or } 31.8\%.$$

TELL ➡ **Conclusion**   Describe what the model says in words and numbers. Be sure to use the names of the variables and their units.

The scatterplot shows a positive, linear relationship and no outliers. The slope of the least squares regression line suggests that cereals have about 2.50 *Calories* more per additional gram of *Sugar*.

The key to interpreting a regression model is to start with the phrase "b_1 *y*-units per *x*-unit," substituting the estimated value of the slope for b_1 and the names of the respective units. The intercept is then a starting or base value.

The intercept predicts that sugar-free cereals would average about 89.5 calories.

R^2 gives the fraction of the variability of **y** accounted for by the linear regression model.

The R^2 says that 31.8% of the variability in *Calories* is accounted for by variation in *Sugar* content.

Find the standard deviation of the residuals, s_e, and compare it to the original s_y.

$s_e = 16.2$ calories. That's smaller than the original SD of 19.5, but still fairly large.

THINK ➡ **Check Again**   Even though we looked at the scatterplot *before* fitting a regression model, a plot of the residuals is essential to any regression analysis because it is the best check for additional patterns and interesting quirks in the data.

AGAIN

TI-*nspire*

Residuals plots. See how the residuals plot changes as you drag points around in a scatterplot.

The residuals show a shapeless form and roughly equal scatter for all predicted values. The linear model appears to be appropriate.

TI Tips REGRESSION LINES AND RESIDUALS PLOTS

By now you will not be surprised to learn that your calculator can do it all: scatterplot, regression line, and residuals plot. Let's try it using the Arizona State tuition data from the last chapter. (TI Tips, p. 149) You should still have that saved in lists named YR and TUIT. First, recreate the scatterplot.

FIND THE EQUATION OF THE REGRESSION LINE. Actually, you already found the line when you used the calculator to get the correlation. But this time we'll be a little fancier so that we can display the line on our scatterplot. We want to tell the calculator to do the regression and save the equation of the model as a graphing variable.

- Under STAT CALC choose LinReg(a + bx).
- Specify that Xlist and Ylist are YR and TUIT, as before, but . . .
- Now add one more specification to store the regression equation. Press VARS, go to the Y-VARS menu, choose 1:Function, and finally(!) choose Y1.
- To Calculate, hit ENTER.

There's the equation. The calculator tells you that the regression line is $\widehat{tuit} = 6440 + 326\,year$. Can you explain what the slope and y-intercept mean?

ADD THE LINE TO THE PLOT. When you entered the LinReg command, the calculator automatically saved the equation as Y1. Just hit GRAPH to see the line drawn across your scatterplot.

CHECK THE RESIDUALS. Remember, you are not finished until you check to see if a linear model is appropriate. That means you need to see if the residuals appear to be randomly distributed. To do that, you need to look at the residuals plot.

This is made easy by the fact that the calculator has already placed the residuals in a list named RESID. Want to see them? Go to STAT EDIT and look through the lists. (If RESID is not already there, go to the first blank list and import the name RESID from your LIST NAMES menu. The residuals should appear.) Every time you have the calculator compute a regression analysis, it will automatically save this list of residuals for you.

NOW CREATE THE RESIDUALS PLOT.
- Set up STAT PLOT Plot2 as a scatterplot with Xlist:YR and Ylist:RESID.

- Before you try to see the plot, go to the Y= screen. By moving the cursor around and hitting ENTER in the appropriate places you can turn off the regression line and Plot1, and turn on Plot2.
- ZoomStat will now graph the residuals plot.

Uh-oh! See the curve? The residuals are high at both ends, low in the middle. Looks like a linear model may not be appropriate after all. Notice that the residuals plot makes the curvature much clearer than the original scatterplot did.

Moral: Always check the residuals plot!

So a linear model might not be appropriate here. What now? The next two chapters provide techniques for dealing with data like these.

Reality Check: Is the Regression Reasonable?

> **Regression: Adjective, Noun, or Verb**
> You may see the term *regression* used in different ways. There are many ways to fit a line to data, but the term "regression line" or "regression" without any other qualifiers always means least squares. People also use *regression* as a verb when they speak of *regressing* a *y*-variable on an *x*-variable to mean fitting a linear model.

Statistics don't come out of nowhere. They are based on data. The results of a statistical analysis should reinforce your common sense, not fly in its face. If the results are surprising, then either you've learned something new about the world or your analysis is wrong.

Whenever you perform a regression, think about the coefficients and ask whether they make sense. Is a slope of 2.5 calories per gram of sugar reasonable? That's hard to say right off. We know from the summary statistics that a typical cereal has about 100 calories and 7 grams of sugar. A gram of sugar contributes some calories (actually, 4, but you don't need to know that), so calories should go up with increasing sugar. The direction of the slope seems right.

To see if the *size* of the slope is reasonable, a useful trick is to consider its order of magnitude. We'll start by asking if shrinking the slope by a factor of 10 seems reasonable. Is 0.25 calories per gram of sugar enough? Then the 7 grams of sugar found in the average cereal would contribute less than 2 calories. That seems too small.

Now let's try inflating the slope by a factor of 10. Is 25 calories per gram reasonable? Then the average cereal would have 175 calories from sugar alone. The average cereal has only 100 calories per serving, though, so that slope seems too big.

We have tried inflating the slope by a factor of 10 and deflating it by 10 and found both to be unreasonable. So, like Goldilocks, we're left with the value in the middle that's just right. And an increase of 2.5 calories per gram of sugar is certainly *plausible*.

It's easy to take something that comes out of a computer at face value and just go with it. The small effort of asking yourself whether the regression equation makes sense is repaid whenever you catch errors or avoid saying something silly or absurd about the data.

WHAT IF ●●● that regression line isn't the one and only?

Remember those engineers investigating the relationship between the age and condition of bridges in the Finger Lakes? Using a random sample of 20 of the over 700 bridges, let's create a regression model that can predict the condition of bridges that are 60 years old.

The relationship looks reasonably linear. With the model

$\widehat{Condition} = 6.35 - 0.023 Age$ we'd estimate that 60-year-old bridges have an average condition rating of 4.97 (out of 7). Because in New York State ratings below 5 mean a bridge will be labeled "deficient", our model suggests that careful attention should be paid to any bridges over 60 years old.

But this model is based on this particular sample, chosen at random. What if we had picked a different sample? Here's another:

Looks linear again (good news), and the model is

$$\widehat{Condition} = 6.40 - 0.022\ Age.$$

This model would have suggested that 60-year-old bridges would rate 5.08, classified as "good" condition.

While it may seem a little strange to think of it this way, we now see that lines are variables. The model we develop depends on the sample we choose. To investigate a bit more, we drew 5 simulated samples and found the 5 regression lines. Without making the plot messy by displaying all the sampled points, here they are:

The good news is that these models are all pretty similar. That's because each is estimating the same underlying relationship between *Age* and *Condition* for all the Finger Lakes bridges. But the differences remind us that samples can only provide insights about what's going on in a population; the truth remains elusive.

Keep this in mind when you write your conclusions. Statements like "60-year-old bridges have a condition rating of 5.08." imply a level of certainty that you simply cannot have. You should be less sure of yourself, and sound it. "My model suggests that the condition ratings of 60-year-old bridges may average about 5.08." It's always better to tell the truth.

WHAT CAN GO WRONG?

There are many ways in which data that appear at first to be good candidates for regression analysis may be unsuitable. And there are ways that people use regression that can lead them astray. Here's an overview of the most common problems. We'll discuss these and more in the next chapter.

- **Don't fit a straight line to a nonlinear relationship.** Linear regression is suited only to relationships that are, well, *linear*. Fortunately, we can often improve the linearity easily by using re-expression. We'll come back to that topic in Chapter 9.

- **Beware of extraordinary points.** Data points can be extraordinary in a regression in two ways: They can have *y*-values that stand off from the linear pattern suggested by the bulk of the data, or extreme *x*-values. Both kinds of extraordinary points require attention.

- **Don't infer that *x* causes *y* just because there is a good linear model for their relationship.** When two variables are strongly correlated, it is often tempting to assume a causal relationship between them. Putting a regression line on a scatterplot tempts us even further, but it doesn't make the assumption of causation any more valid. For example, our regression model predicting hurricane wind speeds from the central pressure was reasonably successful, but the relationship is very complex. It is reasonable to say that low central pressure at the eye is responsible for the high winds because it draws moist, warm air into the center of the storm, where it swirls around, generating the winds. But as is often the case, things aren't quite that simple. The winds themselves also contribute to lowering the pressure at the center of the storm as it becomes a hurricane. Understanding causation requires far more work than just finding a correlation or modeling a relationship.

- **Don't choose a model based on R^2 alone.** Although R^2 measures the *strength* of the linear association, a high R^2 does not demonstrate the *appropriateness* of the regression. Even a relationship that's actually curved could produce a high R^2, as could an outlier even if there's little association in the rest of the data. Or an otherwise strong relationship could have an outlier that makes R^2 misleadingly low. Always look at the scatterplot.

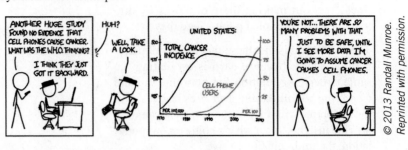

> **Think Variation!**
> R^2 does not mean that protein accounts for 58% of the fat in a BK food item. It is the *variation* in fat content that is accounted for by the linear model.

■ **Don't invert the regression.** The model works in one direction only. If the equation predicts $\hat{y}$ from x, it will not correctly predict $\hat{x}$ from y. This isn't algebra class; you can't solve the equation for the other variable. If you want to make predictions in the other direction, you'll need to create the model that does that.

What Have We Learned?

We've learned that when the relationship between quantitative variables is fairly straight, a linear model can help summarize that relationship and give us insights about it:

■ The regression (best fit) line doesn't pass through all the points, but it is the best compromise in the sense that the sum of squares of the residuals is the smallest possible.

We've learned several things the correlation, r, tells us about the regression:

■ For each SD of x that we are away from the x mean, we expect to be r SDs of y away from the y mean.

■ Because r is always between -1 and $+1$, each predicted y is fewer SDs away from its mean than the corresponding x was, a phenomenon called regression to the mean.

■ The slope of the line is based on the correlation, adjusted for the units of x and y:

$$b_1 = \frac{rs_y}{s_x}$$

We've learned to interpret the slope in context as predicted change in y-units per 1 unit change in x.

We've learned that the residuals and R^2 reveal how well the model works:

■ If a plot of residuals shows a pattern, we should re-examine the data to see why.

■ The standard deviation of the residuals, s_e, quantifies the amount of scatter around the line.

■ The square of the correlation coefficient, R^2, gives us the fraction of the variation of the response variable accounted for by the regression model. The remaining $1 - R^2$ of the variation is left in the residuals and not explained by the model.

■ R^2 is an overall measure of how successful the regression is in linearly relating y to x.

Of course, the linear model makes no sense unless the **Linearity Assumption** is satisfied. We check the **Straight Enough Condition** and **Outlier Condition** with a scatterplot, as we did for correlation, and also with a plot of residuals against either the x or the predicted values. For the standard deviation of the residuals to make sense as a summary, we have to make the **Equal Variance Assumption**. We check it by looking at both the original scatterplot and the residual plot for the **Does the Plot Thicken? Condition**.

Terms

Model	An equation or formula that simplifies and represents reality. (p. 177)
Linear model	A linear model is an equation of a line. To interpret a linear model, we need to know the variables (along with their W's) and their units. (p. 177)
Predicted value	The value of $\hat{y}$ found for a given x-value in the data. A predicted value is found by substituting the x-value in the regression equation. The predicted values are the values on the fitted line; the points $(x, \hat{y})$ all lie exactly on the fitted line. (p. 177)
Residuals	Residuals are the differences between data values and the corresponding values predicted by the regression model—or, more generally, values predicted by any model. (p. 177)
	$$\text{Residual} = \text{observed value} - \text{predicted value} = e = y - \hat{y}$$
Least squares	The least squares criterion specifies the unique line that minimizes the variance of the residuals or, equivalently, the sum of the squared residuals. (p. 178)
Regression to the mean	Because the correlation is always less than 1.0 in magnitude, each predicted $\hat{y}$ tends to be fewer standard deviations from its mean than its corresponding x was from its mean. This is called regression to the mean. (p. 179)
Regression line **Line of best fit**	The particular linear equation $$\hat{y} = b_0 + b_1 x$$ that satisfies the least squares criterion is called the least squares regression line. Casually, we often just call it the regression line, or the line of best fit. (p. 179)
Slope	The slope, b_1, gives a value in "y-units per x-unit." Changes of one unit in x are associated with changes of b_1 units in predicted values of y. The slope can be found by $$b_1 = \frac{rs_y}{s_x}. \quad \text{(p. 181)}$$
Intercept	The intercept, b_0, gives a starting value in y-units. It's the $\hat{y}$-value when x is 0. You can find it from $b_0 = \bar{y} - b_1\bar{x}$. (p. 182)
s_e	The standard deviation of the residuals is found by $s_e = \sqrt{\dfrac{\Sigma e^2}{n - 2}}$. When the assumptions and conditions are met, the residuals can be well described by using this standard deviation and the 68–95–99.7 Rule. (p. 187)
R^2	R^2 (the square of the correlation between y and x) gives the fraction of the variability of y accounted for by the least squares linear regression on x. (p. 189)
Does the Plot Thicken? **Condition**	The scatterplot or residuals plot should show consistent (vertical) spread in y-values. (p. 190)

On the Computer REGRESSION ANALYSIS

All statistics packages make a table of results for a regression. These tables may differ slightly from one package to another, but all are essentially the same—and all include much more than we need to know for now. Every computer regression table includes a section something like the one on the next page (based on data for other fast food items):

R squared

Standard dev of residuals (s_e)

The "dependent," response, or y-variable

```
Dependent variable is: Total Fat
R squared = 69.0%
s = 9.277

Variable    Coefficient  SE(Coeff)  t-ratio  P-value
Intercept   6.83077      2.664      2.56     0.0158
Protein     0.971381     0.1209     8.04     ≤0.0001
```

A S **Finding Least Squares Lines.**
We almost always use technology to find regressions. Practice now—just in time for the exercises.

The "independent," predictor, or x-variable

The slope

The intercept

We'll deal with all of these later in the book. You may ignore them for now.

The slope and intercept coefficient are given in a table such as this one. Usually the slope is labeled with the name of the x-variable, and the intercept is labeled "Intercept" or "Constant." So the regression equation shown here is

$$\widehat{Fat} = 6.83077 + 0.97138\ Protein.$$

It is not unusual for statistics packages to give many more digits of the estimated slope and intercept than could possibly be estimated from the data. (The original data were reported to the nearest gram.) Ordinarily, you should round most of the reported numbers to one digit more than the precision of the data, and the slope to two. We will learn about the other numbers in the regression table later in the book. For now, all you need to be able to do is find the coefficients, the s_e, and the R^2 value.

Exercises

1. **Cereals** For many people, breakfast cereal is an important source of fiber in their diets. Cereals also contain potassium, a mineral shown to be associated with maintaining a healthy blood pressure. An analysis of the amount of fiber (in grams) and the potassium content (in milligrams) in servings of 77 breakfast cereals produced the regression model $\widehat{Potassium} = 38 + 27\ Fiber$. If your cereal provides 9 grams of fiber per serving, how much potassium does the model estimate you will get?

2. **Horsepower** A study that examined the relationship between the fuel economy (mpg) and horsepower for 15 models of cars produced the regression model $\widehat{mpg} = 46.87 - 0.084\ HP$. If the car you are thinking of buying has a 200-horsepower engine, what does this model suggest your gas mileage would be?

3. **More cereal** Exercise 1 describes a regression model that estimates a cereal's potassium content from the amount

of fiber it contains. In this context, what does it mean to say that a cereal has a negative residual?

4. **Horsepower, again** Exercise 2 describes a regression model that uses a car's horsepower to estimate its fuel economy. In this context, what does it mean to say that a certain car has a positive residual?

5. **Another bowl** In Exercise 1, the regression model $\widehat{Potassium} = 38 + 27\ Fiber$ relates fiber (in grams) and potassium content (in milligrams) in servings of breakfast cereals. Explain what the slope means.

6. **More horsepower** In Exercise 2, the regression model $\widehat{mpg} = 46.87 - 0.084\ HP$ relates cars' horsepower to their fuel economy (in mpg). Explain what the slope means.

7. **Cereal again** The correlation between a cereal's fiber and potassium contents is $r = 0.903$. What fraction

of the variability in potassium is accounted for by the amount of fiber that servings contain?

8. Another car The correlation between a car's horsepower and its fuel economy (in mpg) is $r = -0.869$. What fraction of the variability in fuel economy is accounted for by the horsepower?

9. Last bowl! For Exercise 1's regression model predicting potassium content (in milligrams) from the amount of fiber (in grams) in breakfast cereals, $s_e = 30.77$. Explain in this context what that means.

10. Last tank! For Exercise 2's regression model predicting fuel economy (in mpg) from the car's horsepower, $s_e = 3.287$. Explain in this context what that means.

11. Residuals I Tell what each of the residual plots below indicates about the appropriateness of the linear model that was fit to the data.

12. Residuals II Tell what each of the residual plots below indicates about the appropriateness of the linear model that was fit to the data.

13. What slope I? If you create a regression model for predicting the *Weight* of a car (in pounds) from its *Length* (in feet), is the slope most likely to be 3, 30, 300, or 3000? Explain.

14. What slope II? If you create a regression model for estimating the *Height* of a pine tree (in feet) based on the *Circumference* of its trunk (in inches), is the slope most likely to be 0.1, 1, 10, or 100? Explain.

15. True or false If false, explain briefly.
a) We choose the linear model that passes through the most data points on the scatterplot.
b) The residuals are the observed y-values minus the y-values predicted by the linear model.
c) Least squares means that the square of the largest residual is as small as it could possibly be.

16. True or false II If false, explain briefly.
a) Some of the residuals from a least squares linear model will be positive and some will be negative.
b) Least Squares means that some of the squares of the residuals are minimized.
c) We write $\hat{y}$ to denote the predicted values and y to denote the observed values.

17. Bookstore sales revisited Recall the data we saw in Chapter 6, Exercise 3 for a bookstore. The manager wants to predict *Sales* from *Number of Sales People Working*.

Number of Sales People Working	Sales (in $1000)
2	10
3	11
7	13
9	14
10	18
10	20
12	20
15	22
16	22
20	26

Here is the regression analysis of *Sales* vs. *Number of Sales People Working*.

Dependent variable is Sales
R-squared = 93.2%
s = 1.477

Variable	Coefficient
Intercept	8.1006
Num_Workers	0.9134

a) Write the regression equation. Define the variables used in your equation.
b) What does the slope mean in this context?
c) What does the y-intercept mean in this context? Is it meaningful?
d) If 18 people are working, what *Sales* do you predict?
e) If sales for the 18 people are actually $25,000, what is the value of the residual?
f) Have we overestimated or underestimated the sales?

18. Disk drives again In Chapter 6, Exercise 4, we saw some data on hard drives. After correcting for an outlier, these data look like this: we want to predict *Price* from *Capacity*.

Capacity (in TB)	Price (in $)
0.080	29.95
0.120	35.00
0.250	49.95
0.320	69.95
1.0	99.00
2.0	205.00
4.0	449.00

Here is the regression analysis of *Price* vs. *Capacity*.

Dependent variable is Price
R-squared = 98.8%
s = 17.95

Variable	Coefficient
Intercept	18.617
Capacity	103.929

a) Write the regression equation. Define the variables used in your equation.
b) What does the slope mean in this context?
c) What does the y-intercept mean in this context? Is it meaningful?
d) What would you predict for the price of a 3.0 TB drive?
e) You found a 3.0 TB drive for $300. Is this a good buy? How much would you save compared to what you expected to buy?
f) Does the model overestimate or underestimate the price for a 3.0 TB drive?

19. Bookstore sales once more Here are the residuals for a regression of *Sales* on *Number of Sales People Working* for the bookstore Exercise 17:

Number of Sales People Working	Residual
2	0.07
3	0.16
7	−1.49
9	−2.32
10	0.77
10	2.77
12	0.94
15	0.20
16	−0.72
20	−0.37

a) What are the units of the residuals?
b) Which residual contributes the most to the sum that was minimized according to the Least Squares Criterion to find this regression?
c) Which residual contributes least to that sum?

20. Disk drives once more Here are the residuals for a regression of *Price* on *Capacity* for the hard drives of Exercise 18.

Capacity	Residual
0.080	3.02
0.120	3.91
0.250	5.35
0.320	18.075
1.0	−23.55
2.0	−21.475
4.0	14.666

a) Which residual contributes the most to the sum that is minimized by the Least Squares criterion?
b) Two of the residuals are negative. What does that mean about those drives? Be specific and use the correct units.

21. Bookstore sales last time For the regression model for the bookstore of Exercise 17, what is the value of R^2 and what does it mean?

22. Disk drives encore For the hard drive data of Exercise 18, interpret the value of R^2.

23. Residual plots Here are residual plots (residuals plotted against predicted values) for three linear regression models. Indicate which condition appears to be violated (linearity, outlier or equal spread) in each case.

a)

b)

c)
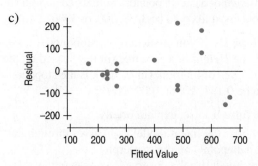

24. Disk drives last time Here is a scatterplot of the residuals from the regression of the hard drive prices on their sizes from Exercise 18.

a) Are any assumptions or conditions violated? If so, which ones?

b) What would you recommend about this regression?

25. Real estate A random sample of records of sales of homes from February 15 to April 30, 1993, from the files maintained by the Albuquerque Board of Realtors gives the *Price* and *Size* (in square feet) of 117 homes. A regression to predict *Price* (in thousands of dollars) from *Size* has an *R*-squared of 71.4%. The residuals plot indicated that a linear model is appropriate.

a) What are the variables and units in this regression?

b) What units does the slope have?

c) Do you think the slope is positive or negative? Explain.

26. Roller coaster The Mitch Hawker poll ranked the Top 10 steel roller coasters in 2011. A table in the previous chapter's exercises shows the length of the initial drop (in feet) and the duration of the ride (in seconds). A regression to predict *Duration* from *Drop* has $R^2 = 15.2\%$.

a) What are the variables and units in this regression?

b) What units does the slope have?

c) Do you think the slope is positive or negative? Explain.

27. Real estate again The regression of *Price* on *Size* of homes in Albuquerque had $R^2 = 71.4\%$, as described in Exercise 25. Write a sentence (in context, of course) summarizing what the R^2 says about this regression.

28. Coasters again Exercise 26 examined the association between the *Duration* of a roller coaster ride and the height of its initial *Drop*, reporting that $R^2 = 15.2\%$. Write a sentence (in context, of course) summarizing what the R^2 says about this regression.

29. Real estate redux The regression of *Price* on *Size* of homes in Albuquerque had $R^2 = 71.4\%$, as described in Exercise 25.

a) What is the correlation between *Size* and *Price*? Explain why you chose the sign (+ or −) you did.

b) What would you predict about the *Price* of a home 1 standard deviation above average in *Size*?

c) What would you predict about the *Price* of a home 2 standard deviations below average in *Size*?

30. Another ride The regression of *Duration* of a roller coaster ride on the height of its initial *Drop*, described in Exercise 26, had $R^2 = 15.2\%$.

a) What is the correlation between *Drop* and *Duration*?

b) What would you predict about the *Duration* of the ride on a coaster whose initial *Drop* was 1 standard deviation below the mean *Drop*?

c) What would you predict about the *Duration* of the ride on a coaster whose initial *Drop* was 3 standard deviations above the mean *Drop*?

31. More real estate Consider the Albuquerque home sales from Exercise 25 again. The regression analysis gives the model $\widehat{Price} = 47.82 + 0.061\ Size$.

a) Explain what the slope of the line says about housing prices and house size.

b) What price would you predict for a 3000-square-foot house in this market?

c) A real estate agent shows a potential buyer a 1200-square-foot home, saying that the asking price is $6000 less than what one would expect to pay for a house of this size. What is the asking price, and what is the $6000 called?

32. Last ride Consider the roller coasters described in Exercise 26 again. The regression analysis gives the model $\widehat{Duration} = 64.232 + 0.180\ Drop$.

a) Explain what the slope of the line says about how long a roller coaster ride may last and the height of the coaster.

b) A new roller coaster advertises an initial drop of 200 feet. How long would you predict the rides last?

c) Another coaster with a 150-foot initial drop advertises a 2-minute ride. Is this longer or shorter than you'd expect? By how much? What's that called?

33. Misinterpretations A Biology student who created a regression model to use a bird's *Height* when perched for predicting its *Wingspan* made these two statements. Assuming the calculations were done correctly, explain what is wrong with each interpretation.

a) My R^2 of 93% shows that this linear model is appropriate.

b) A bird 10 inches tall will have a wingspan of 17 inches.

34. More misinterpretations A Sociology student investigated the association between a country's *Literacy Rate* and *Life Expectancy*, then drew the conclusions listed below. Explain why each statement is incorrect. (Assume that all the calculations were done properly.)

a) The *Literacy Rate* determines 64% of the *Life Expectancy* for a country.

b) The slope of the line shows that an increase of 5% in *Literacy Rate* will produce a 2-year improvement in *Life Expectancy*.

35. ESP People who claim to "have ESP" participate in a screening test in which they have to guess which of several images someone is thinking of. You and a friend both took the test. You scored 2 standard deviations above the mean, and your friend scored 1 standard deviation below the mean. The researchers offer everyone the opportunity to take a retest.

a) Should you choose to take this retest? Explain.

b) Now explain to your friend what his decision should be and why.

36. *SI* jinx Players in any sport who are having great seasons, turning in performances that are much better than anyone might have anticipated, often are pictured on the cover of *Sports Illustrated*. Frequently, their performances then falter somewhat, leading some athletes to believe in a "*Sports Illustrated* jinx." Similarly, it is common for phenomenal rookies to have less stellar second seasons—the so-called "sophomore slump." While fans, athletes, and analysts have proposed many theories about what leads to such declines, a statistician might offer a simpler (statistical) explanation. Explain.

37. Cigarettes Is the nicotine content of a cigarette related to the "tars"? A collection of data (in milligrams) on 29 cigarettes produced the scatterplot, residuals plot, and regression analysis shown:

Dependent variable is: nicotine
R squared = 92.4%

Variable	Coefficient
Constant	0.154030
Tar	0.065052

a) Do you think a linear model is appropriate here? Explain.
b) Explain the meaning of R^2 in this context.

38. Attendance 2010 In the previous chapter, you looked at the relationship between the number of wins by American League baseball teams and the average attendance at their home games for the 2010 season. Here are the scatterplot, the residuals plot, and part of the regression analysis:

Dependent variable is Home Attendance
R-squared = 28.4%

Variable	Coefficient
Constant	−6760.5
Wins	431.22

a) Do you think a linear model is appropriate here? Explain.
b) Interpret the meaning of R^2 in this context.
c) Do the residuals show any pattern worth remarking on?
d) The point in the upper right of the plots is the New York Yankees. What can you say about the residual for the Yankees?

39. Another cigarette Consider again the regression of *Nicotine* content on *Tar* (both in milligrams) for the cigarettes examined in Exercise 37.

a) What is the correlation between *Tar* and *Nicotine*?
b) What would you predict about the average *Nicotine* content of cigarettes that are 2 standard deviations below average in *Tar* content?
c) If a cigarette is 1 standard deviation above average in *Nicotine* content, what do you suspect is true about its *Tar* content?

40. Second inning 2010 Consider again the regression of *Average Attendance* on *Wins* for the baseball teams examined in Exercise 38.

a) What is the correlation between *Wins* and *Average Attendance*?
b) What would you predict about the *Average Attendance* for a team that is 2 standard deviations above average in *Wins*?
c) If a team is 1 standard deviation below average in attendance, what would you predict about the number of games the team has won?

41. Last cigarette Take another look at the regression analysis of tar and nicotine content of the cigarettes in Exercise 37.

a) Write the equation of the regression line.
b) Estimate the *Nicotine* content of cigarettes with 4 milligrams of *Tar.*
c) Interpret the meaning of the slope of the regression line in this context.
d) What does the *y*-intercept mean?

e) If a new brand of cigarette contains 7 milligrams of tar and a nicotine level whose residual is −0.5 mg, what is the nicotine content?

42. Last inning 2010 Refer again to the regression analysis for average attendance and games won by American League baseball teams, seen in Exercise 38.

a) Write the equation of the regression line.

b) Estimate the *Average Attendance* for a team with 50 *Wins.*

c) Interpret the meaning of the slope of the regression line in this context.

d) In general, what would a negative residual mean in this context?

e) The San Francisco Giants, the 2010 World Champions, are not included in these data because they are a National League team. During the 2010 regular season, the Giants won 92 games and averaged 41,736 fans at their home games. Calculate the residual for this team, and explain what it means.

43. Income and housing revisited In Chapter 6, Exercise 32, we learned that the Office of Federal Housing Enterprise Oversight (OFHEO) collects data on various aspects of housing costs around the United States. Here's a scatterplot (by state) of the *Housing Cost Index* (HCI) versus the *Median Family Income* (MFI) for the 50 states. The correlation is $r = 0.65$. The mean HCI is 338.2, with a standard deviation of 116.55. The mean MFI is $46,234, with a standard deviation of $7072.47.

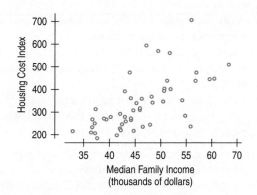

a) Is a regression analysis appropriate? Explain.

b) What is the equation that predicts Housing Cost Index from median family income?

c) For a state with MFI = $44,993, what would be the predicted HCI?

d) Washington, DC, has an MFI of $44,993 and an HCI of 548.02. How far off is the prediction in c) from the actual HCI?

e) If we standardized both variables, what would be the regression equation that predicts standardized HCI from standardized MFI?

f) If we standardized both variables, what would be the regression equation that predicts standardized MFI from standardized HCI?

44. Interest rates and mortgages again In Chapter 6, Exercise 33, we saw a plot of mortgages in the United States (in thousands of dollars) versus the interest rate at various times over the past 26 years. The correlation is $r = -0.86$. The mean mortgage amount is $121.8 thousand and the mean interest rate is 7.74%. The standard deviations are $47.36 thousand for mortgage amounts and 1.79% for the interest rates.

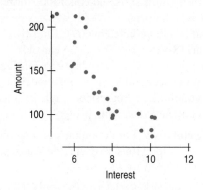

a) Is a regression model appropriate for predicting mortgage amount from interest rates? Explain.

b) What is the equation that predicts mortgage amount from interest rates?

c) What would you predict the mortgage amount would be if the interest rates climbed to 13%?

d) Do you have any reservations about your prediction in part c)?

e) If we standardized both variables, what would be the regression equation that predicts standardized mortgage amount from standardized interest rates?

f) If we standardized both variables, what would be the regression equation that predicts standardized interest rates from standardized mortgage amount?

45. Online clothes An online clothing retailer keeps track of its customers' purchases. For those customers who signed up for the company's credit card, the company also has information on the customer's *Age* and *Income.* A random sample of 500 of these customers shows the following scatterplot of *Total Yearly Purchases* by *Age:*

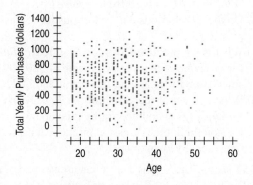

The correlation between *Total Yearly Purchases* and *Age* is $r = 0.037$. Summary statistics for the two variables are:

	Mean	SD
Age	29.67 yrs	8.51 yrs
Total Yearly Purchase	$572.52	$253.62

a) What is the linear regression equation for predicting *Total Yearly Purchase* from *Age*?
b) Do the assumptions and conditions for regression appear to be met?
c) What is the predicted average *Total Yearly Purchase* for an 18-year-old? For a 50-year-old?
d) What percent of the variability in *Total Yearly Purchases* is accounted for by this model?
e) Do you think the regression might be a useful one for the company? Explain.

46. Online clothes II For the online clothing retailer discussed in the previous problem, the scatterplot of *Total Yearly Purchases* by *Income* shows

The correlation between *Total Yearly Purchases* and *Income* is 0.722. Summary statistics for the two variables are:

	Mean	SD
Income	$50,343.40	$16,952.50
Total Yearly Purchase	$572.52	$253.62

a) What is the linear regression equation for predicting *Total Yearly Purchase* from *Income*?
b) Do the assumptions and conditions for regression appear to be met?
c) What is the predicted average *Total Yearly Purchase* for someone with a yearly *Income* of $20,000? For someone with an annual *Income* of $80,000?
d) What percent of the variability in *Total Yearly Purchases* is accounted for by this model?
e) Do you think the regression might be a useful one for the company? Comment.

47. SAT scores The SAT is a test often used as part of an application to college. SAT scores are between 200 and 800, but have no units. Tests are given in both Math and Verbal areas. Doing the SAT-Math problems also involves the ability to

read and understand the questions, but can a person's verbal score be used to predict the math score? Verbal and math SAT scores of a high school graduating class are displayed in the scatterplot, with the regression line added.

Here is the regression analysis of *Math SAT* vs. *Verbal SAT*.

Dependent variable is Math SAT
R-squared = 46.9%
s = 71.75

Variable	Coefficient
Intercept	209.5542
Verbal SAT	0.67507

a) Describe the relationship.
b) Are there any students whose scores do not seem to fit the overall pattern?
c) Find the correlation coefficient and interpret this value in context.
d) Write the equation of the regression line, defining any variables used in the equation.
e) Interpret the slope of this line.
f) Predict the math score of a student with a verbal score of 500.
g) Every year some student scores a perfect 1600 on these two parts of the test. Based on this model, what would be that student's Math score residual?

48. Success in college Colleges use SAT scores in the admissions process because they believe these scores provide some insight into how a high school student will perform at the college level. Regression analysis was computed on using *SAT* to predict *GPA*.

Dependent variable is GPA
R-squared = 22.1%

Variable	Coefficient
Intercept	−1.262
SAT	0.00214

a) Write the equation of the regression line.
b) Explain what the *y*-intercept of the regression line indicates.
c) Interpret the slope of the regression line.
d) Predict the GPA of a freshman who scored a combined 2100.

e) Based upon these statistics, how effective do you think SAT scores would be in predicting academic success during the first semester of the freshman year at this college? Explain.

f) As a student, would you rather have a positive or a negative residual in this context? Explain.

49. **SAT, take 2** Suppose the AP calculus students complained and insisted that we should use SAT math scores to estimate verbal scores (using the same data from exercise 47). Here is the regression analysis of *Math SAT* vs. *Verbal SAT*.

Dependent variable is Verbal
$s = 72.77$

Variable	Coefficient
Intercept	171.333
Math SAT	0.6943

a) What is the correlation?

b) Write the equation of the line of regression predicting verbal scores from math scores.

c) In general, what would a positive residual mean in this context?

d) A person tells you her math score was 500. Predict her verbal score.

e) Using that predicted verbal score and the equation you created in Exercise 47, predict her math score.

f) Why doesn't the result in part e) come out to 500?

50. **Success, part 2** The standard deviation of the residuals in Exercise 48 is 0.275. Interpret this value in context.

51. **Wildfires 2010** The National Interagency Fire Center (www.nifc.gov) reports statistics about wildfires. Here's an analysis of the number of wildfires between 1985 and 2010.

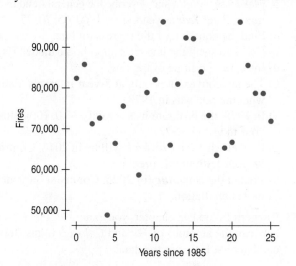

Dependent variable is Fires
R-squared $= 1.9\%$
$s = 11920$

Variable	Coefficient
Intercept	74487.1
Years since 1985	209.728

a) Is a linear model appropriate for these data? Explain.

b) Interpret the slope in this context.

c) Can we interpret the intercept? Why or why not?

d) What does the value of s_e say about the size of the residuals? What does it say about the effectiveness of the model?

e) What does R^2 mean in this context?

52. **Wildfires 2010—sizes** We saw in Exercise 51 that the number of fires was nearly constant. But has the damage they cause remained constant as well? Here's a regression that examines the trend in *Acres per Fire*, (in hundreds of thousands of acres) together with some supporting plots:

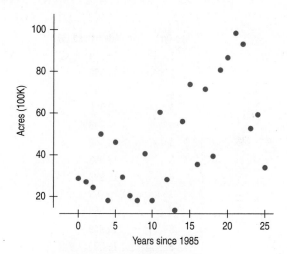

Dependent variable is Acres/fire
R-squared $= 36.6\%$
$s = 20.52$

Variable	Coefficient
Intercept	−3941
Years since 1985	1.997

a) Is the regression model appropriate for these data? Explain.

b) What interpretation (if any) can you give for the R^2 in the regression table?

53. Used cars 2011 Carmax.com lists numerous Toyota Corollas for sale within a 250 mile radius of Redlands, CA. The table below shows the ages of the cars and the advertised prices.

a) Make a scatterplot for these data.
b) Describe the association between *Age* and *Price* of a used Corolla.
c) Do you think a linear model is appropriate?
d) Computer software says that $R^2 = 89.1\%$. What is the correlation between *Age* and *Price*?
e) Explain the meaning of R^2 in this context.
f) Why doesn't this model explain 100% of the variability in the price of a used Corolla?

Age (yr)	Price Advertised ($)
1	17,599
2	14,998
2	15,998
4	13,998
4	14,998
5	14,599
5	13,998
6	11,998
7	9,998
7	11,559
8	10,849
8	10,899
10	9,998

T **54. Drug abuse** In the exercises of the last chapter you examined results of a survey conducted in the United States and 10 countries of Western Europe to determine the percentage of teenagers who had used marijuana and other drugs. Below is the scatterplot. Summary statistics showed that the mean percent that had used marijuana was 23.9%, with a standard deviation of 15.6%. An average of 11.6% of teens had used other drugs, with a standard deviation of 10.2%.

a) Do you think a linear model is appropriate? Explain.
b) For this regression, R^2 is 87.3%. Interpret this statistic in this context.

c) Write the equation you would use to estimate the percentage of teens who use other drugs from the percentage who have used marijuana.
d) Explain in context what the slope of this line means.
e) Do these results confirm that marijuana is a "gateway drug," that is, that marijuana use leads to the use of other drugs?

T **55. More used cars 2011** Use the advertised prices for Toyota Corollas given in Exercise 53 to create a linear model for the relationship between a car's *Age* and its *Price*.

a) Find the equation of the regression line.
b) Explain the meaning of the slope of the line.
c) Explain the meaning of the *y*-intercept of the line.
d) If you want to sell a 7-year-old Corolla, what price seems appropriate?
e) You have a chance to buy one of two cars. They are about the same age and appear to be in equally good condition. Would you rather buy the one with a positive residual or the one with a negative residual? Explain.
f) You see a "For Sale" sign on a 10-year-old Corolla stating the asking price as $8,500. What is the residual?
g) Would this regression model be useful in establishing a fair price for a 25-year-old car? Explain.

T **56. Birthrates 2009** The table shows the number of live births per 1000 women aged 15–44 years in the United States, starting in 1965. (National Center for Health Statistics, www.cdc.gov/nchs/)

Year	1965	1970	1975	1980	1985	1990	1995	2000	2005	2009
Rate	19.4	18.4	14.8	15.9	15.6	16.4	14.8	14.4	14.0	13.5

a) Make a scatterplot and describe the general trend in *Birthrates*. (Enter *Year* as years since 1900: 65, 70, 75, etc.)
b) Find the equation of the regression line.
c) Check to see if the line is an appropriate model. Explain.
d) Interpret the slope of the line.
e) The table gives rates only at 5-year intervals. Estimate what the rate was in 1978.
f) In 1978, the birthrate was actually 15.0. How close did your model come?
g) Predict what the *Birthrate* will be in 2010. Comment on your faith in this prediction.
h) Predict the *Birthrate* for 2025. Comment on your faith in this prediction.

57. Burgers In the last chapter, you examined the association between the amounts of *Fat* and *Calories* in fast-food hamburgers. Here are the data:

Fat (g)	19	31	34	35	39	39	43
Calories	410	580	590	570	640	680	660

a) Create a scatterplot of *Calories* vs. *Fat*.
b) Interpret the value of R^2 in this context.
c) Write the equation of the line of regression.

d) Use the residuals plot to explain whether your linear model is appropriate.

e) Explain the meaning of the y-intercept of the line.

f) Explain the meaning of the slope of the line.

g) A new burger containing 28 grams of fat is introduced. According to this model, its residual for calories is +33. How many calories does the burger have?

58. **Chicken** Chicken sandwiches are often advertised as a healthier alternative to beef because many are lower in fat. Data from tests on 15 different sandwiches randomly selected from the website http://fast-food-nutrition.findthebest.com/d/a/Chicken-Sandwich produced the *Calories* vs. *Fat* scatterplot and the regression analysis below.

Dependent variable is Calories
R-squared = 93.2%
s = 63.73

Variable	Coefficient	SE
Intercept	12.7234	43.910
Fat	21.2171	1.5944

a) Do you think a linear model is appropriate in this situation?

b) Describe the strength of this association.

c) Write the equation of the regression line to estimate calories from the fat content.

d) Explain the meaning of the slope.

e) Explain the meaning of the y-intercept.

f) What does it mean if a certain sandwich has a negative residual?

59. **A second helping of burgers** In Exercise 57 you created a model that can estimate the number of *Calories* in a burger when the *Fat* content is known.

a) Explain why you cannot use that model to estimate the fat content of a burger with 600 calories.

b) Make that estimate using an appropriate model.

60. **A second helping of chicken** In Exercise 58 you created a model to estimate the number of *Calories* in a chicken sandwich when you know the *Fat*.

a) Explain why you cannot use that model to estimate the fat content of a 400-calorie sandwich.

b) Quiznos large mesquite sandwich stands out on the graph with an impressive 53 fat grams and 1190 calories. What effect do you think this value has on the regression equation?

61. **Body fat** It is difficult to determine a person's body fat percentage accurately without immersing him or her in water. Researchers hoping to find ways to make a good estimate immersed 20 male subjects, then measured their waists and recorded their weights.

Waist (in.)	Weight (lb)	Body Fat (%)	Waist (in.)	Weight (lb)	Body Fat (%)
32	175	6	33	188	10
36	181	21	40	240	20
38	200	15	36	175	22
33	159	6	32	168	9
39	196	22	44	246	38
40	192	31	33	160	10
41	205	32	41	215	27
35	173	21	34	159	12
38	187	25	34	146	10
38	188	30	44	219	28

a) Create a model to predict *%Body Fat* from *Weight*.

b) Do you think a linear model is appropriate? Explain.

c) Interpret the slope of your model.

d) Is your model likely to make reliable estimates? Explain.

e) What is the residual for a person who weighs 190 pounds and has 21% body fat?

62. **Body fat again** Would a model that uses the person's *Waist* size be able to predict the *%Body Fat* more accurately than one that uses *Weight*? Using the data in Exercise 61, create and analyze that model.

63. **Heptathlon 2004** We discussed the women's Olympic heptathlon in Chapter 5. The table shows the results from the high jump, 800-meter run, and long jump for the 26 women who successfully completed all three events in the 2004 Olympics.

Name	Country	High Jump (m)	800-m (sec)	Long Jump (m)
Carolina Klüft	SWE	1.91	134.15	6.51
Austra Skujyté	LIT	1.76	135.92	6.30
Kelly Sotherton	GBR	1.85	132.27	6.51
Shelia Burrell	USA	1.70	135.32	6.25
Yelena Prokhorova	RUS	1.79	131.31	6.21
Sonja Kesselschlaeger	GER	1.76	135.21	6.42
Marie Collonville	FRA	1.85	133.62	6.19
Natalya Dobrynska	UKR	1.82	137.01	6.23
Margaret Simpson	GHA	1.79	137.72	6.02
Svetlana Sokolova	RUS	1.70	133.23	5.84
J. J. Shobha	IND	1.67	137.28	6.36
Claudia Tonn	GER	1.82	130.77	6.35
Naide Gomes	POR	1.85	140.05	6.10
Michelle Perry	USA	1.70	133.69	6.02
Aryiro Strataki	GRE	1.79	137.90	5.97

(continued)

Name	Country	High Jump (m)	800-m (sec)	Long Jump (m)
Karin Ruckstuhl	NED	1.85	133.95	5.90
Karin Ertl	GER	1.73	138.68	6.03
Kylie Wheeler	AUS	1.79	137.65	6.36
Janice Josephs	RSA	1.70	138.47	6.21
Tiffany Lott Hogan	USA	1.67	145.10	6.15
Magdalena Szczepanska	POL	1.76	133.08	5.98
Irina Naumenko	KAZ	1.79	134.57	6.16
Yuliya Akulenko	UKR	1.73	142.58	6.02
Soma Biswas	IND	1.70	132.27	5.92
Marsha Mark-Baird	TRI	1.70	141.21	6.22
Michaela Hejnova	CZE	1.70	145.68	5.70

Let's examine the association among these events. Perform a regression to predict high-jump performance from the 800-meter results.

a) What is the regression equation? What does the slope mean?

b) What percent of the variability in high jumps can be accounted for by differences in 800-m times?

c) Do good high jumpers tend to be fast runners? (Be careful—low times are good for running events and high distances are good for jumps.)

d) What does the residuals plot reveal about the model?

e) Do you think this is a useful model? Would you use it to predict high-jump performance? (Compare the residual standard deviation to the standard deviation of the high jumps.)

T **64. Heptathlon 2004 again** We saw the data for the women's 2004 Olympic heptathlon in Exercise 63. Are the two jumping events associated? Perform a regression of the long-jump results on the high-jump results.

a) What is the regression equation? What does the slope mean?

b) What percentage of the variability in long jumps can be accounted for by high-jump performances?

c) Do good high jumpers tend to be good long jumpers?

d) What does the residuals plot reveal about the model?

e) Do you think this is a useful model? Would you use it to predict long-jump performance? (Compare the residual standard deviation to the standard deviation of the long jumps.)

65. Least squares I Consider the four points (10, 10), (20, 50), (40, 20), and (50,80). The least squares line is $\hat{y} = 7.0 + 1.1x$. Explain what "least squares" means, using these data as a specific example.

66. Least squares II Consider the four points (200,1950), (400,1650), (600,1800), and (800,1600). The least squares line is $\hat{y} = 1975 - 0.45x$. Explain what "least squares" means, using these data as a specific example.

 Just Checking ANSWERS

1. You should expect the price to be 0.77 standard deviations above the mean.

2. You should expect the size to be $2(0.77) = 1.54$ standard deviations below the mean.

3. The home is 1.5 standard deviations above the mean in size no matter how size is measured.

4. An increase in home size of 1000 square feet is associated with an increase in price of $94,454, on average.

5. Units are thousands of dollars per thousand square feet.

6. About $188,908, on average

7. No. Even if it were positive, no one wants a house with 0 square feet!

8. Negative; that indicates it's priced lower than a typical home of its size.

9. $280,245

10. $19,755 (positive!)

11. Differences in the size of houses account for about 59.5% of the variation in the house prices.

12. It's positive. The correlation and the slope have the same sign.

13. R^2 would not change, but the slope would. Slope depends on the units used but correlation doesn't.

14. No, the standard deviation of the residuals is 53.79 thousand dollars. We shouldn't be surprised by any residual smaller than 2 standard deviations, and a residual of $50,000 is less than 1 standard deviation.

Regression is used every day throughout the world to predict customer loyalty, numbers of admissions at hospitals, sales of automobiles, and many other things. Because regression is so widely used, it's also widely abused and misinterpreted. This chapter presents examples of regressions in which things are not quite as simple as they may have seemed at first and shows how you can still use regression to discover what the data have to say.

Getting the "Bends": When the Residuals Aren't Straight

No regression analysis is complete without a display of the residuals to check that the linear model is reasonable. Because the residuals are what is "left over" after the model describes the relationship, they often reveal subtleties that were not clear from a plot of the original data. Sometimes these are additional details that help confirm or refine our understanding. Sometimes they reveal violations of the regression conditions that require our attention.

> **Straight Enough?**
>
> We can't *know* whether the Linearity Assumption is true, but we can see if it's *plausible* by checking the Straight Enough Condition.

The fundamental assumption in working with a linear model is that the relationship you are modeling is, in fact, linear. That sounds obvious, but when you fit a regression, you can't take it for granted. Often it's hard to spot non-linearity in the scatterplot before you fit the regression model. Sometimes you can't see a bend in the relationship until you plot the residuals.

Jessica Meir[1] and Paul Ponganis studied emperor penguins at the Scripps Institution of Oceanography's Center for Marine Biotechnology and Biomedicine at the University of California at San Diego. Says Jessica:

> *Emperor penguins are the most accomplished divers among birds, making routine dives of 5–12 minutes, with the longest recorded dive over 27 minutes. These*

---

[1]Since completing this research, Jessica has gone on to loftier things: she's now a NASA astronaut!

birds can also dive to depths of over 500 meters! Since air-breathing animals like penguins must hold their breath while submerged, the duration of any given dive depends on how much oxygen is in the bird's body at the beginning of the dive, how quickly that oxygen gets used, and the lowest level of oxygen the bird can tolerate. The rate of oxygen depletion is primarily determined by the penguin's heart rate. Consequently, studies of heart rates during dives can help us understand how these animals regulate their oxygen consumption in order to make such impressive dives.

The researchers equip emperor penguins with devices that record their heart rates during dives. Here's a scatterplot of the *Dive Heart Rate* (beats per minute) and the *Duration* (minutes) of dives by these high-tech penguins.

Figure 8.1

The scatterplot of *Dive Heart Rate* in beats per minute (bpm) vs. *Duration* (minutes) shows a strong, roughly linear, negative association.

The scatterplot looks fairly linear with a moderately strong negative association ($R^2 = 71.5\%$). The linear regression equation

$$\widehat{DiveHeartRate} = 96.9 - 5.47\,Duration$$

says that for longer dives, the average *Dive Heart Rate* is lower by about 5.47 beats per dive minute, starting from a value of 96.9 beats per minute.

The scatterplot of the *Residuals* against *Duration* is revealing. The Linearity Assumption says we should not see a pattern, but instead there's a bend, starting high on the left, dropping down in the middle of the plot, and rising again at the right. Graphs of residuals often reveal patterns such as this that were easy to miss in the original scatterplot.

Figure 8.2

Plotting the *Residuals* against *Duration* reveals a bend. It was also in the original scatterplot, but here it's easier to see.

Now if you look back at the original scatterplot, you may see that the scatter of points isn't really straight. There's a slight bend to that plot, but that bend is much easier to see in the residuals. Even though it means rechecking the Straight Enough Condition *after* you find the regression, it's always a good idea to check your residuals plot for bends that you might have overlooked in the original scatterplot.

Sifting Residuals for Groups

Figure 8.3

A histogram of the regression residuals shows small modes both above and below the central large mode. These may be worth a second look.

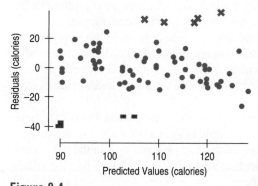

Figure 8.4

A scatterplot of the *Residuals* vs. *Predicted Values* for the cereal regression. The green "x" points are cereals whose calorie content is higher than the linear model predicts. The red "−" points show cereals with fewer calories than the model predicts. Is there something special about these cereals?

In the Step-By-Step analysis in Chapter 7 to predict *Calories* from *Sugar* content in breakfast cereals, we examined a scatterplot of the residuals. Our first impression was that it had no particular structure—a conclusion that supported using the regression model. But let's look again.

Here's a histogram of the residuals. How would you describe its shape? It looks like there might be a small cluster on each side of the central body of the data. One group of cereals seems to stand out as having large negative residuals, with fewer calories than we might have predicted, and another stands out with large positive residuals (more calories). Whenever we suspect multiple modes, we ask whether they are somehow different.

Below the histogram is the residual plot, with the points in those clusters marked. Now we can see that those two groups do stand away from the central pattern in the scatterplot. The high-residual cereals are Just Right Fruit & Nut; Muesli Raisins, Dates & Almonds; Peaches & Pecans; Mueslix Crispy Blend; and Nutri-Grain Almond Raisin. Do these cereals seem to have something in common? They all present themselves as "healthy." This might be surprising, but in fact, "healthy" cereals often contain more fat, and therefore more calories, than we might expect from looking at their sugar content alone.

The low-residual cereals are Puffed Rice, Puffed Wheat, three bran cereals, and Golden Crisps. You might not have grouped these cereals together before. What they have in common is a low calorie count *relative to their sugar content*—even though their sugar contents are quite different.

These observations may not lead us to question the overall linear model, but they do help us understand that other factors may be part of the story. An examination of residuals often leads us to discover groups of observations that are different from the rest.

When we discover that there is more than one group in a regression, we may decide to analyze the groups separately, using a different model for each group. Or we can stick with the original model and simply note that there are groups that are a little different. Either way, the model will be wrong, but useful, so it will improve our understanding of the data.

Subsets

Birds of a Feather?

Here's an important unstated condition for fitting models: All the data must come from the same population.

Cereal manufacturers aim cereals at different segments of the market. Supermarkets and cereal manufacturers try to attract different customers by placing different types of cereals on certain shelves. Cereals for kids tend to be on the "kid's shelf," at their eye level. Toddlers wouldn't be likely to grab a box from this shelf and beg, "Mom, can we please get this All-Bran with Extra Fiber?"

Should we take this extra information into account in our analysis? At the top of the next page you'll see a scatterplot of *Calories* and *Sugar,* colored according to the shelf on which the cereals were found and with a separate regression line fit for each. The top shelf's cereals are clearly different. We might want to report two regressions, one for the top shelf and one for the bottom two shelves.[2]

---

[2]More complex models can take into account both sugar content and shelf information. This kind of *multiple regression* model is a natural extension of the model we're using here. You can learn about such models in Chapter 28 on the DVD.

Figure 8.5

Calories and *Sugar* colored according to the shelf on which the cereal was found in a supermarket, with regression lines fit for each shelf individually. Do these data appear homogeneous? That is, do all the cereals seem to be from the same population of cereals? Or are there different kinds of cereals that we might want to consider separately?

Extrapolation: Reaching Beyond the Data

Case Study: **Predicting Manatee Kills.** Can we use regression to predict the number of manatees that will be killed by power boats this year?

"Prediction is difficult, especially about the future."

—*Niels Bohr, Danish physicist*

Linear models give a predicted value for each case in the data. Put a new x-value into the equation, and it gives a predicted value, $\hat{y}$, to go with it. But when the new x-value lies far from the data we used to build the regression, how trustworthy is the prediction?

The simple answer is that the farther the new x-value is from $\bar{x}$, the less trust we should place in the predicted value. Once we venture into new x territory, such a prediction is called an **extrapolation**. Extrapolations are dubious because they require the very questionable assumption that nothing about the relationship between x and y changes even at extreme values of x and beyond.

Extrapolations can get us into deep trouble. When the x-variable is *Time*, extrapolation becomes an attempt to peer into the future. People have always wanted to see into the future, and it doesn't take a crystal ball to foresee that they always will. In the past, seers, oracles, and wizards were called on to predict the future. Today mediums, fortune-tellers, and Tarot card readers still find many customers.

Those with a more scientific outlook may use a linear model as their digital crystal ball. Linear models are based on the x-values of the data at hand and cannot be trusted beyond that span. Some physical phenomena do exhibit a kind of "inertia" that allows us to guess that current systematic behavior will continue, but regularity can't be counted on in phenomena such as stock prices, sales figures, hurricane tracks, or public opinion.

Extrapolating from current trends is so tempting that even professional forecasters make this mistake, and sometimes the errors are striking. In the mid-1970s, oil prices surged and long lines at gas stations were common. In 1970, oil cost about $17 a barrel (in 2005

When the Data are Years. . .

When the Data are Years. . .

. . . we usually don't enter them as four-digit numbers. Here we used 0 for 1970, 10 for 1980, and so on. Or we may simply enter two digits, using 82 for 1982, for instance. Rescaling years like this often makes calculations easier and equations simpler. We recommend you do it, too. But be careful: If 1982 is 82, then 2004 is 104 (not 4), right?

dollars)—about what it had cost for 20 years or so. But then, within just a few years, the price surged to over $40. In 1975, a survey of 15 top econometric forecasting models (built by groups that included Nobel prize–winning economists) found predictions for 1985 oil prices that ranged from $300 to over $700 a barrel (in 2005 dollars). How close were these forecasts?

Here's a scatterplot of oil prices from 1971 to 1982 (in 2005 dollars).

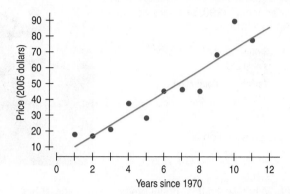

Figure 8.6

The scatterplot shows an average *increase* in the price of a barrel of oil of over $7 per year from 1971 to 1982.

Figure 8.7

This scatterplot of oil prices from 1981 to 1998 shows a fairly constant *decrease* of about $3 per barrel per year.

The regression model

$$\widehat{Price} = -3.68 + 6.90 \, Years \, since \, 1970$$

says that prices had been going up 6.90 dollars per year, or about $69 in 10 years. If you assume that they would *keep going up*, it's not hard to imagine almost any price you want.

So, how did the forecasters do? Well, in the period from 1982 to 1998 oil prices didn't exactly continue that steady increase. In fact, they went down so much that by 1998, prices (adjusted for inflation) were the lowest they'd been since before World War II.

Not one of the experts' models predicted that.

Of course, these decreases clearly couldn't continue, or oil would be free by now. The Energy Information Administration offered two *different* 20-year forecasts for oil prices after 1998, and both called for relatively modest increases in oil prices. So, how accurate have *these* forecasts been? Here's a timeplot of the EIA's predictions and the actual prices (in 2005 dollars).

Figure 8.8

Here are the EIA forecasts with the actual prices from 1981 to 2010. Neither forecast predicted the rapid increase that was soon to occur.

Oops! They seemed to have missed the sharp run-up in oil prices in the early 2000's.

Where do you think oil prices will go in the next decade? *Your* guess may be as good as anyone's!

Of course, knowing that extrapolation is dangerous doesn't stop people. The temptation to see into the future is hard to resist. So our more realistic advice is this:

If you must extrapolate into the future, at least don't believe that the prediction will come true.

For Example EXTRAPOLATION: REACHING BEYOND THE DATA

The U.S. Census Bureau (www.census.gov) reports the median age at first marriage for men and women. Here's a regression of median *Age* (at first marriage) for men against *Year* (since 1890) at every census from 1890 to 1940:

R-squared = 92.6%
s = 0.2417

Variable	Coefficient
Intercept	25.7
Year	−0.04

The regression equation is

$$\widehat{Age} = 25.7 - 0.04\ Year.$$

QUESTION: What would this model predict as the age at first marriage for men in the year 2010?

ANSWER: When *Year* counts from 0 in 1890, the year 2010 is "120." Substituting 120 for *Year*, we find that the model predicts a first marriage *Age* of 25.7 − 0.04×120=20.7 years old in 2010.

QUESTION: In the year 2010, the median *Age* at first marriage for men was 28.2 years. What's gone wrong?

The median age at which men first married fell at the rate of about a year every 25 years from 1890 tp 1940.

ANSWER: It is never safe to extrapolate beyond the data very far. The regression was fit for years up to 1940. To see how absurd a prediction from that period can be when extrapolated into the present look at a scatterplot of the median *Age* at first marriage for men for all the data from 1890 to 2010:

Now we can see why the extrapolation failed. Although the trend in *Age* at first marriage was linear and negative for the first part of the century, after World War II, it leveled off for about 30 years. Since 1980 or so, it has risen steadily. To characterize age and first marriage, we should probably treat these three time periods separately.

Outliers, Leverage, and Influence

Ralph Nader

The outcome of the 2000 U.S. presidential election was determined in Florida amid much controversy. The main race was between George W. Bush and Al Gore, but two minor candidates played a significant role. To the political right of the main party candidates was Pat Buchanan, while to the political left was Ralph Nader. Generally, Nader earned more votes than Buchanan throughout the state. We would expect counties with larger vote totals to give more votes to each candidate. The regression model relating the two candidates' vote totals by county is

$$\widehat{Buchanan} = 50.3 + 0.14\ Nader.$$

It says that, in each county, Buchanan received about 0.14 times (or 14% of) the vote Nader received, starting from a base of 50.3 votes.

Pat Buchanan

This seems like a reasonable regression, with an R^2 of almost 43%. But we've violated all three Rules of Data Analysis by going straight to the regression table without making a picture.[3] Let's have a look.

Here's a scatterplot that shows the vote for Buchanan in each county of Florida plotted against the vote for Nader. The striking **outlier** is Palm Beach County.

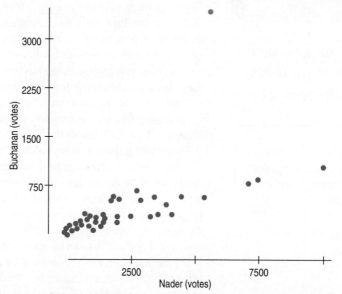

Figure 8.9

Votes received by Buchanan against votes for Nader in all Florida counties in the presidential election of 2000. The red "*x*" point is Palm Beach County, home of the "butterfly ballot."

The so-called "butterfly ballot," used only in Palm Beach County, was a source of controversy. Many claim that the format of this ballot confused voters so that some who intended to vote for the Democrat, Al Gore, punched the wrong hole next to his name and, as a result, voted for Buchanan.

The scatterplot shows a strong, positive, linear association, and one striking point. With Palm Beach removed from the regression, the R^2 jumps from 42.8% to 82.1% and the slope of the line changes to 0.1, suggesting that Buchanan received only about 10% of the vote that Nader received. With more than 82% of the variability of the Buchanan vote accounted for, the model when Palm Beach is omitted certainly fits better. Palm Beach County now stands out, not as a Buchanan stronghold, but rather as a clear violation of the model begging for explanation.

Figure 8.10

The red line shows the effect that one unusual point can have on a regression. Omitting the red point makes the blue line's slope quite different.

A S *Activity*: **Leverage.** You may be surprised to see how sensitive to a single influential point a regression line is.

---

[3] Why didn't you stop us?

"Give me a place to stand and I will move the Earth."

—*Archimedes (287–211 BCE)*

TI-*nspire*

Influential points. Try to make the regression line's slope change dramatically by dragging a point around in the scatterplot.

Figure 8.11

Bozo's extraordinarily large shoes give his data point high leverage in the regression. Wherever Bozo's IQ falls, the regression line will follow.

Figure 8.12

If Bozo's IQ were low, the regression slope would change from positive to negative. A single influential point can change a regression model drastically.

One of the great values of models is that, by establishing an idealized behavior, they help us to see when and how data values are unusual. In regression, a point can stand out in two different ways. First, a data value can have a large residual, as Palm Beach County does in this example. Because they seem to be different from the other cases, points whose residuals are large always deserve special attention.

A data point can also be unusual if its *x*-value is far from the mean of the *x*-values. Such a point is said to have high **leverage**. The physical image of a lever is exactly right. We know the line must pass through $(\bar{x}, \bar{y})$, so you can picture that point as the fulcrum of the lever. Just as sitting farther from the hinge on a see-saw gives you more leverage to pull it your way, points with values far from $\bar{x}$ pull more strongly on the regression line.

A point with high leverage has the potential to change the regression line. But it doesn't always use that potential. If the point lines up with the pattern of the other points, then including it doesn't change our estimate of the line. By sitting so far from $\bar{x}$, though, it may strengthen the relationship, inflating the correlation and R^2. How can you tell if a high-leverage point actually changes the model? Just fit the linear model twice, both with and without the point in question. We say that a point is **influential** if omitting it from the analysis gives a very different model.[4]

Influence depends on both leverage and residual; a case with high leverage whose *y*-value sits right on the line fit to the rest of the data is not influential. Removing that case won't change the slope, even if it does affect R^2. A case with modest leverage but a very large residual (such as Palm Beach County) can be influential. Of course, if a point has enough leverage, it can pull the line right to it. Then it's highly influential, but its residual is small. The only way to be sure is to fit both regressions.

Unusual points in a regression often tell us more about the data and the model than any other points. We face a challenge: The best way to identify unusual points is against the background of a model, but good models are free of the influence of unusual points. Don't give in to the temptation to simply delete points that don't fit the line. You can take points out and discuss what the model looks like with and without them, but arbitrarily deleting points can give a false sense of how well the model fits the data. Your goal should be understanding the data, not making R^2 as big as you can.

In 2000, George W. Bush won Florida (and thus the presidency) by only a few hundred votes, so Palm Beach County's residual is big enough to be meaningful. It's the rare unusual point that determines a presidency, but all are worth examining and trying to understand.

A point with so much influence that it pulls the regression line close to it can make its residual deceptively small. Influential points like that can have a shocking effect on the regression. Here's a plot of *IQ* against *Shoe Size*, again from the fanciful study of intelligence and foot size in comedians we saw in Chapter 6. With Bozo there, $R^2 = 24.8\%$. But almost all of the variance accounted for is due to Bozo. Without him, there is little correlation between *Shoe Size* and *IQ*: the R^2 value is only 0.7%—a very weak linear relationship (as one might expect!).

What would have happened if Bozo hadn't shown his comic genius on IQ tests? Suppose his measured *IQ* had been only 50. The slope of the line would then drop from 0.96 IQ points/shoe size to −0.69 IQ points/shoe size. No matter where Bozo's *IQ* is, the line tends to follow it because his *Shoe Size*, being so far from the mean *Shoe Size*, makes this a high-leverage point.

Even though this example is far fetched, similar situations occur all the time in real life. For example, a regression of sales against floor space for hardware stores that looked primarily at small-town businesses could be dominated in a similar way if The Home Depot were included.

Warning Influential points can hide in plots of residuals. Points with high leverage pull the line close to them, so they often have small residuals. You'll see influential points more easily in scatterplots of the original data or by finding a regression model with and without the points.

---

[4]Some textbooks use the term *influential point* for any observation that influences the slope, intercept, or R^2. We'll reserve the term for points that influence the slope.

Just Checking

Each of these scatterplots shows an unusual point. For each, tell whether the point is a high-leverage point, would have a large residual, or is influential.

Lurking Variables and Causation

> **Causing Change?**
>
> One common way to interpret a regression slope is to say that "a change of 1 unit in x results in a change of b_1 units in Y." This way of saying things encourages causal thinking. Beware.

In Chapter 6, we tried to make it clear that no matter how strong the correlation is between two variables, there's no simple way to show that one variable causes the other. Putting a regression line through a cloud of points just increases the temptation to think and to say that the x-variable *causes* the y-variable. Just to make sure, let's repeat the point again: No matter how strong the association, no matter how large the R^2 value, no matter how straight the line, there is no way to conclude from a regression alone that one variable *causes* the other. There's always the possibility that some third variable is driving both of the variables you have observed. With observational data, as opposed to data from a designed experiment, there is no way to be sure that a **lurking variable** is not the cause of any apparent association.

Here's an example: The scatterplot shows the *Life Expectancy* (average of men and women, in years) for each of 41 countries of the world, plotted against the square root of the number of *Doctors* per person in the country. (The square root is here to make the relationship satisfy the Straight Enough Condition, as we saw back in Chapter 6.)

The strong positive association ($R^2 = 62.4\%$) seems to confirm our expectation that more *Doctors* per person improves healthcare, leading to longer lifetimes and a greater *Life Expectancy*. The strength of the association would *seem* to argue that we should send more doctors to developing countries to increase life expectancy.

That conclusion is about the consequences of a change. Would sending more doctors increase life expectancy? Specifically, do doctors *cause* greater life expectancy? Perhaps, but these are observed data, so there may be another explanation for the association.

On the next page the scatterplot's x-variable is the square root of the number of *Televisions* per person in each country. The positive association in this scatterplot is even *stronger* than the association in the previous plot ($R^2 = 72.3\%$). We can fit the linear model, and quite possibly use the number of TVs as a way to predict life expectancy. Should we conclude that increasing the number of TVs actually extends lifetimes? If so, we should send TVs instead of doctors to developing countries. Not only is the correlation with life expectancy higher, but TVs are much cheaper than doctors.

What's wrong with this reasoning? Maybe we were a bit hasty earlier when we concluded that doctors *cause* longer lives. Maybe there's a lurking variable here. Countries with higher standards of living have both longer life expectancies *and* more doctors (and more TVs). Could higher living standards cause changes in the other variables? If so, then improving living standards might be expected to prolong lives, increase the number of doctors, and increase the number of TVs.

Figure 8.13

The relationship between *Life Expectancy (years)* and availability of Doctors (measured as $\sqrt{doctors/person}$) for countries of the world is strong, positive, and linear.

Figure 8.14

To increase life expectancy, don't send doctors, send TVs; they're cheaper and more fun. Or maybe that's not the right interpretation of this scatterplot of *Life Expectancy* against availability of TVs (as $\sqrt{TVs/person}$).

From this example, you can see how easy it is to fall into the trap of mistakenly inferring causality from a regression. For all we know, doctors (or TVs!) *do* increase life expectancy. But we can't tell that from data like these, no matter how much we'd like to. Resist the temptation to conclude that *x* causes *y* from a regression, no matter how obvious that conclusion seems to you.

Predict Changes? No!

Not only is it incorrect and dangerous to interpret association as causation, but when using regression there's a more subtle danger. Never interpret a regression slope coefficient as predicting how *y* is likely to change if its *x* value in the data were changed. Here's an example: In Chapter 7, we found a regression model relating calories in breakfast cereals to their sugar content as

$$\widehat{Calories} = 89.5 + 2.50\ Sugar.$$

It might be tempting to interpret this slope as implying that adding 1 gram of sugar is expected to lead to an increase of 2.50 calories. We can't say that. The correct interpretation of the slope is that cereals having a gram more sugar in them tend to have about 2.50 more calories per serving. That is, the regression model describes how the cereals differ, but does not tell us how they might change if circumstances were different. As a matter of fact, if there were no other differences in the cereals, simply adding a gram of sugar add 3.90 calories—that's the calorie content of a gram of sugar.

To believe that *y* would change in a certain way if we were to change *x* is to believe the relationship is causal. We can't go there. Regression models describe the data as they are, not as they might be under other circumstances.

Working with Summary Values

Figure 8.15

Weight (lb) against *Height* (in.) for a sample of men. There's a strong, positive, linear association.

Scatterplots of statistics summarized over groups tend to show less variability than we would see if we measured the same variable on individuals. This is because the summary statistics themselves vary less than the data on the individuals do—a fact we will make more specific in coming chapters.

In Chapter 6 we looked at the heights and weights of individual students. There we saw a correlation of 0.644, so R^2 is 41.5%.

Suppose, instead of data on individuals, we knew only the mean weight for each height value. The scatterplot of mean weight by height would show less scatter. And the R^2 would increase to 80.1%.

Figure 8.16

***Mean Weight* (lb) shows a stronger linear association with *Height* than do the weights of individuals.** Means vary less than individual values.

Scatterplots of summary statistics show less scatter than the baseline data on individuals and can give a false impression of how well a line summarizes the data. There's no simple correction for this phenomenon. Once we're given summary data, there's no simple way to get the original values back.

In the life expectancy and TVs example, we have no good measure of exposure to doctors or to TV on an individual basis. But if we did, we should expect the scatterplot to show more variability and the corresponding R^2 to be smaller. The bottom line is that you should be a bit suspicious of conclusions based on regressions of summary data. They may look better than they really are.

For Example USING SEVERAL OF THESE METHODS TOGETHER

Motorcycles designed to run off-road, often known as dirt bikes, are specialized vehicles.

We have data on 104 dirt bikes available for sale in 2005. Some cost as little as $3000, while others are substantially more expensive. Let's investigate how the size and type of engine contribute to the cost of a dirt bike. As always, we start with a scatterplot.

Here's a scatterplot of the manufacturer's suggested retail price (*MSRP*) in dollars against the engine *Displacement*, along with a regression analysis:

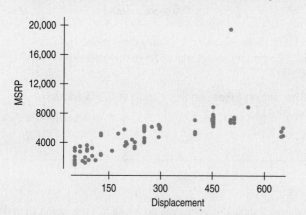

Dependent variable is: MSRP
R-squared = 49.9% s = 1737

Variable	Coefficient
Intercept	2273.67
Displacement	10.0297

QUESTION: What do you see in the scatterplot?

ANSWER: *There is a strong positive association between the engine displacement of dirt bikes and the manufacturer's suggested retail price. One of the dirt bikes is an outlier; its price is more than double that of any other bike.*

The outlier is the Husqvarna TE 510 Centennial. Most of its components are handmade exclusively for this model, including extensive use of carbon fiber throughout. That may explain its $19,500 price tag! Clearly, the TE 510 is not like the other bikes. We'll set it aside for now and look at the data for the remaining dirt bikes.

QUESTION: What effect will removing this outlier have on the regression? Describe how the slope, R^2, and s_e will change.

ANSWER: *The TE 510 was an influential point, tilting the regression line upward. With that point removed, the regression slope will get smaller. With that dirt bike omitted, the pattern becomes more consistent, so the value of R^2 should get larger and the standard deviation of the residuals, s_e, should get smaller.*

With the outlier omitted, here's the new regression and a scatterplot of the residuals:

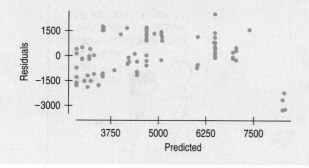

Dependent variable is: MSRP
R-squared = 61.3% s = 1237

Variable	Coefficient
Intercept	2411.02
Displacement	9.05450

(continued)

QUESTION: What do you see in the residuals plot?

ANSWER: *The points at the far right don't fit well with the other dirt bikes. Overall, there appears to be a bend in the relationship, so a linear model may not be appropriate.*

Let's try a re-expression. Here's a scatterplot showing *MSRP* against the cube root of *Displacement* to make the relationship closer to straight. (Since displacement is measured in cubic centimeters, its cube root has the simple units of centimeters.) In addition, we've colored the plot according to the cooling method used in the bike's engine: liquid or air. Each group is shown with its own regression line, as we did for the cereals on different shelves.

QUESTION: What does this plot say about dirt bikes?

ANSWER: *There appears to be a positive, linear relationship between MSRP and the cube root of Displacement. In general, the larger the engine a bike has, the higher the suggested price. Liquid-cooled dirt bikes, however, typically cost more than air-cooled bikes with comparable displacement. A few liquid-cooled bikes appear to be much less expensive than we might expect, given their engine displacements.*

[Jiang Lu, Joseph B. Kadane, and Peter Boatwright, "The Dirt on Bikes: An Illustration of CART Models for Brand Differentiation," provides data on 2005-model bikes.]

WHAT CAN GO WRONG?

This entire chapter has held warnings about things that can go wrong in a regression analysis. So let's just recap. When you make a linear model:

- **Make sure the relationship is straight.** Check the Straight Enough Condition. Always examine the residuals for evidence that the Linearity Assumption has failed. It's often easier to see deviations from a straight line in the residuals plot than in the scatterplot of the original data. Pay special attention to the most extreme residuals because they may have something to add to the story told by the linear model.

- **Be on guard for different groups in your regression.** Check for evidence that the data consist of separate subsets. If you find subsets that behave differently, consider fitting a different linear model to each subset.

- **Beware of extrapolating.** Beware of extrapolation beyond the *x*-values that were used to fit the model. Although it's common to use linear models to extrapolate, the practice is dangerous.

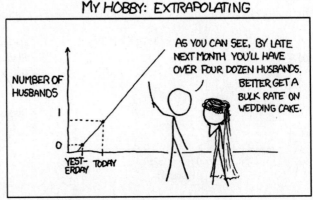

- **Beware especially of extrapolating into the future!** Be especially cautious about extrapolating into the future with linear models. To predict the future, you must assume that future changes will continue at the same rate you've observed in the past. Predicting the future is particularly tempting and particularly dangerous.

- **Look for unusual points.** Unusual points always deserve attention and may well reveal more about your data than the rest of the points combined. Always look for them and try to understand why they stand apart. A scatterplot of the data is a good way to see high-leverage and influential points. A scatterplot of the residuals against the predicted values is a good tool for finding points with large residuals.

- **Beware of high-leverage points and especially of those that are influential.** Influential points can alter the regression model a great deal. The resulting model may say more about one or two points than about the overall relationship.

- **Consider comparing two regressions.** To see the impact of outliers on a regression, it's often wise to run two regressions, one with and one without the extraordinary points, and then to discuss the differences.

- **Treat unusual points honestly.** If you remove enough carefully selected points, you can always get a regression with a high R^2 eventually. But it won't give you much understanding. Some variables are not related in a way that's simple enough for a linear model to fit very well. When that happens, report the failure and stop.

- **Beware of lurking variables.** Think about lurking variables before interpreting a linear model. It's particularly tempting to explain a strong regression by thinking that the x-variable *causes* the y-variable. A linear model alone can never demonstrate such causation, in part because it cannot eliminate the chance that a lurking variable has caused the variation in both x and y.

- **Watch out when dealing with data that are summaries.** Be cautious in working with data values that are themselves summaries, such as means or medians. Such statistics are less variable than the data on which they are based, so they tend to inflate the impression of the strength of a relationship.

- **Don't even *imply* causation.** By now you know (We hope!) that the presence of an association between two variables d`oesn't mean one causes the other, but it's easy to imply causation when interpreting the slope of the regression model. If an analysis shows that automobile fuel efficiency tends to drop about 0.8 miles per gallon for every extra 100 pounds cars weigh, that does not mean your car will get 0.8 mpg less if you give your sister a ride. Be careful that your interpretation of slope does not predict what will happen to y if x changes.

What Have We Learned?

We've learned to be alert to the many ways in which a data set may be unsuitable for a regression analysis.

- Watch out for more than one group hiding in your regression analysis. If you find subsets of the data that behave differently, consider fitting a different regression model to each subset.

- The Straight Enough Condition says that the relationship should be reasonably straight to fit a regression. Somewhat paradoxically, sometimes it's easier to see that the relationship is not straight *after* fitting the regression by examining the residuals. The same is true of outliers.

- The Outlier Condition actually means two things: Points with large residuals or high leverage (especially both) can influence the regression model significantly. It's a good idea to perform the regression analysis with and without such points to see their impact.

And we've learned that even a good regression doesn't mean we should believe that the model says more than it really does.

■ Extrapolation far from $\bar{x}$ can lead to silly and useless predictions.

■ Even an R^2 near 100% doesn't indicate that x causes y (or the other way around). Watch out for lurking variables that may affect both x and y.

■ Be careful when you interpret regressions based on *summaries* of the data sets. These regressions tend to look stronger than the regression based on all the individual data.

Terms

Extrapolation
Although linear models provide an easy way to predict values of y for a given value of x, it is unsafe to predict for values of x far from the ones used to find the linear model equation. Such extrapolation may pretend to see into the future, but the predictions should not be trusted. (p. 212)

Outlier
Any data point that stands away from the others can be called an outlier. In regression, outliers can be extraordinary in two ways: by having a large residual or by having high leverage. (p. 215)

Leverage
Data points whose x-values are far from the mean of x are said to exert leverage on a linear model. High-leverage points pull the line close to them, and so they can have a large effect on the line, sometimes very strongly influencing the slope and intercept. With high enough leverage, their residuals can be deceptively small. (p. 216)

Influential point
If omitting a point from the data results in a regression model with a very different slope, then that point is called an influential point. (p. 216)

Lurking variable
A variable that is not explicitly part of a model but affects the way the variables in the model appear to be related is called a lurking variable. Because we can never be certain that observational data are not hiding a lurking variable that influences both x and y, it is never safe to conclude that a linear model demonstrates a causal relationship, no matter how strong the linear association. (p. 217)

On the Computer REGRESSION DIAGNOSIS

Most statistics technology offers simple ways to check whether your data satisfy the conditions for regression. We have already seen that these programs can make a simple scatterplot. They can also help us check the conditions by plotting residuals.

Exercises

1. Marriage age 2010 Is there evidence that the age at which women get married has changed over the past 100 years? The scatterplot shows the trend in age at first marriage for American women (www.census.gov).

a) Is there a clear pattern? Describe the trend.
b) Is the association strong?
c) Is the correlation high? Explain.
d) Is a linear model appropriate? Explain.

2. Smoking 2009 The Centers for Disease Control and Prevention track cigarette smoking in the United States. How has the percentage of people who smoke changed since the danger became clear during the last half of the 20th century? The scatterplot shows percentages of smokers among men 18–24 years of age, as estimated by surveys, from 1965 through 2009 (http://www.cdc.gov/nchs/).

a) Is there a clear pattern? Describe the trend.
b) Is the association strong?
c) Is a linear model appropriate? Explain.

3. Human Development Index The United Nations Development Programme (UNDP) uses the Human Development Index (HDI) in an attempt to summarize in one number the progress in health, education, and economics of a country. In 2010, the HDI was as high as 0.938 for Norway and as low as 0.14 for Zimbabwe. The gross domestic product per capita (GDPPC), by contrast, is often used to summarize the *overall* economic strength of a country. Is the HDI related to the GDPPC? Here is a scatterplot of *HDI* against *GDPPC*.

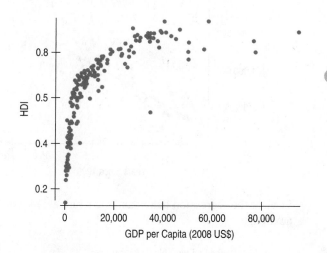

a) Explain why fitting a linear model to these data might be misleading.
b) If you fit a linear model to the data, what do you think a scatterplot of residuals versus predicted *HDI* will look like?

4. HDI revisited The United Nations Development Programme (UNDP) uses the Human Development Index

(HDI) in an attempt to summarize in one number the progress in health, education, and economics of a country. The number of phone subscribers per 100 people is positively associated with economic progress in a country. Can the number of phone subscribers be used to predict the HDI? Here is a scatterplot of HDI against phone subscribers:

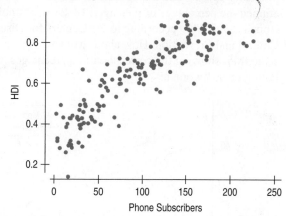

a) Explain why fitting a linear model to these data might be misleading.
b) If you fit a linear model to the data, what do you think a scatterplot of residuals vs. predicted *HDI* will look like?

5. Good model? In justifying his choice of a model, a student wrote, "I know this is the correct model because $R^2 = 99.4\%$."

a) Is this reasoning correct? Explain.
b) Does this model allow the student to make accurate predictions? Explain.

6. Bad model? A student who has created a linear model is disappointed to find that her R^2 value is a very low 13%.

a) Does this mean that a linear model is not appropriate? Explain.
b) Does this model allow the student to make accurate predictions? Explain.

7. Movie dramas Here's a scatterplot of the production budgets (in millions of dollars) vs. the running time (in minutes) for major release movies in 2005. Dramas are plotted as red x's and all other genres are plotted as blue dots. (The re-make of *King Kong* is plotted as a black "-". At the time it was the most expensive movie ever made, and not typical of any genre.) A separate least squares regression line has been fitted to each group. For the following questions, just examine the plot:

a) What are the units for the slopes of these lines?
b) In what way are dramas and other movies similar with respect to this relationship?
c) In what way are dramas different from other genres of movies with respect to this relationship?

T 8. **Smoking 2009, women and men** In Exercise 2, we examined the percentage of men aged 18–24 who smoked from 1965 to 2009 according to the Centers for Disease Control and Prevention. How about women? Here's a scatterplot showing the corresponding percentages for both men and women:

a) In what ways are the trends in smoking behavior similar for men and women?
b) How do the smoking rates for women differ from those for men?
c) Viewed alone, the trend for men may have seemed to violate the Linearity Condition. How about the trend for women? Does the consistency of the two patterns encourage you to think that a linear model for the trend in men might be appropriate? (Note: there is no correct answer to this question; it is raised for you to think about.)

9. **Abalone** Abalones are edible sea snails that include over 100 species. A researcher is working with a model that uses the number of rings in an Abalone's shell to predict its age. He finds an observation that he believes has been miscalculated. After deleting this outlier, he redoes the calculation. Does it appear that this outlier was exerting very much influence?

Before:
Dependent variable is Age
R-squared = 67.5%

Variable	Coefficient
Intercept	1.736
Rings	0.45

After:
Dependent variable is Age
R-squared = 83.9%

Variable	Coefficient
Intercept	1.56
Rings	1.13

10. **Abalone again** The researcher in Exercise 9 is content with the second regression. But he has found a number of shells that have large residuals and is considering removing all of them. Is this good practice?

11. **Skinned knees** There is a strong correlation between the temperature and the number of skinned knees on playgrounds. Does this tell us that warm weather causes children to trip?

12. **Cell phones and life expectancy** The correlation between cell phone usage and life expectancy is very high. Should we buy cell phones to help people live longer?

13. **Grading** A team of Calculus teachers is analyzing student scores on a final exam compared to the midterm scores. One teacher proposes that they already have every teacher's class averages and they should just work with those averages. Explain why this is problematic.

14. **Average GPA** An athletic director proudly states that he has used the average GPAs of the university's sports teams and is predicting a high graduation rate for the teams. Why is this method unsafe?

T 15. **Oakland passengers** The scatterplot below shows the number of passengers departing from Oakland (CA) airport month by month since the start of 1997. Time is shown as years since 1990, with fractional years used to represent each month. (Thus, June of 1997 is 7.5—halfway through the 7th year after 1990.) www.oaklandairport.com

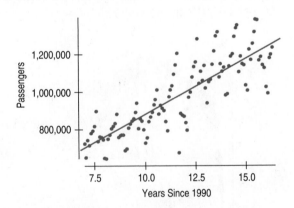

Look at the regression below and the residuals plot on the next page.

Dependent variable is: Passengers
R-squared = 71.1%   s = 104330

Variable	Coefficient
Constant	282584
Year-1990	59704.4

a) Interpret the slope and intercept of the model.
b) What does the value of R^2 say about the model?
c) Interpret s_e in this context.
d) Would you use this model to predict the numbers of passengers in 2010 (*YearsSince1990* = 20) Explain.
e) There's a point near the middle of this time span with a large negative residual. Can you explain this outlier?

16. Tracking hurricanes 2010 In a previous chapter, we saw data on the errors (in nautical miles) made by the National Hurricane Center in predicting the path of hurricanes. The scatterplot below shows the trend in the 24-hour tracking errors since 1970 (www.nhc.noaa.gov).

Dependent variable is Error
R-squared = 68.7%   s = 16.44

Variable	Coefficient
Intercept	132.301
Years - 1970	−2.00662

a) Interpret the slope and intercept of the model.
b) Interpret s_e in this context.
c) The Center would like to achieve an average tracking error of 45 nautical miles by 2015. Will they make it? Defend your response.
d) What if their goal were an average tracking error of 25 nautical miles?
e) What cautions would you state about your conclusion?

17. Unusual points Each of the four scatterplots that follow shows a cluster of points and one "stray" point. For each, answer these questions:

1) In what way is the point unusual? Does it have high leverage, a large residual, or both?
2) Do you think that point is an influential point?

3) If that point were removed, would the correlation become stronger or weaker? Explain.
4) If that point were removed, would the slope of the regression line increase or decrease? Explain.

18. More unusual points Each of the following scatterplots shows a cluster of points and one "stray" point. For each, answer these questions:

1) In what way is the point unusual? Does it have high leverage, a large residual, or both?
2) Do you think that point is an influential point?
3) If that point were removed, would the correlation become stronger or weaker? Explain.
4) If that point were removed, would the slope of the regression line increase or decrease? Explain.

19. The extra point The scatterplot shows five blue data points at the left. Not surprisingly, the correlation for these points is $r = 0$. Suppose *one* additional data point is added at one of the five positions suggested below in green. Match each point (a–e) with the correct new correlation from the list given.

1) −0.90   4) 0.05
2) −0.40   5) 0.75
3) 0.00

20. The extra point revisited The original five points in Exercise 19 produce a regression line with slope 0. Match each of the green points (a–e) with the slope of the line after that one point is added:

1) −0.45   4) 0.05
2) −0.30   5) 0.85
3) 0.00

21. What's the cause? Suppose a researcher studying health issues measures blood pressure and the percentage of body fat for several adult males and finds a strong positive association. Describe three different possible cause-and-effect relationships that might be present.

22. What's the effect? A researcher studying violent behavior in elementary school children asks the children's parents how much time each child spends playing computer games and has their teachers rate each child on the level of aggressiveness they display while playing with other children. Suppose that the researcher finds a moderately strong positive correlation. Describe three different possible cause-and-effect explanations for this relationship.

23. Reading To measure progress in reading ability, students at an elementary school take a reading comprehension test every year. Scores are measured in "grade-level" units; that is, a score of 4.2 means that a student is reading at slightly above the expected level for a fourth grader. The school principal prepares a report to parents that includes a graph showing the mean reading score for each grade. In his comments he points out that the strong positive trend demonstrates the success of the school's reading program.

a) Does this graph indicate that students are making satisfactory progress in reading? Explain.
b) What would you estimate the correlation between *Grade* and *Average Reading Level* to be?
c) If, instead of this plot showing average reading levels, the principal had produced a scatterplot of the reading levels of all the individual students, would you expect the correlation to be the same, higher, or lower? Explain.
d) Although the principal did not do a regression analysis, someone as statistically astute as you might do that. (But don't bother.) What value of the slope of that line would you view as demonstrating acceptable progress in reading comprehension? Explain.

24. Grades A college admissions officer, defending the college's use of SAT scores in the admissions process, produced the graph below. It shows the mean GPAs for last year's freshmen, grouped by SAT scores. How strong is the evidence that *SAT Score* is a good predictor of *GPA*? What concerns you about the graph, the statistical methodology or the conclusions reached?

25. Heating After keeping track of his heating expenses for several winters, a homeowner believes he can estimate the monthly cost ($) from the average daily Fahrenheit temperature with the model $\widehat{Cost} = 133 - 2.13\ Temp$. Here is the residuals plot for his data:

a) Interpret the slope of the line in this context.
b) Interpret the *y*-intercept of the line in this context.
c) During months when the temperature stays around freezing, would you expect cost predictions based on this model to be accurate, too low, or too high? Explain.
d) What heating cost does the model predict for a month that averages 10°?
e) During one of the months on which the model was based, the temperature did average 10°. What were the actual heating costs for that month?
f) Should the homeowner use this model? Explain.
g) Would this model be more successful if the temperature were expressed in degrees Celsius? Explain.

26. Speed How does the speed at which you drive affect your fuel economy? To find out, researchers drove a compact car for 200 miles at speeds ranging from 35 to 75 miles per hour. From their data, they created the model

$$\overline{Fuel\ Efficiency} = 32 - 0.1\ Speed$$ and created this residual plot:

a) Interpret the slope of this line in context.
b) Explain why it's silly to attach any meaning to the *y*-intercept.
c) When this model predicts high *Fuel Efficiency,* what can you say about those predictions?
d) What *Fuel Efficiency* does the model predict when the car is driven at 50 mph?
e) What was the actual *Fuel Efficiency* when the car was driven at 45 mph?
f) Do you think there appears to be a strong association between *Speed* and *Fuel Efficiency*? Explain.
g) Do you think this is the appropriate model for that association? Explain.

T **27. Interest rates** Here's a plot showing the federal rate on 3-month Treasury bills from 1950 to 1980, and a regression model fit to the relationship between the *Rate* (in %) and *Years since 1950* (www.gpoaccess.gov/eop/).

Dependent variable is: Rate
R-squared = 77.4%   s = 1.239

Variable	Coefficient
Intercept	0.640282
Year − 1950	0.247637

a) What is the correlation between *Rate* and *Year*?
b) Interpret the slope and intercept.
c) What does this model predict for the interest rate in the year 2000?

d) Would you expect this prediction to have been accurate? Explain.

T **28. Marriage Age 2010** The graph shows the ages of both men and women at first marriage (www.census.gov).

Clearly, the patterns for men and women are similar. But are the two lines getting closer together?

Here's a timeplot showing the *difference* in average age (men's age − women's age) at first marriage, the regression analysis, and the associated residuals plot.

Dependent variable is Age Difference
R-squared = 75.5%   s = 0.2319

Variable	Coefficient
Intercept	33.396
Year	−0.01571

a) What is the correlation between *Age Difference* and *Year*?
b) Interpret the slope of this line.

c) Predict the average age difference in 2015.
d) Describe reasons why you might not place much faith in that prediction.

29. Interest rates revisited In Exercise 27 you investigated the federal rate on 3-month Treasury bills between 1950 and 1980. The scatterplot below shows that the trend changed dramatically after 1980.

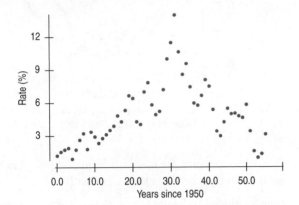

Here's a regression model for the data since 1980.

Dependent variable is: Rate
R-squared = 74.5%   s = 1.630

Variable	Coefficient
Intercept	21.0688
Year − 1950	−0.356578

a) How does this model compare to the one in Exercise 27?
b) What does this model estimate the interest rate to have been in 2000? How does this compare to the rate you predicted in Exercise 27?
c) Do you trust this newer predicted value? Explain.
d) Given these two models, what would you predict the interest rate on 3-month Treasury bills will be in 2020?

30. Ages of couples again Has the trend of decreasing difference in age at first marriage seen in Exercise 28 gotten stronger recently? The scatterplot and residual plot for the data from 1980 through 2010, along with a regression for just those years, are below.

Dependent variable is Men − Women
R-squared = 56.5%   s = 0.212

Variable	Coefficient
Intercept	53.512
Year	−0.0258

a) Is this linear model appropriate for the post-1980 data? Explain.
b) What does the slope say about marriage ages?
c) Explain why it's not reasonable to interpret the y-intercept.

31. Gestation For women, pregnancy lasts about 9 months. In other species of animals, the length of time from conception to birth varies. Is there any evidence that the gestation period is related to the animal's lifespan? The first scatterplot shows *Gestation Period* (in days) vs. *Life Expectancy* (in years) for 18 species of mammals. The highlighted point at the far right represents humans.

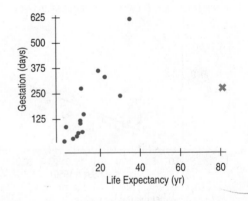

a) For these data, $r = 0.54$, not a very strong relationship. Do you think the association would be stronger or weaker if humans were removed? Explain.
b) Is there reasonable justification for removing humans from the data set? Explain.
c) Here are the scatterplot and regression analysis for the 17 nonhuman species. Comment on the strength of the association.

Dependent variable is: Gestation
R-Squared = 72.2%

Variable	Coefficient
Constant	−39.5172
LifExp	15.4980

d) Interpret the slope of the line.
e) Some species of monkeys have a life expectancy of about 20 years. Estimate the expected gestation period of one of these monkeys.

T **32. Swim the lake 2010** People swam across Lake Ontario 48 times between 1974 and 2010 (www.soloswims.com). We might be interested in whether they are getting any faster or slower. Here are the regression of the crossing *Times* (minutes) against the *Year* of the crossing and the residuals plot:

Dependent variable is Time
R-squared = 2.0%   s = 449.9

Variable	Coefficient
Intercept	−9943.083
Year	5.64227

a) What does the R^2 mean for this regression?
b) Are the swimmers getting faster or slower? Explain.
c) The outlier seen in the residuals plot is a crossing by Vicki Keith in 1987 in which she swam a round trip, north to south, and then back again. Clearly, this swim doesn't belong with the others. Would removing it change the model a lot? Explain.

33. Elephants and hippos We removed humans from the scatterplot in Exercise 31 because our species was an outlier in life expectancy. The resulting scatterplot shows two points that now may be of concern. The point in the upper right corner of this scatterplot is for elephants, and the other point at the far right is for hippos.

a) By removing one of these points, we could make the association appear to be stronger. Which point? Explain.
b) Would the slope of the line increase or decrease?
c) Should we just keep removing animals to increase the strength of the model? Explain.
d) If we remove elephants from the scatterplot, the slope of the regression line becomes 11.6 days per year. Do you think elephants were an influential point? Explain.

T **34. Another swim 2010** In Exercise 32, we saw that Vicki Keith's round-trip swim of Lake Ontario was an obvious outlier among the other one-way times. Here is the new regression after this unusual point is removed:

Dependent variable is Time
R-Squared = 6.0%   s = 326.1

Variable	Coefficient
Intercept	−13123.2
Year	7.21677

a) In this new model, the value of s_e is smaller. Explain what that means in this context.
b) Now would you be willing to say that the Lake Ontario swimmers are getting faster (or slower)?

35. Marriage age 2010 revisited Suppose you wanted to predict the trend in marriage age for American women into the early part of this century.

a) How could you use the data graphed in Exercise 1 to get a good prediction? Marriage ages in selected years starting in 1900 are listed below. Use all or part of these data to create an appropriate model for predicting the average age at which women will first marry in 2020.

1900–1950 (10-yr intervals):	21.9, 21.6, 21.2, 21.3, 21.5, 20.3
1955–2010 (5-yr intervals):	20.3, 20.3, 20.6, 20.8, 21.1, 22.0, 23.3, 23.9, 24.5, 25.1, 25.3, 26.1

b) How much faith do you place in this prediction? Explain.
c) Do you think your model would produce an accurate prediction about your grandchildren, say, 50 years from now? Explain.

36. Unwed births The National Center for Health Statistics reported the data below, showing the percentage of all births that are to unmarried women for selected years between 1980 and 1998. Create a model that describes this trend. Justify decisions you make about how to best use these data.

Year	1980	1985	1990	1991	1992	1993	1994	1995	1996	1997	1998
%	18.4	22.0	28.0	29.5	30.1	31.0	32.6	32.2	32.4	32.4	32.8

37. Life expectancy 2010 Data for 24 Western Hemisphere countries can be used to examine the association between life expectancy and the birth rate (number of births per 1000 population).

Country	Birth Rate (births/ 1000 population)	Life Expectancy
Argentina	18	77
Bahamas, The	16	70
Barbados	13	74
Belize	27	68
Bolivia	26	67
Canada	10	81
Chile	15	77
Colombia	18	74
Costa Rica	17	78
Dominican Republic	22	74
Ecuador	21	75
El Salvador	25	72
Guatemala	28	70
Honduras	26	70
Jamaica	20	74
Mexico	20	76
Nicaragua	23	72
Panama	20	77
Paraguay	28	76
Puerto Rico	12	79
United States	14	78
Uruguay	14	76
Venezuela	21	74
Virgin Islands	12	79

a) Create a scatterplot relating these two variables and describe the association.
b) Find the equation of the regression line.
c) Interpret the value of R^2.
d) Make a plot of the residuals. Are any countries unusual?
e) Is the line an appropriate model?
f) If you conclude that there is an outlier, set it aside and recompute the regression.

g) If government leaders want to increase life expectancy, should they encourage women to have fewer children? Explain.

38. Tour de France 2011 We met the Tour de France data set in Chapter 1 (in Just Checking). One hundred years ago, the fastest rider finished the course at an average speed of about 25.3 kph (around 15.8 mph). In 2005, Lance Armstrong averaged 41.65 kph (25.88 mph) for the fastest average speed in history. (Later, of course, his record was vacated when he admitted doping.)

a) Make a scatterplot of *Avg Speed* against *Year*. Describe the relationship of *Avg Speed* by *Year*, being careful to point out any unusual features in the plot.
b) Find the regression equation of *Avg Speed* on *Year*.
c) Are the conditions for regression met? Comment.

39. Inflation 2011 The Consumer Price Index (CPI) tracks the prices of consumer goods in the United States, as shown in the following table. The CPI is reported monthly, but we can look at selected values. The table shows the January CPI at five-year intervals. It indicates, for example, that the average item costing $17.90 in 1926 cost $220.22 in the year 2011.

Year	JanCPI	Year	JanCPI
1916	10.4	1966	31.8
1921	19.0	1971	39.8
1926	17.9	1976	55.6
1931	15.9	1981	87.0
1936	13.8	1986	109.6
1941	14.1	1991	134.6
1946	18.2	1996	154.4
1951	25.4	2001	175.1
1956	26.8	2006	198.3
1961	29.8	2011	220.223

a) Make a scatterplot showing the trend in consumer prices. Describe what you see.
b) Be an economic forecaster: Project increases in the cost of living over the next decade. Justify decisions you make in creating your model.

40. Second stage 2011 Look once more at the data from the Tour de France. In Exercise 38, we looked at the whole history of the race, but now let's consider just the post-World War II era.

a) Find the regression of *Avg Speed* by *Year* only for years from 1947 to the present. Are the conditions for regression met?
b) Interpret the slope.
c) In 1979, Bernard Hinault averaged 39.8 kph, while in 2005 Lance Armstrong averaged 41.65 kph. Which was the more remarkable performance and why?

41. Bridges covered There is a relationship between the age of a bridge in Tompkins County, New York, and its condition as found by inspection. Below is a graph of data. Two of bridges are considerably older than the rest. The oldest covered bridge in daily use in New York State was built in 1853 and was recently judged to have a condition of 4.523. Another bridge with a rating of 5.4 was built in 1865. Also below is the regression analysis for this relationship using all 196 bridges.

Just Checking ANSWERS

1. Not high leverage, not influential, large residual
2. High leverage, not influential, small residual
3. High leverage, influential, not large residual

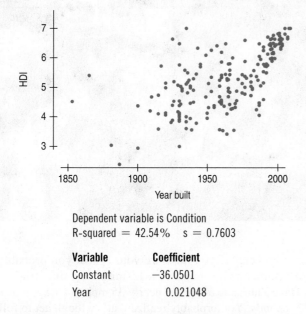

Dependent variable is Condition
R-squared = 42.54%   s = 0.7603

Variable	Coefficient
Constant	−36.0501
Year	0.021048

a) If we use this regression to predict the condition of the oldest bridge, what would its residual be?

We removed the 1853 and 1865 bridges from the analysis and here are the results:

Dependent variable is Condition
R-squared = 46.55%   s = 0.7354

Variable	Coefficient
Constant	−40.7546
Year	0.02343

b) How were these two old bridges affecting the regression?
c) The 1853 bridge was extensively restored in 1972. Does that better explain its condition?

9 Re-expressing Data: Get It Straight!

Beyond $b_0 + b_1 x$
Scan through any Physics book. Most equations have powers, reciprocals, or logs.

How fast can you go on a bicycle? If you measure your speed, you probably do it in miles per hour or kilometers per hour. In a 12-mile-long time trial in the 2005 Tour de France, Dave Zabriskie *averaged* nearly 35 mph (54.7 kph), beating Lance Armstrong by 2 seconds. You probably realize that's a tough act to follow. It's fast. You can tell that at a glance because you have no trouble thinking in terms of distance covered per time.

OK, then, if you averaged 12.5 mph (20.1 kph) for a mile *run,* would *that* be fast? Would it be fast for a 100-m dash? Even if you run the mile often, you probably have to stop and calculate. Running a mile in under 5 minutes (12 mph) is fast. A mile at 16 mph would be a world record (that's a 3-minute, 45-second mile). There's no single *natural* way to measure speed. Sometimes we use time over distance; other times we use the *reciprocal*, distance over time. Neither one is *correct*. We're just used to thinking that way in each case.

So, how does this insight help us understand data? All quantitative data come to us measured in some way, with units specified. But maybe those units aren't the best choice. It's not whether meters are better (or worse) than fathoms or leagues. What we're talking about is a different type of **re-expression**: applying a function, such as a square root, log, or reciprocal to the data. You already use some of them, even though you may not know it. For example, the Richter scale of earthquake strength (logs), the decibel scale for sound intensity (logs), the f/stop scale for camera aperture openings (squares), and the gauges of shotguns (square roots) all include simple functions of this sort.

Why bother? As with speeds, some expressions of the data may be easier to think about. And some may be much easier to analyze with statistical methods. We've seen that symmetric distributions are easier to summarize and straight scatterplots are easier to model with regressions. We often look to re-express our data if doing so makes them more suitable for our methods.

Straight to the Point

We know from common sense and from physics that heavier cars need more fuel, but exactly how does a car's weight affect its fuel efficiency? Here are the scatterplot of *Weight* (in pounds) and *Fuel Efficiency* (in miles per gallon) for 38 cars, and the residuals plot:

Figure 9.1

***Fuel Efficiency* (mpg) vs. *Weight* for 38 cars as reported by *Consumer Reports*.** The scatterplot shows a negative direction, roughly linear shape, and strong relationship. However, the residuals from a regression of *Fuel Efficiency* on *Weight* reveal a bent shape when plotted against the predicted values. Looking back at the original scatterplot, you may be able to see the bend.

Figure 9.2

Extrapolating the regression line gives an absurd answer for vehicles that weigh as little as 6000 pounds.

Hmm . . . . Even though R^2 is 81.6%, the residuals don't show the random scatter we were hoping for. The shape is clearly bent. Looking back at the first scatterplot, you can probably see the slight bending. Think about the regression line through the points. How heavy would a car have to be to have a predicted gas mileage of 0? It looks like the *Fuel Efficiency* would go negative at about 6000 pounds. A Hummer H2 weighs about 6400 pounds. The H2 is hardly known for fuel efficiency, but it does get more than the *minus* 5 mpg this regression predicts. Extrapolation is always dangerous, but it's more dangerous the more the model is wrong, because wrong models tend to do even worse the farther you get from the middle of the data.

The bend in the relationship between *Fuel Efficiency* and *Weight* is the kind of failure to satisfy the conditions for an analysis that we can repair by re-expressing the data. Instead of looking at miles per gallon, we could take the reciprocal and work with gallons per hundred miles.[1]

> **"Gallons Per Hundred Miles—What an Absurd Way to Measure Fuel Efficiency! Who Would Ever Do It That Way?"** Not all re-expressions are easy to understand, but in this case the answer is "Everyone except U.S. drivers." Most of the world measures fuel efficiency in liters per 100 kilometers (L/100 km). This is the same reciprocal form (fuel amount per distance driven) and differs from gallons per 100 miles only by a constant multiple of about 2.38. It has been suggested that most of the world says, "I've got to go 100 km; how much gas do I need?" But Americans say, "I've got 10 gallons in the tank. How far can I drive?" In much the same way, re-expressions "think" about the data differently but don't change what they mean.

Figure 9.3

The reciprocal (1/y) is measured in gallons per mile. Gallons per 100 miles gives more meaningful numbers. The reciprocal is more nearly linear against *Weight* than the original variable, but the re-expression changes the direction of the relationship. The residuals from the regression of *Fuel Consumption* (gal/100 mi) on *Weight* show less of a pattern than before.

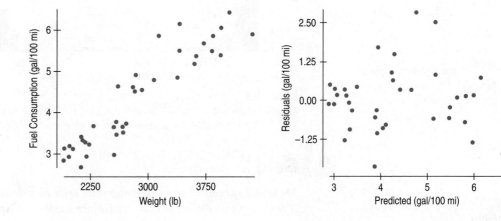

---

[1]Multiplying by 100 to get gallons per 100 miles simply makes the numbers easier to think about: You might have a good idea of how many gallons your car needs to drive 100 miles, but probably a much poorer sense of how much gas you need to go just 1 mile.

The direction of the association is positive now, since we're measuring gas consumption and heavier cars consume more gas per mile. The relationship is much straighter, as we can see from a scatterplot of the regression residuals.

This is more the kind of boring residuals plot (no direction, no particular shape, no outliers, no bends) that we hope to see, so we have reason to think that the Straight Enough Condition is now satisfied. And here's the payoff: What does the reciprocal model say about the Hummer? The regression line fit to *Fuel Consumption* vs. *Weight* predicts somewhere near 9.7 gallons for a car weighing 6400 pounds. What does this mean? It means the car is predicted to use 9.7 gallons for every 100 miles, or in other words,

$$\frac{100 \text{ miles}}{9.7 \text{ gallons}} = 10.3 \text{ mpg.}$$

That's a much more reasonable prediction and very close to the reported value of 11.0 miles per gallon (of course, *your* mileage may vary . . .).

Goals of Re-expression

We re-express data for several reasons. Each of these goals helps make the data more suitable for analysis by our methods.

Goal 1

Make the distribution of a variable (as seen in its histogram, for example) more symmetric. It's easier to summarize the center of a symmetric distribution, and for nearly symmetric distributions, we can use the mean and standard deviation. If the distribution is unimodal, then the resulting distribution may be closer to the Normal model, allowing us to use the 68-95-99.7 Rule.

Here are a histogram, quite skewed, showing the *Assets* of 77 companies selected from the Forbes 500 list (in $100,000) and the more symmetric histogram after taking logs.

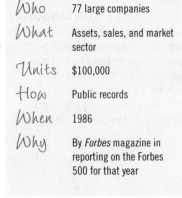

Who	77 large companies
What	Assets, sales, and market sector
Units	$100,000
How	Public records
When	1986
Why	By *Forbes* magazine in reporting on the Forbes 500 for that year

Simulation: **Re-expression in Action.** Slide the re-expression power and watch the histogram change.

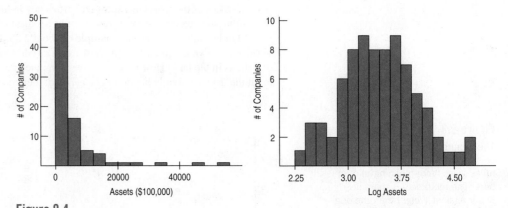

Figure 9.4

The distribution of the *Assets* of large companies is skewed to the right. Data on wealth often look like this. Taking logs makes the distribution more nearly symmetric.

Goal 2

Make the spread of several groups (as seen in side-by-side boxplots) more alike, even if their centers differ. Groups that share a common spread are easier to compare. We'll see methods later in the book that can be applied only to groups with a common standard deviation. We saw an example of re-expression for comparing groups with boxplots in Chapter 4.

Here are the *Assets* of these companies by *Market Sector:*

Figure 9.5

Assets* of large companies by *Market Sector. It's hard to compare centers or spreads, and there seem to be a number of high outliers.

Taking logs makes the individual boxplots more symmetric and gives them spreads that are more nearly equal.

Figure 9.6

After re-expressing by logs, it's much easier to compare across market sectors. The boxplots are more nearly symmetric, most have similar spreads, and the companies that seemed to be outliers before are no longer extraordinary. Two new outliers have appeared in the finance sector. They are the only companies in that sector that are not banks. Perhaps they don't belong there.

Doing this makes it easier to compare assets across market sectors. It can also reveal problems in the data. Some companies that looked like outliers on the high end turned out to be more typical. But two companies in the finance sector now stick out. Unlike the rest of the companies in that sector, they are not banks. They may have been placed in the wrong sector, but we couldn't see that in the original data.

Goal 3

Make the form of a scatterplot more nearly linear. Linear scatterplots are easier to model. We saw an example of scatterplot straightening in Chapter 6. The greater value of re-expression to straighten a relationship is that we can fit a linear model once the relationship is straight.

Here are *Assets* of the companies plotted against the logarithm of *Sales*, clearly bent. Taking logs makes things much more linear.

Figure 9.7

Assets vs. *Log Sales* shows a positive association (bigger sales go with bigger assets) but a bent shape. Note also that the points go from tightly bunched at the left to widely scattered at the right; the plot "thickens." In the second plot, *Log Assets* vs. *Log Sales* shows a clean, positive, linear association. And the variability at each value of *x* is about the same.

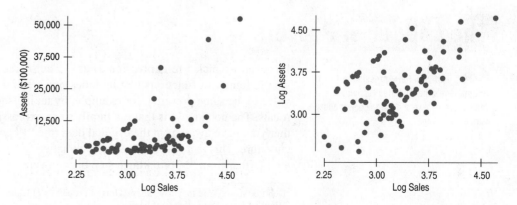

Goal 4

Make the scatter in a scatterplot spread out evenly rather than thickening at one end.
Having consistent scatter is a condition of many methods of Statistics, as we'll see in later chapters. This goal is closely related to Goal 2, but it often comes along with Goal 3. Indeed, a glance back at the scatterplot (Figure 9.7) shows that the plot for *Assets* is much more spread out on the right than on the left, while the plot for log *Assets* has roughly the same variation in log *Assets* for any *x*-value.

For Example RECOGNIZING WHEN A RE-EXPRESSION CAN HELP

In Chapter 8, we saw the awesome ability of emperor penguins to slow their heart rates while diving. Here are three displays relating to the diving heart rates:

(The boxplots show the diving heart rates for each of the 9 penguins whose dives were tracked. The names are those given by the researchers; EP = emperor penguin.)

QUESTION: What features of each of these displays suggest that a re-expression might be helpful?

ANSWER: The scatterplot shows a curved relationship, concave upward, between the duration of the dives and penguins' heart rates. Re-expressing either variable may help to straighten the pattern.

The histogram of heart rates is skewed to the high end. Re-expression often helps to make skewed distributions more nearly symmetric.

The boxplots each show skewness to the high end as well. The medians are low in the boxes, and several show high outliers.

The Ladder of Powers

A S Activity: **Re-expression in Action.** Here's the animated version of the Ladder of Powers. Slide the power and watch the change.

How can we pick a re-expression to use? The secret is to choose a re-expression from a simple family of functions that includes powers and the logarithm.[2] We raise each data value to the same power: $\frac{1}{2}$, for example, by taking square roots. Or -1, by finding reciprocals. The good news is that the family of re-expressions line up in order, so that the farther you move away from the original data (the "1" position), the greater the effect on any curvature. This fact lets you search systematically for a re-expression that works, stepping a bit farther from "1" or taking a step back toward "1" as you see the results.

---

[2]Don't be scared. You may have learned lots of properties of logarithms or done some messy calculations. Relax! You won't need that stuff here.

Where to start? It turns out that certain kinds of data are more likely to be helped by particular re-expressions. Knowing that gives you a good place to start your search, and from there you can look around a bit for a useful re-expression. We call this collection of re-expressions the **Ladder of Powers.**

Power	Name	Comment
2	The square of the data values, y^2.	Try this for unimodal distributions that are skewed to the left.
1	The raw data—no change at all. This is "home base." The farther you step from here up or down the ladder, the greater the effect.	Data that can take on both positive and negative values with no bounds are less likely to benefit from re-expression.
1/2	The square root of the data values, $\sqrt{y}$.	Counts often benefit from a square root re-expression. For counted data, start here.
"0"	Although mathematicians define the "0-th" power differently,[3] for us the place is held by the logarithm. You may feel uneasy about logarithms. Don't worry; the computer or calculator does the work.[4]	Measurements that cannot be negative, and especially values that grow by percentage increases such as salaries or populations, often benefit from a log re-expression. When in doubt, start here. If your data have zeros, try adding a small constant to all values before finding the logs.
−1/2	The (negative) reciprocal square root, $-1/\sqrt{y}$.	An uncommon re-expression, but sometimes useful. Changing the sign to take the *negative* of the reciprocal square root preserves the direction of relationships, making things a bit simpler.
−1	The (negative) reciprocal, $-1/y$.	Ratios of two quantities (miles per hour, for example) often benefit from a reciprocal. (You have about a 50–50 chance that the original ratio was taken in the "wrong" order for simple statistical analysis and would benefit from re-expression.) Often, the reciprocal will have simple units (hours per mile). Change the sign if you want to preserve the direction of relationships. If your data have zeros, try adding a small constant to all values before finding the reciprocal.

TI-*nspire*

Re-expression. See a curved relationship become straighter with each step on the Ladder of Powers.

The Ladder of Powers orders the effects that the re-expressions have on data. If you try, say, taking the square roots of all the values in a variable and it helps, but not enough, then move farther down the ladder to the logarithm or reciprocal root. Those re-expressions will have a similar, but even stronger, effect on your data. If you go too far, you can always back up. But don't forget—when you take a negative power, the *direction* of the relationship will change. That's OK. You can always change the sign of the response variable if you want to keep the same direction. With modern technology, finding a suitable re-expression is no harder than the push of a button.

Just Checking

1. You want to model the relationship between the number of birds counted at a nesting site and the temperature (in degrees Celsius). The scatterplot of counts vs. temperature shows an upwardly curving pattern, with more birds spotted at higher temperatures. What transformation (if any) of the bird counts might you start with?

2. You want to model the relationship between prices for various items in Paris and in Hong Kong. The

scatterplot of Hong Kong prices vs. Parisian prices shows a generally straight pattern with a small amount of scatter. What transformation (if any) of the Hong Kong prices might you start with?

3. You want to model the population growth of the United States over the past 200 years. The scatterplot shows a strongly upwardly curved pattern. What transformation (if any) of the population might you start with?

---

[3]You may remember that for any nonzero number y, $y^0 = 1$. This is not a very exciting transformation for data; every data value would be the same. We use the logarithm in its place.

[4]Your calculator or software package probably gives you a choice between "base 10" logarithms and "natural (base e)" logarithms. Don't worry about that. It doesn't matter at all which you use; they have exactly the same effect on the data. If you want to choose, base 10 logarithms can be a bit easier to interpret.

For Example TRYING A RE-EXPRESSION

RECAP: We've seen curvature in the relationship between emperor penguins' diving heart rates and the duration of the dive. Let's start the process of finding a good re-expression. Heart rate is in beats per minute; maybe heart "speed" in minutes per beat would be a better choice. Here are the corresponding displays for this reciprocal re-expression (as we often do, we've changed the sign to preserve the order of the data values):

QUESTION: Were the re-expressions successful?

ANSWER: The scatterplot bends less than before, but now may be slightly concave downward. The histogram is now slightly skewed to the low end. Most of the boxplots have no outliers. These boxplots seem better than the ones for the raw heart rates.

Overall, it looks like I may have moved a bit "too far" on the ladder of powers. Halfway between "1" (the original data) and "−1" (the reciprocal) is "0," which represents the logarithm. I'd try that for comparison.

Step-by-Step Example RE-EXPRESSING TO STRAIGHTEN A SCATTERPLOT

Standard (monofilament) fishing line comes in a range of strengths, usually expressed as "test pounds." Five-pound test line, for example, can be expected to withstand a pull of up to five pounds without breaking. The convention in selling fishing line is that the price of a spool doesn't vary with strength. Instead, the length of line on the spool varies. Higher test pound line is thicker, though, so spools of fishing line hold about the same amount of material. Some spools hold line that is thinner and longer, some fatter and shorter. Let's look at the *Length* and *Strength* of spools of monofilament line manufactured by the same company and sold for the same price at one store.

Question: How are the *Length* on the spool and the *Strength* related? And what re-expression will straighten the relationship?

THINK ➡ Plan State the problem.	I want to fit a linear model for the length and strength of monofilament fishing line.

(continued)

Variables Identify the variables and report the W's.

I have the *length* and "pound test" *strength* of monofilament fishing line sold by a single vendor at a particular store. Each case is a different strength of line, but all spools of line sell for the same price.

Let *Length* = length (in yards) of fishing line on the spool

   Strength = the test strength (in pounds).

Plot Check that even if there is a curve, the overall pattern does not reach a minimum or maximum and then turn around and go back. An up-and-down curve can't be fixed by re-expression.

The plot shows a negative direction and an association that has little scatter but is not straight.

SHOW ➡ **Mechanics** Try a re-expression.

The lesson of the Ladder of Powers is that if we're moving in the right direction but have not had sufficient effect, we should go farther along the ladder. This example shows improvement, but is still not straight.

(Because *Length* is an amount of something and cannot be negative, we probably should have started with logs. This plot is here in part to illustrate how the Ladder of Powers works.)

Here's a plot of the square root of *Length* against *Strength*:

The plot is less bent, but still not straight.

The scatterplot of the logarithm of *Length* against *Strength* is even less bent:

Stepping from the 1/2 power to the "0" power, we try the logarithm of *Length* against *Strength*.

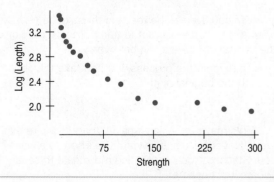

(continued)

The straightness is improving, so we know we're moving in the right direction. But since the plot of the logarithms is not yet straight, we know we haven't gone far enough. To keep the direction consistent, change the sign and re-express to $-1/Length$.

This is much better, but still not straight, so I'll take another step to the "-1" power, or reciprocal.

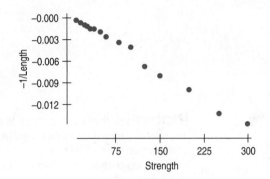

Maybe now I moved too far along the ladder.

We may have to choose between two adjacent re-expressions. For most data analyses, it really doesn't matter which we choose.

A half-step back is the $-1/2$ power: the reciprocal square root.

TELL ➡ **Conclusion** Specify your choice of re-expression. If there's some natural interpretation (as for gallons per 100 miles), give that.

It's hard to choose between the last two alternatives. Either of the last two choices is good enough. I'll choose the $-1/2$ power.

Now that the re-expressed data satisfy the Straight Enough Condition, we can fit a linear model by least squares. We find that

$$\frac{-1}{\sqrt{Length}} = -0.023 - 0.000373 \; Strength.$$

We can use this model to predict the length of a spool of, say, 35-pound test line:

$$\frac{-1}{\sqrt{Length}} = -0.023 - 0.000373 \times 35 = -0.036$$

We could leave the result in these units ($-1/\sqrt{yards}$). Sometimes the new units may be as meaningful as the original, but here we want to transform the predicted value back into yards. Fortunately, each of the re-expressions in the Ladder of Powers can be reversed.

To reverse the process, we first take the reciprocal: $\sqrt{Length} = -1/(-0.036) = 27.778$. Then squaring gets us back to the original units:

$$\widehat{Length} = 27.778^2 = 771.6 \; yards.$$

This may be the most painful part of the re-expression. Getting back to the original units can sometimes be a little work. Nevertheless, it's worth the effort to always consider re-expression. Re-expressions extend the reach of all of your Statistics tools by helping more data to satisfy the conditions they require. Just think how much more useful this course just became!

For Example COMPARING RE-EXPRESSIONS

RECAP: We've concluded that in trying to straighten the relationship between *Diving Heart Rate* and *Dive Duration* for emperor penguins, using the reciprocal re-expression goes a bit "too far" on the ladder of powers. Now we try the logarithm. Here are the resulting displays:

QUESTION: Comment on these displays. Now that we've looked at the original data (rung 1 on the Ladder), the reciprocal (rung −1), and the logarithm (rung 0), which re-expression of *Diving Heart Rate* would you choose?

ANSWER: *The scatterplot is now more linear and the histogram is symmetric. The boxplots are still a bit skewed to the high end, but less so than for the original Diving Heart Rate values. We don't expect real data to cooperate perfectly, and the logarithm seems like the best compromise re-expression, improving several different aspects of the data.*

TI Tips RE-EXPRESSING DATA TO ACHIEVE LINEARITY

Let's revisit the Arizona State tuition data. Recall that back in Chapter 7 when we tried to fit a linear model to the yearly tuition costs, the residuals plot showed a distinct curve. Residuals are high (positive) at the left, low in the middle of the decade, and high again at the right.

This curved pattern indicates that data re-expression may be in order. If you have no clue what re-expression to try, the Ladder of Powers may help. We just used that approach in the fishing line example. Here, though, we can play a hunch. It is reasonable to suspect that tuition increases at a relatively consistent percentage year by year. This suggests that using the logarithm of tuition may help.

- Tell the calculator to find the logs of the tuitions, and store them as a new list. Remember that you must import the name TUIT from the LIST NAMES menu. The command is log(1TUIT) STO L1.
- Check the scatterplot for the re-expressed data by changing your STATPLOT specifications to Xlist:YR and Ylist:L1. (Don't forget to use 9: ZoomStat to resize the window properly.)

The new scatterplot looks quite linear, but it's really the residuals plot that will tell the story. Remember that the TI automatically finds and stores the residuals whenever you ask it to calculate a regression.

(continued)

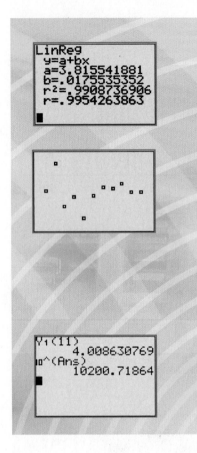

- Perform the regression for the *logarithm of tuition* vs. *year* with the command `LinReg(a + bx)`, setting `Xlist:YR`, `Ylist:L1`, and `RegEQ:Y1` (or on an older calculator, `LinReg(a+bx) 1YR, L1, Y1`). That both creates the residuals and reports details about the model (storing the equation for later use).
- Now that the residuals are stored in `RESID`, set up a new scatterplot, this time specifying `Xlist:YR` and `Ylist:RESID`.

While the residuals for the second and fifth years are comparatively large, the curvature we saw above is gone. The pattern in these residuals seem essentially horizontal and random. This re-expressed model is probably more useful than the original linear model.

Do you know what the model's equation is? Remember, it involves a log re-expression. The calculator does not indicate that; be sure to *Think* when you write your model!

$$\log \widehat{tuit} = 3.816 + 0.018\, yr$$

And you have to *Think* some more when you make an estimate using the calculator's equation. Notice that this model does not actually predict tuition; rather, it predicts the *logarithm* of the tuition.

For example, to estimate the 2001 tuition we must first remember that in entering our data we designated 1990 as year 0. That means we'll use 11 for the year 2001 and evaluate `Y1(11)`.

No, we're not predicting the tuition to be $4! That's the log of the estimated tuition. Since logarithms are exponents, $\log(\widehat{tuit}) = 4$ means $\widehat{tuit} = 10^4$, or about $10,000. When you are working with models that involve re-expressions, you'll often need to "backsolve" like this to find the correct predictions.

Plan B: Attack of the Logarithms

Figure 9.8

Plotting log (*Length*) against log (*Strength*) gives a straighter shape.

The Ladder of Powers is often successful at finding an effective re-expression. Sometimes, though, the curvature is more stubborn, and we're not satisfied with the residual plots. What then?

When none of the data values is zero or negative, logarithms can be a helpful ally in the search for a useful model. Try taking the logs of both the *x*- and *y*-variables. Then re-express the data using some combination of x or $\log(x)$ vs. y or $\log(y)$. You may find that one of these works pretty well.

Model Name	x-axis	y-axis	Comment
Exponential	x	$\log(y)$	This model is the "0" power in the ladder approach, useful for values that grow by percentage increases.
Logarithmic	$\log(x)$	y	A wide range of *x*-values, or a scatterplot descending rapidly at the left but leveling off toward the right, may benefit from trying this model.
Power	$\log(x)$	$\log(y)$	The Goldilocks model: When one of the ladder's powers is too big and the next is too small, this one may be just right.

When we tried to model the relationship between the length of fishing line and its strength, we were torn between the "−1" power and the "−1/2" power. The first showed slight upward curvature, and the second downward. Maybe there's a better power between those values.

The scatterplot shows what happens when we graph the logarithm of *Length* against the logarithm of *Strength*. Technology reveals that the equation of our log–log model is

$$\widehat{\log(Length)} = 4.49 - 1.08 \log(Strength).$$

It's interesting that the slope of this line (-1.08) is a power[5] we didn't try. After all, the ladder can't have every imaginable rung.

A warning, though! Don't expect to be able to straighten every curved scatterplot you find. It may be that there just isn't a very effective re-expression to be had. You'll certainly encounter situations when nothing seems to work the way you wish it would. Don't set your sights too high—you won't find a perfect model. Keep in mind: We seek a *useful* model, not perfection (or even "the best").

TI Tips USING LOGARITHMIC RE-EXPRESSIONS

In Chapter 6 we looked at data showing the relationship between the f/stop of a camera's lens and its shutter speed. Let's use the attack of the logarithms to model this situation.

Shutter speed:	1/1000	1/500	1/250	1/125	1/60	1/30	1/15	1/8
f/stop:	2.8	4	5.6	8	11	16	22	32

- Enter these data into your calculator, shutter *speed* in L1 and *f/stop* in L2.
- Create the scatterplot with Xlist:L1 and Ylist:L2. See the curve?
- Find the logarithms of each variable's values. Keep track of where you store everything so you don't get confused! We put log(*speed*) in L3 and log(*f/stop*) in L4.
- Make three scatterplots:
  ○ *f/stop* vs. log(*speed*) using Xlist:L3 and Ylist:L2
  ○ log(*f/stop*) vs. *speed* using Xlist:L1 and Ylist:L4
  ○ log(*f/stop*) vs. log(*speed*) using Xlist:L3 and Ylist:L4
- Pick your favorite. We liked log(*f/stop*) vs. log(*speed*) a lot! It appears to be very straight. (Don't be misled—this is a situation governed by the laws of Physics. Real data are not so cooperative. Don't expect to achieve this level of perfection often!)
- Remember that before you check the residuals plot, you first have to calculate the regression. In this situation all the errors in the residuals are just round-off errors in the original *f/stops*.
- Use your regression to write the equation of the model. Remember: The calculator does not know there were logarithms involved. You have to Think about that to be sure you write your model correctly.[6]

$$\log(\widehat{f/stop}) = 1.94 + 0.497\log(speed)$$

Why Not Just Use a Curve?

When a clearly curved pattern shows up in the scatterplot, why not just fit a curve to the data? We saw earlier that the association between the *Weight* of a car and its *Fuel Efficiency* was not a straight line. Instead of trying to find a way to straighten the plot, why not find a curve that seems to describe the pattern well?

We can find "curves of best fit" using essentially the same approach that led us to linear models. You won't be surprised, though, to learn that the mathematics and the calculations are considerably more difficult for curved models. Many calculators and

---

[5]For logarithms, $-1.08 \log(Strength) = \log(Strength^{-1.08})$.

[6]See the slope, 0.497? Just about 0.5. That's because the actual relationship involves the square root of shutter speeds. Technically the *f*/stop listed as 2.8 should be $2\sqrt{2} \approx 2.8284$. Rounding off to 2.8 makes sense for photographers, but it's what led to the minor errors you saw in the residuals plot.

computer packages do have the ability to fit curves to data, but this approach has many drawbacks.

Straight lines are easy to understand. We know how to think about the slope and the *y*-intercept, for example. We often want some of the other benefits mentioned earlier, such as making the spread around the model more nearly the same everywhere. In later chapters you will learn more advanced statistical methods for analyzing linear associations.

We give all of that up when we fit a model that is not linear. For many reasons, then, it is usually better to re-express the data to straighten the plot.

TI Tips SOME SHORTCUTS TO AVOID

Your calculator offers many regression options in the STAT CALC menu. There are three that automate fitting simple re-expressions of *y* or *x*:

- 9 : LnReg—fits a logarithmic model ($\hat{y} = a + b\ln x$)
- 0 : ExpReg—fits an exponential model ($\hat{y} = ab^x$)
- A : PwrReg—fits a power model ($\hat{y} = ax^b$)

In addition, the calculator offers two other functions:

- 5 : QuadReg—fits a quadratic model ($\hat{y} = ax^2 + bx + c$)
- 6 : CubicReg—fits a cubic model ($\hat{y} = ax^3 + bx^2 + cx + d$)

These two models have a form we haven't seen, with several *x*-terms. Because x, x^2, and x^3 are likely to be highly correlated with each other, the quadratic and cubic models are almost sure to be unreliable to fit, difficult to understand, and dangerous to use for predictions that are even slight extrapolations. We recommend that you be very wary of models of this type.

Let's try out one of the calculator shortcuts; we'll use the Arizona State tuition data. (For the last time, we promise!) This time, instead of re-expressing *tuition* to straighten the scatterplot, we'll have the calculator do more of the work.

Which model should you use? You could always just play hit-and-miss, but knowing something about the data can save a lot of time. If tuition increases by a consistent percentage each year, then the growth is exponential. The Exp Reg results all look very good: R^2 is high, the curve appears to fit the points quite well, and the residuals plot is acceptably random.

The equation of the model is $\widehat{tuit} = 6539.46(1.041^{year})$.

Notice, though that this is the same residuals plot we saw when we re-expressed the data and fit a line to the logarithm of *tuition*. That's the calculator just did the very same thing. This new equation may look different, but it is equivalent to our earlier model $\log \widehat{tuit} = 3.816 + 0.018\ year$.

Not easy to see that, is it? Here's how it works:

Initially we used a logarithmic re-expression to create a linear model: $\log \hat{y} = a + bx$

Rewrite that equation in exponential form: $\hat{y} = 10^{a+bx}$

Simplify, using the laws of exponents: $\hat{y} = 10^a(10^b)^x$

Let $10^a = a$ and $10^b = b$ (different *a* and *b*!) $\hat{y} = ab^x$

See? Your linear model created by logarithmic re-expression is the same as the calculator model created by ExpReg. In fact, three of the special TI functions correspond to a simple regression model involving re-expression.

Type of Model	Re-expression Equation	Calculator's Curve	
		Command	Equation
Logarithmic	$\hat{y} = a + b\log x$	LnReg	$\hat{y} = a + b\ln x$
Exponential	$\log \hat{y} = a + bx$	ExpReg	$\hat{y} = ab^x$
Power	$\log \hat{y} = a + b\log x$	PwrReg	$\hat{y} = ax^b$

Be careful. It may look like the calculator fit these equations to the data by minimizing the sum of squared residuals, but it really didn't do that. It handles the residuals differently, and the difference matters. If you use a statistics program to fit an "exponential model," it will probably fit the exponential form of the equation and give you a different answer.

You've seen two ways to handle bent relationships:
- straighten the data, then fit a line, or
- use the calculator shortcut to create a curve.

Note that the calculator does not have a shortcut for every model you might want to use—models involving square roots or reciprocals, for instance. And remember: The calculator may be quick, but there are real advantages to finding *linear* models by actually re-expressing the data. That's the approach you should always use.

WHAT CAN GO WRONG?

- **Don't get seduced by ExpReg and its calculator cousins.** Those so-called "curved" regression options look enticing, but don't go there. This course is about *linear* regression. If you see a curve, re-express the data to achieve linearity and then fit a line. Equations of lines are easier to interpret, and will be far easier to work with later on when we do more advanced statistical analyses.

- **Don't expect your model to be perfect.** In Chapter 5 we quoted statistician George Box: "All models are wrong, but some are useful." Be aware that the real world is a messy place and data can be uncooperative. Don't expect to find one elusive re-expression that magically irons out every kink in your scatterplot and produces perfect residuals. You aren't looking for the Right Model, because that mythical creature doesn't exist. Find a useful model and use it wisely.

- **Don't stray too far from the ladder.** It's wise not to stray too far from the powers that we suggest in the Ladder of Powers. Stick to powers between 2 and −2. Even in that interval, you should prefer the simpler powers in the ladder to those in the cracks. A square root is easier to understand than the 0.413 power. That simplicity may compensate for a slightly less straight relationship.

- **Don't choose a model based on R^2 alone.** You've tried re-expressing your data to straighten a curved relationship and found a model with a high R^2. Beware: That doesn't mean the pattern is straight now. On the next page is a plot of a relationship with an R^2 of 98.3%. The R^2 is about as high as we could ask for, but if you look closely, you'll see that there's a consistent bend. Plotting the residuals from the least squares line makes the bend much easier to see.

Occam's Razor If you think that simpler explanations and simpler models are more likely to give a true picture of the way things work, then you should look for opportunities to re-express your data and simplify your analyses.

The general principle that simpler explanations are likely to be the better ones is known as Occam's Razor, after the English philosopher and theologian William of Occam (1284–1347).

Remember the basic rule of data analysis: *Make a picture*. Before you fit a line, always look at the pattern in the scatterplot. After you fit the line, check for linearity again by plotting the residuals.

■ **Beware of multiple modes.** Re-expression can often make a skewed unimodal histogram more nearly symmetric, but it cannot pull separate modes together. A suitable re-expression may, however, make the separation of the modes clearer, simplifying their interpretation and making it easier to separate them to analyze individually.

■ **Watch out for scatterplots that turn around.** Re-expression can straighten many bent relationships but not those that go up and then down or down and then up. You should refuse to analyze such data with methods that require a linear form.

Figure 9.9

The shape of the scatterplot of *Birth Rates* (births per 100,000 women) in the United States shows an oscillation that cannot be straightened by re-expressing the data.

■ **Watch out for zero or negative data values.** It's impossible to re-express negative values by any power that is not a whole number on the Ladder of Powers or to re-express values that are zero for negative powers. One possible cure for zeros and small negative values is to add a constant ($\frac{1}{2}$ and $\frac{1}{6}$ are often used) to bring all the data values above zero.

What Have We Learned?

We've learned that when the conditions for regression are not met, a simple re-expression of the data may help. There are several reasons to consider a re-expression:

■ To make the distribution of a variable more symmetric (as we saw in Chapter 4)
■ To make the spread across different groups more similar
■ To make the form of a scatterplot straighter
■ To make the scatter around the line in a scatterplot more consistent

We've learned that when seeking a useful re-expression, taking logs is often a good, simple starting point. To search further, the Ladder of Powers or the log–log approach can help us find a good re-expression.

We've come to understand that our models won't be perfect, but that re-expression can lead us to a useful model.

Terms

Re-expression   We re-express data by taking the logarithm, the square root, the reciprocal, or some other mathematical operation of all values of a variable. (p. 232)

Ladder of Powers   The Ladder of Powers places in order the effects that many re-expressions have on the data. (p. 237)

On the Computer RE-EXPRESSION

Computers and calculators make it easy to re-express data. Most statistics packages offer a way to re-express and compute with variables. Some packages permit you to specify the power of a re-expression with a slider or other moveable control, possibly while watching the consequences of the re-expression on a plot or analysis. This, of course, is a very effective way to find a good re-expression.

Exercises

1. Residuals Suppose you have fit a linear model to some data and now take a look at the residuals. For each of the following possible residuals plots, tell whether you would try a re-expression and, if so, why.

2. Residuals Suppose you have fit a linear model to some data and now take a look at the residuals. For each of the following possible residuals plots, tell whether you would try a re-expression and, if so, why.

3. Oakland passengers revisited In Chapter 8, Exercise 15, we created a linear model describing the trend in the number of passengers departing from the Oakland (CA) airport each month since the start of 1997. Here's the residual plot, but with lines added to show the order of the values in time:

a) Can you account for the pattern shown here?

b) Would a re-expression help us deal with this pattern? Explain.

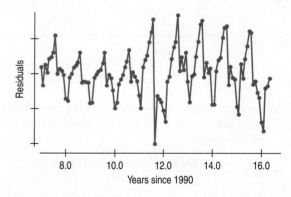

4. Hopkins winds, revisited In Chapter 4, we examined the wind speeds in the Hopkins forest over the course of a year. Here's the scatterplot we saw then:

a) Describe the pattern you see here.

b) Should we try re-expressing either variable to make this plot straighter? Explain.

5. Models For each of the models listed below, predict y when $x = 2$.

a) $\ln \hat{y} = 1.2 + 0.8x$ d) $\hat{y} = 1.2 + 0.8 \ln x$

b) $\sqrt{\hat{y}} = 1.2 + 0.8x$ e) $\log \hat{y} = 1.2 + 0.8 \log x$

c) $\dfrac{1}{\hat{y}} = 1.2 + 0.8x$

6. More models For each of the models listed below, predict y when $x = 2$.

a) $\hat{y} = 1.2 + 0.8 \log x$ d) $\hat{y}^2 = 1.2 + 0.8x$

b) $\log \hat{y} = 1.2 + 0.8x$ e) $\dfrac{1}{\sqrt{\hat{y}}} = 1.2 + 0.8x$

c) $\ln \hat{y} = 1.2 + 0.8 \ln x$

7. Gas mileage As the example in the chapter indicates, one of the important factors determining a car's *Fuel Efficiency* is its *Weight*. Let's examine this relationship again, for 11 cars.

a) Describe the association between these variables shown in the scatterplot.

b) Here is the regression analysis for the linear model. What does the slope of the line say about this relationship?

Dependent variable is: Fuel Efficiency
R-squared = 85.9%

Variable	Coefficient
Intercept	47.9636
Weight	−7.65184

c) Do you think this linear model is appropriate? Use the residuals plot to explain your decision.

8. Crowdedness In a *Chance* magazine article (Summer 2005), Danielle Vasilescu and Howard Wainer used data from the United Nations Center for Human Settlements to investigate aspects of living conditions for several countries. Among the variables they looked at were the country's per capita gross domestic product (*GDP*, in $) and *Crowdedness,* defined as the average number of persons per room living in homes there. This scatterplot displays these data for 56 countries:

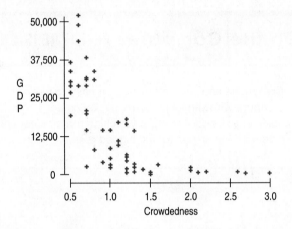

a) Explain why you should re-express these data before trying to fit a model.

b) What re-expression of *GDP* would you try as a starting point?

9. Gas mileage revisited Let's try the re-expressed variable *Fuel Consumption* (gal/100 mi) to examine the fuel efficiency of the 11 cars in Exercise 7. Here are the revised regression analysis and residuals plot:

Dependent variable is: Fuel Consumption
R-squared = 89.2%

Variable	Coefficient
Intercept	0.624932
Weight	1.17791

a) Explain why this model appears to be better than the linear model.

b) Using the regression analysis above, write an equation of this model.

c) Interpret the slope of this line.

d) Based on this model, how many miles per gallon would you expect a 3500-pound car to get?

10. Crowdedness again In Exercise 8 we looked at United Nations data about a country's *GDP* and the average number of people per room (*Crowdedness*) in housing there. For a re-expression, a student tried the reciprocal $-10000/GDP$, representing the number of people per \$10,000 of gross domestic product. Here are the results, plotted against *Crowdedness:*

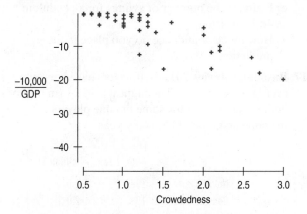

a) Is this a useful re-expression? Explain.

b) What re-expression would you suggest this student try next?

11. GDP The scatterplot shows the *gross domestic product (GDP)* of the United States in billions of dollars plotted against years since 1950.

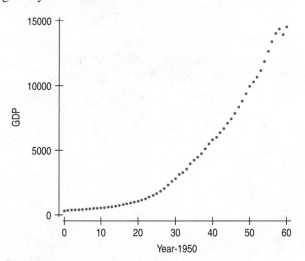

A linear model fit to the relationship looks like this:

Dependent variable is: GDP
R-squared = 87.6%   s = 1597.7456

Variable	Coefficient
Intercept	−2561.3552
Year-1950	237.74577

a) Does the value 87.6% suggest that this is a good model? Explain.

b) Here's a scatterplot of the residuals. Now do you think this is a good model for these data? Explain?

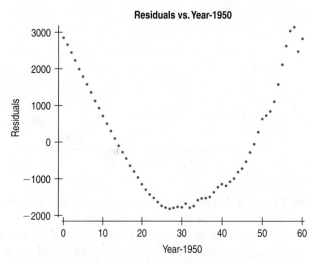

12. Treasury Bills The 3-month Treasury bill interest rate is watched by investors and economists. Here's a scatterplot of the 3-month Treasury bill rate since 1950:

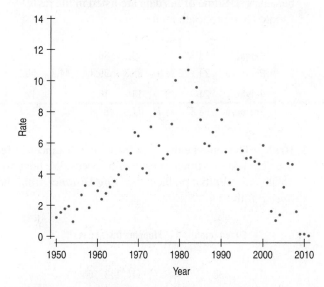

Clearly, the relationship is not linear. Can it be made nearly linear with a re-expression? If so, which one would you suggest? If not, why not?

13. Better GDP model? Consider again the post-1950 trend in U.S. GDP we examined in Exercise 11. Here are a regression and (on the next page) a residual plot when we use the log of GDP in the model. Is this a better model for GDP? Explain.

Dependent variable is: logGDP
R-squared = 98.9%   s = 0.13185

Variable	Coefficient
Intercept	5.6579766
Year-1950	0.070734456

Residuals vs. Year-1950

T **14. Pressure** Scientist Robert Boyle examined the relationship between the volume in which a gas is contained and the pressure in its container. He used a cylindrical container with a moveable top that could be raised or lowered to change the volume. He measured the *Height* in inches by counting equally spaced marks on the cylinder, and measured the *Pressure* in inches of mercury (as in a barometer). Some of his data are listed in the table. Create an appropriate model.

Height	48	44	40	36	32	28
Pressure	29.1	31.9	35.3	39.3	44.2	50.3
Height	24	20	18	16	14	12
Pressure	58.8	70.7	77.9	87.9	100.4	117.6

T **15. Brakes** The table below shows stopping distances in feet for a car tested 3 times at each of 5 speeds. We hope to create a model that predicts *Stopping Distance* from the *Speed* of the car.

Speed (mph)	Stopping Distances (ft)
20	64, 62, 59
30	114, 118, 105
40	153, 171, 165
50	231, 203, 238
60	317, 321, 276

a) Explain why a linear model is not appropriate.
b) Re-express the data to straighten the scatterplot.
c) Create an appropriate model.
d) Estimate the stopping distance for a car traveling 55 mph.
e) Estimate the stopping distance for a car traveling 70 mph.
f) How much confidence do you place in these predictions? Why?

T **16. Pendulum** A student experimenting with a pendulum counted the number of full swings the pendulum made in 20 seconds for various lengths of string. Here are her data.

Length (in.)	6.5	9	11.5	14.5	18	21	24	27	30	37.5
Number of Swings	22	20	17	16	14	13	13	12	11	10

a) Explain why a linear model is not appropriate for using the *Length* of a pendulum to predict the *Number of Swings* in 20 seconds.
b) Re-express the data to straighten the scatterplot.
c) Create an appropriate model.
d) Estimate the number of swings for a pendulum with a 4-inch string.
e) Estimate the number of swings for a pendulum with a 48-inch string.
f) How much confidence do you place in these predictions? Why?

T **17. Baseball salaries 2012** Ballplayers have been signing ever larger contracts. The highest salaries (in millions of dollars per season) for some notable players are given in the table and plotted below by year.

Player	Year	Salary (million $)
Nolan Ryan	1980	1.0
George Foster	1982	2.0
Kirby Puckett	1990	3.0
Jose Canseco	1990	4.7
Roger Clemens	1991	5.3
Ken Griffey, Jr.	1996	8.5
Albert Belle	1997	11.0
Pedro Martinez	1998	12.5
Mike Piazza	1999	12.5
Mo Vaughn	1999	13.3
Kevin Brown	1999	15.0
Carlos Delgado	2001	17.0
Alex Rodriguez	2001	22.0
Manny Ramirez	2004	22.5
Alex Rodriguez	2005	26.0

a) Examine the scatterplot above. Does it look straight? Given what you know about money and inflation, would you expect it to be straight?

Here is the residual plot for regression of *Year* vs. *Salary*.

b) What does this residual plot tell you about using the regression model for *Year* vs. *Salary*?

The log of salary was computed in an attempt to straighten the data. Regression was run on *Year* vs. ln *Salary*. Here are the results and the residual plot.

Dependent variable is: lnSalary
R-squared = 97.2%   s = 0.16478

Variable	Coefficient
Intercept	−251.28738
Year	0.1270045

Residuals vs. Year

c) Was this transformation successful? Use this model to predict the top salary in 2015.

d) After A-Rod's record setting salary, the next three biggest contracts were:

2009	CC Sabathia	$23
2011	Joe Mauer	$23
2012	Albert Pujols	$25.4

Adding the new data to the scatterplot, we see:

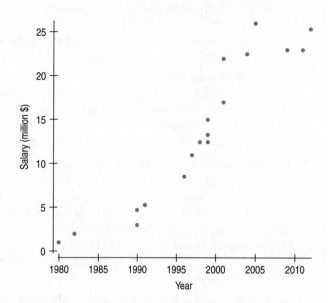

Given this recent trend, how do you feel about your 2015 prediction? Describe how the new data and the graph change your perspective on this model.

T 18. Planet distances and years 2012 At a meeting of the International Astronomical Union (IAU) in Prague in 2006, Pluto was determined not to be a planet, but rather the largest member of the Kuiper belt of icy objects. Let's examine some facts. Here is a table of the 9 sun-orbiting objects formerly known as planets:

Planet	Position Number	Distance from Sun (million miles)	Length of Year (Earth years)
Mercury	1	36	0.24
Venus	2	67	0.61
Earth	3	93	1.00
Mars	4	142	1.88
Jupiter	5	484	11.86
Saturn	6	887	29.46
Uranus	7	1784	84.07
Neptune	8	2796	164.82
Pluto	9	3707	247.68

a) Plot the *Length* of the year against the *Distance* from the sun. Describe the shape of your plot.

b) Re-express one or both variables to straighten the plot. Use the re-expressed data to create a model describing the length of a planet's year based on its distance from the sun.

c) Comment on how well your model fits the data.

T 19. Planet distances and order 2012 Let's look again at the pattern in the locations of the planets in our solar system seen in the table in Exercise 18.

a) Re-express the distances to create a model for the *Distance* from the sun based on the planet's *Position*.

b) Based on this model, would you agree with the International Astronomical Union that Pluto is not a planet? Explain.

20. Planets 2012, part 3 The asteroid belt between Mars and Jupiter may be the remnants of a failed planet. If so, then Jupiter is really in position 6, Saturn is in 7, and so on. Repeat Exercise 19, using this revised method of numbering the positions. Which method seems to work better?

21. Eris: Planets 2012, part 4 In July 2005, astronomers Mike Brown, Chad Trujillo, and David Rabinowitz announced the discovery of a sun-orbiting object, since named Eris,[7] that is 5% larger than Pluto. Eris orbits the sun once every 560 earth years at an average distance of about 6300 million miles from the sun. Based on its *Position,* how does Eris's *Distance* from the sun (re-expressed to logs) compare with the prediction made by your model of Exercise 19?

22. Models and laws: Planets 2012, part 5 The model you found in Exercise 18 is a relationship noted in the 17th century by Kepler as his Third Law of Planetary Motion. It was subsequently explained as a consequence of Newton's Law of Gravitation. The models for Exercises 19–21 relate to what is sometimes called the Titius-Bode "law," a pattern noticed in the 18th century but lacking any scientific explanation.

Compare how well the re-expressed data are described by their respective linear models. What aspect of the model of Exercise 18 suggests that we have found a physical law? In the future, we may learn enough about a planetary system around another star to tell whether the Titius-Bode pattern applies there. If you discovered that another planetary system followed the same pattern, how would it change your opinion about whether this is a real natural "law"? What would you think if the next system we find does not follow this pattern?

23. Logs (not logarithms) The value of a log is based on the number of board feet of lumber the log may contain. (A board foot is the equivalent of a piece of wood 1 inch thick, 12 inches wide, and 1 foot long. For example, a 2″ × 4″ piece that is 12 feet long contains 8 board feet.) To estimate the amount of lumber in a log, buyers measure the diameter inside the bark at the smaller end. Then they look in a table based on the Doyle Log Scale. The table below shows the estimates for logs 16 feet long.

Diameter of Log	8″	12″	16″	20″	24″	28″
Board Feet	16	64	144	256	400	576

a) What model does this scale use?

b) How much lumber would you estimate that a log 10 inches in diameter contains?

c) What does this model suggest about logs 36 inches in diameter?

24. Weightlifting 2012 Listed below are the gold medal–winning men's weight-lifting performances at the 2012 Olympics, followed by some analysis.

Weight Class (kg)	Name (Country)	Weight Lifted (kg)
56	Yun Om (Korea)	293
62	Un Kim (Korea)	327
69	Qingfeng Lin (China)	344
77	Xiaojun Lu (China)	379
85	Adrian Zielinski (Poland)	385
94	Ilya Ilyin (Kazakhstan)	406
105	Oleksiy Torokhtiy (Ukraine)	412

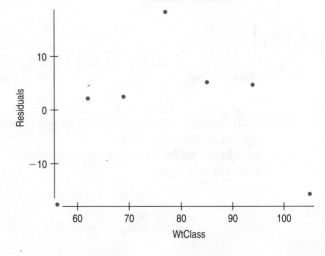

---

[7]Eris is the Greek goddess of warfare and strife who caused a quarrel among the other goddesses that led to the Trojan war. In the astronomical world, Eris stirred up trouble when the question of its proper designation led to the raucous meeting of the IAU in Prague where IAU members voted to demote Pluto and Eris to dwarf-planet status—http://www.gps.caltech.edu/~mbrown/planetlila/#paper.

Dependent variable is: WtLifted
R-sqared = 91.8%   s = 13.7446

Variable	Coefficient
Intercept	176.607
WtClass	2.39

a) What does the residual plot tell you about the need to re-express?

We reexpressed these data two ways, first using the reciprocal and then taking the logs of both variables. Here are the results.

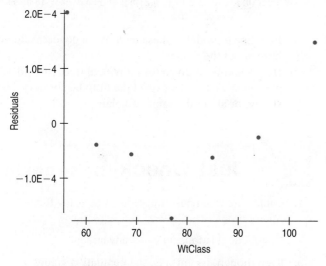

Dependent variable is: 1 / WtLifted
R-squared = 86.8%   s = 0.000143

Variable	Coefficient
Intercept	0.00427
WtClass	−0.000019006

Residuals vs. log10(WtClass)

Dependent variable is: log(WtLifted)
R-squared = 94.1%   s = 0.0144

Variable	Coefficient
Intercept	1.5479
log(WtClass)	0.5360

b) What does the residual plot tell you about the success of the re-expressions?

25. Life expectancy The data in the table below list the *Life Expectancy* for white males in the United States every decade during the last century (1 = 1900 to 1910, 2 = 1911 to 1920, etc.). Create a model to predict future increases in life expectancy. (National Vital Statistics Report)

Decade	1	2	3	4	5	6	7	8	9	10	11
Life exp.	48.6	54.4	59.7	62.1	66.5	67.4	68.0	70.7	72.7	74.9	76.5

26. Lifting record weight 2012 In 2012, Xiaojun Lu from China set a world record in the 77kg weight class with a lift of 379 kg.

a) Use the reciprocal re-expression in Exercise 24 to calculate his residual. Interpret this residual in context.

b) Does the sign of the residual surprise you, given that Xiaojun set a world record?

27. Slower is cheaper? Researchers studying how a car's *Fuel Efficiency* varies with its *Speed* drove a compact car 200 miles at various speeds on a test track. Their data are shown in the table.

Speed (mph)	35	40	45	50	55	60	65	70	75
Fuel Eff. (mpg)	25.9	27.7	28.5	29.5	29.2	27.4	26.4	24.2	22.8

Create a linear model for this relationship and report any concerns you may have about the model.

28. Orange production The table below shows that as the number of oranges on a tree increases, the fruit tends to get smaller. Create a model for this relationship, and express any concerns you may have.

Number of Oranges/Tree	Average Weight/ Fruit (lb)
50	0.60
100	0.58
150	0.56
200	0.55
250	0.53
300	0.52
350	0.50
400	0.49
450	0.48
500	0.46
600	0.44
700	0.42
800	0.40
900	0.38

29. Years to live 2008 Insurance companies and other organizations use actuarial tables to estimate the remaining life spans of their customers. The table below shows the predicted additional years of life for Hispanic females of various ages in the United States, according to a 2008 National Vital Statistics Report. (www.cdc.gov/nchs/deaths.htm)

Age	Years to Live
10	73.9
20	64
30	54.2
40	44.5
50	35.1
60	26.1
70	17.8
80	10.6
90	5.3
100	2.6

Here are the results of a re-expression.

Dependent variable is: sqrt(YrsToLive)
R-squared = 99.4%   s = 0.19068

Variable	Coefficient
Intercept	9.6867
Age	−0.0797

Residuals vs. Age

a) Evaluate the success of the regression.
b) Predict the lifespan of an 18-year-old Hispanic woman.
c) Are you satisfied that your model could predict the life expectancy of a friend of yours?

30. Tree growth A 1996 study examined the growth of grapefruit trees in Texas, determining the average trunk *Diameter* (in inches) for trees of varying *Ages:*

Age (yr)	2	4	6	8	10	12	14	16	18	20
Diameter (in.)	2.1	3.9	5.2	6.2	6.9	7.6	8.3	9.1	10.0	11.4

a) Fit a linear model to these data. What concerns do you have about the model?
b) If data had been given for individual trees instead of averages, would you expect the fit to be stronger, less strong, or about the same? Explain.

Just Checking ANSWERS

1. Counts are often best transformed by using the square root.

2. None. The relationship is already straight.

3. Even though, technically, the population values are counts, you should probably try a stronger transformation like log(population) because populations grow in proportion to their size.

Review of part II

Exploring Relationships Between Variables

Quick Review

You have now survived your second major unit of Statistics. Here's a brief summary of the key concepts and skills:

- We treat data two ways: as categorical and as quantitative.

- To explore relationships in categorical data, check out Chapter 2.

- To explore relationships in quantitative data:

  - Make a picture. Use a scatterplot. Put the explanatory variable on the *x*-axis and the response variable on the *y*-axis.

  - Describe the association between two quantitative variables in terms of direction, form, and strength.

  - The amount of scatter determines the strength of the association.

  - If, as one variable increases so does the other, the association is positive. If one increases as the other decreases, it's negative.

  - If the form of the association is linear, calculate a correlation to measure its strength numerically, and do a regression analysis to model it.

  - Correlations closer to −1 or +1 indicate stronger linear associations. Correlations near 0 indicate weak linear relationships, but other forms of association may still be present.

  - The line of best fit is also called the least squares regression line because it minimizes the sum of the squared residuals.

- The regression line predicts values of the response variable from values of the explanatory variable.

- A residual is the difference between the true value of the response variable and the value predicted by the regression model.

- The slope of the line is a rate of change, best described in "*y*-units" per "*x*-unit."

- R^2 gives the fraction of the variation in the response variable that is accounted for by the model.

- The standard deviation of the residuals measures the amount of scatter around the line.

- Outliers and influential points can distort any of our models.

- If you see a pattern (a curve) in the residuals plot, your chosen model is not appropriate; use a different model. You may, for example, straighten the relationship by re-expressing one of the variables.

- To straighten bent relationships, re-express the data using logarithms or a power (squares, square roots, reciprocals, etc.).

- Always remember that an association is not necessarily an indication that one of the variables causes the other.

Need more help with some of this? Try rereading some sections of Chapters 6 through 9. Starting right here on this very page are more opportunities to review these concepts and skills.

"One must learn by doing the thing; though you think you know it, you have no certainty until you try."

—*Sophocles (495–406 BCE)*

Review Exercises

1. College Every year *US News and World Report* publishes a special issue on many U.S. colleges and universities. The scatterplots below have *Student/Faculty Ratio* (number of students per faculty member) for the colleges and universities on the *y*-axes plotted against 4 other variables. The correct correlations for these scatterplots appear in this list. Match them.

−0.98 −0.71 −0.51 0.09 0.23 0.69

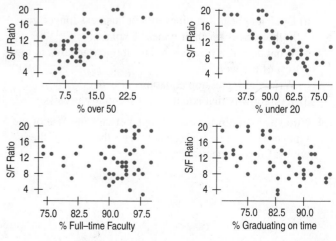

2. **Togetherness** Are good grades in high school associated with family togetherness? A random sample of 142 high school students was asked how many meals per week their families ate together. Their responses produced a mean of 3.78 meals per week, with a standard deviation of 2.2. Researchers then matched these responses against the students' grade point averages (GPAs). The scatterplot appeared to be reasonably linear, so they created a line of regression. No apparent pattern emerged in the residuals plot. The equation of the line was $\widehat{GPA} = 2.73 + 0.11\ Meals$.

a) Interpret the y-intercept in this context.
b) Interpret the slope in this context.
c) What was the mean GPA for these students?
d) If a student in this study had a negative residual, what did that mean?
e) Upon hearing of this study, a counselor recommended that parents who want to improve the grades their children get should get the family to eat together more often. Do you agree with this interpretation? Explain.

3. **Vineyards** Here are the scatterplot and regression analysis for *Case Prices* of 36 wines from vineyards in the Finger Lakes region of New York State and the *Ages* of the vineyards.

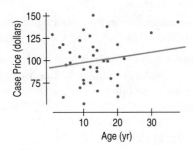

Dependent variable is: Case Price
R-squared = 2.7%

Variable	Coefficient
Constant	92.7650
Age	0.567284

a) Does it appear that vineyards in business longer get higher prices for their wines? Explain.
b) What does this analysis tell us about vineyards in the rest of the world?
c) Write the regression equation.
d) Explain why that equation is essentially useless.

4. **Vineyards again** Instead of *Age*, perhaps the *Size* of the vineyard (in acres) is associated with the price of the wines. Look at the scatterplot:

a) Do you see any evidence of an association?
b) What concern do you have about this scatterplot?
c) If the red "+" data point is removed, would the correlation become stronger or weaker? Explain.
d) If the red "+" data point is removed, would the slope of the line increase or decrease? Explain.

5. **More twins 2009?** As the table shows, the number of twins born in the United States has been increasing. (www.cdc.gov/nchs/births.htm)

Year	Twin Births	Year	Twin Births
1980	68,339	1995	96,736
1981	70,049	1996	100,750
1982	71,631	1997	104,137
1983	72,287	1998	110,670
1984	72,949	1999	114,307
1985	77,102	2000	118,916
1986	79,485	2001	121,246
1987	81,778	2002	125,134
1988	85,315	2003	128,665
1989	90,118	2004	132,219
1990	93,865	2005	133,122
1991	94,779	2006	137,085
1992	95,372	2007	138,961
1993	96,445	2008	138,660
1994	97,064	2009	137,217

a) Find the equation of the regression line for predicting the number of twin births.
b) Explain in this context what the slope means.
c) Predict the number of twin births in the United States for the year 2014. Comment on your faith in that prediction.
d) Comment on the residuals plot.

6. **Dow Jones 2012** The Dow Jones stock index measures the performance of the stocks of America's largest

companies (finance.yahoo.com). A regression of the Dow prices on years 1972–2012 looks like this:

Dependent variable is Dow Index
R-squared $= 83.9\%$ $s = 1659$

Variable	Coefficient
Intercept	−667396
Year	337.605

a) What is the correlation between *Dow Index* and *Year*?
b) Write the regression equation.
c) Explain in this context what the equation says.
d) Here's a scatterplot of the residuals. Which assumption(s) of the regression analysis appear to be violated?

7. Acid rain Biologists studying the effects of acid rain on wildlife collected data from 163 streams in the Adirondack Mountains. They recorded the *pH* (acidity) of the water and the *BCI*, a measure of biological diversity, and they calculated $R^2 = 27\%$. Here's a scatterplot of *BCI* against *pH:*

a) What is the correlation between *pH* and *BCI?*
b) Describe the association between these two variables.
c) If a stream has average *pH,* what would you predict about the *BCI?*
d) In a stream where the *pH* is 3 standard deviations above average, what would you predict about the *BCI?*

T 8. Manatees 2010 Marine biologists warn that the growing number of powerboats registered in Florida threatens the existence of manatees. The data in the table come from the Florida Fish and Wildlife Conservation Commission (myfwc.com/research/manatee/) and the National Marine Manufacturers Association (www.nmma.org/).

Year	Manatees Killed	Power Registrations (in 1000s)	Year	Manatees Killed	Power Registrations (in 1000s)
1982	13	447	1997	53	716
1983	21	460	1998	38	716
1984	24	481	1999	35	716
1985	16	498	2000	49	735
1986	24	513	2001	81	860
1987	20	512	2002	95	923
1988	15	527	2003	73	940
1989	34	559	2004	69	946
1990	33	585	2005	79	974
1992	33	614	2006	92	988
1993	39	646	2007	73	992
1994	43	675	2008	90	932
1995	50	711	2009	97	949
1996	47	719	2010	83	914

a) In this context, which is the explanatory variable?
b) Make a scatterplot of these data and describe the association you see.
c) Find the correlation between *Boat Registrations* and *Manatee Deaths*.
d) Interpret the value of R^2.
e) Does your analysis prove that powerboats are killing manatees?

T 9. A manatee model 2010 Continue your analysis of the manatee situation from the previous exercise.

a) Create a linear model of the association between *Manatee Deaths* and *Powerboat Registrations*.
b) Interpret the slope of your model.
c) Interpret the *y*-intercept of your model.
d) How accurately did your model predict the high number of manatee deaths in 2010?
e) Which is better for the manatees, positive residuals or negative residuals? Explain.
f) What does your model suggest about the future for the manatee?

10. Grades A Statistics instructor created a linear regression equation to predict students' final exam scores from their midterm exam scores. The regression equation was $\widehat{Fin} = 10 + 0.9\ Mid$.

a) If Susan scored a 70 on the midterm, what did the instructor predict for her score on the final?
b) Susan got an 80 on the final. How big is her residual?
c) If the standard deviation of the final was 12 points and the standard deviation of the midterm was 10 points, what is the correlation between the two tests?
d) How many points would someone need to score on the midterm to have a predicted final score of 100?
e) Suppose someone scored 100 on the final. Explain why you can't estimate this student's midterm score from the information given.

f) One of the students in the class scored 100 on the midterm but got overconfident, slacked off, and scored only 15 on the final exam. What is the residual for this student?

g) No other student in the class "achieved" such a dramatic turnaround. If the instructor decides not to include this student's scores when constructing a new regression model, will the R^2 value of the regression increase, decrease, or remain the same? Explain.

h) Will the slope of the new line increase or decrease?

11. **Traffic** Highway planners investigated the relationship between traffic *Density* (number of automobiles per mile) and the average *Speed* of the traffic on a moderately large city thoroughfare. The data were collected at the same location at 10 different times over a span of 3 months. They found a mean traffic *Density* of 68.6 cars per mile (cpm) with standard deviation of 27.07 cpm. Overall, the cars' average *Speed* was 26.38 mph, with standard deviation of 9.68 mph. These researchers found the regression line for these data to be $\widehat{Speed} = 50.55 - 0.352\ Density$.

a) What is the value of the correlation coefficient between *Speed* and *Density?*

b) What percent of the variation in average *Speed* is explained by traffic *Density?*

c) Predict the average *Speed* of traffic on the thoroughfare when the traffic *Density* is 50 cpm.

d) What is the value of the residual for a traffic *Density* of 56 cpm with an observed *Speed* of 32.5 mph?

e) The data set initially included the point *Density* = 125 cpm, *Speed* = 55 mph. This point was considered an outlier and was not included in the analysis. Will the slope increase, decrease, or remain the same if we redo the analysis and include this point?

f) Will the correlation become stronger, weaker, or remain the same if we redo the analysis and include this point (125,55)?

g) A European member of the research team measured the *Speed* of the cars in kilometers per hour (1 km ≈ 0.62 miles) and the traffic *Density* in cars per kilometer. Find the value of his calculated correlation between speed and density.

12. **Cramming** One Thursday, researchers gave students enrolled in a section of basic Spanish a set of 50 new vocabulary words to memorize. On Friday the students took a vocabulary test. When they returned to class the following Monday, they were retested—without advance warning. Here are the test scores for the 25 students.

Fri.	Mon.	Fri.	Mon.	Fri.	Mon.
42	36	48	37	39	41
44	44	43	41	46	32
45	46	45	32	37	36
48	38	47	44	40	31
44	40	50	47	41	32
43	38	34	34	48	39
41	37	38	31	37	31
35	31	43	40	36	41
43	32				

a) What is the correlation between *Friday* and *Monday* scores?

b) What does a scatterplot show about the association between the scores?

c) What does it mean for a student to have a positive residual?

d) What would you predict about a student whose *Friday* score was one standard deviation below average?

e) Write the equation of the regression line.

f) Predict the *Monday* score of a student who earned a 40 on Friday.

13. **Correlations** What factor most explains differences in *Fuel Efficiency* among cars? Below is a correlation matrix exploring that relationship for the car's *Weight, Horsepower,* engine *size (Displacement),* and number of *Cylinders.*

	MPG	Weight	Horse-power	Displace-ment	Cylinders
MPG	1.000				
Weight	−0.903	1.000			
Horsepower	−0.871	0.917	1.000		
Displacement	−0.786	0.951	0.872	1.000	
Cylinders	−0.806	0.917	0.864	0.940	1.000

a) Which factor seems most strongly associated with *Fuel Efficiency?*

b) What does the negative correlation indicate?

c) Explain the meaning of R^2 for that relationship.

14. **Autos revisited** Look again at the correlation table for cars in the previous exercise.

a) Which two variables in the table exhibit the strongest association?

b) Is that strong association necessarily cause-and-effect? Offer at least two explanations why that association might be so strong.

c) Engine displacements for U.S.-made cars are often measured in cubic inches. For many foreign cars, the units are either cubic centimeters or liters. How would changing from cubic inches to liters affect the calculated correlations involving *Displacement?*

d) What would you predict about the *Fuel Efficiency* of a car whose engine *Displacement* is one standard deviation above the mean?

15. **Cars, one more time!** Can we predict the *Horsepower* of the engine that manufacturers will put in a car by knowing the *Weight* of the car? Here are the regression analysis and residuals plot:

Dependent variable is: Horsepower
R-squared = 84.1%

Variable	Coefficient
Intercept	3.49834
Weight	34.3144

a) Write the equation of the regression line.

b) Do you think the car's *Weight* is measured in pounds or thousands of pounds? Explain.

c) Do you think this linear model is appropriate? Explain.

d) The highest point in the residuals plot, representing a residual of 22.5 horsepower, is for a Chevy weighing 2595 pounds. How much horsepower does this car have?

16. Colorblind Although some women are colorblind, this condition is found primarily in men. Why is it wrong to say there's a strong correlation between *Sex* and *Colorblindness*?

(T) 17. Old Faithful There is evidence that eruptions of Old Faithful can best be predicted by knowing the duration of the previous eruption.

a) Describe what you see in the scatterplot of *Intervals* between eruptions vs. *Duration* of the previous eruption.

b) Write the equation of the line of best fit. Here's the regression analysis:

Dependent variable is: Interval
R-squared = 77.0%

Variable	Coefficient
Intercept	33.9668
Duration	10.3582

c) Carefully explain what the slope of the line means in this context.

d) How accurate do you expect predictions based on this model to be? Cite statistical evidence.

e) If you just witnessed an eruption that lasted 4 minutes, how long do you predict you'll have to wait to see the next eruption?

f) So you waited, and the next eruption came in 79 minutes. Use this as an example to define a residual.

(T) 18. Which croc? The ranges inhabited by the Indian gharial crocodile and the Australian saltwater crocodile overlap in Bangladesh. Suppose a very large crocodile skeleton is found there, and we wish to determine the species of the animal. Wildlife scientists have measured the lengths of the heads and the complete bodies of several crocs (in centimeters) of each species, creating the regression analyses below:

Indian Crocodile
Dependent variable is: IBody
R-squared = 97.2%

Variable	Coefficient
Intercept	−69.3693
IHead	7.40004

Australian Crocodile
Dependent variable is: ABody
R-squared = 98.0%

Variable	Coefficient
Intercept	−20.2245
AHead	7.71726

a) Do the associations between the sizes of the heads and bodies of the two species appear to be strong? Explain.

b) In what ways are the two relationships similar? Explain.

c) What is different about the two models? What does that mean?

d) The crocodile skeleton found had a head length of 62 cm and a body length of 380 cm. Which species do you think it was? Explain why.

(T) 19. How old is that tree? One can determine how old a tree is by counting its rings, but that requires cutting the tree down. Can we estimate the tree's age simply from its diameter? A forester measured 27 trees of the same species that had been cut down, and counted the rings to determine the ages of the trees.

Diameter (in.)	Age (yr)	Diameter (in.)	Age (yr)
1.8	4	10.3	23
1.8	5	14.3	25
2.2	8	13.2	28
4.4	8	9.9	29
6.6	8	13.2	30
4.4	10	15.4	30
7.7	10	17.6	33
10.8	12	14.3	34
7.7	13	15.4	35
5.5	14	11.0	38
9.9	16	15.4	38
10.1	18	16.5	40
12.1	20	16.5	42
12.8	22		

a) Find the correlation between *Diameter* and *Age*. Does this suggest that a linear model may be appropriate? Explain.

b) Create a scatterplot and describe the association.

c) Create the linear model.

d) Check the residuals. Explain why a linear model is probably not appropriate.

e) If you used this model, would it generally overestimate or underestimate the ages of very large trees? Explain.

T 20. Improving trees In the last exercise you saw that the linear model had some deficiencies. Let's create a better model.

a) Perhaps the cross-sectional area of a tree would be a better predictor of its age. Since area is measured in square units, try re-expressing the data by squaring the diameters. Does the scatterplot look better?

b) Create a model that predicts *Age* from the square of the *Diameter*.

c) Check the residuals plot for this new model. Is this model more appropriate? Why?

d) Estimate the age of a tree 18 inches in diameter.

21. New homes A real estate agent collects data to develop a model that will use the *Size* of a new home (in square feet) to predict its *Sale Price* (in thousands of dollars). Which of these is most likely to be the slope of the regression line: 0.008, 0.08, 0.8, or 8? Explain.

T 22. Smoking and pregnancy 2006 The Child Trends Data Bank monitors issues related to children. The table shows a 50-state average of the percent of expectant mothers who smoked cigarettes during their pregnancies.

Year	% Smoking While Pregnant	Year	% Smoking While Pregnant
1990	19.2	1999	14.1
1991	18.7	2000	14.0
1992	17.9	2001	13.8
1993	16.8	2002	13.3
1994	16.0	2003	12.7
1995	15.4	2004	10.9
1996	15.3	2005	10.1
1997	14.9	2006	10.0
1998	14.8		

a) Create a scatterplot and describe the trend you see.

b) Find the correlation.

c) How is the value of the correlation affected by the fact that the data are averages rather than percentages for each of the 50 states?

d) Write a linear model and interpret the slope in context.

T 23. No smoking? The downward trend in smoking you saw in the last exercise is good news for the health of babies, but will it ever stop?

a) Explain why you can't use the linear model you created in Exercise 22 to see when smoking during pregnancy will cease altogether.

b) Create a model that could estimate the year in which the level of smoking would be 0%.

c) Comment on the reliability of such a prediction.

24. Tips It's commonly believed that people use tips to reward good service. A researcher for the hospitality industry examined tips and ratings of service quality from 2645 dining parties at 21 different restaurants. The correlation between ratings of service and tip percentages was 0.11. (M. Lynn and M. McCall, "Gratitude and Gratuity." *Journal of Socio-Economics* 29: 203–214)

a) Describe the relationship between *Quality of Service* and *Tip Size*.

b) Find and interpret the value of R^2 in this context.

25. US cities Data from 50 large U.S. cities show the mean *January Temperature* and the *Latitude*. Describe what you see in the scatterplot.

26. Correlations The study of U.S. cities in Exercise 25 found the mean *January Temperature* (degrees Fahrenheit), *Altitude* (feet above sea level), and *Latitude* (degrees north of the equator) for 55 cities. Here's the correlation matrix:

	Jan. Temp	Latitude	Altitude
Jan. Temp	1.000		
Latitude	−0.848	1.000	
Altitude	−0.369	0.184	1.000

a) Which seems to be more useful in predicting *January Temperature—Altitude* or *Latitude*? Explain.

b) If the *Temperature* were measured in degrees Celsius, what would be the correlation between *Temperature* and *Latitude*?

c) If the *Temperature* were measured in degrees Celsius and the *Altitude* in meters, what would be the correlation? Explain.

d) What would you predict about the January *Temperatures* in a city whose *Altitude* is two standard deviations higher than the average *Altitude*?

27. Winter in the city Summary statistics for the data relating the latitude and average January temperature for 55 large U.S. cities are given below.

Variable	Mean	StdDev
Latitude	39.02	5.42
JanTemp	26.44	13.49

Correlation = −0.848

a) What percent of the variation in January *Temperatures* can be explained by variation in *Latitude?*

b) What is indicated by the fact that the correlation is negative?

c) Write the equation of the line of regression for predicting January *Temperature* from *Latitude*.

d) Explain what the slope of the line means.

e) Do you think the *y*-intercept is meaningful? Explain.

f) The latitude of Denver is 40° N. Predict the mean January temperature there.

g) What does it mean if the residual for a city is positive?

28. Depression The September 1998 issue of the *American Psychologist* published an article by Kraut et al. that reported on an experiment examining "the social and psychological impact of the Internet on 169 people in 73 households during their first 1 to 2 years online." In the experiment, 73 households were offered free Internet access for 1 or 2 years in return for allowing their time and activity online to be tracked. The members of the households who participated in the study were also given a battery of tests at the beginning and again at the end of the study. The conclusion of the study made news headlines: Those who spent more time online tended to be more depressed at the end of the experiment. Although the paper reports a more complex model, the basic result can be summarized in the following regression of *Depression* (at the end of the study, in "depression scale units") vs. *Internet Use* (in mean hours per week):

Dependent variable is: Depression
R-squared = 4.6%
s = 0.4563

Variable	Coefficient
Intercept	0.5655
Internet use	0.0199

The news reports about this study clearly concluded that using the Internet causes depression. Discuss whether such a conclusion can be drawn from this regression. If so, discuss the supporting evidence. If not, say why not.

29. Jumps 2012 How are Olympic performances in various events related? The plot shows winning long-jump and high-jump distances, in meters, for the Summer Olympics from 1912 through 2012.

a) Describe the association.

b) Do long-jump performances somehow influence the high-jumpers? How do you account for the relationship you see?

c) The correlation for the plotted data is 0.913. If we converted the jump lengths to centimeters by multiplying by 100, would that make the actual correlation higher or lower?

d) What would you predict about the long jump in a year when the high-jumper jumped one standard deviation better than the average high jump?

30. Modeling jumps 2012 Here are the summary statistics for the Olympic long jumps and high jumps displayed in the previous exercise.

Event	Mean	StdDev
High Jump	2.148	0.1939
Long Jump	8.05	0.5136

Correlation = 0.9125

a) Write the equation of the line of regression for estimating *High Jump* from *Long Jump*.

b) Interpret the slope of the line.

c) In a year when the long jump is 8.9 m, what high jump would you predict?

d) Why can't you use this line to estimate the long jump for a year when you know the high jump was 2.25 m?

e) Write the equation of the line you need to make that prediction.

31. French Consider the association between a student's score on a French vocabulary test and the weight of the student. What direction and strength of correlation would you expect in each of the following situations? Explain.

a) The students are all in third grade.

b) The students are in third through twelfth grades in the same school district.

c) The students are in tenth grade in France.

d) The students are in third through twelfth grades in France.

32. Twins Twins are often born after a pregnancy that lasts less than 9 months. On the next page is a graph from the *Journal of the American Medical Association (JAMA)* showing the rate of preterm twin births in the United States over the past 20 years. In this study, *JAMA* categorized mothers by the level of prenatal medical care they received: inadequate, adequate, or intensive.

a) Describe the overall trend in preterm twin births.

b) Describe any differences you see in this trend, depending on the level of prenatal medical care the mother received.

c) Should expectant mothers be advised to cut back on the level of medical care they seek in the hope of avoiding preterm births? Explain.

Preterm Birth Rate *per 100 live twin births among U.S. twins by intensive, adequate, and less than adequate prenatal care utilization, 1981–1997.* (JAMA *284[2000]: 335–341*)

T **33. Lunchtime** Create and interpret a model for the toddlers' lunchtime data presented below. The table and graph show the number of minutes the kids stayed at the table and the number of calories they consumed.

Calories	Time	Calories	Time
472	21.4	450	42.4
498	30.8	410	43.1
465	37.7	504	29.2
456	33.5	437	31.3
423	32.8	489	28.6
437	39.5	436	32.9
508	22.8	480	30.6
431	34.1	439	35.1
479	33.9	444	33.0
454	43.8	408	43.7

34. Gasoline Since clean-air regulations have dictated the use of unleaded gasoline, the supply of leaded gas in New York state has diminished. The table below was given on the August 2001 New York State Math B exam, a statewide achievement test for high school students.

Year	1984	1988	1992	1996	2000
Gallons (1000's)	150	124	104	76	50

a) Create a linear model and predict the number of gallons that will be available in 2005.
b) The exam then asked students to estimate the year when leaded gasoline will first become unavailable, expecting them to use the model from part a to answer the question. Explain why that method is incorrect.
c) Create a model that *would* be appropriate for that task, and make the estimate.
d) The "wrong" answer from the other model is fairly accurate in this case. *Why?*

T **35. Tobacco and alcohol** Are people who use tobacco products more likely to consume alcohol? Here are data on household spending (in pounds) taken by the British Government on 11 regions in Great Britain. Do tobacco and alcohol spending appear to be related? What questions do you have about these data? What conclusions can you draw?

Region	Alcohol	Tobacco
North	6.47	4.03
Yorkshire	6.13	3.76
Northeast	6.19	3.77
East Midlands	4.89	3.34
West Midlands	5.63	3.47
East Anglia	4.52	2.92
Southeast	5.89	3.20
Southwest	4.79	2.71
Wales	5.27	3.53
Scotland	6.08	4.51
Northern Ireland	4.02	4.56

T **36. Football weights** The Sears Cup was established in 1993 to honor institutions that maintain a broad-based athletic program, achieving success in many sports, both men's and women's. Since its Division III inception in 1995, the cup has been won by Williams College in every year except one. Their football team has a 85.3% winning record under their current coach. Why does the football team win so much? Is it because they're heavier than their opponents? The table shows the average team weights for selected years from 1973 to 1993.

Year	Weight (lb)	Year	Weight (lb)
1973	185.5	1983	192.0
1975	182.4	1987	196.9
1977	182.1	1989	202.9
1979	191.1	1991	206.0
1981	189.4	1993	198.7

a) Fit a straight line to the relationship between *Weight* and *Year*.
b) Does a straight line seem reasonable?

c) Predict the average weight of the team for the year 2003. Does this seem reasonable?

d) What about the prediction for the year 2103? Explain.

e) What about the prediction for the year 3003? Explain.

37. Models Find the predicted value of y, using each model for $x = 10$.

a) $\hat{y} = 2 + 0.8 \ln x$ b) $\log \hat{y} = 5 - 0.23x$

c) $\dfrac{1}{\sqrt{\hat{y}}} = 17.1 - 1.66x$

T 38. Williams vs Texas Here are the average weights of the football team for the University of Texas for various years in the 20th century.

Year	1905	1919	1932	1945	1955	1965
Weight (lb)	164	163	181	192	195	199

a) Fit a straight line to the relationship of *Weight* by *Year* for Texas football players.

b) According to these models, in what year will the predicted weight of the Williams College team from Exercise 36 first be more than the weight of the University of Texas team?

c) Do you believe this? Explain.

39. Vehicle weights The Minnesota Department of Transportation hoped that they could measure the weights of big trucks without actually stopping the vehicles by using a newly developed "weigh-in-motion" scale. After installation of the scale, a study was conducted to find out whether the scale's readings correspond to the true weights of the trucks being monitored. In Exercise 46 of Chapter 6, you examined the scatterplot for the data they collected, finding the association to be approximately linear with $R^2 = 93\%$. Their regression equation is $\widehat{Wt} = 10.85 + 0.64\,Scale$, where both the scale reading and the predicted weight of the truck are measured in thousands of pounds.

a) Estimate the weight of a truck if this scale read 31,200 pounds.

b) If that truck actually weighed 32,120 pounds, what was the residual?

c) If the scale reads 35,590 pounds, and the truck has a residual of −2440 pounds, how much does it actually weigh?

d) In general, do you expect estimates made using this equation to be reasonably accurate? Explain.

e) If the police plan to use this scale to issue tickets to trucks that appear to be overloaded, will negative or positive residuals be a greater problem? Explain.

40. Profit How are a company's profits related to its sales? Let's examine data from 71 large U.S. corporations. All amounts are in millions of dollars.

a) Histograms of *Profits* and *Sales* and histograms of the logarithms of *Profits* and *Sales* appear below. Why are the re-expressed data better for regression?

b) Here are the scatterplot and residuals plot for the regression of logarithm of *Profits* vs. Log(*Sales*). Do you think this model is appropriate? Explain.

c) Here's the regression analysis. Write the equation.

Dependent variable is: Log Profit
R-squared = 48.1%

Variable	Coefficient
Intercept	−0.106259
LogSales	0.647798

d) Use your equation to estimate profits earned by a company with sales of 2.5 billion dollars. (That's 2500 million.)

T 41. Down the drain Most water tanks have a drain plug so that the tank may be emptied when it's to be moved or repaired. How long it takes a certain size of tank to drain depends on the size of the plug, as shown in the table. Create a model.

Plug Dia (in.)	$\frac{3}{8}$	$\frac{1}{2}$	$\frac{3}{4}$	1	$1\frac{1}{4}$	$1\frac{1}{2}$	2
Drain Time (min.)	140	80	35	20	13	10	5

42. Chips A start-up company has developed an improved electronic chip for use in laboratory equipment. The company needs to project the manufacturing cost, so it develops a spreadsheet model that takes into account the purchase of production equipment, overhead, raw materials, depreciation, maintenance, and other business costs. The spreadsheet estimates the cost of producing 10,000 to 200,000 chips per year, as seen in the table. Develop a regression model to predict *Costs* based on the *Level* of production.

Chips Produced (1000s)	Cost per Chip ($)	Chips Produced (1000s)	Cost per Chip ($)
10	146.10	90	47.22
20	105.80	100	44.31
30	85.75	120	42.88
40	77.02	140	39.05
50	66.10	160	37.47
60	63.92	180	35.09
70	58.80	200	34.04
80	50.91		

Practice Exam

I. Multiple Choice

(Questions 1–3) Based on data collected over several sessions, a statistically minded trainer of office typists modeled the linear relationship between the number of hours of training a typist receives and the typist's speed (in words per minute) with the equation $\widehat{speed} = 10.6 + 5.4\,hour$.

1. Which of these statements best interprets this equation?

A) Typists increase their speed by 10.6 wpm for every 5.4 hours of training.

B) Typists increase their speed by 5.4 wpm for every 10.6 hours of training.

C) A typist who trains for an additional hour will benefit with a speed increase of 5.4 wpm.

D) On average, typists tend to increase their speed by roughly 5.4 wpm for every hour of training.

E) For every 5.4 hours of training, typists can increase their speed from 10.6 wpm to faster.

2. Which is the best interpretation of the *y*-intercept for this model?

A) People who can't type need about 10.6 hours of training.

B) Before undergoing this training, typists' average speed was about 10.6 words per minute.

C) The *y*-intercept is meaningless here because no one types at 0 wpm.

D) The *y*-intercept is meaningless here because none of the typists had 0 hours of training.

E) In regression models, the slope has meaning, but not the *y*-intercept.

3. After some training, one of the typists was told that the speed he attained had a residual of 4.3 words per minute. How should he interpret this?

A) He types slower than the model predicted, given the amount of time he spent training.

B) He types faster than the model predicted, given the amount of time he spent training.

C) He can't interpret his residual without also knowing the correlation.

D) He can't interpret his residual without also knowing the size of other people's residuals.

E) He can't interpret his residual without also knowing the standard deviation of the residuals.

4. The Bureau of Labor Statistics looked at the association between students' GPAs in high school (*gpa_HS*) and their freshmen GPAs at a University of California school (*gpa_U*). The resulting least-squares regression equation is $\widehat{gpa\_U} = 0.22 + 0.72 gpa\_HS$. Calculate the residual for a student with a 3.8 in high school who achieved a freshman GPA of 3.5.

A) −0.844   B) −0.544   C) 2.956

D) 0.544   E) 0.844

5. In April of 2012, the Centers for Disease Control and Prevention announced that birth rates for U.S. teenagers reached historic lows. From 2009 to 2010 the rate declined 9%, to a level of 34.3 births per 1000 women aged 15–19. Which of these conclusions is an example of extrapolation in this context?

A) There was a decreasing trend in teenage birth rates at the time of this study.

B) Time is an explanatory variable in the change of teenage birth rates.

C) By 2014, teenage birth rates will be 36% lower and set new records.

D) There is a linear relationship between year and teenage birth rates.

E) None of these is an example of extrapolation.

6. An engineer studying the performance of a certain type of bolt predicts the failure rate (bolts per 1000) from the load (in pounds) using the model $\log(\widehat{fail}) = 1.04 + 0.0013 load$. If these bolts are subjected to a load of 600 pounds, what failure rate should we expect?

A) 0.26   B) 0.60   C) 1.82

D) 6.17   E) 66.07

7. A researcher analyzing some data created a linear model with $R^2 = 94\%$ and having the residuals plot seen here. What should she conclude?

A) The linear model is appropriate, because about the half the residuals are positive and half negative.

B) The linear model is appropriate, because the value of R^2 is quite high.

C) The linear model is not appropriate, because the value of R^2 is not high enough.

D) The linear model is not appropriate, because the residuals plot shows curvature.

E) The linear model is not appropriate, because the residuals plot identifies an outlier.

8. Researchers at UC San Francisco discovered that high plasma levels of vitamins B, C, D, and E are associated with better cognitive performance. "Each standard deviation higher plasma level for these vitamins predicted a global cognitive score 0.28 standard deviations better," the researchers reported. Which value are the researchers interpreting in this statement?

A) the correlation coefficient between plasma level and cognitive score

B) the y-intercept of the regression model predicting cognitive score from plasma level

C) the slope of the regression model predicting cognitive score from plasma level

D) the standard deviation of the regression model's residuals

E) R^2 for the regression model

9. This graph shows the relationship the number of days since the NBA season began and the number of injuries, over the course of three different seasons. In 2011–12, the season was shortened by a labor strike.

Of statements A–D, which of the following is NOT a correct conclusion that can be drawn from this graph?

A) There is a fairly strong linear relationship between days since the start of the season and the number of injuries.

B) At first the rate of injuries was higher during the strike-shortened season.

C) As the strike-shortened season continued the number of injuries became similar to the other two seasons.

D) While the strike-shortened season had more injuries initially, we cannot know for certain if the strike caused the difference or if it was attributable to other variables.

E) All of A-D are correct.

10. Which of statements A–D is true?

A) An influential point always has a large residual.

B) An influential point changes the slope of the regression equation.

C) An influential point decreases the value of R^2.

D) An influential point does not affect the y-intercept.

E) Statements A–D are all false.

(Questions 11–12) The segmented bar charts below depict the data from the NAAL (National Assessment of Adult Literacy) conducted in 2003.

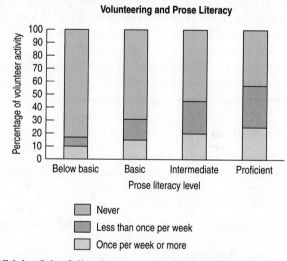

11. Which of the following is greatest?

A) The number of people who volunteer once per week or more and test Below Basic on Prose Literacy.

B) The number of people who volunteer less than once per week and test Basic on Prose Literacy.

C) The number of people who never volunteer and test Proficient on Prose Literacy.

D) The number of people who volunteer less than once per week and test Intermediate in Prose Literacy.

E) It is impossible to determine which is greatest without knowing the actual number of people at each literacy level.

12. Based on the segmented bar graphs, does there appear to be an association between volunteerism and literacy level?

A) Yes, all three bars have the same number of segments.

B) Yes, because all three bars have the same height.

C) Yes, because the corresponding segments of the three bars have different heights.

D) No, because the corresponding segments of the three bars have different heights.

E) No, because the sums of the 3 proportions in each bar are identical.

13. A TV weatherman's end-of-year analysis of the local weather showed that among all the years for which records had been kept, the past year's average temperature had a z-score of 2.8. What does that mean?

A) The past year's average temperature was 2.8° higher than the historical mean.

B) The past year's average temperature was 2.8 standard deviations above the historical mean.

C) The past year's average temperature was 2.8% higher than the historical mean.

D) The past year's temperatures had a standard deviation of 2.8°.

E) The past year had 2.8 times as many days with above average temperatures as is typical for that area.

14. In Statsville there's a city-wide speed limit of 30 mph. If you are caught speeding the fine is $100 plus $10 for every mile per hour you were over the speed limit. For example, if you're ticketed for going 45 mph, your fine is $100 + 10(45 - 30) = 250. Last month all the drivers who were fined for speeding averaged 42 mph with a standard deviation of 7 mph. What were the mean and standard deviation of the fines?

A) $120 and $70   B) $220 and $7
C) $220 and $70   D) $220 and $170
E) $420 and $70

15. Among those Statsville drivers fined for speeding, the fastest 10% were caught exceeding how many miles per hour?

A) 37.0   B) 48.3   C) 51.0   D) 58.3
E) It cannot be determined from the information given.

II. Free Response

1. A diligent statistics student recorded the length of his faithful #2 pencil as he worked away on his homework. He discovered a strong linear relationship between the number of hours that he worked and the length of his pencil. Here is the regression analysis for these data.

Dependent variable: length (cm)
$R^2 = 92.3\%$ $R^2(\text{adj}) = 89.5\%$

	coeff	se	t ratio	p value
constant	17.047	0.128	23.58	<0.0001
time (hr)	−1.914	0.047	35.28	<0.0001

a) Write the equation of the least square regression line.

b) Interpret R^2 in this context.

c) Interpret the equation in this context.

d) This student's girlfriend tried out his model on a pencil she had used for 5 hours, and found a residual of −0.88 cm. How long was her pencil at that time?

e) Should she have expected this model to describe the rate for her pencils? Why or why not?

2. Energy drinks come in different-sized packages: pouches, small bottles, large bottles, twin-packs, 6-packs, and so on. How is the price related to the amount of beverage? Data collected on a variety of packages revealed a mean size of 140.17 ounces with a standard deviation of 78.23 ounces. The packages had an average price of $3.66 with a standard deviation of $1.50, and the correlation between size and price was $r = 0.91$. A scatterplot of these data suggested that the assumptions needed for regression were reasonable.

a) Interpret the value of r in context.

b) Compute the slope of the least-squares regression line for predicting the price of an energy drink. Include the proper units in your answer.

c) Write the equation for the least-squares regression line for these data.

d) For this model the standard deviation of the residuals was $s = 0.26$. Explain what that means in context.

3. The Pew Research Center conducted two surveys, one in December 2011 and another in November 2012, asking people about their reading habits. Pew reported the percentage of people who read at least one e-book in the past year, given that they had read at least one book. Those percentages, broken down by age group, are shown in the table below.

Date of Poll	Age Group				
	16–17	18–29	30–49	50–64	65+
December 2011	13	25	25	19	12
November 2012	28	31	41	23	20

a) Create an appropriate graphical display that allows a comparison of responses between the two years and also among the different age groups.

b) Write a few sentences comparing e-book readership in the two time periods.

c) Is there an association between age and the growth in e-book readership? Use evidence from the table or your graph to justify your answer.

10 Understanding Randomness

We all know what it means for something to be random. Or do we? Many children's games rely on chance outcomes. Rolling dice, spinning spinners, and shuffling cards all select at random. Adult games use randomness as well, from card games to lotteries to Bingo. What's the most important aspect of the randomness in these games? It must be fair.

What is it about random selection that makes it seem fair? It's really two things. First, nobody can guess the outcome before it happens. Second, when we want things to be fair, usually some underlying set of outcomes will be equally likely (although in many games, some combinations of outcomes are more likely than others).

Randomness is not always what we might think of as "at random." Random outcomes have a lot of structure, especially when viewed in the long run. You can't predict how a fair coin will land on any single toss, but you're pretty confident that if you flipped it thousands of times you'd see about 50% heads. As we will see, randomness is an essential tool of Statistics. Statisticians don't think of randomness as the annoying tendency of things to be unpredictable or haphazard. Statisticians use randomness as a tool. In fact, without deliberately applying randomness, we couldn't do most of Statistics, and this book would stop right about here.[1]

But truly random values are surprisingly hard to get. Just to see how fair humans are at selecting, pick a number at random from the top of the next page. Go ahead. Turn the page, look at the numbers quickly, and pick a number at random.

Ready?

Go.

---

[1] Don't get your hopes up.

1 2 3 4

It's Not Easy Being Random

> "The generation of random numbers is too important to be left to chance."
>
> —*Robert R. Coveyou,*
> *Oak Ridge National*
> *Laboratory*

A S **Activity: Random Behavior.**
ActivStats' Random Experiment Tool lets you experiment with truly random outcomes. We'll use it a lot in the coming chapters.

A S **Activity: Truly Random Values on the Internet.** This activity will take you to an Internet site (www.random.org) that generates all the truly random numbers you could want.

An ordinary deck of playing cards, like the ones used in bridge and many other card games, consists of 52 cards. There are numbered cards (2 through 10), and face cards (Jack, Queen, King, Ace) whose value depends on the game you are playing. Each card is also marked by one of four suits (clubs, diamonds, hearts, or spades) whose significance is also game-specific.

Did you pick 3? If so, you've got company. Almost 75% of all people pick the number 3. About 20% pick either 2 or 4. If you picked 1, well, consider yourself a little different. Only about 5% choose 1. Psychologists have proposed reasons for this phenomenon, but for us, it simply serves as a lesson that we've got to find a better way to choose things at random.

So how should we generate **random numbers?** It's surprisingly difficult to get random values even when they're equally likely. Computers, calculators, and even smartphones have become a popular way to generate random numbers. Even though they often do much better than humans, they can't generate truly random numbers either. Start a program or app from the same place, and it will always follow exactly the same path, so such numbers generated are not truly random. Technically, "random" numbers generated this way are *pseudorandom* numbers. Fortunately, pseudorandom values are virtually indistinguishable from truly random numbers, and that's usually good enough.

There *are* ways to generate random numbers so that they are both equally likely and truly random. There are published tables of carefully generated random numbers.[2] Or, we can find genuinely random digits on the Internet. The sites use methods like timing the decay of a radioactive element to generate truly random digits.[3] A string of random digits might look like this:

```
2217726304387410092537086270581997622725849795907032825001108963
3217535822643800292254644943760642389043766557204107354186024508
8906427308645681412198226653885873285801699027843110380420067664
8740522639824530519902027044464984322000946238678577902639002954
8887003319933147508331265192321413908608674496383528968974910533
6944182713168919406022181281304751019321546303870481407676636740
6070204916508913632855535136136104379429342848690946288143179360
7706356513310563210508993624272872250535395513645991015328128202
```

The best ways we know to generate data that give a fair and accurate picture of the world rely on randomness, and the ways in which we draw conclusions from those data depend on the randomness, too. If this sounds familiar to you, it should. The *What If. . .* explorations you've seen in several chapters use simulations based on random numbers. Remember that we promised you'd learn how to create your own simulations later on? Well, "later on" is here!

---

[2] You'll find a table of random digits of this kind in the back of this book.

[3] For example, www.random.org or www.randomnumbers.info.

Aren't You Done Shuffling Yet? Even something as common as card shuffling may not be as random as you might think. If you shuffle cards by the usual method in which you split the deck in half and try to let cards fall roughly alternately from each half, you're doing a "riffle shuffle."

How many times should you shuffle cards to make the deck random? A surprising fact was discovered by statisticians Persi Diaconis, Ronald Graham, and W. M. Kantor. It takes seven riffle shuffles. Fewer than seven leaves order in the deck, but after that, more shuffling does little good. Most people, though, don't shuffle that many times.

When computers were first used to generate hands in bridge tournaments, some professional bridge players complained that the computer was making too many "weird" hands—hands with 10 cards of one suit, for example. Suddenly these hands were appearing more often than players were used to when cards were shuffled by hand. The players assumed that the computer was doing something wrong. But it turns out that it's humans who hadn't been shuffling enough to make the decks really random and have those "weird" hands appear as often as they should.

Let's Simulate!

Suppose a cereal manufacturer puts pictures of famous athletes on cards in boxes of cereal in the hope of boosting sales. The manufacturer announces that 20% of the boxes contain a picture of basketball star LeBron James, 30% a picture of race car driver Danica Patrick, and the rest a picture of tennis champion Serena Williams. You want all three pictures. How many boxes of cereal do you expect to have to buy in order to get the complete set?

How can we answer questions like this? Well, one way is to buy hundreds of boxes of cereal to see what might happen. But let's not. Instead, we'll consider using a random model. Why random? When we pick a box of cereal off the shelf, we don't know what picture is inside. We'll assume that the pictures are randomly placed in the boxes and that the boxes are distributed randomly to stores around the country. Why a model? Because we won't actually buy the cereal boxes. We can't afford all those boxes and we don't want to waste food. So we need an imitation of the real process that we can manipulate and control. In short, we're going to simulate reality.

A **simulation** mimics reality by using random numbers to represent the outcomes of real events. Just as pilots use flight simulators to learn about and practice real situations, we can learn a great deal about the real events by carefully modeling the randomness and analyzing the simulation results.

The question we've asked is how many boxes do you expect to buy to get a complete card collection. But we can't answer our question by completing a card collection just once. We want to understand the *typical* number of boxes to open, how that number varies, and, often, the shape of the distribution. So we'll have to do this over and over. We call each time we obtain a simulated answer to our question a **trial.**

For the sports cards, a trial's outcome is the number of boxes. We'll need at least 3 boxes to get one of each card, but with really bad luck, you could empty the shelves of several supermarkets before finding the card you need to get all 3. So, the possible outcomes of a trial are 3, 4, 5, or lots more. But we can't simply pick one of those numbers at random, because they're not equally likely. We'd be surprised if we only needed 3 boxes to get all the cards, but we'd probably be even more surprised to find that it took exactly 7,359 boxes. In fact, the reason we're doing the simulation is that it's hard to guess how many boxes we'd expect to open.

Building a Simulation

We know how to find equally likely random digits. How can we get from there to simulating the trial outcomes? We know the relative frequencies of the cards: 20% LeBron, 30% Danica, and 50% Serena. So, we can interpret the digits 0 and 1 as finding LeBron; 2, 3, and 4 as finding Danica; and 5 through 9 as finding Serena to simulate opening one box.

Opening one box is the basic building block, called a **component** of our simulation. But the component's outcome isn't the result we want. We need to observe a sequence of components until our card collection is complete. The *trial's* outcome is called the **response variable**; for this simulation that's the *number* of components (boxes) in the sequence.

Let's look at the steps for making a simulation:

Specify how to model a component outcome using equally likely random digits:

1. **Identify the component to be repeated.** In this case, our component is the opening of a box of cereal.
2. **Explain how you will model the component's outcome.** The digits from 0 to 9 are equally likely to occur. Because 20% of the boxes contain LeBron's picture, we'll use 2 of the 10 digits to represent that outcome. Three of the 10 digits can model the 30% of boxes with Danica's cards, and the remaining 5 digits can represent the 50% of boxes with Serena. One possible assignment of the digits, then, is

0, 1 LeBron   2, 3, 4 Danica   5, 6, 7, 8, 9 Serena.

Specify how to simulate trials:

3. **Explain how you will combine the components to model a trial.** We pretend to open boxes (repeat components) until our collection is complete. We do this by looking at each random digit and indicating what picture it represents. We continue until we've found all three.
4. **State clearly what the response variable is.** What are we interested in? We want to find out the number of boxes it might take to get all three pictures.

Put it all together to run the simulation:

5. **Run several trials.** For example, consider the third line of random digits shown earlier (p. 268):

8906427308645681412198226653885873285801699027843110380420067664.

Let's see what happens.

The first random digit, 8, means you get Serena's picture. So the first component's outcome is Serena. The second digit, 9, means Serena's picture is also in the next box. Continuing to interpret the random digits, we get LeBron's picture (0) in the third, Serena's (6) again in the fourth, and finally Danica (4) on the fifth box. Since we've now found all three pictures, we've finished one trial of our simulation. This trial's outcome is 5 boxes.

Now we keep going, running more trials by looking at the rest of our line of random digits:

89064 2730 8645681 41219 822665388587328580 169902 78431 1038 042006 7664.

It's best to create a chart to keep track of what happens:

# of Boxes

Trial Number	Component Outcomes	Trial Outcomes: y = Number of boxes
1	89064 = Serena, Serena, LeBron, Serena, Danica	5
2	2730 = Danica, Serena, Danica, LeBron	4
3	8645681 = Serena, Serena, Danica, . . . , LeBron	7
4	41219 = Danica, LeBron, Danica, LeBron, Serena	5
5	822665388587328580 = Serena, Danica, . . . , LeBron	18
6	169902 = LeBron, Serena, Serena, Serena, LeBron, Danica	6
7	78431 = Serena, Serena, Danica, Danica, LeBron	5
8	1038 = LeBron, LeBron, Danica, Serena	4
9	042006 = LeBron, Danica, Danica, LeBron, LeBron, Serena	6
10	7664 . . . = Serena, Serena, Serena, Danica . . .	?

Analyze the response variable:

6. **Collect and summarize the results of all the trials.** You know how to summarize and display a response variable. You'll certainly want to report the shape, center, and spread, and depending on the question asked, you may want to include more.

7. **State your conclusion,** as always, in the context of the question you wanted to answer. Based on this simulation, we estimate that customers hoping to complete their card collection will need to open a median of 5 boxes, but it could take a lot more.

_**A** **S**_ *Activity:* **Bigger Samples Are Better.** The random simulation tool can generate lots of outcomes with a single click, so you can see more of the long run with less effort.

If you fear that these may not be accurate estimates because we ran only nine trials, you are absolutely correct. The more trials the better, and nine is woefully inadequate. How many is enough? We'll explore that question in this chapter's *What If. . . .*

For Example SIMULATING A DICE GAME

The game of 21 can be played with an ordinary 6-sided die. Competitors each roll the die repeatedly, trying to get the highest total less than or equal to 21. If your total exceeds 21, you lose.

Suppose your opponent has rolled an 18. Your task is to try to beat him by getting more than 18 points without going over 21. How many rolls do you expect to make, and what are your chances of winning?

QUESTION: How will you simulate the components?

ANSWER: A component is one roll of the die. I'll simulate each roll by looking at a random digit from a table or an Internet site. The digits 1 through 6 will represent the results on the die; I'll ignore digits 7–9 and 0.

QUESTION: How will you combine components to model a trial? What's the response variable?

ANSWER: I'll add components until my total is greater than 18, counting the number of rolls. If my total is greater than 21, it is a loss; if not, it is a win. There are two response variables. I'll count the number of times I roll the die, and I'll keep track of whether I win or lose.

QUESTION: How would you use these random digits to run trials? Show your method clearly for two trials.

91129 58757 69274 92380 82464 33089

ANSWER: I've marked the discarded digits in color.

Trial #1:	9	**1**	**1**	**2**	9	**5**	8	7	**5**	7	**6**		
Total:		1	2	4		9			14		20		Outcomes: 6 rolls, won

Trial #2:	9	**2**	7	**4**	9	**2**	**3**	8	0	8	**2**	**4**	**6**	
Total:		2		6		8	11				13	17	23	Outcomes: 7 rolls, lost

QUESTION: Suppose you run 30 trials, getting the outcomes tallied here. What is your conclusion?

ANSWER: Based on my simulation, when competing against an opponent who has a score of 18, I expect my turn to usually last 5 or 6 rolls, and I should win about 70% of the time.

Number of rolls		Result	
4	////	Won	〽〽〽〽 /
5	〽〽	Lost	〽 ////
6	〽〽 /		
7	〽		
8	/		

✓ Just Checking

The baseball World Series consists of up to seven games. The first team to win four games wins the series. The first two are played at one team's home ballpark (Team *A*), the next three at the other team's park (Team *B*), and the final two (if needed) are played back at Team *A*'s park. Records over the past century show that there is a home field advantage; in any game the home team has about a 55% chance of winning. Does the current system of alternating ballparks even out the home field advantage? How often will Team *A*, who begins at home, win the series?

   Let's set up the simulation:

1. What is the component to be repeated?

2. How will you model each component from equally likely random digits?

3. How will you model a trial by combining components?

4. What is the response variable?

5. How will you analyze the response variable?

Step-by-Step Example   SIMULATION

Fifty-seven students participated in a lottery for a particularly desirable dorm room—a triple with a fireplace and private bath in the tower. Twenty of the participants were members of the same varsity team. When all three winners were members of the team, the other students cried foul.

Question: Could an all-team outcome reasonably be expected to happen if everyone had a fair shot at the room?

THINK **Plan**   State the problem. Identify the important parts of your simulation.

> I'll use a simulation to investigate whether it's unlikely that three varsity athletes would get the great room in the dorm if the lottery were fair.

Components   Identify the components.

> A component is the selection of a student.

Outcomes   State how you will model each component using equally likely random digits. You can't just use the digits from 0 to 9 because the outcomes you are simulating are not multiples of 10%.

There are 20 and 37 students in the two groups. This time you must use *pairs* of random digits (and ignore some of them) to represent the 57 students.

> I'll look at two-digit random numbers.
>
> Let 00–19 represent the 20 varsity applicants.
>
> Let 20–56 represent the other 37 applicants.
>
> Skip 57–99. If I get a number in this range, I'll throw it away and go back for another two-digit random number.

Trial   Explain how you will combine the components to simulate a trial. In each of these trials, you can't choose the same student twice, so you'll need to ignore a random number if it comes up a second or third time. Be sure to mention this in describing your simulation.

> Each trial consists of identifying pairs of digits as V (varsity) or N (nonvarsity) until 3 people are chosen, ignoring out-of-range or repeated numbers (X)—I can't put the same person in the room twice.

Response Variable   Define your response variable.

> The response variable is whether or not all three selected students are on the varsity team.

(continued)

SHOW ➡ **Mechanics** Run several trials. Carefully record the random numbers, indicating

1) the corresponding component outcomes (here, varsity, nonvarsity, or ignored number) and

2) the value of the response variable.

Trial Number	Component Outcomes	All Varsity?
1	74 02 94 39 02 77 55 X V X N X X N	No
2	18 63 33 25 V X N N	No
3	05 45 88 91 56 V N X X N	No
4	39 09 07 N V V	No
5	65 39 45 95 43 X N N X N	No
6	98 95 11 68 77 12 17 X X V X X V V	Yes
7	26 19 89 93 77 27 N V X X X N	No
8	23 52 37 N N N	No
9	16 50 83 44 V N X N	No
10	74 17 46 85 09 X V N X V	No

Analyze Summarize the results across all trials to answer the initial question.

"All varsity" occurred once, or 10% of the time.

TELL ➡ **Conclusion** Describe what the simulation shows, and interpret your results in the context of the real world.

In my simulation of "fair" room draws, the three people chosen were all varsity team members only 10% of the time. While this result could happen by chance, it is not particularly likely. I'm suspicious, but I'd need many more trials and a smaller frequency of the all-varsity outcome before I would make an accusation of unfairness.

TI Tips GENERATING RANDOM NUMBERS

Instead of using coins, dice, cards, or tables of random numbers, you may decide to use your calculator for simulations. There are several random number generators offered in the MATH PRB menu.

randInt(is of particular importance. This command will produce any number of random integers in a specified range.

Here are some examples showing how to use randInt for simulations:

- randInt(0,1) randomly chooses a 0 or a 1. This is an effective simulation of a coin toss. You could let 0 represent tails and 1 represent heads.
- randInt(1,6) produces a random integer from 1 to 6, a good way to simulate rolling a die.

(continued)

- `randInt(1,6,2)` simulates rolling *two* dice. To do several rolls in a row, just hit ENTER repeatedly.

- `randInt(0,9,5)` produces five random integers that might represent the pictures in the cereal boxes. Our run gave us two LeBrons (0, 1), no Danicas (2, 3, 4), and three Serenas (5–9).

- `randInt(0,56,3)` produces three random integers between 0 and 56, a nice way to simulate the dorm room lottery. The window shows 6 trials, but we would skip the first one because one student was chosen twice. In none of the remaining 5 trials did three athletes (0–19) win.

WHAT IF ●●● we don't simulate enough trials?

When we showed you how to do simulations, first looking for the cereal box pictures and then checking the fairness of dorm room assignments, we ran just 10 trials. Let's see why that's not really enough. How? With a simulation, of course!

As an easy example, we'll just pretend to open cereal boxes looking for pictures of LeBron James. While he's in 20% of the boxes, that doesn't tell us what we'll actually find as we go box by box. The table shows the results of the first 10 trials.

Remember that the intent of a simulation is to gain insight about situations we don't understand. If we didn't already know that LeBron's picture is in 20% of the boxes, these 10 trials wouldn't tell us that. At best, we might feel comfortable guessing that fewer than half of the boxes contain LeBron's picture, but 10 trials just isn't enough to say anything very definitive.

For the homework[4] exercises we suggest you do 20 trials. How much better is that? Let's open at more cereal boxes. Look at the graph displaying LeBron's percentage after each of the first 10 trials in the table above and for 10 more trials.

Now would we conclude that 20% was correct? Probably not, even though the estimated percentages do appear to be settling down a bit. It appears 20 trials is still too few. So let's go big. We used a computer to run 1000 trials. The graph on the next page shows what happened.

The simulation . . .			Results so far . . .	
Box number	Random digit	Picture found	Number of LeBrons	Percent LeBron
1	8	Serena	0 out of 1	0%
2	3	Danica	0 out of 2	0%
3	3	Danica	0 out of 3	0%
4	0	LeBron	1 out of 4	25%
5	6	Serena	1 out of 5	20%
6	1	LeBron	2 out of 6	33%
7	9	Serena	2 out of 7	28%
8	2	Danica	2 out of 8	25%
9	9	Serena	2 out of 9	22%
10	1	LeBron	3 out of 10	30%

---

[4]Stop making that face. You *knew* there'd be homework.

(continued)

It appears that in this simulation there were quite a few LeBron pictures in the first 100 (or so) boxes, but as the number of trials mounted the percentage drifted toward the true value of 20%. With 1,000 trials we might be able to make a pretty good guess about the cereal boxes. Frankly, though, this is a pretty simple situation. In the real world, simulations are used to explore very complex issues like climate change, election outcomes, and even national defense. Those investigations require tens or even hundreds of thousands of trials![5]

---

[5]We hope that makes you feel better about doing just 20 trials for the homework. See, we're actually being nice to you!

WHAT CAN GO WRONG?

■ **Don't overstate your case.** Let's face it: In some sense, a simulation is *always* wrong. After all, it's not the real thing. We didn't buy any cereal or run a room draw. So beware of confusing what *really* happens with what a simulation suggests *might* happen. Never forget that future results will not match your simulated results exactly.

A S *Activity*: **Estimating Summaries from Random Outcomes.** See how well you can estimate something you can't know just by generating random outcomes.

■ **Model outcome chances accurately.** A common mistake in constructing a simulation is to adopt a strategy that may appear to produce the right kind of results, but that does not accurately model the situation. For example, in our room draw, we could have gotten 0, 1, 2, or 3 team members. Why not just see how often these digits occur in random digits from 0 to 9, ignoring the digits 4 and up?

$$3\ 2\ 1\ 7\ 9\ 0\ 0\ 5\ 9\ 7\ 3\ 7\ 9\ 2\ 5\ 2\ 4\ 1\ 3\ 8$$

$$\mathbf{3\ 2\ 1\ x\ x\ 0\ 0\ x\ x\ x\ 3\ x\ x\ 2\ x\ 2\ x\ 1\ 3\ x}$$

This "simulation" makes it seem fairly likely that three team members would be chosen. There's a big problem with this approach, though: The digits 0, 1, 2, and 3 occur with equal frequency among random digits, making each outcome appear to happen 25% of the time. In fact, the selection of 0, 1, 2, or all 3 team members are not all equally likely outcomes. In our correct simulation, we estimated that all 3 would be chosen only about 10% of the time. If your simulation overlooks important aspects of the real situation, your model will not be accurate.

TI-*nspire*

Simulations. Improve your predictions by running thousands of trials.

■ **Run enough trials.** Simulation is cheap and fairly easy to do. Don't try to draw conclusions based on 5 or 10 trials (even though we did for illustration purposes here). We'll make precise how many trials to use in later chapters. For now, err on the side of large numbers of trials.

What Have We Learned?

We've learned to harness the power of randomness. We've learned that a simulation model can help us investigate a question for which many outcomes are possible, we can't (or don't want to) collect data, and a mathematical answer is hard to calculate. We've learned how to base our simulation on random values generated by a computer, generated by a randomizing device such as a die or spinner, or found on the Internet. Like all models, simulations can provide us with useful insights about the real world.

Terms

Random	An outcome is random if we know the possible values it can have, but not which particular value it takes. (p. 267)
Generating random numbers	Random numbers are hard to generate. Nevertheless, several Internet sites offer an unlimited supply of equally likely random values. (p. 268)
Simulation	A simulation models a real-world situation by using random-digit outcomes to mimic the uncertainty of a response variable of interest. (p. 269)
Simulation component	A component uses equally likely random digits to model simple random occurrences whose outcomes may not be equally likely. (p. 270)
Trial	The sequence of several components representing events that we are pretending will take place. (p. 269)
Response variable	Values of the response variable record the results of each trial with respect to what we were interested in. (p. 270)

On the Computer SIMULATION

Simulations are best done with the help of technology simply because running more trials makes for a better simulation, and computers are fast. There are special computer programs designed for simulation, and most statistics packages and calculators can at least generate random numbers to support a simulation.

All technology-generated random numbers are *pseudorandom*. The random numbers available on the Internet may technically be better, but the differences won't matter for any simulation of modest size. Pseudorandom numbers generate the next random value from the previous one by a specified algorithm. But they have to start somewhere. This starting point is called the "seed." Most programs let you set the seed. There's usually little reason to do this, but if you wish to, go ahead. If you reset the seed to the same value, the programs will generate the same sequence of "random" numbers.

A S **Activity: Creating Random Values.** Learn to use your statistics package to generate random outcomes.

Exercises

1. **Random outcomes** For each of the following scenarios, decide if the outcome is random.
   a) Flip a coin to decide who takes out the trash. Is who takes out the trash random?
   b) A friend asks you to quickly name a professional sports team. Is the sports team named random?
   c) Names are selected out of a hat to decide roommates in a dormitory. Is your roommate for the year random?

2. **More random outcomes** For each of the following scenarios, decide if the outcome is random.
   a) You enter a contest in which the winning ticket is selected from a large drum of entries. Was the winner of the contest random?

   b) When playing a board game, the number of spaces you move is decided by rolling a six-sided die. Is the number of spaces you move random?
   c) Before flipping a coin, your friend asks you to "call it." Is your choice (heads or tails) random?

3. **The lottery** Many states run lotteries, giving away millions of dollars if you match a certain set of winning numbers. How are those numbers determined? Do you think this method guarantees randomness? Explain.

4. **Games** Many kinds of games people play rely on randomness. Cite three different methods commonly used in the attempt to achieve this randomness, and discuss the effectiveness of each.

5. Birth defects The American College of Obstetricians and Gynecologists says that out of every 100 babies born in the United States, 3 have some kind of major birth defect. How would you assign random numbers to conduct a simulation based on this statistic?

6. Colorblind By some estimates, about 10% of all males have some color perception defect, most commonly red–green colorblindness. How would you assign random numbers to conduct a simulation based on this statistic?

7. Geography An elementary school teacher with 25 students plans to have each of them make a poster about two different states. The teacher first numbers the states (in alphabetical order, from 01-Alabama to 50-Wyoming), then uses a random number table to decide which states each kid gets. Here are the random digits:

45921 01710 22892 37076

a) Which two state numbers does the first student get?
b) Which two state numbers go to the second student?

8. Get rich Your state's BigBucks Lottery prize has reached $100,000,000, and you decide to play. You have to pick five numbers between 1 and 60, and you'll win if your numbers match those drawn by the state. You decide to pick your "lucky" numbers using a random number table. Which numbers do you play, based on these random digits?

43680 98750 13092 76561 58712

9. Play the lottery Some people play state-run lotteries by always playing the same favorite "lucky" number. Assuming that the lottery is truly random, is this strategy better, worse, or the same as choosing different numbers for each play? Explain.

10. Play it again, Sam In Exercise 8 you imagined playing the lottery by using random digits to decide what numbers to play. Is this a particularly good or bad strategy? Explain.

11. Bad simulations Explain why each of the following simulations fails to model the real situation properly:
a) Use a random integer from 0 through 9 to represent the number of heads when 9 coins are tossed.
b) A basketball player takes a foul shot. Look at a random digit, using an odd digit to represent a good shot and an even digit to represent a miss.
c) Use random numbers from 1 through 13 to represent the denominations of the cards in a five-card poker hand.

12. More bad simulations Explain why each of the following simulations fails to model the real situation:
a) Use random numbers 2 through 12 to represent the sum of the faces when two dice are rolled.
b) Use a random integer from 0 through 5 to represent the number of boys in a family of 5 children.

c) Simulate a baseball player's performance at bat by letting 0 = an out, 1 = a single, 2 = a double, 3 = a triple, and 4 = a home run.

13. Wrong conclusion A Statistics student properly simulated the length of checkout lines in a grocery store and then reported, "The average length of the line will be 3.2 people." What's wrong with this conclusion?

14. Another wrong conclusion After simulating the spread of a disease, a researcher wrote, "24% of the people contracted the disease." What should the correct conclusion be?

15. Election You're pretty sure that your candidate for class president has about 55% of the votes in the entire school. But you're worried that only 100 students will show up to vote. How often will the underdog (the one with 45% support) win? To find out, you set up a simulation.
a) Describe how you will simulate a component.
b) Describe how you will simulate a trial.
c) Describe the response variable.

16. Two pair or three of a kind? When drawing five cards randomly from a deck, which is more likely, two pairs or three of a kind? A pair is exactly two of the same denomination. Three of a kind is exactly 3 of the same denomination. (Don't count three 8's as a pair—that's 3 of a kind. And don't count 4 of the same kind as two pair—that's 4 of a kind, a very special hand.) How could you simulate 5-card hands? Be careful; once you've picked the 8 of spades, you can't get it again in that hand.
a) Describe how you will simulate a component.
b) Describe how you will simulate a trial.
c) Describe the response variable.

17. Cereal In the chapter's example, 20% of the cereal boxes contained a picture of LeBron James, 30% Danica Patrick, and the rest Serena Williams. Suppose you buy five boxes of cereal. Estimate the probability that you end up with a complete set of the pictures. Your simulation should have at least 20 runs.

18. Cereal, again Suppose you really want the LeBron James picture. How many boxes of cereal do you need to buy to be pretty sure of getting at least one? Your simulation should use at least 10 trials.

19. Multiple choice You take a quiz with 6 multiple choice questions. After you studied, you estimated that you would have about an 80% chance of getting any individual question right. What are your chances of getting them all right? Use at least 20 trials.

20. Lucky guessing? A friend of yours who took the multiple choice quiz in Exercise 19 got all 6 questions right, but now claims to have guessed blindly on every question. If each question offered 4 possible answers, do you believe her? Explain, basing your argument on a simulation involving at least 10 trials.

21. **Beat the lottery** Many states run lotteries to raise money. A Web site advertises that it knows "how to increase YOUR chances of Winning the Lottery." They offer several systems and criticize others as foolish. One system is called *Lucky Numbers*. People who play the *Lucky Numbers* system just pick a "lucky" number to play, but maybe some numbers are luckier than others. Let's use a simulation to see how well this system works.

To make the situation manageable, simulate a simple lottery in which a single digit from 0 to 9 is selected as the winning number. Pick a single value to bet, such as 1, and keep playing it over and over. You'll want to run at least 100 trials. (If you can program the simulations on a computer, run several hundred. Or generalize the questions to a lottery that chooses two- or three-digit numbers—for which you'll need thousands of trials.)

a) What proportion of the time do you expect to win?

b) Would you expect better results if you picked a "luckier" number, such as 7? (Try it if you don't know.) Explain.

22. **Random is as random does** The "beat the lottery" Web site discussed in Exercise 21 suggests that because lottery numbers are random, it is better to select your bet randomly. For the same simple lottery in Exercise 21 (random values from 0 to 9), generate each bet by choosing a separate random value between 0 and 9. Play many games. What proportion of the time do you win?

23. **It evens out in the end** The "beat the lottery" Web site of Exercise 21 notes that in the long run we expect each value to turn up about the same number of times. That leads to their recommended strategy. First, watch the lottery for a while, recording the winners. Then bet the value that has turned up the least, because it will need to turn up more often to even things out. If there is more than one "rarest" value, just take the lowest one (since it doesn't matter). Simulating the simplified lottery described in Exercise 21, play many games with this system. What proportion of the time do you win?

24. **Play the winner?** Another strategy for beating the lottery is the reverse of the system described in Exercise 23. Simulate the simplified lottery described in Exercise 21. Each time, bet the number that just turned up. The Web site suggests that this method should do worse. Does it? Play many games and see.

25. **Driving test** You are about to take the road test for your driver's license. You hear that only 34% of candidates pass the test the first time, but the percentage rises to 72% on subsequent retests.

a) Create a plan for a simulation to estimate the average number of tests drivers take in order to get a license.

b) Here are results of 100 trials of a simulation. Use these results to estimate the average number of tests drivers take in order to get a license.

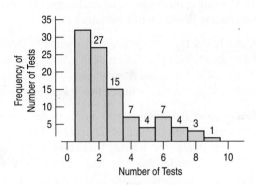

26. **Basketball strategy** Late in a basketball game, the team that is behind often fouls someone in an attempt to get the ball back. Usually the opposing player will get to shoot foul shots "one and one," meaning he gets a shot, and then a second shot only if he makes the first one. Suppose the opposing player has made 72% of his foul shots this season.

a) Create a plan for a simulation to estimate the number of points he will score in a one-and-one situation.

b) Here are the results of 100 trials of a simulation. Use these results to estimate the number of points he will score in a one-and-one situation.

27. **Still learning?** As in Exercise 25, assume that your chance of passing the driver's test is 34% the first time and 72% for subsequent retests. Estimate the percentage of those tested who still do not have a driver's license after two attempts.

28. **Blood donors** A person with type O-positive blood can receive blood only from other type O donors. About 44% of the U.S. population has type O blood. At a blood drive, how many potential donors do you expect to examine in order to get three units of type O blood?

29. **Free groceries** To attract shoppers, a supermarket runs a weekly contest that involves "scratch-off" cards. With each purchase, customers get a card with a black spot obscuring a message. When the spot is scratched away, most of the cards simply say, "Sorry—please try again."

But during the week, 100 customers will get cards that make them eligible for a drawing for free groceries. Ten of the cards say they may be worth $200, 10 others say $100, 20 may be worth $50, and the rest could be worth $20. To register those cards, customers write their names on them and put them in a barrel at the front of the store. At the end of the week the store manager draws cards at random, awarding the lucky customers free groceries in the amount specified on their card. The drawings continue until the store has given away more than $500 of free groceries. Estimate the average number of winners each week.

30. **Find the ace** A technology store holds a contest to attract shoppers. Once an hour, someone at checkout is chosen at random to play in the contest. Here's how it works: An ace and four other cards are shuffled and placed face down on a table. The customer gets to turn over cards one at a time, looking for the ace. The person wins $100 of store credit if the ace is the first card, $50 if it is the second card, and $20, $10, or $5 if it is the third, fourth, or last card chosen. What is the average dollar amount of store credit given away in the contest? Estimate with a simulation.

31. **The family** Many couples want to have both a boy and a girl. If they decide to continue to have children until they have one child of each sex, what would the average family size be? Assume that boys and girls are equally likely.

32. **A bigger family** Suppose a couple will continue having children until they have at least two children of each sex (two boys *and* two girls). How many children might they expect to have?

33. **Dice game** You are playing a children's game in which the number of spaces you get to move is determined by the rolling of a die. You must land exactly on the final space in order to win. If you are 10 spaces away, how many turns might it take you to win?

34. **Parcheesi** You are three spaces from a win in Parcheesi. On each turn, you will roll two dice. To win, you must roll a total of 3 or roll a 3 on one of the dice. How many turns might you expect this to take?

35. **The hot hand** A basketball player with a 65% shooting percentage has just made 6 shots in a row. The announcer says this player "is hot tonight! She's in the zone!" Assume the player takes about 20 shots per game. Is it unusual for her to make 6 or more shots in a row during a game?

36. **The World Series** The World Series ends when a team wins 4 games. Suppose that sports analysts consider one team a bit stronger, with a 55% chance to win any individual game. Estimate the likelihood that the underdog wins the series.

37. **Teammates** Four couples at a dinner party play a board game after the meal. They decide to play as teams of two and to select the teams randomly. All eight people write their names on slips of paper. The slips are thoroughly mixed, then drawn two at a time. How likely is it that every person will be teamed with someone other than the person he or she came to the party with?

38. **Second team** Suppose the couples in Exercise 37 choose the teams by having one member of each couple write their names on the cards and the other people each pick a card at random. How likely is it that every person will be teamed with someone other than the person he or she came with?

39. **Job discrimination?** A company with a large sales staff announces openings for three positions as regional managers. Twenty-two of the current salespersons apply, 12 men and 10 women. After the interviews, when the company announces the newly appointed managers, all three positions go to women. The men complain of job discrimination. Do they have a case? Simulate a random selection of three people from the applicant pool, and make a decision about the likelihood that a fair process would result in hiring all women.

40. **Smartphones** A proud legislator claims that your state's new law banning texting and hand-held phones while driving reduced occurrences of infractions to less than 10% of all drivers. While on a long drive home from your college, you notice a few people seemingly texting. You decide to count everyone using their smartphones illegally who pass you on the expressway for the next 20 minutes. It turns out that 5 out of the 20 drivers were actually using their phones illegally. Does this cast doubt on the legislator's figure of 10%? Use a simulation to estimate the likelihood of seeing at least 5 out of 20 drivers using their phones illegally if the actual usage rate is only 10%. Explain your conclusion clearly.

Just Checking ANSWERS

1. The component is one game.

2. I'll generate random numbers and assign numbers from 00 to 54 to the home team's winning and from 55 to 99 to the visitors' winning.

3. I'll generate components until one team wins 4 games. I'll record which team wins the series.

4. The response is who wins the series.

5. I'll calculate the proportion of wins by Team A (who starts at home).

11 Sample Surveys

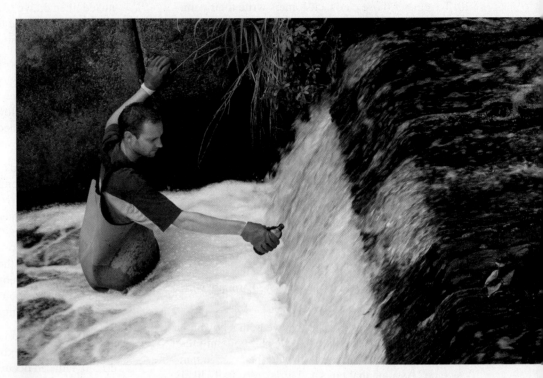

I n December 2010, Pew Research conducted a survey to assess where Americans got their news. Pew sampled 1500 U.S. adults, reaching 1000 of them by landline telephone and another 500 on their cell phones. They report that 66% of respondents say their main source for news is television (down from 74% in 2007). But the timeplot below shows that among 18–29 year olds, the Internet (at 65%) has now passed TV (52%). Pew claimed that these estimates were close to the true percentages that they would have found if they had asked all U.S. adults. That step from a small sample to the entire population is impossible without understanding Statistics. To make business decisions, to do science, to choose wise investments, or to understand how voters think they'll vote in the next election, we need to stretch beyond the data at hand to the world at large.

Figure 11.1

A timeplot showing the responses of 18–29 year olds to Pew polls asking them where they get most of their news about national and international news. Respondents could name up to 2 sources. (www.people-press .org/2011/01/04/internet-gains-on -television-as-publics-main-news -source)

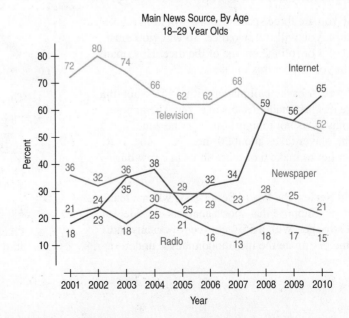

To make that stretch, we need three ideas. You'll find the first one natural. The second may be more surprising. The third is one of the strange but true facts that often confuse those who don't know Statistics.

Idea 1: Examine a Part of the Whole

A S **Activity: Populations and Samples.** Explore the differences between populations and samples.

The W's and Sampling

The population we are interested in is usually determined by the *Why* of our study. The sample we draw will be the *Who*. *When* and *How* we draw the sample may depend on what is practical.

The first idea is to draw a sample. We'd like to know about an entire **population** of individuals, but examining all of them is usually impractical, if not impossible. So we settle for examining a smaller group of individuals—a **sample**—selected from the population.

You do this every day. For example, suppose you wonder how the vegetable soup you're cooking for dinner tonight is going to go over with your friends. To decide whether it meets your standards, you only need to try a small amount. You might taste just a spoonful or two. You certainly don't have to consume the whole pot. You trust that the taste will *represent* the flavor of the entire pot. The idea behind your tasting is that a small sample, if selected properly, can represent the entire population.

It's hard to go a day without hearing about the latest opinion poll. These polls are examples of **sample surveys,** designed to ask questions of a small group of people in the hope of learning something about the entire population. Most likely, you've never been selected to be part of one of these national opinion polls. That's true of most people. So how can the pollsters claim that a sample is representative of the entire population? The answer is that professional pollsters work quite hard to ensure that the "taste"—the sample that they take—represents the population. If not, the sample can give misleading information about the population.

Bias

A S **Video: The *Literary Digest* Poll and the Election of 1936.** Hear the story of one of the most famous polling failures in history.

Selecting a sample to represent the population fairly is more difficult than it sounds. Polls or surveys most often fail because they use a sampling method that tends to over- or underrepresent parts of the population. The method may overlook subgroups that are harder to find (such as the homeless or those who use only cell phones) or favor others (such as Internet users who like to respond to online surveys). Sampling methods that, by their nature, tend to over- or underemphasize some characteristics of the population are said to be **biased.** Bias is the bane of sampling—the one thing above all to avoid. Conclusions based on samples drawn with biased methods are inherently flawed. There is usually no way to fix bias after the sample is drawn and no way to salvage useful information from it.

Here's a famous example of a really dismal failure. By the beginning of the 20th century, it was common for newspapers to ask readers to return "straw" ballots on a variety of topics. (Today's Internet surveys are the same idea, gone electronic.) The earliest known example of such a straw vote in the United States dates back to 1824.

From 1916 to 1936, the magazine *Literary Digest* regularly surveyed public opinion and forecast election results correctly. During the 1936 presidential campaign between Alf Landon and Franklin Delano Roosevelt, it mailed more than 10 million ballots and got back an astonishing 2.4 million. (Polls were still a relatively novel idea, and many people thought it was important to send back their opinions.) The results were clear: Alf Landon would be the next president by a landslide, 57% to 43%. You remember President Landon? No? In fact, Landon carried only two states. Roosevelt won, 62% to 37%, and, perhaps coincidentally, the *Digest* went bankrupt soon afterward.

What went wrong? One problem was that the *Digest*'s sample wasn't representative. Where would *you* find 10 million names and addresses to sample? The *Digest* used the phone book, as many surveys do.[1] But in 1936, at the height of the Great Depression, telephones were a real luxury, so they sampled more rich than poor voters. The campaign of 1936 focused on the economy, and those who were less well off were more likely to vote for the Democrat. So the *Digest*'s sampling method was hopelessly biased.

---

[1]Today phone numbers are computer-generated to make sure that unlisted numbers are included. But even now, cell phones and VOIP Internet phones are often not included.

How do modern polls get their samples to *represent* the entire population? You might think that they'd handpick individuals to sample with care and precision. But in fact, they do something quite different: They select individuals to sample *at random*.

In 1936, a young pollster named George Gallup used a subsample of only 3000 of the 2.4 million responses that the *Literary Digest* received to reproduce the wrong prediction of Landon's victory over Roosevelt. He then used an entirely different sample of 50,000 and predicted that Roosevelt would get 56% of the vote to Landon's 44%. His sample was apparently much more representative of the actual voting populace. The Gallup Organization has gone on to become one of the leading polling companies.

Idea 2: Randomize

Think back to the soup sample. Suppose you add some salt to the pot. If you sample it from the top before stirring, you'll get the misleading idea that the whole pot is salty. If you sample from the bottom, you'll get an equally misleading idea that the whole pot is bland. By stirring, you *randomize* the amount of salt throughout the pot, making each taste more typical of the whole pot.

Not only does randomization protect you against factors that you know are in the data, it can also help protect against factors that you didn't even know were there. Suppose, while you weren't looking, a friend added a handful of peas to the soup. If they're down at the bottom of the pot, and you don't randomize the soup by stirring, your test spoonful won't have any peas. By stirring in the salt, you *also* randomize the peas throughout the pot, making your sample taste more typical of the overall pot *even though you didn't know the peas were there*. So randomizing protects us even in this case.

How do we "stir" people in a survey? We select them at random. **Randomizing** protects us from the influences of *all* the features of our population by making sure that, *on average*, the sample looks like the rest of the population.

Why Not Match the Sample to the Population?

Rather than randomizing, we could try to design our sample so that the people we choose are typical in terms of every characteristic we can think of. We might want the income levels of those we sample to match the population. How about age? Political affiliation? Marital status? Having children? Living in the suburbs? We can't possibly think of all the things that might be important. Even if we could, we wouldn't be able to match our sample to the population for all these characteristics.

A S *Activity:* **Sampling from Some Real Populations.** Draw random samples to see how closely they resemble each other and the population.

For Example IS A RANDOM SAMPLE REPRESENTATIVE?

Here are summary statistics comparing two samples of 8000 drawn at random from a company's database of 3.5 million customers:

Age (yr)	White (%)	Female (%)	# of Children	Income Bracket (1–7)	Wealth Bracket (1–9)	Homeowner? (% Yes)
61.4	85.12	56.2	1.54	3.91	5.29	71.36
61.2	84.44	56.4	1.51	3.88	5.33	72.30

QUESTION: Do you think these samples are representative of the population? Explain.

ANSWER: The two samples look very similar with respect to these seven variables. It appears that randomizing has automatically matched them pretty closely. We can reasonably assume that since the two samples don't differ too much from each other, they don't differ much from the rest of the population either.

Idea 3: It's the Sample Size

How large a random sample do we need for the sample to be reasonably representative of the population? Most people think that we need a large percentage, or *fraction,* of the population, but it turns out that what matters is the *number* of individuals in the sample, not what fraction of the population it is. A random sample of 100 students in a college represents the student body just about as well as a random sample of 100 voters represents the entire electorate of the United States. This is the *third* idea and probably the most surprising one in designing surveys.

How can it be that only the size of the sample, and not the population, matters? Well, let's return one last time to that pot of soup. If you're cooking for a banquet rather than just for a few people, your pot will be bigger, but do you need a bigger spoon to decide how the soup tastes? Of course not. The same-size spoonful is probably enough to make a decision about the entire pot, no matter how large the pot. The *fraction* of the population that you've sampled doesn't matter.[2] It's the **sample size** itself that's important.

How big a sample do you need? That depends on what you're estimating. To get an idea of what's really in the soup, you'll need a large enough taste to get a *representative* sample from the pot. For a survey that tries to find the proportion of the population falling into a category, you'll usually need several hundred respondents to say anything precise enough to be useful.[3]

Larger Is Better.
A friend who knows that you are taking Statistics asks your advice on her study. What can you possibly say that will be helpful? Just say, "If you could just get a larger sample, it would probably improve your study." Even though a larger sample might not be worth the cost, it will almost always make the results more precise.

What Do the Pollsters Do? How do professional polling agencies do their work? The most common polling method today is to contact respondents by telephone. Computers generate random telephone numbers, so pollsters can even call some people with unlisted phone numbers. The person who answers the phone is invited to respond to the survey—if that person qualifies. (For example, only if it's an adult who lives at that address.) If the person answering doesn't qualify, the caller will ask for an appropriate alternative. In phrasing questions, pollsters often list alternative responses (such as candidates' names) in different orders to avoid biases that might favor the first name on the list.

Do these methods work? The Pew Research Center for the People and the Press, reporting on one survey, says that

Across five days of interviewing, surveys today are able to make some kind of contact with the vast majority of households (76%), and there is no decline in this contact rate over the past seven years. But because of busy schedules, skepticism and outright refusals, interviews were completed in just 38% of households that were reached using standard polling procedures.

Nevertheless, studies indicate that those actually sampled can give a good snapshot of larger populations from which the surveyed households were drawn.

Does a Census Make Sense?

Why bother determining the right sample size? Wouldn't it be better to just include everyone and "sample" the entire population? Such a special sample is called a **census.** Although a census would appear to provide the best possible information about the population, there are a number of reasons why it might not.

---

[2]Well, that's not exactly true. If the population is small enough and the sample is more than 10% of the whole population, it *can* matter. It doesn't matter whenever, as usual, our sample is a very small fraction of the population.

[3]Chapter 18 gives the details behind this statement and shows how to decide on a sample size for a survey.

First, it can be difficult to complete a census. Some individuals in the population will be hard (and expensive) to locate. Or a census might just be impractical. If you were a taste tester for the Hostess™ Company, you probably wouldn't want to conduct a census by eating *all* the Twinkies on the production line. Not only might this be life-endangering, but the company wouldn't have any left to sell.

Second, populations rarely stand still. In populations of people, babies are born and folks die or leave the country. In opinion surveys, events may cause a shift in opinion during the survey. A census takes longer to complete and the population changes while you work. A sample surveyed in just a few days may give more accurate information.

Third, taking a census can be more complex than sampling. For example, the U.S. Census records too many college students. Many are counted once with their families and are then counted a second time in a report filed by their schools.

> **The Undercount.** It's particularly difficult to compile a complete census of a population as large, complex, and spread out as the U.S. population. The U.S. Census is known to miss some residents. On occasion, the undercount has been striking. For example, there have been blocks in inner cities in which the number of residents recorded by the Census was smaller than the number of electric meters for which bills were being paid. What makes the problem particularly important is that some groups have a higher probability of being missed than others—undocumented immigrants, the homeless, the poor. The Census Bureau proposed the use of random sampling to estimate the number of residents missed by the ordinary census. Unfortunately, the resulting debate has become more political than statistical.

Populations and Parameters

A study found that teens were less likely to "buckle up." The National Center for Chronic Disease Prevention and Health Promotion reports that 21.7% of U.S. teens never or rarely wear seatbelts. We're sure they didn't take a census, so what *does* the 21.7% mean? We can't know what percentage of all teenagers wear seatbelts. Reality is just too complex. But we can simplify the question by building a model.

Models use mathematics to represent reality. Parameters are the key numbers in those models. A parameter used in a model for a population is sometimes called (redundantly) a **population parameter.**

But let's not forget about the data. We use summaries of the data to estimate the population parameters. As we know, any summary found from the data is a **statistic.** Sometimes you'll see the (also redundant) term **sample statistic.**[4]

We've already met two parameters in Chapter 5: the mean, μ, and the standard deviation, σ. We'll try to keep denoting population model parameters with Greek letters and the corresponding statistics with Latin letters. Usually, but not always, the letter used for the statistic and the parameter correspond in a natural way. So the standard deviation of the data is s, and the corresponding parameter is σ (Greek for s). In Chapter 6, we used r to denote the sample correlation. The corresponding correlation in a model for the population would be called ρ (rho). In Chapter 7, b_1 represented the slope of a linear regression estimated from the data. But when we think about a (linear) *model* for the population, we denote the slope parameter β_1 (beta).

Get the pattern? Good. But now it breaks down. We denote the mean of a population model with μ (because μ is the Greek letter for m). It might make sense to denote the sample mean with m, but long-standing convention is to put a bar over anything when we average it, so we write $\bar{y}$.

---

[4]Where else besides a sample *could* a statistic come from?

What about proportions? Suppose we want to talk about the proportion of teens who don't wear seatbelts. If we use p to denote the proportion from the data, what is the corresponding model parameter? By all rights it should be π. But statements like $\pi = 0.25$ might be confusing because π has been equal to 3.1415926 . . . for so long, and it's worked so *well*. So, once again we violate the rule. We'll use p for the population model parameter and $\hat{p}$ for the proportion from the data (since, like $\hat{y}$ in regression, it's an estimated value).

Here's a table summarizing the notation:

Name	Statistic	Parameter
Mean	$\bar{y}$	μ (mu, pronounced "meeoo," not "moo")
Standard deviation	s	σ (sigma)
Correlation	r	ρ (rho)
Regression coefficient	b	β (beta, pronounced "baytah"[5])
Proportion	$\hat{p}$	p (pronounced "pee"[6])

We draw samples because we can't work with the entire population, but we want the statistics we compute from a sample to reflect the corresponding parameters accurately. A sample that does this is said to be **representative.** A biased sampling methodology tends to over- or underestimate the parameter of interest.

Just Checking

1. Various claims are often made for surveys. Why is each of the following claims not correct?

a) It is always better to take a census than to draw a sample.

b) Stopping students on their way out of the cafeteria is a good way to sample if we want to know about the quality of the food there.

c) We drew a sample of 100 from the 3000 students in a school. To get the same level of precision for a town of 30,000 residents, we'll need a sample of 1000.

d) A poll taken at a statistics support website garnered 12,357 responses. The majority said they enjoy doing statistics homework. With a sample size that large, we can be pretty sure that most Statistics students feel this way, too.

e) The true percentage of all Statistics students who enjoy the homework is called a "population statistic."

Simple Random Samples

How would you select a representative sample? Most people would say that every individual in the population should have an equal chance to be selected, and certainly that seems fair. But it's not sufficient. There are many ways to give everyone an equal chance that still wouldn't give a representative sample. Consider, for example, a school that has equal numbers of males and females. We could sample like this: Flip a coin. If it comes up heads, select 100 female students at random. If it comes up tails, select 100 males at random. Everyone has an equal chance of selection, but every sample is of only a single sex—hardly representative.

We need to do better. Suppose we insist that every possible *sample* of the size we plan to draw has an equal chance to be selected. This ensures that situations like the one just described are not likely to occur and still guarantees that each person has an equal chance of being selected. What's different is that with this method, each *combination* of people

---

[5]If you're from the United States. If you're British or Canadian, it's "beetah."

[6]Just in case you weren't sure.

has an equal chance of being selected as well. A sample drawn in this way is called a **Simple Random Sample,** usually abbreviated **SRS.** An SRS is the standard against which we measure other sampling methods, and the sampling method on which the theory of working with sampled data is based.

To select a sample at random, we first need to define where the sample will come from. The **sampling frame** is a list of individuals from which the sample is drawn. For example, to draw a random sample of students at a college, we might obtain a list of all registered full-time students and sample from that list. In defining the sampling frame, we must deal with the details of defining the population. Are part-time students included? How about those who are attending school elsewhere and transferring credits back to the college?

Once we have a sampling frame, the easiest way to choose an SRS is to assign a random number to each individual in the sampling frame. We then select only those whose random numbers satisfy some rule. Let's look at some ways to do this.

For Example USING RANDOM NUMBERS TO GET AN SRS

There are 80 students enrolled in an introductory Statistics class; you are to select a sample of 5.

QUESTION: How can you select an SRS of 5 students using these random digits found on the Internet: 05166   29305   77482?

ANSWER: First I'll number the students from 00 to 79. Taking the random numbers two digits at a time gives me 05, 16, 62, 93, 05, 77, and 48. I'll ignore 93 because the students were numbered only up to 79. And, so as not to pick the same person twice, I'll skip the repeated number 05. My simple random sample consists of students with the numbers 05, 16, 62, 77, and 48.

Error Okay, Bias No Way! Sampling variability is sometimes referred to as *sampling error,* making it sound like it's some kind of mistake. It's not. We understand that samples will vary, so "sampling error" is to be expected. It's *bias* we must strive to avoid. Bias means our sampling method distorts our view of the population, and that will surely lead to mistakes.

- We can be more efficient when we're choosing a larger sample from a sampling frame stored in a data file. First we assign a random number with several digits (say, from 0 to 10,000) to each individual. Then we arrange the random numbers in numerical order, keeping each name with its number. Choosing the first *n* names from this re-arranged list will give us a random sample of that size.
- Often the sampling frame is so large that it would be too tedious to number everyone consecutively. If our intended sample size is approximately 10% of the sampling frame, we can assign each individual a single random digit 0 to 9. Then we select only those with a specific random digit, say, 5.

Samples drawn at random generally differ one from another. Each draw of random numbers selects *different* people for our sample. These differences lead to different values for the variables we measure. We call these sample-to-sample differences **sampling variability.** Surprisingly, sampling variability isn't a problem; it's an opportunity. In future chapters we'll investigate what the variation in a sample can tell us about its population.

Stratified Sampling

Simple random sampling is not the only fair way to sample. More complicated designs may save time or money or help avoid sampling problems. All statistical sampling designs have in common the idea that chance, rather than human choice, is used to select the sample.

Designs that are used to sample from large populations—especially populations residing across large areas—are often more complicated than simple random samples. Sometimes the population is first sliced into homogeneous groups, called **strata,** before the sample is selected. Then simple random sampling is used within each stratum before the results are combined. This common sampling design is called **stratified random sampling.**

Why would we want to complicate things? Here's an example. Suppose we want to learn how students feel about funding for the football team at a large university. The campus is 60% men and 40% women, and we suspect that men and women have different views on the funding. If we use simple random sampling to select 100 people for the survey, we could end up with 70 men and 30 women or 35 men and 65 women. Our resulting estimates of the level of support for the football funding could vary widely. To help reduce this sampling variability, we can decide to force a representative balance, selecting 60 men at random and 40 women at random. This would guarantee that the proportions of men and women within our sample match the proportions in the population, and that should make such samples more accurate in representing population opinion.

You can imagine the importance of stratifying by race, income, age, and other characteristics, depending on the questions in the survey. Samples taken within a stratum vary less, so our estimates can be more precise. This reduced sampling variability is the most important benefit of stratifying. We'll explore that further in this chapter's What If.

Stratified sampling can also help us notice important differences among groups. As we saw in Chapter 3, if we unthinkingly combine group data, we risk reaching the wrong conclusion, becoming victims of Simpson's paradox.

For Example STRATIFYING THE SAMPLE

RECAP: You're trying to find out what freshmen think of the food served on campus. Food Services believes that men and women typically have different opinions about the importance of the salad bar.

QUESTION: How should you adjust your sampling strategy to allow for this difference?

ANSWER: I will stratify my sample by drawing an SRS of men and a separate SRS of women—assuming that the data from the registrar include information about each person's sex.

Cluster and Multistage Sampling

Suppose we wanted to assess the reading level of this textbook based on the length of the sentences. Simple random sampling could be awkward; we'd have to number each sentence, then find, for example, the 576th sentence or the 2482nd sentence, and so on. Doesn't sound like much fun, does it?

It would be much easier to pick a few *pages* at random and count the lengths of the sentences on those pages. That works if we believe that each page is representative of the entire book in terms of reading level. Splitting the population into representative **clusters** can make sampling more practical. Then we could simply select one or a few clusters at random and perform a census within each of them. This sampling design is called **cluster sampling.**

Clusters are generally selected for reasons of efficiency, practicality, or cost. Ideally, if each cluster represents the full population fairly, cluster sampling will be unbiased. But often we just hope that by choosing a sample of clusters, we can obtain a representative sample of the entire population.

For Example   CLUSTER SAMPLING

RECAP: In trying to find out what freshmen think about the food served on campus, you've considered both an SRS and a stratified sample. Now you have run into a problem: It's simply too difficult and time consuming to track down the individuals whose names were chosen for your sample. Fortunately, freshmen at your school are all housed in 10 freshman dorms.

QUESTION: How could you use this fact to draw a cluster sample? How might that alleviate the problem? What concerns do you have?

ANSWER: *To draw a cluster sample, I would select one or two dorms at random and then try to contact everyone in each selected dorm. I could save time by simply knocking on doors on a given evening and interviewing people. I'd have to assume that freshmen were assigned to dorms pretty much at random and that the people I'm able to contact are representative of everyone in the dorm.*

What's the difference between cluster sampling and stratified sampling? We stratify to ensure that our sample represents different groups in the population, and we sample randomly within each stratum. Strata are internally homogeneous, but differ from one another. By contrast, we select clusters to make sampling more practical or affordable. Clusters can be heterogeneous; we want our randomly selected clusters to provide a representative sample of the population.

Stratified vs. Cluster Sampling

Boston cream pie consists of a layer of yellow cake, a layer of pastry creme, another cake layer, and then a chocolate frosting. Suppose you are a professional taster (yes, there really are such people) whose job is to check your company's pies for quality. You'd need to eat small samples of randomly selected pies, tasting all three components: the cake, the creme, and the frosting.

One approach is to cut a thin vertical slice out of the pie. Such a slice will be a lot like the entire pie, so by eating that slice, you'll learn about the whole pie. This vertical slice containing all the different ingredients in the pie would be a *cluster* sample.

Another approach is to sample in *strata:* Select some tastes of the cake at random, some tastes of creme at random, and some bits of frosting at random. You'll end up with a reliable judgment of the pie's quality.

Many populations you might want to learn about are like this Boston cream pie. You can think of the subpopulations of interest as horizontal strata, like the layers of pie. Cluster samples slice vertically across the layers to obtain clusters, each of which is representative of the entire population. Stratified samples represent the population by drawing some from each layer, reducing variability in the results that could arise because of the differences among the layers.

Sometimes we use a variety of sampling methods together. In trying to assess the reading level of this book, we might worry that it starts out easy and then gets harder as the concepts become more difficult. If so, we'd want to avoid samples that selected heavily from early or from late chapters. To guarantee a fair mix of chapters, we could randomly choose one chapter from each of the seven parts of the book and then randomly select a few pages from each of those chapters. If, altogether, that made too many sentences, we might select a few sentences at random from each of the chosen pages. So, what is our sampling strategy? First we stratify by the part of the book and randomly choose a chapter to represent each stratum. Within each selected chapter, we choose pages as clusters. Finally, we consider an SRS of sentences within each cluster. Sampling schemes that combine several methods are called **multistage samples.** Most surveys conducted by professional polling organizations use some combination of stratified and cluster sampling as well as simple random samples.

For Example MULTISTAGE SAMPLING

RECAP: Having learned that freshmen are housed in separate dorms allowed you to sample their attitudes about the campus food by going to dorms chosen at random, but you're still concerned about possible differences in opinions between men and women. It turns out that these freshmen dorms house the sexes on alternate floors.

QUESTION: How can you design a sampling plan that uses this fact to your advantage?

ANSWER: Now I can stratify my sample by sex. I would first choose one or two dorms at random and then select some dorm floors at random from among those that house men and, separately, from among those that house women. I could then treat each floor as a cluster and interview everyone on that floor.

*Systematic Samples

Some samples select individuals systematically. For example, you might survey every 10th person on an alphabetical list of students. To make it random, you still must start the systematic selection from a randomly selected individual. When the order of the list is not associated in any way with the responses sought, **systematic sampling** can give a representative sample. Systematic sampling can be much less expensive than true random sampling. When you use a systematic sample, you should justify the assumption that the systematic method is not associated with any of the measured variables.

Think about the reading-level sampling example again. Suppose we have chosen a chapter of the book at random, then three pages at random from that chapter, and now we want to select a sample of 10 sentences from the 73 sentences found on those pages. Instead of numbering each sentence so we can pick a simple random sample, it would be easier to sample systematically. A quick calculation shows $73/10 = 7.3$, so we can get our sample by just picking every seventh sentence on the page. But where should you start? At random, of course. We've accounted for $10 \times 7 = 70$ of the sentences, so we'll throw the extra 3 into the starting group and choose a sentence at random from the first 10. Then we pick every seventh sentence after that and record its length.

 Just Checking

2. We need to survey a random sample of the 300 passengers on a flight from San Francisco to Tokyo. Name each sampling method described below.

   a) Pick every 10th passenger as people board the plane, starting with a randomly chosen passenger among the first 10.
   b) From the boarding list, randomly choose 5 people flying first class and 25 of the other passengers.
   c) Randomly generate 30 seat numbers and survey the passengers who sit there.
   d) Randomly select a seat position (right window, right center, right aisle, etc.) and survey all the passengers sitting in those seats.

Step-by-Step Example   SAMPLING

The assignment says, "Conduct your own sample survey to find out how many hours per week students at your school spend watching TV during the school year." Let's see how we might do this step by step. (Remember, though—actually collecting the data from your sample can be difficult and time consuming.)

Question: How would you design this survey?

THINK ➡ **Plan**   State what you want to know.

Population and Parameter
Identify the W's of the study. The *Why* determines the population and the associated sampling frame. The *What* identifies the parameter of interest and the variables measured. The *Who* is the sample we actually draw. The *How, When,* and *Where* are given by the sampling plan.

Often, thinking about the *Why* will help us see whether the sampling frame and plan are adequate to learn about the population.

Sampling Plan   Specify the sampling
method and the sample size, *n*. Specify how the sample was actually drawn. What is the sampling frame? How was the randomization performed?

A good description should be complete enough to allow someone to replicate the procedure, drawing another sample from the same population in the same manner.

> I wanted to design a study to find out how many hours of TV students at my school watch.
>
> The population studied was students at our school. I obtained a list of all students currently enrolled and used it as the sampling frame. The parameter of interest was the number of TV hours watched per week during the school year, which I attempted to measure by asking students how much TV they watched during the previous week.
>
> I decided against stratifying by class or sex because I didn't think TV watching would differ much between males and females or across classes. I selected a simple random sample of students from the list. I obtained an alphabetical list of students, assigned each a random digit between 0 and 9. I randomly decided to select all students assigned a "4." This method generated a sample of 212 students from the population of 2133 students.

SHOW ➡ **Sampling Practice**   Specify *When,*
Where, and *How* the sampling was performed. Specify any other details of your survey, such as how respondents were contacted, what incentives were offered to encourage them to respond, how nonrespondents were treated, and so on.

> The survey was taken over the period Oct. 15 to Oct. 25. Surveys were sent to selected students by e-mail, with the request that they respond by e-mail as well. Students who could not be reached by e-mail were handed the survey in person.

TELL ➡ **Summary and Conclusion**   This
report should include a discussion of all the elements. In addition, it's good practice to discuss any special circumstances. Professional polling organizations report the *When* of their samples

> During the period Oct. 15 to Oct. 25, 212 students were randomly selected, using a simple random sample from a list of all students currently enrolled. The survey they received

(continued)

but will also note, for example, any important news that might have changed respondents' opinions during the sampling process. In this survey, perhaps, a major news story or sporting event might change students' TV viewing behavior.

The question you ask also matters. It's better to be specific ("How many hours did you watch TV last week?") than to ask a general question ("How many hours of TV do you usually watch in a week?").

The report should show a display of the data, provide and interpret the statistics from the sample, and state the conclusions that you reached about the population.

asked the following question: "How many hours did you spend watching television last week?"

Of the 212 students surveyed, 110 responded. It's possible that the nonrespondents differ in the number of TV hours watched from those who responded, but I was unable to follow up on them due to limited time and funds. The 110 respondents reported an average 3.62 hours of TV watching per week. The median was only 2 hours per week. A histogram of the data shows that the distribution is highly right-skewed, indicating that the median might be a more appropriate summary of the typical TV watching of the students.

Most of the students (90%) watch between 0 and 10 hours per week, while 30% reported watching less than 1 hour per week. A few watch much more. About 3% reported watching more than 20 hours per week.

At each step, the group we can study may be constrained further. The *Who* keeps changing, and each constraint can introduce biases. A careful study should address the question of how well each group matches the population of interest. One of the main benefits of simple random sampling is that it never loses its sense of who's *Who*. The *Who* in an SRS is the population of interest from which we've drawn a representative sample. That's not always true for other kinds of samples.

The Valid Survey

It isn't sufficient to just draw a sample and start asking questions. We'll want our survey to be *valid*. A valid survey yields the information we are seeking about the population we are interested in. Before setting out to survey, ask yourself:

- What do I want to know?
- Am I asking the right respondents?
- Am I asking the right questions?
- What would I do with the answers if I had them; would they address the things I want to know?

These questions may sound obvious, but there are a number of pitfalls to avoid.

Know what you want to know. Before considering a survey, understand what you hope to learn and about whom you hope to learn it.

Use the right sampling frame. A valid survey obtains responses from the appropriate respondents. Be sure you have a suitable *sampling frame.* Have you identified the population of interest and sampled from it appropriately? A company might survey customers who returned warranty registration cards, a readily available sampling frame. But if the company wants to know how to make their product more attractive, the most important population is the customers who rejected their product in favor of one from a competitor.

Calvin and Hobbes © 1993 Bill Watterson. Distributed by Universal Uclick. Reprinted with permission. All rights reserved.

Tune your instrument. It is often tempting to ask questions you don't really need, but beware—longer questionnaires yield fewer responses and thus a greater chance of nonresponse bias.

Ask specific rather than general questions. People are not very good at estimating their typical behavior, so it is better to ask "How many hours did you sleep last night?" than "How much do you usually sleep?" Sure, some responses will include some unusual events (My dog was sick; I was up all night.), but overall you'll get better data.

Ask for quantitative results when possible. "How many magazines did you read last week?" is better than "How much do you read: A lot, A moderate amount, A little, or None at all?"

Be careful in phrasing questions. A respondent may not understand the question—or may understand the question differently than the researcher intended it. ("Does anyone in your family ride a motorcycle?" Do you mean just me, my spouse, and my children? Or does "family" include my father, my siblings, and my second cousin once removed? And does a motor scooter count? Respondents are unlikely (or may not have the opportunity) to ask for clarification. A question like "Do you approve of the recent actions of the Secretary of Education?" is likely not to measure what you want if many respondents don't know who the Secretary of Education is or what actions he or she recently made.

Respondents may even lie or shade their responses if they feel embarrassed by the question ("Did you have too much to drink last night?"), are intimidated or insulted by the question ("Could you understand our new *Instructions for Dummies* manual, or was it too difficult for you?"), or if they want to avoid offending the interviewer ("Would you hire a man with a tattoo?" asked by a tattooed interviewer). Also, be careful to avoid phrases that have double or regional meanings. "How often do you go to town?" might be interpreted differently by different people and cultures.

Even subtle differences in phrasing can make a difference. In January 2006, the *New York Times* asked half of the 1229 U.S. adults in their sample the following question:

> *After 9/11, President Bush authorized government wiretaps on some phone calls in the U.S. without getting court warrants, saying this was necessary to reduce the threat of terrorism. Do you approve or disapprove of this?*

A Short Survey Given that the *New York Times* reports[7] that statisticians can earn $125,000 at top companies their first year on the job, do you think this course will be valuable to *you*?

---

[7]www.nytimes.com/2009/08/06/technology/06stats.html

They found that 53% of respondents approved. But when they asked the other half of their sample a question with only slightly different phrasing,

> *After 9/11, George W. Bush authorized government wiretaps on some phone calls in the U.S. without getting court warrants. Do you approve or disapprove of this?*

only 46% approved.

Be careful in phrasing answers. It's often a good idea to offer choices rather than inviting a free response. Open-ended answers can be difficult to analyze. "How did you like the movie?" may start an interesting debate, but it may be better to give a range of possible responses. Be sure to phrase them in a neutral way. When asking "Do you support higher school taxes?" positive responses could be worded "Yes," "Yes, it is important for our children," or "Yes, our future depends on it." But those are not equivalent answers.

THE WIZARD OF ID   parker and hart

By permission of John L. Hart FLP and Creators Syndicate, Inc.

The best way to protect a survey from such unanticipated measurement errors is to perform a pilot survey. A **pilot** is a trial run of the survey you eventually plan to give to a larger group, using a draft of your survey questions administered to a small sample drawn from the same sampling frame you intend to use. By analyzing the results from this smaller survey, you can often discover ways to improve your instrument.

Lots Can Go Wrong: How to Sample Badly

Bad sample designs yield worthless data. Many of the most convenient forms of sampling can be seriously biased. And there is no way to correct for the bias from a bad sample. So it's wise to pay attention to sample design—and to beware of reports based on poor samples.

Mistake 1: Sample Volunteers

One of the most common dangerous sampling methods is a voluntary response sample. In a **voluntary response sample**, a large group of individuals is invited to respond, and those who choose to respond are counted. The respondents, rather than the researcher, decide who will be in the sample. This method is used by call-in shows, 900 numbers, Internet polls, and letters written to members of Congress. Voluntary response samples are almost always biased, so conclusions drawn from them are almost always wrong.

It's often hard to define the sampling frame of a voluntary response study. Practically, the frames are groups such as Internet users who frequent a particular website or those who happen to be watching a particular TV show at the moment. But those sampling frames don't correspond to the population of interest.

Even within the sampling frame, voluntary response samples are often biased toward those with strong opinions or those who are strongly motivated. People with very negative opinions tend to respond more often than those with equally strong positive opinions. The sample is not representative, even though every individual in the population may have been offered the chance to respond. The resulting **voluntary response bias** invalidates the survey.

A S   *Activity:* **Sources of Sampling Bias.** Here's a narrated exploration of sampling bias.

> **If You Had It to Do Over Again, Would You Have Children?**   Ann Landers, the advice columnist, asked parents this question. The overwhelming majority—70% of the more than 10,000 people who wrote in—said no, kids weren't worth it. A more carefully designed survey later showed that about 90% of parents actually are happy with their decision to have children. What accounts for the striking difference in these two results? What parents do you think are most likely to respond to the original question?

For Example   VOLUNTARY RESPONSE SAMPLE

RECAP: You're trying to find out what freshmen think of the food served on campus, and have thought of a variety of sampling methods, all time consuming. A friend suggests that you set up a "Tell Us What You Think" website and invite freshmen to visit the site to complete a questionnaire.

QUESTION: What's wrong with this idea?

ANSWER: *Letting each freshman decide whether to participate makes this a voluntary response survey. Students who were dissatisfied might be more likely to go to the website to record their complaints, and this could give me a biased view of the opinions of all freshmen.*

Mistake 2: Sample Conveniently

> Do you use the Internet?
> Click here ○ for yes
> Click here ○ for no

Another sampling method that doesn't work is convenience sampling. As the name suggests, in **convenience sampling** we simply include the individuals who are convenient for us to sample. Unfortunately, this group may not be representative of the population. Here's an amusing example. Back in 2001, when computer use in the home was not as common as it is today, a survey of 437 potential home buyers in Orange County, California, reached the surprising conclusion that

> *All but 2 percent of the buyers have at least one computer at home, and 62 percent have two or more. Of those with a computer, 99 percent are connected to the Internet (Source: Jennifer Hieger, "Portrait of Homebuyer Household: 2 Kids and a PC," Orange County Register, 27 July 2001).*

Internet Surveys

Internet convenience surveys are worthless. As voluntary response surveys, they have no well-defined sampling frame (all those who use the Internet and visit their site?) and thus report no useful information. Do not believe them.

How was the survey conducted? On the Internet!

Many surveys conducted at shopping malls suffer from the same problem. People in shopping malls are not necessarily representative of the population of interest. Mall shoppers tend to be more affluent and include a larger percentage of teenagers and retirees than the population at large. To make matters worse, survey interviewers tend to select individuals who look "safe," or easy to interview.

For Example   CONVENIENCE SAMPLE

RECAP: To try to gauge freshman opinion about the food served on campus, Food Services suggests that you just stand outside a school cafeteria at lunchtime and stop people to ask them questions.

QUESTION: What's wrong with this sampling strategy?

ANSWER: *This would be a convenience sample, and it's likely to be biased. I would miss people who use the cafeteria for dinner, but not for lunch, and I'd never hear from anyone who hates the food so much that they have stopped coming to the school cafeteria.*

Mistake 3: Use a Bad Sampling Frame

An SRS from an incomplete sampling frame introduces bias because the individuals included may differ from the ones not in the frame. People in prison, homeless people, students, and long-term travelers are all likely to be missed. Professional polling companies now need special procedures to be sure they include in their sampling frame people who can be reached only by cell phone.

Mistake 4: Undercoverage

Many survey designs suffer from **undercoverage**, in which some portion of the population is not sampled at all or has a smaller representation in the sample than it has in the population. Undercoverage can arise for a number of reasons, but it's always a potential source of bias.

Telephone surveys are usually conducted when you are likely to be home, such as dinnertime. If you eat out often, you may be less likely to be surveyed, a possible source of undercoverage.

Nonresponse Bias

A common and serious potential source of bias for most surveys is **nonresponse bias**. No survey succeeds in getting responses from everyone. The problem is that those who don't respond may differ from those who do. And they may differ on just the variables we care about. Rather than sending out a large number of surveys for which the response rate will be low, it is often better to design a smaller randomized survey for which you have the resources to ensure a high response rate. We might offer a small reward for responding, enter respondents in a drawing for a nice prize, or make followup contacts with nonrespondents.

It turns out that the *Literary Digest* survey was wrong on two counts. First, their list of 10 million people was not representative. There was a selection bias in their sampling frame. There was also a nonresponse bias. We know this because the *Digest* also surveyed a *systematic* sample in Chicago, sending the same question used in the larger survey to every third registered voter. They *still* got a result in favor of Landon, even though Chicago voted overwhelmingly for Roosevelt in the election. This suggests that the Roosevelt supporters were less likely to respond to the *Digest* survey. There's a modern version of this problem: It's been suggested that those who screen their calls with caller ID or an answering machine, and so might not talk to a pollster, may differ in wealth or political views from those who just answer the phone.

Response Bias

Response bias[8] refers to anything in the survey design that influences the responses. Response biases include the tendency of respondents to tailor their responses to try to please the interviewer, the natural unwillingness of respondents to reveal personal facts or admit to illegal or unapproved behavior, and the ways in which the wording of the questions can influence responses.

How to Think About Biases

- **Look for biases in any survey you encounter.** If you design one of your own, ask someone else to help look for biases that may not be obvious to you. And do this *before* you collect your data. There's no way to recover from a biased sampling method or a survey that asks biased questions.

    Sorry, it just can't be done.

    A bigger sample size for a biased study just gives you a bigger useless study.

---

[8]Response bias is not the opposite of nonresponse bias. (We don't make these terms up; we just try to explain them.)

What's the Sample?

The population we want to study is determined by asking *why*. When we design a survey, we use the term "sample" to refer to the individuals selected, from whom we hope to obtain responses. Unfortunately, the real sample is just those we can reach to obtain responses—the *who* of the study. These are slightly different uses of the same term *sample*. The context usually makes clear which we mean, but it's important to realize that the difference between the two could undermine even a well-designed study.

A S *Video*: **Biased Question Wording.** Watch a hapless interviewer make every mistake in the book.

A really big sample gives you a really big useless study. (Think of the 2.4 million *Literary Digest* responses.)

- **Spend your time and resources reducing biases.** No other use of resources is as worthwhile as reducing the biases.
- **Think about the members of the population who could have been excluded from your study.** Be careful not to claim that you have learned anything about them.
- **If you can, pilot-test your survey.** Administer the survey in the exact form that you intend to use it to a small sample drawn from the population you intend to sample. Look for misunderstandings, misinterpretation, confusion, or other possible biases. Then refine your survey instrument.
- **Always report your sampling methods in detail.** Others may be able to detect biases where you did not expect to find them.

WHAT IF ●●● we use a stratified sample?

The goal of sampling is to learn something about a population. It's common and straightforward to choose a simple random sample, but when there are subgroups in the population that vary from one another with respect to the question of interest, stratifying is the way to go. What's better about a stratified sample? Let's look at a hypothetical situation, and investigate using (yes, you guessed it) a simulation.

Every year when the National Council of Teachers of Mathematics holds its convention about 10,000 math teachers converge on some lucky city.[9] Suppose we want to conduct a poll to find out what fraction of these teachers approve of the programs and policies of the nation's Secretary of Education. From conference registration records we could learn that 60% of the attendees teach in public schools and 40% in private schools. We might suspect that these two groups have different opinions. In reality, we could not know in advance what that difference might be, but for purposes of our simulation let's assume that the true approval rate is only 30% among the public school teachers and a whopping 90% among the private school crowd. How could this approval gap affect what we might learn from our poll?

We decide to survey 200 randomly selected math teachers. What sampling methodology should we use?

PLAN A: We choose a simple random sample of any 200 teachers.

PLAN B: We stratify our sample by type of school, randomly choosing 120 public school teachers and 80 private school teachers. This makes the respondents' school type ratio the same as the population's.

With that framework, we ran our simulation. In the first SRS 115 teachers expressed approval of the Secretary of Education, a sample proportion of $115/200 = 0.575$. In the next SRS the proportion was $96/200 = 0.46$. And so on; we simulated 100 simple random samples. Then we tried the stratified sampling plan 100 times, too. In our first stratified sample 38 of 120 public school teachers approved, as did 73 of the 80 from private schools. That made this sample's approval proportion $(38 + 73)/200 = 0.555$. One down, 99 more to go.

What did these simulated samples reveal? The boxplots compare the distributions of the simulated sample proportions.

What do you see?

For one thing, the medians are about the same. It appears that both SRSs and stratified samples produce estimates centering around 54% approval.[10] It's good to see that both approaches target the same truth in the NCTM population . . . *on average.*

---

[9]Q: What could be more fun than 10,000 math teachers in one place? (A: Only 5,000?)

[10]Yes, 54% is indeed the true approval rate based on the percentages we used to create this NCTM population. See if you can figure out why. Go ahead, try! We believe in you.

But in the real world we don't have the luxury of choosing 100 samples. We just get one. We'd want to use a sampling method that's more likely to produce a result close to the truth. The boxplots clearly show that when we stratify there's less variation from sample to sample, so we can expect a stratified sample to provide a more accurate estimate.

If there weren't an opinion gap between public and private school teachers, stratifying would be a waste of time; an SRS would work just as well. But when we think there's a difference between subgroups of the population, going to the extra trouble of choosing a stratified sample offers us the important advantage of reducing sampling error.

WHAT CAN GO WRONG?

- **Get a sample that looks like the population.** The principal thing that can go wrong in sampling is that the sample can fail to represent the population. Unfortunately, this can happen in many different ways and for many different reasons. We've considered many of them in the chapter.

- **Make the sample large enough.** It is also an error to draw too small a sample for your needs regardless of how well you draw the sample. However, generally it is more worthwhile to devote effort on improving the quality of your sample than on expanding its size.

- **Avoid bias.** Any of the many types of bias we've discussed can render your results meaningless. Pay attention to all of them.

What Have We Learned?

We've learned that a representative sample can offer us important insights about populations. It's the size of the sample—and not its fraction of the larger population—that determines the precision of the statistics it yields.

We've learned several ways to draw samples, all based on the power of randomness to make them representative of the population of interest:

- A Simple Random Sample (SRS) is our standard. Every possible group of n individuals has an equal chance of being our sample. That's what makes it *simple*.
- Stratified samples can reduce sampling variability by identifying homogeneous subgroups and then randomly sampling within each.
- Cluster samples randomly select among heterogeneous subgroups making our sampling tasks more manageable.
- Systematic samples can work in some situations and are often the least expensive method of sampling. But we still want to start them randomly.
- Multistage samples combine several random sampling methods.

We've learned that bias can destroy our ability to gain insights from our sample:

- Voluntary response samples are almost always biased and should be avoided and distrusted.
- Convenience samples are likely to be flawed for similar reasons.
- Bad sampling frames can lead to samples that don't represent the population of interest.
- Undercoverage occurs when individuals from a subgroup of the population are selected less often than they should be.
- Nonresponse bias can arise when sampled individuals will not or cannot respond.
- Response bias arises when respondents' answers might be affected by external influences, such as question wording or interviewer behavior.

Finally, we've learned to look for biases in any survey we find and to be sure to report our methods whenever we perform a survey so that others can evaluate the fairness and accuracy of our results.

Terms

Population
The entire group of individuals or instances about whom we hope to learn. (p. 281)

Sample
A (representative) subset of a population, examined in hope of learning about the population. (p. 281)

Sample survey
A study that asks questions of a sample drawn from some population in the hope of learning something about the entire population. Polls taken to assess voter preferences are common sample surveys. (p. 281)

Bias
Any systematic failure of a sampling method to represent its population is bias. Biased sampling methods tend to over- or underestimate parameters. It is almost impossible to recover from bias, so efforts to avoid it are well spent. Common errors include

- relying on voluntary response.
- undercoverage of the population.
- nonresponse bias.
- response bias. (p. 281)

Randomization
The best defense against bias is randomization, in which each individual is given a fair, random chance of selection. (p. 282)

Sample size
The number of individuals in a sample. The sample size determines how well the sample represents the population, not the fraction of the population sampled. (p. 283)

Census
A sample that consists of the entire population is called a census. (p. 283)

Population parameter
A numerically valued attribute of a model for a population. We rarely expect to know the true value of a population parameter, but we do hope to estimate it from sampled data. For example, the mean income of all employed people in the country is a population parameter. (p. 284)

Statistic, sample statistic
Statistics are values calculated for sampled data. Those that correspond to, and thus estimate, a population parameter, are of particular interest. For example, the mean income of all employed people in a representative sample can provide a good estimate of the corresponding population parameter. The term "sample statistic" is sometimes used, usually to parallel the corresponding term "population parameter". (p. 284)

Representative
A sample is said to be representative if the statistics computed from it accurately reflect the corresponding population parameters. (p. 285)

Simple random sample (SRS)
A simple random sample of sample size n is a sample in which each set of n elements in the population has an equal chance of selection. (p. 286)

Sampling frame
A list of individuals from whom the sample is drawn is called the sampling frame. Individuals who may be in the population of interest, but who are not in the sampling frame, cannot be included in any sample. (p. 286)

Sampling variability
The natural tendency of randomly drawn samples to differ, one from another. Sometimes, unfortunately, called *sampling error*, sampling variability is no error at all, but just the natural result of random sampling. (p. 286)

Stratified random sample
A sampling design in which the population is divided into several subpopulations, or **strata,** and random samples are then drawn from each stratum. If the strata are homogeneous, but are different from each other, a stratified sample may yield more consistent results than an SRS. (p. 287)

Cluster sample
A sampling design in which entire groups, or **clusters,** are chosen at random. Cluster sampling is usually selected as a matter of convenience, practicality, or cost. Clusters are heterogeneous, and a random sample of clusters should be representative of the population. (p. 287)

Multistage sample	Sampling schemes that combine several sampling methods are called multistage samples. For example, a national polling service may stratify the country by geographical regions, select a random sample of cities from each region, and then interview a cluster of residents in each city. (p. 288)
Systematic sample	A sample drawn by selecting individuals systematically from a sampling frame. When there is no relationship between the order of the sampling frame and the variables of interest, a systematic sample can be representative. (p. 289)
Pilot Study	A small trial run of a survey to check whether questions are clear. A pilot study can reduce errors due to ambiguous questions. (p. 293)
Voluntary response bias	Bias introduced to a sample when individuals can choose on their own whether to participate in the sample. Samples based on voluntary response are always invalid and cannot be recovered, no matter how large the sample size. (p. 294)
Convenience sample	A convenience sample consists of the individuals who are conveniently available. Convenience samples often fail to be representative because every individual in the population is not equally convenient to sample. (p. 294)
Undercoverage	A sampling scheme that biases the sample in a way that gives a part of the population less representation than it has in the population suffers from undercoverage. (p. 295)
Nonresponse bias	Bias introduced when a large fraction of those sampled fails to respond. Those who do respond are likely to not represent the entire population. Voluntary response bias is a form of nonresponse bias, but nonresponse may occur for other reasons. For example, those who are at work during the day won't respond to a telephone survey conducted only during working hours. (p. 295)
Response bias	Anything in a survey design that influences responses falls under the heading of response bias. One typical response bias arises from the wording of questions, which may suggest a favored response. Voters, for example, are more likely to express support of "the president" than support of the particular person holding that office at the moment. (p. 295)

On the Computer SAMPLING

Computer-generated pseudorandom numbers are usually good enough for drawing random samples. But there is little reason not to use the truly random values available on the Internet.

Here's a convenient way to draw an SRS of a specified size using a computer-based sampling frame. The sampling frame can be a list of names or of identification numbers arrayed, for example, as a column in a spreadsheet, statistics program, or database:

1. Generate random numbers of enough digits so that each exceeds the size of the sampling frame list by several digits. This makes duplication unlikely.
2. Assign the random numbers arbitrarily to individuals in the sampling frame list. For example, put them in an adjacent column.
3. Sort the list of random numbers, carrying along the sampling frame list.
4. Now the first n values in the sorted sampling frame column are an SRS of n values from the entire sampling frame.

Exercises

1. **Roper** Through their *Roper Reports Worldwide*, GfK Roper conducts a global consumer survey to help multinational companies understand different consumer attitudes throughout the world. Within 30 countries, the researchers interview 1000 people aged 13–65. Their samples are designed so that they get 500 males and 500 females in each country. (www.gfkamerica.com)

   a) Are they using a simple random sample? Explain.
   b) What kind of design do you think they are using?

2. **Student Center Survey** For their class project, a group of Statistics students decide to survey the student body to assess opinions about the proposed new student center. Their sample of 200 contained 50 first-year students, 50 sophomores, 50 juniors, and 50 seniors.

   a) Do you think the group was using an SRS? Why?
   b) What sampling design do you think they used?

3. **Emoticons** The website www.gamefaqs.com asked, as their question of the day to which visitors to the site were invited to respond, *"Do you ever use emoticons when you type online?"* Of the 87,262 respondents, 27% said that they did not use emoticons. ;-(

   a) What kind of sample was this?
   b) How much confidence would you place in using 27% as an estimate of the fraction of people who use emoticons?

4. **Drug tests** Major League Baseball tests players to see whether they are using performance-enhancing drugs. Officials select a team at random, and a drug-testing crew shows up unannounced to test all 40 players on the team. Each testing day can be considered a study of drug use in Major League Baseball.

   a) What kind of sample is this?
   b) Is that choice appropriate?

5. **Gallup** At its website (www.gallup.com) the Gallup Poll publishes results of a new survey each day. Scroll down to the end, and you'll find a statement that includes words such as these:

   Results are based on telephone interviews with 1,008 national adults, aged 18 and older, conducted January 3–5, 2013. . . . In addition to sampling error, question wording and practical difficulties in conducting surveys can introduce error or bias into the findings of public opinion polls.

   a) For this survey, identify the population of interest.
   b) Gallup performs its surveys by phoning numbers generated at random by a computer program. What is the sampling frame?
   c) What problems, if any, would you be concerned about in matching the sampling frame with the population?

6. **Gallup World** At its website (www.gallupworldpoll .com) the Gallup World Poll describes their methods. After one report they explained:

   Results are based on face-to-face interviews with randomly selected national samples of approximately 1,000 adults, aged 15 and older, who live permanently in each of the 21 sub-Saharan African nations surveyed. Those countries include Angola (areas where land mines might be expected were excluded), Benin, Botswana, Burkina Faso, Cameroon, Ethiopia, Ghana, Kenya, Madagascar (areas where interviewers had to walk more than 20 kilometers from a road were excluded), Mali, Mozambique, Niger, Nigeria, Senegal, Sierra Leone, South Africa, Tanzania, Togo, Uganda (the area of activity of the Lord's Resistance Army was excluded from the survey), Zambia, and Zimbabwe. . . . In all countries except Angola, Madagascar, and Uganda, the sample is representative of the entire population.

   a) Gallup is interested in sub-Saharan Africa. What kind of survey design are they using?
   b) Some of the countries surveyed have large populations. (Nigeria is estimated to have about 130 million people.) Some are quite small. (Togo's population is estimated at 5.4 million.) Nonetheless, Gallup sampled 1000 adults in each country. How does this affect the precision of its estimates for these countries?

7–10. **What did they do?** *For the following reports about statistical studies, identify the following items (if possible). If you can't tell, then say so—this often happens when we read about a survey.*

   a) The population
   b) The population parameter of interest
   c) The sampling frame
   d) The sample
   e) The sampling method, including whether or not randomization was employed
   f) Any potential sources of bias you can detect and any problems you see in generalizing to the population of interest

7. **Medical treatments** Consumers Union, in an attempt to get information about U.S. adults, asked all subscribers whether they had used alternative medical treatments and, if so, whether they had benefited from them. For almost all of the treatments, approximately 20% of those responding reported cures or substantial improvement in their condition. They received replies from 12% of their subscribers.

8. **Snack foods** A company packaging snack foods maintains quality control by randomly selecting 10 cases from each day's production and weighing the bags. Then they open one bag from each case and inspect the contents. For this

exercise, answer the questions both for the bags that are weighed and for the bags that are opened for inspection.

9. **Drinking and driving** In order to determine how adults of legal drinking age in their city feel about whether drinking and driving was a problem, researchers waited outside a bar they had randomly selected from a list of such establishments. They rolled a ten-sided die and it came up 4, so they stopped the fourth person who came out of the bar, then every 10th person after that, and asked whether he or she thought drinking and driving was a serious problem.

10. **Mayoral race** Hoping to learn what issues may resonate with voters in the coming election, the campaign director for a mayoral candidate randomly selects two blocks from each of the city's election districts. Staff members go there and interview all the residents they can find. The residents were asked to select the three most important issues from a prepared list.

11. **Toxic waste** The Environmental Protection Agency took a map of a region near a former industrial waste dump and placed a grid of 552 squares on it. They randomly selected any 16 of those squares from which to collect soil samples and checked each for evidence of toxic chemicals.
    a) What type of sampling did they use?
    b) Is there any sort of bias associated with this sampling procedure?
    c) One researcher suggests that plots closer to the old dump site could contain more contaminants than those farther away. How could the sampling procedure be improved to take this into account?

12. **Social life** A question posted on the gamefaqs.com website on August 1, 2011, asked visitors to the site, "Do you have an active social life outside the Internet?" 22% of the 55,581 respondents said "No" or "Not really, most of my personal contact is online."
    a) Can this survey be used to estimate the proportion of U.S. adults who would say they have an active social life outside the Internet? Why or why not?
    b) Can this survey be used to estimate the proportion of visitors to their site who would say they have an active social life outside the Internet? Why or why not?

13. **Roadblock** State police set up a roadblock to estimate the percentage of cars with up-to-date registration, insurance, and safety inspection stickers. It would be too inconvenient and costly to check every vehicle that passes through a checkpoint, so they decide to stop about 1/20 of the vehicles.
    a) Why would a simple random sample be unreasonable for this situation.
    b) Identify two possible sampling schemes that could be used. Explain how randomization would be used in each.

14. **Milk samples** Dairy inspectors visit farms unannounced and take samples of the milk to test for contamination.

If the milk is found to contain dirt, antibiotics, or other foreign matter, the milk will be destroyed and the farm reinspected until purity is restored.

Would simple random sampling be appropriate for selecting farms for inspection? If so, explain how it would be done. If not, explain why it is not appropriate.

15. **Mistaken poll** A local TV station conducted a "Pulse-Poll" about the upcoming mayoral election. Evening news viewers were invited to phone in their votes, with the results to be announced on the late-night news. Based on the phone calls, the station predicted that Amabo would win the election with 52% of the vote. They were wrong: Amabo lost, getting only 46% of the vote. Do you think the station's faulty prediction is more likely to be a result of bias or sampling error? Explain.

16. **Another mistaken poll** Prior to the mayoral election discussed in Exercise 15, the newspaper also conducted a poll. The paper surveyed a random sample of registered voters stratified by political party, age, sex, and area of residence. This poll predicted that Amabo would win the election with 52% of the vote. The newspaper was wrong: Amabo lost, getting only 46% of the vote. Do you think the newspaper's faulty prediction is more likely to be a result of bias or sampling error? Explain.

17. **Parent opinion, part 1** In a large city school system with 20 elementary schools, the school board is considering the adoption of a new policy that would require elementary students to pass a test in order to be promoted to the next grade. The PTA wants to find out whether parents agree with this plan. Listed below are some of the ideas proposed for gathering data. For each, indicate what kind of sampling strategy is involved and what (if any) biases might result.
    a) Put a big ad in the newspaper asking people to log their opinions on the PTA website.
    b) Randomly select one of the elementary schools and contact every parent by phone.
    c) Send a survey home with every student, and ask parents to fill it out and return it the next day.
    d) Randomly select 20 parents from each elementary school. Send them a survey, and follow up with a phone call if they do not return the survey within a week.

18. **Parent opinion, part 2** Let's revisit the school system described in Exercise 17. Four new sampling strategies have been proposed to help the PTA determine whether parents favor requiring elementary students to pass a test in order to be promoted to the next grade. For each, indicate what kind of sampling strategy is involved and what (if any) biases might result.
    a) Run a poll on the local TV news, asking people to dial one of two phone numbers to indicate whether they favor or oppose the plan.
    b) Hold a PTA meeting at each of the 20 elementary schools, and tally the opinions expressed by those who attend the meetings.

c) Randomly select one class at each elementary school and contact each of those parents.

d) Go through the district's enrollment records, selecting every 40th parent. PTA volunteers will go to those homes to interview the people chosen.

19. **Churches** For your political science class, you'd like to take a survey from a sample of all the Catholic Church members in your city. A list of churches shows 17 Catholic churches within the city limits. Rather than try to obtain a list of all members of all these churches, you decide to pick 3 churches at random. For those churches, you'll ask to get a list of all current members and contact any 100 members selected at random.

a) What kind of design have you used?

b) What could go wrong with your design?

20. **Playground** Some people have been complaining that the children's playground at a municipal park is too small and is in need of repair. Managers of the park decide to survey city residents to see if they believe the playground should be rebuilt. They hand out questionnaires to parents who bring children to the park. Describe possible biases in this sample.

21. **Roller coasters** An amusement park has opened a new roller coaster. It is so popular that people are waiting for up to 3 hours for a 2-minute ride. Concerned about how patrons (who paid a large amount to enter the park and ride on the rides) feel about this, they survey every 10th person on the line for the roller coaster, starting from a randomly selected individual.

a) What kind of sample is this?

b) What is the sampling frame?

c) Is it likely to be representative?

22. **Playground, act two** The survey described in Exercise 20 asked,

Many people believe this playground is too small and in need of repair. Do you think the playground should be repaired and expanded even if that means raising the entrance fee to the park?

Describe two ways this question may lead to response bias.

23. **Wording the survey** Two members of the PTA committee in Exercises 17 and 18 have proposed different questions to ask in seeking parents' opinions.

Question 1: Should elementary school–age children have to pass high-stakes tests in order to remain with their classmates?

Question 2: Should schools and students be held accountable for meeting yearly learning goals by testing students before they advance to the next grade?

a) Do you think responses to these two questions might differ? How? What kind of bias is this?

b) Propose a question with more neutral wording that might better assess parental opinion.

24. **Banning ephedra** An online poll at a website asked:

A nationwide ban of the diet supplement ephedra went into effect recently. The herbal stimulant has been linked to 155 deaths and many more heart attacks and strokes. Ephedra manufacturer NVE Pharmaceuticals, claiming that the FDA lacked proof that ephedra is dangerous if used as directed, was denied a temporary restraining order on the ban yesterday by a federal judge. Do you think that ephedra should continue to be banned nationwide?

65% of 17,303 respondents said "yes." Comment on each of the following statements about this poll:

a) With a sample size that large, we can be pretty certain we know the true proportion of Americans who think ephedra should be banned.

b) The wording of the question is clearly very biased.

c) The sampling frame is all Internet users.

d) Results of this voluntary response survey can't be reliably generalized to any population of interest.

25. **Survey questions** Examine each of the following questions for possible bias. If you think the question is biased, indicate how and propose a better question.

a) Should companies that pollute the environment be compelled to pay the costs of cleanup?

b) Given that 18-year-olds are old enough to vote and to serve in the military, is it fair to set the drinking age at 21?

26. **More survey questions** Examine each of the following questions for possible bias. If you think the question is biased, indicate how and propose a better question.

a) Do you think high school students should be required to wear uniforms?

b) Given humanity's great tradition of exploration, do you favor continued funding for space flights?

27. **Phone surveys** Anytime we conduct a survey, we must take care to avoid undercoverage. Suppose we plan to select 500 names from the city phone book, call their homes between noon and 4 PM, and interview whoever answers, anticipating contacts with at least 200 people.

a) Why is it difficult to use a simple random sample here?

b) Describe a more convenient, but still random, sampling strategy.

c) What kinds of households are likely to be included in the eventual sample of opinion? Excluded?

d) Suppose, instead, that we continue calling each number, perhaps in the morning or evening, until an adult is contacted and interviewed. How does this improve the sampling design?

e) Random-digit dialing machines can generate the phone calls for us. How would this improve our design? Is anyone still excluded?

28. **Cell phone survey** What about drawing a random sample only from cell phone exchanges? Discuss the advantages and disadvantages of such a sampling method compared with surveying randomly generated telephone

numbers from non–cell phone exchanges. Do you think these advantages and disadvantages have changed over time? How do you expect they'll change in the future?

29. Arm length How long is your arm compared with your hand size? Put your right thumb at your left shoulder bone, stretch your hand open wide, and extend your hand down your arm. Put your thumb at the place where your little finger is, and extend down the arm again. Repeat this a third time. Now your little finger will probably have reached the back of your left hand. If the fourth hand width goes past the end of your middle finger, turn your hand sideways and count finger widths to get there.

a) How many hand and finger widths is your arm?

b) Suppose you repeat your measurement 10 times and average your results. What parameter would this average estimate? What is the population?

c) Suppose you now collect arm lengths measured in this way from 9 friends and average these 10 measurements. What is the population now? What parameter would this average estimate?

d) Do you think these 10 arm lengths are likely to be representative of the population of arm lengths in your community? In the country? Why or why not?

30. Fuel economy Occasionally, when I fill my car with gas, I figure out how many miles per gallon my car got. I wrote down those results after 6 fill-ups in the past few months. Overall, it appears my car gets 28.8 miles per gallon.

a) What statistic have I calculated?

b) What is the parameter I'm trying to estimate?

c) How might my results be biased?

d) When the Environmental Protection Agency (EPA) checks a car like mine to predict its fuel economy, what parameter is it trying to estimate?

31. Accounting Between quarterly audits, a company likes to check on its accounting procedures to address any problems before they become serious. The accounting staff processes payments on about 120 orders each day. The next day, the supervisor rechecks 10 of the transactions to be sure they were processed properly.

a) Propose a sampling strategy for the supervisor.

b) How would you modify that strategy if the company makes both wholesale and retail sales, requiring different bookkeeping procedures?

32. Happy workers? A manufacturing company employs 14 project managers, 48 foremen, and 377 laborers. In an effort to keep informed about any possible sources of employee discontent, management wants to conduct job satisfaction interviews with a sample of employees every month.

a) Do you see any potential danger in the company's plan? Explain.

b) Propose a sampling strategy that uses a simple random sample.

c) Why do you think a simple random sample might not provide the representative opinion the company seeks?

d) Propose a better sampling strategy.

e) Listed below are the last names of the project managers. Use random numbers to select two people to be interviewed. Explain your method carefully.

Barrett	Bowman	Chen
DeLara	DeRoos	Grigorov
Maceli	Mulvaney	Pagliarulo
Rosica	Smithson	Tadros
Williams	Yamamoto	

33. Quality control Sammy's Salsa, a small local company, produces 20 cases of salsa a day. Each case contains 12 jars and is imprinted with a code indicating the date and batch number. To help maintain consistency, at the end of each day, Sammy selects three jars of salsa, weighs the contents, and tastes the product. Help Sammy select the sample jars. Today's cases are coded 07N61 through 07N80.

a) Carefully explain your sampling strategy.

b) Show how to use random numbers to pick 3 jars.

c) Did you get a simple random sample of the jars? Explain.

34. A fish story Concerned about reports of discolored scales on fish caught downstream from a newly sited chemical plant, scientists set up a field station in a shoreline public park. For one week they asked fishermen there to bring any fish they caught to the field station for a brief inspection. At the end of the week, the scientists said that 18% of the 234 fish that were submitted for inspection displayed the discoloration. From this information, can the researchers estimate what proportion of fish in the river have discolored scales? Explain.

35. Sampling methods Consider each of these situations. Do you think the proposed sampling method is appropriate? Explain.

a) We want to know what percentage of local doctors accept Medicaid patients. We call the offices of 50 doctors randomly selected from local Yellow Page listings.

b) We want to know what percentage of local businesses anticipate hiring additional employees in the upcoming month. We randomly select a page in the Yellow Pages and call every business listed there.

36. More sampling methods Consider each of these situations. Do you think the proposed sampling method is appropriate? Explain.

a) We want to know if there is neighborhood support to turn a vacant lot into a playground. We spend a Saturday afternoon going door-to-door in the neighborhood, asking people to sign a petition.

b) We want to know if students at our college are satisfied with the selection of food available on campus. We go to the largest cafeteria and interview every 10th person in line.

37. Texas A&M Administrators at Texas A&M University were interested in estimating the percentage of students who are the first in their family to go to college. The A&M student body has about 46,000 members.

a) What problems do you see with asking the following question of students? "Are you the first member of your family to seek higher education?"

b) For each scenario, identify the kind of sample used by the university administrators:

   i. Select several dormitories at random and contact everyone living in the selected dorms.

   ii. Using a computer-based list of registered students, contact 200 freshmen, 200 sophomores, 200 juniors, and 200 seniors selected at random from each class.

   iii. Using a computer-based alphabetical list of registered students, select one of the first 25 on the list by random and then contact the student whose name is 50 names later, and then every 50 names beyond that.

c) A professor teaching a large lecture class of 350 students samples her class by rolling a die. Then, starting with the row number on the die (1 to 6), she passes out a survey to every fourth row of the large lecture hall. She says that this is a Simple Random Sample because everyone had an equal opportunity to sit in any seat and because she randomized the choice of rows. What do you think? Be specific.

d) For each of these proposed survey designs, identify the problem and the effect it would have on the estimate of the percentage of students who are the first in their family to go to college.

   i. Publish an advertisement inviting students to visit a website and answer questions.

   ii. Set up a table in the student union and ask students to stop and answer a survey.

e) The president of the university plans a speech to an alumni group. He plans to talk about the proportion of students who responded in the survey that they are the first in their family to attend college, but the first draft of his speech treats that proportion as the actual proportion of current A&M students who are the first in their families to attend college. Explain to the president the difference between the proportion of respondents who are first attenders and the proportion of the entire student body that are first attenders. Use appropriate statistics terminology.

38. Satisfied workers The managers of a large company wished to know the percentage of employees who feel "extremely satisfied" to work there. The company has roughly 24,000 employees. They contacted a random sample of employees and asked them about their job satisfaction, obtaining 437 completed responses.

a) The company's annual report states, "Our survey shows that 87.34% of our employees are 'very happy' working here." Comment on that claim. Use appropriate statistics terminology.

b) One manager suggested surveying employees by assigning computer-generated random numbers to each employee on a list of all employees and then contacting all those whose assigned random number is divisible by 7. Is this a simple random sample?

c) For each scenario suggested by a different manager, determine the sampling method.

   i. Use the company e-mail directory to contact 150 employees from among those employed for less than 5 years, 150 from among those employed for 5–10 years, and 150 from among those employed for more than 10 years.

   ii. Use the company e-mail directory to contact every 50th employee on the list.

   iii. Select several divisions of the company at random. Within each division, draw an SRS of employees to contact.

d) One manager suggested having the head of each corporate division hold a meeting of their employees to ask whether they are happy on their jobs. They will ask people to raise their hands to indicate whether they are happy. What problems do you see with this plan?

e) For each of these designs proposed by a different manager, identify the problem with the method and the effect it would have on the estimate of the percentage of employees who feel "extremely satisfied" to work there.

   i. Leave a stack of surveys out in the employee cafeteria so people can pick them up and return them.

   ii. Stuff a questionnaire in the mailbox of each employee with the request that they fill it out and return it.

Just Checking ANSWERS

1. a) It can be hard to reach all members of a population, and it can take so long that circumstances change, affecting the responses. A well-designed sample is often a better choice.

   b) This sample is probably biased—students who didn't like the food at the cafeteria might not choose to eat there.

   c) No, only the sample size matters, not the fraction of the overall population.

   d) Students who frequent this website might be more enthusiastic about Statistics than the overall population of Statistics students. A large sample cannot compensate for bias.

   e) It's the population "parameter." "Statistics" describe samples.

2. a) systematic
   b) stratified
   c) simple
   d) cluster

12 Experiments and Observational Studies

Who gets good grades? And, more importantly, why? Is there something schools and parents could do to help weaker students improve their grades? Some people think they have an answer: music! No, not your iPod, but an instrument. In a study conducted at Mission Viejo High School, in California, researchers compared the scholastic performance of music students with that of non-music students. Guess what? The music students had a much higher overall grade point average than the non-music students, 3.59 to 2.91. Not only that: A whopping 16% of the music students had all A's compared with only 5% of the non-music students.

As a result of this study and others, many parent groups and educators pressed for expanded music programs in the nation's schools. They argued that the work ethic, discipline, and feeling of accomplishment fostered by learning to play an instrument also enhance a person's ability to succeed in school. They thought that involving more students in music would raise academic performance. What do you think? Does this study provide solid evidence? Or are there other possible explanations for the difference in grades? Is there any way to really prove such a conjecture?

Observational Studies

This research tried to show an association between music education and grades. But it wasn't a survey. Nor did it assign students to get music education. Instead, it simply observed students "in the wild," recording the choices they made and the outcome. Such studies are called **observational** studies. In observational studies, researchers don't *assign* choices; they simply observe them. In addition, this was a **retrospective study**, because researchers first identified subjects who studied music and then collected data on their past grades.

What's wrong with concluding that music education causes good grades? One high school during one academic year may not be representative of the whole United States. That's true, but the real problem is that the claim that music study *caused* higher grades depends on there being *no other differences* between the groups that could account for the differences in grades, and studying music was not the *only* difference between the two groups of students.

We can think of lots of lurking variables that might cause the groups to perform differently. Students who study music may have better work habits to start with, and this makes them successful in both music and course work. Music students may have more parental support (someone had to pay for all those lessons), and that support may have enhanced their academic performance, too. Maybe they came from wealthier homes and had other advantages. Or it could be that smarter kids just like to play musical instruments.

Observational studies are valuable for discovering trends and possible relationships. They are used widely in public health and marketing. Observational studies that try to discover variables related to rare outcomes, such as specific diseases, are often retrospective. They first identify people with the disease and then look into their history and heritage in search of things that may be related to their condition. But retrospective studies have a restricted view of the world because they are usually restricted to a small part of the entire population. And because retrospective records are based on historical data, they can have errors. (Do you recall *exactly* what you ate even yesterday? How about last Wednesday?)

A somewhat better approach is to observe individuals over time, recording the variables of interest and ultimately seeing how things turn out. For example, we might start by selecting young students who have not begun music lessons. We could then track their academic performance over several years, comparing those who later choose to study music with those who do not. Identifying subjects in advance and collecting data as events unfold would make this a **prospective study.**

Although an observational study may identify important variables related to the outcome we are interested in, there is no guarantee that we have found the right or the most important related variables. Students who choose to study an instrument might still differ from the others in some important way that we failed to observe. It may be this difference—whether we know what it is or not—rather than music itself that leads to better grades. It's just not possible for observational studies, whether prospective or retrospective, to demonstrate a causal relationship.

> **Retrospective Studies Can Give Valuable Clues.** For rare illnesses, it's not practical to draw a large enough sample to see many ill respondents, so the only option remaining is to develop retrospective data. For example, researchers can interview those who have become ill. The likely causes of both legionnaires' disease and HIV were initially identified from such retrospective studies of the small populations who were initially infected. But to confirm the causes, researchers needed laboratory-based experiments.

For Example DESIGNING AN OBSERVATIONAL STUDY

In early 2007, a larger-than-usual number of cats and dogs developed kidney failure; many died. Initially, researchers didn't know why, so they used an observational study to investigate.

QUESTION: Suppose you were called on to plan a study seeking the cause of this problem. Would your design be retrospective or prospective? Explain why.

ANSWER: *I would use a retrospective observational study. Even though the incidence of disease was higher than usual, it was still rare. Surveying all pets would have been impractical. Instead, it makes sense to locate some who were sick and ask about their diets, exposure to toxins, and other possible causes.*

Randomized, Comparative Experiments

> "He that leaves nothing to chance will do few things ill, but he will do very few things."
>
> —*Lord Halifax (1633–1695)*

Is it *ever* possible to get convincing evidence of a cause-and-effect relationship? Well, yes it is, but we would have to take a different approach. We could take a group of third graders, randomly assign half to take music lessons, and forbid the other half to do so. Then we could compare their grades several years later. This kind of study design is called an **experiment.**

Experimental design was advanced in the 19th century by work in psychophysics by Gustav Fechner (1801–1887), the founder of experimental psychology. Fechner designed ingenious experiments that exhibited many of the features of modern designed experiments. Fechner was careful to control for the effects of factors that might affect his results. For example, in his 1860 book *Elemente der Psychophysik* he cautioned readers to group experiment trials together to minimize the possible effects of time of day and fatigue.

An Experiment

manipulates the factor levels to create treatments. *Randomly assigns* subjects to these treatment levels. *Compares* the responses of the subject groups across treatment levels.

The FDA   No drug can be sold in the United States without first showing, in a suitably designed experiment approved by the Food and Drug Administration (FDA), that it's safe and effective. The small print on the booklet that comes with many prescription drugs usually describes the outcomes of that experiment.

An experiment requires a **random assignment** of subjects to treatments. Only an experiment can justify a claim like "Music lessons cause higher grades." Questions such as "Does taking vitamin C reduce the chance of getting a cold?" and "Does working with computers improve performance in Statistics class?" and "Is this drug a safe and effective treatment for that disease?" require a designed experiment to establish cause and effect.

Experiments study the relationship between two or more variables. An experimenter must identify at least one explanatory variable, called a **factor,** to manipulate and at least one **response variable** to measure. What distinguishes an experiment from other types of investigation is that the experimenter actively and deliberately manipulates the factors to control the details of the possible treatments, and assigns the subjects to those treatments *at random.* The experimenter then observes the response variable and *compares* responses for different groups of subjects who have been treated differently. For example, we might design an experiment to see whether the amount of sleep and exercise you get affects your performance.

The individuals on whom or which we experiment are known by a variety of terms. Humans who are experimented on are commonly called **subjects** or **participants.** Other individuals (rats, days, petri dishes of bacteria) are commonly referred to by the more generic term **experimental units.** When we recruit subjects for our sleep deprivation experiment by advertising in Statistics class, we'll probably have better luck if we invite them to be participants than if we advertise that we need experimental units.

The specific values that the experimenter chooses for a factor are called the **levels** of the factor. We might assign our participants to sleep for 4, 6, or 8 hours. Often there are several factors at a variety of levels. (Our subjects will also be assigned to a treadmill for 0 or 30 minutes.) The combination of specific levels from all the factors that an experimental unit receives is known as its **treatment.** (Our experiment has six different treatments—three sleep levels, each at two exercise levels.)

How should we assign our participants to these treatments? Some students prefer 4 hours of sleep, while others need 8. Some exercise regularly; others are couch potatoes. Should we let the students choose the treatments they'd prefer? No. That would not be a good idea. To have any hope of drawing a fair conclusion, we must assign our participants to their treatments *at random.*

It may be obvious to you that we shouldn't let the students choose the treatment they'd prefer, but the need for random assignment is a lesson that was once hard for some to accept. For example, physicians might naturally prefer to assign patients to the therapy that they think best rather than have a random element such as a coin flip determine the treatment. But if anyone knew for sure which treatment was better, we wouldn't be doing an experiment. We've known for more than a century that for the results of an experiment to be valid, we must use deliberate randomization.

The Women's Health Initiative   is a major 15-year research program funded by the National Institutes of Health to address the most common causes of death, disability, and poor quality of life in older women. It consists of an observational study with more than 93,000 participants as well as several randomized comparative experiments. The goals of this study include

- giving reliable estimates of the extent to which known risk factors predict heart disease, cancers, and fractures;
- identifying "new" risk factors for these and other diseases in women;
- comparing risk factors, presence of disease at the start of the study, and new occurrences of disease during the study across all study components; and
- creating a future resource to identify biological indicators of disease, especially substances and factors found in blood.

That is, the study seeks to identify possible risk factors and assess how serious they might be. It seeks to build up data that might be checked retrospectively as the women in the study

(continued)

continue to be followed. There would be no way to find out these things with an experiment because the task includes identifying new risk factors. If we don't know those risk factors, we could never control them as factors in an experiment.

By contrast, one of the clinical trials (randomized experiments) that received much press attention randomly assigned postmenopausal women to take either hormone replacement therapy or an inactive pill. The results published in 2002 and 2004 concluded that hormone replacement with estrogen carried increased risks of stroke.

For Example DETERMINING THE TREATMENTS AND RESPONSE VARIABLE

RECAP: In 2007, deaths of a large number of pet dogs and cats were ultimately traced to contamination of some brands of pet food. The manufacturer now claims that the food is safe, but before it can be released, it must be tested.

QUESTION: In an experiment to test whether the food is now safe for dogs to eat,[1] what would be the treatments and what would be the response variable?

ANSWER: The treatments would be ordinary-size portions of two dog foods: the new one from the company (the *test food*) and one that I was certain was safe (perhaps prepared in my kitchen or laboratory). The response would be a veterinarian's assessment of the health of the test animals.

The Four Principles of Experimental Design

 Video: An Industrial Experiment. Manufacturers often use designed experiments to help them perfect new products. Watch this video about one such experiment.

The deep insight that experiments should use random assignment is quite an old one. It can be attributed to the American philosopher and scientist C. S. Peirce in his experiments with J. Jastrow, published in 1885.

1. **Control.** We control sources of variation other than the factors we are testing by making conditions as similar as possible for all treatment groups. For human subjects, we try to treat them alike. However, there is always a question of degree and practicality. Controlling extraneous sources of variation reduces the variability of the responses, making it easier to detect differences among the treatment groups.

   Making generalizations from the experiment to other levels of the controlled factor can be risky. For example, suppose we test two laundry detergents and carefully control the water temperature at 180°F. This would reduce the variation in our results due to water temperature, but what could we say about the detergents' performance in cold water? Not much.

   Although we control both experimental factors and other sources of variation, we think of them very differently. We control a factor by assigning subjects to different factor levels because we want to see how the response will change at those different levels. We control other sources of variation to *prevent* them from changing and affecting the response variable.

2. **Randomize.** As in sample surveys, randomization allows us to equalize the effects of unknown or uncontrollable sources of variation. It does not eliminate the effects of these sources, but by distributing them equally (on average) across the treatment

---

[1] It may disturb you (as it does us) to think of deliberately putting dogs at risk in this experiment, but in fact that is what is done. The risk is borne by a small number of dogs so that the far larger population of dogs can be kept safe.

levels, it makes comparisons among the treatments fair. Assigning experimental units to treatments at random allows us to use the powerful methods of Statistics to draw conclusions from an experiment. Assigning subjects to treatments at random reduces the risk that an imbalance in some uncontrolled source of variation will produce an apparent, but misleading, effect. (We'll talk more about a different problem called "confounding" later in this chapter.)

Experimenters often control factors that are easy or inexpensive to control. They randomize to protect against the effects of other factors, even factors they haven't thought about. How to choose between the two strategies is best summed up in the old adage that says "control what you can, and randomize the rest."

3. **Replicate.** Drawing conclusions about the world is impossible unless we repeat, or **replicate**, our results. Two kinds of replication show up in comparative experiments. First, we should apply each treatment to a number of subjects. Only with such replication can we estimate the variability of responses. If we have not assessed the variation, the experiment is not complete. The outcome of an experiment on a single subject is an anecdote, not data.

DILBERT © 2002 Scott Adams. Distributed by Universal Uclick. Reprinted with permission. All rights reserved

A second kind of **replication** shows up when the entire experiment is repeated on a different population of experimental units. We may believe that what is true of the students in Psych 101 who volunteered for the sleep experiment is true of all humans, but we'll feel more confident if our results for the experiment are *replicated* in another part of the country, with people of different ages, and at different times of the year. Replication of an entire experiment with the controlled sources of variation at different levels is an essential step in science.

4. **Block.** Randomizing helps us deal with unknown sources of variability, but if we recognize some important difference in our participants, then we may design our experiment to deal with it more directly. For example, suppose the participants available for a study of balance include two members of the varsity girls gymnastics team and 10 other students with no gymnastics experience. Randomizing may place both gymnasts in the same treatment group. In the long run, if we could perform the experiment over and over, it would all equalize. But wouldn't it be better to assign one gymnast to each group (at random) and five of the other students to each group (at random)? By doing this, we would improve fairness in the short run. This approach recognizes the variation due to practice and experience and allocates the participants at random *within* each experience level. When we do this, the variable "Experience" is a blocking variable, and the levels of "Experience" are called blocks.

Sometimes, attributes of the experimental units that we are not studying and that we can't control may nevertheless affect the outcomes of an experiment. If we group similar individuals together and then randomize within each of these **blocks**, we can account for much of the variability due to the difference among the blocks so it won't obscure our comparison of the treatment groups. Blocking is an important compromise between randomization and control. However, unlike the first three principles, blocking is not *required* in an experimental design.

A S *Activity*: **The Three Rules of Experimental Design.** Watch an animated discussion of three rules of design.

A S *Activity*: **Perform an Experiment.** How well can you read pie charts and bar charts? Find out as you serve as the subject in your own experiment.

For Example CONTROL, RANDOMIZE, AND REPLICATE

RECAP: We're planning an experiment to see whether the new pet food is safe for dogs to eat. We'll feed some animals the new food and others a food known to be safe, comparing their health after a period of time.

QUESTION: In this experiment, how will you implement the principles of control, randomization, and replication?

ANSWER: *I'd control the portion sizes eaten by the dogs. To reduce possible variability from factors other than the food, I'd standardize other aspects of their environments—housing the dogs in similar pens and ensuring that each got the same amount of water, exercise, play, and sleep time, for example. I might restrict the experiment to a single breed of dog and to adult dogs to further minimize variation.*

To equalize traits, pre-existing conditions, and other unknown influences, I would assign dogs to the two feed treatments randomly.

I would replicate by assigning more than one dog to each treatment to allow for variability among individual dogs. If I had the time and funding, I might replicate the entire experiment using, for example, a different breed of dog.

Diagrams

An experiment is carried out over time with specific actions occurring in a specified order. A diagram of the procedure can help in thinking about experiments.[2]

The diagram emphasizes the random allocation of subjects to treatment groups, the separate treatments applied to these groups, and the ultimate comparison of results. It's best to specify the responses that will be compared. A good way to start comparing results for the treatment groups is with boxplots.

Step-by-Step Example DESIGNING AN EXPERIMENT

An ad for OptiGro plant fertilizer claims that with this product you will grow "juicier, tastier" tomatoes. You'd like to test this claim, and also wonder whether you might be able to get by with half the specified dose. How can you set up an experiment to check out the claim?

Of course, you'll have to get some tomatoes, try growing some plants with the product and some without, and see what happens. But you'll need a clearer plan than that. How should you design your experiment?

Let's work through the design, step by step. We'll design the simplest kind of experiment, in which each experimental unit is equally likely to end up an any

---

[2]Diagrams of this sort were introduced by David Moore in his textbooks and are still widely used.

A completely randomized experiment is the ideal simple design, just as a *simple random sample* is the ideal simple sample—and for many of the same reasons.

of the treatment groups. This is a **completely randomized experiment** in one factor. Since this is a *design* for an experiment, most of the steps are part of the *Think* stage. The statements in the right column are the kinds of things you would need to say in *proposing* an experiment. You'd need to include them in the "methods" section of a report once the experiment is run.

Question: How would you design an experiment to test OptiGro fertilizer?

THINK ➡ **Plan** State what you want to know.

I want to know whether tomato plants grown with OptiGro yield juicier, tastier tomatoes than plants raised in otherwise similar circumstances but without the fertilizer.

Response Specify the response variable.

I'll evaluate the juiciness and taste of the tomatoes by asking a panel of judges to rate them on a scale from 1 to 10 in juiciness and in taste.

Treatments Specify the factor levels and the treatments.

The factor is fertilizer, specifically OptiGro fertilizer. I'll grow tomatoes at three different factor levels: some with no fertilizer, some with half the specified amount of OptiGro, and some with the full dose of OptiGro. These are the three treatments.

Experimental Units Specify the experimental units.

I'll obtain 24 tomato plants of the same variety from a local garden store.

Experimental Design Observe the principles of design:

Control any sources of variability you know of and can control.

I'll locate the garden plots near each other so that the plants get similar amounts of sun and rain and experience similar temperatures. I will weed the plots equally and otherwise treat the plants alike.

Replicate results by placing more than one plant in each treatment group.

I'll use 8 plants in each treatment group.

Randomly assign experimental units to treatments, to equalize the effects of unknown or uncontrollable sources of variation.

To randomly divide the plants into three groups, first I'll label the plants with numbers 00–23. I'll look at pairs of digits across a random number table. The first 8 plants identified (ignoring numbers 24–99 and any repeats) will get no fertilizer, the next 8 a half dose, and the remaining plants the full amount.

Describe how the randomization will be accomplished.

Make a Picture A diagram of your design can help you think about it clearly.

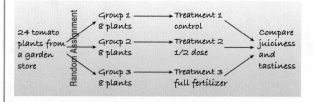

Specify any other experiment details. You must give enough details so that another experimenter could exactly replicate your experiment.

I will grow the plants until the tomatoes are mature, as judged by reaching a standard color.

I'll harvest the tomatoes when ripe and store them for evaluation.

Specify how to measure the response.

I'll set up a numerical scale of juiciness and one of tastiness for the taste testers. Several people will taste slices of tomato and rate them.

(continued)

SHOW ➡	Once you collect the data, you'll need to display them and compare the results for the three treatment groups.	*I will display the results with side-by-side boxplots to compare the three treatment groups.* *I will compare the means of the groups.*
TELL ➡	To answer the initial question, we ask whether the differences we observe in the means of the three groups are meaningful. If so, because this is a randomized experiment, we can attribute significant differences to the treatments. (To do this properly, we'll need methods from what is called "statistical inference," the subject of much of the rest of this book.)	*If the differences in taste and juiciness among the groups are greater than I would expect by knowing the usual variation among tomatoes, I may be able to conclude that these differences can be attributed to treatment with the fertilizer.*

Does the Difference Make a Difference?

A S *Activity*: **Graph the Data.** Do you think there's a significant difference in your perception of pie charts and bar charts? Explore the data from your plot perception experiment.

If the differences among the treatment groups are big enough, we'll attribute the differences to the treatments, but how can we decide whether the differences are big enough?

Would we expect the group means to be identical? Not really. Even if the treatment made no difference at all, there would still be some variation. We assigned the tomato plants to treatments at random. But a different random assignment would have led to different results. Even a repeat of the *same* treatment on a different randomly assigned set of plants would lead to different means. The real question is whether the differences we observed are about as big as we might get just from the randomization alone, or whether they're bigger than that. If we decide that they're bigger, we'll attribute the differences to the treatments. In that case we say the differences are **statistically significant.**

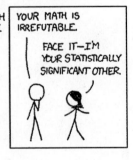

How will we decide if something is different enough to be considered statistically significant? To get some intuition, think about deciding whether a coin is fair. If we flip a fair coin 100 times, we expect, *on average,* to get 50 heads. Suppose we get 54 heads out of 100. That doesn't seem very surprising. It's well within the bounds of ordinary random fluctuations. What if we'd seen 94 heads? That's clearly outside the bounds. We'd be pretty sure that the coin flips were not random. But what about 74 heads? Is that far enough from 50-50 to arouse our suspicions? That's the sort of question we need to ask of our experiment results.

In Statistics terminology, 94 heads would be a statistically significant difference from 50, and 54 heads would not. Whether 74 is *statistically significant* or not would depend on

Figure 12.1

The boxplots in both pairs have medians the same distance apart, but when the spreads are large, that observed difference may be just from random fluctuation.

the chance of getting 74 heads in 100 flips of a fair coin and on our tolerance for believing that rare events can happen to us. Stay tuned: you'll learn to calculate that probability soon. For now, the important point is that an outcome is statistically significant if the probability that it happened just by chance is so low that we're convinced there must be another explanation.

Back at the tomato stand, we ask whether the taste differences we see among the treatment groups could have arisen merely from randomization. A good way to get a feeling for that is to look at how much our results vary among plants that get the *same* treatment. Boxplots of our results by treatment group can give us a general idea.

For example, Figure 12.1 shows two pairs of boxplots whose centers differ by exactly the same amount. In the upper set, that difference appears to be larger than we'd expect just by chance. Why? Because the variation is quite small *within* treatment groups, so the larger difference *between* the groups is unlikely to be just from the randomization. In the bottom pair, that same difference between the centers looks less impressive. There the variation *within* each group swamps the difference *between* the two medians. We'd say the difference is statistically significant in the upper pair and not statistically significant in the lower pair.

In later chapters we'll see statistical tests that quantify this intuition. But we've already seen one way to explore such questions, back in Chapter 4's What If. There we wondered whether the observed difference in speeds between steel and wooden roller coasters was statistically significant. We discovered by simulation that the odds were about 1000:1 against such a large difference arising just by chance, providing strong evidence that steel coasters really are faster. This is a great time to read that What If again. It's on page 93. Go have a look. We'll be here when you get back.

![Just Checking checkmark icon] **Just Checking**

1. At one time, a method called "gastric freezing" was used to treat people with peptic ulcers. An inflatable bladder was inserted down the esophagus and into the stomach, and then a cold liquid was pumped into the bladder. Now you can find the following notice on the Internet site of a major insurance company:

   [Our company] does not cover gastric freezing (intragastric hypothermia) for chronic peptic ulcer disease. . . .

   Gastric freezing for chronic peptic ulcer disease is a non-surgical treatment which was popular about 20 years ago but now is seldom performed. It has been abandoned due to a high complication rate, only temporary improvement experienced by patients, and a lack of effectiveness when tested by double-blind, controlled clinical trials.

   What did that "controlled clinical trial" (experiment) probably look like? (Don't worry about "double-blind"; we'll get to that soon.)

   a) What was the factor in this experiment?
   b) What was the response variable?
   c) What were the treatments?

   d) How did researchers decide which subjects received which treatment?
   e) Were the results statistically significant?

Experiments and Samples

Both experiments and sample surveys use randomization, but they do so in different ways and for different purposes. Sample surveys try to estimate population parameters, so the sample needs to be as representative of the population as possible. By contrast, experiments try to assess the effects of treatments. Experimental units are not always drawn randomly from the population. For example, a medical experiment may deal only with local patients who have the disease being studied. The randomization is in the assignment of their therapy. We want a sample to exhibit the diversity and variability of the population, but for an experiment the more homogeneous the subjects the more easily we'll spot differences in the effects of the treatments.

Unless the experimental units are chosen from the population at random, you should be cautious about generalizing experiment results to larger populations until the experiment has been repeated under different circumstances. Results become more persuasive if they remain the same in completely different settings, such as in a different season, in a different country, or for a different species, to name a few.

Even without choosing experimental units from a population at random, experiments can draw stronger conclusions than surveys. By looking only at the differences across treatment groups, experiments cancel out many sources of variation. For example, the entire pool of subjects may not be representative of the population. (College students may need more sleep, on average, than the general population.) When we assign subjects randomly to treatment groups, all the groups are still unrepresentative, but *in the same way.* When we consider the differences in their responses, this issue cancels out, allowing us to see the *differences* due to treatment effects more clearly.

> **Not a Random Sample!** Experiments are rarely performed on random samples from a population. Don't describe the subjects in an experiment as a random sample unless they really are. More likely, the randomization was in assigning subjects to treatments.

Control Treatments

A S Activity: **Control Groups in Experiments.** Is a control group really necessary?

Suppose you wanted to test a $300 piece of software designed to shorten download times. You could just try it on several files and record the download times, but you probably want to *compare* the speed with what would happen *without* the software installed. Such a baseline measurement is called a **control treatment**, and the experimental units to whom it is applied are called a **control group.**

This is a use of the word "control" in an entirely different context. Previously, we controlled extraneous sources of variation by keeping them constant. Here, we use a control treatment as another *level* of the factor in order to compare the treatment results to a situation in which "nothing happens." That's what we did in the tomato experiment when we used no fertilizer on the 8 tomatoes in Group 1.

Blinding

Humans are notoriously susceptible to errors in judgment.[3] All of us. When we know what treatment was assigned, it's difficult not to let that knowledge influence our assessment of the response, even when we try to be careful.

Suppose you were trying to advise your school on which brand of cola to stock in the school's vending machines. You set up an experiment to see which of the three competing brands students prefer (or whether they can tell the difference at all). But people have brand loyalties. You probably prefer one brand already. So if you knew which brand you were tasting, it might influence your rating. To avoid this problem, it would be better to disguise the brands as much as possible. This strategy is called **blinding** the participants to the treatment.[4]

But it isn't just the subjects who should be blind. Experimenters themselves often subconsciously behave in ways that favor what they believe. Even technicians may treat plants or test animals differently if, for example, they expect them to die. An animal that starts doing a little better than others by showing an increased appetite may get fed a bit more than the experimental protocol specifies.

People are so good at picking up subtle cues about treatments that the best (in fact, the *only*) defense is to keep *anyone* who could affect the outcome or the measurement of the response from knowing which subjects have been assigned to which treatments. So, not only should your cola-tasting subjects be blinded, but also *you,* as the experimenter, shouldn't know which drink is which, either—at least until you're ready to analyze the results.

---

[3]For example, here we are in Chapter 12 and you're still reading the footnotes.

[4]C. S. Peirce, in the same 1885 work in which he introduced randomization, also recommended blinding.

Blinding by Misleading Social science experiments can sometimes blind subjects by misleading them about the purpose of a study. One of the authors participated as an undergraduate volunteer in a (now infamous) psychology experiment using such a blinding method. The subjects were told that the experiment was about three-dimensional spatial perception and were assigned to draw a model of a horse. While they were busy drawing, a loud noise and then groaning were heard coming from the room next door. The *real* purpose of the experiment was to see how people reacted to the apparent disaster. The experimenters wanted to see whether the social pressure of being in groups made people react to the disaster differently. Subjects had been randomly assigned to draw either in groups or alone; that was the treatment. The experimenter had no interest in how well the subjects could draw the horse, but the subjects were blinded to the treatment because they were misled.

There are two main classes of individuals who can affect the outcome of the experiment:

- those who could influence the results (the subjects, treatment administrators, or technicians)
- those who evaluate the results (judges, treating physicians, etc.)

When all the individuals in either one of these classes are blinded, an experiment is said to be **single-blind.** When everyone in *both* classes is blinded, we call the experiment **double-blind.** Even if several individuals in one class are blinded—for example, both the patients and the technicians who administer the treatment—the study would still be just single-blind. If only some of the individuals in a class are blind—for example, if subjects are not told of their treatment, but the administering technician is not blind—there is a substantial risk that subjects can discern their treatment from subtle cues in the technician's behavior or that the technician might inadvertently treat subjects differently. Such experiments cannot be considered truly blind.

In our tomato experiment, we certainly don't want the people judging the taste to know which tomatoes got the fertilizer. That makes the experiment single-blind. We might also not want the people caring for the tomatoes to know which ones were being fertilized, in case they might treat them differently in other ways, too. We can accomplish this double-blinding by having some fake fertilizer for them to put on the other plants. Read on.

For Example BLINDING

RECAP: In our experiment to see if the new pet food is now safe, we're feeding one group of dogs the new food and another group a food we know to be safe. Our response variable is the health of the animals as assessed by a veterinarian.

QUESTION: Should the vet be blinded? Why or why not? How would you do this? (Extra credit: Can this experiment be double-blind? Would that mean that the test animals wouldn't know what they were eating?)

ANSWER: Whenever the response variable involves judgment, it is a good idea to blind the evaluator to the treatments. The veterinarian should not be told which dogs ate which foods.

Extra credit: There is a need for double-blinding. In this case, the workers who care for and feed the animals should not be aware of which dogs are receiving which food. We'll need to make the "safe" food look as much like the "test" food as possible.

Placebos

A S *Activity*: **Blinded Experiments.** This narrated account of blinding isn't a placebo!

Often, simply applying *any* treatment can induce an improvement. Every parent knows the medicinal value of a kiss to make a toddler's scrape or bump stop hurting. Some of the improvement seen with a treatment—even an effective treatment—can be due simply to the act of treating. To separate these two effects, we can use a control treatment that mimics the treatment itself.

> **Active Placebos**   The placebo effect is stronger when placebo treatments are administered with authority or by a figure who appears to be an authority. "Doctors" in white coats generate a stronger effect than salespeople in polyester suits. But the placebo effect is not reduced much even when subjects know that the effect exists. People often suspect that they've gotten the placebo if nothing at all happens. So, recently, drug manufacturers have gone so far in making placebos realistic that they cause the same side effects as the drug being tested! Such "active placebos" usually induce a stronger placebo effect. When those side effects include loss of appetite or hair, the practice may raise ethical questions.

A "fake" treatment that looks just like the treatments being tested is called a **placebo.** Placebos are the best way to blind subjects from knowing whether they are receiving the treatment or not. One common version of a placebo in drug testing is a "sugar pill." Especially when psychological attitude can affect the results, control group subjects treated with a placebo may show an improvement.

The fact is that subjects treated with a placebo sometimes improve. It's not unusual for 20% or more of subjects given a placebo to report reduction in pain, improved movement, or greater alertness, or even to demonstrate improved health or performance. This **placebo effect** highlights both the importance of effective blinding and the importance of comparing treatments with a control. Placebo controls are so effective that you should use them as an essential tool for blinding whenever possible.

The best experiments are usually

- randomized.
- comparative.
- double-blind.
- placebo-controlled.

> **Does Ginkgo Biloba Improve Memory?**   Researchers investigated the purported memory-enhancing effect of ginkgo biloba tree extract (P. R. Solomon, F. Adams, A. Silver, J. Zimmer, R. De Veaux, "Ginkgo for Memory Enhancement. A Randomized Controlled Trial." *JAMA* 288 [2002]: 835–840). In a randomized, comparative, double-blind, placebo-controlled study, they administered treatments to 230 elderly community members. One group received Ginkoba™ according to the manufacturer's instructions. The other received a similar-looking placebo. Thirteen different tests of memory were administered before and after treatment. The placebo group showed greater improvement on 7 of the tests, the treatment group on the other 6. None showed any significant differences. Here are boxplots of one measure.

By permission of John L. Hart FLP and Creators Syndicate, Inc.

Blocking

Suppose we had wanted to use 18 tomato plants of the same variety for our experiment, but the garden store had only 12 plants left. Then we drove down to the nursery and bought 6 more plants of that variety. We worry that the tomato plants from the two stores are different somehow, or have received different care.

How can we design the experiment so that the differences between the stores don't mess up our attempts to see differences among fertilizer levels? We can't measure the effect of a store the same way as we can the fertilizer because we can't assign it at random. You can't tell a tomato what store to come from.

Because stores may vary in the care they give plants or in the sources of their seeds, the plants from either store are likely to be more like each other than they are like the plants from the other store. When groups of experimental units are similar, it's often a good idea to gather them together into **blocks.** By blocking, we isolate the variability attributable to the differences between the blocks, so that we can see the differences caused by the treatments more clearly. Here, we would define the plants from each store to be a block. The randomization is introduced when we randomly assign treatments within each block.

In a completely randomized design, each of the 18 plants would have an equal chance to land in each of the three treatment groups. But we realize that the store may have an effect. To isolate the store effect, we block on store by assigning the plants from each store to treatments at random. So we now have six treatment groups, three for each block. Within each block, we'll randomly assign the same number of plants to each of the three treatments. The experiment is still fair because each treatment is still applied (at random) to the same number of plants and to the same proportion from each store: 4 from store A and 2 from store B. Because the randomization occurs only within the blocks (plants from one store cannot be assigned to treatment groups for the other), we call this a **randomized block design.**

In effect, we conduct two parallel experiments, one for tomatoes from each store, and then combine the results. The picture tells the story:

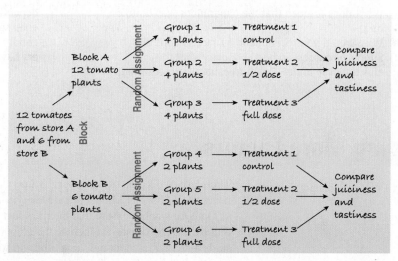

In a retrospective or prospective study, subjects are sometimes paired because they are similar in ways *not* under study. **Matching** subjects in this way can reduce variation in much the same way as blocking. For example, a retrospective study of music education and grades might match each student who studies an instrument with someone of the same sex who is similar in family income but didn't study an instrument. When we compare grades of music students with those of non-music students, the matching would reduce the variation due to income and sex differences.

Blocking in experiments is the same idea as stratifying in sampling. Both methods group together subjects that are similar and randomize within those groups as a way to remove unwanted variation. (But be careful to keep the terms straight. Don't say that we "stratify" an experiment or "block" a sample.) We use blocks to reduce variability so we can see the effects of the factors; we're not usually interested in studying the effects of the blocks themselves.

For Example BLOCKING

RECAP: In 2007, pet food contamination put cats at risk, as well as dogs. Our experiment should probably test the safety of the new food on both animals.

QUESTION: Why shouldn't we randomly assign a mix of cats and dogs to the two treatment groups? What would you recommend instead?

ANSWER: Dogs and cats might respond differently to the foods, and that variability could obscure my results. Blocking by species can remove that superfluous variation. I'd randomize cats to the two treatments (test food and safe food) separately from the dogs. I'd measure their responses separately and look at the results afterward.

 Just Checking

2. Recall the experiment about gastric freezing, an old method for treating peptic ulcers that you read about in the first Just Checking. Doctors would insert an inflatable bladder down the patient's esophagus and into the stomach and then pump in a cold liquid. A major insurance company now states that it doesn't cover this treatment because "double-blind, controlled clinical trials" failed to demonstrate that gastric freezing was effective.

   a) What does it mean that the experiment was double-blind?
   b) Why would you recommend a placebo control?
   c) Suppose that researchers suspected that the effectiveness of the gastric freezing treatment might depend on whether a patient had recently developed the peptic ulcer or had been suffering from the condition for a long time. How might the researchers have designed the experiment?

Adding More Factors

There are two kinds of gardeners. Some water frequently, making sure that the plants are never dry. Others let Mother Nature take her course and leave the watering to her. The makers of OptiGro want to ensure that their product will work under a wide variety of watering conditions. Maybe we should include the amount of watering as part of our experiment. Can we study a second factor at the same time and still learn as much about fertilizer?

We now have two factors (fertilizer at three levels and irrigation at two levels). We combine them in all possible ways to yield six treatments:

	No Fertilizer	Half Fertilizer	Full Fertilizer
No Added Water	1	2	3
Daily Watering	4	5	6

If we allocate the original 12 plants, the experiment now assigns 2 plants to each of these six treatments at random. This experiment is a **completely randomized two-factor experiment** because any plant could end up assigned at random to any of the six treatments (and we have two factors).

It's often important to include several factors in the same experiment in order to see what happens when the factor levels are applied in different *combinations*. For example, we might find out that regular watering allows a full dose of fertilizer to work best, but plants left to the whims of the weather do better on just half the recommended amount. Experiments with more than one factor are both more efficient and provide more information than one-at-a-time experiments. There are many ways to design efficient multifactor experiments. You can take a whole course on the design and analysis of such experiments.

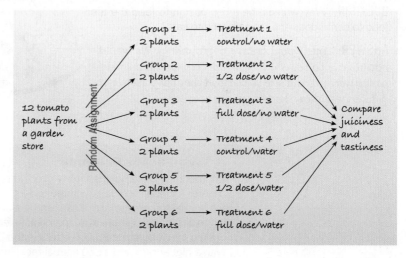

Confounding

Professor Stephen Ceci of Cornell University performed an experiment to investigate the effect of a teacher's classroom style on student evaluations. He taught a class in developmental psychology during two successive terms to a total of 472 students in two very similar classes. He kept everything about his teaching identical (same text, same syllabus, same office hours, etc.) and modified only his style in class. During the fall term, he maintained a subdued demeanor. During the spring term, he lectured with more enthusiasm, varying his vocal pitch and using more hand gestures. He administered a standard student evaluation form at the end of each term.

The students in the fall term class rated him only an average teacher. Those in the spring term class rated him an excellent teacher, praising his knowledge and accessibility, and even the quality of the textbook. On the question "How much did you learn in the course?" the average response changed from 2.93 to 4.05 on a 5-point scale.[5]

How much of the difference he observed was due to his difference in manner, and how much might have been due to the season of the year? Fall term in Ithaca, NY (home of Cornell University), starts out colorful and pleasantly warm but ends cold and bleak. Spring term starts out bitter and snowy and ends with blooming flowers and singing birds. Might students' overall happiness have been affected by the season and reflected in their evaluations?

Unfortunately, there's no way to tell. Nothing in the data enables us to tease apart these two effects, because all the students who experienced the subdued manner did so during the fall term and all who experienced the expansive manner did so during the spring. When the levels of one factor are associated with the levels of another factor, we say that these two factors are **confounded.**

In some experiments, such as this one, it's just not possible to avoid some confounding. Professor Ceci could have randomly assigned students to one of two classes during the same term, but then we might question whether mornings or afternoons were better, or whether he really delivered the same class the second time (after practicing on the first class). Or he could have had another professor deliver the second class, but that would have raised more serious issues about differences in the two professors and concern over more serious confounding.

---

[5]But the two classes performed almost identically well on the final exam.

For Example CONFOUNDING

RECAP: After many dogs and cats suffered health problems caused by contaminated foods, we're trying to find out whether a newly formulated pet food is safe. Our experiment will feed some animals the new food and others a food known to be safe, and a veterinarian will check the response.

QUESTION: Why would it be a bad design to feed the test food to some dogs and the safe food to cats?

ANSWER: This would create confounding. We would not be able to tell whether any differences in animals' health were attributable to the food they had eaten or to differences in how the two species responded.

A Two-Factor Example

Confounding can also arise from a badly designed multifactor experiment. Here's a classic. A credit card bank wanted to test the sensitivity of the market to two factors: the annual fee charged for a card and the annual percentage rate charged. Not wanting to scrimp on sample size, the bank selected 100,000 people at random from a mailing list. It sent out 50,000 offers with a low rate and no fee and 50,000 offers with a higher rate and a $50 annual fee. Guess what happened? That's right—people preferred the low-rate, no-fee card. No surprise. In fact, they signed up for that card at over twice the rate as the other offer. And because of the large sample size, the bank was able to estimate the difference precisely. But the question the bank really wanted to answer was "how much of the change was due to the rate, and how much was due to the fee?" unfortunately, there's simply no way to separate out the two effects. If the bank had sent out all four possible different treatments—low rate with no fee, low rate with $50 fee, high rate with no fee, and high rate with $50 fee—each to 25,000 people, it could have learned about both factors and could have also seen what happens when the two factors occur in combination.

Lurking or Confounding?

Confounding may remind you of the problem of lurking variables we discussed back in Chapters 6 and 8. Confounding variables and lurking variables are alike in that they interfere with our ability to interpret our analyses simply. Each can mislead us, but there are important differences in both how and where the confusion may arise.

A lurking variable creates an association between two other variables that tempts us to think that one may cause the other. This can happen in a regression analysis or an observational study when a lurking variable influences both the explanatory and response variables. Recall that countries with more TV sets per capita tend to have longer life expectancies. We shouldn't conclude it's the TVs "causing" longer life. We suspect instead that a generally higher standard of living may mean that people can afford more TVs and get better health care, too. Our data revealed an association between TVs and life expectancy, but economic conditions were a likely lurking variable. A lurking variable, then, is usually thought of as a variable associated with both y and x that makes it appear that x may be causing y.

Confounding can arise in experiments when some other variable associated with a factor has an effect on the response variable. However, in a designed experiment, the experimenter *assigns* treatments (at random) to subjects rather than just observing them. A confounding variable can't be thought of as causing that assignment. Professor Ceci's choice of teaching styles was not caused by the weather, but because he used one style in the fall and the other in spring, he was unable to tell how much of his students' reactions were attributable to his teaching and how much to the weather. A confounding variable, then, is associated in a noncausal way with a factor and affects the response. Because of the confounding, we find that we can't tell whether any effect we see was caused by our factor or by the confounding variable—or even by both working together.

Both confounding and lurking variables are outside influences that make it harder to understand the relationship we are modeling. However, the nature of the causation is different in the two situations. In regression and observational studies, we can only observe associations between variables. Although we can't demonstrate a causal relationship, we often imagine whether *x* could cause *y*. We can be misled by a lurking variable that influences both. In a designed experiment, we often hope to show that the factor causes a response. Here we can be misled by a confounding variable that's associated with the factor and causes or contributes to the differences we observe in the response.

It's worth noting that the role of blinding in an experiment is to combat a possible source of confounding. There's a risk that knowledge about the treatments could lead the subjects or those interacting with them to behave differently or could influence judgments made by the people evaluating the responses.[6] That means we won't know whether the treatments really do produce different results or if we're being fooled by these confounding influences.

---

[6]If you're thinking that sounds like "bias", good for you. Confounding in an experiment is analogous to bias in sampling. Be sure to use these words in their proper context.

WHAT IF ●●● some of the tomatoes do taste better?

Suppose our backyard tomato experiment has run its course. We've sliced up a nice tomato from each of the 24 plants and had a neighbor who's something of a tomato aficionado taste them. Without knowing which tomatoes came from fertilized plants, she rated each one on a scale from 1 to 10, with 1 representing less than satisfactory taste and 10 meaning absolutely delicious. The data table and boxplots below display her evaluations.

It certainly appears the fertilizer worked. Our tester rated both the tomatoes that got the half dose and those that got the full dose as much tastier than those that went unfertilized. Furthermore, it looks like the full-dose tomatoes were somewhat better than those from plants that got only half the fertilizer. In fact, the mean taste ratings for the three groups were 4.5, 7, and 7.75 out of 10, respectively. But is that ¾-point difference between the two fertilized groups statistically significant? Did the extra fertilizer really help, or could we just be seeing natural variability in tomato plants? To find out, we could randomly split those 16 ratings into two other groups of 8 and see how big a difference might occur by chance. Yes, it's time for another simulation!

In our first random split, Group A (6, 9, 4, 6, 9, 8, 7, 5) had a mean rating of 6.75, and Group B (7, 6, 9, 8, 10, 9, 7, 8) a mean of 8. That's an even bigger difference than what the extra fertilizer appeared to produce. Hmm.

Taste ratings of tomatoes from plants that had ...		
No fertilizer	**Half dose**	**Full dose**
6	6	6
4	9	5
7	7	10
5	6	9
4	8	9
5	4	7
3	7	8
2	9	8

Our simulation repeated this process a total of 500 times. The dotplot below shows the differences in means for the random groupings.

Difference in Mean

That 7.75 to 7 mean taste-rating win for the fully fertilized tomatoes doesn't look very impressive now. In our simulation a difference of 0.75 points happened by chance a whopping 91 times in 500 trials—nearly 20% of the time. Such a difference is far from unusual; in other words, it's not statistically significant. Our neighbor's ratings don't provide evidence that using a full dose of fertilizer instead of only half a dose will produce tomatoes that taste better.

If right now you're thinking that maybe even the apparently large difference between the tomatoes that got fertilizer and those that had none isn't significant either, good for you! Skepticism is the mark of a good statistician. To ease your mind, we checked that out for you. Instead of a simulation, though, we used one of the statistical tests we told you you'd learn about later in this course. That test revealed that a difference as large as the one we see between the mean taste ratings for tomatoes from fertilized vs non-fertilized plants could arise by chance only 1 time in 2000! Now, *that's* statistically significant.

Based on these ratings, then, the OptiGro company would be on solid ground in asserting that its fertilizer produces tastier tomatoes. But the company might also be tempted to warn customers that using only half the recommended amount wouldn't work as well. When you learn to think like a statistician, such claims made without evidence leave a bad taste in your mouth.

WHAT CAN GO WRONG?

- **Don't give up just because you can't run an experiment.** Sometimes we can't run an experiment because we can't identify or control the factors. Sometimes it would simply be unethical to run the experiment. (Consider randomly assigning students to take—and be graded in—a Statistics course deliberately taught to be boring and difficult or one that had an unlimited budget to use multimedia, real-world examples, and field trips to make the subject more interesting.) If we can't perform an experiment, often an observational study is a good choice.

- **Beware of confounding.** Use randomization whenever possible to ensure that the factors not in your experiment are not confounded with your treatment levels. Be alert to confounding that cannot be avoided, and report it along with your results.

■ **Bad things can happen even to good experiments.** Protect yourself by recording additional information. An experiment in which the air conditioning failed for 2 weeks, affecting the results, was saved by recording the temperature (although that was not originally one of the factors) and estimating the effect the higher temperature had on the response.[7]

It's generally good practice to collect as much information as possible about your experimental units and the circumstances of the experiment. For example, in the tomato experiment, it would be wise to record details of the weather (temperature, rainfall, sunlight) that might affect the plants and any facts available about their growing situation. (Is one side of the field in shade sooner than the other as the day proceeds? Is one area lower and a bit wetter?) Sometimes we can use this extra information during the analysis to reduce biases.

■ **Don't spend your entire budget on the first run.** Just as it's a good idea to pretest a survey, it's always wise to try a small pilot experiment before running the full-scale experiment. You may learn, for example, how to choose factor levels more effectively, about effects you forgot to control, and about unanticipated confoundings.

[7]R. D. DeVeaux and M. Szelewski, "Optimizing Automatic Splitless Injection Parameters for Gas Chromatographic Environmental Analysis." *Journal of Chromatographic Science* 27, no. 9 (1989): 513–518.

What Have We Learned?

We've learned to recognize sample surveys, observational studies, and randomized comparative experiments. We know that these methods collect data in different ways and lead us to different conclusions.

We've learned to identify retrospective and prospective observational studies and understand the advantages and disadvantages of each.

We've learned that only well-designed experiments can allow us to reach cause-and-effect conclusions. We manipulate levels of treatments to see if the factor we have identified produces changes in our response variable.

We've learned the principles of experimental design:

■ We want to be sure that variation in the response variable can be attributed to our factor, so we identify and control as many other sources of variability as possible.

■ Because there are many possible sources of variability that we cannot identify, we try to equalize those by randomly assigning experimental units to treatments.

■ We replicate the experiment on as many subjects as possible.

■ We consider blocking to reduce variability from sources we recognize but cannot control.

■ We've learned to recognize the factors, their levels, the treatments, and the response variable in a description of a designed experiment.

We've learned the value of having a control group and of using blinding and placebo controls.

We've learned to recognize the problems posed by confounding variables in experiments and lurking variables in observational studies.

Finally, we are learned the differences between experiments and surveys.

■ Surveys try to estimate facts (parameter) about a population, so they require a representative random sample from that population.

■ Experiments try to estimate the differences in the effects of treatments. They randomize a group of experimental units to treatments, but there is no need for the experimental units to be a representative sample from the population.

Terms

Observational study   A study based on data in which no manipulation of factors has been employed. (p. 305)

Retrospective study   An observational study in which subjects are selected and then their previous conditions or behaviors are determined. Retrospective studies need not be based on random samples and they usually focus on estimating differences between groups or associations between variables. (p. 305)

Prospective study	An observational study in which subjects are followed to observe future outcomes. Because no treatments are deliberately applied, a prospective study is not an experiment. Nevertheless, prospective studies typically focus on estimating differences among groups that might appear as the groups are followed during the course of the study. (p. 306)
Experiment	An experiment *manipulates* factor levels to create treatments, *randomly assigns* subjects to these treatment levels, and then *compares* the responses of the subject groups across treatment levels. (p. 306)
Random assignment	To be valid, an experiment must assign experimental units to treatment groups at random. This is called random assignment. (p. 307)
Factor	A variable whose levels are manipulated by the experimenter. Experiments attempt to discover the effects that differences in factor levels may have on the responses of the experimental units. (p. 307)
Response	A variable whose values are compared across different treatments. In a randomized experiment, large response differences can be attributed to the effect of differences in treatment level. (p. 307)
Experimental units	Individuals on whom an experiment is performed. Usually called **subjects** or **participants** when they are human. (p. 307)
Level	The specific values that the experimenter chooses for a factor are called the levels of the factor. (p. 307)
Treatment	The process, intervention, or other controlled circumstance applied to randomly assigned experimental units. Treatments are the different levels of a single factor or are made up of combinations of levels of two or more factors. (p. 307)
Principles of experimental design	■ **Control** aspects of the experiment that we know may have an effect on the response, but that are not the factors being studied.
	■ **Randomize** subjects to treatments to even out effects that we cannot control.
	■ **Replicate** over as many subjects as possible. Results for a single subject are just anecdotes. If, as often happens, the subjects of the experiment are not a representative sample from the population of interest, replicate the entire study with a different group of subjects, preferably from a different part of the population.
	■ **Block** to reduce the effects of identifiable attributes of the subjects that cannot be controlled. (p. 308)
Completely randomized design	In a completely randomized design, all experimental units have an equal chance of receiving any treatment. (p. 311)
Statistically significant	When an observed difference is too large for us to believe that it is likely to have occurred naturally, we consider the difference to be statistically significant. Subsequent chapters will show specific calculations and give rules, but the principle remains the same. (p. 312)
Control group	The experimental units assigned to a baseline treatment level, typically either the default treatment, which is well understood, or a null, placebo treatment. Their responses provide a basis for comparison. (p. 314)
Blinding	Any individual associated with an experiment who is not aware of how subjects have been allocated to treatment groups is said to be blinded. (p. 314)
Single-blind Double-blind	There are two main classes of individuals who can affect the outcome of an experiment: ■ those who could *influence the results* (the subjects, treatment administrators, or technicians).
	■ those who *evaluate the results* (judges, treating physicians, etc.).
	When every individual in *either* of these classes is blinded, an experiment is said to be single-blind. When everyone in *both* classes is blinded, we call the experiment double-blind. (p. 315)
Placebo	A treatment known to have no effect, administered to one group so that all groups experience the same conditions. Many subjects respond to such a treatment (a response known as a placebo effect). Only by comparing with a placebo can we be sure that the observed effect of a treatment is not due simply to the placebo effect. (p. 316)

Placebo effect	The tendency of many human subjects (often 20% or more of experiment subjects) to show a response even when administered a placebo. (p. 316)
Blocking	When groups of experimental units are similar, it is often a good idea to gather them together into blocks. By blocking, we isolate the variability attributable to the differences between the blocks so that we can see the differences caused by the treatments more clearly. (p. 317)
Randomized block design	In a randomized block design, the subjects are randomly assigned to treatments only within blocks. (p. 317)
Matching	In a retrospective or prospective study, subjects who are similar in ways not under study may be matched and then compared with each other on the variables of interest. Matching, like blocking, reduces unwanted variation. (p. 317)
Confounding	When the levels of one factor are associated with the levels of another factor in such a way that their effects cannot be separated, we say that these two factors are confounded. (p. 319)

On the Computer EXPERIMENTS

Most experiments are analyzed with a statistics package. You should almost always display the results of a comparative experiment with side-by-side boxplots. You may also want to display the means and standard deviations of the treatment groups in a table.

The analyses offered by statistics packages for comparative randomized experiments fall under the general heading of Analysis of Variance, usually abbreviated ANOVA. These analyses are beyond the scope of this chapter.

Exercises

1. **Standardized test scores** For his Statistics class experiment, researcher J. Gilbert decided to study how parents' income affects children's performance on standardized tests like the SAT. He proposed to collect information from a random sample of test takers and examine the relationship between parental income and SAT score.
   a) Is this an experiment? If not, what kind of study is it?
   b) If there is relationship between parental income and SAT score, why can't we conclude that differences in score are caused by differences in parental income?

2. **Heart attacks and height** Researchers who examined health records of thousands of males found that men who died of myocardial infarction (heart attack) tended to be shorter than men who did not.
   a) Is this an experiment? If not, what kind of study is it?
   b) Is it correct to conclude that shorter men are at higher risk for heart attack? Explain.

3. **MS and vitamin D** Multiple sclerosis (MS) is an autoimmune disease that strikes more often the farther people live from the equator. Could vitamin D—which most people get from the sun's ultraviolet rays—be a factor? Researchers compared vitamin D levels in blood samples from 150 U.S. military personnel who have developed MS with blood samples of nearly 300 who have not. The samples were taken, on average, five years before the disease was diagnosed. Those with the highest blood vitamin D levels had a 62% lower risk of MS than those with the lowest levels. (The link was only in whites, not in blacks or Hispanics.)
   a) What kind of study was this?
   b) Is that an appropriate choice for investigating this problem? Explain.
   c) Who were the subjects?
   d) What were the variables?

4. **Super Bowl commercials** When spending large amounts to purchase advertising time, companies want to know

what audience they'll reach. In January 2011, a poll by *The Hollywood Reporter* asked randomly selected American adults whether they planned to watch the upcoming Super Bowl. Men and women were asked separately whether they were looking forward more to the football game or to watching the commercials. Among the men, who planned on watching, 70% were watching for the game. Among women, 60% were looking forward primarily to the game.

a) Was this a stratified sample or a blocked experiment? Explain.

b) Was the design of the study appropriate for the advertisers' questions?

5. **Maggot therapy?** People generally don't think "health" when they hear the word "maggot," but one experiment tested the ability of maggots to remove dead tissue from open wounds that would not heal on their own. Sterile maggots were placed in a small pouch which, in turn, was placed on the wound. 100 men with wounds on their lower limbs were randomly assigned to receive either a traditional surgical treatment or maggot therapy. After eight days, the percentage of dead tissue in the wounds that underwent maggot treatment was 54.5%, compared to 66.5% with the surgical treatment. (The difference decreased with time, and the advantage disappeared by about day 15.) Neither patients nor the doctors evaluating the wounds knew which therapy had been applied. (Patients were blindfolded as bandages were changed.) Surprisingly, the number of patients that reported a crawling sensation in their wound was about the same in both groups! (http://www.myhealthnewsdaily.com/2030-maggots-clean-wounds-faster-surgeons.html)

a) What kind of study was this?

b) Is that an appropriate choice for this investigation?

c) Who were the subjects?

d) Identify the treatment and response variables.

6. **Honesty** Coffee stations in offices often just ask users to leave money in a tray to pay for their coffee, but many people cheat. Researchers at Newcastle University replaced the picture of flowers on the wall behind the coffee station with a picture of staring eyes. They found that the average contribution increased significantly above the well-established standard when people felt they were being watched, even though the eyes were patently not real. (NY *Times* 12/10/06)

a) Was this a survey, an observational study, or an experiment? How can we tell?

b) Identify the variables.

c) What does "increased significantly" mean in a statistical sense?

7–16. What's the design? *Read each brief report of statistical research, and identify*

a) whether it was an observational study or an experiment.

If it was an observational study, identify (if possible)

b) whether it was retrospective or prospective.

c) the subjects studied and how they were selected.

d) the parameter of interest.

e) the nature and scope of the conclusion the study can reach.

If it was an experiment, identify (if possible)

b) the subjects studied.

c) the factor(s) in the experiment and the number of levels for each.

d) the number of treatments.

e) the response variable measured.

f) the design (completely randomized, blocked, or matched).

g) whether it was blind (or double-blind).

h) the nature and scope of the conclusion the experiment can reach.

7. **Superglue** 130 patients with eligible lacerations randomly assigned to have the wound closed either with Octylcyanoacrylate Tissue Adhesive (essentially superglue) or with traditional sutures. When evaluated at the end of the study, the two treatments worked equally well with regard to scarring, and the adhesive was less painful and worked faster.

8. **Truancy** A group of researchers analyzed three observational studies that followed children's attendance and mental health, among other things. They found that students who missed more school tended to have more incidences of depression. One study followed 20,745 secondary students in a random sample of all secondary schools in the United States. Another tracked 2,311 first graders at 18 Baltimore schools who were participating in an intervention program. The third study followed 671 students from first or fifth grade to their senior year in high risk areas of Eugene, OR, who had been randomly assigned to an intervention or to no intervention.

9. **Hypertension** In a test of roughly 200 older men and women, those with moderately high blood pressure (averaging 164/89 mm Hg) did worse on tests of memory and reaction time than those with normal blood pressure. (*Hypertension* 36 [2000]: 1079)

10. **Tossing and turning** Is diet or exercise effective in combating insomnia? Some believe that cutting out desserts can help alleviate the problem, while others recommend exercise. Forty volunteers suffering from insomnia agreed to participate in a month-long test. Half were randomly assigned to a special no-desserts diet; the others continued desserts as usual. Half of the people in each of these groups were randomly assigned to an exercise program, while the others did not exercise. Those who ate no desserts and engaged in exercise showed the most improvement.

11. **Alcohol and estrogen** After menopause, some women take supplemental estrogen. There is some concern that if these women also drink alcohol, their estrogen levels will rise too high. Twelve volunteers who were receiving supplemental estrogen were randomly divided into two groups, as were 12 other volunteers not on estrogen. In each case, one group drank an alcoholic beverage, the other a nonalcoholic beverage. An hour later, everyone's

estrogen level was checked. Only those on supplemental estrogen who drank alcohol showed a marked increase.

12. Dioxin Researchers have linked an increase in the incidence of breast cancer in Italy to dioxin released by an industrial accident in 1976. The study identified 981 women who lived near the site of the accident and were under age 40 at the time. Fifteen of the women had developed breast cancer at an unusually young average age of 45. Medical records showed that they had heightened concentrations of dioxin in their blood and that each tenfold increase in dioxin level was associated with a doubling of the risk of breast cancer. (Science News, Aug. 3, 2002)

13. Boys and girls In 2002 the journal *Science* reported that a study of women in Finland indicated that having sons shortened the lifespans of mothers by about 34 weeks per son, but that daughters helped to lengthen the mothers' lives. The data came from church records from the period 1640 to 1870.

14. Herbal remedy Scientists at a major pharmaceutical firm investigated the effectiveness of an herbal compound to treat the common cold. They exposed each subject to a cold virus, then randomly assigned him or her either the herbal compound or a sugar solution known to have no effect on colds. Several days later they assessed the patient's condition, using a cold severity scale ranging from 0 to 5. They found no evidence of benefits associated with the compound.

15. Depression The May 4, 2000, issue of Science News reported that, contrary to popular belief, depressed individuals cry no more often in response to sad situations than non depressed people. Researchers studied 23 men and 48 women with major depression and 9 men and 24 women with no depression. They showed the subjects a sad film about a boy whose father has died, noting whether or not the subjects cried. Women cried more often than men, but there were no significant differences between the depressed and non depressed groups.

16. Vitamin C doping Some people who race greyhounds give the dogs large doses of vitamin C in the belief that the dogs will run faster. Investigators at the University of Florida tried three different diets in random order on each of five racing greyhounds. They were surprised to find that when the dogs ate high amounts of vitamin C they ran more slowly. (*Science News*, July 20, 2002)

17. Torn ACL Having at least one 15-minute warm-up session per week resulted in a drastic reduction in tears in the *anterior cruciate ligament* (ACL). In a study involving about 4500 adolescent girls' soccer players in Sweden, one group was randomly assigned to warm up with a neuromuscular exercise session. This group had 64% fewer ACL tears than the control group.

a) Is this an experiment or an observational study? Explain why.
b) Identify the treatments in this study. What is the response variable?

c) Give one *statistical* advantage of using only Swedish girls who played soccer in this study.
d) Give one *statistical* disadvantage of using only Swedish girls who played soccer in this study.

18. Losing sleep An article entitled "TV Before Bed May Rob Teens of Sleep" reported on a study published online in *Pediatrics* in January of 2013. The study found that students who watch TV before bedtime tend to go to sleep later than those who engaged in nonscreen sedentary activities before bed. Researchers contacted a nationally representative cross-sectional sample of teens in New Zealand, interviewing participants in person and following up with phone interviews, to look for a relationship between before-bed activities and the length of time before kids go to sleep.

a) Is this an experiment? Explain why or why not.
b) Researchers cautioned that "causality could not be inferred from their cross-sectional study." Explain why this is the case.
c) Comment on the title of the article in light of your answer to part b.

19. Migraines Some people claim they can get relief from migraine headache pain by drinking a large glass of ice water. Researchers plan to enlist several people who suffer from migraines in a test. Participants will be randomly assigned to a standard pain reliever or a placebo. When a participant experiences a migraine headache, he or she will take the pill. Half of each group will also drink ice water. Participants will then report the level of pain relief they experience.

a) Identify the factors and levels in this experiment.
b) Identify the treatments and the response variable.
c) Is there any blinding described in the study?
d) No blocking is described in the study. What might be an appropriate variable on which to block? Clearly explain why you think this variable would be appropriate.

20. Low-cal dog food A dog food company wants to compare a new lower calorie food with their standard dog food to see if it's effective in helping inactive dogs maintain a healthy weight. They have found several dog owners willing to participate in the trial. The dogs have been classified as small, medium, or large breeds, and the company will supply some owners of each size of dog with one of the two foods. The owners have agreed not to feed their dogs anything else for a period of 6 months, after which the dogs' weights will be checked.

a) Identify the treatments, the experimental units, and the response variable.
b) Describe a method of assigning treatments if this is to be a randomized block design with size of the breed as the blocking variable.
c) Is blinding important in this experiment? Double-blinding? How could blinding be conducted?

21. Omega-3 An experiment that showed that high doses of omega-3 fats might be of benefit to people with bipolar disorder involved a control group of subjects who received a placebo. Why didn't the experimenters just give everyone the omega-3 fats to see if they improved?

22. **Insomnia** Exercise 10 describes an experiment showing that exercise helped people sleep better. The experiment involved other groups of subjects who didn't exercise. Why didn't the experimenters just have everyone exercise and see if their ability to sleep improved?

23. **Omega-3 revisited** Exercises 21 describes an experiment investigating a dietary approach to treating bipolar disorder. Researchers randomly assigned 30 subjects to two treatment groups, one group taking a high dose of omega-3 fats and the other a placebo.
    a) Why was it important to randomize in assigning the subjects to the two groups?
    b) What would be the advantages and disadvantages of using 100 subjects instead of 30?

24. **Insomnia again** Exercises 10 and 22 describe an experiment investigating the effectiveness of exercise in combating insomnia. Researchers randomly assigned half of the 40 volunteers to an exercise program.
    a) Why was it important to randomize in deciding who would exercise?
    b) What would be the advantages and disadvantages of using 100 subjects instead of 40?

25. **Omega-3, finis** Exercises 21 and 23 describe an experiment investigating the effectiveness of omega-3 fats in treating bipolar disorder. Suppose some of the 30 subjects were very active people who walked a lot or got vigorous exercise several times a week, while others tended to be more sedentary, working office jobs and watching a lot of TV. Why might researchers choose to block the subjects by activity level before randomly assigning them to the omega-3 and placebo groups?

26. **Insomnia, at last** Exercises 10, 22, and 24 describe an experiment investigating the effectiveness of exercise in combating insomnia. Suppose some of the 40 subjects had maintained a healthy weight, but others were quite overweight. Why might researchers choose to block the subjects by weight level before randomly assigning some of each group to the exercise program?

27. **Tomatoes** Describe a strategy to randomly split the 24 tomato plants into the three groups for the chapter's completely randomized single factor test of OptiGro fertilizer.

28. **Tomatoes II** The chapter also described a completely randomized two-factor experiment testing OptiGro fertilizer in conjunction with two different routines for watering the plants. Describe a strategy to randomly assign the 24 tomato plants to the six treatments.

29. **Shoes** A running-shoe manufacturer wants to test the effect of its new sprinting shoe on 100-meter dash times. The company sponsors 5 athletes who will try out for the 100-meter dash in the 2016 Summer Olympic games.

To test the shoe, it has all 5 runners run the 100-meter dash with a competitor's shoe and then again with their new shoe. The company uses the difference in times as the response variable.
    a) Suggest some improvements to the design.
    b) Why might the shoe manufacturer not be able to generalize the results they find to all runners?

30. **Swimsuits** A swimsuit manufacturer wants to test the speed of its newly designed suit. The company designs an experiment by having 6 randomly selected Olympic swimmers swim as fast as they can with their old swimsuit first and then swim the same event again with the new, expensive swimsuit. The company will use the difference in times as the response variable. Criticize the experiment and point out some of the problems with generalizing the results.

31. **Hamstrings** Athletes who had suffered hamstring injuries were randomly assigned to one of two exercise programs. Those who engaged in static stretching returned to sports activity in a mean of 15.2 days faster than those assigned to a program of agility and trunk stabilization exercises. (Journal of Orthopaedic & Sports Physical Therapy 34 [March 2004]: 3)
    a) Explain why it was important to assign the athletes to the two different treatments randomly.
    b) There was no control group consisting of athletes who did not participate in a special exercise program. Explain the advantage of including such a group.
    c) How might blinding have been used?
    d) One group returned to sports activity in a mean of 37.4 days ($SD = 27.6$ days) and the other in a mean of 22.2 days ($SD = 8.3$ days). Do you think this difference is statistically significant? Explain.

32. **Diet and blood pressure** An experiment that showed that subjects fed the DASH diet were able to lower their blood pressure by an average of 6.7 points compared to a group fed a "control diet." All meals were prepared by dieticians.
    a) Why were the subjects randomly assigned to the diets instead of letting people pick what they wanted to eat?
    b) Why were the meals prepared by dieticians?
    c) Why did the researchers need the control group? If the DASH diet group's blood pressure was lower at the end of the experiment than at the beginning, wouldn't that prove the effectiveness of that diet?
    d) What additional information would you want to know in order to decide whether an average reduction in blood pressure of 6.7 points was statistically significant?

33. **Mozart** Will listening to a Mozart piano sonata make you smarter? In a 1995 study published in the journal *Psychological Science*, Rauscher, Shaw, and Ky reported that when students were given a spatial reasoning section of a standard IQ test, those who listened to Mozart for 10 minutes improved their scores more than those who simply sat quietly.
    a) These researchers said the differences were statistically significant. Explain what that means in context.

b) Steele, Bass, and Crook tried to replicate the original study. In their study, also published in *Psychological Science* (1999), the subjects were 125 college students who participated in the experiment for course credit. Subjects first took the test. Then they were assigned to one of three groups: listening to a Mozart piano sonata, listening to music by Philip Glass, and sitting for 10 minutes in silence. Three days after the treatments, they were retested. Draw a diagram displaying the design of this experiment.

c) These boxplots show the differences in score before and after treatment for the three groups. Did the Mozart group show improvement?

d) Do you think the results prove that listening to Mozart is beneficial? Explain.

34. Full moon It's a common belief that people behave strangely when there's a full moon and that as a result police and emergency rooms are busier than usual. Design a way you could find out whether there is any merit to this belief. Will you use an observational study or an experiment? Why?

35. Wine A 2001 Danish study published in the *Archives of Internal Medicine* casts significant doubt on suggestions that adults who drink wine have higher levels of "good" cholesterol and fewer heart attacks. These researchers followed a group of individuals born at a Copenhagen hospital between 1959 and 1961 for 40 years. Their study found that in this group the adults who drank wine were richer and better educated than those who did not.

a) What kind of study was this?

b) It is generally true that people with high levels of education and high socioeconomic status are healthier than others. How does this call into question the supposed health benefits of wine?

c) Can studies such as these prove causation (that wine helps prevent heart attacks, that drinking wine makes one richer, that being rich helps prevent heart attacks, etc.)? Explain.

36. Swimming Recently, a group of adults who swim regularly for exercise were evaluated for depression. It turned out that these swimmers were less likely to be depressed than the general population. The researchers said the difference was statistically significant.

a) What does "statistically significant" mean in this context?

b) Is this an experiment or an observational study? Explain.

c) News reports claimed this study proved that swimming can prevent depression. Explain why this conclusion is not justified by the study. Include an example of a possible lurking variable.

d) But perhaps it is true. We wonder if exercise can ward off depression, and whether anaerobic exercise (like weight training) is as effective as aerobic exercise (like swimming). We find 120 volunteers not currently engaged in a regular program of exercise. Design an appropriate experiment.

37. Dowsing Before drilling for water, many rural homeowners hire a dowser (a person who claims to be able to sense the presence of underground water using a forked stick). Suppose we wish to set up an experiment to test one dowser's ability. We get 20 identical containers, fill some with water, and ask him to tell which ones they are.

a) How will we randomize this procedure?

b) The dowser correctly identifies the contents of 12 out of 20 containers. Do you think this level of success is statistically significant? Explain.

c) How many correct identifications (out of 20) would the dowser have to make to convince you that the forked-stick trick works? Explain.

38. Healing A medical researcher suspects that giving post-surgical patients large doses of vitamin E will speed their recovery times by helping their incisions heal more quickly. Design an experiment to test this conjecture. Be sure to identify the factors, levels, treatments, response variable, and the role of randomization.

39. Reading Some schools teach reading using phonics (the sounds made by letters) and others using whole language (word recognition). Suppose a school district wants to know which method works better. Suggest a design for an appropriate experiment.

40. Gas mileage Do cars get better gas mileage with premium instead of regular unleaded gasoline? It might be possible to test some engines in a laboratory, but we'd rather use real cars and real drivers in real day-to-day driving, so we get 20 volunteers. Design the experiment.

41. Weekend deaths A study published in the *New England Journal of Medicine* (Aug. 2001) suggests that it's dangerous to enter a hospital on a weekend. During a 10-year period, researchers tracked over 4 million emergency admissions to hospitals in Ontario, Canada. Their findings revealed that patients admitted on weekends had a much higher risk of death than those who went on weekdays.

a) The researchers said the difference in death rates was "statistically significant." Explain in this context what that means.

b) What kind of study was this? Explain.

c) If you think you're quite ill on a Saturday, should you wait until Monday to seek medical help? Explain.

d) Suggest some possible explanations for this troubling finding.

42. Shingles A research doctor has discovered a new ointment that she believes will be more effective than the current medication in the treatment of shingles (a painful skin rash). Eight patients have volunteered to participate in the initial trials of this ointment. You are the statistician hired as a consultant to help design a completely randomized experiment.

a) Describe how you will conduct this experiment.

b) Suppose the eight patients' last names start with the letters A to H. Using the random numbers listed below, show which patients you will assign to each treatment. Explain your randomization procedure clearly.

41098 18329 78458 31685 55259

c) Can you make this experiment double-blind? How?

d) The initial experiment revealed that males and females may respond differently to the ointment. Further testing of the drug's effectiveness is now planned, and many patients have volunteered. What changes in your first design, if any, would you make for this second stage of testing?

43. Beetles Hoping to learn how to control crop damage by a certain species of beetle, a researcher plans to test two different pesticides in small plots of corn. A few days after application of the chemicals, he'll check the number of beetle larvae found on each plant. The researcher wants to know whether either pesticide works and whether there is a significant difference in effectiveness between them. Design an appropriate experiment.

44. SAT Prep Can special study courses actually help raise SAT scores? One organization says that the 30 students they tutored achieved an average gain of 60 points when they retook the test.

a) Explain why this does not necessarily prove that the special course caused the scores to go up.

b) Propose a design for an experiment that could test the effectiveness of the tutorial course.

c) Suppose you suspect that the tutorial course might be more helpful for students whose initial scores were particularly low. How would this affect your proposed design?

45. Safety switch An industrial machine requires an emergency shutoff switch that must be designed so that it can be easily operated with either hand. Design an experiment to find out whether workers will be able to deactivate the machine as quickly with their left hands as with their right hands. Be sure to explain the role of randomization in your design.

46. Washing clothes A consumer group wants to test the effectiveness of a new "organic" laundry detergent and make recommendations to customers about how to best use the product. They intentionally get grass stains on 30 white T-shirts in order to see how well the detergent will clean them. They want to try the detergent in cold water and in hot water on both the "regular" and "delicates" wash cycles. Design an appropriate experiment, indicating the number of factors, levels, and treatments. Explain the role of randomization in your experiment.

47. Skydiving, anyone? A humor piece published in the *British Medical Journal* ("Parachute use to prevent death and major trauma related to gravitational challenge: systematic review of randomized control trials," Gordon, Smith, and Pell, *BMJ*, 2003:327) notes that we can't tell for sure whether parachutes are safe and effective because there has never been a properly randomized, double-blind, placebo-controlled study of parachute effectiveness in skydiving. (Yes, this is the sort of thing statisticians find funny. . . .) Suppose you were designing such a study:

a) What is the factor in this experiment?

b) What experimental units would you propose?[8]

c) What would serve as a placebo for this study?

d) What would the treatments be?

e) What would the response variable be?

f) What sources of variability would you control?

g) How would you randomize this "experiment"?

h) How would you make the experiment double-blind?

_____

[8]Don't include your Statistics instructor!

 Just Checking ANSWERS

1. a. The factor was type of treatment for peptic ulcer.

 b. The response variable could be a measure of relief from gastric ulcer pain or an evaluation by a physician of the state of the disease.

 c. Treatments would be gastric freezing and some alternative control treatment.

 d. Treatments should be assigned randomly.

 e. No. The Web site reports "lack of effectiveness," indicating that no large differences in patient healing were noted.

2. a. Neither the patients who received the treatment nor the doctor who evaluated them afterward knew what treatment they had received.

 b. The placebo is needed to accomplish blinding. The best alternative would be using body-temperature liquid rather than the freezing liquid.

 c. The researchers should block the subjects by the length of time they had had the ulcer, then randomly assign subjects in each block to the freezing and placebo groups.

Review of part III

Gathering Data

Quick Review

Before you can make a boxplot, calculate a mean, describe a distribution, or fit a line, you must have meaningful data to work with. Getting good data is essential to any investigation. No amount of clever analysis can make up for badly collected data. Here's a brief summary of the key concepts and skills:

■ The way you gather data depends both on what you want to discover and on what is practical.

■ To get some insight into what might happen in a real situation, model it with a **simulation** using random numbers.

■ To answer questions about a target population, collect information from a sample with a **survey** or poll.

  ▪ Choose the sample randomly. Random sampling designs include simple, stratified, systematic, cluster, and multistage.

  ▪ A simple random sample draws without restriction from the entire target population.

  ▪ When there are subgroups within the population that may respond differently, use a stratified sample.

  ▪ Avoid bias, a systematic distortion of the results. Sample designs that allow undercoverage or response bias and designs such as voluntary response or convenience samples don't faithfully represent the population.

  ▪ Samples will naturally vary one from another. This sample-to-sample variation is called sampling error. Each sample only approximates the target population.

■ **Observational studies** collect information from a sample drawn from a target population.

  ▪ Retrospective studies examine existing data. Prospective studies identify subjects in advance, then follow them to collect data as the data are created, perhaps over many years.

  ▪ Observational studies can spot associations between variables but cannot establish cause and effect. It's impossible to eliminate the possibility of lurking or confounding variables.

■ To see how different treatments influence a response variable, design an **experiment**.

  ▪ Assign subjects to treatments randomly. If you don't assign treatments randomly, your experiment is not likely to yield valid results.

  ▪ Control known sources of variation as much as possible. Reduce variation that cannot be controlled by using blocking, if possible.

  ▪ Replicate the experiment, assigning several subjects to each treatment level.

  ▪ If possible, replicate the entire experiment with an entirely different collection of subjects.

  ▪ A well-designed experiment can provide evidence that changes in the factors cause changes in the response variable.

Now for more opportunities to review these concepts and skills . . .

Review Exercises

1–18. What design? *Analyze the design of each research example reported. Is it a sample survey, an observational study, or an experiment? If a sample, what are the population, the parameter of interest, and the sampling procedure? If an observational study, was it retrospective or prospective? If an experiment, describe the factors, treatments, randomization, response variable, and any blocking, matching, or blinding that may be present. In each, what kind of conclusions can be reached?*

1. Researchers identified 242 children in the Cleveland area who had been born prematurely (at about 29 weeks). They examined these children at age 8 and again at age 20, comparing them to another group of 233 children not born prematurely. Their report, published in the *New England Journal of Medicine,* said the "preemies" engaged in significantly less risky behavior than the others.

Differences showed up in the use of alcohol and marijuana, conviction of crimes, and teenage pregnancy.

2. The journal *Circulation* reported that among 1900 people who had heart attacks, those who drank an average of 19 cups of tea a week were 44% more likely than nondrinkers to survive at least 3 years after the attack.

3. Researchers at the Purina Pet Institute studied Labrador retrievers for evidence of a relationship between diet and longevity. At 8 weeks of age, 2 puppies of the same sex and weight were randomly assigned to one of two groups—a total of 48 dogs in all. One group was allowed to eat all they wanted, while the other group was fed a diet about 25% lower in calories. The median life span of dogs fed the restricted diet was 22 months longer than that of other dogs. (Source: *Science News* 161, no. 19)

4. The radioactive gas radon, found in some homes, poses a health risk to residents. To assess the level of contamination in their area, a county health department wants to test a few homes. If the risk seems high, they will publicize the results to emphasize the need for home testing. Officials plan to use the local property tax list to randomly choose 25 homes from various areas of the county.

5. Data were collected over a decade from 1021 men and women with a recent history of precancerous colon polyps. Participants were randomly assigned to receive folic acid (a B vitamin) or a placebo, and the study concluded that those receiving the folic acid may actually increase their risk of developing additional precancerous growths. Previous studies suggested that taking folic acid may help to prevent colorectal cancer. (Source: *JAMA* 2007, 297)

6. In the journal *Science,* a research team reported that plants in southern England are flowering earlier in the spring. Records of the first flowering dates for 385 species over a period of 47 years indicate that flowering has advanced an average of 15 days per decade, an indication of climate warming, according to the authors.

7. Fireworks manufacturers face a dilemma. They must be sure that the rockets work properly, but test-firing a rocket essentially destroys it. On the other hand, not testing the product leaves open the danger that they sell a bunch of duds, leading to unhappy customers and loss of future sales. The solution, of course, is to test a few of the rockets produced each day, assuming that if those tested work properly, the others are ready for sale.

8. People who read the last page of a mystery novel first generally like stories better. Researchers recruited 819 college students to read short stories, and for one story, they were given a spoiler paragraph beforehand. On the second and third story, the spoiler was incorporated as the opening paragraph or not given at all. Overall, participants liked the stories best after first reading spoilers. (Source: *Psychological Science*, August 12, 2011)

9. Does keeping a child's lunch in an insulated bag, even with ice packs, protect the food from warming to temperatures where germs can proliferate? Researchers used an electric temperature gun on 235 lunches at preschools 90 minutes before they were to be eaten. Of the lunches with ice packs, over 90% of them were at unsafe temperatures. The study was of particular interest because preschoolers develop up to four times as many foodborne infections as do adults. (Source: *Science News*, August 9, 2011)

10. Some doctors have expressed concern that men who have vasectomies seemed more likely to develop prostate cancer. Medical researchers used a national cancer registry to identify 923 men who had had prostate cancer and 1224 men of similar ages who had not. Roughly one quarter of the men in each group had undergone a vasectomy, many

more than 25 years before the study. The study's authors concluded that there is strong evidence that having the operation presents no long-term risk for developing prostate cancer. (Source: *Science News*, July 20, 2002)

11. Widely used antidepressants may reduce ominous brain plaques associated with Alzheimer's disease. In the study, mice genetically engineered to have large amounts of brain plaque were given a class of antidepressants that boost serotonin in the brain. After a single dose, the plaque levels dropped, and after four months, the mice had about half the brain plaques as the mice that didn't take the drug. (Source: *Proceedings of the National Academy of Sciences*, August 22, 2011)

12. An artisan wants to create pottery that has the appearance of age. He prepares several samples of clay with four different glazes and test fires them in a kiln at three different temperature settings.

13. Tests of gene therapy on laboratory rats have raised hopes of stopping the degeneration of tissue that characterizes chronic heart failure. Researchers at the University of California, San Diego, used hamsters with cardiac disease, randomly assigning 30 to receive the gene therapy and leaving the other 28 untreated. Five weeks after treatment the gene therapy group's heart muscles stabilized, while those of the untreated hamsters continued to weaken. (Source: *Science News*, July 27, 2002)

14. People aged 50 to 71 were initially contacted in the mid-1990s to participate in a study about smoking and bladder cancer. Data were collected from more than 280,000 men and 186,000 women from eight states who answered questions about their health, smoking history, alcohol intake, diet, physical activity, and other lifestyle factors. When the study ended in 2006, about half the bladder cancer cases in adults age 50 and older were traceable to smoking. (Source: *Journal of the American Medical Association*, August 17, 2011)

15. An orange-juice processing plant will accept a shipment of fruit only after several hundred oranges selected from various locations within the truck are carefully inspected. If too many show signs of unsuitability for juice (bruised, rotten, unripe, etc.), the whole truckload is rejected.

16. A soft-drink manufacturer must be sure the bottle caps on the soda are fully sealed and will not come off easily. Inspectors pull a few bottles off the production line at regular intervals and test the caps. If they detect any problems, they will stop the bottling process to adjust or repair the machine that caps the bottles.

17. Older Americans with a college education are significantly more likely to be emotionally well-off than are people in this age group with less education. Among those aged 65 and older, 35% scored 90 or above on the Emotional Health Index, but for those with a college degree, the

percentage rose to 43% (post-graduate degree, 46%). The results are based on phone interviews conducted between January 2010 and July 2011. (Source: gallup.com, August 19, 2011)

18. Does the use of computer software in Introductory Statistics classes lead to better understanding of the concepts? A professor teaching two sections of Statistics decides to investigate. She teaches both sections using the same lectures and assignments, but gives one class statistics software to help them with their homework. The classes take the same final exam, and graders do not know which students used computers during the semester. The professor is also concerned that students who have had calculus may perform differently from those who have not, so she plans to compare software vs. no-software scores separately for these two groups of students.

19. Point spread. When taking bets on sporting events, bookmakers often include a "point spread" that awards the weaker team extra points. In theory, this makes the outcome of the bet a toss-up. Suppose a gambler places a $10 bet and picks the winners of five games. If he's right about fewer than three of the games, he loses. If he gets three, four, or all five correct, he's paid $10, $20, and $50, respectively. Estimate the amount such a bettor might expect to lose over many weeks of gambling.

20. The lottery. Many people spend a lot of money trying to win huge jackpots in state lotteries. Let's play a simplified version using only the numbers from 1 to 20. You bet on three numbers. The state picks five winning numbers. If your three are all among the winners, you are rich!

a) Simulate repeated plays. How long did it take you to win?

b) In real lotteries, there are many more choices (often 54) and you must match all five winning numbers. Explain how these changes affect your chances of hitting the jackpot.

21. Everyday randomness. Aside from casinos, lotteries, and games, there are other situations you encounter in which something is described as "random" in some way. Give three different examples. Describe how randomness is (or is not) achieved in each.

22. Cell phone risks. Researchers at the Washington University School of Medicine randomly placed 480 rats into one of three chambers containing radio antennas. One group was exposed to digital cell phone radio waves, the second to analog cell phone waves, and the third group to no radio waves. Two years later, the rats were examined for signs of brain tumors. In June 2002, the scientists said that differences among the three groups were not statistically significant.

a) Is this a study or an experiment? Explain.

b) Explain in this context what "not statistically significant" means.

c) Comment on the fact that this research was funded by Motorola, a manufacturer of cell phones.

23. Tips. In restaurants, servers rely on tips as a major source of income. Does serving candy after the meal produce larger tips? To find out, two waiters determined randomly whether or not to give candy to 92 dining parties. They recorded the sizes of the tips and reported that guests getting candy tipped an average of 17.8% of the bill, compared with an average tip of only 15.1% from those who got no candy. (Source: "Sweetening the Till: The Use of Candy to Increase Restaurant Tipping." *Journal of Applied Social Psychology* 32, no. 2 [2002]: 300–309)

a) Was this an experiment or an observational study? Explain.

b) Is it reasonable to conclude that the candy caused guests to tip more? Explain.

c) The researchers said the difference was statistically significant. Explain in this context what that means.

24. Tips, take 2. In another experiment to see if getting candy after a meal would induce customers to leave a bigger tip, a waitress randomly decided what to do with 80 dining parties. Some parties received no candy, some just one piece, and some two pieces. Others initially got just one piece of candy, and then the waitress suggested that they take another piece. She recorded the tips received, finding that, in general, the more candy, the higher the tip, but the highest tips (23%) came from the parties who got one piece and then were offered more. (Source: "Sweetening the Till: The Use of Candy to Increase Restaurant Tipping." *Journal of Applied Social Psychology* 32, no. 2 [2002]: 300–309)

a) Diagram this experiment.

b) How many factors are there? How many levels?

c) How many treatments are there?

d) What is the response variable?

e) Did this experiment involve blinding? Double-blinding?

f) In what way might the waitress, perhaps unintentionally, have biased the results?

25. Timing. In August 2011, a Sodahead.com voluntary response poll asked site visitors, "Obama is on Vacation Again: Does He Have the Worst Timing Ever?" 56% of the 629 votes were for "Yes." During the week of the poll, a 5.8 earthquake struck near Washington, D.C., and Hurricane Irene made its way up the East coast. What types of bias may be present in the results of the poll?

26. Laundry. An experiment to test a new laundry detergent, SparkleKleen, is being conducted by a consumer advocate group. They would like to compare its performance with that of a laboratory standard detergent they have used in previous experiments. They can stain 16 swatches of cloth with 2 tsp of a common staining compound and then use a well-calibrated optical scanner to detect the amount of the stain left after washing. To save time in the experiment, several suggestions have been made. Comment on the possible merits and drawbacks of each one.

a) Since data for the laboratory standard detergent are already available from previous experiments, for this experiment wash all 16 swatches with SparkleKleen, and compare the results with the previous data.

b) Use both detergents with eight separate runs each, but to save time, use only a 10-second wash time with very hot water.

c) To ease bookkeeping, first run all of the standard detergent washes on eight swatches, then run all of the SparkleKleen washes on the other eight swatches.

d) Rather than run the experiment, use data from the company that produced SparkleKleen, and compare them with past data from the standard detergent.

27. When to stop? You play a game that involves rolling a die. You can roll as many times as you want, and your score is the total for all the rolls. But . . . if you roll a 6 your score is 0 and your turn is over. What might be a good strategy for a game like this?

a) One of your opponents decides to roll 4 times, then stop (hoping not to get the dreaded 6 before then). Use a simulation to estimate his average score.

b) Another opponent decides to roll until she gets at least 12 points, then stop. Use a simulation to estimate her average score.

c) Propose another strategy that you would use to play this game. Using your strategy, simulate several turns. Do you think you would beat the two opponents?

28. Rivets. A company that manufactures rivets believes the shear strength of the rivets they manufacture follows a Normal model with a mean breaking strength of 950 pounds and a standard deviation of 40 pounds.

a) What percentage of rivets selected at random will break when tested under a 900-pound load?

b) You're trying to improve the rivets and want to examine some that fail. Use a simulation to estimate how many rivets you might need to test in order to find three that fail at 900 pounds (or below).

29. Homecoming. A college Statistics class conducted a survey concerning community attitudes about the college's large homecoming celebration. That survey drew its sample in the following manner: Telephone numbers were generated at random by selecting one of the local telephone exchanges (first three digits) at random and then generating a random four-digit number to follow the exchange. If a person answered the phone and the call was to a residence, then that person was taken to be the subject for interview. (Undergraduate students and those under voting age were excluded, as was anyone

who could not speak English.) Calls were placed until a sample of 200 eligible respondents had been reached.

a) Did every telephone number that could occur in that community have an equal chance of being generated?

b) Did this method of generating telephone numbers result in a simple random sample (SRS) of local residences? Explain.

c) Did this method generate an SRS of local voters? Explain.

d) Is this method unbiased in generating samples of households? Explain.

30. Youthful appearance. *Readers' Digest* (April 2002, p. 152) reported results of several surveys that asked graduate students to examine photographs of men and women and try to guess their ages. Researchers compared these guesses with the number of times the people in the pictures reported having sexual intercourse. It turned out that those who had been more sexually active were judged as looking younger, and that the difference was described as "statistically significant." Psychologist David Weeks, who compiled the research, speculated that lovemaking boosts hormones that "reduce fatty tissue and increase lean muscle, giving a more youthful appearance."

a) What does "statistically significant" mean in this context?

b) Explain in statistical terms why you might be skeptical about Dr. Weeks's conclusion. Propose an alternative explanation for these results.

31. Smoking and Alzheimer's. Medical studies indicate that smokers are less likely to develop Alzheimer's disease than people who never smoked.

a) Does this prove that smoking may offer some protection against Alzheimer's? Explain.

b) Offer an alternative explanation for this association.

c) How would you conduct a study to investigate this?

32. Antacids. A researcher wants to compare the performance of three types of antacid in volunteers suffering from acid reflux disease. Because men and women may react differently to this medication, the subjects are split into two groups, by sex. Subjects in each group are randomly assigned to take one of the antacids or to take a sugar pill made to look the same. The subjects will rate their level of discomfort 30 minutes after eating.

a) What kind of design is this?

b) The experiment uses volunteers rather than a random sample of all people suffering from acid reflux disease. Does this make the results invalid? Explain.

c) How may the use of the placebo confound this experiment? Explain.

33. Sex and violence. Does the content of a television program affect viewers' memory of the products advertised in commercials? Design an experiment to compare the ability of viewers to recall brand names of items featured in commercials during programs with violent content, sexual content, or neutral content.

34. Pubs. In England, a Leeds University researcher said that the local watering hole's welcoming atmosphere helps men get rid of the stresses of modern life and is vital for their psychological well-being. Author of the report, Dr. Colin Gill, said rather than complain, women should encourage men to "pop out for a swift half." "Pub-time allows men to bond with friends and colleagues," he said. "Men need break-out time as much as women and are mentally healthier for it." Gill added that men might feel unfulfilled or empty if they had not been to the pub for a week. The report, commissioned by alcohol-free beer brand Kaliber, surveyed 900 men on their reasons for going to the pub. More than 40% said they went for the conversation, with relaxation and a friendly atmosphere being the other most common reasons.

Only 1 in 10 listed alcohol as the overriding reason.

Let's examine this news story from a statistical perspective.

a) What are the W's: *Who, What, When, Where, Why, How?*

b) What population does the researcher think the study applies to?

c) What is the most important thing about the selection process that the article does *not* tell us?

d) How do *you* think the 900 respondents were selected? (Name a method of drawing a sample that is likely to have been used.)

e) Do you think the report that only 10% of respondents listed alcohol as an important reason for going to the pub might be a biased result? Why?

35. Age and party 2008. The Pew Research Center collected data from national exits polls conducted by *NBC News* after the 2008 presidential election. The following table shows information regarding voter age and party preference:

	Republican	Democrat	Other	Total
18–29	260	390	351	1001
30–44	320	379	300	999
45–64	329	369	300	998
65+	361	392	251	1004
Total	1270	1530	1202	4002

a) What sampling strategy do you think the pollsters used? Explain.

b) What percentage of the people surveyed were Democrats?

c) Do you think this is a good estimate of the percentage of voters in the United States who are registered Democrats? Why or why not?

d) In creating this sample design, what question do you think the pollsters were trying to answer?

36. Bias? Political analyst Michael Barone has written that "conservatives are more likely than others to refuse to respond to polls, particularly those polls taken by media outlets that conservatives consider biased" (Source: *The Weekly Standard*, March 10, 1997). The Pew Research Foundation tested this assertion by asking the same questions in a national survey run by standard methods and in a more rigorous survey that was a true SRS with careful follow-up to encourage participation. The response rate in the "standard survey" was 42%. The response rate in the "rigorous survey" was 71%.

a) What kind of bias does Barone claim may exist in polls?

b) What is the population for these surveys?

c) On the question of political position, the Pew researchers report the following table:

	Standard Survey	Rigorous Survey
Conservative	37%	35%
Moderate	40%	41%
Liberal	19%	20%

What makes you think these results are incomplete?

d) The Pew researchers report that differences between opinions expressed on the two surveys were not statistically significant. Explain what "not statistically significant" means in this context.

37. Save the grapes. Vineyard owners have problems with birds that like to eat the ripening grapes. Some vineyards use scarecrows to try to keep birds away. Others use netting that covers the plants. Owners really would like to know if either method works and, if so, which one is better. One owner has offered to let you use his vineyard this year for an experiment. Propose a design. Carefully indicate how you would set up the experiment, specifying the factor(s) and response variable.

38. Bats. It's generally believed that baseball players can hit the ball farther with aluminum bats than with the traditional wooden ones. Is that true? And, if so, how much farther? Players on your local high school baseball team have agreed to help you find out. Design an appropriate experiment.

39. Acupuncture. Research reported in 2008 brings to light the effectiveness of treating chronic lower back pain with different methods. One-third of nearly 1200 volunteers were administered conventional treatment (drugs, physical therapy, and exercise). The remaining patients got 30-minute acupuncture sessions. Half of these patients were punctured at sites suspected of being useful and half received needles at other spots on their bodies.

Comparable shares of each acupuncture group, roughly 45%, reported decreased back pain for at least six months after their sessions ended. This was almost twice as high as those receiving the conventional therapy, leading the researchers to conclude that results were statistically significant.

a) Why did the researchers feel it was necessary to have some of the patients undergo a "fake" acupuncture?

b) Because patients had to consent to participate in this experiment, the subjects were essentially self-selected— a kind of voluntary response group. Explain why that does not invalidate the findings of the experiment.

c) What does "statistically significant" mean in the context of this experiment?

40. NBA draft lottery. Professional basketball teams hold a "draft" each year in which they get to pick the best available college and high school players. In an effort to promote competition, teams with the worst records get to pick first, theoretically allowing them to add better players. To combat the fear that teams with no chance to make the playoffs might try to get better draft picks by intentionally losing late-season games, the NBA's Board of Governors adopted a weighted lottery system in 1990. Under this system, the 11 teams that did not make the playoffs were eligible for the lottery. The NBA prepared 66 cards, each naming one of the teams. The team with the worst win-loss record was named on 11 of the cards, the second-worst team on 10 cards, and so on, with the team having the best record among the nonplayoff clubs getting only one chance at having the first pick. The cards were mixed, then drawn randomly to determine the order in which the teams could draft players. Suppose there are two exceptional players available in this year's draft and your favorite team had the third-worst record. Use a simulation to find out how likely it is that your team gets to pick first or second. Describe your simulation carefully.

41. Security. There are 20 first-class passengers and 120 coach passengers scheduled on a flight. In addition to the usual security screening, 10% of the passengers will be subjected to a more complete search.

a) Describe a sampling strategy to randomly select those to be searched.

b) Here is the first-class passenger list and a set of random digits. Select two passengers to be searched, carefully demonstrating your process.

65436 71127 04879 41516 20451 02227 94769 23593

Bergman	Cox	Fontana	Perl
Bowman	DeLara	Forester	Rabkin
Burkhauser	Delli-Bovi	Frongillo	Roufaiel
Castillo	Dugan	Furnas	Swafford
Clancy	Febo	LePage	Testut

c) Explain how you would use a random number table to select the coach passengers to be searched.

42. Profiling? Among the 20 first-class passengers on the flight described in Exercise 41, there were four businessmen from the Middle East. Two of them were the two passengers selected to be searched. They complained of profiling, but the airline claims that the selection was random. What do you think? Support your conclusion with a simulation.

43. Par 4. In theory, a golfer playing a par-4 hole tees off, hitting the ball in the fairway, then hits an approach shot onto the green. The first putt (usually long) probably won't go in, but the second putt (usually much shorter) should. Sounds simple enough, but how many strokes might it really take? Use a simulation to estimate a pretty good golfer's score based on these assumptions:

- The tee shot hits the fairway 70% of the time.
- A first approach shot lands on the green 80% of the time from the fairway, but only 40% of the time otherwise.
- Subsequent approach shots land on the green 90% of the time.
- The first putt goes in 20% of the time, and subsequent putts go in 90% of the time.

44. The back nine. Use simulations to estimate more golf scores, similar to the procedure in Exercise 43.

a) On a par 3, the golfer hopes the tee shot lands on the green. Assume that the tee shot behaves like the first approach shot described in Exercise 43.

b) On a par 5, the second shot will reach the green 10% of the time and hit the fairway 60% of the time. If it does not hit the green, the golfer must play an approach shot as described in Exercise 43.

c) Create a list of assumptions that describe your golfing ability, and then simulate your score on a few holes. Explain your simulation clearly.

Practice Exam

Multiple Choice

(Questions 1–2). Do math skills matter when looking for gainful employment? The latest U.S. Department of Education National Assessment of Adult Literacy was conducted in 2003. A random sample of 18,102 adults (aged 16+) living in U.S. households were tested on Quantitative Literacy and asked whether they thought that their math skills limited their job opportunities. Quantitative Literacy was measured as a respondent's ability to identify and perform computations using data embedded in printed materials such as balancing a checkbook, figuring out a tip, completing an order form, or determining the amount of interest on a loan from an advertisement. Respondents were placed into one of four categories (Below Basic, Basic, Intermediate, or Proficient) based upon their scores on various tasks. The table below summarizes the data that were collected.

Number of Adults Who Think Their Math Skills Limit Their Ability to Get a Job					
	Quantitative Proficiency Level				
Response	Below Basic	Basic	Inter- mediate	Proficient	Total
Not at All	1654	4095	4813	2175	12738
A Little	662	807	662	196	2326
Some	786	807	421	49	2062
A Lot	1034	496	120	24	1675
Total	4136	6204	6016	2444	18801

1. What percentage of those with Basic quantitative skills think that their math skills limit their job opportunities a little?
   A) 4%   B) 12%   C) 13%
   D) 33%   E) 35%

2. Adults in which proficiency level are most likely to think that their math skills limit them "A Little" in job opportunities?
   A) Below Basic   B) Basic
   C) Intermediate   D) Proficient
   E) Below Basic and Intermediate have the same high likelihood.

3. The segmented bar graphs below depict data from the NAAL (National Assessment of Adult Literacy) conducted in 2003.
   (*Source: Kutner, M., et al, Literacy in Everyday Life: Results From the 2003 National Assessment of Adult Literacy (NCES 2007–480).U.S. Department of Education. Washington, DC*)

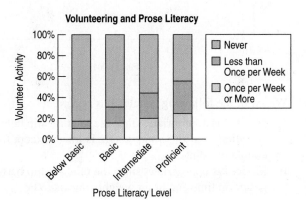

Does there appear to be a relationship between volunteerism and literacy level?
   A) Yes, all three bars have the same number of segments.
   B) Yes, because all three bars have the same height.
   C) Yes, because the corresponding segments of the three bars have different heights.
   D) No, because the corresponding segments of the three bars have different heights.
   E) No, because the sums of the 3 proportions in each bar are identical.

4. Professional basketball scouts are on the lookout for tall players. Ideally, a player makes a great center if they are 7 feet tall or taller. Male height in America is roughly normally distributed with a mean of 69.5″ and a standard deviation of 3″. If there are 4.6 million males in America who are 18 or 19 years old, about how many prospective centers are available?
   A) 3   B) 5   C) 22   D) 48   E) 67

5. A forest ranger researching ecosystem recovery following forest fires collected data on a large growth of young pine trees. The heights of the trees had a mean of 3.2 feet and standard deviation 0.6 feet. A botany journal has asked that the data be expressed in inches. When the data are rescaled, what will be the new mean and standard deviation?
   A) 3.2 and 0.6   B) 3.2 and 7.2   C) 38.4 and 0.6
   D) 38.4 and 7.2   E) 38.4 and 86.4

6. A high school statistics teacher ran an experiment with his classes called "The Barbie Bungee" in which the students tied rubber bands to Barbie's feet and recorded the number of rubber bands used and the distance, in centimeters, of her jump. When they created a scatterplot the relationship appeared to be linear, and the correlation was $r = 0.996$. After discussing these results, the teacher instructed the students to convert the distances to inches. What effect will this have on the correlation between the two variables?
   A) Because 1 cm = 0.393701 inches, the correlation will become .392126.
   B) Because 1 cm = 0.393701 inches, the correlation will become 2.5298.
   C) Because only the length measurements have changed, the correlation will decrease substantially.
   D) Because changing from centimeters to inches does not affect the value of the correlation, the correlation will remain 0.996.
   E) Because inches is a much more common measurement for distance in the United States, the relationship between the data will be stronger and thus the correlation will increase.

Source: http://www.sophia.org/learning-about-linear-regression-with-bungee-jumpi-tutorial

7. A pile of sand on your local beach has no strength and can be knocked down by a toddler with a light kick. Sandstones, however, have a great variety of strength. "Certain horizons in the local Cretaceous Dakota sandstones, can be easily broken and crumbled by hand, while other horizons require a hammer and a good strong blow." A coloration difference indicates a difference in the amount of cementation. The more iron oxide cement the darker and the stronger the sandstone. A geologist collected data to study the relationship between porosity and sandstone strength. Based on those data, the least squares regression line is $\hat{y} = 20560 - 1344.4x$, where x is the percent of porosity and y is unconfined compressive sandstone strength measured in psi (pounds per square inch). Which of the following best describes the meaning of the slope of the least squares regression line?

A) For each increase of 1 psi in strength, the estimated porosity is expected to decrease by 1344.4%.
B) For each increase of 1% in porosity, the estimated strength is expected to increase by 20560 psi.
C) For each increase of 1% in porosity, the estimated strength is expected to increase by 1344.4 psi.
D) For each increase of 1% porosity, the estimated strength is expected to decrease by 1344.4 psi.
E) For each increase of 1% in porosity, the estimated strength is expected to increase by 19,215.6 psi.

8. Can you tell how old a lion is by looking at its nose? A professor at the University of Wisconsin-Madison conducted a study of data taken from 32 lions and observed the relationship between age (in years) and proportion of blackness in the lion's nose. The equation of the least squares regression line was

$$\hat{y} = 0.8790 + 10.6471x$$

where $\hat{y}$ is the predicted age of the lion, measured in years, and x is the proportion of the lion's nose that is black. A lion whose nose was 11% black was known to be 1.9 years old. What is the residual for the age of this lion?

A) −0.15 years   B) 0.15 years
C) 0.88 years   D) 2.05 years
E) 10.65 years
(*Source*: http://www.stat.wisc.edu/~st571-1/15-regression-4.pdf)

9. An analysis of price of crude oil ($/barrel) and gasoline prices at the pump ($/gallon) from 1976 to 2004 found a correlation coefficient of 0.829.
Which of the following is a true statement?

A) Since the correlation coefficient is high, there is a linear relationship between crude oil prices and prices at the pump.
B) Since the correlation coefficient is only moderately high, the relationship between crude oil prices and prices at the pump is probably not linear.

C) For every one dollar increase in crude oil price per barrel, the gasoline price at the pump is expected to increase by $0.829 per gallon.
D) 68.7% of the price of a gallon of gasoline can be explained by crude oil prices.
E) None of the statements A–D is true.

10. Researchers measured gestation time (in days) and brain weight (in grams) for a random sample of 23 unborn infants. They fit two possible regression models to their data.
Here are the equation and residuals plot for **Model I**:

$$\overline{Brain\ weight} = -78.46 + 0.9176\ Gestation$$

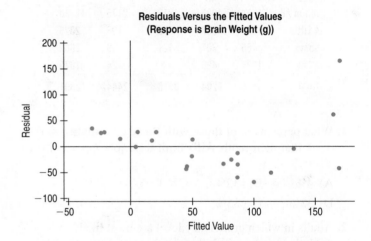

Here are the equation and residuals plot for **Model II**:

$$\log(\overline{Brain\ weight}) = 0.3457 + 0.007371\ Gestation$$

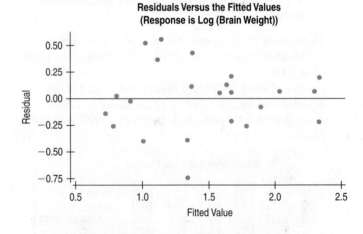

Which of the following conclusions is correct?

A) Model I is appropriate, since the relationship between gestation time and brain weight is linear except for a couple of points.
B) Model I is appropriate, since the relationship between gestation time and brain weight appears to be stronger.

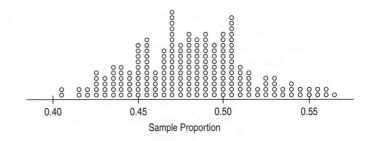

C) Model II is appropriate, since the residuals are smaller.

D) Model II is appropriate, since the relationship between gestation time and brain weight appears to be linear.

E) Model II is appropriate, since the relationship between gestation time and the logarithm of brain weight appears to be linear.

11. An advertising agency is testing the effectiveness of a new commercial. They are trying out the commercial in three different time lengths (15, 30, and 60 seconds) and two formats (animated vs. live action). Each test audience will rate the product after seeing the commercial shown in one way. Which description of this study is correct?

A) An experiment with 2 factors (length and format), one having 3 levels (15, 30, 60) and the other having 2 levels (animated and live) for 6 treatments.

B) An experiment with 2 factors(length and format) at 5 levels (15, 30, 60, animated and live) for 5 treatments.

C) An experiment 1 factor (length) blocked by format with 3 levels (15, 30, 60).

D) A sample survey with two questions and one response.

E) A sample stratified by length and format.

12. A school newspaper is going to investigate students' perceptions regarding the campus security guards. They plan to survey 100 students at the school. The editor-in-chief tells his staff that he is worried about undercoverage bias created by chronically truant students. His concern is. . .

A) unwarranted because some students may refuse to participate and that is inevitable.

B) justified because chronically truant students will be difficult to survey and will probably have a disproportionately negative view of the security guards.

C) unwarranted because students who cannot be surveyed are not a part of the population of interest.

D) unwarranted because students who cannot be surveyed are not a part of the sample.

E) justified because people need to be forced to answer a survey in order for the results to be valid.

13. A poll of 500 randomly selected likely voters predicted that Candidate A would get 51% of the vote. The opposition insists that only 47% of the voters support Candidate A. You test their claim using a simulation, based on the assumption that 47% of the likely voters will vote for Candidate A. You assign digit pairs 00–46 to represent a voter who would pick Candidate A and 47–99 to represent a voter who would not. You do 200 trials, recording the proportion of voters in each of your simulated samples that would vote for Candidate A. The resulting sample proportions are shown in the plot below.

Is the opposition's claim that Candidate A has the support of only 47% of the likely voters plausible?

A) No, the average proportion of voters supporting Candidate A in the samples would be less than 51%, so a sample proportion of 51% would be too unexpected under this model.

B) No, because the majority of sample proportions would be less than 51%, so a sample proportion of 51% would be too unexpected under this model.

C) No, because only 29 of the 200 samples had a proportion of 51% or greater supporting Candidate A, so a sample proportion of 51% would be too unexpected under this model.

D) Yes, because 29 of the 200 samples had a proportion of 51% or greater supporting Candidate A, so a sample proportion of 51% would be plausible under this model.

E) Yes, because a sample proportion of 51% occurred more than once, so it's possible to get a sample proportion of 51% under this model.

14. A small airline runs commuter flights with a plane that holds 10 people. Each ticket-holder has a 10% chance of not showing up, so the airline sells 12 tickets for each flight. Which is an appropriate plan for a simulation that uses a table of random digits to estimate the probability that exactly ten people show up for the flight?

A) Let digit pairs 01–12 represent the 12 tickets. In the table, select pairs of digits, ignoring repeats and pairs that do not represent a ticket. Continue until you get 10 seats filled. Record the number of pairs needed to get 10 seats filled.

B) Let digit pairs 01–12 represent the 12 tickets. In the table, select pairs of digits until you find 10 pairs that represent tickets and record the proportion of trials that required ten or fewer.

C) Let digit pairs 01–12 represent a seat that was filled, and other pairs represent a seat that was not filled. In the table, select 10 pairs of digits, ignoring repeats, and record the number of seats that were filled.

D) Let digit 0 represent a seat that was filled, and 1–9 a seat that was not filled. In the table, select 10 digits and record the number of seats that were filled.

E) Let digit 0 represent a ticket-holder that doesn't show up, and 1–9 a ticket-holder who shows up. Select 12 digits and record the number of passengers who show up.

15. A local news program decides to conduct a poll to see how their viewers feel about their new programming format. At the end of each program during one week they ask viewers to call in and express their opinions. The station gives one number to dial if you like the new format, and another number to dial if you do not like it. Which of the following is a correct characterization of this sampling approach?

A) This sampling approach will be biased because people might dial the number incorrectly, and therefore not everyone will be correctly represented.

B) This sampling approach will be biased because the people who call in will be those who have stronger feelings about the new format, and therefore the sample will not be representative of the population.

C) This sampling approach will be unbiased because people can call in with either opinion, therefore the sample will be representative of the population.

D) This sampling approach will be unbiased because the station doesn't know who will call in, making it a random sample.

E) It is impossible to tell whether this sampling approach will be biased or not because the station cannot predict who will call in.

16. A city in the midwestern United States is considering a plan to add roundabout intersections in some high traffic areas to reduce the number of accidents. One city council member used a phone survey to reach his constituents containing the following question: "Many people object to the city's plan to reduce the unnecessary accidents by turning several intersections into those strange European style roundabouts. Do you also object to it?" He called the homes in his district during his lunch break at his regular job, which is between 12:00 and 12:30 PM. Which of the following is a likely source of bias in his survey?

A) Calling between 12:00 and 12:30 limits his sample to people who will be at home during those hours, which leaves out everyone who regularly works during that time of day. This might influence the estimate of the proportion of people who object to the plan.

B) The phrase "plan to reduce unnecessary accidents" might influence people to voice support for the plan, making the estimate of the proportion of constituents who object to the plan too low.

C) The phrases "Many people object" and "Do you also object?" may influence people to say they object, inflating the estimate of the proportion of constituents who object.

D) The phrase "strange European style roundabouts" may influence people to say they object, inflating the estimate of the proportion of constituents who object.

E) All of these are possible sources of bias for this survey.

17. Which of these is a main difference between experiments and observational studies?

A) There is a response variable in an experiment, but not in an observational study.

B) There is at least one explanatory variable in an experiment, but not in an observational study.

C) An experiment requires blocking, while an observational study does not.

D) An experiment can be used to establish a causal relationship, but an observational study cannot.

E) Observational studies require larger samples than experiments.

18. Statistics teachers often debate the best order in which to teach topics. One group of teachers likes to teach design of studies first. Another group likes to begin with data analysis. To see which order is more effective in preparing students for the AP Exam, an experiment was proposed. A large group of teachers, each of whom teaches two sections of statistics, volunteered to be a part of the experiment. Each teacher will randomly assign one of their classes to begin with design, and the other to begin with data analysis. Which is the correct description of this design?

A) The experimental units are the classrooms, the blocks are the teachers, and the response variable is the difference in average AP Exam score for each teacher's classes.

B) The experimental units are the teachers, there are no blocks, and the response variable is the average AP Exam score for each teacher.

C) The experimental units are the individual students, the blocks are the classrooms, and the response variable is each individual student's AP Exam score.

D) The experimental units are the individual students, the blocks are the teachers, and the response variable is the average AP Exam score for each classroom.

E) The experimental units are the orders of topics, the blocks are the teachers, and the response variable is the average AP Exam score for all students who used each order of topics.

19. In one study, researchers at McGill University recruited 127 people with high cholesterol and split them into two groups. One took a probiotic supplement twice a day for nine weeks, while the second group took a placebo. The probiotic group saw their total cholesterol drop by

9 percent and their LDL, or "bad cholesterol," fall almost 12 percent. In a different study, conducted in Britain, 80 volunteers "were given probiotics for six weeks and then switched, later on, to a placebo. That study found no difference in cholesterol levels when the subjects took the supplement versus the dummy pills."
Which of the following is true:

A) The McGill study is a controlled randomized experiment, while the British study is an observational study.

B) The McGill study is a completely randomized experiment, while the British study is a matched pairs experiment.

C) The results from the McGill study are more valid because it includes blinding while the British study does not.

D) We have reason to believe that the design of one of these experiments was flawed since the results are contradictory.

E) Conclusions from the British study cannot be trusted since there is no control group.

(*Sources: "The Claim: Probiotics Can Lower Cholesterol," NY Times, Feb.19, 2013*. Detail on British study available at: http://www.ncbi.nlm.nih.gov/pmc/articles/PMC2904929/)

20. In a Japanese study, "researchers looked at 35 people with lower back pain who were enrolled in an aquatic exercise program, which included swimming and walking in a pool. Almost all of the patients showed improvements after six months, but the researchers found that those who participated at least twice weekly showed more significant improvement that those who went only once per week."
Which of the following statements is true?

A) This experiment proves that swimming causes a reduction in lower back pain.

B) This study is a poorly designed experiment since there is no control group and no randomization.

C) Conclusions based on this observational study are suspect since participation in the aquatic program may be confounded with other lifestyle behaviors that may cause the improvement in lower back health.

D) Conclusions based on this observational study are suspect since a sample size of 35 is too small.

E) This study is not an observational study since the aquatic program is a treatment.

(*Source: "Ask Well: What are the best swimming strokes to alleviate lower back pain?" New York Times, February 19, 2013, D5.*)

Free Response

1. In 2012, the Washington Nationals had the best record in baseball with 98 wins. The Houston Astros had the worst record with 107 losses. To win, a team needs to score runs. Here are the summary statistics for the runs scored for hitters on each team (excluding pitchers and players with fewer than 50 at bats).

Column	n	Mean	Variance	StdDev	StdErr	Med	Range	Min	Max	Q1	Q3
NatRuns	17	40.7058	1021.22	31.9565	7.7506	25	92	6	98	15	72
AstrosRuns	24	23.5833	317.731	17.8250	3.6385	22	78	2	80	9.5	35

a) Construct parallel boxplots for the runs scored. (Note: the two best hitters for the Astros were Jose Altuve with 80 runs and Justin Maxwell with 46 runs.)

b) Compare the distributions of runs scored.

c) Explain why the difference in means for these distributions is so much larger than the difference in the medians.

2. Over the years Olympic racers have been getting faster in most events, and the women's singles 500-meter kayak race is no exception. A scatterplot displaying the data for years since 1948 (x) and time in seconds (y) suggests that a linear model is appropriate. The equation of the least squares regression line is, $\hat{y} = 144.627 - 0.776x$, and $r^2 = 0.932$.

a) Interpret the value of r^2 in this context.

b) Compute and interpret the value of r in context.

c) The Olympics are held every 4 years. What change in the winning time does this model predict from one Olympics to the next?

d) The residual for the winning time in 1980 was -1.795 seconds. Find this gold medal time.

3. A high school administration wants to collect data on the amount of time the students use computers. There are 1200 students in the school, and they have been assigned to 40 different homerooms, 10 homerooms per grade 9–12, with 30 students in each homeroom. The administration wants a sample of 120 students.

a) Describe a method to select a simple random sample of students.

b) Describe a method to select students using a stratified sample.

c) Describe a method to select students using a cluster sample.

d) Describe any advantages or disadvantages in using the stratified or the cluster sampling method here.

4. The online security firm SecurEnvoy and the research firm OnePoll conducted a survey in October 2012 and found that 60% of Britons admit they don't always understand text message abbreviations they receive. The firms surveyed 1000 British adults.

a) Assuming this was a simple random sample, can you be comfortable that 60% is a good estimate of the percentage of British adults who are sometimes confused by text abbreviations? Explain.

b) Assuming this was a simple random sample, can you be comfortable that 60% is a good estimate of the percentage of American adults who are sometimes confused by text abbreviations? Explain.

c) It seems reasonable to suspect that age may be associated with a person's comfort with text abbreviations. How might the sampling technique be improved by taking this association into account?

d) A blog that reported a story about this poll had a banner at the bottom of the webpage with a multiple choice question:

Do you get confused with abbreviations in text messages?

- *Yes. Sometimes the abbreviations do the opposite of what they're supposed to do.*
- *No. I've been texting for a long time. It's my second language*
- *IDK*

(http://tsminteractive.com/have-you-ever-been-confused-by-text-message-abbreviations-poll/)

What type of sampling is the blog using? Will the results of their survey be likely to match those of the original survey? Explain.

chapter 13 From Randomness to Probability

Early humans saw a world filled with random events. To help them make sense of the chaos around them, they sought out seers, consulted oracles, and read tea leaves. As science developed, we learned to recognize some events as predictable. We can now forecast the change of seasons, tell when eclipses will occur precisely, and even make a reasonably good guess at how warm it will be tomorrow. But many other events are still essentially random. Will the stock market go up or down today? When will the next car pass this corner?

But we have also learned to understand randomness. The surprising fact is that in the long run, many truly random phenomena settle down in a way that's consistent and predictable. It's this property of random phenomena that makes the next steps we're about to take in Statistics possible. The previous three chapters showed that randomness plays a critical role in gathering data. That fact alone makes it important to understand how random events behave. From here on, randomness will be fundamental to how we think about data.

Random Phenomena

Every day you drive through the intersection at College and Main. Even though it may seem that the light is never green when you get there, you know this can't really be true. In fact, if you try really hard, you can recall just sailing through the green light once in a while.

What's random here? The light itself is governed by a timer. Its pattern isn't haphazard. In fact, the light may even be red at precisely the same times each day. It's the time you arrive at the light that is *random*. It isn't that your driving is erratic. Even if you try to leave your house at exactly the same time every day, whether the light is red or green as *you* reach the intersection is a random phenomenon.[1] For us, a **random phenomenon** is a situation

---

[1]If you somehow managed to leave your house at *precisely* the same time every day and there was *no* variation in the time it took you to get to the light, then there wouldn't be any randomness, but that's not very realistic.

in which we know what outcomes can possibly occur, but we don't know which particular outcome will happen. Even though the color of the light is random[2] as you approach it, some *fraction* of the time, the light will be green. How can you figure out what that fraction is?

You might record what happens at the intersection each day and graph the *accumulated percentage* of green lights like this:

Figure 13.1

The accumulated percentage of times the light is green settles down as you see more outcomes.

The first day you recorded the light, it was green. Then on the next five days, it was red, then green again, then green, red, and red. When you plot the percentage of green lights against days, the graph starts at 100% (because the first time, the light was green, so 1 out of 1, for 100%). Then the next day it was red, so the accumulated percentage drops to 50% (1 out of 2). The third day it was green again (2 out of 3, or 67% green), then green (3 out of 4, or 75%), then red twice in a row (3 out of 5, for 60% green, and then 3 out of 6, for 50%), and so on. As you collect a new data value for each day, each new outcome becomes a smaller and smaller fraction of the accumulated experience, so, in the long run, the graph settles down. As it settles down, you can see that, in fact, the light is green about 35% of the time. And here's the really cool thing. No matter when you start recording the data or what the erratic start looks like (perhaps very different from these first few days plotted above), in the long run the percentage of green lights will settle down to this very same value. We'll explore that phenomenon in this chapter's What If.

In general, each occasion upon which we observe a random phenomenon is called a **trial**. At each trial, we note the value of the random phenomenon, and call that the trial's **outcome**. (If this language reminds you of Chapter 10, that's *not* unintentional.)

For the traffic light, there are really three possible outcomes: red, yellow, or green. Often we're more interested in a combination of outcomes rather than in the individual ones. When you see the light turn yellow, what do *you* do? If you race through the intersection, then you treat the yellow more like a green light. If you step on the brakes, you treat it more like a red light. Either way, you might want to group the yellow with one or the other. When we combine outcomes like that, the resulting combination is an **event**.[3] We call the collection of *all possible outcomes* the **sample space**.[4] We'll denote the sample space **S**. (Some books are even fancier and use the Greek letter Ω.) For the traffic light, S = {red, green, yellow}. If you flip a coin once, the sample space is very simple: S = {H, T}. If you flip two coins, it's more complicated because now there are four outcomes, so S = {HH, HT, TH, TT}. If *ABC News* takes a sample of 1023 randomly chosen U.S. adults for a poll, the sample space is incomprehensibly enormous

Day	Light	% Green
1	Green	100
2	Red	50
3	Green	66.7
4	Green	75
5	Red	60
6	Red	50
⋮	⋮	⋮

Trials, Outcomes, and Events

A phenomenon consists of trials. Each trial has an outcome. Outcomes combine to make events.

The Sample Space

For a random phenomenon, the sample space, S, is the set of all possible outcomes of each trial.

---

[2]Even though the randomness here comes from the uncertainty in our arrival time, we can think of the light itself as showing a color at random.

[3]Each individual outcome is also an event.

[4]Mathematicians like to use the term "space" as a fancy name for a set. Sort of like referring to that closet colleges call a dorm room as "living space." But it's really just the set of all outcomes.

because it would list every combination of 1023 adults you could take from the approximately 250 million adults in the United States.

The Law of Large Numbers

"For even the most stupid of men . . . is convinced that the more observations have been made, the less danger there is of wandering from one's goal."

—Jacob Bernoulli, 1713, discoverer of the LLN

What's the *probability* of a green light at College and Main? Based on the graph, it looks like the relative frequency of green lights settles down to about 35%, so saying that the probability is about 0.35 seems like a reasonable answer. But do random phenomena always behave well enough for this to make sense? Might the relative frequency of an event bounce back and forth between two values forever, never settling on just one number?

Fortunately, a principle called the **Law of Large Numbers** (LLN) gives us the guarantee we need. The LLN says that as we repeat a random process over and over, the proportion of times that an event occurs does settle down to one number. We call this number the **probability** of the event. But the law of large numbers requires two key assumptions. First, the random phenomenon we're studying must not change—the outcomes must have the same probabilities for each trial. And, the events must be **independent**.[5] Informally, independence means that the outcome of one trial doesn't affect the outcomes of the others. (We'll see a formal definition of independent events in the next chapter.) The LLN says that as the number of independent trials increases, the long-run *relative frequency* of repeated events gets closer and closer to a single value. We call that the **probability** of the event. If the relative frequency of green lights at that intersection settles down to 35% in the long run, we say that the probability of encountering a green light is 0.35, and we write $P(\text{green}) = 0.35$.

Although he could have said it much more gently, Bernoulli is trying to tell us how intuitive the LLN actually is, even though it wasn't formally proved until the 18th century. Most of us would guess that the law is true from our everyday experiences.

The Nonexistent Law of Averages

Probability

For any event A,

$$P(A) = \frac{\#\text{times A occurs}}{\text{total }\#\text{ of trials}}$$

in the long run.

Even though the LLN seems natural, it is often misunderstood because the idea of the *long run* is hard to grasp. Many people believe, for example, that an outcome of a random event that hasn't occurred in many trials is "due" to occur. Many gamblers bet on numbers that haven't been seen for a while, mistakenly believing that they're likely to come up sooner. A common term for this is the "Law of Averages." After all, we know that in the long run, the relative frequency will settle down to the probability of that outcome, so now we have some "catching up" to do, right?

Wrong. The Law of Large Numbers says nothing about short-run behavior. Relative frequencies even out *only in the long run.* And, according to the LLN, the long run is *really* long (*infinitely* long, in fact).

The so-called Law of Averages doesn't exist at all. But you'll hear people talk about it as if it does. Is a good hitter in baseball who has struck out the last six times *due* for a hit his next time up? If you've been doing particularly well in weekly quizzes in Statistics class, are you *due* for a bad grade? No. This isn't the way random phenomena work. There is *no* Law of Averages for short runs.

"Slump? I ain't in no slump. I just ain't hittin'."

—Yogi Berra

The Law of Averages

Don't let yourself think that there's a Law of Averages that promises short-term compensation for recent deviations from expected behavior. A belief in such a "Law" can lead to money lost in gambling and to poor business decisions.

---

[5]There are stronger forms of the Law that don't require independence, but for our purposes, this form is general enough.

For Example COINS AND THE LAW OF AVERAGES

You've just flipped a fair coin and seen six heads in a row.

QUESTION: Does the coin "owe" you some tails? Suppose you spend that coin and your friend gets it in change. When she starts flipping the coin, should she expect a run of tails?

ANSWER: *Of course not. Each flip is a new event. The coin can't "remember" what it did in the past, so it can't "owe" any particular outcomes in the future.*

Just to see how this works in practice, the authors ran a simulation of 100,000 flips of a fair coin. We collected 100,000 random numbers, letting the numbers 0 to 4 represent heads and the numbers 5 to 9 represent tails. In our 100,000 "flips," there were 2981 streaks of at least 5 heads. The "Law of Averages" suggests that the next flip after a run of 5 heads should be tails more often to even things out. Actually, the next flip was heads more often than tails: 1550 times to 1431 times. That's 51.9% heads. You can perform a similar simulation easily on a computer. Try it!

Of course, sometimes an apparent drift from what we expect means that the probabilities are, in fact, *not* what we thought. If you get 10 heads in a row, maybe the coin has heads on both sides!

The Law of Averages in Everyday Life The advice columnist Abigail Van Buren (Dear Abby) once published a letter from a woman lamenting the fact that her newborn child was another girl – the couple's eighth! While happy that her new daughter was healthy, the woman was really disappointed. Her doctor had told the couple to expect a boy this time, because the Law of Averages was 100 to 1 on their side.

The lesson of the LLN is that sequences of random events don't compensate in the *short* run and don't need to do so to get back to the right long-run probability. If the probability of an outcome doesn't change and the events are independent, the probability of any outcome in another trial is *always* what it was, no matter what has happened in other trials.

Beat the Casino Keno is a simple casino game in which numbers from 1 to 80 are chosen. The numbers, as in most lottery games, are supposed to be equally likely. Payoffs are made depending on how many of those numbers you match on your card. A group of graduate students from a Statistics department decided to take a field trip to Reno. They (*very* discreetly) wrote down the outcomes of the games for a couple of days, then drove back to test whether the numbers were, in fact, equally likely. It turned out that some numbers were *more likely* to come up than others. Rather than bet on the Law of Averages and put their money on the numbers that were "due," the students put their faith in the LLN—and all their (and their friends') money on the numbers that had come up before. After they pocketed more than $50,000, they were escorted off the premises and invited never to show their faces in that casino again.

Just Checking

1. One common proposal for beating the lottery is to note which numbers have come up lately, eliminate those from consideration, and bet on numbers that have not come up for a long time. Proponents of this method argue that in the long run, every number should be selected equally often, so those that haven't come up are due. Explain why this is faulty reasoning.

Modeling Probability

■ NOTATION ALERT

We often use capital letters—and usually from the beginning of the alphabet—to denote events. We *always* use *P* to denote probability. So,

$$P(\mathrm{A}) = 0.35$$

means "the probability of the event A is 0.35."

When being formal, use decimals (or fractions) for the probability values, but sometimes, especially when talking more informally, it's easier to use percentages.

Probability was first studied extensively by a group of French mathematicians who were interested in games of chance.[6] Rather than *experiment* with the games (and risk losing their money), they developed mathematical models. When the probability comes from a mathematical model and not from observation, it is called **theoretical probability**. To make things simple (as we usually do when we build models), they started by looking at games in which the different outcomes were equally likely. Fortunately, many games of chance are like that. Any of 52 cards is equally likely to be the next one dealt from a well-shuffled deck. Each face of a die is equally likely to land up (or at least it *should be*).

It's easy to find probabilities for events that are made up of several *equally likely* outcomes. We just count all the outcomes that the event contains. The probability of the event is the number of outcomes in the event divided by the total number of possible outcomes. We can write

$$P(\mathrm{A}) = \frac{\text{\# outcomes in A}}{\text{\# of possible outcomes}}.$$

For example, the probability of drawing a face card (JQK) from a deck is

$$P(\text{face card}) = \frac{\text{\# face cards}}{\text{\# cards}} = \frac{12}{52} = \frac{3}{13}.$$

How Hard Can Counting Be? Finding the probability of any event when the outcomes are equally likely is straightforward, but not necessarily easy. It gets hard when the number of outcomes in the event (and in the sample space) gets big. Think about flipping two coins. The sample space is $\mathbf{S} = \{\mathrm{HH, HT, TH, TT}\}$ and each outcome is equally likely. So, what's the probability of getting exactly one head and one tail? Let's call that event **A**. Well, there are two outcomes in the event $\mathbf{A} = \{\mathrm{HT, TH}\}$ out of the 4 possible equally likely ones in S, so $P(\mathbf{A}) = \frac{2}{4}$, or $\frac{1}{2}$.

OK, now flip 100 coins. What's the probability of exactly 67 heads? Well, first, how many outcomes are in the sample space? $\mathbf{S} = \{\mathrm{HHHHHHHHHHH \ldots H, HH \ldots T, \ldots}\}$ Um . . . . a lot. In fact, there are 1,267,650,600,228,229,401,496,703,205,376 different outcomes possible when flipping 100 coins. And that's just the denominator of the probability! To answer the question, we'd still have to figure out how many ways there are to get 67 heads. We'll see how in Chapter 16; stay tuned!

Don't get trapped into thinking that random events are always equally likely. The chance of winning a lottery—especially lotteries with very large payoffs—is small. Regardless, people continue to buy tickets. In an attempt to understand why, an interviewer asked someone who had just purchased a lottery ticket, "What do you think your chances are of winning the lottery?" The reply was, "Oh, about 50–50." The shocked interviewer asked, "How do you get that?" to which the response was, "Well, the way I figure it, either I win or I don't!"

The moral of this story is that events are *not* always equally likely.

Personal Probability

What's the probability that your grade in this Statistics course will be an A? You may be able to come up with a number that seems reasonable. Of course, no matter how confident or depressed you feel about your chance of success, your probability should be between 0 and 1. How did you come up with this probability? It can't truly be a probability. For that,

---

[6]Ok, gambling.

you'd have to take the course over and over (and over . . .), and forget everything after each time so the probability of getting an A would stay the same. But people use the word "probability" informally as well.

We use the language of probability in everyday speech to express a degree of uncertainty *without* basing it on long-run relative frequencies or mathematical models. Your personal assessment of your chances of getting an A expresses your uncertainty about the outcome. That uncertainty may be based on how comfortable you're feeling in the course or on your midterm grade, but it can't be based on long-run behavior. We call this informal kind of probability a **subjective** or **personal probability**.

Although personal probabilities may be based on experience, they're not based either on long-run relative frequencies or on equally likely events. So they don't display the kind of consistency that we'll need probabilities to have. For that reason, in Statistics we stick to formally defined probabilities. You should be alert to the difference.

> **Which Kind of Probability?** The line between personal probability and the other two probabilities can be a fuzzy one. When a weather forecaster predicts a 40% probability of rain, is this a personal probability or a relative frequency probability? The National Weather Service bases such claims on decades worth of data showing that when conditions have looked like this it has rained 40% of the time. A local forecaster, though, may be offering a personal opinion based on past experience in your area and a sense of what may happen today. When you hear a probability stated, try to ascertain what kind of probability is intended.

The First Three Rules for Working with Probability

> John Venn (1834–1923) created the Venn diagram. His book on probability, *The Logic of Chance*, was "strikingly original and considerably influenced the development of the theory of Statistics," according to John Maynard Keynes, one of the luminaries of Economics.

1. Make a picture.
2. Make a picture.
3. Make a picture.

We're dealing with probabilities now, not data, but the three rules don't change. The most common kind of picture to make is called a Venn diagram. We'll use Venn diagrams throughout the rest of this chapter. Even experienced statisticians make Venn diagrams to help them think about probabilities of compound and overlapping events. You should, too.

Formal Probability

For some people, the phrase "50/50" means something vague like "I don't know" or "whatever." But when we discuss probabilities of outcomes, it takes on the precise meaning of *equally likely*. Speaking vaguely about probabilities will get us into trouble, so whenever we talk about probabilities, we'll need to be precise.[7] And to do that, we'll need to develop some formal rules[8] about how probability works.

Rule 1. If the probability is 0, the event *never* occurs, and likewise if it has probability 1, it *always* occurs. Even if you think an event is very unlikely, its probability can't be negative, and even if you're sure it will happen, its probability can't be greater than 1. (Think about relative frequencies.) So we require that

A probability is a number between 0 and 1.
For any event A, $0 \leq P(A) \leq 1$.

---

[7]And to be precise, we will be talking only about sample spaces where we can enumerate all the outcomes. Mathematicians call this a countable number of outcomes.

[8]Actually, in mathematical terms, these are axioms—statements that we assume to be true of probability. We'll derive other rules from these in the next chapter.

> **Surprising Probabilities** We've been careful to discuss probabilities only for situations in which the outcomes were finite, or even countably infinite. But if the outcomes can take on *any* numerical value at all (we say they are *continuous*), things can get surprising. For example, what is the probability that a randomly selected child will be *exactly* 3 feet tall? Well, if we mean 3.00000 . . . feet, the answer is zero. No randomly selected child—even one whose height would be recorded as 3 feet, will be *exactly* 3 feet tall (to an infinite number of decimal places). But, if you've grown taller than 3 feet, there must have been a time in your life when you actually *were* exactly 3 feet tall, even if only for a second. So this is an outcome with probability 0 that not only has happened—it has happened to *you*.

Rule 2. If a random phenomenon has only one possible outcome, it's not very interesting (or very random). So we need to distribute the probabilities among all the outcomes a trial can have. How can we do that so that it makes sense? For example, consider what you're doing as you read this book. The possible outcomes might be

A: You read to the end of this chapter before stopping.
B: You finish this section but stop reading before the end of the chapter.
C: You bail out before the end of this section.

When we assign probabilities to these outcomes, the first thing to be sure of is that we distribute all of the available probability. Something always occurs, so the probability of the entire sample space is 1.

Making this more formal gives the **Probability Assignment Rule.**

**The set of all possible outcomes of a trial
must have probability 1.**

$$P(\mathbf{S}) = 1$$

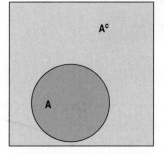

The set **A** and its complement **A**^{C}. Together, they make up the entire sample space **S**.

Rule 3. Suppose the probability that you get to class on time is 0.8. What's the probability that you don't get to class on time? Yes, it's 0.2. The set of outcomes that are *not* in the event **A** is called the **complement** of **A,** and is denoted **A**^{C}. This leads to the **Complement Rule:**

**The probability of an event occurring
is 1 minus the probability that it doesn't occur.**

$$P(\mathbf{A}) = 1 - P(\mathbf{A}^C)$$

For Example APPLYING THE COMPLEMENT RULE

RECAP: We opened the chapter by looking at the traffic light at the corner of College and Main, observing that when we arrive at that intersection, the light is green about 35% of the time.

QUESTION: If $P(\text{green}) = 0.35$, what's the probability the light isn't green when you get to College and Main?

ANSWER: "Not green" is the complement of "green," so $P(\text{not green}) = 1 - P(\text{green})$
$$= 1 - 0.35 = 0.65$$

There's a 65% chance I won't have a green light.

Rule 4. Suppose the probability that (**A**) a randomly selected student is a sophomore is 0.20, and the probability that (**B**) he or she is a junior is 0.30. What is the probability that the student is *either* a sophomore *or* a junior, written $P(\mathbf{A} \cup \mathbf{B})$? If you guessed 0.50, you've deduced the Addition Rule, which says that you can add the probabilities of events that are disjoint. To see whether two events are

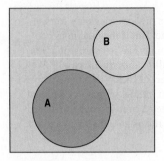

Two disjoint sets, **A** and **B**.

disjoint, we think about whether both can occur at the same time. **Disjoint** (or **mutually exclusive**) events have no outcomes in common. The **Addition Rule** states,

> **For two disjoint events A and B, the probability that one *or* the other occurs is the sum of the probabilities of the two events.**
>
> $P(A \cup B) = P(A) + P(B)$, **provided that A and B are disjoint.**

For Example APPLYING THE ADDITION RULE

RECAP: When you get to the light at College and Main, it's either red, green, or yellow. We know that $P(\text{green}) = 0.35$.

QUESTION: Suppose we find out that $P(\text{yellow})$ is about 0.04. What's the probability the light is red?

ANSWER: To find the probability that the light is green or yellow, I can use the Addition Rule because these are disjoint events: The light can't be both green and yellow at the same time.

$$P(\text{green} \cup \text{yellow}) = 0.35 + 0.04 = 0.39$$

Red is the only remaining alternative, and the probabilities must add up to 1, so

$$P(\text{red}) = P(\text{not }(\text{green} \cup \text{yellow}))$$
$$= 1 - P(\text{green} \cup \text{yellow})$$
$$= 1 - 0.39 = 0.61$$

A S *Activity*: **Addition Rule for Disjoint Events.** Experiment with disjoint events to explore the Addition Rule.

"Baseball is 90% mental. The other half is physical."

—*Yogi Berra*

■ **NOTATION ALERT**

We write $P(A \text{ or } B)$ as $P(A \cup B)$. The symbol $\cup$ means "union," representing the outcomes in event **A** *or* event **B** (or both). The symbol $\cap$ means "intersection," representing outcomes that are in both event **A** *and* event **B**. We write $P(A \text{ and } B)$ as $P(A \cap B)$.

The Addition Rule can be extended to any number of disjoint events, and that's helpful for checking probability assignments. Because individual sample space outcomes are always disjoint, we have an easy way to check whether the probabilities we've assigned to the possible outcomes are legitimate. The Probability Assignment Rule tells us that to be a **legitimate assignment of probabilities**, the sum of the probabilities of all possible outcomes must be exactly 1. No more, no less. For example, if we were told that the probabilities of selecting at random a freshman, sophomore, junior, or senior from all the undergraduates at a school were 0.25, 0.23, 0.22, and 0.20, respectively, we would know that something was wrong. These "probabilities" sum to only 0.90, so this is not a legitimate probability assignment. Either a value is wrong, or we just missed some possible outcomes, like "pre-freshman" or "postgraduate" categories that soak up the remaining 0.10. Similarly, a claim that the probabilities were 0.26, 0.27, 0.29, and 0.30 would be wrong because these "probabilities" sum to more than 1.

But be careful: The Addition Rule doesn't work for events that aren't disjoint. If the probability of owning an MP3 player is 0.50 and the probability of owning a computer is 0.90, the probability of owning either an MP3 player or a computer may be pretty high, but it is *not* 1.40! Why can't you add probabilities like this? Because these events are not disjoint. You *can* own both. In the next chapter, we'll see how to add probabilities for events like these, but we'll need another rule.

Rule 5. Suppose your job requires you to fly from Atlanta to Houston every Monday morning. The airline's website reports that this flight is on time 85% of the time. What's the chance that it will be on time two weeks in a row? That's the same as asking for the probability that your flight is on time this week *and* it's on time again next

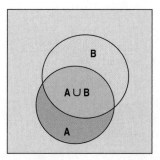

Two sets **A** and **B** that are not disjoint. The event (**A ∩ B**) is their intersection.

A S *Activity*: **Multiplication Rule for Independent Events.** Experiment with independent random events to explore the Multiplication Rule.

A S *Activity*: **Probabilities of Compound Events.** The Random tool also lets you experiment with Compound random events to see if they are independent.

week. For independent events, the answer is very simple. Remember that independence means that the outcome of one event doesn't influence the outcome of the other. What happens with your flight this week doesn't influence whether it will be on time next week, so it's reasonable to assume that those events are independent. The **Multiplication Rule** says that for independent events, to find the probability that both events occur, we just multiply the probabilities together. Formally,

> **For two independent events A and B, the probability that both A *and* B occur is the product of the probabilities of the two events.**
>
> $$P(A \cap B) = P(A) \times P(B), \text{provided that}$$
> **A and B are independent.**

This rule can be extended to more than two independent events. What's the chance of your flight being on time for a month—four Mondays in a row? We can multiply the probabilities of it happening each week:

$$0.85 \times 0.85 \times 0.85 \times 0.85 = 0.522$$

or just over 50–50. Of course, to calculate this probability, we have used the assumption that the four events are independent.

Many Statistics methods require an **Independence Assumption,** but *assuming* independence doesn't make it true. Always *Think* about whether that assumption is reasonable before using the Multiplication Rule.

For Example APPLYING THE MULTIPLICATION RULE (AND OTHERS)

RECAP: We've determined that the probability that we encounter a green light at the corner of College and Main is 0.35, a yellow light 0.04, and a red light 0.61. Let's think about your morning commute in the week ahead.

QUESTION: What's the probability you find the light red both Monday and Tuesday?

ANSWER: *Because the color of the light I see on Monday doesn't influence the color I'll see on Tuesday, these are independent events; I can use the Multiplication Rule:*

$$P(\text{red Monday} \cap \text{red Tuesday}) = P(\text{Red}) \times P(\text{red})$$
$$= (0.61)(0.61)$$
$$= 0.3721$$

There's about a 37% chance I'll hit red lights both Monday and Tuesday mornings.

QUESTION: What's the probability you don't encounter a red light until Wednesday?

ANSWER: *For that to happen, I'd have to see green or yellow on Monday, green or yellow on Tuesday, and then red on Wednesday. I can simplify this by thinking of it as not red on Monday and Tuesday and then red on Wednesday.*

$$P(\text{not red}) = 1 - P(\text{red}) = 1 - 0.61 = 0.39, \text{so}$$

$$P(\text{not red Monday} \cap \text{not red Tuesday} \cap \text{red Wednesday}) = P(\text{not red}) \times P(\text{not red}) \times P(\text{red})$$
$$= (0.39)(0.39)(0.61)$$
$$= 0.092781$$

There's about a 9% chance that this week I'll hit my first red light there on Wednesday morning.

QUESTION: What's the probability that you'll have to stop *at least once* during the week?

ANSWER: *Having to stop at least once means that I have to stop for the light either 1, 2, 3, 4, or 5 times next week. It's easier to think about the complement: never having to stop at a red light. Having to stop at least once means that I didn't make it through the week with no red lights.*

(continued)

P(having to stop at the light at least once in 5 days)

$$= 1 - P(\text{no red lights for 5 days in a row})$$
$$= 1 - P(\text{not red} \cap \text{not red} \cap \text{not red} \cap \text{not red} \cap \text{not red})$$
$$= 1 - (0.39)(0.39)(0.39)(0.39)(0.39)$$
$$= 1 - 0.0090$$
$$= 0.991$$

There's over a 99% chance I'll hit at least one red light sometime this week.

> **At Least**
> Note that the phrase "at least" is often a tip-off to think about the complement. Something that happens *at least once* does happen. Happening at least once is the complement of not happening at all, and that's easier to find.

> **Some: At Least One** In informal English, you may see "some" used to mean "at least one." "What's the probability that some of the eggs in that carton are broken?" means at least one.

Just Checking

2. Opinion polling organizations contact their respondents by telephone. Random telephone numbers are generated, and interviewers try to contact those households. In the 1990s this method could reach about 69% of U.S. households. According to the Pew Research Center for the People and the Press, by 2003 the contact rate had risen to 76%. We can reasonably assume each household's response to be independent of the others. What's the probability that . . .

a) the interviewer successfully contacts the next household on her list?

b) the interviewer successfully contacts both of the next two households on her list?

c) the interviewer's first successful contact is the third household on the list?

d) the interviewer makes at least one successful contact among the next five households on the list?

Step-by-Step Example PROBABILITY

The five rules we've seen can be used in a number of different combinations to answer a surprising number of questions. Let's try one to see how we might go about it.

M&M's® Milk Chocolate candies now come in 7 colors, but they've changed over time. In 1995, Americans voted to change tan M&M's (which had replaced violet in 1949) to blue. In 2002, Mars™, the parent company of M&M's, used the Internet to solicit global opinion for a seventh color. To decide which color to add, Mars surveyed kids in nearly every country of the world and asked them to vote among purple, pink, and teal. The global winner was purple!

In the United States, 42% of those who voted said purple, 37% said teal, and only 19% said pink. But in Japan the percentages were 38% pink, 36% teal, and only 16% purple. Let's use Japan's percentages to ask some questions:

1. What's the probability that a Japanese M&M's survey respondent selected at random preferred either pink or teal?
2. If we pick two respondents at random, what's the probability that they both selected purple?
3. If we pick three respondents at random, what's the probability that *at least one* preferred purple?

(continued)

THINK ➡️ The probability of an event is its long-term relative frequency. It can be determined in several ways: by looking at many replications of an event, by deducing it from equally likely events, or by using some other information. Here, we are told the relative frequencies of the three responses.

The M&M's Website reports the proportions of Japanese votes by color. These give the probability of selecting a voter who preferred each of the colors:

$$P(\text{pink}) = 0.38$$
$$P(\text{teal}) = 0.36$$
$$P(\text{purple}) = 0.16$$

Make sure the probabilities are legitimate. Here, they're not. Either there was a mistake, or the other voters must have chosen a color other than the three given. A check of the reports from other countries shows a similar deficit, so probably we're seeing those who had no preference or who wrote in another color.

Each is between 0 and 1, but they don't all add up to 1. The remaining 10% of the voters must have not expressed a preference or written in another color. I'll put them together into "no preference" and add $P(\text{no preference}) = 0.10$.

Now, I have a legitimate assignment of probabilities.

Question 1: What's the probability that a Japanese M&M's survey respondent selected at random preferred either pink or teal?

THINK ➡️ **Plan** Decide which rules to use and check the conditions they require.

The events "Pink" and "Teal" are individual outcomes (a respondent can't choose both colors), so they are disjoint. I can apply the Addition Rule.

SHOW ➡️ **Mechanics** Show your work.

$$P(\text{pink} \cup \text{teal}) = P(\text{pink}) + P(\text{teal})$$
$$= 0.38 + 0.36 = 0.74$$

TELL ➡️ **Conclusion** Interpret your results in the proper context.

The probability that the respondent said pink or teal is 0.74.

Question 2: If we pick two respondents at random, what's the probability that they both said purple?

THINK ➡️ **Plan** The word "both" suggests we want $P(\textbf{A} \text{ and } \textbf{B})$, which calls for the Multiplication Rule. Think about the assumption.

✓ **Independence Assumption:** It's unlikely that the choice made by one random respondent affected the choice of the other, so the events seem to be independent. I can use the Multiplication Rule.

SHOW ➡️ **Mechanics** Show your work.
For both respondents to pick purple, each one has to pick purple.

$P(\text{both purple})$

$$= P(\text{first purple} \cap \text{second purple})$$
$$= P(\text{first purple}) \times P(\text{second purple})$$
$$= 0.16 \times 0.16 = 0.0256$$

(continued)

TELL ➡ **Conclusion**   Interpret your results in the proper context.

The probability that both respondents pick purple is 0.0256.

Question 3:   If we pick three respondents at random, what's the probability that at least one preferred purple?

THINK ➡ **Plan**   The phrase "at least . . ." often flags a question best answered by looking at the complement, and that's the best approach here. The complement of "At least one preferred purple" is "None of them preferred purple."

$P(\text{at least one purple})$
$= P(\{\text{none purple}\}^C)$
$= 1 - P(\text{none purple}).$
$= 1 - P(\text{not purple} \cap \text{not purple} \cap \text{not purple}).$

Think about the assumption.

✓ **Independence Assumption:** These are independent events because they are choices by three random respondents. I can use the Multiplication Rule.

SHOW ➡ **Mechanics**   First we find $P(\text{not purple})$ with the Complement Rule.

$P(\text{not purple}) = 1 - P(\text{purple})$
$= 1 - 0.16 = 0.84$

Next we calculate $P(\text{none picked purple})$ by using the Multiplication Rule.

$P(\text{at least one picked purple})$
$= 1 - P(\text{none purple})$
$= 1 - P(\text{not purple} \cap \text{not purple} \cap \text{not purple})$
$= 1 - (0.84)(0.84)(0.84)$

Then we can use the Complement Rule to get the probability we want.

$= 1 - 0.5927$
$= 0.4073$

TELL ➡ **Conclusion**   Interpret your results in the proper context.

There's about a 40.7% chance that at least one of the respondents picked purple.

WHAT IF ●●● we test the Law of Large Numbers?

When you collected data on that traffic light at the corner of College and Main, over the first few days the relative frequency of green lights bounced around a lot. After a while, though, it seemed to settle down to around 35%. (If you don't remember what that timeplot looked like, take another peek at Fig 13.1 on page 344.) But that only shows what happened for this one driver during this particular sequence of days. If other people replicated this investigation, what conclusions might they have reached?

(continued)

Our simulation imagines 5 different drivers going through this intersection independently. Since each arrives there at different times, some will hit the light when it's red, others when it's green. Here's a graph summarizing their commuting experiences for the first 5 days:

At the end of this week, these 5 drivers would form very different perceptions about the behavior of the traffic light. The man in the blue car thinks it's green most of the time. The people in the green and purple cars have hit a green light just over half the time. The lady in the red car only got one green light, while the guy in the yellow car saw red lights all week long. He's thinking, "I'm due for a green one."

At the end of the month, they've each commuted through the intersection 20 times.

That guy who was thinking that after 5 reds in a row he'd surely see a green light soon, well, it didn't happen until Day 13! The other 4 drivers, though, are starting to form the impression that the light is green 30–40% of the time.

We continued this simulation to model 2 years of daily commuting—that's 500 days.

And now we see that, despite the very different experiences the drivers had for the first several weeks, over time all 5 of them are seeing green lights on about 35% of the days. Even that guy who began with 13 red lights in a row. If you're thinking the "Law of Averages" kicked in and he got a few extra greens to make up for that bad start, sorry. From then on he hit green lights on only 32.4% of the remaining days, raising his overall rate to 31.6%. Like all the other drivers, he's *approaching* 35%.

That's how the Law of Large Numbers works. Even after 500 days the percentages are not all the same. But, remember, the Law of Large Numbers is not about what happens in 5 days, or 20 days, or even 500 days. It's about what happens in the long run. *The very long run.*

WHAT CAN GO WRONG?

■ **Beware of probabilities that don't add up to 1.** To be a legitimate probability assignment, the sum of the probabilities for all possible outcomes must total 1. If the sum is less than 1, you may need to add another category ("other") and assign the remaining probability to that outcome. If the sum is more than 1, check that the outcomes are disjoint. If they're not, then you can't assign probabilities by just counting relative frequencies.

■ **Don't add probabilities of events if they're not disjoint.** Events must be disjoint to use the Addition Rule. The probability of being under 80 *or* a female is not the probability of being under 80 *plus* the probability of being female. That sum may be more than 1.

■ **Don't multiply probabilities of events if they're not independent.** The probability of selecting a student at random who is over 6′10″ tall *and* on the basketball team is *not* the probability the student is over 6′10″ tall *times* the probability he's on the basketball team. Knowing that the student is over 6′10″ changes the probability of his being on the basketball team. You can't multiply these probabilities. The multiplication of probabilities of events that are not independent is one of the most common errors people make in dealing with probabilities.

■ **Don't confuse disjoint and independent.** Disjoint events *can't* be independent. If **A** = {you get an A in this class} and **B** = {you get a B in this class}, **A** and **B** are disjoint. Are they independent? If you find out that **A** is true, does that change the probability of **B?** You bet it does! So they can't be independent. We'll return to this issue in the next chapter.

What Have We Learned?

We've learned that the probability of an event is its long-run frequency of occurrence. We understand that the Law of Large Numbers speaks only of long-run (very long) behavior. We've learned not to fall victim to the short-run false reasoning called the "Law of Averages."

We've learned some basic probability rules, and how to apply them.

■ A probability is a number between 0 and 1.

■ The sum of the probabilities for all outcomes must be 1.

■ The **Complement Rule** says that $P(\text{not } A) = P(A^C) = 1 - P(A)$.

■ The **Addition Rule** says that $P(A \cup B) = P(A) + P(B)$, provided events **A** and **B** are disjoint.

■ The **Multiplication Rule** says that $P(A \cap B) = P(A) \cdot P(B)$, provided events **A** and **B** are independent.

Terms

Random phenomenon   A phenomenon is random if we know what outcomes could happen, but not which particular values will happen. (p. 343)

Trial   A single attempt or realization of a random phenomenon. (p. 344)

Outcome   The outcome of a trial is the value measured, observed, or reported for an individual instance of that trial. (p. 344)

Event   A collection of outcomes. Usually, we identify events so that we can attach probabilities to them. We denote events with bold capital letters such as **A, B,** or **C**. (p. 344)

Sample Space   The collection of all possible outcome values. The sample space has a probability of 1. (p. 344)

Law of Large Numbers	The Law of Large Numbers states that the long-run *relative frequency* of repeated independent events gets closer and closer to the *true* relative frequency as the number of trials increases. (p. 345)
Probability	The probability of an event is a number between 0 and 1 that reports the long-run frequency of that event's occurrence. We write $P(\textbf{A})$ for the probability of the event **A**. (p. 345)
Independence (informally)	Two events are *independent* if learning that one event occurs does not change the probability that the other event occurs. (p. 345)
Theoretical probability	When a probability is based on a model (such as equally likely outcomes), it is called a theoretical probability. (p. 347)
Personal probability	When a probability is subjective and represents your personal degree of belief, it is called a personal probability. (p. 348)
The Probability Assignment Rule	The probability of the entire sample space must be 1. $P(\textbf{S}) = 1$. (p. 349)
Complement Rule	The probability of an event occurring is 1 minus the probability that it doesn't occur. (p. 349) $$P(\textbf{A}) = 1 - P(\textbf{A}^C)$$
Disjoint (Mutually exclusive)	Two events are disjoint if they share no outcomes in common. If **A** and **B** are disjoint, then knowing that **A** occurs tells us that **B** cannot occur. Disjoint events are also called "mutually exclusive". (p. 350)
Addition Rule	If **A** and **B** are disjoint events, then the probability of **A** *or* **B** is $$P(\textbf{A} \cup \textbf{B}) = P(\textbf{A}) + P(\textbf{B}). \quad \text{(p. 350)}$$
Legitimate probability assignment	An assignment of probabilities to outcomes is legitimate if ■ each probability is between 0 and 1 (inclusive). ■ the sum of the probabilities is 1. (p. 350)
Multiplication Rule	If **A** and **B** are independent events, then the probability of **A** *and* **B** is $$P(\textbf{A} \cap \textbf{B}) = P(\textbf{A}) \times P(\textbf{B}). \quad \text{(p. 351)}$$
Independence Assumption	We often require events to be independent. (So you should think about whether this assumption is reasonable). (p. 351)

Exercises

1. **Sample spaces** For each of the following, list the sample space and tell whether you think the events are equally likely:
   a) Toss 2 coins; record the order of heads and tails.
   b) A family has 3 children; record the number of boys.
   c) Flip a coin until you get a head or 3 consecutive tails; record each flip.
   d) Roll two dice; record the larger number.

2. **Sample spaces** For each of the following, list the sample space and tell whether you think the events are equally likely:
   a) Roll two dice; record the sum of the numbers.
   b) A family has 3 children; record each child's sex in order of birth.

   c) Toss four coins; record the number of tails.
   d) Toss a coin 10 times; record the length of the longest run of heads.

3. **Roulette** A casino claims that its roulette wheel is truly random. What should that claim mean?

4. **Rain** The weather reporter on TV makes predictions such as a 25% chance of rain. What do you think is the meaning of such a phrase?

5. **Winter** Comment on the following quotation:

   "What I think is our best determination is it will be a colder than normal winter," said Pamela Naber Knox, a Wisconsin state climatologist. "I'm basing that on a couple of different things. First, in looking at the past

few winters, there has been a lack of really cold weather. Even though we are not supposed to use the law of averages, we are due." (Associated Press, fall 1992, quoted by Schaeffer et al.)

6. Snow After an unusually dry autumn, a radio announcer is heard to say, "Watch out! We'll pay for these sunny days later on this winter." Explain what he's trying to say, and comment on the validity of his reasoning.

7. Cold streak A batter who had failed to get a hit in seven consecutive times at bat then hits a game-winning home run. When talking to reporters afterward, he says he was very confident that last time at bat because he knew he was "due for a hit." Comment on his reasoning.

8. Crash Commercial airplanes have an excellent safety record. Nevertheless, there are crashes occasionally, with the loss of many lives. In the weeks following a crash, airlines often report a drop in the number of passengers, probably because people are afraid to risk flying.

   a) A travel agent suggests that since the law of averages makes it highly unlikely to have two plane crashes within a few weeks of each other, flying soon after a crash is the safest time. What do you think?
   b) If the airline industry proudly announces that it has set a new record for the longest period of safe flights, would you be reluctant to fly? Are the airlines due to have a crash?

9. Auto insurance Insurance companies collect annual payments from drivers in exchange for paying for the cost of accidents.

   a) Why should you be reluctant to accept a $1500 payment from your neighbor to cover his automobile accidents in the next year?
   b) Why can the insurance company make that offer?

10. Jackpot On February 11, 2009, the AP news wire released the following story:

   (*LAS VEGAS, Nev.*)—A man in town to watch the NCAA basketball tournament hit a $38.7 million jackpot on Friday, the biggest slot machine payout ever. The 25-year-old software engineer from Los Angeles, whose name was not released at his request, won after putting three $1 coins in a machine at the Excalibur hotel-casino, said Rick Sorensen, a spokesman for slot machine maker International Game Technology.

   a) How can the Excalibur afford to give away millions of dollars on a $3 bet?
   b) Why was the maker willing to make a statement? Wouldn't most businesses want to keep such a huge loss quiet?

11. Wardrobe In your dresser are five blue shirts, three red shirts, and two black shirts.

   a) What is the probability of randomly selecting a red shirt?
   b) What is the probability that a randomly selected shirt is not black?

12. Playlists Your list of favorite songs contains 10 rock songs, 7 rap songs, and 3 country songs.

   a) What is the probability that a randomly played song is a rap song?
   b) What is the probability that a randomly played song is not country?

13. Cell phones and surveys A 2010 study conducted by the National Center for Health Statistics found that 25% of U.S. households had no landline service. This raises concerns about the accuracy of certain surveys, as they depend on random-digit dialing to households via landlines. We are going to pick five U.S. households at random:

   a) What is the probability that all five of them have a landline?
   b) What is the probability that at least one of them does not have a landline?
   c) What is the probability that at least one of them does have a landline?

14. Cell phones and surveys II The survey by the National Center for Health Statistics further found that 49% of adults ages 25–29 had only a cell phone and no landline. We randomly select four 25–29-year-olds:

   a) What is the probability that all of these adults have a only a cell phone and no landline?
   b) What is the probability that none of these adults have only a cell phone and no landline?
   c) What is the probability that at least one of these adults has only a cell phone and no landline?

15. Spinner The plastic arrow on a spinner for a child's game stops rotating to point at a color that will determine what happens next. Which of the following probability assignments are possible?

	Probabilities of . . .			
	Red	**Yellow**	**Green**	**Blue**
a)	0.25	0.25	0.25	0.25
b)	0.10	0.20	0.30	0.40
c)	0.20	0.30	0.40	0.50
d)	0	0	1.00	0
e)	0.10	0.20	1.20	−1.50

16. Scratch off Many stores run "secret sales": Shoppers receive cards that determine how large a discount they get, but the percentage is revealed by scratching off that black stuff (what *is* that?) only after the purchase has been totaled at the cash register. The store is required to reveal (in the fine print) the distribution of discounts available. Which of these probability assignments are legitimate?

	Probabilities of . . .			
	10% off	**20% off**	**30% off**	**50% off**
a)	0.20	0.20	0.20	0.20
b)	0.50	0.30	0.20	0.10
c)	0.80	0.10	0.05	0.05
d)	0.75	0.25	0.25	−0.25
e)	1.00	0	0	0

17. Electronics Suppose that 46% of families living in a certain county own a computer and 18% own an HDTV. The Addition Rule might suggest, then, that 64% of families own either a computer or an HDTV. What's wrong with that reasoning?

18. Homes Funding for many schools comes from taxes based on assessed values of local properties. People's homes are assessed higher if they have extra features such as garages and swimming pools. Assessment records in a certain school district indicate that 37% of the homes have garages and 3% have swimming pools. The Addition Rule might suggest, then, that 40% of residences have a garage or a pool. What's wrong with that reasoning?

19. Speeders Traffic checks on a certain section of highway suggest that 60% of drivers are speeding there. Since $0.6 \times 0.6 = 0.36$, the Multiplication Rule might suggest that there's a 36% chance that two vehicles in a row are both speeding. What's wrong with that reasoning?

20. Lefties Although it's hard to be definitive in classifying people as right- or left-handed, some studies suggest that about 14% of people are left-handed. Since $0.14 \times 0.14 = 0.0196$, the Multiplication Rule might suggest that there's about a 2% chance that a brother and a sister are both lefties. What's wrong with that reasoning?

21. College admissions For high school students graduating in 2007, college admissions to the nation's most selective schools were the most competitive in memory. (*The New York Times*, "A Great Year for Ivy League Schools, but Not So Good for Applicants to Them," April 4, 2007). Harvard accepted about 9% of its applicants, Stanford 10%, and Penn 16%. Jorge has applied to all three. Assuming that he's a typical applicant, he figures that his chances of getting into both Harvard and Stanford must be about 0.9%.
a) How has he arrived at this conclusion?
b) What additional assumption is he making?
c) Do you agree with his conclusion?

22. College admissions II In Exercise 21, we saw that in 2007 Harvard accepted about 9% of its applicants,

Stanford 10%, and Penn 16%. Jorge has applied to all three. He figures that his chances of getting into at least one of the three must be about 35%.
a) How has he arrived at this conclusion?
b) What assumption is he making?
c) Do you agree with his conclusion?

23. Car repairs A consumer organization estimates that over a 1-year period 17% of cars will need to be repaired once, 7% will need repairs twice, and 4% will require three or more repairs. What is the probability that a car chosen at random will need
a) no repairs?
b) no more than one repair?
c) some repairs?

24. Family Music Your family has an iPod filled with music. It has many thousands of songs. You figure that roughly 60% of the songs are music you like, 25% of the music is annoying songs from your little sister, and the rest is stuff from the 80s that only your parents still think is cool. Driving across town, your Mom puts the iPod on shuffle and you all listen to whatever randomness produces. What is the probability that the first song is
a) an 80s song?
b) a song picked by one of the kids?
c) not one of your songs?

25. More repairs Consider again the auto repair rates described in Exercise 23. If you own two cars, what is the probability that
a) neither will need repair?
b) both will need repair?
c) at least one car will need repair?

26. More music You listen to two songs, as described in Exercise 24. What is the probability that
a) neither will be songs that you like?
b) both will be annoying songs from your sister?
c) at least one will be a song of your Mom's choice?

27. Repairs, again You used the Multiplication Rule to calculate repair probabilities for your cars in Exercise 25.
a) What must be true about your cars in order to make that approach valid?
b) Do you think this assumption is reasonable? Explain.

28. Coda You used the Multiplication Rule to calculate probabilities about the music choices of your iPod in Exercise 26.
a) What must be true about the songs to make that approach valid?
b) Do you think this assumption is reasonable? Explain.

29. Energy 2011 A Gallup Poll in March 2011 asked 1012 U.S. adults whether increasing domestic energy production or protecting the environment should be given a higher priority. Here are the results:

Response	Number
Increase Production	511
Protect Environment	419
Equally Important	41
No Opinion	41
Total	**1012**

If we select a person at random from this sample of 1012 adults,

a) what is the probability that the person responded "Increase production"?

b) what is the probability that the person responded "Equally important" or had no opinion?

30. Failing fathers? A Pew Research poll in 2011 asked 2005 U.S. adults whether being a father today is harder than it was a generation ago. Here's how they responded:

Response	Number
Easier	501
Same	802
Harder	682
No Opinion	20
Total	**2005**

If we select a respondent at random from this sample of 2005 adults,

a) what is the probability that the selected person responded "Harder"?

b) what is the probability that the person responded the "Same" or "Easier"?

31. More energy Exercise 29 shows the results of a Gallup Poll about energy. Suppose we select three people at random from this sample.

a) What is the probability that all three responded "Protect the environment"?

b) What is the probability that none responded "Equally important"?

c) What assumption did you make in computing these probabilities?

d) Explain why you think that assumption is reasonable.

32. Fathers, revisited Consider again the results of the poll about fathering discussed in Exercise 30. If we select two people at random from this sample,

a) what is the probability that both think that being a father is easier today?

b) what is the probability that neither thinks being a father is easier today?

c) what is the probability that the first person thinks being a father is easier today and the second one doesn't?

d) what assumption did you make in computing these probabilities?

e) explain why you think that assumption is reasonable.

33. Polling As mentioned in the chapter, opinion-polling organizations contact their respondents by sampling random telephone numbers. Although interviewers now can reach about 62% of U.S. households, the percentage of those contacted who agree to cooperate with the survey has fallen from 43% in 1997 to only 14% in 2012 (Pew Research Center for the People and the Press). Each household, of course, is independent of the others.

a) What is the probability that the next household on the list will be contacted but will refuse to cooperate?

b) What was the probability (in 2012) of failing to contact a household or of contacting the household but not getting them to agree to the interview?

c) Show another way to calculate the probability in part b.

34. Polling, part II According to Pew Research, the contact rate (probability of contacting a selected household) was 90% in 1997 and 62% in 2012. However, the cooperation rate (probability of someone at the contacted household agreeing to be interviewed) was 43% in 1997 and dropped to 14% in 2012.

a) What was the probability (in 2012) of obtaining an interview with the next household on the sample list? (To obtain an interview, an interviewer must both contact the household and then get agreement for the interview.)

b) Was it more likely to obtain an interview from a randomly selected household in 1997 or in 2012?

35. M&M's The Masterfoods company says that before the introduction of purple, yellow candies made up 20% of their plain M&M's, red another 20%, and orange, blue, and green each made up 10%. The rest were brown.

a) If you pick an M&M at random, what was the probability that
   1) it is brown?
   2) it is yellow or orange?
   3) it is not green?
   4) it is striped?

b) If you pick three M&M's in a row, what is the probability that
   1) they are all brown?
   2) the third one is the first one that's red?
   3) none are yellow?
   4) at least one is green?

36. Blood The American Red Cross says that about 45% of the U.S. population has Type O blood, 40% Type A, 11% Type B, and the rest Type AB.

a) Someone volunteers to give blood. What is the probability that this donor
 1) has Type AB blood?
 2) has Type A or Type B?
 3) is not Type O?

b) Among four potential donors, what is the probability that
 1) all are Type O?
 2) no one is Type AB?
 3) they are not all Type A?
 4) at least one person is Type B?

37. Disjoint or independent? In Exercise 35 you calculated probabilities of getting various M&M's. Some of your answers depended on the assumption that the outcomes described were *disjoint;* that is, they could not both happen at the same time. Other answers depended on the assumption that the events were *independent;* that is, the occurrence of one of them doesn't affect the probability of the other. Do you understand the difference between disjoint and independent?

a) If you draw one M&M, are the events of getting a red one and getting an orange one disjoint, independent, or neither?

b) If you draw two M&M's one after the other, are the events of getting a red on the first and a red on the second disjoint, independent, or neither?

c) Can disjoint events ever be independent? Explain.

38. Disjoint or independent? In Exercise 36 you calculated probabilities involving various blood types. Some of your answers depended on the assumption that the outcomes described were *disjoint;* that is, they could not both happen at the same time. Other answers depended on the assumption that the events were *independent;* that is, the occurrence of one of them doesn't affect the probability of the other. Do you understand the difference between disjoint and independent?

a) If you examine one person, are the events that the person is Type A and that the person is Type B disjoint, independent, or neither?

b) If you examine two people, are the events that the first is Type A and the second Type B disjoint, independent, or neither?

c) Can disjoint events ever be independent? Explain.

39. Dice You roll a fair die three times. What is the probability that

a) you roll all 6's?
b) you roll all odd numbers?
c) none of your rolls gets a number divisible by 3?
d) you roll at least one 5?
e) the numbers you roll are not all 5's?

40. Slot machine A slot machine has three wheels that spin independently. Each has 10 equally likely symbols: 4 bars, 3 lemons, 2 cherries, and a bell. If you play, what is the probability that

a) you get 3 lemons?
b) you get no fruit symbols?
c) you get 3 bells (the jackpot)?
d) you get no bells?
e) you get at least one bar (an automatic loser)?

41. Champion bowler A certain bowler can bowl a strike 70% of the time. What's the probability that she

a) goes three consecutive frames without a strike?
b) makes her first strike in the third frame?
c) has at least one strike in the first three frames?
d) bowls a perfect game (12 consecutive strikes)?

42. The train To get to work, a commuter must cross train tracks. The time the train arrives varies slightly from day to day, but the commuter estimates he'll get stopped on about 15% of work days. During a certain 5-day work week, what is the probability that he

a) gets stopped on Monday and again on Tuesday?
b) gets stopped for the first time on Thursday?
c) gets stopped every day?
d) gets stopped at least once during the week?

43. Voters Suppose that in your city 37% of the voters are registered as Democrats, 29% as Republicans, and 11% as members of other parties (Liberal, Right to Life, Green, etc.). Voters not aligned with any official party are termed "Independent." You are conducting a poll by calling registered voters at random. In your first three calls, what is the probability you talk to

a) all Republicans?
b) no Democrats?
c) at least one Independent?

44. Religion Census reports for a city indicate that 62% of residents classify themselves as Christian, 12% as Jewish, and 16% as members of other religions (Muslims, Buddhists, etc.). The remaining residents classify themselves as nonreligious. A polling organization seeking information about public opinions wants to be sure to talk with people holding a variety of religious views, and makes random phone calls. Among the first four people they call, what is the probability they reach

a) all Christians?
b) no Jews?
c) at least one person who is nonreligious?

45. Tires You bought a new set of four tires from a manufacturer who just announced a recall because 2% of those tires are defective. What is the probability that at least one of yours is defective?

46. Pepsi For a sales promotion, the manufacturer places winning symbols under the caps of 10% of all Pepsi bottles. You buy a six-pack. What is the probability that you win something?

47. 9/11? On September 11, 2002, the first anniversary of the terrorist attack on the World Trade Center, the New York State Lottery's daily number came up 9–1–1. An interesting coincidence or a cosmic sign?

a) What is the probability that the winning three numbers match the date on any given day?
b) What is the probability that a whole year passes without this happening?
c) What is the probability that the date and winning lottery number match at least once during any year?
d) If every one of the 50 states has a three-digit lottery, what is the probability that at least one of them will come up 9–1–1 on September 11?

48. Red cards You shuffle a deck of cards and then start turning them over one at a time. The first one is red. So is the second. And the third. In fact, you are surprised to get 10 red cards in a row. You start thinking, "The next one is due to be black!"

a) Are you correct in thinking that there's a higher probability that the next card will be black than red? Explain.
b) Is this an example of the Law of Large Numbers? Explain.

 Just Checking ANSWERS

1. The LLN works only in the long run, not in the short run. The random methods for selecting lottery numbers have no memory of previous picks, so there is no change in the probability that a certain number will come up.

2. **a)** 0.76

   b) $0.76(0.76) = 0.5776$

   c) $(1 - 0.76)^2(0.76) = 0.043776$

   d) $1 - (1 - 0.76)^5 = 0.9992$

14 Probability Rules!

Pull a bill from your wallet or pocket without looking at it. An outcome of this trial is the kind of bill you select. The sample space is all the bills in circulation: **S** = { $1 bill, $2 bill, $5 bill, $10 bill, $20 bill, $50 bill, $100 bill}.[1] These are *all* the possible outcomes. (In spite of what you may have seen in bank robbery movies, there are no $500 or $1000 bills.)

We can combine the outcomes in different ways to make many different events. For example, the event **A** = { $1, $5, $10} represents selecting a $1, $5, or $10 bill. The event **B** = {a bill that does not have a president on it} is the collection of outcomes (Don't look! Can you name them?): { $10 (Hamilton), $100 (Franklin)}. The event **C** = {enough money to pay for a $12 meal with one bill} is the set of outcomes { $20, $50, $100}.

Notice that these outcomes are not equally likely. You'd no doubt be more surprised (and pleased) to pull out a $100 bill than a $1 bill—it's not as likely, though. You probably carry many more $1 than $100 bills, but without information about the probability of each outcome, we can't calculate the probability of an event.

The probability of the event **C** (getting a bill worth more than $12) is *not* 3/7. There are 7 possible outcomes, and 3 of them exceed $12, but they are not *equally likely*.

The General Addition Rule

Now look at the bill in your hand. There are images of famous buildings in the center of the backs of all but two bills in circulation. The $1 bill has the word ONE in the center, and the $2 bill shows the signing of the Declaration of Independence.

---

[1] Well, technically, the sample space is all the bills in your pocket. You may be quite sure there isn't a $100 bill in there, but *we* don't know that, so humor us that it's at least *possible* that any legal bill could be there.

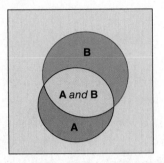

Events **A** and **B** and their intersection.

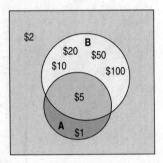

Denominations of bills that are odd (**A**) or that have a building on the reverse side (**B**). The two sets both include the $5 bill, and both exclude the $2 bill.

What's the probability of randomly selecting A = {a bill with an odd-numbered value} *or* B = {a bill with a building on the reverse}? We know **A** = {$1, $5} and **B** = {$5, $10, $20, $50, $100}. But $P(\mathbf{A} \text{ or } \mathbf{B})$ is not simply the sum $P(\mathbf{A}) + P(\mathbf{B})$, because the events **A** and **B** are not disjoint. The $5 bill is in both sets. So what can we do? We'll need a new probability rule.

As the diagrams show, we can't get the right answer by just adding the two probabilities because the events are not disjoint; they overlap. There's an outcome (the $5 bill) in the *intersection* of **A** and **B**. The Venn diagram represents the sample space. Notice that the $2 bill has neither a building nor an odd denomination, so it sits outside both circles.

The $5 bill plays a crucial role here because it is both odd *and* has a building on the reverse. It's in both **A** and **B**, which places it in the *intersection* of the two circles. The reason we can't simply add the probabilities of **A** and **B** is that we'd count the $5 bill twice.

If we did add the two probabilities, we could compensate by *subtracting* out the probability of that $5 bill. So,

P(odd number value or building)
= P(odd number value) + P(building) − P(odd number value *and* building)
= P($1, $5) + P($5, $10, $20, $50, $100) − P($5).

This method works in general. We add the probabilities of two events and then subtract out the probability of their intersection. This approach gives us the **General Addition Rule,** which does not require disjoint events:

$$P(\mathbf{A} \cup \mathbf{B}) = P(\mathbf{A}) + P(\mathbf{B}) - P(\mathbf{A} \cap \mathbf{B}).$$

For Example USING THE GENERAL ADDITION RULE

A survey of college students found that 56% live in a campus residence hall, 62% participate in a campus meal program, and 42% do both.

QUESTION: What's the probability that a randomly selected student either lives or eats on campus?

ANSWER:

Let **L** = {student lives on campus} and **M** = {student has a campus meal plan}.

P(a student either lives or eats on campus) = $P(\mathbf{L} \cup \mathbf{M})$
$= P(\mathbf{L}) + P(\mathbf{M}) - P(\mathbf{L} \cap \mathbf{M})$
$= 0.56 + 0.62 - 0.42$
$= 0.76$

There's a 76% chance that a randomly selected college student either lives or eats on campus.

Would You Like Dessert or Coffee?

Natural language can be ambiguous. In this question, is the answer one of the two alternatives, or simply "yes"? Must you decide between them, or may you have both? That kind of ambiguity can confuse our probabilities.

Suppose we had been asked a different question: What is the probability that the bill we draw has *either* an odd value *or* a building but *not both*? Which bills are we talking about now? The set we're interested in would be {$1, $10, $20, $50, $100}. We don't include the $5 bill in the set because it has both characteristics.

Why isn't this the same answer as before? The problem is that when we say the word "or," we usually mean *either* one *or* both. We don't usually mean the *exclusive* version of "or" as in, "Would you like the steak *or* the vegetarian entrée?" Ordinarily when we ask for the probability that **A** or **B** occurs, we mean **A** *or* **B** or both. And we know *that* probability is $P(\mathbf{A}) + P(\mathbf{B}) - P(\mathbf{A}\ and\ \mathbf{B})$. The General Addition Rule subtracts the probability of the outcomes in **A** and **B** because we've counted those outcomes *twice*. But they're still there.

If we really mean **A** or **B** but NOT both, we have to get rid of the outcomes in $\{\mathbf{A}\ and\ \mathbf{B}\}$. So $P(\mathbf{A}\ or\ \mathbf{B}\ but\ not\ both) = P(\mathbf{A} \cup \mathbf{B}) - P(\mathbf{A} \cap \mathbf{B}) = P(\mathbf{A}) + P(\mathbf{B}) - 2 \times P(\mathbf{A} \cap \mathbf{B})$. Now we've subtracted $P(\mathbf{A} \cap \mathbf{B})$ twice—once because we don't want to double-count these events and a second time because we really didn't want to count them at all.

Confused? *Make a picture.* It's almost always easier to think about such situations by looking at a Venn diagram.

For Example USING VENN DIAGRAMS

RECAP: We return to our survey of college students: 56% live on campus, 62% have a campus meal program, and 42% do both.

QUESTION: Based on a Venn diagram, what is the probability that a randomly selected student

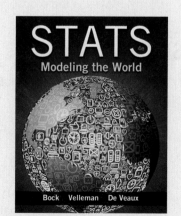

a) lives off campus and doesn't have a meal program?
b) lives in a residence hall but doesn't have a meal program?

ANSWER: Let $\mathbf{L} = \{$student lives on campus$\}$ and $\mathbf{M} = \{$student has a campus meal plan$\}$. In the Venn diagram, the intersection of the circles is $P(\mathbf{L} \cap \mathbf{M}) = 0.42$. Since $P(\mathbf{L}) = 0.56$, $P(\mathbf{L} \cap \mathbf{M}^C) = 0.56 - 0.42 = 0.14$. Also, $P(\mathbf{L}^C \cap \mathbf{M}) = 0.62 - 0.42 = 0.20$. Now, $0.14 + 0.42 + 0.20 = 0.76$, leaving $1 - 0.76 = 0.24$ for the region outside both circles.

Now ... $P(\text{off campus and no meal program}) = P(\mathbf{L}^C \cap \mathbf{M}^C) = 0.24$
$\qquad\qquad P(\text{on campus and no meal program}) = P(\mathbf{L} \cap \mathbf{M}^C) = 0.14$

 # Just Checking

1. We sampled some pages of this book at random to see whether they held an equation, a graph, or other data display and found the following:
   48% of pages had some kind of data display,
   27% of pages had an equation, and
   7% of pages had both a data display and an equation.

   a) Display these results in a Venn diagram.
   b) What is the probability that a randomly selected sample page had neither a data display nor an equation?
   c) What is the probability that a randomly selected sample page had a data display but no equation?

Step-by-Step Example USING THE GENERAL ADDITION RULE

Police report that 78% of drivers stopped on suspicion of drunk driving are given a breath test, 36% a blood test, and 22% both tests.

Question: What is the probability that a randomly selected DWI suspect is given

1. a test?
2. a blood test or a breath test, but not both?
3. neither test?

THINK ➡ **Plan** Define the events we're interested in. There are no conditions to check; the General Addition Rule works for any events!

Plot Make a picture, and use the given probabilities to find the probability for each region.

The blue region represents **A** but not **B**. The green intersection region represents **A** *and* **B**. Note that since $P(\mathbf{A}) = 0.78$ and $P(\mathbf{A} \cap \mathbf{B}) = 0.22$, the probability of **A** but not **B** must be $0.78 - 0.22 = 0.56$.

The yellow region is **B** but not **A**.

The gray region outside both circles represents the outcome neither **A** nor **B**. All the probabilities must total 1, so you can determine the probability of that region by subtraction.

Now, figure out what you want to know. The probabilities can come from the diagram or a formula. Sometimes translating the words to equations is the trickiest step.

Let $\mathbf{A} = \{suspect\ is\ given\ a\ breath\ test\}$.
Let $\mathbf{B} = \{suspect\ is\ given\ a\ blood\ test\}$.

I know that $P(\mathbf{A}) = 0.78$
$$P(\mathbf{B}) = 0.36$$
$$P(\mathbf{A} \cap \mathbf{B}) = 0.22$$

So $P(\mathbf{A} \cap \mathbf{B}^{C}) = 0.78 - 0.22 = 0.56$
$$P(\mathbf{B} \cap \mathbf{A}^{C}) = 0.36 - 0.22 = 0.14$$
$$P(\mathbf{A}^{C} \cap \mathbf{B}^{C}) = 1 - (0.56 + 0.22 + 0.14)$$
$$= 0.08$$

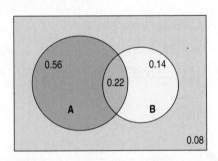

Question 1: What is the probability that the suspect is given a test?

SHOW ➡ **Mechanics** The probability the suspect is given a test is $P(\mathbf{A} \cup \mathbf{B})$. We can use the General Addition Rule, or we can add the probabilities seen in the diagram.

$$P(\mathbf{A} \cup \mathbf{B}) = P(\mathbf{A}) + P(\mathbf{B}) - P(\mathbf{A} \cap \mathbf{B})$$
$$= 0.78 + 0.36 - 0.22$$
$$= 0.92$$

OR

$$P(\mathbf{A} \cup \mathbf{B}) = 0.56 + 0.22 + 0.14 = 0.92$$

TELL ➡ **Conclusion** Don't forget to interpret your result in context.

92% of all DWI suspects are given a test.

(continued)

Question 2: What is the probability that the suspect gets either a blood test or a breath test but NOT both?

SHOW ➡ **Mechanics** We can use the rule, or just add the appropriate probabilities seen in the Venn diagram.

$$P(\mathbf{A} \text{ or } \mathbf{B} \text{ but NOT both}) = P(\mathbf{A} \cup \mathbf{B}) - P(\mathbf{A} \cap \mathbf{B})$$
$$= 0.92 - 0.22 = 0.70$$

OR

$$P(\mathbf{A} \text{ or } \mathbf{B} \text{ but NOT both}) = P(\mathbf{A} \cap \mathbf{B}^c) + P(\mathbf{B} \cap \mathbf{A}^c)$$
$$= 0.56 + 0.14 = 0.70$$

TELL ➡ **Conclusion** Interpret your result in context.

70% of the suspects get exactly one of the tests.

Question 3: What is the probability that the suspect gets neither test?

SHOW ➡ **Mechanics** Getting neither test is the complement of getting one or the other. Use the Complement Rule or just notice that "neither test" is represented by the region outside both circles.

$$P(\text{neither test}) = 1 - P(\text{either test})$$
$$= 1 - P(\mathbf{A} \cup \mathbf{B})$$
$$= 1 - 0.92 = 0.08$$

OR

$$P(\mathbf{A}^c \cap \mathbf{B}^c) = 0.08$$

TELL ➡ **Conclusion** Interpret your result in context.

Only 8% of the suspects get no test.

It Depends . . .

Two psychologists surveyed 478 children in grades 4, 5, and 6 in elementary schools in Michigan. They stratified their sample, drawing roughly 1/3 from rural, 1/3 from suburban, and 1/3 from urban schools. Among other questions, they asked the students whether their primary goal was to get good grades, to be popular, or to be good at sports. The researchers wondered whether boys and girls at this age had similar goals.

Here's a *contingency table* giving counts of the students by their goals and sex:

Table 14.1

The distribution of goals for boys and girls.

		Goals			
		Grades	**Popular**	**Sports**	**Total**
Sex	**Boy**	117	50	60	**227**
	Girl	130	91	30	**251**
	Total	**247**	**141**	**90**	**478**

Using these 478 students as our sample space, some probabilities aren't hard to find:

- What's the probability that a person selected at random from these students is a girl? There are 251 girls among the 478 students, so $P(\text{Girl}) = \dfrac{251}{478} = 0.525$.

- The probability that a randomly selected student's goal is to excel in sports is
$$P(\text{Sports}) = \dfrac{90}{478} = 0.188.$$

- What's the probability that a randomly selected student is a girl who hopes to be good at sports? Well, 30 girls named sports as their goal, so the probability is
$$P(\text{Girl} \cap \text{Sports}) = \dfrac{30}{478} = 0.063.$$

But how do the goals of boys and girls compare?

We looked at contingency tables and graphed *conditional distributions* back in Chapter 2. These pie charts show the *relative frequencies* with which boys and girls named the three goals.

We see that girls are much less likely to say their goal is to excel at sports than are boys. When we restrict our focus to girls, we look only at the girls' row of the table. Of the 251 girls, only 30 of them said their goal was to excel at sports.

We write the probability that a selected student wants to excel at sports *given that we have selected a girl* as

$$P(\text{Sports}|\text{Girl}) = 30/251 = 0.120$$

What about boys? Look at the top row of the table. There, of the 227 boys, 60 said their goal was to excel at sports. So, $P(\text{Sports}|\text{Boy}) = 60/227 = 0.264$, more than twice the girls's probability.

In general, when we want the probability of an event from a *conditional* distribution, we write $P(\mathbf{B}|\mathbf{A})$ and pronounce it "the probability of **B** *given* **A**." A probability that takes into account a given *condition* such as this is called a **conditional probability.**

Let's look at what we did. We worked with the counts, but we could work with the probabilities just as well. There were 30 students who both were girls and had sports as their goal, and there are 251 girls. So we found the probability to be 30/251. To find the probability of the event **B** *given* the event **A**, we restrict our attention to the outcomes in **A**. We then find in what fraction of *those* outcomes **B** also occurred. Formally, we write:

$$P(\mathbf{B}|\mathbf{A}) = \dfrac{P(\mathbf{A} \cap \mathbf{B})}{P(\mathbf{A})}.$$

Thinking this through, we can see that it's just what we've been doing, but now with probabilities rather than with counts. Look back at the girls for whom sports was the goal. How did we calculate $P(\text{Sports}|\text{Girl})$?

The rule says to use probabilities. It says to find $P(\mathbf{A} \cap \mathbf{B})/P(\mathbf{A})$. The result is the same whether we use counts or probabilities because the total number in the sample cancels out:

$$P(\text{Sports}|\text{Girl}) = \dfrac{P(\text{Sports} \cap \text{Girl})}{P(\text{Girl})} = \dfrac{30/478}{251/478} = \dfrac{30}{251}.$$

Be careful, though. When we work with conditional probabilities, we must be sure to pay attention to which event is given. We've just seen that the probability a girl hopes to excel in sports is $\dfrac{30}{251}$, or about 0.12. That's $P(\text{Sports}|\text{Girl})$. What about $P(\text{Girl}|\text{Sports})$? First, lets think about what that means. $P(\text{Girl}|\text{Sports})$ is the probability that a student with a goal of being good at sports is a girl. Now we restrict our thinking to the 90 students who said

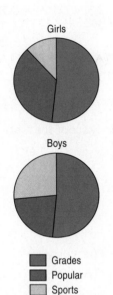

Girls

Boys

■ Grades
■ Popular
□ Sports

Figure 14.1
The distribution of goals for boys and girls.

A S *Activity*: **Birthweights and Smoking.** Does smoking increase the chance of having a baby with low birth weight?

they wanted to excel in sports. Among them, 30 are girls, so $P(\text{Girl}|\text{Sports}) = \dfrac{30}{90} = 0.33$, a much different result. Using our conditional probability formula, we'd find

A S **Activity: Conditional Probability.** Simulation is great for seeing conditional probabilities at work.

$$P(\text{Girl}|\text{Sports}) = \frac{P(\text{Girl} \cap \text{Sports})}{P(\text{Sports})} = \frac{30/478}{90/478} = \frac{30}{90} \approx 0.33$$

So remember, $P(\mathbf{A}|\mathbf{B})$ is not the same as $P(\mathbf{B}|\mathbf{A})$. Whenever you're finding a conditional probability, always *Think* about which event is given.

For Example FINDING A CONDITIONAL PROBABILITY

RECAP: Our survey found that 56% of college students live on campus, 62% have a campus meal program, and 42% do both.

QUESTION: While dining in a campus facility open only to students with meal plans, you meet someone interesting. What is the probability that your new acquaintance lives on campus?

ANSWER: Let **L** = {student lives on campus} and **M** = {student has a campus meal plan}.

$$
\begin{aligned}
P(\text{student lives on campus given that the student has a meal plan}) &= P(\mathbf{L}|\mathbf{M})\\
&= \frac{P(\mathbf{L} \cap \mathbf{M})}{P(\mathbf{M})}\\
&= \frac{0.42}{0.62}\\
&\approx 0.677
\end{aligned}
$$

There's a probability of about 0.677 that a student with a meal plan lives on campus. Notice that this is higher than the probability for all students.

Independence

Independence

If we had to pick one idea in this chapter that you should understand and remember, it's the definition and meaning of independence. We'll need this idea in every one of the chapters that follow.

It's time to return to the question of just what it means for events to be independent. We've said informally that what we mean by independence is that the outcome of one event does not influence the probability of the other. With our new notation for conditional probabilities, we can write a formal definition: Events **A** and **B** are **independent** whenever

$$P(\mathbf{B}|\mathbf{A}) = P(\mathbf{B}).$$

In other words, the probability that **B** happens is the same whether **A** happened or not.

Let's look again at the study about the goals of 4th, 5th, and 6th grade children. Is wanting to excel in sports independent of a student's sex? We've already looked at the two probabilities we need:

A S **Activity: Independence.** Are *Smoking and Low Birthweight* independent?

- $P(\text{Sports}) = \dfrac{90}{478} = 0.188$

- $P(\text{Sports}|\text{Girl}) = \dfrac{30}{251} = 0.120$

While nearly 19% of all students had being good in sports as their goal, only about 12% of the girls did. Apparently, girls attach less importance to sports. Because these probabilities aren't equal, choosing success in sports as a goal is *not* independent of the student's sex.

	Goals			
	Grades	**Popular**	**Sports**	**Total**
Boy	117	50	60	**227**
Girl	130	91	30	**251**
Total	**247**	**141**	**90**	**478**

(The left edge of the table is labeled vertically: Sex)

Table 14.2

The distributions of goals for boys and girls (again).

What about grades? Is the probability of having good grades as a goal independent of the sex of the responding student? We need to check whether

$$P(\text{Grades}|\text{Girl}) = P(\text{Grades})$$

$$\frac{130}{251} = 0.52 \overset{?}{=} \frac{247}{478} = 0.52$$

To two decimal place accuracy, it looks like we can consider choosing good grades as a goal to be independent of sex.

For Example CHECKING FOR INDEPENDENCE

RECAP: Our survey told us that 56% of college students live on campus, 62% have a campus meal program, and 42% do both.

QUESTION: Are living on campus and having a meal plan independent? Are they disjoint?

ANSWER: Let L = {student lives on campus} and M = {student has a campus meal plan}. If these events are independent, then knowing that a student lives on campus doesn't affect the probability that he or she has a meal plan. I'll check to see if $P(M|L) = P(M)$:

$$P(M|L) = \frac{P(L \cap M)}{P(L)}$$

$$= \frac{0.42}{0.56}$$

$$= 0.75, \quad \text{but } P(M) = 0.62.$$

Because $0.75 \neq 0.62$, the events are not independent; students who live on campus are more likely to have meal plans. Living on campus and having a meal plan are not disjoint either; in fact, 42% of college students do both.

Independent ≠ Disjoint

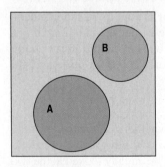

Figure 14.2

Because these events are mutually exclusive, learning that **A** happened tells us that **B** didn't. The probability of **B** has changed from whatever it was to zero. So the disjoint events **A** and **B** are not independent.

Are disjoint events independent? These concepts seem to have similar ideas of separation and distinctness about them, but in fact disjoint events *cannot* be independent.[2] Let's see why. Consider the two disjoint events {you get an A in this course} and {you get a B in this course}. They're disjoint because they have no outcomes in common. Suppose you learn that you *did* get an A in the course. Now what is the probability that you got a B? You can't get both grades, so it must be 0.

Think about what that means. Knowing that the first event (getting an A) occurred changed your probability for the second event (down to 0). So these events aren't independent.

Disjoint events can't be independent. They have no outcomes in common, so if one occurs, the other doesn't. A common error is to treat disjoint (mutually exclusive) events as if they were independent and apply the Multiplication Rule for independent events. Don't make that mistake.

Let's summarize:

- Two events could be either independent or disjoint, *but not both.*
- And they could be *neither* disjoint nor independent.

---

[2]Well, technically two disjoint events *can* be independent, but only if the probability of one of the events is 0. For practical purposes, though, we can ignore this case. After all, as statisticians we don't anticipate having data about things that never happen.

One spring day your high school baseball team will be playing for the league championship right after school. You and your friends are planning to go, but when you look outside at lunchtime it's raining. Consider the events **R**: rain at lunchtime and **B**: baseball after school. They're not disjoint: even though it's raining now, the game may still go on as scheduled. And they're not independent either: you thought you were going to the game, but now that you see the rain, you're not so sure anymore. Knowing **R** changes the probability of **B**.

 ## Just Checking

2. A company's office of Human Resources reports a breakdown of employees by job type and sex as seen in this table.

   a) Are being male and having a supervision job disjoint events?
   b) Is having a supervisor's job independent of the sex of the employee?

	Sex	
Job Type	Male	Female
Management	7	6
Supervision	8	12
Production	45	72

Check for Independence

In earlier chapters we said informally that two events were independent if learning that one occurred didn't change what you thought about the other occurring. Now we can be more formal. Events **A** and **B** are independent if (and only if) the probability of **A** is the same when we are given that **B** has occurred. That is, $P(\mathbf{A}) = P(\mathbf{A}|\mathbf{B})$.

Although sometimes your intuition is enough, now that we have the formal rule, use it whenever you can.

Depending on Independence   A note of caution: People often estimate the probability of a compound event by multiplying probabilities together without thinking about whether those probabilities are independent.

For example, experts have assured us that the probability of a major commercial nuclear plant failure is so small that we should not expect such a failure to occur even in a span of hundreds of years. Yet in only a few decades of commercial nuclear power, the world has seen three failures (Chernobyl, Three Mile Island, and Fukushima). How could the estimates have been so wrong?

One simple part of the failure calculation is to test a particular valve and determine that valves such as this one fail only once in, say, 100 years of normal use. For a coolant failure to occur, several valves must fail. So we need the compound probability, P(valve 1 fails *and* valve 2 fails *and* . . .). A simple risk assessment might multiply the small probability of one valve failure together as many times as needed. But if the valves all came from the same manufacturer, a flaw in one might be found in the others. And maybe when the first fails, it puts additional pressure on the next one in line. In either case, the events aren't independent and so we can't simply multiply the probabilities.

Whenever you see probabilities multiplied together, stop and ask whether you think they are really independent before you believe the result.

Tables, Venn Diagrams, and Probability

One of the easiest ways to think about conditional probabilities is with contingency tables. We did that earlier in the chapter when we began our discussion. But sometimes we're given probabilities without a table. You can often construct a simple table to correspond to the probabilities.

For instance, in the drunk driving example, we were told that 78% of suspect drivers get a breath test, 36% a blood test, and 22% both. That's enough information. Translating percentages to probabilities, what we know looks like this:

		Breath Test		
		Yes	No	Total
Blood Test	Yes	0.22		**0.36**
	No			
	Total	**0.78**		**1.00**

Notice that the 0.78 and 0.36 are *marginal* probabilities and so they go into the *margins*. The 0.22 is the probability of getting both tests—a breath test *and* a blood test—so that's a *joint* probability. Those belong in the interior of the table.

Because the cells of the table show disjoint events, the probabilities always add to the marginal totals going across rows or down columns. So, filling in the rest of the table is quick:

		Breath Test		
		Yes	No	Total
Blood Test	Yes	0.22	0.14	**0.36**
	No	0.56	0.08	**0.64**
	Total	**0.78**	**0.22**	**1.00**

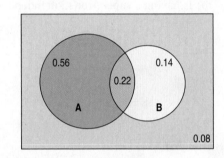

Compare this with the Venn diagram. Notice which entries in the table match up with the sets in this diagram. Whether a Venn diagram or a table is better to use will depend on what you are given and the questions you're being asked. Try both.

Step-by-Step Example   ARE THE EVENTS DISJOINT? INDEPENDENT?

Let's take another look at the drunk driving situation. Police report that 78% of drivers are given a breath test, 36% a blood test, and 22% both tests.

Question: 1. Are giving a DWI suspect a blood test and a breath test mutually exclusive?
2. Are giving the two tests independent?

THINK ➡ **Plan** Define the events we're interested in. State the given probabilities.

Let **A** = {suspect is given a breath test}
Let **B** = {suspect is given a blood test}.

I know that $P(\mathbf{A}) = 0.78$
$P(\mathbf{B}) = 0.36$
$P(\mathbf{A} \cap \mathbf{B}) = 0.22$

(continued)

Question 1: Are giving a DWI suspect a blood test and a breath test mutually exclusive?

SHOW ➡ **Mechanics** Disjoint events cannot *both* happen at the same time, so check to see if $P(\mathbf{A} \cap \mathbf{B}) = 0$.

$P(A \cap B) = 0.22$. Since some suspects are given both tests, $P(A \cap B) \neq 0$. The events are not mutually exclusive.

TELL ➡ **Conclusion** State your conclusion in context.

22% of all suspects get both tests, so a breath test and a blood test are not disjoint events.

Question 2: Are the two tests independent?

THINK ➡ **Plan** Make a table.

		Breath Test		
		Yes	No	Total
Blood Test	Yes	0.22	0.14	**0.36**
	No	0.56	0.08	**0.64**
	Total	**0.78**	**0.22**	**1.00**

SHOW ➡ **Mechanics** Does getting a breath test change the probability of getting a blood test? That is, does $P(\mathbf{B}|\mathbf{A}) = P(\mathbf{B})$?

Because the two probabilities are *not* the same, the events are not independent.

$$P(\mathbf{B}|\mathbf{A}) = \frac{P(\mathbf{A} \cap \mathbf{B})}{P(\mathbf{A})} = \frac{0.22}{0.78} \approx 0.28$$

$$P(\mathbf{B}) = 0.36$$

$$P(\mathbf{B}|\mathbf{A}) \neq P(\mathbf{B})$$

TELL ➡ **Conclusion** Interpret your results in context.

Overall, 36% of the drivers get blood tests, but only 28% of those who get a breath test do. Since suspects who get a breath test are less likely to have a blood test, the two events are not independent.

 Just Checking

3. Remember our sample of pages in this book from the earlier Just Checking. . . ?

48% of pages had a data display.
27% of pages had an equation, and
7% of pages had both a data display and an equation.

a) Make a contingency table for the variables *display* and *equation*.

b) What is the probability that a randomly selected sample page with an equation also had a data display?

c) Are having an equation and having a data display disjoint events?

d) Are having an equation and having a data display independent events?

The General Multiplication Rule

A Little Algebra

We know:

$$P(\mathbf{B}|\mathbf{A}) = \frac{P(\mathbf{A} \cap \mathbf{B})}{P(\mathbf{A})}$$

Multiply both sides of the equation by $P(\mathbf{A})$ to get:

$$P(\mathbf{A} \cap \mathbf{B}) = P(\mathbf{A}) \cdot P(\mathbf{B}|\mathbf{A})$$

 A S *Activity:* **The General Multiplication Rule.** The best way to understand the General Multiplication Rule is with an experiment.

Remember the Multiplication Rule for the probability of **A** *and* **B**? It said

$$P(\mathbf{A} \cap \mathbf{B}) = P(\mathbf{A}) \times P(\mathbf{B}) \text{ when } \mathbf{A} \text{ and } \mathbf{B} \text{ are independent.}$$

Now we can write a more general rule that doesn't require independence. In fact, we've *already* written it down. We just need to rearrange the equation a bit.

The equation in the definition for conditional probability contains the probability of **A** *and* **B.** Rewriting the equation gives

$$P(\mathbf{A} \cap \mathbf{B}) = P(\mathbf{A}) \times P(\mathbf{B}|\mathbf{A}).$$

This is a **General Multiplication Rule** for compound events that does not require the events to be independent. Better than that, it even makes sense. The probability that two events, **A** and **B,** *both* occur is the probability that event **A** occurs multiplied by the probability that event **B** *also* then occurs—that is, by the probability that event **B** occurs *given* that event **A** has occurred.

Notice that this General Multiplication Rule works regardless of whether the events are independent. If they are, then $P(\mathbf{B}|\mathbf{A}) = P(\mathbf{B})$, so $P(\mathbf{A} \cap \mathbf{B}) = P(\mathbf{A}) \cdot P(\mathbf{B}|\mathbf{A}) = P(\mathbf{A}) \cdot P(\mathbf{B})$ for independent events. We hope that looks familiar.

For Example USING THE GENERAL MULTIPLICATION RULE

A factory produces two types of batteries, regular and rechargeable. Quality inspection tests show that 2% of the regular batteries come off the manufacturing line with a defect while only 1% of the rechargeable batteries have a defect. Rechargeable batteries make up 25% of the company's production.

QUESTION: What's the probability that if we choose one of the company's batteries at random we get

    a) a defective rechargeable battery?
    b) a regular battery and it's not defective?

ANSWER: Let R = rechargeable and B = a regular battery. It's given that $P(R) = 0.25$, so $P(B) = 0.75$.

Let D = defective. It's given that $P(D|B) = 0.02$ and $P(D|R) = 0.01$.

Now: $P(R \cap D) = P(R) \times P(D|R) = (0.25)(0.01) = 0.0025$

If 2% of the regular batteries are defective, then the other 98% aren't; in other words: $P(D^C|B) = 0.98$. So:

$$P(B \cap D^C) = P(B) \times P(D^C|B) = (0.75)(0.98) = 0.735$$

Only $\frac{1}{4}$ of 1% of the company's batteries are rechargeable and defective, while 73.5% are nondefective regular batteries.

Drawing Without Replacement

Room draw is a process for assigning dormitory rooms to students who live on a college campus. When it's time for you and your friend to draw, there are 12 rooms left. Three are in Gold Hall, a very desirable dorm. You get to draw first, and then your friend will draw. Naturally, you would both like to score rooms in Gold. What are your chances? In particular, what's the chance that you *both* can get rooms in Gold?

When you go first, the chance that *you* will draw one of the Gold rooms is 3/12. Suppose you do. Now, with you clutching your prized room assignment, what chance does your friend have? At this point there are only 11 rooms left and just 2 left in Gold, so your friend's chance is now 2/11.

Using our notation, we write

$$P(\text{friend draws Gold} \mid \text{you draw Gold}) = 2/11.$$

The reason the denominator changes is that we draw these rooms *without replacement*. That is, once one is drawn, it doesn't go back into the pool.

We often sample without replacement. When we draw from a very large population, the change in the denominator is too small to worry about. But when there's a small population to draw from, as in this case, we need to take note of the changing probabilities.

What are the chances that *both* of you will luck out? Well, now we've calculated the two probabilities we need for the General Multiplication Rule, so we can write:

$$P(\text{you draw Gold} \cap \text{friend draws Gold})$$
$$= P(\text{you draw Gold}) \times P(\text{friend draws Gold} \mid \text{you draw Gold})$$
$$= 3/12 \times 2/11 = 1/22 = 0.045$$

In this instance, it doesn't matter who went first, or even if the rooms were drawn simultaneously. Even if the room draw was accomplished by shuffling cards containing the names of the dormitories and then dealing them out to 12 applicants (rather than by each student drawing a room in turn), we can still *think* of the calculation as having taken place in two steps:

Picturing conditional probabilities this way leads to a more general way of helping us think with pictures—tree diagrams.

For Example DRAWING WITHOUT REPLACEMENT

You just bought a small bag of Skittles. Not that you could know this, but inside are 20 candies: 7 green, 5 orange, 4 red, 3 yellow, and only 1 brown. You tear open one corner of the package and begin eating them by shaking out one at a time.[3]

QUESTION: What's the probability that your first 2 Skittles are both orange? That none of your first 3 candies is green?

ANSWER: *Getting two orange candies in a row means I draw an orange one first* **and** *then another one second, with one orange candy already missing from the bag:*

$$P(2 \text{ orange}) = P(\text{orange first} \cap \text{orange second})$$
$$= P(\text{orange first}) \cdot P(\text{orange second} \mid \text{orange first})$$
$$= \frac{5}{20} \cdot \frac{4}{19} = \frac{1}{19}$$

There's a 1 in 19 chance (just over 5%) that I'd shake out 2 orange Skittles in a row.

(continued)

---
[3]Wow—what self-control!

Not getting any green ones means all of the first 3 Skittles were among the colors (13 candies):

$$P(3 \text{ non-greens}) = P(\text{green}^c \cap \text{green}^c \cap \text{green}^c)$$

$$= \frac{13}{20} \cdot \frac{12}{19} \cdot \frac{11}{18} \approx 0.25$$

There's about a 25% chance I won't get any green Skittles among the first 3 I shake out of the bag.

Just Checking

4. Think some more about that bag of Skittles described in the previous *For Example* (7 green, 5 orange, 4 red, 3 yellow, 1 brown). Write out the fractions you'd multiply together to find the probabilities of these outcomes. (Don't bother multiplying them together—unless you're curious.)

   a) The first two are both red.
   b) You eat three without seeing a yellow one.

   c) The fourth candy out of the bag is the brown one.

Tree Diagrams

"Why," said the Dodo, "the best way to explain it is to do it."

—*Lewis Carroll*

A recent Maryland highway safety study found that in 77% of all accidents the driver was wearing a seat belt. Accident reports indicated that 92% of those drivers escaped serious injury (defined as hospitalization or death), but only 63% of the nonbelted drivers were so fortunate. Overall, what's the probability that a driver involved in an accident was seriously injured?

The best way to organize information like this is—you guessed it—to make a picture.

Here we'll use a **tree diagram,** because it shows sequences of events, like those we had in the room draw, as paths that look like branches of a tree. It is a good idea to make a tree diagram almost any time you have conditional probabilities and plan to use the General Multiplication Rule. The number of different paths we can take can get large, so we usually draw the tree starting from the left and growing vine-like across the page.

The first branch of our tree separates drivers who had accidents according to whether they wore a seat belt. We label each branch of the tree with a possible outcome and its corresponding probability.

Figure 14.3

The first branches. We can diagram the two seat belt possibilities and indicate their respective probabilities with a simple tree diagram.

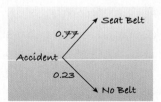

Notice that because we cover all possible outcomes with the branches, the probabilities add up to one. But we're also interested in injuries. The probability of being seriously injured *depends* on one's seat belt behavior.

Because the probabilities are *conditional,* we draw those outcomes separately on each branch of the tree:

Figure 14.4

The second branches. Extending the tree diagram, we can show both seat belt and injury outcomes. The injury probabilities are conditional on the seat belt outcomes, and they change depending on which branch we follow.

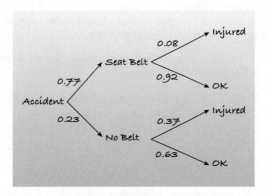

On each of the second set of branches, we write the possible outcomes associated with having a car accident (being seriously injured or not) and the associated probability. These probabilities are different because they are *conditional* depending on the driver's seat belt behavior. (It shouldn't be too surprising that those who don't wear their seat belts have a higher probability of serious injury or death.) Each set of probabilities add up to one, because given the outcome on the first branch, these outcomes cover all the possibilities.

Looking back at the General Multiplication Rule, we can see how the tree helps with the calculation.

To find the probability that in a randomly selected accident the driver was wearing a seat belt and was seriously injured, we follow the top branches, multiplying as we go:

$$P(\text{Seat Belt} \cap \text{Injured}) = P(\text{Seat Belt}) \times P(\text{Injured} \mid \text{Seat Belt})$$
$$= (0.77)(0.08)$$
$$= 0.0616$$

And we do the same for each combination of outcomes:

Figure 14.5

The completed tree. We can find the probabilities of compound events by multiplying the probabilities along the branch of the tree that leads to the event, just the way the General Multiplication Rule specifies.

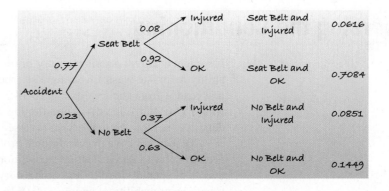

Which Picture?
- A tree diagram is usually best for working with conditional probabilities.
- Venn diagrams are great for intersections and unions (*and*s and *or*s).
- A carefully designed table can handle any situation, so try that if you get stuck.

Here's a hint to help you check your work creating a tree diagram. All the outcomes at the far right are disjoint and they are *all* the possibilities, so the final probabilities must add up to one. Always check!

And now (at last!) we can answer our original question: Overall, what's the probability that a driver involved in an accident was seriously injured? We simply find all the branches that lead to a serious injury; there are two (the first and the third). Adding the probabilities of those disjoint outcomes we find that

$$P(\text{Injured}) = 0.0616 + 0.0851 = 0.1467.$$

For Example TREE DIAGRAMS

RECAP: Let's revisit the battery factory. Remember, it produced both regular and rechargeable batteries, with 25% of them rechargeable. Past history indicates that 2% of the regular batteries and 1% of the rechargeable batteries have some kind of defect.

QUESTION: What's the probability that a battery chosen at random from a shipment of this factory's products turns out to be defective?

ANSWER: *First, I'll create a tree diagram. Batteries are either regular (B) or rechargeable (R), and each type may be defective (D) or OK.*

Check to see if the probabilities of all the possible outcomes add up to 1:

$$0.0150 + 0.7350 + 0.0025 + 0.2475 = 1.0000 \ (Hooray!)$$

And now, the probability that a randomly chosen battery is defective is:

$$P(D) = P(B \cap D) + P(R \cap D) = 0.0150 + 0.0025 = 0.0175$$

Overall, 1.75% of the batteries produced at this factory are defective.

Reversing the Conditioning

If someone suffers a serious injury or dies in a Maryland auto accident, what's the probability he or she wasn't wearing a seat belt? That's an interesting question, but we can't just read it from the tree in Figure 14.5. The tree gives us $P(\text{Injured} \mid \text{No Belt})$, but we want $P(\text{No Belt} \mid \text{Injured})$—conditioning in the other direction. The two probabilities are definitely *not* the same. We have reversed the conditioning.

We may not have the conditional probability we want, but we do know everything we need to know to find it. To find a conditional probability, we need the probability that both events happen divided by the probability that the given event occurs. We have already found the probability of an injury: $0.0616 + 0.0851 = 0.1467$. Also, the joint probability that a person was not wearing a seat belt and was injured is found on the third branch of the tree: $P(\text{No Belt} \cap \text{Injured}) = 0.0851$. With those two pieces of information, we can now find the conditional probability.

$$P(\text{No Belt} \mid \text{Injured}) = \frac{P(\text{No Belt} \cap \text{Injured})}{P(\text{Injured})} = \frac{0.0851}{0.1467} = 0.58$$

Think about that. Even though only 23% all drivers weren't wearing their seat belts, they accounted for 58% of all serious injuries and deaths![4]

---

[4]Just some advice from your friends, the authors: *Please buckle up*. We want you to finish this course.

Step-by-Step Example REVERSING THE CONDITIONING

When the authors were in college, there were only three requirements for graduation that were the same for all students: You had to be able to tread water for 2 minutes, you had to learn a foreign language, and you had to be free of tuberculosis. For the last requirement, all freshmen had to take a TB screening test that consisted of a nurse jabbing what looked like a corncob holder into your forearm. You were then expected to report back in 48 hours to have it checked. If you were healthy and TB-free, your arm was supposed to look as though you'd never had the test.

Sometime during the 48 hours, one of us had a reaction. When he finally saw the nurse, his arm was about 50% bigger than normal and a very unhealthy red. Did he have TB? The nurse had said that the test was about 99% effective, so it seemed that the chances must be pretty high that he had TB. How high do you think the chances were? Go ahead and guess. Guess low.

We'll call TB the event of actually having TB and + the event of testing positive. To start a tree, we need to know $P(\textbf{TB})$, the probability of having TB.[5] We also need to know the conditional probabilities $P(+\,|\,\textbf{TB})$ and $P(+\,|\,\textbf{TB}^{\textbf{C}})$. Diagnostic tests can make two kinds of errors. They can give a positive result for a healthy person (a *false positive*) or a negative result for a sick person (a *false negative*). Being 99% accurate usually means a false-positive rate of 1%. That is, someone who doesn't have the disease has a 1% chance of testing positive anyway. We can write $P(+\,|\,\textbf{TB}^{\textbf{C}}) = 0.01$.

Since a false negative is more serious (because a sick person might not get treatment), tests are usually constructed to have a lower false-negative rate. We don't know exactly, but let's assume a 0.1% false-negative rate. So only 0.1% of sick people test negative. We can write $P(-\,|\,\textbf{TB}) = 0.001$.

THINK ➡		

THINK ➡ Plan Define the events we're interested in and their probabilities.

Figure out what you want to know in terms of the events. Use the notation of conditional probability to write the event whose probability you want to find.

Let $\textbf{TB} = \{\text{having } \textbf{TB}\}$ and $\textbf{TB}^{C} = \{\text{no } \textbf{TB}\}$
   $+ = \{\text{testing positive}\}$ and
   $- = \{\text{testing negative}\}$

I know that $P(+\,|\,\textbf{TB}^{C}) = 0.01$ and $P(-\,|\,\textbf{TB}) = 0.001$. I also know that $P(\textbf{TB}) = 0.00005$.

I'm interested in the probability that the author had TB given that he tested positive: $P(\textbf{TB}\,|\,+)$.

SHOW ➡ Plot Draw the tree diagram. When probabilities are very small like these are, be careful to keep all the significant digits.

To finish the tree we need $P(\textbf{TB}^{\textbf{C}})$, $P(-\,|\,\textbf{TB}^{\textbf{C}})$, and $P(-\,|\,\textbf{TB})$. We can find each of these from the Complement Rule:

$$P(\textbf{TB}^{\textbf{C}}) = 1 - P(\textbf{TB}) = 0.99995$$
$$P(-\,|\,\textbf{TB}^{\textbf{C}}) = 1 - P(+\,|\,\textbf{TB}^{\textbf{C}})$$
$$= 1 - 0.01 = 0.99 \text{ and}$$
$$P(+\,|\,\textbf{TB}) = 1 - P(-\,|\,\textbf{TB})$$
$$= 1 - 0.01 = 0.999$$

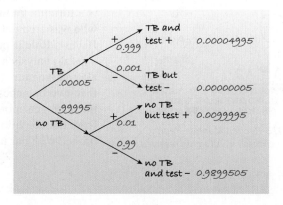

---

[5]This isn't given, so we looked it up. Although TB is a matter of serious concern to public health officials, it is a fairly uncommon disease, with an incidence of about 5 cases per 100,000 in the United States (see *http://www.cdc.gov/tb/default.htm*).

Mechanics Multiply along the branches to find the probabilities of the four possible outcomes. Check your work by seeing if they total 1.

(Check: 0.00004995 + 0.00000005 + 0.0099995 + 0.98995050 = 1)

Add up the probabilities corresponding to the condition of interest—in this case, testing positive. We can add because the tree shows disjoint events.

$$P(+) = P(\text{TB} \cap +) + P(\text{TB}^C \cap +)$$
$$P = 0.00004995 + 0.0099995$$
$$= 0.01004945$$

Divide the probability of both events occurring (here, having TB and a positive test) by the probability of satisfying the condition (testing positive).

$$P(\text{TB} \mid +) = \frac{P(\text{TB} \cap +)}{P(+)}$$
$$= \frac{0.00004995}{0.01004945}$$
$$= 0.00497$$

TELL ➡ **Conclusion** Interpret your result in context.

The chance of having TB after you test positive is less than 0.5%.

When we reverse the order of conditioning, we change the *Who* we are concerned with. With events of low probability, the result can be surprising. That's the reason patients who test positive for HIV, for example, are always told to seek medical counseling. They may have only a small chance of actually being infected. That's why global drug or disease testing can have unexpected consequences if people interpret *testing* positive as *being* positive.

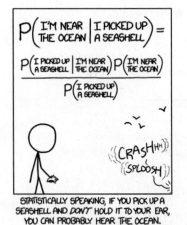

STATISTICALLY SPEAKING, IF YOU PICK UP A SEASHELL AND *DON'T* HOLD IT TO YOUR EAR, YOU CAN PROBABLY HEAR THE OCEAN.

Bayes's Rule

When we have $P(A \mid B)$ but want the *reverse* probability $P(B \mid A)$, we need to find $P(A \cap B)$ and $P(A)$. A tree is often a convenient way of finding these probabilities. It can work even when we have more than two possible events, as we saw in the seat belt example. Instead of using the tree, we *could* write the calculation algebraically, showing exactly how we found the quantities that we needed: $P(A \cap B)$ and $P(A)$. The result is a formula known as Bayes's Rule, after the Reverend Thomas Bayes (1702?–1761), who was credited with the rule after his death, when he could no longer defend himself. Bayes's Rule is quite important in Statistics and is the foundation of an approach to Statistical analysis known as Bayesian Statistics. Although the simple rule deals with two alternative outcomes, the rule can be extended to the situation in which there are more than two branches to the first split of the tree. The principle remains the same (although the math gets more difficult). Bayes's Rule is just a formula[6] for reversing the probability from the conditional probability that you're originally given, the same feat we accomplished with our tree diagram.

---

[6]Bayes's Rule for two events says that $P(B \mid A) = \dfrac{P(A \mid B)P(B)}{P(A \mid B)P(B) + P(A \mid B^C)P(B^C)}$.

Masochists may wish to try it with the TB testing probabilities. (It's easier to just draw the tree, isn't it?)

For Example REVERSING THE CONDITIONING

RECAP: Remember the battery factory that produced 25% regular batteries and the rest the regular kind? A small number of its batteries had defects: 2% of the regular batteries and only 1% of the rechargeable ones.

QUESTION: A quality control inspector inspects some batteries selected at random from each shipment. If one of those batteries turns out to be defective, what's the probability it's the rechargeable type?

ANSWER: In the tree diagram, **B** = regular and **R** = rechargeable; **D** = defective.

$$P(R|D) = \frac{P(R \cap D)}{P(D)}$$

$$= \frac{0.0025}{0.0150 + 0.0025}$$

$$= 0.143$$

Only 14.3% of all defective batteries are the rechargeable kind.

WHAT IF ●●● basketball is really just random events?

You've seen the headlines: ***Redhot LeBron Leads Team to Victory.*** And you've read the article explaining how he hit 9 straight shots. Or heard sportscasters marvel that a player somehow reached a new level of performance. He was "in the zone," they say. Fans, coaches, and players alike believe that for occasional periods of time athletes defy the odds with runs of extraordinary success, proving that for those magic moments they've become even better players. But Tom Gilovich and fellow researchers[7] looked at years of data for many NBA players and found no evidence that anything beyond simple random events (like tossing a coin) ever happens. How can that be?

A S *Video*: **Is There a Hot Hand in Basketball?** Most coaches and fans believe that basketball players sometimes get "hot" and make more of their shots. What do the conditional probabilities say?

A S *Activity*: **Hot Hand Simulation.** Can you tell the difference between real and simulated sequences of basketball shot hits and misses?

Let's do a simulation. Based on a recent season we can expect a great player like LeBron James to attempt about 1500 shots and hit about 55% of them. We randomly generated a sequence of 1's and 0's to mimic this shooting percentage. It started with the following string of "shots":

[7]Gilovich, Thomas, Robert Vallone, and Amos Tversky. The *Hot Hand In Basketball: On The Misperception of Random Sequences*. *Psych.cornell.edu*. Academic Press, Inc. (1985), pg. 295.

01111011000001111110101 0101.

That sequence might represent one game in which he took 25 shots. He missed the first one, but then hit 4 in a row. At one point later on he missed 5 in a row, then bounced back with a streak when he made 6 consecutive shots. Can't you almost hear the breathless announcers talking about this dramatic turnaround?

Sure, six baskets in a row in that game seems like a lot. Sometimes, though, players do even better. We generated an entire simulated season of 1500 shot attempts. The table shows how many times streaks of various lengths occurred.

During this simulated season, our player hit 6 baskets in a row 6 different times. And had 10 streaks that were even longer. Once he even made 13 shots in a row *purely by random chance*. Imagine how impressed fans and announcers—and teammates—would be. But this is not a hot streak. It's just a set of random outcomes based on the same constant shooting percentage. He's not "in the zone"; he just had a run of good luck, no different from you tossing a coin and occasionally getting several heads in a row.

In fact, though it doesn't show in this table, there was one stretch during that season when our simulated player *missed* 11 shots in a row. Boy, was he "cold" that night!

Maybe you're thinking this particular outcome was unusual? Nope. We repeated our simulation to create a total of 15 seasons—a whole career for most players. It turns out that long streaks like this are actually commonplace. In only 2 of the 15 seasons did our player not have a streak of at least 10 consecutive baskets. He made 13 or more in a row (like this season) 5 different times, including one amazing stretch where he hit 17 baskets in a row! Just imagine what sportscasters would say about that . . .

Length of Streak	Number of Times
1	173
2	96
3	61
4	23
5	5
6	6
7	4
8	3
9	1
10	1
13	1

WHAT CAN GO WRONG?

- **Don't use a simple probability rule where a general rule is appropriate.** Don't assume independence without reason to believe it. Don't assume that outcomes are disjoint without checking that they are. Remember that the general rules always apply, even when outcomes are in fact independent or disjoint.

- **Don't find probabilities for samples drawn without replacement as if they had been drawn with replacement.** Remember to adjust the denominator of your probabilities. This warning applies only when we draw from small populations or draw a large fraction of a finite population. When the population is very large relative to the sample size, the adjustments make very little difference, and we ignore them.

- **Don't reverse conditioning naively.** As we have seen, the probability of **A** *given* **B** may not, and, in general does not, resemble the probability of **B** *given* **A.** The true probability may be counterintuitive.

- **Don't confuse "disjoint" with "independent."** Disjoint events *cannot* happen at the same time. When one happens, you know the other did not, so $P(\mathbf{B}|\mathbf{A}) = 0$. Independent events *must* be able to happen at the same time. When one happens, you know it has no effect on the other, so $P(\mathbf{B}|\mathbf{A}) = P(\mathbf{B})$.

What Have We Learned?

We've learned the general rules of probability and how to apply them:

- The General Addition Rule says that $P(\mathbf{A} \cup \mathbf{B}) = P(\mathbf{A}) + P(\mathbf{B}) - P(\mathbf{A} \cap \mathbf{B})$.
- The General Multiplication Rule says that $P(\mathbf{A} \cap \mathbf{B}) = P(\mathbf{A}) \cdot P(\mathbf{B}|\mathbf{A})$.

We've learned to work with conditional probabilities, and have seen that $P(\mathbf{B}|\mathbf{A}) = \dfrac{P(\mathbf{A} \cap \mathbf{B})}{P(\mathbf{A})}$.

We've learned to think clearly about independence:

- We can use conditional probability to determine whether events are independent and to work with events that are not independent.
- Events **A** and **B** are independent if the occurrence of one does not effect the probability that the other occurs: $P(\mathbf{B}|\mathbf{A}) = P(\mathbf{B})$.
- Events that are mutually exclusive (disjoint) cannot be independent.

We've learned to organize our thinking about probability with Venn diagrams, tables, and tree diagrams, and to use tree diagrams to solve problems about reverse conditioning.

Terms

General Addition Rule For any two events, **A** and **B**, the probability of **A** or **B** is

$$P(\mathbf{A} \cup \mathbf{B}) = P(\mathbf{A}) + P(\mathbf{B}) - P(\mathbf{A} \cap \mathbf{B}). \quad \text{(p. 364)}$$

Conditional probability $P(\mathbf{B}|\mathbf{A}) = \dfrac{P(\mathbf{A} \cap \mathbf{B})}{P(\mathbf{A})}$; $P(\mathbf{B}|\mathbf{A})$ is read "the probability of **B** *given* **A**". (p. 368)

Independence (used formally) Events **A** and **B** are independent when $P(\mathbf{B}|\mathbf{A}) = P(\mathbf{B})$. (p. 369)

General Multiplication Rule For any two events, **A** and **B**, the probability of **A** and **B** is

$$P(\mathbf{A} \cap \mathbf{B}) = P(\mathbf{A}) \times P(\mathbf{B}|\mathbf{A}). \quad \text{(p. 374)}$$

Tree diagram A display of conditional events or probabilities that is helpful in thinking through conditioning. (p. 376)

Exercises

1. **Pet ownership** Suppose that 25% of people have a dog, 29% of people have a cat, and 12% of people own both. What is the probability that someone owns a dog or a cat?

2. **Cooking and shopping** Forty-five percent of Americans like to cook and 59% of Americans like to shop, while 23% enjoy both activities. What is the probability that a randomly selected American either enjoys cooking or shopping or both?

3. **Sports** What is the probability that a person likes to watch football, given that she also likes to watch basketball?

	Football	No Football
Basketball	27	13
No Basketball	38	22

4. **Sports again** From Exercise 3, if someone doesn't like to watch basketball, what is the probability that she will be a football fan?

5. **Late to the train** A student always catches his train if class ends on time. However, 30% of classes run late and then there's a 45% chance he'll miss it. What is the probability that he misses the train today?

6. **Field goals** A nervous kicker usually makes 70% of his first field goal attempts. If he makes his first attempt, his success rate rises to 90%. What is the probability that he makes his first two kicks?

7. **Titanic** On the *Titanic*, the probability of survival was 0.323. Among first class passengers, it was 0.625. Were *survival* and *ticket class* independent? Explain.

8. **Births** If the sex of a child is independent of all other births, is the probability of a woman giving birth to a girl after having four boys greater than it was on her first birth? Explain.

9. **Facebook** Facebook reports that 70% of its users are from outside the United States and that 50% of its users log on to Facebook every day. Suppose that 20% of its users are U.S. users who log on every day. Make a probability table.

10. **Online banking** A national survey indicated that 30% of adults conduct their banking online. It also found that 40% are younger than 50, and that 25% are younger than 50 and conduct their banking online. Make a probability table.

11. **Phones** Recent research suggests that 73% of Americans have a home phone, 83% have a cell phone, and 58% of people have both. What is the probability that an American has

    a) a home or cell phone?
    b) neither a home phone nor a cell phone?
    c) a cell phone but no home phone?

12. **Travel** Suppose the probability that a U.S. resident has traveled to Canada is 0.18, to Mexico is 0.09, and to both countries is 0.04. What's the probability that an American chosen at random has

    a) traveled to Canada but not Mexico?
    b) traveled to either Canada or Mexico?
    c) not traveled to either country?

13. **Amenities** A check of dorm rooms on a large college campus revealed that 38% had refrigerators, 52% had TVs, and 21% had both a TV and a refrigerator. What's the probability that a randomly selected dorm room has

    a) a TV but no refrigerator?
    b) a TV or a refrigerator, but not both?
    c) neither a TV nor a refrigerator?

14. **Workers** Employment data at a large company reveal that 72% of the workers are married, that 44% are college graduates, and that half of the college grads are married. What's the probability that a randomly chosen worker

    a) is neither married nor a college graduate?
    b) is married but not a college graduate?
    c) is married or a college graduate?

15. **Global survey** The marketing research organization GfK Custom Research North America conducts a yearly survey on consumer attitudes worldwide. They collect demographic information on the roughly 1500 respondents from each country that they survey. Here is a table showing the number of people with various levels of education in five countries:

Educational Level by Country					
	Post-graduate	College	Some high school	Primary or less	No answer	Total
China	7	315	671	506	3	1502
France	69	388	766	309	7	1539
India	161	514	622	227	11	1535
U.K.	58	207	1240	32	20	1557
U.S.	84	486	896	87	4	1557
Total	379	1910	4195	1161	45	7690

If we select someone at random from this survey,

a) what is the probability that the person is from the United States?
b) what is the probability that the person completed his or her education before college?
c) what is the probability that the person is from France *or* did some post-graduate study?
d) what is the probability that the person is from France *and* finished only primary school or less?

16. **Birth order** A survey of students in a large Introductory Statistics class asked about their birth order (1 = oldest or only child) and which college of the university they were enrolled in. Here are the results:

		Birth Order		
		1 or only	2 or more	Total
College	**Arts & Sciences**	34	23	57
	Agriculture	52	41	93
	Human Ecology	15	28	43
	Other	12	18	30
	Total	113	110	223

Suppose we select a student at random from this class. What is the probability that the person is

a) a Human Ecology student?
b) a firstborn student?
c) firstborn *and* a Human Ecology student?
d) firstborn *or* a Human Ecology student?

17. **Cards** You draw a card at random from a standard deck of 52 cards. Find each of the following conditional probabilities:

a) The card is a heart, given that it is red.
b) The card is red, given that it is a heart.
c) The card is an ace, given that it is red.
d) The card is a queen, given that it is a face card.

18. **Pets** In its monthly report, the local animal shelter states that it currently has 24 dogs and 18 cats available for adoption. Eight of the dogs and 6 of the cats are male.

Find each of the following conditional probabilities if an animal is selected at random:

a) The pet is male, given that it is a cat.
b) The pet is a cat, given that it is female.
c) The pet is female, given that it is a dog.

19. Health The probabilities that an adult American man has high blood pressure and/or high cholesterol are shown in the table.

	Blood Pressure	
Cholesterol	High	OK
High	0.11	0.21
OK	0.16	0.52

What's the probability that

a) a man has both conditions?
b) a man has high blood pressure?
c) a man with high blood pressure has high cholesterol?
d) a man has high blood pressure if it's known that he has high cholesterol?

20. Immigration The table shows the political affiliations of U.S. voters and their positions on supporting stronger immigration enforcement.

	Stronger Immigration Enforcement		
Party	Favor	Oppose	No Opinion
Republican	0.30	0.04	0.03
Democrat	0.22	0.11	0.02
Other	0.16	0.07	0.05

a) What's the probability that
   i) a randomly chosen voter favors stronger immigration enforcement?
   ii) a Republican favors stronger enforcement?
   iii) a voter who favors stronger enforcement is a Democrat?
b) A candidate thinks she has a good chance of gaining the votes of anyone who is a Republican or in favor of stronger enforcement of immigration policy. What proportion of voters is that?

21. Global survey, take 2 Look again at the table summarizing the Roper survey in Exercise 15.

a) If we select a respondent at random, what's the probability we choose a person from the United States who has done post-graduate study?
b) Among the respondents who have done post-graduate study, what's the probability the person is from the United States?
c) What's the probability that a respondent from the United States has done post graduate study?

d) What's the probability that a respondent from China has only a primary-level education?
e) What's the probability that a respondent with only a primary-level education is from China?

22. Birth order, take 2 Look again at the data about birth order of Intro Stats students and their choices of colleges shown in Exercise 16.

a) If we select a student at random, what's the probability the person is an Arts and Sciences student who is a second child (or more)?
b) Among the Arts and Sciences students, what's the probability a student was a second child (or more)?
c) Among second children (or more), what's the probability the student is enrolled in Arts and Sciences?
d) What's the probability that a first or only child is enrolled in the Agriculture College?
e) What is the probability that an Agriculture student is a first or only child?

23. Sick kids Seventy percent of kids who visit a doctor have a fever, and 30% of kids with a fever have sore throats. What's the probability that a kid who goes to the doctor has a fever and a sore throat?

24. Sick cars Twenty percent of cars that are inspected have faulty pollution control systems. The cost of repairing a pollution control system exceeds $100 about 40% of the time. When a driver takes her car in for inspection, what's the probability that she will end up paying more than $100 to repair the pollution control system?

25. Cards You are dealt a hand of three cards, one at a time. Find the probability of each of the following.

a) The first heart you get is the third card dealt.
b) Your cards are all red (that is, all diamonds or hearts).
c) You get no spades.
d) You have at least one ace.

26. Another hand You pick three cards at random from a deck. Find the probability of each event described below.

a) You get no aces.
b) You get all hearts.
c) The third card is your first red card.
d) You have at least one diamond.

27. Batteries A junk box in your room contains a dozen old batteries, five of which are totally dead. You start picking batteries one at a time and testing them. Find the probability of each outcome.

a) The first two you choose are both good.
b) At least one of the first three works.
c) The first four you pick all work.
d) You have to pick 5 batteries to find one that works.

28. Shirts The soccer team's shirts have arrived in a big box, and people just start grabbing them, looking for the right size. The box contains 4 medium, 10 large, and 6 extra-large shirts. You want a medium for you and one for your sister. Find the probability of each event described.

a) The first two you grab are the wrong sizes.
b) The first medium shirt you find is the third one you check.
c) The first four shirts you pick are all extra-large.
d) At least one of the first four shirts you check is a medium.

29. **Eligibility** A university requires its biology majors to take a course called BioResearch. The prerequisite for this course is that students must have taken either a Statistics course or a computer course. By the time they are juniors, 52% of the Biology majors have taken Statistics, 23% have had a computer course, and 7% have done both.

a) What percent of the junior Biology majors are ineligible for BioResearch?
b) What's the probability that a junior Biology major who has taken Statistics has also taken a computer course?
c) Are taking these two courses disjoint events? Explain.
d) Are taking these two courses independent events? Explain.

30. **Benefits** Fifty-six percent of all American workers have a workplace retirement plan, 68% have health insurance, and 49% have both benefits. We select a worker at random.

a) What's the probability he has neither employer-sponsored health insurance nor a retirement plan?
b) What's the probability he has health insurance if he has a retirement plan?
c) Are having health insurance and a retirement plan independent events? Explain.
d) Are having these two benefits mutually exclusive? Explain.

31. **Cell phones in the home** A survey found that 73% of Americans have a home phone, 83% have a cell phone and 58% of people have both.

a) If a person has a home phone, what's the probability that they have a cell phone also?
b) Are having a home phone and a cell phone independent events? Explain.
c) Are having a home phone and a cell phone mutually exclusive? Explain.

32. **On the road again** According to Exercise 12, the probability that a U.S. resident has traveled to Canada is 0.18, to Mexico is 0.09, and to both countries is 0.04.

a) What's the probability that someone who has traveled to Mexico has visited Canada too?
b) Are traveling to Mexico and to Canada disjoint events? Explain.
c) Are traveling to Mexico and to Canada independent events? Explain.

33. **Cards** If you draw a card at random from a well-shuffled deck, is getting an ace independent of the suit? Explain.

34. **Pets again** The local animal shelter in Exercise 18 reported that it currently has 24 dogs and 18 cats available for adoption; 8 of the dogs and 6 of the cats are male. Are the species and sex of the animals independent? Explain.

35. **Unsafe food** Early in 2010, *Consumer Reports* published the results of an extensive investigation of broiler chickens purchased from food stores in 23 states. Tests for bacteria in the meat showed that 62% of the chickens were contaminated with campylobacter, 14% with salmonella, and 9% with both.

a) What's the probability that a tested chicken was not contaminated with either kind of bacteria?
b) Are contamination with the two kinds of bacteria disjoint? Explain.
c) Are contamination with the two kinds of bacteria independent? Explain.

36. **Birth order, finis** In Exercises 16 and 22 we looked at the birth orders and college choices of some Intro Stats students. For these students:

a) Are enrolling in Agriculture and Human Ecology disjoint? Explain.
b) Are enrolling in Agriculture and Human Ecology independent? Explain.
c) Are being firstborn and enrolling in Human Ecology disjoint? Explain.
d) Are being firstborn and enrolling in Human Ecology independent? Explain.

37. **Men's health, again** Given the table of probabilities from Exercise 19, are high blood pressure and high cholesterol independent? Explain.

		Blood Pressure	
		High	OK
Cholesterol	High	0.11	0.21
	OK	0.16	0.52

38. **Politics** Given the table of probabilities from Exercise 20, are party affiliation and position on immigration independent? Explain.

		Stronger Immigration Enforcement		
		Favor	Oppose	No Opinion
Party	Republican	0.30	0.04	0.03
	Democrat	0.22	0.11	0.02
	Other	0.16	0.07	0.05

39. Phone service According to estimates from the federal government's 2010 National Health Interview Survey, based on face-to-face interviews in 16,676 households, approximately 63.6% of U.S. adults have both a landline in their residence and a cell phone, 25.4% have only cell phone service but no landline, and 1.8% have no telephone service at all.

a) Polling agencies won't phone cell phone numbers because customers object to paying for such calls. What proportion of U.S. households can be reached by a landline call?

b) Are having a cell phone and having a landline independent? Explain.

40. Snoring After surveying 995 adults, 81.5% of whom were over 30, the National Sleep Foundation reported that 36.8% of all the adults snored. 32% of the respondents were snorers over the age of 30.

a) What percent of the respondents were under 30 and did not snore?

b) Is snoring independent of age? Explain.

41. Gender A 2009 poll conducted by Gallup classified respondents by sex and political party, as shown in the table. Is party affiliation independent of the respondents' sex? Explain.

	Democrat	Republican	Independent
Male	32	28	34
Female	41	25	26

42. Cars A random survey of autos parked in student and staff lots at a large university classified the brands by country of origin, as seen in the table. Is country of origin independent of type of driver?

		Driver	
		Student	Staff
Origin	American	107	105
	European	33	12
	Asian	55	47

43. Luggage Leah is flying from Boston to Denver with a connection in Chicago. The probability her first flight leaves on time is 0.15. If the flight is on time, the probability that her luggage will make the connecting flight in Chicago is 0.95, but if the first flight is delayed, the probability that the luggage will make it is only 0.65.

a) Are the first flight leaving on time and the luggage making the connection independent events? Explain.

b) What is the probability that her luggage arrives in Denver with her?

44. Graduation A private college report contains these statistics:

70% of incoming freshmen attended public schools.
75% of public school students who enroll as freshmen eventually graduate.
90% of other freshmen eventually graduate.

a) Is there any evidence that a freshman's chances to graduate may depend upon what kind of high school the student attended? Explain.

b) What percent of freshmen eventually graduate?

45. Late luggage Remember Leah (Exercise 43)? Suppose you pick her up at the Denver airport, and her luggage is not there. What is the probability that Leah's first flight was delayed?

46. Graduation, part II What percent of students who graduate from the college in Exercise 44 attended a public high school?

47. Absenteeism A company's records indicate that on any given day about 1% of their day-shift employees and 2% of the night-shift employees will miss work. Sixty percent of the employees work the day shift.

a) Is absenteeism independent of shift worked? Explain.

b) What percent of employees are absent on any given day?

48. E-readers In 2011, 12% of Americans owned an electronic reader of some sort. Suppose that 43% of people with e-readers read at least 3 books last year, while among people without an e-reader, only 11% read 3 or more books in the course of the year.

a) Explain how these statistics indicate that owning an e-reader and reading 3 or more books are not independent.

b) What's the probability that a randomly selected American has read 3 or more books?

49. Absenteeism, part II At the company described in Exercise 47, what percent of the absent employees are on the night shift?

50. E-readers II Given the e-reader data presented in Exercise 48, if a randomly selected American has read 3 or more books, what's the probability that he or she owns an e-reader?

51. Drunks Police often set up sobriety checkpoints—roadblocks where drivers are asked a few brief questions to allow the officer to judge whether or not the person may have been drinking. If the officer does not suspect a problem, drivers are released to go on their way. Otherwise, drivers are detained for a Breathalyzer test that will determine whether or not they will be arrested. The police say that based on the brief initial stop, trained officers can make the right decision 80% of the time. Suppose the police operate a sobriety checkpoint after 9 P.M. on a Saturday night, a time when national traffic safety experts suspect that about 12% of drivers have been drinking.

a) You are stopped at the checkpoint and, of course, have not been drinking. What's the probability that you are detained for further testing?

b) What's the probability that any given driver will be detained?

c) What's the probability that a driver who is detained has actually been drinking?

d) What's the probability that a driver who was released had actually been drinking?

52. No-shows An airline offers discounted "advance-purchase" fares to customers who buy tickets more than 30 days before travel and charges "regular" fares for tickets purchased during those last 30 days. The company has noticed that 60% of its customers take advantage of the advance-purchase fares. The "no-show" rate among people who paid regular fares is 30%, but only 5% of customers with advance-purchase tickets are no-shows.

a) What percent of all ticket holders are no-shows?

b) What's the probability that a customer who didn't show had an advance-purchase ticket?

c) Is being a no-show independent of the type of ticket a passenger holds? Explain.

53. Dishwashers Dan's Diner employs three dishwashers. Al washes 40% of the dishes and breaks only 1% of those he handles. Betty and Chuck each wash 30% of the dishes, and Betty breaks only 1% of hers, but Chuck breaks 3% of the dishes he washes. (He, of course, will need a new job soon. . . .) You go to Dan's for supper one night and hear a dish break at the sink. What's the probability that Chuck is on the job?

54. Parts A company manufacturing electronic components for home entertainment systems buys electrical connectors from three suppliers. The company prefers to use supplier A because only 1% of those connectors prove to be defective, but supplier A can deliver only 70% of the connectors needed. The company must also purchase connectors from two other suppliers, 20% from supplier B and the rest from supplier C. The rates of defective connectors from B and C are 2% and 4%, respectively. You buy one of these components, and when you try to use it you find that the connector is defective. What's the probability that your component came from supplier A?

55. HIV testing In July 2005 the journal *Annals of Internal Medicine* published a report on the reliability of HIV testing. Results of a large study suggested that among people with HIV, 99.7% of tests conducted were (correctly) positive, while for people without HIV 98.5% of the tests were (correctly) negative. A clinic serving an at-risk population offers free HIV testing, believing that 15% of the patients may actually carry HIV. What's the probability that a patient testing negative is truly free of HIV?

56. Polygraphs Lie detectors are controversial instruments, barred from use as evidence in many courts. Nonetheless, many employers use lie detector screening as part of their hiring process in the hope that they can avoid hiring people who might be dishonest. There has been some research, but no agreement, about the reliability of polygraph tests. Based on this research, suppose that a polygraph can detect 65% of lies, but incorrectly identifies 15% of true statements as lies.

A certain company believes that 95% of its job applicants are trustworthy. The company gives everyone a polygraph test, asking, "Have you ever stolen anything from your place of work?" Naturally, all the applicants answer "No," but the polygraph identifies some of those answers as lies, making the person ineligible for a job. What's the probability that a job applicant rejected under suspicion of dishonesty was actually trustworthy?

 Just Checking ANSWERS

1. a)

b) 0.32   c) 0.41

2. a) No; 8 people are both male and supervisors.

b) Yes. $P(F|S) = \dfrac{12}{20} = 0.6$ and $P(F) = \dfrac{90}{150} = 0.6$, so $P(F|S) = P(F)$.

3. a)

		Equation		
		Yes	No	Total
Display	Yes	0.07	0.41	0.48
	No	0.20	0.32	0.52
	Total	0.27	0.73	1.00

b) $P(D|Eq) = P(D \cap Eq)/P(Eq) = 0.07/0.27$
$= 0.259$

c) No, pages can (and 7% do) have both.

d) To be independent, we'd need $P(D|Eq) = P(D)$. $P(D|Eq) = 0.259$, but $P(D) = 0.48$. Overall, 48% of pages have data displays, but only about 26% of pages with equations do. They do not appear to be independent.

4. a) $\dfrac{4}{20} \cdot \dfrac{3}{19}$ b) $\dfrac{17}{20} \cdot \dfrac{16}{19} \cdot \dfrac{15}{18}$ c) $\dfrac{19}{20} \cdot \dfrac{18}{19} \cdot \dfrac{17}{18} \cdot \dfrac{1}{17}$

15 Random Variables

Insurance companies make bets. They bet that you're going to live a long life. You bet that you're going to die sooner. Both you and the insurance company want the company to stay in business, so it's important to find a "fair price" for your bet. Of course, the right price for *you* depends on many factors, and nobody can predict exactly how long you'll live. But when the company averages over enough customers, it can make reasonably accurate estimates of the amount it can expect to collect on a policy before it has to pay its benefit.

Here's a simple example. An insurance company offers a "death and disability" policy that pays $10,000 when you die or $5000 if you are permanently disabled. It charges a premium of only $50 a year for this benefit. Is the company likely to make a profit selling such a plan? To answer this question, the company needs to know the *probability* that its clients will die or be disabled in any year. From actuarial information like this, the company can calculate the expected value of this policy.

Expected Value: Center

We'll want to build a probability model in order to answer the questions about the insurance company's risk. First we need to define a few terms. The amount the company pays out on an individual policy is called a **random variable** because its numeric value is based on the outcome of a random event. We use a capital letter, like X, to denote a random variable. We'll denote a particular value that it can have by the corresponding lowercase letter, in this case x. For the insurance company, x can be $10,000 (if you die that year), $5000 (if you are disabled), or $0 (if neither occurs). Because we can list all the outcomes, we might formally call this random variable a **discrete random variable**. Otherwise,

A S   *Activity*: **Random Variables.**
Learn more about random variables from
this animated tour.

Table 15.1

The probability model shows all
the possible values of the random variable and their associated
probabilities.

we'd call it a **continuous random variable**. The collection of all the possible values and
the probabilities that they occur is called the **probability model** for the random variable.

Suppose, for example, that the death rate in any year is 1 out of every 1000 people,
and that another 2 out of 1000 suffer some kind of disability. Then we can display the
probability model for this insurance policy in a table like this:

Policyholder Outcome	Payout x	Probability $P(x)$
Death	10,000	$\frac{1}{1000}$
Disability	5000	$\frac{2}{1000}$
Neither	0	$\frac{997}{1000}$

Given these probabilities, what should the insurance company expect to pay out? They
can't know exactly what will happen in any particular year, but they can calculate what to
expect in the long run. In a probability model, we call that the **expected value** and denote
in two ways: $E(X)$ and μ. $E(X)$ is just short hand for expected value of x. We use μ when
we want to emphasize that it is a parameter of a model. To understand the calculation
for the expected value, imagine that the company insures exactly 1000 people. Further
imagine that, in perfect accordance with the probabilities, 1 of the policyholders dies,
2 are disabled, and the remaining 997 survive the year unscathed. The company would pay
$10,000 to one client and $5000 to each of 2 clients. That's a total of $20,000, or an aver-
age of 20000/1000 = $20 per policy. Since it is charging people $50 for the policy, the
company expects to make a profit of $30 per customer. Not bad!

Let's look at this expected value calculation more closely. We imagined that we have
exactly 1000 clients. Of those, exactly 1 died and 2 were disabled, corresponding to what
the probabilities say. The average payout is:

$$\mu = E(X) = \frac{10,000(1) + 5000(2) + 0(997)}{1000} = \$20 \text{ per policy.}$$

Instead of writing the expected value as one big fraction, we can rewrite it as separate
terms with a common denominator of 1000.

$$E(X) = \$10,000\left(\frac{1}{1000}\right) + \$5000\left(\frac{2}{1000}\right) + \$0\left(\frac{997}{1000}\right)$$
$$= \$20.$$

How convenient! See the probabilities? For each policy, there's a 1/1000 chance that we'll
have to pay $10,000 for a death and a 2/1000 chance that we'll have to pay $5000 for a
disability. Of course, there's a 997/1000 chance that we won't have to pay anything.

Take a good look at the expression now. It's easy to calculate the **expected value** of
a (discrete) random variable—just multiply each possible value by the probability that it
occurs, and find the sum:

$$\mu = E(X) = \sum xP(x).$$

Be sure that every possible outcome is included in the sum. And verify that you have a
valid probability model to start with—the probabilities should each be between 0 and 1
and should sum to one.

For Example   LOVE AND EXPECTED VALUES

On Valentine's Day the *Quiet Nook* restaurant offers a *Lucky Lovers Special* that could save couples money on their romantic dinners. When the waiter brings the check, he'll also bring the four aces from a deck of cards. He'll shuffle them and lay them out face down on the table. The couple will then get to turn one card over. If it's a black ace, they'll owe the full amount, but if it's the ace of hearts, the waiter will give them a $20 Lucky Lovers discount. If they first turn over the ace of diamonds (hey—at least it's red!), they'll then get to turn over one of the remaining cards, earning a $10 discount for finding the ace of hearts this time.

QUESTION: Based on a probability model for the size of the Lucky Lovers discounts the restaurant will award, what's the expected discount for a couple?

ANSWER: Let $X =$ the Lucky Lovers discount. The probabilities of the three outcomes are:

$$P(X = 20) = P(A\heartsuit) = \frac{1}{4}$$

$$P(X = 10) = P(A\diamondsuit, \text{ then } A\heartsuit) = P(A\diamondsuit) \cdot P(A\heartsuit | A\diamondsuit)$$

$$= \frac{1}{4} \cdot \frac{1}{3} = \frac{1}{12}$$

$$P(X = 0) = P(X \neq 20 \text{ or } 10) = 1 - \left(\frac{1}{4} + \frac{1}{12} \right) = \frac{2}{3}.$$

My probability model is:

Outcome	A♥	A♦, then A ♥	Black Ace
x	20	10	0
P(X = x)	$\frac{1}{4}$	$\frac{1}{12}$	$\frac{2}{3}$

$$E(X) = 20 \cdot \frac{1}{4} + 10 \cdot \frac{1}{12} + 0 \cdot \frac{2}{3} = \frac{70}{12} \approx 5.83$$

Couples dining at the Quiet Nook can expect an average discount of $5.83.

 ## Just Checking

1. One of the authors took his minivan in for repair because the air conditioner was cutting out intermittently. The mechanic identified the problem as dirt in a control unit. He said that in about 75% of such cases, drawing down and then recharging the coolant a couple of times cleans up the problem—and costs only $60. If that fails, then the control unit must be replaced at an additional cost of $100 for parts and $40 for labor.

a) Define the random variable and construct the probability model.
b) What is the expected value of the cost of this repair?
c) What does that mean in this context?

Oh—in case you were wondering—the $60 fix worked!

First Center, Now Spread . . .

Of course, this expected value (or mean) is not what actually happens to any *particular* policyholder. No individual policy actually costs the company $20. We are dealing with random events, so some policyholders receive big payouts, others nothing. Because the insurance company must anticipate this variability, it needs to know the *standard deviation* of the random variable.

For data, we calculated the **standard deviation** by first computing the deviation from the mean and squaring it. We do that with (discrete) random variables as well. First, we find the deviation of each payout from the mean (expected value):

Table 15.2

Deviations from the mean.

Policyholder Outcome	Payout x	Probability $P(X = x)$	Deviation $(x - \mu)$
Death	10,000	$\frac{1}{1000}$	$(10,000 - 20) = 9980$
Disability	5000	$\frac{2}{1000}$	$(5000 - 20) = 4980$
Neither	0	$\frac{997}{1000}$	$(0 - 20) = -20$

Next, we square each deviation. The **variance** is the expected value of those squared deviations, so we multiply each by the appropriate probability and sum those products. That gives us the variance of X. Here's what it looks like:

$$Var(X) = 9980^2\left(\frac{1}{1000}\right) + 4980^2\left(\frac{2}{1000}\right) + (-20)^2\left(\frac{997}{1000}\right) = 149,600.$$

Finally, we take the square root to get the standard deviation:

$$SD(X) = \sqrt{149,600} \approx \$386.78.$$

The insurance company can expect an average payout of $20 per policy, with a standard deviation of $386.78.

Think about that. The company charges $50 for each policy and expects to pay out $20 per policy. Sounds like an easy way to make $30. In fact, most of the time (probability 997/1000) the company pockets the entire $50. But would you consider selling your neighbor such a policy? The problem is that occasionally the company loses big. With probability 1/1000, it will pay out $10,000, and with probability 2/1000, it will pay out $5000. That's probably more risk than you're willing to take on in a single policy with your neighbor. The standard deviation of $386.78 gives an indication that the outcome is no sure thing. That's a pretty big spread (and risk) for an average profit of $30.

Here are the formulas for what we just did. Because these are parameters of our probability model, the variance and standard deviation can also be written as σ^2 and σ. You should recognize both kinds of notation.

Variance and Standard Deviation

$\sigma^2 = Var(X)$

$\quad = \sum (x - \mu)^2 \, P(X)$

$\sigma = SD(X) = \sqrt{Var(X)}$

$$\sigma^2 = Var(X) = \sum (x - \mu)^2 P(x)$$
$$\sigma = SD(X) = \sqrt{Var(X)}$$

For Example FINDING THE STANDARD DEVIATION

RECAP: Here's the probability model for the Lucky Lovers restaurant discount.

Outcome	A♥	A♦, then A♥	Black Ace
x	20	10	0
$P(X = x)$	$\dfrac{1}{4}$	$\dfrac{1}{12}$	$\dfrac{2}{3}$

We found that couples can expect an average discount of $\mu = \$5.83$.

QUESTION: What's the standard deviation of the discounts?

ANSWER: First find the variance:

$$Var(X) = \sum (x - \mu)^2 \cdot P(x)$$

$$= (20 - 5.83)^2 \cdot \frac{1}{4} + (10 - 5.83)^2 \cdot \frac{1}{12} + (0 - 5.83)^2 \cdot \frac{2}{3}$$

$$\approx 74.306.$$

So, $SD(X) = \sqrt{74.306} \approx \$8.62$

Couples can expect the Lucky Lovers discounts to average $\$5.83$, with a standard deviation of $\$8.62$.

Step-by-Step Example EXPECTED VALUES AND STANDARD DEVIATIONS FOR DISCRETE RANDOM VARIABLES

As the head of inventory for Knowway computer company, you were thrilled that you had managed to ship 2 computers to your biggest client the day the order arrived. You are horrified, though, to find out that someone had restocked refurbished computers in with the new computers in your storeroom. The shipped computers were selected randomly from the 15 computers in stock, but 4 of those were actually refurbished.

If your client gets 2 new computers, things are fine. If the client gets one refurbished computer, it will be sent back at your expense—$100—and you can replace it. However, if both computers are refurbished, the client will cancel the order this month and you'll lose a total of $1000.

Question: What's the expected value and the standard deviation of the company's loss?

THINK ➡ **Plan** State the problem.	I want to find the company's expected loss for shipping refurbished computers and the standard deviation.
Variable Define the random variable.	Let X = amount of loss.

(continued)

Plot Make a picture. This is another job for tree diagrams.

If you prefer calculation to drawing, find $P(\mathbf{NN})$ and $P(\mathbf{RR})$, then use the Complement Rule to find $P(\mathbf{NR}\ or\ \mathbf{RN})$.

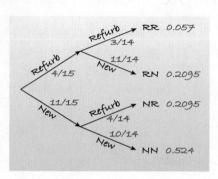

Model List the possible values of the random variable, and determine the probability model.

Outcome	x	$P(X=x)$
Two refurbs	1000	$P(\mathbf{RR}) = 0.057$
One refurb	100	$P(\mathbf{NR} \cup \mathbf{RN}) = 0.2095$ $+ 0.2095 = 0.419$
New/new	0	$P(\mathbf{NN}) = 0.524$

SHOW ➡ **Mechanics** Find the expected value.

$$E(X) = 0(0.524) + 100(0.419) + 1000(0.057)$$
$$= \$98.90$$

Find the variance.

$$\begin{aligned} Var(X) = &(0 - 98.90)^2(0.524) \\ &+ (100 - 98.90)^2(0.419) \\ &+ (1000 - 98.90)^2(0.057) \\ &= 51,408.79 \end{aligned}$$

Find the standard deviation.

$$SD(X) = \sqrt{51,408.79} = \$226.735$$

TELL ➡ **Conclusion** Interpret your results in context.

REALITY CHECK ➡ Both numbers seem reasonable. The expected value of $98.90 is between the extremes of $0 and $1000, and there's great variability in the outcome values.

I expect this mistake to cost the firm $98.90, with a standard deviation of $226.74. The large standard deviation reflects the fact that there's a pretty large range of possible losses.

TI Tips FINDING THE MEAN AND SD OF A RANDOM VARIABLE

You can easily calculate means and standard deviations for a random variable with your TI. Let's do the Knowway computer example.
- Enter the values of the variable in a list, say, L1: 0, 100, 1000.
- Enter the probability model in another list, say, L2. Notice that you can enter the probabilities as fractions. For example, multiplying along the top branches of the

(continued)

```
1-Var Stats L1,L
2■
```

```
1-Var Stats
x̄=99.04761905
Σx=99.04761905
Σx²=61333.3333
Sx=
σx=226.986569
↓n=1
■
```

tree gives the probability of a $1000 loss to be $\frac{4}{15} \cdot \frac{3}{14}$. When you enter that, the TI will automatically calculate the probability as a decimal!

- Under the STAT CALC menu, choose 1-Var Stats with List:L1, FreqList:L2, then go to Calculate and hit ENTER. (OR on an older calculator just ask for 1-Var Stats L1,L2.)

Now you see the mean and standard deviation (along with some other things). Don't fret that the calculator's mean and standard deviation aren't precisely the same as the ones we found. Such minor differences can arise whenever we round off probabilities to do the work by hand.

Beware: Although the calculator knows enough to call the standard deviation σ, it uses $\bar{x}$ where it should say μ. Make sure you don't make that mistake!

More About Means and Variances

Our insurance company expected to pay out an average of $20 per policy, with a standard deviation of about $387. If we take the $50 premium into account, we see the company makes a profit of $50 - 20 = \$30$ per policy. Suppose the company lowers the premium by $5 to $45. It's pretty clear that the expected profit also drops an average of $5 per policy, to $45 - 20 = \$25$.

What about the standard deviation? The differences among payouts hasn't changed. We know that adding or subtracting a constant from data shifts the mean but doesn't change the variance or standard deviation. The same is true of random variables.[1]

$$E(X \pm c) = E(X) \pm c \qquad Var(X \pm c) = Var(X).$$

For Example ADDING A CONSTANT

RECAP: We've determined that couples dining at the *Quiet Nook* can expect Lucky Lovers discounts averaging $5.83 with a standard deviation of $8.62. Suppose that for several weeks the restaurant has also been distributing coupons worth $5 off any one meal (one discount per table).

QUESTION: If every couple dining there on Valentine's Day brings a coupon, what will be the mean and standard deviation of the total discounts they'll receive?

ANSWER: Let D = total discount (Lucky Lovers plus the coupon); then $D = X + 5$.

$$E(D) = E(X + 5) = E(X) + 5 = 5.83 + 5 = \$10.83$$

$$Var(D) = Var(X + 5) = Var(X) = 8.62^2$$

$$SD(D) = \sqrt{Var(X)} = \$8.62$$

Couples with the coupon can expect total discounts averaging $10.83. The standard deviation is still $8.62.

Back to insurance . . . What if the company decides to double all the payouts—that is, pay $20,000 for death and $10,000 for disability? This would double the average payout per policy and also increase the variability in payouts. We have seen that multiplying or dividing all data values by a constant changes both the mean and the standard deviation by the same factor. Variance, being the square of standard deviation, changes by the square of the constant. The same is true of random variables. In general, multiplying each value of

---

[1]The rules in this section are true for both discrete *and* continuous random variables.

a random variable by a constant multiplies the mean by that constant and the variance by the *square* of the constant.

$$E(aX) = aE(X) \qquad Var(aX) = a^2Var(X)$$

For Example DOUBLE THE LOVE

RECAP: On Valentine's Day at the *Quiet Nook*, couples may get a Lucky Lovers discount averaging $5.83 with a standard deviation of $8.62. When two couples dine together on a single check, the restaurant doubles the discount offer—$40 for the ace of hearts on the first card and $20 on the second.

QUESTION: What are the mean and standard deviation of discounts for such foursomes?

ANSWER:

$$E(2X) = 2E(X) = 2(5.83) = \$11.66$$
$$Var(2x) = 2^2 Var(x) = 2^2 \cdot 8.62^2 = 297.2176$$
$$SD(2X) = \sqrt{297.2176} = \$17.24$$

If the restaurant doubles the discount offer, two couples dining together can expect to save an average of $11.66 with a standard deviation of $17.24.

An insurance company sells policies to more than just one person. How can the company find the total expected value (and standard deviation) of policies taken over all policyholders? Consider a simple case: just two customers, Mr. Ecks and Ms. Wye. With an expected payout of $20 on each policy, we might predict a total of $20 + $20 = $40 to be paid out on the two policies. Nothing surprising there. The expected value of the sum is the sum of the expected values.

$$E(X + Y) = E(X) + E(Y).$$

The variability is another matter. Is the risk of insuring two people the same as the risk of insuring one person for twice as much? We wouldn't expect both clients to die or become disabled in the same year. Because we've spread the risk, the standard deviation should be smaller. Indeed, this is the fundamental principle behind insurance. By spreading the risk among many policies, a company can keep the standard deviation quite small and predict costs more accurately.

But how much smaller is the standard deviation of the sum? It turns out that, if the random variables are independent, there is a simple **Addition Rule for Variances:** *The variance of the sum of two independent random variables is the sum of their individual variances.*

For Mr. Ecks and Ms. Wye, the insurance company can expect their outcomes to be independent, so (using *X* for Mr. Ecks's payout and *Y* for Ms. Wye's)

$$Var(X + Y) = Var(X) + Var(Y)$$
$$= 149{,}600 + 149{,}600$$
$$= 299{,}200.$$

If they had insured only Mr. Ecks for twice as much, there would only be one outcome rather than two *independent* outcomes, so the variance would have been

$$Var(2X) = 2^2Var(X) = 4 \times 149{,}600 = 598{,}400, \text{ or}$$

twice as big as with two independent policies.

Of course, variances are in squared units. The company would prefer to know standard deviations, which are in dollars. The standard deviation of the payout for two independent policies is $\sqrt{299{,}200} = \$546.99$. But the standard deviation of the payout for a single policy of twice the size is $\sqrt{598{,}400} = \$773.56$, or about 40% more.

If the company has two customers, then, it will have an expected annual total payout of $40 with a standard deviation of about $547.

For Example ADDING THE DISCOUNTS

RECAP: The Valentine's Day Lucky Lovers discount for couples averages $5.83 with a standard deviation of $8.62. We've seen that if the restaurant doubles the discount offer for two couples dining together on a single check, they can expect to save $11.66 with a standard deviation of $17.24. Some couples decide instead to get separate checks and pool their two discounts.

QUESTION: You and your amour go to this restaurant with another couple and agree to share any benefit from this promotion. Does it matter whether you pay separately or together?

ANSWER: Let X_1 and X_2 represent the two separate discounts, and T the total; then $T = X_1 + X_2$.

$$E(T) = E(X_1 + X_2) = E(X_1) + E(X_2) = 5.83 + 5.83 = \$11.66,$$

so the expected saving is the same either way.

The cards are reshuffled for each couple's turn, so the discounts couples receive are independent. It's okay to add the variances:

$$Var(T) = Var(X_1 + X_2) = Var(X_1) + Var(X_2) = 8.62^2 + 8.62^2 = 148.6088$$
$$SD(T) = \sqrt{148.6088} = \$12.19$$

When two couples get separate checks, there's less variation in their total discount. The standard deviation is $12.19, compared to $17.24 for couples who play for the double discount on a single check. It does, therefore, matter whether they pay separately or together.

In general,

- *The mean of the sum of two random variables is the sum of the means.*
- *The mean of the difference of two random variables is the difference of the means.*
- *If the random variables are independent, the variance of their sum or difference is always the sum of the variances.*

$$E(X \pm Y) = E(X) \pm E(Y) \qquad Var(X \pm Y) = Var(X) + Var(Y)$$

Wait a minute! Is that third part correct? Do we always *add* variances? Yes. Think about the two insurance policies. Suppose we want to know the mean and standard deviation of the *difference* in payouts to the two clients. Since each policy has an expected payout of $20, the expected difference is $20 - 20 = \$0$. If we also subtract variances, we get $0, too, and that surely doesn't make sense. Note that if the outcomes for the two clients are independent, the difference in payouts could range from $10,000 - \$0 = \$10,000$ to $\$0 - \$10,000 = -\$10,000$, a spread of $20,000. The variability in differences increases as much as the variability in sums. If the company has two customers, the difference in payouts has a mean of $0 and a standard deviation of about $547 (again).

Pythagorean Theorem of Statistics

We often use the standard deviation to measure variability, but when we add independent random variables, we use their variances. Think of the Pythagorean Theorem. In a right triangle (only), the *square* of the length of the hypotenuse is the sum of the *squares* of the lengths of the other two sides:

$$c^2 = a^2 + b^2.$$

For independent random variables (only), the *square* of the standard deviation of their sum is the sum of the *squares* of their standard deviations:

$$SD^2(X + Y) = SD^2(X) + SD^2(Y).$$

It's simpler to write this with *variances:*

For independent random variables, X and Y, $Var(X + Y) = Var(X) + Var(Y)$.

For Example WORKING WITH DIFFERENCES

RECAP: The Lucky Lovers discount at the *Quiet Nook* averages $5.83 with a standard deviation of $8.62. Just up the street, the *Wise Fool* restaurant has a competing Lottery of Love promotion. There a couple can select a specially prepared chocolate from a large bowl and unwrap it to learn the size of their discount. The restaurant's manager says the discounts vary with an average of $10.00 and a standard deviation of $15.00.

QUESTION: How much more can you expect to save at the *Wise Fool*? With what standard deviation?

ANSWER: Let W = *discount at the Wise Fool*, X = *the discount at the Quiet Nook*, and D = *the difference:* $D = W - X$. These are different promotions at separate restaurants, so the outcomes are independent.

$$E(W - X) = E(W) - E(X) = 10.00 - 5.83 = \$4.17$$
$$\begin{aligned} SD(W - X) &= \sqrt{Var(W - X)} \\ &= \sqrt{Var(W) + Var(X)} \\ &= \sqrt{15^2 + 8.62^2} \\ &\approx \$17.30 \end{aligned}$$

Discounts at the Wise Fool will average $4.17 more than at the Quiet Nook, with a standard deviation of $17.30.

For Random Variables, Does $X + X + X = 3X$?

Maybe, but be careful. As we've just seen, insuring one person for $30,000 is not the same risk as insuring three people for $10,000 each. When each instance represents a different outcome for the same random variable, it's easy to fall into the trap of writing all of them with the same symbol. Don't make this common mistake. Make sure you write each instance as a *different* random variable. Just because each random variable describes a similar situation doesn't mean that each random outcome will be the same.

These are *random* variables, not the variables you saw in Algebra. Being random, they take on different values each time they're evaluated. So what you really mean is $X_1 + X_2 + X_3$. Written this way, it's clear that the sum shouldn't necessarily equal 3 times *anything*. We'll explore this important issue in this chapter's What If.

For Example SUMMING A SERIES OF OUTCOMES

RECAP: The *Quiet Nook*'s Lucky Lovers promotion offers couples discounts averaging $5.83 with a standard deviation of $8.62. The restaurant owner is planning to serve 40 couples on Valentine's Day.

QUESTION: What's the expected total of the discounts the owner will give? With what standard deviation?

ANSWER: Let $X_1, X_2, X_3, \ldots, X_{40}$ represent the discounts to the 40 couples, and T the total of all the discounts. Then:

$$T = X_1 + X_2 + X_3 + \cdots + X_{40}$$
$$\begin{aligned} E(T) &= E(X_1 + X_2 + X_3 + \cdots + X_{40}) \\ &= E(X_1) + E(X_2) + E(X_3) + \cdots + E(X_{40}) \\ &= 5.83 + 5.83 + 5.83 + \cdots + 5.83 \\ &= \$233.20 \end{aligned}$$

Reshuffling cards between couples makes the discounts independent, so:

$$\begin{aligned} SD(T) &= \sqrt{Var(X_1 + X_2 + X_3 + \cdots + X_{40})} \\ &= \sqrt{Var(X_1) + Var(X_2) + Var(X_3) + \cdots + Var(X_{40})} \\ &= \sqrt{8.62^2 + 8.62^2 + 8.62^2 + \cdots + 8.62^2} \\ &\approx \$54.52 \end{aligned}$$

The restaurant owner can expect the 40 couples to win discounts totaling $233.20, with a standard deviation of $54.52.

Just Checking

2. Suppose the time it takes a customer to get and pay for seats at the ticket window of a baseball park is a random variable with a mean of 100 seconds and a standard deviation of 50 seconds. When you get there, you find only two people in line in front of you.
   a) How long do you expect to wait for your turn to get tickets?
   b) What's the standard deviation of your wait time?
   c) What assumption did you make about the two customers in finding the standard deviation?

Step-by-Step Example HITTING THE ROAD: MEANS AND VARIANCES

You're planning to spend next year wandering through the mountains of Kyrgyzstan. You plan to sell your used SUV so you can purchase an off-road Honda motor scooter when you get there. Used SUVs of the year and mileage of yours are selling for a mean of $6940 with a standard deviation of $250. Your research shows that scooters in Kyrgyzstan are going for about 65,000 Kyrgyzstan som with a standard deviation of 500 som. One U.S. dollar is worth about 38.5 Kyrgyzstan som (38 som and 50 tylyn).

Question: How much cash can you expect to pocket after you sell your SUV and buy the scooter?

THINK ➡ **Plan** State the problem.	I want to model how much money I'd have (in som) after selling my SUV and buying the scooter.
Variables Define the random variables.	Let A = sale price of my SUV (in dollars), B = price of a scooter (in som), and D = profit (in som)
Write an appropriate equation.	$D = 38.5A - B$
Think about the assumptions.	✓ **Independence Assumption:** The prices are independent.

(*continued*)

SHOW ➡ **Mechanics** Find the expected value, using the appropriate rules.

$$E(D) = E(38.5A - B)$$
$$= 38.5E(A) - E(B)$$
$$= 38.5(6,940) - (65,000)$$
$$E(D) = 202,190 \text{ som}$$

Find the variance, using the appropriate rules. Be sure to check the assumptions first!

Since sale and purchase prices are independent,

$$Var(D) = Var(38.5A - B)$$
$$= Var(38.5A) + Var(B)$$
$$= (38.5)^2 Var(A) + Var(B)$$
$$= 1482.25(250)^2 + (500)^2$$
$$Var(D) = 92,890,625$$

Find the standard deviation.

$$SD(D) = \sqrt{92,890,625} = 9637.98 \text{ som}$$

TELL ➡ **Conclusion** Interpret your results in context. (Here that means talking about dollars.)

I can expect to clear about 202,190 som ($5252) with a standard deviation of 9638 som ($250).

REALITY CHECK➡ Given the initial cost estimates, the mean and standard deviation seem reasonable.

Continuous Random Variables

A S *Activity*: **Numeric Outcomes.** You've seen how to simulate discrete random outcomes. There's a tool for simulating continuous outcomes, too.

A company manufactures home theater systems. At the end of the production line, the systems are packaged and prepared for shipping. Stage 1 of this process is called "packing." Workers must collect all the system components (a subwoofer, four speakers, a power cord, some cables, and a remote control), put each in plastic bags, and then place everything inside a protective styrofoam form. The packed form then moves on to Stage 2, called "boxing." There, workers place the form and a packet of instructions in a cardboard box, close it, then seal and label the box for shipping.

The company says that times required for the packing stage can be described by a Normal model with a mean of 9 minutes and standard deviation of 1.5 minutes. The times for the boxing stage can also be modeled as Normal, with a mean of 6 minutes and standard deviation of 1 minute.

This is a common way to model events. Do our rules for random variables apply here? What's different? We no longer have a list of discrete outcomes, with their associated probabilities. Instead, we have **continuous random variables** that can take on any value. Now any single value won't have a probability. We saw this back in Chapter 5 when we first saw the Normal model (although we didn't talk then about "random variables" or "probability"). We know that the probability that $z = 1.5$ doesn't make sense, but we *can* talk about the probability that z lies *between* 0.5 and 1.5. For a Normal random variable, the probability that it falls within an interval is just the area under the Normal curve over that interval.

A S *Activity*: **Means of Random Variables.** Experiment with continuous random variables to learn how their expected values behave.

Some continuous random variables have Normal models; others may be skewed, uniform, or bimodal. Regardless of shape, all continuous random variables have means (which we also call *expected values*) and variances. In this book we won't worry about how to calculate them, but we can still work with models for continuous random variables when we're given these parameters.

The good news is that nearly everything we've said about how discrete random variables behave is true of continuous random variables, as well. When two independent continuous random variables have Normal models, so does their sum or difference. This simple fact is a special property of Normal models and is very important. It allows us to apply our knowledge of Normal probabilities to questions about the sum or difference of independent random variables.

Step-by-Step Example PACKAGING STEREOS

Consider the company that manufactures and ships home theater systems that we just discussed.

Recall that times required to pack the systems can be described by a Normal model with a mean of 9 minutes and standard deviation of 1.5 minutes. The times for the boxing stage can also be modeled as Normal, with a mean of 6 minutes and standard deviation of 1 minute.

Questions:

1. What is the probability that packing two consecutive systems takes over 20 minutes?

2. What percentage of the theater systems take longer to pack than to box?

Question 1: What is the probability that packing two consecutive systems takes over 20 minutes?

THINK ➡ **Plan** State the problem.	I want to estimate the probability that packing two consecutive systems takes over 20 minutes.
	Let P_1 = time for packing the first system
	$\qquad P_2$ = time for packing the second
Variables Define your random variables.	$\qquad T$ = total time to pack two systems
Write an appropriate equation.	$T = P_1 + P_2$
Think about the assumptions. Sums of independent Normal random variables follow a Normal model. Such simplicity isn't true in general.	✓ **Normal Model Assumption:** We are told that both random variables follow Normal models.
	✓ **Independence Assumption:** We can reasonably assume that the two packing times are independent.
SHOW ➡ **Mechanics** Find the expected value.	$E(T) = E(P_1 + P_2)$
	$\qquad\quad = E(P_1) + E(P_2)$
	$\qquad\quad = 9 + 9 = 18$ minutes
For sums of independent random variables, variances add. (We don't need the variables to be Normal for this to be true—just independent.)	Since the times are independent,
	$Var(T) = Var(P_1 + P_2)$
	$\qquad\quad = Var(P_1) + Var(P_2)$
	$\qquad\quad = 1.5^2 + 1.5^2$
	$Var(T) = 4.50$
Find the standard deviation.	$SD(T) = \sqrt{4.50} \approx 2.12$ minutes

(continued)

Now we use the fact that these independent random variables both follow Normal models to say that their sum is also Normal.

I'll model T with $N(18, 2.12)$.

Sketch a picture of the Normal model for the total time, shading the region representing over 20 minutes.

Find the z-score for 20 minutes.

$$z = \frac{20 - 18}{2.12} = 0.94$$

Use technology or Table Z to find the probability.

$$P(T > 20) = P(z > 0.94) = 0.1736$$

TELL ➡ **Conclusion** Interpret your result in context.

There's a little more than a 17% chance that it will take a total of over 20 minutes to pack two consecutive home theater systems.

Question 2: What percent of the home theater systems take longer to pack than to box?

THINK ➡ **Plan** State the question.

I want to estimate the percentage of the systems that take longer to pack than to box.

Variables Define your random variables.

Let P = time for packing a system

B = time for boxing a system

D = difference in times to pack and box a system

Write an appropriate equation.

$D = P - B$

What are we trying to find? Notice that we can tell which of two quantities is greater by subtracting and asking whether the difference is positive or negative.

The probability that it takes longer to pack than to box a system is the probability that the difference $P - B$ is greater than zero.

Don't forget to think about the assumptions.

✓ **Normal Model Assumption:** We are told that both random variables follow Normal models.

✓ **Independence Assumption:** We can assume that the times it takes to pack and to box a system are independent.

SHOW ➡ **Mechanics** Find the expected value.

$$E(D) = E(P - B)$$
$$= E(P) - E(B)$$
$$= 9 - 6 = 3 \text{ minutes}$$

(continued)

For the difference of independent random variables, variances add.

Since the times are independent,

$$Var(D) = Var(P - B)$$
$$= Var(P) + Var(B)$$
$$= 1.5^2 + 1^2$$
$$Var(D) = 3.25$$

$$SD(D) = \sqrt{3.25} \approx 1.80 \text{ minutes}$$

Find the standard deviation.

State what model you will use.

I'll model D with $N(3, 1.80)$

Sketch a picture of the Normal model for the difference in times, and shade the region representing a difference greater than zero.

Find the z-score for 0 minutes, then use Table Z or technology to find the probability.

$$z = \frac{0 - 3}{1.80} = -1.67$$
$$P(D > 0) = P(z > -1.67) = 0.9525$$

TELL ➡ **Conclusion**   Interpret your result in context.

About 95% of all the home theater systems will require more time for packing than for boxing.

WHAT IF ●●● we confuse $X_1 + X_2 + X_3$ with $3X$?

When we work with random variables, some situations call for us to add the results of several random outcomes together. At other times we need to multiply one random outcome by some constant. It's critical that we recognize the difference between these, yet they often seem confusingly similar. That confusion is forgivable, because in algebra class it's certainly true that $x + x + x = 3x$. Always. So why don't random variables in Statistics behave themselves like this? We'll investigate by (you knew this was coming . . .) simulation.

Imagine that you get to roll dice for bonus points on your next test.[2] Your teacher offers you two options.

 Plan A: Roll one die and get 3 bonus points for every dot that shows.

 Plan B: Roll 3 dice and add them up to get your extra points.

Which option would *you* pick? Seriously. That's not an easy question to answer. Stop reading for a moment and think about what you would do.

OK, made your decision? Let's see if you chose wisely.

––––––––––
[2]Go ahead, pitch this idea to your teacher. We dare you.

We simulated each plan 1000 times; these histograms compare the results:

So, which way would you be better off? Well, that's up to you. Plan A gives you a much better shot at getting a 15- or 18-point bonus than Plan B, but Plan B makes it far less likely you'll end up with only a few extra points. It's your call.

The histograms show the most important lesson here: multiplying one die's outcome by 3 ($3D$) is clearly *not* the same as summing the results on 3 dice ($D_1 + D_2 + D_3$).

The good news is that we can work with both. Let's do the math. Let random variable D = the outcome on one die. Because we know you can figure these out, we'll just tell you the expected value is $E(D) = 3.5$ and the standard deviation $SD(D) = 1.708$.[3]

So, for Plan A: $\quad E(3D) = 3E(D) = 3(3.5) = 10.5$

$$SD(3D) = \sqrt{Var(3D)} = \sqrt{3^2 Var(D)} = \sqrt{9(1.708)^2} = 5.124$$

Simulations are designed to mimic theoretical situations, so ours should have come out close to this. And in fact, the mean for our 1000 Plan A trials was 10.557 with standard deviation 5.053. Nice.

What about Plan B? Let's start with the mean.

$$E(D1 + D2 + D3) = E(D1) + E(D2) + E(D3) = 3.5 + 3.5 + 3.5 = 10.5$$

You're thinking, "That's the same as before." Yes, but it's important that now we think of it as adding, not multiplying. The distinction becomes critical when we turn our attention to the standard deviation. Because the three rolls are independent, we can use to the Pythagorean Theorem of Statistics and add the variances:

$$SD(D1 + D2 + D3) = \sqrt{Var(D1) + Var(D2) + Var(D3)}$$
$$= \sqrt{1.708^2 + 1.708^2 + 1.708^2}$$
$$= 2.958$$

And in our simulation? The 1000 Plan B trials had a mean of 10.583 with standard deviation 2.866.[4]

Here's our last bit of advice. When working with random variables you must be careful to distinguish between such similar-sounding scenarios. How? Ask yourself: *How many things are happening at random?* In Plan A, you roll one die. In Plan B, you get to roll 3. That tells you Plan B involves addition, not multiplication.

---

[3]Check our work. It'll be easy for you, because you read this chapter. Right?

[4]*Don't you love it when things turn out the way they're supposed to?*

WHAT CAN GO WRONG?

- **Probability models are still just models.** Models can be useful, but they are not reality. Think about the assumptions behind your models. Are your dice really perfectly fair? (They are probably pretty close.) But when you hear that the probability of a nuclear accident is 1/10,000,000 per year, is that likely to be a precise value? Question probabilities as you would data.

- **If the model is wrong, so is everything else.** Before you try to find the mean or standard deviation of a random variable, check to make sure the probability model is reasonable. As a start, the probabilities in your model should add up to 1. If not, you may have calculated a probability incorrectly or left out a value of the random variable. For instance, in the insurance example, the description mentions only death and disability. Good health is by far the most likely outcome, not to mention the best for both you and the insurance company (who gets to keep your money). Don't overlook that.

- **Don't assume everything's Normal.** Just because a random variable is continuous or you happen to know a mean and standard deviation doesn't mean that a Normal model will be useful. You must *Think* about whether the Normality Assumption is justified. Using a Normal model when it really does not apply will lead to wrong answers and misleading conclusions.

  To find the expected value of the sum or difference of random variables, we simply add or subtract means. Center is easy; spread is trickier. Watch out for some common traps.

- **Watch out for variables that aren't independent.** You can add expected values of *any* two random variables, but you can only add variances of independent random variables. Suppose a survey includes questions about the number of hours of sleep people get each night and also the number of hours they are awake each day. From their answers, we find the mean and standard deviation of hours asleep and hours awake. The expected total must be 24 hours; after all, people are either asleep or awake.[5] The means still add just fine. Since all the totals are exactly 24 hours, however, the standard deviation of the total will be 0. We can't add variances here because the number of hours you're awake depends on the number of hours you're asleep. Be sure to check for independence before adding variances.

- **Don't forget: Variances of independent random variables add. Standard deviations don't.**

- **Don't forget: Variances of independent random variables add, even when you're looking at the difference between them.**

- **Don't write independent instances of a random variable with notation that looks like they are the same variables.** Make sure you write each instance as a different random variable. Just because each random variable describes a similar situation doesn't mean that each random outcome will be the same. These are *random* variables, not the variables you saw in Algebra. Write $X_1 + X_2 + X_3$ rather than $X + X + X$.

---

[5]Although some students do manage to attain a state of consciousness somewhere between sleeping and wakefulness during Statistics class.

What Have We Learned?

We've learned to work with random variables. We can use the probability model for a discrete random variable to find its expected value and its standard deviation.

We've learned that the mean of the sum or difference of two random variables, discrete or continuous, is just the sum or difference of their means. And we've learned the Pythagorean Theorem of Statistics: *For independent random variables,* the variance of their sum or difference is always the *sum* of their variances.

Finally, we've learned that Normal models are once again special. Sums or differences of normally distributed independent random variables also follow Normal models.

Terms

Random variable
A random variable assumes any of several different numeric values as a result of some random event. Random variables are denoted by a capital letter such as X. (p. 389)

Discrete random variable
A random variable that can take one of a finite number[6] of distinct outcomes is called a discrete random variable. (p. 389)

Continuous random variable
A random variable that can take any numeric value within an interval of values is called a continuous random variable. The interval may be infinite or bounded at either or both ends. (p. 390)

Probability model
The probability model is a function that associates a probability P with each value of a discrete random variable X, denoted $P(X = x)$, or with any interval of values of a continuous random variable. (p. 390)

Expected value
The expected value of a random variable is its theoretical long-run average value, the center of its model. Denoted μ or $E(X)$, it is found (if the random variable is discrete) by summing the products of variable values and probabilities:

$$\mu = E(X) = \sum xP(x). \quad \text{(p. 390)}$$

Variance
The variance of a random variable is the expected value of the squared deviation from the mean. For discrete random variables, it can be calculated as:

$$\sigma^2 = Var(X) = \sum (x - \mu)^2 P(x). \quad \text{(p. 392)}$$

Standard deviation
The standard deviation of a random variable describes the spread in the model, and is the square root of the variance:

$$\sigma = SD(X) = \sqrt{Var(X)}. \quad \text{(p. 392)}$$

Changing a random variable by a constant
$$E(X \pm c) = E(X) \pm c \qquad Var(X \pm c) = Var(X). \quad \text{(p. 395)}$$
$$E(aX) = aE(X) \qquad\qquad Var(aX) = a^2 Var(X). \quad \text{(p. 396)}$$

Addition Rule for Expected Values of Random Variables
$$E(X \pm Y) = E(X) \pm E(Y). \quad \text{(p. 396)}$$

Addition Rule for Variances of Random Variables
If X and Y are *independent*: $Var(X \pm Y) = Var(X) + Var(Y)$,
$$\text{and } SD(X \pm Y) = \sqrt{Var(X) + Var(Y)}.$$

Pythagorean Theorem of Statistics (p. 396)

---

[6]Actually, there could be an infinite number of outcomes, as long as they're *countable*. Essentially that means we can imagine listing them all in order, like the counting numbers 1, 2, 3, 4, 5, . . .

On the Computer RANDOM VARIABLES

Statistics packages deal with data, not with random variables. Nevertheless, the calculations needed to find means and standard deviations of random variables are little more than weighted means. Most packages can manage that, but then they are just being overblown calculators. For technological assistance with these calculations, we recommend you pull out your calculator.

Exercises

1. **Expected value** Find the expected value of each random variable:

a)
x	10	20	30
$P(X = x)$	0.3	0.5	0.2

b)
x	2	4	6	8
$P(X = x)$	0.3	0.4	0.2	0.1

2. **Expected value** Find the expected value of each random variable:

a)
x	0	1	2
$P(X = x)$	0.2	0.4	0.4

b)
x	100	200	300	400
$P(X = x)$	0.1	0.2	0.5	0.2

3. **Oranges** A citrus farmer has observed the following distribution for the number of oranges per tree. How many oranges does he expect on average?

Oranges	25	30	35	40
Probability	0.10	0.40	0.30	0.20

4. **Caffeinated** A coffee shop tracks sales and has observed the distribution in the following table. What is the average daily sales that it can expect?

# of Sales	145	150	155	160	170
Probability	0.15	0.22	0.37	0.19	0.07

5. **Oranges again** What is the standard deviation for Exercise 3?

6. **Caffeinated again** What is the standard deviation for Exercise 4?

7. **Pick a card, any card** You draw a card from a deck. If you get a red card, you win nothing. If you get a spade, you win $5. For any club, you win $10 plus an extra $20 for the ace of clubs.

a) Create a probability model for the amount you win.
b) Find the expected amount you'll win.
c) What would you be willing to pay to play this game?

8. **You bet!** You roll a die. If it comes up a 6, you win $100. If not, you get to roll again. If you get a 6 the second time, you win $50. If not, you lose.

a) Create a probability model for the amount you win.
b) Find the expected amount you'll win.
c) What would you be willing to pay to play this game?

9. **Kids** A couple plans to have children until they get a girl, but they agree that they will not have more than three children even if all are boys. (Assume boys and girls are equally likely.)

a) Create a probability model for the number of children they might have.
b) Find the expected number of children.
c) Find the expected number of boys they'll have.

10. **Carnival** A carnival game offers a $100 cash prize for anyone who can break a balloon by throwing a dart at it. It costs $5 to play, and you're willing to spend up to $20 trying to win. You estimate that you have about a 10% chance of hitting the balloon on any throw.

a) Create a probability model for this carnival game.
b) Find the expected number of darts you'll throw.
c) Find your expected winnings.

11. **Software** A small software company bids on two contracts. It anticipates a profit of $60,000 if it gets the larger contract and a profit of $20,000 on the smaller contract. The company estimates there's a 30% chance it will get the larger contract and a 60% chance it will get the smaller contract. Assuming the contracts will be awarded independently, what's the expected profit?

12. Racehorse A man buys a racehorse for $20,000 and enters it in two races. He plans to sell the horse afterward, hoping to make a profit. If the horse wins both races, its value will jump to $100,000. If it wins one of the races, it will be worth $50,000. If it loses both races, it will be worth only $10,000. The man believes there's a 20% chance that the horse will win the first race and a 30% chance it will win the second one. Assuming that the two races are independent events, find the man's expected profit.

13. Variation 1 Find the standard deviations of the random variables in Exercise 1.

14. Variation 2 Find the standard deviations of the random variables in Exercise 2.

15. Pick another card Find the standard deviation of the amount you might win drawing a card in Exercise 7.

16. The die Find the standard deviation of the amount you might win rolling a die in Exercise 8.

17. Kids again Find the standard deviation of the number of children the couple in Exercise 9 may have.

18. Darts Find the standard deviation of your winnings throwing darts in Exercise 10.

19. Repairs The probability model below describes the number of repair calls that an appliance repair shop may receive during an hour.

Repair Calls	0	1	2	3
Probability	0.1	0.3	0.4	0.2

a) How many calls should the shop expect per hour?
b) What is the standard deviation?

20. Red lights A commuter must pass through five traffic lights on her way to work and will have to stop at each one that is red. She estimates the probability model for the number of red lights she hits, as shown below.

X = # of Red	0	1	2	3	4	5
P(X = x)	0.05	0.25	0.35	0.15	0.15	0.05

a) How many red lights should she expect to hit each day?
b) What's the standard deviation?

21. Defects A consumer organization inspecting new cars found that many had appearance defects (dents, scratches, paint chips, etc.). While none had more than three of these defects, 7% had three, 11% two, and 21% one defect. Find the expected number of appearance defects in a new car and the standard deviation.

22. Insurance An insurance policy costs $100 and will pay policyholders $10,000 if they suffer a major injury

(resulting in hospitalization) or $3000 if they suffer a minor injury (resulting in lost time from work). The company estimates that each year 1 in every 2000 policyholders may have a major injury, and 1 in 500 a minor injury only.

a) Create a probability model for the profit on a policy.
b) What's the company's expected profit on this policy?
c) What's the standard deviation?

23. Cancelled flights Mary is deciding whether to book the cheaper flight home from college after her final exams, but she's unsure when her last exam will be. She thinks there is only a 20% chance that the exam will be scheduled after the last day she can get a seat on the cheaper flight. If it is and she has to cancel the flight, she will lose $150. If she can take the cheaper flight, she will save $100.

a) If she books the cheaper flight, what can she expect to gain, on average?
b) What is the standard deviation?

24. Day trading An option to buy a stock is priced at $200. If the stock closes above 30 on May 15, the option will be worth $1000. If it closes below 20, the option will be worth nothing, and if it closes between 20 and 30 (inclusively), the option will be worth $200. A trader thinks there is a 50% chance that the stock will close in the 20–30 range, a 20% chance that it will close above 30, and a 30% chance that it will fall below 20 on May 15.

a) Should she buy the stock option?
b) How much does she expect to gain?
c) What is the standard deviation of her gain?

25. Contest You play two games against the same opponent. The probability you win the first game is 0.4. If you win the first game, the probability you also win the second is 0.2. If you lose the first game, the probability that you win the second is 0.3.

a) Are the two games independent? Explain.
b) What's the probability you lose both games?
c) What's the probability you win both games?
d) Let random variable X be the number of games you win. Find the probability model for X.
e) What are the expected value and standard deviation?

26. Contracts Your company bids for two contracts. You believe the probability you get contract #1 is 0.8. If you get contract #1, the probability you also get contract #2 will be 0.2, and if you do not get #1, the probability you get #2 will be 0.3.

a) Are the two contracts independent? Explain.
b) Find the probability you get both contracts.
c) Find the probability you get no contract.
d) Let X be the number of contracts you get. Find the probability model for X.
e) Find the expected value and standard deviation.

27. Batteries In a group of 10 batteries, 3 are dead. You choose 2 batteries at random.

    a) Create a probability model for the number of good batteries you get.

    b) What's the expected number of good ones you get?

    c) What's the standard deviation?

28. Kittens In a litter of seven kittens, three are female. You pick two kittens at random.

    a) Create a probability model for the number of male kittens you get.

    b) What's the expected number of males?

    c) What's the standard deviation?

29. Random variables Given independent random variables with means and standard deviations as shown, find the mean and standard deviation of:

    a) $3X$

    b) $Y + 6$

    c) $X + Y$

    d) $X - Y$

    e) $X_1 + X_2$

	Mean	SD
X	10	2
Y	20	5

30. Random variables Given independent random variables with means and standard deviations as shown, find the mean and standard deviation of:

    a) $X - 20$

    b) $0.5Y$

    c) $X + Y$

    d) $X - Y$

    e) $Y_1 + Y_2$

	Mean	SD
X	80	12
Y	12	3

31. Random variables Given independent random variables with means and standard deviations as shown, find the mean and standard deviation of:

    a) $0.8Y$

    b) $2X - 100$

    c) $X + 2Y$

    d) $3X - Y$

    e) $Y_1 + Y_2$

	Mean	SD
X	120	12
Y	300	16

32. Random variables Given independent random variables with means and standard deviations as shown, find the mean and standard deviation of:

    a) $2Y + 20$

    b) $3X$

    c) $0.25X + Y$

    d) $X - 5Y$

    e) $X_1 + X_2 + X_3$

	Mean	SD
X	80	12
Y	12	3

33. Salary An employer pays a mean salary for a 5-day workweek of $1250 with a standard deviation of $129. On the weekends, his salary expenses have a mean of $450 with a standard deviation of $57. What is the mean and standard deviation of his total weekly salaries?

34. Golf scores A golfer keeps track of his score for playing nine holes of golf (half a normal golf round). His mean score is 85 with a standard deviation of 11. Assuming

that the second 9 has the same mean and standard deviation, what is the mean and standard deviation of his total score if he plays a full 18 holes?

35. Eggs A grocery supplier believes that in a dozen eggs, the mean number of broken ones is 0.6 with a standard deviation of 0.5 eggs. You buy 3 dozen eggs without checking them.

    a) How many broken eggs do you expect to get?

    b) What's the standard deviation?

    c) What assumptions did you have to make about the eggs in order to answer this question?

36. Garden A company selling vegetable seeds in packets of 20 estimates that the mean number of seeds that will actually grow is 18, with a standard deviation of 1.2 seeds. You buy 5 different seed packets.

    a) How many good seeds do you expect to get?

    b) What's the standard deviation?

    c) What assumptions did you make about the seeds? Do you think that assumption is warranted? Explain.

37. Eggs again In Exercise 35 you bought 3 dozen eggs.

    a) How many good eggs do you expect?

    b) What's the standard deviation of the number of good eggs?

    c) Why does it make sense for the standard deviation not to change?

38. Garden grows In Exercise 36 you bought 5 seed packets.

    a) How many bad seeds do you expect?

    b) What is the standard deviation of the bad seed count?

    c) Why does it make sense that the standard deviation is not different?

39. SAT or ACT revisited Remember back in Chapter 5 when we used the equation $SAT = 40 \times ACT + 150$ to convert an ACT score into a SAT score. Let's use this transformation again, now with random variables.

    a) Suppose your school has a mean ACT score of 29. What would its equivalent mean SAT score be?

    b) If your school has a standard deviation of 5 ACT points, what is the standard of its equivalent SAT score?

40. Colder? We used the formula $°F = 9/5°C + 32$ to convert Celsius to Fahrenheit in Chapter 5. Let's put it to use in a random variable setting.

    a) Suppose your town has a mean January temperature of 11°C. What is the mean temperature in °F?

    b) Fortunately your local weatherman has recently taken a statistics course and is keen to show off his newfound knowledge. He reports that January has a standard deviation of 6°C. What is the standard deviation in °F?

41. Repair calls Suppose that the appliance shop in Exercise 19 plans an 8-hour day.

    a) Find the mean and standard deviation of the number of repair calls they should expect in a day.

b) What assumption did you make about the repair calls?

c) Use the mean and standard deviation to describe what a typical 8-hour day will be like.

d) At the end of a day, a worker comments "Boy, I'm tired. Today was sure unusually busy!" How many repair calls would justify such an observation?

42. Stop! Suppose the commuter in Exercise 20 has a 5-day workweek.

a) Find the mean and standard deviation of the number of red lights the commuter should expect to hit in her week.

b) What assumption did you make about the days?

c) Use the mean and standard deviation to describe a typical week.

d) Upon arriving home on Friday, the commuter remarks, "Wow! My commute was quick all week." How many red lights would it take to deserve a feeling of good luck?

43. Tickets A delivery company's trucks occasionally get parking tickets, and based on past experience, the company plans that the trucks will average 1.3 tickets a month, with a standard deviation of 0.7 tickets.

a) If they have 18 trucks, what are the mean and standard deviation of the total number of parking tickets the company will have to pay this month?

b) What assumption did you make in answering?

44. Donations Organizers of a televised fundraiser know from past experience that most people donate small amounts ($10–$25), some donate larger amounts ($50–$100), and a few people make very generous donations of $250, $500, or more. Historically, pledges average about $32 with a standard deviation of $54.

a) If 120 people call in pledges, what are the mean and standard deviation of the total amount raised?

b) What assumption did you make in answering this question?

45. Fire! An insurance company estimates that it should make an annual profit of $150 on each homeowner's policy written, with a standard deviation of $6000.

a) Why is the standard deviation so large?

b) If it writes only two of these policies, what are the mean and standard deviation of the annual profit?

c) If it writes 10,000 of these policies, what are the mean and standard deviation of the annual profit?

d) Is the company likely to be profitable? Explain.

e) What assumptions underlie your analysis? Can you think of circumstances under which those assumptions might be violated? Explain.

46. Casino A casino knows that people play the slot machines in hopes of hitting the jackpot but that most of them lose their dollar. Suppose a certain machine pays out an average of $0.92, with a standard deviation of $120.

a) Why is the standard deviation so large?

b) If you play 5 times, what are the mean and standard deviation of the casino's profit?

c) If gamblers play this machine 1000 times in a day, what are the mean and standard deviation of the casino's profit?

d) Is the casino likely to be profitable? Explain.

47. Cereal The amount of cereal that can be poured into a small bowl varies with a mean of 1.5 ounces and a standard deviation of 0.3 ounces. A large bowl holds a mean of 2.5 ounces with a standard deviation of 0.4 ounces. You open a new box of cereal and pour one large and one small bowl.

a) How much more cereal do you expect to be in the large bowl?

b) What's the standard deviation of this difference?

c) If the difference follows a Normal model, what's the probability the small bowl contains more cereal than the large one?

d) What are the mean and standard deviation of the total amount of cereal in the two bowls?

e) If the total follows a Normal model, what's the probability you poured out more than 4.5 ounces of cereal in the two bowls together?

f) The amount of cereal the manufacturer puts in the boxes is a random variable with a mean of 16.3 ounces and a standard deviation of 0.2 ounces. Find the expected amount of cereal left in the box and the standard deviation.

48. Pets The American Veterinary Association claims that the annual cost of medical care for dogs averages $100, with a standard deviation of $30, and for cats averages $120, with a standard deviation of $35.

a) What's the expected difference in the cost of medical care for dogs and cats?

b) What's the standard deviation of that difference?

c) If the costs can be described by Normal models, what's the probability that medical expenses are higher for someone's dog than for her cat?

d) What concerns do you have?

49. More cereal In Exercise 47 we poured a large and a small bowl of cereal from a box. Suppose the amount of cereal that the manufacturer puts in the boxes is a random variable with mean 16.2 ounces and standard deviation 0.1 ounces.

a) Find the expected amount of cereal left in the box.

b) What's the standard deviation?

c) If the weight of the remaining cereal can be described by a Normal model, what's the probability that the box still contains more than 13 ounces?

50. More pets You're thinking about getting two dogs and a cat. Assume that annual veterinary expenses are independent and have a Normal model with the means and standard deviations described in Exercise 48.

a) Define appropriate variables and express the total annual veterinary costs you may have.

b) Describe the model for this total cost. Be sure to specify its name, expected value, and standard deviation.

c) What's the probability that your total expenses will exceed $400?

51. Medley In the 4 × 100 medley relay event, four swimmers swim 100 yards, each using a different stroke. A college team preparing for the conference championship looks at the times their swimmers have posted and creates a model based on the following assumptions:

- The swimmers' performances are independent.
- Each swimmer's times follow a Normal model.
- The means and standard deviations of the times (in seconds) are as shown:

Swimmer	Mean	SD
1 (backstroke)	50.72	0.24
2 (breaststroke)	55.51	0.22
3 (butterfly)	49.43	0.25
4 (freestyle)	44.91	0.21

a) What are the mean and standard deviation for the relay team's total time in this event?

b) The team's best time so far this season was 3:19.48. (That's 199.48 seconds.) Do you think the team is likely to swim faster than this at the conference championship? Explain.

52. Bikes Bicycles arrive at a bike shop in boxes. Before they can be sold, they must be unpacked, assembled, and tuned (lubricated, adjusted, etc.). Based on past experience, the shop manager makes the following assumptions about how long this may take:

- The times for each setup phase are independent.
- The times for each phase follow a Normal model.
- The means and standard deviations of the times (in minutes) are as shown:

Phase	Mean	SD
Unpacking	3.5	0.7
Assembly	21.8	2.4
Tuning	12.3	2.7

a) What are the mean and standard deviation for the total bicycle setup time?

b) A customer decides to buy a bike like one of the display models but wants a different color. The shop has one, still in the box. The manager says they can have it ready in half an hour. Do you think the bike will be set up and ready to go as promised? Explain.

53. Farmers' market A farmer has 100 lb of apples and 50 lb of potatoes for sale. The market price for apples (per pound) each day is a random variable with a mean of 0.5 dollars and a standard deviation of 0.2 dollars. Similarly, for a pound of potatoes, the mean price is 0.3 dollars and the standard deviation is 0.1 dollars. It also costs him 2 dollars to bring all the apples and potatoes to the market. The market is busy with eager shoppers, so we can assume that he'll be able to sell all of each type of produce at that day's price.

a) Define your random variables, and use them to express the farmer's net income.

b) Find the mean.

c) Find the standard deviation of the net income.

d) Do you need to make any assumptions in calculating the mean? How about the standard deviation?

54. Bike sale The bicycle shop in Exercise 52 will be offering 2 specially priced children's models at a sidewalk sale. The basic model will sell for $120 and the deluxe model for $150. Past experience indicates that sales of the basic model will have a mean of 5.4 bikes with a standard deviation of 1.2, and sales of the deluxe model will have a mean of 3.2 bikes with a standard deviation of 0.8 bikes. The cost of setting up for the sidewalk sale is $200.

a) Define random variables and use them to express the bicycle shop's net income.

b) What's the mean of the net income?

c) What's the standard deviation of the net income?

d) Do you need to make any assumptions in calculating the mean? How about the standard deviation?

55. Coffee and doughnuts At a certain coffee shop, all the customers buy a cup of coffee; some also buy a doughnut. The shop owner believes that the number of cups he sells each day is normally distributed with a mean of 320 cups and a standard deviation of 20 cups. He also believes that the number of doughnuts he sells each day is independent of the coffee sales and is normally distributed with a mean of 150 doughnuts and a standard deviation of 12.

a) The shop is open every day but Sunday. Assuming day-to-day sales are independent, what's the probability he'll sell over 2000 cups of coffee in a week?

b) If he makes a profit of 50 cents on each cup of coffee and 40 cents on each doughnut, can he reasonably expect to have a day's profit of over $300? Explain.

c) What's the probability that on any given day he'll sell a doughnut to more than half of his coffee customers?

56. Weightlifting The Atlas BodyBuilding Company (ABC) sells "starter sets" of barbells that consist of one bar, two 20-pound weights, and four 5-pound weights. The bars weigh an average of 10 pounds with a standard deviation of 0.25 pounds. The weights average the specified amounts, but the standard deviations are 0.2 pounds for the 20-pounders and 0.1 pounds for the 5-pounders. We can assume that all the weights are normally distributed.

a) ABC ships these starter sets to customers in two boxes: The bar goes in one box and the six weights go in another. What's the probability that the total weight in that second box exceeds 60.5 pounds? Define your variables clearly and state any assumptions you make.

b) It costs ABC $0.40 per pound to ship the box containing the weights. Because it's an odd-shaped package, though, shipping the bar costs $0.50 a pound plus a $6.00 surcharge. Find the mean and standard deviation of the company's total cost for shipping a starter set.

c) Suppose a customer puts a 20-pound weight at one end of the bar and the four 5-pound weights at the other end. Although he expects the two ends to weigh the same, they might differ slightly. What's the probability the difference is more than a quarter of a pound?

Just Checking ANSWERS

1. a)

Outcome	X = cost	Probability
Recharging works	$60	0.75
Replace control unit	$200	0.25

b) $60(0.75) + 200(0.25) = \$95$

c) Car owners with this problem will spend an average of $95 to get it fixed.

2. a) $100 + 100 = 200$ seconds

b) $\sqrt{50^2 + 50^2} = 70.7$ seconds

c) The times for the two customers are independent.

16 Probability Models

Suppose a cereal manufacturer puts pictures of famous athletes on cards in boxes *of cereal, in the hope of increasing sales. The manufacturer announces that 20% of the boxes contain a picture of LeBron James, 30% a picture of Damica Patrick, and the rest a picture of Serena Williams.*

Sound familiar? In Chapter 10 we simulated to find the number of boxes we'd need to open to get one of each card. That's a fairly complex question and one well suited for simulation. But many important questions can be answered more directly by using simple probability models.

Searching for LeBron: Bernoulli Trials

Suppose you're a huge LeBron James fan. You don't care about completing the whole sports card collection, but you've just *got* to have the LeBron's picture. How many boxes do you expect you'll have to open before you find him? This isn't the same question that we asked before, but this situation is simple enough for a probability model.

We'll keep the assumption that pictures are distributed at random and we'll trust the manufacturer's claim that 20% of the cards are LeBron. So, when you open the box, the probability that you succeed in finding LeBron is 0.20. Now we'll call the act of opening *each* box a trial, and note that:

- There are only two possible outcomes (called *success* and *failure*) on each trial. Either you get LeBron's picture (success), or you don't (failure).
- In advance, the probability of success, denoted p, is the same on every trial. Here $p = 0.20$ for each box.
- As we proceed, the trials are independent. Finding LeBron in the first box does not change what might happen when you reach for the next box.

Daniel Bernoulli (1700–1782) was the nephew of Jacob, whom you saw in Chapter 14. He was the first to work out the mathematics for what we now call Bernoulli trials.

A S   *Activity*: **Bernoulli Trials.** Guess what! We've been generating Bernoulli trials all along. Look at the Random Simulation Tool in a new way.

Situations like this occur often and are called **Bernoulli trials.** Common examples of Bernoulli trials include tossing a coin, looking for defective products rolling off an assembly line, or even shooting free throws in a basketball game. Just as we found equally likely random digits to be the building blocks for our simulation, we can use Bernoulli trials to build a wide variety of useful probability models.

Calvin and Hobbes © 1993 Bill Watterson. Distributed by Universal Uclick. Reprinted with permission. All rights reserved.

Back to finding LeBron. We want to know how many boxes we'll need to open to find his card. Let's call this random variable $Y = $ # boxes, and build a probability model for it. What's the probability you find his picture in the first box of cereal? It's 20%, of course. We could write $P(Y = 1) = 0.20$.

How about the probability that you don't find LeBron until the second box? Well, that means you fail on the first trial and then succeed on the second. With the probability of success 20%, the probability of failure, denoted q, is $1 - 0.2 = 80\%$. Since the trials are independent, the probability of getting your first success on the second trial is $P(Y = 2) = (0.8)(0.2) = 0.16$.

Of course, you could have a run of bad luck. Maybe you won't find LeBron until the fifth box of cereal. What are the chances of that? You'd have to fail 4 straight times and then succeed, so $P(Y = 5) = (0.8)^4(0.2) = 0.08192$.

How many boxes might you expect to have to open? We could reason that since LeBron's picture is in 20% of the boxes, or 1 in 5, we expect to find his picture, on average, in the fifth box; that is, $E(Y) = \frac{1}{0.2} = 5$ boxes. That's correct, but not easy to prove.[1]

The 10% "Rule"

One of the important requirements for Bernoulli trials is that the trials be independent. Sometimes that's a reasonable assumption—when tossing a coin or rolling a die, for example. But that becomes a problem when (often!) we're looking at situations involving samples chosen without replacement. We said that whether we find a LeBron James card in one box has no effect on the probabilities in other boxes. This is *almost* true. Technically, if exactly 20% of the boxes have LeBron cards, then when you find one, you've reduced the number of remaining LeBron cards. With a few million boxes of cereal, though, the difference is hardly worth mentioning. But if you knew there were 2 LeBron James cards hiding in the 10 boxes of cereal on the market shelf, then finding one in the first box you try would clearly change your chances of finding his picture in the next box.

If we had an infinite number of boxes, there wouldn't be a problem. It's selecting from a finite population that causes the probabilities to change, making the trials not independent. Obviously, taking 2 out of 10 boxes changes the probability. Taking even a few hundred out of millions, though, makes very little difference. Fortunately, it turns out that if we look at less than 10% of the population, we can pretend that the trials are independent and still calculate probabilities that are quite accurate. That's our 10% rule of thumb:

---

[1] See the Math Box, coming soon to a textbook near you.

The 10% Condition: Bernoulli trials must be independent. If that assumption is violated, it is still okay to proceed as long as we randomly sample fewer than 10% of the population.[2]

Just Checking

1. Think about each of these situations. Are these random variables based on Bernoulli trials? If you don't think so, explain why not.

   a) The waitstaff at a small restaurant consists of 5 males and 8 females. They write their names on slips of paper and the boss chooses 4 people at random to work overtime on a holiday weekend. We count the number of females who are chosen.

   b) In the United States about 1 in every 90 pregnant women gives birth to twins. We count the number of twins born to a group of pregnant women who work in the same office.

   c) We count the number of times a woman who has been pregnant 3 times gave birth to twins.

   d) We pick 40 M&M's at random from a large bag, counting how many of each color we get.

   e) A small town's merchant's association says that 26% of all businesses there are owned by women. You call 15 businesses randomly chosen from the 77 listed in the local Yellow Pages, counting the number owned by women.

The Geometric Model: Waiting for Success

TI-*nspire*

Geometric probabilities.
See what happens to a geometric model as you change the probability of success.

We want to model how long it will take to achieve the first success in a series of Bernoulli trials. The model that tells us this probability is called the **Geometric probability model.** Geometric models are completely specified by one parameter, p, the probability of success, and are denoted Geom(p). Since achieving the first success on trial number x requires first experiencing $x - 1$ failures, the probabilities are easily expressed by a formula.

■ **NOTATION ALERT**

Now we have two more reserved letters. Whenever we deal with Bernoulli trials, p represents the probability of success, and q the probability of failure. (Of course, $q = 1 - p$.)

> **Geometric Probability Model for Bernoulli Trials: Geom(p)**
>
> p = probability of success (and $q = 1 - p$ = probability of failure)
> X = number of trials until the first success occurs
>
> $$P(X = x) = q^{x-1}p$$
>
> Expected value: $E(X) = \mu = \dfrac{1}{p}$ *Standard deviation: $\sigma = \sqrt{\dfrac{q}{p^2}}$

For Example SPAM AND THE GEOMETRIC MODEL

Postini is a global company specializing in communications security. The company monitors over 1 billion Internet messages per day and recently reported that 91% of e-mails are spam!

Let's assume that your e-mail is typical—91% spam. We'll also assume you aren't using a spam filter, so every message gets dumped in your inbox. And, since spam comes from many different sources, we'll consider your messages to be independent.

(continued)

---

[2]There is a formula that can adjust for even larger samples, called the finite population correction, but it's beyond the scope of this course.

QUESTION: Overnight your inbox collects email. When you first check your email in the morning, about how many spam emails should you expect to have to wade through and discard before you find a real message? What's the probability that the 4th message in your inbox is the first one that isn't spam?

ANSWER: When I check my emails one by-one:

- There are two possible outcomes each time: a real message (success) or spam (failure).
- Since 91% of all emails are spam, the probability of success is

$$p = 1 - 0.91 = 0.09.$$

- My messages arrive in random order from many different sources and are far fewer than 10% of all email messages. I can treat them as independent.

Let $X =$ the number of emails I'll check until I find a real message. I can use the model Geom(0.09).

$$E(X) = \frac{1}{p} = \frac{1}{0.09} = 11.1$$

$$P(X = 4) = (0.91)^3(0.09) = 0.0678$$

On average, I expect to have to check just over 11 emails before I find a real message. There's slightly less than a 7% chance that my first real message will be the 4th one I check.

Note that this probability calculation isn't new. It's simply Chapter 13's Multiplication Rule used to find $P(\text{spam} \cap \text{spam} \cap \text{spam} \cap \text{real})$.

Math Box

We want to find the mean (expected value) of random variable X, using a geometric model with probability of success p.

First, write the probabilities:

x	1	2	3	4	$\cdots$
$P(X = x)$	p	qp	q^2p	q^3p	$\cdots$

The expected value is:	$E(X) = 1p + 2qp + 3q^2p + 4q^3p + \cdots$
Let $p = 1 - q$:	$= (1 - q) + 2q(1 - q) + 3q^2(1 - q) + 4q^3(1 - q) + \cdots$
Simplify:	$= 1 - q + 2q - 2q^2 + 3q^2 - 3q^3 + 4q^3 - 4q^4 + \cdots$
That's an infinite geometric series, with first term 1 and common ratio q:	$= 1 + q + q^2 + q^3 + \cdots$
	$= \dfrac{1}{1 - q}$
So, finally . . .	$E(X) = \dfrac{1}{p}.$

Step-by-Step Example WORKING WITH A GEOMETRIC MODEL

People with O-negative blood are called "universal donors" because O-negative blood can be given to anyone else, regardless of the recipient's blood type. Only about 6% of people have O-negative blood.

Questions:

1. If donors line up at random for a blood drive, how many do you expect to examine before you find someone who has O-negative blood?

2. What's the probability that the first O-negative donor found is one of the first four people in line?

THINK ➡ **Plan** State the questions.		*I want to estimate how many people I'll need to check to find an O-negative donor, and the probability that 1 of the first 4 people is O-negative.*
	Check to see that these are Bernoulli trials.	✓ *There are two outcomes:* *success = O-negative* *failure = other blood types* ✓ *The probability of success for each person is p = 0.06, because they lined up randomly.* ✓ **10% Condition:** *Trials aren't independent because the population is finite, but the donors lined up are fewer than 10% of all possible donors.*
	Variable Define the random variable.	*Let X = number of donors until one is O-negative.*
	Model Specify the model.	*I can model X with Geom(0.06).*

SHOW ➡ **Mechanics** Find the mean.

$$E(X) = \frac{1}{0.06} \approx 16.7$$

Calculate the probability of success on one of the first four trials. That's the probability that $X = 1, 2, 3,$ or 4.

$$P(X \le 4) = P(X = 1) + P(X = 2) + P(X = 3) + P(X = 4)$$
$$= (0.06) + (0.94)(0.06) + (0.94)^2(0.06) + (0.94)^3(0.06)$$
$$\approx 0.2193$$

TELL ➡ **Conclusion** Interpret your results in context.

Blood drives such as this one expect to examine an average of 16.7 people to find a universal donor. About 22% of the time there will be one within the first 4 people in line.

TI Tips FINDING GEOMETRIC PROBABILITIES

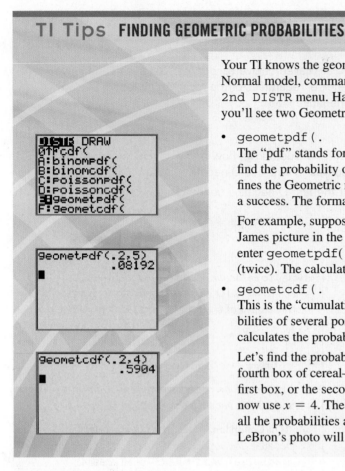

Your TI knows the geometric model. Just as you saw back in Chapter 5 with the Normal model, commands to calculate probability distributions are found in the 2nd DISTR menu. Have a look. After many others (Yes, there's still more to learn!) you'll see two Geometric probability functions at the bottom of the list.

- geometpdf(.
  The "pdf" stands for "probability density function." This command allows you to find the probability of any *individual* outcome. You need only specify *p*, which defines the Geometric model, and *x*, which indicates the number of trials until you get a success. The format is geometpdf(p,x).

  For example, suppose we want to know the probability that we find our first LeBron James picture in the fifth box of cereal. Since LeBron is in 20% of the boxes, we enter geometpdf(with p:0.2, x value:5, then go to Paste and hit ENTER (twice). The calculator says there's about an 8% chance.

- geometcdf(.
  This is the "cumulative density function," meaning that it finds the sum of the probabilities of several possible outcomes. In general, the command geometcdf(p,x) calculates the probability of finding the first success *on or before* the *x*th trial.

  Let's find the probability of getting a LeBron James picture by the time we open the fourth box of cereal—in other words, the probability our first success comes on the first box, or the second, or the third, or the fourth. Again we specify *p* = 0.2, and now use *x* = 4. The command geometcdf with p:0.2, x value:4 calculates all the probabilities and sums them. There's about a 59% chance that our quest for a LeBron's photo will succeed by the time we open the fourth box.

The Binomial Model: Counting Successes

We can use the Bernoulli trials to answer other common questions. Suppose you buy 5 boxes of cereal. What's the probability you get *exactly* 2 pictures of LeBron James? Before, we asked how long it would take until our first success. Now we want to find the probability of getting 2 successes among the 5 trials. We are still talking about Bernoulli trials, but we're asking a different question.

This time we're interested in the *number of successes* in the 5 trials, so we'll call it X = number of successes. We want to find $P(X = 2)$. This is an example of a **Binomial probability.** It takes two parameters to define this **Binomial model:** the number of trials, n, and the probability of success, p. We denote this model Binom(n, p). Here, n = 5 trials, and p = 0.2, the probability of finding a LeBron James card in any trial.

Exactly 2 successes in 5 trials means 2 successes and 3 failures. It seems logical that the probability should be $(0.2)^2(0.8)^3$. Too bad: It's not that easy. That calculation would give you the probability of finding LeBron in the first 2 boxes and not in the next 3—*in that order.* But you could find LeBron in the third and fifth boxes and still have 2 successes. The probability of those outcomes in that particular order is $(0.8)(0.8)(0.2)(0.8)(0.2)$. That's also $(0.2)^2(0.8)^3$. In fact, the probability will always be the same, no matter what order the successes and failures occur in. Anytime we get 2 successes in 5 trials, regardless of the order, the probability will be $(0.2)^2(0.8)^3$. We just need to count all the possible orders in which the outcomes can occur.

Fortunately, these possible orders are *disjoint.* (For example, if your two successes came on the first two trials, they couldn't come on the last two.) So we could use the Addition Rule to add up the probabilities, but since they're are all the same, we really only need to know how many orders are possible. For small *n*'s, we can just make a tree diagram and count the branches. For larger numbers this isn't practical: fortunately, there's a formula for that.

Each different order in which we can have k successes in n trials is called a "combination." The total number of ways that can happen is written $\binom{n}{k}$ or $_nC_k$ and pronounced "n choose k."

$$\binom{n}{k} = \frac{n!}{k!(n-k)!} \text{ where } n! \text{ (pronounced "}n\text{ factorial")} = n \times (n-1) \times \cdots \times 1$$

For 2 successes in 5 trials,

$$\binom{5}{2} = \frac{5!}{2!(5-2)!} = \frac{5 \times 4 \times 3 \times 2 \times 1}{2 \times 1 \times 3 \times 2 \times 1} = \frac{5 \times 4}{2 \times 1} = 10.$$

So there are 10 ways to get 2 LeBron pictures in 5 boxes, and the probability of each is $(0.2)^2(0.8)^3$. Now we can find what we wanted:

$$P(\#success = 2) = 10(0.2)^2(0.8)^3 = 0.2048$$

In general, the probability of exactly k successes in n trials is $\binom{n}{k}p^kq^{n-k}$.

It's not hard to find the expected value for a binominal random variable. If we have 5 boxes, and LeBron's picture is in 20% of them, then we would expect to have $5(0.2) = 1$ success. If we had 100 trials with probability of success 0.2, how many successes would you expect? Can you think of any reason not to say 20? It seems so simple that most people wouldn't even stop to think about it. You just multiply the probability of success by n. In other words, $E(X) = np$. Not fully convinced? We prove it in the next Math Box.

The standard deviation is less obvious; you can't just rely on your intuition. Fortunately, the formula for the standard deviation also boils down to something simple: $SD(X) = \sqrt{npq}$. (If you're curious about where that comes from, it's in the Math Box too!) In 100 boxes of cereal, we expect to find 20 LeBron James cards, with a standard deviation of $\sqrt{100 \times 0.8 \times 0.2} = 4$ pictures.

Time to summarize. A Binomial probability model describes the number of successes in a specified number of trials. It takes two parameters to specify this model: the number of trials n and the probability of success p.

NOTATION ALERT

Now punctuation! Throughout mathematics $n!$, pronounced "n factorial," is the product of all the integers from 1 to n. For example, $4! = 4 \cdot 3 \cdot 2 \cdot 1 = 24$.

TI-*nspire*

Binomial probabilities. Do-it-yourself Binomial models! Watch the probabilities change as you control n and p.

Binomial Probability Model for Bernoulli Trials: Binom(n, p)

n = number of trials
p = probability of success (and $q = 1 - p$ = probability of failure)
X = number of successes in n trials

$$P(X = x) = {}_nC_x\, p^xq^{n-x}, \text{ where } {}_nC_x = \frac{n!}{x!(n-x)!}$$

Mean: $\mu = np$
Standard Deviation: $\sigma = \sqrt{npq}$

Math Box

To derive the formulas for the mean and standard deviation of a Binomial model we start with the most basic situation.

Consider a single Bernoulli trial with probability of success p. Let's find the mean and variance of the number of successes.

Here's the probability model for the number of successes:

x	0	1
$P(X = x)$	q	p

(continued)

Find the expected value:

$$E(X) = 0q + 1p$$
$$E(X) = p$$

And now the variance:

$$Var(X) = (0 - p)^2 q + (1 - p)^2 p$$
$$= p^2 q + q^2 p$$
$$= pq(p + q)$$
$$= pq(1)$$
$$Var(X) = pq$$

What happens when there is more than one trial, though? A Binomial model simply counts the number of successes in a series of n independent Bernoulli trials. That makes it easy to find the mean and standard deviation of a binomial random variable, Y.

$$\text{Let } Y = X_1 + X_2 + X_3 + \cdots + X_n$$
$$E(Y) = E(X_1 + X_2 + X_3 + \cdots + X_n)$$
$$= E(X_1) + E(X_2) + E(X_3) + \cdots + E(X_n)$$
$$= p + p + p + \cdots + p \text{ (There are } n \text{ terms.)}$$

So, as we thought, the mean is $E(Y) = np$.

And since the trials are independent, the Pythagorean Theorem of Statistics tells us that the variances add:

$$Var(Y) = Var(X_1 + X_2 + X_3 + \cdots + X_n)$$
$$= Var(X_1) + Var(X_2) + Var(X_3) + \cdots + Var(X_n)$$
$$= pq + pq + pq + \cdots + pq \text{ (Again, } n \text{ terms.)}$$
$$Var(Y) = npq$$

Voilà! The standard deviation is $SD(Y) = \sqrt{npq}$.

For Example SPAM AND THE BINOMIAL MODEL

RECAP: The communications monitoring company *Postini* has reported that 91% of email messages are spam. Suppose your inbox contains 25 messages.

QUESTIONS: What are the mean and standard deviation of the number of real messages you should expect to find in your inbox? What's the probability that you'll find only 1 or 2 real messages?

ANSWER: I assume that messages arrive independently and at random, with the probability of success (a real message) $p = 1 - 0.91 = 0.09$. Let $X =$ the number of real messages among 25. I can use the model Binom(25, 0.09).

$$E(X) = np = 25(0.09) = 2.25$$
$$SD(X) = \sqrt{npq} = \sqrt{25(0.09)(0.91)} = 1.43$$
$$P(X = 1 \text{ or } 2) = P(X = 1) + P(X = 2)$$
$$= \binom{25}{1}(0.09)^1(0.91)^{24} + \binom{25}{2}(0.09)^2(0.91)^{23}$$
$$= 0.2340 + 0.2777$$
$$= 0.5117$$

Among 25 email messages, I expect to find an average of 2.25 that aren't spam, with a standard deviation of 1.43 messages. There's just over a 50% chance that 1 or 2 of my 25 emails will be real messages.

Step-by-Step Example WORKING WITH A BINOMIAL MODEL

Suppose 20 donors come to a blood drive. Recall that 6% of people are "universal donors."

Question:

1. What are the mean and standard deviation of the number of universal donors among them?
2. What is the probability that there are 2 or 3 universal donors?

THINK ➡ **Plan** State the question.

I want to know the mean and standard deviation of the number of universal donors among 20 people, and the probability that there are 2 or 3 of them.

Check to see that these are Bernoulli trials.

✓ There are two outcomes:

$$\text{success} = O\text{-negative}$$
$$\text{failure} = \text{other blood types}$$

✓ $p = 0.06$, because people have lined up at random.

✓ **10% Condition:** Trials are not independent, because the population is finite, but fewer than 10% of all possible donors are lined up.

Variable Define the random variable.

Let $X =$ number of O-negative donors among $n = 20$ people.

Model Specify the model.

I can model X with Binom(20, 0.06).

SHOW ➡ **Mechanics** Find the expected value and standard deviation.

Calculate the probability

$$E(X) = np = 20(0.06) = 1.2$$
$$SD(X) = \sqrt{npq} = \sqrt{20(0.06)(0.94)} \approx 1.06$$
$$P(X = 2 \text{ or } 3) = P(X = 2) + P(X = 3)$$
$$= \binom{20}{2}(0.06)^2(0.94)^{18}$$
$$+ \binom{20}{3}(0.06)^3(0.94)^{17}$$
$$\approx 0.2246 + 0.0860$$
$$= 0.3106$$

TELL ➡ **Conclusion** Interpret your results in context.

In groups of 20 randomly selected blood donors, I expect to find an average of 1.2 universal donors, with a standard deviation of 1.06. About 31% of the time, I'd find 2 or 3 universal donors among the 20 people.

Not-So-Random Assignment

To compare the effects of 2 different energy drinks a trainer plans to engage the first 40 people who arrive at his gym one morning in an experiment. He'll give half of them one drink and half the other, then compare the length of time they work out. He knows he has to assign the drinks at random, and he has a simple idea. As people walk in the door (and agree to participate), he'll flip a coin: heads they get Drink A, tails and it's Drink B. When one of the drinks has been assigned to 20 people, he'll simply give the other one to the rest so that there are 20 in that group, too.

Seems like a good idea, doesn't it? The coin tosses are unpredictable, and this plan would be easy to implement. *But it's not really random.*

For assignments to be truly random, there must be a 50-50 chance that any 2 volunteers will end up in the same group. While that's true for the first two people who arrive at the gym, think about the last two of the 40. They will get the same drink unless the first 38 coin tosses come out 19 heads and 19 tails. The binomial probability of 19 heads in 38 tosses is 0.13. That means there's an 87% chance that the last two volunteers will be assigned the same drink. There's a 25% chance that the first 3 people will all get the same drink, but a 74% chance that the last 3 will. That's not random.

Could this really matter? Absolutely! Maybe the first people in the door came early because they plan to work out longer. Some people who arrive together may be friends with similar workout plans. Or there could be many other reasons why failing to completely randomize will undermine this experiment.

(Note that such a heads-tails assignment technique is random if the trainer actually flips the coin for all 40 people with no stop-at-20 rule. Sure, he may—and probably will—end up with groups of different sizes, but that's okay. Failing to randomize is not.)

TI Tips FINDING BINOMIAL PROBABILITIES

Remember how the calculator handles Geometric probabilities? Well, the commands for finding Binomial probabilities are essentially the same. Again you'll find them in the 2nd DISTR menu.

- `binompdf(`
  This probability density function allows you to find the probability of an *individual* outcome. You need to define the Binomial model by specifying n and p, and then indicate the desired number of successes, x. The format is `binompdf(n,p,X)`.

For example, recall that LeBron James's picture is in 20% of the cereal boxes. Suppose that we want to know the probability of finding LeBron exactly twice among 5 boxes of cereal. We enter `binompdf(` with `trials:5`, `p:0.2`, `x value:2`, then go to `Paste` and hit `ENTER` (twice). There's about a 20% chance of getting two pictures of LeBron in five boxes of cereal.

- `binomcdf(`
  Need to add several Binomial probabilities? To find the total probability of getting x or fewer successes among the n trials use the cumulative Binomial density function `binomcdf(n,p,X)`.

For example, suppose we have ten boxes of cereal and wonder about the probability of finding up to 4 pictures of LeBron. That's the probability of 0, 1, 2, 3 or 4 successes, so we specify the command `binomcdf(` with `trials:10`, `p:0.2`, `x value:2`. Pretty likely!

Of course "up to 4" allows for the possibility that we end up with none. What's the probability we get at least 4 pictures of LeBron in 10 boxes? Well, "at least 4" means "not 3 or fewer." That's the complement of 0, 1, 2, or 3 successes. Have your TI evaluate `1-binomcdf(10,.2,3)`. There's about a 12% chance we'll find at least 4 pictures of LeBron in 10 boxes of cereal.

Just Checking

2. The Pew Research Center reports that they are only able to contact 76% of randomly selected households drawn for telephone surveys. Suppose a pollster has a list of 12 calls to make.

a) Why can these phone calls be considered Bernoulli trials.
b) Find the probability that the fourth call is the first one that makes contact.
c) Find the expected number of successful calls out of the 12.
d) Find the standard deviation of the number of successful calls.
e) Find the probability that exactly 9 of the 12 calls are successful.
f) Find the probability that at least 9 of the calls are successful.

The Normal Model to the Rescue!

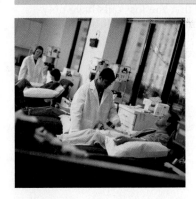

A S *Activity: Normal Approximation.* Binomial probabilities can be hard to calculate. With the Simulation Tool you'll see how well the Normal model can approximate the Binomial—a much easier method.

Suppose the Tennessee Red Cross anticipates the need for at least 1850 units of O-negative blood this year. It estimates that it will collect blood from 32,000 donors. How great is the risk that the Tennessee Red Cross will fall short of meeting its need? We've just learned how to calculate such probabilities. We can use the Binomial model with $n = 32,000$ and $p = 0.06$. The probability of getting *exactly* 1850 units of O-negative blood from 32,000 donors is $\binom{32000}{1850} \times 0.06^{1850} \times 0.94^{30150}$. No calculator on earth can calculate that first term (it has more than 100,000 digits).[3] And that's just the beginning. The problem said *at least* 1850, so we have to do it again for 1851, for 1852, and all the way up to 32,000. No thanks.

When we're dealing with a large number of trials like this, making direct calculations of the probabilities becomes tedious (or outright impossible). Here an old friend—the Normal model—comes to the rescue.

The Binomial model has mean $np = 1920$ and standard deviation $\sqrt{npq} \approx 42.48$. We could try approximating its distribution with a Normal model, using the same mean and standard deviation. Remarkably enough, that turns out to be a very good approximation. (We'll see why in the next chapter.) With that approximation, we can find the *probability:*

$$P(X < 1850) = P\left(z < \frac{1850 - 1920}{42.48}\right) \approx P(z < -1.65) \approx 0.05$$

There seems to be about a 5% chance that this Red Cross chapter will run short of O-negative blood.

Can we always use a Normal model to make estimates of Binomial probabilities? No. Consider the LeBron James situation—pictures in 20% of the cereal boxes. If we buy five boxes, the actual Binomial probabilities that we get 0, 1, 2, 3, 4, or 5 pictures of LeBron are 33%, 41%, 20%, 5%, 1%, and 0.03%, respectively. The first histogram shows that this probability model is skewed. That makes it clear that we should not try to estimate these probabilities by using a Normal model.

Now suppose we open 50 boxes of this cereal and count the number of LeBron pictures we find. The second histogram shows this probability model. It is centered at $np = 50(0.2) = 10$ pictures, as expected, and it appears to be fairly symmetric around that center. Let's have a closer look.

---

[3]If your calculator *can* find Binom(32000,0.06), then it's smart enough to use an approximation. Read on to see how you can, too.

0   5   10   15   20

TI-*nspire*

How close to Normal? How well does a Normal curve fit a Binomial model? Check out the Success/Failure Condition for yourself.

The third histogram again shows Binom(50, 0.2), this time magnified somewhat and centered at the expected value of 10 pictures of LeBron. It looks close to Normal, for sure. With this larger sample size, it appears that a Normal model might be a useful approximation.

A Normal model, then, is a close enough approximation only for a large enough number of trials. And what we mean by "large enough" depends on the probability of success. We'd need a larger sample if the probability of success were very low (or very high). It turns out that a Normal model works pretty well if we expect to see at least 10 successes and 10 failures. That is, we check the **Success/Failure Condition.**

The Success/Failure Condition: A Binomial model is approximately Normal if we expect at least 10 successes and 10 failures:

$$np \geq 10 \text{ and } nq \geq 10.$$

Math Box

Let's see where the magic number 10 comes from. You just need to remember how Normal models work. The problem is that a Normal model extends infinitely in both directions. But a Binomial model must have between 0 and n successes, so if we use a Normal to approximate a Binomial, we have to cut off its tails. That's not very important if the center of the Normal model is so far from 0 and n that the lost tails have only a negligible area. More than three standard deviations should do it, because a Normal model has little probability past that.

So the mean needs to be at least 3 standard deviations from 0 and at least 3 standard deviations from n. Let's look at the 0 end.

We require:	$\mu - 3\sigma > 0$
Or in other words:	$\mu > 3\sigma$
For a Binomial, that's:	$np > 3\sqrt{npq}$
Squaring yields:	$n^2p^2 > 9npq$
Now simplify:	$np > 9q$
Since $q \leq 1$, we can require:	$np > 9$

For simplicity, we usually require that np (and nq for the other tail) be at least 10 to use the Normal approximation, the Success/Failure Condition.

For Example   SPAM AND THE NORMAL APPROXIMATION TO THE BINOMIAL

RECAP: The communications monitoring company *Postini* has reported that 91% of email messages are spam. Recently, you installed a spam filter. You observe that over the past week it okayed only 151 of 1422 emails you received, classifying the rest as junk. Should you worry that the filtering is too aggressive?

QUESTION: What's the probability that no more than 151 of 1422 emails is a real message?

ANSWER: I assume that messages arrive randomly and independently, with a probability of success (a real message) $p = 0.09$. The model Binom(1422, 0.09) applies, but will be hard to work with. Checking conditions for the Normal approximation, I see that:

✓ These messages represent less than 10% of all email traffic.

✓ I expect $np = (1422)(0.09) = 127.98$ real messages and $nq = (1422)(0.91) = 1294.02$ spam messages, both far greater than 10.

(continued)

It's okay to approximate this binomial probability by using a Normal model.

$$\mu = np = 1422(0.09) = 127.98$$
$$\sigma = \sqrt{npq} = \sqrt{1422(0.09)(0.91)} \approx 10.79$$
$$P(x \le 151) = P\left(z \le \frac{151 - 127.98}{10.79}\right)$$
$$= P(z \le 2.13)$$
$$= 0.9834$$

Among my 1422 emails, there's over a 98% chance that no more than 151 of them were real messages, so the filter may be working properly.

A Word About Continuous Random Variables

There's a problem with approximating a Binomial model with a Normal model. The Binomial is discrete, giving probabilities for specific counts, but the Normal models a **continuous** random variable that can take on *any value*. For continuous random variables, we can no longer list all the possible outcomes and their probabilities, as we could for discrete random variables.[4]

As we saw in the previous chapter, models for continuous random variables give probabilities for *intervals* of values. So, when we use the Normal model, we no longer calculate the probability that the random variable equals a *particular* value, but only that it lies *between* two values. We won't calculate the probability of getting exactly 1850 units of blood, but we have no problem approximating the probability of getting 1850 *or more*, which was, after all, what we really wanted.[5]

 # Just Checking

3. Let's think about the Pew Research pollsters one more time. They tell us they are successful in contacting 76% of the households randomly selected for telephone surveys. When surveying public opinion, they hope to poll at least 1000 adults. Suppose Pew has compiled a list of 1300 phone numbers to call. What's the probability that they'll reach enough people?

a) Despite the fact that pollsters sample people without replacement, can we think of these calls as independent trials?
b) Find the mean and standard deviation of X = the number of adults Pew may successfully contact.
c) We want to find $P(X \ge 1000)$. Can we use a Normal model to approximate this Binomial probability?
d) Find the approximate probability Pew is able to contact at least 1000 voters on their list.

---

[4]In fact, some people use an adjustment called the "continuity correction" to help with this problem. It's related to the suggestion we make in the next footnote and is discussed in more advanced textbooks.

[5]If we really had been interested in a single value, we might have approximated it by finding the probability of getting between 1849.5 and 1850.5 units of blood.

Should I Be Surprised? A First Look at Statistical Significance

You watch a friend toss a coin 100 times and get 67 heads. That's more heads than you'd expect, but is it enough more that you should think she might be cheating somehow? You probably wouldn't consider 52 or 53 heads instead of a "perfect" 50 to be unusual, but if she tossed heads 90 times out of 100 you'd be really suspicious. How about 67? After all, random outcomes do vary, sometimes ending up higher or lower than expected. Is 67 heads too strange to be explained away as just random chance?

For quite a while now we've been thinking about *statistical significance*. The results of an experiment or a sample are said to be **statistically significant** if it's not reasonable to believe they occurred just by chance.

Let's think about your friend's 67 heads in 100 tosses. Coin tosses are Bernoulli trials; here there are $n = 100$ trials with probability of success $p = 0.5$. We can model the random variable X = number of heads with Binom(100,0.5). For our model, the mean is $np = 100(0.5) = 50$. OK, on average we expect 50 heads (duh!), but we know it won't be *exactly* 50 every time. The standard deviation is $\sqrt{npq} = \sqrt{100(0.5)(0.5)} = 5$ heads, and that's our clue about how much variation is reasonable.

Add one more key insight and we're ready to go: since we expect more than 10 successes (50) and more than 10 failures (nq is also 50), a Normal model is useful here. Her 67 heads is 17 more than we expected. Because the SD = 5, we know her results are over 3 standard deviations above the mean.

(To be exact, $z = \dfrac{67 - 50}{5} = 3.4$) Remember the 68-95-99.7 Rule?[6] More than 99.7% of the time, the result should be within 3 standard deviations of the mean, but hers isn't. If her coin-tossing method is fair, this would be an exceedingly rare outcome. Such an unusual result is statistically significant—friend or not, we should be very suspicious.

This is a real breakthrough! (Drumroll, please!) For the first time we've been able to decide whether what we've observed is just a chance occurrence or is strong evidence that something unusual is afoot. We'll explore this kind of reasoning in greater detail in the chapters ahead. For now it's enough to recognize that when a Normal model is useful,[7] outcomes more than 2 standard deviations from the expected value should be considered surprising.

Step-by-Step Example LOOKING FOR STATISTICAL SIGNIFICANCE

Before a blood drive, a local Red Cross agency puts out a plea for universal donors, hoping that they'll get more than the usual 6% among the donors who show up. That day they collected 202 units of blood, and among them 17 units were Type O-negative.

Question: Does this suggest that making a public plea is an effective way to get more O-negative donors to come to blood drives?

(continued)

---

[6]No? Well, it was a long time ago. Take a quick peek at page 115.

[7]*Always* check the conditions to be sure!

THINK ➡️ **Plan** State the question.

I expect 6% of all blood donors to be O-negative. I want to decide whether getting 17 O-negative donors among 202 people is statistically significant evidence that the Red Cross's public plea may have worked.

Variable Define the random variable.

X = number of O-negative donors

Check the conditions We've already confirmed that these are Bernoulli trials (p. 353), but it's critical to be sure that a Normal model applies.

✓ **10% Condition:** 202 < 10% of all possible donors.

✓ *Success/Failure Condition:* Among 202 donors with $P = 0.06$ I expect:
$np = (202)(0.06) = 12.12$ successes, and
$nq = (202)(0.94) = 189.88$ failures. Both are at least 10.

Model Name your model.

OK to use a Normal model.

SHOW ➡️ **Mechanics** Find the mean and standard deviation.

$$n = 202 \quad p = 0.06$$
$$E(X) = np = 202(0.6) = 12.12$$
$$SD(X) = \sqrt{npq} = \sqrt{202(0.06)(0.94)} = 3.375$$

Find the *z*-score for the observed result.

$$z = \frac{17 - 12.12}{3.375} = 1.45$$

Use the 68-95-99.7 Rule to think about whether that *z*-score seems unusual. We shouldn't be surprised unless the outcome is more than 2 standard deviations above or below the mean.

This doesn't look unusual; it's within 2 standard deviations of the mean.

TELL ➡️ **Conclusion** Explain (in context, of course) whether or not you consider the outcome to be statistically significant.

Although it was a good turnout, getting 17 Type O-negative donors among 202 people is only about 1.5 standard deviations more than expected. This could have been just random chance, so it's not strong evidence that the Red Cross's public plea raised the number of universal donors who came to the blood drive.

WHAT CAN GO WRONG?

- **Be sure you have Bernoulli trials.** Be sure to check the requirements first: two possible outcomes per trial ("success" and "failure"), a constant probability of success, and independence. Remember to check the 10% Condition when sampling without replacement.

- **Don't confuse Geometric and Binomial models.** Both involve Bernoulli trials, but the issues are different. If you are repeating trials until your first success, that's a Geometric probability. You don't know in advance how many trials you'll need—theoretically, it could take forever. If you are counting the number of successes in a specified number of trials, that's a Binomial probability.

- **Don't use the Normal approximation with small _n_.** To use a Normal approximation in place of a Binomial model, there must be at least 10 expected successes and 10 expected failures.

What Have We Learned?

We've learned that Bernoulli trials show up in lots of places. Depending on the random variable of interest, we can use one of three models to estimate probabilities for Bernoulli trials:

- a Geometric model when we're interested in the number of Bernoulli trials until the next success;
- a Binomial model when we're interested in the number of successes in a certain number of Bernoulli trials;
- a Normal model to approximate a Binomial model when we expect at least 10 successes and 10 failures.

We've learned (yet again) the importance of checking assumptions and conditions before proceeding.

And we've learned to use a Normal model to help us think about statistical significance. We consider observations more than 2 standard deviations from what's expected to be unusual.

Terms

Bernoulli trials, if . . .
1. there are two possible outcomes.
2. the probability of success is constant.
3. the trials are independent. (p. 414)

10% Condition
When sampling without replacement, trials are not independent. It's still okay to proceed as long as the random sample is smaller than 10% of the population. (p. 415)

Geometric probability model
A Geometric model is appropriate for a random variable that counts the number of Bernoulli trials until the first success.

$$X = \text{the number of trials until the first success}$$

$$P(x) = q^{x-1}p, \text{ where } p = \text{the probability of success and } q = 1 - p$$

$$E(X) = \frac{1}{p} \quad \text{(p. 418)}$$

Binomial probability model	A Binomial model is appropriate for a random variable that counts the number of successes in a fixed number of Bernoulli trials. X = the number of successes in n trials $$P(x) = \binom{n}{x} p^x q^{n-x}, \text{ where } p = \text{ the probability of success and } q = 1 - p$$ $$\mu = E(X) = np \text{ and } \sigma = SD(X) = \sqrt{npq} \quad \text{(p. 418)}$$
Success/Failure Condition	For a Normal model to be a good approximation of a Binomial model, we must expect at least 10 successes and 10 failures. That is, $np \geq 10$ and $nq \geq 10$. (p. 424)
Statistically significant	The results of a study are considered statistically significant if there's a very low probability they could have occurred by chance. (p. 426)

On the Computer THE BINOMIAL MODEL

Most statistics packages offer functions that compute Binomial probabilities. Some technology solutions automatically use the Normal approximation for the Binomial when the exact calculations become unmanageable.

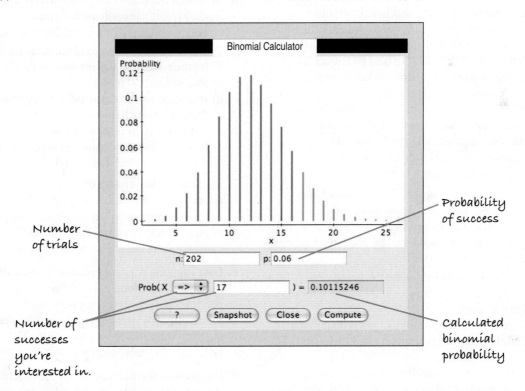

Exercises

1. **Bernoulli** Do these situations involve Bernoulli trials? Explain.
   a) We roll 50 dice to find the distribution of the number of spots on the faces.
   b) How likely is it that in a group of 120 the majority may have Type A blood, given that Type A is found in 43% of the population?
   c) We deal 7 cards from a deck and get all hearts. How likely is that?
   d) We wish to predict the outcome of a vote on the school budget, and poll 500 of the 3000 likely voters to see how many favor the proposed budget.
   e) A company realizes that about 10% of its packages are not being sealed properly. In a case of 24, is it likely that more than 3 are unsealed?

2. **Bernoulli 2** Do these situations involve Bernoulli trials? Explain.
   a) You are rolling 5 dice and need to get at least two 6's to win the game.
   b) We record the distribution of eye colors found in a group of 500 people.
   c) A manufacturer recalls a doll because about 3% have buttons that are not properly attached. Customers return 37 of these dolls to the local toy store. Is the manufacturer likely to find any dangerous buttons?
   d) A city council of 11 Republicans and 8 Democrats picks a committee of 4 at random. What's the probability they choose all Democrats?
   e) A 2002 Rutgers University study found that 74% of high school students have cheated on a test at least once. Your local high school principal conducts a survey in homerooms and gets responses that admit to cheating from 322 of the 481 students.

3. **Simulating the model** Think about the LeBron James picture search again. You are opening boxes of cereal one at a time looking for his picture, which is in 20% of the boxes. You want to know how many boxes you might have to open in order to find LeBron.
   a) Describe how you would simulate the search for LeBron using random numbers.
   b) Run at least 30 trials.
   c) Based on your simulation, estimate the probabilities that you might find your first picture of LeBron in the first box, the second, etc.
   d) Calculate the actual probability model.
   e) Compare the distribution of outcomes in your simulation to the probability model.

4. **Simulation II** You are one space short of winning a child's board game and must roll a 1 on a die to claim victory. You want to know how many rolls it might take.
   a) Describe how you would simulate rolling the die until you get a 1.
   b) Run at least 30 trials.
   c) Based on your simulation, estimate the probabilities that you might win on the first roll, the second, the third, etc.
   d) Calculate the actual probability model.
   e) Compare the distribution of outcomes in your simulation to the probability model.

5. **LeBron again** Let's take one last look at the LeBron James picture search. You know his picture is in 20% of the cereal boxes. You buy five boxes to see how many pictures of LeBron you might get.
   a) Describe how you would simulate the number of pictures of LeBron you might find in five boxes of cereal.
   b) Run at least 30 trials.
   c) Based on your simulation, estimate the probabilities that you get no pictures of LeBron, 1 picture, 2 pictures, etc.
   d) Find the actual probability model.
   e) Compare the distribution of outcomes in your simulation to the probability model.

6. **Seatbelts** Suppose 75% of all drivers always wear their seatbelts. Let's investigate how many of the drivers might be belted among five cars waiting at a traffic light.
   a) Describe how you would simulate the number of seatbelt-wearing drivers among the five cars.
   b) Run at least 30 trials.
   c) Based on your simulation, estimate the probabilities there are no belted drivers, exactly one, two, etc.
   d) Find the actual probability model.
   e) Compare the distribution of outcomes in your simulation to the probability model.

7. **On time** A Department of Transportation report about air travel found that, nationwide, 76% of all flights are on time. Suppose you are at the airport and your flight is one of 50 scheduled to take off in the next two hours. Can you consider these departures to be Bernoulli trials? Explain.

8. **Lost luggage** A Department of Transportation report about air travel found that airlines misplace about 5 bags per 1000 passengers. Suppose you are traveling with a group of people who have checked 22 pieces of luggage on your flight. Can you consider the fate of these bags to be Bernoulli trials? Explain.

9. **Hoops** A basketball player has made 80% of his foul shots during the season. Assuming the shots are independent, find the probability that in tonight's game he
   a) misses for the first time on his fifth attempt.
   b) makes his first basket on his fourth shot.
   c) makes his first basket on one of his first 3 shots.

10. **Chips** Suppose a computer chip manufacturer rejects 2% of the chips produced because they fail presale testing.
    a) What's the probability that the fifth chip you test is the first bad one you find?
    b) What's the probability you find a bad one within the first 10 you examine?

11. **More hoops** For the basketball player in Exercise 9,
    a) What's the expected number of shots until he misses?
    b) If the player shoots 10 foul shots in the fourth quarter, how many shots do you expect him to make?
    c) What is the standard deviation of the 10 shots?

12. **Chips ahoy** For the computer chips described in Exercise 10,
    a) How many do you expect to test before finding a bad one?
    b) In a random sample of 400 chips, what is the mean number of chips that are expected to fail?
    c) What is the standard deviation of that same sample?

13. **Customer center operator** Raaj works at the customer service call center of a major credit card bank. Cardholders call for a variety of reasons, but regardless of their reason for calling, if they hold a platinum card, Raaj is instructed to offer them a double-miles promotion. About 10% of all cardholders hold platinum cards, and about 50% of those will take the double-miles promotion. On average, how many calls will Raaj have to take before finding the first cardholder to take the double-miles promotion?

14. **Cold calls** Justine works for an organization committed to raising money for Alzheimer's research. From past experience, the organization knows that about 20% of all potential donors will agree to give something if contacted by phone. They also know that of all people donating, about 5% will give $100 or more. On average, how many potential donors will she have to contact until she gets her first $100 donor?

15. **Blood** Only 4% of people have Type AB blood.
    a) On average, how many donors must be checked to find someone with Type AB blood?
    b) What's the probability that there is a Type AB donor among the first 5 people checked?
    c) What's the probability that the first Type AB donor will be found among the first 6 people?
    d) What's the probability that we won't find a Type AB donor before the 10th person?

16. **Colorblindness** About 8% of males are colorblind. A researcher needs some colorblind subjects for an experiment and begins checking potential subjects.
    a) On average, how many men should the researcher expect to check to find one who is colorblind?
    b) What's the probability that she won't find anyone colorblind among the first 4 men she checks?
    c) What's the probability that the first colorblind man found will be the sixth person checked?

d) What's the probability that she finds someone who is colorblind before checking the 10th man?

17. **Smartphones** According to a September 2012 Nielsen study, 58% of teenagers (age 13–18) have a smartphone. If we select 8 teenagers at random, find the probability of each outcome described below.
    a) The first smartphone owner is the fourth person chosen.
    b) There is at least 1 smartphone among the 8 people.
    c) The first smartphone owner is the second or third person chosen.
    d) There are exactly 6 smartphones in the group.
    e) There are at least 6 smartphones in the group.
    f) There are no more than 6 smartphones in the group.

18. **Arrows** An Olympic archer is able to hit the bull's-eye 80% of the time. Assume each shot is independent of the others. If she shoots 6 arrows, what's the probability of each of the following results?
    a) Her first bull's-eye comes on the third arrow.
    b) She misses the bull's-eye at least once.
    c) Her first bull's-eye comes on the fourth or fifth arrow.
    d) She gets exactly 4 bull's-eyes.
    e) She gets at least 4 bull's-eyes.
    f) She gets at most 4 bull's-eyes.

19. **Smartphones redux** Consider our group of 8 people from Exercise 17.
    a) How many smartphones do you expect in the group?
    b) With what standard deviation?
    c) If we keep picking people until we find a smartphone, how long do you expect it will take until we find one?

20. **More arrows** Consider our archer from Exercise 18.
    a) How many bull's-eyes do you expect her to get?
    b) With what standard deviation?
    c) If she keeps shooting arrows until she hits the bull's-eye, how long do you expect it will take?

21. **Still more smartphones** Suppose we choose 20 people instead of the 8 chosen in Exercise 17.
    a) Find the mean and standard deviation of the number of non–smartphone owners in the group.
    b) What's the probability that
        i) they're not all smartphone owners?
        ii) there are no more than 15 smartphones?
        iii) there are exactly 10 of each?
        iv) the majority don't have a smartphone?

22. **Still more arrows** Suppose our archer from Exercise 18 shoots 10 arrows.
    a) Find the mean and standard deviation of the number of bull's-eyes she may get.
    b) What's the probability that
        i) she never misses?
        ii) there are no more than 8 bull's-eyes?
        iii) there are exactly 8 bull's-eyes?
        iv) she hits the bull's-eye more often than she misses?

23. **Vision** It is generally believed that nearsightedness affects about 12% of all children. A school district tests the vision of 169 incoming kindergarten children. How many would you expect to be nearsighted? With what standard deviation?

24. **International students** At a certain college, 6% of all students come from outside the United States. Incoming students there are assigned at random to freshman dorms, where students live in residential clusters of 40 freshmen sharing a common lounge area. How many international students would you expect to find in a typical cluster? With what standard deviation?

25. **Tennis, anyone?** A certain tennis player makes a successful first serve 70% of the time. Assume that each serve is independent of the others. If she serves 6 times, what's the probability she gets
    a) all 6 serves in?
    b) exactly 4 serves in?
    c) at least 4 serves in?
    d) no more than 4 serves in?

26. **Frogs** A wildlife biologist examines frogs for a genetic trait he suspects may be linked to sensitivity to industrial toxins in the environment. Previous research had established that this trait is usually found in 1 of every 8 frogs. He collects and examines a dozen frogs. If the frequency of the trait has not changed, what's the probability he finds the trait in
    a) none of the 12 frogs?
    b) at least 2 frogs?
    c) 3 or 4 frogs?
    d) no more than 4 frogs?

27. **Second serve** Consider the tennis player in Exercise 25 who successfully serves 70% of the first time.
    a) What are the four conditions that need to be met to justify your answers for Exercise 25?
    b) Do you think those conditions are valid? Explain.

28. **Easy being green** The biologist in Exercise 26 studied frogs with a 1 in 8 chance of having a certain trait.
    a) What are the conditions that must be met to justify your answers for Exercise 26?
    b) Do you think those conditions are satisfied? Explain.

29. **And more tennis** Suppose the tennis player in Exercise 25 serves 80 times in a match.
    a) What are the mean and standard deviation of the number of good first serves expected?
    b) Verify that you can use a Normal model to approximate the distribution of the number of good first serves.
    c) Use the 68–95–99.7 Rule to describe this distribution.
    d) What's the probability she makes at least 65 first serves?

30. **More arrows** The archer in Exercise 18 will be shooting 200 arrows in a large competition.

a) What are the mean and standard deviation of the number of bull's-eyes she might get?
b) Is a Normal model appropriate here? Explain.
c) Use the 68–95–99.7 Rule to describe the distribution of the number of bull's-eyes she may get.
d) Would you be surprised if she made only 140 bull's-eyes? Explain.

31. **Apples** An orchard owner knows that he'll have to use about 6% of the apples he harvests for cider because they will have bruises or blemishes. He expects a tree to produce about 300 apples.
    a) Describe an appropriate model for the number of cider apples that may come from that tree. Justify your model.
    b) Find the probability there will be no more than a dozen cider apples.
    c) Is it likely there will be more than 50 cider apples? Explain.

32. **Frogs, part III** Based on concerns raised by his preliminary research, the biologist in Exercise 26 decides to collect and examine 150 frogs.
    a) Assuming the frequency of the trait is still 1 in 8, determine the mean and standard deviation of the number of frogs with the trait he should expect to find in his sample.
    b) Verify that he can use a Normal model to approximate the distribution of the number of frogs with the trait.
    c) He found the trait in 22 of his frogs. Do you think this proves that the trait has become more common? Explain.

33. **Lefties** A lecture hall has 200 seats with folding arm tablets, 30 of which are designed for left-handers. The typical size of classes that meet there is 188, and we can assume that about 13% of students are left-handed. What's the probability that a right-handed student in one of these classes is forced to use a lefty arm tablet?

34. **No-shows** An airline, believing that 5% of passengers fail to show up for flights, overbooks (sells more tickets than there are seats). Suppose a plane will hold 265 passengers, and the airline sells 275 tickets. What's the probability the airline will not have enough seats, so someone gets bumped?

35. **Annoying phone calls** A newly hired telemarketer is told he will probably make a sale on about 12% of his phone calls. The first week he called 200 people, but only made 10 sales. Should he suspect he was misled about the true success rate? Explain.

36. **The euro** Shortly after the introduction of the euro coin in Belgium, newspapers around the world published articles claiming the coin is biased. The stories were based on reports that someone had spun the coin 250 times and gotten 140 heads—that's 56% heads. Do you think this is evidence that spinning a euro is unfair? Explain.

37. Seatbelts II Police estimate that 80% of drivers now wear their seatbelts. They set up a safety roadblock, stopping cars to check for seatbelt use.

a) How many cars do they expect to stop before finding a driver whose seatbelt is not buckled?
b) What's the probability that the first unbelted driver is in the 6th car stopped?
c) What's the probability that the first 10 drivers are all wearing their seatbelts?
d) If they stop 30 cars during the first hour, find the mean and standard deviation of the number of drivers expected to be wearing seatbelts.
e) If they stop 120 cars during this safety check, what's the probability they find at least 20 drivers not wearing their seatbelts?

38. Rickets Vitamin D is essential for strong, healthy bones. Our bodies produce vitamin D naturally when sunlight falls upon the skin, or it can be taken as a dietary supplement. Although the bone disease rickets was largely eliminated in England during the 1950s, some people there are concerned that this generation of children is at increased risk because they are more likely to watch TV or play computer games than spend time outdoors. Recent research indicated that about 20% of British children are deficient in vitamin D. Suppose doctors test a group of elementary school children.

a) What's the probability that the first vitamin D–deficient child is the 8th one tested?
b) What's the probability that the first 10 children tested are all okay?
c) How many kids do they expect to test before finding one who has this vitamin deficiency?
d) They will test 50 students at the third-grade level. Find the mean and standard deviation of the number who may be deficient in vitamin D.
e) If they test 320 children at this school, what's the probability that no more than 50 of them have the vitamin deficiency?

39. ESP Scientists wish to test the mind-reading ability of a person who claims to "have ESP." They use five cards with different and distinctive symbols (square, circle, triangle, line, squiggle). Someone picks a card at random and thinks about the symbol. The "mind reader" must correctly identify which symbol was on the card. If the test consists of 100 trials, how many would this person need to get right in order to convince you that ESP may actually exist? Explain.

40. True-False A true-false test consists of 50 questions. How many does a student have to get right to convince you that he is not merely guessing? Explain.

41. Hot hand A basketball player who ordinarily makes about 55% of his free throw shots has made 4 in a row.

Is this evidence that he has a "hot hand" tonight? That is, is this streak so unusual that it means the probability he makes a shot must have changed? Explain.

42. New bow Our archer in Exercise 18 purchases a new bow, hoping that it will improve her success rate to more than 80% bull's-eyes. She is delighted when she first tests her new bow and hits 6 consecutive bull's-eyes. Do you think this is compelling evidence that the new bow is better? In other words, is a streak like this unusual for her? Explain.

43. Hotter hand Our basketball player in Exercise 41 has new sneakers, which he thinks improve his game. Over his past 40 shots, he's made 32—much better than the 55% he usually shoots. Do you think his chances of making a shot really increased? In other words, is making at least 32 of 40 shots really unusual for him? (Do you think it's his sneakers?)

44. New bow, again The archer in Exercise 42 continues shooting arrows, ending up with 45 bull's-eyes in 50 shots. Now are you convinced that the new bow is better? Explain.

Just Checking ANSWERS

1. a) No; the probability of choosing a female changes with each name drawn.
    b) Yes.
    c) No; women who have had twins are more likely to have them again.
    d) No; there are more than two possible outcomes (colors).
    e) No; the sample is more than 10% of the population.
2. a) There are 2 outcomes (contact or not); $p = 0.26$; fewer than 10% of the population are being contacted randomly.
    b) $(0.24)^3(0.76) = 0.011$
    c) $\mu = np = 12(0.76) = 9.12$
    d) $\sigma = \sqrt{npq} = \sqrt{12(0.76)(0.24)} \approx 1.48$
    e) $\binom{12}{9}(0.76)^9(0.24)^3 \approx 0.26$
    f) $\binom{12}{9}(0.76)^9(0.24)^3 + \binom{12}{10}(0.76)^{10}(0.24)^2 + \binom{12}{11}(0.76)^{11}(0.24)^1 + (0.76)^{12} \approx 0.68$
3. a) $1300 < 10\%$ of all households.
    b) $\mu = 1300(0.76) = 988$; $\sigma = \sqrt{1300(0.76)(0.24)} = 15.4$
    c) Yes; $np = 1300(0.76) = 988$ and $nq = 1300(0.24) = 312$ are both at least 10.
    d) $P(z > 0.78) = 0.22$

Review of part IV

Randomness and Probability

Quick Review

Here's a brief summary of the key concepts and skills in probability and probability modeling:

- The Law of Large Numbers says that the more times we try something, the closer the results will come to theoretical perfection.

  - Don't mistakenly misinterpret the Law of Large Numbers as the "Law of Averages." There's no such thing.

- Basic rules of probability can handle most situations:

  - To find the probability that an event OR another event happens, add their probabilities and subtract the probability that both happen.

  - To find the probability that an event AND another independent event both happen, multiply probabilities.

  - Conditional probabilities tell you how likely one event is to happen, knowing that another event has happened.

  - Mutually exclusive events (also called "disjoint") cannot both happen at the same time.

- Two events are independent if the occurrence of one doesn't change the probability that the other happens.

- A probability model for a random variable describes the theoretical distribution of outcomes.

  - The mean of a random variable is its expected value.

  - For sums or differences of independent random variables, variances add.

  - To estimate probabilities involving quantitative variables, you may be able to use a Normal model—but only if the distribution of the variable is unimodal and symmetric.

  - To estimate the probability you'll get your first success on a certain trial, use a Geometric model.

  - To estimate the probability you'll get a certain number of successes in a specified number of independent trials, use a Binomial model.

Ready? Here are some opportunities to check your understanding of these ideas.

Review Exercises

1. **Quality control** A consumer organization estimates that 29% of new cars have a cosmetic defect, such as a scratch or a dent, when they are delivered to car dealers. This same organization believes that 7% have a functional defect—something that does not work properly—and that 2% of new cars have both kinds of problems.

   a) If you buy a new car, what's the probability that it has some kind of defect?
   b) What's the probability it has a cosmetic defect but no functional defect?
   c) If you notice a dent on a new car, what's the probability it has a functional defect?
   d) Are the two kinds of defects disjoint events? Explain.
   e) Do you think the two kinds of defects are independent events? Explain.

2. **Workers** A company's human resources officer reports a breakdown of employees by job type and sex shown in the table.

		Sex	
		Male	Female
Job Type	Management	7	6
	Supervision	8	12
	Production	45	72

   a) What's the probability that a worker selected at random is
   i) female?
   ii) female or a production worker?
   iii) female, if the person works in production?
   iv) a production worker, if the person is female?
   b) Do these data suggest that job type is independent of being male or female? Explain.

3. **Airfares** Each year a company must send 3 officials to a meeting in China and 5 officials to a meeting in France. Airline ticket prices vary from time to time, but the company purchases all tickets for a country at the same price. Past experience has shown that tickets to China have a mean price of $1000, with a standard deviation of $150, while the mean airfare to France is $500, with a standard deviation of $100.

   a) Define random variables and use them to express the total amount the company will have to spend to send these delegations to the two meetings.
   b) Find the mean and standard deviation of this total cost.
   c) Find the mean and standard deviation of the difference in price of a ticket to China and a ticket to France.
   d) Do you need to make any assumptions in calculating these means? How about the standard deviations?

4. Autism Psychiatrists estimate that about 1 in 100 adults has autism. What's the probability that in a city of 20,000, there are more than 300 people with this condition? Be sure to verify that a Normal model can be used here.

5. A game To play a game, you must pay $5 for each play. There is a 10% chance you will win $5, a 40% chance you will win $7, and a 50% chance you will win only $3.

a) What are the mean and standard deviation of your net winnings?

b) You play twice. Assuming the plays are independent events, what are the mean and standard deviation of your total winnings?

6. Emergency switch Safety engineers must determine whether industrial workers can operate a machine's emergency shutoff device. Among a group of test subjects, 66% were successful with their left hands, 82% with their right hands, and 51% with either hand.

a) What percent of these workers could not operate the switch with either hand?

b) Are success with right and left hands independent events? Explain.

c) Are success with right and left hands mutually exclusive? Explain.

7. Facebook According to Pew Research, 50% of adults and 75% of teenagers were using a social networking site in early 2012. Most of that activity was on Facebook. Let's assume these probabilities apply strictly to Facebook (after all, MySpace is empty, right?) Among a group of 10 people, what's the probability that

a) at least one of the people was not on Facebook if they were all adults?

b) at least one of the people was not on Facebook if they were all teenagers?

c) at least one of the people was not on Facebook if half were teenagers?

8. Deductible A car owner may buy insurance that will pay the full price of repairing the car after an at-fault accident, or save $12 a year by getting a policy with a $500 deductible. Her insurance company says that about 0.5% of drivers in her area have an at-fault auto accident during any given year. Based on this information, should she buy the policy with the deductible or not? How does the value of her car influence this decision?

9. More Facebook Using the percentages from Exercise 7, suppose there is a group of 5 teenagers. What's the probability that

a) all will be on Facebook?

b) exactly 1 will be on Facebook?

c) at least 3 will be on Facebook?

10. At fault The car insurance company in Exercise 8 believes that about 0.5% of drivers have an at-fault accident during a given year. Suppose the company insures 1355 drivers in that city.

a) What are the mean and standard deviation of the number who may have at-fault accidents?

b) Can you describe the distribution of these accidents with a Normal model? Explain.

11. Friend me? One hundred fifty-eight teenagers are standing in line for a big movie premier night. (See Exercise 7.)

a) What are the mean and standard deviation of the number of Facebook users we might expect to find among this group of teens?

b) Can we use a Normal model in this situation?

c) What the probability that no more than 110 of the teenagers are Facebook users?

12. Child's play In a board game you determine the number of spaces you may move by spinning a spinner and rolling a die. The spinner has three regions: Half of the spinner is marked "5," and the other half is equally divided between "10"and "20." The six faces of the die show 0, 0, 1, 2, 3, and 4 spots. When it's your turn, you spin and roll, adding the numbers together to determine how far you may move.

a) Create a probability model for the outcome on the spinner.

b) Find the mean and standard deviation of the spinner results.

c) Create a probability model for the outcome on the die.

d) Find the mean and standard deviation of the die results.

e) Find the mean and standard deviation of the number of spaces you get to move.

13. Language Neurological research has shown that in about 80% of people, language abilities reside in the brain's left side. Another 10% display right-brain language centers, and the remaining 10% have two-sided language control. (The latter two groups are mainly left-handers; *Science News*, 161 no. 24 [2002].)

a) Assume that a freshman composition class contains 25 randomly selected people. What's the probability that no more than 15 of them have left-brain language control?

b) In a randomly chosen group of 5 of these students, what's the probability that no one has two-sided language control?

c) In the entire freshman class of 1200 students, how many would you expect to find of each type?

d) What are the mean and standard deviation of the number of these freshmen who might be right-brained in language abilities?

e) If an assumption of Normality is justified, use the 68–95–99.7 Rule to describe how many students in the freshman class might have right-brain language control.

14. Play again If you land in a "penalty zone" on the game board described in Exercise 12, your move will be determined by subtracting the roll of the die from the result on the spinner. Now what are the mean and standard deviation of the number of spots you may move?

15. Beanstalks In some cities tall people who want to meet and socialize with other tall people can join Beanstalk Clubs. To qualify, a man must be over 6′2″ tall, and a woman over 5′10″. According to the National Health Survey, heights of adults may have a Normal model with mean heights of 69.1″ for men and 64.0″ for women. The respective standard deviations are 2.8″ and 2.5″.

a) You're probably not surprised to learn that men are generally taller than women, but what does the greater standard deviation for men's heights indicate?

b) Are men or women more likely to qualify for Beanstalk membership?

c) Beanstalk members believe that height is an important factor when people select their spouses. To investigate, we select at random a married man and, independently, a married woman. Define two random variables, and use them to express how many inches taller the man is than the woman.

d) What's the mean of this difference?

e) What's the standard deviation of this difference?

f) What's the probability that the man is taller than the woman (that the difference in heights is greater than 0)?

g) Suppose a survey of married couples reveals that 92% of the husbands were taller than their wives. Based on your answer to part f, do you believe that people's choice of spouses is independent of height? Explain.

16. Stocks Since the stock market began in 1872, stock prices have risen in about 73% of the years. Assuming that market performance is independent from year to year, what's the probability that

a) the market will rise for 3 consecutive years?

b) the market will rise 3 years out of the next 5?

c) the market will fall during at least 1 of the next 5 years?

d) the market will rise during a majority of years over the next decade?

17. Multiple choice A multiple choice test has 50 questions, with 4 answer choices each. You must get at least 30 correct to pass the test, and the questions are very difficult.

a) Are you likely to be able to pass by guessing on every question? Explain.

b) Suppose, after studying for a while, you believe you have raised your chances of getting each question right to 70%. How likely are you to pass now?

c) Assuming you are operating at the 70% level and the instructor arranges questions randomly, what's the probability that the third question is the first one you get right?

18. Stock strategy Many investment advisors argue that after stocks have declined in value for 2 consecutive years, people should invest heavily because the market rarely declines 3 years in a row.

a) Since the stock market began in 1872, there have been two consecutive losing years eight times. In six of those cases, the market rose during the following year. Does this confirm the advice?

b) Overall, stocks have risen in value during 95 of the 130 years since the market began in 1872. How is this fact relevant in assessing the statistical reasoning of the advisors?

19. Insurance A 65-year-old woman takes out a $100,000 term life insurance policy. The company charges an annual premium of $520. Estimate the company's expected profit on such policies if mortality tables indicate that only 2.6% of women age 65 die within a year.

20. Teen smoking The Centers for Disease Control say that about 30% of high school students smoke tobacco (down from a high of 38% in 1997). Suppose you randomly select high school students to survey them on their attitudes toward scenes of smoking in the movies. What's the probability that

a) none of the first 4 students you interview is a smoker?

b) the first smoker is the sixth person you choose?

c) there are no more than 2 smokers among 10 people you choose?

21. Passing stats Molly's college offers two sections of Statistics 101. From what she has heard about the two professors listed, Molly estimates that her chances of passing the course are 0.80 if she gets Professor Scedastic and 0.60 if she gets Professor Kurtosis. The registrar uses a lottery to randomly assign the 120 enrolled students based on the number of available seats in each class. There are 70 seats in Professor Scedastic's class and 50 in Professor Kurtosis's class.

a) What's the probability that Molly will pass Statistics?

b) At the end of the semester, we find out that Molly failed. What's the probability that she got Professor Kurtosis?

22. Teen smoking II Suppose that, as reported by the Centers for Disease Control, about 30% of high school students smoke tobacco. You randomly select 120 high school students to survey them on their attitudes toward scenes of smoking in the movies.

a) What's the expected number of smokers?

b) What's the standard deviation of the number of smokers?

c) The number of smokers among 120 randomly selected students will vary from group to group. Explain why that number can be described with a Normal model.

d) Using the 68–95–99.7 Rule, create and interpret a model for the number of smokers among your group of 120 students.

23. **Random variables** Given independent random variables with means and standard deviations as shown, find the mean and standard deviation of each of these variables:

a) $X + 50$
b) $10Y$
c) $X + 0.5Y$
d) $X - Y$
e) $X_1 + X_2$

	Mean	SD
X	50	8
Y	100	6

24. **Merger** Explain why the facts you know about variances of independent random variables might encourage two small insurance companies to merge. (*Hint:* Think about the expected amount and potential variability in payouts for the separate and the merged companies.)

25. **Youth survey** According to a recent Gallup survey, 93% of teens use the Internet, but there are differences in how teen boys and girls say they use computers. The telephone poll found that 77% of boys had played computer games in the past week, compared with 65% of girls. On the other hand, 76% of girls said they had e-mailed friends in the past week, compared with only 65% of boys.

a) For boys, the cited percentages are 77% playing computer games and 65% using e-mail. That total is 142%, so there is obviously a mistake in the report. No? Explain.
b) Based on these results, do you think playing games and using e-mail are mutually exclusive? Explain.
c) Do you think whether a child e-mails friends is independent of being a boy or a girl? Explain.
d) Suppose that in fact 93% of the teens in your area do use the Internet. You want to interview a few who do not, so you start contacting teenagers at random. What is the probability that it takes you 5 interviews until you find the first person who does not use the Internet?

26. **Meals** A college student on a seven-day meal plan reports that the amount of money he spends daily on food varies with a mean of $13.50 and a standard deviation of $7.

a) What are the mean and standard deviation of the amount he might spend in two consecutive days?
b) What assumption did you make in order to find that standard deviation? Are there any reasons you might question that assumption?
c) Estimate his average weekly food costs, and the standard deviation.
d) Do you think it likely he might spend less than $50 in a week? Explain, including any assumptions you make in your analysis.

27. **Travel to Kyrgyzstan** Your pocket copy of *Kyrgyzstan on 4237 ± 360 Som a Day* claims that you can expect to spend about 4237 som each day with a standard deviation of 360 som. How well can you estimate your expenses for the trip?

a) Your budget allows you to spend 90,000 som. To the nearest day, how long can you afford to stay in Kyrgyzstan, on average?
b) What's the standard deviation of your expenses for a trip of that duration?
c) You doubt that your total expenses will exceed your expectations by more than two standard deviations. How much extra money should you bring? On average, how much of a "cushion" will you have per day?

28. **Picking melons** Two stores sell watermelons. At the first store the melons weigh an average of 22 pounds, with a standard deviation of 2.5 pounds. At the second store the melons are smaller, with a mean of 18 pounds and a standard deviation of 2 pounds. You select a melon at random at each store.

a) What's the mean difference in weights of the melons?
b) What's the standard deviation of the difference in weights?
c) If a Normal model can be used to describe the difference in weights, what's the probability that the melon you got at the first store is heavier?

29. **Home, sweet home 2009** According to the 2009 Census, 67% of U.S. households own the home they live in. A mayoral candidate conducts a survey of 820 randomly selected homes in your city and finds only 523 owned by the current residents. The candidate then attacks the incumbent mayor, saying that there is an unusually low level of homeownership in the city. Do you agree? Explain.

30. **Buying melons** The first store in Exercise 28 sells watermelons for 32 cents a pound. The second store is having a sale on watermelons—only 25 cents a pound. Find the mean and standard deviation of the difference in the price you may pay for melons randomly selected at each store.

31. **Who's the boss?** The 2000 Census revealed that 26% of all firms in the United States are owned by women. You call some firms doing business locally, assuming that the national percentage is true in your area.

a) What's the probability that the first 3 you call are all owned by women?
b) What's the probability that none of your first 4 calls finds a firm that is owned by a woman?
c) Suppose none of your first 5 calls found a firm owned by a woman. What's the probability that your next call does?

32. **Jerseys** A Statistics professor comes home to find that all four of his children got white team shirts from soccer camp this year. He concludes that this year, unlike other

years, the camp must not be using a variety of colors. But then he finds out that in each child's age group there are 4 teams, only 1 of which wears white shirts. Each child just happened to get on the white team at random.

a) Why was he so surprised? If each age group uses the same 4 colors, what's the probability that all four kids would get the same-color shirt?

b) What's the probability that all 4 would get white shirts?

c) We lied. Actually, in the oldest child's group there are 6 teams instead of the 4 teams in each of the other three groups. How does this change the probability you calculated in part b?

33. When to stop? In Exercise 27 of the Review Exercises for Part III, we posed this question:

You play a game that involves rolling a die. You can roll as many times as you want, and your score is the total for all the rolls. But . . . if you roll a 6, your score is 0 and your turn is over. What might be a good strategy for a game like this?

You attempted to devise a good strategy by simulating several plays to see what might happen. Let's try calculating a strategy.

a) On what roll would you expect to get a 6 for the first time?

b) So, roll *one time less* than that. Assuming all those rolls were not 6's, what's your expected score?

c) What's the probability that you can roll that many times without getting a 6?

34. Plan B Here's another attempt at developing a good strategy for the dice game in Exercise 33. Instead of stopping after a certain number of rolls, you could decide to stop when your score reaches a certain number of points.

a) How many points would you expect a roll to *add* to your score?

b) In terms of your current score, how many points would you expect a roll to *subtract* from your score?

c) Based on your answers in parts a and b, at what score will another roll "break even"?

d) Describe the strategy this result suggests.

35. Technology on campus 2005 Every 5 years, the Conference Board of the Mathematical Sciences surveys college math departments. In 2005, the board reported that 51%

of all undergraduates taking Calculus I were in classes that used graphing calculators and 21% were in classes that used computer assignments. Suppose that 10% used both calculators and computers.

a) What percent used neither kind of technology?

b) What percent used calculators but not computers?

c) What percent of the calculator users had computer assignments?

d) Based on this survey, do calculator and computer use appear to be independent events? Explain.

36. Dogs A census by the county dog control officer found that 18% of homes kept one dog as a pet, 4% had two dogs, and 1% had three or more. If a salesman visits two homes selected at random, what's the probability he encounters

a) no dogs?

b) some dogs?

c) dogs in each home?

d) more than one dog in each home?

37. Socks In your sock drawer you have 4 blue socks, 5 grey socks, and 3 black ones. Half asleep one morning, you grab 2 socks at random and put them on. Find the probability you end up wearing

a) 2 blue socks.

b) no grey socks.

c) at least 1 black sock.

d) a green sock.

e) matching socks.

38. Gym Time A local gym charges a $150 one-time fee to join. After joining, there is a $40 per month charge to continue your membership.

a) The gym has tracked its members over time and estimates that the customers have a mean membership length of 21 months. What is the mean total of the money collected?

b) The standard deviation of the length of stay is 5 months. What is the standard deviation of the total money collected?

39. Coins A coin is to be tossed 36 times.

a) What are the mean and standard deviation of the number of heads?

b) Suppose the resulting number of heads is unusual, two standard deviations above the mean. How many "extra" heads were observed?

c) If the coin were tossed 100 times, would you still consider the same number of extra heads unusual? Explain.

d) In the 100 tosses, how many extra heads would you need to observe in order to say the results were unusual?

e) Explain how these results refute the "Law of Averages" but confirm the Law of Large Numbers.

40. The Drake equation In 1961 astronomer Frank Drake developed an equation to try to estimate the number of extraterrestrial civilizations in our galaxy that might be able to communicate with us via radio transmissions. Now largely accepted by the scientific community, the Drake equation has helped spur efforts by radio astronomers to search for extraterrestrial intelligence. Here is the equation:

$$N_C = N \cdot f_p \cdot n_e \cdot f_l \cdot f_i \cdot f_c \cdot f_L$$

OK, it looks a little messy, but here's what it means:

Factor	What It Represents	Possible Value
N	Number of stars in the Milky Way Galaxy	200–400 billion
f_p	Probability that a star has planets	20%–50%
n_e	Number of planets in a solar system capable of sustaining earth-type life	1? 2?
f_l	Probability that life develops on a planet with a suitable environment	1%–100%
f_i	Probability that life evolves intelligence	50%?
f_c	Probability that intelligent life develops radio communication	10%–20%
f_L	Fraction of the planet's life for which the civilization survives	$\dfrac{1}{1,000,000}$?
N_c	Number of extraterrestrial civilizations in our galaxy with which we could communicate	?

So, how many ETs are out there? That depends; values chosen for the many factors in the equation depend on ever-evolving scientific knowledge and one's personal guesses. But now, some questions.

a) What quantity is calculated by the first product, $N \cdot f_p$?
b) What quantity is calculated by the product, $N \cdot f_p \cdot n_e \cdot f_l$?
c) What probability is calculated by the product $f_l \cdot f_i$?
d) Which of the factors in the formula are conditional probabilities? Restate each in a way that makes the condition clear.

Note: A quick Internet search will find you a site where you can play with the Drake equation yourself.

41. Recalls In a car rental company's fleet, 70% of the cars are American brands, 20% are Japanese, and the rest are German. The company notes that manufacturers' recalls seem to affect 2% of the American cars, but only 1% of the others.
a) What's the probability that a randomly chosen car is recalled?
b) What's the probability that a recalled car is American?

42. Pregnant? Suppose that 70% of the women who suspect they may be pregnant and purchase an in-home pregnancy test are actually pregnant. Further suppose that the test is 98% accurate. What's the probability that a woman whose test indicates that she is pregnant actually is?

43. Door prize You are among 100 people attending a charity fundraiser at which a large-screen TV will be given away as a door prize. To determine who wins, 99 white balls and 1 red ball have been placed in a box and thoroughly mixed. The guests will line up and, one at a time, pick a ball from the box. Whoever gets the red ball wins the TV, but if the ball is white, it is returned to the box. If none of the 100 guests gets the red ball, the TV will be auctioned off for additional benefit of the charity.
a) What's the probability that the first person in line wins the TV?
b) You are the third person in line. What's the probability that you win the TV?
c) What's the probability that the charity gets to sell the TV because no one wins?
d) Suppose you get to pick your spot in line. Where would you want to be in order to maximize your chances of winning?
e) After hearing some protest about the plan, the organizers decide to award the prize by not returning the white balls to the box, thus ensuring that 1 of the 100 people will draw the red ball and win the TV. Now what position in line would you choose in order to maximize your chances?

Practice Exam

Multiple Choice

1. Researchers who were interested in the types of movies preferred by children of different age groups asked students student in the sixth grade and in the eighth grade if they would prefer to see an animated feature like "The Lion King" or an action feature like "The Avengers." The results are summarized in the table:

	Animated	Action
Sixth Grade	45	35
Eighth Grade	40	60

Which proportions represent the conditional distribution of grade for children who preferred an action feature?
A) 0.194 and 0.333
B) 0.368 and 0.632
C) 0.40 and 0.60
D) 0.444 and 0.556
E) 0.529 and 0.471

2. A bookstore asked a sample of adults how many e-books they had downloaded to their e-book readers during the last three months. Some of the data are shown below:

Values below Q1	Q1	Median	Q3	Values above Q3
8, 11, 14	16	18	22	25, 28, 30, 33

Which values will a boxplot identify as outliers?

A) None   B) 8 only   C) 33 only
D) 8 and 33 only   E) 8, 30, and 33

3. A company that supplies LP gas for heating keeps data on the low temperature for each day of each month. It summarizes these data by finding the mean, median, standard deviation and interquartile range. The company assumes that people use the LP gas for heat when the temperature is below 70°, so it creates a second set of data by subtracting 70 from each daily low temperature. Which of the four summary statistics will change for this second data set?

A) mean and median only
B) mean and standard deviation only
C) median and IQR only
D) standard deviation and IQR only
E) mean, median, standard deviation, and IQR

4. Luis is a star high jumper for his high school track team. His Statistics teacher tells him that his jumps last year had an average z-score of +1.8, while this year his jumps have an average z-score of +2.1. Which of these statements can Luis know for certain?

A) He is jumping higher this year than last year.
B) The competing jumpers are not jumping as high this year as they did last year.
C) He is winning the high jump competitions more often this year.
D) His jumps average 0.3 feet higher this year.
E) His jumps this year are higher relative to the other jumpers.

5. A forest ranger has data on the heights of a large growth of young pine trees. The mean height is 3.2 feet and the standard deviation 0.6 feet. A histogram shows that the distribution of heights is approximately normal. Approximately what fraction of the trees should we expect to be between 4.0 and 4.4 feet tall?

A) 2%   B) 7%   C) 9%
D) 91%   E) 98%

6. A study conducted by students in an AP Psychology class at South Kent School in CT discovered a correlation of −0.38 between students' hours of sleep (x) and GPAs (y). The scatterplot below displays the relationship. What conclusion can be reasonably drawn from these data?

A) Students who get more sleep tend to earn higher GPAs.
B) If a student wishes to earn higher grades, he or she should get more sleep.
C) Based on the pattern in the scatterplot and the correlation, these data indicate that as students sleep longer, they tend to earn lower GPAs.
D) The scatterplot and the correlation coefficient contradict one another.
E) The scatterplot shows two influential points that affect the value of the correlation.

7. In 2009 the Organization for Economic Cooperation and Development (OECD) conducted a study of 34 member countries called the Programme for International Student Assessment (PISA). The OECD looked at reading scores among 15-year-olds in each country as well as the students' socioeconomic backgrounds. PISA reported that higher socioeconomic status was associated with higher reading scores, with their socioeconomic index explaining 14% of the variability in reading scores.
(Source: Sahlberg, Pasi, "A Model Lesson: Finland Shows Us What Equal Opportunity Looks Like," *American Educator* Spring 2012: 25.)

Which of the following can be correctly concluded from this information?

A) For every additional point in the mean reading score in an OECD country, the PISA socioeconomic index is expected to increase by 14% on average.
B) For every additional one percent increase in the PISA socioeconomic index, the mean reading score is expected to increase by 14% on average.
C) If the U.S. government wants to increase the mean reading score for American 15-year-olds, it would be wise to institute programs to improve students' socioeconomic status.
D) The correlation between the PISA socioeconomic index and reading scores for these OECD countries is 0.374.
E) 14% of the variation in reading scores is caused by student socioeconomic status.

8. Once a month a local theater group stages a live theater production. The group is able to sell enough tickets so that the theater is almost full each month. However, the number of adult tickets and children's tickets that are sold vary depending on the play being performed. For these productions, which statement best describes the correlation between the number of adult tickets and the number of children's tickets sold?

A) The correlation will be exactly 1.
B) The correlation will be negative.
C) The correlation will be 0.
D) The correlation will be positive and less than 1.
E) The correlation cannot be described based on the information given.

9. A set of paired data has a least squares regression line with equation $\hat{y} = 0.50x + 2.0$ and a correlation coefficient of $r = 0.80$. Suppose we convert the data for each variable to z-scores and then compute the new regression line. What will the equation be?

A) $\hat{z}_y = 0.50z_x$ B) $\hat{z}_y = 0.64z_x$
C) $\hat{z}_y = 0.80z_x$ D) $\hat{z}_y = 0.50z_x + 20$
E) $\hat{z}_y = 0.80z_x + 20$

10. A researcher examined a sample of rainbow trout taken from the Spokane River in Washington State, recording their lengths (mm) and weights (grams). For example, one trout was 360 mm long and weighed 469 g. Because a scatterplot using length to predict weight showed an exponential relationship, the researcher took the log of weight and successfully linearized the relationship. Use his regression model $\log(\widehat{weight}) = 1.491 + 0.00331length$ to predict the weight of a 400 mm rainbow trout.

A) 2.815 g   B) 16.7 g   C) 509 g
D) 598 g   E) 653 g

11. In 2000 two organizations conducted surveys to ascertain the public's opinion on banning gay men from serving in leadership roles in the Boy Scouts.

■ A Pew poll asked respondents whether they agreed with "the recent decision by the Supreme Court" that "the Boy Scouts of America have a constitutional right to block gay men from becoming troop leaders."

■ A Los Angeles Times poll asked respondents whether they agreed with the following statement: "A Boy Scout leader should be removed from his duties as a troop leader if he is found out to be gay, even if he is considered by the Scout organization to be a model Boy Scout leader."

Source: http://www.gallup.com/poll/9916/Homosexuality.aspx#5

One of these polls found 36% agreement; the other found 56% agreement. Which of the following statements is true?

A) The Pew poll found 36% agreement, and the Los Angeles Times poll found 56% agreement.
B) The Pew poll includes a leading question, while the Los Angeles Times poll uses neutral wording.
C) The Los Angeles Times Poll includes a leading question, while the Pew poll uses neutral wording.
D) Both polls use neutral wording, so the different results reflect sampling error.
E) Both polls contained leading questions, which may explain the very different results.

12. In 2012 American Idol completed its 11th season with Jennifer Lopez, Steven Tyler, and Randy Jackson as judges. Although the judges express their opinions about the contestants' performances, viewers vote to keep their favorites in the competition. During each episode, host Ryan Seacrest explains how viewers can vote via text, phone call, or online. Nothing prevents a viewer from submitting multiple votes. Each week the contestant with the lowest number of votes is removed from the competition. Why are the results not a true depiction of all Americans' musical preferences?

A) Undercoverage bias
B) Voluntary response bias
C) The judges' comments influence voters' opinions.
D) Since some voters may vote multiple times, one vote does not correspond to one person.
E) All of the above.

13. Suppose you wish to compare the ages at inauguration of Democratic and Republican presidents. Which is the most appropriate type of technique for gathering the needed data?

A) Census
B) Sample survey
C) Experiment
D) Prospective observational study
E) None of these methods is appropriate.

14. Which of these is the best description of a block?

A) A random sample of a population who serve as subjects in an experiment
B) Any of the different groups randomly selected in a stratified sample
C) Any subgroup of the subjects in an experiment
D) A subgroup of experimental subjects randomly assigned the same treatment
E) A subgroup of experimental subjects that are the same with regard to some source of variation

15. The weight of large bags of one brand of potato chips is a random variable with a mean of 18 ounces and a standard deviation of 0.7 ounces. The amount of chips a professional caterer pours into a certain size serving bowl can be described as a random variable with a mean of 4 ounces and a standard deviation of 0.3 ounces. The mean amount left in the bag after the caterer has filled two bowls is 10 ounces. Assuming these two variables are independent, what is the standard deviation of this remaining weight?

A) 0.10 oz.   B) 0.36 oz.   C) 0.56 oz.
D) 0.82 oz.   E) 0.92 oz.

16. A quality control procedure at a manufacturing facility involves selecting 5 items at random from a large batch, and then accepting the entire batch if at least 3 of selected items pass inspection. If in reality 80% of all items produced would individually pass inspection, what is the probability that the batch will be accepted?

A) 0.2048   B) 0.7373   C) 0.8000
D) 0.9421   E) 0.9488

17. The bar graph below summarizes information a New York City Survey of Elementary School Children (NYC DOHMH and DOE 2003). In a random sample of 400 Hispanic NYC schoolchildren, what is the expected number who are obese and the standard deviation?

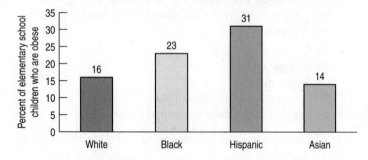

Obesity in NYC elementary schools is highest among Hispanic children — nearly 1 in 3

A) mean 124; standard deviation 7.7
B) mean 148; standard deviation 7.7
C) mean 124; standard deviation 9.25
D) mean 148; standard deviation is 9.66
E) mean 124; standard deviation 11.1355

18. The Substance Abuse and Mental Health Services Administration reports that in 2011, 20,783 people were treated for medical emergencies related to energy drinks, largely attributable to the very high doses of caffeine in drinks of this type. The table below shows the age distribution for these patients. What is the probability that a person who visited the emergency department for an energy drink–related emergency was under 40 given that the person was at least 18 years old?

A) 0.676
B) 0.729
C) 0.806
D) 0.904
E) 0.928

Age range	Number of patients
12–17	1,499
18–25	7,322
26–39	6,729
40 or older	5,233

Source: http://www.samhsa.gov/data/2k13/DAWN126/sr126-energy-drinks-use.pdf

19. As reported in the *New York Times* on August 14, 2012, "University of Michigan researchers analyzed data on more than 21,000 children observed in cars at gas stations, fast-food restaurants, recreation centers, and child care centers from 2007–2009. Twenty-one percent of children younger than 4 years old were not sitting in car seats as recommended." What is the probability that researchers found their first unrestrained child four years old or younger by the fifth car they checked?

A) 0.002   B) 0.082   C) 0.210
D) 0.692   E) 0.999

(Questions 20–21) A local bookseller carefully collects data on the customers that enter her store. She uses the term "unit" to describe either a customer that comes in alone or a group of customers that come in together. Based on past experience, she estimates that 8% of the units who enter her store will make some type of purchase. It appears that the units are independent.

20. What is the probability that three units in a row will make a purchase?

A) 0.000512   B) 0.068   C) 0.08
D) 0.203   E) 0.24

21. In one hour, ten units enter her store. What is the probability that at least one of them makes a purchase?

A) 0.038   B) 0.378   C) 0.434
D) 0.566   E) 0.812

22. A pet store sells fancy handcrafted nametags to go on dogs' and cats' collars. The store's profit is $6 for each dog nametag and $5 for each cat nametag sold. Each week the store sells an average of 12 dog tags with a standard deviation of 4 tags, and an average of 15 cat tags with a standard deviation of 3. The store's expected weekly profit on these products is $147. Assuming sales are independent, what's the standard deviation of this weekly profit?

A) $11.87   B) $14.80   C) $28.30
D) $39   E) $55

23. An electronics retailer is developing a model for insurance policies on new cell phone purchases. It estimates that 60% of customers never make a claim, 25% of customers require a small repair costing an average of $50, and 15% of customers request a full refund costing $200. What is the long-term average cost the retailer should expect to pay to its customers per claim?

A) $42.50   B) $50   C) $83.33
D) $125   E) None of these

24. A university reports that 80% of its students enroll there as freshmen, while the rest transfer in from other 2- or 4-year colleges. The eventual graduation rate is 85% among the transfers, but only 70% among those who arrived as freshmen. What's the probability that a former student who never graduated was a transfer student?

A) 0.03   B) 0.11   C) 0.15
D) 0.27   E) 0.77

25. Iron is an essential nutrient. Iron deficiency has been linked with symptoms such as anemia, rapid heartbeat, increased risk of infections, and lightheadedness. At the other end of the spectrum is iron overload, described in an August 2012 *New York Times* article. Excess iron is deposited in the liver, heart, and pancreas and can cause cirrhosis, liver cancer, cardiac arrhythmias, and diabetes. According to a Framingham Heart Study researcher, "About one person in 250 inherits a genetic disorder called hemochromatosis that increases iron absorption and results in a gradual, organ-damaging buildup of stored iron." Suppose we have a random sample of 1000 adults, and want to find the probability that at least 5 of them have this disorder. Which of these statements is true?

I. We would expect 4 of these people to have hemochromatosis.
II. We can calculate this probability using a Binomial model.
III. We can approximate this probability using a Normal model.

A) None   B) I only   C) I and II only
D) I and III only   E) I, II, and III

Free Response

1. In the National Football League every team must submit an injury list prior to each game. The list contains the names of the players who are "out", meaning they will not play in the game. There are also three other categories: "Doubtful", "Questionable", and "Probable". The guidelines state that "Doubtful" should mean about a 25% chance that the player will play, and "Questionable" means about a 50% chance that the player will play. The table below shows what happened with the three categories of players for a particular week.

	Doubtful	Questionable	Probable
Played	4	16	52
Did Not Play	16	20	12

a) What percent of these players who did play had been listed as probable?
b) What percent of these players were listed as probable and did play?
c) Create a graph that compares players' chances of getting into the game for the three status categories.
d) Based on this information, is whether or not a player gets into the game independent of his pre-game status? Explain.

2. In the United States, homes are measured by the total number of square feet of area of all floors of the house. Data from www.census.gov show how the median size of a home in the Northeast changed from 1973 through 2010. Here are the regression results and a plot of the residuals against predicted values.

Dependent variable is: NEsqft
No Selector
R squared =94.8%   R squared (adjusted) = 94.6%
S = 72.07 with 38 − 2 = 36 degrees of freedom

Source	Sum of Squares	df	Mcan Square	F-ratio
Regression	3403782	1	3403782	655
Residual	186979	36	5193.87	

Variable	Coefficient	s.e. of Coeff	t-ratio	prob
Constant	−52410.9	2123	−24.7	≤ 0.0001
Year	27.2927	1.066	25.6	≤ 0.0001

a) Write the equation of the regression line.
b) Do you think this linear model is appropriate? Explain.
c) What does this residual plot tell you about estimates based on this model?
d) According to this model how has median home size changed per decade?
e) Estimate the *actual* median home size in 2009.

3. Fifty patients suffering from knee pain have volunteered to participate in an experiment to examine the effects of pain relief medication. Half will be assigned to take 400 mg of the pain reliever every six hours, and the other half to take 200 mg every 3 hours. The patients will report their pain levels on a 1–10 scale at the beginning of the experiment and again after 8 days on the medication.

   a) Define the experimental units and the response variable.
   b) What are the treatments?
   c) Describe an appropriate method to assign the volunteers to treatments.
   d) An alternate design could include a control group who take a placebo. Describe an advantage of adding a control group.

4. In 2012 the Centers for Disease Control and Prevention reported that the rate of autism in children has risen to 1 in 50. A study conducted in Norway suggests that the risk of autism may be significantly lower if women take folic acid supplements during pregnancy. Suppose a large medical center enlists 900 expectant mothers in a trial; the women agree to take the folic acid for the duration of their pregnancies. The researchers will track the children after birth to see how many develop autism.

   a) Explain why the number of autism cases is a binomial random variable.

   b) If folic acid supplements actually have no effect, what is the mean and standard deviation of the number of autism cases that these researchers might find?
   c) How low would the number of autism cases among these 900 children have to be in order to convince you that the folic acid supplements may be effective? Explain your reasoning.

5. An online retailer sells its $25 gift cards at supermarkets. The retailer knows that 20% of people who purchase these cards spend the full value, 70% leave a balance of $5 that is never spent, and the rest never use the cards at all. Because people pay cash for the cards, the amounts that are not spent are pure profit for the retailer.

   a) If you buy three of these cards for three friends at Christmas, what is the probability that none of the three cards will be completely used?
   b) What is the online retailer's expected profit per card?
   c) What is the standard deviation of the retailer's profit per card?
   d) A service club makes a proposal to the retailer. The club wants to buy 100 of the cards for only $2000, so it can sell them as a fundraiser. Should the retailer worry that it might suffer a loss if it sells the club these cards at this discount? Explain.

17 Sampling Distribution Models

n May 2007, the Gallup poll surveyed 1007 U.S. adults by telephone. When asked about "Evolution, that is, the idea that human beings developed over millions of years from less advanced forms of life," 43% responded that this was either definitely or probably true. At roughly the same time, the Baylor Religion Survey (administered by the Gallup organization) collected responses from 1594 randomly selected U.S. adults. They asked if respondents agreed with the statement "Humans evolved from other primates over millions of years," and found that 668, or 41.9%, of them agreed or "strongly agreed" with the statement. Should we be surprised to find that we can get different values from properly selected random samples drawn from the same population?

The Sampling Distribution of a Proportion

Imagine
We see only the sample that we actually drew, but by simulating or modeling, we can *imagine* what we might have seen had we drawn other possible random samples.

If we (or Gallup) had surveyed every possible sample of 1007 U.S. adults, we could find the proportion $\hat{p}$ of each sample who believe the theory of evolution. Then we could see how much those sample proportions varied. Of course, we can't do that, but we can imagine drawing those samples. In fact, we can do better than that. We can simulate what the distribution will look like. For our simulation, we first have to pick a particular value to be the "true" population proportion. Recall that we denote the true population proportion with the letter p. We'll pretend for now that Gallup got it exactly right and the true proportion is 43%. Then we'll draw 1007 simulated responses.

■ **NOTATION ALERT**

The letter p is our choice for the *parameter* of the model for proportions. It violates our "Greek letters for parameters" rule, but if we stuck to that, our natural choice would be π. We could use π to be perfectly consistent, but then we'd have to write statements like $\pi = 0.46$. That just seems a bit weird to us. After all, we've known that $\pi = 3.1415926 \ldots$ since the Greeks, and it's a hard habit to break.

So, we'll use p for the model parameter (the probability of a success) and $\hat{p}$ for the observed proportion in a sample. We'll also use q for the probability of a failure ($q = 1 - p$) and $\hat{q}$ for its observed value.

But be careful. We've already used capital P for a general probability. And we'll soon see another use of P in the next chapter! There are a lot of p's in this course; you'll need to think clearly about the context to keep them straight.

TI-*nspire*

Sample Proportions. Generate sample after sample to see how the proportions vary.

A S *Activity:* **Sampling Distribution of a Proportion.** You don't have to imagine—you can simulate.

The first three times we drew the 1007 yes/no values, we got 424, 415, and 422 evolution supporters (for $\hat{p} = 0.421, 0.412$, and 0.419 respectively). When we ran the simulation 2000 times, a histogram of all the sample proportions looked like this:

Figure 17.1

The distribution of sample proportions for 2000 simulated samples of 1007 adults drawn from a population with proportion, $p = 0.43$, of evolution supporters.

It should be no surprise that we don't get the same proportion for each sample we draw, even though the underlying true value is the same for the population. Each $\hat{p}$ comes from a different simulated sample. The histogram above is a simulation of what we'd get if we could see *all the proportions from all possible samples of this size*. That distribution has a special name. It is called the **sampling distribution** of the sample proportions.[1]

The histogram is unimodal, symmetric, and centered at p. That's probably not surprising. But, it's an amazing and fortunate fact that a Normal model is just the right one for the histogram of sample proportions. This was proved mathematically nearly 300 years ago by the French mathematician Abraham De Moivre.

There is no reason you should guess that the Normal model would work here,[2] and, indeed, the importance of De Moivre's result was not immediately understood. But (unlike De Moivre's contemporaries in 1718) we know how useful the Normal model can be.

Modeling how sample statistics, such as proportions or means, vary from sample to sample is one of the most powerful ideas we'll see in this course. A **sampling distribution model** for how a statistic varies from sample to sample allows us to quantify that variation and to make statements about where we think the corresponding population parameter is.

For Example   USING THE SAMPLING DISTRIBUTION MODEL FOR A PROPORTION

According to the Centers for Disease Control and Prevention, about 18% of U.S. adults still smoke.

QUESTION: How much would we expect the proportion of smokers in a sample of size 1000 to vary from sample to sample?

ANSWER: We'll simulate by taking random integers between 1 and 100. We'll assign values between 1 and 18 to represent a smoker, and the rest to represent a nonsmoker. Then we'll repeat the process, drawing samples of size 1000 over and over and recording the sample proportion of smokers for each

(continued)

---

[1] A word of caution. Until now, we've been plotting the *distribution of the sample*, a display of the actual data that were collected in that one sample. But now we've plotted the *sampling distribution*; a display of summary statistics ($\hat{p}$'s, for example) for many different samples. "Sample distribution" and "sampling distribution" sound a lot alike, but they refer to very different things. (Sorry about that—we didn't make up the terms.) And the distinction is critical. Whenever you read or write something about one of these, choose your terms carefully.

[2] Well, the fact that we spent much of Chapter 5 on the Normal model might have been a hint.

sample. If we did this 10,000 times, we might get a histogram of sample proportions that looks like this:

The mean of this distribution is located at the population proportion, 0.18. The standard deviation of these 10,000 sample proportions is 0.0122 or 1.22%. Because the shape is symmetric and roughly Normal, the 68–95–99.7 Rule should work quite well. For example, we would expect about 95% of these samples to have proportions within 2 SDs (2.44%) of 18%. In fact, 9541 or 95.41% of them did.

Looking at lots of sample proportions is a big shift in our point of view. Up to now, you've thought of the sample proportion as a fixed value that comes from one particular sample. But now we're trying to understand how proportions from different random samples might behave. Each random sample has its own proportion and each of those proportions may be different, so we really have to think of proportions as random quantities that vary from sample to sample.

Abraham de Moivre (1667–1754)

A S *Simulation:* **Simulating Sampling Distributions.** Watch the Normal model appear from random proportions.

Figure 17.2

A Normal model centered at p with a standard deviation of $\sqrt{\dfrac{pq}{n}}$ is a good model for a collection of proportions found for many random samples of size n from a population with success probability p.

■ **NOTATION ALERT**

In Chapter 7, we introduced $\hat{y}$ as the predicted value for y. The "hat" here plays a similar role. It indicates that $\hat{p}$—the observed proportion in our data—is our *estimate* of the parameter p.

To use a Normal model, we need to specify two parameters: its mean and standard deviation. The center of the histogram is naturally at p, so that's what we'll use for the mean.

What about the standard deviation? There's a special fact about proportions that makes the standard deviation easy to find. Once we know the mean, p, we automatically also know the standard deviation. We saw in the last chapter that for a Binomial model the standard deviation of the *number* of successes is $\sqrt{npq}$. Now we want the standard deviation of the *proportion* of successes, $\hat{p}$. The sample proportion $\hat{p}$ is the number of successes divided by the number of trials, n, so the standard deviation is also divided by n:

$$\sigma(\hat{p}) = SD(\hat{p}) = \frac{\sqrt{npq}}{n} = \sqrt{\frac{pq}{n}}.$$

When we draw simple random samples of n individuals, the proportions we find will vary from sample to sample. As long as n is reasonably large,[3] we can model the distribution of these sample proportions with a probability model that is

$$N\left(p, \sqrt{\frac{pq}{n}}\right).$$

Although we'll never know the true proportion of adults who believe in evolution, for this investigation we're supposing it to be 43%. Once we put the center at $p = 0.43$, the standard deviation for the Gallup poll is

$$SD(\hat{p}) = \sqrt{\frac{pq}{n}} = \sqrt{\frac{(0.43)(0.57)}{1007}} = 0.0156, \text{ or } 1.56\%.$$

---

[3]For smaller n, we could use a Binomial model (see Chapter 16).

Here's a picture of the Normal model for our simulation histogram:

Figure 17.3

Using 0.43 for *p* gives this Normal model for Figure 17.1's histogram of the sample proportions of adults believing in evolution (*n* = 1007).

0.3832	0.3988	0.4144	0.4300	0.4456	0.4612	0.4768
−3σ	−2σ	−1σ	*p*	1σ	2σ	3σ

Let's see how well the theory works. In our simulation of smokers (in the For Example, page 447), the SD of $\hat{p}$ was 0.0122. The formula says it actually should be[4]

$$SD(\hat{p}) = \sqrt{\frac{pq}{n}} = \sqrt{\frac{(0.18)(0.82)}{1000}} = 0.0121.$$

That works.

Because we have a Normal model, we can use the 68–95–99.7 Rule or look up exact probabilities using a table or technology. For example, we know that 95% of Normally distributed values are within roughly two standard deviations of the mean, so we should not be surprised if various polls that may appear to ask the same question report a variety of results. Such sample-to-sample variation is sometimes called **sampling error**. It's not really an *error* at all, but just *variability* you'd expect to see from one sample to another. A better term would be **sampling variability**.[5]

A S *Simulation:* **The Standard Deviation of a Proportion.** Do you believe this formula for standard deviation? Don't just take our word for it—convince yourself with an experiment.

Can We Always Use a Normal Model?

Figure 17.4

Proportions from samples of size 2 can take on only three possible values. A Normal model does not work well.

De Moivre claimed that the sampling distribution can be modeled well by a Normal model. But, does it always work? Well, no. For example, if we drew samples of size 2, the only possible proportion values would be 0, 0.5, and 1. There's no way a histogram consisting only of those values could look like a Normal model.

In fact, De Moivre's claim is only approximately true. (But that's OK. After all, models are only supposed to be approximately true.) The Normal model becomes a better and better representation of the sampling distribution as the size of the samples gets bigger.[6] Samples of size 1 or 2 won't work at all, but the distributions of sample proportions of larger samples are modeled well by the Normal model as long as *p* isn't too close to 0 or 1.

Populations with a true proportion, *p*, close to 0 or 1 can be a problem. Suppose a basketball coach surveys students to see how many male high school seniors are over 6'6". What will the proportions of samples of size 1000 look like? If the true proportion of students that tall is 0.001, then the coach is likely to get only a few seniors over 6'6" in any random sample of 1000. Most samples will have proportions of 0/1000, 1/1000, 2/1000, 3/1000, 4/1000 with only a very few samples having a higher proportion. A simulation of 2000 surveys of size 1000 with *p* = 0.001 shows a sampling

---

[4]The standard deviation is 0.012149 or 1.21%. Remember that the standard deviation always has the same units as the data. Here our units are %, or "percentage points." The standard deviation isn't 1.21% of anything, it is just 1.21 percentage points. If that's confusing, try writing the units out as "percentage points" instead of using the symbol %. Many polling agencies now do that too.

[5]Of course, polls differ for many reasons in addition to sampling variability. These can include minor differences in question wording (Gallup's question was about evolving from "less advanced forms of life" but Baylor described evolution "from other primates") and in the specific ways that the random samples are selected. But polling agencies report their "margin of error" based only on sampling variability. We'll see how they find that value in Chapter 18.

[6]Formally, we say the claim is true in the limit as *n* grows.

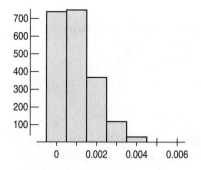

Figure 17.5
The distribution of sample proportions for 2000 samples of size 1000 with $p = 0.001$. Because the true proportion is so small, the sampling distribution is skewed to the right and the Normal model won't work well.

Successes and Failures

The terms "success" and "failure" for the outcomes that have probability p and q are common in Statistics. But they are completely arbitrary labels. When we say that a disease occurs with probability p, we certainly don't mean that getting sick is a "success" in the ordinary sense of the word.

---

distribution for $\hat{p}$ that's skewed to the right because p is so close to 0. (Had p been very close to 1, it would have been skewed to the left.) So, even though n is large, p is too small, and so the Normal model still won't work well.

Assumptions and Conditions

When does the Normal model with mean p and standard deviation $\sqrt{\dfrac{pq}{n}}$ work well as a model for the sampling distribution of a sample proportion? Before proceeding we must check the following assumptions and conditions:

The Independence Assumption: The individuals in the sample must be independent of each other.

You can't know if an assumption is true or not, but you *can* check conditions to see whether the data were collected in a way that makes this assumption plausible.

Randomization Condition: If your data come from an experiment, subjects should have been randomly assigned to treatments. If you have a survey, your sample should be a simple random sample of the population. If some other sampling design was used, be sure the sampling method was not biased and that the data are representative of the population. If they aren't, then your inferences may not be valid.

10% Condition: It can also be a problem to sample too much of the population. Once you've sampled more than about 10% of the population, the remaining individuals are no longer really independent of each other.[7]

We've seen that a Normal model becomes more useful for describing the sampling distribution of $\hat{p}$ as the sample size increases. That's the basis for another assumption we must think about.

The Sample Size Assumption: The sample size, n, must be large enough. How large? That depends on the value of p; the closer p is to 0 or 1, the larger n must be. We check a familiar condition:

Success/Failure Condition: The sample size must be large enough that we expect to see at least 10 successes and at least 10 failures. (We check that $np \geq 10$ and $nq \geq 10$.)

These last two conditions seem to conflict with each other. The **Success/Failure Condition** wants sufficient data. How much depends on p. If p is near 0.5, we need a sample of only 20 or so. If p is only 0.01, however, we'd need 1000. But the **10% Condition** says that a sample should be no larger than 10% of the population. If you're thinking, "Wouldn't a larger sample be better?" you're right of course. It's just that if the sample were more than 10% of the population, we'd need to use different methods to analyze the data. Fortunately, this isn't usually a problem in practice. Often, as in polls that sample from all U.S. adults or industrial samples from a day's production, the populations are much larger than 10 times the sample size.

A Sampling Distribution Model for a Proportion

We've arrived at a key moment in this course (Drumroll, please!) We have changed our point of view in a very important way. No longer is a proportion something we just compute for a set of data. We now see it as a random variable quantity that has a probability distribution, and thanks to De Moivre we have a model for that distribution. We call that the **sampling distribution model** for the proportion, and we'll make good use of it.

---

[7]There are special formulas that you can use to adjust for sampling a large part of a population but, as the saying goes, they are "beyond the scope of this course."

A S *Simulation:* **Simulate the Sampling Distribution Model of a Proportion.** You probably don't want to work through the formal mathematical proof; a simulation is far more convincing!

The Sampling Distribution Model for a Sample Proportion

Provided that the sampled values are independent and the sample size is large enough, the sampling distribution of $\hat{p}$ is modeled by a Normal model with mean $\mu(\hat{p}) = p$ and standard deviation $SD(\hat{p}) = \sqrt{\dfrac{pq}{n}}$.

Without sampling distribution models, the rest of Statistics just wouldn't exist. Sampling models are what make Statistics work. They inform us about the amount of variation we should expect when we sample. A sampling distribution model tells us how surprising a sample statistic is and enables us to make informed decisions about how precise our estimate of the true value of a parameter might be. That's exactly what we'll be doing for the rest of this book.

Sampling distribution models enable us to say something about the population when all we have is data from a sample. This is the huge leap of Statistics. By imagining what *might* happen if we were to draw many, many samples from the same population, we can learn a lot about how close the statistics computed from our one particular sample may be to the corresponding population parameters they estimate. That's the path to the *margin of error* you hear about in polls and surveys. We'll see how to determine that in the next chapter.

For Example USING THE SAMPLING DISTRIBUTION MODEL FOR PROPORTIONS

The Centers for Disease Control and Prevention report that 22% of 18-year-old women in the United States have a body mass index (BMI)[8] of 25 or more—a value considered by the National Heart Lung and Blood Institute to be associated with increased health risks.

As part of a routine health check at a large college, the physical education department usually requires students to come in to be measured and weighed. This year, the department decided to try out a self-report system. It asked 200 randomly selected female students to report their heights and weights (from which their BMIs could be calculated). Only 31 of these students had BMIs greater than 25.

QUESTION: Is this proportion of high-BMI students unusually small?

ANSWER: First, check the conditions:

✓ Randomization Condition: The department drew a random sample, so the respondents should be independent and randomly selected from the population.

✓ 10% Condition: 200 respondents is less than 10% of all the female students at a "large college."

✓ Success/Failure Condition: The department expected $np = 200(0.22) = 44$ "successes" and $nq = 200(0.78) = 156$ "failures," both at least 10.

It's okay to use a Normal model to describe the sampling distribution of the proportion of respondents with BMIs above 25.

The phys ed department observed $\hat{p} = \dfrac{31}{200} = 0.155$.

The department expected $E(\hat{p}) = p = 0.22$, with $SD(\hat{p}) = \sqrt{\dfrac{pq}{n}} = \sqrt{\dfrac{(0.22)(0.78)}{200}} = 0.029$,

so $z = \dfrac{\hat{p} - p}{SD(\hat{p})} = \dfrac{0.155 - 0.22}{0.029} = -2.24$.

By the 68–95–99.7 Rule, I know that values more than 2 standard deviations below the mean of a Normal model show up less than 2.5% of the time. Perhaps women at this college differ from the general population, or self-reporting may not provide accurate heights and weights.

---

[8]BMI = weight in kg/(height in m)^2.

Just Checking

1. You want to poll a random sample of 100 students at a large university to see if they are in favor of the proposed location for the new student center. Of course, you'll get just one number, your sample proportion, $\hat{p}$. But if you imagined all the possible samples of 100 students you could draw and imagined the histogram of all the sample proportions from these samples, what shape would it have?

2. Where would the center of that histogram be?

3. If you think that about half the students are in favor of the plan, what would the standard deviation of the sample proportions be?

Step-by-Step Example WORKING WITH SAMPLING DISTRIBUTION MODELS FOR PROPORTIONS

Suppose that about 13% of the population is left-handed.[9] A 200-seat school auditorium has been built with 15 "lefty seats," seats that have the built-in desk on the left rather than the right arm of the chair. (For the right-handed readers among you, have you ever tried to take notes in a chair with the desk on the left side?)

Question: In a class of 90 students, what's the probability that there will not be enough seats for the left-handed students?

THINK ➡ **Plan** State what we want to know.

I want to find the probability that in a group of 90 students, more than 15 will be left-handed. Since 15 out of 90 is 16.7%, I need the probability of finding more than 16.7% left-handed students out of a sample of 90 if the proportion of lefties in the population is 13%.

Model Think about the assumptions and check the conditions.

You might be able to think of cases where the **Independence Assumption** is not plausible—for example, if the students are all related, or if they were selected for being left- or right-handed. Sampling randomly is the key to independence.

✓ **Independence Assumption:** It is reasonable to assume that the probability that one student is left-handed is not changed by the fact that another student is right- or left-handed.

✓ **Randomization Condition:** The 90 students in the class can be thought of as a random sample of students.

✓ **10% Condition:** 90 is surely less than 10% of the population of all students. (Even if the school itself is small, I'm thinking of the population of all *possible* students who could have gone to the school.)

---

[9]Actually, it's quite difficult to get an accurate estimate of the proportion of lefties in the population. Estimates range from 8% to 15%.

✓ **Success/Failure Condition:**

$np = 90(0.13) = 11.7 \geq 10$

$nq = 90(0.87) = 78.3 \geq 10$

State the parameters and the sampling distribution model.

The population proportion is $p = 0.13$. The conditions are satisfied, so I'll model the sampling distribution of $\hat{p}$ with a Normal model with mean 0.13 and a standard deviation of

$$SD(\hat{p}) = \sqrt{\frac{pq}{n}} = \sqrt{\frac{(0.13)(0.87)}{90}} \approx 0.035$$

My model for $\hat{p}$ is $N(0.13, 0.035)$.

SHOW ➡ **Plot**   Make a picture. Sketch the model and shade the area we're interested in, in this case the area to the right of 16.7%.

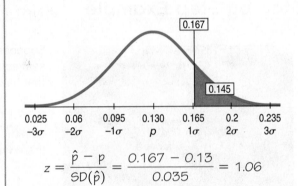

Mechanics   Use the standard deviation as a ruler to find the z-score of the cutoff proportion. We see that 16.7% lefties would be just over one standard deviation above the mean.

$$z = \frac{\hat{p} - p}{SD(\hat{p})} = \frac{0.167 - 0.13}{0.035} = 1.06$$

Find the resulting probability from a table of Normal probabilities, a computer program, or a calculator.

$$P(\hat{p} > 0.167) = P(z > 1.06) = 0.1446$$

TELL ➡ **Conclusion**   Interpret the probability in the context of the question.

There is about a 14.5% chance that there will not be enough seats for the left-handed students in the class.

The Sampling Distributions of Other Statistics

The sampling distribution for the sample proportion is especially useful because the Normal model provides such a good approximation. But, it might be useful to know the sampling distribution for *any* statistic that we can calculate, not just the sample proportion. (Chapter 5's What If first explored this.) Is the Normal model a good model for all statistics? Would you expect that the sampling distribution of the minimum or the maximum or the variance of a sample to be Normally distributed? What about the median?

Simulating the Sampling Distributions of Other Statistics

A study of body measurements of 250 men found the median weight to be 176.125 lbs and the variance to be 730.9 lbs^2. Treating these 250 men as a population, we can draw repeated random samples of 10 and compute the median, the variance, and the minimum for each sample.

Each of these histograms depicts the sampling distribution of its respective statistic. And it is easy to see that they aren't all the same. The sampling distribution of the medians is unimodal and symmetric. The sampling distribution of the variances is skewed to the right. And the sampling distribution of the minimums is, well, messy.

We can simulate to get a look at the sampling distribution of *any* statistic we like: the maximum, the IQR, the 37th percentile, anything. Both the proportion and the mean (as we'll see in the next section) have sampling distributions that are well approximated by a Normal model. That's good news, since these are the two summary statistics that we use most often.

Simulating the Sampling Distribution of a Mean

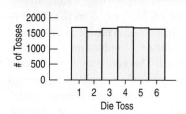

Figure 17.6

Here's a simple simulation. Let's start with one fair die. If we toss this die 10,000 times, what should the histogram of the numbers on the face of the die look like? Figure 17.6 shows the results of a simulated 10,000 tosses.

Figure 17.7

Now let's toss a *pair* of dice and record the average of the two. If we repeat this (or at least simulate repeating it) 10,000 times, recording the average of each pair, what will the histogram of the 10,000 averages look like? Before you look, think a minute. Is getting an average of 1 on *two* dice as likely as getting an average of 3 or 3.5? Not at all, right? We're much more likely to get an average near 3.5 than we are to get one near 1 or 6. After all, the *only* way to get an average of 1 is to get two 1's. To get a total of 7 (for an average of 3.5), though, there are many more possibilities. This distribution even has a name: the *triangular* distribution, as seen in Figure 17.7.

Figure 17.8

What if we average 3 dice? We simulated 10,000 tosses of 3 dice and took their average. Figure 17.8 shows the result.

What's happening? First notice that it's getting harder to have averages near the ends. Getting an average of 1 or 6 with 3 dice requires all three to come up 1 or 6, respectively. That's less likely than for 2 dice to come up both 1 or both 6. The distribution is being pushed toward the middle.

Figure 17.9

Let's continue this simulation to see what happens with larger samples. Figure 17.9 shows a histogram of the averages for 10,000 tosses of 5 dice.

The pattern is becoming clearer. Two things continue to happen. The first fact we knew already from the Law of Large Numbers. It says that as the sample size (number of dice) gets larger, each sample average is more likely to be closer to the population mean. So, we see the distribution continuing to tighten around 3.5. But the shape is the surprising part. It's approaching the Normal model.

Figure 17.10

Let's skip ahead and try 20 dice. The histogram of averages for 10,000 throws of 20 dice is Figure 17.10.

Now we see the Normal shape clearly (and notice how much smaller the spread is). But can we count on this happening for situations other than dice throws? It turns out that Normal models work well amazingly often.

The Fundamental Theorem of Statistics

A S **Activity: The Sampling Distribution Model for Means.** Don't just sit there reading about the simulation—do it yourself.

"The theory of probabilities is at bottom nothing but common sense reduced to calculus."

—Pierre-Simon *Laplace,* in Théorie analytique des probabilités, *1812*

The Greatness of Laplace

Laplace was one of the greatest scientists and mathematicians of his time. In addition to his contributions to probability and statistics, he published many new results in mathematics, physics, and astronomy (where his nebular theory was one of the first to describe the formation of the solar system in much the way it is understood today). He also played a leading role in establishing the metric system of measurement.

TI-*nspire*

The Central Limit Theorem. See the sampling distribution of sample means take shape as you choose sample after sample.

What we saw with dice is true for means of repeated samples for almost every situation. When we looked at the sampling distribution of a proportion, we had to check only a few conditions. For means, the result is even more remarkable. *There are almost no conditions at all.*

Let's say that again: The sampling distribution of *any* mean becomes more nearly Normal as the sample size grows. All we need is for the observations to be independent and collected with randomization. We don't even care about the shape of the population distribution![10] This surprising fact is the result Laplace proved in a fairly general form in 1810. At the time, Laplace's theorem caused quite a stir (at least in mathematics circles) because it is so unintuitive. Laplace's result is called the **Central Limit Theorem**[11] (CLT).

Why should the Normal model show up again for the sampling distribution of means as well as proportions? We're not going to try to persuade you that it is obvious, clear, simple, or straightforward. In fact, the CLT is surprising and a bit weird. Not only does the distribution of means of many random samples get closer and closer to a Normal model as the sample size grows, *this is true regardless of the shape of the population distribution!* Even if we sample from a skewed or bimodal population, the Central Limit Theorem tells us that means of repeated random samples will tend to follow a Normal model as the sample size grows. If the sample is small it works better the closer the population distribution is to a Normal model. If the data come from a population that's exactly Normal to start with, then the observations themselves are Normal. If we take samples of size 1, their "means" are just the observations—so, of course, they have Normal sampling distribution. But even if the population distribution is very skewed (like the CEO compensations from Chapter 4, for example) the CLT still works, although it may take a sample size of dozens or even hundreds of observations for the Normal model to work well. We'll explore this further in this chapter's What If.

What about a really bimodal population, one that consists of only 0's and 1's? The CLT says that even means of samples from this population will follow a Normal sampling distribution model. But wait. Suppose we have a categorical variable and we assign a 1 to each individual in the category and a 0 to each individual not in the category. And then we find the mean of these 0's and 1's. That's the same as counting the number of individuals who are in the category and dividing by *n*. That mean will be . . . the *sample proportion, $\hat{p}$,* of individuals who are in the category (a "success"). So maybe it wasn't so surprising after all that proportions, like means, have Normal sampling distribution models; they are actually just a special case of Laplace's remarkable theorem. Of course, for such an extremely bimodal population, we'll need a reasonably large sample size—and that's where the special conditions for proportions come in.

The Central Limit Theorem (CLT)

The mean of a random sample is a random variable whose sampling distribution can be approximated by a Normal model. The larger the sample, the better the approximation will be.

---

[10]OK, one technical condition. The data must come from a population with a finite variance. You probably can't imagine a population with an infinite variance, but statisticians can construct such things, so we have to discuss them in footnotes like this. It really makes no difference in how you think about the important stuff, so you can just forget we mentioned it.

[11]The word "central" in the name of the theorem means "fundamental." It doesn't refer to the center of a distribution.

Assumptions and Conditions

A S **Activity: The Central Limit Theorem.** Does it really work for samples from non-Normal populations?

The CLT requires essentially the same assumptions as we saw for modelling proportions:

> **Independence Assumption:** The sampled values must be independent of each other.
> **Sample Size Assumption:** The sample size must be sufficiently large.

We can't check these directly, but we can think about whether the **Independence Assumption** is plausible. We can also check some related conditions:

> **Randomization Condition:** The data values must be sampled randomly, or the concept of a sampling distribution makes no sense.

> **10% Condition:** When the sample is drawn without replacement (as is usually the case), the sample size, *n*, should be no more than 10% of the population.

> **Large Enough Sample Condition:** Although the CLT tells us that a Normal model is useful in thinking about the behavior of sample means when the sample size is large enough, it doesn't tell us how large a sample we need. The truth is, it depends; there's no one-size-fits-all rule. If the population is unimodal and symmetric, even a fairly small sample is okay. If the population is strongly skewed, like the compensation for CEOs we looked at in Chapter 4, it can take a pretty large sample to allow use of a Normal model to describe the distribution of sample means. For now you'll just need to think about your sample size in the context of what you know about the population, and then tell whether you believe the Large Enough Sample Condition has been met.

But Which Normal?

A S **Activity: The Standard Deviation of Means.** Experiment to see how the variability of the mean changes with the sample size.

A S **Activity: The Sampling Distribution of the Mean.** The CLT tells us what to expect. In this activity you can work with the CLT or simulate it if you prefer.

The CLT says that the sampling distribution of any mean or proportion is approximately Normal. But which Normal model? Any Normal model is specified by its mean and standard deviation. For proportions, the sampling distribution is centered at the population proportion. For means, it's centered at the population mean. What else would we expect?

What about the standard deviations, though? We noticed in our dice simulation that the histograms got narrower as we averaged more and more dice together. This shouldn't be surprising. Means vary less than the individual observations. Think about it for a minute. Which would be more surprising, having *one* person in your Statistics class who is over 6′6″ tall or having the *mean* of all students taking the course be over 6′6″? The first event is fairly rare.[12] You may have seen somebody this tall in one of your classes sometime. But finding a whole class whose mean height is over 6′6″ tall just won't happen. Why? Because *means have smaller standard deviations than individuals.*

How much smaller? Well, we have good news and bad news. The good news is that the standard deviation of $\bar{y}$ falls as the sample size grows. The bad news is that it doesn't drop as fast as we might like. Like proportions, it only goes down by the *square root* of the sample size. Why? The Math Box will show you that the Normal model for the sampling distribution of the mean has a standard deviation equal to

$$SD(\bar{y}) = \frac{\sigma}{\sqrt{n}}$$

where σ is the standard deviation of the population. To emphasize that this is a standard deviation *parameter* of the sampling distribution model for the sample mean, $\bar{y}$, we write $SD(\bar{y})$ or $\sigma(\bar{y})$.

---

[12]If students are a random sample of adults, fewer than 1 out of 10,000 should be taller than 6′9″. Why might students not really be a random sample with respect to height? Even if it's not a perfectly random sample, choosing a student over 6′9″ tall is still rare.

> ### The Sampling Distribution Model for a Sample Mean (CLT)
>
> When a random sample of size n is drawn from any population with mean μ and standard deviation σ, its sample mean, $\bar{y}$, has a sampling distribution with the same mean μ but whose *standard deviation* is $\dfrac{\sigma}{\sqrt{n}}$ (and we write $\sigma(\bar{y}) = SD(\bar{y}) = \dfrac{\sigma}{\sqrt{n}}$).
>
> No matter what population the random sample comes from, the *shape* of the sampling distribution is approximately Normal as long as the sample size is large enough. The larger the sample used, the more closely the Normal model approximates the sampling distribution for the mean.

Math Box

We know that $\bar{y}$ is a sum divided by n:

$$\bar{y} = \frac{y_1 + y_2 + y_3 + \cdots + y_n}{n}.$$

As we saw in Chapter 15, when a random variable is divided by a constant its variance is divided by the *square* of the constant:

$$Var(\bar{y}) = \frac{Var(y_1 + y_2 + y_3 + \cdots + y_n)}{n^2}.$$

To get our sample, we draw the y's randomly, ensuring they are independent. For independent random variables, variances add:

$$Var(\bar{y}) = \frac{Var(y_1) + Var(y_2) + Var(y_3) + \cdots + Var(y_n)}{n^2}.$$

All n of the y's were drawn from our population, so they all have the same variance, σ^2:

$$Var(\bar{y}) = \frac{\sigma^2 + \sigma^2 + \sigma^2 + \cdots + \sigma^2}{n^2} = \frac{n\sigma^2}{n^2} = \frac{\sigma^2}{n}.$$

The standard deviation of $\bar{y}$ is the square root of this variance:

$$SD(\bar{y}) = \sqrt{\frac{\sigma^2}{n}} = \frac{\sigma}{\sqrt{n}}.$$

We now have two closely related sampling distribution models that we can use when the appropriate assumptions and conditions are met. Which one we use depends on which kind of data we have:

- When we have categorical data, we calculate a sample proportion, $\hat{p}$; the sampling distribution of this random variable has a Normal model with a mean at the true proportion ("Greek letter") p and a standard deviation of $SD(\hat{p}) = \sqrt{\dfrac{pq}{n}} = \dfrac{\sqrt{pq}}{\sqrt{n}}$.

We'll use this model in Chapters 18 through 21.

- When we have quantitative data, we calculate a sample mean, $\bar{y}$; the sampling distribution of this random variable has a Normal model with a mean at the true mean, μ, and a standard deviation of $SD(\bar{y}) = \dfrac{\sigma}{\sqrt{n}}$. We'll use this model in Chapters 22, 23, and 24.

The means of these models are easy to remember, so all you need to be careful about is the standard deviations. Remember that these are standard deviations of the *statistics* $\hat{p}$ and $\bar{y}$. They both have a square root of n in the denominator. That tells us that the larger the sample, the less either statistic will vary.

■ **NOTATION ALERT**

To avoid confusion over what standard deviations we're talking about, when working with sampling models we always write:

- $SD(\hat{p})$ for categorical data;
- $SD(\bar{y})$ for quantitative data.

For Example USING THE CLT FOR MEANS

RECAP: A college physical education department asked a random sample of 200 female students to self-report their heights and weights, but the percentage of students with body mass indexes over 25 seemed suspiciously low. One possible explanation may be that the respondents "shaded" their weights down a bit. The CDC reports that the mean weight of 18-year-old women is 143.74 lb, with a standard deviation of 51.54 lb, but these 200 randomly selected women reported a mean weight of only 140 lb.

QUESTION: Based on the Central Limit Theorem and the 68–95–99.7 Rule, does the mean weight in this sample seem exceptionally low, or might this just be random sample-to-sample variation?

ANSWER: The conditions check out okay:

✓ Randomization Condition: The women were a random sample and their weights can be assumed to be independent.

✓ 10% Condition: They sampled fewer than 10% of all women at the college.

✓ Large Enough Sample Condition: The distribution of college women's weights is likely to be unimodal and reasonably symmetric, so the CLT applies to means of even small samples; 200 values is plenty.

The sampling model for sample means is approximately Normal with $E(\bar{y}) = 143.7$ and

$$SD(\bar{y}) = \frac{\sigma}{\sqrt{n}} = \frac{51.54}{\sqrt{200}} = 3.64.$$ The expected distribution of sample means is:

The 68–95–99.7 Rule suggests that although the reported mean weight of 140 pounds is somewhat lower than expected, it does not appear to be unusual. Such variability is not all that extraordinary for samples of this size.

Step-by-Step Example WORKING WITH THE SAMPLING DISTRIBUTION MODEL FOR A MEAN

The Centers for Disease Control and Prevention reports that the mean weight of adult men in the United States is 190 lb with a standard deviation of 59 lb.[13]

Question: An elevator in our building has a weight limit of 10 persons or 2500 lb. What's the probability that if 10 men get on the elevator, they will overload its weight limit?

THINK ➡ Plan State what we want to know.

Asking the probability that the total weight of a sample of 10 men exceeds 2500 pounds is equivalent to asking the probability that their mean weight is greater than 250 pounds.

(continued)

[13]Cynthia L. Ogden, Cheryl D. Fryar, Margaret D. Carroll, and Katherine M. Flegal, *Mean Body Weight, Height, and Body Mass Index, United States 1960–2002, Advance Data from Vital and Health Statistics Number 347*, Oct. 27, 2004. http://www.cdc.gov/nchs

Model Think about the assumptions and check the conditions.

✓ **Independence Assumption:** It's reasonable to think that the weights of 10 randomly sampled men will be independent of each other. (But there could be exceptions—for example, if they were all from the same family or if the elevator were in a building with a diet clinic!)

✓ **Randomization Condition:** I'll assume that the 10 men getting on the elevator are a random sample from the population.

✓ **10% Condition:** 10 men is surely less than 10% of the population of possible elevator riders.

Note that if the sample were larger we'd be less concerned about the shape of the distribution of all weights.

✓ **Large Enough Sample Condition:** I suspect the distribution of population weights is roughly unimodal and symmetric, so my sample of 10 men seems large enough.

State the parameters and the sampling model.

The mean for all weights is $\mu = 190$ and the standard deviation is $\sigma = 59$ pounds. Since the conditions are satisfied, the CLT says that the sampling distribution of $\bar{y}$ has a Normal model with mean 190 and standard deviation

$$SD(\bar{y}) = \frac{\sigma}{\sqrt{n}} = \frac{59}{\sqrt{10}} \approx 18.66$$

Show ➡ **Plot** Make a picture. Sketch the model and shade the area we're interested in. Here the mean weight of 250 pounds appears to be far out on the right tail of the curve.

134.0 152.7 171.3 190.0 208.7 227.3 246.0 $\bar{y}$

Mechanics Use the standard deviation as a ruler to find the z-score of the cutoff mean weight. We see that an average of 250 pounds is more than 3 standard deviations above the mean.

$$z = \frac{\bar{y} - \mu}{SD(\bar{y})} = \frac{250 - 190}{18.66} = 3.21$$

Find the resulting probability from a table of Normal probabilities such as Table Z, a computer program, or a calculator.

$$P(\bar{y} > 250) = P(z > 3.21) = 0.0007$$

Tell ➡ **Conclusion** Interpret your result in the proper context, being careful to relate it to the original question.

The chance that a random collection of 10 men will exceed the elevator's weight limit is only 0.0007. So, if they are a random sample, it is quite unlikely that 10 people will exceed the total weight allowed on the elevator.

About Variation

"The n's justify the means."

—*Apocryphal
statistical saying*

Means vary less than individual data values. That makes sense. If the same test is given to many sections of a large course and the class average is, say, 80%, some students may score 95% because individual scores vary a lot. But we'd be shocked (and pleased!) if the *average* score of the students in any section was 95%. Averages are much less variable. Not only do group averages vary less than individual values, but common sense suggests that averages should be more consistent for larger groups. The Central Limit Theorem confirms this hunch; the fact that both $SD(\hat{p}) = \sqrt{\dfrac{pq}{n}}$ and $SD(\bar{y}) = \dfrac{\sigma}{\sqrt{n}}$ have n in the denominator shows that the variability of sample means decreases as the sample size increases. There's a catch, though. The standard deviation of the sampling distribution declines only with the square root of the sample size and not, for example, with $1/n$.

The mean of a random sample of 4 has half $\left(\dfrac{1}{\sqrt{4}} = \dfrac{1}{2}\right)$ the standard deviation of an individual data value. To cut the standard deviation in half again, we'd need a sample of 16, and a sample of 64 to halve it once more.

If only we had a much larger sample, we could get the standard deviation of the sampling distribution *really* under control so that the sample mean could tell us still more about the unknown population mean, but larger samples cost more and take longer to survey. And while we're gathering all those extra data, the population itself may change, or a news story may alter opinions. There are practical limits to most sample sizes. As we shall see, that nasty square root limits how much we can make a sample tell about the population. This is an example of something that's known as the Law of Diminishing Returns.

A Billion Dollar Misunderstanding? In the late 1990s the Bill and Melinda Gates Foundation began funding an effort to encourage the breakup of large schools into smaller schools. Why? It had been noticed that smaller schools were more common among the best-performing schools than one would expect. In time, the Annenberg Foundation, the Carnegie Corporation, the Center for Collaborative Education, the Center for School Change, Harvard's Change Leadership Group, the Open Society Institute, Pew Charitable Trusts, and the U.S. Department of Education's Smaller Learning Communities Program all supported the effort. Well over a billion dollars was spent to make schools smaller.

But was it all based on a misunderstanding of sampling distributions? Statisticians Howard Wainer and Harris Zwerling[14] looked at the mean test scores of schools in Pennsylvania. They found that indeed 12% of the top-scoring 50 schools were from the smallest 3% of Pennsylvania schools—substantially more than the 3% we'd naively expect. But then they looked at the *bottom* 50. There they found that 18% were small schools! The explanation? Mean test scores are, well, means. We are looking at a rough real-world simulation in which each school is a trial. Even if all Pennsylvania schools were equivalent, we'd expect their mean scores to vary. How much? The CLT tells us that means of test scores vary according to $\dfrac{\sigma}{\sqrt{n}}$. Smaller schools have (by definition) smaller n's, so the sampling distributions of their mean scores naturally have larger standard deviations. It's natural, then, that small schools have both higher and lower mean scores.

On October 26, 2005, *The Seattle Times* reported:

> [T]he Gates Foundation announced last week it is moving away from its emphasis on converting large high schools into smaller ones and instead giving grants to specially

(continued)

---

[14]Wainer, H. and Zwerling, H., "Legal and empirical evidence that smaller schools do not improve student achievement," *The Phi Delta Kappan* 2006 87:300–303. Discussed in Howard Wainer, "The Most Dangerous Equation," *American Scientist*, May–June 2007, pp. 249–256; also at www.Americanscientist.org.

> *selected school districts with a track record of academic improvement and effective leadership. Education leaders at the Foundation said they concluded that improving classroom instruction and mobilizing the resources of an entire district were more important first steps to improving high schools than breaking down the size.*

The Real World and the Model World

Be careful. We have been slipping smoothly between the real world, in which we draw random samples of data, and a magical mathematical model world, in which we describe how the sample means and proportions we observe in the real world behave as random variables in all the random samples that we might have drawn. Now we have *two* distributions to deal with. The first is the real-world distribution of the sample, which we might display with a histogram (for quantitative data) or with a bar chart or table (for categorical data). The second is the math world *sampling distribution model* of the statistic, a Normal model based on the Central Limit Theorem. Don't confuse the two.

For example, don't mistakenly think the CLT says that the *data* are Normally distributed as long as the sample is large enough. In fact, as samples get larger, we expect the distribution of the data to look more and more like the population from which they are drawn—skewed, bimodal, whatever—but not necessarily Normal. You can collect a sample of CEO salaries for the next 1000 years,[15] but the histogram will never look Normal. It will be skewed to the right. The Central Limit Theorem doesn't talk about the distribution of the data from the sample. It talks about the sample *means* and sample *proportions* of many different random samples drawn from the same population. Of course, the CLT does require that the sample be big enough when the population shape is not unimodal and symmetric, but the fact that, even then, a Normal model is useful is still a very surprising and powerful result.

 ## Just Checking

4. Human gestation times have a mean of about 266 days, with a standard deviation of about 10 days. If we record the gestation times of a sample of 100 women, do we know that a histogram of the times will be well modeled by a Normal model?

5. Suppose we look at the *average* gestation times for a sample of 100 women. If we imagined all the possible random samples of 100 women we could take and looked at the histogram of all the sample means, what shape would it have?

6. Where would the center of that histogram be?

7. What would be the standard deviation of that histogram?

Sampling Distribution Models

Let's summarize what we've learned about sampling distributions. At the heart is the idea that *the statistic itself is a random variable*. We can't know what our statistic will be because it comes from a random sample. It's just one instance of something that happened for our

---

[15]Don't forget to adjust for inflation.

particular random sample. A different random sample would have given a different result. This sample-to-sample variability is what generates the sampling distribution. The sampling distribution shows us the distribution of possible values that the statistic could have had.

We could simulate that distribution by pretending to take lots of samples. Fortunately, for the mean and the proportion, the CLT tells us that we can model their sampling distribution directly with a Normal model.

The two basic truths about sampling distributions are:

Simulation: **The CLT for Real Data.** Why settle for a picture when you can see it in action?

1. Sampling distributions arise because samples vary. Each random sample will contain different cases and, so, a different value of the statistic.
2. Although we can always simulate a sampling distribution, the Central Limit Theorem saves us the trouble for means and proportions.

Here's a picture showing the process going into the sampling distribution model:

Figure 17.11

We start with a population model, which can have any shape. It can even be bimodal or skewed (as this one is). We label the mean of this model μ and its standard deviation, σ.

We draw one real sample (solid line) of size n and show its histogram and summary statistics. We imagine (or simulate) drawing many other samples the same size (dotted lines), which have their own histograms and summary statistics.

We (imagine) gathering all the means into a histogram.

The CLT tells us we can model the shape of this histogram with a Normal model. The mean of this Normal is μ, and the standard deviation is

$$SD(\bar{y}) = \frac{\sigma}{\sqrt{n}}.$$

WHAT IF ●●● the population is far from Normal?

The great power of the Central Limit Theorem is that it applies to proportions or means of samples drawn from *any* population. The farther the population distribution is from normal, the larger the sample we'll need. To see this in action, let's revisit the compensations of the Fortune 500 CEOs we saw back in Chapter 4. Here's the distribution; the boxplot below helps illustrate just how highly skewed these data are.

If we sample from this distribution, there's a good chance that we'll get an exceptionally large value. So some samples will have sample means much larger than others. Here is the simulated sampling distribution of the means from 1000 samples of size 10:

This distribution is not *as* skewed as the population's distribution, but still strongly right skewed.

What happens if we take a larger sample? Here is the simulated sampling distribution of means from samples of size 50:

This distribution is less skewed than the corresponding distribution from smaller samples and its mean is again near 10,000.

Will this continue as we increase the sample size? Let's try samples of size 100:

Now the simulated sampling distribution of sample means is even more symmetric, but still skewed enough that we would not want to apply a Normal model.

As we take larger samples, the distribution of means becomes more and more symmetric.

By the time we get to samples of size 200, the distribution is quite symmetrical and, of course, has a mean quite close to 10,000. Now a Normal model will work!

We hope you're amazed. Even though the compensations are extremely skewed, these simulations demonstrate that with large enough samples a Normal model can help us understand the behavior of sample means. The Central Limit Theorem deserves its reputation as one of the most stunning results in all of mathematics.

WHAT CAN GO WRONG?

■ **Don't confuse the sampling distribution with the distribution of the sample.** It's important to be clear on which of these you're thinking about. The distribution of the data in a sample and the sampling distribution model of a sample statistic are two completely different things. For example, let's have a look using the CEO data once more. On the left is the distribution of the compensations of a sample of 200 randomly selected CEOs. On the right is the simulated sampling distribution of sample means that we just saw in the What If.

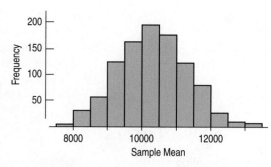

Obviously, they're very different. Be careful not to confuse these two important ideas:

- The larger the sample, the more the sample should look like the population—symmetric, skewed, bimodal, whatever.
- The larger the sample, the more the sampling distribution of sample proportions or means will look like a Normal model.

■ **Beware of observations that are not independent.** The CLT depends crucially on the assumption of independence. If our elevator riders are related, are all from the same school (for example, an elementary school), or in some other way aren't a random sample, then the statements we try to make about the mean are going to be wrong. Unfortunately, this isn't something you can check in your data. You have to think about how the data were gathered. Good sampling practice and well-designed randomized experiments ensure independence.

■ **Watch out for small samples from skewed populations.** The CLT assures us that the sampling distribution model is Normal if n is large enough. If the population is nearly Normal, even small samples (like our 10 elevator riders) work. If the population is very skewed, then n will have to be large before the Normal model will work well. If we sampled 15 or even 20 CEOs and used $\bar{y}$ to make a statement about the mean of all CEOs' compensation, we'd likely get into trouble because the underlying data distribution is so skewed. Unfortunately, there's no good rule of thumb.[16] It just depends on how skewed the data distribution is. Always plot the data to check.

---

[16]For proportions, of course, there is a rule: the **Success/Failure Condition.** That works for proportions because the standard deviation of a proportion is linked to its mean.

What Have We Learned?

We've learned to model the variation in statistics from sample to sample with a sampling distribution.

- The Central Limit Theorem tells us that the sampling distributions of both the sample proportion and the sample mean are approximately Normal for large enough samples.

We've learned that, usually, the mean of a sampling distribution is the value of the parameter estimated.

- For the sampling distribution of $\hat{p}$, the mean is p.
- For the sampling distribution of $\bar{y}$, the mean is μ.

We've learned about the standard deviation of a sampling distribution.

- The standard deviation of a sampling model is the most important information about it.

- The standard deviation of the sampling distribution of a proportion is $\sqrt{\dfrac{pq}{n}}$, where $q = 1 - p$.

- The standard deviation of the sampling distribution of a mean is $\dfrac{\sigma}{\sqrt{n}}$, where σ is the population standard deviation.

We've learned about the Central Limit Theorem, the most important theorem in Statistics.

- The sampling distribution of a sample mean tends toward Normal, *no matter what the underlying distribution of the data is.*
- The CLT says that this happens in the limit, as the sample size grows. The Normal model applies sooner when sampling from a unimodal, symmetric population and more gradually when the population is very non-Normal.

Terms

Sampling distribution	Different random samples give different values for a statistic. The distribution of the statistics over all possible samples is called the sampling distribution. The sampling distribution model shows the behavior of the statistic over all the possible samples for the same size *n*. (p. 446)
Sampling distribution model	Because we can never see all possible samples, we often use a model as a practical way of describing the theoretical sampling distribution. (p. 446)
Sampling variability (sampling error)	The variability we expect to see from one random sample to another. It is sometimes called sampling error, but sampling variability is the better term. (p. 448)
Sampling distribution model for a sample proportion	If assumptions of independence and random sampling are met, and we expect at least 10 successes and 10 failures, then the sampling distribution of a sample proportion is modeled by a Normal model with a mean equal to the true proportion value, p, and a standard deviation equal to $\sqrt{\dfrac{pq}{n}}$. (p. 450)
Central Limit Theorem	The Central Limit Theorem (CLT) states that the sampling distribution model of the sample mean (and proportion) from a random sample is approximately Normal for large *n*, *regardless of the distribution of the population, as long as the observations are independent.* (p. 454)
Sampling distribution model for a sample mean	If assumptions of independence and random sampling are met, and the sample size is large enough, the sampling distribution of the sample mean is modeled by a Normal model with a mean equal to the population mean, μ, and a standard deviation equal to $\dfrac{\sigma}{\sqrt{n}}$. (p. 456)

Exercises

1. **Send money** When they send out their fundraising letter, a philanthropic organization typically gets a return from about 5% of the people on their mailing list. To see what the response rate might be for future appeals, they did a simulation using samples of size 20, 50, 100, and 200. For each sample size, they simulated 1000 mailings with success rate $p = 0.05$ and constructed the histogram of the 1000 sample proportions. Explain how these four histograms demonstrate what the Central Limit Theorem says about the sampling distribution model for sample proportions. Be sure to talk about shape, center, and spread.

Samples of Size 100

Samples of Size 200

2. Character recognition An automatic character recognition device can successfully read about 85% of handwritten credit card applications. To estimate what might happen when this device reads a stack of applications, the company did a simulation using samples of size 20, 50, 75, and 100. For each sample size, they simulated 1000 samples with success rate $p = 0.85$ and constructed the histogram of the 1000 sample proportions, shown here. Explain how these four histograms demonstrate what the Central Limit Theorem says about the sampling distribution model for sample proportions. Be sure to talk about shape, center, and spread.

Samples of Size 20

Samples of Size 50

Samples of Size 75

Samples of Size 100

3. AP Exam A small class of five statistics students received the following scores on their AP Exam: 5, 4, 4, 3, 1.

a) Calculate the mean and standard deviation of these five scores.

b) List all possible sets of size 2 that could be chosen from this class. (There are $_5C_2 = 10$ such sets.)

c) Calculate the mean of each of these sets of 2 scores and make a dotplot of the sampling distribution of the sample mean.

d) Calculate the mean and standard deviation of this sampling distribution. How do they compare to those of the individual scores? Is the sample mean an unbiased estimator of the population mean?

4. AP Exam II For the small class described in Exercise 3 (mean $= 3.4$, sd $= 1.517$), you will be selecting samples of size 3 (sampling without replacement).

a) List all possible samples of size 3 that could be chosen from this class.

b) Construct the sampling distribution of the sample mean for samples of size 3.

c) How do the mean and standard deviation of this sampling distribution compare to the mean and standard deviation of the population? To the mean and standard deviation of the sampling distribution of the sample mean from Exercise 3 part d?

5. Marriage According to a Pew Research survey, about 27% of American adults are pessimistic about the future of marriage and the family. This is based on a sample, but assume that this percentage is correct for all American adults.

a) Using a binomial model, what is the probability that, in a sample of 20 American adults, 25% or fewer of the people in the sample are pessimistic about the future of marriage and family?

b) Now use a Normal model to compute that probability. How does this compare to your answer from part a?

c) Using a Binomial model, what is the probability that, in a sample of 700 American adults, 25% or fewer of the people of the people in the sample are pessimistic about the future of marriage and family?

d) Now use a Normal model to compute that probability. How does this compare to your answer from part c?

e) What do these answers tell you about the importance of checking that np and nq are both at least 10?

6. Wow. Just, wow According to a 2013 poll from Public Policy Polling, 4% of American voters believe that shape-shifting reptilian people control our world by taking on human form and gaining power. Yes, you read that correctly! (This was a poll about conspiracy theories.) Assume that's the actual proportion of Americans who hold that belief.

a) Use a binomial model to calculate the probability that, in a random sample of 100 people, at least 6% of those in the sample believe the thing about reptilian people controlling our world.

b) Use a Normal model to calculate the same probability. How does this compare with the answer in part a?

c) That same poll found that 51% of American voters believe there was a larger conspiracy responsible for the assassination of President Kennedy. Use a binomial model to calculate the probability that, in a random sample of 100 people, at least 57% of those in the sample believe in the JFK conspiracy theory.

d) Use a normal model to calculate the same probability. How does this compare with the answer in part c?

e) What do these answers tell you about the importance of checking that np and nq are both at least 10?

7. **Send money, again** The philanthropic organization in Exercise 1 expects about a 5% success rate when they send fundraising letters to the people on their mailing list. In Exercise 1 you looked at the histograms showing distributions of sample proportions from 1000 simulated mailings for samples of size 20, 50, 100, and 200. The sample statistics from each simulation were as follows:

n	mean	st. dev.
20	0.0497	0.0479
50	0.0516	0.0309
100	0.0497	0.0215
200	0.0501	0.0152

a) According to the Central Limit Theorem, what should the theoretical mean and standard deviations be for these sample sizes?

b) How close are those theoretical values to what was observed in these simulations?

c) Looking at the histograms in Exercise 1, at what sample size would you be comfortable using the Normal model as an approximation for the sampling distribution?

d) What does the Success/Failure Condition say about the choice you made in part c?

8. **Character recognition, again** The automatic character recognition device discussed in Exercise 2 successfully reads about 85% of handwritten credit card applications. In Exercise 2 you looked at the histograms showing distributions of sample proportions from 1000 simulated samples of size 20, 50, 75, and 100. The sample statistics from each simulation were as follows:

n	mean	st. dev.
20	0.8481	0.0803
50	0.8507	0.0509
75	0.8481	0.0406
100	0.8488	0.0354

a) According to the Central Limit Theorem, what should the theoretical mean and standard deviations be for these sample sizes?

b) How close are those theoretical values to what was observed in these simulations?

c) Looking at the histograms in Exercise 2, at what sample size would you be comfortable using the Normal model as an approximation for the sampling distribution?

d) What does the Success/Failure Condition say about the choice you made in part c?

9. **Sample maximum** The distribution of scores on a Statistics test for a particular class is skewed to the left. The professor wants to predict the maximum score and so wants to understand the distribution of the sample maximum. She simulates the distribution of the maximum of the test for 30 different tests (with $n = 5$). The histogram below shows a simulated sampling distribution of the sample maximum from these tests.

a) Would a Normal model be a useful model for this sampling distribution? Explain.

b) The mean of this distribution is 46.3 and the SD is 3.5. Would you expect about 95% of the samples to have their maximums within 7 of 46.3? Why or why not?

10. **Soup** A machine is supposed to fill cans with 16 oz of soup. Of course there, will be some variation in the amount actually dispensed, and measurement errors are often approximately normally distributed. The manager would like to understand the variability of the variances of the samples, so he collects information from the last 250 batches of size 10 and plots a histogram of the variances:

a) Would a Normal model be a useful model for this sampling distribution? Explain.

b) The mean of this distribution is 0.009 and the SD is 0.004. Would you expect about 95% of the samples to have their variances within 0.008 of 0.009? Why or why not?

11. **Coin tosses** In a large class of introductory Statistics students, the professor has each person toss a coin 16 times and calculate the proportion of his or her tosses that were heads. The students then report their results, and the professor plots a histogram of these several proportions.

a) What shape would you expect this histogram to be? Why?

b) Where do you expect the histogram to be centered?

c) How much variability would you expect among these proportions?

d) Explain why a Normal model should not be used here.

12. **M&M's** The candy company claims that 16% of the Milk Chocolate M&M's it produces are green. Suppose that the candies are thoroughly mixed and then packaged in small bags containing about 50 M&M's. A class of elementary school students learning about percents opens several bags, counts the various colors of the candies, and calculates the proportion that are green.

a) If we plot a histogram showing the proportions of green candies in the various bags, what shape would you expect it to have?

b) Can that histogram be approximated by a Normal model? Explain.

c) Where should the center of the histogram be?

d) What should the standard deviation of the sampling distribution be?

13. **More coins** Suppose the class in Exercise 11 repeats the coin-tossing experiment.

a) The students toss the coins 25 times each. Use the 68–95–99.7 Rule to describe the sampling distribution model.

b) Confirm that you can use a Normal model here.

c) They increase the number of tosses to 64 each. Draw and label the appropriate sampling distribution model. Check the appropriate conditions to justify your model.

d) Explain how the sampling distribution model changes as the number of tosses increases.

14. **Bigger bag** Suppose the class in Exercise 12 buys bigger bags of candy, with 200 M&M's each. Again the students calculate the proportion of green candies they find.

a) Explain why it's appropriate to use a Normal model to describe the distribution of the proportion of green M&M's they might expect.

b) Use the 68–95–99.7 Rule to describe how this proportion might vary from bag to bag.

c) How would this model change if the bags contained even more candies?

15. **Just (un)lucky?** One of the students in the introductory Statistics class in Exercise 13 claims to have tossed her coin 200 times and found only 42% heads. What do you think of this claim? Explain.

16. **Too many green ones?** In a really large bag of M&M's, the students in Exercise 14 found 500 candies, and 18% of them were green. Is this an unusually large proportion of green M&M's? Explain.

17. **Speeding** State police believe that 70% of the drivers traveling on a major interstate highway exceed the speed limit. They plan to set up a radar trap and check the speeds of 80 cars.

a) Using the 68–95–99.7 Rule, draw and label the distribution of the proportion of these cars the police will observe speeding.

b) Do you think the appropriate conditions necessary for your analysis are met? Explain.

18. **Smoking** Public health statistics for 2009 indicate that 20.6% of American adults smoke cigarettes. Using the 68–95–99.7 Rule, describe the sampling distribution model for the proportion of smokers among a randomly selected group of 50 adults. Be sure to discuss your assumptions and conditions.

19. **Vision** It is generally believed that nearsightedness affects about 12% of all children. A school district has registered 170 incoming kindergarten children.

a) Can you apply the Central Limit Theorem to describe the sampling distribution model for the sample proportion of children who are nearsighted? Check the conditions and discuss any assumptions you need to make.

b) Sketch and clearly label the sampling model, based on the 68–95–99.7 Rule.

c) How many of the incoming students might the school expect to be nearsighted? Explain.

20. **Mortgages** In July 2010, Lender Processing Services reported that homeowners were defaulting in record numbers; 12.4% of mortgages were delinquent or in foreclosure. Suppose a large bank holds 1731 adjustable-rate mortgages.

a) Can you apply the Central Limit Theorem to describe the sampling distribution model for the sample proportion of foreclosures? Check the conditions and discuss any assumptions you need to make.

b) Sketch and clearly label the sampling model, based on the 68–95–99.7 Rule.

c) How many of these homeowners might the bank expect will default on their mortgages? Explain.

21. **Loans** Based on past experience, a bank believes that 7% of the people who receive loans will not make payments on time. The bank has recently approved 200 loans.

a) What are the mean and standard deviation of the proportion of clients in this group who may not make timely payments?

b) What assumptions underlie your model? Are the conditions met? Explain.

c) What's the probability that over 10% of these clients will not make timely payments?

22. **Teens with phones** Pew Research reported that, in 2013, 78% of all teens had a cell phone. Assume this estimate is correct.

a) We randomly pick 100 teens. Let $\hat{p}$ represent the proportion of teens in this sample who own a cell phone. What's the appropriate model for the distribution of $\hat{p}$? Specify the name of the distribution, the mean, and the standard deviation. Be sure to verify that the conditions are met.

b) What's the approximate probability that less than three fourths of this sample own a cell phone?

23. Back to school? Best known for its testing program, ACT, Inc., also compiles data on a variety of issues in education. In 2004 the company reported that the national college freshman-to-sophomore retention rate held steady at 74% over the previous four years. Consider random samples of 400 freshmen who took the ACT. Use the 68–95–99.7 Rule to describe the sampling distribution model for the percentage of those students we expect to return to that school for their sophomore years. Do you think the appropriate conditions are met?

24. Binge drinking A national study found that 44% of college students engage in binge drinking (5 drinks at a sitting for men, 4 for women). Use the 68–95–99.7 Rule to describe the sampling distribution model for the proportion of students in a randomly selected group of 200 college students who engage in binge drinking. Do you think the appropriate conditions are met?

25. Back to school, again Based on the 74% national retention rate described in Exercise 23, does a college where 522 of the 603 freshman returned the next year as sophomores have a right to brag that it has an unusually high retention rate? Explain.

26. Binge sample After hearing of Exercise 24's national result that 44% of students engage in binge drinking (5 drinks at a sitting for men, 4 for women), a professor surveyed a random sample of 244 students and found that 96 of them admitted to binge drinking in the past week. Should he be surprised at this result? Explain.

27. Polling Just before a referendum on a school budget, a local newspaper polls 400 voters in an attempt to predict whether the budget will pass. Suppose that the budget actually has the support of 52% of the voters. What's the probability the newspaper's sample will lead them to predict defeat? Be sure to verify that the assumptions and conditions necessary for your analysis are met.

28. Seeds Information on a packet of seeds claims that the germination rate is 92%. What's the probability that more than 95% of the 160 seeds in the packet will germinate? Be sure to discuss your assumptions and check the conditions that support your model.

29. Gaydar Exercise 10 in Chapter 1 describes a study that showed that heterosexual women, during ovulation, were significantly better at correctly identifying the sexual orientation of a man from a photograph of his face than women who were not ovulating. In other words, ovulation improves a woman's "gaydar." Near ovulation, on average women correctly identified the orientation of about 65% of the 100 men shown to them. If this is the probability of correctly identifying the orientation of a man in any given photograph, what is the probability a woman would correctly classify 80 or more of the men (as two women in the study did)?

30. Genetic defect It's believed that 4% of children have a gene that may be linked to juvenile diabetes. Researchers hoping to track 20 of these children for several years test 732 newborns for the presence of this gene. What's the probability that they find enough subjects for their study?

31. "No Children" section Some restaurant owners, at the request of some of their less tolerant customers, have stopped allowing children into their restaurant. This, naturally, outrages other customers. One restaurateur hopes to please both sets of customers by having a "no children" section. She estimates that in her 120-seat restaurant, about 30% of her seats, on average, are taken by families with children. How many seats should be in the "children allowed" area in order to be very sure of having enough seating there? Comment on the assumptions and conditions that support your model, and explain what "very sure" means to you.

32. Meals A restauranteur anticipates serving about 180 people on a Friday evening, and believes that about 20% of the patrons will order the chef's steak special. How many of those meals should he plan on serving in order to be pretty sure of having enough steaks on hand to meet customer demand? Justify your answer, including an explanation of what "pretty sure" means to you.

33. Sampling A sample is chosen randomly from a population that can be described by a Normal model.
  a) What's the sampling distribution model for the sample mean? Describe shape, center, and spread.
  b) If we choose a larger sample, what's the effect on this sampling distribution model?

34. Sampling, part II A sample is chosen randomly from a population that was strongly skewed to the left.
  a) Describe the sampling distribution model for the sample mean if the sample size is small.
  b) If we make the sample larger, what happens to the sampling distribution model's shape, center, and spread?
  c) As we make the sample larger, what happens to the expected distribution of the data in the sample?

35. Waist size A study measured the *Waist Size* of 250 men, finding a mean of 36.33 inches and a standard deviation of 4.02 inches. Here is a histogram of these measurements

a) Describe the histogram of *Waist Size*.
b) To explore how the mean might vary from sample to sample, they simulated by drawing many samples of size 2, 5, 10, and 20, with replacement, from the 250 measurements. Here are histograms of the sample means for each simulation. Explain how these histograms demonstrate what the Central Limit Theorem says about the sampling distribution model for sample means.

36. CEO compensation In Chapter 5 we saw the distribution of the total compensation of the chief executive officers (CEOs) of the 800 largest U.S. companies (the Fortune 800). The average compensation (in thousands of dollars) is 10,307.31 and the standard deviation is 17,964.62. Here is a histogram of their annual compensations (in $1000):

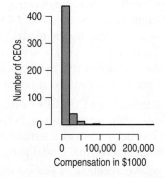

a) Describe the histogram of *Total Compensation*. A research organization simulated sample means by drawing samples of 30, 50, 100, and 200, with replacement, from the 800 CEOs. The histograms show the distributions of means for many samples of each size.

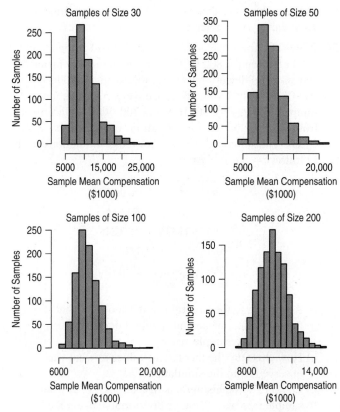

b) Explain how these histograms demonstrate what the Central Limit Theorem says about the sampling distribution model for sample means. Be sure to talk about shape, center, and spread.
c) Comment on an oft-cited "rule of thumb" that "With a sample size of at least 30, the sampling distribution of the mean is Normal."

37. Waist size revisited Researchers measured the *Waist Sizes* of 250 men in a study on body fat. The true mean and standard deviation of the *Waist Sizes* for the 250 men are 36.33 in and 4.019 inches, respectively. In Exercise 35 you looked at the histograms of simulations that drew samples of sizes 2, 5, 10, and 20 (with replacement). The summary statistics for these simulations were as follows:

n	mean	st. dev.
2	36.314	2.855
5	36.314	1.805
10	36.341	1.276
20	36.339	0.895

a) According to the Central Limit Theorem, what should the theoretical mean and standard deviation be for each of these sample sizes?
b) How close are the theoretical values to what was observed in the simulation?
c) Looking at the histograms in Exercise 35, at what sample size would you be comfortable using the Normal model as an approximation for the sampling distribution?

d) What about the shape of the distribution of *Waist Size* explains your choice of sample size in part c?

38. CEOs revisited In Exercise 36 you looked at the annual compensation for 800 CEOs, for which the true mean and standard deviation were (in thousands of dollars) 10,307.31 and 17,964.62, respectively. A simulation drew samples of sizes 30, 50, 100, and 200 (with replacement) from the total annual compensations of the Fortune 800 CEOs. The summary statistics for these simulations were as follows:

n	mean	st. dev.
30	10,251.73	3359.64
50	10,343.93	2483.84
100	10,329.94	1779.18
200	10,340.37	1230.79

a) According to the Central Limit Theorem, what should the theoretical mean and standard deviation be for each of these sample sizes?
b) How close are the theoretical values to what was observed from the simulation?
c) Looking at the histograms in Exercise 36, at what sample size would you be comfortable using the Normal model as an approximation for the sampling distribution?
d) What about the shape of the distribution of *Total Compensation* explains your answer in part c?

39. GPAs A college's data about the incoming freshmen indicates that the mean of their high school GPAs was 3.4, with a standard deviation of 0.35; the distribution was roughly mound-shaped and only slightly skewed. The students are randomly assigned to freshman writing seminars in groups of 25. What might the mean GPA of one of these seminar groups be? Describe the appropriate sampling distribution model—shape, center, and spread—with attention to assumptions and conditions. Make a sketch using the 68–95–99.7 Rule.

40. Home values Assessment records indicate that the value of homes in a small city is skewed right, with a mean of $140,000 and standard deviation of $60,000. To check the accuracy of the assessment data, officials plan to conduct a detailed appraisal of 100 homes selected at random. Using the 68–95–99.7 Rule, draw and label an appropriate sampling model for the mean value of the homes selected.

Ⓣ **41. Lucky Spot?** A reporter working on a story about the New York lottery contacted one of the authors of this book, wanting help analyzing data to see if some ticket sales outlets were more likely to produce winners. His data for each of the 966 New York lottery outlets are graphed below; the scatterplot shows the ratio *TotalPaid/TotalSales* vs. *TotalSales* for the state's "instant winner" games for all of 2007.

The reporter thinks that by identifying the outlets with the highest fraction of bets paid out, players might be able to increase their chances of winning. (Typically—but not always—instant winners are paid immediately (instantly) at the store at which they are purchased. However, the fact that tickets may be scratched off and then cashed in at any outlet may account for some outlets paying out more than they take in. The few with very low payouts may be on interstate highways where players may purchase cards but then leave.)

a) Explain why the plot has this funnel shape.
b) Explain why the reporter's idea wouldn't have worked anyway.

42. Safe cities Allstate Insurance Company identified the 10 safest and 10 least-safe U.S. cities from among the 200 largest cities in the United States, based on the mean number of years drivers went between automobile accidents. The cities on both lists were all smaller than the 10 largest cities. Using facts about the sampling distribution model of the mean, explain why this is not surprising.

43. Pregnancy Assume that the duration of human pregnancies can be described by a Normal model with mean 266 days and standard deviation 16 days.

a) What percentage of pregnancies should last between 270 and 280 days?
b) At least how many days should the longest 25% of all pregnancies last?
c) Suppose a certain obstetrician is currently providing prenatal care to 60 pregnant women. Let $\bar{y}$ represent the mean length of their pregnancies. According to the Central Limit Theorem, what's the distribution of this sample mean, $\bar{y}$? Specify the model, mean, and standard deviation.
d) What's the probability that the mean duration of these patients' pregnancies will be less than 260 days?

44. Rainfall Statistics from Cornell's Northeast Regional Climate Center indicate that Ithaca, NY, gets an average of 35.4″ of rain each year, with a standard deviation of 4.2″. Assume that a Normal model applies.

a) During what percentage of years does Ithaca get more than 40″ of rain?
b) Less than how much rain falls in the driest 20% of all years?
c) A Cornell University student is in Ithaca for 4 years. Let $\bar{y}$ represent the mean amount of rain for those

4 years. Describe the sampling distribution model of this sample mean, $\bar{y}$.

d) What's the probability that those 4 years average less than 30″ of rain?

45. Pregnant again The duration of human pregnancies may not actually follow the Normal model described in Exercise 37.

a) Explain why it may be somewhat skewed to the left.

b) If the correct model is in fact skewed, does that change your answers to parts a, b, and c of Exercise 43? Explain why or why not for each.

46. At work Some business analysts estimate that the length of time people work at a job has a mean of 6.2 years and a standard deviation of 4.5 years.

a) Explain why you suspect this distribution may be skewed to the right.

b) Explain why you could estimate the probability that 100 people selected at random had worked for their employers an average of 10 years or more, but you could not estimate the probability that an individual had done so.

47. Dice and dollars You roll a die, winning nothing if the number of spots is odd, $1 for a 2 or a 4, and $10 for a 6.

a) Find the expected value and standard deviation of your prospective winnings.

b) You play twice. Find the mean and standard deviation of your total winnings.

c) You play 40 times. What's the probability that you win at least $100?

48. New game You pay $10 and roll a die. If you get a 6, you win $50. If not, you get to roll again. If you get a 6 this time, you get your $10 back.

a) Create a probability model for this game.

b) Find the expected value and standard deviation of your prospective winnings.

c) You play this game five times. Find the expected value and standard deviation of your average winnings.

d) 100 people play this game. What's the probability the person running the game makes a profit?

49. AP Stats 2012 The College Board reported the score distribution shown in the table for all students who took the 2012 AP Statistics exam.

Score	Percent of Students
5	12.5
4	21.1
3	25.6
2	18.0
1	22.8

a) Find the mean and standard deviation of the scores.

b) If we select a random sample of 40 AP Statistics students, would you expect their scores to follow a Normal model? Explain.

c) Consider the mean scores of random samples of 40 AP Statistics students. Describe the sampling model for these means (shape, center, and spread).

50. Museum membership A museum offers several levels of membership, as shown in the table.

Member Category	Amount of Donation ($)	Percent of Members
Individual	50	41
Family	100	37
Sponsor	250	14
Patron	500	7
Benefactor	1000	1

a) Find the mean and standard deviation of the donations.

b) During their annual membership drive, they hope to sign up 50 new members each day. Would you expect the distribution of the donations for a day to follow a Normal model? Explain.

c) Consider the mean donation of the 50 new members each day. Describe the sampling model for these means (shape, center, and spread).

51. AP Stats 2012, again An AP Statistics teacher had 63 students preparing to take the AP exam discussed in Exercise 49. Though they were obviously not a random sample, he considered his students to be "typical" of all the national students. What's the probability that his students will achieve an average score of at least 3?

52. Joining the museum One of the museum's phone volunteers sets a personal goal of getting an average donation of at least $100 from the new members she enrolls during the membership drive described in Exercise 50. If she gets 80 new members and they can be considered a random sample of all the museum's members, what is the probability that she can achieve her goal?

53. Pollution Carbon monoxide (CO) emissions for a certain kind of car vary with mean 2.9 g/mi and standard deviation 0.4 g/mi. A company has 80 of these cars in its fleet. Let $\bar{y}$ represent the mean CO level for the company's fleet.

a) What's the approximate model for the distribution of $\bar{y}$? Explain.

b) Estimate the probability that $\bar{y}$ is between 3.0 and 3.1 g/mi.

c) There is only a 5% chance that the fleet's mean CO level is greater than what value?

54. Potato chips The weight of potato chips in a medium-size bag is stated to be 10 ounces. The amount that the packaging machine puts in these bags is believed to have a Normal model with mean 10.2 ounces and standard deviation 0.12 ounces.

a) What fraction of all bags sold are underweight?
b) Some of the chips are sold in "bargain packs" of 3 bags. What's the probability that none of the 3 is underweight?
c) What's the probability that the mean weight of the 3 bags is below the stated amount?
d) What's the probability that the mean weight of a 24-bag case of potato chips is below 10 ounces?

55. Tips A waiter believes the distribution of his tips has a model that is slightly skewed to the right, with a mean of $9.60 and a standard deviation of $5.40.

a) Explain why you cannot determine the probability that a given party will tip him at least $20.
b) Can you estimate the probability that the next 4 parties will tip an average of at least $15? Explain.
c) Is it likely that his 10 parties today will tip an average of at least $15? Explain.

56. Groceries A grocery store's receipts show that Sunday customer purchases have a skewed distribution with a mean of $32 and a standard deviation of $20.

a) Explain why you cannot determine the probability that the next Sunday customer will spend at least $40.
b) Can you estimate the probability that the next 10 Sunday customers will spend an average of at least $40? Explain.
c) Is it likely that the next 50 Sunday customers will spend an average of at least $40? Explain.

57. More tips The waiter in Exercise 55 usually waits on about 40 parties over a weekend of work.

a) Estimate the probability that he will earn at least $500 in tips.
b) How much does he earn on the best 10% of such weekends?

58. More groceries Suppose the store in Exercise 56 had 312 customers this Sunday.

a) Estimate the probability that the store's revenues were at least $10,000.
b) If, on a typical Sunday, the store serves 312 customers, how much does the store take in on the worst 10% of such days?

59. IQs Suppose that IQs of East State University's students can be described by a Normal model with mean 130 and standard deviation 8 points. Also suppose that IQs of students from West State University can be described by a Normal model with mean 120 and standard deviation 10.

a) We select a student at random from East State. Find the probability that this student's IQ is at least 125 points.
b) We select a student at random from each school. Find the probability that the East State student's IQ is at least 5 points higher than the West State student's IQ.
c) We select 3 West State students at random. Find the probability that this group's average IQ is at least 125 points.
d) We also select 3 East State students at random. What's the probability that their average IQ is at least 5 points higher than the average for the 3 West Staters?

60. Milk Although most of us buy milk by the quart or gallon, farmers measure daily production in pounds. Ayrshire cows average 47 pounds of milk a day, with a standard deviation of 6 pounds. For Jersey cows, the mean daily production is 43 pounds, with a standard deviation of 5 pounds. Assume that Normal models describe milk production for these breeds.

a) We select an Ayrshire at random. What's the probability that she averages more than 50 pounds of milk a day?
b) What's the probability that a randomly selected Ayrshire gives more milk than a randomly selected Jersey?
c) A farmer has 20 Jerseys. What's the probability that the average production for this small herd exceeds 45 pounds of milk a day?
d) A neighboring farmer has 10 Ayrshires. What's the probability that his herd average is at least 5 pounds higher than the average for part c's Jersey herd?

 Just Checking ANSWERS

1. A Normal model (approximately).

2. At the actual proportion of all students who are in favor.

3. $SD(\hat{p}) = \sqrt{\dfrac{(0.5)(0.5)}{100}} = 0.05$

4. No, this is a histogram of individuals. It may or may not be approximately Normal, but we can't tell from the information provided.

5. A Normal model (approximately).

6. 266 days

7. $\dfrac{10}{\sqrt{100}} = 1.0$ day

18 Confidence Intervals for Proportions

oral reef communities are home to one quarter of all marine plants and animals worldwide. These reefs support large fisheries by providing breeding grounds and safe havens for young fish of many species. Coral reefs are seawalls that protect shorelines against tides, storm surges, and hurricanes, and are sand "factories" that produce the limestone and sand of which beaches are made. Beyond the beach, these reefs are major tourist attractions for snorkelers and divers, driving a tourist industry worth tens of billions of dollars.

But marine scientists say that 10% of the world's reef systems have been destroyed in recent times. At current rates of loss, 70% of the reefs could be gone in 40 years. Pollution, global warming, outright destruction of reefs, and increasing acidification of the oceans are all likely factors in this loss.

Dr. Drew Harvell's lab studies corals and the diseases that affect them. They sampled sea fans[1] at 19 randomly selected reefs along the Yucatan peninsula and diagnosed whether the animals were affected by the disease *aspergillosis*.[2] In specimens collected at a depth of 40 feet at the Las Redes Reef in Akumal, Mexico, these scientists found that 54 of 104 sea fans sampled were infected with that disease.

Of course, we care about much more than these particular 104 sea fans. We care about the health of coral reef communities throughout the Caribbean. What can this study tell us about the prevalence of the disease among sea fans?

We have a sample proportion, which we write as $\hat{p}$, of 54/104, or 51.9%. Our first guess might be that this observed proportion is close to the population proportion, p. But we also know that because of natural sampling variability, if the researchers had drawn

Who	Sea fans
What	Percent infected
When	June 2000
Where	Las Redes Reef, Akumal, Mexico, 40 feet deep
Why	Research

---

[1]That's a sea fan in the picture. Although they look like trees, they are actually colonies of genetically identical animals.

[2]K. M. Mullen, C. D. Harvell, A. P. Alker, D. Dube, E. Jordán-Dahlgren, J. R. Ward, and L. E. Petes, "Host range and resistance to aspergillosis in three sea fan species from the Yucatan," *Marine Biology* (2006), Springer-Verlag.

a second sample of 104 sea fans at roughly the same time, the proportion infected from that sample probably wouldn't have been exactly 51.9%.

What *can* we say about the population proportion, p? To start to answer this question, think about how different the sample proportion might have been if we'd taken another random sample from the same population. But wait. Remember—we aren't actually going to take more samples. We just want to *imagine* how the sample proportions might vary from sample to sample. In other words, we want to know about the *sampling distribution* of the sample proportion of infected sea fans.

A Confidence Interval

A S *Activity*: **Confidence Intervals and Sampling Distributions.** Simulate the sampling distribution, and see how it gives a confidence interval.

Let's look at our model for the sampling distribution. What do we know about it? We know it's approximately Normal (under certain assumptions, which we must be careful to check) and that its mean is the proportion of all infected sea fans on the Las Redes Reef. Is the infected proportion of *all* sea fans 51.9%? No, that's just $\hat{p}$, our estimate. We don't know the proportion, p, of all the infected sea fans; that's what we're trying to find out. We do know, though, that the sampling distribution model of $\hat{p}$ is centered at p, and we know that the standard deviation of the sampling distribution is $\sqrt{\dfrac{pq}{n}}$.

■ NOTATION ALERT

Remember that $\hat{p}$ is our sample-based estimate of the true proportion p. Recall also that q is just shorthand for $1 - p$, and $\hat{q} = 1 - \hat{p}$.

When we use $\hat{p}$ to estimate the standard deviation of the sampling distribution model, we call that the standard error and write $SE(\hat{p}) = \sqrt{\dfrac{\hat{p}\hat{q}}{n}}$.

Now we have a problem: Since we don't know p, we can't find the true standard deviation of the sampling distribution model. We do know the observed proportion, $\hat{p}$, so, of course we just use what we know, and we estimate. That may not seem like a big deal, but it gets a special name. Whenever we estimate the standard deviation of a sampling distribution, we call it a **standard error**.[3] For a sample proportion, $\hat{p}$, the standard error is

$$SE(\hat{p}) = \sqrt{\dfrac{\hat{p}\hat{q}}{n}}.$$

For the sea fans, then:

$$SE(\hat{p}) = \sqrt{\dfrac{\hat{p}\hat{q}}{n}} = \sqrt{\dfrac{(0.519)(0.481)}{104}} = 0.049 = 4.9\%.$$

Now we know that the sampling model for $\hat{p}$ should look like this:

Figure 18.1

The sampling distribution model for $\hat{p}$ is Normal with a mean of p and a standard deviation we estimate to be 0.049.

Great. What does that tell us? Well, because it's Normal, it says that about 68% of all samples of 104 sea fans will have $\hat{p}$'s within 1 *SE*, 0.049, of p. And about 95% of all these samples will be within $p \pm 2$ *SE*s. But where is *our* sample proportion in this picture? And what value does p have? We still don't know!

We do know that for 95% of random samples, $\hat{p}$ will be no more than 2 *SE*s away from p. So let's look at this from $\hat{p}$'s point of view. If I'm $\hat{p}$, there's a 95% chance that p is no more than 2 *SE*s away from me. If I reach out 2 *SE*s, or 2 × 0.049, away from me on both sides, I'm 95% sure that p will be within my grasp. Now I've got him! Probably.

---

[3]This isn't such a great name because it isn't standard and nobody made an error. But it's much shorter and more convenient than saying, "the estimated standard deviation of the sampling distribution of the sample statistic."

Of course, even if my interval does catch p, I still don't know its true value. The best I can do is to produce an interval, and even then I can't be positive it contains p.

Figure 18.2

Reaching out 2 SEs on either side of $\hat{p}$ makes us 95% confident that we'll trap the true proportion, p.

ACME p-trap: Guaranteed*
to capture p.

*with 95% confidence

$\hat{p} - 2\,SE$ $\hat{p}$ $\hat{p} + 2\,SE$

 Activity: Can We Estimate a Parameter? Consider these four interpretations of a confidence interval by simulating to see whether they could be right.

So what can we really say about p? Here's a list of things we'd like to be able to say, in order of strongest to weakest and the reasons we can't say most of them:

1. **"51.9% of *all* sea fans on the Las Redes Reef are infected."** It would be nice to be able to make absolute statements about population values with certainty, but we just don't have enough information to do that. There's no way to be sure that the population proportion is the same as the sample proportion; in fact, it almost certainly isn't. Observations vary. Another sample would almost certainly yield a different sample proportion.

2. **"It is *probably* true that 51.9% of all sea fans on the Las Redes Reef are infected."** No. In fact, we can be pretty sure that whatever the true proportion is, it's not exactly 51.900%. So the statement is not true.

3. **"We don't know exactly what proportion of sea fans on the Las Redes Reef is infected, but we *know* that it's within the interval 51.9% $\pm$ 2 $\times$ 4.9%. That is, it's between 42.1% and 61.7%."** This is getting closer, but we still can't be certain. We can't know *for sure* that the true proportion is in this interval—or in any particular interval.

4. **"We don't know exactly what proportion of sea fans on the Las Redes Reef is infected, but the interval from 42.1% to 61.7% *probably* contains the true proportion."** We've now fudged twice—first by giving an interval and second by admitting that we only think the interval "probably" contains the true value. And this statement is true.

"Far better an approximate answer to the right question, . . . than an exact answer to the wrong question."

—*John W. Tukey*

That last statement may be true, but it's a bit wishy-washy. We can tighten it up a bit by quantifying what we mean by "probably." We saw that 95% of the time when we reach out 2 *SE*s from $\hat{p}$ we capture p, so we can be 95% confident that this is one of those times. After putting a number on the probability that this interval covers the true proportion, we've given our best guess of where the parameter is and how certain we are that it's within some range.

5. **"We are 95% confident that between 42.1% and 61.7% of Las Redes sea fans are infected."** Statements like these describe **confidence intervals.** They're the best we can do.

Each confidence interval discussed in the book has a name. You'll see many different kinds of confidence intervals in the following chapters. Some will be about more than *one* sample, some will be about statistics other than *proportions,* and some will use models other than the Normal. The interval calculated and interpreted here is sometimes called a **one-proportion z-interval.**[4]

---

[4]In fact, this confidence interval is so standard for a single proportion that you may see it simply called a "confidence interval for the proportion."

Just Checking

A Pew Research study regarding cell phones asked questions about cell phone experience. One growing concern is unsolicited advertising in the form of text messages. Pew asked cell phone owners, "Have you ever received unsolicited text messages on your cell phone from advertisers?" and 17% reported that they had. Pew estimates a 95% confidence interval to be 0.17 ± 0.04, or between 13% and 21%.

Are the following statements about people who have cell phones correct? Explain.

1. In Pew's sample, somewhere between 13% and 21% of respondents reported that they had received unsolicited advertising text messages.

2. We can be 95% confident that 17% of U.S. cell phone owners have received unsolicited advertising text messages.

3. We are 95% confident that between 13% and 21% of all U.S. cell phone owners

have received unsolicited advertising text messages.

4. We know that between 13% and 21% of all U.S. cell phone owners have received unsolicited advertising text messages.

5. 95% of all U.S. cell phone owners have received unsolicited advertising text messages.

What Does "95% Confidence" Really Mean?

What do we mean when we say we have 95% confidence that our interval contains the true proportion? Formally, what we mean is that "95% of samples of this size will produce confidence intervals that capture the true proportion." This is correct, but a little long winded, so we sometimes say, "we are 95% confident that the true proportion lies in our interval." Our uncertainty is about whether the particular sample we have at hand is one of the successful ones or one of the 5% that fail to produce an interval that captures the true value.

Back in Chapter 17 we saw that proportions vary from sample to sample. If other researchers select their own samples of sea fans, they'll also find some infected by the disease, but each person's sample proportion will almost certainly differ from ours. When they each try to estimate the true rate of infection in the entire population, they'll center *their* confidence intervals at the proportions they observed in their own samples. Each of us will end up with a different interval.

Our interval guessed the true proportion of infected sea fans to be between about 42% and 62%. Another researcher whose sample contained more infected fans than ours did might guess between 46% and 66%. Still another who happened to collect fewer infected fans might estimate the true proportion to be between 23% and 43%. And so on. Every possible sample would produce yet another confidence interval. Although wide intervals like these can't pin down the actual rate of infection very precisely, we expect that most of them should be winners, capturing the true value. Nonetheless, some will be duds, missing the population proportion entirely.

On the next page you'll see confidence intervals produced by simulating 20 different random samples. The red dots are the proportions of infected fans in each sample, and the blue segments show the confidence intervals found for each. The green line represents the true rate of infection in the population, so you can see that most of the intervals caught it—but a few missed. (And notice again that it is the *intervals* that vary from sample to sample; the green line doesn't move.)

Figure 18.3

The horizontal green line shows the true percentage of all sea fans that are infected. Most of the 20 simulated samples produced confidence intervals that captured the true value, but a few missed.

Of course, there's a huge number of possible samples that *could* be drawn, each with its own sample proportion. These are just some of them. Each sample proportion can be used to make a confidence interval. That's a large pile of possible confidence intervals, and ours is just one of those in the pile. Did *our* confidence interval "work"? We can never be sure, because we'll never know the true proportion of all the sea fans that are infected. However, the Central Limit Theorem assures us that 95% of the intervals in the pile are winners, covering the true value, and only 5% are duds. *That's* why we're 95% confident that our interval is a winner!

So, What *Can* I Say?

Technically, we should say, "I am 95% confident that the interval from 42.1% and 61.7% captures the true proportion of sea fans on the Las Redes Reef that are infected." That formal phrasing emphasizes that *our confidence (and our uncertainty) is about the interval, not the true proportion.* But you may choose a more casual phrasing like "I am 95% confident that between 42.1% and 61.7% of Las Redes sea fans are infected." Because you've made it clear that the uncertainty is yours and you didn't suggest that the randomness is in the true proportion, this is OK. Keep in mind that it's the interval that's random and is the focus of both our confidence and doubt.

For Example POLLS AND MARGIN OF ERROR

In April and May 2011, the Yale Project on Climate Change Communication and the George Mason University Center for Climate Change Communication interviewed 1010 U.S. adults about American's global warming beliefs and attitudes.[5]

QUESTION: It is standard among pollsters to use a 95% confidence level unless otherwise stated. Given that, what do these researchers mean by their confidence interval in this context?

ANSWER: *If this polling were done repeatedly, 95% of all random samples would yield confidence intervals that contain the true proportion of all U.S. adults who believe that there's a lot of disagreement among scientists about global warming.*

---

[5]Among their questions, they asked what respondents thought were the views of scientists. Among respondents, 40% agreed with the alternative "There is a lot of disagreement among scientists about whether or not global warming is happening." The investigators provide a confidence interval from 37% to 43%.

Margin of Error: Certainty vs. Precision

We've just claimed that with a certain confidence we've captured the true proportion of all infected sea fans. Our confidence interval had the form

$$\hat{p} \pm 2\,SE(\hat{p}).$$

The extent of the interval on either side of $\hat{p}$ is called the **margin of error** (*ME*).

We'll want to use the same approach for many other situations besides estimating proportions. In fact, almost any population parameter—a proportion, a mean, or a regression slope, for example—can be estimated with some margin of error. The margin of error is a way to describe our uncertainty in estimating the population value. We'll see how to find a margin of error for each of these values and for others.

For all of those statistics, regardless of how we calculate the margin of error, we'll be able to construct a confidence interval that looks like this:

$$\textit{Estimate} \pm \textit{ME}.$$

The margin of error for our 95% confidence interval was 2 *SE*. What if we wanted to be more confident? To be more confident, we'll need to capture *p* more often, and to do that we'll need to make the interval wider. For example, if we want to be 99.7% confident, the margin of error will have to be 3 *SE*.

Figure 18.4

Reaching out 3 SEs on either side of $\hat{p}$ makes us 99.7% confident we'll trap the true proportion *p*. Compare with Figure 18.2.

NEW!! ACME *p*-trap: Guaranteed* to capture *p*. *Now with 99.7% confidence! IMPROVED!!

$\hat{p} - 3\,SE$ $\hat{p}$ $\hat{p} + 3\,SE$

A S *Activity*: **Balancing Precision and Certainty.** What percent of parents expect their kids to pay for college with a student loan? Investigate the balance between the precision and the certainty of a confidence interval.

The more confident we want to be, the larger the margin of error must be. We can be 100% confident that the proportion of infected sea fans is between 0% and 100%, but this isn't likely to be very useful. On the other hand, we could give a confidence interval from 51.8% to 52.0%, but we can't be very confident about a precise statement like this. Every confidence interval is a balance between certainty and precision.

The tension between certainty and precision is always there. Fortunately, in most cases we can be both sufficiently certain and sufficiently precise to make useful statements. There is no simple answer to the conflict. You must choose a confidence level yourself. The data can't do it for you. The choice of confidence level is somewhat arbitrary. The most commonly chosen confidence levels are 90%, 95%, and 99%, but any percentage can be used. (In practice, though, using something like 92.9% or 97.2% is likely to make people think you're up to something.)

Garfield © 1999 Paws, Inc. Distributed by Universal Uclick. Reprinted with permission. All rights reserved.

For Example FINDING THE MARGIN OF ERROR (TAKE 1)

RECAP: An April 2011 Yale/George Mason poll of 1010 U.S. adults asking questions about current topics reported a margin of error of 3%. It is a convention among pollsters to use a 95% confidence level and to report the "worst case" margin of error, based on $p = 0.5$.

QUESTION: How did the researchers calculate their margin of error?

ANSWER: Assuming $p = 0.5$, for random samples of $n = 1010$,

$$SD(\hat{p}) = \sqrt{\frac{pq}{n}} = \sqrt{\frac{(0.5)(0.5)}{1010}} = 0.0157.$$

For a 95% confidence level, $ME = 2(0.0157) = 0.031$, so their margin of error is just a bit over 3%.

Critical Values

■ **NOTATION ALERT**

We'll put an asterisk on a letter to indicate a critical value, so z^* is always a critical value from a normal model.

In our sea fans example we used $2SE$ to give us a 95% confidence interval. To change the confidence level, we'd need to change the *number* of SEs so that the size of the margin of error corresponds to the new level. This number of SEs is called the **critical value.** Here it's based on the Normal model, so we denote it z^*. For any confidence level, we can find the corresponding critical value from a computer, a calculator, or a Normal probability table, such as Table Z.

For a 95% confidence interval, you'll find the precise critical value is $z^* = 1.96$. That is, 95% of a Normal model is found within ± 1.96 standard deviations of the mean. We've been using $z^* = 2$ from the 68–95–99.7 Rule because it's easy to remember.

Figure 18.5

For a 90% confidence interval, the critical value is 1.645, because, for a Normal model, 90% of the values are within 1.645 standard deviations from the mean.

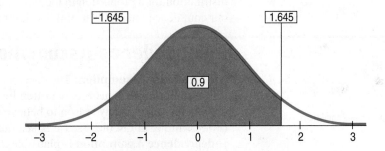

For Example FINDING THE MARGIN OF ERROR (TAKE 2)

RECAP: In April 2011, a Yale/George Mason poll of 1010 U.S. adults found that 40% of the respondents believed that scientists disagreed about whether global warming exists. They reported a 95% confidence interval with a margin of error of 3%.

QUESTIONS: Using the critical value of z and the standard error based on the observed proportion, what would be the margin of error for a 90% confidence interval? What's good and bad about this change?

$$\text{With } n = 1010 \text{ and } \hat{p} = 0.40, \quad SE(\hat{p}) = \sqrt{\frac{\hat{p}\hat{q}}{n}} = \sqrt{\frac{(0.40)(0.60)}{1010}} = 0.0154$$

For a 90% confidence level, $z^* = 1.645$, so $ME = 1.645(0.0154) = 0.0254$.

ANSWER: Now the margin of error is only about 2.5%, producing a narrower interval. What's good about the change is that we now have a smaller interval, but what's bad is that we are less certain that the interval actually contains the true proportion of adults who think that scientists disagree about global warming.

Just Checking

Think some more about the 95% confidence interval originally created for the proportion of U.S. adults who believe that scientists disagree about whether global warming exists.

6. If the researchers wanted to be 98% confident, would their confidence interval need to be wider or narrower?

7. The study's margin of error was about 3%. If the researchers wanted to reduce it to 2%, would their level of confidence be higher or lower?

8. If the researchers had polled more people, would the interval's margin of error have been larger or smaller?

Assumptions and Conditions

We've just made some pretty sweeping statements about sea fans. Those statements were possible because we used a Normal model for the sampling distribution. But is that model appropriate?

We've also said that the same basic ideas will work for other statistics. One difference for those statistics is that some will have sampling distribution models that are different than the Normal. But the background theory (which we won't bother with) is so similar for all of those models that they share the same basic assumptions and conditions about independence and sample size. Then for each statistic, we usually tack on a special-case assumption or two. We'll deal with those as we get to them in later chapters.

We saw the assumptions and conditions for using the Normal to model the sampling distribution for a proportion in the last chapter. Because they are so crucial to making sure our confidence interval is useful, we'll repeat them here.

Independence Assumption

A S *Activity*: **Assumptions and Conditions.** Here's an animated review of the assumptions and conditions.

Independence Assumption: The data values must be independent. To think about whether this assumption is plausible, we often look for reasons to suspect that it fails. We wonder whether there is any reason to believe that the data values somehow affect each other. (For example, might the disease in sea fans be contagious?) Whether you decide that the **Independence Assumption** is plausible depends on your knowledge of the situation. It's not one you can check by looking at the data.

However, now that we have data, there are two conditions that we can check:

Randomization Condition: Were the data sampled at random or generated from a properly randomized experiment? Proper randomization can help ensure independence.

10% Condition: If you sample more than 10% of a population, the formula for the standard error won't be quite right. There is a special formula (found in advanced books) that corrects for this, but it isn't a common problem unless your population is small.

Sample Size Assumption

The model we use for inference for proportions is based on the Central Limit Theorem. We need to know whether the sample is large enough to make the sampling model for the sample proportions approximately Normal. It turns out that we need more data as the proportion gets closer and closer to either extreme (0 or 1). That's why we check the:

Success/Failure Condition: We must expect at least 10 "successes" and at least 10 "failures." Recall that by tradition we arbitrarily label one alternative (usually the outcome being counted) as a "success" even if it's something bad (like getting a disease). The other alternative is, of course, then a "failure."

Activity: A Confidence Interval for p. View the video story of pollution in Chesapeake Bay, and make a confidence interval for the analysis with the interactive tool.

> **One-Proportion z-Interval**
>
> When the conditions are met, we are ready to find a level C confidence interval for the population proportion, p. The confidence interval is $\hat{p} \pm z^* \times SE(\hat{p})$ where the standard deviation of the proportion is estimated by $SE(\hat{p}) = \sqrt{\dfrac{\hat{p}\hat{q}}{n}}$ and the critical value, z^*, specifies the number of SEs needed for C% of random samples to yield confidence intervals that capture the true parameter value.

Step-by-Step Example A CONFIDENCE INTERVAL FOR A PROPORTION

In October 2010, the Gallup Poll[6] asked 510 randomly sampled adults the question "Generally speaking, do you believe the death penalty is applied fairly or unfairly in this country today?" Of these, 58% answered "Fairly," 36% said "Unfairly," and 7% said they didn't know. (Percentages add up to 101% due to rounding).

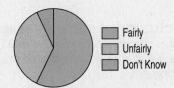

Who	Adults in the United States
What	Response to a question about the death penalty
When	October 2010
Where	United States
How	510 adults were randomly sampled and asked by the Gallup Poll
Why	Public opinion research

Question: From this survey, what can we conclude about the opinions of *all* adults?

To answer this question, we'll build a confidence interval for the proportion of all U.S. adults who believe the death penalty is applied fairly. There are four steps to building a confidence interval for proportions: Plan, Model, Mechanics, and Conclusion.

THINK ➡ **Plan** State the problem and the W's.

Identify the *parameter* you wish to estimate.

Identify the *population* about which you wish to make statements.

Choose and state a confidence level.

Model Think about the assumptions and check the conditions.

I want to find an interval that is likely, with 95% confidence, to contain the true proportion, p, of U.S. adults who think the death penalty is applied fairly. I have a random sample of 510 U.S. adults.

✓ **Randomization Condition:** Gallup drew a random sample from all U.S. adults. I can be confident that the respondents are independent.

✓ **10% Condition:** The sample is certainly less than 10% of the population.

✓ **Success/Failure Condition:**
$n\hat{p} = 510(58\%) = 296 \geq 10$ and
$n\hat{q} = 510(42\%) = 214 \geq 10$,
so the sample appears to be large enough to use the Normal model.

State the sampling distribution model for the statistic.

Choose your method.

The conditions are satisfied, so I can use a Normal model to find a **one-proportion z-interval**.

[6]www.gallup.com/poll/1606/death-penalty.aspx

SHOW ➡ **Mechanics** Construct the confidence interval.

First find the standard error. (Remember: It's called the "standard error" because we don't know p and have to use $\hat{p}$ instead.)

Next find the margin of error. We could informally use 2 for our critical value, but 1.96 (found from a table or technology) is more accurate.

Write the confidence interval (CI).

REALITY CHECK ➡ The CI is centered at the sample proportion and the width seems reasonable for a sample of 500.

$n = 510, \hat{p} = 0.58$, so

$$SE(\hat{p}) = \sqrt{\frac{\hat{p}\hat{q}}{n}} = \sqrt{\frac{(0.58)(0.42)}{510}} = 0.022.$$

Because the sampling model is Normal, for a 95% confidence interval, the critical value $z^* = 1.96$.

The margin of error is

$$ME = z^* \times SE(\hat{p}) = 1.96(0.022) = 0.043.$$

So the 95% confidence interval is

$$0.58 \pm 0.043 \text{ or } (0.537, 0.623).$$

TELL ➡ **Conclusion** Interpret the confidence interval in the proper context. We're 95% confident that our interval captured the true proportion.

I am 95% confident that between 53.7% and 62.3% of all U.S. adults think that the death penalty is applied fairly.

TI Tips FINDING CONFIDENCE INTERVALS

It will come as no surprise that your TI can calculate a confidence interval for a population proportion. Remember the sea fans? Of 104 sea fans, 54 were diseased. To find the resulting confidence interval, we first take a look at a whole new menu.

- Under STAT go to the TESTS menu. Quite a list! Commands are found here for the inference procedures you will learn through the coming chapters.

- We're using a Normal model to find a confidence interval for a proportion based on one sample. Scroll down the list and select 1-PropZInt.
- Enter the number of successes observed and the sample size.
- Specify a confidence level and then Calculate.

And there it is! Note that the TI calculates the sample proportion for you, but the important result is the interval itself, 42% to 62%. The calculator did the easy part—just Show. Tell is harder. It's your job to interpret that interval correctly.

Beware: You may run into a problem. When you enter the value of x, you need a *count*, not a percentage. Suppose the marine scientists had reported that 52% of the 104 sea fans were infected. You can enter x:.52*104, and the calculator will evaluate that as 54.08. Wrong. Unless you fix that result, you'll get an error message. Think about it—the number of infected sea fans must have been a whole number, evidently 54. When the scientists reported the results, they rounded off the actual percentage $(54 \div 104 = 51.923\%)$ to 52%. Simply change the value of x to 54 and you should be able to Calculate the correct interval.

Choosing Your Sample Size

The question of how large a sample to take is an important step in planning any study. We weren't ready to make that calculation when we first looked at study design in Chapter 11, but now we can—and we always should.

Suppose a candidate is planning a poll and wants to estimate voter support within 3% with 95% confidence. How large a sample does she need?

Let's look at the margin of error:

$$ME = z^* \sqrt{\frac{\hat{p}\hat{q}}{n}}$$

$$0.03 = 1.96 \sqrt{\frac{\hat{p}\hat{q}}{n}}.$$

We want to find n, the sample size. To find n we need a value for $\hat{p}$. We don't know $\hat{p}$ because we don't have a sample yet, but we can probably guess a value. The worst case— the value that makes $\hat{p}\hat{q}$ (and therefore n) largest—is 0.50, so if we use that value for $\hat{p}$, we'll certainly be safe. Our candidate probably expects to be near 50% anyway.

Our equation, then, is

$$0.03 = 1.96 \sqrt{\frac{(0.5)(0.5)}{n}}.$$

To solve for n, we first multiply both sides of the equation by $\sqrt{n}$ and then divide by 0.03:

$$0.03\sqrt{n} = 1.96\sqrt{(0.5)(0.5)}$$

$$\sqrt{n} = \frac{1.96\sqrt{(0.5)(0.5)}}{0.03} \approx 32.67$$

Notice that evaluating this expression tells us the *square root* of the sample size. We need to square that result to find n:

$$n \approx (32.67)^2 \approx 1067.1$$

To be safe, we round up and conclude that we need at least 1068 respondents to keep the margin of error as small as 3% with a confidence level of 95%.

> **What Do I Use Instead of $\hat{p}$?**
>
> Often we have an estimate of the population proportion based on experience or perhaps a previous study. If so, use that value as $\hat{p}$ in calculating what size sample you need. If not, the cautious approach is to use $p = 0.5$ in the sample size calculation; that will determine the largest sample necessary regardless of the true proportion.

For Example CHOOSING A SAMPLE SIZE

RECAP: The Yale/George Mason poll that estimated that 40% of all voters believed that scientists disagree about whether global warming exists had a margin of error of ±3%. Suppose an environmental group planning a follow-up survey of voters' opinions on global warming wants to determine a 95% confidence interval with a margin of error of no more than ±2%.

QUESTION: How large a sample do they need? (You could take $p = 0.5$, but we have data that indicate $p = 0.40$, so we can use that.)

ANSWER:

$$ME = z^* \sqrt{\frac{\hat{p}\hat{q}}{n}}$$

$$0.02 = 1.96 \sqrt{\frac{(0.40)(0.60)}{n}}$$

$$\sqrt{n} = \frac{1.96\sqrt{(0.40)(0.60)}}{0.02} \approx 48.01$$

$$n = 48.01^2 = 2304.96$$

The environmental group's survey will need at least 2305 respondents.

How Big Should a Margin of Error Be?
Public opinion polls often sample 1000 people, which gives an ME of 3% when $p = 0.5$. But businesses and nonprofit organizations typically use much larger samples to estimate the proportion who will accept a direct mail offer. Why? Because that proportion is very low— often far below 5%. An ME of 3% wouldn't be precise enough. An ME like 0.1% would be more useful, and that requires a very large sample size.

Unfortunately, bigger samples cost more money and more effort. Because the standard error declines only with the *square root* of the sample size, to cut the standard error (and thus the ME) in half, we must *quadruple* the sample size.

Generally a margin of error of 5% or less is acceptable, but different circumstances call for different standards. For a pilot study, a margin of error of 10% may be fine, so a sample of 100 will do quite well. In a close election, a polling organization might want to get the margin of error down to 2%. Drawing a large sample to get a smaller ME, however, can run into trouble. It takes time to survey 2400 people, and a survey that extends over a week or more may be trying to hit a target that moves during the time of the survey. An important event can change public opinion in the middle of the survey process.

Keep in mind that the sample size for a survey is the number of respondents, not the number of people to whom questionnaires were sent or whose phone numbers were dialed. And also keep in mind that a low response rate turns any study essentially into a voluntary response study, which is of little value for inferring population values. It's almost always better to spend resources on increasing the response rate than on surveying a larger group. A full or nearly full response by a modest-size sample can yield useful results.

Surveys are not the only place where proportions pop up. Banks sample huge mailing lists to estimate what proportion of people will accept a credit card offer. Even pilot studies may mail offers to over 50,000 customers. Most don't respond; that doesn't make the sample smaller—they simply said "No thanks." Those who do respond want the card. To the bank, the response rate[7] is $\hat{p}$. With a typical success rate around 0.5%, the bank needs a very small margin of error—often as low as 0.1%—to make a sound business decision. That calls for a large sample, and the bank must take care in estimating the size needed. For our election poll calculation we used $p = 0.5$, both because it's safe and because we honestly believed p to be near 0.5. If the bank used 0.5, they'd get an absurd answer. Instead, they base their calculation on a proportion closer to the one they expect to find.

For Example SAMPLE SIZE REVISITED

A credit card company is about to send out a mailing to test the market for a new credit card. From that sample, they want to estimate the true proportion of people who will sign up for the card nationwide. A pilot study suggests that about 0.5% of the people receiving the offer will accept it.

QUESTION: To be within a tenth of a percentage point (0.001) of the true rate with 95% confidence, how big does the test mailing have to be?

ANSWER: Using the estimate $\hat{p} = 0.5\%$:

$$ME = 0.001 = z^* \sqrt{\frac{\hat{p}\hat{q}}{n}} = 1.96 \sqrt{\frac{(0.005)(0.995)}{n}}$$

$$(0.001)^2 = 1.96^2 \frac{(0.005)(0.995)}{n} \Rightarrow n = \frac{1.96^2(0.005)(0.995)}{(0.001)^2}$$

$$= 19,111.96 \text{ or } 19,112$$

That's a lot, but it's actually a reasonable size for a trial mailing such as this. Note, however, that if they had assumed 0.50 for the value of p, they would have found

$$ME = 0.001 = z^* \sqrt{\frac{pq}{n}} = 1.96 \sqrt{\frac{(0.5)(0.5)}{n}}$$

$$(0.001)^2 = 1.96^2 \frac{(0.5)(0.5)}{n} \Rightarrow n = \frac{1.96^2(0.5)(0.5)}{(0.001)^2} = 960,400.$$

Quite a different (and unreasonable) result.

---

[7]In marketing studies every mailing yields a response—"yes" or "no"—and "response rate" means the proportion of customers who accept an offer. That's not the way we use the term for survey response.

WHAT IF ●●● a confidence interval is "wrong"?

Before every election pollsters try to predict what will happen when the voters finally make their decision. They predict the percentage of the vote that a candidate will get, and they state the poll's margin of error. It's impressive that they're usually right. But when they're wrong, commentators speculate about why the poll was mistaken. They suggest that voters changed their minds at the last minute. Or that the voter turnout was lower (or higher) than anticipated. Or that the poll failed to contact some group of voters. And so on. It's rare to hear anyone recognize that the pollsters could have done everything correctly and still been wrong in their prediction.

Let's examine the process by—yes—simulation. To start, we created an imaginary population of 10,000 voters, of whom 52% will vote for Candidate A. Of course, that outcome won't actually be known until Election Day.

We imagine trying to predict what will happen in advance by polling 600 randomly selected voters. In our first simulated sample, 331 of the 600 people chosen favored Candidate A. That predicts A will receive 55% of the vote, with a margin of error of $\pm 4\%$.

On Election Day, Candidate A gets only 52% of the vote, but does win. So was this poll right or wrong? What we care about as statisticians is that the 95% confidence interval of 55% $\pm$ 4% (from 51% to 59%) correctly captured the actual outcome of 52%.

We tried this repeatedly, for 100 simulated samples. Most turned out like the first one, leading to intervals that work. But not all. Here's a table showing a few of the results.

Sample #	Respondents favoring A	Predicted percentage	Confidence interval (95%)	Did it capture the true 52%?
1	331	55.2%	51.2% to 59.1%	Yes
2	302	50.3%	46.3% to 54.3%	Yes
23	286	47.7%	43.7% to 51.7%	**No**
81	291	48.5%	44.5% to 52.5%	Yes
83	339	56.5%	52.5% to 60.5%	**No**

Newspeople and the public at large would be critical of results like those generated by simulated samples 23 and 81. In each of those cases the poll suggests that Candidate A should get less than 50% of the vote, and thus lose. When A wins on Election Day, those folks will say the poll was wrong. But we're statisticians, so we look at the margin of error. Sample #81 actually produced a successful confidence interval, because the true result of 52% is between 44.5% and 52.5%. But Sample #23's interval did miss the truth. And so did Sample #83. Sure, that one picked Candidate A as the winner, but its confidence interval does not contain the actual 52% of the vote that A got.

What went wrong in these samples? *Nothing!* Our simulation did everything properly: we selected each sample at random, the responses were independent, and the sample was large but less than 10% of the population. Nonetheless, two of the simulated confidence intervals shown in the table plus 2 others we didn't show you failed to capture the population parameter of 52%. That's only 4 failures in 100 attempts. Why "only"? Because the intervals we created have a 95% level of confidence, in 100 samples we'd expect 95 successes and 5 failures.

The bottom line: Usually confidence intervals work, but not always. Will a confidence interval *you* create be right? Probably. But you can't know for sure. You can only be 95% confident.

WHAT CAN GO WRONG?

Confidence intervals are powerful tools. Not only do they tell what we know about the parameter value, but—more important—they also tell what we *don't* know. In order to use confidence intervals effectively, you must be clear about what you say about them.

Don't Misstate What the Interval Means

- **Don't suggest that the parameter varies.** A statement like "There is a 95% chance that the true proportion is between 42.7% and 51.3%" sounds as though you think the population proportion wanders around and sometimes happens to fall between 42.7% and 51.3%. When you interpret a confidence interval, make it clear that *you* know that the population parameter is fixed and that it is the interval that varies from sample to sample.

- **Don't claim that other samples will agree with yours.** Keep in mind that the confidence interval makes a statement about the true population proportion. An interpretation such as "In 95% of samples of U.S. adults, the proportion who think marijuana should be decriminalized will be between 42.7% and 51.3%" is just wrong. The interval isn't about sample proportions but about the population proportion.

- **Don't be certain about the parameter.** Saying "Between 42.1% and 61.7% of sea fans are infected" asserts that the population proportion cannot be outside that interval. Of course, we can't be absolutely certain of that. (Just pretty sure.)

- **Don't forget: It's about the parameter.** Don't say, "I'm 95% confident that $\hat{p}$ is between 42.1% and 61.7%." Of course you are—in fact, we calculated that $\hat{p} = 51.9\%$ of the fans in our sample were infected. So we already *know* the sample proportion. The confidence interval is about the (unknown) population parameter, p.

- **Don't claim to know too much.** Don't say, "I'm 95% confident that between 42.1% and 61.7% of all the sea fans in the world are infected." You didn't sample from all 500 species of sea fans found in coral reefs around the world. Just those of this type on the Las Redes Reef.

- **Do take responsibility.** Confidence intervals are about *uncertainty. You* are the one who is uncertain, not the parameter. You have to accept the responsibility and consequences of the fact that not all the intervals you compute will capture the true value. In fact, about 5% of the 95% confidence intervals you find will fail to capture the true value of the parameter. You *can* say, "I am 95% confident that between 42.1% and 61.7% of the sea fans on the Las Redes Reef are infected."[8]

- **Do treat the whole interval equally.** Although a confidence interval is a set of plausible values for the parameter, don't think that the values in the middle of a confidence interval are somehow "more plausible" than the values near the edges. Your interval provides no information about where in your current interval (if at all) the parameter value is most likely to be hiding.

Beware of a Margin of Error Too Large to Be Useful

We know we can't be exact, but how precise do we need to be? A confidence interval that says that the percentage of infected sea fans is between 10% and 90% wouldn't be of much use. Most likely, you have some sense of how large a margin of error you can tolerate. What can you do?

One way to make the margin of error smaller is to reduce your level of confidence. But that may not be a useful solution. It's a rare study that reports confidence levels lower than 80%. Levels of 95% or 99% are more common.

The time to think about whether your margin of error is small enough to be useful is when you design your study. Don't wait until you compute your confidence interval. To get a narrower interval without giving up confidence, you need to have less variability in your sample proportion. How can you do that? Choose a larger sample.

---

[8]When we are being very careful we say, "95% of samples of this size will produce confidence intervals that capture the true proportion of infected sea fans on the Las Redes Reef."

Look for Violations of Assumptions

Confidence intervals and margins of error are often reported along with poll results and other analyses. But it's easy to misuse them and wise to be aware of other ways things can go wrong.

■ **Watch out for biased sampling.** Don't forget about the potential sources of bias in surveys that we discussed in Chapter 12. Just because we have more statistical machinery now doesn't mean we can forget what we've already learned. A questionnaire that finds that 85% of people enjoy filling out surveys still suffers from nonresponse bias even though now we're able to put confidence intervals around this (biased) estimate.

■ **Think about independence.** The assumption that the values in our sample are mutually independent is one that we usually cannot check. It always pays to think about it, though. For example, the disease affecting the sea fans might be contagious, so that fans growing near a diseased fan are more likely themselves to be diseased. Such contagion would violate the Independence Assumption and could severely affect our sample proportion. It could be that the proportion of infected sea fans on the entire reef is actually quite small, and the researchers just happened to find an infected area. To avoid this, the researchers should be careful to sample sites far enough apart to make contagion unlikely.

What Have We Learned?

We've learned to construct a confidence interval for a proportion, p, as the statistic, $\hat{p}$, plus and minus a margin of error.
- ■ The margin of error consists of a critical value based on the sampling model times a standard error based on the sample.
- ■ The critical value is found from the Normal model.
- ■ The standard error of a sample proportion is calculated as $\sqrt{\dfrac{\hat{p}\hat{q}}{n}}$.

We've learned to interpret a confidence interval correctly.
- ■ You can claim to have the specified level of confidence that the interval you have computed actually covers the true value.

We've come to understand the relationship of the sample size, n, to both the certainty (confidence level) and precision (margin of error).
- ■ For the same sample size and true population proportion, more certainty means less precision (wider interval) and more precision (narrower interval) implies less certainty.

We've learned to check the assumptions and conditions for finding and interpreting confidence intervals.
- ■ Independence Assumption or Randomization Condition
- ■ 10% Condition
- ■ Success/Failure Condition

We've learned to find the sample size required, given a proportion, a confidence level, and a desired margin of error.

Terms

Standard error   When we estimate the standard deviation of a sampling distribution using statistics found from the data, the estimate is called a standard error.

$$SE(\hat{p}) = \sqrt{\dfrac{\hat{p}\hat{q}}{n}}. \quad \text{(p. 474)}$$

Confidence interval   A level C confidence interval for a model parameter is an interval of values usually of the form

estimate ± *margin of error*

found from data in such a way that C% of all random samples will yield intervals that capture the true parameter value. (p. 475)

One-proportion z-interval   A confidence interval for the true value of a proportion. The confidence interval is

$$\hat{p} \pm z^*SE(\hat{p}),$$

where z^* is a critical value from the Standard Normal model corresponding to the specified confidence level. (p. 475)

Margin of error   In a confidence interval, the extent of the interval on either side of the observed statistic value is called the margin of error. A margin of error is typically the product of a critical value from the sampling distribution and a standard error from the data. A small margin of error corresponds to a confidence interval that pins down the parameter precisely. A large margin of error corresponds to a confidence interval that gives relatively little information about the estimated parameter. For a proportion,

$$ME = z^*\sqrt{\frac{\hat{p}\hat{q}}{n}}.\quad \text{(p. 478)}$$

Critical value   The number of standard errors to move away from the sample statistic to specify an interval that corresponds to the specified level of confidence. The critical value, denoted z^*, is usually found from a table or with technology. (p. 479)

On the Computer **CONFIDENCE INTERVALS FOR PROPORTIONS**

Confidence intervals for proportions are so easy and natural that many statistics packages don't offer special commands for them. Most statistics programs want the "raw data" for computations. For proportions, the raw data are the "success" and "failure" status for each case. Usually, these are given as 1 or 0, but they might be category names like "yes" and "no." Other software and graphing calculators allow you to create confidence intervals from summaries of the data—all you need to enter are the number of successes and the sample size.

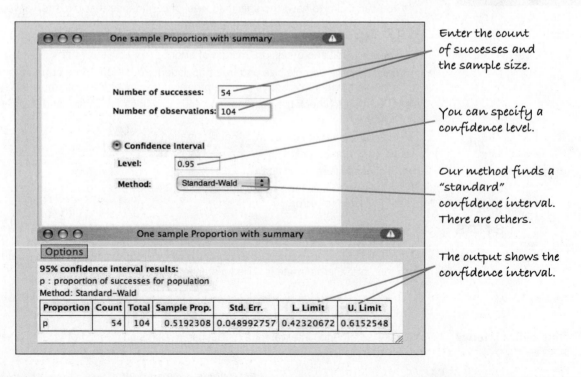

Enter the count of successes and the sample size.

You can specify a confidence level.

Our method finds a "standard" confidence interval. There are others.

The output shows the confidence interval.

95% confidence interval results:
p : proportion of successes for population
Method: Standard–Wald

Proportion	Count	Total	Sample Prop.	Std. Err.	L. Limit	U. Limit
p	54	104	0.5192308	0.048992757	0.42320672	0.6152548

Exercises

1. **Margin of error** A TV newscaster reports the results of a poll of voters, and then says, "The margin of error is plus or minus 4%." Explain carefully what that means.

2. **Margin of error** A medical researcher estimates the percentage of children exposed to lead-based paint, adding that he believes his estimate has a margin of error of about 3%. Explain what the margin of error means.

3. **Conditions** For each situation described below, identify the population and the sample, explain what p and $\hat{p}$ represent, and tell whether the methods of this chapter can be used to create a confidence interval.

    a) Police set up an auto checkpoint at which drivers are stopped and their cars inspected for safety problems. They find that 14 of the 134 cars stopped have at least one safety violation. They want to estimate the percentage of all cars that may be unsafe.

    b) A TV talk show asks viewers to register their opinions on prayer in schools by logging on to a website. Of the 602 people who voted, 488 favored prayer in schools. We want to estimate the level of support among the general public.

    c) A school is considering requiring students to wear uniforms. The PTA surveys parent opinion by sending a questionnaire home with all 1245 students; 380 surveys are returned, with 228 families in favor of the change.

    d) A college admits 1632 freshmen one year, and four years later 1388 of them graduate on time. The college wants to estimate the percentage of all their freshman enrollees who graduate on time.

4. **More conditions** Consider each situation described. Identify the population and the sample, explain what p and $\hat{p}$ represent, and tell whether the methods of this chapter can be used to create a confidence interval.

    a) A consumer group hoping to assess customer experiences with auto dealers surveys 167 people who recently bought new cars; 3% of them expressed dissatisfaction with the salesperson.

    b) What percent of college students have cell phones? 2883 students were asked as they entered a football stadium, and 243 said they had phones with them.

    c) 240 potato plants in a field in Maine are randomly checked, and only 7 show signs of blight. How severe is the blight problem for the U.S. potato industry?

    d) 12 of the 309 employees of a small company suffered an injury on the job last year. What can the company expect in future years?

5. **Conclusions** A catalog sales company promises to deliver orders placed on the Internet within 3 days. Follow-up calls to a few randomly selected customers show that

a 95% confidence interval for the proportion of all orders that arrive on time is 88% $\pm$ 6%. What does this mean? Are these conclusions correct? Explain.

    a) Between 82% and 94% of all orders arrive on time.

    b) 95% of all random samples of customers will show that 88% of orders arrive on time.

    c) 95% of all random samples of customers will show that 82% to 94% of orders arrive on time.

    d) We are 95% sure that between 82% and 94% of the orders placed by the sampled customers arrived on time.

    e) On 95% of the days, between 82% and 94% of the orders will arrive on time.

6. **More conclusions** In January 2002, two students made worldwide headlines by spinning a Belgian euro 250 times and getting 140 heads—that's 56%. That makes the 90% confidence interval (51%, 61%). What does this mean? Are these conclusions correct? Explain.

    a) Between 51% and 61% of all euros are unfair.

    b) We are 90% sure that in this experiment this euro landed heads on between 51% and 61% of the spins.

    c) We are 90% sure that spun euros will land heads between 51% and 61% of the time.

    d) If you spin a euro many times, you can be 90% sure of getting between 51% and 61% heads.

    e) 90% of all spun euros will land heads between 51% and 61% of the time.

7. **Confidence intervals** Several factors are involved in the creation of a confidence interval. Among them are the sample size, the level of confidence, and the margin of error. Which statements are true?

    a) For a given sample size, higher confidence means a smaller margin of error.

    b) For a specified confidence level, larger samples provide smaller margins of error.

    c) For a fixed margin of error, larger samples provide greater confidence.

    d) For a given confidence level, halving the margin of error requires a sample twice as large.

8. **Confidence intervals, again** Several factors are involved in the creation of a confidence interval. Among them are the sample size, the level of confidence, and the margin of error. Which statements are true?

    a) For a given sample size, reducing the margin of error will mean lower confidence.

    b) For a certain confidence level, you can get a smaller margin of error by selecting a bigger sample.

    c) For a fixed margin of error, smaller samples will mean lower confidence.

    d) For a given confidence level, a sample 9 times as large will make a margin of error one third as big.

9. Cars What fraction of cars is made in Japan? The computer output below summarizes the results of a random sample of 50 autos. Explain carefully what it tells you.

z-Interval for proportion
With 90.00% confidence,
0.29938661 < p(japan) < 0.46984416

10. Parole A study of 902 decisions made by the Nebraska Board of Parole produced the following computer output. Assuming these cases are representative of all cases that may come before the Board, what can you conclude?

z-Interval for proportion
With 95.00% confidence,
0.56100658 < p(parole) < 0.62524619

11. Mislabeled seafood In December 2011, Consumer Reports published their study of labeling of seafood sold in New York, New Jersey, and Connecticut. They purchased 190 pieces of seafood from various kinds of food stores and restaurants in the three states and genetically compared the pieces to standard gene fragments that can identify the species. Laboratory results indicated that 22% of these packages of seafood were mislabeled, incompletely labeled, or misidentified by store or restaurant employees.

a) Construct a 95% confidence interval for the proportion of all seafood packages in those three states that are mislabeled or misidentified.

b) Explain what your confidence interval says about seafood sold in these three states.

c) A 2009 report by the Government Accountability Board says that the Food and Drug Administration has spent very little time recently looking for seafood fraud. Suppose an official said, "That's only190 packages out of the billions of pieces of seafood sold in a year. With the small number tested, I don't know that one would want to change one's buying habits." (An official was quoted similarly in a different but similar context). Is this argument valid? Explain.

12. Mislabeled seafood, second course The December 2011 Consumer Reports study described in Exercise 11 also found that 12 of the 22 "red snapper" packages tested were a different kind of fish.

a) Are the conditions for creating a confidence interval satisfied? Explain.

b) Construct a 95% confidence interval.

c) Explain what your confidence interval says about "red snapper" sold in these three states.

13. Baseball fans In a national poll taken in February 2008, Gallup asked 1006 adults whether they were baseball fans; 43% said they were. Two months previously, in December 2007, 40% of a similar-size sample had reported being baseball fans.

a) Find the margin of error for the 2008 poll if we want 90% confidence in our estimate of the percent of national adults who are baseball fans.

b) Explain what that margin of error means.

c) If we wanted to be 99% confident, would the margin of error be larger or smaller? Explain.

d) Find that margin of error.

e) In general, if all other aspects of the situation remain the same, will smaller margins of error produce greater or less confidence in the interval?

14. Lying about age Pew Research, in November 2011, polled a random sample of 799 U.S. teens about Internet use. 49% of those teens reported that they had misrepresented their age online to gain access to websites and online services.

a) Find the margin of error for this poll if we want 95% confidence in our estimate of the percent of American teens who have misrepresented their age online.

b) Explain what that margin of error means.

c) If we only need to be 90% confident, will the margin of error be larger or smaller? Explain.

d) Find that margin of error.

e) In general, if all other aspects of the situation remain the same, would smaller samples produce smaller or larger margins of error?

15. Contributions, please The Paralyzed Veterans of America is a philanthropic organization that relies on contributions. They send free mailing labels and greeting cards to potential donors on their list and ask for a voluntary contribution. To test a new campaign, they recently sent letters to a random sample of 100,000 potential donors and received 4781 donations.

a) Give a 95% confidence interval for the true proportion of their entire mailing list who may donate.

b) A staff member thinks that the true rate is 5%. Given the confidence interval you found, do you find that percentage plausible?

16. Take the offer First USA, a major credit card company, is planning a new offer for their current cardholders. The offer will give double airline miles on purchases for the next 6 months if the cardholder goes online and registers for the offer. To test the effectiveness of the campaign, First USA recently sent out offers to a random sample of 50,000 cardholders. Of those, 1184 registered.

a) Give a 95% confidence interval for the true proportion of those cardholders who will register for the offer.

b) If the acceptance rate is only 2% or less, the campaign won't be worth the expense. Given the confidence interval you found, what would you say?

17. Teenage drivers An insurance company checks police records on 582 accidents selected at random and notes that teenagers were at the wheel in 91 of them.

a) Create a 95% confidence interval for the percentage of all auto accidents that involve teenage drivers.

b) Explain what your interval means.

c) Explain what "95% confidence" means.

d) A politician urging tighter restrictions on drivers' licenses issued to teens says, "In one of every five auto accidents, a teenager is behind the wheel." Does your confidence interval support or contradict this statement? Explain.

18. Junk mail Direct mail advertisers send solicitations (a.k.a. "junk mail") to thousands of potential customers in the hope that some will buy the company's product. The acceptance rate is usually quite low. Suppose a company wants to test the response to a new flyer, and sends it to 1000 people randomly selected from their mailing list of over 200,000 people. They get orders from 123 of the recipients.

a) Create a 90% confidence interval for the percentage of people the company contacts who may buy something.

b) Explain what this interval means.

c) Explain what "90% confidence" means.

d) The company must decide whether to now do a mass mailing. The mailing won't be cost-effective unless it produces at least a 5% return. What does your confidence interval suggest? Explain.

19. Safe food Some food retailers propose subjecting food to a low level of radiation in order to improve safety, but sale of such "irradiated" food is opposed by many people. Suppose a grocer wants to find out what his customers think. He has cashiers distribute surveys at checkout and ask customers to fill them out and drop them in a box near the front door. He gets responses from 122 customers, of whom 78 oppose the radiation treatments. What can the grocer conclude about the opinions of all his customers?

20. Local news The mayor of a small city has suggested that the state locate a new prison there, arguing that the construction project and resulting jobs will be good for the local economy. A total of 183 residents show up for a public hearing on the proposal, and a show of hands finds only 31 in favor of the prison project. What can the city council conclude about public support for the mayor's initiative?

21. Death penalty, again In the survey on the death penalty you read about in the chapter, the Gallup Poll actually split the sample at random, asking 510 respondents the question quoted earlier, "Generally speaking, do you believe the death penalty is applied fairly or unfairly in this country today?" The other 510 were asked "Generally speaking, do you believe the death penalty is applied unfairly or fairly in this country today?" Seems like the same question, but sometimes the order of the choices matters. Asked the first question, 58% said the death penalty was fairly applied; only 54% said so with the second wording.

a) What kind of bias may be present here?

b) If we combine them, considering the overall group to be one larger random sample of 1020 respondents, what is a 95% confidence interval for the proportion of the general public that thinks the death penalty is being fairly applied?

c) How does the margin of error based on this pooled sample compare with the margins of error from the separate groups? Why?

22. Gambling A city ballot includes a local initiative that would legalize gambling. The issue is hotly contested, and two groups decide to conduct polls to predict the outcome. The local newspaper finds that 53% of 1200 randomly selected voters plan to vote "yes," while a college Statistics class finds 54% of 450 randomly selected voters in support. Both groups will create 95% confidence intervals.

a) Without finding the confidence intervals, explain which one will have the larger margin of error.

b) Find both confidence intervals.

c) Which group concludes that the outcome is too close to call? Why?

23. Rickets Vitamin D, whether ingested as a dietary supplement or produced naturally when sunlight falls on the skin, is essential for strong, healthy bones. The bone disease rickets was largely eliminated in England during the 1950s, but now there is concern that a generation of children more likely to watch TV or play computer games than spend time outdoors is at increased risk. A recent study of 2700 children randomly selected from all parts of England found 20% of them deficient in vitamin D.

a) Find a 98% confidence interval.

b) Explain carefully what your interval means.

c) Explain what "98% confidence" means.

24. Teachers A 2011 Gallup poll found that 76% of Americans believe that high achieving high school students should be recruited to become teachers. This poll was based on a random sample of 1002 Americans.

a) Find a 90% confidence interval for the proportion of Americans who would agree with this.

b) Interpret your interval in this context.

c) Explain what "90% confidence" means.

d) Do these data refute a pundit's claim that 2/3 of Americans believe this statement? Explain.

25. Payments In a May 2007 Experian/Gallup Personal Credit Index poll of 1008 U.S. adults aged 18 and over, 8% of respondents said they were very uncomfortable with their ability to make their monthly payments on their current debt during the next three months. A more detailed poll surveyed 1288 adults, reporting similar overall results and also noting differences among four age groups: 18–29, 30–49, 50–64, and 65+.

a) Do you expect the 95% confidence interval for the true proportion of all 18- to 29-year-olds who are worried to be wider or narrower than the 95% confidence interval for the true proportion of all U.S. consumers? Explain.

b) Do you expect this second poll's overall margin of error to be larger or smaller than the Experian/Gallup poll's? Explain.

26. Back to campus In 2004 ACT, Inc., reported that 74% of 1644 randomly selected college freshmen returned to college the next year. The study was stratified by type of college—public or private. The retention rates were 71.9% among 505 students enrolled in public colleges and 74.9% among 1139 students enrolled in private colleges.
   a) Will the 95% confidence interval for the true national retention rate in private colleges be wider or narrower than the 95% confidence interval for the retention rate in public colleges? Explain.
   b) Do you expect the margin of error for the overall retention rate to be larger or smaller? Explain.

27. Deer ticks Wildlife biologists inspect 153 deer taken by hunters and find 32 of them carrying ticks that test positive for Lyme disease.
   a) Create a 90% confidence interval for the percentage of deer that may carry such ticks.
   b) If the scientists want to cut the margin of error in half, how many deer must they inspect?
   c) What concerns do you have about this sample?

28. Back to campus again Suppose ACT, Inc. wants to update their information from Exercise 26 on the percentage of freshmen that return for a second year of college.
   a) They want to cut the stated margin of error in half. How many college freshmen must be surveyed?
   b) Do you have any concerns about this sample? Explain.

29. Graduation It's believed that as many as 25% of adults over 50 never graduated from high school. We wish to see if this percentage is the same among the 25 to 30 age group.
   a) How many of this younger age group must we survey in order to estimate the proportion of non-grads to within 6% with 90% confidence?
   b) Suppose we want to cut the margin of error to 4%. What's the necessary sample size?
   c) What sample size would produce a margin of error of 3%?

30. Hiring In preparing a report on the economy, we need to estimate the percentage of businesses that plan to hire additional employees in the next 60 days.
   a) How many randomly selected employers must we contact in order to create an estimate in which we are 98% confident with a margin of error of 5%?
   b) Suppose we want to reduce the margin of error to 3%. What sample size will suffice?
   c) Why might it not be worth the effort to try to get an interval with a margin of error of only 1%?

31. Graduation, again As in Exercise 29, we hope to estimate the percentage of adults aged 25 to 30 who never graduated from high school. What sample size would allow us to increase our confidence level to 95% while reducing the margin of error to only 2%?

32. Better hiring info Editors of the business report in Exercise 30 are willing to accept a margin of error of 4% but want 99% confidence. How many randomly selected employers will they need to contact?

33. Pilot study A state's environmental agency worries that many cars may be violating clean air emissions standards. The agency hopes to check a sample of vehicles in order to estimate that percentage with a margin of error of 3% and 90% confidence. To gauge the size of the problem, the agency first picks 60 cars and finds 9 with faulty emissions systems. How many should be sampled for a full investigation?

34. Another pilot study During routine screening, a doctor notices that 22% of her adult patients show higher than normal levels of glucose in their blood—a possible warning signal for diabetes. Hearing this, some medical researchers decide to conduct a large-scale study, hoping to estimate the proportion to within 4% with 98% confidence. How many randomly selected adults must they test?

35. Approval rating A newspaper reports that the governor's approval rating stands at 65%. The article adds that the poll is based on a random sample of 972 adults and has a margin of error of 2.5%. What level of confidence did the pollsters use?

36. Amendment A TV news reporter says that a proposed constitutional amendment is likely to win approval in the upcoming election because a poll of 1505 likely voters indicated that 52% would vote in favor. The reporter goes on to say that the margin of error for this poll was 3%.
   a) Explain why the poll is actually inconclusive.
   b) What confidence level did the pollsters use?

 Just Checking ANSWERS

1. While true, we know that in the sample 17% said "yes"; there's no need for a margin of error.

2. No, we are 95% confident that the percentage falls in some interval, not exactly on a particular value.

3. Yes. That's what the confidence interval means.

4. No. We don't know for sure that's true; we are only 95% confident.

5. No. That's our level of confidence, not the proportion of people receiving unsolicited text messages. The sample suggests the proportion is much lower.

6. Wider.

7. Lower.

8. Smaller.

19 Testing Hypotheses About Proportions

A **S** *Activity:* **Testing a Claim.** Can we really draw a reasonable conclusion from a random sample? Run this simulation before you read the chapter, and you'll gain a solid sense of what we're doing here.

"Half the money I spend on advertising is wasted; the trouble is I don't know which half."

—*John Wanamaker (attributed)*

ngots are huge pieces of metal, often weighing more than 20,000 pounds, made in a giant mold. The metal, used for making parts for cars and planes, must be cast in one large piece. If it cracks while being made, the crack can ruin the part. Airplane manufacturers insist that metal for their planes be defect-free, so the ingot must be made over if any cracking is detected, a process costing thousands of dollars.

Metal manufacturers would like to avoid cracking if at all possible. But the casting process is complicated and not everything is completely under control. In one plant that specializes in very large (over 30,000 lb) ingots designed for the airplane industry, about 20% of the ingots have had some kind of crack. Hoping to reduce cracking, the plant engineers and chemists recently tried out some changes in the casting process. Since then, 400 ingots have been cast and only 17% of them have cracked. Should management declare victory? Has the cracking rate really decreased, or was 17% just due to luck?

We can treat the 400 ingots cast with the new method as a random sample. We know that each random sample will have a somewhat different proportion of cracked ingots. Is the 17% we observe merely a result of natural sampling variability, or is this lower cracking rate strong enough evidence to assure management that the true cracking rate now is really below 20%?

People want answers to questions like these all the time. Has the president's approval rating changed since last month? Has teenage smoking decreased in the past five years? Is the global temperature increasing? Did the Super Bowl ad we bought actually increase sales? To answer such questions, we test *hypotheses* about models.

Hypotheses

We want to know if the changes made by the engineers have lowered the cracking rate from 20%. Humans are natural skeptics, so to test whether the changes have worked, we'll assume that they haven't. We'll make the **hypothesis** that the cracking rate is still 20% and

Hypothesis n.; pl. {Hypotheses}. A supposition; a proposition or principle which is supposed or taken for granted, in order to draw a conclusion or inference for proof of the point in question; something not proved, but assumed for the purpose of argument.
—*Webster's Unabridged Dictionary, 1913*

■ NOTATION ALERT

Capital H is the standard letter for hypotheses. H_0 always labels the null hypothesis, and H_A labels the alternative hypothesis.

see if the data convince us otherwise. Hypotheses are models that we adopt temporarily—until we can test them once we have data. This starting hypothesis to be tested is called the **null hypothesis**—null because it assumes that nothing has changed. We denote it H_0. It specifies a parameter value—here that the cracking rate is 20%—which we usually write in the form H_0: *parameter = hypothesized value*. So, for the ingots we would write: H_0: $p = 0.20$.

The **alternative hypothesis**, which we denote H_A, contains the values of the parameter that we consider plausible if we reject the null hypothesis. Our null hypothesis is that $p = 0.20$. What's the alternative? Management is interested in *reducing* the cracking rate, so their alternative is H_A: $p < 0.20$.

What would convince you that the cracking rate had actually gone down? If only 4 out of the next 400 ingots crack (for a rate of 1%), most folks would conclude that the changes helped. But if the sample cracking rate is 19.8% instead of 20%, you should be skeptical. After all, observed proportions do vary, so we wouldn't be surprised to see some difference. How much smaller must the cracking rate be before we *are* convinced that it has changed? That's the crucial question in a hypothesis test. As usual in statistics, when we think about how big the change has to be, we think of using the standard deviation as the ruler to measure that change. So let's start by finding the standard deviation of the sample cracking rate.

Since the company changed the process, 400 new ingots have been cast of which 68 have visible surface cracks, for a sample proportion of 17%. The sample size of 400 is big enough to satisfy the **Success/Failure Condition**. (We expect $0.20 \times 400 = 80$ ingots to crack.) Although not a random sample, the engineers think that whether an ingot cracks should be independent from one ingot to the next, so the Normal sampling distribution model should work well. The standard deviation of the sampling model is

$$SD(\hat{p}) = \sqrt{\frac{pq}{n}} = \sqrt{\frac{(0.20)(0.80)}{400}} = 0.02.$$

Why Is This a Standard Deviation and Not a Standard Error?
Remember that we reserve the term standard error for the *estimate* of the standard deviation of the sampling distribution. But we're not estimating here—we have a value of p from our null hypothesis model. To remind us that the parameter value comes from the null hypothesis, it is sometimes written as p_0 and the standard deviation as $SD(\hat{p}) = \sqrt{p_0 q_0/n}$. That's different than when we found a confidence interval for p. In that case we couldn't assume that we knew its value, so we estimated the standard deviation from the sample value $\hat{p}$.

Now we know both parameters of the Normal sampling distribution model: $p = 0.20$ and $SD(\hat{p}) = 0.02$, so we can find out how likely it would be to see the observed value of $\hat{p} = 0.17$. Since we are using a Normal model, we find the *z*-score:

$$z = \frac{0.17 - 0.20}{0.02} = -1.5$$

Figure 19.1

How likely is a *z*-score of -1.5 (or lower)? This is what it looks like. The red area is 0.067 of the total area under the curve.

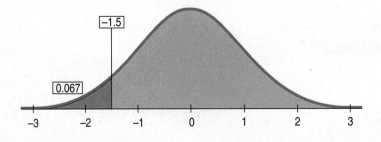

Then we ask, "How likely is it to observe a value at least 1.5 standard deviations below the mean of a Normal model?" The answer (from a calculator, computer program, or the Normal table) is about 0.067. This is the probability of observing a cracking rate of 17% or less in a sample of 400 if the null hypothesis is true and the true cracking rate is still 20%. Management must now decide whether an event that would happen 6.7% of the time by chance is strong enough evidence to conclude that the true cracking proportion has decreased.

A Trial as a Hypothesis Test

Does the reasoning of hypothesis tests seem backward? That could be because we usually prefer to think about getting things right rather than getting them wrong. You have seen this reasoning before because it's the logic of jury trials.

Let's suppose a defendant has been accused of robbery. In British common law and those systems derived from it (including U.S. law), the null hypothesis is that the defendant is innocent. Instructions to juries are quite explicit about this.

How is the null hypothesis tested? The prosecution first collects evidence. ("If the defendant were innocent, wouldn't it be remarkable that the police found him at the scene of the crime with a bag full of money in his hand, a mask on his face, and a getaway car parked outside?") For us, the data is the evidence.

The next step is to judge the evidence. Evaluating the evidence is the responsibility of the jury in a trial, but it falls on your shoulders in hypothesis testing. The jury considers the evidence in light of the *presumption* of innocence and judges whether the evidence against the defendant would be plausible *if the defendant were in fact innocent.*

Like the jury, you ask, "Could these data plausibly have happened by chance if the null hypothesis were true?" If they are very unlikely to have occurred, then the evidence raises a reasonable doubt about the null hypothesis.

Ultimately, you must make a decision. The standard of "beyond a reasonable doubt" is wonderfully ambiguous because it leaves the jury to decide the degree to which the evidence contradicts the hypothesis of innocence. Juries don't explicitly use probability to help them decide whether to reject that hypothesis. But when you ask the same question of your null hypothesis, you have the advantage of being able to quantify exactly how surprising the evidence would be if the null hypothesis were true.

How unlikely is unlikely? Some people set rigid standards, like 1 time out of 20 (0.05) or 1 time out of 100 (0.01). But if *you* have to make the decision, you must judge for yourself in each situation whether the probability of observing your data is small enough to constitute "reasonable doubt."

A S *Activity:* **The Reasoning of Hypothesis Testing.** Our reasoning is based on a rule of logic that dates back to ancient scholars. Here's a modern discussion of it.

P-Values: Are We Surprised?

Beyond a Reasonable Doubt We ask whether the data were unlikely beyond a reasonable doubt. We've just calculated that probability. The probability that the observed statistic value (or an even more extreme value) could occur if the null model were true—in this case, 0.067—is the P-value.

The fundamental step in our reasoning is the question "Are these data surprising, given the null hypothesis?" The key calculation is to determine exactly how likely the data we observed would be if the null hypothesis were a true model of the world. Specifically, we want to find the *probability* of seeing data like these (or something even more extreme) *given* that the null hypothesis is true. This probability tells us how surprised we'd be to see the data we collected if the null hypothesis is true. It's so important that it gets a special name: it's called the **P-value**.[1]

When the P-value *is* small enough, it says that we are very surprised. It means that it's very unlikely we'd observe data like these if our null hypothesis were true. The model we started with (the null hypothesis) and the data we collected are at odds with each other, so we have to make a choice. Either the null hypothesis is correct and we've just seen something remarkable, or the null hypothesis is wrong, and we were wrong to use it as the basis

---

[1]You'd think if it were that special it would have a better name, but "P-value" is about as creative as statisticians get.

"If the People fail to satisfy their burden of proof, you must find the defendant not guilty."

—*NY state jury instructions*

Don't "Accept" the Null Hypothesis

Think about the null hypothesis that H_0: All swans are white. Does collecting a sample of 100 white swans prove the null hypothesis? The data are *consistent* with this hypothesis and seem to lend support to it, but they don't *prove* it. In fact, all we can do is disprove the null hypothesis—for example, by finding just one non-white swan.

for computing our P-value. On the other hand, if you believe in data more than in assumptions, then, given that choice, you should reject the null hypothesis.

When the P-value is high, we haven't seen anything unlikely or surprising at all. Events that have a high probability of happening happen often. The data are consistent with the model from the null hypothesis, and we have no reason to reject the null hypothesis. But many other similar hypotheses could also account for the data we've seen, so *we haven't proven that the null hypothesis is true.* The most we can say is that it doesn't appear to be false. Formally, we "fail to reject" the null hypothesis. That's a pretty weak conclusion, but it's all we can do with a high P-value.

What to Do with an "Innocent" Defendant

If the evidence is not strong enough to reject the defendant's presumption of innocence, what verdict does the jury return? They say "not guilty." Notice that they do not say that the defendant is innocent. All they say is that they have not seen sufficient evidence to convict, to reject innocence. The defendant may, in fact, be innocent, but the jury has no way to be sure.

Said statistically, the jury's null hypothesis is H_0: innocent defendant. If the evidence is too unlikely given this assumption—if the P-value is too small—the jury rejects the null hypothesis and finds the defendant guilty. But—and this is an important distinction—if there is *insufficient evidence* to convict the defendant, the jury does not decide that H_0 is true and declare the defendant innocent. Juries can only *fail to reject* the null hypothesis and declare the defendant "not guilty."

In the same way, if the data are not particularly unlikely under the assumption that the null hypothesis is true, then the most we can do is to "fail to reject" our null hypothesis. We never declare the null hypothesis to be true (or "accept" the null), because we simply do not know whether it's true or not. (After all, more evidence may come along later.)

In the trial, the burden of proof is on the prosecution. In a hypothesis test, the burden of proof is on the unusual claim. The null hypothesis is the ordinary state of affairs, so it's the alternative to the null hypothesis that we consider unusual and for which we must marshal evidence.

Imagine a clinical trial testing the effectiveness of a new headache remedy. In Chapter 12, we saw the value of comparing such treatments to a placebo. The null hypothesis, then, is that the new treatment is no more effective than the placebo. This is important because some patients will improve even when administered the placebo treatment. If we use only six people to test the drug, the results are likely *not to be clear* and we'll be unable to reject the hypothesis. Does this mean the drug doesn't work? Of course not. It simply means that we don't have enough evidence to reject our assumption. That's why we don't start by assuming that the drug *is more effective.* If we were to do that, then we could test just a few people, find that the results aren't clear, and claim that since we've been unable to reject our original assumption the drug must be effective. The FDA is unlikely to be impressed by that argument.

 Just Checking

1. A research team wants to know if aspirin helps to thin blood. The null hypothesis says that it doesn't. They test 12 patients, observe the proportion with thinner blood, and get a P-value of 0.32. They proclaim that aspirin doesn't work. What would you say?

2. An allergy drug has been tested and found to give relief to 75% of the patients in a large clinical trial.

Now the scientists want to see if the new, improved version works even better. What would the null hypothesis be?

3. The new drug is tested and the P-value is 0.0001. What would you conclude about the new drug?

The Reasoning of Hypothesis Testing

> "The null hypothesis is never proved or established, but is possibly disproved, in the course of experimentation. Every experiment may be said to exist only in order to give the facts a chance of disproving the null hypothesis."
>
> —*Sir Ronald Fisher,*
> **The Design of Experiments**

Hypothesis tests follow a carefully structured path. To avoid getting lost as we navigate down it, we divide that path into four distinct sections.

1. Hypotheses

First, we state the null hypothesis. That's usually the skeptical claim that nothing's different. Are we considering a (New! Improved!) possibly better method? The null hypothesis says, "Oh yeah? Convince me!" To convert a skeptic, we must pile up enough evidence against the null hypothesis that we can reasonably reject it.

In statistical hypothesis testing, hypotheses are almost always about model parameters. To assess how unusual our data may be, we need a null model. The null hypothesis specifies a particular parameter value to use in our model. In the usual shorthand, we write $H_0: parameter = hypothesized\ value$. The alternative hypothesis, H_A, contains the values of the parameter we consider plausible when we reject the null.

For Example WRITING HYPOTHESES

A large city's Department of Motor Vehicles claimed that 80% of candidates pass driving tests, but a newspaper reporter's survey of 90 randomly selected local teens who had taken the test found only 68 who passed.

QUESTION: Does this finding suggest that the passing rate for teenagers is lower than the DMV reported? Write appropriate hypotheses.

ANSWER: I'll assume that the passing rate for teenagers is the same as the DMV's overall rate of 80%, unless there's strong evidence that it's lower.

$$H_O: p = 0.80$$
$$H_A: p < 0.80$$

How to Say It
You might think that the 0 in H_0 should be pronounced as "zero" or "0," but it's actually pronounced "naught" as in "all is for naught."

2. Model

To plan a statistical hypothesis test, specify the *model* you will use to test the null hypothesis and the parameter of interest. Of course, all models require assumptions, so you will need to state them and check any corresponding conditions.

Your Model step should end with a statement such as

Because the conditions are satisfied, I can model the sampling distribution of the sample proportion with a Normal model.

Watch out, though. Your Model step could end with

Because the conditions are not satisfied, I can't proceed with the test.

If that's the case, stop and reconsider.

Each test in the book has a name that you should include in your report. We'll see many tests in the chapters that follow. Some will be about more than one sample, some will involve statistics other than proportions, and some will use models other than the Normal (and so will not use z-scores). The test about proportions is called a **one-proportion z-test**.[2]

When the Conditions Fail . . .
You might proceed with caution, explicitly stating your concerns. Or you may need to do the analysis with and without an outlier, or on different subgroups, or after re-expressing the response variable. Or you may not be able to proceed at all.

---

[2]It's also called the "one-sample test for a proportion."

A **S** *Activity:* **Was the Observed Outcome Unlikely?** Complete the test you started in the first activity for this chapter. The narration explains the steps of the hypothesis test.

One-Proportion z-Test

The conditions for the one-proportion z-test are the same as for the one-proportion z-interval. We test the hypothesis $H_0: p = p_0$ using the statistic $z = \dfrac{(\hat{p} - p_0)}{SD(\hat{p})}$. We use the hypothesized proportion to find the standard deviation, $SD(\hat{p}) = \sqrt{\dfrac{p_0 q_0}{n}}$.

When the conditions are met and the null hypothesis is true, this statistic follows the standard Normal model, so we can use that model to obtain a P-value.

For Example CHECKING THE CONDITIONS

RECAP: A large city's DMV claimed that 80% of candidates pass driving tests. A reporter has results from a survey of 90 randomly selected local teens who had taken the test.

QUESTION: Are the conditions for inference satisfied?

✓ **Randomization Condition:** The 90 teens surveyed were a random sample of local teenage driving candidates.

✓ **10% Condition:** 90 is fewer than 10% of the teenagers who take driving tests in a large city.

✓ **Success/Failure Condition:** We expect $np_0 = 90(0.80) = 72$ successes and $nq_0 = 90(0.20) = 18$ failures. Both are at least 10.

ANSWER: *The conditions are satisfied, so it's okay to use a Normal model and perform a one-proportion z-test.*

Conditional Probability

Did you notice that a P-value is a conditional probability? It's the probability that the observed results could have happened *if (or given that) the null hypothesis were true.*

3. Mechanics

Under "Mechanics," we place the actual calculation of our test statistic from the data. Different tests we encounter will have different formulas and different test statistics. Usually, the mechanics are handled by a statistics program or calculator, but it's good to have the formulas recorded for reference and to know what's being computed. The ultimate goal of the calculation is to find out how surprising our data would be if the null hypothesis were true. We measure this by the P-value—the probability that the observed statistic value (or an even more extreme value) occurs if the null model is correct. If the P-value is small enough, we'll reject the null hypothesis.

For Example FINDING A P-VALUE

RECAP: A large city's DMV claimed that 80% of candidates pass driving tests, but a survey of 90 randomly selected local teens who had taken the test found only 68 who passed.

QUESTION: What's the P-value for the one-proportion z-test?

ANSWER: I have $n = 90$, $x = 68$, and a hypothesized $p = 0.80$.

$$\hat{p} = \frac{68}{90} \approx 0.756$$

$$SD(\hat{p}) = \sqrt{\frac{p_0 q_0}{n}} = \sqrt{\frac{(0.8)(0.2)}{90}} \approx 0.042$$

$$z = \frac{\hat{p} - p_0}{SD(\hat{p})} = \frac{0.756 - 0.800}{0.042} \approx -1.05$$

$$\text{P-value} = P(z < -1.05) = 0.146$$

4. Conclusion

"... They make things admirably plain,
 But one hard question will remain:
 If one hypothesis you lose,
 Another in its place you choose ..."

—*James Russell Lowell*,
Credidimus Jovem Regnare

The conclusion in a hypothesis test is always a statement about the null hypothesis. The conclusion must state either that we reject or that we fail to reject the null hypothesis. And, as always, the conclusion should be stated in context.

Your conclusion about the null hypothesis should never be the end of a testing procedure. Often, there are actions to take or policies to change. In our ingot example, management must decide whether to continue the changes proposed by the engineers. The decision always includes the practical consideration of whether the new method is worth the cost. Suppose management decides to reject the null hypothesis of 20% cracking in favor of the alternative that the percentage has been reduced. They must still evaluate how much the cracking rate has been reduced and how much it cost to accomplish the reduction. The *size of the effect* is always a concern when we test hypotheses. A good way to look at the **effect size** is to examine a confidence interval.

For Example STATING THE CONCLUSION

RECAP: A large city's DMV claimed that 80% of candidates pass driving tests. Data from a reporter's survey of randomly selected local teens who had taken the test produced a P-value of 0.146.

QUESTION: What can the reporter conclude? And how might the reporter explain what the P-value means for the newspaper story?

ANSWER: Because the P-value of 0.146 is so large, I fail to reject the null hypothesis. These survey data do not provide sufficient evidence to convince us that the passing rate for teenagers taking the driving test is lower than 80%.

If the passing rate for teenage driving candidates were actually 80%, we'd expect to see success rates this low in about 1 in 7 (14.6%) samples of this size. This seems too likely to happen just by chance to assert that the DMV's stated success rate does not apply to teens.

> **How Much Does it Cost?**
> Formal tests of a null hypothesis base the decision of whether to reject the null hypothesis solely on the size of the P-value. But in real life, we want to evaluate the costs of our decisions as well. How much would you be willing to pay for a faster computer? Shouldn't your decision depend on how much faster? And on how much more it costs? Costs are not just monetary either. Would you use the same standard of proof for testing the safety of an airplane as for the speed of your new computer?

Alternative Alternatives

A **S** *Activity:* **The Alternative Hypotheses.** This interactive tool provides easy ways to visualize how one- and two-tailed alternative hypotheses work.

Tests on the ingot data can be viewed in two different ways. We know the old cracking rate is 20%, so the null hypothesis is

$$H_0: p = 0.20.$$

But we have a choice of alternative hypotheses. A metallurgist working for the company might be interested in *any* change in the cracking rate due to the new process. Even if the rate got worse, she might learn something useful from it. In that case, she's interested in possible changes on both sides of the null hypothesis. So she would write her alternative hypothesis as

$$H_A: p \neq 0.20.$$

An alternative hypothesis such as this is known as a **two-sided alternative**[3] because we are equally interested in deviations on either side of the null hypothesis value. For two-sided alternatives, the P-value is the probability of deviating in *either* direction from the null hypothesis value.

But management is really interested only in *lowering* the cracking rate below 20%. The scientific value of knowing how to *increase* the cracking rate may not appeal to them. The only alternative of interest to them is that the cracking rate *decreases*. So, they would write their alternative hypothesis as

$$H_A: p < 0.20.$$

An alternative hypothesis that focuses on deviations from the null hypothesis value in only one direction is called a **one-sided alternative**.

---

[3]It is also called a **two-tailed alternative**, because the probabilities we care about are found in both tails of the sampling distribution.

For a hypothesis test with a one-sided alternative, the P-value is the probability of deviating *only in the direction of the alternative* away from the null hypothesis value. For the same data, the one-sided P-value is half the two-sided P-value. So, a one-sided test will reject the null hypothesis more often. If you aren't sure which to use, a two-sided test is always more conservative. Be sure you can justify the choice of a one-sided test from the *Why* of the situation.

Step-by-Step Example TESTING A HYPOTHESIS

Advances in medical care such as prenatal ultrasound examination now make it possible to determine a child's sex early in a pregnancy. There is a fear that in some cultures some parents may use this technology to select the sex of their children. A study from Punjab, India (E. E. Booth, M. Verma, and R. S. Beri, "Fetal Sex Determination in Infants in Punjab, India: Correlations and Implications," *BMJ* 309 [12 November 1994]: 1259–1261), reports that, in 1993, in one hospital, 56.9% of the 550 live births that year were boys. It's a medical fact that male babies are slightly more common than female babies. The study's authors report a baseline for this region of 51.7% male live births.

Question: Is there evidence that the proportion of male births is different for this hospital?

THINK ➡ **Plan** State what we want to know.

Define the variables and discuss the W's.

Hypotheses The null hypothesis makes the claim of no difference from the baseline.

Before seeing the data, we were interested in any change in male births, so the alternative hypothesis is two-sided.

Model Think about the assumptions and check the appropriate conditions.

I want to know whether the proportion of male births in this hospital is different from the established baseline of 51.7%. The data are the recorded sexes of the 550 live births from a hospital in Punjab, India, in 1993, collected for a study on fetal sex determination. The parameter of interest, p, is the proportion of male births:

$$H_0: p = 0.517$$
$$H_A: p \neq 0.517$$

✓ **Independence Assumption:** There is no reason to think that the sex of one baby can affect the sex of other babies, so births can reasonably be assumed to be independent with regard to the sex of the child.

✓ **Randomization Condition:** The 550 live births are not a random sample, so I must be cautious about any general conclusions. I hope that this is a representative year, and I think that the births at this hospital may be typical of this area of India.

✓ **10% Condition:** I would like to be able to make statements about births at similar hospitals in India. These 550 births are fewer than 10% of all of those births.

(continued)

For testing proportions, the conditions are the same ones we had for making confidence intervals, except that we check the **Success/Failure Condition** with the *hypothesized* proportions rather than with the *observed* proportions.

Specify the sampling distribution model.
Tell what test you plan to use.

✓ **Success/Failure Condition:** Both
$np_0 = 550(0.517) = 284.35$ and
$nq_0 = 550(0.483) = 265.65$ are greater than 10; I expect the births of at least 10 boys and at least 10 girls, so the sample is large enough.

The conditions are satisfied, so I can use a Normal model and perform a **one-proportion z-test**.

SHOW ➡ **Mechanics** The null model gives us the mean, and (because we are working with proportions) the mean gives us the standard deviation.

The null model is a Normal distribution with a mean of 0.517 and a standard deviation of

$$SD(\hat{p}) = \sqrt{\frac{p_0 q_0}{n}} = \sqrt{\frac{(0.517)(1 - 0.517)}{550}}$$
$$= 0.0213.$$

The observed proportion, $\hat{p}$, is 0.569, so

We find the z-score for the observed proportion to find out how many standard deviations it is from the hypothesized proportion.

$$z = \frac{\hat{p} - p_0}{SD(\hat{p})} = \frac{0.569 - 0.517}{0.0213} = 2.44.$$

The sample proportion lies 2.44 standard deviations above the mean.

Make a picture. Sketch a Normal model centered at $p_0 = 0.517$. Shade the region to the right of the observed proportion, and because this is a two-tail test, also shade the corresponding region in the other tail.

$$P = 2P(z > 2.44) = 2(0.0073) = 0.0146$$

From the z-score, we can find the P-value, which tells us the probability of observing a value that extreme (or more). Because this is a two-tail test, the P-value is the probability of observing an outcome more than 2.44 standard deviations from the mean of a Normal model *in either direction*. We must therefore *double* the probability we find in the upper tail.

TELL ➡ **Conclusion** State your conclusion in context.

This P-value is roughly 1 time in 70. That's clearly significant, but don't jump to other conclusions. We can't be sure how this deviation came about. For instance, we don't know whether this hospital is typical, or whether the time period studied was selected at random. And we certainly can't conclude that ultrasound played any role.

The P-value of 0.0146 says that if the true proportion of male babies were still at 51.7%, then an observed proportion as different as 56.9% male babies would occur at random only about 15 times in 1000. With a P-value this small, I reject H_0. This is strong evidence that the proportion of boys is not equal to the baseline for the region. It appears that the proportion of boys may be larger.

TI Tips   TESTING A HYPOTHESIS

By now probably nothing surprises you about your calculator. Of course it can help you with the mechanics of a hypothesis test. But that's not much. It cannot write the correct hypotheses, check the appropriate conditions, interpret the results, or state a conclusion. You still have to do the tough stuff!

Let's do the mechanics of the Step-By-Step example about the post-ultrasound male birthrate. Based on historical evidence, we hypothesized that 51.7% of babies would be males, but one year at one hospital the rate was 56.9% among 550 births.

```
1-PropZTest
 p0:.517
 x:550*.569■
 n:0
 prop≠p0 <p0 >p0
 Calculate Draw
```

- Go to the STAT TESTS menu. Scroll down and select 1-PropZTest.
- Specify the hypothesized proportion p_0
- Enter x, the observed number of males. Since you don't know the actual count, enter 550*.569 there and then round the resulting 312.95 off to a whole number.

```
1-PropZTest
 p0:.517
 x:313
 n:550
 prop≠p0 <p0 >p0
 Calculate Draw
```

- Specify the sample size.
- Since this is a two-tailed test, indicate that you want to see if the observed proportion is significantly different ($\neq$) from what was hypothesized.
- Calculate the result.

```
1-PropZTest
 prop≠.517
 z=2.444693651
 p=.0144975283
 p̂=.5690909091
 n=550
 ■
```

Okay, the rest is up to you. The calculator reports a z-score of 2.445 and a P-value of 0.0145. Such a small P-value indicates that this higher rate of male births is unlikely to be just sampling error. Be careful how you state your conclusion.

P-Values and Decisions: What to Tell About a Hypothesis Test

TELL ➡
MORE

Don't We Want to Reject the Null? Often the folks who collect the data or perform the experiment hope to reject the null. (They hope the new drug is better than the placebo, or the new ad campaign is better than the old one.) But when we practice Statistics, we can't allow that hope to affect our decision. The essential attitude for a hypothesis tester is skepticism. Until we become convinced otherwise, we cling to the null's assertion that there's nothing unusual, no effect, no difference, etc. As in a jury trial, the burden of proof rests with the alternative hypothesis—innocent until proven guilty. When you test a hypothesis, you must act as judge and jury, but you are not the prosecutor.

Hypothesis tests are particularly useful when we must make a decision. Is the defendant guilty or not? Should we choose print advertising or television? The absolute nature of the hypothesis test decision, however, makes some people (including the authors) uneasy. Whenever possible, it's a good idea to report a confidence interval for the parameter of interest as well.

How small should the P-value be to reject the null hypothesis? A jury needs enough evidence to show the defendant guilty "beyond a reasonable doubt." How does that translate to P-values? The answer is that there is no good, universal answer. How small the P-value has to be to reject the null hypothesis is highly context-dependent. When we're screening for a disease and want to be sure we treat all those who are sick, we may be willing to reject the null hypothesis of no disease with a P-value as large as 0.10. That would mean that 10% of the healthy people would be treated as sick and subjected to further testing. We might rather treat (or recommend further testing for) the occasional healthy person than fail to treat someone who was really sick. But a long-standing hypothesis, believed by many to be true, needs stronger evidence (and a correspondingly small P-value) to reject it.

See if you require the same P-value to reject each of the following null hypotheses:

- A renowned musicologist claims that she can distinguish between the works of Mozart and Haydn simply by hearing a randomly selected 20 seconds of music from any work by either composer. What's the null hypothesis? If she's just guessing, she'll get 50% of the pieces correct, on average. So our null hypothesis is that

"Extraordinary claims require extraordinary proof."

—*Marcello Truzzi*

p equals 50%. If she's for real, she'll get more than 50% correct. Now, we present her with 10 pieces of Mozart or Haydn chosen at random. She gets 9 out of 10 correct. It turns out that the P-value associated with that result is 0.011. (In other words, if you tried to just guess, you'd get at least 9 out of 10 correct only about 1% of the time.) What would *you* conclude? Most people would probably reject the null hypothesis and be convinced that she has some ability to do as she claims. Why? Because the P-value is small and we don't have any particular reason to doubt the alternative.

- On the other hand, imagine a student who bets that he can make a flipped coin land the way he wants just by thinking hard. To test him, we flip a fair coin 10 times. Suppose he gets 9 out of 10 right. This also has a P-value of 0.011. Are you willing now to reject this null hypothesis? Are you convinced that he's not just lucky? What amount of evidence *would* convince you? We require more evidence if rejecting the null hypothesis would contradict long-standing beliefs or other scientific results. Of course, with sufficient evidence we would revise our opinions (and scientific theories). That's how science makes progress.

Another factor in choosing a P-value is the importance of the issue being tested. Consider the following two tests:

- A researcher claims that the proportion of college students who hold part-time jobs now is higher than the proportion known to hold such jobs a decade ago. You might be willing to believe the claim (and reject the null hypothesis of no change) with a P-value of 0.05.
- An engineer claims that even though there were several problems with the rivets holding the wing on an airplane in their fleet, they've retested the proportion of faulty rivets and now the P-value is small enough to reject the null hypothesis that the proportion is the same. What P-value would be small enough to get you to fly on that plane?

Your conclusion about any null hypothesis should always be accompanied by the P-value of the test. Don't just declare the null hypothesis rejected or not rejected. Report the P-value to show the strength of the evidence against the hypothesis and the effect size. This will let each reader decide whether or not to reject the null hypothesis and whether or not to consider the result important if it is statistically significant.

To complete your analysis, follow your test with a confidence interval for the parameter of interest, to report the size of the effect.

 Activity: Hypothesis Tests for Proportions. You've probably noticed that the tools for confidence intervals and for hypothesis tests are similar. See how tests and intervals for proportions are related—and an important way in which they differ.

A S **Activity: Practice with Testing Hypotheses About Proportions.** Here's an interactive tool that makes it easy to see what's going on in a hypothesis test.

✔ Just Checking

4. A bank is testing a new method for getting delinquent customers to pay their past-due credit card bills. The standard way was to send a letter (costing about $0.40) asking the customer to pay. That worked 30% of the time. They want to test a new method that involves sending a DVD to customers encouraging them to contact the bank and set up a payment plan. Developing and sending the video costs about $10.00 per customer. What is the parameter of interest? What are the null and alternative hypotheses?

5. The bank sets up an experiment to test the effectiveness of the DVD. They mail it out to several randomly selected delinquent customers and keep track of how many actually do contact the bank to arrange payments. The bank's statistician calculates a P-value of 0.003. What does this P-value suggest about the DVD?

6. The statistician tells the bank's management that the results are clear and that they should switch to the DVD method. Do you agree? What else might you want to know?

Step-by-Step Example TESTS AND INTERVALS

Anyone who plays or watches sports has heard of the "home field advantage." Tournaments in many sports are designed to try to neutralize the advantage of the home team or player. Most people believe that teams tend to win more often when they play at home. But do they?

If there were no home field advantage, the home teams would win about half of all games played. To test this, we'll use the games in the Major League Baseball 2012 season. That year, there were 2430 regular-season games. It turns out that the home team won 1295 of the 2430 games, or 53.29% of the time.

Question: Could this deviation from 50% be explained just from natural sampling variability, or is it evidence to suggest that there really is a home field advantage, at least in professional baseball?

THINK ➡ **Plan** State what we want to know.

Define the variables and discuss the W's.

I want to know whether the home team in professional baseball is more likely to win. The data are all 2430 games from the 2012 Major League Baseball season. The variable is whether or not the home team won. The parameter of interest is the proportion of home team wins. If there's no advantage, I'd expect that proportion to be 0.50.

$$H_0: p = 0.50$$
$$H_A: p > 0.50$$

Hypotheses The null hypothesis makes the claim of no difference from the baseline. Here, that means no home field advantage.

We are interested only in a home field *advantage*, so the alternative hypothesis is one-sided.

Model Think about the assumptions and check the appropriate conditions. This is not a random sample. If we wanted to talk only about this season there would be no inference. So, we view the 2430 games here not as a random sample, but as a representative collection of games. Our inference is about all years of Major League Baseball.

✓ **Independence Assumption:** Generally, the outcome of one game has no effect on the outcome of another game. But this may not be strictly true. For example, if a key player is injured, the probability that the team will win in the next couple of games may decrease slightly, but independence is still roughly true. The data come from one entire season, but I expect other seasons to be similar.

I'm not just interested in 2012, and those games, while not randomly selected, should be a reasonable representative sample of all Major League Baseball games in the recent past and near future.

✓ **10% Condition:** We are interested in home field advantage for Major League Baseball for all seasons. While not a random sample, these 2430 games are fewer than 10% of all games played over the years.

(continued)

✓ **Success/Failure Condition:** Both
$np_0 = 2430(0.50) = 1215$ and
$nq_0 = 2430(0.50) = 1215$ are at
least 10.

Specify the sampling distribution model.
State what test you plan to use.

Because the conditions are satisfied, I'll use a
Normal model for the sampling distribution of the
sample proportion and do a one-proportion z-test.

SHOW ➡ **Mechanics** The null model gives us
the mean, and (because we are working with
proportions) the mean gives us the standard
deviation.

The null model is a Normal distribution with a mean
of 0.50 and a standard deviation of

$$SD(\hat{p}) = \sqrt{\frac{p_0 q_0}{n}} = \sqrt{\frac{(0.5)(1 - 0.5)}{2430}}$$

$$= 0.010143.$$

The observed proportion is

$$\hat{p} = \frac{1295}{2430} = 0.5329.$$

Next, we find the z-score for the observed
proportion, to find out how many standard
deviations it is from the hypothesized proportion.

So the z-value is

$$z = \frac{0.5329 - 0.5}{0.010143} = 3.246.$$

The sample proportion lies 3.246 standard
deviations above the mean.

From the z-score, we can find the P-value,
which tells us the probability of observing a
value that extreme (or more so).

The probability of observing a value 3.246 or
more standard deviations above the mean of
a Normal model can be found by computer,
calculator, or table to be 0.0006.

The corresponding P-value is 0.0006.

TELL ➡ **Conclusion** State your conclusion about
the parameter—in context, of course!

The P-value of 0.0006 says that if the true proportion of home team wins were 0.50, then an
observed value of 0.5329 (or larger) would occur
only 6 times in 10,000. With a P-value so small,
I reject H_0. I have reasonable evidence that the
true proportion of home team wins is greater than
50%. It appears there is a home field advantage in
Major League Baseball.

(continued)

Question: OK, but how big a difference are we talking about? Just knowing that there is an effect is only part of the answer. Let's find a confidence interval for the home field advantage.

THINK ➡ **Model** Think about the assumptions and check the conditions.

The conditions are identical to those for the hypothesis test, with one difference: Now we are not given a hypothesized proportion, p_0, so we must instead work with the observed results.

Specify the sampling distribution model.

Tell what method you plan to use.

✓ **Success/Failure Condition:** There were 1295 home team wins and 1135 losses, both at least 10.

The conditions are satisfied, so I can model the sampling distribution of the sample proportion with a Normal model and find a **one-proportion z-interval.**

SHOW ➡ **Mechanics** We can't find the sampling model standard deviation from the null model proportion. (In fact, we've just rejected it.) Instead, we find the standard error of $\hat{p}$ from the *observed* proportions. Other than that substitution, the calculation looks the same as for the hypothesis test.

With this large a sample size, the difference is negligible, but in smaller samples, it could make a bigger difference.

$$SE(\hat{p}) = \sqrt{\frac{\hat{p}\hat{q}}{n}} = \sqrt{\frac{(0.5329)(1 - 0.5329)}{2430}}$$
$$= 0.01012$$

The sampling model is Normal, so for a 95% confidence interval, the critical value $z^* = 1.96$.

The margin of error is

$$ME = z^* \times SE(\hat{p}) = 1.96 \times 0.01012$$
$$= 0.0198.$$

So the 95% confidence interval is

$$0.5329 \pm 0.0198 \text{ or } (0.5131, 0.5527).$$

TELL ➡ **Conclusion** Confidence intervals help us think about the size of the effect. Here we can see that the home field advantage may affect enough games to make a real difference.

I am 95% confident that, in professional baseball, home teams win between 51.3% and 55.3% of the games.

In a season of 162 games, the low end of this interval, 51.3% of the 81 home games is about one extra home victory, on average. The upper end, 55.3%, is just over 4 extra wins.

WHAT IF ●●● we don't reject the null?

Dice can be "loaded" to make one face come up more often by inserting extra weight under the spots on the opposite side. If you're playing a game and your opponent seems to be having an unusual amount of luck rolling 6's to beat you, you might become suspicious of his die. So you try rolling it 60 times. If it's really a fair die, you'd expect to see face 6 about 10 times, but in your test you actually get 14 of them. Wondering if that's statistically significant, you test the hypothesis that his die is fair, $H_o:p = \frac{1}{6}$. You do a two-tailed test, because a die is unfair if a face will show up either more often or less often than it should. The P-value of your one-proportion z-test comes out 0.166, too high to reject the null. Were you to therefore *accept* the null hypothesis, then you'd be concluding that his die is fair. But do you really know that? Let's simulate.

Suppose your opponent's die actually has been tweaked so that 6's will come up 25% of the time, enough to give him an unfair advantage in the game. What might happen if you rolled a bogus die like this 60 times? We simulated that, running 500 trials. The histogram summarizes what happened.

Our simulation suggests a die loaded to favor 6's to the tune of 25% certainly could behave like the one you rolled. After all, in 220 of the 500 trials (44%), our simulated die came up a 6 in no more than 14 out of 60 rolls.

So what about the one your opponent is rolling? Is it fair? With so high a P-value you can't reject the null. But you can't accept it either. That die *could* be loaded. All you can say is that there's not enough evidence to accuse your opponent of cheating. That does not mean he isn't . . .

WHAT CAN GO WRONG?

Hypothesis tests are so widely used—and so widely misused—that we've devoted all of the next chapter to discussing the pitfalls involved, but there are a few issues that we can talk about already.

■ **Don't base your null hypothesis on what you see in the data.** You are not allowed to look at the data first and then adjust your null hypothesis so that it will be rejected. When your sample value turns out to be $\hat{p} = 51.8\%$, with a standard deviation of 1%, don't form a null hypothesis like $H_0: p = 49.8\%$, knowing that you can reject it. You should always *Think* about the situation you are investigating and make your null hypothesis describe the "nothing interesting" or "nothing has changed" scenario. No peeking at the data!

■ **Don't base your alternative hypothesis on the data, either.** Again, you need to *Think* about the situation. Are you interested only in knowing whether something has *increased*? Then write a one-sided (upper-tail) alternative. Or would you be equally interested in a change in either direction? Then you want a two-sided alternative. You should decide whether to do a one- or two-sided test based on what results would be of interest to you, not what you see in the data.

■ **Don't make your null hypothesis what you want to show to be true.** Remember, the null hypothesis is the status quo, the nothing-is-strange-here position a skeptic would take. You wonder whether the data cast doubt on that. You can reject the null hypothesis, but you can never "accept" or "prove" the null.

■ **Don't forget to check the conditions.** The reasoning of inference depends on randomization. No amount of care in calculating a test result can recover from biased sampling. The probabilities we compute depend on the independence assumption. And the sample must be large enough to justify the use of a Normal model.

■ **Don't accept the null hypothesis.** You may not have found enough evidence to reject it, but you surely have *not* proven it's true!

■ **If you fail to reject the null hypothesis, don't think that a bigger sample would be more likely to lead to rejection.** If the results you looked at were "almost" significant, it's enticing to think that because you would have rejected the null had these same observations come from a larger sample, then a larger sample would surely lead to rejection. Don't be misled. Remember, each sample is different, and a larger sample won't necessarily duplicate your current observations. Indeed, the Central Limit Theorem tells us that statistics will vary *less* in larger samples. We should therefore expect such results to be less extreme. Maybe they'd be statistically significant but maybe (perhaps even probably) not. Even if you fail to reject the null hypothesis, it's a good idea to examine a confidence interval. If none of the plausible parameter values in the interval would matter to you (for example, because none would be *practically* significant), then even a larger study with a correspondingly smaller standard error is unlikely to be worthwhile.

What Have We Learned?

We've learned to use what we see in a random sample to test a hypothesis about the world. Hypothesis tests go hand in hand with confidence intervals.

■ A hypothesis test makes a decision about the plausibility of a parameter value.
■ A confidence interval estimates a range of plausible values for the parameter.

We've learned that testing a hypothesis requires four important steps:

■ writing **hypotheses**;
■ determining what **test** to use by checking appropriate assumptions and conditions;
■ completing the **mechanics** of the test by finding a z-score and a P-value; and
■ stating our **conclusion** in the proper context.

We've learned to formulate appropriate hypotheses.

■ The null hypothesis specifies the parameter of a model we'll test using our data. It has the form $H_0: p = p_0$.
■ The alternative hypothesis states what we'll have evidence for if we reject the null. It can be one-sided or two-sided, depending on what we want to investigate.

We've learned to confirm that we can use a Normal model, and to name the test we'll perform.

■ We check the Independence Assumption, the Randomization Condition, the 10% Condition, and the Success/Failure Condition.
■ If all of these check out, we use a Normal model to perform a one-proportion z-test.

We've learned to complete the mechanics of the test.

■ Based on our assumption that the null hypothesis is true, we find the standard deviation of the sampling model for the sample proportion: $SD(\hat{p}) = \sqrt{\dfrac{p_0 q_0}{n}}$.

■ We calculate the test statistic $z = \dfrac{\hat{p} - p_0}{SD(\hat{p})}$ and use a Normal model to find the P-value. The P-value is the probability of observing an outcome at least as extreme as ours if the null hypothesis is actually true.

We've learned to state an appropriate conclusion.

- If the P-value is large, then it's plausible that the results we've observed may be just sampling error. We'll fail to reject the null hypothesis and conclude there's not enough evidence to suggest that hypothesis is false.

- If the P-value is very small, then it's highly unlikely we'd observe results like ours if the null hypothesis were true. We'll reject the null hypothesis and conclude there's strong evidence to suggest that hypothesis is false.

Terms

Null hypothesis
The claim being assessed in a hypothesis test is called the null hypothesis. Usually, the null hypothesis is a statement of "no change from the traditional value," "no effect," "no difference," or "no relationship." For a claim to be a testable null hypothesis, it must specify a value for some population parameter that can form the basis for assuming a sampling distribution for a test statistic. (p. 494)

Alternative hypothesis
The alternative hypothesis proposes what we should conclude if we find the null hypothesis to be unlikely. (p. 494)

Two-sided alternative (Two-tailed alternative)
An alternative hypothesis is two-sided ($H_A: p \neq p_0$) when we are interested in deviations in *either* direction away from the hypothesized parameter value. (p. 500)

One-sided alternative (One-tailed alternative)
An alternative hypothesis is one-sided (e.g., $H_A: p > p_0$ or $H_A: p < p_0$) when we are interested in deviations in *only one* direction away from the hypothesized parameter value. (p. 500)

P-value
The probability of observing a value for a test statistic at least as far from the hypothesized value as the statistic value actually observed if the null hypothesis is true. A small P-value indicates either that the observation is improbable or that the probability calculation was based on incorrect assumptions. The assumed truth of the null hypothesis is the assumption under suspicion. (p. 495)

One-proportion z-test
A test of the null hypothesis that the proportion of a single sample equals a specified value ($H_0: p = p_0$) by referring the statistic $z = \dfrac{\hat{p} - p_0}{SD(\hat{p})}$ to a Standard Normal model. (p. 497)

Effect size
The difference between the null hypothesis value and the actual value of population parameter. (p. 499)

On the Computer HYPOTHESIS TESTS FOR A PROPORTION

You can conduct a hypothesis test for a proportion using a graphing calculator or a statistics software package on a computer. Often all you need to do is enter information about the hypotheses, the observed number of successes, and the sample size. Some programs want the original data, in which success and failure may be coded as 1 and 0 or "yes" and "no." The technology will report the z-score and the P-value.

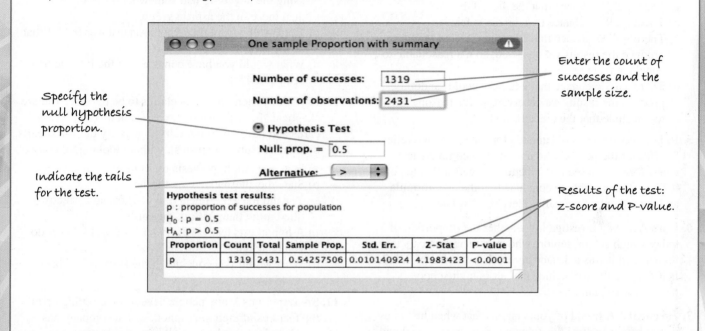

Specify the null hypothesis proportion.

Enter the count of successes and the sample size.

Indicate the tails for the test.

Results of the test: z-score and P-value.

Exercises

1. **Hypotheses** Write the null and alternative hypotheses you would use to test each of the following situations:

   a) A governor is concerned about his "negatives"—the percentage of state residents who express disapproval of his job performance. His political committee pays for a series of TV ads, hoping that they can keep the negatives below 30%. They will use follow-up polling to assess the ads' effectiveness.

   b) Is a coin fair?

   c) Only about 20% of people who try to quit smoking succeed. Sellers of a motivational tape claim that listening to the recorded messages can help people quit.

2. **More hypotheses** Write the null and alternative hypotheses you would use to test each situation.

   a) In the 1950s only about 40% of high school graduates went on to college. Has the percentage changed?

   b) 20% of cars of a certain model have needed costly transmission work after being driven between 50,000 and 100,000 miles. The manufacturer hopes that a redesign of a transmission component has solved this problem.

   c) We field-test a new-flavor soft drink, planning to market it only if we are sure that over 60% of the people like the flavor.

3. **Negatives** After the political ad campaign described in Exercise 1a, pollsters check the governor's negatives. They test the hypothesis that the ads produced no change against the alternative that the negatives are now below 30% and find a P-value of 0.22. Which conclusion is appropriate? Explain.

   a) There's a 22% chance that the ads worked.

   b) There's a 78% chance that the ads worked.

   c) There's a 22% chance that their poll is correct.

d) There's a 22% chance that natural sampling variation could produce poll results like these if there's really no change in public opinion.

4. Dice The seller of a loaded die claims that it will favor the outcome 6. We don't believe that claim, and roll the die 200 times to test an appropriate hypothesis. Our P-value turns out to be 0.03. Which conclusion is appropriate? Explain.

a) There's a 3% chance that the die is fair.

b) There's a 97% chance that the die is fair.

c) There's a 3% chance that a loaded die could randomly produce the results we observed, so it's reasonable to conclude that the die is fair.

d) There's a 3% chance that a fair die could randomly produce the results we observed, so it's reasonable to conclude that the die is loaded.

5. Relief A company's old antacid formula provided relief for 70% of the people who used it. The company tests a new formula to see if it is better and gets a P-value of 0.27. Is it reasonable to conclude that the new formula and the old one are equally effective? Explain.

6. Cars A survey investigating whether the proportion of today's high school seniors who own their own cars is higher than it was a decade ago finds a P-value of 0.017. Is it reasonable to conclude that more high-schoolers have cars? Explain.

7. He cheats! A friend of yours claims that when he tosses a coin he can control the outcome. You are skeptical and want him to prove it. He tosses the coin, and you call heads; it's tails. You try again and lose again.

a) Do two losses in a row convince you that he really can control the toss? Explain.

b) You try a third time, and again you lose. What's the probability of losing three tosses in a row if the process is fair?

c) Would three losses in a row convince you that your friend cheats? Explain.

d) How many times in a row would you have to lose in order to be pretty sure that this friend really can control the toss? Justify your answer by calculating a probability and explaining what it means.

8. Candy Someone hands you a box of a dozen chocolate-covered candies, telling you that half are vanilla creams and the other half peanut butter. You pick candies at random and discover the first three you eat are all vanilla.

a) If there really were 6 vanilla and 6 peanut butter candies in the box, what is the probability that you would have picked three vanillas in a row?

b) Do you think there really might have been 6 of each? Explain.

c) Would you continue to believe that half are vanilla if the fourth one you try is also vanilla? Explain.

9. Better than aspirin? A very large study showed that aspirin reduced the rate of first heart attacks by 44%. A pharmaceutical company thinks they have a drug that will be more effective than aspirin, and plans to do a randomized clinical trial to test the new drug.

a) What is the null hypothesis the company will use?

b) What is their alternative hypothesis?
   They conducted the study and found that the group using the new drug had somewhat fewer heart attacks than those in the aspirin group.

c) The P-value from the hypothesis test was 0.28. What do you conclude?

d) What would you have concluded if the P-value had been 0.004?

10. Psychic A friend of yours claims to be psychic. You are skeptical. To test this you take a stack of 100 playing cards and have your friend try to identify the suit (hearts, diamonds, clubs, or spades), without looking, of course!

a) State the null hypothesis for your experiment.

b) State the alternative hypothesis.
   You did the experiment and your friend correctly identified more than 25% of the cards.

c) A hypothesis test gave a P-value of 0.014. What do you conclude?

d) What would you conclude if the P-value had been 0.245?

11. Smartphones Many people have trouble setting up all the features of their smartphones, so a company has developed what it hopes will be easier instructions. The goal is to have at least 96% of customers succeed. The company tests the new system on 200 people, of whom 188 were successful. Is this strong evidence that the new system fails to meet the company's goal? A student's test of this hypothesis is shown. How many mistakes can you find?

$H_0: \hat{p} = 0.96$

$H_A: \hat{p} \neq 0.96$

SRS, $0.96(200) > 10$

$\frac{188}{200} = 0.94;$ $SD(\hat{p}) = \sqrt{\frac{(0.94)(0.06)}{200}} = 0.017$

$z = \frac{0.96 - 0.94}{0.017} = 1.18$

$P = P(z > 1.18) = 0.12$

There is strong evidence the new instructions don't work.

12. Obesity 2008 In 2008, the Centers for Disease Control and Prevention reported that 34% of adults in the United States are obese. A county health service planning a new awareness campaign polls a random sample of 750 adults living there. In this sample, 228 people were found to be obese based on their answers to a health questionnaire.

Do these responses provide strong evidence that the 34% figure is not accurate for this region? Correct the mistakes you find in a student's attempt to test an appropriate hypothesis.

$H_0: \hat{p} = 0.34$

$H_A: \hat{p} < 0.34$

SRS, $750 \geq 10$

$\frac{228}{750} = 0.304; \quad SD(\hat{p}) = \sqrt{\frac{(0.304)(0.696)}{750}} = 0.017$

$z = \frac{0.304 - 0.34}{0.017} = -2$

$P = P(z > -2) = 0.977$

There is more than a 97% chance that the stated percentage is correct for this region.

13. **Dowsing** In a rural area, only about 30% of the wells that are drilled find adequate water at a depth of 100 feet or less. A local man claims to be able to find water by "dowsing"—using a forked stick to indicate where the well should be drilled. You check with 80 of his customers and find that 27 have wells less than 100 feet deep. What do you conclude about his claim?
   a) Write appropriate hypotheses.
   b) Check the necessary assumptions.
   c) Perform the mechanics of the test. What is the P-value?
   d) Explain carefully what the P-value means in context.
   e) What's your conclusion?

14. **Abnormalities** In the 1980s it was generally believed that congenital abnormalities affected about 5% of the nation's children. Some people believe that the increase in the number of chemicals in the environment has led to an increase in the incidence of abnormalities. A recent study examined 384 children and found that 46 of them showed signs of an abnormality. Is this strong evidence that the risk has increased?
   a) Write appropriate hypotheses.
   b) Check the necessary assumptions.
   c) Perform the mechanics of the test. What is the P-value?
   d) Explain carefully what the P-value means in context.
   e) What's your conclusion?
   f) Do environmental chemicals cause congenital abnormalities?

15. **Absentees** The National Center for Education Statistics monitors many aspects of elementary and secondary education. Their 1996 numbers are often used as a baseline to assess changes. In 1996 34% of students had not been absent from school even once during the previous month. In the 2000 survey, responses from 8302 randomly chosen students showed that this figure

had slipped to 33%. Do these figures give evidence of a change in student attendance?
   a) Write appropriate hypotheses.
   b) Check the assumptions and conditions.
   c) Perform the test and find the P-value.
   d) State your conclusion.
   e) Do you think this difference is meaningful? Explain.

16. **Educated mothers** The National Center for Education Statistics monitors many aspects of elementary and secondary education. Their 1996 numbers are often used as a baseline to assess changes. In 1996, 31% of students reported that their mothers had graduated from college. In 2000, responses from 8368 randomly chosen students found that this figure had grown to 32%. Is this evidence of a change in education level among mothers?
   a) Write appropriate hypotheses.
   b) Check the assumptions and conditions.
   c) Perform the test and find the P-value.
   d) State your conclusion.
   e) Do you think this difference is meaningful? Explain.

17. **Contributions, please, part II** In Chapter 18, Exercise 15, you learned that the Paralyzed Veterans of America is a philanthropic organization that relies on contributions. They send free mailing labels and greeting cards to potential donors on their list and ask for a voluntary contribution. To test a new campaign, the organization recently sent letters to a random sample of 100,000 potential donors and received 4781 donations. They've had a contribution rate of 5% in past campaigns, but a staff member worries that the rate will be lower if they run this campaign as currently designed.
   a) What are the hypotheses?
   b) Are the assumptions and conditions for inference met?
   c) Do you think the rate would drop? Explain.

18. **Take the offer, part II** In Chapter 18, Exercise 16, you learned that First USA, a major credit card company, is planning a new offer for their current cardholders. First USA will give double airline miles on purchases for the next 6 months if the cardholder goes online and registers for this offer. To test the effectiveness of this campaign, the company recently sent out offers to a random sample of 50,000 cardholders. Of those, 1184 registered. A staff member suspects that the success rate for the full campaign will be comparable to the standard 2% rate that they are used to seeing in similar campaigns. What do you predict?
   a) What are the hypotheses?
   b) Are the assumptions and conditions for inference met?
   c) Do you think the rate would change if they use this fundraising campaign? Explain.

19. **Law School** According to the Law School Admission Council, in the fall of 2007, 66% of law school applicants

were accepted to some law school.[4] The training program *LSATisfaction* claims that 163 of the 240 students trained in 2006 were admitted to law school. You can safely consider these trainees to be representative of the population of law school applicants. Has *LSATisfaction* demonstrated a real improvement over the national average?

a) What are the hypotheses?

b) Check the conditions and find the P-value.

c) Would you recommend this program based on what you see here? Explain.

20. Med School 2011 According to the Association of American Medical Colleges, only 46% of medical school applicants were admitted to a medical school in the fall of 2011.[5] Upon hearing this, the trustees of Striving College expressed concern that only 77 of the 180 students in their class of 2011 who applied to medical school were admitted. The college president assured the trustees that this was just the kind of year-to-year fluctuation in fortunes that is to be expected and that, in fact, the school's success rate was consistent with the national average. Who is right?

a) What are the hypotheses?

b) Check the conditions and find the P-value.

c) Are the trustees right to be concerned, or is the president correct? Explain.

21. Pollution A company with a fleet of 150 cars found that the emissions systems of 7 out of the 22 they tested failed to meet pollution control guidelines. Is this strong evidence that more than 20% of the fleet might be out of compliance? Test an appropriate hypothesis and state your conclusion. Be sure the appropriate assumptions and conditions are satisfied before you proceed.

22. Scratch and dent An appliance manufacturer stockpiles washers and dryers in a very large warehouse for shipment to retail stores. Sometimes in handling them the appliances get damaged. Even though the damage may be minor, the company must sell those machines at drastically reduced prices. The company goal is to keep the level of damaged machines below 2%. One day an inspector randomly checks 60 washers and finds that 5 of them have scratches or dents. Is this strong evidence that the warehouse is failing to meet the company goal? Test an appropriate hypothesis and state your conclusion. Be sure the appropriate assumptions and conditions are satisfied before you proceed.

23. Twins In 2009 a national vital statistics report indicated that about 3% of all births produced twins. Is the rate of twin births the same among very young mothers? Data from a large city hospital found that only 7 sets of twins were born to 469 teenage girls. Test an appropriate

hypothesis and state your conclusion. Be sure the appropriate assumptions and conditions are satisfied before you proceed.

24. Football 2010 During the 2010 season, the home team won 143 of the 246 regular-season National Football League games. Is this strong evidence of a home field advantage in professional football? Test an appropriate hypothesis and state your conclusion. Be sure the appropriate assumptions and conditions are satisfied before you proceed.

25. WebZine A magazine is considering the launch of an online edition. The magazine plans to go ahead only if it's convinced that more than 25% of current readers would subscribe. The magazine contacted a simple random sample of 500 current subscribers, and 137 of those surveyed expressed interest. What should the company do? Test an appropriate hypothesis and state your conclusion. Be sure the appropriate assumptions and conditions are satisfied before you proceed.

26. Seeds A garden center wants to store leftover packets of vegetable seeds for sale the following spring, but the center is concerned that the seeds may not germinate at the same rate a year later. The manager finds a packet of last year's green bean seeds and plants them as a test. Although the packet claims a germination rate of 92%, only 171 of 200 test seeds sprout. Is this evidence that the seeds have lost viability during a year in storage? Test an appropriate hypothesis and state your conclusion. Be sure the appropriate assumptions and conditions are satisfied before you proceed.

27. Women executives A company is criticized because only 13 of 43 people in executive-level positions are women. The company explains that although this proportion is lower than it might wish, it's not surprising given that only 40% of all its employees are women. What do you think? Test an appropriate hypothesis and state your conclusion. Be sure the appropriate assumptions and conditions are satisfied before you proceed.

28. Jury Census data for a certain county show that 19% of the adult residents are Hispanic. Suppose 72 people are called for jury duty and only 9 of them are Hispanic. Does this apparent underrepresentation of Hispanics call into question the fairness of the jury selection system? Explain.

29. Dropouts Some people are concerned that new tougher standards and high-stakes tests adopted in many states have driven up the high school dropout rate. The National Center for Education Statistics reported that the high school dropout rate for the year 2004 was 10.3%. One school district whose dropout rate has always been very close to the national average reports that 210 of their 1782 high school students dropped out last year. Is this evidence that their dropout rate may be increasing? Explain.

---

[4]As reported by the Cornell office of career services in their Class of 2007 Postgraduate Report.

[5]www.aamc.org/data/facts/applicantmatriculant/

30. Acid rain A study of the effects of acid rain on trees in the Hopkins Forest shows that 25 of 100 trees sampled exhibited some sort of damage from acid rain. This rate seemed to be higher than the 15% quoted in a recent *Environmetrics* article on the average proportion of damaged trees in the Northeast. Does the sample suggest that trees in the Hopkins Forest are more susceptible than trees from the rest of the region? Comment, and write up your own conclusions based on an appropriate confidence interval as well as a hypothesis test. Include any assumptions you made about the data.

31. Lost luggage An airline's public relations department says that the airline rarely loses passengers' luggage. It further claims that on those occasions when luggage is lost, 90% is recovered and delivered to its owner within 24 hours. A consumer group that surveyed a large number of air travelers found that only 103 of 122 people who lost luggage on that airline were reunited with the missing items by the next day. Does this cast doubt on the airline's claim? Explain.

32. TV ads A start-up company is about to market a new computer printer. It decides to gamble by running commercials during the Super Bowl. The company hopes that name recognition will be worth the high cost of the ads. The goal of the company is that over 40% of the public recognize its brand name and associate it with computer equipment. The day after the game, a pollster contacts 420 randomly chosen adults and finds that 181 of them know that this company manufactures printers. Would you recommend that the company continue to advertise during Super Bowls? Explain.

33. John Wayne Like a lot of other Americans, John Wayne died of cancer. But is there more to this story? In 1955 Wayne was in Utah shooting the film *The Conqueror*. Across the state line, in Nevada, the United States military was testing atomic bombs. Radioactive fallout from those tests drifted across the filming location. A total of 46 of the 220 people working on the film eventually died of cancer. Cancer experts estimate that one would expect only about 30 cancer deaths in a group this size.

a) Is the death rate among the movie crew unusually high?

b) Does this prove that exposure to radiation increases the risk of cancer?

34. AP Stats The College Board reported that 58.7% of all students who took the 2010 AP Statistics exam earned scores of 3 or higher. One teacher wondered if the performance of her school was better. She believed that year's students to be typical of those who will take AP Stats at that school and was pleased when 34 of her 54 students achieved scores of 3 or better. Can she claim that her school is better? Explain.

 Just Checking ANSWERS

1. You can't conclude that the null hypothesis is true. You can conclude only that the experiment was unable to reject the null hypothesis. They were unable, on the basis of 12 patients, to show that aspirin was effective.

2. The null hypothesis is $H_0: p = 0.75$.

3. With a P-value of 0.0001, this is very strong evidence against the null hypothesis. We can reject H_0 and conclude that the improved version of the drug gives relief to a higher proportion of patients.

4. The parameter of interest is the proportion, p, of all delinquent customers who will pay their bills. $H_0: p = 0.30$ and $H_A: p > 0.30$.

5. The very low P-value leads us to reject the null hypothesis. There is strong evidence that the DVD is more effective in getting people to start paying their debts than just sending a letter had been.

6. All we know is that there is strong evidence to suggest that $p > 0.30$. We don't know how much higher than 30% the new proportion is. We'd like to see a confidence interval to see if the new method is worth the cost.

chapter 20 More About Tests and Intervals

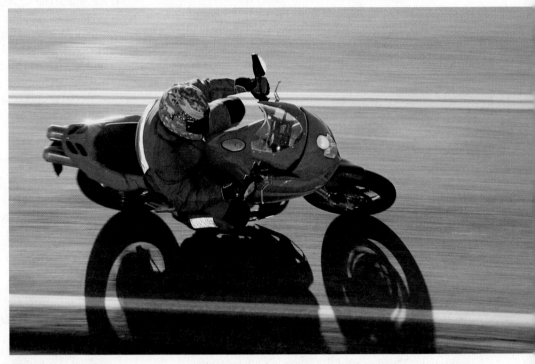

Who	Florida motorcycle riders aged 20 and younger involved in motorcycle accidents
What	% wearing helmets
When	2001–2003
Where	Florida
Why	Assessment of injury rates commissioned by the National Highway Traffic Safety Administration (NHTSA)

In 2000 Florida changed its motorcycle helmet law. No longer are riders 21 and older required to wear helmets. Under the new law, those under 21 still must wear helmets, but a report by the Preusser Group (www.preussergroup.com) suggests that helmet use may have declined in this group, too.

It isn't practical to survey young motorcycle riders. (For example, how can you construct a sampling frame? If you contacted licensed riders, would they admit to riding illegally without a helmet?) The researchers adopted a different strategy. Police reports of motorcycle accidents record whether the rider wore a helmet and give the rider's age. Before the change in the helmet law, 60% of youths involved in a motorcycle accident had been wearing their helmets. The Preusser study looked at accident reports during 2001–2003, the three years following the law change, considering these riders to be a representative sample of the larger population. They observed 781 young riders who were involved in accidents. Of these, 396 (or 50.7%) were wearing helmets. Is this evidence of a decline in helmet-wearing, or just the natural fluctuation of such statistics?

Zero In on the Null

How do we choose the null hypothesis? The appropriate null arises directly from the context of the problem. It is dictated, not by the data, but by the situation. One good way to identify both the null and alternative hypotheses is to think about why the study is being done and what we hope to learn from the test. Typical null hypotheses might be that the proportion of patients recovering after receiving a new drug is the same as we would expect of patients receiving a placebo or that the mean strength attained by athletes training with new equipment is the same as with the old equipment. The alternative hypotheses would be that the new drug cures a higher proportion of patients or that the new equipment results in a greater mean strength.

To write a null hypothesis, identify a parameter and choose a null value that relates to the question at hand. Even though the null usually means no difference or no change, you can't automatically interpret "null" to mean zero. A claim that "nobody" wears a motorcycle helmet would be absurd. The null hypothesis for the Florida study is that the true rate of helmet use remained the same at $p = 0.60$ among young riders after the law changed. The alternative is that the proportion has decreased. Both the value for the parameter in the null hypothesis and the nature of the alternative arise from the context of the problem.

There is a temptation to state your *claim* as the null hypothesis. As we have seen, however, you cannot prove a null hypothesis true any more than a trial proves a defendant innocent. So, it makes more sense to use what you want to show as the *alternative*. This way, if you reject the null, you are left with what you want to show.

For Example WRITING HYPOTHESES

The diabetes drug Avandia® was approved to treat Type 2 diabetes in 1999. But in 2007 an article in the *New England Journal of Medicine* (*NEJM*)[1] raised concerns that the drug might carry an increased risk of heart attack. This study combined results from a number of other separate studies to obtain an overall sample of 4485 diabetes patients taking Avandia. People with Type 2 diabetes are known to have about a 20.2% chance of suffering a heart attack within a seven-year period. According to the article's author, Dr. Steven E. Nissen,[2] the risk found in the *NEJM* study was equivalent to a 28.9% chance of heart attack over seven years. The FDA is the government agency responsible for relabeling Avandia to warn of the risk if it is judged to be unsafe. Although the statistical methods they use are more sophisticated, we can get an idea of their reasoning with the tools we have learned.

QUESTION: What null hypothesis and alternative hypothesis about seven-year heart attack risk would you test? Explain.

ANSWER:
$$H_0: p = 0.202$$
$$H_A: p > 0.202$$

The parameter of interest is the proportion of diabetes patients suffering a heart attack in seven years. The FDA is concerned only with whether Avandia *increases* the seven-year risk of heart attacks above the baseline value of 20.2%, so a one-sided upper-tail test is appropriate.

I CAN'T BELIEVE SCHOOLS ARE STILL TEACHING KIDS ABOUT THE NULL HYPOTHESIS.

I REMEMBER READING A BIG STUDY THAT CONCLUSIVELY DISPROVED IT *YEARS* AGO.

One-Sided or Two? In the 1930s, a series of experiments was performed at Duke University in an attempt to see whether humans were capable of extrasensory perception, or ESP. Psychologist Karl Zener designed a set of cards with 5 symbols, later made infamous in the movie *Ghostbusters*:

In the experiment, the "sender" selects one of the 5 cards at random from a deck and then concentrates on it. The "receiver" tries to determine which card it is. If we let p be the proportion of correct responses, what's the null hypothesis? The null hypothesis is that ESP makes no difference. Without ESP, the receiver would just be guessing, and since there are 5 possible responses, there would be a 20% chance of guessing each card correctly. So, H_0 is $p = 0.20$. What's the alternative? It seems that it should be $p > 0.20$, a one-sided alternative. But some ESP researchers have expressed the claim that if the proportion guessed were much *lower* than expected, that would show an "interference" and should be considered evidence for ESP as well. So they argue for a two-sided alternative.

---

[1]Steven E. Nissen, M.D., and Kathy Wolski, M.P.H., "Effect of Rosiglitazone on the Risk of Myocardial Infarction and Death from Cardiovascular Causes," *NEJM* 2007; 356.

[2]Interview reported in the *New York Times* (May 26, 2007).

How to Think About P-Values

Which Conditional?

Suppose that as a political science major you are offered the chance to be a White House intern. There would be a very high probability that next summer you'd be in Washington, D.C. That is, $P(\text{Washington} \mid \text{Intern})$ would be high. But if we find a student in Washington, D.C., is it likely that he's a White House intern? Almost surely not; $P(\text{Intern} \mid \text{Washington})$ is low. You can't switch around conditional probabilities. The P-value is $P(\text{data} \mid H_0)$. We might wish we could report $P(H_0 \mid \text{data})$, but these two quantities are NOT the same.

A P-value is a conditional probability. It tells us the probability of getting results at least as unusual as the observed statistic, *given* that the null hypothesis is true. We can write P-value $= P(\text{observed statistic value [or even more extreme]} \mid H_0)$.

Writing the P-value this way helps to make clear that the P-value is *not* the probability that the null hypothesis is true. It is a probability about the data. Let's say that again:

The P-value is not the probability that the null hypothesis is true.

The P-value is not even the conditional probability that the null hypothesis is true given the data. We would write that probability as $P(H_0 \mid \text{observed statistic value})$. This is a conditional probability but in reverse. It would be nice to know this probability, but we can't. As we saw in Chapter 14, reversing the order in a conditional probability is difficult, and the results can be counterintuitive.

We can find the P-value, $P(\text{observed statistic value} \mid H_0)$, because H_0 gives the parameter values that we need to calculate the required probability. But there's no direct way to find $P(H_0 \mid \text{observed statistic value})$.[3] As tempting as it may be to say that a P-value of 0.03 means there's a 3% chance that the null hypothesis is true, that just isn't right. All we can say is that, given the null hypothesis, there's a 3% chance of observing the statistic value that we have actually observed (or one more unlike the null value).

What to Do with a Small P-Value

We know that a small P-value means that the result we just observed is unlikely to occur if the null hypothesis is true. So we have evidence against the null hypothesis. An even smaller P-value implies stronger evidence against the null hypothesis, but it doesn't mean that the null hypothesis is "less true" (see "How Guilty Is the Suspect" on page 520).

How small the P-value has to be for you to reject the null hypothesis depends on a lot of things, not all of which can be precisely quantified. Your belief in the null hypothesis will influence your decision. Your trust in the data, in the experimental method if the data come from a planned experiment, in the survey protocol if the data come from a designed survey, all influence your decision. The P-value should serve as a measure of the strength of the evidence against the null hypothesis, but should never serve as a hard and fast rule for decisions. You have to take that responsibility on yourself.

As a review, let's look at the helmet law example from the chapter opener. Did helmet wearing among young riders decrease after the law allowed older riders to ride without helmets? What is the evidence?

Step-by-Step Example   ANOTHER ONE-PROPORTION z-TEST

Question: Has helmet use in Florida declined among riders under the age of 21 subsequent to the change in the helmet laws?

---

[3]The approach to statistical inference known as Bayesian Statistics addresses the question in just this way, but it requires more advanced mathematics and more assumptions. See p. 380 for more about the founding father of this approach.

THINK ➡ **Plan** State the problem and discuss the variables and the W's.

Hypotheses The null hypothesis is established by the rate set before the change in the law. The study was concerned with safety, so they'll want to know of any decline in helmet use, making this a lower-tail test.

I want to know whether the rate of helmet wearing among Florida's motorcycle riders under the age of 21 decreased after the law changed to allow older riders to go without helmets. The proportion before the law was passed was 60% so I'll use that as my null hypothesis value. The alternative is one-sided because I'm interested only in seeing if the rate decreased. I have data from accident records showing 396 of 781 young riders were wearing helmets.

$$H_O: p = 0.60$$
$$H_A: p < 0.60$$

SHOW ➡ **Model** Check the conditions.

✓ **Independence Assumption:** The data are for riders involved in accidents during a three-year period. Individuals are independent of one another.

✗ **Randomization Condition:** No randomization was applied, but we are considering these riders involved in accidents to be a representative sample of all riders. We should take care in generalizing our conclusions.

✓ **10% Condition:** These 781 riders are a small sample of a larger population of all young motorcycle riders.

✓ **Success/Failure Condition:** We'd expect $np = 781(0.6) = 468.6$ helmeted riders and $nq = 781(0.4) = 312.4$ non-helmeted. Both are at least 10.

Specify the sampling distribution model and name the test.

The conditions are satisfied, so I can use a Normal model and perform a one-proportion z-test.

SHOW ➡ **Mechanics** Find the standard deviation of the sampling model using the hypothesized proportion.

There were 396 helmet wearers among the 781 accident victims.

$$\hat{p} = \frac{396}{781} = 0.507$$

$$SD(\hat{p}) = \sqrt{\frac{p_0 q_0}{n}} = \sqrt{\frac{(0.60)(0.40)}{781}} = 0.0175$$

Find the z-score for the observed proportion.

$$z = \frac{\hat{p} - p_0}{SD(\hat{p})} = \frac{0.507 - 0.60}{0.0175} = -5.31$$

(continued)

Make a picture. Sketch a Normal model centered at the hypothesized helmet rate of 60%. This is a lower-tail test, so shade the region to the left of the observed rate.

0.507   0.60 $\hat{p}$

Given this z-score, the P-value is obviously very low.

The observed helmet rate is 5.31 standard deviations below the former rate. The corresponding P-value is less than 0.001.

TELL ➡ **Conclusion** Link the P-value to your decision about the null hypothesis, and then state your conclusion in context.

The very small P-value says that if the true rate of helmet-wearing among riders under 21 were still 60%, the probability of observing a rate no higher than 50.7% in a sample like this is less than 1 chance in 1000, so I reject the null hypothesis. There is strong evidence that there has been a decline in helmet use among riders under 21.

The P-value in the helmet example is quite small—less than 0.001. That's strong evidence to suggest that the rate has decreased since the law was changed. But it doesn't say that it was "a lot lower." To answer that question, you'd need to construct a confidence interval:

$$\hat{p} \pm z^* \sqrt{\frac{\hat{p}\hat{q}}{n}} = 0.507 \pm 1.96(0.0175) = (0.472, 0.542)$$

(using 95% confidence).

There is strong evidence that the rate is no longer 60%, but the small P-value by itself says nothing about how much lower the rate might be. The confidence interval provides that information; the rate seems to be closer to 50% now. Whether a change from 60% to 50% makes an important difference in safety is a judgment that depends on the situation, but not on the P-value. Not coincidentally, on July 1, 2008, Florida required a motorcycle "endorsement" for all motorcycle riders. For riders under 21, that requires a motorcycle safety course. Although only about 70% of motorcycle riders are endorsed, the percentage of unendorsed riders involved in crashes dropped considerably after 2008.[4]

How Guilty Is the Suspect?

We might like to know $P(H_0 | \text{data})$, but when you think about it, we can't talk about the probability that the null hypothesis is true. The null is not a random event, so either it is true or it isn't. The data, however, are random in the sense that if we were to repeat a randomized experiment or draw another random sample, we'd get different data and expect to find a different statistic value. So we can talk about the probability of the data given the null hypothesis, and that's the P-value.

But it does make sense that the smaller the P-value, the more confident we can be in declaring that we doubt the null hypothesis. Think again about the jury trial. Our null hypothesis is that the defendant is innocent. Then the evidence starts rolling in.

(continued)

A car the same color as his was parked in front of the bank. Well, there are lots of cars that color. The probability of that happening (given his innocence) is pretty high, so we're not persuaded that he's guilty. The bank's security camera showed the robber was male and about the defendant's height and weight. Hmmm. Could that be a coincidence? If he's innocent, then it's a little less likely that the car and description would *both* match, so our P-value goes down. We're starting to question his innocence a little. Witnesses said the robber wore a blue jacket just like the one the police found in a garbage can behind the defendant's house. Well, if he's innocent, then that doesn't seem very likely, does it? If he's really innocent, the probability that all of these could have happened is getting pretty low. Now our P-value may be small enough to be called "beyond a reasonable doubt" and lead to a conviction. Each new piece of evidence strains our skepticism a bit more. The more compelling the evidence—the more *unlikely* it would be were he innocent—the more convinced we become that he's guilty.

But even though it may make *us* more confident in declaring him guilty, additional evidence does not make *him* any guiltier. Either he robbed the bank or he didn't. Additional evidence (like the teller picking him out of a police lineup) just makes us more confident that we did the right thing when we convicted him. The lower the P-value, the more comfortable we feel about our decision to reject the null hypothesis, but the null hypothesis doesn't get any more false.

> "The wise man proportions his belief to the evidence."
>
> —*David Hume,*
> *"Enquity Concerning Human Understanding," 1748*

For Example THINKING ABOUT THE P-VALUE

RECAP: A *New England Journal of Medicine* paper reported that the seven-year risk of heart attack in diabetes patients taking the drug Avandia was increased from the baseline of 20.2% to an estimated risk of 28.9% and said the P-value was 0.03.

QUESTION: How should the P-value be interpreted in this context?

ANSWER: The $P\text{-value} = P(\hat{p} \geq 28.9\% \mid p = 20.2\%)$. That is, it's the probability of seeing such a high heart attack rate among the people studied if, in fact, taking Avandia really didn't increase the risk at all.

What to Do with a High P-Value

A S *Video*: **Is There Evidence for Therapeutic Touch?** This video shows the experiment and tells the story.

A S *Activity*: **Testing Therapeutic Touch.** Perform the one-proportion z-test using *ActivStats* technology. The test in *ActivStats* is two-sided. Do you think this is the appropriate choice?

Therapeutic touch (TT), taught in many schools of nursing, is a therapy in which the practitioner moves her hands near, but does not touch, a patient in an attempt to manipulate a "human energy field." Therapeutic touch practitioners believe that by adjusting this field they can promote healing. However, no instrument has ever detected a human energy field, and no experiment has ever shown that TT practitioners can detect such a field.

In 1998, the *Journal of the American Medical Association* published a paper reporting work by a then nine-year-old girl.[5] She had performed a simple experiment in which she challenged 15 TT practitioners to detect whether her unseen hand was hovering over their left or right hand (selected by the flip of a coin).

The practitioners "warmed up" with a period during which they could see the experimenter's hand, and each said that they could detect the girl's human energy field. Then a screen was placed so that the practitioners could not see the girl's hand, and they attempted 10 trials each. Overall, of 150 trials, the TT practitioners were successful only 70 times—a success proportion of 46.7%.

The null hypothesis here is that the TT practitioners were just guessing. If that were the case, since the hand was chosen using a coin flip, the practitioners would guess correctly 50% of the time. So the null hypothesis is that $p = 0.5$ and the alternative that they could actually detect a human energy field is (one-sided) $p > 0.5$.

---

[5]L. Rosa, E. Rosa, L. Sarner, and S. Barrett, "A Close Look at Therapeutic Touch," *JAMA* 279(13) [1 April 1998]: 1005–1010.

What would constitute evidence that they weren't guessing? Certainly, a very high proportion of correct guesses out of 150 would convince most people. Exactly how high the proportion of correct guesses has to be for you to reject the null hypothesis depends on how small a P-value you need to be convinced (which, in turn, depends on how often you're willing to make mistakes—a topic we'll discuss later in the chapter).

But let's look again at the TT practitioners' proportion. Does it provide any evidence that they weren't guessing? The proportion of correct guesses is 46.7%—that's *less* than the hypothesized value, not greater! When we find $SD(\hat{p}) = 0.041$ (or 4.1%) we can see that 46.7% is almost 1 SD *below* the hypothesized proportion:

$$SD(\hat{p}) = \sqrt{\frac{p_0 q_0}{n}} = \sqrt{\frac{(0.5)(0.5)}{150}} \approx 0.041$$

The observed proportion, $\hat{p}$, is 0.467.

$$z = \frac{\hat{p} - p_0}{SD(\hat{p})} = \frac{0.467 - 0.5}{0.041} = -0.805$$

0.467 0.50 $\hat{p}$

The observed success rate is 0.805 standard deviations below the hypothesized mean.

$$\text{P-value} = p(z > -0.805) = 0.790$$

If the practitioners had been highly successful, we would have seen a low P-value. In that case, we would then have concluded that they could actually detect a human energy field.

But that's not what happened. What we observed was a $\hat{p} = 0.467$ success rate. The P-value for this proportion is greater than 0.5 because the observed value is on the "wrong" side of the null hypothesis value. To convince us, the practitioners should be doing better than guessing, not worse!

Obviously, we won't be rejecting the null hypothesis; for us to reject it, the P-value would have to be quite small.

Big P-values just mean that what we've observed isn't surprising. That is, the results are in line with our assumption that the null hypothesis models the world, so we have no reason to reject it. A big P-value doesn't prove that the null hypothesis is true, but it certainly offers no evidence that it's *not* true. When we see a large P-value, all we can say is that we "don't reject the null hypothesis."

For Example MORE ABOUT P-VALUES

RECAP: The question of whether the diabetes drug Avandia increased the risk of heart attack was raised by a study in the *New England Journal of Medicine*. This study estimated the seven-year risk of heart attack to be 28.9% and reported a P-value of 0.03 for a test of whether this risk was higher than the baseline seven-year risk of 20.2%. An earlier study (the ADOPT study) had estimated the seven-year risk to be 26.9% and reported a P-value of 0.27.

QUESTION: Why did the researchers in the ADOPT study not express alarm about the increased risk they had seen?

ANSWER: A P-value of 0.27 means that a heart attack rate at least as high as the one they observed could be expected in 27% of similar experiments even if, in fact, there were no increased risk from taking Avandia. That's not remarkable enough to reject the null hypothesis. In other words, the ADOPT study wasn't convincing.

Alpha Levels

A S *Activity*: **Rejecting the Null Hypothesis.** See alpha levels at work in the animated hypothesis-testing tool.

■ **NOTATION ALERT**

The first Greek letter, α, is used in Statistics for the threshold value of a hypothesis test. You'll hear it referred to as the alpha level. Common values are 0.10, 0.05, 0.01, and 0.001.

Sir Ronald Fisher (1890–1962) was one of the founders of modern Statistics.

It Could Happen to You! Of course, if the null hypothesis *is* true, no matter what alpha level you choose, you still have a probability α of rejecting the null hypothesis by mistake. This is the rare event we want to protect ourselves against. When we do reject the null hypothesis, no one ever thinks that *this* is one of those rare times. As statistician Stu Hunter notes, '*The statistician says 'rare events do happen— but not to me!*''

Sometimes we need to make a firm decision about whether or not to reject the null hypothesis. A jury must decide whether the evidence reaches the level of "beyond a reasonable doubt." A business must select a Web design. You need to decide which section of Statistics to enroll in.

When the P-value is small, it tells us that our data are rare, *given the null hypothesis*. As humans, we are suspicious of rare events. If the data are "rare enough," we just don't think that could have happened due to chance. Since the data *did* happen, something must be wrong. All we can do now is reject the null hypothesis.

But how rare is "rare"?

We can define "rare event" arbitrarily by setting a threshold for our P-value. If our P-value falls below that point, we'll reject the null hypothesis, deeming the results statistically significant. The threshold is called an **alpha level.** Not surprisingly, it's labeled with the Greek letter α. Common α levels are 0.10, 0.05, and 0.01. You have the option— almost the *obligation*—to consider your alpha level carefully and choose an appropriate one for the situation. If you're assessing the safety of air bags, you'll want a low alpha level; even 0.01 might not be low enough. If you're just wondering whether folks prefer their pizza with or without pepperoni, you might be happy with $\alpha = 0.10$. It can be hard to justify your choice of α, though, so often people arbitrarily choose 0.05. Note, however: You must select the alpha level *before* you look at the data. Otherwise you can be accused of cheating by tuning your alpha level to suit the data.

> **Where Did the Value 0.05 Come From?** In 1931, in a famous book called *The Design of Experiments*, Sir Ronald Fisher discussed the amount of evidence needed to reject a null hypothesis. He said that it was *situation dependent*, but remarked, somewhat casually, that for many scientific applications, 1 out of 20 *might be* a reasonable value. Since then, some people—indeed some entire disciplines—have treated the number 0.05 as sacrosanct.

The alpha level is also called the **significance level.** When we reject the null hypothesis, we say that the test is "significant at that level." For example, we might say that we reject the null hypothesis "at the 5% level of significance."

What can you say if the P-value does not fall below α?

When you have not found sufficient evidence to reject the null according to the standard you have established, you should say that "The data have failed to provide sufficient evidence to reject the null hypothesis." Don't say that you "accept the null hypothesis." You certainly haven't proven or established it; it was merely assumed to begin with. Say that you've failed to reject it.

The automatic nature of the reject/fail-to-reject decision when we use an alpha level may make you uncomfortable. If your P-value falls just slightly above your alpha level, you're not allowed to reject the null. Yet a P-value just barely below the alpha level leads to rejection. If this bothers you, you're in good company. Many statisticians think it better to report the P-value than to base a decision on an arbitrary alpha level.

> **It's in the Stars**
>
> Some disciplines carry the idea further and code P-values by their size. In this scheme, a P-value between 0.05 and 0.01 gets highlighted by *. A P-value between 0.01 and 0.001 gets **, and a P-value less than 0.001 gets ***. This can be a convenient summary of the weight of evidence against the null hypothesis if it's not taken too literally. But we warn you against taking the distinctions too seriously and against making a black-and-white decision near the boundaries. The boundaries are a matter of tradition, not science; there is nothing special about 0.05. A P-value of 0.051 should be looked at very seriously and not casually thrown away just because it's larger than 0.05, and one that's 0.009 is not very different from one that's 0.011.

When you decide to declare a verdict, it's always a good idea to report the P-value as an indication of the strength of the evidence. Sometimes it's best to report that the conclusion is not yet clear and to suggest that more data be gathered. (In a trial, a jury may "hang" and be unable to return a verdict.) In these cases, the P-value is the best summary we have of what the data say or fail to say about the null hypothesis.

Practical vs. Statistical Significance

What do we mean when we say that a test is statistically significant? All we mean is that the test statistic had a P-value lower than our alpha level. Don't be lulled into thinking that statistical significance carries with it any sense of practical importance or impact.

For large samples, even small, unimportant ("insignificant") deviations from the null hypothesis can be statistically significant. On the other hand, if the sample is not large enough, even large financially or scientifically "significant" differences may not be statistically significant.

It's good practice to report the magnitude of the difference between the observed statistic value and the null hypothesis value (in the data units) along with the P-value on which we base statistical significance.

> **Statistically Significant Yes. But Is It Important?** A large insurance company mined its data and found a statistically significant ($P = 0.04$) difference between the mean value of policies sold in 2001 and 2002. The difference in the mean values was $9.83. Even though it was statistically significant, management did not see this as an important difference when a typical policy sold for more than $1000. On the other hand, even a clinically important improvement of 10% in cure rate with a new treatment is not likely to be statistically significant in a study of fewer than 225 patients. A small clinical trial would probably not be conclusive.

Confidence Intervals and Hypothesis Tests

For the motorcycle helmet example, a 95% confidence interval would give $0.507 \pm 1.96 \times 0.0179 = (0.472, 0.542)$, or 47.2% to 54.2%. If the previous rate of helmet compliance had been, say, 50%, we would not have been able to reject the null hypothesis because 50% is in the interval, so it's a plausible value. Indeed, *any* hypothesized value for the true proportion of helmet wearers in this interval is consistent with the data. Any value outside the confidence interval would make a null hypothesis that we would reject, but we'd feel more strongly about values far outside the interval.

Confidence intervals and hypothesis tests are built from the same calculations.[6] They have the same assumptions and conditions. As we have just seen, you can approximate a hypothesis test by examining the confidence interval. As an alternative to finding a P-value. You can just ask whether the null hypothesis value is consistent with a confidence interval for the parameter at the corresponding confidence level. Because confidence intervals are naturally two-sided, they correspond to two-sided tests. For example, a 95% confidence interval corresponds to a two-sided hypothesis test at $\alpha = 5\%$. In general, a confidence interval with a confidence level of $C\%$ corresponds to a two-sided hypothesis test with an α level of $100 - C\%$.

The relationship between confidence intervals and one-sided hypothesis tests is more complicated. For a one-sided test with $\alpha = 5\%$, the corresponding confidence interval has a confidence level of 90%—that's 5% in each tail. In general, a one-sided significance level α corresponds to a $(100 - 2\alpha)\%$ confidence interval.

---

[6]As we saw in Chapter 19, this is not *exactly* true for proportions. For a confidence interval, we estimate the standard deviation of $\hat{p}$ from $\hat{p}$ itself. Because we estimate it from the data, we have a *standard error*. For the corresponding hypothesis test, we use the model's standard deviation for $\hat{p}$, based on the null hypothesis value p_0. When $\hat{p}$ and p_0 are close, these calculations give similar results. When they differ, you're likely to reject H_0 (because the observed proportion is far from your hypothesized value). In that case, you're better off building your confidence interval with a standard error estimated from the data.

For Example MAKING A DECISION BASED ON A CONFIDENCE INTERVAL

RECAP: The baseline seven-year risk of heart attacks for diabetics is 20.2%. In 2007 a *NEJM* study reported a 95% confidence interval equivalent to 20.8% to 40.0% for the risk among patients taking the diabetes drug Avandia.

QUESTION: What did this confidence interval suggest to the FDA about the safety of the drug?

ANSWER: *The FDA could be 95% confident that the interval from 20.8% to 40.0% included the true risk of heart attack for diabetes patients taking Avandia. Because the lower limit of this interval was higher than the baseline risk of 20.2%, there was evidence of an increased risk.*

Just Checking

1. An experiment to test the fairness of a roulette wheel gives a *z*-score of 0.62. What would you conclude?

2. In the last chapter we encountered a bank that wondered if it could get more customers to make payments on delinquent balances by sending them a DVD urging them to set up a payment plan. Well, the bank just got back the results on their test of this strategy. A 90% confidence interval for the success rate is (0.29, 0.45). Their old send-a-letter method had worked 30% of the time. Can you reject the null hypothesis that the proportion is still 30% at $\alpha = 0.05$? Explain.

3. Given the confidence interval the bank found in their trial of DVDs, what would you recommend that they do? Should they scrap the DVD strategy?

Step-by-Step Example WEAR THAT SEATBELT!

Teens are at the greatest risk of being killed or injured in traffic crashes. According to the National Highway Traffic Safety Administration, 65% of young people killed were not wearing a safety belt. In 2001, a total of 3322 teens were killed in motor vehicle crashes, an average of 9 teenagers a day. Because many of these deaths could easily be prevented by the use of safety belts, several states have begun "Click It or Ticket" campaigns in which increased enforcement and publicity have resulted in significantly higher seatbelt use. Overall use in Massachusetts quickly increased from 51% in 2002 to 64.8% in 2006, with a goal of surpassing the national average of 82%. Recently, a local newspaper reported that a roadblock resulted in 23 tickets to drivers who were unbelted out of 134 stopped for inspection.

Question: Does this provide evidence that the goal of over 82% compliance was met?

Let's use a confidence interval to test this hypothesis.

THINK ➡ **Plan** State the problem and discuss the variables and the W's.

Hypotheses The null hypothesis is that the compliance rate is only 82%. The alternative is that it is now higher. It's clearly a one-sided test, so if we use a confidence interval, we'll have to be careful about what level we use.

The data come from a local newspaper report that tells the number of tickets issued and number of drivers stopped at a recent roadblock. I want to know whether the rate of compliance with the seatbelt law is greater than 82%.

$$H_0: p = 0.82$$
$$H_A: p > 0.82$$

(continued)

Model Think about the assumptions and check the conditions.

✓ **Independence Assumption:** Drivers are not likely to influence one another when it comes to wearing a seatbelt.

✓ **Randomization Condition:** This wasn't a random sample, but I assume these drivers are representative of the driving public.

✓ **10% Condition:** The police stopped fewer than 10% of all drivers.

We are finding a confidence interval, so we work from the data rather than the null model.

✓ **Success/Failure Condition:** There were 111 successes and 23 failures, both at least 10. The sample is large enough.

State your method.

Under these conditions, the sampling model is Normal. I'll create a one-proportion z-interval.

SHOW ➡ **Mechanics** Write down the given information, and determine the sample proportion.

$n = 134$, so

$$\hat{p} = \frac{111}{134} = 0.828 \text{ and}$$

To use a confidence interval, we need a confidence level that corresponds to the alpha level of the test. If we use $\alpha = 0.05$, we should construct a 90% confidence interval, because this is a one-sided test.

$$SE(\hat{p}) = \sqrt{\frac{\hat{p}\hat{q}}{n}} = \sqrt{\frac{(0.828)(0.172)}{134}} = 0.033$$

That will leave 5% on *each* side of the observed proportion. Determine the standard error of the sample proportion and the margin of error. The critical value is $z^* = 1.645$.

$$ME = z^* \times SE(\hat{p})$$
$$= 1.645(0.033) = 0.054$$

The confidence interval is

estimate $\pm$ margin of error.

The 90% confidence interval is

$$0.828 \pm 0.054 \text{ or}$$
$$(0.774, 0.882).$$

TELL ➡ **Conclusion** Link the confidence interval to your decision about the null hypothesis, and then state your conclusion in context.

I am 90% confident that between 77.4% and 88.2% of all drivers wear their seatbelts. Because the hypothesized rate of 82% is within this interval, I do not reject the null hypothesis. There is insufficient evidence to conclude that the campaign was truly effective and now more than 82% of all drivers are wearing seatbelts.

The upper limit of the confidence interval shows it's possible that the campaign is quite successful, but the small sample size makes the interval too wide to be very specific.

*A 95% Confidence Interval for Small Samples

When the **Success/Failure Condition** fails, all is not lost. A simple adjustment to the calculation lets us make a 95% confidence interval anyway.

All we do is add four *phony* observations—two to the successes, two to the failures. So instead of the proportion $\hat{p} = \dfrac{y}{n}$, we use the adjusted proportion $\tilde{p} = \dfrac{y + 2}{n + 4}$ and, for convenience, we write $\tilde{n} = n + 4$. We modify the interval by using these adjusted values:

$$\tilde{p} \pm z^* \sqrt{\frac{\tilde{p}(1 - \tilde{p})}{\tilde{n}}}.$$

Called a "plus-four" interval, this adjusted form gives better performance overall[7] and works much better for proportions near 0 or 1. It has the additional advantage that we no longer need to check the **Success/Failure Condition** that $n\hat{p}$ and $n\hat{q}$ are greater than 10. Because of these properties, the use of plus-four intervals is becoming increasingly common in modern statistics.

For Example   AN AGRESTI-COULL "PLUS-FOUR" INTERVAL

Surgeons examined their results to compare two methods for a surgical procedure used to alleviate pain on the outside of the wrist. A new method was compared with the traditional "freehand" method for the procedure. Of 45 operations using the "freehand" method, three were unsuccessful, for a failure rate of 6.7%. With only 3 failures, the data don't satisfy the **Success/Failure Condition**, so we can't use a standard confidence interval.

QUESTION: What's the confidence interval using the "plus-four" method?

ANSWER: There were 42 successes and 3 failures. Adding 2 "pseudo-successes" and 2 "pseudo-failures," we find

$$\tilde{p} = \frac{3 + 2}{45 + 4} = 0.102$$

A 95% confidence interval is then

$$0.102 \pm 1.96 \sqrt{\frac{0.102(1 - 0.102)}{49}} = 0.102 \pm 0.085 \text{ or } (0.017, 0.187).$$

Notice that although the observed failure rate of 0.067 is contained in the interval, it is not at the center of the interval—something we haven't seen with any of the other confidence intervals we've considered.

Making Errors

Nobody's perfect. Even with lots of evidence, we can still make the wrong decision. In fact, when we perform a hypothesis test, we can make mistakes in *two* ways:

I. The null hypothesis is true, but we mistakenly reject it.
II. The null hypothesis is false, but we fail to reject it.

These two types of errors are known as **Type I** and **Type II errors.** One way to keep

---

[7]By "better performance," we mean that a 95% confidence interval has more nearly a 95% chance of covering the true population proportion. Simulation studies have shown that our original, simpler confidence interval in fact is less likely than 95% to cover the true population proportion when the sample size is small or the proportion very close to 0 or 1. The original idea for this method can be attributed to E. B. Wilson. The simpler approach discussed here was proposed by Agresti and Coull (A. Agresti and B. A. Coull, "Approximate Is Better Than 'Exact' for Interval Estimation of Binomial Proportions," *The American Statistician*, 52[1998]: 119–129).

A S *Activity*: **Hypothesis Tests Are Random.** Simulate hypothesis tests and watch Type I errors occur. When you conduct real hypothesis tests you'll never know, but simulation can tell you when you've made an error.

■ NOTATION ALERT

In Statistics, α is almost always saved for the alpha level. But β is also used for the parameters of a linear model.

Finding β The null hypothesis specifies a single value for the parameter. So it's easy to calculate the probability of a Type I error. But the alternative gives a whole range of possible values, and one could find a β for any alternative parameter value.

What n Do We Need? We have seen ways to find a sample size by specifying the margin of error. Choosing the sample size to achieve a specified β (for a particular alternative value) is sometimes more appropriate, but the calculation is more complex and lies beyond the scope of this book.

the names straight is to remember that we start by assuming the null hypothesis is true, so a Type I error is the first kind of error we could make.

In medical disease testing, the null hypothesis is usually the assumption that a person is healthy. The alternative is that he or she has the disease we're testing for. So a Type I error is a *false positive:* A healthy person is diagnosed with the disease. A Type II error, in which an infected person is diagnosed as disease free, is a *false negative.*

Which type of error is more serious depends on the situation. In the jury trial, a Type I error occurs if the jury convicts an innocent person. A Type II error occurs if the jury lets a guilty person go free. Which seems more serious? In medical diagnosis, a false negative could mean that a sick patient goes untreated. A false positive might mean that the person must undergo further tests. In a Statistics final exam (with H_0: the student has learned only 60% of the material), a Type I error would be passing a student who in fact learned less than 60% of the material, while a Type II error would be failing a student who knew enough to pass. Which of these errors seems more serious? It depends on the situation, the cost, and your point of view.

Here's an illustration of the possibilities:

How often will a Type I error occur? It happens when the null hypothesis is true but we've had the bad luck to draw an unusual sample. To reject H_0, the P-value must fall below α. When H_0 is true, that happens *exactly* with probability α. So when you choose level α, you're setting the probability of a Type I error to α.

What if H_0 is not true? Then we can't possibly make a Type I error. You can't get a false positive from a sick person. A Type I error can happen only when H_0 is true.

When H_0 is false but we fail to reject it, we have made a Type II error. We assign the letter β to the probability of this mistake. What's the value of β? That's harder to assess than α because it depends on what the value of the parameter really is, and we don't know that.

We could reduce β for *all* alternative parameter values by increasing α. By making it easier to reject the null, we'd be more likely to reject it whether it's true or not. So we'd reduce β, the chance that we fail to reject a false null—but we'd make more Type I errors. This tension between Type I and Type II errors is inevitable. In the political arena, think of the ongoing debate between those who favor provisions to reduce Type I errors in the courts (supporting Miranda rights, requiring warrants for wiretaps, providing legal representation for those who can't afford it) and those who advocate changes to reduce Type II errors (admitting into evidence confessions made when no lawyer is present, eavesdropping on conferences with lawyers, restricting paths of appeal, etc.).

The only way to reduce *both* types of error is to collect more evidence or, in statistical terms, to collect more data. Too often, studies fail because their sample sizes are too small to detect the change they are looking for.

Of course, what we really want to do is to detect a false null hypothesis. When H_0 is false and we reject it, we have done the right thing. A test's ability to detect a false null hypothesis is called the **power** of the test. In a jury trial, power is the ability of the criminal justice system to convict people who are guilty—a good thing! We'll have a lot more to say about power soon.

For Example THINKING ABOUT ERRORS

RECAP: A published study found the risk of heart attack to be increased in patients taking the diabetes drug Avandia. The issue of the *New England Journal of Medicine* (*NEJM*) in which that study appeared also included an editorial that said, in part, "A few events either way might have changed the findings for myocardial infarction[8] or for death from cardiovascular causes. In this setting, the possibility that the findings were due to chance cannot be excluded."

QUESTION: What kind of error would the researchers have made if, in fact, their findings were due to chance? What could be the consequences of this error?

ANSWER: *The null hypothesis said the risk didn't change, but the researchers rejected that model and claimed evidence of a higher risk. If these findings were just due to chance, they rejected a true null hypothesis—a Type I error.*

If, in fact, Avandia carried no extra risk, then patients might be deprived of its benefits for no good reason.

Power

When we failed to reject the null hypothesis about TT practitioners, did we prove that they were just guessing? No, it could be that they actually *can* discern a human energy field but we just couldn't tell. For example, suppose they really have the ability to get 53% of the trials right but just happened to get only 47% in our experiment. Our confidence interval shows that with these data we wouldn't have rejected the null even though the true proportion was actually greater than 50%. That means we would have made a Type II error because we failed to detect their ability.

Remember, we can never prove a null hypothesis true. We can only fail to reject it. But when we fail to reject a null hypothesis, it's natural to wonder whether we looked hard enough. Might the null hypothesis actually be false and our test too weak to tell?

When the null hypothesis actually *is* false, we hope our test is strong enough to reject it. We'd like to know how likely we are to succeed. The power of the test gives us a way to think about that. The **power** of a test is the probability that it correctly rejects a false null hypothesis. When the power is high, we can be confident that we've looked hard enough. We know that β is the probability that a test *fails* to reject a false null hypothesis, so the power of the test is the probability that it *does* reject: $1 - \beta$.

Whenever a study fails to reject its null hypothesis, the test's power comes into question. Was the sample size big enough to detect an effect had there been one? Might we have missed an effect large enough to be interesting just because we failed to gather sufficient data or because there was too much variability in the data we could gather? The therapeutic touch experiment failed to reject the null hypothesis that the TT practitioners were just guessing. Might the problem be that the experiment simply lacked adequate power to detect their ability?

For Example ERRORS AND POWER

RECAP: The study of Avandia published in the *NEJM* combined results from 47 different trials—a method called *meta-analysis.* The drug's manufacturer, GlaxoSmithKline (GSK), issued a statement that pointed out, "Each study is designed differently and looks at unique questions: For example, individual studies vary in size and length, in the type of patients who participated, and in the outcomes they investigate." Nevertheless, by combining data from many studies, meta-analyses can achieve a much larger sample size.

(continued)

---

[8]Doctorese for "heart attack."

> **QUESTION:** How could this larger sample size help?
>
> **ANSWER:** If Avandia really did increase the seven-year heart attack rate, doctors needed to know. To overlook that would have been a Type II error (failing to detect a false null hypothesis), resulting in patients being put at greater risk. Increasing the sample size could increase the power of the analysis, making it more likely that researchers will detect the danger if there is one.

Effect Size

 A S *Activity:* **The Power of a Test.** Power is a concept that's much easier to understand when you can visualize what's happening.

When we think about power, we imagine that the null hypothesis is false. The value of the power depends on how far the truth lies from the null hypothesis value. We call the distance between the null hypothesis value, p_0, and the truth, p, the **effect size.** The power of a test depends directly on the effect size. It's easier to see larger effects, so the farther p_0 is from p, the greater the power. If the therapeutic touch practitioners were in fact able to detect human energy fields 90% of the time, it should be easy to see that they aren't guessing. With an effect size this large, we'd have a powerful test.

Small effects are more difficult to detect. If the true success rate of the TT practitioner's were only 53%, we'd probably miss that, committing a Type II error. To have high power to detect such a small effect size (and reject H_0), we'd need a much larger sample size.

Whenever a hypothesis test fails to reject the null, the question of power can arise. For example, in the therapeutic touch experiment, with a sample size of 150, if the researchers took an increase to 75% as a reasonably interesting effect size (keeping in mind that 50% is the level of guessing) and used an α-level of 0.05, they could determine that the TT experiment would have been able to detect such an ability with a power of 99.99%. So there is only a very small chance that their study would have failed to detect a practitioner's ability at that level, had it existed. If, on the other hand, they thought that even an increase to 51% was interesting, then they would need a sample size of over 20,000 trials to have a 90% chance of detecting it!

✔ Just Checking

4. Remember our bank that's sending out DVDs to try to get customers to make payments on delinquent loans? It is looking for evidence that the costlier DVD strategy produces a higher success rate than the letters it has been sending. Explain what a Type I error is in this context and what the consequences would be to the bank.

5. What's a Type II error in the bank experiment context, and what would the consequences be?

6. For the bank, which situation has higher power: a strategy that works really well, actually getting 60% of people to pay off their balances, or a strategy that barely increases the payoff rate to 32%? Explain briefly.

A Picture Worth $\dfrac{1}{P(z > 3.09)}$ Words

It makes intuitive sense that the larger the effect size, the easier it should be to see it. Obtaining a larger sample size decreases the probability of a Type II error, so it increases the power. It also makes sense that the more willing we are to accept a Type I error, the less likely we will be to make a Type II error.

Figure 20.1

The power of a test is the probability that it rejects a false null hypothesis. The upper figure shows the null hypothesis model. We'd reject the null in a one-sided test if we observed a value of $\hat{p}$ in the red region to the right of the critical value, p^*. The lower figure shows the true model. If the true value of p is greater than p_0, then we're more likely to observe a value that exceeds the critical value and make the correct decision to reject the null hypothesis. The power of the test is the green region on the right of the lower figure. Of course, even drawing samples whose observed proportions are distributed around p, we'll sometimes get a value in the red region on the left and make a Type II error of failing to reject the null.

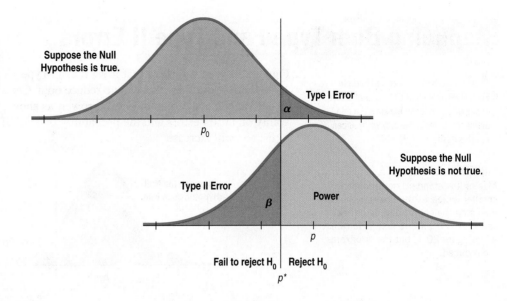

■ **NOTATION ALERT**

We've attached symbols to many of the p's. Let's keep them straight. p is a true proportion parameter. p_0 is a hypothesized value of p. $\hat{p}$ is an observed proportion. p^* is a critical value of a proportion corresponding to a specified α.

Fisher and $\alpha = 0.05$ Why did Sir Ronald Fisher suggest 0.05 as a criterion for testing hypotheses? It turns out that he had in mind small initial studies. Small studies have relatively little power. Fisher was concerned that they might make too many Type II errors—failing to discover an important effect—if too strict a criterion were used. Once a test failed to reject a null hypothesis, it was unlikely that researchers would return to that hypothesis to try again.

On the other hand, the increased risk of Type I errors arising from a generous criterion didn't concern him as much for exploratory studies because these are ordinarily followed by a replication or a larger study. The probability of a Type I error is α—in this case, 0.05. The probability that two independent studies would both make Type I errors is $0.05 \times 0.05 = 0.0025$, so Fisher was confident that Type I errors in initial studies were not a major concern.

The widespread use of the relatively generous 0.05 criterion even in large studies is most likely not what Fisher had in mind.

Figure 20.1 shows a good way to visualize the relationships among these concepts. Suppose we are testing $H_0: p = p_0$ against the alternative $H_A: p > p_0$. We'll reject the null if the observed proportion, $\hat{p}$, is big enough. By big enough, we mean $\hat{p} > p^*$ for some critical value, p^* (shown as the red region in the right tail of the upper curve). For example, we might be willing to believe the ability of therapeutic touch practitioners if they were successful in 65% of our trials. This is what the upper model shows. It's a picture of the sampling distribution model for the proportion if the null hypothesis were true. We'd make a Type I error whenever the sample gave us $\hat{p} > p^*$, because we would reject the (true) null hypothesis. And unusual samples like that would happen only with probability α.

In reality, though, the null hypothesis is rarely *exactly* true. The lower probability model supposes that H_0 is not true. In particular, it supposes that the true value is p, not p_0. (Perhaps the TT practitioner really can detect the human energy field 72% of the time.) It shows a distribution of possible observed $\hat{p}$ values around this true value. Because of sampling variability, sometimes $\hat{p} < p^*$ and we fail to reject the (false) null hypothesis. Suppose a TT practitioner with a true ability level of 72% is actually successful on fewer than 65% of our tests. Then we'd make a Type II error. The area under the curve to the left of p^* in the bottom model represents how often this happens. The probability is β. In this picture, β is less than half, so most of the time we *do* make the right decision. The *power* of the test—the probability that we make the right decision—is shown as the green region to the right of p^*. It's $1 - \beta$.

How often we correctly reject H_0 when it's *false* depends on the effect size. We can see from the picture that if the effect size were larger (the true proportion were farther above the hypothesized value), the bottom curve would shift to the right, making the power greater.

We can see several important relationships from this figure:

- Power = $1 - \beta$.
- Reducing α to lower the chance of committing a Type I error will move the critical value, p^*, to the right (in this example). This will have the effect of increasing β, the probability of a Type II error, and correspondingly reducing the power.
- The larger the real difference between the hypothesized value, p_0, and the true population value, p, the smaller the chance of making a Type II error and the greater the power of the test. If the two proportions are very far apart, the two models will barely overlap, and we will not be likely to make any Type II errors at all—but then, we are unlikely to really need a formal hypothesis-testing procedure to see such an obvious difference. If the TT practitioners were successful almost all the time, we'd be able to see that with even a small experiment.

Reducing Both Type I and Type II Errors

A **S** *Activity*: **Power and Sample Size.** Investigate how the power of a test changes with the sample size. The interactive tool is really the only way you can see this easily.

Figure 20.2

Making the standard deviations smaller increases the power without changing the corresponding critical value. The means are just as far apart as in Figure 20.1, but the error rates are reduced.

Figure 20.1 seems to show that if we reduce Type I error, we automatically must increase Type II error. But there is a way to reduce both. Can you think of it?

If we can make both curves narrower, as shown in Figure 20.2, then both the probability of Type I errors and the probability of Type II errors will decrease, and the power of the test will increase.

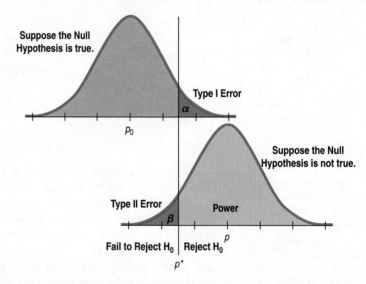

TI-*nspire*

Errors and power. Explore the relationships among Type I and Type II errors, sample size, effect size, and the power of a test.

How can we accomplish that? The only way is to reduce the standard deviations by increasing the sample size. (Remember, these are pictures of sampling distribution models, not of data.) Increasing the sample size works regardless of the true population parameters. But recall the curse of diminishing returns. The standard deviation of the sampling distribution model decreases only as the *square root* of the sample size, so to halve the standard deviations we must *quadruple* the sample size.

For Example SAMPLE SIZE, ERRORS, AND POWER

RECAP: The meta-analysis of the risks of heart attacks in patients taking the diabetes drug Avandia combined results from 47 smaller studies. In its rebuttal the drug's manufacturer, pointed out, "Data from the ADOPT clinical trial did show a small increase in reports of myocardial infarction among the *Avandia*-treated group . . . however, the number of events is too small to reach a reliable conclusion about the role any of the medicines may have played in this finding."

QUESTION: Why would this smaller study have been less likely to detect the difference in risk? What are the appropriate statistical concepts for comparing the smaller studies?

ANSWER: A small study is subject to greater sampling variability; that is, the sampling distribution of the sample proportion has a larger standard deviation. That gives small studies less power: They're be less able to discern whether an apparently higher risk was merely the result of chance variation or evidence of real danger. The FDA doesn't want to restrict the use of a drug that's safe and effective (Type I error), nor does it want patients to continue taking a medication that puts them at risk (Type II error). Larger sample sizes can reduce the risk of both kinds of error. Greater power (the probability of rejecting a false null hypothesis) means a better chance of spotting a genuinely higher risk of heart attacks.

WHAT IF ●●● we test many hypotheses?

With a significance level of $\alpha = 0.05$, there's 1 chance in 20 that we'll commit a Type I error and conclude that we've found something significant when there's nothing more going on than sampling error. By now you must be assuming that What Ifs always use a simulation to say something more. Not this time. If you see what's funny about this cartoon, then you understand the concept.

WHAT CAN GO WRONG?

- **Don't interpret the P-value as the probability that H_0 is true.** The P-value is about the data, not the hypothesis. It's the probability of observing data this unusual, *given* that H_0 is true, not the other way around.

- **Don't believe too strongly in arbitrary alpha levels.** There's not really much difference between a P-value of 0.051 and a P-value of 0.049, but sometimes it's regarded as the difference between night (having to refrain from rejecting H_0) and day (being able to shout to the world that your results are "statistically significant"). It may just be better to report the P-value and a confidence interval and let the world decide along with you.

- **Don't confuse practical and statistical significance.** A large sample size can make it easy to discern even a trivial change from the null hypothesis value. On the other hand, an important difference can be missed if your test lacks sufficient power.

- **Don't forget that in spite of all your care, you might make a wrong decision.** We can never reduce the probability of a Type I error (α) or of a Type II error (β) to zero (but increasing the sample size helps).

What Have We Learned?

We've learned that there's a lot more to hypothesis testing than a simple yes/no decision.

- We've learned that the P-value can indicate evidence against the null hypothesis when it's small, but it does not tell us the probability that the null hypothesis is true.
- We've learned that the alpha level of the test establishes the level of proof we'll require. That determines the critical value of z that will lead us to reject the null hypothesis.
- We've also learned more about the connection between hypothesis tests and confidence intervals; they're really two ways of looking at the same question. The hypothesis test gives us the answer to a decision about a parameter; the confidence interval tells us the plausible values of that parameter.

We've learned about the two kinds of errors we might make, and we've seen why in the end we're never sure we've made the right decision.

- If the null hypothesis is really true and we reject it, that's a Type I error; the alpha level of the test is the probability that this could happen.
- If the null hypothesis is really false but we fail to reject it, that's a Type II error.
- The power of the test is the probability that we reject the null hypothesis when it's false. The larger the size of the effect we're testing for, the greater the power of the test to detect it.
- We've seen that tests with a greater likelihood of Type I error have more power and less chance of a Type II error. We can increase power while reducing the chances of both kinds of error by increasing the sample size.

Terms

Alpha level	The threshold P-value that determines when we reject a null hypothesis. If we observe a statistic whose P-value based on the null hypothesis is less than α, we reject that null hypothesis. (p. 523)
Statistically significant	When the P-value falls below the alpha level, we say that the test is "statistically significant" at that alpha level. (p. 523)
Significance level	The alpha level is also called the significance level, most often in a phrase such as a conclusion that a particular test is "significant at the 5% significance level.". (p. 523)
Type I error	The error of rejecting a null hypothesis when in fact it is true (also called a "false positive"). The probability of a Type I error is α. (p. 528)
Type II error	The error of failing to reject a null hypothesis when in fact it is false (also called a "false negative"). The probability of a Type II error is commonly denoted β and depends on the effect size. (p. 528)

| Power | The probability that a hypothesis test will correctly reject a false null hypothesis is the power of the test. To find power, we must specify a particular alternative parameter value as the "true" value. For any specific value in the alternative, the power is $1 - \beta$. (p. 528) |
| Effect size | The difference between the null hypothesis value and true value of a model parameter is called the effect size. (p. 530) |

On the Computer HYPOTHESIS TESTS

Reports about hypothesis tests generated by technologies don't follow a standard form. Most will name the test and provide the test statistic value, its standard deviation, and the P-value. But these elements may not be labeled clearly. For example, the expression "Prob > |z|" is a fancy (and not very clear) way of saying two-tailed P-value. In some packages, you can specify that the test be one-sided. Others might report three P-values, covering the ground for both one-sided tests and the two-sided test.

Sometimes a confidence interval and hypothesis test are automatically given together. The CI ought to be for the corresponding confidence level: $1 - \alpha$ for 2-tailed tests, $1 - 2\alpha$ for 1-tailed tests.

Often, the standard deviation of the statistic is called the "standard error," and usually that's appropriate because we've had to estimate its value from the data. That's not the case for proportions, however: We get the standard deviation for a proportion from the null hypothesis value. Nevertheless, you may see the standard deviation called a "standard error" even for tests with proportions.

It's common for statistics packages and calculators to report more digits of "precision" than could possibly have been found from the data. You can safely ignore them. Round values such as the standard deviation to one digit more than the number of digits reported in your data.

The example of results shown below is not from any program or calculator we know of, but it displays some of the things you might see in typical computer output.

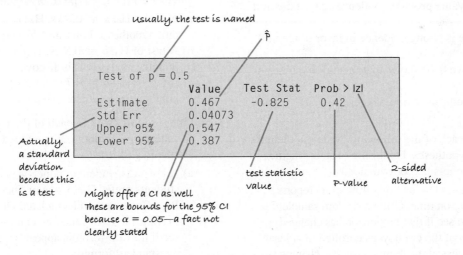

Exercises

1. Parameters and hypotheses For each of the following situations, define the parameter (proportion or mean) and write the null and alternative hypotheses in terms of parameter values. Example: We want to know if the proportion of up days in the stock market is 50%. Answer: Let p = the proportion of up days. $H_0: p = 0.5$ vs. $H_A: p \neq 0.5$.

a) A casino wants to know if their slot machine really delivers the 1 in 100 win rate that it claims.

b) A pharmaceutical company wonders if their new drug has a cure rate different from the 30% reported by the placebo.

c) A bank wants to know if the percentage of customers using their website has changed from the 40% that used it before their system crashed last week.

2. **Hypotheses and parameters** As in Exercise 1, for each of the following situations, define the parameter and write the null and alternative hypotheses in terms of parameter values.

a) Seat-belt compliance in Massachusetts was 65% in 2008. The state wants to know if it has changed.

b) Last year, a survey found that 45% of the employees were willing to pay for on-site day care. The company wants to know if that has changed.

c) Regular card customers have a default rate of 6.7%. A credit card bank wants to know if that rate is different for their Gold card customers.

3. **P-values** Which of the following are true? If false, explain briefly.

a) A very high P-value is strong evidence that the null hypothesis is false.

b) A very low P-value proves that the null hypothesis is false.

c) A high P-value shows that the null hypothesis is true.

d) A P-value below 0.05 is always considered sufficient evidence to reject a null hypothesis.

4. **More P-values** Which of the following are true? If false, explain briefly.

a) A very low P-value provides evidence against the null hypothesis.

b) A high P-value is strong evidence in favor of the null hypothesis.

c) A P-value above 0.10 shows that the null hypothesis is true.

d) If the null hypothesis is true, you can't get a P-value below 0.01.

5. **Hypotheses** For each of the following, write out the null and alternative hypotheses, being sure to state whether the alternative is one-sided or two-sided.

a) A company knows that last year 40% of its reports in accounting were on time. Using a random sample this year, it wants to see if that proportion has changed.

b) Last year, 42% of the employees enrolled in at least one wellness class at the company's site. Using a survey, it wants to see whether a greater percentage is planning to take a wellness class this year.

c) A political candidate wants to know from recent polls if she's going to garner a majority of votes in next week's election.

6. **More hypotheses** For each of the following, write out the alternative hypothesis, being sure to indicate whether it is one-sided or two-sided.

a) Consumer Reports discovered that 20% of a certain computer model had warranty problems over the first three months. From a random sample, the manufacturer wants to know if a new model has improved that rate.

b) The last time a philanthropic agency requested donations, 4.75% of people responded. From a recent pilot mailing, they wonder if that rate has increased.

c) A student wants to know if other students on her campus prefer Coke or Pepsi.

7. **Errors** For each of the following situations, state whether a Type I, a Type II, or neither error has been made. Explain briefly.

a) A bank wants to see if the enrollment on their website is above 30% based on a small sample of customers. It tests $H_0: p = 0.3$ vs. $H_A: p > 0.3$ and rejects the null hypothesis. Later the bank finds out that actually 28% of all customers enrolled.

b) A student tests 100 students to determine whether other students on her campus prefer Coke or Pepsi and finds no evidence that preference for Coke is not 0.5. Later, a marketing company tests all students on campus and finds no difference.

c) A pharmaceutical company tests whether a drug lifts the headache relief rate from the 25% achieved by the placebo. The test fails to reject the null hypothesis because the P-value is 0.465. Further testing shows that the drug actually relieves headaches in 38% of people.

8. **More errors** For each of the following situations, state whether a Type I, a Type II, or neither error has been made.

a) A test of $H_0: p = 0.8$ vs. $H_A: p < 0.8$ fails to reject the null hypothesis. Later it is discovered that $p = 0.9$.

b) A test of $H_0: p = 0.5$ vs. $H_A: p \neq 0.5$ rejects the null hypothesis. Later is it discovered that $p = 0.65$.

c) A test of $H_0: p = 0.7$ vs. $H_A: p < 0.7$ fails to reject the null hypothesis. Later is it discovered that $p = 0.6$.

9. **One sided or two?** In each of the following situations, is the alternative hypothesis one-sided or two-sided? What are the hypotheses?

a) A business student conducts a taste test to see whether students prefer Diet Coke or Diet Pepsi.

b) PepsiCo recently reformulated Diet Pepsi in an attempt to appeal to teenagers. The company runs a taste test to see if the new formula appeals to more teenagers than the standard formula.

c) A budget override in a small town requires a two-thirds majority to pass. A local newspaper conducts a poll to see if there's evidence it will pass.

d) One financial theory states that the stock market will go up or down with equal probability. A student collects data over several years to test the theory.

10. **Which alternative?** In each of the following situations, is the alternative hypothesis one-sided or two-sided? What are the hypotheses?

a) A college dining service conducts a survey to see if students prefer plastic or metal cutlery.

b) In recent years, 10% of college juniors have applied for study abroad. The dean's office conducts a survey to see if that's changed this year.

c) A pharmaceutical company conducts a clinical trial to see if more patients who take a new drug experience headache relief than the 22% who claimed relief after taking the placebo.

d) At a small computer peripherals company, only 60% of the hard drives produced passed all their performance tests the first time. Management recently invested a lot of resources into the production system and now conducts a test to see if it helped.

11. P-value A medical researcher tested a new treatment for poison ivy against the traditional ointment. He concluded that the new treatment is more effective. Explain what the P-value of 0.047 means in this context.

12. Another P-value Have harsher penalties and ad campaigns increased seat-belt use among drivers and passengers? Observations of commuter traffic failed to find evidence of a significant change compared with three years ago. Explain what the study's P-value of 0.17 means in this context.

13. Alpha A researcher developing scanners to search for hidden weapons at airports has concluded that a new device is significantly better than the current scanner. He made this decision based on a test using $\alpha = 0.05$. Would he have made the same decision at $\alpha = 0.10$? How about $\alpha = 0.01$? Explain.

14. Alpha again Environmentalists concerned about the impact of high-frequency radio transmissions on birds found that there was no evidence of a higher mortality rate among hatchlings in nests near cell towers. They based this conclusion on a test using $\alpha = 0.05$. Would they have made the same decision at $\alpha = 0.10$? How about $\alpha = 0.01$? Explain.

15. Significant? Public health officials believe that 90% of children have been vaccinated against measles. A random survey of medical records at many schools across the country found that, among more than 13,000 children, only 89.4% had been vaccinated. A statistician would reject the 90% hypothesis with a P-value of $P = 0.011$.

a) Explain what the P-value means in this context.

b) The result is statistically significant, but is it important? Comment.

16. Significant again? A new reading program may reduce the number of elementary school students who read below grade level. The company that developed this program supplied materials and teacher training for a large-scale test involving nearly 8500 children in several different school districts. Statistical analysis of the results showed that the percentage of students who did not meet the grade-level goal was reduced from 15.9% to 15.1%. The hypothesis that the new reading program produced no improvement was rejected with a P-value of 0.023.

a) Explain what the P-value means in this context.

b) Even though this reading method has been shown to be significantly better, why might you not recommend that your local school adopt it?

17. Groceries In January 2011, Yahoo surveyed 2400 U.S. men. 1224 of the men identified themselves as the primary grocery shopper in their household.

a) Estimate the percentage of all American males who identify themselves as the primary grocery shopper. Use a 98% confidence interval. Check the conditions first.

b) A grocery store owner believed that only 45% of men are the primary grocery shopper for their family, and targets his advertising accordingly. He wishes to conduct a hypothesis test to see if the fraction is in fact higher than 45%. What does your confidence interval indicate? Explain.

c) What is the level of significance of this test? Explain.

18. Is the Euro fair? Soon after the Euro was introduced as currency in Europe, it was widely reported that someone had spun a Euro coin 250 times and gotten heads 140 times. We wish to test a hypothesis about the fairness of spinning the coin.

a) Estimate the true proportion of heads. Use a 95% confidence interval. Don't forget to check the conditions.

b) Does your confidence interval provide evidence that the coin is unfair when spun? Explain.

c) What is the significance level of this test? Explain.

19. Approval 2011 In November 2011, Barack Obama's approval rating stood at 45% in Rasmussen's daily tracking poll of 1500 randomly surveyed U.S. adults.

a) Make a 95% confidence interval for his approval rating by all U.S. adults.

b) Based on the confidence interval, test the null hypothesis that Obama's approval rating was no worse than his November 2009 approval rating of 50%.

20. Hard times In June 2010, a random poll of 800 working men found that 9% had taken on a second job to help pay the bills. (www.careerbuilder.com)

a) Estimate the true percentage of men that are taking on second jobs by constructing a 95% confidence interval.

b) A pundit on a TV news show claimed that only 6% of working men had a second job. Use your confidence interval to test whether his claim is plausible given the poll data.

21. Dogs Canine hip dysplasia is a degenerative disease that causes pain in many dogs. Sometimes advanced warning signs appear in puppies as young as 6 months. A veterinarian checked 42 puppies whose owners brought them to a vaccination clinic, and she found 5 with early hip dysplasia. She considers this group to be a random sample of all puppies.

a) Explain we cannot use this information to construct a confidence interval for the rate of occurrence of early hip dysplasia among all 6-month-old puppies.

*b) Construct a "plus-four" confidence interval and interpret it in this context.

22. Fans A survey of 81 randomly selected people standing in line to enter a football game found that 73 of them were home team fans.

a) Explain why we cannot use this information to construct a confidence interval for the proportion of all people at the game who are fans of the home team.

*b) Construct a "plus-four" confidence interval and interpret it in this context.

23. Loans Before lending someone money, banks must decide whether they believe the applicant will repay the loan. One strategy used is a point system. Loan officers assess information about the applicant, totaling points they award for the person's income level, credit history, current debt burden, and so on. The higher the point total, the more convinced the bank is that it's safe to make the loan. Any applicant with a lower point total than a certain cutoff score is denied a loan.

We can think of this decision as a hypothesis test. Since the bank makes its profit from the interest collected on repaid loans, their null hypothesis is that the applicant will repay the loan and therefore should get the money. Only if the person's score falls below the minimum cutoff will the bank reject the null and deny the loan. This system is reasonably reliable, but, of course, sometimes there are mistakes.

a) When a person defaults on a loan, which type of error did the bank make?

b) Which kind of error is it when the bank misses an opportunity to make a loan to someone who would have repaid it?

c) Suppose the bank decides to lower the cutoff score from 250 points to 200. Is that analogous to choosing a higher or lower value of α for a hypothesis test? Explain.

d) What impact does this change in the cutoff value have on the chance of each type of error?

24. Spam Spam filters try to sort your e-mails, deciding which are real messages and which are unwanted. One method used is a point system. The filter reads each incoming e-mail and assigns points to the sender, the subject, key words in the message, and so on. The higher the point total, the more likely it is that the message is unwanted. The filter has a cutoff value for the point total; any message rated lower than that cutoff passes through to your inbox, and the rest, suspected to be spam, are diverted to the junk mailbox.

We can think of the filter's decision as a hypothesis test. The null hypothesis is that the e-mail is a real message and should go to your inbox. A higher point total provides evidence that the message may be spam; when there's sufficient evidence, the filter rejects the null, classifying the message as junk. This usually works pretty well, but, of course, sometimes the filter makes a mistake.

a) When the filter allows spam to slip through into your inbox, which kind of error is that?

b) Which kind of error is it when a real message gets classified as junk?

c) Some filters allow the user (that's you) to adjust the cutoff. Suppose your filter has a default cutoff of 50 points, but you reset it to 60. Is that analogous to choosing a higher or lower value of α for a hypothesis test? Explain.

d) What impact does this change in the cutoff value have on the chance of each type of error?

25. Second loan Exercise 23 describes the loan score method a bank uses to decide which applicants it will lend money. Only if the total points awarded for various aspects of an applicant's financial condition fail to add up to a minimum cutoff score set by the bank will the loan be denied.

a) In this context, what is meant by the power of the test?

b) What could the bank do to increase the power?

c) What's the disadvantage of doing that?

26. More spam Consider again the points-based spam filter described in Exercise 24. When the points assigned to various components of an e-mail exceed the cutoff value you've set, the filter rejects its null hypothesis (that the message is real) and diverts that e-mail to a junk mailbox.

a) In this context, what is meant by the power of the test?

b) What could you do to increase the filter's power?

c) What's the disadvantage of doing that?

27. Homeowners 2009 In 2009, the U.S. Census Bureau reported that 67.4% of American families owned their homes. Census data reveal that the ownership rate in one small city is much lower. The city council is debating a plan to offer tax breaks to first-time home buyers in order to encourage people to become homeowners. They decide to adopt the plan on a 2-year trial basis and use the data they collect to make a decision about continuing the tax breaks. Since this plan costs the city tax revenues, they will continue to use it only if there is strong evidence that the rate of home ownership is increasing.

a) In words, what will their hypotheses be?

b) What would a Type I error be?

c) What would a Type II error be?

d) For each type of error, tell who would be harmed.

e) What would the power of the test represent in this context?

28. Alzheimer's Testing for Alzheimer's disease can be a long and expensive process, consisting of lengthy tests and medical diagnosis. Recently, a group of researchers (Solomon et al., 1998) devised a 7-minute test to serve as a quick screen for the disease for use in the general population of senior citizens. A patient who tested positive would then go through the more expensive battery of tests and medical diagnosis. The authors reported a false positive rate of 4% and a false negative rate of 8%.

a) Put this in the context of a hypothesis test. What are the null and alternative hypotheses?

b) What would a Type I error mean?

c) What would a Type II error mean?

d) Which is worse here, a Type I or Type II error? Explain.

e) What is the power of this test?

29. Testing cars A clean air standard requires that vehicle exhaust emissions not exceed specified limits for various pollutants. Many states require that cars be tested annually to be sure they meet these standards. Suppose state regulators double-check a random sample of cars that a suspect repair shop has certified as okay. They will revoke the shop's license if they find significant evidence that the shop is certifying vehicles that do not meet standards.

a) In this context, what is a Type I error?
b) In this context, what is a Type II error?
c) Which type of error would the shop's owner consider more serious?
d) Which type of error might environmentalists consider more serious?

30. Quality control Production managers on an assembly line must monitor the output to be sure that the level of defective products remains small. They periodically inspect a random sample of the items produced. If they find a significant increase in the proportion of items that must be rejected, they will halt the assembly process until the problem can be identified and repaired.

a) In this context, what is a Type I error?
b) In this context, what is a Type II error?
c) Which type of error would the factory owner consider more serious?
d) Which type of error might customers consider more serious?

31. Cars again As in Exercise 29, state regulators are checking up on repair shops to see if they are certifying vehicles that do not meet pollution standards.

a) In this context, what is meant by the power of the test the regulators are conducting?
b) Will the power be greater if they test 20 or 40 cars? Why?
c) Will the power be greater if they use a 5% or a 10% level of significance? Why?
d) Will the power be greater if the repair shop's inspectors are only a little out of compliance or a lot? Why?

32. Production Consider again the task of the quality control inspectors in Exercise 30.

a) In this context, what is meant by the power of the test the inspectors conduct?
b) They are currently testing 5 items each hour. Someone has proposed that they test 10 instead. What are the advantages and disadvantages of such a change?
c) Their test currently uses a 5% level of significance. What are the advantages and disadvantages of changing to an alpha level of 1%?
d) Suppose that, as a day passes, one of the machines on the assembly line produces more and more items that are defective. How will this affect the power of the test?

33. Equal opportunity? A company is sued for job discrimination because only 19% of the newly hired candidates were minorities when 27% of all applicants were minorities. Is this strong evidence that the company's hiring practices are discriminatory?

a) Is this a one-tailed or a two-tailed test? Why?
b) In this context, what would a Type I error be?
c) In this context, what would a Type II error be?

d) In this context, what is meant by the power of the test?
e) If the hypothesis is tested at the 5% level of significance instead of 1%, how will this affect the power of the test?
f) The lawsuit is based on the hiring of 37 employees. Is the power of the test higher than, lower than, or the same as it would be if it were based on 87 hires?

34. Stop signs Highway safety engineers test new road signs, hoping that increased reflectivity will make them more visible to drivers. Volunteers drive through a test course with several of the new- and old-style signs and rate which kind shows up the best.

a) Is this a one-tailed or a two-tailed test? Why?
b) In this context, what would a Type I error be?
c) In this context, what would a Type II error be?
d) In this context, what is meant by the power of the test?
e) If the hypothesis is tested at the 1% level of significance instead of 5%, how will this affect the power of the test?
f) The engineers hoped to base their decision on the reactions of 50 drivers, but time and budget constraints may force them to cut back to 20. How would this affect the power of the test? Explain.

35. Dropouts A Statistics professor has observed that for several years about 13% of the students who initially enroll in his Introductory Statistics course withdraw before the end of the semester. A salesman suggests that he try a statistics software package that gets students more involved with computers, predicting that it will cut the dropout rate. The software is expensive, and the salesman offers to let the professor use it for a semester to see if the dropout rate goes down significantly. The professor will have to pay for the software only if he chooses to continue using it.

a) Is this a one-tailed or two-tailed test? Explain.
b) Write the null and alternative hypotheses.
c) In this context, explain what would happen if the professor makes a Type I error.
d) In this context, explain what would happen if the professor makes a Type II error.
e) What is meant by the power of this test?

36. Ads A company is willing to renew its advertising contract with a local radio station only if the station can prove that more than 20% of the residents of the city have heard the ad and recognize the company's product. The radio station conducts a random phone survey of 400 people.

a) What are the hypotheses?
b) The station plans to conduct this test using a 10% level of significance, but the company wants the significance level lowered to 5%. Why?
c) What is meant by the power of this test?
d) For which level of significance will the power of this test be higher? Why?
e) They finally agree to use $\alpha = 0.05$, but the company proposes that the station call 600 people instead of

the 400 initially proposed. Will that make the risk of Type II error higher or lower? Explain.

37. Dropouts, part II Initially, 203 students signed up for the Stats course in Exercise 35. They used the software suggested by the salesman, and only 11 dropped out of the course.

a) Should the professor spend the money for this software? Support your recommendation with an appropriate test.

b) Explain what your P-value means in this context.

38. Testing the ads The company in Exercise 36 contacts 600 people selected at random, and only 133 remember the ad.

a) Should the company renew the contract? Support your recommendation with an appropriate test.

b) Explain what your P-value means in this context.

39. Two coins In a drawer are two coins. They look the same, but one coin produces heads 90% of the time when spun while the other one produces heads only 30% of the time. You select one of the coins. You are allowed to spin it *once* and then must decide whether the coin is the 90%- or the 30%-head coin. Your null hypothesis is that your coin produces 90% heads.

a) What is the alternative hypothesis?

b) Given that the outcome of your spin is tails, what would you decide? What if it were heads?

c) How large is α in this case?

d) How large is the power of this test? (*Hint:* How many possibilities are in the alternative hypothesis?)

e) How could you lower the probability of a Type I error and increase the power of the test at the same time?

40. Faulty or not? You are in charge of shipping computers to customers. You learn that a faulty disk drive was put into some of the machines. There's a simple test you can perform, but it's not perfect. All but 4% of the time, a good disk drive passes the test, but unfortunately, 35% of the bad disk drives pass the test, too. You have to decide on the basis of one test whether the disk drive is good or bad. Make this a hypothesis test.

a) What are the null and alternative hypotheses?

b) Given that a computer fails the test, what would you decide? What if it passes the test?

c) How large is α for this test?

d) What is the power of this test? (*Hint:* How many possibilities are in the alternative hypothesis?)

41. Hoops A basketball player with a poor foul-shot record practices intensively during the off-season. He tells the coach that he has raised his proficiency from 60% to 80%. Dubious, the coach asks him to take 10 shots, and is surprised when the player hits 9 out of 10. Did the player prove that he has improved?

a) Suppose the player really is no better than before—still a 60% shooter. What's the probability he can hit at least 9 of 10 shots anyway? (*Hint:* Use a Binomial model.)

b) If that is what happened, now the coach thinks the player has improved when he has not. Which type of error is that?

c) If the player really can hit 80% now, and it takes at least 9 out of 10 successful shots to convince the coach, what's the power of the test?

d) List two ways the coach and player could increase the power to detect any improvement.

42. Pottery An artist experimenting with clay to create pottery with a special texture has been experiencing difficulty with these special pieces. About 40% break in the kiln during firing. Hoping to solve this problem, she buys some more expensive clay from another supplier. She plans to make and fire 10 pieces and will decide to use the new clay if at most one of them breaks.

a) Suppose the new, expensive clay really is no better than her usual clay. What's the probability that this test convinces her to use it anyway? (*Hint:* Use a Binomial model.)

b) If she decides to switch to the new clay and it is no better, what kind of error did she commit?

c) If the new clay really can reduce breakage to only 20%, what's the probability that her test will not detect the improvement?

d) How can she improve the power of her test? Offer at least two suggestions.

Just Checking ANSWERS

1. With a z-score of 0.62, you can't reject the null hypothesis. The experiment shows no evidence that the wheel is not fair.

2. At $\alpha = 0.05$, you can't reject the null hypothesis because 0.30 is contained in the 90% confidence interval—it's plausible that sending the DVDs is no more effective than just sending letters.

3. The confidence interval is from 29% to 45%. The DVD strategy is more expensive and may not be worth it. We can't distinguish the success rate from 30% given the results of this experiment, but 45% would represent a large improvement. The bank should consider another trial, increasing their sample size to get a narrower confidence interval.

4. A Type I error would mean deciding that the DVD success rate is higher than 30% when it really isn't. They would adopt a more expensive method for collecting payments that's no better than the less expensive strategy.

5. A Type II error would mean deciding that there's not enough evidence to say that the DVD strategy works when in fact it does. The bank would fail to discover an effective method for increasing their revenue from delinquent accounts.

6. 60%; the larger the effect size, the greater the power. It's easier to detect an improvement to a 60% success rate than to a 32% rate.

21 Comparing Two Proportions

who	6971 male drivers
what	Seatbelt use
why	Highway safety
when	2007
where	Massachusetts

Do men take more risks than women? Psychologists have documented that in many situations men choose riskier behavior than women do. But what is the effect of having a woman by their side? A seatbelt observation study in Massachusetts[1] found that, not surprisingly, male drivers wear seatbelts less often than women do. The study also noted that men's belt-wearing jumped more than 16 percentage points when they had a female passenger. Seatbelt use was recorded at 161 locations in Massachusetts, using random-sampling methods developed by the National Highway Traffic Safety Administration (NHTSA). Female drivers wore belts more than 70% of the time, regardless of the sex of their passengers. Of 4208 male drivers with female passengers, 2777 (66.0%) were belted. But among 2763 male drivers with male passengers only, 1363 (49.3%) wore seatbelts. This was only a random sample, but it suggests there may be a shift in men's risk-taking behavior when women are present. What would we estimate the true size of that gap to be?

Comparisons between two percentages are much more common than questions about isolated percentages. And they are more interesting. We often want to know how two groups differ, whether a treatment is better than a placebo control, or whether this year's results are better than last year's.

Another Ruler

We know the difference between the proportions of men wearing seatbelts seen in the *sample*. It's 16.7%. But what's the *true* difference for all men? We know that our estimate probably isn't exactly right. To be able to say more, we need a new ruler—the standard deviation of the sampling distribution model for the difference in the sample proportions. Now we have two proportions, and each will vary from sample to sample. We are interested in the difference between them. So what is the correct standard deviation?

---

[1]Massachusetts Traffic Safety Research Program (June 2007).

The answer comes to us from Chapter 15. Remember the Pythagorean Theorem of Statistics?

The variance of the sum or difference of two independent random variables is the sum of their variances.

This is such an important (and powerful) idea in Statistics that it's worth pausing a moment to review the reasoning. Here's some intuition about why variation increases even when we subtract two random quantities.

Grab a full box of cereal. The box claims to contain 16 ounces of cereal. We know that's not exact: There's some small variation from box to box. Now pour a bowl of cereal. Of course, your 2-ounce serving will not be exactly 2 ounces. There'll be some variation there, too. How much cereal would you guess was left in the box? Do you think your guess will be as close as your guess for the full box? *After* you pour your bowl, the amount of cereal in the box is still a random quantity (with a smaller mean than before), but it is even *more variable* because of the additional variation in the amount you poured.

According to our rule, the variance of the amount of cereal left in the box would now be the *sum* of the two *variances*.

We want a standard deviation, not a variance, but that's just a square root away. We can write symbolically what we've just said:

$$Var(X - Y) = Var(X) + Var(Y), \text{ so}$$

$$SD(X - Y) = \sqrt{SD^2(X) + SD^2(Y)} = \sqrt{Var(X) + Var(Y)}.$$

> **Remember**
>
> For *independent* random variables, variances add.

Be careful, though—this simple formula applies only when X and Y are independent. Just as the Pythagorean Theorem[2] works only for right triangles, our formula works only for independent random variables. Always check for independence before using it.

The Standard Deviation of the Difference Between Two Proportions

> **Variation Grows**
>
> Combining independent random quantities always *increases* the overall variation, so even for *differences* of independent random variables, **variances add**.

Fortunately, proportions observed in independent random samples *are* independent, so we can put those two proportions in for X and Y and add their variances. We just need to use careful notation to keep things straight.

When we have two samples, each can have a different size and proportion value, so we keep them straight with subscripts. Often we choose subscripts that remind us of the groups. For our example, we might use "$_M$" and "$_F$", but generically we'll just use "$_1$" and "$_2$". We will represent the two sample proportions as $\hat{p}_1$ and $\hat{p}_2$, and the two sample sizes as n_1 and n_2.

The standard deviations of the sample proportions are $SD(\hat{p}_1) = \sqrt{\dfrac{p_1 q_1}{n_1}}$ and $SD(\hat{p}_2) = \sqrt{\dfrac{p_2 q_2}{n_2}}$, so the variance of the difference in the proportions is

$$Var(\hat{p}_1 - \hat{p}_2) = \left(\sqrt{\frac{p_1 q_1}{n_1}}\right)^2 + \left(\sqrt{\frac{p_2 q_2}{n_2}}\right)^2 = \frac{p_1 q_1}{n_1} + \frac{p_2 q_2}{n_2}.$$

The standard deviation is the square root of that variance:

$$SD(\hat{p}_1 - \hat{p}_2) = \sqrt{\frac{p_1 q_1}{n_1} + \frac{p_2 q_2}{n_2}}.$$

---

[2]If you don't remember the formula, don't rely on the Scarecrow's version from *The Wizard of Oz*. He may have a brain and have been awarded his Th.D. (Doctor of Thinkology), but he gets the formula wrong.

We usually don't know the true values of p_1 and p_2. When we have the sample proportions in hand from the data, we use them to estimate the variances. So the standard error is

$$SE(\hat{p}_1 - \hat{p}_2) = \sqrt{\frac{\hat{p}_1\hat{q}_1}{n_1} + \frac{\hat{p}_2\hat{q}_2}{n_2}}.$$

For Example FINDING THE STANDARD ERROR OF A DIFFERENCE IN PROPORTIONS

A recent survey of 886 randomly selected teenagers (aged 12–17) found that more than half of them had online profiles.[3] Some researchers and privacy advocates are concerned about the possible access to personal information about teens in public places on the Internet. There appear to be differences between boys and girls in their online behavior. Among teens aged 15–17, 57% of the 248 boys had posted profiles, compared to 70% of the 256 girls. Let's start the process of estimating how large the true gender gap might be.

QUESTION: What's the standard error of the difference in sample proportions?

ANSWER: *Because the boys and girls were selected at random, it's reasonable to assume their behaviors are independent, so it's okay to use the Pythagorean Theorem of Statistics and add the variances:*

$$SE(\hat{p}_{boys}) = \sqrt{\frac{0.57 \times 0.43}{248}} = 0.0314 \qquad SE(\hat{p}_{girls}) = \sqrt{\frac{0.70 \times 0.30}{256}} = 0.0286$$

$$SE(\hat{p}_{girls} - \hat{p}_{boys}) = \sqrt{0.0314^2 + 0.0286^2} = 0.0425$$

Assumptions and Conditions

As always, we need to check assumptions and conditions. The first one is new, and very important.

Independence Assumptions

Because we are comparing two groups, we need a new Independence Assumption. In fact, this is the most important of these assumptions. If it is violated, these methods just won't work.

Independent Groups Assumption: The two groups, we're comparing must be independent *of each other*. Usually, the independence of the groups from each other is evident from the way the data were collected.

Why is the Independent Groups Assumption so important? If we compare husbands with their wives, or a group of subjects before and after some treatment, we can't just add the variances. Subjects' performance after a treatment might very well be related to their performance before the treatment. That means the proportions are not independent and the Pythagorean-style variance formula does not hold. We'll see a way to compare a common kind of nonindependent samples in a later chapter.

You'll recognize the rest of the assumptions and conditions.

Independence Assumption: Within each group, the data should be based on results for independent individuals. We can't check that for certain, but we *can* check the following:

Randomization Condition: The data in each group should be drawn independently and at random from a homogeneous population or generated by a randomized comparative experiment.

The 10% Condition: If the data are sampled without replacement, the sample should not exceed 10% of the population.

---

[3]Princeton Survey Research Associates International for the Pew Internet & American Life Project.

Sample Size Condition

Each of the groups must be big enough. As with individual proportions, we need larger groups to estimate proportions that are near 0% or 100%. We usually check the Success/Failure Condition for each group.

Success/Failure Condition: Both groups are big enough that at least 10 successes and at least 10 failures have been observed in each.

For Example CHECKING ASSUMPTIONS AND CONDITIONS

RECAP: Among randomly sampled teens aged 15–17, 57% of the 248 boys had posted online profiles, compared to 70% of the 256 girls.

QUESTION: Can we use these results to make inferences about all 15–17-year-olds?

ANSWER:

✓ **Randomization Condition:** The sample of boys and the sample of girls were both chosen randomly.

✓ **10% Condition:** 248 boys and 256 girls are each less than 10% of all teenage boys and girls.

✓ **Independent Groups Assumption:** Because the samples were selected at random, it's reasonable to believe the boys' online behaviors are independent of the girls' online behaviors.

✓ **Success/Failure Condition:** Among the boys, 248(0.57) = 141 had online profiles and the other 248(0.43) = 107 did not. For the girls, 256(0.70) = 179 successes and 256(0.30) = 77 failures. All counts are at least 10.

Because all the assumptions and conditions are satisfied, it's okay to proceed with inference for the difference in proportions.

(Note that when we find the *observed counts* of successes and failures, we round off to whole numbers. We're using the reported percentages to recover the actual counts.)

A Confidence Interval

We're almost there. We just need one more fact about proportions. We already know that for large enough samples, each of our sample proportions has an approximately Normal sampling distribution. The same is true of their difference.

Why Normal?
In Chapter 15 you learned that sums and differences of Independent Normal random variables also follow a Normal model. That's the reason we use a Normal model for the difference of two independent sample proportions.

The Sampling Distribution Model for a Difference Between Two Independent Sample Proportions
Provided that the sampled values are independent, the samples are independent, and the sample sizes are large enough, the sampling distribution of $\hat{p}_1 - \hat{p}_2$ is modeled by a Normal model with mean $\mu = p_1 - p_2$ and standard deviation

$$SD(\hat{p}_1 - \hat{p}_2) = \sqrt{\frac{p_1 q_1}{n_1} + \frac{p_2 q_2}{n_2}}.$$

The sampling distribution model and the standard deviation give us all we need to find a margin of error for the difference in proportions—or at least they would if we knew the true proportions, p_1 and p_2. However, we don't know the true values, so we'll work with the observed proportions, $\hat{p}_1$ and $\hat{p}_2$, and use $SE(\hat{p}_1 - \hat{p}_2)$ to estimate the standard deviation. The rest is just like a one-proportion z-interval.

A Two-Proportion z-Interval

When the conditions are met, we are ready to find the confidence interval for the difference of two proportions, $p_1 - p_2$. The confidence interval is

$$(\hat{p}_1 - \hat{p}_2) \pm z^* \times SE(\hat{p}_1 - \hat{p}_2)$$

where we find the standard error of the difference,

$$SE(\hat{p}_1 - \hat{p}_2) = \sqrt{\frac{\hat{p}_1 \hat{q}_1}{n_1} + \frac{\hat{p}_2 \hat{q}_2}{n_2}},$$

from the observed proportions.

The critical value z^* depends on the particular confidence level, C, that we specify.

For Example   FINDING A TWO-PROPORTION z-INTERVAL

RECAP: Among randomly sampled teens aged 15–17, 57% of the 248 boys had posted online profiles, compared to 70% of the 256 girls. We calculated the standard error for the difference in sample proportions to be $SE(\hat{p}_{girls} - \hat{p}_{boys}) = 0.0425$ and found that the assumptions and conditions required for inference checked out okay.

QUESTION: What does a confidence interval say about the difference in online behavior?

ANSWER: A 95% confidence interval for $p_{girls} - p_{boys}$ is $(\hat{p}_{girls} - \hat{p}_{boys}) \pm z^* SE(\hat{p}_{girls} - \hat{p}_{boys})$

$$(0.70 - 0.57) \pm 1.96(0.0425)$$
$$0.13 \pm 0.083$$
$$(4.7\%, 21.3\%)$$

We can be 95% confident that among teens aged 15–17, the proportion of girls who post online profiles is between 4.7 and 21.3 percentage points higher than the proportion of boys who do. It seems clear that teen girls are more likely to post profiles than are boys the same age.

Step-by-Step Example   A TWO-PROPORTION z-INTERVAL

Now we are ready to be more precise about the passenger-based gap in male drivers' seatbelt use. We'll estimate the difference with a confidence interval using a method called the two-proportion z-interval and follow the four confidence interval steps.

Question: How much difference is there in the proportion of male drivers who wear seatbelts when sitting next to a male passenger and the proportion who wear seatbelts when sitting next to a female passenger?

THINK ➡ **Plan** State what you want to know. Discuss the variables and the W's.

Identify the parameter you wish to estimate. (It usually doesn't matter in which direction we subtract, so, for convenience, we usually choose the direction with a positive difference.)

I want to know the true difference in the population proportion, p_M, of male drivers who wear seatbelts when sitting next to a man and p_F, the proportion who wear seatbelts when sitting next to a woman. The data are from a random sample of drivers in Massachusetts in 2007, observed according to procedures developed by the NHTSA. The parameter of interest is the difference $p_F - p_M$.

(continued)

Choose and state a confidence level.

Model Think about the assumptions and check the conditions.

I will find a 95% confidence interval for this parameter.

✓ **Independent Groups Assumption:** There's no reason to believe that seatbelt use among drivers with male passengers and those with female passengers are not independent.

✓ **Independence Assumption:** Driver behavior was independent from car to car.

✓ **Randomization Condition:** The NHTSA methods are more complex than an SRS, but they result in a suitable random sample.

✓ **10% Condition:** The samples include far fewer than 10% of all male drivers accompanied by male or by female passengers.

The Success/Failure Condition must hold for each group.

✓ **Success Failure Condition:** Among male drivers with female passengers, 2777 wore seatbelts and 1431 did not; of those driving with male passengers, 1363 wore seatbelts and 1400 did not. Each group contained far more than 10 successes and 10 failures.

State the sampling distribution model for the statistic.

Choose your method.

Under these conditions, the sampling distribution of the difference between the sample proportions is approximately Normal, so I'll find a **two-proportion z-interval**.

SHOW ➡ **Mechanics** Construct the confidence interval.

I know

$$n_F = 4208, n_M = 2763.$$

The observed sample proportions are

$$\hat{p}_F = \frac{2777}{4208} = 0.660, \hat{p}_M = \frac{1363}{2763} = 0.493$$

As often happens, the key step in finding the confidence interval is estimating the standard deviation of the sampling distribution model of the statistic. Here the statistic is the difference between the sample proportion of men who wear seatbelts when they have a female passenger and the sample proportion who do so with a male passenger.

I'll estimate the SD of the difference with

$$SE(\hat{p}_F - \hat{p}_M) = \sqrt{\frac{\hat{p}_F \hat{q}_F}{n_F} + \frac{\hat{p}_M \hat{q}_M}{n_M}}$$

$$= \sqrt{\frac{(0.660)(0.340)}{4208} + \frac{(0.493)(0.507)}{2763}}$$

$$= 0.012$$

The sampling distribution is Normal, so the critical value for a 95% confidence interval, z^*, is 1.96. The margin of error is the critical value times the SE.

$$ME = z^* \times SE(\hat{p}_F - \hat{p}_M)$$
$$= 1.96(0.012) = 0.024$$

(continued)

The confidence interval is the statistic $\pm$ ME.	The observed difference in proportions is $\hat{p}_F - \hat{p}_M = 0.660 - 0.493 = 0.167$, so the 95% confidence interval is $$0.167 \pm 0.024$$ or 14.3% to 19.1%
TELL ➡ **Conclusion** Interpret your confidence interval in the proper context. (Remember: We're 95% confident that our interval captured the true difference.)	I am 95% confident that the proportion of male drivers who wear seatbelts when driving next to a female passenger is between 14.3 and 19.1 percentage points higher than the proportion who wear seatbelts when driving next to a male passenger.

This is an interesting result—but be careful not to try to say too much! In Massachusetts, overall seatbelt use is lower than the national average, so we can't be certain that these results generalize to other states. And these were two different groups of men, so we can't say that, individually, men are more likely to buckle up when they have a woman passenger. You can probably think of several alternative explanations; we'll suggest just a couple. Perhaps age is a lurking variable: Maybe older men are more likely to wear seatbelts and also more likely to be driving with their wives. Or maybe men who don't wear seatbelts have trouble attracting women!

TI Tips FINDING THE CONFIDENCE INTERVAL

You can use a routine in the STAT TESTS menu to create confidence intervals for the difference of two proportions. Remember, the calculator can do only the mechanics—checking conditions and writing conclusions are still up to you.

A Gallup Poll asked whether the attribute "intelligent" described men in general. The poll revealed that 28% of 506 men thought it did, but only 14% of 520 women agreed. We want to estimate the true size of the gender gap by creating a 95% confidence interval.

- Go to the STAT TESTS menu. Scroll down the list and select 2-PropZInt.

- Enter the observed number of males: .28*506. Remember that the actual number of males must be a whole number, so be sure to round off.
- Enter the sample size: 506 males.
- Repeat those entries for women: .14*520 agreed, and the sample size was 520.

- Specify the desired confidence level.
- Calculate the result.

And now explain what you see: We are 95% confident that the proportion of men who think the attribute "intelligent" describe males in general is between 9 and 19 percentage points higher than the proportion of women who think so.

Just Checking

A public broadcasting station plans to launch a special appeal for additional contributions from current members. Unsure of the most effective way to contact people, they run an experiment. They randomly select two groups of current members. They send the same request for donations to everyone, but it goes to one group by e-mail and to the other group by regular mail. The station was successful in getting contributions from 26% of the members they e-mailed but only from 15% of those who received the request by regular mail. A 90% confidence interval estimated the difference in donation rates to be 11% ± 7%.

1. Interpret the confidence interval in this context.

2. Based on this confidence interval, what conclusion would we reach if we tested the hypothesis that there's no difference in the response rates to the two methods of fundraising? Explain.

Will I Snore When I'm 64?

Who	Randomly selected U.S. adults over age 18
What	Proportion who snore, categorized by age (less than 30, 30 or older)
When	2001
Where	United States
Why	To study sleep behaviors of U.S. adults

The National Sleep Foundation asked a random sample of 1010 U.S. adults questions about their sleep habits. The sample was selected in the fall of 2001 from random telephone numbers, stratified by region and sex, guaranteeing that an equal number of men and women were interviewed (2002 Sleep in America Poll, National Sleep Foundation, Washington, DC).

One of the questions asked about snoring. Of the 995 respondents, 37% of adults reported that they snored at least a few nights a week during the past year. Would you expect that percentage to be the same for all age groups? Split into two age categories, 26.1% of the 184 people under 30 snored, compared with 39.2% of the 811 in the older group. Is this difference of 13% real, or due only to natural fluctuations in the sample we've chosen?

The question calls for a hypothesis test. Now the parameter of interest is the true *difference* between the snoring rates of the two age groups.

What's the appropriate null hypothesis? That's easy here. We hypothesize that there is no difference in the proportions. This is such a natural null hypothesis that we rarely consider any other. But instead of writing $H_0: p_1 = p_2$, we usually express it in a slightly different way. To make it relate directly to the *difference*, we hypothesize that the difference in proportions is zero:

$$H_0: p_1 - p_2 = 0.$$

We'll reject the null hypothesis if we see a difference in our sample proportions that's so large it's unlikely to be sampling error. How can we decide if $\hat{p}_1 - \hat{p}_2$ is unusually large? The same way we always do: we'll use a standard error to find a z-score.

Everyone into the Pool

Our hypothesis is about a new parameter: the *difference* in proportions. We need a standard error for that statistic. Wait—don't we know that SE already? Yes and no. We know that the standard error of the difference in sample proportions is

$$SE(\hat{p}_1 - \hat{p}_2) = \sqrt{\frac{\hat{p}_1 \hat{q}_1}{n_1} + \frac{\hat{p}_2 \hat{q}_2}{n_2}},$$

Rounding

When we have only proportions and not the counts, we have to reconstruct the number of successes by multiplying the sample sizes by the proportions:

$Success_1 = n_1 \hat{p}_1$ and $Success_2 = n_2 \hat{p}_2$.

If these calculations don't come out to whole numbers, round them first. There must have been a whole number of successes, after all. (This is the *only* time you should round values in the middle of a calculation.)

and we could just plug in the given numbers, but that presents a logical dilemma. Our null hypothesis says we're assuming the two proportions p_1 and p_2 are equal. Why would we then substitute the unequal values $\hat{p}_1$ and $\hat{p}_2$ as estimates? That's like saying, "I think Tony and Maria are the same age. I'd guess he's 15 and she's 18." To avoid such a contradiction, we need to come up with a single value for $\hat{p}$ in the SE formula.

Let's see how to do this for the snoring example. If the null hypothesis is true, then the two groups aren't really different, so we can think of our two samples as really just parts of one bigger sample of the combined population. Overall, we saw $48 + 318 = 366$ snorers out of a total of $184 + 811 = 995$ adults who responded to this question. The overall proportion of snorers was $366/995 = 0.3678$.

Combining the counts like this to get an estimated proportion is called **pooling.** Whenever we have data from different sources or different groups but we believe that they really came from the same underlying population, we pool them to get better estimates.

Using the counts for each group, we can find the pooled proportion as

$$\hat{p}_{pooled} = \frac{Success_1 + Success_2}{n_1 + n_2},$$

where $Success_1$ is the number of successes in group 1 and $Success_2$ is the number of successes in group 2. That's our best estimate of the population proportion of success.

We then put this pooled value into the formula, substituting it for *both* sample proportions in the standard error formula:

$$SE_{pooled}(\hat{p}_1 - \hat{p}_2) = \sqrt{\frac{\hat{p}_{pooled}\,\hat{q}_{pooled}}{n_1} + \frac{\hat{p}_{pooled}\,\hat{q}_{pooled}}{n_2}}$$

$$= \sqrt{\frac{0.3678 \times (1 - 0.3678)}{184} + \frac{0.3678 \times (1 - 0.3678)}{811}}.$$

This comes out to 0.039.

Improving the Success/Failure Condition

The vaccine Gardasil® was introduced to prevent the strains of human papillomavirus (HPV) that are responsible for almost all cases of cervical cancer. In randomized placebo-controlled clinical trials,[4] only 1 case of HPV was diagnosed among 7897 women who received the vaccine, compared with 91 cases diagnosed among 7899 who received a placebo. The one observed HPV case ("success") doesn't meet the at-least-10-successes criterion. Surely, though, we should not refuse to test the effectiveness of the vaccine just because it failed so rarely; that would be absurd.

For that reason, in a two-proportion z-test, the proper Success/Failure test uses the *expected* frequencies, which we can find from the pooled proportion. In this case,

$$\hat{p}_{pooled} = \frac{91 + 1}{7899 + 7897} = 0.0058$$

$$n_1 \hat{p}_{pooled} = 7899(0.0058) = 46$$

$$n_2 \hat{p}_{pooled} = 7897(0.0058) = 46,$$

so we can proceed with the hypothesis test.

Often it is easier just to check the observed numbers of successes and failures. If they are both greater than 10, you don't need to look further. But keep in mind that the correct test uses the expected frequencies rather than the observed ones.

---

[4]*Quadrivalent Human Papillomavirus Vaccine: Recommendations of the Advisory Committee on Immunization Practices (ACIP)*, National Center for HIV/AIDS, Viral Hepatitis, STD and TB Prevention [May 2007].

The Two-Proportion z-Test

At last, we're ready to test a hypothesis about the difference of two proportions. We use the pooled estimate of the population proportion to find the standard error. That provides the yardstick we need to first calculate a z-score and then determine the P-value that allows us to decide whether the difference we see in the sample proportions is statistically significant.

> **Two-Proportion z-Test**
>
> The conditions for the two-proportion z-test are the same as for the two-proportion z-interval. We are testing the hypothesis
>
> $$H_0: p_1 - p_2 = 0.$$
>
> Because we hypothesize that the proportions are equal, we pool the groups to find
>
> $$\hat{p}_{pooled} = \frac{Success_1 + Success_2}{n_1 + n_2}$$
>
> and use that pooled value to estimate the standard error:
>
> $$SE_{pooled}(\hat{p}_1 - \hat{p}_2) = \sqrt{\frac{\hat{p}_{pooled}\hat{q}_{pooled}}{n_1} + \frac{\hat{p}_{pooled}\hat{q}_{pooled}}{n_2}}.$$
>
> Now we find the test statistic,
>
> $$z = \frac{(\hat{p}_1 - \hat{p}_2) - 0}{SE_{pooled}(\hat{p}_1 - \hat{p}_2)}.$$
>
> When the conditions are met and the null hypothesis is true, this statistic follows the standard Normal model, so we can use that model to obtain a P-value.

Step-by-Step Example A TWO-PROPORTION z-TEST

Question: Are the snoring rates of the two age groups really different?

THINK ➡ **Plan** State what you want to know. Discuss the variables and the W's.	I want to know whether snoring rates differ for those under and over 30 years old. The data are from a random sample of 1010 U.S. adults surveyed in the 2002 Sleep in America Poll. Of these, 995 responded to the question about snoring, indicating whether or not they had snored at least a few nights a week in the past year.

(continued)

Hypotheses The study simply broke down the responses by age, so there is no sense that either alternative was preferred. A two-sided alternative hypothesis is appropriate.

Model Think about the assumptions and check the conditions.

H_O: There is no difference in snoring rates in the two age groups:

$$p_{old} - p_{young} = 0.$$

H_A: The rates are different: $p_{old} - p_{young} \neq 0$.

✓ **Independent Groups Assumption:** The two groups are independent of each other because the sample was selected at random.

✓ **Independence Assumption:** The National Sleep Foundation selected respondents at random, so they should be independent.

✓ **Randomization Condition:** The respondents were randomly selected by telephone number and stratified by sex and region.

✓ **10% Condition:** The number of adults surveyed in each age group is certainly far less than 10% of that population.

✓ **Success/Failure Condition:** In the younger age group, 48 snored and 136 didn't. In the older group, 318 snored and 493 didn't. The observed numbers of both successes and failures are much more than 10 for both groups.[5]

State the null model.

Choose your method.

Because the conditions are satisfied, I'll use a Normal model and perform a **two-proportion z-test.**

SHOW ➡ Mechanics

The hypothesis is that the proportions are equal, so pool the sample data.

Use the pooled SE to estimate $SD(\hat{p}_{old} - \hat{p}_{young})$.

$n_{young} = 184, \quad y_{young} = 48, \quad \hat{p}_{young} = 0.261$

$n_{old} = 811, \quad y_{old} = 318, \quad \hat{p}_{old} = 0.392$

$\hat{p}_{pooled} = \dfrac{y_{old} + y_{young}}{n_{old} + n_{young}} = \dfrac{318 + 48}{811 + 184} = 0.3678$

$SE_{pooled}(\hat{p}_{old} - \hat{p}_{young})$

$= \sqrt{\dfrac{\hat{p}_{pooled}\,\hat{q}_{pooled}}{n_{old}} + \dfrac{\hat{p}_{pooled}\,\hat{q}_{pooled}}{n_{young}}}$

$= \sqrt{\dfrac{(0.3678)(0.6322)}{811} + \dfrac{(0.3678)(0.6322)}{184}}$

≈ 0.039375

The observed difference in sample proportions is $\hat{p}_{old} - \hat{p}_{young} = 0.392 - 0.261 = 0.131$

(continued)

[5]This is one of those situations in which the traditional term "success" seems a bit weird. A success here could be that a person snores. "Success" and "failure" are arbitrary labels left over from studies of gambling games.

Make a picture. Sketch a Normal model centered at the hypothesized difference of 0. Shade the region to the right of the observed difference, and because this is a two-tailed test, also shade the corresponding region in the other tail.

Find the z-score for the observed difference in proportions, 0.131.

Find the P-value using Table Z or technology. Because this is a two-tailed test, we must *double* the probability we find in the upper tail.

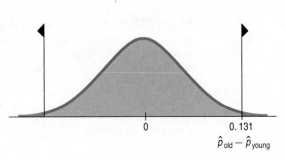

$$z = \frac{(\hat{p}_{old} - \hat{p}_{young}) - 0}{SE_{pooled}(\hat{p}_{old} - \hat{p}_{young})} = \frac{0.131 - 0}{0.039375} = 3.33$$

$$P = 2P(z \geq 3.33) = 0.0008$$

TELL ➡ **Conclusion** Link the P-value to your decision about the null hypothesis, and state your conclusion in context.

The P-value of 0.0008 says that if there really were no difference in snoring rates between the two age groups, then the difference observed in this study would happen only 8 times in 10,000. This is so small that I reject the null hypothesis of no difference and conclude that there is evidence of a difference in the rate of snoring between older adults and younger adults. It appears that older adults are more likely to snore.

TI Tips TESTING THE HYPOTHESIS

Yes, of course, there's a STAT TESTS routine to test a hypothesis about the difference of two proportions. Let's do the mechanics for the test about snoring. Of 811 people over 30 years old, 318 snored, while only 48 of the 184 people under 30 did.

• In the STAT TESTS menu select 2-PropZTest.

• Enter the observed numbers of snorers and the sample sizes for both groups.
• Since this is a two-tailed test, indicate that you want to see if the proportions are unequal. When you choose this option, the calculator will automatically include both tails as it determines the P-value.
• Calculate the result. Don't worry; for this procedure the calculator will pool the proportions automatically.

Now it is up to you to interpret the result and state a conclusion. We see a z-score of 3.33 and the P-value is 0.0008. Such a small P-value indicates that the observed difference is unlikely to be sampling error. What does that mean about snoring and age? Here's a great opportunity to follow up with a confidence interval so you can Tell even more!

Just Checking

3. A June 2004 public opinion poll asked 1000 randomly selected adults whether the United States should decrease the amount of immigration allowed; 49% of those responding said "yes." In June of 1995, a random sample of 1000 had found that 65% of adults thought immigration should be curtailed. To see if that percentage has decreased, why can't we just use a one-proportion z-test of H_0: $p = 0.65$ and see what the P-value for $\hat{p} = 0.49$ is?

4. For opinion polls like this, which has more variability: the percentage of respondents answering "yes" in either year or the difference in the percentages between the two years?

For Example ANOTHER 2-PROPORTION z-TEST

RECAP: One concern of the study on teens' online profiles was safety and privacy. In the random sample, girls were less likely than boys to say that they are easy to find online from their profiles. Only 19% (62 girls) of 325 teen girls with profiles say that they are easy to find, while 28% (75 boys) of the 268 boys with profiles say the same.

QUESTION: Are these results evidence of a real difference between boys and girls? Perform a two-proportion z-test and discuss what you find.

ANSWER:

$$H_O: p_{boys} - p_{girls} = 0$$
$$H_A: p_{boys} - p_{girls} \neq 0$$

✓ **Randomization Condition:** The sample of boys and the sample of girls were both chosen randomly.

✓ **10% Condition:** 268 boys and 325 girls are each less than 10% of all teenage boys and girls with online profiles.

✓ **Independent Groups Assumption:** Because the samples were selected at random, it's reasonable to believe the boys' perceptions are independent of the girls'.

✓ **Success/Failure Condition:** Among the girls, there were 62 "successes" and 263 failures, and among boys, 75 successes and 193 failures. These counts are at least 10 for each group.

Because all the assumptions and conditions are satisfied, it's okay to do a **two-proportion z-test**:

$$\hat{p}_{pooled} = \frac{75 + 62}{268 + 325} = 0.231$$

$$SE_{pooled}(\hat{p}_{boys} - \hat{p}_{girls}) = \sqrt{\frac{0.231 \times 0.769}{268} + \frac{0.231 \times 0.769}{325}} = 0.0348$$

$$z = \frac{(0.28 - 0.19) - 0}{0.0348} = 2.59$$

$$P(z > 2.59) = 0.0048$$

This is a two-tailed test, so the P-value = $2(0.0048) = 0.0096$. Because this P-value is very small, I reject the null hypothesis. This study provides strong evidence that there really is a difference in the proportions of teen girls and boys who say they are easy to find online.

WHAT IF ●●● we test a hypothesis by simulation?

It's estimated that world-wide 50,000 pregnant women die each year of eclampsia, a condition involving high blood pressure and seizures. In 2002 the medical journal *Lancet* reported on an experiment that involved nearly 10,000 at-risk women at 175 hospitals in 33 countries. The good news: researchers found that treating women with magnesium sulfate significantly reduced the occurrence of eclampsia, an important advance in women's health.

However, there was one cause for concern. Unfortunately, 27.5% women (11 out of 40) who developed eclampsia despite receiving the magnesium sulfate treatment died. In the placebo group, only 20.8% of the women (20 out of 96) who developed eclampsia died. Does this indicate that even though this treatment dramatically reduces the occurrence of eclampsia, women who develop the condition anyway may face a greater risk of death?

You now know how to do a 2-proportion z-test to see if the observed difference in mortality rates is statistically significant, so we won't do that.[6] Instead, we'll attack the question through simulation.

We'll do what's called a permutation test. The actual experiment saw a total of 31 deaths among 136 women. We wonder if the higher mortality rate in the magnesium sulfate group could have arisen by chance, or was a result of the treatment. To find out we create a list of 136 subjects denoting 31 as having died and 105 as having survived. Then we randomly divide them into a group of 40 and a group of 96, and look at the difference in mortality rates. That gives us a peek at what sort of difference could arise just because of the random allocation of subjects to treatments that actually had the same effect.

In our first simulation trial, 9 of the deaths randomly landed in the magnesium sulfate group and the other 22 in the placebo group. The simulated mortality rates came out $\frac{9}{40} = 22.5\%$ and $\frac{22}{96} = 22.9\%$, almost identical. Maybe the actual experiment's observed difference of $27.5 - 20.8 = 6.7\%$ is significant?

As you know, though, one trial does not a simulation make. In our next trial the random assignment resulted in a simulated death rate of $\frac{12}{40} = 30.0\%$ in the magnesium sulfate group and only $\frac{19}{96} = 19.8\%$ in the placebo group. This difference of $30.0 - 19.8 = 10.2\%$ is even larger than the 6.7% actually observed, and it arose purely by chance. Could something like that happen often? To find out, we simulated 500 trials.

Here's a histogram of the differences in sample proportions that arose by chance. In over 25% of our trials (127 out of 500) the simulated difference in mortality rates was at least as large as that seen in the actual experiment. Because such an apparently large difference isn't unusual, it's not evidence of a heightened risk for women receiving preventative treatment for eclampsia.

Permutation tests like this are pretty cool. Although they're not a required topic in this course,[7] the now widespread use of computer simulations makes them an increasingly common analytical tool. In fact, a permutation test is the correct way to analyze the results of an experiment. Fortunately, a Normal model and the 2-proportion z-test that we use for sample data provide a very good approximation for experiments.[8]

---

[6]Try it, though. It's good practice, and you'll be able to compare that result to the clever alternative we're showing you here.

[7]Yes, you can exhale. It won't be on the test.

[8]And *way* easier to do on a calculator!

WHAT CAN GO WRONG?

■ **Don't use two-sample proportion methods when the samples aren't independent.** These methods give wrong answers when this assumption of independence is violated. Good random sampling is usually the best insurance of independent groups. Make sure there is no relationship between the two groups. For example, you can't compare the proportion of respondents who own SUVs with the proportion of those same respondents who think the tax on gas should be eliminated. The responses are not independent because you've asked the same people. To use these methods to estimate or test the difference, you'd need to survey two different groups of people.

Alternatively, if you have a random sample, you can split your respondents according to their answers to one question and treat the two resulting groups as independent samples. So, you could test whether the proportion of SUV owners who favored eliminating the gas tax was the same as the corresponding proportion among non-SUV owners.

■ **Don't apply inference methods where there was no randomization.** If the data do not come from representative random samples or from a properly randomized experiment, then the inference about the differences in proportions will be wrong.

■ **Don't interpret a significant difference in proportions causally.** It turns out that people with higher incomes are more likely to snore. Does that mean money affects sleep patterns? Probably not. We have seen that older people are more likely to snore, and they are also likely to earn more. In a prospective or retrospective study, there is always the danger that other lurking variables not accounted for are the real reason for an observed difference. Be careful not to jump to conclusions about causality.

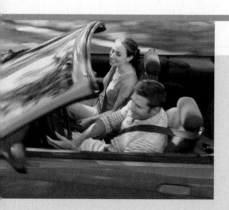

What Have We Learned?

We've learned how to extend our understanding of statistical inference to create confidence intervals and test hypotheses about the difference in two proportions.

■ We've learned that inference for the difference in two proportions is based on Normal models. In addition to the usual assumptions and conditions, we've learned to check the assumption that the groups are independent so that we can use the Pythagorean Theorem of Statistics to find the standard error for the difference in the two proportions.

■ We've learned that when the null hypothesis assumes the proportions are equal we must pool the sample data to estimate the true proportion. We don't pool for confidence intervals because there's no such assumption.

Perhaps most important, we've learned that the concepts, reasoning, and interpretations of statistical inference remain the same; only the mechanics change.

Terms

Variances of independent random variables add
The variance of a sum or difference of independent random variables is the sum of the variances of those variables. (p. 542)

Sampling distribution of the difference between two sample proportions
The sampling distribution of $\hat{p}_1 - \hat{p}_2$ is, under appropriate assumptions, modeled by a Normal model with mean $\mu = p_1 - p_2$ and standard deviation $SD(\hat{p}_1 - \hat{p}_2) = \sqrt{\dfrac{p_1 q_1}{n_1} + \dfrac{p_2 q_2}{n_2}}$. (p. 541)

Two-proportion z-interval
A two-proportion z-interval gives a confidence interval for the true difference in proportions, $p_1 - p_2$, in two independent groups.

The confidence interval is $(\hat{p}_1 - \hat{p}_2) \pm z^* \times SE(\hat{p}_1 - \hat{p}_2)$, where z^* is a critical value from the standard Normal model corresponding to the specified confidence level. (p. 545)

Pooling	When we believe a proportion is the same in two different groups, we can get a better estimate of this common proportion by combining the data from our two samples.

$$\hat{p}_{pooled} = \frac{x_1 + x_2}{n_1 + n_2}$$

The resulting standard error is based on more data and hence more reliable (if the null hypothesis is true). (p. 549)

Two-proportion z-test	Test the null hypothesis $H_0: p_1 - p_2 = 0$ by referring the statistic

$$z = \frac{\hat{p}_1 - \hat{p}_2}{SE_{pooled}(\hat{p}_1 - \hat{p}_2)}$$

to a standard Normal model. (p. 551)

On the Computer INFERENCES FOR THE DIFFERENCE BETWEEN TWO PROPORTIONS

It is so common to test against the null hypothesis of no difference between the two true proportions that most statistics programs simply assume this null hypothesis. And most will automatically use the pooled standard deviation. If you wish to test a different null (say, that the true difference is 0.3), you may have to search for a way to do it. Here's some typical computer software output for confidence intervals or hypothesis tests for the difference in proportions from two independent groups.

Identify the two groups.

Observed difference in the proportions

Hypothesis test results:
p_1 : proportion of successes for FemalePassenger
p_2 : proportion of successes for MalePassenger
$p_1 - p_2$: difference in proportions
$H_0: p_1 - p_2 = 0$
$H_A: p_1 - p_2 \neq 0$

Difference	Count1	Total1	Count2	Total2	Sample Diff.	Std. Err.	Z-Stat	P-value
$p_1 - p_2$	2777	4208	1363	2763	0.16662908	0.012025254	13.856596	<0.0001

Specify the hypotheses.

Number of successes and sample size for each group

The z-score and the P-value of the test. When the P-value is very small we don't need to know the exact value.

Note that often statistics packages don't offer special commands for inference for differences between proportions. As with inference for single proportions, many statistics programs want the "success" and "failure" status for each case. Computer packages don't usually deal with summary statistics. Calculators typically do a better job.

Exercises

1. **Canada** Suppose an advocacy organization surveys 960 Canadians and 192 of them reported being born in another country (www.unitednorthamerica.org/simdiff .htm). Similarly, 170 out of 1250 Americans reported being foreign-born. Find the standard error of the difference in sample proportions.

2. **Non-profits** Do people who work for non-profit organizations differ from those who work at for-profit companies when it comes to personal job satisfaction? Separate random samples were collected by a polling agency to investigate the difference. Data collected from 422 employees at non-profit organizations revealed that 377 of them were "highly satisfied." From the for-profit companies, 431 out 518 employees reported the same level of satisfaction. Find the standard error of the difference in sample proportions.

3. **Canada, deux** The information in Exercise 1 was used to create a 95% confidence interval for the difference between the proportions of Canadians and Americans who were born in foreign countries.
   a) Interpret this interval with a sentence in context. 95% confidence interval for
      $p_{Canadians} - p_{Americans}$ is $(3.24\%, 9.56\%)$
   b) For this interval, explain what "95% confidence" means.

4. **Non-profits, part 2** The researchers from Exercise 2 created a 95% confidence interval for the difference in proportions who are "highly satisfied" among people who work at non-profits versus people who work at for-profit companies.
   a) Interpret the interval with a sentence in context. 95% confidence interval for
      $p_{non-profits} - p_{for-profits} = (1.77\%, 10.50\%)$
   b) For this interval, explain what "95% confidence" means.

5. **How do you get online?** A 2013 Pew Research survey of 802 teens found that, among teens aged 12–17, girls were significantly more likely to access the internet mostly through their smartphone. 29% of the girls surveyed said they access the internet mostly on their phone, compared to 20% of the boys. What does it mean to say that the difference in proportions is "significant"?

6. **Social network news** In 2012, Pew Research noted that 34% of young adults age 18 to 24 got some news from a social networking site the previous day. This is not a significant difference from the 32% of 25- to 29-year-olds who did so. What does it mean to say that the difference is not significant?

7. **Name recognition** A political candidate runs a weeklong series of TV ads designed to attract public attention to his campaign. Polls taken before and after the ad campaign show some increase in the proportion of voters who now recognize this candidate's name, with a P-value of 0.033. Is it reasonable to believe the ads may be effective?

8. **Origins** In a 1993 Gallup poll, 47% of the respondents agreed with the statement "*God created human beings pretty much in their present form at one time within the last 10,000 years or so.*" When Gallup asked the same question in 2008, only 44% of those respondents agreed. Is it reasonable to conclude that there was a change in public opinion given that the P-value is 0.17? Explain.

9. **Revealing information** 886 randomly sampled teens were asked which of several personal items of information they thought it okay to share with someone they had just met. 44% said it was okay to share their e-mail addresses, but only 29% said they would give out their cell phone numbers. A researcher claims that a two-proportion z-test could tell whether there was a real difference among all teens. Explain why that test would not be appropriate for these data.

10. **Regulating access** When a random sample of 935 parents were asked about rules in their homes, 77% said they had rules about the kinds of TV shows their children could watch. Among the 790 of those parents whose teenage children had Internet access, 85% had rules about the kinds of Internet sites their teens could visit. That looks like a difference, but can we tell? Explain why a two-sample z-test would not be appropriate here.

11. **Gender gap** A presidential candidate fears he has a problem with women voters. His campaign staff plans to run a poll to assess the situation. They'll randomly sample 300 men and 300 women, asking if they have a favorable impression of the candidate. Obviously, the staff can't know this, but suppose the candidate has a positive image with 59% of males but with only 53% of females.
   a) What sampling design is his staff planning to use?
   b) What difference would you expect the poll to show?
   c) Of course, sampling error means the poll won't reflect the difference perfectly. What's the standard deviation for the difference in the proportions?
   d) Sketch a sampling model for the size difference in proportions of men and women with favorable impressions of this candidate that might appear in a poll like this.
   e) Could the campaign be misled by the poll, concluding that there really is no gender gap? Explain.

12. Buy it again? A consumer magazine plans to poll car owners to see if they are happy enough with their vehicles that they would purchase the same model again. They'll randomly select 450 owners of American-made cars and 450 owners of Japanese models. Obviously, the actual opinions of the entire population couldn't be known, but suppose 76% of owners of American cars and 78% of owners of Japanese cars would purchase another.

a) What sampling design is the magazine planning to use?

b) What difference would you expect their poll to show?

c) Of course, sampling error means the poll won't reflect the difference perfectly. What's the standard deviation for the difference in the proportions?

d) Sketch a sampling model for the difference in proportions that might appear in a poll like this.

e) Could the magazine be misled by the poll, concluding that owners of American cars are much happier with their vehicles than owners of Japanese cars? Explain.

13. Arthritis The Centers for Disease Control and Prevention reported a survey of randomly selected Americans age 65 and older, which found that 411 of 1012 men and 535 of 1062 women suffered from some form of arthritis.

a) Are the assumptions and conditions necessary for inference satisfied? Explain.

b) Create a 95% confidence interval for the difference in the proportions of senior men and women who have this disease.

c) Interpret your interval in this context.

d) Does this confidence interval suggest that arthritis is more likely to afflict women than men? Explain.

14. Graduation In October 2000 the U.S. Department of Commerce reported the results of a large-scale survey on high school graduation. Researchers contacted more than 25,000 randomly chosen Americans aged 24 years to see if they had finished high school; 84.9% of the 12,460 males and 88.1% of the 12,678 females indicated that they had high school diplomas.

a) Are the assumptions and conditions necessary for inference satisfied? Explain.

b) Create a 95% confidence interval for the difference in graduation rates between males and females.

c) Interpret your confidence interval.

d) Does this provide strong evidence that girls are more likely than boys to complete high school? Explain.

15. Pets Researchers at the National Cancer Institute released the results of a study that investigated the effect of weed-killing herbicides on house pets. They examined 827 dogs from homes where an herbicide was used on a regular basis, diagnosing malignant lymphoma in 473 of them. Of the 130 dogs from homes where no herbicides were used, only 19 were found to have lymphoma.

a) What's the standard error of the difference in the two proportions?

b) Construct a 95% confidence interval for this difference.

c) State an appropriate conclusion.

16. Carpal tunnel The painful wrist condition called carpal tunnel syndrome can be treated with surgery or less invasive wrist splints. In September 2002, *Time* magazine reported on a study of 176 patients. Among the half that had surgery, 80% showed improvement after three months, but only 54% of those who used the wrist splints improved.

a) What's the standard error of the difference in the two proportions?

b) Construct a 95% confidence interval for this difference.

c) State an appropriate conclusion.

17. Prostate cancer There has been debate among doctors over whether surgery can prolong life among men suffering from prostate cancer, a type of cancer that typically develops and spreads very slowly. Recently, *The New England Journal of Medicine* published results of some Scandinavian research. Men diagnosed with prostate cancer were randomly assigned to either undergo surgery or not. Among the 347 men who had surgery, 16 eventually died of prostate cancer, compared with 31 of the 348 men who did not have surgery.

a) Was this an experiment or an observational study? Explain.

b) Create a 95% confidence interval for the difference in rates of death for the two groups of men.

c) Based on your confidence interval, is there evidence that surgery may be effective in preventing death from prostate cancer? Explain.

18. Race and smoking 2010 Data collected in 2010 by the Behavioral Risk Factor Surveillance System revealed that in the state of New Jersey, 15.7% of whites and 15.9% of blacks were cigarette smokers. Suppose these proportions were based on samples of 2449 whites and 464 blacks.

a) Create a 90% confidence interval for the difference in the percentage of smokers between black and white adults in New Jersey.

b) Does this survey indicate a race-based difference in smoking among New Jersey adults? Explain, using your confidence interval to test an appropriate hypothesis.

c) What alpha level did your test use?

19. Ear infections A new vaccine was recently tested to see if it could prevent the painful and recurrent ear infections that many infants suffer from. *The Lancet*, a medical journal, reported a study in which babies about a year old were randomly divided into two groups. One group received vaccinations; the other did not. During the following year, only 333 of 2455 vaccinated children had ear infections, compared to 499 of 2452 unvaccinated children in the control group.

a) Are the conditions for inference satisfied?

b) Find a 95% confidence interval for the difference in rates of ear infection.

c) Use your confidence interval to explain whether you think the vaccine is effective.

20. **Anorexia** The *Journal of the American Medical Association* reported on an experiment intended to see if the drug Prozac® could be used as a treatment for the eating disorder anorexia nervosa. The subjects, women being treated for anorexia, were randomly divided into two groups. Of the 49 who received Prozac, 35 were deemed healthy a year later, compared to 32 of the 44 who got the placebo.

a) Are the conditions for inference satisfied?

b) Find a 95% confidence interval for the difference in outcomes.

c) Use your confidence interval to explain whether you think Prozac is effective.

21. **Another ear infection** In Exercise 19 you used a confidence interval to examine the effectiveness of a vaccine against ear infections in babies. Suppose that instead you had conducted a hypothesis test. (Answer these questions *without* actually doing the test.)

a) What hypotheses would you test?

b) State a conclusion based on your confidence interval.

c) What alpha level did your test use?

d) If that conclusion is wrong, which type of error did you make?

e) What would be the consequences of such an error?

22. **Anorexia again** In Exercise 20 you used a confidence interval to examine the effectiveness of Prozac in treating anorexia nervosa. Suppose that instead you had conducted a hypothesis test. (Answer these questions *without* actually doing the test.)

a) What hypotheses would you test?

b) State a conclusion based on your confidence interval.

c) What alpha level did your test use?

d) If that conclusion is wrong, which type of error did you make?

e) What would be the consequences of such an error?

23. **Teen smoking, part I** A Vermont study published by the American Academy of Pediatrics examined parental influence on teenagers' decisions to smoke. A group of students who had never smoked were questioned about their parents' attitudes toward smoking. These students were questioned again two years later to see if they had started smoking. The researchers found that, among the 284 students who indicated that their parents disapproved of kids smoking, 54 had become established smokers. Among the 41 students who initially said their parents were lenient about smoking, 11 became smokers. Do these data provide strong evidence that parental attitude influences teenagers' decisions about smoking?

a) What kind of design did the researchers use?

b) Write appropriate hypotheses.

c) Are the assumptions and conditions necessary for inference satisfied?

d) Test the hypothesis and state your conclusion.

e) Explain in this context what your P-value means.

f) If that conclusion is actually wrong, which type of error did you commit?

24. **Depression** A study published in the *Archives of General Psychiatry* examined the impact of depression on a patient's ability to survive cardiac disease. Researchers identified 450 people with cardiac disease, evaluated them for depression, and followed the group for 4 years. Of the 361 patients with no depression, 67 died. Of the 89 patients with minor or major depression, 26 died. Among people who suffer from cardiac disease, are depressed patients more likely to die than non-depressed ones?

a) What kind of design was used to collect these data?

b) Write appropriate hypotheses.

c) Are the assumptions and conditions necessary for inference satisfied?

d) Test the hypothesis and state your conclusion.

e) Explain in this context what your P-value means.

f) If your conclusion is actually incorrect, which type of error did you commit?

25. **Teen smoking, part II** Consider again the Vermont study discussed in Exercise 23.

a) Create a 95% confidence interval for the difference in the proportion of children who may smoke and have lenient parents and those who may smoke and have disapproving parents.

b) Interpret your interval in this context.

c) Carefully explain what "95% confidence" means.

26. **Depression revisited** Consider again the study of the association between depression and cardiac disease survivability in Exercise 24.

a) Create a 95% confidence interval for the difference in survival rates.

b) Interpret your interval in this context.

c) Carefully explain what "95% confidence" means.

27. **Pregnancy** In 1998, a San Diego reproductive clinic reported 42 live births to 157 women under the age of 38, but only 7 live births for 89 clients aged 38 and older. Is this strong evidence of a difference in the effectiveness of the clinic's methods for older women?

a) Was this an experiment? Explain.

b) Test an appropriate hypothesis and state your conclusion in context.

c) If you concluded there was a difference, estimate that difference with a confidence interval and interpret your interval in context.

28. **Birthweight** In 2003 the *Journal of the American Medical Association* reported a study examining the possible impact of air pollution caused by the 9/11 attack

on New York's World Trade Center on the weight of babies. Researchers found that 8% of 182 babies born to mothers who were exposed to heavy doses of soot and ash on September 11 were classified as having low birth weight. Only 4% of 2300 babies born in another New York City hospital whose mothers had not been near the site of the disaster were similarly classified. Does this indicate a possibility that air pollution might be linked to a significantly higher proportion of low-weight babies?

a) Was this an experiment? Explain.

b) Test an appropriate hypothesis and state your conclusion in context.

c) If you concluded there is a difference, estimate that difference with a confidence interval and interpret that interval in context.

29. **Political scandal!** One month before the election, a poll of 630 randomly selected voters showed 54% planning to vote for a certain candidate. A week later, it became known that he had tweeted inappropriate pictures of himself, and a new poll showed only 51% of 1010 voters supporting him. Do these results indicate a decrease in voter support for his candidacy?

a) Test an appropriate hypothesis and state your conclusion.

b) If your conclusion turns out to be wrong, did you make a Type I or Type II error?

c) If you concluded there was a difference, estimate that difference with a confidence interval and interpret your interval in context.

30. **Shopping** A survey of 430 randomly chosen adults found that 21% of the 222 men and 18% of the 208 women had purchased books online.

a) Is there evidence that men are more likely than women to make online purchases of books? Test an appropriate hypothesis and state your conclusion in context.

b) If your conclusion in fact proves to be wrong, did you make a Type I or Type II error?

c) Estimate this difference with a confidence interval.

d) Interpret your interval in context.

31. **Twins** In 2001, one county reported that, among 3132 white women who had babies, 94 were multiple births. There were also 20 multiple births to 606 black women. Does this indicate any racial difference in the likelihood of multiple births?

a) Test an appropriate hypothesis and state your conclusion in context.

b) If your conclusion is incorrect, which type of error did you commit?

32. **Mammograms** A 9-year study in Sweden compared 21,088 women who had mammograms with 21,195 who did not. Of the women who underwent screening, 63 died of breast cancer, compared to 66 deaths among the control group. (*The New York Times*, Dec 9, 2001)

a) Do these results support the effectiveness of regular mammograms in preventing deaths from breast cancer?

b) If your conclusion is incorrect, what kind of error have you committed?

33. **Pain** Researchers comparing the effectiveness of two pain medications randomly selected a group of patients who had been complaining of a certain kind of joint pain. They randomly divided these people into two groups, then administered the pain killers. Of the 112 people in the group who received medication A, 84 said this pain reliever was effective. Of the 108 people in the other group, 66 reported that pain reliever B was effective.

a) Write a 95% confidence interval for the percent of people who may get relief from this kind of joint pain by using medication A. Interpret your interval.

b) Write a 95% confidence interval for the percent of people who may get relief by using medication B. Interpret your interval.

c) Do the intervals for A and B overlap? What do you think this means about the comparative effectiveness of these medications?

d) Find a 95% confidence interval for the difference in the proportions of people who may find these medications effective. Interpret your interval.

e) Does this interval contain zero? What does that mean?

f) Why do the results in parts c and e seem contradictory? If we want to compare the effectiveness of these two pain relievers, which is the correct approach? Why?

34. **Gender gap** Candidates for political office realize that different levels of support among men and women may be a crucial factor in determining the outcome of an election. One candidate finds that 52% of 473 men polled say they will vote for him, but only 45% of the 522 women in the poll express support.

a) Write a 95% confidence interval for the percent of male voters who may vote for this candidate. Interpret your interval.

b) Write and interpret a 95% confidence interval for the percent of female voters who may vote for him.

c) Do the intervals for males and females overlap? What do you think this means about the gender gap?

d) Find a 95% confidence interval for the difference in the proportions of males and females who will vote for this candidate. Interpret your interval.

e) Does this interval contain zero? What does that mean?

f) Why do the results in parts c and e seem contradictory? If we want to see if there is a gender gap among voters with respect to this candidate, which is the correct approach? Why?

35. **Food preference** GfK Roper Consulting gathers information on consumer preferences around the world to help companies monitor attitudes about health, food, and

healthcare products. They asked people in many different cultures how they felt about the following statement:

> *I have a strong preference for regional or traditional products and dishes from where I come from.*

In a random sample of 800 respondents, 417 of 646 people who live in urban environments agreed (either completely or somewhat) with that statement, compared to 78 out of 154 people who live in rural areas.

Based on this sample, is there evidence that the percentage of people agreeing with the statement about regional preferences differs between all urban and rural dwellers?

36. **Fast food** The global survey we learned about in Exercise 35 also asked respondents how they felt about the statement "I try to avoid eating fast foods." The random sample of 800 included 411 people 35 years old or younger, and of those, 197 agreed (completely or somewhat) with the statement. Of the 389 people over 35 years old, 246 people agreed with the statement. Is there evidence that the percentage of people avoiding fast food is different in the two age groups?

37. **Online activity checks** Are more parents checking up on their teen's online activities? A Pew survey in 2004 found that 33% of 868 randomly sampled teens said that their parents checked to see what Web sites they visited. In 2006 the same question posed to 811 teens found 41% reporting such checks. Do these results provide evidence that more parents are checking?

38. **Computer gaming** Who plays online or electronic games? A survey in 2006 found that 69% of 223 boys aged 12–14 said they "played computer or console games like Xbox or PlayStation . . . or games online." Of 248 boys aged 15–17, only 62% played these games. Is this evidence of a real age-based difference?

 Just Checking ANSWERS

1. We're 90% confident that if members are contacted by e-mail, the donation rate will be between 4 and 18 percentage points higher than if they received regular mail.

2. Since a difference of 0 is not in the confidence interval, we'd reject the null hypothesis. There is evidence that more members will donate if contacted by e-mail.

3. The proportion from the sample in 1995 has variability, too. If we do a one-proportion z-test, we won't take that variability into account and our P-value will be incorrect.

4. The difference in the proportions between the two years has more variability than either individual proportion. The variance of the difference is the sum of the two variances.

Review of part V

From the Data at Hand to the World at Large

Quick Review

What do samples really tell us about the populations from which they are drawn? Are the results of an experiment meaningful, or are they just sampling error? Statistical inference based on our understanding of sampling models can help answer these questions. Here's a brief summary of the key concepts and skills:

- Sampling models describe the variability of sample statistics using a remarkable result called the Central Limit Theorem.

  - When the number of trials is sufficiently large, proportions found in different samples vary according to an approximately Normal model.

  - When samples are sufficiently large, the means of different samples vary, with an approximately Normal model.

  - The variability of sample statistics decreases as sample size increases.

  - Statistical inference procedures are based on the Central Limit Theorem.

  - No inference procedure is valid unless the underlying assumptions are true. Always check the conditions before proceeding.

- A confidence interval uses a sample statistic (such as a proportion) to estimate a range of plausible values for the parameter of a population model.

  - All confidence intervals involve an estimate of the parameter, a margin of error, and a level of confidence.

  - For confidence intervals based on a given sample, the greater the margin of error, the higher the confidence.

  - At a given level of confidence, the larger the sample, the smaller the margin of error.

- A hypothesis test proposes a model for the population, then examines the observed statistics to see if that model is plausible.

- A null hypothesis suggests a parameter value for the population model. Usually, we assume there is nothing interesting, unusual, or different about the sample results.

- The alternative hypothesis states what we will believe if the sample results turn out to be inconsistent with our null model.

- We compare the difference between the statistic and the hypothesized value with the standard deviation of the statistic. It's the sampling distribution of this ratio that gives us a P-value.

- The P-value of the test is the conditional probability that the null model could produce results at least as extreme as those observed in the sample or the experiment just as a result of sampling error.

- A low P-value indicates evidence against the null model. If it is sufficiently low, we reject the null model.

- A high P-value indicates that the sample results are not inconsistent with the null model, so we cannot reject it. However, this does not prove the null model is true.

- Sometimes we will mistakenly reject the null hypothesis even though it's actually true—that's called a Type I error. If we fail to reject a false null hypothesis, we commit a Type II error.

- The power of a test measures its ability to detect a false null hypothesis.

- You can lower the risk of a Type I error by requiring a higher standard of proof (lower P-value) before rejecting the null hypothesis. But this will raise the risk of a Type II error and decrease the power of the test.

- The only way to increase the power of a test while decreasing the chance of committing either error is to design a study based on a larger sample.

And now for some opportunities to review these concepts and skills . . .

Review Exercises

1. **Crohn's disease.** Omega-3 fatty acids have been tested as a means to prevent relapse of Crohn's disease. Two large, randomized, placebo-controlled studies have shown no such benefit from omega-3 fatty acids. Suppose you are asked to design an experiment to further study this claim. Imagine that you have collected data on Crohn's relapses in subjects who have used these omega-3 fatty acids and similar subjects who have not used them and that you can measure incidences of relapse for these subjects. State the null and alternative hypotheses you would use in your study.

2. **Colorblind.** Medical literature says that about 8% of males are colorblind. A university's introductory psychology course is taught in a large lecture hall. Among the

students, there are 325 males. Each semester when the professor discusses visual perception, he shows the class a test for colorblindness. The percentage of males who are colorblind varies from semester to semester.

a) Is the sampling distribution model for the sample proportion likely to be Normal? Explain.

b) What are the mean and standard deviation of this sampling distribution model?

c) Sketch the sampling model, using the 68–95–99.7 Rule.

d) Write a few sentences explaining what the model says about this professor's class.

3. Birth days. During a 2-month period in 2002, 72 babies were born at the Tompkins Community Hospital in up-state New York. The table shows how many babies were born on each day of the week.

a) If births are uniformly distributed across all days of the week, how many would you expect on each day?

b) Only 7 births occurred on a Monday. Does this indicate that women might be less likely to give birth on a Monday? Explain.

c) Are the 17 births on Tuesdays unusually high? Explain.

Day	Births
Mon.	7
Tues.	17
Wed.	8
Thurs.	12
Fri.	9
Sat.	10
Sun.	9

d) Can you think of any reasons why births may not occur completely at random?

4. Polling 2004. In the 2004 U.S. presidential election, the official results showed that George W. Bush received 50.7% of the vote and John Kerry received 48.3%. Ralph Nader, running as a third-party candidate, picked up only 0.4%. After the election, there was much discussion about exit polls, which had initially indicated a different result. Suppose you had taken a random sample of 1000 voters in an exit poll and asked them for whom they had voted.

a) Would you always get 507 votes for Bush and 483 for Kerry?

b) In 95% of such polls, your sample proportion of voters for Bush should be between what two values?

c) In 95% of such polls, your sample proportion of voters for Nader should be between what two numbers?

d) Would you expect the sample proportion of Nader votes to vary more, less, or about the same as the sample proportion of Bush votes? Why?

5. Leaky gas tanks. Nationwide, it is estimated that 40% of service stations have gas tanks that leak to some extent. A new program in California is designed to lessen the prevalence of these leaks. We want to assess the effectiveness of the program by seeing if the percentage of service stations whose tanks leak has decreased. To do this, we randomly sample 27 service stations in

California and determine whether there is any evidence of leakage. In our sample, only 7 of the stations exhibit any leakage. Is there evidence that the new program is effective?

a) What are the null and alternative hypotheses?

b) Check the assumptions necessary for inference.

c) Test the null hypothesis.

d) What do you conclude (in plain English)?

e) If the program actually works, have you made an error? What kind?

f) What two things could you do to decrease the probability of making this kind of error?

g) What are the advantages and disadvantages of taking those two courses of action?

6. Surgery and germs. Joseph Lister (for whom Listerine is named!) was a British physician who was interested in the role of bacteria in human infections. He suspected that germs were involved in transmitting infection, so he tried using carbolic acid as an operating room disinfectant. In 75 amputations, he used carbolic acid 40 times. Of the 40 amputations using carbolic acid, 34 of the patients lived. Of the 35 amputations without carbolic acid, 19 patients lived. The question of interest is whether carbolic acid is effective in increasing the chances of surviving an amputation.

a) What kind of a study is this?

b) What do you conclude? Support your conclusion by testing an appropriate hypothesis.

c) What reservations do you have about the design of the study?

7. Scrabble. Using a computer to play many simulated games of Scrabble, researcher Charles Robinove found that the letter "A" occurred in 54% of the hands. This study had a margin of error of $\pm 10\%$. (*Chance*, 15, no. 1 [2002])

a) Explain what the margin of error means in this context.

b) Why might the margin of error be so large?

c) Probability theory predicts that the letter "A" should appear in 63% of the hands. Does this make you concerned that the simulation might be faulty? Explain.

8. Dice. When one die is rolled, the number of spots showing has a mean of 3.5 and a standard deviation of 1.7. Suppose you roll 10 dice. What's the approximate probability that your total is between 30 and 40 (that is, the average for the 10 dice is between 3 and 4)? Specify the model you use and the assumptions and conditions that justify your approach.

9. Net-Newsers. In June 2008, the Pew Research Foundation sampled 3615 U.S. adults and asked about their choice of news sources. They identified 13% as "Net-Newsers" who regularly get their news from online sources rather than TV or newspapers.

a) Pew reports a margin of error of $\pm 2\%$ for this result. Explain what the margin of error means.

b) Pew's survey included 2802 respondents contacted by landline and 813 contacted by cell phone. If the percentage of Net-Newsers is the same in both groups and Pew estimated those percentages separately, which group would have the larger margin of error? Explain.

c) Pew reports that 82% of the 470 Net-Newsers in their survey get news during the course of the day, far more than other respondents. Find a 95% confidence interval for this statistic.

d) How does the margin of error for your confidence interval compare with the values in parts a and b? Explain why.

10. Gay marriage. In May 2012, a CNN/ORC Poll asked a random sample of 1009 U.S. adults this question:

Do you think marriages between gay and lesbian couples should or should not be recognized by the law as valid, with the same rights as traditional marriages?

Of those polled, 54% said they favored marriage equality, the highest percentage to date.

a) Create a 95% confidence interval for the percentage of all American adults who support marriage equality.

b) Based on your confidence interval, can you conclude that a majority of U.S. adults support legally recognizing gay marriages? Explain.

c) If pollsters wanted to follow up on this poll with another survey that could determine the level of support for gay marriage to within 2% with 98% confidence, how many people should they poll?

11. Bimodal. We are sampling randomly from a distribution known to be bimodal.

a) As our sample size increases, what's the expected shape of the sample's distribution?

b) What's the expected value of our sample's mean? Does the size of the sample matter?

c) How is the variability of sample means related to the standard deviation of the population? Does the size of the sample matter?

d) How is the shape of the sampling distribution model affected by the sample size?

12. Vitamin D 2012. In 2012, the *American Journal of Clinical Nutrition* reported that 31% of Australian adults over age 25 have a vitamin D deficiency. The data came from the AusDiab study of 11,218 Australians.

a) Do these data meet the assumptions necessary for inference? What would you like to know that you don't?

b) Create a 95% confidence interval.

c) Interpret the interval in this context.

d) Explain in this context what "95% confidence" means.

13. Archery. A champion archer can generally hit the bull's-eye 80% of the time. Suppose she shoots 200 arrows during competition. Let $\hat{p}$ represent the percentage of bull's-eyes she gets (the sample proportion).

a) What are the mean and standard deviation of the sampling distribution model for $\hat{p}$?

b) Is a Normal model appropriate here? Explain.

c) Sketch the sampling model, using the 68–95–99.7 Rule.

d) What's the probability that she gets at least 85% bull's-eyes?

14. Occupy Wall Street. In 2011, the Occupy Wall Street movement protested the concentration of wealth and power in the United States. A 2012 University of Delaware survey asked a random sample of 901 American adults whether they agreed or disagreed with the following statement:

The Occupy Wall Street protesters offered new insights on social issues.

Of those asked, 59.9% said they strongly or somewhat agreed with this statement. We know that if we could ask the entire population of American adults, we would not find that exactly 59.9% think that Wall Street workers would be willing to break the law to make money. Construct a 95% confidence interval for the true percentage of American adults who agree with the statement.

15. Twins. There is some indication in medical literature that doctors may have become more aggressive in inducing labor or doing preterm cesarean sections when a woman is carrying twins. Records at a large hospital show that, of the 43 sets of twins born in 1990, 20 were delivered before the 37th week of pregnancy. In 2000, 26 of 48 sets of twins were born preterm. Does this indicate an increase in the incidence of early births of twins? Test an appropriate hypothesis and state your conclusion.

16. Eclampsia. It's estimated that 50,000 pregnant women worldwide die each year of eclampsia, a condition involving elevated blood pressure and seizures. A research team from 175 hospitals in 33 countries investigated the effectiveness of magnesium sulfate in preventing the occurrence of eclampsia in at-risk patients. Results are summarized below. (*Lancet*, June 1, 2002)

		Reported side effects	Developed eclampsia	Deaths	Total Subjects
Treatment	Magnesium sulfate	1201	40	11	**4999**
	Placebo	228	96	20	**4993**

a) Write a 95% confidence interval for the increase in the proportion of women who may develop side effects from this treatment. Interpret your interval.

b) Is there evidence that the treatment may be effective in preventing the development of eclampsia? Test an appropriate hypothesis and state your conclusion.

17. Eclampsia. Refer again to the research summarized in Exercise 16. Is there any evidence that when eclampsia does occur, the magnesium sulfate treatment may help prevent the woman's death?

a) Write an appropriate hypothesis.
b) Check the assumptions and conditions.
c) Find the P-value of the test.
d) What do you conclude about the magnesium sulfate treatment?
e) If your conclusion is wrong, which type of error have you made?
f) Name two things you could do to increase the power of this test.
g) What are the advantages and disadvantages of those two options?

18. Eggs. The ISA Babcock Company supplies poultry farmers with hens, advertising that a mature B300 Layer produces eggs with a mean weight of 60.7 grams. Suppose that egg weights follow a Normal model with standard deviation 3.1 grams.

a) What fraction of the eggs produced by these hens weigh more than 62 grams?
b) What's the probability that a dozen randomly selected eggs average more than 62 grams?
c) Using the 68–95–99.7 Rule, sketch a model of the total weights of a dozen eggs.

19. Polling disclaimer. A newspaper article that reported the results of an election poll included the following explanation:

The Associated Press poll on the 2012 presidential campaign is based on telephone interviews with 798 randomly selected registered voters from all states except Alaska and Hawaii.

The results were weighted to represent the population by demographic factors such as age, sex, region, and education.

No more than 1 time in 20 should chance variations in the sample cause the results to vary by more than 4 percentage points from the answers that would be obtained if all Americans were polled.

The margin of sampling error is larger for responses of subgroups, such as income categories or those in political parties. There are other sources of potential error in polls, including the wording and order of questions.

a) Did they describe the 5 W's well?
b) What kind of sampling design could take into account the several demographic factors listed?
c) What was the margin of error of this poll?
d) What was the confidence level?
e) Why is the margin of error larger for subgroups?
f) Which kinds of potential bias did they caution readers about?

20. Enough eggs? One of the important issues for poultry farmers is the production rate—the percentage of days on which a given hen actually lays an egg. Ideally, that would be 100% (an egg every day), but realistically, hens tend to lay eggs on about 3 of every 4 days. ISA Babcock wants to advertise the production rate for the B300 Layer (see Exercise 18) as a 95% confidence interval with a margin of error of $\pm 2\%$. How many hens must they collect data on?

21. Teen deaths. Traffic accidents are the leading cause of death among people aged 15 to 20. In May 2002, the National Highway Traffic Safety Administration reported that even though only 6.8% of licensed drivers are between 15 and 20 years old, they were involved in 14.3% of all fatal crashes. Insurance companies have long known that teenage boys were high risks, but what about teenage girls? One insurance company found that the driver was a teenage girl in 44 of the 388 fatal accidents they investigated. Is this strong evidence that the accident rate is lower for girls than for teens in general?

a) Test an appropriate hypothesis and state your conclusion.
b) Explain what your P-value means in this context.

22. Perfect pitch. A recent study of perfect pitch tested students in American music conservatories. It found that 7% of 1700 non-Asian and 32% of 1000 Asian students have perfect pitch. A test of the difference in proportions resulted in a P-value of <0.0001.

a) What are the researchers' null and alternative hypotheses?
b) State your conclusion.
c) Explain in this context what the P-value means.
d) The researchers claimed that the data prove that genetic differences between the two populations cause a difference in the frequency of occurrence of perfect pitch. Do you agree? Why or why not?

23. Largemouth bass. Organizers of a fishing tournament believe that the lake holds a sizable population of largemouth bass. They assume that the weights of these fish have a model that is skewed to the right with a mean of 3.5 pounds and a standard deviation of 2.2 pounds.

a) Explain why a skewed model makes sense here.
b) Explain why you cannot determine the probability that a largemouth bass randomly selected ("caught") from the lake weighs over 3 pounds.
c) Each fisherman in the contest catches 5 fish each day. Can you determine the probability that someone's catch averages over 3 pounds? Explain.
d) The 12 fishermen competing each caught the limit of 5 fish. What's the probability that the total catch of 60 fish averaged more than 3 pounds?

24. Cheating. A Rutgers University study released in 2002 found that many high school students cheat on tests. The researchers surveyed a random sample of 4500 high school students nationwide; 74% of them said they had cheated at least once.

a) Create a 90% confidence interval for the level of cheating among high school students. Don't forget to check the appropriate conditions.

b) Interpret your interval.

c) Explain what "90% confidence" means.

d) Would a 95% confidence interval be wider or narrower? Explain without actually calculating the interval.

25. Language. Neurological research has shown that in about 80% of people language abilities reside in the brain's left side. Another 10% display right-brain language centers, and the remaining 10% have two-sided language control. (The latter two groups are mainly left-handers.) (*Science News*, 161, no. 24 [2002])

a) We select 60 people at random. Is it reasonable to use a Normal model to describe the possible distribution of the proportion of the group that has left-brain language control? Explain.

b) What's the probability that our group has at least 75% left-brainers?

c) If the group had consisted of 100 people, would that probability be higher, lower, or about the same? Explain why, without actually calculating the probability.

d) How large a group would almost certainly guarantee at least 75% left-brainers? Explain.

26. Cigarettes 2009. In 1999, the Centers for Disease Control and Prevention estimated that about 34.8% of high school students smoked cigarettes. They established a national health goal of reducing that figure to 16% by the year 2010. To that end, they would be on track if they achieved a reduction to 17.7% by 2009. In 2009, they released a research study in which 19.5% of a random sample of 5080 high school students said they were current smokers. Is this evidence that progress toward the goal is off track?

a) Write appropriate hypotheses.

b) Verify that the appropriate assumptions are satisfied.

c) Find the P-value of this test.

d) Explain what the P-value means in this context.

e) State an appropriate conclusion.

f) Of course, your conclusion may be incorrect. If so, which kind of error did you commit?

27. Crohn's disease. In 2002 the medical journal *The Lancet* reported that 335 of 573 patients suffering from Crohn's disease responded positively to injections of the arthritis-fighting drug infliximab.

a) Create a 95% confidence interval for the effectiveness of this drug.

b) Interpret your interval in context.

c) Explain carefully what "95% confidence" means in this context.

28. Teen smoking 2009. The Centers for Disease Control and Prevention say that about 19.5% of teenagers smoke tobacco (down from a high of 38% in 1997). A college has 522 students in its freshman class. Is it likely that more than 25% of them are smokers? Explain.

29. Alcohol abuse. Growing concern about binge drinking among college students has prompted one large state university to conduct a survey to assess the size of the problem on its campus. The university plans to randomly select students and ask how many have been drunk during the past week. If the school hopes to estimate the true proportion among all its students with 90% confidence and a margin of error of $\pm 4\%$, how many students must be surveyed?

30. Errors. An auto parts company advertises that its special oil additive will make the engine "run smoother, cleaner, longer, with fewer repairs." An independent laboratory decides to test part of this claim. It arranges to use a taxicab company's fleet of cars. The cars are randomly divided into two groups. The company's mechanics will use the additive in one group of cars but not in the other. At the end of a year the laboratory will compare the percentage of cars in each group that required engine repairs.

a) What kind of a study is this?

b) Will they do a one-tailed or a two-tailed test?

c) Explain in this context what a Type I error would be.

d) Explain in this context what a Type II error would be.

e) Which type of error would the additive manufacturer consider more serious?

f) If the cabs with the additive do indeed run significantly better, can the company conclude it is an effect of the additive? Can they generalize this result and recommend the additive for all cars? Explain.

31. Preemies. Among 242 Cleveland-area children born prematurely at low birth weights between 1977 and 1979, only 74% graduated from high school. Among a comparison group of 233 children of normal birth weight, 83% were high school graduates. ("Outcomes in Young Adulthood for Very-Low-Birth-Weight Infants," *New England Journal of Medicine*, 346, no. 3 [2002])

a) Create a 95% confidence interval for the difference in graduation rates between children of normal and children of very low birth weights. Be sure to check the appropriate assumptions and conditions.

b) Does this provide evidence that premature birth may be a risk factor for not finishing high school? Use your confidence interval to test an appropriate hypothesis.

c) Suppose your conclusion is incorrect. Which type of error did you make?

32. Safety. Observers in Texas watched children at play in eight communities. Of the 814 children seen biking, roller skating, or skateboarding, only 14% wore a helmet.

a) Create and interpret a 95% confidence interval.

b) What concerns do you have about this study that might make your confidence interval unreliable?

c) Suppose we want to do this study again, picking various communities and locations at random, and hope to end up with a 98% confidence interval having a margin of error of $\pm 4\%$. How many children must we observe?

33. Fried PCs. A computer company recently experienced a disastrous fire that ruined some of its inventory. Unfortunately, during the panic of the fire, some of the damaged computers were sent to another warehouse, where they were mixed with undamaged computers. The engineer responsible for quality control would like to check out each computer in order to decide whether it's undamaged or damaged. Each computer undergoes a series of 100 tests. The number of tests it fails will be used to make the decision. If it fails more than a certain number, it will be classified as damaged and then scrapped. From past history, the distribution of the number of tests failed is known for both undamaged and damaged computers. The relative frequencies of each outcome are listed in the table below:

Number of tests failed	0	1	2	3	4	5	>5
Undamaged (%)	80	13	2	4	1	0	0
Damaged (%)	0	10	70	5	4	1	10

The table indicates, for example, that 80% of the undamaged computers have no failures, while 70% of the damaged computers have 2 failures.

a) To the engineers, this is a hypothesis-testing situation. State the null and alternative hypotheses.

b) Someone suggests classifying a computer as damaged if it fails any of the tests. Discuss the advantages and disadvantages of this test plan.

c) What number of tests would a computer have to fail in order to be classified as damaged if the engineers want to have the probability of a Type I error equal to 5%?

d) What's the power of the test plan in part c?

e) A colleague points out that by increasing α just 2%, the power can be increased substantially. Explain.

34. Power. We are replicating an experiment. How will each of the following changes affect the power of our test? Indicate whether it will increase, decrease, or remain the same, assuming that all other aspects of the situation remain unchanged.

a) We increase the number of subjects from 40 to 100.

b) We require a higher standard of proof, changing from $\alpha = 0.05$ to $\alpha = 0.01$.

35. Approval 2008. Of all the post–World War II presidents, Richard Nixon had the highest disapproval rating near the end of his presidency. His disapproval rating peaked at 66% in July 1974, just before he resigned. This percentage has been considered by some pundits as a high water mark for presidential disapproval. However, in April 2008, George W. Bush's disapproval rating peaked at 69%, according to a Gallup poll of 1016 voters. Pundits started discussing whether his rating was discernibly worse than the previous high water mark of 66%. What do you think?

36. Grade inflation 2012. In 1996, 20% of the students at a major university had an overall grade point average of 3.5 or higher (on a scale of 4.0). In 2012, a random sample of 1100 student records found that 25% had a GPA of 3.5 or higher. Is this evidence of grade inflation?

37. Name recognition. An advertising agency won't sign an athlete to do product endorsements unless it is sure the person is known to more than 25% of its target audience. The agency always conducts a poll of 500 people to investigate the athlete's name recognition before offering a contract. Then it tests $H_0: p = 0.25$ against $H_A: p > 0.25$ at a 5% level of significance.

a) Why does the company use upper tail tests in this situation?

b) Explain what Type I and Type II errors would represent in this context, and describe the risk that each error poses to the company.

c) The company is thinking of changing its test to use a 10% level of significance. How would this change the company's exposure to each type of risk?

38. Name recognition, part II. The advertising company described in Exercise 37 is thinking about signing a WNBA star to an endorsement deal. In its poll, 27% of the respondents could identify her.

a) Fans who never took Statistics can't understand why the company did not offer this WNBA player an endorsement contract even though the 27% recognition rate in the poll is above the 25% threshold. Explain it to them.

b) Suppose that further polling reveals that this WNBA star really is known to about 30% of the target audience. Did the company initially commit a Type I or Type II error in not signing her?

c) Would the power of the company's test have been higher or lower if the player were more famous? Explain.

39. NIMBY. In March 2007, the Gallup Poll split a sample of 1003 randomly selected U.S. adults into two groups at random. Half ($n = 502$) of the respondents were asked,

"Overall, do you strongly favor, somewhat favor, somewhat oppose, or strongly oppose the use of nuclear energy as one of the ways to provide electricity for the U.S.?"

They found that 53% were either "somewhat" or "strongly" in favor. The other half ($n = 501$) were asked,

> *"Overall, would you strongly favor, somewhat favor, somewhat oppose, or strongly oppose the construction of a nuclear energy plant in your area as one of the ways to provide electricity for the U.S.?"*

Only 40% were somewhat or strongly in favor. This difference is an example of the *NIMBY* (Not In My BackYard) phenomenon and is a serious concern to policy makers and planners. How large is the difference between the proportion of American adults who think nuclear energy is a good idea and the proportion who would be willing to have a nuclear plant in their area? Construct and interpret an appropriate confidence interval.

40. Women. The U.S. Census Bureau reports that 26% of all U.S. businesses are owned by women. A Colorado consulting firm surveys a random sample of 410 businesses in the Denver area and finds that 115 of them have women owners. Should the firm conclude that its area is

unusual? Test an appropriate hypothesis and state your conclusion.

41. Skin cancer. In February 2012, MedPage Today reported that researchers used vemurafenib to treat metastatic melanoma (skin cancer). Out of 152 patients, 53% had a partial or complete response to vemurafenib.

a) Write a 95% confidence interval for the proportion helped by the treatment, and interpret it in this context.

b) If researchers subsequently hope to produce an estimate (with 95% confidence) of treatment effectiveness for metastatic melanoma that has a margin of error of only 6%, how many patients should they study?

42. Streams. Researchers in the Adirondack Mountains collect data on a random sample of streams each year. One of the variables recorded is the substrate of the stream—the type of soil and rock over which they flow. The researchers want to estimate the proportion of streams that have a substrate of shale to within a margin of error of 7% (with 95% confidence). How many streams must they sample?

Practice Exam

I. Multiple Choice

1. A teacher gives a test and the distribution of scores turns out to be bimodal. One-third of the class earned scores between 60% and 75%, while the rest of the class scored between 88% and 98%. Which of the following are the most plausible estimates of the mean and the median?

A) The median is about 80 and the mean is about 85.
B) The median and the mean are both about 85.
C) The median is about 85 and the mean is about 90.
D) The median is about 90 and the mean is about 85.
E) The median and the mean are both about 90.

2. A researcher is reporting characteristics of the subjects she used in a recent study. Two of the variables are hair color and age. Which of these are appropriate choices to summarize these data?

A) Bar charts for both hair color and age.
B) Histograms for both hair color and age.
C) A bar chart for hair color and a histogram for age.
D) A histogram for hair color and a bar chart for age.
E) Either bar charts or histograms are good choices for both hair color and age.

3. An employer is ready to give her employees a raise and is considering two plans. Plan A is to give each person an $8.00 per day increase. Plan B is to give each person a 10% increase. Data on the current pay of these employees shows that the median pay is $80 per day with an interquartile range of $10. Which of the following data sets will have the same median?

A) The current pay and Plan A only
B) The current pay and Plan B only
C) Plan A and Plan B only
D) The current pay, Plan A, and Plan B
E) None; the current pay, Plan A, and Plan B will all have different medians.

4. Using the same context as question 3, which of the following data sets will have the same interquartile range?

A) The current pay and Plan A only
B) The current pay and Plan B only
C) Plan A and Plan B only
D) The current pay, Plan A, and Plan B
E) None; the current pay, Plan A, and Plan B will all have different IQRs.

5. A study conducted by students in an AP Psychology class at South Kent School in Connecticut discovered a correlation of –0.38 between hours of sleep (x) and GPA (y). If we change the variable on the horizontal axis to hours awake ($24 - x$), but make no change to the GPA data, which of the following would be true about the new scatterplot?

A) It slopes down, and the correlation is -0.38.
B) It slopes down, and the correlation is $+0.38$.
C) It slopes up, and the correlation is -0.38.
D) It slopes up, and the correlation is $+0.38$.
E) None of the above choices is correct.

(Source: http://www.cardinalnewsnetwork.org/ south-kent-community/sleep-study-sks-style/)

6. An entomologist observes that the relationship between the antennae length of a certain species of grasshopper and its age appears to be linear, and calculates an R^2 of 78%. This value tells us that:

A) Errors in predicting the age of a grasshopper by using its antenna length will not exceed 22%.
B) 78% of a grasshopper's antennae length can be explained by its age.
C) The regression model makes accurate predictions for 78% of all grasshoppers.
D) The variability in antenna length depends on the age 78% of the time.
E) There is a moderately strong relationship between antennae length and age.

7. In a study of association between age (x) and the ability to recall a list of animal names (y), the residual for the data point (5, 13) is 2, and the residual for the data point (7, 20) is -1. What is the slope of the least-squares regression line?

A) 1   B) 2   C) 3   D) 4   E) 5

8. Researchers studying the association between (x) the number of daily hours of intense artificial light and (y) the rate of plant growth (cm/month) found $s_x = 1.5$, $s_y = 3.0$, and $r = 0.80$. The regression model they created predicts growth of 5 cm/month with 2 hours of this light per day. What's the predicted growth rate for 3 hours of artificial light daily?

A) 5.4 cm/mo
B) 5.8 cm/mo
C) 6.6 cm/mo
D) 7.0 cm/mo
E) 7.8 cm/mo

9. Based on data from a study of the association between miles driven and gallons of gasoline used, let r_1 represent the correlation coefficient for (miles, gallons) data points, and let r_2 represent the correlation coefficient for (gallons, miles) data points. Which of the following must be true?

A) $r_1 + r_2 = 0$
B) $r_1 - r_2 = 0$
C) $r_1 + r_2 = 1$
D) $r_1 \cdot r_2 = 1$
E) $r_1 \cdot r_2 = -1$

10. Which of these is the main reason for using a placebo in an experiment?

A) blinding
B) randomization
C) reducing bias
D) blocking
E) reducing within-treatment variability

11. The staff of a school newspaper plans to investigate students' opinions regarding the school's security guards. They decide to survey 100 students at the school. The editor-in-chief recommends that they survey 25 randomly selected students from each class (freshmen, sophomores, juniors, seniors). This is an example of a

A) blocked random sample.
B) multi-stage random sample.
C) simple random sample.
D) stratified random sample.
E) random cluster sample.

12. A company's human resources director randomly selected 100 employees to complete a confidential survey about the effectiveness of a new incentive program. Only 70 of those selected returned the survey. Should the director be concerned about making a conclusion about the incentive program based on the results of the survey?

A) No, the employees were selected randomly.
B) No, 70 is a large sample.
C) No, it was not an experiment.
D) Yes, if the decision to not return the survey is related to one's opinion about the program.
E) Yes, if 70 is more than 10% of the company's employees.

13. A conclusion that differences in the explanatory variable actually causes differences in the response variable would require a design that uses

A) a random sample.
B) random samples taken from at least two different populations.
C) a random sample of individuals who are then randomly assigned to treatment groups.
D) a control group.
E) random assignment of individuals to treatment groups.

14. In an experiment to investigate a medication's effect, researchers randomly assign 60 volunteers to receive 10 mg, 20 mg, or 30 mg of the medication. What is one of the limitations of this experiment?

A) The conclusions will be questionable because the design has no blocking.
B) The conclusions will be questionable because no group gets a placebo.
C) The conclusions cannot be generalized to a population because the subjects are volunteers.
D) The conclusions will be questionable because there are only 20 subjects in each group.
E) The conclusions will be questionable because each subject is not given all of the doses.

15. The Bureau of Justice conducted a study of 272,111 former inmates released from prisons in 15 states in 1994. A classification of the former prisoners by "most serious offense for which released" yielded the following: 22.5% violent; 33.5% property; 32.6% drug; 11.4% other. Of those prisoners whose most serious crime was violent, 61.7% were rearrested within 3 years of release. For the

other categories, the recidivism rate (defined as rearrested within 3 years of release) were: property 73.8%, drug 66.7%, and other 62.6%. If a released prisoner from these 15 states is rearrested within 3 years of release, what is the probability that this prisoner's most serious offense was violent?

A) 0.139   B) 0.206   C) 0.223

D) 0.617   E) 0.690

(Source: Levine, D and Mangan, P. Recidivism of Prisoners Released in 1994; NCJ193427; Bureau of Justice; File name rpr94bxl.csv http://bjs.ojp.usdoj.gov/index.cfm?ty=pbdetail&iid=1134)

16. A study of over 3000 network and cable programs found that nearly 60% featured violence. Suppose you want to simulate a collection of five randomly selected programs. Which of the following assignments of the digits 0 through 9 would be appropriate for modeling whether individual programs feature violence for each trial of 5 programs?

A) Assign the digits 0, 1, 2, 3, 4, 5 as featuring violence and 6, 7, 8, and 9 as not featuring violence.
B) Assign the digits 0, 1, 2, 3, 4, 5 as featuring violence and 6, 7, 8, and 9 as not featuring violence, and ignore repeats.
C) Assign 6 as featuring violence and the digits 0, 1, 2, 3, 4, 5, 7, 8, and 9 as not featuring violence.
D) Assign the digits 1, 2, 3, 4, 5, 6 as featuring violence and 7, 8, and 9 as not featuring violence, ignoring 0.
E) Assign 0 as featuring violence, the digits 1, 2, 3, 4, and 5 as not featuring violence, and ignore 6, 7, 8 and 9.

17. Researchers collected data on spending habits of freshmen students versus senior students at five different high schools. They asked students if they spent less than $10 on fast food in week, $10–$20, or $20 or more. Students were told to count only the money they spent out of their own personal funds, not food purchased for them by their parents or other adults. Here is a summary of the data:

	< $10	$10–$20	> $20	Total
Freshmen	23	12	8	43
Seniors	25	19	28	72
Total	48	31	36	115

If one of these students is selected at random, what is the probability that person is a senior or spent less than $10?

A) 0.217   B) 0.347   C) 0.521

D) 0.783   E) 0.826

18. Airport security personnel screen carry-on luggage for items that are not allowed on planes. Most bags pass through without a problem, but occasionally one is pulled aside for further inspection. Those inspections actually find some item that must removed in 2 of every 5 bags that are pulled aside. If 7 of the people on a certain flight had luggage that was subjected to further inspection, what's the probability that at least 2 of them had items removed?

A) 0.159   B) 0.261   C) 0.420

D) 0.580   E) 0.841

19. A chemist knows that when he uses his 50 mL pipet there will be some measurement error. The distribution of the errors is very close to a normal distribution with a mean of 0.05 mL and a standard deviation of 0.01 mL. If the chemist measures out 150 mL by using this pipet three times and combining the quantities, the mean error will be 0.15 mL. What's the standard deviation of that error?

A) 0.010   B) 0.017   C) 0.030

D) 0.087   E) 0.090

20. Over 50% of family households in the U.S. have no children under 18 living at home and fewer than 10% have 3 or more, so the distribution of the number of children living in family households is skewed to the right. The mean is 0.96 children, with a standard deviation of 1.26 children. What's the probability that in a random sample of 250 family households the mean number of children living in those homes will be greater than 1?

A) 0.056   B) 0.308   C) 0.487

D) It cannot be determined, because the sample is not large enough.
E) It cannot be determined, because the population distribution is not normal.

21. In describing the findings of a 2010 Public Religion Research Institute poll, the researchers reported a 3% margin of error at a 95% confidence level. If instead they had used a 99% confidence level, their margin of error would have been:

A) still 3%, because the same sample is used.
B) more than 3%, because they reported greater confidence.
C) less than 3%, because they reported greater confidence.
D) more than 3%, because this sample is too small.
E) less than 3%, because this sample is too small.

22. Mitt Romney was a presidential candidate in the November 2012 U.S. presidential elections and he is a Mormon. The Public Religion Research Institute, during its 2011 American Values Survey, asked 1019 random Americans if they would be comfortable with a Mormon serving as President. Fifty three percent reported that they would be somewhat or very comfortable with this. Which of the following hypotheses should we test to determine whether the majority of Americans would be somewhat or very comfortable with a Mormon serving as president?

A) $H_0: p = 0.50$ $H_A: p \neq 0.50$
B) $H_0: p = 0.50$ $H_A: p < 0.50$
C) $H_0: p = 0.50$ $H_A: p > 0.50$
D) $H_0: p = 0.53$ $H_A: p > 0.53$
E) $H_0: \hat{p} = 0.53$ $H_A: \hat{p} \neq 0.53$

(Source: http://publicreligion.org/research/2011/11/
2011-american-values-survey/)

23. Bank officers are considering targeting young parents with an offer to start a college savings plan. They have decided to pursue this new marketing program only if at least 30% of such parents might be interested in this kind of plan. They surveyed 250 randomly selected parents, of whom 38% indicated some interest. Which formula below correctly computes the z-score they should use to see if this survey result provides evidence that the true proportion is over 30%?

A) $z = \dfrac{0.38 - 0.30}{\sqrt{\dfrac{(0.30)(0.70)}{250}}}$

B) $z = \dfrac{0.30 - 0.38}{\sqrt{\dfrac{(0.30)(0.70)}{250}}}$

C) $z = \dfrac{0.38 - 0.30}{\sqrt{\dfrac{(0.38)(0.62)}{250}}}$

D) $z = \dfrac{0.38 - 0.30}{\sqrt{250(0.30)(0.70)}}$

E) $z = \dfrac{0.30 - 0.38}{\sqrt{250(0.38)(0.62)}}$

24. In 2012 the *New York Times* surveyed adults about whether they had gone back to school for additional training and, if so, whether they felt this training helped them get new jobs or promotions. In the article that presented the results, the newspaper included the following information: "The nationwide telephone poll was conducted May 31 to June 3 with 976 adults, of whom 229 said they went to school in the last five years. Margin of sampling error for all adults is plus or minus 3 percentage points; for adults who went back to school, plus or minus 6 percentage points. Of those who went back to school, 84% reported that the training was a good investment of time and money." What is the most logical reason for the larger margin of error that was indicated for results about adults who went back to school?

A) The researchers used a smaller confidence level.
B) The researchers used a larger confidence level.
C) The sample size was smaller.
D) The sample size was larger.
E) Adults who returned to school were more variable in terms of employment.

(Source: Connelly, M., Stefan, M. and Kayda, A. *Is it Worth it?* Education Life. New York Times. July 22, 2012: 31.)

25. A survey of 1025 teens found that 20% of students aged 14 to 18 plan to borrow no money to pay for college. What's the margin of error for a 90% confidence interval for the proportion of all students aged 14 to 18 who plan to borrow no money to pay for college?

A) $\pm 1.25\%$ B) $\pm 1.60\%$ C) $\pm 2.06\%$
D) $\pm 2.45\%$ E) $\pm 21.1\%$

26. Researchers conducted an experiment to compare two treatments for high blood pressure. Their null hypothesis was that the proportion of people whose blood pressure would improve is the same for both treatments. The two-sided test resulted in a P-value of 0.07. Which of the following statements is true?

A) There is a 7% chance that the drugs are equally effective.
B) There is a 7% chance the drugs are not equally effective.
C) The null hypothesis should be rejected at the 0.05 level.
D) 0 would be contained in the 95% confidence interval for the difference in proportions.
E) 0 would be contained in the 90% confidence interval for the difference in proportions.

27. Sometimes a drug that cures a disease turns out to have a nasty side effect. For example, some antidepressant drugs may cause suicidal thoughts in younger patients. A researcher conducts a study of such a drug to look for evidence of such side effects. He tests the hypothesis that there's no side effect at the 0.05 significance level and finds a P-value of 0.03. Which of the following statements is true?

A) He could have committed a Type I error by concluding there's no such side effect if, in fact, there really is.
B) He could have committed a Type II error by concluding there's no such side effect if, in fact, there really is.
C) He could have committed a Type I error by concluding there is such a side effect if, in fact, there really isn't.
D) He could have committed a Type II error by concluding there is such a side effect if, in fact, there really isn't.
E) With a P-value as low as 0.03 he could not have committed either type of error.

28. For her final project, Stacy plans on surveying a random sample of 50 students on whether they plan to leave town for Spring Break. Based on past years, she guesses that about 85% of the students will go somewhere. Is it appropriate for her to use a Normal model for the sampling distribution of the sample proportion?

A) Yes, because $np = 42.5$, which is greater than 10.

B) Yes, because a Normal model always applies to sample proportions.

C) No, because a Normal model is not appropriate for a binomial situation.

D) No, because $nq = 7.5$, which is less than 10.

E) We don't know, because we don't know the shape of the population distribution.

29. A marketing researcher for a phone company conducted a survey of 500 people and then constructed a confidence interval for the proportion of customers who are likely to switch providers when their contract expires. After seeing this interval, the company CEO insisted that the researcher repeat the survey and provide an interval with a margin of error half as large. What sample size should the researcher use for the new survey?

A) 250   B) 354   C) 708

D) 1000   E) 2000

30. In which of the following situations is a one-sided alternative hypothesis appropriate?

A) A business student conducts a blind taste test to see whether students prefer Diet Coke or Diet Pepsi.

B) A budget override in a small town requires a two-thirds majority to pass. A local newspaper conducts a poll to see if there's evidence it will pass.

C) In recent years, 10% of college juniors have applied for study abroad. The dean's office conducts a preliminary survey to see if that has changed this year.

D) A taxi company checks gasoline usage records to see if a recent change in the brand of tires installed on the cars has any impact on their fleet's fuel economy.

E) The Centers for Disease Control and Prevention wants to see if a controversial new advertising campaign has had any effect on teen smoking rates.

II. Free Response

1. The table below summarizes combined city/highway fuel economy for the 2012 cars classified as "large" models.

N	MEAN	STDEV	SEMEAN	TRMEAN	MIN	Q1	MEDIAN	Q3	MAX
65	19.6	3.36	0.417	19.5	14	17	19	21	28

a) What do the summary statistics suggest about the overall shape of the distribution? Explain.

b) Are there any outliers in the data? Explain how you determined this.

c) In advertising for its own vehicle, a competitor described one large car that gets 17 mpg as exceptionally gas-thirsty. Based on the summary statistics above, comment on that characterization.

2. Based on data collected at a store where tomatoes are sold by the pound, the association between the number of tomatoes purchased and the total cost (in dollars) appeared to be linear. The output below shows the regression analysis.

Dependent variable is Cost
R-squared = 92.4%

Variable	Coefficient
Intercept	0.03
Tomatoes	0.47

a) Predict the cost of 12 tomatoes.

b) Interpret the value of R^2 in this context.

c) Interpret the meaning of the slope in this context.

d) Is there any meaningful interpretation of the y-intercept in this context?

3. The management team at a company sent out the following email to all 75 employees:

"Several complaints have come in that morale in this company is low, and that people are not happy with the working conditions. Naturally, as managers we are concerned about this. Please reply to this email with an answer to the following question: Are you happy with the working conditions here?"

Based on answers received from 24 employees who replied to the email, the managers concluded there is no morale problem at their company.

a) Explain how bias could have been introduced by the wording of the survey, and how that might mislead the company managers.

b) Explain how bias could be introduced by conducting the survey by email, and how that might mislead the company managers.

c) Explain how bias could have been introduced by the fact that only 24 of the 75 employees responded, and how that might mislead the company managers.

4. Do American's views of the war in Iraq vary by religion? A study conducted in 2007 by The Baylor Department of Sociology asked, "Was going to Iraq the right decision?" The table summarizes responses from a 1571 randomly selected adults.

	Religious Affiliation					
		Protestant	Catholic	Jewish	Other	None	Total
Opinion	Disagree	413	188	18	55	147	821
	Agree	422	135	12	37	29	635
	Undecided	82	24	1	4	4	115
	Total	917	347	31	96	180	1571

(Source: Baylor Religion Survey, Wave 2, 2007)

a) What percentage of respondents agreed that going to Iraq was the right thing to do?

b) What percentage of respondents classified their religion as "None" and agreed that the war in Iraq was the right decision?

c) What percentage of the respondents who classified their religion as "None" agreed that the war in Iraq was the right decision?

d) What percentage of the respondents who agreed that the war in Iraq was the right thing to do classified their religion as "None"?

e) Were respondents' opinions about the war in Iraq independent of whether or not they were religious? Justify your decision using your answers from 2 of the prior questions.

f) Among the respondents who identified a religion, was there an association between the religious affiliation and opinion about the war in Iraq? Explain, citing appropriate statistics to support your conclusion.

5. In 1986 a catastrophic nuclear accident in the Ukranian city of Chernobyl exposed the surrounding area to large amounts of radiation. Since then, scientists have been investigating the aftereffects on plants, animals, and humans. A 2007 study examined 841 barn swallows there, noting the incidence of deformed toes and beaks, abnormal tail feathers, deformed eyes, tumors, and albinism (a condition that involves discoloration of the birds' plumage). Previous studies involving thousands of barn swallows in Spain, Italy, and Denmark had seen albinism in 4.35% of this species, yet 112 of the Chernobyl swallows displayed the condition.

a) Is there evidence that the albinism rate is higher among Chernobyl barn swallows?

b) What does this study tell us about the effect of nuclear contamination on barn swallows?

6. Members of the We-Luv-Plants club are interested in comparing the tendency of males vs. females to choose a vegetarian diet. They randomly sampled 75 males and 75 females at their school. The data showed that 23 of the girls were vegetarians whereas only 12 of the boys could stay away from a tasty burger.

a) Construct and interpret a 95% confidence interval for the difference in the proportions between girls and boys who are vegetarians.

b) Does this confidence interval provide evidence that girls are more likely than boys to choose a vegetarian diet? Explain.

c) To what population(s) do your conclusions in parts a and b apply?

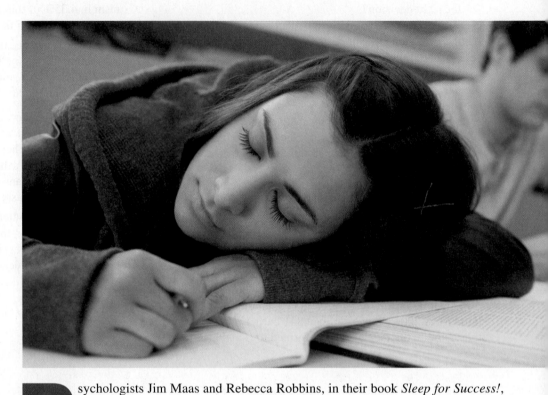

Psychologists Jim Maas and Rebecca Robbins, in their book *Sleep for Success!*, say that

> In general, high school and college students are the most pathologically sleep-deprived segment of the population. Their alertness during the day is on par with that of untreated narcoleptics and those with untreated sleep apnea. Not surprisingly, teens are also 71 percent more likely to drive drowsy and/or fall asleep at the wheel compared to other age groups. (Males under the age of twenty-six are particularly at risk.)

They report that adults require between 7 and 9 hours of sleep each night and claim that college students require 9.25 hours of sleep to be fully alert. They note that "There is a 19 percent memory deficit in sleep-deprived individuals" (p. 35).

A student surveyed students at a small school in the northeast U.S. and asked, among other things, how much they had slept the previous night. Here's a histogram and the data for 25 of the students selected at random from the survey.

6	6	6
7	6	7
6	6	8
7	8	5
6	8	8
7	8	7
7	5	6
7	4	7
8		

We're interested both in estimating the mean amount slept by college students and in testing whether it is less than the minimum recommended amount of 7 hours. These data were collected in a suitably randomized survey so we can treat them as representative of students at that college, and possibly as representative of college students in general.

These data differ from data on proportions in one important way. Proportions are summaries of individual responses, which had two possible values such as "yes" and "no," "male" and "female," or "1" and "0." Quantitative data, though, report a quantitative value for each individual. When you have quantitative data, you should remember the three rules of data analysis and plot the data, as we have done here.

Getting Started: The Central Limit Theorem (Again)

You've learned how to create confidence intervals and test hypotheses about proportions. We always center confidence intervals at our best guess of the unknown parameter. Then we add and subtract a margin of error. For proportions, we write $\hat{p} \pm ME$.

We found the margin of error as the product of the standard error, $SE(\hat{p})$, and a critical value, z^*, from the Normal table. So we had $\hat{p} \pm z^*SE(\hat{p})$.

We knew we could use z because the Central Limit Theorem told us (back in Chapter 17) that the sampling distribution model for proportions is Normal.

Now we want to do exactly the same thing for means, and fortunately, the Central Limit Theorem (still in Chapter 17) told us that a Normal model also works as the sampling distribution for means.

> **The Central Limit Theorem**
>
> When a random sample is drawn from any population with mean μ and standard deviation σ, its sample mean, $\bar{y}$, has a sampling distribution with the same *mean* μ but whose *standard deviation* is $\dfrac{\sigma}{\sqrt{n}}$ $\left(\text{and we write } \sigma(\bar{y}) = SD(\bar{y}) = \dfrac{\sigma}{\sqrt{n}}\right)$.
>
> No matter what population the random sample comes from, the *shape* of the sampling distribution is approximately Normal as long as the sample size is large enough. The larger the sample used, the more closely the Normal approximates the sampling distribution for the mean.

For Example USING THE CENTRAL LIMIT THEOREM (AS IF WE KNEW σ)

Based on weighing thousands of animals, the American Angus Association reports that mature Angus cows have a mean weight of 1309 pounds with a standard deviation of 157 pounds. This result was based on a very large sample of animals from many herds over a period of 15 years, so let's assume that these summaries are the population parameters and that the distribution of the weights was unimodal and reasonably symmetric.

QUESTION: What does the Central Limit Theorem (CLT) predict about the mean weight seen in random samples of 100 mature Angus cows?

ANSWER: It's given that weights of all mature Angus cows have $\mu = 1309$ and $\sigma = 157$ pounds. Because $n = 100$ animals is a fairly large sample, I can apply the Central Limit Theorem. I expect the resulting sample means $\bar{y}$ will average 1309 pounds and have a standard deviation of

$$SD(\bar{y}) = \frac{\sigma}{\sqrt{n}} = \frac{157}{\sqrt{100}} = 15.7 \text{ pounds.}$$

(continued)

The CLT also says that the distribution of sample means follows a Normal model, so the 68–95–99.7 Rule applies. I'd expect that

✓ in 68% of random samples of 100 mature Angus cows, the mean weight will be between 1309 − 15.7 = 1293.3 and 1309 + 15.7 = 1324.7 pounds;

✓ in 95% of such samples, 1277.6 ≤ ȳ ≤ 1340.4 pounds;

✓ in 99.7% of such samples, 1261.9 ≤ ȳ ≤ 1356.1 pounds.

Standard Error

Because we estimate the standard deviation of the sampling distribution model from the data, it's a *standard error*. So we use the $SE(\bar{y})$ notation. Remember, though, that it's just the estimated standard deviation of the sampling distribution model for means.

A S *Activity*: **Estimating the Standard Error.** What's the average age at which people have heart attacks? A confidence interval gives a good answer, but we must estimate the standard deviation from the data to construct the interval.

The CLT says that all we need to model the sampling distribution of $\bar{y}$ is a random sample of quantitative data.

And the true population standard deviation, σ.

Uh oh. That's a big problem. How are we supposed to know σ? With proportions, we had a link between the proportion value and the standard deviation of the sample proportion: $SD(\hat{p}) = \sqrt{\dfrac{pq}{n}}$. And there was an obvious way to estimate the standard deviation from the data: $SE(\hat{p}) = \sqrt{\dfrac{\hat{p}\hat{q}}{n}}$. But for means, $SD(\bar{y}) = \dfrac{\sigma}{\sqrt{n}}$, so knowing $\bar{y}$ doesn't tell us anything about $SD(\bar{y})$. We know n, the sample size, but the population standard deviation, σ, could be *anything*. So what should we do? We do what any sensible person would do: We estimate the population parameter σ with s, the sample standard deviation based on the data. The resulting standard error is $SE(\bar{y}) = \dfrac{s}{\sqrt{n}}$.

A century ago, people used this standard error with the Normal model, assuming it would work. And for large sample sizes it *did* work pretty well. But they began to notice problems with smaller samples. A sample standard deviation, s, like any other statistic, varies from sample to sample. And this extra variation in the standard error was messing up the P-values and margins of error.

William S. Gosset is the man who investigated this fact. He realized that not only do we need to allow for the extra variation with larger margins of error and P-values, but we even need a new sampling distribution model. In fact, we need a whole *family* of models, depending on the sample size, n. These models are unimodal, symmetric, bell-shaped models, but the smaller our sample, the more we must stretch out the tails. Gosset's work transformed Statistics, but most people who use his work don't even know his name.

Gosset's *t*

Gosset had a job that made him the envy of many. He was the chief Experimental Brewer for the Guinness Brewery in Dublin, Ireland. The brewery was a pioneer in scientific brewing and Gosset's job was to meet the demands of the brewery's many discerning customers by developing the best stout (a thick, dark beer) possible.

Gosset's experiments often required as much as a day to make the necessary chemical measurements or a full year to grow a new crop of hops. For these reasons, not to mention his health, his samples sizes were small—often as small as 3 or 4.

When he calculated means of these small samples, Gosset wanted to compare them to a target mean to judge the quality of the batch. To do so, he followed common statistical practice of the day, which was to calculate z-scores and compare them to the Normal model. But Gosset noticed that with samples of this size, his tests weren't quite right. He knew this because when the batches that he rejected were sent back to the laboratory for more extensive testing, too often they turned out to be OK. In fact, about 3 times more often than he expected. Gosset knew something was wrong, and it bugged him.

Guinness granted Gosset time off to earn a graduate degree in the emerging field of Statistics, and naturally he chose this problem to work on. He figured out that when he used the standard error, $\dfrac{s}{\sqrt{n}}$, as an estimate of the standard deviation of the mean, the shape of the sampling model changed. He even figured out what the new model should be.

How Gosset Did His Homework   To find the sampling distribution of $\dfrac{\bar{y}-\mu}{s/\sqrt{n}}$,

Gosset simulated it *by hand*. He drew 750 samples of size 4 by shuffling 3000 cards on which he'd written the heights of some prisoners and computed the means and standard deviations with a mechanically cranked calculator. (He knew μ because he was simulating and knew the population from which his samples were drawn.) Today you could repeat in seconds on a computer the experiment that took him over a year. Gosset's work was so meticulous that not only did he get the shape of the new histogram approximately right, but he even figured out the exact *formula* for it from his sample. The formula was not confirmed mathematically until many years later by Sir R. A. Fisher.

The Guinness Company may have been ahead of its time in using statistical methods to manage quality, but they also had a policy that forbade their employees to publish. Gosset pleaded that his results were of no specific value to brewers and was eventually allowed to publish under the pseudonym "Student." This important result is still widely known as **Student's *t***.

Gosset's sampling distribution model is always bell-shaped, but the details change with different sample sizes. When the sample size is very large, the model is nearly Normal, but when it's small the tails of the distribution are much fatter than the Normal. That means that values far from the mean are more common, especially for small samples (see Figure 22.1). So the Student's *t*-models form a whole *family* of related distributions that depend on a parameter known as **degrees of freedom**. The degrees of freedom of a distribution represent the number of independent quantities that are left after we've estimated the parameters. Here it's simply the number of data values, *n*, minus the number of estimated parameters. For means, that's just $n - 1$. We often denote degrees of freedom as *df* and the model as t_{df}, with the degrees of freedom as a subscript.

A S *Activity*: **Student's Distributions.** Interact with Gosset's family of *t*-models. Watch the shape of the model change as you slide the degrees of freedom up and down.

Figure 22.1

The *t*-model (solid curve) on 2 degrees of freedom has fatter tails than the Normal model (dashed curve). So the 68–95–99.7 Rule doesn't work for *t*-models with only a few degrees of freedom. It may not look like a big difference, but a *t* with 2 df is more than 4 times as likely to have a value greater than 2 than a standard Normal.

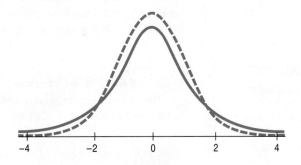

Don't divide by *n*

Some calculators offer an alternative button for standard deviation that divides by *n* instead of by $n - 1$. Why don't you stick a wad of gum over the "*n*" button so you won't be tempted to use it? Use $n - 1$.

Degrees of Freedom   The formula for standard deviation divides by $n - 1$ rather than by *n*. Our simulation in Chapter 3's What If confirmed that $n - 1$ is better, and we promised to explain why later. The reason is closely tied to the concept of degrees of freedom.

If only we knew the true population mean, μ, we would use it in our formula for the sample standard deviation:

$$s = \sqrt{\dfrac{\sum (y - \mu)^2}{n}}$$

But we don't know μ, so we naturally use $\bar{y}$ in its place. And that causes a small problem. For any sample, the data values will generally be closer to their own sample mean, $\bar{y}$, than to the true population mean, μ. Why is that? Imagine that we take a random sample of 10 high school seniors. The mean SAT verbal score is 500 in the United States. But the sample mean, $\bar{y}$, for *these* 10 seniors won't be exactly 500. Are the 10 seniors' scores closer to 500 or $\bar{y}$? They'll

(continued)

always be closer to their own average $\bar{y}$. So, when we calculate s using $\sum (y - \bar{y})^2$ instead of $\sum (y - \mu)^2$, our standard deviation estimate is too small. The amazing mathematical fact is that we can fix it by dividing by $n - 1$ instead of by n. This difference is much more important when n is small than when the sample size is large. The t-distribution inherits this same number and we call $n - 1$ the degrees of freedom.

A Confidence Interval for Means

■ NOTATION ALERT
Ever since Gosset, t has been reserved in Statistics for his distribution.

To make confidence intervals or test hypotheses for means, we need to use Gosset's model. Which one? Well, for means, it turns out the right value for degrees of freedom is $df = n - 1$.

> **A Practical Sampling Distribution Model for Means**
> When certain assumptions and conditions[1] are met, the standardized sample mean,
> $$t = \frac{\bar{y} - \mu}{SE(\bar{y})},$$
> follows a Student's t-model with $n - 1$ degrees of freedom. We estimate the standard deviation with
> $$SE(\bar{y}) = \frac{s}{\sqrt{n}}.$$

When Gosset corrected the model to take account of the extra uncertainty, the margin of error got bigger, as you might guess. When you use Gosset's model instead of the Normal model, your confidence intervals will be just a bit wider and your P-values just a bit larger. That's the correction you need. By using the t-model, you've compensated for the extra variability in precisely the right way.[2]

■ NOTATION ALERT
When we found critical values from a Normal model, we called them z^*. When we use a Student's t-model, we'll denote the critical values t^*.

> **One-Sample t-Interval for the Mean**
> When the assumptions and conditions[3] are met, we are ready to find the confidence interval for the population mean, μ. The confidence interval is
> $$\bar{y} \pm t^*_{n-1} \times SE(\bar{y}),$$
> where the standard error of the mean is $SE(\bar{y}) = \frac{s}{\sqrt{n}}$.
>
> The critical value t^*_{n-1} depends on the particular confidence level, C, that you specify and on the number of degrees of freedom, $n - 1$, which we get from the sample size.

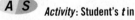 **Activity: Student's *t* in Practice.** Use a statistics package to find a t-based confidence interval; that's how it's almost always done.

---

[1] You can probably guess what they are. We'll see them in the next section.

[2] Gosset, as the first to recognize the consequence of using s rather than σ, was also the first to give the sample standard deviation, s, a different letter than the population standard deviation, σ.

[3] Yes, the same ones, and they're still coming in the next section.

For Example A ONE-SAMPLE *t*-INTERVAL FOR THE MEAN

In 2004, a team of researchers published a study of contaminants in farmed salmon.[4] Fish from many sources were analyzed for 14 organic contaminants. The study expressed concerns about the level of contaminants found. One of those was the insecticide mirex, which has been shown to be carcinogenic and is suspected to be toxic to the liver, kidneys, and endocrine system. One farm in particular produced salmon with very high levels of mirex. After those outliers are removed, summaries for the mirex concentrations (in parts per million) in the rest of the farmed salmon are:

$$n = 150 \qquad \bar{y} = 0.0913 \text{ ppm} \qquad s = 0.0495 \text{ ppm}.$$

QUESTION: What does a 95% confidence interval say about mirex?

ANSWER: $df = 150 - 1 = 149$

$$SE(\bar{y}) = \frac{s}{\sqrt{n}} = \frac{0.0495}{\sqrt{150}} = 0.0040$$

$t^*_{149} \approx 1.977$ (from Table T, using 140 *df*)
(actually, $t^*_{149} \approx 1.976$ from technology)

So the confidence interval for μ is $\bar{y} \pm t^*_{149} \times SE(\bar{y}) = 0.0913 \pm 1.977(0.0040)$

$$= 0.0913 \pm 0.0079$$
$$= (0.0834, 0.0992)$$

I'm 95% confident that the mean level of mirex concentration in farm-raised salmon is between 0.0834 and 0.0992 parts per million.

TI-*nspire*

The *t*-models. See how *t*-models change as you change the degrees of freedom.

Student's *t*-models are unimodal, symmetric, and bell-shaped, just like the Normal. But *t*-models with only a few degrees of freedom have longer tails and a larger standard deviation than the Normal. (That's what makes the margin of error bigger.) As the degrees of freedom increase, the *t*-models look more and more like the standard Normal. In fact, the *t*-model with infinite degrees of freedom is exactly Normal.[5] This is great news if you happen to have an infinite number of data values, but that's not likely. Fortunately, above a few hundred degrees of freedom it's very hard to tell the difference. Of course, in the rare situation that we *know* σ, it would be foolish not to use that information. And if we don't have to estimate σ, we can use a Normal model.

---

z or t?

If you know σ, use *z*. (That's rare!) Whenever you use *s* to estimate σ, use *t*.

When σ is known Administrators of a hospital were concerned about the prenatal care given to mothers in their part of the city. To study this, they examined the gestation times of babies born there. They drew a sample of 25 babies born in their hospital in the previous 6 months. Human gestation times for healthy pregnancies are thought to be well-modeled by a Normal with a mean of 280 days and a standard deviation of 10 days. The hospital administrators wanted to test the mean gestation time of their sample of babies against the known standard. For this test, they should use the established value for the standard deviation, 10 days, rather than estimating the standard deviation from their sample. Because they use the model parameter value for σ, they should base their test on the Normal model rather than Student's *t*.

---

[4]Ronald A. Hites, Jeffery A. Foran, David O. Carpenter, M. Coreen Hamilton, Barbara A. Knuth, and Steven J. Schwager, "Global Assessment of Organic Contaminants in Farmed Salmon," *Science* 9 January 2004: Vol. 303, no. 5655, pp. 226–229.

[5]Formally, in the limit as *n* goes to infinity.

TI Tips FINDING *t*-MODEL PROBABILITIES AND CRITICAL VALUES

FINDING PROBABILITIES You already know how to use your TI to find probabilities for Normal models using *z*-scores and normalcdf. What about *t*-models? Yes, the calculator can work with them, too.

You know from your experience with confidence intervals that $z = 1.645$ cuts off the upper 5% in a Normal model. Use the TI to check that. From the DISTR menu, enter normalcdf(1.645, 99). Only 0.04998? Close enough for statisticians!

We might wonder about the probability of observing a *t*-value greater than 1.645, but we can't find that. There's only one Normal model, but there are many *t*-models, depending on the number of degrees of freedom. We need to be more specific.

Let's find the probability of observing a *t*-value greater than 1.645 when there are 12 degrees of freedom. That we can do. Look in the DISTR menu again. See it? Yes, tcdf. That function works essentially like normalcdf. Use tcdf(with lower:1.645, upper:99, df:12, then choose Paste and hit ENTER twice (or on an older calculator, enter the command tcdf(1.645,99,12)).

The upper tail probability for t_{12} is 0.063, higher than the Normal model's 0.05. That should make sense to you—remember, *t*-models are a bit fatter in the tails, so more of the distribution lies beyond the 1.645 cutoff. (That means we'll have to go a little wider to make a 90% confidence interval.)

Check out what happens when there are more degrees of freedom, say, 25. The command tcdf(1.645,99,25) yields a probability of 0.056. That's closer to 0.05, for a good reason: *t*-models look more and more like the Normal model as the number of degrees of freedom increases.

FINDING CRITICAL VALUES Your calculator can also determine the critical value of *t* that cuts off a specified percentage of the distribution, using invT. It works just like invNorm, but for *t* we also have to specify the number of degrees of freedom (of course).

Suppose we have 6 degrees of freedom and want to create a 98% confidence interval. A confidence level of 98% leaves 1% in each tail of our model, so we need to find the value of *t* corresponding to the 99th percentile. If a Normal model were appropriate, we'd use $z = 2.33$. (Try it: invNorm(.99)). Now think. How should the critical value for *t* compare?

If you thought, "It'll be larger, because *t*-models are more spread out," you're right. Check with your TI. Use invT(with area:0.99, df:6, then choose Paste and hit ENTER twice (or on an older calculator, enter the command invT(.99,6)). Were you surprised, though, that the critical value of *t* is so much larger?

So think once more. How would the critical value of *t* differ if there were 60 degrees of freedom instead of only 6? When you think you know, check it out on your TI.

UNDERSTANDING *t* Use your calculator to play around with tcdf and invT a bit. Try to develop a clear understanding of how *t*-models compare to the more familiar Normal model. That will help you as you learn to use *t*-models to make inferences about means.

Assumptions and Conditions

Sir Ronald Fisher (1890–1962) was one of the founders of modern Statistics.

Gosset initially found the *t*-model by simulation and then made calculations to derive its form. He made some assumptions so his math would work out. Years later, when Sir Ronald A. Fisher showed mathematically that Gosset was right, he confirmed that Gosset's assumptions were necessary for the math to work. These are the assumptions we need to use the Student's *t*-models.

Independence Assumption

The data values should be mutually independent. There's really no way to check independence of the data by looking at the sample, but you should think about whether the assumption is reasonable.

Randomization Condition: This condition is satisfied if the data arise from a random sample or suitably randomized experiment. Randomly sampled data—and especially data from a Simple Random Sample—are almost surely independent. If the data don't satisfy the **Randomization Condition** then you should think about whether the values are likely to be independent for the variables you are concerned with and whether the sample you have is likely to be representative of the population you wish to learn about.

When we sample without replacement (essentially always), technically the selections are not independent. We should confirm that we have not sampled so much of the population that it matters. We check the

10% Condition: The sample is less than 10% of the population.

For means, though, we tend to use samples that are smaller than those we need for proportions, so we often don't mention this. The independence problem arises only if the population is small, and then there's a correction formula we could use.[6] And if we're dealing with a randomized experiment, this issue is irrelevant, because there's no sampling at all.

Normal Population Assumption

Student's *t*-models won't work for small samples that are badly skewed. How skewed is too skewed? Well, formally, we assume that the data are from a population that follows a Normal model. Practically speaking, we can't be sure this is true.

And it's almost certainly *not* true. Models are idealized; real data are, well, real— *never* exactly Normal. The good news, however, is that even for small samples, it's sufficient to check the . . .

Nearly Normal Condition: The data come from a distribution that appears to be unimodal and symmetric, without strong skewness or outliers.

Check this condition by making a histogram or Normal probability plot. Normality is less important for larger sample sizes. Just our luck: It matters most when it's hardest to check.[7]

For very small samples ($n < 15$ or so), the data should follow a Normal model pretty closely. Of course, with so little data, it can be hard to tell as you'll see in this chapter's What If. But if you do find outliers or strong skewness, don't use these methods.

For moderate sample sizes (n between 15 and 40 or so), the *t* methods will work well as long as the data are unimodal and reasonably symmetric. Make a histogram.

When the sample size is larger than 40 or 50, the *t* methods are safe to use unless the data are extremely skewed. Be sure to make a histogram. If you find outliers in the data, it's always a good idea to perform the analysis twice, once with and once without the

---

We Don't *Want* to Stop

We check conditions hoping that we can make a meaningful analysis of our data. The conditions serve as *disqualifiers*—we keep going unless there's a serious problem. If we find minor issues, we note them and express caution about our results.

- If the sample is not an SRS but we believe it's representative of some populations, we limit our conclusions accordingly.
- If there are outliers, rather than stop, we perform the analysis both with and without them.
- If the sample looks bimodal, we try to analyze subgroups separately.

Only when there's major trouble—like a strongly skewed small sample or an obviously nonrepresentative sample—are we unable to proceed at all.

---

[6]Let's not get into that here. If you're just dying to know, take a second Statistics course.

[7]There are formal tests of Normality, but they don't really help. When we have a small sample—just when we really care about checking Normality—these tests are not very effective. So it doesn't make much sense to use them in deciding whether to perform a *t*-test. We don't recommend that you use them.

outliers, even for large samples. Outliers may well hold additional information about the data, but you may decide to give them individual attention and then summarize the rest of the data. If you find multiple modes, you may well have different groups that should be analyzed and understood separately.

For Example CHECKING ASSUMPTIONS AND CONDITIONS FOR STUDENT'S *t*

RECAP: Researchers purchased whole farmed salmon from 51 farms in eight regions in six countries. The histogram shows the concentrations of the insecticide mirex in 150 farmed salmon.

QUESTION: Are the assumptions and conditions for inference satisfied?

ANSWER:

✓ **Independence Assumption:** The fish were not a random sample because no simple population existed to sample from. But they were raised in many different places, and samples were purchased independently from several sources, so they were likely to be independent and to represent the population of farmed salmon worldwide.

✓ **Nearly Normal Condition:** The histogram of the data is unimodal. Although it may be somewhat skewed to the right, this is not a concern with a sample size of 150.

It's okay to use these data for inference about farm-raised salmon.

✓ Just Checking

Every 10 years, the United States takes a census. The census tries to count every resident. There have been two forms, known as the "short form," answered by most people, and the "long form," slogged through by about one in six or seven households chosen at random. (For the 2010 Census, the long form was replaced by the American Community Survey.) According to the Census Bureau (www.census.gov), ". . . each estimate based on the long form responses has an associated confidence interval."

1. Why does the Census Bureau need a confidence interval for long-form information but not for the questions that appear on both the long and short forms?

2. Why must the Census Bureau base these confidence intervals on *t*-models?

The Census Bureau goes on to say, "These confidence intervals are wider . . . for geographic areas with smaller populations and for characteristics that occur less frequently in the area being examined (such as the proportion of people in poverty in a middle-income neighborhood)."

3. Why is this so? For example, why should a confidence interval for the mean amount families spend monthly on housing be wider for a sparsely populated area of farms in the Midwest than for a densely populated area of an urban center? How does the formula show this will happen?

To deal with this problem, the Census Bureau reports long-form data only for ". . . geographic areas from which about two hundred or more long forms were completed—which are large enough to produce good quality estimates. If smaller weighting areas had been used, the confidence intervals around the estimates would have been significantly wider, rendering many estimates less useful. . . ."

4. Suppose the Census Bureau decided to report on areas from which only 50 long forms were completed. What effect would that have on a 95% confidence interval for, say, the mean cost of housing? Specifically, which values used in the formula for the margin of error would change? Which would change a lot and which would change only slightly?

5. Approximately how much wider would that confidence interval based on 50 forms be than the one based on 200 forms?

Step-by-Step Example A ONE-SAMPLE *t*-INTERVAL FOR THE MEAN

Let's build a 90% confidence interval for the mean amount that college students sleep in a night.

Question: What can we say about the mean amount of sleep that college students get?

THINK ➡ **Plan** State what we want to know. Identify the parameter of interest.

Identify the variables and review the W's.

Make a picture. Check the distribution shape and look for skewness, multiple modes, and outliers.

REALITY CHECK ➡ The histogram centers around 7 hours, and the data lie between 4 and 9 hours. We'd expect a confidence interval to place the population mean within an hour or so of 7.

I want to find a 90% confidence interval for the mean, μ, of hours slept by college students. I have data on the number of hours that 25 students slept.

Here's a histogram of the 25 observed amounts that students slept.

Model Think about the assumptions and check the conditions.

Because this was a randomized survey, we check the randomization condition.

✓ **Randomization Condition:** These are data from a randomized survey, so respondents are likely to be independent.

✓ **10% Condition:** These 25 students are fewer than 10% of all college students.

✓ **Nearly Normal Condition:** The histogram of the *Hours of Sleep* is unimodal and slightly skewed, but not enough to be a concern.

State the sampling distribution model for the statistic.

The conditions are satisfied, so I will use a Student's *t*-model with

$$n - 1 = 24 \text{ degrees of freedom}$$

Choose your method.

and find a **one-sample *t*-interval for the mean.**

SHOW ➡ **Mechanics** Construct the confidence interval.

Be sure to include the units along with the statistics.

Calculating from the data (see page 574):

$n = 25$ students

$\bar{y} = 6.64$ hours

$s = 1.075$ hours

(continued)

The critical value we need to make a 90% interval comes from a Student's t table such as Table T at the back of the book, a computer program, or a calculator. We have $25 - 1 = 24$ degrees of freedom. The selected confidence level says that we want 90% of the probability to be caught in the middle, so we exclude 5% in *each* tail, for a total of 10%. The degrees of freedom and 5% tail probability are all we need to know to find the critical value.

REALITY CHECK ➡ The result looks plausible and in line with what we thought.

TELL ➡ **Conclusion** Interpret the confidence interval in the proper context.

When we construct confidence intervals in this way, we expect 90% of them to cover the true mean and 10% to miss the true value. That's what "90% confident" means.

The standard error of $\bar{y}$ is

$$SE(\bar{y}) = \frac{s}{\sqrt{n}} = \frac{1.075}{\sqrt{25}} = 0.215 \text{ hours.}$$

The 90% critical value is $t^*_{24} = 1.711$, so the margin of error is

$$ME = t^*_{24} \times SE(\bar{y})$$
$$= 1.711(0.215)$$
$$= 0.368 \text{ hours.}$$

The 90% confidence interval for the mean number of sleep hours is 6.64 ± 0.368 hours $=$ (6.272, 7.008) hours.

I am 90% confident that the interval from 6.272 to 7.008 hours contains the true mean number of hours that college students sleep.

Figure 22.2
A Normal probability plot of sleep hours is reasonably straight. The data are integers. That is why there are flat regions of the plot. Nevertheless, the plot looks reasonably straight.

Higher Degrees of Freedom and the t-Table

As degrees of freedom increase, the shape of Student's t-models changes more gradually. Table T at the back of the book includes degrees of freedom between 100 and 1000 selected so that you can pin down the P-value for just about any df. If your df's aren't listed, take the cautious approach by using the next lower value, or use technology

Make a Picture, Make a Picture, Make a Picture

The only reasonable way to check the Nearly Normal Condition is with graphs of the data. In the Step-by-Step we made a histogram of the data to verify that its distribution is unimodal and symmetric and that it has no outliers. You could also make a Normal probability plot to see that it's reasonably straight. You'll be able to spot deviations from the Normal model more easily with a Normal probability plot, but it's easier to understand the particular nature of the deviations from a histogram.

Using Table T to Find *t*-Values

Usually we find critical values and margins of error for Student's t-based intervals and tests with technology. But you can also use tables such as Table T at the back of this book. The tables run down the page for as many degrees of freedom as can fit. As the degrees of freedom increase, the t-model gets closer and closer to the Normal, so the tables give a final row with the critical values from the Normal model and label it "∞ df."

Here's the part of the Student's t table that gives the critical value we needed for the Step-by-Step confidence interval. To find a critical value, locate the row of the table corresponding to the degrees of freedom and the column corresponding to the probability you want. Our 90% confidence interval leaves 5% of the values on either side, so look for 0.05 at the top of the column or 90% at the bottom. The value in the table at that intersection is the critical value we need: 1.711.

A S *Activity*: **Building *t*-Intervals with the *t*-Table.** Interact with an animated version of Table T.

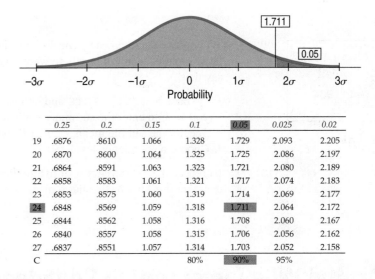

	0.25	0.2	0.15	0.1	0.05	0.025	0.02
19	.6876	.8610	1.066	1.328	1.729	2.093	2.205
20	.6870	.8600	1.064	1.325	1.725	2.086	2.197
21	.6864	.8591	1.063	1.323	1.721	2.080	2.189
22	.6858	.8583	1.061	1.321	1.717	2.074	2.183
23	.6853	.8575	1.060	1.319	1.714	2.069	2.177
24	.6848	.8569	1.059	1.318	1.711	2.064	2.172
25	.6844	.8562	1.058	1.316	1.708	2.060	2.167
26	.6840	.8557	1.058	1.315	1.706	2.056	2.162
27	.6837	.8551	1.057	1.314	1.703	2.052	2.158
C				80%	90%	95%	

For confidence intervals, the values in the table are usually enough to cover most cases of interest. If you can't find a row for the df you need, just use the next smaller df in the table.[8] Of course, you can also create the confidence interval with computer software or a calculator.

TI Tips FINDING A CONFIDENCE INTERVAL FOR A MEAN

Yes, your calculator can create a confidence interval for a mean. And it's so easy we'll do two!

FIND A CONFIDENCE INTERVAL GIVEN A SET OF DATA
• Enter the data for the number of hours the 25 students slept in L1. Go ahead; we'll wait.

67676 77786 66888 54678 58767

• Set up a STATPLOT to create a histogram of the data using Xscl = 1 so you can check the nearly Normal condition. Looks okay—unimodal and roughly symmetric.
• Under STAT TESTS choose TInterval.

• Choose Inpt:Data, then specify that your data is List:L1.
• For these data the frequency is 1. (If your data have a frequency distribution stored in another list, you would specify that.)
• Choose the confidence level you want.
• Calculate the interval.

There's the 90% confidence interval. That was easy—but remember, the calculator only does the *Show*. Now you have to *Tell* what it means.

NO DATA? FIND A CONFIDENCE INTERVAL GIVEN THE SAMPLE'S MEAN AND STANDARD DEVIATION
Sometimes instead of the original data you just have the summary statistics. For instance, suppose a random sample of 53 lengths of fishing line had a mean strength of 83 pounds and standard deviation of 4 pounds. Let's make a 95% confidence interval for the mean strength of this kind of fishing line.

(continued)

---
[8]You can also find tables on the Internet. Search for terms like "statistical tables *z t*."

Without the data you can't check the Nearly Normal Condition. But 53 is a moderately large sample, so assuming there were no outliers, it's okay to proceed. You need to say that.

- Go back to STAT TESTS and choose TInterval again. This time indicate that you wish to enter the summary statistics. To do that, select Stats, then hit ENTER.
- Specify the sample mean, standard deviation, and sample size.
- Choose a confidence level and Calculate the interval.

If (repeat, IF . . .) strengths of fishing lines follow a Normal model, we are 95% confident that this kind of line has a mean strength between 81.9 and 84.1 pounds.

Be Careful When Interpreting Confidence Intervals

A S *Activity*: Intuition for *t*-based Intervals. A narrated review of Student's *t*.

So What *Should* We Say?

Since 90% of random samples yield an interval that captures the true mean, we *should* say, "I am 90% confident that the interval from 6.272 and 7.008 hours per night contains the mean amount that students sleep." It's also okay to say something less formal: "I am 90% confident that the average amount that students sleep is between 6.272 and 7.008 hours per night." Remember: *Our uncertainty is about the interval, not the true mean*. The interval varies randomly. The true mean sleep is neither variable nor random—just unknown.

Confidence intervals for means offer new, tempting, wrong interpretations. Here are some things you *shouldn't* say:

- ***Don't say***, "*90% of all students* sleep between 6.272 and 7.008 hours per night." The confidence interval is about the *mean* sleep, not about the sleep of *individual* students.
- ***Don't say***, "We are 90% confident that *a randomly selected student* will sleep between 6.272 and 7.008 hours per night." This false interpretation is also about individual students rather than about the *mean*. We are 90% confident that the *mean* amount of sleep is between 6.272 and 7.008 hours per night.
- ***Don't say***, "The mean amount students sleep is 6.64 hours 90% of the time." That's about means, but still wrong. It implies that the true mean varies, when in fact it is the confidence interval that would have been different had we gotten a different sample.
- Finally, ***don't say***, "*90% of all samples* will have mean sleep between 6.272 and 7.008 hours per night." That statement suggests that *this* interval somehow sets a standard for every other interval. In fact, this interval is no more (or less) likely to be correct than any other. You could say that 90% of all possible samples will produce intervals that actually do contain the true mean sleep. (The problem is that, because we'll never know where the true mean sleep really is, we can't know if our sample was one of those 90%.)
- ***Do say***, "90% of intervals that could be found in this way would cover the true value." Or make it more personal and say, "I am 90% confident that the true mean amount that students sleep is between 6.272 and 7.008 hours per night."

A Hypothesis Test for the Mean

We are told that adults need between 7 and 9 hours of sleep. Do college students get what they need? Can we say that the mean amount college students sleep is at least 7 hours? A question like this calls for a hypothesis test called the **one-sample *t*-test for the mean**.

You already know enough to construct this test. The test statistic looks just like the others we've seen. It compares the difference between the observed statistic and a hypothesized value to the standard error of the observed statistic. We've seen that, for means, the appropriate probability model to use for P-values is Student's t with $n - 1$ degrees of freedom.

Activity: A *t*-Test for Wind Speed. Watch the video in the preceding activity, and then use the interactive tool to test whether there's enough wind for electricity generation at a site under investigation.

One-Sample *t*-Test for the Mean

The assumptions and conditions for the one-sample *t*-test for the mean are the same as for the one-sample *t*-interval. We test the hypothesis $H_0: \mu = \mu_0$ using the statistic

$$t_{n-1} = \frac{\bar{y} - \mu_0}{SE(\bar{y})}.$$

The standard error of $\bar{y}$ is $SE(\bar{y}) = \dfrac{s}{\sqrt{n}}$.

When the conditions are met and the null hypothesis is true, this statistic follows a Student's *t*-model with $n - 1$ degrees of freedom. We use that model to obtain a P-value.

For Example A ONE-SAMPLE *t*-TEST FOR THE MEAN

RECAP: Researchers tested 150 farm-raised salmon for organic contaminants. They found the mean concentration of the carcinogenic insecticide mirex to be 0.0913 parts per million, with standard deviation 0.0495 ppm. As a safety recommendation to recreational fishers, the Environmental Protection Agency's (EPA) recommended "screening value" for mirex is 0.08 ppm.

QUESTION: Are farmed salmon contaminated beyond the level permitted by the EPA?

ANSWER: *(We've already checked the conditions; see page 582.)*

$$H_0: \mu = 0.08$$
$$H_A: \mu > 0.08$$

These data satisfy the conditions for inference; I'll do a one-sample t-test for the mean:

$n = 150, df = 149$

$\bar{y} = 0.0913, s = 0.0495$

$SE(\bar{y}) = \dfrac{0.0495}{\sqrt{150}} = 0.0040$

$t_{149} = \dfrac{0.0913 - 0.08}{0.0040} = 2.825$

$P(t_{149} > 2.825)$
$\quad = 0.0027$ *(from technology)*

With a P-value that low, I reject the null hypothesis and conclude that, in farm-raised salmon, the mirex contamination level does exceed the EPA screening value.

Step-by-Step Example A ONE-SAMPLE *t*-TEST FOR THE MEAN

Let's apply the one-sample *t*-test to the student sleep survey. It is clear at a glance that students don't get as much sleep as Maas and Robbins recommend. Do they even make it to the minimum for adults? The Sleep Foundation (www.sleepfoundation.org) says that adults should get at least 7 hours of sleep each night.

Question: Is the mean amount that college students sleep at least as much as the 7-hour minimum recommended for adults?

(continued)

THINK ➡ **Plan** State what we want to test. Make clear what the population and parameter are.

Identify the variables and review the W's.

I want to know whether the mean amount that college students sleep meets or exceeds the recommended minimum of 7 hours per night. I have a sample of 25 student reports of their sleep amounts.

H_0: Mean sleep, $\mu = 7$ hours

H_A: Mean sleep, $\mu < 7$ hours

Hours of Sleep

Hypotheses The null hypothesis is that the true mean sleep is equal to the minimum recommended. Because we're interested in whether students get enough sleep, the alternative is one-sided.

Make a picture. Check the distribution for skewness, multiple modes, and outliers.

REALITY CHECK ➡ The histogram of the observed hours of sleep is clustered around a value less than 7. But is this enough evidence to suggest that the mean for all college students is less than 7?

Model Think about the assumptions and check the conditions.

Since 25 students is such a tiny fraction of all college students, we won't bother checking the 10% condition.

State the sampling distribution model. (Be sure to include the degrees of freedom.)

Choose your method.

✓ **Randomization Condition:** The students were sampled in a randomized survey, so the amounts they sleep are likely to be mutually independent.

✓ **Nearly Normal Condition:** The histogram of the speeds is unimodal and reasonably symmetric.

The conditions are satisfied, so I'll use a Student's *t*-model with $(n - 1) = 24$ degrees of freedom to do a **one-sample *t*-test for the mean**.

SHOW ➡ **Mechanics** Be sure to include the units when you write down what you know from the data.

We use the null model to find the P-value. Make a picture of the *t*-model centered at $\mu = 7$. Since this is a lower-tail test, shade the region to the left of the observed mean speed.

The *t*-statistic calculation is just a standardized value, like *z*. We subtract the hypothesized mean and divide by the standard error.

The P-value is the probability of observing a sample mean as small as 6.64 (or smaller) *if* the true mean were 7, as the null hypothesis states. We can find this P-value from a table, calculator, or computer program.

REALITY CHECK ➡ The *t*-statistic is negative because the observed mean is below the hypothesized value. That makes sense because we suspect students get too little sleep, not too much.

From the data,

$$n = 25 \text{ students}$$
$$\bar{y} = 6.64 \text{ hours}$$
$$s = 1.075 \text{ hours}$$
$$SE(\bar{y}) = \frac{s}{\sqrt{n}} = \frac{1.075}{\sqrt{25}} = 0.215 \text{ hours.}$$
$$t = \frac{\bar{y} - \mu_0}{SE(\bar{y})} = \frac{6.64 - 7.0}{0.215} = -1.67$$

The observed mean is 1.67 standard errors below the hypothesized value.

$$\text{P-value} = P(t_{24} < -1.67) = 0.054.$$

(continued)

TELL ➡ **Conclusion** Link the P-value to your decision about H_0, and state your conclusion in context.

The confidence interval will often offer additional insights.

The P-value of 0.054 says that if the true mean student nightly sleep amount were 7 hours, samples of 25 students can be expected to have an observed mean of 6.64 hours or less by random chance about 54 times in 1000. If we use 0.05 as the cutoff value, then this P-value is not quite small enough to provide reasonable evidence to reject the hypothesis that the true mean is at least 7 hours. The 90% confidence interval (6.272, 7.008) shows the plausible values for the mean number of hours that college students get, and it does include 7. Although we are unable to reject the hypothesis that the mean is 7 hours, the P-value is very close to 0.05 and the confidence interval shows that there are plausible values well below 7. If those values would result in negative health impacts to students, it might be worth collecting more data to reduce the margin of error.

Significant and/or Important?

Remember that "statistically significant" does not mean "actually important" or "meaningful," even though it sort of sounds that way. A large insurance company found a statistically significant difference in the mean value of policies sold in 2001 and 2002. The difference was $9.83 per policy. Because a typical policy sold for over $1000, management did not see this as an important difference even though it was significant. On the other hand, even a clinically important improvement of 10% in cure rate with a new treatment may not show up as statistically significant in a study involving fewer than 225 patients. Large sample sizes offer the power to detect small differences, and often those are meaningful. Supplementing a hypothesis test with a confidence interval can help us think about whether any change is significant and also whether it's important.

TI Tips TESTING A HYPOTHESIS ABOUT A MEAN

TESTING A HYPOTHESIS GIVEN A SET OF DATA Still have the student sleep data in L1? Good. Let's use the TI to see if the mean is significantly lower than 7 hours (you've already checked the histogram to verify the nearly Normal condition, of course).

- Go to the STAT TESTS menu, and choose T-Test.
- Tell it you want to use the stored Data.
- Enter the mean of the null model, and indicate where the data are.
- Since this is a lower tail test, choose the $< \mu_0$, option.
- Calculate.

There's everything you need to know: the summary statistics, the calculated value of t, and the P-value of 0.054. (t and P differ slightly from the values in our worked example because when we did it by hand we rounded off the mean and standard deviation. No harm done.)

As always, the *Tell* is up to you.

(continued)

TESTING A HYPOTHESIS GIVEN THE SAMPLE'S MEAN AND STANDARD DEVIATION Don't have the actual data? Just summary statistics? No problem, assuming you can verify the necessary conditions. In the last TI Tips we created a confidence interval for the strength of fishing line. We had test results for a random sample of 53 lengths of line showing a mean strength of 83 pounds and a standard deviation of 4 pounds. Is there evidence that this kind of fishing line exceeds the "80-lb test" as labeled on the package?

We bet you know what to do even without our help. Try it before you read on.

- Go back to T-Test.
- You're entering Stats this time.
- Specify the hypothesized mean and the sample statistics.
- Choose the alternative being tested (upper tail here).
- Calculate.

The results of the calculator's mechanics show a large t and a really small P-value (0.0000007). We have very strong evidence that the mean breaking strength of this kind of fishing line is over the 80 pounds claimed by the manufacturer.

Intervals and Tests

The 90% confidence interval for the mean number of sleep hours was 6.64 hours $\pm$ 0.368, or (6.272 to 7.008 hours). If someone hypothesized that the mean sleep amount was really 8 hours, how would you feel about it? How about 6 hours?

Because the confidence interval includes 7 hours, it certainly looks like 7 hours might be a plausible value for the true mean sleep time. A hypothesized mean of 7 hours lies *within the confidence interval*. It's only one of the plausible values for the mean.

Confidence intervals and significance tests are built from the same calculations. In fact, they are complementary ways of looking at the same question. Here's the connection: The confidence interval contains all the null hypothesis values we can't reject with these data.

How is the confidence level related to the P-value? To be precise, a level C confidence interval contains *all* of the plausible null hypothesis values that would *not* be rejected if you use a (two-sided) P-value of $(1 - C)$ as the cutoff for deciding to reject H_0.

Confidence intervals are naturally two-sided, so they correspond to two-sided P-values. When, as in our example, the hypothesis is one-sided, the interval contains values that would not be rejected using a cutoff P-value of $(1 - C)/2$.

> **Fail to reject** Our 90% confidence interval was 6.272 to 7.008 hours. If any of these values had been the null hypothesis for the mean, then the corresponding hypothesis test using a P-value cutoff of 0.05 (because $\dfrac{1 - 0.90}{2} = 0.05$) would not have been able to reject the null. So, we would not reject any hypothesized value between 6.272 and 7.008 hours.

Confidence intervals and hypothesis tests look at the same problem from two different perspectives. A hypothesis test starts with a *proposed parameter value* and asks if the *data* are consistent with that value. If the observed statistic is too far from the proposed parameter value, that makes it less plausible that the proposed value is the truth. So we reject the null hypothesis. By contrast, a confidence interval starts with the *data* and finds an interval of plausible values for where the parameter may lie.

Just Checking

In discussing estimates based on the long-form samples, the Census Bureau notes, "The disadvantage . . . is that . . . estimates of characteristics that are also reported on the short form will not match the [long-form estimates]."

The short-form estimates are values from a complete census, so they are the "true" values—something we don't usually have when we do inference.

6. Suppose we use long-form data to make 95% confidence intervals for the mean age of residents for each of 100 of the Census-defined areas. How many of these 100 intervals should we expect will fail to include the true mean age (as determined from the complete short-form Census data)?

7. Based only on the long-form sample, we might test the null hypothesis about the mean household income in a region. Would the power of the test increase or decrease if we used an area with more long forms?

Choosing the Sample Size

A S *Activity*: **The Real Effect of Small Sample Size.** We know that smaller sample sizes lead to wider confidence intervals, but is that just because they have fewer degrees of freedom?

How large a sample do we need? The simple answer is "more." But more data costs money, effort, and time, so how much is enough? Suppose your computer just took half an hour to download a movie you wanted to watch. You're not happy. You hear about a program that claims to download movies in less than 15 minutes. You're interested enough to spend $29.95 for it, but only if it really delivers. So you get the free evaluation copy and test it by downloading that movie 5 different times. Of course, the mean download time is not exactly 15 minutes as claimed. Observations vary. If the margin of error were 8 minutes, you might find it hard to decide whether the software is worth the money. Doubling the sample size would require another 5 hours of testing and would reduce your margin of error to just less than 6 minutes. You'll need to decide whether that's worth the effort.

Armed with the *ME* that you decide is necessary to draw a conclusion and confidence level, you can find the sample size you need. Approximately.

For a mean, $ME = t^*_{n-1} \times SE(\bar{y})$ and $SE(\bar{y}) = \dfrac{s}{\sqrt{n}}$, so to find the sample size, solve this equation for n (see the margin note).

$$ME = t^*_{n-1}\dfrac{s}{\sqrt{n}}$$

But the equation needs s and before we collect the data, we don't know s. A guess is often good enough, but if you have no idea what the standard deviation might be, or if the sample size really matters (for example, because each additional individual is very expensive to sample or experiment on), a small *pilot study* can provide some information about the standard deviation.

That's not all. Without knowing n, you don't know the degrees of freedom so we can't find the critical value, t^*_{n-1}. One common approach is to use the corresponding z^* value from the Normal model. If you've chosen a 95% confidence level, then just use 2, following the 68–95–99.7 Rule. If your estimated sample size is, say, 60 or more, it's probably okay—z^* was a good guess. If it's smaller than that, you may want to add a step, using z^* at first, finding n, and then replacing z^* with the corresponding t^*_{n-1} and calculating the sample size once more.

Sample size calculations are *never* exact. But it's always a good idea to know whether your sample size is large enough to give you a good chance of being able to tell you what you want to know before you collect your data.

For Example FINDING SAMPLE SIZE

A company claims its program will allow your computer to download movies quickly. We'll test the free evaluation copy by downloading a movie several times, hoping to estimate the mean download time with a margin of error of only 8 minutes. We think the standard deviation of download times is about 10 minutes.

QUESTION: How many trial downloads must we run if we want 95% confidence in our estimate with a margin of error of only 8 minutes?

ANSWER: Using $z^* = 2$ (from the 68–95–99.7 Rule), solve

$$8 = 2\frac{10}{\sqrt{n}}$$

$$\sqrt{n} = \frac{20}{8} = 2.5$$

$$n = 2.5^2 = 6.25.$$

That's a small sample size, so I'll use $6 - 1 = 5$ degrees of freedom[9] to substitute an appropriate t^* value. At 95%, $t_5^* = 2.571$. Solving the equation one more time:

$$8 = 2.571\frac{10}{\sqrt{n}}$$

$$\sqrt{n} = \frac{2.571 \times 10}{8} \approx 3.214$$

$$n = (3.214)^2 \approx 10.33$$

To make sure the *ME* is no larger, we'll round *up*, which gives $n = 11$ runs. So, to get an *ME* of 8 minutes, we'll find the downloading times for 11 movies.

WHAT IF ●●● the sample is small?

The Normality Assumption is about the population. The Central Limit Theorem tells us that the sampling distribution of sample means approaches a Normal model regardless of the shape of the population. That may seem like a free pass to apply the CLT everywhere, but then the theorem goes on to describe this Normal model based on the population standard deviation, which we never know. In the real world we have to use the sample's standard deviation as an estimate. The good news is that Gossett's *t*-models describe the sampling distribution we can use. The bad news is that now the shape of the population matters: the mathematics underlying *t*-models assumes the population is normal.

Of course, we don't (and can't) know that. We have to think about whether that assumption is reasonable, so we check the Nearly Normal Condition: we look at a histogram of the sample data. We're not looking to see if the sample data are normally distributed (a common misconception). We're looking to see if it's reasonable to believe that our sample could have come from a population that's Normal. And that's where things get messy.

When the sample is fairly large, the shape of the distribution of the sample data should look a lot like the population. If it's unimodal and symmetric, that's a good sign. The trouble is, small samples don't paint a very clear picture. A simulation will show you what we mean.

---

[9]Ordinarily, we'd round the sample size *up*. But at this stage of the calculation, rounding *down* is the safer choice. Can you see why?

First we created a Normal population with mean 100 and standard deviation 10. Then we drew 9 random samples of size 12. Here are the histograms of the sample data.

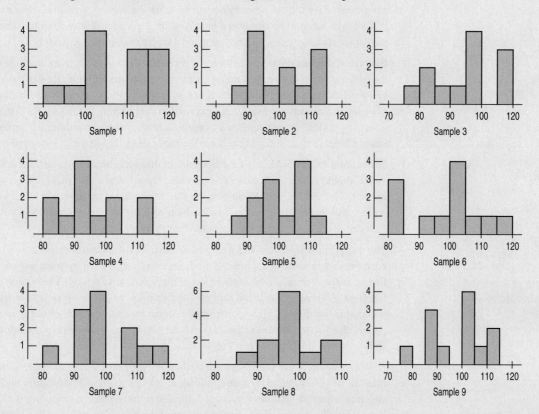

When we look at these samples, we certainly don't immediately think, "Aha, Normal!" The bottom line is that we can't expect our sample to look perfect. Its shape will almost certainly be ragged, and smaller samples could look even worse than these. We have to ask whether the sample's shape is so distorted that we should worry about the population.

If you see histograms like these, they're mostly fine. After all even Sample 3, which is quite skewed to the left, did come from our Normal population. Unless the sample's histogram screams out that population wasn't Normal, our major concern is really that the skewness (or outliers) in the sample may make it unwise to use the sample mean and standard deviation. Those are the statistics we need to use to do inference, and that's the reason we might not proceed with Sample 3.

WHAT CAN GO WRONG?

The most fundamental issue you face is knowing when to use Student's *t* methods.

- **Don't confuse proportions and means.** When you treat your data as categorical, counting successes and summarizing with a sample proportion, make inferences using the Normal model methods you learned about in Chapters 18 and 19. When you treat your data as quantitative, summarizing with a sample mean, make your inferences using Student's *t* methods.

Student's *t* methods work only when the Normality Assumption is true. Naturally, many of the ways things can go wrong turn out to be different ways that the Normality Assumption can fail. It's always a good idea to look for the most common kinds of failure. It turns out that you can even fix some of them.

- **Beware of multimodality.** The Nearly Normal Condition clearly fails if a histogram of the data has two or more modes. When you see this, look for the possibility that your data come from two groups. If so, your best bet is to try to separate the data into different groups. (Use the variables to help distinguish the modes, if possible. For example, if the modes seem to be composed mostly of men in one and women in the other, split the data according to sex.) Then you could analyze each group separately.

- **Beware of skewed data.** Make a Normal probability plot and a histogram of the data. If the data are very skewed, you might try re-expressing the variable. Re-expressing may yield a distribution that is unimodal and symmetric, more appropriate for Student's *t* inference methods for means. Re-expression cannot help if the sample distribution is not unimodal. Some people may object to re-expressing the data, but unless your sample is very large, you just can't use the methods of this chapter on skewed data.

- **Set outliers aside.** Student's *t* methods are built on the mean and standard deviation, so we should beware of outliers when using them. When you make a histogram to check the Nearly Normal Condition, be sure to check for outliers as well. If you find some, consider doing the analysis twice, both with the outliers excluded and with them included in the data, to get a sense of how much they affect the results.

The suggestion that you can perform an analysis with outliers removed may be controversial in some disciplines. Setting aside outliers is seen by some as "cheating." But an analysis of data with outliers left in place is *always* wrong. The outliers violate the Nearly Normal Condition and also the implicit assumption of a homogeneous population, so they invalidate inference procedures. An analysis of the nonoutlying points, along with a separate discussion of the outliers, is often much more informative and can reveal important aspects of the data.

How can you tell whether there are outliers in your data? The "outlier nomination rule" of boxplots can offer some guidance, but it's just a rule of thumb and not an absolute definition. The best practical definition is that a value is an outlier if removing it substantially changes your conclusions about the data. You won't want a single value to determine your understanding of the world unless you are very, very sure that it is correct and similar in nature to the other cases in your data. Of course, when the outliers affect your conclusion, this can lead to the uncomfortable state of not really knowing what to conclude. Such situations call for you to use your knowledge of the real world and your understanding of the data you are working with.[10]

Of course, Normality issues aren't the only risks you face when doing inferences about means. Remember to *Think* about the usual suspects.

- **Watch out for bias.** Measurements of all kinds can be biased. If your observations differ from the true mean in a systematic way, your confidence interval may not capture the true mean. And there is no sample size that will save you. A bathroom scale that's 5 pounds off will be 5 pounds off even if you weigh yourself 100 times and take the average. We've seen several sources of bias in surveys, and measurements can be biased, too. Be sure to think about possible sources of bias in your measurements.

- **Make sure cases are independent.** Student's *t* methods also require the sampled values to be mutually independent. We check for random sampling. You should also think hard about whether there are likely violations of independence in the data collection method. If there are, be very cautious about using these methods.

- **Make sure that data are from an appropriately randomized sample.** Ideally, all data that we analyze are drawn from a simple random sample or generated by a randomized experiment. When they're not, be careful about making inferences from them. You may still compute a confidence interval correctly, or get the mechanics of the P-value right, but this might not save you from making a serious mistake in inference.

---

[10]An important reason for *you* to know Statistics rather than let someone else analyze your data.

Don't Ignore Outliers

As tempting as it is to get rid of annoying values, you can't just throw away outliers and not discuss them. It isn't appropriate to lop off the highest or lowest values just to improve your results.

■ **Interpret your confidence interval correctly.** Many statements that sound tempting are, in fact, misinterpretations of a confidence interval for a mean. You might want to have another look at some of the common mistakes, explained on page 000. Keep in mind that a confidence interval is about the mean of the population, not about the means of samples, individuals in samples, or individuals in the population.

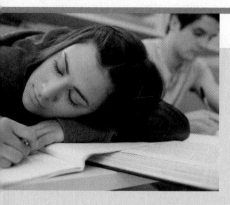

What Have We Learned?

We first learned to create confidence intervals for and test hypotheses about proportions. Now we have extended these ideas to means and seen that the concepts remain the same; only the model and mechanics have changed.

■ We've learned to check the usual assumptions and conditions. The Normality Assumption is of special importance when using a *t*-model. We've learned to check the Nearly Normal Condition to see if it's plausible that our sample could have come from a population that's normally distributed. A histogram of the sample data may appear only roughly unimodal and symmetric, but we should not proceed with inference if the sample is strongly skewed or there are outliers present. This concern is very critical in small samples, but less important with sample sizes of 30 to 40 or more.

■ We've learned that to use the Central Limit Theorem for the mean in practical applications we must estimate the standard deviation with a *standard error*:

$$SE(\bar{y}) = \frac{s}{\sqrt{n}}.$$

■ We've learned that a Student's *t*-model with $n - 1$ degrees of freedom accounts for the extra uncertainty that arises from using the *SE*.

■ We've learned to construct confidence intervals for the true mean μ, in the form

$$\bar{y} \pm ME \text{ where } ME = t_{df}^* SE(\bar{y}).$$

■ We've learned to find the sample size needed to produce a desired margin of error.

■ We've learned to test hypotheses about the mean, using the test statistic $t_{df} = \dfrac{\bar{y} - \mu}{\dfrac{s}{\sqrt{n}}}$.

■ We've learned to write clear conclusions, interpreting a confidence interval or the results of a hypothesis test in context.

Above all, we've learned that the reasoning of inference remains the same regardless of whether we are investigating means or proportions.

Terms

Student's *t*
A family of distributions indexed by its degrees of freedom. The *t*-models are unimodal, symmetric, and bell shaped, but have fatter tails and a narrower center than the Normal model. As the degrees of freedom increase, *t*-distributions approach the Normal. (p. 577)

Degrees of freedom for Student's *t*-distribution
For the *t*-distribution, the degrees of freedom are equal to $n - 1$, where *n* is the sample size. (p. 577)

Nearly Normal Condition
Using a *t*-model requires the Normality Assumption. We look at a histogram of the sample data to see if it's plausible that our sample could have come from a population that's normally distributed. It may appear only roughly unimodal and symmetric, but we should not proceed with inference if the sample is strongly skewed or there are outliers present. This concern is very critical in small samples, but less important with sample sizes of 30 to 40 or more. (p. 581)

| One-sample *t*-interval for the mean | A one-sample *t*-interval for the population mean is (p. 583) |

$$\bar{y} \pm t_{n-1} \times SE(\bar{y}), \text{ where } SE(\bar{y}) = \frac{s}{\sqrt{n}}.$$

The critical value t^*_{n-1} depends on the particular confidence level, C, that you specify and on the number of degrees of freedom, $n - 1$.

| One-sample *t*-test for the mean | The one-sample *t*-test for the mean tests the hypothesis $H_0: \mu = \mu_0$ using the statistic (p. 586) |

$$t_{n-1} = \frac{\bar{y} - \mu_0}{SE(\bar{y})}.$$

The standard error of $\bar{y}$ is

$$SE(\bar{y}) = \frac{s}{\sqrt{n}}.$$

On the Computer INFERENCE FOR MEANS

Statistics packages offer convenient ways to make histograms of the data. Even better for assessing near-Normality is a Normal probability plot. When you work on a computer, there is simply no excuse for skipping the step of plotting the data to check that it is nearly Normal.

Any standard statistics package can compute a hypothesis test. Here's what the package output might look like in general (although no package we know gives the results in exactly this form):[11]

A S *Activity:* **Student's *t* in Practice.** We almost always use technology to do inference with Student's *t*. Here's a chance to do that as you investigate several questions.

The package computes the sample mean and sample standard deviation of the variable and finds the P-value from the *t*-distribution based on the appropriate number of degrees of freedom. All modern statistics packages report P-values. The package may also provide additional information such as the sample mean, sample standard deviation, *t*-statistic value, and degrees of freedom. These are useful for interpreting the resulting P-value and telling the difference between a meaningful result and one that is merely statistically significant. Statistics packages that report the estimated standard deviation of the sampling distribution usually label it "standard error" or "SE."

(continued)

---

[11]Many statistics packages keep as many as 16 digits for all intermediate calculations. If we had kept as many, our results in the Step-by-Step section would have been closer to these.

Inference results are also sometimes reported in a table. You may have to read carefully to find the values you need. Often, test results and the corresponding confidence interval bounds are given together. And often you must read carefully to find the alternative hypotheses. Here's an example of that kind of output:

Exercises

1. **Salmon** A specialty food company sells whole King Salmon to various customers. The mean weight of these salmon is 35 pounds with a standard deviation of 2 pounds. The company ships them to restaurants in boxes of 4 salmon, to grocery stores in cartons of 16 salmon, and to discount outlet stores in pallets of 100 salmon. To forecast costs, the shipping department needs to estimate the standard deviation of the mean weight of the salmon in each type of shipment
   a) Find the standard deviations of the mean weight of the salmon in each type of shipment.
   b) The distribution of the salmon weights turns out to be skewed to the high end. Would the distribution of shipping weights be better characterized by a Normal model for the boxes or pallets? Explain.

2. **LSAT** The LSAT (a test taken for law school admission) has a mean score of 151 with a standard deviation of 9 and a unimodal, symmetric distribution of scores. A test preparation organization teaches small classes of 9 students at a time. A larger organization teaches classes of 25 students at a time. Both organizations publish the mean scores of all their classes.
   a) What would you expect the distribution of mean class scores to be for each organization?

   b) If either organization has a graduating class with a mean score of 160, they'll take out a full-page ad in the local school paper to advertise. Which organization is more likely to have that success? Explain.
   c) Both organizations advertise that if any class has an average score below 145, they'll pay for everyone to retake the LSAT. Which organization is at greater risk to have to pay?

3. **t-models, part I** Using the t tables, software, or a calculator, estimate
   a) the critical value of t for a 90% confidence interval with df = 17.
   b) the critical value of t for a 98% confidence interval with df = 88.
   c) the P-value for $t \geq 2.09$ with 4 degrees of freedom.
   d) the P-value for $|t| > 1.78$ with 22 degrees of freedom.

4. **t-models, part II** Using the t tables, software, or a calculator, estimate
   a) the critical value of t for a 95% confidence interval with df = 7.
   b) the critical value of t for a 99% confidence interval with df = 102.

c) the P-value for $t \leq 2.19$ with 41 degrees of freedom.

d) the P-value for $|t| > 2.33$ with 12 degrees of freedom.

5. *t-models, part III* Describe how the shape, center, and spread of *t*-models change as the number of degrees of freedom increases.

6. *t-models, part IV (last one!)* Describe how the critical value of *t* for a 95% confidence interval changes as the number of degrees of freedom increases.

7. **Cattle** Livestock are given a special feed supplement to see if it will promote weight gain. Researchers report that the 77 cows studied gained an average of 56 pounds, and that a 95% confidence interval for the mean weight gain this supplement produces has a margin of error of ±11 pounds. Some students wrote the following conclusions. Did anyone interpret the interval correctly? Explain any misinterpretations.

a) 95% of the cows studied gained between 45 and 67 pounds.

b) We're 95% sure that a cow fed this supplement will gain between 45 and 67 pounds.

c) We're 95% sure that the average weight gain among the cows in this study was between 45 and 67 pounds.

d) The average weight gain of cows fed this supplement will be between 45 and 67 pounds 95% of the time.

e) If this supplement is tested on another sample of cows, there is a 95% chance that their average weight gain will be between 45 and 67 pounds.

8. **Teachers** Software analysis of the salaries of a random sample of 288 Nevada teachers produced the confidence interval shown below. Which conclusion is correct? What's wrong with the others?

t-Interval for μ: with 90.00% Confidence,
$38944 < \mu(\text{TchPay}) < 42893$

a) If we took many random samples of 288 Nevada teachers, about 9 out of 10 of them would produce this confidence interval.

b) If we took many random samples of Nevada teachers, about 9 out of 10 of them would produce a confidence interval that contained the mean salary of all Nevada teachers.

c) About 9 out of 10 Nevada teachers earn between $38,944 and $42,893.

d) About 9 out of 10 of the teachers surveyed earn between $38,944 and $42,893.

e) We are 90% confident that the average teacher salary in the United States is between $38,944 and $42,893.

9. **Meal plan** After surveying students at Dartmouth College, a campus organization calculated that a 95% confidence interval for the mean cost of food for one term (of three in the Dartmouth trimester calendar) is ($1102, $1290). Now the organization is trying to write its report and is considering the following interpretations. Comment on each.

a) 95% of all students pay between $1102 and $1290 for food.

b) 95% of the sampled students paid between $1102 and $1290.

c) We're 95% sure that students in this sample averaged between $1102 and $1290 for food.

d) 95% of all samples of students will have average food costs between $1102 and $1290.

e) We're 95% sure that the average amount all students pay is between $1102 and $1290.

10. **Snow** Based on meteorological data for the past century, a local TV weather forecaster estimates that the region's average winter snowfall is 23″, with a margin of error of ±2 inches. Assuming he used a 95% confidence interval, how should viewers interpret this news? Comment on each of these statements:

a) During 95 of the last 100 winters, the region got between 21″ and 25″ of snow.

b) There's a 95% chance the region will get between 21″ and 25″ of snow this winter.

c) There will be between 21″ and 25″ of snow on the ground for 95% of the winter days.

d) Residents can be 95% sure that the area's average snowfall is between 21″ and 25″.

e) Residents can be 95% confident that the average snowfall during the last century was between 21″ and 25″ per winter.

11. **Pulse rates** A medical researcher measured the pulse rates (beats per minute) of a sample of randomly selected adults and found the following Student's *t*-based confidence interval:

With 95.00% Confidence,
$70.887604 < \mu(\text{Pulse}) < 74.497011$

a) Explain carefully what the software output means.

b) What's the margin of error for this interval?

c) If the researcher had calculated a 99% confidence interval, would the margin of error be larger or smaller? Explain.

12. **Crawling** Data collected by child development scientists produced this confidence interval for the average age (in weeks) at which babies begin to crawl:

t-Interval for μ
(95.00% Confidence): $29.202 < \mu(\text{age}) < 31.844$

a) Explain carefully what the software output means.

b) What is the margin of error for this interval?

c) If the researcher had calculated a 90% confidence interval, would the margin of error be larger or smaller? Explain.

13. **Home sales** The housing market has recovered slowly from the economic crisis of 2008. Recently, in one large community, realtors randomly sampled 36 bids from potential buyers to estimate the average loss in home value. The sample showed the average loss was $9560 with a standard deviation of $1500.

a) What assumptions and conditions must be checked before finding a confidence interval? How would you check them?

b) Find a 95% confidence interval for the mean loss in value per home.

c) Interpret this interval and explain what 95% confidence means in this context.

14. Home sales again In the previous exercise, you found a 95% confidence interval to estimate the average loss in home value.

a) Suppose the standard deviation of the losses had been $3000 instead of $1500. What would the larger standard deviation do to the width of the confidence interval (assuming the same level of confidence)?

b) Your classmate suggests that the margin of error in the interval could be reduced if the confidence level were changed to 90% instead of 95%. Do you agree with this statement? Why or why not?

c) Instead of changing the level of confidence, would it be more statistically appropriate to draw a bigger sample?

15. CEO compensation A sample of 20 CEOs from the Forbes 500 shows total annual compensations ranging from a minimum of $0.1 to $62.24 million. The average for these 20 CEOs is $7.946 million. Here's a histogram:

Total Compensation in $ Million

Based on these data, a computer program found that a 95% confidence interval for the mean annual compensation of all Forbes 500 CEOs is (1.69, 14.20) $ million. Why should you be hesitant to trust this confidence interval?

16. Credit card charges A credit card company takes a random sample of 100 cardholders to see how much they charged on their card last month. Here's a histogram.

March 2005 Charges

A computer program found that the resulting 95% confidence interval for the mean amount spent in March 2005 is (−$28366.84, $90691.49). Explain why the analysts didn't find the confidence interval useful, and explain what went wrong.

17. Normal temperature The researcher described in Exercise 11 also measured the body temperatures of that randomly selected group of adults. Here are summaries of the data he collected. We wish to estimate the average (or "normal") temperature among the adult population.

Summary	Temperature
Count	52
Mean	98.285
Median	98.200
MidRange	98.600
StdDev	0.6824
Range	2.800
IntQRange	1.050

Body Temperature (°F)

a) Check the conditions for creating a *t*-interval.

b) Find a 98% confidence interval for mean body temperature.

c) Explain the meaning of that interval.

d) Explain what "98% confidence" means in this context.

e) 98.6°F is commonly assumed to be "normal." Do these data suggest otherwise? Explain.

18. Parking Hoping to lure more shoppers downtown, a city builds a new public parking garage in the central business district. The city plans to pay for the structure through parking fees. During a two-month period (44 weekdays), daily fees collected averaged $126, with a standard deviation of $15.

a) What assumptions must you make in order to use these statistics for inference?

b) Write a 90% confidence interval for the mean daily income this parking garage will generate.

c) Interpret this confidence interval in context.

d) Explain what "90% confidence" means in this context.

e) The consultant who advised the city on this project predicted that parking revenues would average $130 per day. Based on your confidence interval, do you think the consultant was correct? Why?

19. Normal temperatures, part II Consider again the statistics about human body temperature in Exercise 17.

a) Would a 90% confidence interval be wider or narrower than the 98% confidence interval you calculated before? Explain. (Don't compute the new interval.)

b) What are the advantages and disadvantages of the 98% confidence interval?

c) If we conduct further research, this time using a sample of 500 adults, how would you expect the 98% confidence interval to change? Explain.

d) How large a sample might allow you to estimate the mean body temperature to within 0.1 degrees with 98% confidence?

20. Parking II Suppose that, for budget planning purposes, the city in Exercise 18 needs a better estimate of the mean daily income from parking fees.

a) Someone suggests that the city use its data to create a 95% confidence interval instead of the 90% interval first created. How would this interval be better for the city? (You need not actually create the new interval.)

b) How would the 95% interval be worse for the planners?

c) How could they achieve an interval estimate that would better serve their planning needs?

d) How many days' worth of data should they collect to have 95% confidence of estimating the true mean to within $3?

21. Speed of light In 1882 Michelson measured the speed of light (usually denoted c as in Einstein's famous equation $E = mc^2$). His values are in km/sec and have 299,000 subtracted from them. He reported the results of 23 trials with a mean of 756.22 and a standard deviation of 107.12.

a) Find a 95% confidence interval for the true speed of light from these statistics.

b) State in words what this interval means. Keep in mind that the speed of light is a physical constant that, as far as we know, has a value that is true throughout the universe.

c) What assumptions must you make in order to use your method?

22. Better light After his first attempt to determine the speed of light (described in Exercise 21), Michelson conducted an "improved" experiment. In 1897 he reported results of 100 trials with a mean of 852.4 and a standard deviation of 79.0.

a) What is the standard error of the mean for these data?

b) Without computing it, how would you expect a 95% confidence interval for the second experiment to differ from the confidence interval for the first? Note at least three specific reasons why they might differ, and indicate the ways in which these differences would change the interval.

c) According to Stigler (who reports these values), the true speed of light is 299,710.5 km/sec, corresponding to a value of 710.5 for Michelson's

1897 measurements. What does this indicate about Michelson's two experiments? Explain, using your confidence interval.

23. Departures 2011 What are the chances your flight will leave on time? The U.S. Bureau of Transportation Statistics of the Department of Transportation publishes information about airline performance. Here are a histogram and summary statistics for the percentage of flights departing on time each month from 1995 thru September 2011. (www.transtats.bts.gov/HomeDrillChart.asp)

There is no evidence of a trend over time.

a) Check the assumptions and conditions for inference.

b) Find a 90% confidence interval for the true percentage of flights that depart on time.

c) Interpret this interval for a traveler planning to fly.

24. Arrivals 2011 Will your flight get you to your destination on time? The U.S. Bureau of Transportation Statistics reported the percentage of flights that were late each month from 1995 through September of 2011. Here's a histogram, along with some summary statistics:

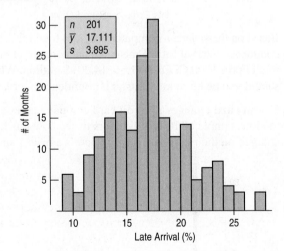

We can consider these data to be a representative sample of all months. There is no evidence of a time trend ($r = 0.07$).

a) Check the assumptions and conditions for inference about the mean.

b) Find a 99% confidence interval for the true percentage of flights that arrive late.

c) Interpret this interval for a traveler planning to fly.

25. Home prices In 2011, the average home in the region of the country studied in Exercise 13 lost $9010. Was the community studied in Exercise 13 unusual? Use a *t*-test to decide if the average loss observed was significantly different from the regional average.

26. Home prices II Suppose the standard deviation of home price losses had been $3000, as in Exercise 14? What would your conclusion be then?

T 27. For Example, 2nd look This chapter's For Examples looked at mirex contamination in farmed salmon. We first found a 95% confidence interval for the mean concentration to be 0.0834 to 0.0992 parts per million. Later we rejected the null hypothesis that the mean did not exceed the EPA's recommended safe level of 0.08 ppm based on a P-value of 0.0027. Explain how these two results are consistent. Your explanation should discuss the confidence level, the P-value, and the decision.

28. Hot Dogs A nutrition lab tested 40 hot dogs to see if their mean sodium content was less than the 325 mg upper limit set by regulations for "reduced sodium" franks. The lab failed to reject the hypothesis that the hot dogs did not meet this requirement, with a P-value of 0.142. A 90% confidence interval estimated the mean sodium content for this kind of hot dog at 317.2 to 326.8 mg. Explain how these two results are consistent. Your explanation should discuss the confidence level, the P-value, and the decision.

29. Pizza A researcher tests whether the mean cholesterol level among those who eat frozen pizza exceeds the value considered to indicate a health risk. She gets a P-value of 0.07. Explain in this context what the "7%" represents.

30. Golf balls The United States Golf Association (USGA) sets performance standards for golf balls. For example, the initial velocity of the ball may not exceed 250 feet per second when measured by an apparatus approved by the USGA. Suppose a manufacturer introduces a new kind of ball and provides a sample for testing. Based on the mean speed in the test, the USGA comes up with a P-value of 0.34. Explain in this context what the "34%" represents.

31. TV safety The manufacturer of a metal stand for home TV sets must be sure that its product will not fail under the weight of the TV. Since some larger sets weigh nearly 300 pounds, the company's safety inspectors have set a standard of ensuring that the stands can support an average of over 500 pounds. Their inspectors regularly subject a random sample of the stands to increasing weight until they fail. They test the hypothesis $H_0: \mu = 500$ against $H_A: \mu > 500$, using the level of significance $\alpha = 0.01$. If the sample of stands fails to pass this safety test, the inspectors will not certify the product for sale to the general public.

a) Is this an upper-tail or lower-tail test? In the context of the problem, why do you think this is important?

b) Explain what will happen if the inspectors commit a Type I error.

c) Explain what will happen if the inspectors commit a Type II error.

32. Catheters During an angiogram, heart problems can be examined via a small tube (a catheter) threaded into the heart from a vein in the patient's leg. It's important that the company that manufactures the catheter maintain a diameter of 2.00 mm. (The standard deviation is quite small.) Each day, quality control personnel make several measurements to test $H_0: \mu = 2.00$ against $H_A: \mu \neq 2.00$ at a significance level of $\alpha = 0.05$. If they discover a problem, they will stop the manufacturing process until it is corrected.

a) Is this a one-sided or two-sided test? In the context of the problem, why do you think this is important?

b) Explain in this context what happens if the quality control people commit a Type I error.

c) Explain in this context what happens if the quality control people commit a Type II error.

33. TV safety revisited The manufacturer of the metal TV stands in Exercise 31 is thinking of revising its safety test.

a) If the company's lawyers are worried about being sued for selling an unsafe product, should they increase or decrease the value of α? Explain.

b) In this context, what is meant by the power of the test?

c) If the company wants to increase the power of the test, what options does it have? Explain the advantages and disadvantages of each option.

34. Catheters again The catheter company in Exercise 32 is reviewing its testing procedure.

a) Suppose the significance level is changed to $\alpha = 0.01$. Will the probability of a Type II error increase, decrease, or remain the same?

b) What is meant by the power of the test the company conducts?

c) Suppose the manufacturing process is slipping out of proper adjustment. As the actual mean diameter of the catheters produced gets farther and farther above the desired 2.00 mm, will the power of the quality control test increase, decrease, or remain the same?

d) What could they do to improve the power of the test?

35. Marriage In 1960, census results indicated that the age at which American men first married had a mean of 23.3 years. It is widely suspected that young people today are waiting longer to get married. We want to find out if the mean age of first marriage has increased during the past 40 years.

a) Write appropriate hypotheses.

b) We plan to test our hypothesis by selecting a random sample of 40 men who married for the first time last year. Do you think the necessary assumptions for inference are satisfied? Explain.

c) Describe the approximate sampling distribution model for the mean age in such samples.

d) The men in our sample married at an average age of 24.2 years, with a standard deviation of 5.3 years. What's the P-value for this result?

e) Explain (in context) what this P-value means.

f) What's your conclusion?

36. Saving gas Congress regulates corporate fuel economy and sets an annual gas mileage for cars. A company with a large fleet of cars hopes to meet the 2011 goal of 30.2 mpg or better for their fleet of cars. To see if the goal is being met, they check the gasoline usage for 50 company trips chosen at random, finding a mean of 32.12 mpg and a standard deviation of 4.83 mpg. Is this strong evidence that they have attained their fuel economy goal?

a) Write appropriate hypotheses.

b) Are the necessary assumptions to make inferences satisfied?

c) Describe the sampling distribution model of mean fuel economy for samples like this.

d) Find the P-value.

e) Explain what the P-value means in this context.

f) State an appropriate conclusion.

37. Ruffles Students investigating the packaging of potato chips purchased 6 bags of Lay's Ruffles marked with a net weight of 28.3 grams. They carefully weighed the contents of each bag, recording the following weights (in grams): 29.3, 28.2, 29.1, 28.7, 28.9, 28.5.

a) Do these data satisfy the assumptions for inference? Explain.

b) Find the mean and standard deviation of the weights.

c) Create a 95% confidence interval for the mean weight of such bags of chips.

d) Explain in context what your interval means.

e) Comment on the company's stated net weight of 28.3 grams.

38. Doritos Some students checked 6 bags of Doritos marked with a net weight of 28.3 grams. They carefully weighed the contents of each bag, recording the following weights (in grams): 29.2, 28.5, 28.7, 28.9, 29.1, 29.5.

a) Do these data satisfy the assumptions for inference? Explain.

b) Find the mean and standard deviation of the weights.

c) Create a 95% confidence interval for the mean weight of such bags of chips.

d) Explain in context what your interval means.

e) Comment on the company's stated net weight of 28.3 grams.

39. Popcorn Yvon Hopps ran an experiment to test optimum power and time settings for microwave popcorn. His goal was to find a combination of power and time that would deliver high-quality popcorn with less than 10% of the kernels left unpopped, on average. After experimenting

with several bags, he determined that power 9 at 4 minutes was the best combination.

a) He concluded that this popping method achieved the 10% goal. If it really does not work that well, what kind of error did Hopps make?

b) To be sure that the method was successful, he popped 8 more bags of popcorn (selected at random) at this setting. All were of high quality, with the following percentages of uncooked popcorn: 7, 13.2, 10, 6, 7.8, 2.8, 2.2, 5.2. Does this provide evidence that he met his goal of an average of no more than 10% uncooked kernels? Explain.

40. Ski wax Bjork Larsen was trying to decide whether to use a new racing wax for cross-country skis. He decided that the wax would be worth the price if he could average less than 55 seconds on a course he knew well, so he planned to test the wax by racing on the course 8 times.

a) Suppose that he eventually decides not to buy the wax, but it really would lower his average time to below 55 seconds. What kind of error would he have made?

b) His 8 race times were 56.3, 65.9, 50.5, 52.4, 46.5, 57.8, 52.2, and 43.2 seconds. Should he buy the wax? Explain.

41. Chips Ahoy In 1998, as an advertising campaign, the Nabisco Company announced a "1000 Chips Challenge," claiming that every 18-ounce bag of their Chips Ahoy cookies contained at least 1000 chocolate chips. Dedicated Statistics students at the Air Force Academy (no kidding) purchased some randomly selected bags of cookies, and counted the chocolate chips. Some of their data are given below. (*Chance*, 12, no. 1[1999])

1219	1214	1087	1200	1419	1121	1325	1345
1244	1258	1356	1132	1191	1270	1295	1135

a) Check the assumptions and conditions for inference. Comment on any concerns you have.

b) Create a 95% confidence interval for the average number of chips in bags of Chips Ahoy cookies.

c) What does this evidence say about Nabisco's claim? Use your confidence interval to test an appropriate hypothesis and state your conclusion.

42. Yogurt *Consumer Reports* tested 11 brands of vanilla yogurt and found these numbers of calories per serving:

130	160	150	120	120	110	170	160	110	130	90

a) Check the assumptions and conditions for inference.

b) Create a 95% confidence interval for the average calorie content of vanilla yogurt.

c) A diet guide claims that you will get an average of 120 calories from a serving of vanilla yogurt. What does this evidence indicate? Use your confidence interval to test an appropriate hypothesis and state your conclusion.

43. Jelly A consumer advocate wants to collect a sample of jelly jars and measure the actual weight of the product in the container. He needs to collect enough data to construct a confidence interval with a margin of error of no more than 2 grams with 99% confidence. The standard deviation of these jars is usually 4 grams. What do you recommend for his sample size?

44. A good book An English professor is attempting to estimate the mean number of novels that the student body reads during their time in college. He is conducting an exit survey with seniors. He hopes to have a margin of error of 3 books with 95% confidence. From reading previous studies, he expects a large standard deviation and is going assume it is 10. How many students should he survey?

45. Maze Psychology experiments sometimes involve testing the ability of rats to navigate mazes. The mazes are classified according to difficulty, as measured by the mean length of time it takes rats to find the food at the end. One researcher needs a maze that will take rats an average of about one minute to solve. He tests one maze on several rats, collecting the data shown.

Time (sec)	
38.4	57.6
46.2	55.5
62.5	49.5
38.0	40.9
62.8	44.3
33.9	93.8
50.4	47.9
35.0	69.2
52.8	46.2
60.1	56.3
55.1	

a) Plot the data. Do you think the conditions for inference are satisfied? Explain.
b) Test the hypothesis that the mean completion time for this maze is 60 seconds. What is your conclusion?
c) Eliminate the outlier, and test the hypothesis again. What is your conclusion?
d) Do you think this maze meets the "one-minute average" requirement? Explain.

46. Braking A tire manufacturer is considering a newly designed tread pattern for its all-weather tires. Tests have indicated that these tires will provide better gas mileage and longer tread life. The last remaining test is for braking effectiveness. The company hopes the tire will allow a car traveling at 60 mph to come to a complete stop within an average of 125 feet after the brakes are applied. They will adopt the new tread pattern unless there is strong evidence that the tires do not meet this objective. The distances (in feet) for 10 stops on a test track were 129, 128, 130, 132, 135, 123, 102, 125, 128, and 130. Should the company adopt the new tread pattern? Test an appropriate hypothesis and state your conclusion. Explain how you dealt with the outlier and why you made the recommendation you did.

47. Arrows A team of anthropologists headed by researcher Nicole Waguespack studied the difference between stone-tipped and wooden-tipped arrows. Stone arrow tips are tougher, but also take longer to make. Many cultures used both types of arrow tips, including the Apache in North America and the Tiwi in Australia. The researchers set up a compound bow with 60 lbs. of force. They shot arrows of both types into a hide-covered ballistics gel. (Waguespack, Nicole, et. al., *Antiquity*, 2009)

a) Here are the data for seven shots at the target with a wooden tip. They measured the penetration depth in mm. Find and interpret a 95% confidence interval for the penetration depth.

216 211 192 208 203 210 203

b) Here are the penetration depths (mm) for seven shots with a stone tip. Find and interpret a 95% confidence interval for the penetration depth.

240 208 213 225 232 214 240

48. Accuracy The researchers in the previous problem also measured the accuracy of the two types of tips. The bow was aimed at a target and the distance was measured from the center.

a) Here are the data from the six wooden-tipped shots. Find and interpret a 95% confidence interval for the measure of accuracy (measured in cm).

9.3 16.7 7.1 14 1 1.2

b) Here are the data from the six stone-tipped shots. Find and interpret a 95% interval for the measure of accuracy (measured in cm).

4.9 21.1 7 1.8 5.4 8.6

49. Sue me! Business professor Richard Posthuma examined the number of lawsuits filed in all 50 states from 1996 to 2003. He collected data on lawsuits filed in federal court regarding employment issues. Some states had a few hundred lawsuits, while other states had thousands. Here are the summary statistics. (*Business Horizons*, 2012, 55)

n	50
MEAN	5734.18
SD	6387.56
MED	3626
MIN	178
Q1	1303
Q3	7897
MAX	2631

a) Find and interpret a 90% confidence interval for the mean number of employment-related lawsuits that states might expect.
b) What are the shortcomings of this interval?

50. Sued again Dr. Posthuma (see Exercise 49) also tabulated the total amount of the lawsuits, in 1000's of dollars. Here are the statistics.

n	50
MEAN	67.1674
SD	110.237
MED	35.105
MIN	1.13
Q1	16.14
Q3	65.35
MAX	624.86

a) Find and interpret a 90% confidence interval for the expected average cost of lawsuits for states.
b) What are risks associated with using this confidence interval?

51. Driving distance 2011 How far do professional golfers drive a ball? (For non-golfers, the drive is the shot hit from a tee at the start of a hole and is typically the longest shot.) The next page shows a histogram of the average driving distances of the 186 leading professional golfers by end of November 2011 along with summary statistics (www.pgatour.com).

a) Find a 95% confidence interval for the mean drive distance.
b) Interpreting this interval raises some problems. Discuss.
c) The data are the mean driving distance for each golfer. Is that a concern in interpreting the interval? (*Hint:* Review the What Can Go Wrong warnings of Chapter 8. Chapter 8?! Yes, Chapter 8.)

T **52. Wind power** Should you generate electricity with your own personal wind turbine? That depends on whether you have enough wind on your site. To produce enough energy, your site should have an annual average wind speed above 8 miles per hour, according to the Wind Energy Association. One candidate site was monitored for a year, with wind speeds recorded every 6 hours. A total of 1114 readings of wind speed averaged 8.019 mph with a standard deviation of 3.813 mph. You've been asked to make a statistical report to help the landowner decide whether to place a wind turbine at this site.

a) Discuss the assumptions and conditions for using Student's *t* inference methods with these data. Here are some plots that may help you decide whether the methods can be used:

b) What would you tell the landowner about whether this site is suitable for a small wind turbine? Explain.

 Just Checking ANSWERS

1. Questions on the short form are answered by everyone in the population. This is a census, so means or proportions *are* the true population values. The long forms are given just to a sample of the population. When we estimate parameters from a sample, we use a confidence interval to take sample-to-sample variability into account.

2. They don't know the population standard deviation, so they must use the sample SD as an estimate. The additional uncertainty is taken into account by *t*-models.

3. The margin of error for a confidence interval for a mean depends, in part, on the standard error,

$$SE(\bar{y}) = \frac{s}{\sqrt{n}}.$$

Since *n* is in the denominator, smaller sample sizes lead to larger SEs and correspondingly wider intervals. Long forms returned by one in every six or seven households in a less populous area will be a smaller sample.

4. The critical values for *t* with fewer degrees of freedom would be slightly larger. The $\sqrt{n}$ part of the standard error changes a lot, making the SE much larger. Both would increase the margin of error.

5. The smaller sample is one fourth as large, so the confidence interval would be roughly twice as wide.

6. We expect 95% of such intervals to cover the true value, so 5 of the 100 intervals might be expected to miss.

7. The power would increase if we have a larger sample size.

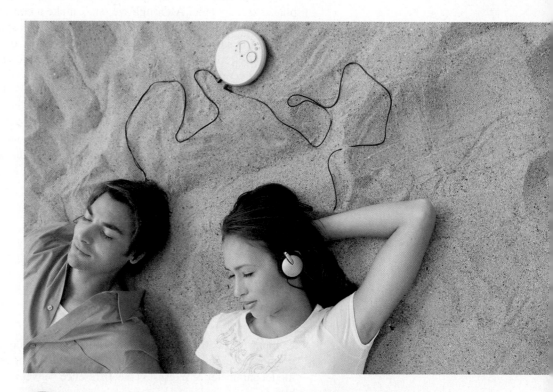

Who	AA alkaline batteries
What	Length of battery life while playing a CD continuously
Units	Minutes
Why	Class project
When	1998

A S *Video:* **Can Diet Prolong Life?**
Watch a video that tells the story of an experiment. We'll analyze the data later in this chapter.

---

chapter
23
Comparing Means

Should you buy generic rather than brand-name batteries? A Statistics student designed a study to test battery life. He wanted to know whether there was any real difference between brand-name batteries and a generic brand. To estimate the difference in mean lifetimes, he kept a battery-powered CD player[1] continuously playing the same CD[2], with the volume control fixed at 5, and measured the time until no more music was heard through the headphones. (He ran an initial trial to find out approximately how long that would take so that he didn't have to spend the first 3 hours of each run listening to the same CD.) For his trials he used six sets of AA alkaline batteries from two major battery manufacturers: a well-known brand name and a generic brand. He measured the time in minutes until the sound stopped. To account for changes in the CD player's performance over time, he randomized the run order by choosing sets of batteries at random. The table shows his data (times in minutes).

Brand Name	Generic
194.0	190.7
205.5	203.5
199.2	203.5
172.4	206.5
184.0	222.5
169.5	209.4

Studies that compare two groups are common throughout both science and industry. We might want to compare the effects of a new drug with the traditional therapy, the fuel efficiency of two car engine designs, or the sales of new products in two different test cities. In fact, battery manufacturers do research like this on their products and competitors' products themselves.

---

[1] Once upon a time, not so very long ago, there were no iPods. At the turn of the century, people actually carried CDs around—and devices to play them. We bet you can find one in your parents' closet.

[2] Even CDs are becoming outdated. We do hope you remember them.

Plot the Data

Figure 23.1

Boxplots comparing the brand-name and generic batteries suggest a difference in duration.

The natural display for comparing two groups is boxplots of the data for the two groups, placed side by side. Although we can't make a confidence interval or test a hypothesis from the boxplots themselves, you should always start with boxplots when comparing groups. Let's look at the boxplots of the battery test data.

It sure looks like the generic batteries lasted longer. And we can see that they were also more consistent. But is the difference large enough to change our battery-buying behavior? Can we be confident that the difference is more than just random fluctuation? That's why we need statistical inference.

The boxplot for the generic data identifies two possible outliers. That's interesting, but with only six measurements in each group, the outlier nomination rule is not very reliable. Both of the extreme values are plausible results, and the range of the generic values is smaller than the range of the brand-name values, even with the outliers. So we're probably better off just leaving these values in the data.

Comparing Two Means

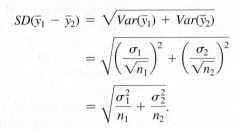

The Pythagorean Theorem of Statistics

Comparing two means is not very different from comparing two proportions. In fact, it's not different in concept from any of the methods we've seen. Now the population model parameter of interest is the difference between the *mean* battery lifetimes of the two brands, $\mu_1 - \mu_2$.

The rest is the same as before. The statistic of interest is the difference in the two observed means, $\bar{y}_1 - \bar{y}_2$. We'll start with this statistic to build our confidence interval, but we'll need to know its standard deviation and its sampling model. Then we can build confidence intervals and find P-values for hypothesis tests.

We know that, for independent random variables, the variance of their *difference* is the *sum* of their individual variances, $Var\,(Y - X) = Var\,(Y) + Var\,(X)$. To find the standard deviation of the difference between the two independent sample means, we add their variances and then take a square root:

$$SD(\bar{y}_1 - \bar{y}_2) = \sqrt{Var(\bar{y}_1) + Var(\bar{y}_2)}$$

$$= \sqrt{\left(\frac{\sigma_1}{\sqrt{n_1}}\right)^2 + \left(\frac{\sigma_2}{\sqrt{n_2}}\right)^2}$$

$$= \sqrt{\frac{\sigma_1^2}{n_1} + \frac{\sigma_2^2}{n_2}}.$$

Of course, we still don't know the true standard deviations of the two groups, σ_1 and σ_2, so as usual, we'll use the estimates, s_1 and s_2. Using the estimates gives us the *standard error*:

$$SE(\bar{y}_1 - \bar{y}_2) = \sqrt{\frac{s_1^2}{n_1} + \frac{s_2^2}{n_2}}.$$

We'll use the standard error to see how big the difference really is. Because we are working with means and estimating the standard error of their difference using the data, we shouldn't be surprised that the sampling model is a Student's *t*.

For Example FINDING THE STANDARD ERROR OF THE DIFFERENCE IN INDEPENDENT SAMPLE MEANS

Can you tell how much you are eating from how full you are? Or do you need visual cues? Researchers[3] constructed a table with two ordinary 18 oz soup bowls and two identical-looking bowls that had been modified to slowly, imperceptibly, refill as they were emptied. They assigned experiment participants to the bowls randomly and served them tomato soup. Those eating from the ordinary bowls had their bowls refilled by ladle whenever they were one-quarter full. If people judge their portions by internal cues, they should eat about the same amount. How big a difference was there in the amount of soup consumed? The table summarizes their results.

	Ordinary bowl	Refilling bowl
n	27	27
$\bar{y}$	8.5 oz	14.7 oz
s	6.1 oz	8.4 oz

QUESTION: How much variability do we expect in the difference between the two means? Find the standard error.

ANSWER: Participants were randomly assigned to bowls, so the two groups should be independent. It's okay to add variances.

$$SE(\bar{y}_{refill} - \bar{y}_{ordinary}) = \sqrt{\frac{s_r^2}{n_r} + \frac{s_o^2}{n_o}} = \sqrt{\frac{8.4^2}{27} + \frac{6.1^2}{27}} = 2.0 \ oz.$$

The confidence interval we build is called a **two-sample *t*-interval** (for the difference in means). The corresponding hypothesis test is called a **two-sample *t*-test.** The interval looks just like all the others we've seen—the statistic plus or minus an estimated margin of error:

$$(\bar{y}_1 - \bar{y}_2) \pm ME$$
$$\text{where } ME = t^* \times SE(\bar{y}_1 - \bar{y}_2).$$

This formula is almost the same as the one for the confidence interval for the difference of two proportions we saw in Chapter 21. It's just that here we use a Student's *t*-model instead of a Normal model to find the critical t^*-value corresponding to our chosen confidence level.

What are we missing? Only the degrees of freedom for the Student's *t*-model. Unfortunately, *that* formula is strange.

The deep, dark secret is that the sampling model isn't *really* Student's *t*, but only something close. The trick is that by using a special, adjusted degrees-of-freedom value, we can make it so close to a Student's *t*-model that nobody can tell the difference. The adjustment formula is straightforward but doesn't help our understanding much, so we leave it to the computer or calculator. (If you are curious and really want to see the formula, look in the footnote.[4])

> **z or t?**
>
> If you know σ, use z. (That's rare!) Whenever you use s to estimate σ, use t.

---

[3]Brian Wansink, James E. Painter, and Jill North, "Bottomless Bowls: Why Visual Cues of Portion Size May Influence Intake," *Obesity Research*, Vol. 13, No. 1, January 2005.

[4]
$$df = \frac{\left(\frac{s_1^2}{n_1} + \frac{s_2^2}{n_2}\right)^2}{\frac{1}{n_1 - 1}\left(\frac{s_1^2}{n_1}\right)^2 + \frac{1}{n_2 - 1}\left(\frac{s_2^2}{n_2}\right)^2}$$

Are you sorry you looked? This formula usually doesn't even give a whole number. If you are using a table, you'll need a whole number, so round down to be safe. If you are using technology, it's even easier. Computers and calculators deal with degrees of freedom automatically.

An Easier Rule?
The formula for the degrees of freedom of the sampling distribution of the difference in sample means is long, but the number of degrees of freedom is always at *least* the smaller of the two *n*'s, minus 1. Using that easier value is a poor choice because it can give fewer than *half* the degrees of freedom you're entitled to from the correct formula.

A Sampling Distribution for the Difference Between Two Sample Means
When the conditions are met, the sampling distribution of the standardized sample difference between the means of two independent groups,

$$t = \frac{(\bar{y}_1 - \bar{y}_2) - (\mu_1 - \mu_2)}{SE(\bar{y}_1 - \bar{y}_2)},$$

can be modeled by a Student's *t*-model with a number of degrees of freedom found with a special formula. We estimate the standard error with

$$SE(\bar{y}_1 - \bar{y}_2) = \sqrt{\frac{s_1^2}{n_1} + \frac{s_2^2}{n_2}}.$$

Assumptions and Conditions

Now we've got everything we need. Before we can make a two-sample *t*-interval or perform a two-sample *t*-test, though, we have to check the assumptions and conditions.[5]

Independent Groups Assumption

Independent Groups Assumption: To use the two-sample *t* methods, the two groups we are comparing must be independent of each other. In fact, this test is sometimes called the two *independent samples t*-test. No statistical test can verify this assumption. You have to think about how the data were collected. The assumption would be violated, for example, if one group consisted of husbands and the other group their wives. Whatever we measure on couples might naturally be related. Similarly, if we compared subjects' performances before some treatment with their performances afterward, we'd expect a relationship of each "before" measurement with its corresponding "after" measurement. In cases such as these, where the observational units in the two groups are related or matched, *the two-sample methods of this chapter can't be applied*. When this happens, we need a different procedure that we'll see in the next chapter.

Independence Assumption

Independence Assumption: The data in each group must be drawn independently and at random, or generated by a randomized comparative experiment. Without randomization of some sort, there are no sampling distribution models and no inference. We can check two conditions:

Randomization Condition: Were the data collected with suitable randomization? For surveys, are they a representative random sample? For experiments, was the experiment randomized?

10% Condition: We often don't check this condition for differences of means. We'll check it only if we have a very small population or an extremely large sample. We needn't worry about it at all for randomized experiments.

Normal Population Assumption

As we did before with Student's *t*-models, we should check the assumption that the underlying populations are *each* Normally distributed. We check the . . .

Nearly Normal Condition: We must check this for *both* groups; a violation by either one violates the condition. As we saw for single sample means, the Normality Assumption

---

[5]No surprise there, eh?

matters most when sample sizes are small. For samples of $n < 15$ in either group, you should not use these methods if the histogram or Normal probability plot shows severe skewness. For n's closer to 40, a mildly skewed histogram is OK, but you should remark on any outliers you find and not work with severely skewed data. When both groups are bigger than 40, the Central Limit Theorem starts to kick in no matter how the data are distributed, so the Nearly Normal Condition for the data matters less. Even in large samples, however, you should still be on the lookout for outliers, extreme skewness, and multiple modes.

For Example CHECKING ASSUMPTIONS AND CONDITIONS

RECAP: Researchers randomly assigned people to eat soup from one of two bowls: 27 got ordinary bowls that were refilled by ladle, and 27 others bowls that secretly refilled slowly as the people ate.

QUESTION: Can the researchers use their data to make inferences about the role of visual cues in determining how much people eat?

ANSWER:

✓ **Independent Groups Assumption:** Randomization to treatment groups guarantees this.

✓ **Independence Assumption:** The amount consumed by one person should be independent of the amount consumed by others.

✓ **Randomization Condition:** Subjects were randomly assigned to the treatments.

✓ **Nearly Normal Condition:** The histograms for both groups look unimodal but somewhat skewed to the right. I believe both groups are large enough (27) to allow use of t-methods.

It's okay to construct a two-sample t-interval for the difference in means.

Note: When you check the Nearly Normal Condition it's important that you include the graphs you looked at (histograms or Normal probability plots).

Two-Sample *t*-Interval for the Difference Between Means

When the conditions are met, we are ready to find the confidence interval for the difference between means of two independent groups, $\mu_1 - \mu_2$. The confidence interval is

$$(\bar{y}_1 - \bar{y}_2) \pm t^*_{df} \times SE(\bar{y}_1 - \bar{y}_2),$$

where the standard error of the difference of the means

$$SE(\bar{y}_1 - \bar{y}_2) = \sqrt{\frac{s_1^2}{n_1} + \frac{s_2^2}{n_2}}.$$

The critical value t^*_{df} depends on the particular confidence level, C, that you specify and on the number of degrees of freedom, which we get from the sample sizes and a special formula.

For Example FINDING A CONFIDENCE INTERVAL FOR THE DIFFERENCE IN SAMPLE MEANS

RECAP: Researchers studying the role of internal and visual cues in determining how much people eat conducted an experiment in which some people ate soup from bowls that secretly refilled. The results are summarized in the table.

    We've already checked the assumptions and conditions, and have found the standard error for the difference in means to be $SE(\bar{y}_{refill} - \bar{y}_{ordinary}) = 2.0$ oz.

	Ordinary bowl	Refilling bowl
n	27	27
$\bar{y}$	8.5 oz	14.7 oz
s	6.1 oz	8.4 oz

QUESTION: What does a 95% confidence interval say about the difference in mean amounts eaten?

ANSWER: The observed difference in means is $\bar{y}_{refill} - \bar{y}_{ordinary} = (14.7 - 8.5) = 6.2$ oz

$$\text{From technology: } df = 47.46 \quad t^*_{47.46} = 2.011$$

$$ME = t^* \times SE(\bar{y}_{refill} - \bar{y}_{ordinary}) = 2.011(2.0) = 4.02 \text{ oz}$$

The 95% confidence interval for $\mu_{refill} - \mu_{ordinary}$ is 6.2 ± 4.02, or $(2.18, 10.22)$ oz.

I am 95% confident that people eating from a subtly refilling bowl will eat an average of between 2.18 and 10.22 more ounces of soup than those eating from an ordinary bowl.

Step-by-Step Example A TWO-SAMPLE *t*-INTERVAL

Judging from the boxplot, the generic batteries seem to have lasted about 20 minutes longer than the brand-name batteries. Before we change our buying habits, what should we expect to happen with the next batteries we buy?

Question: How much longer might the generic batteries last?

THINK ➡ **Plan** State what we want to know.

Identify the *parameter* you wish to estimate. Here our parameter is the difference in the means, not the individual group means.

Identify the *population(s)* about which you wish to make statements. We hope to make decisions about purchasing batteries, so we're interested in all the AA batteries of these two brands.

Identify the variables and review the W's.

REALITY CHECK ➡ From the boxplots, it appears our confidence interval should be centered near a difference of 20 minutes.

I have measurements of the lifetimes (in minutes) of 6 sets of generic and 6 sets of brand-name AA batteries from a randomized experiment. I want to find an interval that is likely, with 95% confidence, to contain the true difference $\mu_G - \mu_B$ between the mean lifetime of the generic AA batteries and the mean lifetime of the brand-name batteries.

Model Think about the appropriate assumptions and check the conditions to be sure that a Student's *t*-model for the sampling distribution is appropriate.

✓ **Independent Groups Assumption:** Batteries manufactured by two different companies and purchased in separate packages should be independent.

✓ **Independence Assumption:** The batteries were packaged together, so they may not be independent. For example, a storage problem might affect all the batteries in the same pack. Repeating the study for several different packs of batteries would make the conclusions stronger.

For very small samples like these, we often don't worry about the 10% Condition.

✓ **Randomization Condition:** The batteries were selected at random from those available for sale. Not exactly an SRS, but a reasonably representative random sample.

✓ **Nearly Normal Condition:** The samples are small, but the histograms look unimodal and symmetric:

Make a picture. Boxplots are the display of choice for comparing groups, but now we want to check the *shape* of distribution of each group. Histograms or Normal probability plots do a better job there.

State the sampling distribution model for the statistic. The degrees of freedom come from that messy approximation formula.

Under these conditions, it's okay to use a Student's *t*-model.

Specify your method.

I'll use a **two-sample *t*-interval**.

SHOW ➡ **Mechanics** Construct the confidence interval.

I know $n_G = 6$ $\qquad n_B = 6$

Be sure to include the units along with the statistics. Use meaningful subscripts to identify the groups.

$\bar{y}_G = 206.0$ min $\quad \bar{y}_B = 187.4$ min

$s_G = 10.3$ min $\quad s_B = 14.6$ min

Use the sample standard deviations to find the standard error of the sampling distribution.

The groups are independent, so

$$SE(\bar{y}_G - \bar{y}_B) = \sqrt{SE^2(\bar{y}_G) + SE^2(\bar{y}_B)}$$

$$= \sqrt{\frac{s_G^2}{n_G} + \frac{s_B^2}{n_B}}$$

$$= \sqrt{\frac{10.3^2}{6} + \frac{14.6^2}{6}}$$

$$= \sqrt{\frac{106.09}{6} + \frac{213.16}{6}}$$

$$= \sqrt{53.208}$$

$$= 7.29 \text{ min.}$$

(continued)

The computer or calculator automatically uses the approximation formula for df. This gives a fractional degree of freedom (here df = 8.98).

Technology will use the fractional df to find the critical value for the confidence interval. Here it's $t^* = 2.263$, but your computer or calculator probably won't tell you that. When showing your work, it's okay to just leave t^* in the formulas and not try to find the actual value yourself.

df (from technology[6]) = 8.98

The corresponding critical value for a 95% confidence level is $t^* = 2.263$.

So the margin of error is

$$ME = t^* \times SE(\bar{y}_G - \bar{y}_B)$$
$$= 2.263(7.29)$$
$$= 16.50 \text{ min.}$$

The 95% confidence interval is

$$(206.0 - 187.4) \pm 16.5 \text{ min.}$$
$$\text{or } 18.6 \pm 16.5 \text{ min.}$$
$$= (2.1, 35.1) \text{ min.}$$

TELL ➡ **Conclusion** Interpret the confidence interval in the proper context.

Less formally, you could say, "I'm 95% confident that generic batteries last an average of 2.1 to 35.1 minutes longer than brand-name batteries."

I am 95% confident that the interval from 2.1 minutes to 35.1 minutes captures the mean amount of time by which generic batteries outlast brand-name batteries for this task. If generic batteries are cheaper, there seems little reason not to use them. If it is more trouble or costs more to buy them, then I'd consider whether the additional performance is worth it.

Another One Just Like the Other Ones?

A S *Activity*: **Find Two-Sample *t*-Intervals.** Who wants to deal with that ugly df formula? We usually find these intervals with a statistics package. Learn how here.

Yes. That's been our point all along. Once again we see a statistic plus or minus the margin of error. And the ME is just a critical value times the standard error. Just look out for that crazy degrees of freedom formula.

TI Tips CREATING THE CONFIDENCE INTERVAL

If you have been successful using your TI to make confidence intervals for proportions and 1-sample means, then you can probably already use the 2-sample function just fine. But humor us while we do one. Please?

FIND A CONFIDENCE INTERVAL FOR THE DIFFERENCE IN MEANS, GIVEN DATA FROM TWO INDEPENDENT SAMPLES

- Let's do the batteries. Always think about whether the samples are independent. If not, stop right here. These procedures are appropriate only for independent groups.

(continued)

---

[6]If you try to find the degrees of freedom with that messy approximation formula (We dare you! It's in the footnote on page 607) using the values above, you'll get 8.99. The minor discrepancy is because we rounded the standard deviations to the nearest 10th.

- Enter the data into two lists.

NameBrand in L1:	194.0	205.5	199.2	172.4	184.0	169.5
Generic in L2:	190.7	203.5	203.5	206.5	222.5	209.4

- Make histograms of the data to check the Nearly Normal Condition. We see that L1's histogram doesn't look so good. But remember—this is a very small data set. The bars represent only one or two values each. It's not unusual for the histogram to look a little ragged. Try resetting the WINDOW to plot an interval of 160 to 220 with Xscl=20, and Ymax=4. Redraw the GRAPH. Looks better.
- It's your turn to try this. Check L2. Go on, do it.
- Under STAT TESTS choose 2-SampTInt.
- Specify that you are using the Data in L1 and L2, specify 1 for both frequencies, and choose the confidence level you want.
- Pooled? We'll discuss this issue later in the chapter, but the easy advice is: Just Say No.
- To Calculate the interval, you need to scroll down one more line.

Now you have the 95% confidence interval. See df? The calculator did that messy degrees of freedom calculation for you. You have to love that!

Notice that the interval bounds are negative. That's because the TI is doing $\mu_1 - \mu_2$, and the generic batteries (L2) lasted longer. No harm done—you just need to be careful to interpret that result correctly when you *Tell* what the confidence interval means.

NO DATA? FIND A CONFIDENCE INTERVAL USING THE SAMPLE STATISTICS In many situations we don't have the original data, but must work with the summary statistics from the two groups. As we saw in the last chapter, you can still have your TI create the confidence interval with 2-SampTInt by choosing the Inpt:Stats option. Enter both means, standard deviations, and sample sizes, then Calculate. We show you the details in the next TI Tips.

 ## Just Checking

Carpal tunnel syndrome (CTS) causes pain and tingling in the hand, sometimes bad enough to keep sufferers awake at night and restrict their daily activities. Researchers studied the effectiveness of two alternative surgical treatments for CTS (Mackenzie, Hainer, and Wheatley, *Annals of Plastic Surgery*, 2000). Patients were randomly assigned to have endoscopic or open-incision surgery. Four weeks later the endoscopic surgery patients demonstrated a mean pinch strength of 9.1 kg compared to 7.6 kg for the open-incision patients.

1. Why is the randomization of the patients into the two treatments important?

2. A 95% confidence interval for the difference in mean strength is about (0.04 kg, 2.96 kg). Explain what this interval means.

3. Why might we want to examine such a confidence interval in deciding between these two surgical procedures?

4. Why might you want to see the data before trusting the confidence interval?

A Test for the Difference Between Two Means

If you bought a used camera in good condition from a friend, would you pay the same as you would if you bought the same item from a stranger? A researcher at Cornell University[7] wanted to know how friendship might affect simple sales such as this. She randomly divided subjects into two groups and gave each group descriptions of items they might want to buy. One group was told to imagine buying from a friend whom they expected to see again. The other group was told to imagine buying from a stranger.

Here are the prices they offered for a used camera in good condition:

Amount Offered ($)

Buy from Friend   Buy from Stranger

Who	University students
What	Prices offered for a used camera
Units	$
Why	Study of the effects of friendship on transactions
When	1990s
Where	U.C. Berkeley

Price Offered for a Used Camera ($)	
Buying from a Friend	**Buying from a Stranger**
275	260
300	250
260	175
300	130
255	200
275	225
290	240
300	

The researcher who designed this study had a specific concern. Previous theories had doubted that friendship had a measurable effect on pricing. She hoped to find an effect of friendship. This calls for a hypothesis test—in this case a **two-sample *t*-test for the difference between means.**[8]

You already know enough to construct this test. The test statistic looks just like the others we've seen. It finds the difference between the observed group means and compares this with a hypothesized value for that difference. We'll call that hypothesized difference Δ_0 ("delta naught"). It's so common for that hypothesized difference to be zero that we often just assume $\Delta_0 = 0$. We then compare the difference in the means with the standard error of that difference. We already know that for a difference between independent means, we can find P-values from a Student's *t*-model on that same special number of degrees of freedom.

A S *Activity*: **The Two-Sample *t*-Test.** How different are beef hot dogs and chicken hot dogs? Test whether measured differences are statistically significant.

Two-Sample *t*-Test for the Difference Between Means

The conditions for the two-sample *t*-test for the difference between the means of two independent groups are the same as for the two-sample *t*-interval. We test the hypothesis

$$H_0: \mu_1 - \mu_2 = \Delta_0$$

(continued)

---

[7]J. J. Halpern, "The Transaction Index: A Method for Standardizing Comparisons of Transaction Characteristics Across Different Contexts," *Group Decision and Negotiation*, 6: 557–572.

[8]Because it is performed so often, this test is usually just called a "two-sample *t*-test."

Δ_0—delta naught—isn't so standard that you can assume everyone will understand it. We use it because it's the Greek letter (good for a parameter) "D" for "difference." You should say "delta naught" rather than "delta zero"—that's standard for parameters associated with null hypotheses.

where the hypothesized difference is almost always 0, using the statistic

$$t = \frac{(\bar{y}_1 - \bar{y}_2) - \Delta_0}{SE(\bar{y}_1 - \bar{y}_2)}.$$

The standard error of $\bar{y}_1 - \bar{y}_2$ is

$$SE(\bar{y}_1 - \bar{y}_2) = \sqrt{\frac{s_1^2}{n_1} + \frac{s_2^2}{n_2}}.$$

When the conditions are met and the null hypothesis is true, this statistic can be closely modeled by a Student's t-model with a number of degrees of freedom given by a special formula. We use that model to obtain a P-value.

Step-by-Step Example A TWO-SAMPLE t-TEST FOR THE DIFFERENCE BETWEEN TWO MEANS

The usual null hypothesis is that there's no difference in means. That's just the right null hypothesis for the camera purchase prices.

Question: Is there a difference in the price people would offer a friend rather than a stranger?

THINK ➡ **Plan** State what we want to know.

Identify the *parameter* you wish to estimate. Here our parameter is the difference in the means, not the individual group means.

Identify the variables and check the W's.

Hypotheses State the null and alternative hypotheses. The research claim is that friendship changes what people are willing to pay.[9] The natural null hypothesis is that friendship makes no difference.

We didn't start with any knowledge of whether friendship might increase or decrease the price, so we choose a two-sided alternative.

Model Think about the assumptions and check the conditions. (Note that, because this is

I want to know whether people are likely to offer a different amount for a used camera when buying from a friend than when buying from a stranger. I wonder whether the difference between mean amounts is zero. I have bid prices from 8 subjects buying from a friend and 7 buying from a stranger, found in a randomized experiment.

H_O: The difference in mean price offered to friends and the mean price offered to strangers is zero:

$$\mu_F - \mu_S = 0.$$

H_A: The difference in mean prices is not zero:

$$\mu_F - \mu_S \neq 0.$$

✓ **Independent Groups Assumption:** Randomizing the experiment gives independent groups.

✓ **Independence Assumption:** This is an experiment, so there is no need for the subjects to be randomly selected from any particular population. All we need to check is whether they were assigned randomly to treatment groups.

(continued)

---
[9]This claim is a good example of what is called a "research hypothesis" in many social sciences. The only way to check it is to deny that it's true and see where the resulting null hypothesis leads us.

Note that because this is a randomized experiment, we haven't sampled at all, so the 10% Condition does not apply.

✓ **Randomization Condition:** The experiment was randomized. Subjects were assigned to treatment groups at random.

✓ **Nearly Normal Condition:** Histograms of the two sets of prices are roughly unimodal and symmetric:

Make a picture. Boxplots are the display of choice for comparing groups, as seen on page 614. We also want to check the shapes of the distribution. Histograms or Normal probability plots do a better job for that.

Buy from Friend

Buy from Stranger

State the sampling distribution model.

Specify your method.

The assumptions are reasonable and the conditions are okay, so I'll use a Student's t-model to perform a **two-sample t-test.**

SHOW ➡ **Mechanics** List the summary statistics. Be sure to use proper notation.

From the data:

$$n_F = 8 \qquad n_S = 7$$
$$\bar{y}_F = \$281.88 \qquad \bar{y}_S = \$211.43$$
$$s_F = \$18.31 \qquad s_S = \$46.43$$

Use the null model to find the P-value. First determine the standard error of the difference between sample means.

For independent groups,

$$SE(\bar{y}_F - \bar{y}_S) = \sqrt{SE^2(\bar{y}_F) + SE^2(\bar{y}_S)}$$
$$= \sqrt{\frac{s_F^2}{n_F} + \frac{s_S^2}{n_S}}$$
$$= \sqrt{\frac{18.31^2}{8} + \frac{46.43^2}{7}}$$
$$= 18.70$$

The observed difference is

$$(\bar{y}_F - \bar{y}_S) = 281.88 - 211.43 = \$70.45$$

Find the t-value.

$$t = \frac{(\bar{y}_F - \bar{y}_S) - (0)}{SE(\bar{y}_F - \bar{y}_S)} = \frac{70.45}{18.70} = 3.77$$

Make a picture. Sketch the t-model centered at the hypothesized difference of zero. Because this is a two-tailed test, shade the region to the right of the t value for the observed difference and the corresponding region in the other tail.

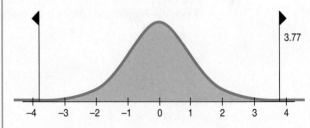

A statistics program or graphing calculator finds the P-value using the fractional degrees of freedom from the approximation formula.

$df = 7.62$ (from technology)

$$\text{P-value} = 2P(t_{7.62} > 3.77) = 0.006$$

(continued)

TELL ➡ **Conclusion** Link the P-value to your decision about the null hypothesis, and state the conclusion in context.

Be cautious about generalizing to items whose prices are outside the range of those in this study.

If there were no difference in the mean prices, a difference this large would occur only 6 times in 1000. That's too rare to believe, so I reject the null hypothesis and conclude that people are likely to offer a friend more than they'd offer a stranger for a used camera (and possibly for other, similar items).

TI Tips TESTING A HYPOTHESIS ABOUT A DIFFERENCE IN MEANS

Now let's use the TI to do a hypothesis test for the difference of two means—independent, of course! (Have we said that enough times yet?)

TEST A HYPOTHESIS WHEN YOU KNOW THE SAMPLE STATISTICS We'll demonstrate by using the statistics from the camera-pricing example. A sample of 8 people suggested they'd sell the camera to a friend for an average price of $281.88 with standard deviation $18.31. An independent sample of 7 other people would charge a stranger an average of $211.43 with standard deviation $46.43. Does this represent a significant difference in prices?

- From the STAT TESTS menu select 2-SampTTest.
- Specify Inpt:Stats, and enter the appropriate sample statistics.

- You have to scroll down to complete the specifications. This is a two-tailed test, so choose alternative $\neq \mu 2$.
- Pooled? Just say No. (We did promise to explain that and we will, coming up next.)
- Ready . . . set . . . Calculate!

The TI reports a calculated value of $t = 3.77$ and a P-value of 0.006. It's hard to tell who your real friends are.

BY NOW WE PROBABLY DON'T HAVE TO TELL YOU HOW TO DO A 2-SampTTEST STARTING WITH DATA IN LISTS
So we won't.

 ## Just Checking

Recall the experiment comparing patients 4 weeks after surgery for carpal tunnel syndrome. The patients who had endoscopic surgery demonstrated a mean pinch strength of 9.1 kg compared to 7.6 kg for the open-incision patients.

5. What hypotheses would you test?

6. The P-value of the test was less than 0.05. State a brief conclusion.

7. The study reports work on 36 "hands," but there were only 26 patients. In fact, 7 of the endoscopic surgery patients had both hands operated on, as did 3 of the open-incision group. Does this alter your thinking about any of the assumptions? Explain.

For Example A TWO-SAMPLE *t*-TEST

Many office "coffee stations" collect voluntary payments for the food consumed. Researchers at the University of Newcastle upon Tyne performed an experiment to see whether the image of eyes watching would change employee behavior.[10] They alternated pictures (seen here) of eyes looking at the viewer with pictures of flowers each week on the cupboard behind the "honesty box." They measured the consumption of milk to approximate the amount of food consumed and recorded the contributions (in £) each week per liter of milk. The table summarizes their results.

QUESTION: Do these results provide evidence that there really is a difference in honesty even when it's only photographs of eyes that are "watching"?

ANSWER: $H_O: \mu_{eyes} - \mu_{flowers} = 0$
$H_A: \mu_{eyes} - \mu_{flowers} \neq 0$

	Eyes	Flowers
n (# weeks)	5	5
$\bar{y}$	0.417 £/*l*	0.151 £/*l*
s	0.1811 £/*l*	0.067 £/*l*

✓ **Independent Groups Assumption:** The same workers were recorded each week, but week-to-week independence is plausible.

✓ **Independence Assumption:** The amount paid by one person should be independent of the amount paid by others.

✓ **Randomization Condition:** This study was observational. Treatments alternated a week at a time and were applied to the same group of office workers.

✓ **Nearly Normal Condition:** I don't have the data to check, but it seems unlikely there would be outliers in either group. I could be more certain if I could see histograms for both groups.

It's okay to do a two-sample *t*-test for the difference in means:

$$SE(\bar{y}_{eyes} - \bar{y}_{flowers}) = \sqrt{\frac{s^2_{eyes}}{n_{eyes}} + \frac{s^2_{flowers}}{n_{flowers}}} = \sqrt{\frac{0.1811^2}{5} + \frac{0.067^2}{5}} = 0.0864$$

$$df = 5.07$$

$$t_5 = \frac{(\bar{y}_{eyes} - \bar{y}_{flowers}) - 0}{SE(\bar{y}_{eyes} - \bar{y}_{flowers})} = \frac{0.417 - 0.151}{0.0864} = 3.08$$

$$P(|t_5| > 3.08) = 0.027$$

Assuming the data were free of outliers, the very low P-value leads me to reject the null hypothesis. This study provides evidence that people will leave higher average voluntary payments for food if pictures of eyes are "watching."

(Note: In Table T we can see that at 5 df, *t* = 3.08 lies between the critical values for *P* = 0.02 and *P* = 0.05, so we could report *P* < 0.05.)

---

[10]Melissa Bateson, Daniel Nettle, and Gilbert Roberts, "Cues of Being Watched Enhance Cooperation in a Real-World Setting," *Biol. Lett. doi*:10.1098/rsbl.2006.0509.

Back into the Pool?

Remember that when we know a proportion, we know its standard deviation. When we tested the null hypothesis that two proportions were equal, that link meant we could assume their variances were equal as well. This led us to pool our data to estimate a standard error for the hypothesis test.

For means, there is also a pooled *t*-test. Like the two-proportions *z*-test, this test assumes that the variances in the two groups are equal. But be careful: Knowing the mean of some data doesn't tell you anything about their variance. And knowing that two means are equal doesn't say anything about whether their variances are equal. If we were willing to *assume* that their variances are equal, we could pool the data from two groups to estimate the common variance. We'd estimate this pooled variance from the data, so we'd still use a Student's *t*-model. This test is called a **pooled *t*-test (for the difference between means)**.

Pooled *t*-tests have a couple of advantages. They often have a few more degrees of freedom than the corresponding two-sample test and a much simpler degrees of freedom formula. But these advantages come at a price: You have to pool the variances and think about another assumption. The assumption of equal variances is a strong one, is often not true, and is difficult to check. For these reasons, we recommend that you use a two-sample *t*-test instead. It's never wrong *not* to pool.

*The Pooled *t*-Test (In Case You're Curious)

Termites cause billions of dollars of damage each year, to homes and other buildings, but some tropical trees seem to be able to resist termite attack. A researcher extracted a compound from the sap of one such tree and tested it by feeding it at two different concentrations to randomly assigned groups of 25 termites.[11] After 5 days, 8 groups fed the lower dose had an average of 20.875 termites alive, with a standard deviation of 2.23. But 6 groups fed the higher dose had an average of only 6.667 termites alive, with a standard deviation of 3.14. Is this a large enough difference to declare the sap compound effective in killing termites? In order to use the pooled *t*-test, we must make the **Equal Variance Assumption** that the variances of the two populations from which the samples have been drawn are equal. That is, $\sigma_1^2 = \sigma_2^2$. (Of course, we could think about the standard deviations being equal instead.) The corresponding **Similar Spreads Condition** really just consists of looking at the boxplots to check that the spreads are not wildly different. We were going to make boxplots anyway, so there's really nothing new here.

Once we decide to pool, we estimate the common variance by combining numbers we already have:

$$s_{pooled}^2 = \frac{(8-1)2.23^2 + (6-1)3.14^2}{(8-1) + (6-1)} = 7.01$$

$$s_{pooled}^2 = \frac{(n_1 - 1)s_1^2 + (n_2 - 1)s_2^2}{(n_1 - 1) + (n_2 - 1)}.$$

(If the two sample sizes are equal, this is just the average of the two variances.)

Now we just substitute this pooled variance in place of each of the variances in the standard error formula.

$$SE_{pooled}(\bar{y}_1 - \bar{y}_2) = \sqrt{\frac{7.01}{8} + \frac{7.01}{6}} = 1.43$$

$$SE_{pooled}(\bar{y}_1 - \bar{y}_2) = \sqrt{\frac{s_{pooled}^2}{n_1} + \frac{s_{pooled}^2}{n_2}}.$$

[11]Adam Messer, Kevin McCormick, Sunjaya, H. H. Hagedorm, Ferny Tumbel, and J. Meinwald, "Defensive role of tropical tree resins: antitermitic sesquiterpenes from Southeast Asian Dipterocarpaceae," *J Chem Ecology*, 16:122, pp. 3333–3352.

$df = 8 + 6 - 2 = 12$

$t = \dfrac{20.875 - 6.667}{1.43} = 9.935$

The formula for degrees of freedom for the Student's t-model is simpler, too. It was so complicated for the two-sample t that we stuck it in a footnote.[12] Now it's just $df = n_1 + n_2 - 2$.

Substitute the pooled-t estimate of the standard error and its degrees of freedom into the steps of the confidence interval or hypothesis test, and you'll be using the pooled-t method. For the termites, $\bar{y}_1 - \bar{y}_2 = 14.208$, giving a t-value $= 9.935$ with 12 df and a P-value ≤ 0.0001.

Of course, if you decide to use a pooled-t method, you must defend your assumption that the variances of the two groups are equal.

---

*Pooled t-Test and Confidence Interval for Means The conditions for the pooled t-test for the difference between the means of two independent groups (commonly called a "pooled t-test") are the same as for the two-sample t-test with the additional assumption that the variances of the two groups are the same. We test the hypothesis

$$H_0: \mu_1 - \mu_2 = \Delta_0$$

where the hypothesized difference, Δ_0, is almost always 0, using the statistic

$$t = \frac{(\bar{y}_1 - \bar{y}_2) - \Delta_0}{SE_{pooled}(\bar{y}_1 - \bar{y}_2)}.$$

The standard error of $\bar{y}_1 - \bar{y}_2$ is

$$SE_{pooled}(\bar{y}_1 - \bar{y}_2) = \sqrt{\frac{s_{pooled}^2}{n_1} + \frac{s_{pooled}^2}{n_2}}$$

where the pooled variance is

$$s_{pooled}^2 = \frac{(n_1 - 1)s_1^2 + (n_2 - 1)s_2^2}{(n_1 - 1) + (n_2 - 1)}.$$

When the conditions are met, we can model this statistic's sampling distribution with a Student's t-model with $(n_1 - 1) + (n_2 - 1)$ degrees of freedom. We use that model to obtain a P-value for a test or a margin of error for a confidence interval.

The corresponding confidence interval is

$$(\bar{y}_1 - \bar{y}_2) \pm t_{df}^* \times SE_{pooled}(\bar{y}_1 - \bar{y}_2).$$

A S *Activity*: **The Pooled t-Test.** It's those hot dogs again. The same interactive tool can handle a pooled t-test, too. Take it for a spin here.

Is the Pool All Wet?

We're testing whether the means are equal, so we admit that we don't *know* whether they are equal. Doesn't it seem a bit much to just *assume* that the variances are equal? Well, yes—but there are some special cases to consider. So when *should* you use pooled-t methods rather than two-sample t methods?

Never.

What, never?

Well, hardly ever.

You see, when the variances of the two groups are in fact equal, the two methods give pretty much the same result. (For the termites, the two-sample t statistic is barely different—9.436 with 8 df—and the P-value is still < 0.001.) Pooled methods have a small advantage (slightly narrower confidence intervals, slightly more powerful tests) mostly because they usually have a few more degrees of freedom, but the advantage is slight.

---

[12]But not this one. See page 607.

> **When Should I use the Pooled t-Test?**
>
> Because the advantages of pooling are small, and you are allowed to pool only rarely (when the Equal Variances Assumption is met), *don't.*
>
> It's never wrong *not* to pool.

When the variances are *not* equal, the pooled methods are just not valid and can give poor results. You have to use the two-sample methods instead.

Pooling may make sense in a randomized comparative experiment. We start by assigning our experimental units to treatments at random. We know that at the start of the experiment each treatment group is a random sample from the same population,[13] so each treatment group begins with the same population variance. When we test whether the true means are equal, we may be willing to go a bit farther and say that the treatments made no difference *at all*. Then it's not much of a stretch to assume that the variances have remained equal. It's still an assumption, and there are conditions that need to be checked (make the boxplots, make the boxplots, make the boxplots), but at least it's a plausible assumption.

This line of reasoning is important. The methods used to analyze comparative experiments *do* pool variances in exactly this way and defend the pooling with a version of this argument. The chapter on Analysis of Variance on the DVD introduces these methods.

---

[13]That is, the population of experimental subjects. Remember that to be valid, experiments do not need a representative sample drawn from a population because we are not trying to estimate a population model parameter.

WHAT IF ●●● we simulate differences in means?

In Chapter 21's What If we used a permutation test to see if the difference of two proportions was statistically significant. Now let's try that on means. Here once again are the data showing how many minutes brand name and generic batteries lasted in an antique CD player.

Brand name	194.0	205.5	199.2	172.4	184.0	169.5
Generic	190.7	203.5	203.5	206.5	222.5	209.4

These 6 generic batteries lasted an average of 18.6 minutes longer than the 6 brand name batteries. Is that a big enough difference to indicate that generic batteries really are better, or could it just be typical random variation we might reasonably expect to see in very small samples like this? Yes, indeed, it's simulation time!

We'll do a permutation test. Here's a good way to think about how it works. Imagine that you and an opponent are playing a card game. There are 12 cards; each has one of the battery lifetimes written on it. Your opponent shuffles the cards and deals 6 to you and 6 to himself. The prize will be determined by how much higher the winner's 6-card average is than the loser's. Of course, if he were to deal you the 6 lowest cards while he got the 6 highest, you'd be pretty suspicious that the game wasn't all that fair. But a clever cheater wouldn't be that brazen. Oh, he'd be sure he won, and by quite a bit, of course; just not by enough to attract attention.

So it's game on! He shuffles, he deals, you both calculate your averages, and. . . you lose. You lose by 18.6 points. Seems like a lot. Do you think he cheated? How likely is it that a difference this large would happen just by chance?

Our simulation plays that very card game. It plays fair. It randomly divides the cards into two hands, computes the average for each hand, and sees how big the difference is. And then it does it again. And again. 200 trials in all. Here's the histogram of the differences.

We see that if the two hands are randomly (fairly) created, usually the difference in means would be pretty small. In fact, only 6 times in 200 "games" would it be at least as large as 18.6 in either person's favor. That makes your 18.6-point loss seem pretty unusual.

(continued)

Back to the batteries. The generic batteries were dealt the winning hand. This permutation test shows that if both types of batteries really perform equally well on average, a difference in means this large would happen by chance only 6 times in 200. Think: What *is* that proportion?

We hope you recognize we've tested the hypothesis that the mean lifespans of the two types of batteries are the same, and that our simulation estimated a P-value of $6/200 = 0.03$. Now grab your calculator and run this chapter's 2-sample *t*-test for these data. Check that P-value. Go ahead; it'll take about one minute. We think you'll be impressed with the permutation test.

WHAT CAN GO WRONG?

- **Watch out for paired data.** The Independent Groups Assumption deserves special attention. If the samples are not independent, you can't use these two-sample methods. This is probably the main thing that can go wrong when using these two-sample methods. The methods of this chapter can be used *only* if the observations in the two groups are *independent*. Matched-pairs designs in which the observations are deliberately related arise often and are important. The next chapter deals with them.

- **Look at the plots.** The usual (by now) cautions about checking for outliers and non-Normal distributions apply, of course. The simple defense is to make and examine boxplots. You may be surprised how often this simple step saves you from the wrong or even absurd conclusions that can be generated by a single undetected outlier. You don't want to conclude that two methods have very different means just because one observation is atypical.

- **Be cautious if you apply inference methods where there was no randomization.** If the data do not come from representative random samples or from a properly randomized experiment, then the inference about the differences between the groups may be wrong.

- **Don't interpret a significant difference in proportions or means causally.** Studies find that people with higher incomes are more likely to snore. Would surgery to increase your snoring be a wise investment? Probably not. It turns out that older people are more likely to snore, and they are also likely to earn more. In a prospective or retrospective study, there is always the danger that other lurking variables not accounted for are the real reason for an observed difference. Be careful not to jump to conclusions about causality.

Do What We Say, Not What We Do . . . Precision machines used in industry often have a bewildering number of parameters that have to be set, so experiments are performed in an attempt to try to find the best settings. Such was the case for a hole-punching machine used by a well-known computer manufacturer to make printed circuit boards. The data were analyzed by one of the authors, but because he was in a hurry, he didn't look at the boxplots first and just performed *t*-tests on the experimental factors. When he found extremely small P-values even for factors that made no sense, he plotted the data. Sure enough, there was one observation 1,000,000 times bigger than the others. It turns out that it had been recorded in microns (millionths of an inch), while all the rest were in inches.

What Have We Learned?

Are the means of two groups the same? If not, how different are they? We've learned to use statistical inference to compare the means of two independent groups.

- We've seen that confidence intervals and hypothesis tests about the difference between two means, like those for an individual mean, use *t*-models.
- Once again we've seen the importance of checking assumptions that tell us whether our method will work.
- We've seen that, as when comparing proportions, finding the standard error for the difference in sample means depends on believing that our data come from independent groups. Unlike proportions, however, pooling is usually not the best choice here.
- And we've seen once again that we can add variances of independent random variables to find the standard deviation of the difference in two independent means.
- Finally, we've learned that the reasoning of statistical inference remains the same; only the mechanics change.

Terms

Two-sample *t* methods

Two-sample *t* methods allow us to draw conclusions about the difference between the means of two independent groups. The two-sample methods make relatively few assumptions about the underlying populations, so they are usually the method of choice for comparing two sample means. However, the Student's *t*-models are only approximations for their true sampling distribution. To make that approximation work well, the two-sample *t* methods have a special rule for estimating degrees of freedom. (p. 608)

Two-sample *t*-interval for the difference between means

A confidence interval for the difference between the means of two independent groups found as

$$(\bar{y}_1 - \bar{y}_2) \pm t^*_{df} \times SE(\bar{y}_1 - \bar{y}_2)$$

where

$$SE(\bar{y}_1 - \bar{y}_2) = \sqrt{\frac{s_1^2}{n_1} + \frac{s_2^2}{n_2}}$$

and the number of degrees of freedom is given by a special formula (see footnote 4 on page 607).

Two-sample *t*-test for the difference between means

A hypothesis test for the difference between the means of two independent groups. It tests the null hypothesis

$$H_0: \mu_1 - \mu_2 = \Delta_0,$$

where the hypothesized difference, Δ_0, is almost always 0, using the statistic

$$t_{df} = \frac{(\bar{y}_1 - \bar{y}_2) - \Delta_0}{SE(\bar{y}_1 - \bar{y}_2)},$$

with the number of degrees of freedom given by the special formula. (p. 614)

*Pooled-*t* methods

Pooled-*t* methods provide inferences about the difference between the means of two independent populations under the assumption that both populations have the same standard deviation. When the assumption is justified, pooled-*t* methods generally produce slightly narrower confidence intervals and more powerful significance tests than two-sample *t* methods. When the assumption is not justified, they generally produce worse results—sometimes substantially worse.

We recommend that you use unpooled two-sample *t* methods instead. (p. 619)

On the Computer INFERENCE FOR THE DIFFERENCE OF MEANS

Here's some typical computer software output for confidence intervals or hypothesis tests for the difference in means based on two independent groups.

Specify the confidence level.

Identify the two groups.

Remember—when software or your calculator asks if you want to pool variances, Just Say No.

95% confidence interval results:
μ_1 : mean of Friend$
μ_2 : mean of Stranger$
$\mu_1 - \mu_2$: mean difference
(without pooled variances)

Difference	Sample Mean	Std. Err.	DF	L. Limit	U. Limit
$\mu_1 - \mu_2$	70.44643	18.70566	7.622948	26.936884	113.95597

The observed difference in means

Many programs give far too many decimal places. Just ignore the extra digits.

The software calculates the degrees of freedom.

The confidence interval

Identify the two groups.

The hypotheses. This is a 2-tail test

Hypothesis test results:
μ_1 : mean of Generic
μ_2 : mean of Brand Name
$\mu_1 - \mu_2$: mean difference
$H_0 : \mu_1 - \mu_2 = 0$
$H_A : \mu_1 - \mu_2 \neq 0$
(without pooled variances)

Just say No.

Difference	Sample Mean	Std. Err.	DF	T-Stat	P-value
$\mu_1 - \mu_2$	18.583334	7.298451	8.9862795	2.5462024	0.0314

The observed difference in sample means

Degrees of freedom, done automatically.

The t-score and P-value of the test.

Exercises

1. **Dogs and calories** In July 2007, *Consumer Reports* examined the calorie content of two kinds of hot dogs: meat (usually a mixture of pork, turkey, and chicken) and all beef. The researchers purchased samples of several different brands. The meat hot dogs averaged 111.7 calories, compared to 135.4 for the beef hot dogs. A test of the null hypothesis that there's no difference in mean calorie content yields a P-value of 0.124. Would a 95% confidence interval for $\mu_{Meat} - \mu_{Beef}$ include 0? Explain.

2. **Dogs and sodium** The *Consumer Reports* article described in Exercise 1 also listed the sodium content (in mg) for the various hot dogs tested. A test of the null hypothesis that beef hot dogs and meat hot dogs don't differ in the mean amounts of sodium yields a P-value of 0.11. Would a 95% confidence interval for $\mu_{Meat} - \mu_{Beef}$ include 0? Explain.

3. **Dogs and fat** The *Consumer Reports* article described in Exercise 1 also listed the fat content (in grams) for samples of beef and meat hot dogs. The resulting 90% confidence interval for $\mu_{Meat} - \mu_{Beef}$ is $(-6.5, -1.4)$.
   a) The endpoints of this confidence interval are negative numbers. What does that indicate?
   b) What does the fact that the confidence interval does not contain 0 indicate?
   c) If we use this confidence interval to test the hypothesis that $\mu_{Meat} - \mu_{Beef} = 0$, what's the corresponding alpha level?

4. **Washers** In June 2007, *Consumer Reports* examined top-loading and front-loading washing machines, testing samples of several different brands of each type. One of the variables the article reported was "cycle time", the number of minutes it took each machine to wash a load of clothes. Among the machines rated good to excellent, the 98% confidence interval for the difference in mean cycle time ($\mu_{Top} - \mu_{Front}$) is $(-40, -22)$.
   a) The endpoints of this confidence interval are negative numbers. What does that indicate?
   b) What does the fact that the confidence interval does not contain 0 indicate?
   c) If we use this confidence interval to test the hypothesis that $\mu_{Top} - \mu_{Front} = 0$, what's the corresponding alpha level?

5. **Dogs and fat, second helping** In Exercise 3, we saw a 90% confidence interval of $(-6.5, -1.4)$ grams for $\mu_{Meat} - \mu_{Beef}$, the difference in mean fat content for meat vs. all-beef hot dogs. Explain why you think each of the following statements is true or false:
   a) If I eat a meat hot dog instead of a beef dog, there's a 90% chance I'll consume less fat.

   b) 90% of meat hot dogs have between 1.4 and 6.5 grams less fat than a beef hot dog.
   c) I'm 90% confident that meat hot dogs average 1.4–6.5 grams less fat than the beef hot dogs.
   d) If I were to get more samples of both kinds of hot dogs, 90% of the time the meat hot dogs would average 1.4–6.5 grams less fat than the beef hot dogs.
   e) If I tested many samples, I'd expect about 90% of the resulting confidence intervals to include the true difference in mean fat content between the two kinds of hot dogs.

6. **Second load of wash** In Exercise 4, we saw a 98% confidence interval of $(-40, -22)$ minutes for $\mu_{Top} - \mu_{Front}$, the difference in time it takes top-loading and front-loading washers to do a load of clothes. Explain why you think each of the following statements is true or false:
   a) 98% of top loaders are 22 to 40 minutes faster than front loaders.
   b) If I choose the laundromat's top loader, there's a 98% chance that my clothes will be done faster than if I had chosen the front loader.
   c) If I tried more samples of both kinds of washing machines, in about 98% of these samples I'd expect the top loaders to be an average of 22 to 40 minutes faster.
   d) If I tried more samples, I'd expect about 98% of the resulting confidence intervals to include the true difference in mean cycle time for the two types of washing machines.
   e) I'm 98% confident that top loaders wash clothes an average of 22 to 40 minutes faster than front-loaders.

7. **Learning math** The Core Plus Mathematics Project (CPMP) is an innovative approach to teaching Mathematics that engages students in group investigations and mathematical modeling. After field tests in 36 high schools over a three-year period, researchers compared the performances of CPMP students with those taught using a traditional curriculum. In one test, students had to solve applied Algebra problems using calculators. Scores for 320 CPMP students were compared to those of a control group of 273 students in a traditional Math program. Computer software was used to create a confidence interval for the difference in mean scores. (*Journal for Research in Mathematics Education,* 31, no. 3[2000])

Conf level: 95%	Variable: Mu(CPMP) – Mu(Ctrl)
Interval: (5.573, 11.427)	

   a) What's the margin of error for this confidence interval?
   b) If we had created a 98% CI, would the margin of error be larger or smaller?

c) Explain what the calculated interval means in context.

d) Does this result suggest that students who learn Mathematics with CPMP will have significantly higher mean scores in Algebra than those in traditional programs? Explain.

8. Stereograms Stereograms appear to be composed entirely of random dots. However, they contain separate images that a viewer can "fuse" into a three-dimensional (3D) image by staring at the dots while defocusing the eyes. An experiment was performed to determine whether knowledge of the form of the embedded image affected the time required for subjects to fuse the images. One group of subjects (group NV) received no information or just verbal information about the shape of the embedded object. A second group (group VV) received both verbal information and visual information (specifically, a drawing of the object). The experimenters measured how many seconds it took for the subject to report that he or she saw the 3D image.

> 2-Sample t-Interval for $\mu 1 - \mu 2$
> Conf level = 90%   df = 70
> μ(NV) $-$ μ(VV) interval: (0.55, 5.47)

a) Interpret your interval in context.

b) Does it appear that viewing a picture of the image helps people "see" the 3D image in a stereogram?

c) What's the margin of error for this interval?

d) Explain what the 90% confidence level means.

e) Would you expect a 99% confidence level to be wider or narrower? Explain.

f) Might that change your conclusion in part b? Explain.

9. CPMP, again During the study described in Exercise 7, students in both CPMP and traditional classes took another Algebra test that did not allow them to use calculators. The table below shows the results. Are the mean scores of the two groups significantly different?

Math Program	n	Mean	SD
CPMP	312	29.0	18.8
Traditional	265	38.4	16.2

Performance on Algebraic Symbolic Manipulation Without Use of Calculators

a) Write an appropriate hypothesis.

b) Do you think the assumptions for inference are satisfied? Explain.

c) Here is computer output for this hypothesis test. Explain what the P-value means in this context.

> 2-Sample t-Test of $\mu 1 - \mu 2 \neq 0$
> t-Statistic = -6.451 w/574.8761 df
> P < 0.0001

d) State a conclusion about the CPMP program.

10. CPMP and word problems The study of the new CPMP Mathematics methodology described in Exercise 7 also tested students' abilities to solve word problems. This table shows how the CPMP and traditional groups performed. What do you conclude?

Math Program	n	Mean	SD
CPMP	320	57.4	32.1
Traditional	273	53.9	28.5

11. Cost of shopping Do consumers spend more on a trip to Walmart or Target? Suppose researchers interested in this question collected a systematic sample from 85 Walmart customers and 80 Target customers by asking customers for their purchase amount as they left the stores. The data collected is summarized in the table below.

	Walmart	Target
n	85	80
$\bar{y}$	$45	$53
s	$21	$19

To perform inference on these two samples, what conditions must be met? Are they? Explain.

12. Athlete ages A sports reporter suggests that professional baseball players must, on average, be older than professional football players, since football is a contact sport and players are more susceptible to concussions and serious injuries (www.sports.yahoo.com). One player was selected at random from each team in both professional baseball (MLB) and professional football (NFL). The data are summarized below.

	MLB	NFL
n	30	32
$\bar{y}$	27.5	26.16
s	3.94	2.78

To perform inference on these two samples, what conditions must be met? Are they? Explain.

13. Cost of shopping, again Using the summary statistics provided in Exercise 11, researchers calculated a 95% confidence interval for the mean difference between Walmart and Target purchase amounts. The interval was $(-\$14.15, -\$1.85)$. Explain in context what this interval means.

14. Athlete ages, again Using the summary statistics provided in Exercise 12, the sports reporter calculated the following 95% confidence interval for the mean difference between major league baseball players and professional football players. The 95% interval for $\mu_{MLB} - \mu_{NFL}$ was $(-0.41, 3.09)$. Summarize in context what the interval means.

15. Commuting A man who moves to a new city sees that there are two routes he could take to work. A neighbor who has lived there a long time tells him Route A will average 5 minutes faster than Route B. The man decides to experiment. Each day he flips a coin to determine which way to go, driving each route 20 days. He finds that Route A takes an average of 40 minutes, with standard deviation 3 minutes, and Route B takes an average of 43 minutes, with standard deviation 2 minutes. Histograms of travel times for the routes are roughly symmetric and show no outliers.

   a) Find a 95% confidence interval for the difference in average commuting time for the two routes.
   b) Should the man believe the old-timer's claim that he can save an average of 5 minutes a day by always driving Route A? Explain.

16. Pulse rates A researcher wanted to see whether there is a significant difference in resting pulse rates for men and women. The data she collected are displayed in the boxplots and summarized below.

	Sex	
	Male	**Female**
Count	28	24
Mean	72.75	72.625
Median	73	73
StdDev	5.37225	7.69987
Range	20	29
IQR	9	12.5

   a) What do the boxplots suggest about differences between male and female pulse rates?
   b) Is it appropriate to analyze these data using the methods of inference discussed in this chapter? Explain.
   c) Create a 90% confidence interval for the difference in mean pulse rates.
   d) Does the confidence interval confirm your answer to part a? Explain.

17. Cereal Here are the data for the sugar content (as a percentage of weight) of several national brands of children's and adults' cereals. Create and interpret a 95% confidence interval for the difference in mean sugar content. Be sure to check the necessary assumptions and conditions.

Children's cereals: 40.3, 55, 45.7, 43.3, 50.3, 45.9, 53.5, 43, 44.2, 44, 47.4, 44, 33.6, 55.1, 48.8, 50.4, 37.8, 60.3, 46.6

Adults' cereals: 20, 30.2, 2.2, 7.5, 4.4, 22.2, 16.6, 14.5, 21.4, 3.3, 6.6, 7.8, 10.6, 16.2, 14.5, 4.1, 15.8, 4.1, 2.4, 3.5, 8.5, 10, 1, 4.4, 1.3, 8.1, 4.7, 18.4

T 18. Egyptians Some archaeologists theorize that ancient Egyptians interbred with several different immigrant populations over thousands of years. To see if there is any indication of changes in body structure that might have resulted, they measured 30 skulls of male Egyptians dated from 4000 B.C.E and 30 others dated from 200 B.C.E. (A. Thomson and R. Randall-Maciver, *Ancient Races of the Thebaid,* Oxford: Oxford University Press, 1905)

   a) Are these data appropriate for inference? Explain.
   b) Create a 95% confidence interval for the difference in mean skull breadth between these two eras.
   c) Do these data provide evidence that the mean breadth of males' skulls changed over this period? Explain.

Maximum Skull Breadth (mm)			
4000 B.C.E.		**200** B.C.E.	
131	131	141	131
125	135	141	129
131	132	135	136
119	139	133	131
136	132	131	139
138	126	140	144
139	135	139	141
125	134	140	130
131	128	138	133
134	130	132	138
129	138	134	131
134	128	135	136
126	127	133	132
132	131	136	135
141	124	134	141

T 19. Reading An educator believes that new reading activities for elementary school children will improve reading comprehension scores. She randomly assigns third graders to an eight-week program in which some will use these activities and others will experience traditional teaching methods. At the end of the experiment, both groups take a reading comprehension exam. Their scores are shown in the back-to-back stem-and-leaf display. Do these results suggest that the new activities are better? Test an appropriate hypothesis and state your conclusion.

New Activities		Control
	1	07
4	2	068
3	3	377
96333	4	12222238
9876432	5	355
721	6	02
1	7	
	8	5

T 20. Streams Researchers collected samples of water from streams in the Adirondack Mountains to look for any differences in the effects of acid rain. They measured the pH (acidity) of the water and classified the streams with respect to the kind of substrate (type of rock over which they flow). A lower pH means the water is more acidic. Here is a plot of the pH of the streams by substrate (limestone, mixed, or shale):

Here are selected parts of a software analysis comparing the pH of streams with limestone and shale substrates:

2-Sample t-Test of $\mu 1 - \mu 2$
Difference Between Means = 0.735
t-Statistic = 16.30 w/133 df
$p \leq 0.0001$

a) State the null and alternative hypotheses for this test.
b) From the information you have, do the assumptions and conditions appear to be met?
c) What conclusion would you draw?

21. Baseball American League baseball teams play their games with the designated hitter rule, meaning that pitchers do not bat. The league believes that replacing the pitcher, traditionally a weak hitter, with another player in the batting order produces more runs and generates more interest among fans. Below are the average numbers of home runs hit per game in American League and National League stadiums for the 2011 season.

American	National	American	National
1.500	1.354	0.913	0.948
1.267	1.314	0.903	0.941
1.230	1.160	0.880	0.919
1.186	1.110	0.789	0.862
1.144	1.095	0.786	0.799
1.060	1.062	0.708	0.774
1.037	0.987		0.735
0.987	0.950		0.596

a) Create an appropriate display of these data. What do you see?
b) With a 95% confidence interval, estimate the mean number of home runs hit in American League games.

c) Coors Field, in Denver, stands a mile above sea level, an altitude far greater than that of any other major league ball park. Some believe that the thinner air makes it harder for pitchers to throw curve balls and easier for batters to hit the ball a long way. Do you think the 1.354 home runs hit per game at Coors is unusual? Explain.

22. Handy A factory hiring people to work on an assembly line gives job applicants a test of manual agility. This test counts how many strangely shaped pegs the applicant can fit into matching holes in a one-minute period. The table below summarizes the data by sex of the job applicant. Assume that all conditions necessary for inference are met.

	Male	Female
Number of subjects	50	50
Pegs placed:		
Mean	19.39	17.91
SD	2.52	3.39

a) Find 95% confidence intervals for the average number of pegs that males and females can each place.
b) Those intervals overlap. What does this suggest about any sex-based difference in manual agility?
c) Find a 95% confidence interval for the difference in the mean number of pegs that could be placed by men and women.
d) What does this interval suggest about any difference in manual agility between men and women?
e) The two results seem contradictory. Which method is correct: doing two-sample inference or doing one-sample inference twice?
f) Why don't the results agree?

23. Double header Look again at the data in Exercise 21.
a) Explain why you should not use two separate confidence intervals to decide whether the two leagues differ in average number of home runs per game.
b) Using a 95% confidence interval, estimate the difference between the mean number of home runs hit in American and National League games.
c) Interpret your interval.
d) Does this interval suggest that the two leagues may differ in average number of home runs hit per game?

T 24. Hard water In an investigation of environmental causes of disease, data were collected on the annual mortality rate (deaths per 100,000) for males in 61 large towns in England and Wales. In addition, the water hardness was recorded as the calcium concentration (parts per million, ppm) in the drinking water. The data set also notes, for each town, whether it was south or north of Derby. Is there a significant difference in mortality rates in the two regions? Here are the summary statistics.

Summary of:	mortality		
For categories in:	Derby		
Group	**Count**	**Mean**	**Median**	**StdDev**
---	---	---	---	---
North	34	1631.59	1631	138.470
South	27	1388.85	1369	151.114

a) Test appropriate hypotheses and state your conclusion.
b) The boxplots of the two distributions show an outlier among the data north of Derby. What effect might that have had on your test?

25. Job satisfaction A company institutes an exercise break for its workers to see if this will improve job satisfaction, as measured by a questionnaire that assesses workers' satisfaction. Scores for 10 randomly selected workers before and after implementation of the exercise program are shown. The company wants to assess the effectiveness of the exercise program. Explain why you can't use the methods discussed in this chapter to do that. (Don't worry, we'll give you another chance to do this the right way.)

Worker Number	Job Satisfaction Index	
	Before	After
1	34	33
2	28	36
3	29	50
4	45	41
5	26	37
6	27	41
7	24	39
8	15	21
9	15	20
10	27	37

26. Summer school Having done poorly on their math final exams in June, six students repeat the course in summer school, then take another exam in August. If we consider these students representative of all students who might attend this summer school in other years, do these results provide evidence that the program is worthwhile?

June	54	49	68	66	62	62
Aug.	50	65	74	64	68	72

27. Sex and violence The *Journal of Applied Psychology* reported on a study that examined whether the content of TV shows influenced the ability of viewers to recall brand names of items featured in the commercials. The researchers randomly assigned volunteers to watch one of three programs, each containing the same nine commercials. One of the programs had violent content,

another sexual content, and the third neutral content. After the shows ended, the subjects were asked to recall the brands of products that were advertised. Here are summaries of the results:

	Program Type		
	Violent	Sexual	Neutral
No. of subjects	108	108	108
Brands recalled			
Mean	2.08	1.71	3.17
SD	1.87	1.76	1.77

a) Do these results indicate that viewer memory for ads may differ depending on program content? A test of the hypothesis that there is no difference in ad memory between programs with sexual content and those with violent content has a P-value of 0.136. State your conclusion.
b) Is there evidence that viewer memory for ads may differ between programs with sexual content and those with neutral content? Test an appropriate hypothesis and state your conclusion.

28. Ad campaign You are a consultant to the marketing department of a business preparing to launch an ad campaign for a new product. The company can afford to run ads during one TV show, and has decided not to sponsor a show with sexual content. You read the study described in Exercise 27, then use a computer to create a confidence interval for the difference in mean number of brand names remembered between the groups watching violent shows and those watching neutral shows.

TWO-SAMPLE T
95% CI FOR MUviol − MUneut: (−1.578, −0.602)

a) At the meeting of the marketing staff, you have to explain what this output means. What will you say?
b) What advice would you give the company about the upcoming ad campaign?

29. Sex and violence II In the study described in Exercise 27, the researchers also contacted the subjects again, 24 hours later, and asked them to recall the brands advertised. Results are summarized below.

	Program Type		
	Violent	Sexual	Neutral
No. of subjects	101	106	103
Brands recalled			
Mean	3.02	2.72	4.65
SD	1.61	1.85	1.62

a) Is there a significant difference in viewers' abilities to remember brands advertised in shows with violent vs. neutral content?
b) Find a 95% confidence interval for the difference in mean number of brand names remembered between

the groups watching shows with sexual content and those watching neutral shows. Interpret your interval in this context.

30. **Ad recall** In Exercises 27 and 29, we see the number of advertised brand names people recalled immediately after watching TV shows and 24 hours later. Strangely enough, it appears that they remembered more about the ads the next day. Should we conclude this is true in general about people's memory of TV ads?

   a) Suppose one analyst conducts a two-sample hypothesis test to see if memory of brands advertised during violent TV shows is higher 24 hours later. If his P-value is 0.00013, what might he conclude?

   b) Explain why his procedure was inappropriate. Which of the assumptions for inference was violated?

   c) How might the design of this experiment have tainted the results?

   d) Suggest a design that could compare immediate brand-name recall with recall one day later.

31. **View of the water** How much extra is having a water-front property worth? A student took a random sample of 170 recently sold properties in Upstate New York to examine the question. Here are her summaries and boxplots of the two groups of prices:

Non-Waterfront Properties		Waterfront Properties	
n	100	n	70
$\bar{y}$	$219,896.60	$\bar{y}$	$319,906.40
s	$94,627.15	s	$153,303.80

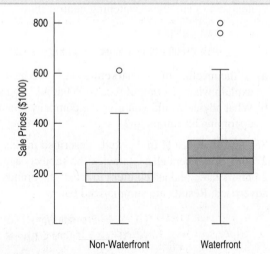

Construct and interpret a 95% confidence interval for the mean additional amount that waterfront property is worth. (From technology, df = 105.48.)

32. **New construction** The sample of house sales we looked at in Exercise 31 also listed whether homes were new construction or not. Find and interpret a 95% confidence interval for how much more an agent can expect to sell a new home for. (From technology, df = 197.8.) Here are the summaries and boxplots of the *Sale Prices*:

Old Construction		New Construction	
n	100	n	100
$\bar{y}$	$201,707.50	$\bar{y}$	$267,878.10
s	$96,116.88	s	$93,302.18

33. **More arrows** In Chapter 22, Exercise 47, we looked at the penetration of stone-tipped versus wooden-tipped arrows. Do these data suggest that stone-tipped arrows penetrate further than wooden-tipped?

   | Wooden (mm): | 216 | 211 | 192 | 208 | 203 | 210 | 203 |
   | Stone (mm): | 240 | 208 | 213 | 225 | 232 | 214 | 240 |

   a) Perform a complete hypothesis test.
   b) Does the difference in penetration seem worth the extra time, effort and cost? Find a 95% confidence interval to measure the average difference in penetration.

34. **Final Shot** In Exercise 48 of Chapter 22, we considered the accuracy of stone-tipped versus wooden-tipped arrows. Do these data give statistically significant evidence of a difference in accuracy between the two types?

   | Wooden: | 9.3 | 16.7 | 7.1 | 14 | 1 | 1.2 |
   | Stone: | 4.9 | 21.1 | 7 | 1.8 | 5.4 | 8.6 |

35. **Fast or not?** In 2009, *Antiquity* published isotope analyses of the human skeletal tissue excavated from 1957 to 1967 at Whithorn Cathedral Priory, Scotland. These analyses sought to use new isotope methods to test common assumptions about the lifestyle and diet of the bishops and clerics compared to lay individuals buried at the same site.

   Specifically, Dr. Muldner (and others) tested an isotope that would indicate whether or not individuals ate seafood regularly. It is believed that the bishops and priests had access to seafood for fast days and other holy days, while the lay individuals did not. A higher measurement of collagen (%) indicates more seafood in a person's diet. Here are the data:

Percent Collagen	
Priests/Bishops	Lay Individuals
9.6	6.1
7.3	8.1
8.1	5.6
7.5	5.3
7.8	5.2
9.8	3.5
9.8	5.8

Is there statistically significant evidence of a higher collagen level among priests and bishops?

36. I can relate Professor Jody Gittell analyzed workers in the health care system. Specifically, she was interested to see if relational coordination (communicating and relating for the purpose of task integration) makes a worker more resilient in response to external threats that require a coordinated response across multiple roles in the organization. She rated whether physicians and nurses were selected for teamwork qualities, on a scale of 0 to 2. Do these summaries show evidence that teamwork qualities are valued differently in doctors than in nurses? Assume the conditions for inference have been met. (*Journal of Applied Behavioral Science*, March 2008)

Position	Mean	SD	n
Physicians	0.44	0.88	9
Nurses	1.44	0.73	9

37. Hungry? Researchers investigated how the size of a bowl affects how much ice cream people tend to scoop when serving themselves.[14] At an "ice cream social," people were randomly given either a 17 oz or a 34 oz bowl (both large enough that they would not be filled to capacity). They were then invited to scoop as much ice cream as they liked. Did the bowl size change the selected portion size? Here are the summaries:

Small Bowl		Large Bowl	
n	26	n	22
$\bar{y}$	5.07 oz	$\bar{y}$	6.58 oz
s	1.84 oz	s	2.91 oz

Test an appropriate hypothesis and state your conclusions. Assume any assumptions and conditions that you cannot test are sufficiently satisfied to proceed.

38. Thirsty? Researchers randomly assigned participants either a tall, thin "highball" glass or a short, wide "tumbler," each of which held 355 ml. Participants were asked to pour a shot (1.5 oz = 44.3 ml) into their glass. Did the shape of the glass make a difference in how much liquid they poured?[15] Here are the summaries:

highball		tumbler	
n	99	n	99
$\bar{y}$	42.2 ml	$\bar{y}$	60.9 ml
s	16.2 ml	s	17.9 ml

Test an appropriate hypothesis and state your conclusions. Assume any assumptions and conditions that you cannot test are sufficiently satisfied to proceed.

39. Lower scores? Newspaper headlines recently announced a decline in science scores among high school seniors. In 2000, a total of 15,109 seniors tested by The National Assessment in Education Program (NAEP) scored a mean of 147 points. Four years earlier, 7537 seniors had averaged 150 points. The standard error of the difference in the mean scores for the two groups was 1.22.

a) Have the science scores declined significantly? Cite appropriate statistical evidence to support your conclusion.

b) The sample size in 2000 was almost double that in 1996. Does this make the results more convincing or less? Explain.

40. The Internet The NAEP report described in Exercise 39 compared science scores for students who had home Internet access to the scores of those who did not, as shown in the graph. They report that the differences are statistically significant.

a) Explain what "statistically significant" means in this context.

b) If their conclusion is incorrect, which type of error did the researchers commit?

c) Does this prove that using the Internet at home can improve a student's performance in science?

41. Running heats In Olympic running events, preliminary heats are determined by random draw, so we should expect that the abilities of runners in the various heats to be about the same, on average. Here are the times (in seconds) for the 400-m women's run in the 2004 Olympics in Athens for preliminary heats 2 and 5. Is there any evidence that the mean time to finish is different for randomized heats? Explain. Be sure to include a discussion of assumptions and conditions for your analysis.

[14]Brian Wansink, Koert van Ittersum, and James E. Painter, "Ice Cream Illusions: Bowls, Spoons, and Self-Served Portion Sizes," *Am J Prev Med* 2006.

[15]Brian Wansink and Koert van Ittersum, "Shape of Glass and Amount of Alcohol Poured: Comparative Study of Effect of Practice and Concentration," *BMJ* 2005;331;1512–1514.

Country	Name	Heat	Time
USA	HENNAGAN Monique	2	51.02
BUL	DIMITROVA Mariyana	2	51.29
CHA	NADJINA Kaltouma	2	51.50
JAM	DAVY Nadia	2	52.04
BRA	ALMIRAO Maria Laura	2	52.10
FIN	MYKKANEN Kirsi	2	52.53
CHN	BO Fanfang	2	56.01
BAH	WILLIAMS-DARLING Tonique	5	51.20
BLR	USOVICH Svetlana	5	51.37
UKR	YEFREMOVA Antonina	5	51.53
CMR	NGUIMGO Mireille	5	51.90
JAM	BECKFORD Allison	5	52.85
TOG	THIEBAUD-KANGNI Sandrine	5	52.87
SRI	DHARSHA K V Damayanthi	5	54.58

(T) **42. Swimming heats** In Exercise 41 we looked at the times in two different heats for the 400-m women's run from the 2004 Olympics. Unlike track events, swimming heats are *not* determined at random. Instead, swimmers are seeded so that better swimmers are placed in later heats. Here are the times (in seconds) for the women's 400-m freestyle from heats 2 and 5. Do these results suggest that the mean times of seeded heats are not equal? Explain. Include a discussion of assumptions and conditions for your analysis.

Country	Name	Heat	Time
ARG	BIAGIOLI Cecilia Elizabeth	2	256.42
SLO	CARMAN Anja	2	257.79
CHI	KOBRICH Kristel	2	258.68
MKD	STOJANOVSKA Vesna	2	259.39
JAM	ATKINSON Janelle	2	260.00
NZL	LINTON Rebecca	2	261.58
KOR	HA Eun-Ju	2	261.65
UKR	BERESNYEVA Olga	2	266.30
FRA	MANAUDOU Laure	5	246.76
JPN	YAMADA Sachiko	5	249.10
ROM	PADURARU Simona	5	250.39
GER	STOCKBAUER Hannah	5	250.46
AUS	GRAHAM Elka	5	251.67
CHN	PANG Jiaying	5	251.81
CAN	REIMER Brittany	5	252.33
BRA	FERREIRA Monique	5	253.75

43. Tees Does it matter what kind of tee a golfer places the ball on? The company that manufactures "Stinger" tees claims that the thinner shaft and smaller head will lessen drag, reducing spin and allowing the ball to travel farther. In August 2003, Golf Laboratories, Inc., compared the distance traveled by golf balls hit off regular wooden tees to those hit off Stinger tees. All the balls were struck by

the same golf club using a robotic device set to swing the club head at approximately 95 miles per hour. Summary statistics from the test are shown in the table. Assume that 6 balls were hit off each tee and that the data were suitable for inference.

		Total Distance (yards)	Ball Velocity (mph)	Club Velocity (mph)
Regular tee	Avg.	227.17	127.00	96.17
	SD	2.14	0.89	0.41
Stinger tee	Avg.	241.00	128.83	96.17
	SD	2.76	0.41	0.52

Is there evidence that balls hit off the Stinger tees would have a higher initial velocity?

44. Golf again Given the test results on golf tees described in Exercise 43, is there evidence that balls hit off Stinger tees would travel farther? Again, assume that 6 balls were hit off each tee and that the data were suitable for inference.

45. Crossing Ontario Between 1954 and 2003, swimmers have crossed Lake Ontario 43 times. Both women and men have made the crossing. Here are some plots (we've omitted a crossing by Vikki Keith, who swam a round trip—North to South to North—in 3390 minutes):

The summary statistics are:

		Count	Mean	StdDev
Group	F	22	1271.59	261.111
	M	20	1196.75	304.369

How much difference is there between the mean amount of time (in minutes) it would take female and male swimmers to swim the lake?

a) Construct and interpret a 95% confidence interval for the difference between female and male times.

b) Comment on the assumptions and conditions.

46. Music and memory Is it a good idea to listen to music when studying for a big test? In a study conducted by some Statistics students, 62 people were randomly assigned to listen to rap music, music by Mozart, or no music while attempting to memorize objects pictured on a page. They were then asked to list all the objects they could remember. Here are summary statistics:

	Rap	Mozart	No Music
Count	29	20	13
Mean	10.72	10.00	12.77
SD	3.99	3.19	4.73

a) Does it appear that it is better to study while listening to Mozart than to rap music? Test an appropriate hypothesis and state your conclusion.

b) Create a 90% confidence interval for the mean difference in memory score between students who study to Mozart and those who listen to no music at all. Interpret your interval.

47. Rap Using the results of the experiment described in Exercise 46, does it matter whether one listens to rap music while studying, or is it better to study without music at all?

a) Test an appropriate hypothesis and state your conclusion.

b) If you concluded there is a difference, estimate the size of that difference with a confidence interval and explain what your interval means.

48. Cuckoos Cuckoos lay their eggs in the nests of other (host) birds. The eggs are then adopted and hatched by the host birds. But the potential host birds lay eggs of different sizes. Does the cuckoo change the size of her eggs for different foster species? The numbers in the table are lengths (in mm) of cuckoo eggs found in nests of three different species of other birds. The data are drawn from the work of O.M. Latter in 1902 and were used in a fundamental textbook on statistical quality control by L.H.C. Tippett (1902–1985), one of the pioneers in that field.

Cuckoo Egg Length (MM)		
Foster Parent Species		
Sparrow	Robin	Wagtail
20.85	21.05	21.05
21.65	21.85	21.85
22.05	22.05	21.85
22.85	22.05	21.85
23.05	22.05	22.05
23.05	22.25	22.45
23.05	22.45	22.65
23.05	22.45	23.05
23.45	22.65	23.05
23.85	23.05	23.25
23.85	23.05	23.45
23.85	23.05	24.05
24.05	23.05	24.05
25.05	23.05	24.05
	23.05	24.85
	23.25	
	23.85	

Investigate the question of whether the mean length of cuckoo eggs is the same for different species, and state your conclusion.

Just Checking ANSWERS

1. Randomization should balance unknown sources of variability in the two groups of patients and helps us believe the two groups are independent.

2. We can be 95% confident that after 4 weeks endoscopic surgery patients will have a mean pinch strength between 0.04 kg and 2.96 kg higher than open-incision patients.

3. The lower bound of this interval is close to 0, so the difference may not be great enough that patients could actually notice the difference. We may want to consider other issues such as cost or risk in making a recommendation about the two surgical procedures.

4. Without data, we can't check the Nearly Normal Condition.

5. H_0: Mean pinch strength is the same after both surgeries. ($\mu_E - \mu_O = 0$)

   H_A: Mean pinch strength is different after the two surgeries. ($\mu_E - \mu_O \neq 0$)

6. With a P-value this low, we reject the null hypothesis. We can conclude that mean pinch strength differs after 4 weeks in patients who undergo endoscopic surgery vs. patients who have open-incision surgery. Results suggest that the endoscopic surgery patients may be stronger, on average.

7. If some patients contributed two hands to the study, then the groups may not be internally independent. It is reasonable to assume that two hands from the same patient might respond in similar ways to similar treatments.

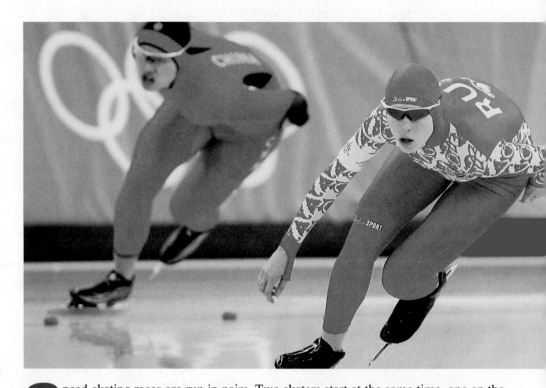

Who	Olympic speed-skaters
What	Time for women's 1500 m
Units	Seconds
When	2006
Where	Torino, Italy
Why	To see whether one lane is faster than the other

Speed-skating races are run in pairs. Two skaters start at the same time, one on the inner lane and one on the outer lane. Halfway through the race, they cross over, switching lanes so that each will skate the same distance in each lane. Even though this seems fair, at the 2006 Olympics some fans thought there might have been an advantage to starting on the outside. After all, the winner, Cindy Klassen, started on the outside and skated a remarkable 1.47 seconds faster than the silver medalist.

Here are the data for the women's 1500-m race:

Inner Lane		Outer Lane	
Name	**Time**	**Name**	**Time**
OLTEAN Daniela	129.24	(no competitor)	
ZHANG Xiaolei	125.75	NEMOTO Nami	122.34
ABRAMOVA Yekaterina	121.63	LAMB Maria	122.12
REMPEL Shannon	122.24	NOH Seon Yeong	123.35
LEE Ju-Youn	120.85	TIMMER Marianne	120.45
ROKITA Anna Natalia	122.19	MARRA Adelia	123.07
YAKSHINA Valentina	122.15	OPITZ Lucille	122.75
BJELKEVIK Hedvig	122.16	HAUGLI Maren	121.22
ISHINO Eriko	121.85	WOJCICKA Katarzyna	119.96
RANEY Catherine	121.17	BJELKEVIK Annette	121.03
OTSU Hiromi	124.77	LOBYSHEVA Yekaterina	118.87
SIMIONATO Chiara	118.76	JI Jia	121.85
ANSCHUETZ THOMS Daniela	119.74	WANG Fei	120.13
BARYSHEVA Varvara	121.60	van DEUTEKOM Paulien	120.15
GROENEWOLD Renate	119.33	GROVES Kristina	116.74
RODRIGUEZ Jennifer	119.30	NESBITT Christine	119.15
FRIESINGER Anni	117.31	KLASSEN Cindy	115.27
WUST Ireen	116.90	TABATA Maki	120.77

We can view this skating event as an experiment testing whether the lanes were equally fast. Skaters were assigned to lanes randomly. The boxplots of times recorded in the inner and outer lanes don't show much difference. But that's not the right way to compare these times. Conditions can change during the day. The question raised is about the difference between starting lanes, so it is those differences we should examine.

Figure 24.1

Using boxplots to compare times in the inner and outer lanes shows little because it ignores the fact that the skaters raced in pairs.

Paired Data

Data such as these are called **paired**. We have the times for skaters in each lane for each race. We want to compare the inner and outer lanes across all the races, so what we're interested in is the *differences* in times for each racing pair.

Paired data arise in a number of ways. Perhaps the most common way is to compare subjects with themselves before and after a treatment. When pairs arise from an experiment, the pairing is a type of *blocking*. When they arise from an observational study, it is a form of *matching*.

For Example IDENTIFYING PAIRED DATA

Do flexible schedules reduce the demand for resources? The Lake County, Illinois, Health Department experimented with a flexible four-day workweek. For a year, the department recorded the mileage driven by 11 field workers on an ordinary five-day workweek. Then it changed to a flexible four-day workweek and recorded mileage for another year.[1] The data are shown.

Name	5-Day Mileage	4-Day Mileage
Jeff	2798	2914
Betty	7724	6112
Roger	7505	6177
Tom	838	1102
Aimee	4592	3281
Greg	8107	4997
Larry G.	1228	1695
Tad	8718	6606
Larry M.	1097	1063
Leslie	8089	6392
Lee	3807	3362

QUESTION: Why are these data paired?

ANSWER: The mileage data are paired because each driver's mileage is measured before and after the change in schedule. I'd expect drivers who drove more than others before the schedule change to continue to drive more afterwards, so the two sets of mileages can't be considered independent.

---

[1]Charles S. Catlin, "Four-day Work Week Improves Environment," *Journal of Environmental Health*, Denver, 59:7.

Paired, or Independent?

It matters. A lot. *You cannot use paired-t methods when the groups are independent, nor 2-sample t-methods when the data are paired.* And the data can't tell you what to do; it's all about the study design. Think about how the data were collected. Is there some connection between the two variables that links each value of one to a value of the other? Would it be okay to rearrange the order of the values in just one of the data sets without rearranging the other data set, too?

A S   *Activity*: **Differences in Means of Paired Groups.** Are married couples typically the same age, or do wives tend to be younger than their husbands, on average?

Pairing isn't a problem; it's an opportunity. If you know the data are paired, you can take advantage of that fact—in fact, you *must* take advantage of it. You *may not* use the two-sample methods of the previous chapter when the data are paired. Remember: Those methods rely on the Pythagorean Theorem of Statistics, and that requires the two samples be independent. Paired data aren't. There is no test that can tell you whether the data are paired. You must determine that from understanding how they were collected and what they mean (check the W's).

Once we recognize that the speed-skating data are matched pairs, it makes sense to consider the difference in times for each two-skater race. So we look at the *pairwise* differences:

Skating Pair	Inner Time	Outer Time	Inner − Outer
1	129.24		·
2	125.75	122.34	3.41
3	121.63	122.12	−0.49
4	122.24	123.35	−1.11
5	120.85	120.45	0.40
6	122.19	123.07	−0.88
7	122.15	122.75	−0.60
8	122.16	121.22	0.94
9	121.85	119.96	1.89
10	121.17	121.03	0.14
11	124.77	118.87	5.90
12	118.76	121.85	−3.09
13	119.74	120.13	−0.39
14	121.60	120.15	1.45
15	119.33	116.74	2.59
16	119.30	119.15	0.15
17	117.31	115.27	2.04
18	116.90	120.77	−3.87

The first skater raced alone, so we'll omit that race. Because it is the *differences* we care about, we'll treat them as if *they* were the data, ignoring the original two columns. Now that we have only one column of values to consider, we can use a one-sample *t*-test. Mechanically, a **paired *t*-test** is just a one-sample *t*-test for the means of these pairwise differences. The sample size is the number of pairs.

So you've already seen the *Show*.

Assumptions and Conditions

Paired Data Condition

Paired Data Condition: The data must be paired. You can't just decide to pair data when in fact the samples are independent. When you have two groups with the same number of observations, it may be tempting to match them up. Don't, unless you are prepared to justify your claim that there is a reason to pair them.

On the other hand, be sure to recognize paired data when you have them. Remember, two-sample *t* methods aren't valid without independent groups, and paired groups aren't independent. Although this is a strictly required assumption, it is one that can be easy to check if you understand how the data were collected.

Independence Assumption

Independence Assumption: If the data are paired, the *groups* are not independent. For these methods, it's the *differences* that must be independent of each other. There's no reason to believe that the difference in speeds of one pair of races could affect the difference in speeds for another pair.

    Randomization Condition: Randomness can arise in many ways. The pairs may be a random sample. In an experiment, the order of the two treatments may be randomly assigned, or the treatments may be randomly assigned to one member of each pair. In a before-and-after study, we may believe that the observed differences are a representative sample from a population of interest. In our example, skaters were assigned to the lanes at random.

    10% Condition: We're thinking of the speed-skating data as an experiment testing the difference between lanes. The 10% Condition doesn't apply to randomized experiments, where no sampling takes place.

> **10% of What?** A fringe benefit of checking the 10% Condition is that it forces us to think about what population we're hoping to make inferences about.

Normal Population Assumption

We need to assume that the population of *differences* follows a Normal model. We don't need to check the individual groups.

    Nearly Normal Condition: This condition can be checked with a histogram or Normal probability plot of the *differences*—but not of the individual groups. As with the one-sample *t*-methods, this assumption matters less the more pairs we have to consider. You may be pleasantly surprised when you check this condition. Even if your original measurements are skewed or bimodal, the *differences* may be nearly Normal. After all, the individual who was way out in the tail on an initial measurement is likely to still be out there on the second one, giving a perfectly ordinary difference.

For Example CHECKING ASSUMPTIONS AND CONDITIONS

RECAP: Field workers for a health department compared driving mileage on a five-day work schedule with mileage on a new four-day schedule. To see if the new schedule changed the amount of driving they did, we'll look at paired differences in mileages before and after.

QUESTION: Is it okay to use these data to test whether the new schedule changed the amount of driving?

ANSWER:

✓ **Paired Data Condition:** The data are paired because each value is the mileage driven by the same person before and after a change in work schedule.

✓ **Independence Assumption:** The driving behavior of any individual worker is independent of the others, so the differences are mutually independent.

✓ **Randomization Condition:** The mileages are the sums of many individual trips, each of which experienced random events that arose while driving. Repeating the experiment in two new years would give randomly different values.

✓ **Nearly Normal Condition:** The histogram of the mileage differences is unimodal and symmetric:

Since the assumptions and conditions are satisfied, it's okay to use paired *t* methods for these data.

Name	5-Day Mileage	4-Day Mileage	Difference
Jeff	2798	2914	−116
Betty	7724	6112	1612
Roger	7505	6177	1328
Tom	838	1102	−264
Aimee	4592	3281	1311
Greg	8107	4997	3110
Larry G.	1228	1695	−467
Tad	8718	6606	2112
Larry M.	1097	1063	34
Leslie	8089	6392	1697
Lee	3807	3362	445

The steps in testing a hypothesis for paired differences are very much like the steps for a one-sample *t*-test for a mean.

The Paired *t*-Test

When the conditions are met, we are ready to test whether the mean of paired differences is significantly different from zero. We test the hypothesis

$$H_0: \mu_d = \Delta_0,$$

where the *d*'s are the pairwise differences and Δ_0 is almost always 0.
    We use the statistic

$$t_{n-1} = \frac{\bar{d} - \Delta_0}{SE(\bar{d})},$$

where $\bar{d}$ is the mean of the pairwise differences, *n* is the number of *pairs*, and

$$SE(\bar{d}) = \frac{s_d}{\sqrt{n}}.$$

$SE(\bar{d})$ is the ordinary standard error for the mean, applied to the differences.
    When the conditions are met and the null hypothesis is true, we can model the sampling distribution of this statistic with a Student's *t*-model with $n - 1$ degrees of freedom, and use that model to obtain a P-value.

Step-by-Step Example   A PAIRED *t*-TEST

Question: Was there a difference in speeds between the inner and outer speed-skating lanes at the 2006 Winter Olympics?

THINK ➡ **Plan** State what we want to know.	I want to know whether there really was a difference in the speeds of the two lanes for speed skating at the 2006 Olympics. I have data for 17 pairs of racers at the women's 1500-m race.
Identify the *parameter* we wish to estimate. Here our parameter is the mean difference in race times.	
Identify the variables and check the W's.	
Hypotheses State the null and alternative hypotheses.	H_0: Neither lane offered an advantage: $$\mu_d = 0.$$
Although fans suspected one lane was faster, we can't use the data we have to specify the direction of a test. We (and Olympic officials) would be interested in a difference in either direction, so we test a two-sided alternative.	H_A: The mean difference is different from zero: $$\mu_d \neq 0.$$

(continued)

Model Think about the assumptions and check the conditions.

State why you think the data are paired. Simply having the same number of individuals in each group and displaying them in side-by-side columns doesn't make them paired.

Think about what we hope to learn and where the randomization comes from. Here, the randomization comes from the racer pairings and lane assignments.

Make a picture—just one. Don't plot separate distributions of the two groups—that entirely misses the pairing. For paired data, it's the Normality of the *differences* that we care about. Treat those paired differences as you would a single variable, and check the Nearly Normal Condition with a histogram (or a Normal probability plot).

✓ **Paired Data Condition:** The data are paired because racers compete in pairs.

✓ **Independence Assumption:** Each race is independent of the others, so the differences are mutually independent.

✓ **Randomization Condition:** Skaters are assigned to lanes at random.

✓ **Nearly Normal Condition:** The histogram of the differences is unimodal and symmetric:

Specify the sampling distribution model.

Choose the method.

The conditions are met, so I'll use a Student's *t*-model with $(n - 1) = 16$ degrees of freedom, and perform a **paired *t*-test**.

SHOW ➡ Mechanics

n is the number of *pairs*—in this case, the number of races.

$\bar{d}$ is the mean difference.

s_d is the standard deviation of the differences.

Find the standard error and the *t*-score of the observed mean difference. There is nothing new in the mechanics of the paired *t* methods. These are the mechanics of the *t*-test for a mean applied to the differences.

The data give

$$n = 17 \text{ pairs}$$
$$\bar{d} = 0.499 \text{ seconds}$$
$$s_d = 2.333 \text{ seconds.}$$

I estimate the standard deviation of $\bar{d}$ using

$$SE(\bar{d}) = \frac{s_d}{\sqrt{n}} = \frac{2.333}{\sqrt{17}} = 0.5658$$

$$\text{So } t_{16} = \frac{\bar{d} - 0}{SE(\bar{d})} = \frac{0.499}{0.5658} = 0.882$$

Make a picture. Sketch a *t*-model centered at the hypothesized mean of 0. Because this is a two-tail test, shade both the region to the right of the calculated *t*-value and the corresponding region in the lower tail.

Find the P-value, using technology.

$$\text{P-value} = 2P(t_{16} > 0.882) = 0.39$$

REALITY CHECK ➡ The mean difference is 0.499 seconds. That may not seem like much, but a smaller difference determined the Silver and Bronze medals. The standard error is about this big, so a *t*-value less than 1.0 isn't surprising. Nor is a large P-value.

(continued)

TELL ➡ **Conclusion** Link the P-value to your decision about H_0, and state your conclusion in context.

The P-value is large. Events that happen more than a third of the time are not remarkable. So, even though there is an observed difference between the lanes, I can't conclude that it isn't due simply to random chance. It appears the fans may have interpreted a random fluctuation in the data as favoring one lane. There's insufficient evidence to declare any lack of fairness.

For Example DOING A PAIRED *t*-TEST

RECAP: We want to test whether a change from a five-day workweek to a four-day workweek could change the amount driven by field workers of a health department. We've already confirmed that the assumptions and conditions for a paired *t*-test are met.

QUESTION: Is there evidence that a four-day workweek would change how many miles workers drive?

ANSWER: H_0: The change in the health department workers' schedules didn't change the mean mileage driven; the mean difference is zero:

$$\mu_d = 0.$$

H_A: The mean difference is different from zero:

$$\mu_d \neq 0.$$

The conditions are met, so I'll use a Student's *t*-model with $(n - 1) = 10$ degrees of freedom and perform a paired *t*-test.

The data give

$$n = 11 \text{ pairs}$$
$$\bar{d} = 982 \text{ miles}$$
$$s_d = 1139.6 \text{ miles}.$$

$$SE(\bar{d}) = \frac{s_d}{\sqrt{n}} = \frac{1139.6}{\sqrt{11}} = 343.6$$

$$\text{So } t_{10} = \frac{\bar{d} - 0}{SE(\bar{d})} = \frac{982.0}{343.6} = 2.86$$

$$\text{P-value} = 2P(t_{10} > 2.86) = 0.017$$

The P-value is small, so I reject the null hypothesis and conclude that the change in workweek did lead to a change in average driving mileage. It appears that changing the work schedule may reduce the mileage driven by workers.

Note: We should propose a course of action, but it's hard to tell from the hypothesis test whether the reduction matters. Is the difference in mileage important in the sense of reducing air pollution or costs, or is it merely statistically significant? To help make that decision, we should look at a confidence interval. If the difference in mileage proves to be large in a practical sense, then we might recommend a change in schedule for the rest of the department.

TI Tips TESTING A HYPOTHESIS WITH PAIRED DATA

Since the inference procedures for matched data are essentially just the one-sample t procedures, you already know what to do . . . once you have the list of paired differences, that is. That list is not hard to create.

TEST A HYPOTHESIS ABOUT THE MEAN OF PAIRED DIFFERENCES

- Think: Are the samples independent or paired. Independent? Go back to the last chapter! Paired? Read on.
- Enter the driving data from page 588 into two lists, say *5-Day mileage* in L1, *4-Day mileage* in L2.
- Create a list of the differences. We want to take each value in L1, subtract the corresponding value in L2, and store the paired difference in L3. The command is L1−L2 → L3. (The arrow is the STO button.) Now take a look at L3. See—it worked!
- Make a histogram of the differences, L3, to check the nearly Normal condition. Notice that we do not look at the histograms of the *5-day mileage* or the *4-day mileage*. Those are not the data that we care about now that we are using a paired procedure. Note also that the calculator's first histogram is not close to Normal. More work to do . . .
- As you have seen before, small samples often produce ragged histograms, and these may look very different after a change in bar width. Reset the WINDOW to Xmin = −3000, Xmax = 4500, and Xscl = 1500. The new histogram looks okay.
- Under STAT TESTS simply use T-Test, as you've done before for hypothesis tests about a mean.
- Specify that the hypothesized difference is 0, you're using the Data in L3, and it's a two-tailed test.
- Calculate.

The small P-value shows strong evidence that on average the change in the workweek reduces the number of miles workers drive.

Confidence Intervals for Matched Pairs

In developed countries, the average age of women is generally higher than that of men. After all, women tend to live longer. But if we look at *married couples*, husbands tend to be slightly older than wives. How much older, on average, are husbands? We have data from a random sample of 200 British couples, the first 7 of which are shown below. Only 170 couples provided ages for both husband and wife, so we can work only with that many pairs. Let's form a confidence interval for the mean difference of husband's and wife's ages for these 170 couples. Here are the first 7 pairs:

Who	170 randomly sampled couples
What	Ages
Units	Years
When	Recently
Where	Britain

Wife's Age	Husband's Age	Difference (husband − wife)
43	49	6
28	25	−3
30	40	10
57	52	−5
52	58	6
27	32	5
52	43	−9
⋮	⋮	⋮

Clearly, these data are paired. The survey selected *couples* at random, not individuals. We're interested in the mean age difference within couples. How would we construct a confidence interval for the true mean difference in ages?

Paired *t*-Interval

When the conditions are met, we are ready to find the confidence interval for the mean of the paired differences. The confidence interval is

$$\bar{d} \pm t^*_{n-1} \times SE(\bar{d}),$$

where the standard error of the mean difference is $SE(\bar{d}) = \dfrac{s_d}{\sqrt{n}}$.

The critical value t^* from the Student's *t*-model depends on the particular confidence level, *C*, that you specify and on the degrees of freedom, $n - 1$, which is based on the number of pairs, *n*.

Making confidence intervals for matched pairs follows exactly the steps for a one-sample *t*-interval.

Step-by-Step Example A PAIRED *t*-INTERVAL

Question: How big a difference is there, on average, between the ages of husbands and wives?

THINK ➡ **Plan** State what we want to know.

Identify the variables and check the W's.

Identify the parameter you wish to estimate. For a paired analysis, the parameter of interest is the mean of the differences. The population of interest is the population of differences.

I want to estimate the mean difference in age between husbands and wives. I have a random sample of 200 British couples, 170 of whom provided both ages.

Model Think about the assumptions and check the conditions.

✓ **Paired Data Condition:** *The data are paired because they are on members of married couples.*

✓ **Independence Assumption:** *The data are from a randomized survey, so couples should be independent of each other.*

✓ **Randomization Condition:** *These couples were randomly sampled.*

✓ **10% Condition:** *The sample is far less than 10% of all married couples in Britain.*

(continued)

Make a picture. We focus on the differences, so a histogram or Normal probability plot is best here.

REALITY CHECK ➤ The histogram shows husbands are often older than wives (because most of the differences are greater than 0). The mean difference seen here of about 2 years is reasonable.

✓ **Nearly Normal Condition:** The histogram of the husband – wife differences is unimodal and symmetric:

Husband's Age – Wife's Age (yr)

State the sampling distribution model.

Choose your method.

The conditions are met, so I can use a Student's *t*-model with $(n - 1) = 169$ degrees of freedom and find a **paired *t*-interval**.

SHOW ➡ Mechanics

n is the number of *pairs*, here, the number of couples.

$\bar{d}$ is the mean difference.

s_d is the standard deviation of the differences.

Be sure to include the units along with the statistics.

$n = 170$ couples

$\bar{d} = 2.2$ years

$s_d = 4.1$ years

I estimate the standard error of $\bar{d}$ as

$$SE(\bar{d}) = \frac{s_d}{\sqrt{n}} = \frac{4.1}{\sqrt{170}} = 0.31 \text{ years.}$$

The df for the *t*-model is $n - 1 = 169$.

The critical value we need to make a 95% interval comes from a Student's *t* table, a computer program, or a calculator.

The 95% critical value for t_{169} is 1.97.

Find the margin of error and the confidence interval.

The margin of error is

$$ME = t^*_{169} \times SE(\bar{d}) = 1.97(0.31) = 0.61.$$

So the 95% confidence interval is

$$2.2 \pm 0.6 \text{ years,}$$

or an interval of $(1.6, 2.8)$ years.

REALITY CHECK ➤ This result makes sense. Our everyday experience confirms that an average age difference of about 2 years is reasonable.

TELL ➡ Conclusion   Interpret the confidence interval in context.

I am 95% confident that British husbands are, on average, 1.6 to 2.8 years older than their wives.

TI Tips   CREATING A CONFIDENCE INTERVAL

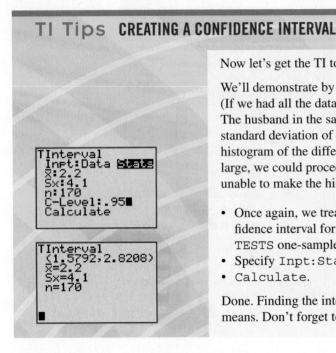

Now let's get the TI to create a confidence interval for the mean of paired differences.

We'll demonstrate by using the statistics about the ages of the British married couples. (If we had all the data, we could enter that, of course. All 170 couples? Um, no thanks.) The husband in the sample were an average of 2.2 years older than their wives, with a standard deviation of 4.1 years. We've already seen that the data are paired and that a histogram of the differences satisfies the Nearly Normal Condition. (With a sample this large, we could proceed with inference even if we didn't have the actual data and were unable to make the histogram.)

- Once again, we treat the paired differences just like data from one sample. A confidence interval for the mean difference, then, like that for a mean, uses the STAT TESTS one-sample procedure TInterval.
- Specify Inpt: Stats, and enter the statistics for the paired differences.
- Calculate.

Done. Finding the interval was the easy part. Now it's time for you to *Tell* what it means. Don't forget to talk about married couples in Britain.

Effect Size

When we examined the speed-skating times, we failed to reject the null hypothesis, so we couldn't be certain whether there really was a difference between the lanes. Maybe there wasn't any difference, or maybe whatever difference there might have been was just too small to matter at all. Were the fans right to be concerned?

We can't tell from the hypothesis test, but using the same summary statistics, we can find that the corresponding 95% confidence interval for the mean difference is $(-0.70 < \mu_d < 1.70)$ seconds.

A confidence interval is a good way to get a sense for the size of the effect we're trying to understand. That gives us a plausible range of values for the true mean difference in lane times. If differences of 1.7 seconds were too small to matter in 1500-m Olympic speed skating, we'd be pretty sure there was no need for concern.

But in fact, except for the Gold − Silver gap, the successive gaps between each skater and the next-faster one were *all* less than the high end of this interval, and most were right around the middle of the interval.

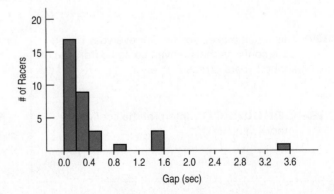

So even though we were unable to discern a real difference, the confidence interval shows that the effects we're considering may be big enough to be important. We may want to continue this investigation by checking out other races on this ice and being alert for possible differences at other venues.

For Example LOOKING AT EFFECT SIZE WITH A PAIRED *t* CONFIDENCE INTERVAL

RECAP: We know that, on average, the switch from a five-day workweek to a four-day workweek reduced the amount driven by field workers in that Illinois health department. However, finding that there is a significant difference doesn't necessarily mean that difference is meaningful or worthwhile. To assess the size of the effect, we need a confidence interval. We already know the assumptions and conditions are met.

QUESTION: By how much, on average, might a change in workweek schedule reduce the amount driven by workers?

ANSWER:
$$\bar{d} = 982 \text{ mi} \quad SE(\bar{d}) = 343.6 \quad t^*_{10} = 2.228 \text{ (for 95\%)}$$
$$ME = t^*_{10} \times SE(\bar{d}) = 2.228(343.6) = 765.54$$

So the 95% confidence interval for μ_d is 982 ± 765.54 or $(216.46, 1747.54)$ fewer miles.

With 95% confidence, I estimate that by switching to a four-day workweek employees would drive an average of between 216 and 1748 fewer miles per year. With high gas prices, this could save a lot of money.

Blocking

Figure 24.2
This display is worthless. It does no good to compare all the wives as a group with all the husbands. We care about the paired differences.

Because the sample of British husbands and wives includes both older and younger couples, there's a lot of variation in the ages of the men and in the ages of the women. In fact, that variation is so great that a boxplot of the two groups would show little difference. But that would be the wrong plot. It's the *difference* we care about. Pairing isolates the extra variation and allows us to focus on the individual differences. In Chapter 12, we saw how to design an experiment with blocking to isolate the variability between identifiable groups of subjects. Blocking makes it easier to see the variability among treatment groups that is attributable to their responses to the treatment. A paired design is an example of blocking.

When we pair, we have roughly half the degrees of freedom of a two-sample test. You may see discussions that suggest that in "choosing" a paired analysis we "give up" these degrees of freedom. This isn't really true, though. If the data are paired, then there never were additional degrees of freedom, and we have no "choice." The fact of the pairing determines how many degrees of freedom are available.

Matching pairs generally removes so much extra variation that it more than compensates for having only half the degrees of freedom, so it is usually a good choice when you design a study. Of course, inappropriate matching when the groups are in fact independent (say, by matching on the first letter of the last name of subjects) would cost degrees of freedom without the benefit of reducing the variance. When you design a study or experiment, you should consider using a paired design if possible.

Just Checking

Think about each of the 5 situations described here.

- Would you use a two-sample *t* or paired *t* method (or neither)? Why? or Why not?
- Would you perform a hypothesis test or find a confidence interval?

1. Random samples of 50 men and 50 women are asked to imagine buying a birthday present for their best friend. We want to estimate the difference in how much they are willing to spend.

2. Mothers of twins were surveyed and asked how often in the past month strangers had asked whether the twins were identical.

(continued)

3. Are parents equally strict with boys and girls? In a random sample of families, researchers asked a brother and sister from each family to rate how strict their parents were.

4. Forty-eight overweight subjects are randomly assigned to either aerobic or stretching exercise programs. They are weighed at the beginning and at the end of the experiment to see how much weight they lost.

a) We want to estimate the mean amount of weight lost by those doing aerobic exercise.

b) We want to know which program is more effective at reducing weight.

5. Couples at a dance club were separated and each person was asked to rate the band. Do men or women like this band more?

WHAT CAN GO WRONG?

- **Don't use a two-sample *t*-test when you have paired data.** See the What Can Go Wrong? discussion in Chapter 23.

- **Don't use a paired *t* method when the samples aren't paired.** Just because two groups have the same number of observations doesn't mean they can be paired, even if they are shown side by side in a table. We might have 25 men and 25 women in our study, but they could be completely independent of one another. If they were siblings or spouses, we might consider them paired. Remember that you cannot *choose* which method to use based on your preferences. If the data are from two independent samples, use two-sample *t* methods. If the data are from an experiment in which observations were paired, you must use a paired method. If the data are from an observational study, you must be able to defend your decision to use matched pairs or independent groups.

- **Don't forget to look for outliers.** The outliers we care about now are in the differences. A subject who is extraordinary both before and after a treatment may still have a perfectly typical difference. But one outlying difference can completely distort your conclusions. Be sure to plot the differences (even if you also plot the data).

- **Don't look for the difference between the means of paired groups with side-by-side boxplots or histograms.** The point of the paired analysis is to remove extra variation. Separate displays of each group still contain that variation. Comparing them is likely to be misleading.

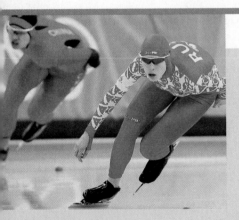

What Have We Learned?

When we looked at study designs in Chapters 11 and 12 we saw that pairing can be a very effective strategy. Because pairing helps control variability between individual subjects, paired methods are usually more powerful than methods that compare independent groups. Now we've learned that analyzing data from matched pairs requires different inference procedures.

- We've learned to think about the design of the study that collected the data to recognize when data are paired or matched, and when they are not. Paired data cannot be analyzed using independent *t*-procedures.

- We've learned (again) the importance of checking the appropriate assumptions and conditions before proceeding with inference.

- We've learned that paired *t*-methods look at pairwise differences, analyzing them using the mechanics of one-sample *t*-methods we saw in Chapter 22.

- We've learned to construct a confidence interval for the mean difference based on paired data.

- We've learned to test a hypothesis about the mean difference based on paired data.

And once again we've learned that the reasoning of inference and the proper interpretation of confidence intervals and P-values remain the same.

Terms

Paired data Data are paired when the observations are collected in pairs or the observations in one group are naturally related to observations in the other. The simplest form of pairing is to measure each subject twice—often before and after a treatment is applied. More sophisticated forms of pairing in experiments are a form of blocking and arise in other contexts. Pairing in observational and survey data is a form of matching. (p. 635)

Paired *t*-test A hypothesis test for the mean of the pairwise differences of two groups. It tests the null hypothesis

$$H_0: \mu_d = \Delta_0,$$

where the hypothesized difference is almost always 0, using the statistic

$$t = \frac{\bar{d} - \Delta_0}{SE(\bar{d})}$$

with $n - 1$ degrees of freedom, where $SE(\bar{d}) = \dfrac{s_d}{\sqrt{n}}$, and n is the number of pairs. (p. 636)

Paired *t* confidence interval A confidence interval for the mean of the pairwise differences between paired groups found as

$$\bar{d} \pm t^{*}_{n-1} \times SE(\bar{d}), \text{ where } SE(\bar{d}) = \frac{s_d}{\sqrt{n}} \text{ and } n \text{ is the number of pairs.} \quad \text{(p. 642)}$$

On the Computer PAIRED *t* INFERENCE

Most statistics programs can compute paired *t* analyses. The computer, of course, cannot verify that the variables are naturally paired. Most programs will automatically omit any pair that is missing a value for either variable (as we did with the British couples). You must look carefully to see whether that has happened.

As we've seen with other inference results, some packages pack a lot of information into a simple table, but you must locate what you want for yourself. Here's a generic example with comments:

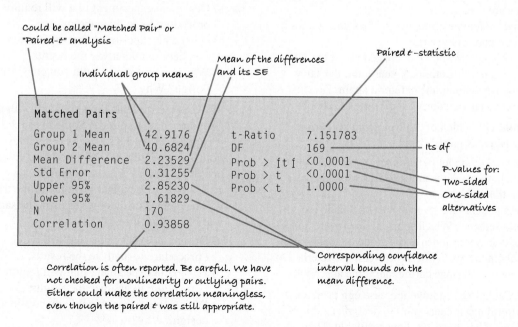

Could be called "Matched Pair" or "Paired-t" analysis

Individual group means

Mean of the differences and its SE

Paired t-statistic

```
Matched Pairs

Group 1 Mean      42.9176      t-Ratio      7.151783
Group 2 Mean      40.6824      DF           169
Mean Difference    2.23529     Prob > |t|   <0.0001
Std Error          0.31255     Prob > t     <0.0001
Upper 95%          2.85230     Prob < t      1.0000
Lower 95%          1.61829
N                170
Correlation        0.93858
```

Its df

P-values for:
Two-sided
One-sided alternatives

Corresponding confidence interval bounds on the mean difference.

Correlation is often reported. Be careful. We have not checked for nonlinearity or outlying pairs. Either could make the correlation meaningless, even though the paired *t* was still appropriate.

(continued)

Other packages try to be more descriptive. It may be easier to find the results, but you may get less information from the output table.

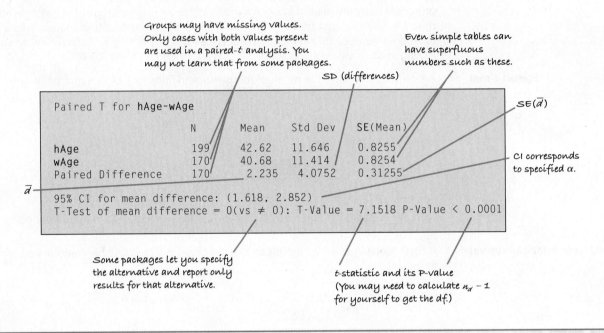

Groups may have missing values. Only cases with both values present are used in a paired-t analysis. You may not learn that from some packages.

Even simple tables can have superfluous numbers such as these.

SD (differences)

SE($\bar{d}$)

```
Paired T for hAge-wAge

                  N      Mean   Std Dev   SE(Mean)
hAge             199     42.62   11.646    0.8255
wAge             170     40.68   11.414    0.8254
Paired Difference 170    2.235   4.0752    0.31255

95% CI for mean difference: (1.618, 2.852)
T-Test of mean difference = 0(vs ≠ 0): T-Value = 7.1518 P-Value < 0.0001
```

$\bar{d}$

CI corresponds to specified α.

Some packages let you specify the alternative and report only results for that alternative.

t-statistic and its P-value (You may need to calculate $n_d - 1$ for yourself to get the df.)

Exercises

1. **Which method?** Which of the following scenarios should be analyzed as paired data?

   a) Students take a MCAT prep course. Their before and after scores are compared.

   b) 20 male and 20 females students in class take a midterm. We compare their scores.

   c) A group of college freshmen are asked about the quality of the university cafeteria. A year later, the same students are asked about the cafeteria again. Do student's opinions change during their time at school?

2. **Which method II?** Which of the following scenarios should be analyzed as paired data?

   a) Spouses are asked about the number of hours of sleep they get each night. We want to see if husbands get more sleep than wives.

   b) 50 insomnia patients are given a placebo and 50 are given a mild sedative. Which subjects sleep more hours?

   c) A group of college freshmen and a group of sophomores are asked about the quality of the university cafeteria. Do students' opinions change during their time at school?

3. **More eggs?** Can a food additive increase egg production? Agricultural researchers want to design an experiment to find out. They have 100 hens available. They

have two kinds of feed: the regular feed and the new feed with the additive. They plan to run their experiment for a month, recording the number of eggs each hen produces.

   a) Design an experiment that will require a two-sample t procedure to analyze the results.

   b) Design an experiment that will require a matched-pairs t procedure to analyze the results.

   c) Which experiment would you consider the stronger design? Why?

4. **MTV** Some students do homework with the TV on. (Anyone come to mind?) Some researchers want to see if people can work as effectively with as without distraction. The researchers will time some volunteers to see how long it takes them to complete some relatively easy crossword puzzles. During some of the trials, the room will be quiet; during other trials in the same room, a TV will be on, tuned to MTV.

   a) Design an experiment that will require a two-sample t procedure to analyze the results.

   b) Design an experiment that will require a matched-pairs t procedure to analyze the results.

   c) Which experiment would you consider the stronger design? Why?

5. **Sex sells?** Ads for many products use sexual images to try to attract attention to the product. But do these ads bring people's attention to the item that was being advertised? We want to design an experiment to see if the presence of sexual images in an advertisement affects people's ability to remember the product.
   a) Describe an experimental design requiring a matched-pairs *t* procedure to analyze the results.
   b) Describe an experimental design requiring an independent sample procedure to analyze the results.

6. **Freshman 15?** Many people believe that students gain weight as freshmen. Suppose we plan to conduct a study to see if this is true.
   a) Describe a study design that would require a matched-pairs *t* procedure to analyze the results.
   b) Describe a study design that would require a two-sample *t* procedure to analyze the results.

7. **Women** Values for the labor force participation rate of women (LFPR) are published by the U.S. Bureau of Labor Statistics. We are interested in whether there was a difference between female participation in 1968 and 1972, a time of rapid change for women. We check LFPR values for 19 randomly selected cities for 1968 and 1972. Shown below is software output for two possible tests:

   Paired t-Test of $\mu(1-2)$
   Test Ho: $\mu(1972\text{-}1968) = 0$ vs Ha: $\mu(1972-1968) \neq 0$
   Mean of Paired Differences = 0.0337
   t-Statistic = 2.458 w/18 df   p = 0.0244

   2-Sample t-Test of $\mu1 - \mu2$
   Ho: $\mu1 - \mu2 = 0$ vs Ha: $\mu1 - \mu2 \neq 0$
   Test Ho: $\mu(1972) - \mu(1968) = 0$ vs
   Ha: $\mu(1972) - \mu(1968) \neq 0$
   Difference Between Means = 0.0337
   t-Statistic = 1.496 w/35 df   p = 0.1434

   a) Which of these tests is appropriate for these data? Explain.
   b) Using the test you selected, state your conclusion.

8. **Rain** Simpson, Alsen, and Eden (*Technometrics* 1975) report the results of trials in which clouds were seeded and the amount of rainfall recorded. The authors report on 26 seeded and 26 unseeded clouds in order of the amount of rainfall, largest amount first. Here are two possible tests to study the question of whether cloud seeding works. Which test is appropriate for these data? Explain your choice. Using the test you select, state your conclusion.

   Paired t-Test of $\mu(1-2)$
   Mean of Paired Differences = −277.39615
   t-Statistic = −3.641 w/25 df   p = 0.0012

   2-Sample t-Test of $\mu1 - \mu2$
   Difference Between Means = −277.4
   t-Statistic = −1.998 w/33 df   p = 0.0538

   a) Which of these tests is appropriate for these data? Explain.
   b) Using the test you selected, state your conclusion.

9. **Friday the 13th, I** In 1993 the *British Medical Journal* published an article titled, "Is Friday the 13th Bad for Your Health?" Researchers in Britain examined how Friday the 13th affects human behavior. One question was whether people tend to stay at home more on Friday the 13th. The data below are the number of cars passing Junctions 9 and 10 on the M25 motorway for consecutive Fridays (the 6th and 13th) for five different periods.

Year	Month	6th	13th
1990	July	134,012	132,908
1991	September	133,732	131,843
1991	December	121,139	118,723
1992	March	124,631	120,249
1992	November	117,584	117,263

   Here are summaries of two possible analyses:

   Paired t-Test of mu$(1-2) = 0$ vs. mu$(1-2) > 0$
   Mean of Paired Differences: 2022.4
   t-Statistic = 2.9377 w/4 df   P = 0.0212

   2-Sample t-Test of mu1 = mu2 vs. mu1 > mu2
   Difference Between Means: 2022.4
   t-Statistic = 0.4273 w/7.998 df   P = 0.3402

   a) Which of the tests is appropriate for these data? Explain.
   b) Using the test you selected, state your conclusion.
   c) Are the assumptions and conditions for inference met?

10. **Friday the 13th, II:** The researchers in Exercise 9 also examined the number of people admitted to emergency rooms for vehicular accidents on 12 Friday evenings (6 each on the 6th and 13th).

Year	Month	6th	13th
1989	October	9	13
1990	July	6	12
1991	September	11	14
1991	December	11	10
1992	March	3	4
1992	November	5	12

    Based on these data, is there evidence that more people are admitted, on average, on Friday the 13th? Here and on the next page are two possible analyses:

    Paired t-Test of mu$(2-1) = 0$ vs. mu$(2-1) > 0$
    Mean of Paired Differences = 3.333
    t-Statistic = 2.7116 w/5 df   P = 0.0211

2-Sample t-Test of mu2 = mu1 vs. mu2 > mu1
Difference Between Means = 3.333
t-Statistic = 1.6644 w/9.940 df   P = 0.0636

a) Which of these tests is appropriate for these data? Explain.
b) Using the test you selected, state your conclusion.
c) Are the assumptions and conditions for inference met?

11. Online insurance I After seeing countless commercials claiming one can get cheaper car insurance from an online company, a local insurance agent was concerned that he might lose some customers. To investigate, he randomly selected profiles (type of car, coverage, driving record, etc.) for 10 of his clients and checked online price quotes for their policies. The comparisons are shown in the table below. His statistical software produced the following summaries (where *PriceDiff* = *Local* − *Online*):

Variable	Count	Mean	StdDev
Local	10	799.200	229.281
Online	10	753.300	256.267
PriceDiff	10	45.9000	175.663

Local	Online	PriceDiff
568	391	177
872	602	270
451	488	−37
1229	903	326
605	677	−72
1021	1270	−249
783	703	80
844	789	55
907	1008	−101
712	702	10

At first, the insurance agent wondered whether there was some kind of mistake in this output. He thought the Pythagorean Theorem of Statistics should work for finding the standard deviation of the price differences—in other words, that $SD(Local - Online) = \sqrt{SD^2(Local) + SD^2(Online)}$. But when he checked, he found that $\sqrt{(229.281)^2 + (256.267)^2} = 343.864$, not 175.663 as given by the software. Tell him where his mistake is.

12. Windy, part I To select the site for an electricity-generating wind turbine, wind speeds were recorded at several potential sites every 6 hours for a year. Two sites not far from each other looked good. Each had a mean wind speed high enough to qualify, but we should choose the site with a higher average daily wind speed. Because the sites are near each other and the wind speeds were recorded at the same times, we should view the speeds as paired. Here are the summaries of the speeds (in miles per hour):

Variable	Count	Mean	StdDev
site2	1114	7.452	3.586
site4	1114	7.248	3.421
site2–site4	1114	0.204	2.551

Is there a mistake in this output? Why doesn't the Pythagorean Theorem of Statistics work here? In other words, shouldn't
$SD(site2 - site4) = \sqrt{SD^2(site2) + SD^2(site4)}$?
But $\sqrt{(3.586)^2 + (3.421)^2} = 4.956$, not 2.551 as given by the software. Explain why this happened.

13. Online insurance II In Exercise 11, we saw summary statistics for 10 drivers' car insurance premiums quoted by a local agent and an online company. Here are displays for each company's quotes and for the difference (*Local – Online*):

a) Which of the summaries would help you decide whether the online company offers cheaper insurance? Why?
b) The standard deviation of *PriceDiff* is quite a bit smaller than the standard deviation of prices quoted by either the local or online companies. Discuss why.
c) Using the information you have, discuss the assumptions and conditions for inference with these data.

14. Windy, part II In Exercise 12, we saw summary statistics for wind speeds at two sites near each other, both being considered as locations for an electricity-generating wind turbine. The data, recorded every 6 hours for a year, showed each of the sites had a mean wind speed high enough to qualify, but how can we tell which site is best? Here are some displays:

site2–site4 (mph)

a) The boxplots show outliers for each site, yet the histogram shows none. Discuss why.

b) Which of the summaries would you use to select between these sites? Why?

c) Using the information you have, discuss the assumptions and conditions for paired *t* inference for these data. (*Hint:* Think hard about the independence assumption in particular.)

15. Online insurance 3 Exercises 11 and 13 give summaries and displays for car insurance premiums quoted by a local agent and an online company. Test an appropriate hypothesis to see if there is evidence that drivers might save money by switching to the online company.

16. Windy, part III Exercises 12 and 14 give summaries and displays for two potential sites for a wind turbine. Test an appropriate hypothesis to see if there is evidence that either of these sites has a higher average wind speed.

17. Cars and trucks We have data on the city and highway fuel efficiency of 316 cars and 316 trucks.

a) Would it be appropriate to use paired *t* methods to compare the cars and the trucks?

b) Would it be appropriate to use paired *t* methods to compare the city and highway fuel efficiencies of these vehicles?

c) A histogram of the differences (highway – city) is shown. Are the conditions for inference satisfied?

18. Weighing trucks One kind of scale for weighing trucks can measure their weight as they drive across a plate. Is this method consistent with the traditional method of static weighing? Are the conditions for

matched pairs inference satisfied? Weights are in 1000s of pounds.

Weight-in-Motion	Static Weight	Diff (Static − Motion)
26.0	27.9	−1.9
29.9	29.1	0.8
39.5	38.0	1.5
25.1	27.0	−1.9
31.6	30.3	1.3
36.2	34.5	1.7
25.1	27.8	−2.7
31.0	29.6	1.4
35.6	33.1	2.5
40.2	35.5	4.7

19. Cars and trucks again In Exercise 17, after deleting an outlying value of −27, the mean difference in fuel efficiencies for the 632 vehicles was 7.37 mpg with a standard deviation of 2.52 mpg. Find a 95% confidence interval for this difference and interpret it in context.

20. Weighing trucks II Find a 98% confidence interval of the weight differences in Exercise 18. Interpret this interval in context.

21. Blocking cars and trucks Thinking about the data on fuel efficiency in Exercise 17, why is the blocking accomplished by a matched pairs analysis particularly important for a sample that has both cars and trucks?

22. Weighing trucks III Consider the weights from Exercise 18. The side-by-side boxplots below show little difference between the two groups. Should this be sufficient to draw a conclusion about the accuracy of the weigh-in-motion scale?

Weight-in-Motion   Static Weight
Variable

23. Temperatures The table on the next page gives the average high temperatures in January and July for several European cities. Write a 90% confidence interval for the mean temperature difference between summer and winter in Europe. Be sure to check conditions for inference, and clearly explain what your interval means.

	Mean High Temperatures (°F)	
City	Jan.	July
Vienna	34	75
Copenhagen	36	72
Paris	42	76
Berlin	35	74
Athens	54	90
Rome	54	88
Amsterdam	40	69
Madrid	47	87
London	44	73
Edinburgh	43	65
Moscow	21	76
Belgrade	37	84

T 24. NY Marathon 2011 The table below shows the winning times (in minutes) for men and women in the New York City Marathon between 1978 and 2011. Assuming that performances in the Big Apple resemble performances elsewhere, we can think of these data as a sample of performance in marathon competitions. Create a 90% confidence interval for the mean difference in winning times for male and female marathon competitors. (www.nycmarathon.org)

Year	Men	Women	Year	Men	Women
1978	132.2	152.5	1995	131.0	148.1
1979	131.7	147.6	1996	129.9	148.3
1980	129.7	145.7	1997	128.2	148.7
1981	128.2	145.5	1998	128.8	145.3
1982	129.5	147.2	1999	129.2	145.1
1983	129.0	147.0	2000	130.2	145.8
1984	134.9	149.5	2001	127.7	144.4
1985	131.6	148.6	2002	128.1	145.9
1986	131.1	148.1	2003	130.5	142.5
1987	131.0	150.3	2004	129.5	143.2
1988	128.3	148.1	2005	129.5	144.7
1989	128.0	145.5	2006	130.0	145.1
1990	132.7	150.8	2007	129.1	143.2
1991	129.5	147.5	2008	128.7	143.9
1992	129.5	144.7	2009	129.3	148.9
1993	130.1	146.4	2010	128.3	148.3
1994	131.4	147.6	2011	125.1	143.3

T 25. Push-ups Every year the students at Gossett High School take a physical fitness test during their gym classes. One component of the test asks them to do as many push-ups as they can. Results for one class are shown below, separately for boys and girls. Assuming that students at Gossett are assigned to gym classes at random, create a 90% confidence interval for how many more push-ups boys can do than girls, on average, at that high school.

Boys	17	27	31	17	25	32	28	23	25	16	11	34
Girls	24	7	14	16	2	15	19	25	10	27	31	8

T 26. Brain waves An experiment was performed to see whether sensory deprivation over an extended period of time has any effect on the alpha-wave patterns produced by the brain. To determine this, 20 subjects, inmates in a Canadian prison, were randomly split into two groups. Members of one group were placed in solitary confinement. Those in the other group were allowed to remain in their own cells. Seven days later, alpha-wave frequencies were measured for all subjects, as shown in the following table. (P. Gendreau et al., "Changes in EEG Alpha Frequency and Evoked Response Latency During Solitary Confinement," *Journal of Abnormal Psychology* 79 [1972]: 54–59)

Nonconfined	Confined
10.7	9.6
10.7	10.4
10.4	9.7
10.9	10.3
10.5	9.2
10.3	9.3
9.6	9.9
11.1	9.5
11.2	9.0
10.4	10.9

a) What are the null and alternative hypotheses? Be sure to define all the terms and symbols you use.
b) Are the assumptions necessary for inference met?
c) Perform the appropriate test, indicating the formula you used, the calculated value of the test statistic, the df, and the P-value.
d) State your conclusion.

T 27. Job satisfaction (When you first read about this exercise break plan in Chapter 23, you did not have an inference method that would work. Try again now.) A company institutes an exercise break for its workers to see if it will improve job satisfaction, as measured by a questionnaire that assesses workers' satisfaction. Scores for 10 randomly selected workers before and after the implementation of the exercise program are shown in the table.

a) Identify the procedure you would use to assess the effectiveness of the exercise program, and check to see if the conditions allow the use of that procedure.
b) Test an appropriate hypothesis and state your conclusion.

c) If your conclusion turns out to be incorrect, what kind of error did you commit?

Worker Number	Job Satisfaction Index	
	Before	**After**
1	34	33
2	28	36
3	29	50
4	45	41
5	26	37
6	27	41
7	24	39
8	15	21
9	15	20
10	27	37

28. Summer school (When you first read about the summer school issue in Chapter 23 you did not have an inference method that would work. Try again now.) Having done poorly on their Math final exams in June, six students repeat the course in summer school and take another exam in August.

June	54	49	68	66	62	62
August	50	65	74	64	68	72

a) If we consider these students to be representative of all students who might attend this summer school in other years, do these results provide evidence that the program is worthwhile?

b) This conclusion, of course, may be incorrect. If so, which type of error was made?

29. Yogurt Is there a significant difference in calories between servings of strawberry and vanilla yogurt? Based on the data shown in the table, test an appropriate hypothesis and state your conclusion. Don't forget to check assumptions and conditions!

		Calories per Serving	
		Strawberry	**Vanilla**
Brand	America's Choice	210	200
	Breyer's Lowfat	220	220
	Columbo	220	180
	Dannon Light 'n Fit	120	120
	Dannon Lowfat	210	230
	Dannon la Crème	140	140
	Great Value	180	80
	La Yogurt	170	160
	Mountain High	200	170
	Stonyfield Farm	100	120
	Yoplait Custard	190	190
	Yoplait Light	100	100

30. Gasoline Many drivers of cars that can run on regular gas actually buy premium in the belief that they will get better gas mileage. To test that belief, we use 10 cars from a company fleet in which all the cars run on regular gas. Each car is filled first with either regular or premium gasoline, decided by a coin toss, and the mileage for that tankful is recorded. Then the mileage is recorded again for the same cars for a tankful of the other kind of gasoline. We don't let the drivers know about this experiment.

Here are the results (miles per gallon):

Car #	1	2	3	4	5	6	7	8	9	10
Regular	16	20	21	22	23	22	27	25	27	28
Premium	19	22	24	24	25	25	26	26	28	32

a) Is there evidence that cars get significantly better fuel economy with premium gasoline?

b) How big might that difference be? Check a 90% confidence interval.

c) Even if the difference is significant, why might the company choose to stick with regular gasoline?

d) Suppose you had done a "bad thing." (We're sure you didn't.) Suppose you had mistakenly treated these data as two independent samples instead of matched pairs. What would the significance test have found? Carefully explain why the results are so different.

31. Braking test A tire manufacturer tested the braking performance of one of its tire models on a test track. The company tried the tires on 10 different cars, recording the stopping distance for each car on both wet and dry pavement. Results are shown in the table.

	Stopping Distance (ft)	
Car #	**Dry Pavement**	**Wet Pavement**
1	150	201
2	147	220
3	136	192
4	134	146
5	130	182
6	134	173
7	134	202
8	128	180
9	136	192
10	158	206

a) Write a 95% confidence interval for the mean dry pavement stopping distance. Be sure to check the appropriate assumptions and conditions, and explain what your interval means.

b) Write a 95% confidence interval for the mean increase in stopping distance on wet pavement. Be sure to check the appropriate assumptions and conditions, and explain what your interval means.

32. Braking test 2 For another test of the tires in Exercise 31, a car made repeated stops from 60 miles per hour. The test was run on both dry and wet pavement, with results as shown in the table. (Note that actual *braking distance*, which takes into account the driver's reaction time, is much longer, typically nearly 300 feet at 60 mph!)

a) Write a 95% confidence interval for the mean dry pavement stopping distance. Be sure to check the appropriate assumptions and conditions, and explain what your interval means.

b) Write a 95% confidence interval for the mean increase in stopping distance on wet pavement. Be sure to check the appropriate assumptions and conditions, and explain what your interval means.

Stopping Distance (ft)	
Dry Pavement	**Wet Pavement**
145	211
152	191
141	220
143	207
131	198
148	208
126	206
140	177
135	183
133	223

33. Tuition 2011 How much more do public colleges and universities charge out-of-state students for tuition per year? A random sample of 19 public colleges and universities listed at www.collegeboard.com yielded the following data.

Institution	Resident	Nonresident
Univ of Akron (OH)	9545	17468
Athens State (AL)	5340	9930
Ball State (IN)	8558	22538
Bloomsburg U (PA)	8082	17442
UC Irvine (CA)	13122	36000
Central State (OH)	5672	12648
Clarion U (PA)	8828	15068
Dakota State	7621	9336
Fairmont State (WV)	5326	11230
Johnson State (VT)	9468	19908
Lock Haven U (PA)	8239	15599
New College of Florida	6060	29088
Oakland U (MI)	9938	23190
U Pittsburgh	16132	25540
Savannah State (GA)	6032	17646
SE Louisiana	4634	14139
W Liberty Univ (WV)	5266	13140
W Texas College	2370	3750
Worcester State (MA)	7653	13733

a) Create a 90% confidence interval for the mean difference in cost. Be sure to justify your procedure.

b) Interpret your interval in context.

c) A national magazine claims that public institutions charge state residents an average of $7000 less than out-of-staters for tuition each year. What does your confidence interval indicate about this assertion?

34. Sex sells, part II In Exercise 11 you considered the question of whether sexual images in ads affected people's abilities to remember the item being advertised. To investigate, a group of Statistics students cut ads out of magazines. They were careful to find two ads for each of 10 similar items, one with a sexual image and one without. They arranged the ads in random order and had 39 subjects look at them for one minute. Then they asked the subjects to list as many of the products as they could remember. Their data are shown in the table. Is there evidence that the sexual images mattered?

Subject Number	Ads Remembered		Subject Number	Ads Remembered	
	Sexual Image	**No Sex**		**Sexual Image**	**No Sex**
1	2	2	21	2	3
2	6	7	22	4	2
3	3	1	23	3	3
4	6	5	24	5	3
5	1	0	25	4	5
6	3	3	26	2	4
7	3	5	27	2	2
8	7	4	28	2	4
9	3	7	29	7	6
10	5	4	30	6	7
11	1	3	31	4	3
12	3	2	32	4	5
13	6	3	33	3	0
14	7	4	34	4	3
15	3	2	35	2	3
16	7	4	36	3	3
17	4	4	37	5	5
18	1	3	38	3	4
19	5	5	39	4	3
20	2	2			

35. Strikes Advertisements for an instructional video claim that the techniques will improve the ability of Little League pitchers to throw strikes and that, after undergoing the training, players will be able to throw strikes on at least 60% of their pitches. To test this claim, we have 20 Little Leaguers throw 50 pitches each, and we record the number of strikes. After the players participate in the training program, we repeat the test. The table shows the number of strikes each player threw before and after the training.

a) Is there evidence that after training players can throw strikes more than 60% of the time?

b) Is there evidence that the training is effective in improving a player's ability to throw strikes?

Number of Strikes (out of 50)		Number of Strikes (out of 50)	
Before	After	Before	After
28	35	33	33
29	36	33	35
30	32	34	32
32	28	34	30
32	30	34	33
32	31	35	34
32	32	36	37
32	34	36	33
32	35	37	35
33	36	37	32

36. Freshman 15, revisited In Exercise 6 you thought about how to design a study to see if it's true that students tend to gain weight during their first year in college. Well, Cornell Professor of Nutrition David Levitsky did just that. He recruited students from two large sections of an introductory health course. Although they were volunteers, they appeared to match the rest of the freshman class in terms of demographic variables such as sex and ethnicity. The students were weighed during the first week of the semester, then again 12 weeks later. Based on Professor Levitsky's data, estimate the mean weight gain in first-semester freshmen and comment on the "freshman 15." (Weights are in pounds.)

Subject Number	Initial Weight	Terminal Weight	Subject Number	Initial Weight	Terminal Weight
1	171	168	35	148	150
2	110	111	36	164	165
3	134	136	37	137	138
4	115	119	38	198	201
5	150	155	39	122	124
6	104	106	40	146	146
7	142	148	41	150	151
8	120	124	42	187	192
9	144	148	43	94	96
10	156	154	44	105	105
11	114	114	45	127	130
12	121	123	46	142	144
13	122	126	47	140	143
14	120	115	48	107	107
15	115	118	49	104	105

Subject Number	Initial Weight	Terminal Weight	Subject Number	Initial Weight	Terminal Weight
16	110	113	50	111	112
17	142	146	51	160	162
18	127	127	52	134	134
19	102	105	53	151	151
20	125	125	54	127	130
21	157	158	55	106	108
22	119	126	56	185	188
23	113	114	57	125	128
24	120	128	58	125	126
25	135	139	59	155	158
26	148	150	60	118	120
27	110	112	61	149	150
28	160	163	62	149	149
29	220	224	63	122	121
30	132	133	64	155	158
31	145	147	65	160	161
32	141	141	66	115	119
33	158	160	67	167	170
34	135	134	68	131	131

37. Wheelchair marathon 2010 The Boston Marathon has had a wheelchair division since 1977. Who do you think is typically faster, the men's marathon winner on foot or the women's wheelchair marathon winner? Because the conditions differ from year to year, and speeds have improved over the years, it seems best to treat these as paired measurements. Here are summary statistics for the pairwise differences in finishing time (in minutes): (www.boston.com/sports/marathon/history/champions/)

Summary of wheelchrF − runM
N = 34
Mean = −3.57
SD = 35.083

a) Comment on the assumptions and conditions.

b) Assuming that these times are representative of such races and the differences appeared acceptable for inference, construct and interpret a 95% confidence interval for the mean difference in finishing times.

c) Would a hypothesis test at $\alpha = 0.05$ reject the null hypothesis of no difference? What conclusion would you draw?

38. Marathon start-up years 2010 When we considered the Boston Marathon in Exercise 37, we were unable to check the Nearly Normal Condition. Here's a histogram of the differences:

whF-rnM

Those three large differences are the first three years of wheelchair competition: 1977, 1978, and 1979. Often the start-up years of new events are different; later on, more athletes train and compete. If we omit those three years, the summary statistics change as follows:

Summary of wheelchrF − runM
N = 31
Mean = −12.794
SD = 18.64

a) Comment on the assumptions and conditions.

b) Assuming that these times are representative of such races, construct and interpret a 95%

confidence interval for the mean difference in finishing time.

c) Would a hypothesis test at $\alpha = 0.05$ reject the null hypothesis of no difference? What conclusion would you draw?

Just Checking ANSWERS

1. These are independent groups sampled at random, so use a two-sample t confidence interval to estimate the size of the difference.

2. There is only one sample. Use a one-sample t-interval.

3. A brother and sister from the same family represent a matched pair. The question calls for a paired t-test.

4. a) A before-and-after study calls for paired t methods. To estimate the loss, find a confidence interval for the before–after differences.

   b) The two treatment groups were assigned randomly, so they are independent. Use a two-sample t-test to assess whether the mean weight losses differ.

5. Sometimes it just isn't clear. Most likely, couples would discuss the band or even decide to go to the club because they both like a particular band. If we think that's likely, then these data are paired. But maybe not. If we asked them their opinions of, say, the decor or furnishings at the club, the fact that they were couples might not affect the independence of their answers.

Review of VI part

Learning About the World

Quick Review

We continue to explore how to answer questions about the statistics we get from samples and experiments. In this part, those questions have been about means—means of one sample, two independent samples, or matched pairs. Here's a brief summary of the key concepts and skills:

- A confidence interval uses a sample statistic to estimate a range of possible values for a parameter of interest.

- A hypothesis test proposes a model, then examines the plausibility of that model by seeing how surprising our observed data would be if the model were true.

- Statistical inference procedures for proportions are based on the Central Limit Theorem. We can make inferences about a single proportion or the difference of two proportions using Normal models.

- Statistical inference procedures for means are also based on the Central Limit Theorem, but we don't usually know the population standard deviation. Student's t-models take into account the additional uncertainty of independently estimating the standard deviation.

  - We can make inferences about one mean, the difference of two independent means, or the mean of paired differences using t-models.

  - No inference procedure is valid unless the underlying assumptions are true. Always check the conditions before proceeding.

- Because t-models assume that samples are drawn from Normal populations, data in the sample should appear to be nearly Normal. Skewness and outliers are particularly problematic, especially for small samples.

- When there are two variables, you must think carefully about how the data were collected. You may use two-sample t procedures only if the groups are independent.

- Unless there is some obvious reason to suspect that two independent populations have the same standard deviation, you should not pool the variances. It is never wrong to use unpooled t procedures.

- If the two groups are somehow paired, the data are *not* from independent groups. You must use matched-pairs t procedures.

Now for some opportunities to review these concepts. Be careful. You have a lot of thinking to do. These review exercises mix questions about proportions and means. You have to determine which of our inference procedures is appropriate in each situation. Then you have to check the proper assumptions and conditions. Keeping track of those can be difficult, so first we summarize the many procedures with their corresponding assumptions and conditions on the next page. Look them over carefully . . . then, on to the Exercises!

Quick Guide to Inference

Inference about?	One group or two?	Procedure	Model	Parameter	Estimate	SE	Chapter
			Think		**Show**		**Tell?**
Proportions	One sample	1-Proportion z-Interval	z	p	$\hat{p}$	$\sqrt{\dfrac{\hat{p}\hat{q}}{n}}$	19
		1-Proportion z-Test				$\sqrt{\dfrac{p_0 q_0}{n}}$	20, 21
	Two independent groups	2-Proportion z-Interval	z	$p_1 - p_2$	$\hat{p}_1 - \hat{p}_2$	$\sqrt{\dfrac{\hat{p}_1\hat{q}_1}{n_1} + \dfrac{\hat{p}_2\hat{q}_2}{n_2}}$	22
		2-Proportion z-Test				$\sqrt{\dfrac{\hat{p}\hat{q}}{n_1} + \dfrac{\hat{p}\hat{q}}{n_2}}, \hat{p} = \dfrac{y_1 + y_2}{n_1 + n_2}$	22
Means	One sample	t-Interval t-Test	t df $= n - 1$	μ	$\bar{y}$	$\dfrac{s}{\sqrt{n}}$	23
	Two independent groups	2-Sample t-Test 2-Sample t-Interval	t df from technology	$\mu_1 - \mu_2$	$\bar{y}_1 - \bar{y}_2$	$\sqrt{\dfrac{s_1^2}{n_1} + \dfrac{s_2^2}{n_2}}$	24
	Matched pairs	Paired t-Test Paired t-Interval	t df $= n - 1$	μ_d	$\bar{d}$	$\dfrac{s_d}{\sqrt{n}}$	25

Assumptions for Inference	**And the Conditions That Support or Override Them**

Proportions (z)
- **One sample**
    1. Individuals are independent.
    2. Sample is sufficiently large.

    1. SRS and $n < 10\%$ of the population.
    2. Successes and failures each ≥ 10.

- **Two groups**
    1. Groups are independent.
    2. Data in each group are independent.
    3. Both groups are sufficiently large.

    1. (Think about how the data were collected.)
    2. Both are SRSs and $n < 10\%$ of populations OR random allocation.
    3. Successes and failures each ≥ 10 for both groups.

Means (t)
- **One sample** (df $= n - 1$)
    1. Individuals are independent.
    2. Population has a Normal model.

    1. SRS and $n < 10\%$ of the population.
    2. Histogram is unimodal and symmetric.*

- **Matched pairs** (df $= n - 1$)
    1. Data are matched.
    2. Individuals are independent.
    3. Population of differences is Normal.

    1. (Think about the design.)
    2. SRS and $n < 10\%$ OR random allocation.
    3. Histogram of differences is unimodal and symmetric.*

- **Two independent groups** (df from technology)
    1. Groups are independent.
    2. Data in each group are independent.
    3. Both populations are Normal.

    1. (Think about the design.)
    2. SRSs and $n < 10\%$ OR random allocation.
    3. Both histograms are unimodal and symmetric.*

    (*less critical as n increases)

Review Exercises

1. **Crawling** A study found that babies born at different times of the year may develop the ability to crawl at different ages! The author of the study suggested that these differences may be related to the temperature at the time the infant is 6 months old. (Benson and Janette, *Infant Behavior and Development* [1993])

    a) The study found that 32 babies born in January crawled at an average age of 29.84 weeks, with a standard deviation of 7.08 weeks. Among 21 July babies, crawling ages averaged 33.64 weeks, with a standard deviation of 6.91 weeks. Is this difference significant?

    b) For 26 babies born in April the mean and standard deviation were 31.84 and 6.21 weeks, while for 44 October babies the mean and standard deviation of crawling ages were 33.35 and 7.29 weeks. Is this difference significant?

    c) Are these results consistent with the researcher's conjecture?

2. **Mazes and smells** Can pleasant smells improve learning? Researchers timed 21 subjects as they tried to complete paper-and-pencil mazes. Each subject attempted a maze both with and without the presence of a floral aroma. Subjects were randomized with respect to whether they did the scented trial first or second. Is there any evidence that the floral scent improved the subjects' ability to complete the mazes? (A. R. Hirsch and L. H. Johnston, "Odors and Learning." Chicago: Smell and Taste Treatment and Research Foundation)

Time to Complete the Maze (sec)	
Unscented	Scented
25.7	30.2
41.9	56.7
51.9	42.4
32.2	34.4
64.7	44.8
31.4	42.9
40.1	42.7
43.2	24.8
33.9	25.1
40.4	59.2
58.0	42.2
61.5	48.4
44.6	32.0
35.3	48.1
37.2	33.7
39.4	42.6
77.4	54.9
52.8	64.5
63.6	43.1
56.6	52.8
58.9	44.3

3. **Women** The U.S. Census Bureau reports that 26% of all U.S. businesses are owned by women. A Colorado consulting firm surveys a random sample of 410 businesses in the Denver area and finds that 115 of them have women owners. Should the firm conclude that its area is unusual? Test an appropriate hypothesis and state your conclusion.

4. **Drugs** In a full-page ad that ran in many U.S. newspapers in August 2002, a Canadian discount pharmacy listed costs of drugs that could be ordered from a Web site in Canada. The table compares prices (in US$) for commonly prescribed drugs.

	Cost per 100 Pills		
Drug Name	United States	Canada	Percent savings
Cardizem	131	83	37
Celebrex	136	72	47
Cipro	374	219	41
Pravachol	370	166	55
Premarin	61	17	72
Prevacid	252	214	15
Prozac	263	112	57
Tamoxifen	349	50	86
Vioxx	243	134	45
Zantac	166	42	75
Zocor	365	200	45
Zoloft	216	105	51

    a) Give a 95% confidence interval for the average savings in dollars.

    b) Give a 95% confidence interval for the average savings in percent.

    c) Which analysis is more appropriate? Why?

    d) In small print the newspaper ad says, "Complete list of all 1500 drugs available on request." How does this comment affect your conclusions above?

5. **Pottery** Archaeologists can use the chemical composition of clay found in pottery artifacts to determine whether different sites were populated by the same ancient people. They collected five samples of Romano–British pottery from each of two sites in Great Britain and measured the percentage of aluminum oxide in each. Based on these data, do you think the same people used these two kiln sites? Base your conclusion on a 95% confidence interval for the difference in aluminum oxide content of pottery made at the sites. (A. Tubb, A. J. Parker, and G. Nickless, "The Analysis of Romano–British Pottery by Atomic Absorption Spectrophotometry." *Archaeometry*, 22[1980]:153–171)

Ashley Rails	19.1	14.8	16.7	18.3	17.7
New Forest	20.8	18.0	18.0	15.8	18.3

6. **Streams** Researchers in the Adirondack Mountains collect data on a random sample of streams each year. One of the variables recorded is the substrate of the streams—the type of soil and rock over which they flow. The researchers found that 69 of the 172 sampled streams had a substrate of shale. Construct a 95% confidence interval for the proportion of Adirondack streams with a shale substrate. Clearly interpret your interval in context.

7. Gehrig Ever since Lou Gehrig developed amyotrophic lateral sclerosis (ALS), this deadly condition has been commonly known as Lou Gehrig's disease. Some believe that ALS is more likely to strike athletes or the very fit. Columbia University neurologist Lewis P. Rowland recorded personal histories of 431 patients he examined between 1992 and 2002. He diagnosed 280 as having ALS; 38% of them had been varsity athletes. The other 151 had other neurological disorders, and only 26% of them had been varsity athletes. (*Science News*, Sept. 28 [2002])

a) Is there evidence that ALS is more common among athletes?

b) What kind of study is this? How does that affect the inference you made in part a?

8. Teen drinking A study of the health behavior of school-aged children asked a sample of 15-year-olds in several different countries if they had been drunk at least twice. The results are shown in the table, by gender. Give a 95% confidence interval for the difference in the rates for males and females. Be sure to check the assumptions that support your chosen procedure, and explain what your interval means. (*Health and Health Behavior Among Young People*. Copenhagen: World Health Organization, 2000)

Percent of 15-Year-OldsDrunk at Least Twice		
Country	Female	Male
Denmark	63	71
Wales	63	72
Greenland	59	58
England	62	51
Finland	58	52
Scotland	56	53
No. Ireland	44	53
Slovakia	31	49
Austria	36	49
Canada	42	42
Sweden	40	40
Norway	41	37
Ireland	29	42
Germany	31	36
Latvia	23	47
Estonia	23	44
Hungary	22	43
Poland	21	39
USA	29	34
Czech Rep.	22	36
Belgium	22	36
Russia	25	32
Lithuania	20	32
France	20	29
Greece	21	24
Switzerland	16	25
Israel	10	18

9. Babies The National Perinatal Statistics Unit of the Sydney Children's Hospital reports that the mean birth weight of all babies born in Australia in 1999 was 3361 grams—about 7.41 pounds. A Missouri hospital reports that the average weight of 112 babies born there last year was 7.68 pounds, with a standard deviation of 1.31 pounds. If we believe the Missouri babies fairly represent American newborns, is there any evidence that U.S. babies and Australian babies do not weigh the same amount at birth?

10. Petitions To get a voter initiative on a state ballot, petitions that contain at least 250,000 valid voter signatures must be filed with the Elections Commission. The board then has 60 days to certify the petitions. A group wanting to create a statewide system of universal health insurance has just filed petitions with a total of 304,266 signatures. As a first step in the process, the Board selects an SRS of 2000 signatures and checks them against local voter lists. Only 1772 of them turn out to be valid.

a) What percent of the sample signatures were valid?

b) What percent of the petition signatures submitted must be valid in order to have the initiative certified by the Elections Commission?

c) What will happen if the Elections Commission commits a Type I error?

d) What will happen if the Elections Commission commits a Type II error?

e) Does the sample provide evidence in support of certification? Explain.

f) What could the Elections Commission do to increase the power of the test?

11. Feeding fish In the midwestern United States, a large aquaculture industry raises largemouth bass. Researchers wanted to know whether the fish would grow better if fed a natural diet of fathead minnows or an artificial diet of food pellets. They stocked six ponds with bass fingerlings weighing about 8 grams. For one year, the fish in three of the ponds were fed minnows, and the others were fed the commercially prepared pellets. The fish were then harvested, weighed, and measured. The bass fed a natural food source had a higher average length (19.6 cm) and weight (95.9 g) than those fed the commercial fish food (17.3 cm and 72.0 g, respectively). The researchers reported P-values for differences in both measurements to be less than 0.001.

a) Explain to someone who has not studied Statistics what the P-values mean here.

b) What advice should the researchers give the people who raise largemouth bass?

c) If that advice turns out to be incorrect, what type of error occurred?

12. Risk A study of auto safety determined the number of driver deaths per million vehicle sales, classified by type of vehicle. The data on the next page are for 6 midsize models and 6 SUVs. Wondering if there is evidence that drivers of SUVs are safer, we hope to create a 95% confidence interval for the difference in driver death rates for the two types of vehicles. Are these data appropriate for this inference? Explain. (Ross and Wenzel, *An Analysis of Traffic Deaths by Vehicle Type and Model*, March 2002)

Midsize	47	54	64	76	88	97
SUV	55	60	62	76	91	109

13. Age In a study of how depression may affect one's ability to survive a heart attack, the researchers reported the ages of the two groups they examined. The mean age of 2397 patients without cardiac disease was 69.8 years (SD = 8.7 years), while for the 450 patients with cardiac disease, the mean and standard deviation of the ages were 74.0 and 7.9, respectively.

a) Create a 95% confidence interval for the difference in mean ages of the two groups.
b) How might an age difference confound these research findings about the relationship between depression and ability to survive a heart attack?

14. Smoking In the depression and heart attack research described in Exercise 13, 32% of the diseased group were smokers, compared with only 23.7% of those free of heart disease.

a) Create a 95% confidence interval for the difference in the proportions of smokers in the two groups.
b) Is this evidence that the two groups in the study were different? Explain.
c) Could this be a problem in analyzing the results of the study? Explain.

15. Computer use A Gallup telephone poll of 1240 teens conducted in 2001 found that boys were more likely than girls to play computer games, by a margin of 77% to 65%. Equal numbers of boys and girls were surveyed.

a) What kind of sampling design was used?
b) Give a 95% confidence interval for the difference in game playing by gender.
c) Does your confidence interval suggest that among all teens a higher percentage of boys than girls play computer games?

16. Recruiting In September 2002, CNN reported on a method of grad student recruiting by the Haas School of Business at U.C.-Berkeley. The school notifies applicants by formal letter that they have been admitted, and also e-mails the accepted students a link to a Web site that greets them with personalized balloons, cheering, and applause. The director of admissions says this extra effort at recruiting has really worked well. The school accepts 500 applicants each year, and the percentage that actually choose to enroll at Berkeley increased from 52% the year before the Web greeting to 54% this year.

a) Create a 95% confidence interval for the change in enrollment rates.
b) Based on your confidence interval, are you convinced that this new form of recruiting has been effective? Explain.

17. Bimodal We are sampling randomly from a distribution known to be bimodal.

a) As our sample size increases, what's the expected shape of the sample's distribution?
b) What's the expected value of our sample's mean? Does the size of the sample matter?
c) How is the variability of sample means related to the standard deviation of the population? Does the size of the sample matter?
d) How is the shape of the sampling distribution model affected by the sample size?

18. Eggs The ISA Babcock Company supplies poultry farmers with hens, advertising that a mature B300 Layer produces eggs with a mean weight of 60.7 grams. Suppose that egg weights follow a Normal model with standard deviation 3.1 grams.

a) What fraction of the eggs produced by these hens weigh more than 62 grams?
b) What's the probability that a dozen randomly selected eggs average more than 62 grams?
c) Using the 68–95–99.7 Rule, sketch a model of the total weights of a dozen eggs.

T **19. Hearing** Fitting someone for a hearing aid requires assessing the patient's hearing ability. In one method of assessment, the patient listens to a tape of 50 English words. The tape is played at low volume, and the patient is asked to repeat the words. The patient's hearing ability score is the number of words perceived correctly. Four tapes of equivalent difficulty are available so that each ear can be tested with more than one hearing aid. These lists were created to be equally difficult to perceive in silence, but hearing aids must work in the presence of background noise. Researchers had 24 subjects with normal hearing compare two of the tapes when a background noise was present, with the order of the tapes randomized. Is it

Subject	List A	List B
1	24	26
2	32	24
3	20	22
4	14	18
5	32	24
6	22	30
7	20	22
8	26	28
9	26	30
10	38	16
11	30	18
12	16	34
13	36	32
14	32	34
15	38	32
16	14	18
17	26	20
18	14	20
19	38	40
20	20	26
21	14	14
22	18	14
23	22	30
24	34	42

reasonable to assume that the two lists are still equivalent for purposes of the hearing test when there is background noise? Base your decision on a confidence interval for the mean difference in the number of words people might misunderstand. (Faith Loven, *A Study of the Interlist Equivalency of the CID W-22 Word List Presented in Quiet and in Noise.* University of Iowa [1981])

20. Cesareans Some people fear that differences in insurance coverage can affect healthcare decisions. A survey of several randomly selected hospitals found that 16.6% of 223 recent births in Vermont involved cesarean deliveries, compared to 18.8% of 186 births in New Hampshire. Is this evidence that the rate of cesarean births in the two states is different?

T 21. Newspapers Who reads the newspaper more, men or women? Eurostat, an agency of the European Union (EU), conducts surveys on several aspects of daily life in EU countries. Recently, the agency asked samples of 1000 respondents in each of 14 European countries whether they read the newspaper on a daily basis. The table on the next page shows the data.

Country	% Reading a Newspaper Daily	
	Men	Women
Belgium	56.3	45.5
Denmark	76.8	70.3
Germany	79.9	76.8
Greece	22.5	17.2
Spain	46.2	24.8
Ireland	58.0	54.0
Italy	50.2	29.8
Luxembourg	71.0	67.0
Netherlands	71.3	63.0
Austria	78.2	74.1
Portugal	58.3	24.1
Finland	93.0	90.0
Sweden	89.0	88.0
UK	32.6	30.4

a) Examine the differences in the percentages for each country. Which of these countries seem to be outliers? What do they have in common?
b) After eliminating the outliers, is there evidence that in Europe men are more likely than women to read the newspaper?

T 22. Meals A college student is on a "meal program." His budget allows him to spend an average of $10 per day for the semester. He keeps track of his daily food expenses for 2 weeks; the data are given in the table. Is there strong evidence that he will overspend his food allowance? Explain.

Date	Cost ($)	Date	Cost ($)
7/29	15.20	8/5	8.55
7/30	23.20	8/6	20.05
7/31	3.20	8/7	14.95
8/1	9.80	8/8	23.45
8/2	19.53	8/9	6.75
8/3	6.25	8/10	0
8/4	0	8/11	9.01

23. Occupy Wall Street In 2011, the Occupy Wall Street movement protested the concentration of wealth and power in the United States. A 2012 University of Delaware survey asked a random sample of 901 American adults whether they agreed or disagreed with the following statement:

> The Occupy Wall Street protesters offered new insights on social issues.

Of those asked, 59.9% said they strongly or somewhat agreed with this statement. We know that if we could ask the entire population of American adults, we would not find that exactly 59.9% think that Wall Street workers would be willing to break the law to make money. Construct a 95% confidence interval for the true percentage of American adults who agree with the statement.

24. Power We are replicating an experiment. How will each of the following changes affect the power of our test? Indicate whether it will increase, decrease, or remain the same, assuming that all other aspects of the situation remain unchanged.

a) We increase the number of subjects from 40 to 100.
b) We require a higher standard of proof, changing from $\alpha = 0.05$ to $\alpha = 0.01$.

25. Herbal cancer A report in the *New England Journal of Medicine* notes growing evidence that the herb *Aristolochia fangchi* can cause urinary tract cancer in those who take it. Suppose you are asked to design an experiment to study this claim. Imagine that you have data on urinary tract cancers in subjects who have used this herb and similar subjects who have not used it and that you can measure incidences of cancer and precancerous lesions in these subjects. State the null and alternative hypotheses you would use in your study.

26. Free throws 2011 During the 2010–2011 NBA season, Stephen Curry led the league by making 212 of 227 free throws, for a success rate of 93.39%. But Chauncey Billups was close behind, with 384 of 419 (91.65%).

a) Find a 95% confidence interval for the difference in their free throw percentages.
b) Based on your confidence interval, is it certain that Curry is better than Billups at making free throws?

27. Rain and fire At the University of California, Riverside, Dr. Richard Minnich collected data on the rainfall in the areas east of Los Angeles. He noted that an increase in rainfall was responsible for an increase in wildfires over these years. Here is the rainfall data from three regions in the area. (*SMCMA Quarterly* 42(3), Richard A. Minnich)

Annual Precipitation (cm)			
Year	Victorville	29 Palms	Mitchell's Cavern
1976–77	166	140	65
1977–78	257	228	194
1978–79	180	137	113
1979–80	191	221	270
1980–81	58	111	92
1981–82	120	52	90
1982–83	257	220	178
1983–84	74	215	119
1984–85	135	179	187

a) Create and interpret a 95% confidence interval for the mean rainfall difference between Victorville and 29 Palms.

b) Create and interpret a 95% confidence interval for the mean rainfall difference between Victorville and Mitchell's Cavern.

c) Create and interpret a 95% confidence interval for the mean rainfall difference between 29 Palms and Mitchell's Cavern.

d) Does it appear, based on these intervals, that these regions receive different average levels of rainfall?

28. Teach for America Several programs attempt to address the shortage of qualified teachers by placing uncertified instructors in schools with acute needs—often in inner cities. A 1999–2000 study compared students taught by certified teachers to others taught by uncertified teachers in the same schools. Reading scores of the students of certified teachers averaged 35.62 points with standard deviation 9.31. The scores of students instructed by uncertified teachers had mean 32.48 points with standard deviation 9.43 points on the same test. There were 44 students in each group. The appropriate *t* procedure has 86 degrees of freedom. Is there evidence of lower scores with uncertified teachers? Discuss. (*The Effectiveness of "Teach for America" and Other Under-certified Teachers on Student Academic Achievement: A Case of Harmful Public Policy.* Education Policy Analysis Archives [2002])

T **29. Legionnaires' disease** In 1974, the Bellevue-Stratford Hotel in Philadelphia was the scene of an outbreak of what later became known as legionnaires' disease. The cause of the disease was finally discovered to be bacteria that thrived in the air-conditioning units of the hotel. Owners of the Rip Van Winkle Motel, hearing about the Bellevue-Stratford, replaced their air-conditioning system. The following data are the bacteria counts in the air of eight rooms, before and after a new air-conditioning system was installed (measured in colonies per cubic foot of air). Has the new system succeeded in lowering the bacterial count? Base your analysis on a confidence interval. Be sure to list all your assumptions, methods, and conclusions.

Room Number	Before	After
121	11.8	10.1
163	8.2	7.2
125	7.1	3.8
264	14	12
233	10.8	8.3
218	10.1	10.5
324	14.6	12.1
325	14	13.7

30. Teach for America, Part II The study described in Exercise 22 also looked at scores in mathematics and language. Here are software outputs for the appropriate tests. Explain what they show.

Mathematics
T-TEST OF Mu(1) − Mu(2) = 0
Mu(Cert) − Mu(NoCert) = 4.53   t(86) = 2.95   p = 0.002

Language
T-TEST OF Mu(1) − Mu(2) = 0
Mu(Cert) − Mu(NoCert) = 2.13   t(84) = 1.71   p = 0.045

31. Bipolar kids The June 2002 *American Journal of Psychiatry* reported that researchers used medication and psychotherapy to treat children aged 7 to 16 who exhibit bipolar symptoms. After 2 years, symptoms had cleared up in only 26 of the 89 children involved in the study.

a) Write a 95% confidence interval; interpret it in context.

b) If researchers subsequently hope to produce an estimate of treatment effectiveness for bipolar disorder that has a margin of error of only 6%, how many patients should they study?

T **32. Online testing** The Educational Testing Service is now administering several of its standardized tests online—the CLEP and GMAT exams, for example. Since taking a test on a computer is different from taking a test with pencil and paper, one wonders if the scores will be the same. To investigate this question, researchers created two versions of an SAT-type test and got 20 volunteers to participate in an experiment. Each volunteer took both versions of the test, one with pencil and paper and the other online. Subjects were randomized with respect to the order in which they sat for the tests (online/paper) and which form they took (Test A, Test B) in which environment. The scores (out of a possible 20) are summarized in the table.

a) Were the two forms (A/B) of the test equivalent in terms of difficulty? Test an appropriate hypothesis and state your conclusion.

b) Is there evidence that the testing environment (paper/online) matters? Test an appropriate hypothesis and state your conclusion.

Subject	Paper Test A	Online Test B	Subject	Paper Test A	Online Test B
1	14	13	11	8	13
2	10	13	12	11	13
3	16	8	13	15	17
4	15	14	14	11	13
5	17	16	15	13	14
6	14	11	16	9	9
7	9	12	17	15	9
8	12	12	18	14	15
9	16	16	19	16	12
10	7	14	10	8	10

33. Bread Clarksburg Bakery is trying to predict how many loaves of bread to bake. In the last 100 days, the bakery has sold between 95 and 140 loaves per day. Here are a histogram and the summary statistics for the number of loaves sold for the last 100 days.

Summary of Sales

Mean	103
Median	100
SD	9.000
Min	95
Max	140
Q_1	97
Q_3	105.5

a) Can you use these data to estimate the number of loaves sold on the busiest 10% of all days? Explain.
b) Explain why you can use these data to construct a 95% confidence interval for the mean number of loaves sold per day.
c) Calculate a 95% confidence interval and carefully interpret what that confidence interval means.
d) If the bakery would have been satisfied with a confidence interval whose margin of error was twice as wide, how many days' data could they have used?
e) When the bakery opened, the owners estimated that they would sell an average of 100 loaves per day. Does your confidence interval provide strong evidence that this estimate was incorrect? Explain.

34. Irises Can measurements of the petal length of flowers be of value when you need to determine the species of a certain flower? Here are the summary statistics from measurements of the petals of two species of irises. (R. A. Fisher, "The Use of Multiple Measurements in Axonomic Problems." *Annals of Eugenics 7* [1936]: 179–188)

	Species	
	Versicolor	Virginica
Count	50	50
Mean	55.52	43.22
Median	55.50	44.00
SD	5.519	5.362
Min	45	30
Max	69	56
Lower Quartile	51	40
Upper Quartile	59	47

a) Make parallel boxplots of petal lengths for the two species.
b) Describe the differences seen in the boxplots.
c) Write a 95% confidence interval for the difference in petal length.
d) Explain what your interval means.
e) Based on your confidence interval, is there evidence of a difference in petal length? Explain.

35. Insulin and diet A study published in the *Journal of the American Medical Association* examined people to see if they showed any signs of IRS (insulin resistance syndrome) involving major risk factors for Type 2 diabetes and heart disease. Among 102 subjects who consumed dairy products more than 35 times per week, 24 were identified with IRS. In comparison, IRS was identified in 85 of 190 individuals with the lowest dairy consumption, fewer than 10 times per week.

a) Is this strong evidence that IRS risk is different in people who frequently consume dairy products than in those who do not?
b) Does this indicate that diary consumption influences the development of IRS? Explain.

36. Speeding A newspaper report in August 2002 raised the issue of racial bias in the issuance of speeding tickets. The following facts were noted:
- 16% of drivers registered in New Jersey are black.
- Of the 324 speeding tickets issued in one month on a 65-mph section of the New Jersey Turnpike, 25% went to black drivers.

a) Is the percentage of speeding tickets issued to blacks unusually high compared to registrations?
b) Does this suggest that racial profiling may be present?
c) What other statistics would you like to know about this situation?

37. Rainmakers? In an experiment to determine whether seeding clouds with silver iodide increases rainfall, researchers randomly assigned clouds to be seeded or not. The table summarizes the resulting rainfall (in acre-feet). Create a 95% confidence interval for the average amount of additional rain created by seeding clouds. Explain what your interval means.

	Unseeded Clouds	Seeded Clouds
Count	26	26
Mean	164.588	441.985
Median	44.200	221.600
SD	278.426	650.787
IntQRange	138.600	337.600
25 %ile	24.400	92.400
75 %ile	163	430

38. Fritos As a project for an introductory Statistics course, students checked 6 bags of Fritos marked with a net weight of 35.4 grams. They carefully weighed the contents of each bag, recording the following weights (in grams): 35.5, 35.3, 35.1, 36.4, 35.4, 35.5. Is there evidence that the mean weight of bags of Fritos is less than advertised?

a) Write appropriate hypotheses.
b) Check the assumptions for inference.
c) Test your hypothesis using all 6 weights.
d) Retest your hypothesis with the one unusually high weight removed.
e) What would you conclude about the stated weight?

T 39. Color or text? In an experiment, 32 volunteer subjects are briefly shown seven cards, each displaying the name of a color printed in a different color (example: **red, blue,** and so on). The subject is asked to perform one of two tasks: memorize the order of the words or memorize the order of the colors. Researchers record the number of cards remembered correctly. Then the cards are shuffled and the subject is asked to perform the other task. The table displays the results for each subject. Is there any evidence that either the color or the written word dominates perception?

a) What role does randomization play in this experiment?
b) Test appropriate hypotheses and state your conclusion.

Subject	Color	Word	Subject	Color	Word
1	4	7	17	4	3
2	1	4	18	7	4
3	5	6	19	4	3
4	1	6	20	0	6
5	6	4	21	3	3
6	4	5	22	3	5
7	7	3	23	7	3
8	2	5	24	3	7
9	7	5	25	5	6
10	4	3	26	3	4
11	2	0	27	3	5
12	5	4	28	1	4
13	6	7	29	2	3
14	3	6	30	5	3
15	4	6	31	3	4
16	4	7	32	6	7

40. And it means? Every statement about a confidence interval contains two parts: the level of confidence and the interval. Suppose that an insurance agent estimating the mean loss claimed by clients after home burglaries created the 95% confidence interval ($1644, $2391).

a) What's the margin of error for this estimate?
b) Carefully explain what the interval means.
c) Carefully explain what the confidence level means.

41. Batteries We work for the "Watchdog for the Consumer" consumer advocacy group. We've been asked to look at a battery company that claims its batteries last an average of 100 hours under normal use. There have been several complaints that the batteries don't last that long, so we decide to test them. To do this, we select 16 batteries and run them until they die. They lasted a mean of 97 hours, with a standard deviation of 12 hours.

a) One of the editors of our newsletter (who does not know statistics) says that 97 hours is a lot less than the advertised 100 hours, so we should reject the company's claim. Explain to him the problem with doing that.
b) What are the null and alternative hypotheses?
c) What assumptions must we make in order to proceed with inference?
d) At a 5% level of significance, what do you conclude?
e) Suppose that, in fact, the average life of the company's batteries is only 98 hours. Has an error been made in part d? If so, what kind?

42. Hamsters How large are hamster litters? Among 47 golden hamster litters recorded, there were an average of 7.72 baby hamsters, with a standard deviation of 2.5.

a) Create and interpret a 90% confidence interval.
b) Would a 98% confidence interval have a larger or smaller margin of error? Explain.
c) How many litters must be used to estimate the average litter size to within 1 baby hamster with 95% confidence?

T 43. Cramming Students in two basic Spanish classes were required to learn 50 new vocabulary words. One group of 45 students received the list on Monday and studied the words all week. Statistics summarizing this group's scores on Friday's quiz are given. The other group of 25 students did not get the vocabulary list until Thursday. They also took the quiz on Friday, after "cramming" Thursday night. Then, when they returned to class the following Monday, they were retested—without advance warning. Both sets of test scores for these students are shown.

Group 1
Fri.
Number of students = 45
Mean = 43.2 (of 50)
StDev = 3.4
Students passing (score ≥ 40) = 33%

Group 2							
Fri.	Mon.	Fri.	Mon.	Fri.	Mon.	Fri.	Mon.
42	36	50	47	35	31	40	31
44	44	34	34	43	32	41	32
45	46	38	31	48	37	48	39
48	38	43	40	43	41	37	31
44	40	39	41	45	32	36	41
43	38	46	32	47	44		
41	37	37	36				

a) Did the week-long study group have a mean score significantly higher than that of the overnight crammers?
b) Was there a significant difference in the percentages of students who passed the quiz on Friday?
c) Is there any evidence that when students cram for a test, their "learning" does not last for 3 days?

Practice Exam

I. Multiple Choice

1. On a college admissions test where the scores were approximately normally distributed, Amy's score of 31 was at the 98th percentile. If the mean of the test scores was 20, the standard deviation was approximately

A) 4.7   B) 5.4   C) 5.6   D) 6.7   E) 9.7

2. Let Q1 represent the first quartile, Q2 represent the second quartile (or median), and Q3 represent the third quartile. Which of the following computes an important value when considering which data values may be outliers?

A) Q2 + (Q3 − Q1)
B) Q2 + 1.5(Q3 − Q1)
C) Q3 + (Q2 − Q1)
D) Q3 + 1.5(Q2 − Q1)
E) Q3 + 1.5(Q3 − Q1)

3. The distribution of a large set of temperatures is approximately Normal with a mean of 60° and a standard deviation of 5. Estimate the interquartile range for these data.

A) 6.7°   B) 7.5°   C) 10.0°
D) 13.4°   E) 15.0°

4. Cam and Denise wish to check the data from their lab experiment to see if the distribution is approximately Normal. Which of the following would be most useful for assessing normality?

A) boxplot
B) stem and leaf plot
C) bar graph
D) scatterplot
E) residuals plot

5. A researcher re-expressed a data set of scientific measurements. In the display shown, the upper box and whisker plot summarizes original data and the lower box and whisker plot summarizes the re-expressed data. Which re-expression might the researcher have used?

A) Subtract 5 from each data value
B) Add 5 to each data value
C) Multiply each data value by 5
D) Divide each data value by 5
E) Convert the data values to z-scores

6. One day in gym class students took a physical fitness test by doing push-ups and sit-ups. The standard deviation of the number of sit-ups they were able to do was 7 and the standard deviation of the number of push-ups was 2. A Statistics student used these data to create a least squares regression line to predict the number of sit-ups a student was able to do based on the number of push-ups the student did. Which of the following could NOT be the slope of that line?

A) −2   B) −0.5   C) 1
D) 3   E) 4

7. The least-squares regression line for a set of (*Age, Skill_Score*) data is $\hat{y} = 5.0x + 0.7$. The data point for age 6 has residual −1.4. What is the skill score for age 6?

A) −4.6   B) 4.6   C) 29.3
D) 30.7   E) 32.1

8. When working with bivariate data, which of these are useful when deciding whether it's appropriate to use a linear model?

   I. the scatterplot
  II. the residuals plot
 III. the correlation coefficient

A) I only
B) II only
C) III only
D) I and II only
E) I, II, and III

9. Josie transformed bivariate data by taking the square root of the y values and found the least-squares regression line $\sqrt{\hat{y}} = 7.6 - 0.4x$ to be a useful model. Predict y for $x = 9$.

   A) 2   B) 3.5   C) 4   D) 11.5   E) 16

10. The least-squares regression line for predicting the price (in cents) of boxes of cereal from their net weight (in ounces)has a slope of 20. The predicted price for 10-ounce boxes is $4.50. What is the predicted price for 12-ounce boxes?

    A) $4.70   B) $4.74   C) $4.90
    D) $5.40   E) $6.90

11. Collecting data from all persons in the population of interest is called

    A) a block.   B) a census.   C) a cluster.
    D) a sample.   E) a stratum.

12. Joanne needs to test boxes of pasta to see if they contain the correct amount of product. Each of the 500 boxes in a recent batch has a unique serial number from 1001 to 1500 stamped on it. She will pick a random sample of 50 boxes by generating a random integer between 1001 and 1011 to select the first box, and then selecting every tenth number in sequence. Her method is called

    A) cluster sampling.
    B) convenience sampling.
    C) multistage sampling.
    D) stratified sampling.
    E) systematic sampling.

13. A two-factor experiment will investigate the best way to make cookies by trying 3 different oven temperatures and 4 different baking times. How many different treatments are there?

    A) 2   B) 7   C) 9   D) 12   E) 20

14. Which of these describes an advantage of using blocking in an experiment?

    A) It increases the sample size.
    B) It ensures that treatments are assigned randomly.
    C) It reduces the variability due to differences between blocks.
    D) It allows a conclusion of cause and effect.
    E) It allows the results to be generalized to a larger group.

15. Two Statistics classes took a practice exam. This computer output shows the summary statistics:

Group	Count	Mean	Median	StdDev
Class 1	32	80.4	78.5	6.1
Class 2	24	76.3	74.2	7.0

    What is the overall mean for all of the students on this exam?

    A) 76.35   B) 76.657   C) 77.28
    D) 78.35   E) 78.643

16. Researchers surveyed samples of freshmen students and senior students about their spending habits. Among several questions, they asked the students how much they spent on fast food during the past week. Here is a summary of the responses:

	<$10	$10–20	>$20	Total
Freshmen	23	12	8	43
Seniors	25	19	28	72
Totals	48	31	36	115

    Which best describes the two variables in this study, money spent on fast food and high school class?

    A) Disjoint and independent
    B) Disjoint but not independent
    C) Independent but not disjoint
    D) Neither independent nor disjoint
    E) Independence cannot be determined because the sample sizes are unequal.

17. Based on data a coffee shop owner has collected, she believes that 12% of her customers will buy a cookie to go with their coffee and that these purchases are independent. One day as she's getting ready to close, 6 customers enter the shop and she has only 2 cookies left. What is the probability that no more than 2 of these last 6 customers will want a cookie?

    A) 0.026   B) 0.130   C) 0.156
    D) 0.610   E) 0.974

18. Every scale has some measurement error, and such errors are roughly normally distributed. A certain deli scale is correctly calibrated, but the standard deviation of the errors is 0.15 ounces. What is the probability that a measurement on this scale is within 0.30 ounces of the correct weight?

    A) 0.118   B) 0.236   C) 0.477
    D) 0.954   E) 0.977

19. The bar chart shown below summarizes U.S. Census data regarding 1999 household incomes for New York City residents, by ethnic group.

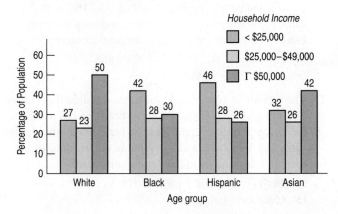

Based on what you see here, which of these statements are true?

I. The distribution of White household incomes is skewed to the left.

II. Approximately the same number of Black households and Hispanic households have incomes between $25,000 and $49,000.

III. Household income and ethnicity appear to be independent.

A) None of I, II, or III
B) I only
C) II only
D) III only
E) II and III only

20. A store had a sale on a popular soft drink, with a limit of 5 packs per customer. The table shows the probability model for the random variable $X =$ number of packs a customer purchases.

X	1	2	3	4	5
P(X)	0.20	0.16	0.10	0.24	0.30

What is the expected number of packs a customer purchases?

A) 2.50   B) 3.00   C) 3.28
D) 3.54   E) 4.00

21. In December 2001 the research firm GfK surveyed baby boomers (Americans born between 1946 and 1964) with $100,000 or more investable assets. GfK contacted 1006 randomly selected baby boomers by cell phone or landline. The vast majority, 71%, reported that they have provided support to their adult children in the form of helping them pay for college tuition or loans. GfK reported a margin of error of plus or minus 3%. What confidence level were the pollsters using?
Source: New York Times, May 5, 2012.

A) 90%   B) 95%   C) 96%
D) 98%   E) 99%

22. For the survey described in Question 21, if GfK had wanted a smaller margin of error of only 1%, with the same confidence level, how many randomly selected baby boomers would the pollsters need to survey?

A) 96   B) 336   C) 1865
D) 3010   E) 9054

23. The weight of a jar of mild salsa has a standard deviation of 5 gm and the weight of a jar of hot salsa has a standard deviation of 6 gm. A combination pack has 2 jars of mild salsa and 3 jars of hot salsa. What is the variance of the total weight of salsa in the combination pack?

A) 28   B) 61   C) 74
D) 158   E) 424

24. A sociologist is comparing the proportion of teenage boys who receive a text message at least once an hour to the percentage of teenage girls who do. She has collected data from 150 men and 150 women. Which of the following is NOT a condition that the sociologist should check before creating a confidence interval for the difference in population proportions?

A) The samples are each approximately Normal.
B) There at least 10 successes and 10 failures in each sample.
C) The male sample and female sample are independent.
D) The people in each sample were selected at random.
E) No more than 10% of each population was sampled.

25. A random sample of students at a college shows that 54 of 200 students had part-time jobs. Which of the following is the correct formula for a 90% confidence interval for the proportion of all students at this college with part-time jobs?

A) $0.27 \pm 1.28\sqrt{\dfrac{(0.27)(0.73)}{200}}$

B) $0.27 \pm 1.28\sqrt{\dfrac{(0.5)(0.5)}{200}}$

C) $0.27 \pm 1.645\sqrt{\dfrac{(0.27)(0.73)}{200}}$

D) $0.27 \pm 1.645\sqrt{\dfrac{(0.5)(0.5)}{200}}$

E) $0.27 \pm 1.96\sqrt{\dfrac{(0.27)(0.73)}{200}}$

26. Fred is constructing a 95% confidence interval to estimate the average length (in minutes) of movies he watches. His random sample of 15 movies averaged 114 minutes long with a standard deviation of 11 minutes. What critical value and standard error of the mean should he use?

A) $t^* = 2.131, SE = 2.84$
B) $t^* = 2.131, SE = 2.94$
C) $t^* = 2.131, SE = 11$
D) $t^* = 2.145, SE = 2.84$
E) $t^* = 2.145, SE = 2.94$

27. In Question 26, Fred constructed a 95% confidence interval to estimate the average length (in minutes) of the movies he watches, with a random sample of 15 movies. He plans to continue collecting data until he has a random sample of 40 movies, and then create a new confidence interval. Which of these statements is most accurate?

A) Both the standard error of the mean and the critical value will probably decrease for the larger sample.

B) The critical value will increase for the larger sample, but the standard error will probably decrease.

C) Since 40 is a large sample, he will be able to use the Normal distribution.

D) With this larger sample he can be more sure that his 95% confidence interval captures the true mean length of all movies he watches.

E) He should not perform inference at all, because he will probably continue watching movies for many years to come.

28. Consider the following scenarios.

I. A group of musicians count how many times they can snap their fingers in 10 seconds with their dominant hands and with their nondominant hands. They will test to see if musicians can snap faster with dominant hands than nondominant hands.

II. A group of musicians is trying to measure the effectiveness of practice time. Half of them practice a difficult piece of music for one hour each day while the other half practice two hours each day. After a week, each musician plays the piece as a judge counts the number of errors. They will test to see if more practice results in fewer errors.

Which are the proper choices for the hypothesis tests?

A) Both scenarios require a 2-sample *t*-test.

B) Both scenarios require a matched pairs *t*-test.

C) Scenario I calls for a 2-sample *t*-test, and Scenario II calls for a matched pairs *t*-test.

D) Scenario I calls for a matched pairs *t*-test, and Scenario II calls for a 2-sample *t*-test.

E) The proper tests cannot be determined until we see the actual data that are collected.

29. Proponents of Neuro-Linguistic Programming (NLP) claim that certain eye movements are reliable indicators of lying. According to this notion, a person looking up to their right suggests a lie whereas looking up to their left is indicative of truth telling. In 2012, British researchers tested the claim that you can spot a lie by watching a person's eyes. They recruited 50 male volunteers and randomly assigned some of them to be trained in using the NLP technique. Then all participants watched 32 video clips that each showed two interviews, one including a true statement and the other a lie. After each clip, the participants identified the interviewee they thought was lying. The 21 NLP-trained participants averaged 16.33 correct decisions with a standard deviation of 3.53, compared to the 29 participants in the control group who averaged 16.59 correct with standard deviation 3.84.

What type of test should be conducted to answer the question, "Do these data provide evidence that you can spot a lie by watching a person's eyes?"

A) One proportion *z*-test

B) One sample *t*-test

C) Matched pairs *t*-test

D) *Z*-test for the difference of two proportions

E) Two sample *t*-test for the difference of two means

Source: Wiseman R, Watt C, ten Brinke L, Porter S, Couper S-L, et al. (2012) The Eyes Don't Have It: Lie Detection and Neuro-Linguistic Programming. PLoS ONE 7(7): e40259.

30. Using the data collected in the British study of NLP training as described in Question 29, the formula for the critical value of the test statistic is:

A) $z = \dfrac{0.5103 - 0.5}{\sqrt{\dfrac{(0.5)(0.5)}{21}}}$

B) $z = \dfrac{0.5103 - 0.5184}{\sqrt{(0.515)(0.485)\left(\dfrac{1}{21} + \dfrac{1}{29}\right)}}$

C) $t = \dfrac{16.33 - 16}{\dfrac{3.53}{\sqrt{32}}}$

D) $z = \dfrac{16.33 - 16.59}{\dfrac{3.84}{\sqrt{21}}}$

E) $t = \dfrac{16.33 - 16.59}{\sqrt{\dfrac{3.53^2}{21} + \dfrac{3.84^2}{29}}}$

31. A clothing store has randomly selected sales receipts from 20 female and 20 male customers. They wish to test their belief that on average females' purchases are larger than males'. Which alternate hypothesis is correct?

A) $H_A: \mu_d > 0$; the paired difference in female minus male spending is greater than zero.

B) $H_A: \mu_f - \mu_m > 0$; the average female purchase minus the average male purchase is greater than zero.

C) $H_A: \mu_f \neq \mu_m$; the average female purchase is different from the average male purchase.

D) $H_A: \bar{x}_f > \bar{x}_m$; the average purchase for the 20 females is greater than the average for the 20 males.

E) $H_A: \bar{x}_f - \bar{x}_m \neq 0$; the average purchase for the 20 females is different than the average for the 20 males.

32. A medical study compares the time it takes for a wound to heal with two different medications. At the end of the study, the conclusion is stated as a 95% confidence interval. The average difference between the two treatments (in days) was reported to be $(-2.45, 3.86)$. From this interval we can conclude that

A) One of the medications is statistically significantly better than the other, because zero is contained in the interval.

B) The experiment must have made a Type I error, because zero is in the interval.

C) Because zero is in the interval, there is no statistically significant difference between the two medications.

D) Because more of the interval is above zero than below it, there must be a slight statistical difference between the treatments.

E) The researcher made a mistake because the difference should not result in a negative number of days.

33. The senior Class of 2012 at Rancho High strongly believes that they have better grades than their predecessors. The class president, a Statistics student, calculates the mean GPA for the 725 seniors to be 2.97 with a standard deviation of 0.67. The Class of 2011 had a mean of 2.83 with a standard deviation of 0.81. Which of the following is correct?

A) The means should be compared with a one sample *t*-test, using 2011's mean as the null hypothesis.

B) The means should be compared using a two sample *t*-test.

C) The means should be compared using a confidence interval for the difference of means.

D) The means should be compared by a matched-pair *t*-test on the mean difference in GPAs.

E) Inference is not necessary since the data are for both populations.

34. An ecologist who analyzes water samples tests the null hypothesis that any contaminants in the water are below dangerous concentrations. Because he uses $\alpha = 0.05$, a set of samples from a small lake that produced a P-value of 0.07 led him to conclude that the evidence did not point to unsafe water conditions. Which is true?

A) There's a 7% chance the lake's water really is safe.

B) There's a 93% chance the lake's water really is safe.

C) There's a 7% chance his sampling would have shown as much contamination as it did even if the lake's water really is safe.

D) If the lake's water really is unsafe, there's a 5% chance he wouldn't notice.

E) If he had taken more samples, he probably would have rejected the null and concluded that the water was unsafe.

35. A torn meniscus is a common type of knee injury. In the case of minor tears, there's some question about whether initial surgery followed by physical therapy (PT) results in a better outcome than just physical therapy alone. To find out, experimenters will randomly assign some subjects with this type of injury to have surgery followed by PT and others to just do PT, and then compare the recovery experiences of the tow groups. Which of statements A–D is false?

A) If the researchers mistakenly conclude that the surgery was beneficial, they will commit a Type I error.

B) The power of this test is its ability to detect that there really is a benefit of surgery.

C) The more participants they use in this experiment, the higher the power of the test will be.

D) Demanding stronger evidence by using $\alpha = 0.01$ instead of $\alpha = 0.05$ would give the researchers greater power.

E) None; statements A–D are all true.

II. Free Response

1. Summary statistics for data relating the latitude (°North) and average January low temperature (°F) for 55 large U.S. cities are given below. A scatterplot suggests a linear model is appropriate.

	Latitude	Avg Jan Low
Mean	39.02	26.44
StDev	5.42	13.49
Correlation	−0.848	

a) Write the equation of the least squares regression line that predicts a U.S. city's average January low temperature based on its latitude.

b) Interpret the slope in context.

c) Do you think the *y*-intercept is meaningful? Explain.

d) For a certain city the residual is −4.2°. Explain what that means.

2. Customers using the drive up window at fast food restaurants are sometimes greeted by a message encouraging them to purchase an item that's currently "on special." Marketing researchers at one restaurant chain want to test the effectiveness of such a message. They have selected 10 of their restaurants in various locations for an experiment, and are considering two different designs.

Design I: 5 of the restaurants will use the message and the other 5 will not.
Design II: Each restaurant will alternate between playing or not playing the message as customers arrive at the drive-thru order station.

a) Describe a method of assigning the restaurants to the groups for Design I using this list of random digits: 08530 08629 32279 29478 50228

b) Which do you think is the better design? Explain why.

c) For the design you chose in part b, describe the data you would collect and the method of analysis you would use to determine whether there is evidence that the message improves sales of the special item.

3. One of the ways contestants on a television game show can win cash is by playing "Bucks-in-the-Box." In this game the host offers the contestant 2 boxes that appear to be identical. The contestant reaches into one of the boxes and draws out an envelope, winning the money inside it. While the contestant can't tell which box is which, the contents are quite different. One box has 5 envelopes; 4 contain $100 and the other contains $1000. The other

box has only 4 envelopes, with $100 in one of them and $1000 in each of the other 3.

a) What is the probability that a Bucks-in-the-Box contestant will win $1000?

b) If a contestant wins $1000, what's the probability it was the only $1000 envelope in that box?

c) What are Bucks-in-the-Box contestants' expected winnings?

4. A bank is considering a marketing campaign that would urge parents of preschool children to start college savings program, targeting families with annual incomes above $40,000. Bank officials will proceed with the campaign only if it appears that at least 10% of such families might be interested. When a pilot test presented the campaign materials to a random sample of 200 parents, 26 expressed some interest in the college savings plan.

a) At the 5% level of significance, do these sample results provide evidence that the bank should pursue the marketing campaign?

b) Describe two ways that the bank could increase the power of the test, and explain a disadvantage of each.

5. Bicycle frames can be made from carbon, steel, or other materials. Carbon frames are much more expensive than steel, but they are also much lighter. Could this lighter frame have a significant impact on speed? In 2010 a doctor in England, who commutes daily to work 27 miles roundtrip on a bicycle, ran his own randomized experiment to investigate.

Each day for several months Dr. Groves flipped a coin to determine whether he would ride his steel frame bicycle or his carbon frame bicycle to work and back. He rode a total of 30 journeys on the steel frame bike and 26 journeys on his carbon frame bicycle, recording his total commuting time for each day. The summary statistics are shown below:

	Steel	Carbon
Commutes	30	26
Mean (min)	107.80	109.35
SD (min)	4.90	6.25

Source: Groves, Jeremy. "Bicycle Weight and Commuting Time: A randomized Trial," Significance: Statistics Making Sense, *June 2011, Volume 8 Issue 2: 95–97.*

a) Why would you prefer to see the actual data before proceeding with inference?

b) Construct and interpret a 95% confidence interval for the difference in average commuting times.

c) Explain in this context what "95% confidence" means.

d) Based on your confidence interval, do the doctor's data provide evidence that there's a difference in average commuting time for the two bikes? Explain.

6. Hank likes to simulate baseball seasons, playing the manager of his favorite team. In real life the team won 97 of 162 games, but in Hank's simulation his team won 104 games. Hank claims his success is due to his excellent ability as manager, but his friend Joe says he was just lucky. In attempts to settle the argument, Joe finds a z-score and Hank runs a computer simulation.

a) What's Joe's best estimate of the probability this team can win a game?

b) What did Joe calculate to be the z-score for 104-win season?

c) What conclusion did Joe reach about Hank's managerial prowess?

d) Hank ran 50 simulated seasons letting the computer's randomness act as the team's manager. The number and frequency of games won are shown in the histogram below.

Does the histogram provide Hank with evidence that his 104-win season was due to more than luck? Explain.

The name suggests a certain badness-of-grammar, but it is quite standard. After all, we are asking whether the model that births are uniformly distributed over the signs fits the data good, . . . er, well.

Goodness-of-fit involves testing a hypothesis. We have specified a model for the distribution and want to know whether it fits. There is no single parameter to estimate, so a confidence interval wouldn't make much sense. A one-proportion *z*-test won't work because we have 12 hypothesized proportions, one for each sign. We need a test that considers all of them together and gives an overall idea of whether the observed distribution differs from the hypothesized one.

For Example FINDING EXPECTED COUNTS

Birth month may not be related to success as a CEO, but what about on the ball field? It has been proposed by some researchers that children who are the older ones in their class at school naturally perform better in sports and that these children then get more coaching and encouragement. Could that make a difference in who makes it to the professional level in sports?

Month	Ballplayer Count	National Birth %	Month	Ballplayer Count	National Birth %
1	137	8%	7	102	9%
2	121	7%	8	165	9%
3	116	8%	9	134	9%
4	121	8%	10	115	9%
5	126	8%	11	105	8%
6	114	8%	12	122	9%
			Total	**1478**	**100%**

Baseball is a remarkable sport, in part because so much data are available, including the birth date of every player who ever played in a major league game. Since the effect we're suspecting may be due to relatively recent policies (and to keep the sample size moderate), we'll consider the birth months of 1478 major league players born since 1975. We can also look up the national demographic statistics to find what percentage of people were born in each month. Let's test whether the observed distribution of ballplayers' birth months shows just random fluctuations or whether it represents a real deviation from the national pattern.

QUESTION: How can we find the expected counts?

ANSWER: *There are 1478 players in this set of data. I found the national percentage of births in each month. Based on the national birth percentages, I'd expect 8% of players to have been born in January, and 1478(0.08) = 118.24. I won't round off, because expected "counts" needn't be integers. Multiplying 1478 by each of the birth percentages gives the expected counts shown in the table in the margin.*

Month	Expected	Month	Expected
1	118.24	7	133.02
2	103.46	8	133.02
3	118.24	9	133.02
4	118.24	10	133.02
5	118.24	11	118.24
6	118.24	12	133.02

 ## Just Checking

Some people can roll their tongues, like the little girl in the picture. (Can you?) Some people's earlobes are attached to their necks, while others dangle freely. You wouldn't think these two traits have anything to do with each other, but they're actually controlled by the same gene!

(continued)

Genetic theory predicts that people will have neither, one, or both of these traits in the ratio 1:3:3:9, as described in the table.

1. The 124 students in a college Biology class plan to collect data on themselves about tongue rolling and earlobes. How many people should they expect to find in each of the four groups?

Tongue	Earlobes	Predicted Fraction
Non-curling	Attached	1/16
Non-curling	Free	3/16
Curling	Attached	3/16
Curling	Free	9/16

Assumptions and Conditions

These data are organized in tables as we saw in Chapter 2, and the assumptions and conditions reflect that. Rather than having an observation for each individual, we typically work with summary counts in categories. In our example, we don't see the birth signs of each of the 256 executives, only the totals for each sign.

Counted Data Condition

The values in each **cell** must be *counts* for the categories of a categorical variable. This might seem a simplistic, even silly condition. But we can't apply these methods to proportions, percentages, or measurements just because they happen to be organized in a table.

Independence Assumption

The counts in the cells should be independent of each other. The easiest case is when the individuals who are counted in the cells are sampled independently from some population. That's what we'd like to have if we want to draw conclusions about that population. Randomness can arise in other ways, though. For example, these Fortune 400 executives are not a random sample of company executives, but it's reasonable to think that their birth dates should be randomly distributed throughout the year.

If we want to generalize to a large population, we should check two more conditions:

- **Randomization Condition:** The individuals who have been counted should be a random sample from the population of interest.
- **10% Condition:** Our sample is less than 10% of the population.

Sample Size Assumption

We must have enough data for the methods to work, so we usually check the

- **Expected Cell Frequency Condition.** We should expect to see at least 5 individuals in each cell.

This is quite similar to the condition that np and nq be at least 10 when we tested proportions. In our astrology example, assuming equal births in each zodiac sign leads us to expect 21.3 births per sign, so the condition is easily met here.

For Example CHECKING ASSUMPTIONS AND CONDITIONS

RECAP: Are professional baseball players more likely to be born in some months than in others? We have observed and expected counts for the 1478 players born since 1975.

QUESTION: Are the assumptions and conditions met for performing a goodness-of-fit test?

ANSWER:

✓ **Counted Data Condition:** I have month-by-month counts of ballplayer births.
✓ **Independence Assumption:** These births were independent.

(continued)

✓ **Randomization Condition:** Although they are not a random sample, we can take these players to be representative of players past and future.

✓ **10% Condition:** These 1478 players are less than 10% of the population of 16,804 players who have ever played (or will play) major league baseball.

✓ **Expected Cell Frequency Condition:** The expected counts extend from 103.46 to 133.02, all much greater than 5.

It's okay to use these data for a goodness-of-fit test.

Just Checking

A Biology class of 124 students collected data on themselves to check the genetic theory about the frequency of tongue-rolling and free-hanging earlobes.

Free

Attached

Their results are summarized in the table.

Tongue	Earlobes	Observed Count	Expected Count
Non-curling	Attached	12	7.75
Non-curling	Free	22	23.25
Curling	Attached	31	23.25
Curling	Free	59	69.75

2. Is it okay to proceed with inference? Check the assumptions and conditions.

Calculations

Are the discrepancies between what we observed and what we expected just natural sampling variability, or are they so large that they indicate something important? It's natural to look at the *differences* between these observed and expected counts, denoted $(Obs - Exp)$. Just adding up these differences won't work because some are positive; others negative. We've been in this predicament before, and we handle it same way now: We square them. That gives us positive values and focuses attention on any cells with large differences from what we expected. Because the differences between observed and expected counts generally get larger the more data we have, we also need to get an idea of the *relative* sizes of the differences. To do that, we divide each squared difference by the expected count for that cell.

The test statistic, called the **chi-square** (or chi-squared) **statistic**, is found by adding up the sum of the squares of the deviations between the observed and expected counts divided by the expected counts:

$$\chi^2 = \sum_{all\ cells} \frac{(Obs - Exp)^2}{Exp}.$$

The chi-square statistic is denoted χ^2, where χ is the Greek letter chi (pronounced "ky" as in "sky"). It refers to a family of sampling distribution models we have not seen before called (remarkably enough) the **chi-square models**.

This family of models, like the Student's *t*-models, differ only in the number of degrees of freedom. The number of degrees of freedom for a goodness-of-fit test is $n - 1$. Here, however, *n* is *not* the sample size, but instead is the number of categories. For the zodiac example, we have 12 signs, so our χ^2 statistic has 11 degrees of freedom.

■ **NOTATION ALERT**

We compare the counts *observed* in each cell with the counts we *expect* to find. The usual notation uses O's and E's or abbreviations such as those we've used here. The method for finding the expected counts depends on the model.

■ **NOTATION ALERT**

In Statistics the Greek letter χ (chi) is used to represent both a test statistic and the associated sampling distribution. This is another violation of our "rule" that Greek letters represent population parameters. Here we are using a Greek letter to name a family of distribution models and a statistic.

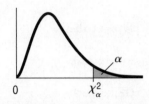

If the observed counts perfectly matched the expected, the χ^2 value would be 0. The greater the differences, positive or negative, the larger χ^2 becomes. If the calculated value is large enough, we'll reject the null hypothesis. What's "large enough" depends on the degrees of freedom. We use technology to find a P-value.

But what does rejecting the null tell us? Since squaring the differences makes all the deviations positive whether the observed counts were higher or lower than expected, there's no direction to the rejection of the null. Because we only worry about unexpectedly large χ^2 values, this behaves like a one-sided test, but it's really *many*-sided. With so many proportions, there are many ways the null model can be wrong. All we know is that it doesn't fit.

For Example   DOING A GOODNESS-OF-FIT TEST

RECAP: The birth months data for major league baseball players are appropriate for performing a χ^2 test.

QUESTIONS: What are the hypotheses, and what does the test show?

ANSWER: H_0: The distribution of birth months for major league ballplayers is the same as that for the general population.

H_A: The distribution of birth months for major league ballplayers differs from that of the rest of the population.

$$df = 12 - 1 = 11$$

$$\chi^2 = \sum \frac{(Obs - Exp)^2}{Exp}$$

$$= \frac{(137 - 118.24)^2}{118.24} + \frac{(121 - 103.46)^2}{103.46} + \cdots$$

$$= 26.48 \text{ (by technology)}$$

$$\text{P-value} = P(\chi^2_{11} \geq 26.48) = 0.0055 \text{ (by technology)}$$

Because of the small P-value, I reject H_0; there's evidence that birth months of major league ballplayers have a different distribution from the rest of us.

Step-by-Step Example   A CHI-SQUARE TEST FOR GOODNESS-OF-FIT

We have counts of 256 executives in 12 zodiac sign categories. The natural null hypothesis is that birth dates of executives are divided equally among all the zodiac signs.

Question: Are CEOs more likely to be born under some zodiac signs than others?

(continued)

THINK ➡ **Plan** State what you want to know.

Identify the variables and check the W's.

Hypotheses State the null and alternative hypotheses. For χ^2 tests, it's usually easier to do that in words than in symbols.

Model Make a picture. The null hypothesis is that the frequencies are equal, so a bar chart (with a line at the hypothesized "equal" value) is a good display.

Think about the assumptions and check the conditions.

Specify the sampling distribution model.

Name the test you will use.

I want to know whether births of successful people are uniformly distributed across the signs of the zodiac. I have counts of 256 Fortune 400 executives, categorized by their birth sign.

H_0: Births are uniformly distributed over zodiac signs.[1]

H_A: Births are not uniformly distributed over zodiac signs.

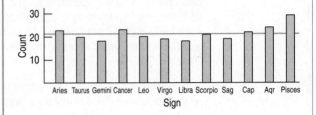

The bar chart shows some variation from sign to sign, and Pisces is the most frequent. But it is hard to tell whether the variation is more than I'd expect from random variation.

✓ **Counted Data Condition:** I have counts of the number of executives in 12 categories.

✓ **Independence Assumption:** The birth dates of executives should be independent of each other.

✓ **Randomization Condition.** This is a convenience sample of executives. It is random, but not a SRS. However, there's no reason to suspect bias.

✓ **Expected Cell Frequency Condition:** The null hypothesis expects that 1/12 of the 256 births, or 21.333, should occur in each sign. These expected values are all at least 5, so the condition is satisfied.

The conditions are satisfied, so I'll use a χ^2 model with $12 - 1 = 11$ degrees of freedom and do a **chi-square goodness-of-fit test**.

(continued)

---

[1]It may seem that we have broken our rule of thumb that null hypotheses should specify parameter values. If you want to get formal about it, the null hypothesis is that

$$p_{\text{Aries}} = p_{\text{Taurus}} = \cdots = p_{\text{Pisces}}.$$

That is, we hypothesize that the true proportions of births of CEOs under each sign are equal. The role of the null hypothesis is to specify the model so that we can compute the test statistic. That's what this one does.

SHOW ➡ **Mechanics** Each cell contributes an $\frac{(Obs - Exp)^2}{Exp}$ value to the chi-square sum. We add up these components for each zodiac sign. If you do it by hand, it can be helpful to arrange the calculation in a table. We show that after this Step-by-Step Example.

The expected value for each zodiac sign is 21.333.

$$\chi^2 = \sum \frac{(Obs - Exp)^2}{Exp} = \frac{(23 - 21.333)^2}{21.333}$$

$$+ \frac{(20 - 21.333)^2}{21.333} + \cdots$$

$$= 5.094 \text{ for all 12 signs.}$$

The χ^2 models are skewed to the high end, and change shape depending on the degrees of freedom. The P-value considers only the right tail. Large χ^2 statistic values correspond to small P-values, which lead us to reject the null hypothesis.

The P-value is the area in the upper tail of the χ^2 model above the computed χ^2 value.

$$P\text{-value} = P(\chi^2_{11} > 5.094) = 0.926$$

TELL ➡ **Conclusion** Link the P-value to your decision. Remember to state your conclusion in terms of what the data mean, rather than just making a statement about the distribution of counts.

The P-value of 0.926 says that if the zodiac signs of executives were in fact distributed uniformly, an observed chi-square value of 5.09 or higher would occur about 93% of the time. This certainly isn't unusual, so I fail to reject the null hypothesis, and conclude that these data show virtually no evidence that executives are more likely to have certain zodiac signs.

The Chi-Square Calculation

A S *Activity: Calculating Standardized Residuals.* The incident of the earlier ActivStats activity in which children were placed at risk, also put women at risk. Standardized residuals help us understand the relative risks.

Let's make the chi-square procedure very clear. Here are the steps:

1. **Find the expected values.** These come from the null hypothesis model. Every model gives a hypothesized proportion for each cell. The expected value is the product of the total number of observations times this proportion.

   For our example, the null model hypothesizes *equal* proportions. With 12 signs, 1/12 of the 256 executives should be in each category. The expected number for each sign is 21.333.

2. **Compute the residuals.** Once you have expected values for each cell, find the residuals, *Observed − Expected*.

3. **Square the residuals.**

4. **Compute the components.** Now find the **component,** $\frac{(Observed - Expected)^2}{Expected}$, for each cell.

A S *Activity: The Chi-Square Test.* This animation completes the calculation of the chi-square statistic and the hypothesis test based on it.

5. **Find the sum of the components.** That's the chi-square statistic.

6. **Find the degrees of freedom.** It's equal to the number of cells minus one. For the zodiac signs, that's $12 - 1 = 11$ degrees of freedom.

7. **Test the hypothesis.** Large chi-square values mean lots of deviation from the hypothesized model, so they give small P-values. Look up the critical value from a table of chi-square values, or use technology to find the P-value directly.

The steps of the chi-square calculations are often laid out in tables using one row for each category, with columns for observed counts, expected counts, residuals, squared residuals, and the contributions to the chi-square total like this:

Sign	Observed	Expected	Residual = (Obs − Exp)	(Obs − Exp)²	Component = $\frac{(Obs - Exp)^2}{Exp}$
Aries	23	21.333	1.667	2.778889	0.130262
Taurus	20	21.333	−1.333	1.776889	0.083293
Gemini	18	21.333	−3.333	11.108889	0.520737
Cancer	23	21.333	1.667	2.778889	0.130262
Leo	20	21.333	−1.333	1.776889	0.083293
Virgo	19	21.333	−2.333	5.442889	0.255139
Libra	18	21.333	−3.333	11.108889	0.520737
Scorpio	21	21.333	−0.333	0.110889	0.005198
Sagittarius	19	21.333	−2.333	5.442889	0.255139
Capricorn	22	21.333	0.667	0.444889	0.020854
Aquarius	24	21.333	2.667	7.112889	0.333422
Pisces	29	21.333	7.667	58.782889	2.755491
					$\sum$ = 5.094

A **S** *Lesson*: The Chi-Square Family
of Curves. (Not an activity like the others,
but there's no better way to see how
x^2 changes with more df.) Click on the
Lesson Book's Resources tab and open the
chi-square table. Watch the curve at the
top as you click on a row and scroll down
the degrees-of-freedom column.

*How Big Is Big?

When we calculated χ^2 for the zodiac sign example, we got 5.094. That value would
have been big for z or t, leading us to reject the null hypothesis. Not here, though. Were
you surprised that $\chi^2 = 5.094$ had a huge P-value of 0.926? What *is* big for a χ^2 statistic,
anyway?

Think about how χ^2 is calculated. In every cell, any deviation from the expected count
contributes to the sum. Large deviations generally contribute more, but if there are a lot of
cells, even small deviations can add up to make the χ^2 value large. So the more cells there are,
the higher the value of χ^2 has to get before it becomes noteworthy. For χ^2, then, the decision
about how big is big depends on the number of degrees of freedom.

Unlike the Normal and t families, χ^2 models are skewed. Here, for example, are the χ^2
curves for 5 and 9 degrees of freedom.

Notice that the value $\chi^2 = 10$ might seem somewhat extreme when there are 5 degrees of
freedom, but appears to be rather ordinary for 9 degrees of freedom. Here are two simple facts
to help you think about χ^2 models:

- The mode is at $\chi^2 = df - 2$. (Look back at the curves; their peaks are at 3 and 7, see?)
- The expected value (mean) of a χ^2 model is its number of degrees of freedom. That's a bit to
  the right of the mode—as we would expect for a skewed distribution.

Our test for zodiac birthdays had 11 df, so the relevant χ^2 curve peaks at 9 and has a mean
of 11. Knowing that, we might have easily guessed that the calculated χ^2 value of 5.094
wasn't going to be significant.

TI Tips TESTING GOODNESS OF FIT

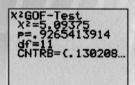

As always, the TI makes doing the mechanics of a goodness-of-fit test pretty easy, but it does take a little work to set it up. Let's use the zodiac data to run through the steps for a χ^2 GOF-Test.

- Enter the counts of executives born under each star sign in L1.

  Those counts were: 23 20 18 23 20 19 18 21 19 22 24 29

- Enter the expected percentages (or fractions, here 1/12) in L2. In this example they are all the same value, but that's not always the case.

- Convert the expected percentages to expected counts by multiplying each of them by the total number of observations. We use the calculator's summation command in the LIST MATH menu to find the total count for the data summarized in L1 and then multiply that sum by the percentages stored in L2 to produce the expected counts. The command is sum(L1)*L2 → L2. (We don't ever need the percentages again, so we can replace them by storing the expected counts in L2 instead.)

- Choose χ^2 GOF-Test from the STATS TESTS menu.
- Specify the lists where you stored the observed and expected counts, and enter the number of degrees of freedom, here 11.
- Ready, set, Calculate...

- ... and there are the calculated value of χ^2 and your P-value.
- Notice, too, there's a list of values called CNTRB. You can scroll across them, or use LIST NAMES to display them as a data list (as seen on the next page). Those are the cell-by-cell components of the χ^2 calculation. We aren't very interested in them this time, because our data failed to provide evidence that the zodiac sign mattered. However, in a situation where we rejected the null hypothesis, we'd want to look at the components to see where the biggest effects occurred. You'll read more about doing that later in this chapter.

BY HAND? If there are only a few cells, you may find that it's just as easy to write out the formula and then simply use the calculator to help you with the arithmetic. After you have found $\chi^2 = 5.09375$ you can use your TI to find the P-value, the probability of observing a χ^2 value at least as high as the one you calculated from your data. As you probably expect, that process is akin to normalcdf and tcdf. You'll find what you need in the DISTR menu at χ^2cdf. Just specify the left and right boundaries and the number of degrees of freedom.

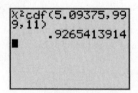

- Enter χ^2cdf(5.09375,999,11), as shown. (Why 999? Unlike t and z, chi-square values can get pretty big, especially when there are many cells. You may need to go a long way to the right to get to where the curve's tail becomes essentially meaningless. You can see what we mean by looking at Table C, showing chi-square values.)

And there's the P-value, a whopping 0.93! There's nothing at all unusual about these data. (So much for the zodiac's predictive power.)

Just Checking

Here's the table summarizing the frequency of two traits in that Biology class. Students are checking the genetic theory that the ratio of people with none, one, or both traits is 1:3:3:9.

3. Write the null and alternative hypothesis.

4. How many degrees of freedom are there?

5. Calculate the component of χ^2 for the bottom cell.

6. For these data $\chi^2 = 6.64$. What's the P-value?

7. What should the students conclude?

Tongue	Earlobes	Observed Count	Expected Count
Non-curling	Attached	12	7.75
Non-curling	Free	22	23.25
Curling	Attached	31	23.25
Curling	Free	59	69.75

The Trouble with Goodness-of-Fit Tests: What's the Alternative?

Goodness-of-fit tests are likely to be performed by people who have a theory of what the proportions *should* be in each category and who believe their theory to be true. Unfortunately, the only *null* hypothesis available for a goodness-of-fit test is that the theory is true. And as we know, the hypothesis-testing procedure allows us only to *reject* the null or *fail to reject* it. We can never confirm that a theory is in fact true, which is often what people want to do.

And we can't fix the problem by turning things around. Suppose we try to make our favored hypothesis the alternative. Then it is impossible to pick a single null. For example, suppose, as a doubter of astrology, you want to prove that the distribution of executive births is uniform. If you choose uniform as the null hypothesis, you can only *fail* to reject it. So you'd like uniformity to be your alternative hypothesis. Which particular violation of equally distributed births would you choose as your null? The problem is that the model can be wrong in many, many ways. There's no way to frame a null hypothesis the other way around. There's just no way to prove that a favored model is true.

Homogeneity: Comparing Observed Distributions

Many universities survey graduating classes to determine the plans of the graduates. We might wonder whether the plans of students are the same at different colleges within a large university. Here's a **two-way table** for Class of 2011 graduates from several colleges at one such university. Each cell of the table shows how many students from a particular college made a certain choice.

Table 25.1

Postgraduation activities of the class of 2011 for several colleges of a large university. (Note: ILR is the School of Industrial and Labor Relations.)

		College				
		Agriculture	Arts & Sciences	Engineering	ILR	Total
Choice	Employed	209	198	177	101	**685**
	Grad School	104	171	158	33	**466**
	Other	135	115	39	16	**305**
	Total	**448**	**484**	**374**	**150**	**1456**

Because the colleges' sizes are so different, we can see differences better by examining the proportions within each college rather than the counts:

Table 25.2

Activities of graduates as a percentage of respondents from each college.

	Agriculture	Arts & Sciences	Engineering	ILR	Combined
Employed	46.7%	40.9%	47.3%	67.3%	**47.0%**
Grad School	23.2%	35.3%	42.2%	22.0%	**32.0%**
Other	30.1%	23.8%	10.4%	10.7%	**20.9%**
Total	100.0%	100.0%	100.0%	100.0%	100.0%

Who   Graduates from four colleges at an upstate New York university

What   Postgraduation activities

When   2011

Why   Survey for general information

A S *Video*: **The Incident.** You may have guessed which famous incident put women and children at risk. Here you can view the story complete with rare film footage.

We already know how to test whether *two* proportions are the same. For example, we could use a two-proportion *z*-test to see whether the proportion of students choosing graduate school is the same for Agriculture students as for Engineering students. But now we have more than two groups. We want to test whether the students' choices are the same across all four colleges. The *z*-test for two proportions generalizes to a **chi-square test of homogeneity**.

Chi-square again? It turns out that the mechanics of this test are *identical* to the chi-square test for goodness-of-fit that we just saw. (How similar can you get?) Why a different name, then? The tests are really quite different. The goodness-of-fit test compared counts with a theoretical model. But here we're asking whether the distribution of choices is the same among different groups, so we find the expected counts for each category directly from the data. As a result, we count the degrees of freedom slightly differently as well.

The term "homogeneity" means that things are the same. Here, we ask whether the postgraduation choices made by students are the *same* for these four colleges. The homogeneity test comes with a built-in null hypothesis: We hypothesize that the distribution does not change from group to group. The test looks for differences too large to reasonably arise from random sample-to-sample variation. It can reveal a large deviation in a single category or small, but persistent, differences over all the categories—or anything in between.

Assumptions and Conditions

The assumptions and conditions are the same as for the chi-square test for goodness-of-fit. The **Counted Data Condition** says that these data must be counts. You can't do a test of homogeneity on proportions, so you have to work with the counts of graduates given in the first table. Also, you can't do a chi-square test on measurements. For example, if we had recorded GPAs for these same groups, we wouldn't be able to determine whether the mean GPAs were different using this test.[2]

Ideally, when we compare the proportions across several groups, we would like the cases within each group to be selected randomly. We need to know whether the **Independence Assumption** both within and across groups is reasonable. As usual, check the **Randomization Condition** and the **10% Condition** to make the assumption plausible.

We still must be sure we have enough data for this method to work. The **Expected Cell Frequency Condition** says that the expected count in each cell must be at least 5. We'll confirm that as we do the calculations.

Homogeneity Calculations

The null hypothesis says that the distribution of the proportions of graduates choosing each alternative is the same for all four colleges, so we can estimate those overall proportions by pooling our data from the four colleges together. Within each college, the expected proportion for each choice is just the overall proportion of all students making

---

[2]To do that, you'd use a method called Analysis of Variance (Chapter 24 on the DVD).

Table 25.3

Expected values for the 2011 graduates.

that choice. The expected counts are those proportions applied to the number of students graduating from each college.

For example, overall, 685, or about 47.0%, of the 1456 students who responded to the survey were employed. If the distributions are homogeneous (as the null hypothesis asserts), then 47% of the 448 Agriculture school graduates (or about 210.76 students) should be employed. Similarly, 47% of the 374 Engineering grads (or about 175.95) should be employed.

Working in this way, we (or, more likely, the computer) can fill in expected values for each cell. Because these are theoretical values, they don't have to be integers. The expected values look like this:

EXPECTED	Agriculture	Arts & Sciences	Engineering	ILR	Total
Employed	210.769	227.706	175.955	70.570	**685**
Grad School	143.385	154.907	119.701	48.008	**466**
Other	93.846	101.387	78.345	31.422	**305**
Total	**448**	**484**	**374**	**150**	**1456**

Now check the **Expected Cell Frequency Condition**. Indeed, there are at least 5 individuals expected in each cell.

Following the pattern of the goodness-of-fit test, we compute the component for each cell of the table. For the highlighted cell, employed students graduating from the Ag school, that's

$$\frac{(Obs - Exp)^2}{Exp} = \frac{(209 - 210.769)^2}{210.769} = 0.0148.$$

Summing these components across all cells gives

$$\chi^2 = \sum_{all\ cells} \frac{(Obs - Exp)^2}{Exp} = 93.66.$$

How about the degrees of freedom? We don't really need to calculate all the expected values in the table. We know there is a total of 685 employed students, so once we find the expected values for three of the colleges, we can determine the expected number for the fourth by just subtracting. Similarly, we know how many students graduated from each college, so after filling in two rows, we can find the expected values for the remaining row by subtracting. To fill out the table, we need to know the counts in only $R - 1$ rows and $C - 1$ columns. So the table has $(R - 1)(C - 1)$ degrees of freedom.

In our example, we need to calculate only 2 choices in each column and counts for 3 of the 4 colleges, for a total of $2 \times 3 = 6$ degrees of freedom. We'll need the degrees of freedom to find a P-value for the chi-square statistic.

■ **NOTATION ALERT**

For a contingency table, R represents the number of rows and C the number of columns.

Step-by-Step Example A CHI-SQUARE TEST FOR HOMOGENEITY

We have reports from four colleges on the postgraduation activities of their 2011 graduating classes.

Question: Are the distributions of students' choices of postgraduation activities the same across all the colleges?

(continued)

THINK ➡ **Plan** State what you want to know.

Identify the variables and check the W's.

I want to test whether postgraduation choices are the same for students from each of four colleges. I have a table of counts classifying each college's Class of 2011 respondents according to their activities.

Hypotheses State the null and alternative hypotheses.

H_0: Students' postgraduation activities are distributed in the same way for all four colleges.

H_A: Students' plans *do not* have the same distribution.

Model Make a picture: A side-by-side bar chart shows the four distributions of postgraduation activities. Plot column percents to remove the effect of class size differences. A split bar chart would also be an appropriate choice.

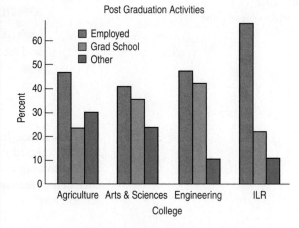

Post Graduation Activities

A side-by-side bar chart shows how the distributions of choices differ across the four colleges.

Think about the assumptions and check the conditions. Since we don't want to make inferences about other colleges or other classes, there is no need to check for a random sample.

✓ **Counted Data Condition:** I have counts of the number of students in categories.

✓ **Independence Assumption:** Even though this isn't a random sample, student plans should be largely independent of each other. The occasional friends who decide to join Teach for America together or couples who make grad school decisions together are too rare to affect this analysis.

✓ **Expected Cell Frequency Condition:** The expected values (shown below) are all at least 5.

State the sampling distribution model and name the test you will use.

The conditions seem to be met, so I can use a χ^2 model with $(3 - 1) \times (4 - 1) = 6$ degrees of freedom and do a **chi-square test of homogeneity**.

SHOW ➡ **Mechanics** Show the expected counts for each cell of the data table. You could make separate tables for the observed and expected counts, or put both counts in each cell as shown here. While observed counts must be whole numbers, expected counts rarely are—don't be tempted to round those off.

	Ag	A&S	Eng	ILR
Empl.	209	198	177	101
	210.769	227.706	175.955	70.570
Grad School	104	171	158	33
	143.385	154.907	119.701	48.008
Other	135	115	39	16
	93.846	101.387	78.345	31.422

(continued)

Calculate χ^2.

$$\chi^2 = \sum_{all\ cells} \frac{(Obs - Exp)^2}{Exp}$$
$$= \frac{(209 - 210.769)^2}{210.769} + \dots$$
$$= 93.66$$

The shape of a χ^2 model depends on the degrees of freedom. A χ^2 model with 6 df is skewed to the high end.

The P-value considers only the right tail. Here, the calculated value of the χ^2 statistic is off the scale, so the P-value is quite small.

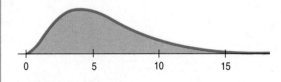

P-value = $P(\chi^2_6 > 93.66) < 0.0001$

TELL ➡ **Conclusion** State your conclusion in the context of the data. You should specifically talk about whether the distributions for the groups appear to be different.

The P-value is very small, so I reject the null hypothesis and conclude that there's evidence that the postgraduation activities of students from these four colleges don't have the same distribution.

If you find that simply rejecting the hypothesis of homogeneity is a bit unsatisfying, you're in good company. OK, so the postgraduation plans are different. What we'd really like to know is what the differences are, where they're the greatest, and where they're smallest. The test for homogeneity doesn't answer these interesting questions, but it does provide some evidence that can help us.

Examining the Residuals

Whenever we reject the null hypothesis, it's a good idea to examine the differences between observed and expected counts—the residuals. (We don't need to do that when we fail to reject because when the χ^2 value is small, all of its components must have been small.) For chi-square tests, we want to compare residuals for cells that may have very different counts. So we're better off standardizing the residuals. We know the mean residual is zero,[3] but we need to know each residual's standard deviation. When we tested proportions, we saw a link between the expected proportion and its standard deviation. For counts, there's a similar link. To standardize a cell's residual, we just divide by the square root of its expected value:

$$c = \frac{(Obs - Exp)}{\sqrt{Exp}}.$$

Notice that these **standardized residuals** are just the square roots of the **components** we calculated for each cell, and their sign indicates whether we observed more cases than we expected, or fewer.

The standardized residuals give us a chance to think about the underlying patterns and to consider the ways in which the distribution of postgraduation plans may differ from college to college. Now that we've subtracted the mean (the residual, $Obs - Exp$, has mean 0,

---

[3]Residual = *Observed* − *Expected*. Because the total of the expected values is set to be the same as the observed total, the residuals must sum to zero.

so it's already subtracted) and divided by their standard deviations, these are z-scores. If the null hypothesis was true, we could even appeal to the Central Limit Theorem, think of the Normal model, and use the 68–95–99.7 Rule to judge how extraordinary the large ones are.

Here are the standardized residuals for the Class of 2011 data:

Table 25.4

Standardized residuals can help show how the table differs from the null hypothesis pattern.

RESIDUALS	Ag	A&S	Eng	ILR
Employed	−0.121866	−1.96860	0.078805	3.62235
Grad School	−3.28909	1.29304	3.50062	−2.16607
Other	4.24817	1.35192	−4.44511	−2.75117

The most extreme standardized residuals, both positive and negative, attract our attention. It appears that ILR graduates are most likely to be employed. Engineering college graduates seem more likely to go on to graduate work and very unlikely to take time off for "volunteering and travel, among other activities" (as the "Other" category is explained). By contrast, Ag school graduates seem to favor those other pursuits and be less likely to start grad school immediately after college.

For Example LOOKING AT χ^2 RESIDUALS

RECAP: Some people suggest that schoolchildren who are the older ones in their class naturally perform better in sports and therefore get more coaching and encouragement. To see if there's any evidence for this, we looked at major league baseball players born since 1975. A goodness-of-fit test found their birth months to have a distribution that's significantly different from the rest of us. The table shows the standardized residuals.

Month	Residual	Month	Residual
1	1.73	7	−2.69
2	1.72	8	2.77
3	−0.21	9	0.08
4	0.25	10	−1.56
5	0.71	11	−1.22
6	−0.39	12	−0.96

QUESTION: What's different about the distribution of birth months among major league ballplayers?

ANSWER: It appears that, compared to the general population, fewer ballplayers than expected were born in July and more than expected in August. Either month would make them the younger kids in their grades in school, so these data don't offer support for the conjecture that being older is an advantage in terms of a career as a professional baseball player.

 ## Just Checking

Tiny black potato flea beetles can damage potato plants in a vegetable garden. These pests chew holes in the leaves, causing the plants to wither or die. They can be killed with an insecticide, but a canola oil spray has been suggested as a nonchemical "natural" method of controlling the beetles. To conduct an experiment to test the effectiveness of the natural spray, we gather 500 beetles and place them in three Plexiglas® containers. Two hundred beetles go in the first container, where we spray them with the canola oil mixture. Another 200 beetles go in the second container; we spray them with the insecticide. The remaining 100 beetles in the last container serve as a control group; we simply spray them with water. Then we wait 6 hours and count the number of surviving beetles in each container.

(continued)

8. Why do we need the control group?

9. What would our null hypothesis be?

10. After the experiment is over, we could summarize the results in a table as shown. How many degrees of freedom does our χ^2 test have?

	Natural Spray	Insecticide	Water	Total
Survived				
Died				
Total	200	200	100	500

11. Suppose that, all together, 125 beetles survived. (That's the first-row total.) What's the expected count in the first cell—survivors among those sprayed with the natural spray?

12. If it turns out that only 40 of the beetles in the first container survived, what's the calculated component of χ^2 for that cell?

13. If the total calculated value of χ^2 for this table turns out to be around 10, would you expect the P-value of our test to be large or small? Explain.

Chi-Square Test of Independence

A study from the University of Texas Southwestern Medical Center examined 626 people being treated for non–blood-related diseases to see whether the risk of hepatitis C was related to whether people had tattoos and to where they got their tattoos. Hepatitis C causes about 10,000 deaths each year in the United States, but often goes undetected for years after infection.

The data from this study can be summarized in a two-way table, as follows:

Table 25.5

Counts of patients classified by their hepatitis C test status according to whether they had a tattoo from a tattoo parlor or from another source, or had no tattoo.

	Hepatitis C	No Hepatitis C	Total
Tattoo, Parlor	17	35	52
Tattoo, Elsewhere	8	53	61
None	22	491	513
Total	47	579	626

Who	Patients being treated for non–blood-related disorders
What	Tattoo status and hepatitis C status
When	1991, 1992
Where	Texas

A S **Activity: Independence and Chi-Square.** This unusual simulation shows how independence arises (and fails) in contingency tables.

These data differ from the kinds of data we've considered before in this chapter because they categorize subjects from a single group on two categorical variables rather than on only one. The categorical variables here are *Hepatitis C Status* ("Hepatitis C" or "No Hepatitis C") and *Tattoo Status* ("Parlor," "Elsewhere," "None"). We've seen counts classified by two categorical variables displayed like this in Chapter 2, so we know such tables are called contingency tables. **Contingency tables** categorize counts on two (or more) variables so that we can see whether the distribution of counts on one variable is contingent on the other.

The natural question to ask of these data is whether the chance of having hepatitis C is *independent* of tattoo status. Recall that for events **A** and **B** to be independent $P(\mathbf{A})$ must equal $P(\mathbf{A}|\mathbf{B})$. Here, this means the probability that a randomly selected patient has hepatitis C should be the same regardless of the patient's tattoo status. If *Hepatitis Status* is independent of tattoos, we'd expect the proportion of people testing positive for hepatitis to be the same for the three levels of *Tattoo Status*. Of course, for real data we won't expect them to be exactly the same, so we look to see how close they are. This sounds a lot like the test of homogeneity. In fact, the mechanics of the calculation are identical. The difference is that now we have two categorical variables measured on a single population. For the homogeneity test, we had a single categorical variable measured independently on two or more populations. But now we ask a different question: "Are the variables independent?" rather than "Are the groups homogeneous?" These are subtle differences, but they are important when we state hypotheses and draw conclusions. When we ask whether two variables measured on the same population are independent we're performing a **chi-square test of independence**.

> ### Look at the Design
> Homogeneity? Or independence? Look at how the data were collected. When our data are categories of a single variable gathered from more than one group, we wonder whether the groups are the same; the question is homogeneity. When our data cross-categorize two variables gathered from a single group, we wonder whether there's an association; the question is independence.

For Example WHICH χ^2 TEST?

Many states and localities now collect data on traffic stops regarding the race of the driver. The initial concern was that Black drivers were being stopped more often (the "crime" ironically called "Driving While Black"). With more data in hand, attention has turned to other issues. For example, data from 2533 traffic stops in Cincinnati[4] report the race of the driver (Black, White, or Other) and whether the traffic stop resulted in a search of the vehicle.

		Race		
		Black	White	Other	Total
Search	No	787	594	27	1408
	Yes	813	293	19	1125
	Total	1600	887	46	2533

QUESTION: Which test would be appropriate to examine whether race is a factor in vehicle searches? What are the hypotheses?

ANSWER: *These data represent one group of traffic stops in Cincinnati, categorized on two variables, Race and Search. I'll do a chi-square test of independence.*

H_0: *Whether or not police search a vehicle is independent of the race of the driver.*

H_A: *Decisions to search vehicles are not independent of the driver's race.*

_A S_ *Activity:* **Chi-Square Tables.** Work with *ActivStats'* interactive chi-square table to perform a hypothesis test.

Assumptions and Conditions

Of course, we still need counts and enough data so that the expected values are at least 5 in each cell.

If we're interested in the independence of variables, we usually want to generalize from the data to some population. In that case, we'll need to check that the data are a representative random sample from, and fewer than 10% of, that population.

---

[4]John E. Eck, Lin Liu, and Lisa Growette Bostaph, Police Vehicle Stops in Cincinnati, Oct. 1, 2003, available at www.cincinnati-oh.gov. Data for other localities can be found by searching from www.racialprofilinganalysis.neu.edu.

Step-by-Step Example A CHI-SQUARE TEST FOR INDEPENDENCE

We have counts of 626 individuals categorized according to their "tattoo status" and their "hepatitis status."

Question: Are tattoo status and hepatitis status independent?

THINK ➡ **Plan** State what you want to know.

Identify the variables and check the W's.

Hypotheses State the null and alternative hypotheses.

We perform a test of independence when we suspect the variables may not be independent. We are on the familiar ground of making a claim (in this case, that knowing *Tattoo Status* will change probabilities for *Hepatitis C Status*) and testing the null hypothesis that it is *not* true.

Model Make a picture. Because these are only two categories—Hepatitis C and No Hepatitis C—a simple bar chart of the distribution of tattoo sources for Hep C patients shows all the information.

Think about the assumptions and check the conditions.

I want to test whether the categorical variables *Tattoo Status* and *Hepatitis Status* are statistically independent. I have a contingency table of 626 Texas patients under treatment for a non–blood-related disease.

H_O: *Tattoo Status* and *Hepatitis Status* are independent.[5]

H_A: *Tattoo Status* and *Hepatitis Status* are not independent.

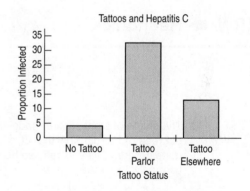

The bar chart suggests strong differences in Hepatitis C risk based on tattoo status.

✓ **Counted Data Condition:** I have counts of individuals categorized on two variables.

✓ **Independence Assumption:** The people in this study are likely to be independent of each other.

✓ **Randomization Condition:** These data are from a retrospective study of patients being treated for something unrelated to hepatitis. Although they are not an SRS, they were selected to avoid biases.

(continued)

---

[5]Once again, parameters are hard to express. The hypothesis of independence itself tells us how to find expected values for each cell of the contingency table. That's all we need.

✓ **10% Condition:** These 626 patients are far fewer than 10% of all those with tattoos or hepatitis.

✗ **Expected Cell Frequency Condition:** The expected values do not meet the condition that all are at least 5.

This table shows both the observed and expected counts for each cell. The expected counts are calculated exactly as they were for a test of homogeneity; in the first cell, for example, we expect $\frac{52}{626}$ (that's 8.3%) of 47.

Warning: Be wary of proceeding when there are small expected counts. If we see expected counts that fall far short of 5, or if many cells violate the condition, we should not use χ^2. (We will soon discuss ways you can fix the problem.) If you do continue, always check the residuals to be sure those cells did not have a major influence on your result.

Specify the model.

Name the test you will use.

	Hepatitis C	No Hepatitis C	Total
Tattoo, Parlor	17	35	52
	3.904	48.096	
Tattoo, Elsewhere	8	53	61
	4.580	56.420	
None	22	491	513
	38.516	474.484	
Total	47	579	626

Although the Expected Cell Frequency Condition is not satisfied, the values are close to 5. I'll go ahead, but I'll check the residuals carefully. I'll use a χ^2 model with $(3 - 1) \times (2 - 1) = 2$ df and do a **chi-square test of independence**.

SHOW ➡ **Mechanics**   Calculate χ^2.

$$\chi^2 = \sum_{all\ cells} \frac{(Obs - Exp)^2}{Exp}$$

$$= \frac{(17 - 3.094)^2}{3.094} + \cdots = 57.91$$

The shape of a chi-square model depends on its degrees of freedom. With 2 df, the model looks quite different, as you can see here. We still care only about the right tail.

$$\text{P-Value} = P(\chi^2_2 > 57.91) < 0.0001$$

TELL ➡ **Conclusion**   Link the P-value to your decision. State your conclusion about the independence of the two variables.

(We should be wary of this conclusion because of the small expected counts. A complete solution must include the additional analysis, recalculation, and final conclusion discussed in the following section.)

The P-value is very small, so I reject the null hypothesis and conclude that *Hepatitis Status* is not independent of *Tattoo Status.* Because the Expected Cell Frequency Condition was violated, I need to check that the two cells with small expected counts did not influence this result too greatly.

For Example CHI-SQUARE MECHANICS

RECAP: We have data that allow us to investigate whether police searches of vehicles they stop are independent of the driver's race.

QUESTIONS: What are the degrees of freedom for this test? What is the expected frequency of searches for the Black drivers who were stopped? What's that cell's component in the χ^2 computation? And how is the standardized residual for that cell computed?

		Race		
		Black	**White**	**Other**	**Total**
Search	**No**	787	594	27	**1408**
	Yes	813	293	19	**1125**
	Total	**1600**	**887**	**46**	**2533**

ANSWER: This is a 2 × 3 contingency table, so $df = (2 - 1)(3 - 1) = 2$.
Overall, 1125 of 2533 vehicles were searched. If searches are conducted independent of race,

then I'd expect $\dfrac{1125}{2533}$ of the 1600 Black drivers to have been searched: $\dfrac{1125}{2533} \times 1600 \approx 710.62$.

That cell's term in the χ^2 calculation is $\dfrac{(Obs - Exp)^2}{Exp} = \dfrac{(813 - 710.62)^2}{710.62} = 14.75$

The standardized residual for that cell is $\dfrac{Obs - Exp}{\sqrt{Exp}} = \dfrac{813 - 710.62}{\sqrt{710.62}} = 3.84$

Examine the Residuals

Each cell of the contingency table contributes a term to the chi-square sum. As we did earlier, we should examine the residuals because we have rejected the null hypothesis. In this instance, we have an additional concern that the cells with small expected frequencies not be the ones that make the chi-square statistic large.

Our interest in the data arises from the potential for improving public health. If patients with tattoos are more likely to test positive for hepatitis C, perhaps physicians should be advised to suggest blood tests for such patients.

The standardized residuals look like this:

Table 25.6

Standardized residuals for the hepatitis and tattoos data. Are any of them particularly large in magnitude?

	Hepatitis C	No Hepatitis C
Tattoo, Parlor	6.628	−1.888
Tattoo, Elsewhere	1.598	−0.455
None	−2.661	0.758

The chi-square value of 57.91 is the sum of the squares of these six values. The cell for people with tattoos obtained in a tattoo parlor who have hepatitis C is large and positive, indicating there are more people in that cell than the null hypothesis of independence would predict. Maybe tattoo parlors are a source of infection or maybe those who go to tattoo parlors also engage in risky behavior.

The second-largest component is a negative value for those with no tattoos who test positive for hepatitis C. A negative value says that there are fewer people in this cell than independence would expect. That is, those who have no tattoos are less likely to be infected with hepatitis C than we might expect if the two variables were independent.

What about the cells with small expected counts? The formula for the chi-square standardized residuals divides each residual by the square root of the expected frequency. Too small an expected frequency can arbitrarily inflate the residual and lead to an inflated chi-square statistic. Any expected count close to the arbitrary minimum of 5 calls for checking that cell's standardized residual to be sure it is not particularly large. In this case, the standardized residual for the "Hepatitis C and Tattoo, Elsewhere" cell is not particularly large, but the standardized residual for the "Hepatitis C and Tattoo, Parlor" cell is large.

We might choose not to report the results because of concern with the small expected frequency. Alternatively, we could include a warning along with our report of the results. Yet another approach is to combine categories to get a larger category total and correspondingly larger expected frequencies, if there are some categories that can be appropriately combined. Here, we might naturally combine the two rows for tattoos, obtaining a 2 × 2 table:

Table 25.7

Combining the two tattoo categories gives a table with all expected counts greater than 5.

	Hepatitis C	No Hepatitis C	Total
Tattoo	25	88	113
None	22	491	513
Total	47	579	626

This table has expected values of at least 5 in every cell, and a chi-square value of 42.42 on 1 degree of freedom. The corresponding P-value is <0.0001.

We conclude that *Tattoo Status* and *Hepatitis C Status* are not independent. The data *suggest* that tattoo parlors may be a particular problem, but we don't have enough data to draw that conclusion.

For Example WRITING CONCLUSIONS FOR χ^2 TESTS

RECAP: We're looking at Cincinnati traffic stop data to see if police decisions about searching cars show evidence of racial bias. With 2 df, technology calculates $\chi^2 = 73.25$, a P-value less than 0.0001, and these standardized residuals:

		Race		
		Black	White	Other
Search	No	−3.43	4.55	0.28
	Yes	3.84	−5.09	−0.31

QUESTION: What's your conclusion?

ANSWER: *The very low P-value leads me to reject the null hypothesis. There's strong evidence that police decisions to search cars at traffic stops are associated with the driver's race.*

The largest residuals are for White drivers, who are searched less often than independence would predict. It appears that Black drivers' cars are searched more often.

TI Tips TESTING HOMOGENEITY OR INDEPENDENCE

Yes, the TI will do chi-square tests of homogeneity and independence. Let's use the tattoo data. Here goes.

TEST A HYPOTHESIS OF HOMOGENEITY OR INDEPENDENCE

Stage 1: You need to enter the data as a matrix. A "matrix" is just a formal mathematical term for a table of numbers.
- Push the MATRIX button, and choose to EDIT matrix [A].
- First specify the dimensions of the table, rows × columns.
- Enter the appropriate counts, one cell at a time. The calculator automatically asks for them row by row.

Stage 2: Do the test.
- In the STAT TESTS menu choose χ^2-Test.
- The TI now confirms that you have placed the observed frequencies in [A]. It also tells you that when it finds the expected frequencies it will store those in [B] for you. Now Calculate the mechanics of the test.

The TI reports a calculated value of $\chi^2 = 57.91$ and an exceptionally small P-value.

Stage 3: Check the expected counts.
- Go back to MATRIX EDIT and choose [B].

Notice that two of the cells fail to meet the condition that expected counts be at least 5. This problem enters into our analysis and conclusions.

Stage 4: And now some bad news. There's no easy way to calculate the standardized residuals. Look at the two matrices, [A] and [B]. Large residuals will happen when the corresponding entries differ greatly, especially when the expected count in [B] is small (because you will divide by the square root of the entry in [B]). The first cell is a good candidate, so we show you the calculation of its standardized residual.

A residual of over 6 is pretty large—possibly an indication that you're more likely to get hepatitis in a tattoo parlor, but the expected count is smaller than 5. We're pretty sure that hepatitis status is not independent of having a tattoo, but we should be wary of saying anything more. Probably the best approach is to combine categories to get cells with expected counts above 5.

Chi-Square and Causation

Chi-square tests are common. Tests for independence are especially widespread. Unfortunately, many people interpret a small P-value as proof of causation. We know better. Just as correlation between quantitative variables does not demonstrate causation, a failure of independence between two categorical variables does not show a cause-and-effect relationship between them, nor should we say that one variable *depends* on the other.

The chi-square test for independence treats the two variables symmetrically. There is no way to differentiate the direction of any possible causation from one variable to the other. In our example, it is unlikely that having hepatitis causes one to crave a tattoo, but other examples are not so clear.

In this case, it's easy to imagine that lurking variables are responsible for the observed lack of independence. Perhaps the lifestyles of some people include both tattoos and behaviors that put them at increased risk of hepatitis C, such as body piercings or even drug use. Even a small subpopulation of people with such a lifestyle among those with tattoos might be enough to create the observed result. After all, we observed only 25 patients with both tattoos and hepatitis.

In some sense, a failure of independence between two categorical variables is less impressive than a strong, consistent, linear association between quantitative variables. Two categorical variables can fail the test of independence in many ways, including ways that show no consistent pattern of failure. Examination of the chi-square standardized residuals can help you think about the underlying patterns.

Just Checking

Which of the three chi-square tests—goodness-of-fit, homogeneity, or independence—would you use in each of the following situations?

14. A restaurant manager wonders whether customers who dine on Friday nights have the same preferences among the four "chef's special" entrées as those who dine on Saturday nights. One weekend he has the wait staff record which entrées were ordered each night. Assuming these customers to be typical of all weekend diners, he'll compare the distributions of meals chosen Friday and Saturday.

15. Company policy calls for parking spaces to be assigned to everyone at random, but you suspect that may not be so. There are three lots of equal size: lot A,

next to the building; lot B, a bit farther away; and lot C, on the other side of the highway. You gather data about employees at middle management level and above to see how many were assigned parking in each lot.

16. Is a student's social life affected by where the student lives? A campus survey asked a random sample of students whether they lived in a dormitory, in off-campus housing, or at home, and whether they had been out on a date 0, 1–2, 3–4, or 5 or more times in the past two weeks.

WHAT IF ●●● we take a closer look at χ^2 residuals?

After the last few chapters, you probably thought we'd use this What If to do a permutation test for a table of categorical counts, didn't you? Well, we could. But we already did that, way back in Chapter 2![6] (Go have a look.) Instead, let's think some more about residuals.

Imagine rolling a die 96 times and counting the number of times each face shows up. We chose 96 because then we'd expect 16 of each, or thereabouts, and dividing by $\sqrt{16}$ makes finding the residuals easy. We simulated the 96 rolls, and got the results shown in the table.

Face	Count
1	21
2	15
3	21
4	18
5	12
6	9

If you rolled a real die and got results like this, would you be surprised? Might it seem that the die could be unfair, biased against rolling higher numbers?

The second table shows the standardized residuals. The largest is $\dfrac{9-16}{\sqrt{16}} = -1.75$, for face 6. Is that large for a residual?

Face	Resid
1	1.25
2	−0.25
3	1.25
4	0.50
5	−1.00
6	−1.75

---

[6]We didn't actually call it a "permutation test" then. You were too young to understand.

We investigated by repeating this simulation. We ran 100 trials, rolling 96 dice each time. For each trial we calculated the residuals, just as you see here. That gave us a total of 600 residuals in all. Here's a histogram displaying their distribution:

Hmm. . . centered at 0, most values between −1 and +1, very few outside the interval −2 to +2. . . Look familiar? We hope so. Sure, it looks a bit ragged after only 100 trials. And it may be slightly skewed, but we rolled the die only 96 times. In the long run, the sampling distribution of the standardized residuals approaches a Normal model, and that means it's helpful to think of the residuals as z-scores. A z-score of −1.75 isn't strange enough to raise concern.[7]

With this in mind, we see that each component in a chi-square calculation is simply a z^2 term. This insight reveals that a chi-square model describes the sampling distribution of the sum of the squares of a bunch of z-scores.

Look back at our path through inference. We used a Normal model for proportions. We tweaked that a bit when we used a t-model for means, but those mechanics still looked pretty similar. Now we see that even though this chapter's procedures appear to be very different, we really didn't stray very far from that amazing Normal curve after all!

---

[7]The goodness-of-fit test has a P-value of 0.186, no reason to question a die's fairness.

WHAT CAN GO WRONG?

A S Simulation: **Sample Size and Chi-Square.** Chi-square statistics have a peculiar problem. They don't respond to increasing the sample size in quite the same way you might expect.

■ **Don't use chi-square methods unless you have counts.** All three of the chi-square tests apply only to counts. Other kinds of data can be arrayed in two-way tables. Just because numbers are in a two-way table doesn't make them suitable for chi-square analysis. Data reported as proportions or percentages can be suitable for chi-square procedures, *but only after they are converted to counts.* If you try to do the calculations without first finding the counts, your results will be wrong.

■ **Beware large samples.** Beware *large* samples?! That's not the advice you're used to hearing. The chi-square tests, however, are unusual. Be wary of chi-square tests performed on very large samples. No hypothesized distribution fits perfectly, no two groups are exactly homogeneous, and two variables are rarely perfectly independent. The degrees of freedom for chi-square tests don't grow with the sample size. With a sufficiently large sample size, a chi-square test can always reject the null hypothesis. But we have no measure of how far the data are from the null model. There are no confidence intervals to help us judge the effect size.

■ **Don't say that one variable "depends" on the other just because they're not independent.** Dependence suggests a pattern and implies causation, but variables can fail to be independent in many different ways. When variables fail the test for independence, you might just say they are "associated."

What Have We Learned?

We've learned how to test hypotheses about categorical variables. We use one of three related methods. All look at counts of data in categories and rely on chi-square models, a new family indexed by degrees of freedom.

We've learned to look at the study design to see which test is appropriate.

- Goodness-of-fit tests compare the observed distribution of a single categorical variable to an expected distribution based on a theory or a model.
- Tests of homogeneity compare the observed distributions in several groups for a single categorical variable.
- Tests of independence examine observed counts from a single group for evidence of an association between two categorical variables.

We've learned to write appropriate hypotheses for each test, and especially to understand how these differ for tests of homogeneity and independence.

We've learned again to check assumptions and conditions before proceeding with inference.

- **Counted Data Condition:** we have observed counts for the categories.
- **Independence Assumption:** randomization makes independence more plausible.
- **Expected Cell Frequency Condition:** expect at least 5 observations in each cell.
- **10% Condition:** The sample is less than 10% of the population.

We've learned that mechanically the three tests are almost identical. We've learned to find the expected counts, determine the number of degrees of freedom, calculate the value of χ^2, and determine the *P*-value.

We've learned to interpret the results of the inference test.

- These tests are conceptually many-sided, because there are many ways that observed counts can deviate significantly from what we hypothesized.
- Failing to reject the null hypothesis does not confirm that the data do fit the model.
- If we reject the null hypothesis, we can examine the standardized residuals (or components) to better understand what cells were responsible.

Terms

Chi-square test of goodness-of-fit
A test of whether the distribution of counts in one categorical variable matches the distribution predicted by a model is called a test of goodness-of-fit. In a chi-square goodness-of-fit test, the expected counts come from the predicting model. The test finds a P-value from a chi-square model with $n - 1$ degrees of freedom, where n is the number of categories in the categorical variable. (p. 673)

Cell
A cell is one element of a table corresponding to a specific row and a specific column. Table cells can hold counts, percentages, or measurements on other variables. Or they can hold several values. (p. 674)

Chi-square statistic
The chi-square statistic can be used to test whether the observed counts in a frequency distribution or contingency table match the counts we would expect according to some model. It is calculated as

$$\chi^2 = \sum_{all\ cells} \frac{(Obs - Exp)^2}{Exp}.$$

Chi-square statistics differ in how expected counts are found, depending on the question asked. (p. 675)

Chi-square model
Chi-square models are skewed to the right. They are parameterized by their degrees of freedom and become less skewed with increasing degrees of freedom. (p. 675)

Chi-square component
The components of a chi-square calculation are

$$\frac{(Observed - Expected)^2}{Expected},$$

found for each cell of the table. (pp. 678, 685)

| Two-way table | Each *cell* of a two-way table shows counts of individuals. One way classifies a sample according to a categorical variable. The other way can classify different groups of individuals according to the same variable or classify the same individuals according to a different categorical variable. (p. 681) |

Chi-square test of homogeneity — A test comparing the distribution of counts for *two or more groups* on the same categorical variable. A chi-square test of homogeneity finds expected counts based on the overall frequencies, adjusted for the totals in each group under the (null hypothesis) assumption that the distributions are the same for each group. We find a P-value from a chi-square distribution with (#*Rows* − 1) × (#*Cols* − 1) degrees of freedom, where #*Rows* gives the number of categories and #*Cols* gives the number of independent groups. (p. 682)

Standardized residual — In each cell of a two-way table, a standardized residual is the square root of the chi-square component for that cell with the sign of the *Observed−Expected* difference:

$$\frac{(Obs - Exp)}{\sqrt{Exp}}.$$

When we reject a chi-square test, an examination of the standardized residuals can sometimes reveal more about how the data deviate from the null model. (p. 685)

Contingency table — A two-way table that classifies individuals according to two categorical variables. (p. 687)

Chi-square test of independence — A test of whether two categorical variables are independent examines the distribution of counts for *one group of individuals* classified according to both variables. A chi-square test of *independence* finds expected counts by assuming that knowing the marginal totals tells us the cell frequencies, assuming that there is no association between the variables. This turns out to be the same calculation as a test of homogeneity. We find a P-value from a chi-square distribution with (#*Rows* − 1) × (#*Cols* − 1) degrees of freedom, where #*Rows* gives the number of categories in one variable and #*Cols* gives the number of categories in the other. (p. 688)

On the Computer CHI-SQUARE

Most statistics packages perform chi-square tests on contingency tables. It's up to you to properly interpret them as tests of homogeneity or independence. Some packages want to create the contingency table from the actual data, while others allow you to enter the summary counts. Goodness-of-fit tests may be missing.

Some software packages display the standardized residuals, others the components.

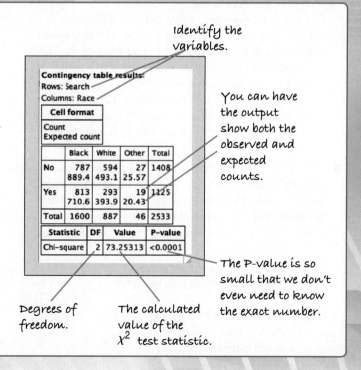

Identify the variables.

You can have the output show both the observed and expected counts.

The P-value is so small that we don't even need to know the exact number.

Degrees of freedom.

The calculated value of the χ^2 test statistic.

Exercises

1. **Which test?** For each of the following situations, state whether you'd use a chi-square goodness-of-fit test, a chi-square test of homogeneity, a chi-square test of independence, or some other statistical test:

   a) A brokerage firm wants to see whether the type of account a customer has (Silver, Gold, or Platinum) affects the type of trades that customer makes (in person, by phone, or on the Internet). It collects a random sample of trades made for its customers over the past year and performs a test.

   b) That brokerage firm also wants to know if the type of account affects the size of the account (in dollars). It performs a test to see if the mean size of the account is the same for the three account types.

   c) The academic research office at a large community college wants to see whether the distribution of courses chosen (Humanities, Social Science, or Science) is different for its residential and nonresidential students. It assembles last semester's data and performs a test.

2. **Which test again?** For each of the following situations, state whether you'd use a chi-square goodness-of-fit test, a chi-square test of homogeneity, a chi-square test of independence, or some other statistical test:

   a) Is the quality of a car affected by what day it was built? A car manufacturer examines a random sample of the warranty claims filed over the past two years to test whether defects are randomly distributed across days of the work week.

   b) A medical researcher wants to know if blood cholesterol level is related to heart disease. She examines a database of 10,000 patients, testing whether the cholesterol level (in milligrams) is related to whether or not a person has heart disease.

   c) A student wants to find out whether political leaning (liberal, moderate, or conservative) is related to choice of major. He surveys 500 randomly chosen students and performs a test.

3. **Dice** After getting trounced by your little brother in a children's game, you suspect the die he gave you to roll may be unfair. To check, you roll it 60 times, recording the number of times each face appears. Do these results cast doubt on the die's fairness?

   a) If the die is fair, how many times would you expect each face to show?

   b) To see if these results are unusual, will you test goodness-of-fit, homogeneity, or independence?

Face	Count
1	11
2	7
3	9
4	15
5	12
6	6

   c) State your hypotheses.

   d) Check the conditions.

   e) How many degrees of freedom are there?

   f) Find χ^2 and the P-value.

   g) State your conclusion.

4. **M&M's** As noted in an earlier chapter, the Master-foods Company says that until very recently yellow candies made up 20% of its milk chocolate M&M's, red another 20%, and orange, blue, and green 10% each. The rest are brown. On his way home from work the day he was writing these exercises, one of the authors bought a bag of plain M&M's. He got 29 yellow ones, 23 red, 12 orange, 14 blue, 8 green, and 20 brown. Is this sample consistent with the company's stated proportions? Test an appropriate hypothesis and state your conclusion.

   a) If the M&M's are packaged in the stated proportions, how many of each color should the author have expected to get in his bag?

   b) To see if his bag was unusual, should he test goodness-of-fit, homogeneity, or independence?

   c) State the hypotheses.

   d) Check the conditions.

   e) How many degrees of freedom are there?

   f) Find χ^2 and the P-value.

   g) State a conclusion.

5. **Human births** If there is no seasonal effect on human births, we would expect equal numbers of children to be born in each season (winter, spring, summer, and fall). A student takes a census of her statistics class and finds that of the 120 students in the class, 25 were born in winter, 35 in spring, 32 in summer, and 28 in fall. She wonders if the excess in the spring is an indication that births are not uniform throughout the year.

   a) What is the expected number of births in each season if there is no "seasonal effect" on births?

   b) Compute the χ^2 statistic.

   c) How many degrees of freedom does the χ^2 statistic have?

   d) Find the $\alpha = 0.05$ critical value for the χ^2 distribution with the appropriate number of df.

   e) Using the critical value, what do you conclude about the null hypothesis at $\alpha = 0.05$?

6. **Bank cards** At a major credit card bank, the percentages of people who historically apply for the Silver, Gold, and Platinum cards are 60%, 30%, and 10%, respectively. In a recent sample of customers responding to a promotion, of 200 customers, 110 applied for Silver, 55 for Gold, and 35 for Platinum. Is there evidence to suggest that the percentages for this promotion may be different from the historical proportions?

a) What is the expected number of customers applying for each type of card in this sample if the historical proportions are still true?

b) Compute the χ^2 statistic.

c) How many degrees of freedom does the χ^2 statistic have?

d) Find the $\alpha = 0.05$ critical value for the χ^2 distribution with the appropriate number of df.

e) Using the critical value, what do you conclude about the null hypothesis at $\alpha = 0.05$?

7. Nuts A company says its premium mixture of nuts contains 10% Brazil nuts, 20% cashews, 20% almonds, and 10% hazelnuts, and the rest are peanuts. You buy a large can and separate the various kinds of nuts. Upon weighing them, you find there are 112 grams of Brazil nuts, 183 grams of cashews, 207 grams of almonds, 71 grams of hazelnuts, and 446 grams of peanuts. You wonder whether your mix is significantly different from what the company advertises.

a) Explain why the chi-square goodness-of-fit test is not an appropriate way to find out.

b) What might you do instead of weighing the nuts in order to use a χ^2 test?

8. Mileage A salesman who is on the road visiting clients thinks that, on average, he drives the same distance each day of the week. He keeps track of his mileage for several weeks and discovers that he averages 122 miles on Mondays, 203 miles on Tuesdays, 176 miles on Wednesdays, 181 miles on Thursdays, and 108 miles on Fridays. He wonders if this evidence contradicts his belief in a uniform distribution of miles across the days of the week. Explain why it is not appropriate to test his hypothesis using the chi-square goodness-of-fit test.

9. NYPD and race Census data for New York City indicate that 29.2% of the under-18 population is white, 28.2% black, 31.5% Latino, 9.1% Asian, and 2% other ethnicities. The New York Civil Liberties Union points out that, of 26,181 police officers, 64.8% are white, 14.5% black, 19.1% Latino and 1.4% Asian. Do the police officers reflect the ethnic composition of the city's youth? Test an appropriate hypothesis and state your conclusion.

10. Violence against women 2009 In its study *When Men Murder Women: An Analysis of 2009 Homicide Data*, 2011, the Violence Policy Center (www.vpc.org) reported that 1818 women were murdered by men in 2009. Of these victims, a weapon could be identified for 1654 of them. Of those for whom a weapon could be identified, 861 were killed by guns, 364 by knives or other cutting instruments, 214 by other weapons, and 215 by personal attack (battery, strangulation, etc.). The FBI's Uniform Crime Report says that, among all murders nationwide, the weapon use rates were as follows: guns 63.4%, knives 13.1%, other weapons 16.8%, personal attack 6.7%. Is there evidence that violence against women involves

different weapons than other violent attacks in the United States?

11. Fruit flies Offspring of certain fruit flies may have yellow or ebony bodies and normal wings or short wings. Genetic theory predicts that these traits will appear in the ratio 9:3:3:1 (9 yellow, normal: 3 yellow, short: 3 ebony, normal: 1 ebony, short). A researcher checks 100 such flies and finds the distribution of the traits to be 59, 20, 11, and 10, respectively.

a) Are the results this researcher observed consistent with the theoretical distribution predicted by the genetic model?

b) If the researcher had examined 200 flies and counted exactly twice as many in each category—118, 40, 22, 20—what conclusion would he have reached?

c) Why is there a discrepancy between the two conclusions?

12. Pi Many people know the mathematical constant π is approximately 3.14. But that's not exact. To be more precise, here are 20 decimal places: 3.14159265358979323846. Still not exact, though. In fact, the actual value is irrational, a decimal that goes on forever without any repeating pattern. But notice that there are no 0's and only one 7 in the 20 decimal places above. Does that pattern persist, or do all the digits show up with equal frequency? The table shows the number of times each digit appears in the first million digits. Test the hypothesis that the digits 0 through 9 are uniformly distributed in the decimal representation of π.

The first million digits of π	
Digit	Count
0	99,959
1	99,758
2	100,026
3	100,229
4	100,230
5	100,359
6	99,548
7	99,800
8	99,985
9	100,106

13. Hurricane frequencies The National Hurricane Center provides data that list the numbers of large (category 3, 4, or 5) hurricanes that have struck the United States, by decade since 1851 (www.nhc.noaa.gov/dcmi.shtml). The data are summarized below.

Decade	Count	Decade	Count
1851–1860	6	1931–1940	8
1861–1870	1	1941–1950	10
1871–1880	7	1951–1960	9
1881–1890	5	1961–1970	6
1891–1900	8	1971–1980	4
1901–1910	4	1981–1990	4
1911–1920	7	1991–2000	5
1921–1930	5	2001–2010	7

Recently, there's been some concern that perhaps the number of large hurricanes has been increasing. The natural null hypothesis would be that the frequency of such hurricanes has remained constant.

a) With 96 large hurricanes observed over the 16 periods, what are the expected value(s) for each cell?

b) What kind of chi-square test would be appropriate?

c) State the null and alternative hypotheses.

d) How many degrees of freedom are there?

e) The value of χ^2 is 12.67. What's the P-value?

f) State your conclusion.

14. **Lottery numbers** The fairness of the South African lottery was recently challenged by one of the country's political parties. The lottery publishes historical statistics at its Website (http://www.nationallottery.co.za/lotto/statistics.aspx). Here is a table of the number of times each of the 49 numbers has been drawn in the main lottery and as the "bonus ball" number as of June 2007:

Number	Count	Bonus	Number	Count	Bonus
1	81	14	26	78	12
2	91	16	27	83	16
3	78	14	28	76	7
4	77	12	29	76	12
5	67	16	30	99	16
6	87	12	31	78	10
7	88	15	32	73	15
8	90	16	33	81	14
9	80	9	34	81	13
10	77	19	35	77	15
11	84	12	36	73	8
12	68	14	37	64	17
13	79	9	38	70	11
14	90	12	39	67	14
15	82	9	40	75	13
16	103	15	41	84	11
17	78	14	42	79	8
18	85	14	43	74	14
19	67	18	44	87	14
20	90	13	45	82	19
21	77	13	46	91	10
22	78	17	47	86	16
23	90	14	48	88	21
24	80	8	49	76	13
25	65	11			

We wonder if all the numbers are equally likely to be the "bonus ball."

a) What kind of test should we perform?

b) There are 655 bonus ball observations. What are the appropriate expected value(s) for the test?

c) State the null and alternative hypotheses.

d) How many degrees of freedom are there?

e) The value of χ^2 is 34.5. What's the P-value?

f) State your conclusion.

15. **Childbirth, part 1** There is some concern that if a woman has an epidural to reduce pain during childbirth, the drug can get into the baby's bloodstream, making the baby sleepier and less willing to breastfeed. In December 2006, the *International Breastfeeding Journal* published results of a study conducted at Sydney University. Researchers followed up on 1178 births, noting whether the mother had an epidural and whether the baby was still nursing after 6 months. Here are their results:

		Epidural?		
		Yes	No	Total
Breastfeeding at 6 months?	Yes	206	498	704
	No	190	284	474
	Total	396	782	1178

a) What kind of test would be appropriate?

b) State the null and alternative hypotheses.

16. **Does your doctor know?** A survey[8] of articles from the *New England Journal of Medicine* (*NEJM*) classified them according to the principal statistics methods used. The articles recorded were all noneditorial articles appearing during the indicated years. Let's just look at whether these articles used statistics at all.

	Publication Year			
	1978–79	1989	2004–05	Total
No stats	90	14	40	144
Stats	242	101	271	614
Total	332	115	311	758

Has there been a change in the use of Statistics?

a) What kind of test would be appropriate?

b) State the null and alternative hypotheses.

17. **Childbirth, part 2** In Exercise 15, the table shows results of a study investigating whether aftereffects of epidurals administered during childbirth might interfere with successful breastfeeding. We're planning to do a chi-square test.

a) How many degrees of freedom are there?

b) The smallest expected count will be in the epidural/ no breastfeeding cell. What is it?

c) Check the assumptions and conditions for inference.

18. **Does your doctor know? (part 2)** The table in Exercise 16 shows whether *NEJM* medical articles during various time periods included statistics or not. We're planning to do a chi-square test.

a) How many degrees of freedom are there?

b) The smallest expected count will be in the 1989/No cell. What is it?

c) Check the assumptions and conditions for inference.

---

[8]Suzanne S. Switzer and Nicholas J. Horton, "What Your Doctor Should Know about Statistics (but Perhaps Doesn't)" *Chance*, 20:1, 2007.

19. Childbirth, part 3 In Exercises 15 and 17, we've begun to examine the possible impact of epidurals on successful breastfeeding.

    a) Calculate the component of chi-square for the epidural/ no breastfeeding cell.

    b) For this test, $\chi^2 = 14.87$. What's the P-value?

    c) State your conclusion.

20. Does your doctor know? (part 3) In Exercises 16 and 18, we've begun to examine whether the use of statistics in *NEJM* medical articles has changed over time.

    a) Calculate the component of chi-square for the 1989/ No cell.

    b) For this test, $\chi^2 = 25.28$. What's the P-value?

    c) State your conclusion.

21. Childbirth, part 4 In Exercises 15, 17, and 19, we've tested a hypothesis about the impact of epidurals on successful breastfeeding. The table shows the test's residuals.

		Epidural?	
		Yes	**No**
Breastfeeding at 6 months?	**Yes**	−1.99	1.42
	No	2.43	−1.73

    a) Show how the residual for the epidural/no breastfeeding cell was calculated.

    b) What can you conclude from the standardized residuals?

22. Does your doctor know? (part 4) In Exercises 16, 18, and 20, we've tested a hypothesis about whether the use of statistics in *NEJM* medical articles has changed over time. The table shows the test's residuals.

	1978–79	1989	2004–05
No stats	3.39	−1.68	−2.48
Stats	−1.64	0.81	1.20

    a) Show how the residual for the 1989/No cell was calculated.

    b) What can you conclude from the patterns in the standardized residuals?

23. Childbirth, part 5 In Exercises 15, 17, 19, and 21, we've looked at a study examining epidurals as one factor that might inhibit successful breastfeeding of newborn babies. Suppose a broader study included several additional issues, including whether the mother drank alcohol, whether this was a first child, and whether the parents occasionally supplemented breastfeeding with bottled formula. Why would it not be appropriate to use chi-square methods on the 2 × 8 table with yes/no columns for each potential factor?

24. Does your doctor know? (part 5) In Exercises 16, 18, 20, and 22, we considered data on articles in the *NEJM*. The original study listed 23 different Statistics methods. (The

list read: *t*-tests, contingency tables, linear regression, ...) Why would it not be appropriate to use a chi-square test on the 23 × 3 table with a row for each method?

25. Internet use poll A Pew Research poll in April 2009 from a random sample of U.S. adults asked the questions "Did you use the Internet yesterday?" and "Are you White, Black, or Hispanic/Other?" Is the response to the question about the Internet independent of race?

		Did You Use the Internet Yesterday?	
		Yes	**No**
Ethnicity	**White**	2546	856
	Black	314	146
	Hispanic/Other	431	174

    a) Under the null hypothesis, what are the expected values?

    b) Compute the χ^2 statistic.

    c) How many degrees of freedom does it have?

    d) Find the P-value.

    e) What do you conclude?

26. Internet use poll, II The same poll as in Exercise 25 also asked the questions "Did you use the Internet yesterday?" and "What is your educational level?" Is the response to the question about the internet independent of educational level?

		Did You Use the Internet Yesterday?	
		Yes	**No**
Education	**Less Than High School**	209	131
	High School	932	550
	Some College	958	346
	College Grad	1447	247

    a) Under the null hypothesis, what are the expected values?

    b) Compute the χ^2 statistic.

    c) How many degrees of freedom does it have?

    d) Find the P-value.

    e) What do you conclude?

27. Titanic Here is a table we first saw in Chapter 2 showing who survived the sinking of the *Titanic* based on whether they were crew members, or passengers booked in first-, second-, or third-class staterooms:

	Crew	First	Second	Third	Total
Alive	212	202	118	178	**710**
Dead	673	123	167	528	**1491**
Total	**885**	**325**	**285**	**706**	**2201**

    a) If we draw an individual at random, what's the probability that we will draw a member of the crew?

    b) What's the probability of randomly selecting a third-class passenger who survived?

c) What's the probability of a randomly selected passenger surviving, given that the passenger was a first-class passenger?

d) If someone's chances of surviving were the same regardless of their status on the ship, how many members of the crew would you expect to have lived?

e) State the null and alternative hypotheses.

f) Give the degrees of freedom for the test.

g) The chi-square value for the table is 187.8, and the corresponding P-value is barely greater than 0. State your conclusions about the hypotheses.

28. NYPD and sex discrimination The table below shows the rank attained by male and female officers in the New York City Police Department (NYPD). Do these data summaries indicate that men and women are equitably represented at all levels of the department?

		Sex	
		Male	Female
Rank	Officer	21,900	4,281
	Detective	4,058	806
	Sergeant	3,898	415
	Lieutenant	1,333	89
	Captain	359	12
	Higher ranks	218	10

a) What's the probability that a person selected at random from the NYPD is a female?

b) What's the probability that a person selected at random from the NYPD is a detective?

c) Assuming no bias in promotions, how many female detectives would you expect the NYPD to have?

d) To see if there is evidence of differences in ranks attained by males and females, will you test goodness-of-fit, homogeneity, or independence?

e) State the hypotheses.

f) Test the conditions.

g) How many degrees of freedom are there?

h) The chi-square value for the table is 290.1 and the P-value is less than 0.0001. State your conclusion about the hypotheses.

29. Titanic again Examine and comment on this table of the standardized residuals for the chi-square test you looked at in Exercise 27.

	Crew	First	Second	Third
Alive	−4.35	9.49	2.72	−3.30
Dead	3.00	−6.55	−1.88	2.27

30. NYPD again Examine and comment on this table of the standardized residuals for the chi-square test you looked at in Exercise 28.

	Sex	
	Male	Female
Officer	−2.34	5.57
Detective	−1.18	2.80
Sergeant	3.84	−9.14
Lieutenant	3.58	−8.52
Captain	2.46	−5.86
Higher ranks	1.74	−4.14

31. Cranberry juice It's common folk wisdom that drinking cranberry juice can help prevent urinary tract infections in women. In 2001 the *British Medical Journal* reported the results of a Finnish study in which three groups of 50 women were monitored for these infections over 6 months. One group drank cranberry juice daily, another group drank a lactobacillus drink, and the third drank neither of those beverages, serving as a control group. In the control group, 18 women developed at least one infection, compared to 20 of those who consumed the lactobacillus drink and only 8 of those who drank cranberry juice. Does this study provide supporting evidence for the value of cranberry juice in warding off urinary tract infections?

a) Is this a survey, a retrospective study, a prospective study, or an experiment? Explain.

b) Will you test goodness-of-fit, homogeneity, or independence?

c) State the hypotheses.

d) Test the conditions.

e) How many degrees of freedom are there?

f) Find χ^2 and the P-value.

g) State your conclusion.

h) If you concluded that the groups are not the same, analyze the differences using the standardized residuals of your calculations.

32. Cars A random survey of autos parked in the student lot and the staff lot at a large university classified the brands by country of origin, as seen in the table. Are there differences in the national origins of cars driven by students and staff?

		Driver	
		Student	Staff
Origin	American	107	105
	European	33	12
	Asian	55	47

a) Is this a test of independence or homogeneity?

b) Write appropriate hypotheses.

c) Check the necessary assumptions and conditions.

d) Find the P-value of your test.

e) State your conclusion and analysis.

33. Montana A poll conducted by the University of Montana classified respondents by whether they were male or female and political party, as shown in the table. We wonder if there is evidence of an association between being male or female and party affiliation.

	Democrat	Republican	Independent
Male	36	45	24
Female	48	33	16

a) Is this a test of homogeneity or independence?
b) Write an appropriate hypothesis.
c) Are the conditions for inference satisfied?
d) Find the P-value for your test.
e) State a complete conclusion.

34. Fish diet Medical researchers followed 6272 Swedish men for 30 years to see if there was any association between the amount of fish in their diet and prostate cancer. ("Fatty Fish Consumption and Risk of Prostate Cancer," *Lancet*, June 2001)

Fish Consumption	Total Subjects	Prostate Cancers
Never/Seldom	124	14
Small Part of Diet	2621	201
Moderate Part	2978	209
Large Part	549	42

a) Is this a survey, a retrospective study, a prospective study, or an experiment? Explain.
b) Is this a test of homogeneity or independence?
c) Do you see evidence of an association between the amount of fish in a man's diet and his risk of developing prostate cancer?
d) Does this study prove that eating fish does not prevent prostate cancer? Explain.

35. Montana revisited The poll described in Exercise 33 also investigated the respondents' party affiliations based on what area of the state they lived in. Test an appropriate hypothesis about this table and state your conclusions.

	Democrat	Republican	Independent
West	39	17	12
Northeast	15	30	12
Southeast	30	31	16

36. Working parents In April 2009, Gallup published results from data collected from a large sample of adults in the 27 European Union member states. One of the questions asked was, "Which is the most practicable and realistic option for child care, taking into account the need to earn a living?" The counts below are representative of the entire collection of responses.

	Male	Female
Both Parents Work Full Time	161	140
One Works Full Time, Other Part Time	259	308
One Works Full Time, Other Stays Home for Kids	189	161
Both Parents Work Part Time	49	63
No Opinion	42	28

Source: www.gallup.com/poll/117358/Work-Life-Balance-Tilts-Against-Women-Single-Parents.aspx

a) Is this a survey, a retrospective study, a prospective study, or an experiment?
b) Will you test goodness-of-fit, homogeneity, or independence?
c) Based on these results, do you think men and women have differing opinions when it comes to raising children?

37. Maryland lottery In the Maryland Pick-3 Lottery, three random digits are drawn each day. A fair game depends on every value (0 to 9) being equally likely to show up in all three positions. If not, someone who detects a pattern could take advantage of that. The table shows how many times each of the digits was drawn during a recent 32-week period, and some of them—4 and 7, for instance—seem to come up a lot. Could this just be a result of randomness, or is there evidence the digits aren't equally likely to occur?

Digit	Count
0	62
1	55
2	66
3	64
4	75
5	57
6	71
7	74
8	69
9	61

38. Stock market Some investors believe that stock prices show weekly patterns, claiming for example that Fridays are more likely to be "up" days. From the trading sessions since October 1, 1928 we selected a random sample of 1000 days on which the Dow Jones Industrial Average (DJIA) showed a gain in stock prices. The table shows how many of these fell on each day of the week. Sure enough, more of them are Fridays—and Tuesday looks like a bad day to own stocks. Can this be explained as just randomness, or is there evidence here to help an investor?

Day of the Week	Number of "up" Days
Mon	192
Tues	189
Wed	202
Thu	199
Fri	218

39. Grades Two different professors teach an introductory Statistics course. The table shows the distribution of final grades they reported. We wonder whether one of these professors is an "easier" grader.

	Prof. Alpha	Prof. Beta
A	3	9
B	11	12
C	14	8
D	9	2
F	3	1

a) Will you test goodness-of-fit, homogeneity, or independence?
b) Write appropriate null hypotheses.
c) Find the expected counts for each cell, and explain why the chi-square procedures are not appropriate.

40. Full moon Some people believe that a full moon elicits unusual behavior in people. The table shows the number of arrests made in a small town during weeks of six full moons and six other randomly selected weeks in the same year. We wonder if there is evidence of a difference in the types of illegal activity that take place.

	Full Moon	Not Full
Violent (murder, assault, rape, etc.)	2	3
Property (burglary, vandalism, etc.)	17	21
Drugs/Alcohol	27	19
Domestic Abuse	11	14
Other Offenses	9	6

a) Will you test goodness-of-fit, homogeneity, or independence?
b) Write appropriate null hypotheses.
c) Find the expected counts for each cell, and explain why the chi-square procedures are not appropriate.

41. Grades again In some situations where the expected cell counts are too small, as in the case of the grades given by Professors Alpha and Beta in Exercise 39, we can complete an analysis anyway. We can often proceed after combining cells in some way that makes sense and also produces a table in which the conditions are satisfied. Here we create a new table displaying the same data, but calling D's and F's "Below C":

	Prof. Alpha	Prof. Beta
A	3	9
B	11	12
C	14	8
Below C	12	3

a) Find the expected counts for each cell in this new table, and explain why a chi-square procedure is now appropriate.
b) With this change in the table, what has happened to the number of degrees of freedom?
c) Test your hypothesis about the two professors, and state an appropriate conclusion.

42. Full moon, next phase In Exercise 40 you found that the expected cell counts failed to satisfy the conditions for inference.

a) Find a sensible way to combine some cells that will make the expected counts acceptable.
b) Test a hypothesis about the full moon and state your conclusion.

43. Racial steering A subtle form of racial discrimination in housing is "racial steering." Racial steering occurs when real estate agents show prospective buyers only homes in neighborhoods already dominated by that family's race. This violates the Fair Housing Act of 1968. According to an article in *Chance* magazine (Vol. 14, no. 2 [2001]), tenants at a large apartment complex recently filed a lawsuit alleging racial steering. The complex is divided into two parts: Section A and Section B. The plaintiffs claimed that white potential renters were steered to Section A, while African-Americans were steered to Section B. The table describes the data that were presented in court to show the locations of recently rented apartments. Do you think there is evidence of racial steering?

	New Renters		
	White	Black	Total
Section A	87	8	95
Section B	83	34	117
Total	170	42	212

44. *Titanic*, redux Newspaper headlines at the time, and traditional wisdom in the succeeding decades, have held that women and children escaped the *Titanic* in greater proportions than men. Here's a summary of the relevant data. Do you think that survival was independent of whether the person was male or female? Explain.

		Sex		
		Female	Male	Total
Survival	Alive	343	367	710
	Dead	127	1364	1491
	Total	470	1731	2201

45. Steering revisited You could have checked the data in Exercise 43 for evidence of racial steering using two-proportion z procedures.

a) Find the z-value for this approach, and show that when you square your z-value, you get the value of χ^2 you calculated in Exercise 37.
b) Show that the resulting P-values are the same.

46. Survival on the *Titanic*, one more time In Exercise 44 you could have checked for a difference in the chances of survival for men and women using two-proportion z procedures.

a) Find the *z*-value for this approach.

b) Show that the square of your calculated value of *z* is the value of χ^2 you calculated in Exercise 44.

c) Show that the resulting P-values are the same.

47. Pregnancies Most pregnancies result in live births, but some end in miscarriages or stillbirths. A June 2001 National Vital Statistics Report examined those outcomes in the United States during 1997, broken down by the age of the mother. The table shows counts consistent with that report. Is there evidence that the distribution of outcomes is not the same for these age groups?

		Outcome	
		Live Births	Fetal Losses
Age of Mother	Under 20	49	13
	20–29	201	41
	30–34	88	21
	35 or over	49	21

48. Education by age Use the survey results in the table to investigate differences in education level attained among different age groups in the United States.

		Age Group				
		25–34	35–44	45–54	55–64	≥ 65
Education	Not HS grad	27	50	52	71	101
	HS	82	19	88	83	59
	1–3 years college	43	56	26	20	20
	≥ 4 years college	48	75	34	26	20

Just Checking ANSWERS

1. 7.75, 23.25, 23.25, 69.75

2. Yes. These are counts; student traits are independent of each other; although not a random sample, these Biology students should be representative of all students; 124 is fewer than 10% of all students; all expected counts are at least 5.

3. H_0: The traits occur in the predicted proportions 1:3:3:9.
   H_A: The proportions of some of the traits are not as predicted.

4. 3 df

5. 1.66

6. 0.084

7. There's no evidence to suggest the genetic theory is incorrect.

8. We need to know how well beetles can survive 6 hours in a Plexiglas® box so that we have a baseline to compare the treatments.

9. There's no difference in survival rate in the three groups.

10. $(2 - 1)(3 - 1) = 2$ df

11. 50

12. 2

13. The mean value for a χ^2 with 2 df is 2, so 10 seems pretty large. The P-value is probably small.

14. This is a test of homogeneity. The clue is that the question asks whether the distributions are alike.

15. This is a test of goodness-of-fit. We want to test the model of equal assignment to all lots against what actually happened.

16. This is a test of independence. We have responses on two variables for the same individuals.

chapter 26 Inferences for Regression

Three percent of a man's body is essential fat. (For a woman, the percentage is closer to 12.5%.) As the name implies, essential fat is necessary for a normal, healthy body. Fat is stored in small amounts throughout your body. Too much body fat, however, can be dangerous to your health. For men between 18 and 39 years old, a healthy percent body fat ranges from 8% to 19%. (For women of the same age, it's 21% to 32%.)

Measuring body fat can be tedious and expensive. The "standard reference" measurement is by dual-energy X-ray absorptiometry (DEXA), which involves two low-dose X-ray generators and takes from 10 to 20 minutes.

How close can we get to a useable prediction of body fat from easily measurable variables such as *Height*, *Weight*, or *Waist* size? Here's a scatterplot of *%Body Fat* plotted against *Waist* size for a sample of 250 males of various ages.

Who	250 male subjects
What	Body fat and waist size
Units	% Body fat and inches
When	1990s
Where	United States
Why	Scientific research

Figure 26.1

Percent Body Fat vs. *Waist* size for 250 men of various ages. The scatterplot shows a strong, positive, linear relationship.

Back in Chapter 7, we modeled relationships like this by fitting a least squares line. The plot is clearly straight, so we can find that line. The equation of the least squares line for these data is

$$\widehat{\%Body\ Fat} = -42.7 + 1.7\ Waist.$$

The slope says that, on average, *%Body Fat* is greater by 1.7 percent for each additional inch around the waist.

How useful is this model? When we fit linear models before, we used them to describe the relationship between the variables and we interpreted the slope and intercept as descriptions of the data. Now we'd like to understand what the regression model can tell us beyond the 250 men in this study. To do that, we'll want to make confidence intervals and test hypotheses about the slope and intercept of the regression line.

The Population and the Sample

When we found a confidence interval for a mean, we could imagine a single, true underlying value for the mean. When we tested whether two means or two proportions were equal, we imagined a true underlying difference. But what does it mean to do inference for regression? We know better than to think that the data would line up perfectly in a straight line even if we knew every population value. After all, even in our sample, not all men who have 38-inch waists have the same *%Body Fat*. In fact, there's a whole distribution of *%Body Fat* for these men:

Figure 26.2

The distribution *of %Body Fat* for men with a *Waist* size of 38 inches is unimodal and symmetric.

This is true at each *Waist* size. In fact, we could depict the distribution of *%Body Fat* at different *Waist* sizes like this:

Figure 26.3

There's a distribution of *%Body Fat* for each value of *Waist* size. We'd like the means of these distributions to line up.

But we want to *model* the relationship between *%Body Fat* and *Waist* size for all men. To do that, we imagine an idealized regression line. The model assumes that the *means* of the distributions of *%Body Fat* for each *Waist* size fall along the line, even though the individuals are scattered around it. We know that this model is not a perfect description of how the variables are associated, but it may be useful for predicting *%Body Fat* and for understanding how it's related to *Waist* size.

If only we had all the values in the population, we could find the slope and intercept of this *idealized regression line* explicitly by using least squares. Following our usual conventions, we write the idealized line with Greek letters and consider the coefficients (the slope and intercept) to be *parameters*: β_0 is the intercept and β_1 is the slope. Corresponding to our fitted line of $\hat{y} = b_0 + b_1 x$, we write

$$\mu_y = \beta_0 + \beta_1 x.$$

Why μ_y instead of $\hat{y}$? Because this is a model. There is a distribution of *%Body Fat* for each *Waist* size. The model places the *means* of the distributions of *%Body Fat* for each *Waist* size on the same straight line.

Of course, not all the individual y's are at these means. (In fact, the line will miss most—and quite possibly all—of the plotted points.) Some individuals lie above and some below the line, so, like all models, this one makes **errors**. Lots of them. In fact, one at each point. These errors are random and, of course, can be positive or negative. They are model errors, so we use a Greek letter and denote them by ε.

When we put the errors into the equation, we can account for each individual y:

$$y = \beta_0 + \beta_1 x + \varepsilon.$$

This equation is now true for each data point (since there is an ε to soak up the deviation), so the model gives a value of y for any value of x.

For the body fat data, an idealized model such as this provides a summary of the relationship between *%Body Fat* and *Waist* size. Like all models, it simplifies the real situation. We know there is more to predicting body fat than waist size alone. But the advantage of a model is that the simplification might help us to think about the situation and assess how well *%Body Fat* can be predicted from simpler measurements.

We estimate the β's by finding a regression line, $\hat{y} = b_0 + b_1 x$, as we did in Chapter 7. The residuals, $e = y - \hat{y}$, are the sample-based versions of the errors, ε. We'll use them to help us assess the regression model.

We know that least squares regression will give reasonable estimates of the parameters of this model from a random sample of data. Our challenge is to account for our uncertainty in how well they do. For that, we need to make some assumptions about the model and the errors.

Assumptions and Conditions

A S *Activity:* **Conditions for Regression Inference.** View an illustrated discussion of the conditions for regression inference.

Back in Chapter 7 when we fit lines to data, we needed to check only the Straight Enough Condition. Now, when we want to make inferences about the coefficients of the line, we'll have to make more assumptions. Fortunately, we can check conditions to help us judge whether these assumptions are reasonable for our data. And as we did before, we'll need to wait to make some checks until *after* we find the regression equation.

Also, we need to be careful about the order in which we check conditions. If our initial assumptions are not true, it makes no sense to check the later ones. So now we number the assumptions to keep them in order.

1. Linearity Assumption

Check the Scatterplot

The shape must be linear or we can't use linear regression at all.

If the true relationship is far from linear and we use a straight line to fit the data, our entire analysis will be useless, so we always check this first.

The **Straight Enough Condition** is satisfied if a scatterplot looks straight. It's generally not a good idea to draw a line through the scatterplot when checking. That can fool your eyes into seeing the plot as more straight. Sometimes it's easier to see violations of the Straight Enough Condition by looking at a scatterplot of the residuals against x or against the predicted values, $\hat{y}$. That plot will have a horizontal direction and should have no pattern if the condition is satisfied.

If the scatterplot is straight enough, we can go on to some assumptions about the errors. If not, stop here, or consider re-expressing the data (see Chapter 9) to make the scatterplot more nearly linear. For the *%Body Fat* data, the scatterplot is beautifully linear.

2. Independence Assumption

Figure 26.4
The residuals show only random scatter when plotted against *Waist* size.

Independence Assumption: The errors in the true underlying regression model (the ε's) must be independent of each other. As usual, there's no way to be sure that the Independence Assumption is true, but there are some things we can think about.

If we hope to apply our regression model to a larger population we can check the **Randomization Condition** that the individuals are a random sample from that population. We may settle for a sample that we perceive to be representative.

We can also check the **Random Residuals Condition** by looking at the residuals plot for evidence of patterns, trends, or clumping, any of which would suggest a failure of independence. In the special case when the x-variable is related to time, a common violation of the Independence Assumption is for the errors to be correlated. (The error our model makes today may be similar to the one it made for yesterday.) This violation can be checked by plotting the residuals against the x-variable and looking for patterns.

The *%Body Fat* data were collected on a sample of men taken to be representative. The subjects were not related in any way, so we can be pretty sure that their measurements are independent. The residuals plot shows no pattern.

3. Equal Variance Assumption

Figure 26.5
A scatterplot of residuals against predicted values can reveal plot thickening. In this plot of the residuals from a regression of diamond prices on carat weight, we see that larger diamonds have more price variation than smaller diamonds. When the Equal Spread Assumption is violated, we can't summarize how the residuals vary with a single number.

The variability of y should be about the same for all values of x. In Chapter 7, we looked at the standard deviation of the residuals (s_e) to measure the size of the scatter. Now we'll need this standard deviation to build confidence intervals and test hypotheses. The standard deviation of the residuals is the building block for the standard errors of all the regression parameters. But it makes sense only if the scatter of the residuals is the same everywhere. In effect, the standard deviation of the residuals "pools" information across all of the individual distributions at each x-value, and pooled estimates are appropriate only when they combine information for groups with the same variance.

Practically, what we can check is the **Does the Plot Thicken? Condition**. The scatterplot offers a visual check: Make sure the spread around the line is nearly constant. Often it is better to look at the residuals plot: Be alert for a "fan" shape or other tendency for the variation to grow or shrink in one part of the scatterplot (as in the plot on the left). That's not a problem for the body fat data, where the residuals plot (Fig. 26.4) shows the spread of *%Body Fat* around the line is remarkably constant across *Waist* sizes from 30 inches to about 45 inches.

If the plot is straight enough, the data are independent, and the plot doesn't thicken, you can now move on to the final assumption.

4. Normal Population Assumption

We assume the errors around the idealized regression line at each value of x follow a Normal model. We need this assumption so that we can use a Student's t-model for inference.

As we have at other times when we've used Student's t, we'll settle for the residuals satisfying the **Nearly Normal Condition** and the **Outlier Condition**. Look at a histogram or Normal probability plot of the residuals.[1]

---

[1]*This* is why we have to check the conditions in order. We have to check that the residuals are independent and that the variation is the same for all x's so that we can lump all the residuals together for a single check of the Nearly Normal Condition.

Check a Histogram of the Residuals
The distribution of the residuals should be unimodal and symmetric.

Figure 26.6

A histogram of the residuals is one way to check whether they are Nearly Normal. Alternatively, we can look at a Normal probability plot.

The histogram of residuals in the *%Body Fat* regression certainly looks nearly Normal. As we have noted before, the Normality Assumption becomes less important as the sample size grows, because the model is about means and the Central Limit Theorem takes over.

If all four assumptions were true, the idealized regression model would look like this:

Figure 26.7

The regression model has a distribution of *y*-values for each *x*-value. These distributions follow a Normal model with means lined up along the line and with the same standard deviations.

At each value of *x*, there is a distribution of *y*-values that follows a Normal model, and each of these Normal models is centered on the line and has the same standard deviation. Of course, we don't expect the assumptions to be exactly true, and we know that all models are wrong, but the linear model is often close enough to be very useful.

For Example   CHECKING ASSUMPTIONS AND CONDITIONS

Look at the moon with binoculars or a telescope, and you'll see craters formed by thousands of impacts. The earth, being larger, has been hit even more often. Meteor Crater in Arizona was the first recognized impact crater and was identified as such only in the 1920s. With the help of satellite images, more and more craters have been identified; now more than 180 are known. These, of course, are only a small sample of all the impacts the earth has experienced: Only 29% of earth's surface is land, and many craters have been covered or eroded away. Astronomers have recognized a roughly 35 million–year cycle in the frequency of cratering, although the cause of this cycle is not fully understood. Here's a scatterplot of the known impact craters from the most recent 35 million years.[2] We've taken logs of both age (in millions of years ago) and diameter (km) to make the relationship simpler. (See Chapter 9.)

Who	39 impact craters
What	Diameter and age
Units	km and millions of years ago
When	Past 35 million years
Where	Worldwide
Why	Scientific research

(continued)

[2]Data, pictures, and much more information at the Earth Impact Database found at www.passc.net/EarthImpactDatabase/index.html

QUESTION: Are the assumptions and conditions satisfied for fitting a linear regression model to these data?

ANSWER:

✓ **Linearity Assumption:** The scatterplot satisfies the Straight Enough Condition.

✓ **Independence Assumption:** Sizes of impact craters are likely to be generally independent.

✗ **Randomization Condition:** These are the only known craters, and may differ from others that may have disappeared or not yet been found. I'll be careful not to generalize my conclusions too broadly.

✓ **Random Residuals Condition:** The residuals appear to be randomly scattered.

✓ **Does the Plot Thicken? Condition:** After fitting a linear model, I find the residuals shown.

Two points seem to give the impression that the residuals may be more variable for higher predicted values than for lower ones, but this doesn't seem to be a serious violation of the Equal Variance Assumption.

✓ **Nearly Normal Condition:** A histogram suggests a bit of skewness in the distribution of residuals.

There are no violations severe enough to stop my regression analysis, but I'll be cautious about my conclusions.

Which Come First: The Conditions or the Residuals?

"Truth will emerge more readily from error than from confusion."

—*Francis Bacon (1561–1626)*

In regression, there's a little catch. The best way to check many of the conditions is with the residuals, but we get the residuals only *after* we compute the regression. Before we compute the regression, however, we should check at least one of the conditions.

So we work in this order:

1. Make a scatterplot of the data to check the Straight Enough Condition. (If the relationship is curved, try re-expressing the data. Or stop.)
2. If the data are straight enough, fit a regression and find the predicted values, $\hat{y}$, and the residuals, e.
3. Make a scatterplot of the residuals against x or against the predicted values. This plot should have no pattern. Check in particular for any bend (which would suggest that the data weren't all that straight after all), for any thickening (or thinning), and, of course, for any outliers. (If there are outliers, and you can correct them or justify removing them, do so and go back to step 1, or consider performing two regressions—one with and one without the outliers.)
4. If the data are measured over time, plot the residuals against time to check for evidence of patterns that might suggest they are not independent.
5. If the scatterplots look OK, then make a histogram (or Normal probability plot) of the residuals to check the Nearly Normal Condition.
6. If all the conditions seem to be reasonably satisfied, go ahead with inference.

Step-by-Step Example REGRESSION INFERENCE

If our data can jump through all these hoops, we're ready to do regression inference. Let's see how much more we can learn about body fat and waist size from a regression model.

Questions: What is the relationship between *%Body Fat* and *Waist* size in men? What model best predicts body fat from waist size, and how well does it do the job?

THINK ➡ **Plan** Specify the question of interest.

Name the variables and report the W's.

Identify the parameters you want to estimate.

Model Think about the assumptions and check the conditions.

Make pictures. For regression inference, you'll need a scatterplot, a residuals plot, and either a histogram or a Normal probability plot of the residuals.

I have quantitative body measurements on 250 adult males from the BYU Human Performance Research Center. I want to understand the relationship between *%Body Fat* and *Waist* size.

✓ **Straight Enough Condition:** There's no obvious bend in the original scatterplot of the data or in the residuals plot.

✓ **Independence Assumption:** These data are not collected over time, and there's no reason to think that the *%Body Fat* of one man influences the *%Body Fat* of another.

✗ **Randomization Condition:** These data are not from a random sample, but it's reasonable to assume these men are representative of the population.

✓ **Random Residuals Condition:** The residuals appear to be randomly scattered.

✓ **Does the Plot Thicken? Condition:** The residuals plot shows no evidence of change in the spread about the line.

(continued)

You could also check the Nearly Normal Condition by making a Normal probability plot; it's quite straight.

Choose your method.

✓ **Nearly Normal Condition, Outlier Condition:** A histogram of the residuals is unimodal and symmetric, without outliers. A Normal model is reasonable for the errors.

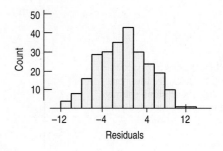

Under these conditions a **regression model** is appropriate.

SHOW ➡ **Mechanics** Let's just "push the button" and see what the regression looks like.

The formula for the regression equation can be found in Chapter 7, and the standard error formulas will be shown a bit later, but regressions are almost always computed with a computer program or calculator.

Write the regression equation.

Dependent variable is %BF
R-squared = 67.8%
$s = 4.713$ with $250 - 2 = 248$ degrees of freedom

Variable	Coeff	SE(Coeff)	t-Ratio	P-Value
Intercept	−42.734	2.717	−15.7	<0.0001
Waist	1.70	0.0743	22.9	<0.0001

The estimated regression equation is

$$\widehat{\%Body\ Fat} = -42.73 + 1.70\ Waist.$$

TELL ➡ **Conclusion** Interpret your results in context.

More Interpretation We haven't worked it out in detail yet, but the output gives us numbers labeled as *t*-statistics and corresponding P-values, and we have a general idea of what those mean. Now it's time to learn more about regression inference.

The R^2 for the regression is 67.8%. Waist size seems to account for about 2/3 of the %Body Fat variation in men. The slope of the regression says that %Body Fat increases by about 1.7 percentage points per inch of Waist size, on average.

The standard error of 0.07 for the slope is much smaller than the slope itself, so it looks like the estimate is reasonably precise.

Intuition About Regression Inference

Wait a minute! We've just pulled a fast one. We've pushed the "regression button" on our computer or calculator but haven't discussed where the standard errors for the slope or intercept come from. We know that if we had collected similar data on a different random sample of men, the slope and intercept would be different. Each sample would have produced its own regression line, with slightly different b_0's and b_1's. This sample-to-sample variation is what generates the sampling distributions for the coefficients.

There's only one regression model; each sample regression is trying to estimate the same parameters, β_0 and β_1. We expect any sample to produce a b_1 whose expected value is the true slope, β_1. What about its standard deviation? What aspects of the data affect how much the slope (and intercept) vary from sample to sample?

- **Spread around the line.** Here are two situations in which we might do regression. Which situation would be more likely to yield a consistent slope? That is, if we were to sample over and over from the two underlying populations that these samples come from and compute all the slopes, which group of slopes would vary less?

Figure 26.8
Which of these scatterplots shows a situation that would give the more consistent regression slope estimate if we were to sample repeatedly from its underlying population?

Clearly, data like those in the left plot will give more consistent slopes estimates.

Less scatter around the line means the slope will be more consistent from sample to sample. The spread around the line is measured with the **residual standard deviation**, s_e. You can always find s_e in the regression output, often just labeled s. It estimates the typical distance true values lie from values predicted by the model. Also, s_e estimates the standard deviation of the y-values at any specific value of x. You're not likely to calculate the residual standard deviation by hand. When we first saw this formula in Chapter 7, we said that it looks a lot like the standard deviation of y, only subtracting the predicted values rather than the mean and dividing by $n - 2$ instead of $n - 1$:

$$s_e = \sqrt{\frac{\sum (y - \hat{y})^2}{n - 2}}.$$

The less scatter around the line, the smaller the residual standard deviation and the stronger the relationship between x and y.

Some people prefer to assess the strength of a regression by looking at s_e rather than R^2. After all, s_e has the same units as y, and because it's the standard deviation of the errors around the line, it tells you how close the data are to our model. By contrast, R^2 is the proportion of the variation of y accounted for by x. We say, why not look at both?

- **Spread of the x's.** Here are two more situations. Which of these would yield more consistent slope estimates?

n − 2?

For standard deviation (in Chapter 3), we divided by $n - 1$ because we didn't know the true mean and had to estimate it. Now it's later in the course and there's even more we don't know. Here we don't know *two* things: the slope and the intercept. If we knew them both, we'd divide by n and have n degrees of freedom. When we *estimate* both, however, we adjust by subtracting 2, so we divide by $n - 2$ and (as we will see soon) have 2 fewer degrees of freedom.

Figure 26.9
Which of these scatterplots shows a situation that would give the more consistent regression slope estimates if we were to sample repeatedly from the underlying population?

A S *Simulation:* **x-Variance and Slope Variance.** You don't have to just imagine how the variability of the slope depends on the spread of the x's.

A plot like the one on the right has a broader range of x-values. We'd expect the slopes in samples from populations like that to vary less from sample to sample. If s_x, the standard deviation of x is large, it provides a more stable regression.

• **Sample size.** What about these two?

Figure 26.10

Which of these scatterplots shows a situation that would give the more consistent regression slope estimates if we were to sample repeatedly from the underlying population?

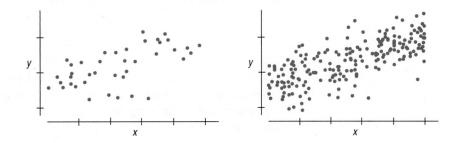

It shouldn't be a surprise that having a larger sample size, *n*, gives more consistent estimates from sample to sample.

Standard Error for the Slope

Three aspects of the scatterplot, then, affect the standard error of the regression slope:

• Spread around the line: s_e
• Spread of *x* values: s_x
• Sample size: *n*

These are in fact the *only* things that affect the standard error of the slope. Although you'll probably never have to calculate it by hand, the formula for the standard error is

$$SE(b_1) = \frac{s_e}{\sqrt{n-1}\, s_x}.$$

A S *Activity*: **Regression Slope Standard Error.** See how $SE(b_1)$ is constructed and where the values used in the formula are found in the regression output table.

The error standard deviation, s_e, is in the *numerator*, since spread around the line *increases* the slope's standard error. The denominator has both a sample size term, $\sqrt{n-1}$, and s_x, because increasing either of these *decreases* the slope's standard error.

We know the b_1s vary from sample to sample. As you'd expect, their sampling distribution model is centered at β_1, the slope of the idealized regression line. Now we can estimate its standard deviation with $SE(b_1)$. What about its shape? Here the Central Limit Theorem and "Wild Bill" Gosset come to the rescue again. When we standardize the slopes by subtracting the model mean and dividing by their standard error, we get a Student's *t*-model, this time with $n-2$ degrees of freedom:

Don't confuse the standard deviation of the residuals, s_e, with the standard error of the slope, $SE(b_1)$. The first, s_e, measures the scatter around the line, telling us how reliably we can estimate *y*-values. The second, $SE(b_1)$, indicates how much slopes vary from sample to sample, telling us how reliably we can estimate the true slope.

A Sampling Distribution for Regression Slopes

When the conditions are met, the standardized estimated regression slope,

$$t = \frac{b_1 - \beta_1}{SE(b_1)},$$

follows a Student's *t*-model with $n-2$ degrees of freedom. We estimate the standard deviation with

$$SE(b_1) = \frac{s_e}{\sqrt{n-1}\, s_x},$$

where $s_e = \sqrt{\dfrac{\sum (y - \hat{y})^2}{n-2}}.$

What About the Intercept?

The same reasoning applies for the intercept. We could write

$$\frac{b_0 - \beta_0}{SE(b_0)} \sim t_{n-2}$$

and use it to construct confidence intervals and test hypotheses, but often the value of the intercept isn't something we care about—or have a natural null hypothesis for. The intercept usually isn't interesting. Most hypothesis tests and confidence intervals for regression are about the slope.

Regression Inference

TI-*nspire*

Regression Inference. How big must a slope be in order to be considered statistically significant? See for yourself by exploring the natural sample-to-sample variability in slopes.

What if the Slope Were 0? If $b_1 = 0$, our prediction is $\hat{y} = b_0 + 0x$. The equation collapses to just $\hat{y} = b_0$. Now x is nowhere in sight, so y doesn't depend on x at all.

And b_0 would turn out to be $\bar{y}$. Why? We know that $b_0 = \bar{y} - b_1\bar{x}$, but when $b_1 = 0$, that becomes simply $b_0 = \bar{y}$. It turns out, then, that when the slope is 0, the equation is just $\hat{y} = \bar{y}$; at every value of x, we always predict the mean value for y.

Now that we have the standard error of the slope and its sampling distribution, we can test a hypothesis about it and make confidence intervals. The usual null hypothesis about the slope is that it's equal to 0. Why? Well, a slope of zero would say that y doesn't tend to change linearly when x changes—in other words, that there is no linear association between the two variables. If the slope were zero, there wouldn't be much left of our regression equation.

So a null hypothesis of a zero slope questions the entire claim of a linear relationship between the two variables—and often that's just what we want to know. In fact, every software package or calculator that does regression simply assumes that you want to test the null hypothesis that the slope is really zero.

To test $H_0: \beta_1 = 0$, we find

$$t_{n-2} = \frac{b_1 - 0}{SE(b_1)}.$$

This is just like every t-test we've seen: a difference between the statistic and its hypothesized value, divided by its standard error.

For our body fat data, the computer found the slope (1.7), its standard error (0.0743), and the ratio of the two: $\dfrac{1.7 - 0}{0.0743} = 22.9$ (see p. 713). Nearly 23 standard errors from the hypothesized value certainly seems big. The P-value (<0.0001) confirms that a t-ratio this large would be very unlikely to occur if the true slope were zero.

Maybe the standard null hypothesis isn't all that interesting here. Did you have any doubts that *%Body Fat* is related to *Waist* size? A more sensible use of these same values might be to make a confidence interval for the slope instead.

We can build a confidence interval in the usual way, as an estimate plus or minus a margin of error. As always, the margin of error is just the product of the standard error and a critical value. Here the critical value comes from the t-distribution with $n - 2$ degrees of freedom, so a 95% **confidence interval for β** is

$$b_1 \pm t^*_{n-2} \times SE(b_1).$$

For the body fat data, $t^*_{248} = 1.970$, so that comes to $1.7 \pm 1.97 \times 0.074$, or an interval from 1.55 to 1.85 *%Body Fat* per inch of *Waist* size.

For Example INTERPRETING A REGRESSION MODEL

RECAP: On a log scale, there seems to be a linear relationship between the diameter and the age of recent terrestrial impact craters. We have regression output from statistics software:

Dependent variable is LogDiam
R-squared = 63.6%
$s = 0.6362$ with $39 - 2 = 37$ degrees of freedom

Variable	Coefficient	Se(coeff)	t-Ratio	P-Value
Intercept	0.358262	0.1106	3.24	0.0025
LogAge	0.526674	0.0655	8.05	≤0.0001

QUESTION: What's the regression model, and what can it tell us?

ANSWER: *For terrestrial impact craters younger than 35 million years, the logarithm of Diameter grows linearly with the logarithm of Age:*

$$\widehat{logDiam} = 0.358 + 0.527 \, logAge.$$

The P-value for each coefficient's t-statistic is very small, so I'm quite confident that neither coefficient is zero. Based on my model, I conclude that, on average, the older a crater is, the larger it tends to be. This model accounts for 63.6% of the variation in logDiam.

Although it is possible that impacts (and their craters) are getting smaller, it is more likely that I'm seeing the effects of age on craters. Small craters are probably more likely to erode or become buried or otherwise be difficult to find as they age. Larger craters may survive the huge expanses of geologic time more successfully.

Just Checking

Researchers in Food Science studied how big people's mouths tend to be. They measured mouth volume by pouring water into the mouths of subjects who lay on their backs. Unless this is your idea of a good time, it would be helpful to have a model to estimate mouth volume more simply. Fortunately, mouth volume is related to height. (Mouth volume is measured in cubic centimeters and height in meters.)

The data were checked and deemed suitable for regression. Take a look at the computer output to answer these questions:

1. What does the *t*-ratio of 3.27 for the slope tell about this relationship? How does the P-value help your understanding?

2. Would you say that measuring a person's height could reliably be used as a substitute for the wetter method of determining how big a person's mouth is? What numbers in the output helped you reach that conclusion?

3. What does the value of s_e add to this discussion?

4. Interpret the value of the standard error of the slope in this context.

Summary of	Mouth Volume
Mean	60.2704
StdDev	16.8777

Dependent variable is Mouth Volume
R-squared = 15.3%
$s = 15.66$ with $61 - 2 = 59$ degrees of freedom

Variable	Coefficient	SE(coeff)	t-Ratio	P-Value
Intercept	−44.7113	32.16	−1.39	0.1697
Height	61.3787	18.77	3.27	0.0018

Another Example: Breaking Up Is Hard to Predict

A S   *Activity:* **A Hypothesis Test for the Regression Slope.** View an animated discussion of testing the standard null hypothesis for slope.

Every spring, Nenana, Alaska, hosts a contest in which participants try to guess the exact minute that a wooden tripod placed on the frozen Tanana River will fall through the breaking ice. The contest started in 1917 as a diversion for railroad engineers, with a jackpot of $800 for the closest guess. It has grown into an event in which hundreds of thousands of entrants enter their guesses on the Internet[3] and vie for as much as $300,000.

Because so much money and interest depends on the time of breakup, it has been recorded to the nearest minute with great accuracy ever since 1917. And because a standard measure of breakup has been used throughout this time, the data are consistent. An article in *Science*[4] used the data to investigate global warming—whether greenhouse gasses and other human actions have been making the planet warmer. Others might just want to make a good prediction of next year's breakup time.

Of course, we can't use regression to tell the *causes* of any change. But we can estimate the *rate* of change (if any) and use it to make better predictions.

Here are some of the data:

Who	Years
What	Year, day, and hour of ice breakup
Units	x is in years since 1900.
	y is in days after midnight Dec. 31.
When	1917–present
Where	Nenana, Alaska
Why	Wagering, but proposed to look at global warming

Year (since 1900)	Breakup Date (days after Jan. 1)	Year (since 1900)	Breakup Date (days after Jan. 1)
17	119.4792	27	127.7938
18	130.3979	28	129.3910
19	122.6063	29	121.4271
20	131.4479	30	127.8125
21	130.2792	31	119.5882
22	131.5556	32	134.5639
23	128.0833	33	120.5403
24	131.6319	34	131.8361
25	126.7722	35	125.8431
26	115.6688	36	118.5597

Step-by-Step Example   A REGRESSION SLOPE *t*-TEST

The slope of the regression gives the change in Nenana ice breakup date per year.

Questions: Is there sufficient evidence to claim that ice breakup times are changing? If so, how rapid is the change?

THINK ➡ **Plan**   State what you want to know.

Identify the *parameter* you wish to estimate. Here our parameter is the slope.

Identify the variables and review the W's.

Hypotheses   Write your null and alternative hypotheses.

I wonder whether the date of ice breakup in Nenana has changed over time. The slope of that change might indicate climate change. I have the date of ice breakup annually for 95 years starting in 1917, recorded as the number of days and fractions of a day until the ice breakup.

H_0: *There is no change in the date of ice breakup:* $\beta_1 = 0$
H_A: *Yes, there is:* $\beta_1 \neq 0$

(continued)

---

[3]www.nenanaakiceclassic.com

[4]"Climate Change in Nontraditional Data Sets." *Science* 294 [26 October 2001]: 811.

Model Think about the assumptions and check the conditions.

Make pictures. Because the scatterplot seems straight enough, we can find and plot the residuals.

✓ **Straight Enough Condition:** I have quantitative data with no obvious bend in the scatterplot.

✓ **Independence Assumption:** These data are a time series, which raises my suspicions that they may not be independent. To check, here's a plot of the residuals against time.

Usually, we check for suggestions that the Independence Assumption fails by plotting the residuals against the predicted values. Patterns and clusters in that plot raise our suspicions. But when the data are measured over time, it is always a good idea to plot residuals against time to look for trends and oscillations.

✓ **Random Residuals Condition:** I see a hint that the data oscillate up and down, which suggests some failure of independence, but not so strongly that I can't proceed with the analysis.

✗ **Randomization Condition:** These data are not a random sample, so I'm reluctant to extend my conclusions beyond this river and these years.

✓ **Does the Plot Thicken? Condition:** The residuals plot shows no obvious changes in the spread.

✓ **Nearly Normal Condition, Outlier Condition:** A histogram of the residuals is unimodal and symmetric.

(continued)

State the sampling distribution model.

Under these conditions, the sampling distribution of the regression slope can be modeled by a Student's t-model with $(n - 2) = 93$ degrees of freedom.

Choose your method.

*I'll do a **regression slope t-test**.*

SHOW ➡ Mechanics The regression equation can be found from the formulas in Chapter 7, but regressions are almost always found from a computer program or calculator.

The P-values given in the regression output table are from the Student's t-distribution on $(n - 2) = 93$ degrees of freedom. They are appropriate for two-sided alternatives.

Here's the computer output for this regression:

Dependent variable is Breakup Date
R-squared = 11.1%
s = 5.600 with 95 − 2 = 93 degrees of freedom

Variable	Coefficient	SE(Coeff)	t-Ratio	P-Value
Intercept	128.732	1.459	88.2	≤0.0001
Year Since 1900	−0.071436	0.0210	−3.41	0.0010

The estimated regression equation is

$$\widehat{Date} = 128.732 - 0.071\ YearSince1900.$$

TELL ➡ Conclusion Link the P-value to your decision and state your conclusion in the proper context.

The P-value of 0.0010 means that the association we see in the data is unlikely to have occurred by chance even though the R^2 is not particularly strong. I reject the null hypothesis, and conclude that there is strong evidence that, on average, the ice breakup is occurring earlier each year. But the hint of an oscillation pattern in the residuals raises concerns.

SHOW MORE ➡ Create a confidence interval for the true slope

A 95% confidence interval for β_1 is

$$b_1 \pm t^*_{93} \times SE(b_1)$$
$$-0.071 \pm (1.986)(0.0210)$$
$$or\ (-0.11, -0.03)\ days\ per\ year.$$

TELL MORE ➡ Interpret the interval Simply rejecting the standard null hypothesis doesn't guarantee that the size of the effect is large enough to be important. Whether we want to know the breakup time to the nearest minute or are interested in global warming, a change measured in hours each year is big enough to be interesting.

I am 95% confident that the ice has been breaking up, on average, between 0.03 days (about 40 minutes) and 0.11 days (about 3 hours) earlier each year since 1900.

But is it global warming? So the ice is breaking up earlier. Temperatures are higher. Must be global warming, right?

Maybe.

An article challenging the original analysis of the Nenana data proposed a possible confounding variable. It noted that the city of Fairbanks is upstream from Nenana and suggested that the growth of Fairbanks could have warmed the river. So maybe it's not global warming.

Or maybe global warming is a lurking variable, leading more people to move to a now balmier Fairbanks and also leading to generally earlier ice breakup in Nenana.

Or maybe there's some other variable or combination of variables at work. We can't set up an experiment, so we may never really know.

Only one thing is for sure. When you try to explain an association by claiming cause and effect, you're bound to be on thin ice.[5]

TI Tips DOING REGRESSION INFERENCE

The TI will easily do almost everything you need for inference for regression: scatterplots, residual plots, histograms of residuals, and *t*-tests and confidence intervals for the slope of the regression line. OK, it won't tell you *SE(b)*, but it will give you enough information to easily figure it out for yourself. Not bad.

As an example we'll use data from *Chance* magazine (Vol. 12, No. 4, 1999) for 11 of the top performances in women's marathons during the 1990s. Let's examine the influence of temperature on the times for elite runners.

°F	44	46	47	50	51	52	54	55	57	60	65
Min	142.7	142.1	143.4	143.6	144.0	143.4	142.4	143.1	143.7	143.4	143.4

TEST A HYPOTHESIS ABOUT THE ASSOCIATION

- Enter the temperatures (nearest degree Fahrenheit) in L1 and the runners' times (nearest tenth of a minute) in L2.
- Check the scatterplot. It's not obviously nonlinear, so go ahead.
- Under STAT TESTS choose LinRegTTest.
- Specify the two data lists (with Freq:1).
- Choose the two-tailed option. (We are interested in whether higher temperatures enhance or interfere with a runner's performance.)
- Tell it to store the regression equation in Y1 (VARS, Y-VARS, Function... remember?), then Calculate.

The TI creates so much information you have to scroll down to look at it all! See:
- The calculated value of t and the P-value.
- The coefficients of the regression equation, a and b.
- The value of s, our sample estimate of the common standard deviation of errors around the true line.
- The values of r² and r.

Wait, where's *SE(b)*? It's not there. No problem—if you need it, you can figure it out. Remember that the *t*-value is *b* divided by *SE(b)*. So *SE(b)* must be *b* divided by *t*. Here $SE(b) = 0.0325 \div 1.1358 = 0.0286$.

```
LinRegTTest
Xlist:L1
Ylist:L2
Freq:1
β & ρ:≠0 <0 >0
RegEQ:Y1
Calculate
```

```
LinRegTTest
y=a+bx
β≠0 and ρ≠0
t=1.135800072
p=.2853799627
df=9
↓a=141.4824942
■
```

```
LinRegTTest
y=a+bx
β≠0 and ρ≠0
↑b=.032517321
s=.5680158733
r²=.1253679846
r=.354073417
■
```

(continued)

---

[5]How *do* scientists sort out such messy situations? Even though they can't conduct an experiment, they *can* look for replications elsewhere. A number of studies of ice on other bodies of water have also shown earlier ice breakup times in recent years. That suggests they need an explanation that's more comprehensive than just Fairbanks and Nenana.

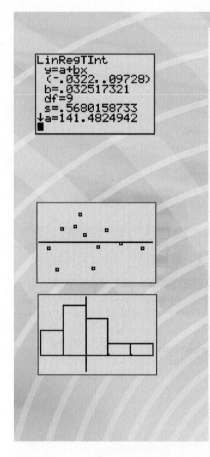

CREATE A CONFIDENCE INTERVAL FOR THE SLOPE
- Back to STAT TEST; this time you want LinRegTInt.
- The specifications for the data lists and the regression equation remain what you entered for the hypothesis test.
- Choose a confidence level, say 95%, and Calculate.

(A question for you: How is this confidence interval consistent with the P-value for the hypothesis test?)

CHECK THE CONDITIONS Beware!!! Before you try to interpret any of this, you must check the conditions to see if inference for regression is allowed.
- We already looked at the scatterplot; it was reasonably linear.
- To create the residuals plot, set up another scatterplot with RESID (from LIST NAMES) as your Ylist. OK, it looks fairly random.
- The residuals plot may show a slight hint of diminishing scatter, but with so few data values it's not very clear.

- The histogram of the residuals is unimodal and roughly symmetric.

WHAT DOES IT ALL MEAN? Because the conditions check out okay, we can try to summarize what we have learned. With a P-value over 28%, it's quite possible that any perceived relationship could be just sampling error. The confidence interval suggests the slope could be positive or negative, so it's possible that as temperatures increase, women marathoners may run faster—or slower. Based on these 11 races there appears to be little evidence of a linear association between temperature and women's performances in the marathon.

*Standard Errors for Predicted Values

Once we have a useful regression, how can we indulge our natural desire to predict, without being irresponsible? We know how to compute predicted values of y for any value of x. We first did that in Chapter 7. This predicted value would be our best estimate, but it's still just an informed guess.

Now, however, we have standard errors. We can use those to construct a confidence interval for the predictions and to report our uncertainty honestly.

From our model of *%Body Fat* and *Waist* size, we might want to use *Waist* size to get a reasonable estimate of *%Body Fat*. A confidence interval can tell us how precise that prediction will be. The precision depends on the question we ask, however, and there are two questions: Do we want to know the mean *%Body Fat* for *all* men with a *Waist* size of, say, 38 inches? Or do we want to estimate the *%Body Fat* for a particular man with a 38-inch *Waist* without making him climb onto the X-ray table?

What's the difference between the two questions? The predicted *%Body Fat* is the same, but one question leads to an answer much more precise than the other. We can predict the *mean %Body Fat* for *all* men whose *Waist* size is 38 inches with a lot more precision than we can predict the *%Body Fat* of a *particular individual* whose *Waist* size happens to be 38 inches. Both are interesting questions.

We start with the same prediction in both cases. We are predicting the value for a new individual, one that was not part of the original data set. To emphasize this, we'll call his x-value "x sub new" and write it x_ν.[6] Here, x_ν is 38 inches. The regression equation predicts *%Body Fat* as $\hat{y}_\nu = b_0 + b_1 x_\nu$.

---

[6]Yes, this is a bilingual pun. The Greek letter ν is called "nu." Don't blame me; my co-author suggested this.

Now that we have the predicted value, we construct both intervals around this same number. Both intervals take the form

$$\hat{y}_{\nu} \pm t^{*}_{n-2} \times SE.$$

Even the t^{*} value is the same for both. It's the critical value (from Table T or technology) for $n - 2$ degrees of freedom and the specified confidence level. The intervals differ because they have different standard errors. Our choice of ruler depends on which interval we want.

Several factors contribute uncertainty—that is, variability—to an estimate. When there's more spread around the line, the standard error will be larger. If we're less certain of the slope, we'll be less certain of our predictions. If we have more data (n larger), our predictions will be more precise. Finally, predictions farther from the mean of x will have more variability than predictions closer to it. This last factor is new but makes intuitive sense: It's a lot easier to predict a data point near the middle of the data set than far from the center.

Because these factors are independent of each other, we can add their variances to find the total variability. The resulting formula for the standard error of the predicted *mean* value explicitly takes them all into account:

$$SE(\hat{\mu}_{\nu}) = \sqrt{SE^{2}(b_1) \times (x_{\nu} - \bar{x})^2 + \frac{s_e^2}{n}}.$$

Individual values vary more than means, so the standard error for a single predicted value has to be larger than the standard error for the mean. In fact, the standard error of a single predicted value has an *extra* source of variability: the variation of individuals around the predicted mean. That appears as the extra variance term, s_e^2, at the end under the square root:

$$SE(\hat{y}_{\nu}) = \sqrt{SE^{2}(b_1) \times (x_{\nu} - \bar{x})^2 + \frac{s_e^2}{n} + s_e^2}.$$

Remember to keep this distinction between the two kinds of standard errors when looking at computer output. The smaller one is for the predicted *mean* value, and the larger one is for a predicted *individual* value.

Mean vs. Individual Predictions For the Nenana Ice Classic, someone who planned to place a bet would want to predict this year's breakup time. By contrast, scientists studying global warming are likely to be more interested in the mean breakup time. If you want to gamble, be sure to take into account that the variability is greater when predicting for a single year.

For Example *FINDING CONFIDENCE INTERVALS FOR PREDICTED VALUES*

Let's use our analysis to create confidence intervals for predictions about *% Body Fat*. From the data and the regression output we know:

$$n = 250 \qquad \bar{x} = 36.3 \qquad s_e = 4.713 \qquad SE(b_1) = 0.074$$

QUESTION 1: What's a 95% confidence interval for the mean *% Body Fat* for all men with 38-inch waists?

ANSWER: For $x_{\nu} = 38$ the regression model predicts

$$\hat{y}_{\nu} = -42.7 + 1.7(38) = 21.9\%.$$

The standard error is

$$SE(\hat{\mu}_{\nu}) = \sqrt{0.074^2(38 - 36.3)^2 + \frac{4.713^2}{250}} = 0.32\%.$$

With $250 - 2 = 248$ df, for 95% confidence $t^{*} = 1.97$.

Putting it all together, the 95% confidence interval is: $21.9\% \pm 1.97(0.32)$
$$21.9\% \pm 0.63\%, \text{ or } (21.27, 22.53)$$

I'm 95% confidence that the mean body fat level for all men with 38-inch waists is between 21.3% and 22.5% body fat.

(continued)

QUESTION 2: What's a 95% prediction interval for the *%Body Fat* of an individual man with a 38-inch waist?

ANSWER: The standard error is

$$SE(\hat{y}_\nu) = \sqrt{0.074^2(38 - 36.3)^2 + \frac{4.713^2}{250} + 4.713^2} = 4.72\%.$$

The *prediction interval* is: 21.9% ± 1.97(4.72)
$$21.9\% \pm 9.3\%, \text{ or } (12.6, 31.2)$$

I'm 95% confident that a randomly selected man with a 38-inch waist will have between 12.6% and 31.2% body fat.

Notice how much wider this interval is than the first one. As we've known since Chapter 18, the mean is such less variable than a randomly selected individual value.

Figure 26.11

A scatterplot of *%Body Fat* vs. *Waist* size with a least squares regression line. The solid green lines near the regression line show the extent of the 95% confidence intervals for mean *%Body Fat* at each *Waist* size. The dashed red lines show the prediction intervals. Most of the points are contained within the prediction intervals, but not within the confidence intervals.

---

*Math Box

So where do those messy formulas for standard errors of predicted values come from? They're based on many of the ideas we've studied so far. Start with regression, add random variables, then throw in the Pythagorean Theorem, the Central Limit Theorem, and a dose of algebra. Mix well. . . .

We begin our quest with an equation of the regression line. Usually we write the line in the form $\hat{y} = b_0 + b_1 x$. Mathematicians call that the "slope-intercept" form; in your algebra class you wrote it as $y = mx + b$. In that algebra class you also learned another way to write equations of lines. When you know that a line with slope m passes through the point (x_1, y_1), the "point-slope" form of its equation is $y - y_1 = m(x - x_1)$.

We know the regression line passes through the mean-mean point $(\bar{x}, \bar{y})$ with slope b_1, so we can write its equation in point-slope form as $\hat{y} - \bar{y} = b_1(x - \bar{x})$. Solving for $\hat{y}$ yields $\hat{y} = b_1(x - \bar{x}) + \bar{y}$. This equation predicts the mean y-value for a specific x_ν:

$$\hat{\mu}_y = b_1(x_\nu - \bar{x}) + \bar{y}.$$

To create a confidence interval for the mean value we need to measure the variability in this prediction:

$$Var(\hat{\mu}_y) = Var(b_1(x_\nu - \bar{x}) + \bar{y}).$$

We now call on the Pythagorean Theorem of Statistics once more: the slope, b_1, and mean, $\bar{y}$, should be independent, so their variances add:

$$Var(\hat{\mu}_y) = Var(b_1(x_\nu - \bar{x})) + Var(\bar{y}).$$

The horizontal distance from our specific x-value to the mean, $x_\nu - \bar{x}$, is a constant:

$$Var(\hat{\mu}_y) = (Var(b_1))(x_\nu - \bar{x})^2 + Var(\bar{y}).$$

Let's write that equation in terms of standard deviations:

$$SD(\hat{\mu}_y) = \sqrt{(SD^2(b_1))(x_\nu - \bar{x})^2 + SD^2(\bar{y})}.$$

Because we'll need to estimate these standard deviations using samples statistics, we're really dealing with standard errors:

$$SE(\hat{\mu}_y) = \sqrt{(SE^2(b_1))(x_\nu - \bar{x})^2 + SE^2(\bar{y})}.$$

(continued)

The Central Limit Theorem tells us that the standard deviation of $\bar{y}$ is $\dfrac{\sigma}{\sqrt{n}}$. Here we'll estimate σ using s_e, which describes the variability in how far the line we drew through our sample mean may lie above or below the true mean:

$$SE(\hat{\mu}_y) = \sqrt{(SE^2(b_1))(x_\nu - \bar{x})^2 + \left(\frac{s_e}{\sqrt{n}}\right)^2}$$

$$= \sqrt{(SE^2(b_1))(x_\nu - \bar{x})^2 + \frac{s_e^2}{n}}.$$

And there it is—the standard error we need to create a confidence interval for a predicted mean value.

When we try to predict an individual value of y, we must also worry about how far the true point may lie above or below the regression line. We represent that uncertainty by adding another term, *e*, to the original equation:

$$y = b_1(x_\nu - \bar{x}) + \bar{y} + e.$$

To make a long story short (and the equation a wee bit longer), that additional term simply adds one more standard error to the sum of the variances:

$$SE(\hat{y}) = \sqrt{(SE^2(b_1))(x_\nu - \bar{x}) + \frac{s_e^2}{n} + s_e^2}.$$

WHAT IF ●●● we simulate slopes?

You've had your blood pressure taken. The measurement gave you two readings, such as 118/72.[7] The bigger number (systolic) is the pressure on your artery walls when your heart beats, pushing blood through your body. The smaller number (diastolic) is the pressure when your heart has relaxed between beats. Blood pressures vary from person to person, and even yours varies somewhat throughout each day, so the two numbers change. It's reasonable to ask how the systolic and diastolic pressures are related.

To investigate, we'd need to get readings from several different people, make a scatterplot, and examine the relationship. And then we'd face our perpetual dilemma: "OK, I see what's happening in *these* sample data. But what does that tell me about the true relationship in the whole population?"

Let's explore that a bit. We start with a data set containing 1406 blood pressure readings. Rather than analyze *them*, let's pretend they are our entire population. We'll play the Statistics game: draw a sample from this population and use it to guess at the truth. So, here's what we see in our first random sample of 25.

The scatterplot shows an association that's straight enough for regression. The slope of the line suggests we might expect systolic pressure to rise about 1.11 points for each 1-point increase in diastolic pressure.

$\widehat{\text{Systolic}} = 51 + 1.11\,\text{Diastolic}$

---

[7]The units are "millimeters of mercury," indicating how high the fluid would rise in a glass tube in the now old-fashioned sphygmomanometers (Yeah, that's the word!) used to measure blood pressure.

That seems pretty straightforward. But keep in mind, that's just where *this* sample leads us. Had we randomly selected a different group of 25 people, we might reach a different conclusion, right? Let's see.

Here's our second random sample. Hmmm. The relationship still looks linear, but this sample suggests the slope is 2.38 systolic points per diastolic point. That seems quite different. So what can a sample tell us about the true slope?

In the Real World we generally get only one sample, but the beauty of a What If is that we get to play around some more. We simulated this process 200 times. Here's a histogram showing the slopes of the 200 regression lines generated by those samples.

Now, stop and think. What important concept is on display here? Don't read farther until you know what this histogram represents.

Got it? This is a simulated sampling distribution of sample slopes.

We've encountered a lot of sampling distributions in this course. They form the basis for all statistical inference. The sampling distribution of sample proportions approaches a Normal model. The Central Limit Theorem (yes, *that* theorem again!) tells us the sampling distribution of sample means is also Normal (but not knowing the population standard deviation, we must use a t-model to do inference). In the last What If we saw that the sampling distribution of the sum of squares of z-scores calls for a χ^2 model. And now…what? Let's think about the sampling distribution of sample slopes in the usual way: shape, center, and spread.

- **Shape:** The histogram above looks roughly unimodal and symmetric. If you're thinking "Normal," good for you. Once again, though, we don't know the population standard deviation, so (as we told you earlier in this chapter), we'll need a t-model. This one has 23 degrees of freedom, right?
- **Center:** Would you say a bit over 1.5? Good guess. The mean of these 200 sample slopes is 1.59. And now, because we're playing what-if, we get to cheat. We get to actually peek at the population. The true slope for all 1406 BPs is (wait for it…) 1.563. Random samples are producing slopes that target the right number. That's good news: sample slope (b_1) is an unbiased estimator of population slope (β_1).
- **Spread:** For small samples like this, slopes can apparently vary quite a bit. The standard deviation of these 200 sample slopes is 0.355, providing a good estimate of how all sample slopes would vary. Any individual sample allows us to make a similar estimate. You know it as $SE(b_1)$, the standard error of sample slopes.

When we're looking at only one sample, its slope (b_1) is our best guess for the true slope (β_1). And then we attach a margin of error. How big should that be? Big enough to capture the true slope starting with 95% of all samples. Look carefully at the histogram and see if you agree that the middle 95% of our 200 samples have slopes varying from about 0.9 to 2.3. That suggests a typical margin of error should be about

$$\frac{1}{2}(2.3 - 0.9) = 0.7.$$

How well would our first two samples have worked? For Sample #1, the confidence interval would have been $1.11 \pm 0.7 = (0.41, 1.81)$, capturing the true slope of 1.563. Hooray! A hit!

For Sample #2, $2.38 \pm 0.7 = (1.68, 3.08)$. With this sample we struck out.

But in Statistics, overall we bat 0.950! Hall of Fame, baby!

WHAT CAN GO WRONG?

In this chapter we've added inference to the regression explorations that we did in Chapters 7 and 8. Everything covered in those chapters that could go wrong with regression can still go wrong. It's probably a good time to review Chapter 8. Take your time; we'll wait.

With inference, we've put numbers on our estimates and predictions, but these numbers are only as good as the model. Here are the main things to watch out for:

- **Don't fit a linear regression to data that aren't straight.** This is the most fundamental assumption. If the relationship between x and y isn't approximately linear, there's no sense in fitting a straight line to it.

- **Watch out for the plot thickening.** The common part of confidence and prediction intervals is the estimate of the error standard deviation, the spread around the line. If it changes with x, the estimate won't make sense. Imagine making a prediction interval for these data.

  When x is small, we can predict y precisely, but as x gets larger, it's much harder to pin y down. Unfortunately, if the spread changes, the single value of s_e won't pick that up. The prediction interval will use the average spread around the line, with the result that we'll be too pessimistic about our precision for low x-values and too optimistic for high x-values. A re-expression of y is often a good fix for changing spread.

- **Make sure the errors are Normal.** When we make a prediction interval for an individual, the Central Limit Theorem can't come to our rescue. For us to believe the prediction interval, the errors must be from the Normal model. Check the histogram and Normal probability plot of the residuals to see if this assumption looks reasonable.

- **Watch out for extrapolation.** It's tempting to think that because we have prediction *intervals*, they'll take care of all our uncertainty so we don't have to worry about extrapolating. Wrong. The interval is only as good as the model. The uncertainty our intervals predict is correct only if our model is true. There's no way to adjust for wrong models. That's why it's always dangerous to predict for x-values that lie far from the center of the data.

- **Watch out for influential points and outliers.** We always have to be on the lookout for a few points that have undue influence on our estimated model—and regression is certainly no exception.

- **Watch out for one-tailed tests.** Because tests of hypotheses about regression coefficients are usually two-tailed, software packages report two-tailed P-values. If you are using software to conduct a one-tailed test about slope, you'll need to divide the reported P-value in half.

What Have We Learned?

In Chapters 6, 7, and 8 we learned to examine the relationship between two quantitative variables by looking at a scatterplot. We've learned (if it's linear) to quantify the strength and direction with a correlation and model it with least squares regression. Now we have learned to apply inference to these regression models.

We've learned that we can interpret the standard deviation of the residuals, s_e, two ways:

- s_e estimates the standard deviation of the y's at any value of x;
- s_e estimates the typical error between predicted values and actual values.

We've learned that the standard error of the slope, $SE(b_1)$, estimates the sample-to-sample variability in slopes of regression lines.

We've learned the assumptions for inference (and how to check them, in order):

- the Linearity Assumption (Straight Enough Condition);
- the Independence Assumption (Random Residuals Condition);
- the Equal Variances Assumption (Does The Plot Thicken? Condition);
- the Normality Assumption (Nearly Normal and Outlier Conditions).

We've learned that when these assumptions are met, the sampling distribution for the slope of a regression line can be described by a t-model with $n - 2$ degrees of freedom.

We've learned to use this model to test the hypothesis that there's no linear association between two quantitative variables, $H_0: \beta_1 = 0$.

And we've learned to construct and interpret a confidence interval around b_1 for the true slope β_1.

Terms

Conditions for inference in regression (and checks for some of them)

- **Straight Enough Condition** for linearity. (Check that the scatterplot of y against x has linear form and that the scatterplot of residuals against predicted values has no obvious pattern.) (p. 708)
- **Independence Assumption.** (Think about the nature of the data.) (p. 709)
- **Randomization Condition** for data that come from a random (or at least representative) sample of the population. (p. 709)
- **Random Residuals Condition** for any evidence of patterns, trends, or clumping. (p. 709)
- **Does the Plot Thicken? Condition** for constant variance. (Check that the scatterplot shows consistent spread across the range of the x-variable, and that the residuals plot has constant variance, too. A common problem is increasing spread with increasing predicted values—the *plot thickens!*) (p. 709)
- **Nearly Normal Condition** for Normality of the residuals. (Check a histogram of the residuals.) (p. 709)

Residual standard deviation

The spread of the data around the regression line is measured with the residual standard deviation, s_e: (p. 714)

$$s_e = \sqrt{\frac{\sum (y - \hat{y})^2}{n - 2}} = \sqrt{\frac{\sum e^2}{n - 2}}.$$

Standard error for the slope

The standard error of b_1 estimates the standard deviation of the sampling distribution model for slopes of regression lines. (p. 715)

$$SE(b_1) = \frac{s_e}{\sqrt{n - 1}\, s_x}$$

t-test for the regression slope

When the assumptions are satisfied, we can perform a test for the slope coefficient. We usually test the null hypothesis that the true value of the slope is zero against the alternative that it is not. A zero slope would indicate a complete absence of linear relationship between y and x. (p. 715)

To test $H_0: \beta_1 = 0$, we find the P-value from the Student's t-model with $n - 2$ degrees of freedom where

$$t = \frac{b_1 - 0}{SE(b_1)}$$

Confidence interval for the regression slope (β)

When the assumptions are satisfied, we can find a confidence interval for the slope parameter from $b_1 \pm t^*_{n-2} \times SE(b_1)$. The critical value, t^*_{n-2}, depends on the confidence level specified and on Student's t-model with $n - 2$ degrees of freedom. (p. 716)

On the Computer REGRESSION ANALYSIS

All statistics packages make a table of results for a regression. These tables differ slightly from one package to another, but all are essentially the same. We've seen other examples of such tables already.

All packages offer analyses of the residuals. With some, you must request plots of the residuals as you request the regression. Others let you find the regression first and then analyze the residuals afterward. Either way, your analysis is not complete if you don't check the residuals with a histogram or Normal probability plot and a scatterplot of the residuals against x or the predicted values.

You should, of course, always look at the scatterplot of your two variables before computing a regression.

Regressions are almost always found with a computer or calculator. The calculations are too long to do conveniently by hand for data sets of any reasonable size. No matter how the regression is computed, the results are usually presented in a table that has a standard form. Here's a portion of a typical regression results table, along with annotations showing where the numbers come from:

A **S** **Activity: Regression on the Computer.** How fast is the universe expanding? And how old is it? A prominent astronomer used regression to astound the scientific community. Read the story, analyze the data, and interactively learn about each of the numbers in a typical computer regression output table.

The regression table gives the coefficients (once you find them in the middle of all this other information), so we can see that the regression equation is

$$\widehat{\%BF} = -42.73 + 1.7 \, Waist$$

and that the R^2 for the regression is 67.8%. (Is accounting for 68% of the variation in *%Body Fat* good enough to be useful? Is a prediction *ME* of more than 9% good enough? Health professionals might not be satisfied.)

The column of t-ratios gives the test statistics for the respective null hypotheses that the true values of the coefficients are zero. The corresponding P-values are also usually reported.

Exercises

1. Graduation rates A prestigious college is interested in factors that might be associated with better graduation rates. The administrators wonder whether there is a relationship between acceptance rates and graduation rates. Before proceeding with their regression inference, what conditions and assumptions must be satisfied? Are they? (Source: *US News and World Report* National University Rankings October 2011)

Dependent variable is 6-Year Graduation Rate
R-squared = 69.6%
s = 2.86 with 50 − 2 = 48 degrees of freedom

Variable	Coefficient	SE(Coeff)	t-Ratio	P-Value
Intercept	97.0663	0.84406	114.999	<0.0001
Acceptance Rate	−0.25101	0.023951	−10.480	<0.0001

2. Shoot to score A college hockey coach collected data from the 2010–2011 National Hockey League season. He hopes to convince his players that the number of shots taken has an effect on the number of goals scored. The coach performed a preliminary analysis, using the scoring statistics from 293 offensive players who play professional hockey. He predicts *Goals* from number of *Shots* (taken for the season). Discuss each of the conditions and assumptions required for him to proceed with the regression analysis. (www.nhl.com)

Dependent variable is Goals
R-squared = 63.4%
s = 5.13 with 293 − 2 = 291 degrees of freedom

Variable	Coefficient	SE(Coeff)	t-Ratio	P-Value
Intercept	−1.77095	0.9087	−1.9488	0.052
Shots	0.11938	0.0053	22.460	<0.0001

3. **Graduation rates, part II** Using the regression output in Exercise 1, identify the standard deviation of the residuals and explain what it means in the context of the problem.

4. **Shoot to score another one** Using the regression output from Exercise 2, identify the standard deviation of the residuals and explain its meaning with a sentence in context.

5. **Graduation rates, part III** Continuing with the regression of Exercise 1, write a sentence that explains the meaning of the standard error of the slope of the regression line, $SE(b_1) = 0.0240$.

6. **Shoot to score, hat trick** Returning to the results of Exercise 2, write a sentence to explain the meaning of the standard error of the slope of the regression line, $SE(b_1) = 0.0053$.

7. **Graduation, part IV** The college administrators in Exercise 1 tested the hypotheses $H_0: \beta_1 = 0$ vs. $H_A: \beta_1 \neq 0$ and rejected the null hypothesis because the P-value was less than 0.0001. What can they conclude about the relationship between admission rates and graduation rates?

8. **Shoot to score, number four** What can the hockey coach in Exercise 2 conclude about shooting and scoring goals from the fact that the P-value < 0.0001 for the slope of the regression line? Write a sentence in context.

9. **Graduation, part V** The college administrators in Exercise 1 constructed a 95% confidence interval for the slope of their regression line. Interpret the meaning of their interval $(-0.299\%, -0.203\%)$ within the context of the problem.

10. **Shoot to score, overtime** The coach in Exercise 2 found a 95% confidence interval for the slope of his regression line. Recall that he is trying to predict total goals scored based on shots taken. Interpret with a sentence the meaning of the interval 0.12 ± 0.01.

11. **Tracking hurricanes 2010** In Chapter 6, we looked at data from the National Oceanic and Atmospheric Administration about their success in predicting hurricane tracks. Below is a scatterplot of the error (in nautical miles) for predicting hurricane locations 24 hours in the future vs. the year in which the prediction (and the hurricane) occurred.

In Chapter 6, we could describe this relationship only in general terms. Now we can learn more. Here is the regression analysis:

Dependent variable is 24Error
R-squared = 68.7%
s = 16.44 with 41 − 2 = 39 degrees of freedom

Variable	Coefficient	SE(Coeff)	t-Ratio	P-Value
Intercept	132.3	5.043	26.2	≤0.0001
Years Since 1970	−2.01	0.217	−9.25	≤0.0001

a) Explain in words and numbers what the regression says.
b) State the hypothesis about the slope (both numerically and in words) that describes how hurricane prediction quality has changed.
c) Assuming that the assumptions for inference are satisfied, perform the hypothesis test and state your conclusion. Be sure to state it in terms of prediction errors and years.
d) Explain what the R-squared means in terms of this regression.

12. **Drug use** The *2011 World Drug Report* investigated the prevalence of drug use as a percentage of the population aged 15 to 64. Data from 22 European countries are shown in the following scatterplot and regression analysis. (Source: *World Drug Report*, 2011. www.unodc.org/unodc/en/data-and-analysis/WDR-2011.html)

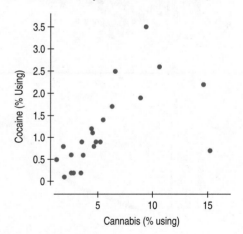

Dependent variable is Cocaine
R-squared = 38.1%
s = 0.724 with 22 − 2 = 20 degrees of freedom

Variable	Coefficient	SE(Coeff)	t-Ratio	P-Value
Intercept	0.35707	0.2757	1.295	0.21
Cannabis%	0.14264	0.0406	3.512	0.002

a) Explain in context what the regression says.
b) State the hypothesis about the slope (both numerically and in words) that describes how use of marijuana is associated with other drugs.
c) Assuming that the assumptions for inference are satisfied, perform the hypothesis test and state your conclusion in context.
d) Explain what R-squared means in context.
e) Do these results indicate that marijuana use leads to the use of harder drugs? Explain.

13. Movie budgets How does the cost of a movie depend on its length? Data on the cost (millions of dollars) and the running time (minutes) for major release films of 2005 are summarized in these plots and computer output:

Dependent variable is: Budget($M)
R squared = 15.4%
s = 32.95 with 120 − 2 = 118 degrees of freedom

Variable	Coefficient	SE(Coeff)	t-ratio	P-value
Intercept	−31.3869	17.12	−1.83	0.0693
Run Time	0.714400	0.1541	4.64	≤0.0001

a) Explain in context what the regression says.
b) The intercept is negative. Discuss its value, taking note of the P-value.
c) The output reports s = 32.95. Explain what that means in this context.
d) What's the value of the standard error of the slope of the regression line?
e) Explain what that means in this context.

14. House prices How does the price of a house depend on its size? Data from Saratoga, New York, on 1064 randomly selected houses that had been sold include data on price ($1000's) and size (1000's ft²), producing the following graphs and computer output:

Dependent variable is: Price
R squared = 59.5%
s = 53.79 with 1064 − 2 = 1062 degrees of freedom

Variable	Coefficient	SE(Coeff)	t-ratio	P-value
Intercept	−3.11686	4.688	−0.665	0.5063
Size	94.4539	2.393	39.5	≤0.0001

a) Explain in context what the regression says.
b) The intercept is negative. Discuss its value, taking note of its P-value.
c) The output reports s = 53.79. Explain what that means in this context.
d) What's the value of the standard error of the slope of the regression line?
e) Explain what that means in this context.

15. Movie budgets: the sequel Exercise 13 shows computer output examining the association between the length of a movie and its cost.

a) Check the assumptions and conditions for inference.
b) Find a 95% confidence interval for the slope and interpret it in context.

16. Second home Exercise 14 shows computer output examining the association between the sizes of houses and their sale prices.

a) Check the assumptions and conditions for inference.

b) Find a 95% confidence interval for the slope and interpret it in context.

17. Hot dogs Healthy eating probably doesn't include hot dogs, but if you are going to have one, you'd probably hope it's low in both calories and sodium. In its July 2007 issue, *Consumer Reports* listed the number of calories and sodium content (in milligrams) for 13 brands of all-beef hot dogs it tested. Examine the association, assuming that the data satisfy the conditions for inference.

Dependent variable is: Sodium
R squared = 60.5%
s = 59.66 with 13 − 2 = 11 degrees of freedom

Variable	Coefficient	SE(Coeff)	t-ratio	P-value
Constant	90.9783	77.69	1.17	0.2663
Calories	2.29959	0.5607	4.10	0.0018

a) State the appropriate hypotheses about the slope.

b) Test your hypotheses and state your conclusion in the proper context.

18. Cholesterol Does a person's cholesterol level tend to change with age? Data collected from 1406 adults aged 45 to 62 produced the regression analysis shown. Assuming that the data satisfy the conditions for inference, examine the association between age and cholesterol level.

Dependent variable is: Chol
s = 46.16

Variable	Coefficient	SE(Coeff)	t-ratio	P-value
Intercept	194.232	13.55	14.3	≤0.0001
Age	0.771639	0.2574	3.00	0.0027

a) State the appropriate hypothesis for the slope.

b) Test your hypothesis and state your conclusion in the proper context.

19. Second frank Look again at Exercise 17's regression output for the calorie and sodium content of hot dogs.

a) The output reports $s = 59.66$. Explain what that means in this context.

b) What's the value of the standard error of the slope of the regression line?

c) Explain what that means in this context.

20. More cholesterol Look again at Exercise 18's regression output for age and cholesterol level.

a) The output reports $s = 46.16$. Explain what that means in this context.

b) What's the value of the standard error of the slope of the regression line?

c) Explain what that means in this context.

21. Last dog Based on the regression output seen in Exercise 17, create a 95% confidence interval for the slope of the regression line and interpret your interval in context.

22. Cholesterol, finis Based on the regression output seen in Exercise 18, create a 95% confidence interval for the slope of the regression line and interpret it in context.

23. Marriage age 2010 The scatterplot below suggests a decrease in the difference in ages at first marriage for men and women since 1975. We want to examine the regression to see if this decrease is significant.

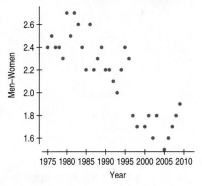

Dependent variable is Men-Women
R-squared = 72.6%
s = 0.186 with 33 − 2 = 31 degrees of freedom

Variable	Coefficient	SE(Coeff)	t-Ratio	P-Value
Intercept	60.055	6.396	9.39	≤0.0001
Year	−0.029	0.0032	−9.05	≤0.0001

a) Write appropriate hypotheses.

b) Here is the residuals plot and a histogram of the residuals. Do you think the conditions for inference are satisfied? Explain.

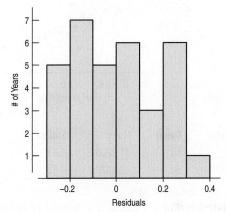

c) Test the hypothesis and state your conclusion about the trend in age at first marriage.

24. Used cars 2010 Vehix.com offered several used Toyota Corollas for sale. The following table displays the ages of the cars and the advertised prices.

Age (yr)	Price ($)	Age (yr)	Price ($)
1	15988	6	9995
1	13988	6	11988
2	14488	7	8990
3	10995	8	9488
3	13998	8	8995
4	13622	9	5990
4	12810	10	4100
5	9988	12	2995

a) Make a scatterplot for these data.
b) Do you think a linear model is appropriate? Explain.
c) Find the equation of the regression line.
d) Check the residuals to see if the conditions for inference are met.

25. Marriage age 2010, again Based on the analysis of marriage ages since 1975 given in Exercise 23, find a 95% confidence interval for the rate at which the age gap is closing. Explain what your confidence interval means.

26. Used cars 2010, again Based on the analysis of used car prices you did for Exercise 24, create a 95% confidence interval for the slope of the regression line and explain what your interval means in context.

27. Fuel economy A consumer organization has reported test data for 50 car models. We will examine the association between the weight of the car (in thousands of pounds) and the fuel efficiency (in miles per gallon). Here are the scatterplot, summary statistics, and regression analysis:

Variable	Count	Mean	StdDev
MPG	50	25.0200	4.83394
wt/1000	50	2.88780	0.511656

Dependent variable is: MPG
R-squared = 75.6%
s = 2.413 with 50 − 2 = 48 df

Variable	Coefficient	SE(Coeff)	t-ratio	P-value
Intercept	48.7393	1.976	24.7	≤0.0001
Weight	−8.21362	0.6738	−12.2	≤0.0001

a) Is there strong evidence of an association between the weight of a car and its gas mileage? Write an appropriate hypothesis.
b) Are the assumptions for regression satisfied?
c) Test your hypothesis and state your conclusion.

28. SAT scores How strong was the association between student scores on the Math and Verbal sections of the old SAT? Scores on each ranged from 200 to 800 and were widely used by college admissions offices. Here are summaries and plots of the scores for a graduating class at Ithaca High School:

Variable	Count	Mean	Median	StdDev	Range	IntQRange
Verbal	162	596.296	610	99.5199	490	140
Math	162	612.099	630	98.1343	440	150

Dependent variable is: Math
R-squared = 46.9%
s = 71.75 with 162 − 2 = 160 df

Variable	Coefficient	SE(Coeff)	t-ratio	P-value
Intercept	209.554	34.35	6.10	≤0.0001
Verbal	0.675075	0.0568	11.9	≤0.0001

a) Is there evidence of an association between Math and Verbal scores? Write an appropriate hypothesis.
b) Discuss the assumptions for inference.
c) Test your hypothesis and state an appropriate conclusion.

29. Fuel economy, part II Consider again the data in Exercise 27 about the gas mileage and weights of cars.

a) Create a 95% confidence interval for the slope of the regression line.
b) Explain in this context what your confidence interval means.

30. SATs, part II Consider the high school SAT scores data from Exercise 28.

a) Find a 90% confidence interval for the slope of the true line describing the association between Math and Verbal scores.
b) Explain in this context what your confidence interval means.

31. *Fuel economy, part III Consider again the data in Exercise 27 about the gas mileage and weights of cars.

a) Create a 95% confidence interval for the average fuel efficiency among cars weighing 2500 pounds, and explain what your interval means.
b) Create a 95% prediction interval for the gas mileage you might get driving your new 3450-pound SUV, and explain what that interval means.

32. *SATs again Consider the high school SAT scores data from Exercise 28 once more.

a) Find a 90% confidence interval for the mean SAT-Math score for all students with an SAT-Verbal score of 500.
b) Find a 90% prediction interval for the Math score of the senior class president if you know she scored 710 on the Verbal section.

33. Cereal A healthy cereal should be low in both calories and sodium. Data for 77 cereals were examined and judged acceptable for inference. The 77 cereals had between 50 and 160 calories per serving and between 0 and 320 mg of sodium per serving. Here's the regression analysis:

Dependent variable is: Sodium
R-squared = 9.0%
s = 80.49 with 77 − 2 = 75 degrees of freedom

Variable	Coefficient	SE(Coeff)	t-ratio	P-value
Intercept	21.4143	51.47	0.416	0.6786
Calories	1.29357	0.4738	2.73	0.0079

a) Is there an association between the number of calories and the sodium content of cereals? Explain.
b) Do you think this association is strong enough to be useful? Explain.

34. Brain size Does your IQ depend on the size of your brain? A group of female college students took a test that measured their verbal IQs and also underwent an MRI scan to measure the size of their brains (in 1000s of pixels). The scatterplot and regression analysis are shown, and the assumptions for inference were satisfied.

Dependent variable is: IQ_Verbal
R-squared = 6.5%   s = 21.5291   df = 18

Variable	Coefficient	SE(Coeff)
Intercept	24.1835	76.38
Size	0.098842	0.0884

a) Test an appropriate hypothesis about the association between brain size and IQ.
b) State your conclusion about the strength of this association.

35. Another bowl Further analysis of the data for the breakfast cereals in Exercise 33 looked for an association between *Fiber* content and *Calories* by attempting to construct a linear model. Here are several graphs. Which of the assumptions for inference are violated? Explain.

36. Winter The output shows an attempt to model the association between average *January Temperature* (in degrees Fahrenheit) and *Latitude* (in degrees north of the equator) for 59 U.S. cities. Which of the assumptions for inference do you think are violated? Explain.

37. Acid rain Biologists studying the effects of acid rain on wildlife collected data from 163 streams in the Adirondack Mountains. They recorded the *pH* (acidity) of the water and the *BCI*, a measure of biological diversity. Here's a scatterplot of *BCI* against *pH*:

And here is part of the regression analysis:

Dependent variable is: BCI
R-squared = 27.1%
s = 140.4 with 163 − 2 = 161 degrees of freedom

Variable	Coefficient	SE(Coeff)
Intercept	2733.37	187.9
pH	−197.694	25.57

a) State the null and alternative hypotheses under investigation.
b) Assuming that the assumptions for regression inference are reasonable, find the *t*- and P-values.
c) State your conclusion.

38. Climate change and CO₂ 2011 Data collected from around the globe show that the earth is getting warmer. The most common theory relates climate change to an increase in atmospheric levels of carbon dioxide (CO_2), a greenhouse gas. The mean annual CO_2 concentration in the atmosphere (parts per million) is measured at the top of Mauna Loa in Hawaii, away from any local contaminants. (Available at ftp://ftp.cmdl.noaa.gov/ccg/co2/trends/co2_annmean_mlo.tx)

The mean surface air temperature is recorded as the change in °C relative to a base period of 1951 to 1980. (Available at data.giss.nasa.gov/gistemp/graphs_v3/)

Here are a scatterplot and regression for the years from 1959 to 2011:

Dependent variable is Temp
R-squared = 82.3%
s = 0.0985 with 53 − 2 = 51 degrees of freedom

Variable	Coefficient	SE(Coeff)	t-Ratio	P-Value
Intercept	−2.98861	0.2086	−14.3	≤0.0001
CO_2	0.0092	0.0006	15.4	≤0.0001

a) Write the equation of the regression line.
b) Is there evidence of an association between CO_2 level and global temperature?
c) Do you think predictions made by this regression will be very accurate? Explain.
d) Does this regression prove that increasing CO_2 levels are causing global warming? Discuss.

39. Ozone The Environmental Protection Agency is examining the relationship between the ozone level (in parts per million) and the population (in millions) of U.S. cities. Part of the regression analysis is shown.

Dependent variable is: Ozone
R-squared = 84.4%
s = 5.454 with 16 − 2 = 14 df

Variable	Coefficient	SE(Coeff)
Intercept	18.892	2.395
Pop	6.650	1.910

a) We suspect that the greater the population of a city, the higher its ozone level. Is the relationship significant? Assuming the conditions for inference are satisfied, test an appropriate hypothesis and state your conclusion in context.
b) Do you think that the population of a city is a useful predictor of ozone level? Use the values of both R^2 and s in your explanation.

40. Sales and profits A business analyst was interested in the relationship between a company's sales and its profits. She collected data (in millions of dollars) from a random sample of Fortune 500 companies and created the regression analysis and summary statistics shown. The assumptions for regression inference appeared to be satisfied.

	Profits	Sales
Count	79	79
Mean	209.839	4178.29
Variance	635,172	49,163,000
Std Dev	796.977	7011.63

Dependent variable is: Profits
R-squared = 66.2%   s = 466.2

Variable	Coefficient	SE(Coeff)
Intercept	−176.644	61.16
Sales	0.092498	0.0075

a) Is there a significant association between sales and profits? Test an appropriate hypothesis and state your conclusion in context.
b) Do you think that a company's sales serve as a useful predictor of its profits? Use the values of both R^2 and s in your explanation.

41. Ozone, again Consider again the relationship between the population and ozone level of U.S. cities that you analyzed in Exercise 39.

a) Give a 90% confidence interval for the approximate increase in ozone level associated with each additional million city inhabitants.
*b) For the cities studied, the mean population was 1.7 million people. The population of Boston is approximately 0.6 million people. Predict the mean ozone level for cities of that size with an interval in which you have 90% confidence.

42. More sales and profits Consider again the relationship between the sales and profits of Fortune 500 companies that you analyzed in Exercise 40.

a) Find a 95% confidence interval for the slope of the regression line. Interpret your interval in context.
*b) Last year the drug manufacturer Eli Lilly, Inc., reported gross sales of $9 billion (that's $9,000 million). Create a 95% prediction interval for the company's profits, and interpret your interval in context.

43. Tablet computers In October 2011, cnet.com listed the battery life (in hours) and luminous intensity (i.e., screen brightness, in cd/m^2) for a sample of tablet computers. We want to know if brighter screens drain the battery more quickly.

Dependent variable is Video battery life (in hours)
R-squared = 27.9%
s = 1.913 with 23 − 2 = 21 degrees of freedom

Variable	Coefficient	SE(Coeff)	t-Ratio	P-Value
Intercept	2.8467073	1.6628386	1.7119566	0.10163
Brightness	0.014080549	0.0049373503	2.851843	0.00955

a) How many tablet computers were tested?
b) Are the conditions for inference satisfied? Explain.
c) Is there evidence of an association between maximum brightness of the screen and battery life? Test an appropriate hypothesis and state your conclusion.
d) Is the association strong? Explain.
e) What is the equation of the regression line?
f) Create a 90% confidence interval for the slope of the true line.
g) Interpret your interval in this context.

44. Crawling Researchers at the University of Denver Infant Study Center wondered whether temperature might influence the age at which babies learn to crawl. Perhaps the extra clothing that babies wear in cold weather would restrict movement and delay the age at which they started crawling. Data were collected on 208 boys and 206 girls. Parents reported the month of the baby's birth and the age (in weeks) at which their child first crawled. The table gives the average *Temperature* (°F) when the babies were 6 months old and average *Crawling Age* (in weeks) for each month of the year. Make the plots and compute the analyses necessary to answer the following questions.

Birth Month	6-Month Temperature	Average Crawling Age
Jan.	66	29.84
Feb.	73	30.52
Mar.	72	29.70
April	63	31.84
May	52	28.58
June	39	31.44
July	33	33.64
Aug.	30	32.82
Sept.	33	33.83
Oct.	37	33.35
Nov.	48	33.38
Dec.	57	32.32

a) Would this association appear to be weaker, stronger, or the same if data had been plotted for individual babies instead of using monthly averages? Explain.

b) Is there evidence of an association between *Temperature* and *Crawling Age*? Test an appropriate hypothesis and state your conclusion. Don't forget to check the assumptions.
c) Create and interpret a 95% confidence interval for the slope of the true relationship.

45. Body fat Do the data shown in the table below indicate an association between *Waist* size and *%Body Fat*?

a) Test an appropriate hypothesis and state your conclusion.
*b) Give a 95% confidence interval for the mean *%Body Fat* found in people with 40-inch *Waists*.

Waist (in.)	Weight (lb)	Body Fat (%)	Waist (in.)	Weight (lb)	Body Fat (%)
32	175	6	33	188	10
36	181	21	40	240	20
38	200	15	36	175	22
33	159	6	32	168	9
39	196	22	44	246	38
40	192	31	33	160	10
41	205	32	41	215	27
35	173	21	34	159	12
38	187	25	34	146	10
38	188	30	44	219	28

46. Body fat, again Use the data from Exercise 45 to examine the association between *Weight* and *%Body Fat*.

a) Find a 90% confidence interval for the slope of the regression line of *%Body Fat* on *Weight*.
b) Interpret your interval in context.
*c) Give a 95% prediction interval for the *%Body Fat* of an individual who weighs 165 pounds.

47. Grades The data set below shows midterm scores from an Introductory Statistics course.

First Name	Midterm 1	Midterm 2	Homework
Timothy	82	30	61
Karen	96	68	72
Verena	57	82	69
Jonathan	89	92	84
Elizabeth	88	86	84
Patrick	93	81	71
Julia	90	83	79
Thomas	83	21	51
Marshall	59	62	58
Justin	89	57	79
Alexandra	83	86	78
Christopher	95	75	77
Justin	81	66	66
Miguel	86	63	74

(continued)

First Name	Midterm 1	Midterm 2	Homework
Brian	81	86	76
Gregory	81	87	75
Kristina	98	96	84
Timothy	50	27	20
Jason	91	83	71
Whitney	87	89	85
Alexis	90	91	68
Nicholas	95	82	68
Amandeep	91	37	54
Irena	93	81	82
Yvon	88	66	82
Sara	99	90	77
Annie	89	92	68
Benjamin	87	62	72
David	92	66	78
Josef	62	43	56
Rebecca	93	87	80
Joshua	95	93	87
Ian	93	65	66
Katharine	92	98	77
Emily	91	95	83
Brian	92	80	82
Shad	61	58	65
Michael	55	65	51
Israel	76	88	67
Iris	63	62	67
Mark	89	66	72
Peter	91	42	66
Catherine	90	85	78
Christina	75	62	72
Enrique	75	46	72
Sarah	91	65	77
Thomas	84	70	70
Sonya	94	92	81
Michael	93	78	72
Wesley	91	58	66
Mark	91	61	79
Adam	89	86	62
Jared	98	92	83
Michael	96	51	83
Kathryn	95	95	87
Nicole	98	89	77
Wayne	89	79	44
Elizabeth	93	89	73
John	74	64	72
Valentin	97	96	80
David	94	90	88
Marc	81	89	62
Samuel	94	85	76
Brooke	92	90	86

a) Fit a model predicting the second midterm score from the first.

b) Comment on the model you found, including a discussion of the assumptions and conditions for regression. Is the coefficient for the slope statistically significant?

c) A student comments that because the P-value for the slope is very small, Midterm 2 is very well predicted from Midterm 1. So, he reasons, next term the professor can give just one midterm. What do you think?

T 48. Grades? The professor teaching the Introductory Statistics class discussed in Exercise 47 wonders whether performance on homework can accurately predict midterm scores.

a) To investigate it, she fits a regression of the sum of the two midterms scores on homework scores. Fit the regression model.

b) Comment on the model including a discussion of the assumptions and conditions for regression. Is the coefficient for the slope "statistically significant"?

c) Do you think she can accurately judge a student's performance without giving the midterms? Explain.

T 49. Strike two Remember the Little League instructional video discussed in Chapter 24? Ads claimed it would improve the performances of Little League pitchers. To test this claim, 20 Little Leaguers threw 50 pitches each, and we recorded the number of strikes. After the players participated in the training program, we repeated the test. The table shows the number of strikes each player threw before and after the training. A test of paired differences failed to show that this training improves ability to throw strikes. Is there any evidence that the effectiveness of the video (*After – Before*) depends on the player's initial ability to throw strikes (*Before*)? Test an appropriate hypothesis and state your conclusion. Propose an explanation for what you find.

Number of Strikes (out of 50)			
Before	After	Before	After
28	35	33	33
29	36	33	35
30	32	34	32
32	28	34	30
32	30	34	33
32	31	35	34
32	32	36	37
32	34	36	33
32	35	37	35
33	36	37	32

T 50. All the efficiency money can buy 2011 A sample of 84 model-2011 cars from an online information service was examined to see how fuel efficiency (as highway mpg) relates to the cost (Manufacturer's Suggested Retail Price in dollars) of cars. Here are displays and computer output:

Dependent variable is MPG
R-squared = 0.0216%
s = 3.54

Variable	Coefficient	SE(Coeff)	t-Ratio	P-Value
Intercept	36.514	1.496	24.406	<0.0001
Slope	−8.089E −6	6.439E −5	−0.1256	0.900

a) State what you want to know, identify the variables, and give the appropriate hypotheses.
b) Check the assumptions and conditions.
c) If the conditions are met, complete the analysis.

51. Education and mortality The software output below is based on the mortality rate (deaths per 100,000 people) and the education level (average number of years in school) for 58 U.S. cities.

Variable	Count	Mean	StdDev
Mortality	58	942.501	61.8490
Education	58	11.0328	0.793480

Dependent variable is: Mortality
R-squared = 41.0%
s = 47.92 with 58 − 2 = 56 degrees of freedom

Variable	Coefficient	SE(Coeff)
Intercept	1493.26	88.48
Education	−49.9202	8.000

a) Comment on the assumptions for inference.
b) Is there evidence of a strong association between the level of *Education* in a city and the *Mortality* rate? Test an appropriate hypothesis and state your conclusion.
c) Can we conclude that getting more education is likely (on average) to prolong your life? Why or why not?
d) Find a 95% confidence interval for the slope of the true relationship.
e) Explain what your interval means.
*f) Find a 95% confidence interval for the average *Mortality* rate in cities where the adult population completed an average of 12 years of school.

52. Property assessments The software outputs below provide information about the *Size* (in square feet) of 18 homes in Ithaca, New York, and the city's assessed *Value* of those homes.

Variable	Count	Mean	StdDev	Range
Size	18	2003.39	264.727	890
Value	18	60946.7	5527.62	19710

Dependent variable is: Value
R-squared = 32.5%
s = 4682 with 18 − 2 = 16 degrees of freedom

Variable	Coefficient	SE(Coeff)
Intercept	37108.8	8664
Size	11.8987	4.290

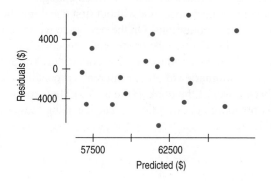

a) Explain why inference for linear regression is appropriate with these data.
b) Is there a significant association between the *Size* of a home and its assessed *Value*? Test an appropriate hypothesis and state your conclusion.
c) What percentage of the variability in assessed *Value* is explained by this regression?
d) Give a 90% confidence interval for the slope of the true regression line, and explain its meaning in the proper context.
e) From this analysis, can we conclude that adding a room to your house will increase its assessed *Value*? Why or why not?
*f) The owner of a home measuring 2100 square feet files an appeal, claiming that the $70,200 assessed *Value* is too high. Do you agree? Explain your reasoning.

Just Checking ANSWERS

1. A high *t*-ratio of 3.27 indicates that the slope is different from zero—that is, that there is a linear relationship between height and mouth size. The small P-value says that a slope this large would be very unlikely to occur by chance if, in fact, there was no linear relationship between the variables.

2. Not really. The R^2 for this regression is only 15.3%, so height doesn't account for very much of the variability in mouth size.

3. The value of *s* tells the standard deviation of the residuals. Mouth sizes have a mean of 60.3 cubic centimeters. A standard deviation of 15.7 in the residuals indicates that the errors made by this regression model can be quite large relative to what we are estimating. Errors of 15 to 30 cubic centimeters would be common.

4. In repeated sampling, we estimate slopes of regression lines predicting mouth volume would vary with a standard deviation of 18.77 cubic centimeters per meter of height.

Review of part VII

Inference When Variables Are Related

Quick Review

With these last two chapters, you have added important analytical tools to your ways of looking at data. Here's a brief summary of those key concepts and skills, as well as an overview of statistical inference:

- Inferences about distributions of counts use chi-square models.

  - To see if an observed distribution is consistent with a proposed model, use a goodness-of-fit test.
  - To see if two or more observed distributions could have arisen from populations with the same model, use a test of homogeneity.

- Inference about association between two variables tests the hypothesis that it is plausible to consider the variables independent.

  - If the variables are categorical, display the data in a contingency table and use a chi-square test of independence.
  - If the variables are quantitative, display them with a scatterplot. You may use a linear regression *t*-test if there appears to be a linear association for which the residuals are random, consistent in terms of spread, and approximately Normal.

- You can now use statistical inference to answer questions about means, proportions, distributions, and associations.

  - No inference procedure is valid unless the underlying assumptions are true. Always check the conditions before proceeding. Many of those checks should be made by examining a graph.
  - You can make inferences about a single proportion or the difference of two proportions using Normal models.

  - You can make inferences about one mean, the difference of two independent means, or the mean of paired differences using *t*-models.
  - You can make inferences about distributions using chi-square models.
  - You can make inferences about associations between categorical variables using chi-square models.
  - You can make inferences about linear associations between quantitative variables using *t*-models.

If you look back at where we've been in this book, you'll see that statistical inference relies on almost everything we've seen. In Chapters 11 and 12 we learned techniques of collecting data using randomization—that's what makes inference possible at all. In Chapters 2, 3, and 6 we learned to plot our data and to look for the patterns and relationships we use to check the conditions that allow inference. In Chapters 2, 4, and 7 we learned about the summary statistics we use to do the mechanics of inference. We use our knowledge of randomness and probability from Chapters 10, 13, and 14 to help us think clearly about uncertainty, and the probability models of Chapters 5, 15, and 16 to measure our uncertainty precisely. Ultimately, the Central Limit Theorem of Chapter 17 makes all of inference possible.

Remember (have we said this often enough yet?): Never use any inference procedure without first checking the assumptions and conditions. On the next page we summarize the new types of inference procedures, the corresponding formulas, and the assumptions and conditions. You'll find complete summaries of all our inference procedures inside the back cover of the book. Have a look. Then you'll be ready for more opportunities to practice using these concepts and skills. . . .

Quick Guide to Inference

Think				Show				Tell?
Inference about?	One group or two?	Procedure	Model	Parameter	Estimate	SE		Chapter
Distributions (one categorical variable)	One sample	Goodness-of-Fit	χ^2 df $= cells - 1$			$\sum \dfrac{(\text{obs} - \text{exp})^2}{\text{exp}}$		26
	Many independent groups	Homogeneity χ^2 Test	χ^2 df $= (r - 1)(c - 1)$					
Independence (two categorical variables)	One sample	Independence χ^2 Test						
Association (two quantitative variables)	One sample	Linear Regression t-Test or Confidence Interval for β	t df $= n - 2$	β_1	b_1	$\dfrac{s_e}{s_x \sqrt{n - 1}}$ (compute with technology)		27
		*Confidence Interval for μ_ν		μ_ν	$\hat{y}_\nu$	$\sqrt{SE^2(b_1) \cdot (x_\nu - \bar{x})^2 + \dfrac{s_e^2}{n}}$		
		*Prediction Interval for y_ν		y_ν	$\hat{y}_\nu$	$\sqrt{SE^2(b_1) \cdot (x_\nu - \bar{x})^2 + \dfrac{s_e^2}{n} + s_e^2}$		

Assumptions for Inference	And the Conditions That Support or Override Them

Distributions/Association (χ^2)

- **Goodness-of-fit** [df $=$ # of cells -1; one variable, one sample compared with population model]

  1. Data are counts.
  2. Data in sample are independent.
  3. Sample is sufficiently large.

  1. (Are they?)
  2. SRS and $n < 10\%$ of the population.
  3. All expected counts ≥ 5.

- **Homogeneity** [df $= (r - 1)(c - 1)$; samples from many populations compared on one variable]

  1. Data are counts.
  2. Data in groups are independent.
  3. Groups are sufficiently large.

  1. (Are they?)
  2. SRSs and $n < 10\%$ OR random allocation.
  3. All expected counts ≥ 5.

- **Independence** [df $= (r - 1)(c - 1)$; sample from one population classified on two variables]

  1. Data are counts.
  2. Data are independent.
  3. Sample is sufficiently large.

  1. (Are they?)
  2. SRSs and $n < 10\%$ of the population.
  3. All expected counts ≥ 5.

Regression (t, df $= n - 2$)

- **Association** between two quantitative variables ($\beta = 0$?)

  1. Form of relationship is linear.
  2. Errors are independent.
  3. Variability of errors is constant.
  4. Errors have a Normal model.

  1. Scatterplot looks approximately linear.
  2. No apparent pattern in residuals plot.
  3. Residuals plot has consistent spread.
  4. Histogram of residuals is approximately unimodal and symmetric or Normal probability plot reasonably straight.*

(*less critical as n increases)

Review Exercises

1. Genetics Two human traits controlled by a single gene are the ability to roll one's tongue and whether one's ear lobes are free or attached to the neck. Genetic theory says that people will have neither, one, or both of these traits in the ratio 1:3:3:9 (1 attached, noncurling; 3 attached, curling; 3 free, noncurling; 9 free, curling). An Introductory Biology class of 122 students collected the data shown. Are they consistent with the genetic theory? Test an appropriate hypothesis and state your conclusion.

	Trait			
	Attached, noncurling	Attached, curling	Free, noncurling	Free, curling
Count	10	22	31	59

2. Tableware Nambe Mills manufactures plates, bowls, and other tableware made from an alloy of several metals. Each item must go through several steps, including polishing. To better understand the production process and its impact on pricing, the company checked the polishing time (in minutes) and the retail price (in US$) of these items. The regression analysis is shown below. The scatterplot showed a linear pattern, and residuals were deemed suitable for inference.

Dependent variable is: Price
R-squared = 84.5%
s = 20.50 with 59 − 2 = 57 degrees of freedom

Variable	Coefficient	SE(Coeff)
Intercept	−2.89054	5.730
Time	2.49244	0.1416

a) How many different products were included in this analysis?
b) What fraction of the variation in retail price is explained by the polishing time?
c) Create a 95% confidence interval for the slope of this relationship.
d) Interpret your interval in this context.

3. Hard water In an investigation of environmental causes of disease, data were collected on the annual mortality rate (deaths per 100,000) for males in 61 large towns in England and Wales. In addition, the water hardness was recorded as the calcium concentration (parts per million, or ppm) in the drinking water. Here are the scatterplot and regression analysis of the relationship between mortality and calcium concentration.

Dependent variable is: mortality
R-squared = 43%
s = 143.0 with 61 − 2 = 59 degrees of freedom

Variable	Coefficient	SE(Coeff)
Intercept	1676	29.30
calcium	−3.23	0.48

a) Is there an association between the hardness of the water and the mortality rate? Write the appropriate hypothesis.
b) Assuming the assumptions for regression inference are met, what do you conclude?
c) Create a 95% confidence interval for the slope of the true line relating calcium concentration and mortality.
d) Interpret your interval in context.

4. Mutual funds In July 2011, the *Wall Street Journal Online* reported the rate of return for the top 20 large-cap mutual funds over the last 10 years. ("Large cap" refers to companies worth over $10 billion.) Among other results, the *Journal* listed the 3-year and 5-year returns. (online.wsj.com)

a) Create a 95% confidence interval for the difference in rate of return for the 3- and 5-year periods covered by these data. Clearly explain what your interval means.
b) It's common for advertisements to carry the disclaimer "Past returns may not be indicative of future performance," but do these data indicate that there was an association between 3-year and 5-year rates of return?

Annualized Returns (%)		
Fund Name	3-year	5-year
Yact man Focused	18.48	11.63
Yackt man	17.5	10.64
CGM Focus	−19.78	2.3
Fairholme	4.76	5.89
Mass Mutual	10.66	6.87
Amana Trust Income	5.26	7.56
Amana Trust Growth	5.01	7.34
Columbia Strategic	2.37	4.69
Columbia Masico	−0.62	2.54
Marsico 21st Century	−0.87	2.03
Wasatch	1.17	4.47

(continued)

Fund Name	3-year	5-year
Fidelity Contrafund	2.06	5.25
Gabelli Asset	6.04	6.6
Parnassus Equity	4.93	7.2
Balck Rock Equity	1.72	4.46
CGM Mutual	−5.32	4.63
Eaton Vance	−5.46	3.66
Gabelli Equity	4.79	5.17
Auxier Focus	7.18	5.87
Oppenheimer Equity	9.15	5.82

5. Resume fraud In 2002 the Veritas Software company found out that its chief financial officer did not actually have the MBA he had listed on his resume. They fired him, and the value of the company's stock dropped 19%. Kroll, Inc., a firm that specializes in investigating such matters, said that they believe as many as 25% of background checks might reveal false information. How many such random checks would they have to do to estimate the true percentage of people who misrepresent their backgrounds to within ±5% with 98% confidence?

T **6. Paper airplanes** In preparation for a regional paper airplane competition, a student tried out her latest design. The distances her plane traveled (in feet) in 11 trial flights are given here. (The world record is an astounding 193.01 feet!) The data were 62, 52, 68, 23, 34, 45, 27, 42, 83, 56, and 40 feet. Here are some summaries:

Count	11
Mean	48.3636
Median	45
StdDev	18.0846
StdErr	5.45273
IntQRange	25
25th %tile	35.5000
75th %tile	60.5000

a) Construct a 95% confidence interval for the true distance.
b) Based on your confidence interval, is it plausible that the mean distance is 40 ft? Explain.
c) How would a 99% confidence interval for the true distance differ from your answer in part a? Explain briefly, without actually calculating a new interval.

d) How large a sample size would the student need to get a confidence interval half as wide as the one you got in part a, at the same confidence level?

T **7. Back to Montana** The respondents to the Montana poll described in Exercise 33 in Chapter 25 were also classified by income level: low (under $20,000), middle ($20,000–$35,000), or high (over $35,000). Is there any evidence that party enrollment there is associated with income? Test an appropriate hypothesis based on this table, and state your conclusions.

	Democrat	Republican	Independent
Low	30	16	12
Middle	28	24	22
High	26	38	6

T **8. Wild horses** Large herds of wild horses can become a problem on some federal lands in the West. Researchers hoping to improve the management of these herds collected data to see if they could predict the number of foals that would be born based on the size of the current herd. Their attempt to model this herd growth is summarized in the graphs and output shown.

Variable	Count	Mean	StdDev
Adults	38	110.237	71.1809
Foals	38	15.3947	11.9945

Dependent variable is: Foals
R-squared = 83.5%
s = 4.941 with 38 − 2 = 36 degrees of freedom

Variable	Coefficient	SE(Coeff)	t-ratio	P-value
Intercept	−1.57835	1.492	−1.06	0.2970
Adults	0.153969	0.0114	13.5	≤0.0001

a) How many herds of wild horses were studied?
b) Are the conditions necessary for inference satisfied? Explain.
c) Create a 95% confidence interval for the slope of this relationship.
d) Explain in this context what that slope means.
e) Suppose that a new herd with 80 adult horses is located. Estimate, with a 90% prediction interval, the number of foals that may be born.

9. **Lefties and music** In an experiment to see if left- and right-handed people have different abilities in music, subjects heard a tone and were then asked to identify which of several other tones matched the first. Of 76 right-handed subjects, 38 were successful in completing this test, compared with 33 of 53 lefties. Is this strong evidence of a difference in musical abilities based on handedness?

T 10. **AP Statistics scores 2010** In 2010, almost 130,000 Statistics students nationwide took the Advanced Placement Examination in Statistics. The national distribution of scores and the results at Ithaca High School are shown in the table.

		Ithaca High School	
Score	National Distribution	Number of Boys	Number of Girls
5	12.8%	13	13
4	22.4%	21	15
3	23.5%	6	13
2	18.2%	7	3
1	23.1%	4	2

a) Is the distribution of scores at this high school significantly different from the national results?
b) Was there a significant different between the performances of boys and girls at this school?

T 11. **Polling** How accurate are pollsters in predicting the outcomes of congressional elections? The table shows the actual number of Democratic party seats in the House of Representatives and the number predicted by the Gallup organization for nonpresidential election years between World War II and 1998.

Democratic Party Congressmen		
Year	Predicted	Actual
1946	190	188
1950	235	234
1954	232	232
1958	272	283
1962	259	258
1966	247	248
1970	260	255
1974	292	291
1978	277	277
1982	275	269
1986	264	258
1990	260	267
1994	201	204
1998	211	211

a) Is there a significant difference between the number of seats predicted for the Democrats and the number they actually held? Test an appropriate hypothesis and state your conclusions.
b) Is there a strong association between the pollsters' predictions and the outcomes of the elections? Test an appropriate hypothesis and state your conclusions.

T 12. **Twins** In 2000 The *Journal of the American Medical Association* published a study that examined a sample of pregnancies that resulted in the birth of twins. Births were classified as preterm with intervention (induced labor or cesarean), preterm without such procedures, or term or postterm. Researchers also classified the pregnancies by the level of prenatal medical care the mother received (inadequate, adequate, or intensive). The data, from the years 1995–1997, are summarized in the table below. Figures are in thousands of births. (*JAMA* 284 [2000]: 335–341)

		Twin Births, 1995–1997 (in Thousands)			
		Preterm (induced or Cesarean)	Preterm (without procedures)	Term or postterm	Total
Level of Prenatal Care	Intensive	18	15	28	61
	Adequate	46	43	65	154
	Inadequate	12	13	38	63
	Total	76	71	131	278

Is there evidence of an association between the duration of the pregnancy and the level of care received by the mother?

13. Twins, again After reading of the *JAMA* study in Exercise 12, a large city hospital examined their records of twin births for several years and found the data summarized in the table below. Is there evidence that the way the hospital deals with pregnancies involving twins may have changed?

Outcome of Pregnancy	1990	1995	2000
Preterm (induced or cesarean)	11	13	19
Preterm (without procedures)	13	14	18
Term or postterm	27	26	32

14. Preemies Do the effects of being born prematurely linger into adulthood? Researchers examined 242 Cleveland-area children born prematurely between 1977 and 1979, and compared them with 233 children of normal birth weight; 24 of the "preemies" and 12 of the other children were described as being of "subnormal height" as adults. Is this evidence that babies born with a very low birth weight are more likely to be smaller than normal adults? ("Outcomes in Young Adulthood for Very-Low-Birth-Weight Infants," *New England Journal of Medicine*, 346, no. 3 [January 2002])

15. LA rainfall The Los Angeles Almanac website reports recent annual rainfall (in inches), as shown in the table.

Year	Rain (in.)	Year	Rain (in.)
1980	8.96	1991	21.00
1981	10.71	1992	27.36
1982	31.28	1993	8.14
1983	10.43	1994	24.35
1984	12.82	1995	12.46
1985	17.86	1996	12.40
1986	7.66	1997	31.01
1987	12.48	1998	9.09
1988	8.08	1999	11.57
1989	7.35	2000	17.94
1990	11.99	2001	4.42

a) Create a 90% confidence interval for the mean annual rainfall in LA.

b) If you wanted to estimate the mean annual rainfall with a margin of error of only 2 inches, how many years' data would you need?

c) Do these data suggest any change in annual rainfall as time passes? Check for an association between rainfall and year.

16. Age and party 2011 The Pew Research Center conducted a representative telephone survey during 2011. Among the reported results was the following table concerning the preferred political party affiliation of respondents and their ages for white voters. Is there evidence of age-based differences in party affiliation in the United States for white voters?

	Leaning Republican	Leaning Democrat	Other	Total
18–29	274	216	36	526
30–49	888	581	146	1615
50–64	1173	962	211	2346
65+	1062	812	209	2083
Total	3397	2571	602	6570

a) Will you conduct a test of homogeneity or independence? Why?

b) Test an appropriate hypothesis.

c) State your conclusion, including an analysis of differences you find (if any).

17. Birth days During a 2-month period, 72 babies were born at the Tompkins Community Hospital in upstate New York. The table shows how many babies were born on each day of the week.

Day	Births
Mon.	7
Tues.	17
Wed.	8
Thurs.	12
Fri.	9
Sat.	10
Sun.	9

a) If births are uniformly distributed across all days of the week, how many would you expect on each day?

b) Test the hypothesis that babies are equally likely to be born on any of the days of the week.

c) Given the results of part b, do you think that the 7 births on Monday and 17 births on Tuesday indicate that women might be less likely to give birth on Monday, or more likely to give birth on Tuesday?

d) Can you think of any reasons why births may not occur completely at random?

18. Wealth distribution Based on their responses to a June 2011 Gallup Poll, Americans were classified as high income (over $75,000), middle income ($30k–$75k), or low income (less than $30k). Those polled were asked for their views on redistributing U.S. wealth by heavily taxing the rich. The data are summarized in the table on the next page.

	Should Redistribute Wealth	Should Not	No Opinion
High Income	170	371	9
Middle Income	306	282	12
Low Income	362	184	29

Is there any evidence that income level is associated with feelings toward the wealth distribution in the United States? Test an appropriate hypothesis about this table, and state your conclusions.

T **19. Eye and hair color** A survey of 1021 school-age children was conducted by randomly selecting children from several large urban elementary schools. Two of the questions concerned eye and hair color. In the survey, the following codes were used:

Hair Color	Eye Color
1 = Blond	1 = Blue
2 = Brown	2 = Green
3 = Black	3 = Brown
4 = Red	4 = Grey
5 = Other	5 = Other

The Statistics students analyzing the data were asked to study the relationship between eye and hair color.

a) One group of students produced the output shown below. What kind of analysis is this? What are the null and alternative hypotheses? Is the analysis appropriate? If so, summarize the findings, being sure to include any assumptions you've made and/or limitations to the analysis. If it's not an appropriate analysis, state explicitly why not.

Dependent variable is: Eyes
R-squared = 3.7%
s = 1.112 with 1021 − 2 = 1019 degrees of freedom

Variable	Coefficient	SE(Coeff)	t-ratio	P-value
Intercept	1.99541	0.08346	23.9	≤0.0001
Hair	0.211809	0.03372	0.28	≤0.0001

b) A second group of students used the same data to produce the output shown below. The table displays counts and standardized residuals in each cell. What kind of analysis is this? What are the null and

alternative hypotheses? Is the analysis appropriate? If so, summarize the findings, being sure to include any assumptions you've made and/or limitations to the analysis. If it's not an appropriate analysis, state explicitly why not.

	Eye Color				
Hair Color	**1**	**2**	**3**	**4**	**5**
1	143	30	58	15	12
	7.67540	0.41799	−5.88169	−0.63925	−0.31451
2	90	45	215	30	20
	−2.57141	0.29019	1.72235	0.49189	−0.08246
3	28	15	190	10	10
	−5.39425	−2.34780	6.28154	−1.76376	−0.80382
4	30	15	10	10	5
	2.06116	2.71589	−4.05540	2.37402	0.75993
5	10	5	15	5	5
	−0.52195	0.33262	−0.94192	1.36326	2.07578

$$\sum \frac{(Observed - Expected)^2}{Expected} = 223.6 \quad \text{P-value} < 0.00001$$

20. Depression and the Internet The September 1998 issue of the *American Psychologist* published an article reporting on an experiment examining "the social and psychological impact of the Internet on 169 people in 73 households during their first 1 to 2 years online." In the experiment, a sample of households was offered free Internet access for one or two years in return for allowing their time and activity online to be tracked. The members of the households who participated in the study were also given a battery of tests at the beginning and again at the end of the study. One of the tests measured the subjects' levels of depression on a 4-point scale, with higher numbers meaning the person was more depressed. Internet usage was measured in average number of hours per week. The regression analysis examines the association between the subjects' depression levels and the amounts of Internet use. The conditions for inference were satisfied.

Dependent variable is: Depression After
R-squared = 4.6%
s = 0.4563 with 162 − 2 = 160 degrees of freedom

Variable	Coefficient	SE(coeff)	t-ratio	Prob
Constant	0.565485	0.0399	14.2	≤0.0001
Intr_use	0.019948	0.0072	2.76	0.0064

a) Do these data indicate that there is an association between Internet use and depression? Test an appropriate hypothesis and state your conclusion clearly.
b) One conclusion of the study was that those who spent more time online tended to be more depressed at the end of the experiment. News headlines said that too much time on the Internet can lead to depression. Does the study support this conclusion? Explain.

c) As noted, the subjects' depression levels were tested at both the beginning and the end of this study; higher scores indicated the person was more depressed. Results are summarized in the table. Is there evidence that the depression level of the subjects changed during this study?

Depression Level
162 subjects

Variable	Mean	StdDev
DeprBfore	0.730370	0.487817
DeprAfter	0.611914	0.461932
Difference	−0.118457	0.552417

21. Pregnancy In 1998 a San Diego reproductive clinic reported 42 live births to 157 women under the age of 38, but only 7 successes for 89 clients aged 38 and older. Is this evidence of a difference in the effectiveness of the clinic's methods for older women?

a) Test the appropriate hypotheses, using the two-proportion z-procedure.
b) Repeat the analysis, using an appropriate chi-square procedure.
c) Explain how the two results are equivalent.

22. Eating in front of the TV Roper Reports asked a random sample of people in 30 countries whether they agreed with the statement "I like to nibble while reading or watching TV." Allowable responses were "Agree completely", "Agree somewhat", "Neither disagree nor agree", "Disagree somewhat", "Disagree completely", and "I Don't Know/No Response." Does a person's age influence their response? The table summarizes data from 3792 respondents in the 2006 sample of five countries (China, India, France, United Kingdom, and United States) for three age groups (Teens, 30's (30–39) and Over 60):

	Agree Completely	Agree Somewhat	Neither Disagree Nor Agree	Disagree Somewhat	Disagree Completely
Teen	369	540	299	175	106
30's	272	522	325	229	170
60+	93	207	153	154	178

a) Make an appropriate display of these data.
b) Does a person's age seem to affect their response to the question about nibbling?

T 23. Old Faithful As you saw in an earlier chapter, Old Faithful isn't all that faithful. Eruptions do not occur at uniform intervals and may vary greatly. Can we improve our chances of predicting the time of the next eruption if we know how long the previous eruption lasted?

a) Describe what you see in this scatterplot.

b) Write an appropriate hypothesis.
c) Here are a histogram of the residuals and the residuals plot. Do you think the assumptions for inference are met? Explain.

d) State a conclusion based on this regression analysis:

Dependent variable is: Interval
R-squared = 77.0%
s = 6.159 with 222 − 2 = 220 degrees of freedom

Variable	Coefficient	SE(Coeff)	t-ratio	P-value
Intercept	33.9668	1.428	23.8	≤0.0001
Duration	10.3582	0.3822	27.1	≤0.0001

Variable	Mean	StdDev
Duration	3.57613	1.08395
Interval	71.0090	12.7992

e) The second table shows the summary statistics for the two variables. Create a 95% confidence interval for the mean length of time that will elapse following a 2-minute eruption.
f) You arrive at Old Faithful just as an eruption ends. Witnesses say it lasted 4 minutes. Create a 95% prediction interval for the length of time you will wait to see the next eruption.

24. Togetherness Are good grades in high school associated with family togetherness? A simple random sample of 142 high-school students was asked how many meals per week their families ate together. Their responses produced a mean of 3.78 meals per week, with a standard deviation of 2.2. Researchers then matched these responses against the students' grade point averages. The scatterplot appeared to be reasonably linear, so they went ahead with the regression analysis, seen below. No apparent pattern emerged in the residuals plot.

Dependent variable: GPA
R-squared = 11.0%
s = 0.6682 with 142 − 2 = 140 df

Variable	Coefficient	SE(Coeff)
Intercept	2.7288	0.1148
Meals/wk	0.1093	0.0263

a) Is there evidence of an association? Test an appropriate hypothesis and state your conclusion.
b) Do you think this association would be useful in predicting a student's grade point average? Explain.
c) Are your answers to parts a and b contradictory? Explain.

25. **Learning math** Developers of a new math curriculum called "Accelerated Math" compared performances of students taught by their system with control groups of students in the same schools who were taught using traditional instructional methods and materials. Statistics about pretest and posttest scores are shown in the table. (J. Ysseldyke and S. Tardrew, *Differentiating Math Instruction*, Renaissance Learning, 2002)

a) Did the groups differ in average math score at the start of this study?
b) Did the group taught using the Accelerated Math program show a significant improvement in test scores?
c) Did the control group show a significant improvement in test scores?
d) Were gains significantly higher for the Accelerated Math group than for the control group?

		Instructional Method	
		Acc. math	Control
Number of students		231	245
Pretest	Mean	560.01	549.65
	St. Dev	84.29	74.68
Post-test	Mean	637.55	588.76
	St. Dev	82.9	83.24
Individual gain	Mean	77.53	39.11
	St. Dev.	78.01	66.25

26. **Pesticides** A study published in 2002 in the journal *Environmental Health Perspectives* examined the gender ratios of children born to workers exposed to dioxin in Russian pesticide factories. The data covered the years from 1961 to 1988 in the city of Ufa, Bashkortostan, Russia. Of 227 children born to workers exposed to dioxin, only 40% were male. Overall in the city of Ufa,

the proportion of males was 51.2%. Is this evidence that human exposure to dioxin may result in the birth of more girls? (An interesting note: It appeared that paternal exposure was most critical; 51% of babies born to mothers exposed to the chemical were boys.)

27. **Dairy sales** Peninsula Creameries sells both cottage cheese and ice cream. The CEO recently noticed that in months when the company sells more cottage cheese, it seems to sell more ice cream as well. Two of his aides were assigned to test whether this is true or not. The first aide's plot and analysis of sales data for the past 12 months (in millions of pounds for cottage cheese and for ice cream) appear below.

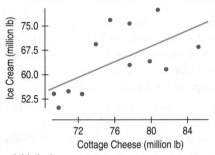

Dependent variable is: Ice cream
R-squared = 36.9%
s = 8.320 with 12 − 2 = 10 degrees of freedom

Variable	Coefficient	SE(Coeff)	t-ratio	P-value
Constant	−26.5306	37.68	−0.704	0.4975
Cottage C . . .	1.19334	0.4936	2.42	0.0362

The other aide looked at the differences in sales of ice cream and cottage cheese for each month and created the following output:

Cottage Cheese − Ice Cream

Count	12
Mean	11.8000
Median	15.3500
StdDev	7.99386
IntQRange	14.3000
25th %tile	3.20000
75th %tile	17.5000

Test H_0: $\mu(CC − IC) = 0$ vs H_a: $\mu(CC − IC) \neq 0$
Sample Mean = 11.800000 t-Statistic = 5.113 w/11 df
Prob = 0.0003
Lower 95% bound = 6.7209429
Upper 95% bound = 16.879057

a) Which analysis would you use to answer the CEO's question? Why?

b) What would you tell the CEO?

c) Which analysis would you use to test whether the company sells more cottage cheese or ice cream in a typical year? Why?

d) What would you tell the CEO about this other result?

e) What assumptions are you making in the analysis you chose in part a? What assumptions are you making in the analysis in part c?

f) Next month's cottage cheese sales are 82 million pounds. Ice cream sales are not yet available. How much ice cream do you predict Peninsula Creameries will sell?

g) Give a 95% confidence interval for the true slope of the regression equation of ice cream sales by cottage cheese sales.

h) Explain what your interval means.

28. **Infliximab** In an article appearing in the journal *The Lancet* in 2002, medical researchers reported on the experimental use of the arthritis drug infliximab in treating Crohn's disease. In a trial, 573 patients were given initial 5-mg injections of the drug. Two weeks later, 335 had responded positively. These patients were then randomly assigned to three groups. Group I received continued injections of a placebo, Group II continued with 5 mg of infliximab, and Group III received 10 mg of the drug. After 30 weeks, 23 of 110 Group I patients were in remission, compared with 44 of 113 Group II and 50 of 112 Group III patients. Do these data indicate that continued treatment with infliximab is of value for Crohn's disease patients who exhibit a positive initial response to the drug?

Ⓣ 29. **Weight loss** A weight loss clinic advertises that its program of diet and exercise will allow clients to lose 10 pounds in one month. A local reporter investigating weight reduction gets permission to interview a randomly selected sample of clients who report the given weight losses during their first month in this program. Create a confidence interval to test the clinic's claim that the typical weight loss is 10 pounds.

Pounds Lost	
9.5	9.5
13	9
9	8
10	7.5
11	10
9	7
5	8
9	10.5
12.5	10.5
6	9

Ⓣ 30. **Education vs. income** The following displays examine the median income and education level (years in school) for several U.S. cities.

Variable	Count	Mean	StdDev
Education	57	10.9509	0.848344
Income	57	32742.6	3618.01

Dependent variable is: Income
R-squared = 32.9%
s = 2991 with 57 − 2 = 55 degrees of freedom

Variable	Coefficient	SE(Coeff)	t-ratio	P-value
Intercept	5970.05	5175	1.15	0.2537
Education	2444.79	471.2	5.19	≤0.0001

a) Do you think the assumptions for inference are met? Explain.

b) Does there appear to be an association between education and income levels in these cities?

c) Would this association appear to be weaker, stronger, or the same if data were plotted for individual people rather than for cities in aggregate? Explain.

d) Create and interpret a 95% confidence interval for the slope of the true line that describes the association between income and education.

e) Predict the median income for cities where residents spent an average of 11 years in school. Describe your estimate with a 90% confidence interval, and interpret that result.

Ⓣ 31. **Diet** Thirteen overweight women volunteered for a study to determine whether eating specially prepared crackers before a meal could help them lose weight. The subjects were randomly assigned to eat crackers with different types of fiber (bran fiber, gum fiber, both, and a control cracker). Unfortunately, some of the women developed uncomfortable bloating and upset stomachs. Researchers suspected that some of the crackers might be at fault. The contingency table of "Cracker" versus "Bloat" shows the relationship between the four different types of crackers and the reported bloating. The study was paid for by the manufacturers of the gum fiber. What would you recommend to them about the prospects for marketing their new diet cracker?

		Bloat	
		Little/None	Moderate/Severe
Cracker	Bran	11	2
	Gum	4	9
	Combo	7	6
	Control	8	4

T **32. Cramming** Students in two basic Spanish classes were required to learn 50 new vocabulary words. One group of 45 students received the list on Monday and studied the words all week. Statistics summarizing this group's scores on Friday's quiz are given. The other group of 25 students did not get the vocabulary list until Thursday. They also took the quiz on Friday, after "cramming" Thursday night. Then, when they returned to class the following Monday, they were retested—without advance warning. Both sets of test scores for these students are shown.

Group 1

Fri.
Number of students = 45
Mean = 43.2 (of 50)
StDev = 3.4
Students passing (score $\geq$ 40) = 33%

Group 2

Fri.	Mon.	Fri.	Mon.
42	36	50	47
44	44	34	34
45	46	38	31
48	38	43	40
44	40	39	41
43	38	46	32
41	37	37	36
35	31	40	31
43	32	41	32
48	37	48	39
43	41	37	31
45	32	36	41
47	44		

a) On Friday, did the week-long study group have a mean score significantly higher than that of the overnight crammers?
b) Was there a significant difference in the percentages of students who passed the quiz on Friday?
c) Is there any evidence that when students cram for a test, their "learning" does not last for 3 days?
d) Use a 95% confidence interval to estimate the mean number of words that might be forgotten by crammers.
e) Is there any evidence that how much students forget depends on how much they "learned" to begin with?

Practice Exam

Multiple Choice

1. To make a decision about closing time, a fast food restaurant has recorded data on the number of customers using their drive up window between 10:00 and 11:00 PM on Sunday through Thursday. A summary is shown in the table.

Number of customers	Relative frequency
1	0.08
2	0.10
3	0.14
4	0.12
5	0.15
6	0.10
7	0.20
8	0.05
9	0.06

What is the inter-quartile range for these data?

A) 3   B) 4   C) 5   D) 6   E) 7

2. Factory supervisors conducting a quality control study sampled twenty production batches and recorded the number of items in each sample that failed to meet standards. The results are summarized in the cumulative relative frequency histogram below.

What is the probability that a sample had exactly five items that didn't meet standards?

A) 0.167   B) 0.15   C) 0.20
D) 0.65   E) 0.85

3. In a symmetric distribution, which of the following should be true?

I.   Maximum value − Median ≈ Median − Minimum value
II. Q3 − Median ≈ Median − Q1
III. Maximum value − Q3 ≈ Q1 − Minimum value

A) I only   B) II only   C) III only
D) II and III only   E) I, II, and III

4. Logs to be sawed into boards are first rough-cut to be a bit over 8 feet long. Measurements found the actual mean length to be 98.5 inches and 20% of the logs were longer than 100 inches. Assuming the log lengths to be normally distributed, approximately what percent of such logs should be between 97 and 100 inches long?

A) 20%   B) 40%   C) 60%
D) 68%   E) 80%

5. For a Statistics class project at a large high school, a senior asked a random sample of students how often they use social networking sites. The relative frequencies of responses by male and female students are displayed in the bar chart below.

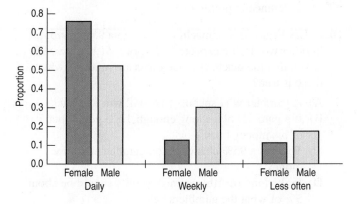

Which of the following statements can be supported by the display?

A) More females use social networking sites than males.
B) In general, females use the sites more often than males.
C) Approximately the same number of males and females were surveyed.
D) About 43% of all students surveyed said they use social networking sites weekly.
E) There appears to be no association between gender and frequency of use of social networking sites.

6. A zoologist studying a certain species of rats investigated the relationship between the weights of adult rats (in grams) and the lengths of their tails (in centimeters). The regression analysis for his data is shown below.

Response variable: weight

Variable	Coef	Std Error	t-ratio	p-value
Constant	−323.5	3.89	6.04	0.0001
Length	72.1	0.24	8.79	0.0001

$s = 14.31$ R-Sq $= 75.4\%$

By how much do weight estimates produced by this model typically differ from the actual weights of the rats?

A) 0.24 g   B) 3.89 g   C) 14.31 g
D) 323.5 g   E) 24.6%

7. A researcher discovered a data error that led her to eliminate the outlier seen in the lower left of her original scatterplot below.

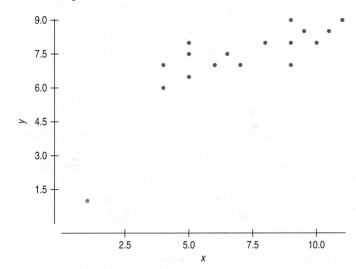

What effect did removing that point have on the slope of the regression line and the value of R^2 she had previously calculated?

A) The slope decreased and R^2 increased.
B) The slope increased and R^2 decreased.
C) The slope decreased, but R^2 remained the same.
D) Both the slope and R^2 increased.
E) Both the slope and R^2 decreased.

8. An analysis of laboratory data collected with the goal of modeling the weight (in grams) of a bacterial culture after several hours of growth produced the least squares regression line $\overline{\log(weight)} = 0.25 + 0.61 hours$. Estimate the weight of the culture after 3 hours.

A) 0.32 g   B) 2.08 g   C) 8.0 g
D) 67.9 g   E) 120.2 g

9. Which is the appropriate interpretation of the slope of the bacterial growth model in Question 8 above?

A) For every additional 0.61 hours of growth, the culture is predicted to weigh 1 more gram.
B) For every additional hour of growth, the culture is predicted to weigh 0.61 more grams.
C) For every additional hour of growth, the logarithm of the culture's weight in grams is predicted to increase by 0.61.

D) At the beginning of the experiment, the logarithm of the culture's weight in grams was approximately 0.61 ounces.

E) 61% of variability in weight is explained by the number of hours the culture has been growing.

10. The price of video games often starts high when initial demand is high and then decreases as time goes on. A scatterplot of data collected for a sample of games suggests a linear model is reasonable, and an analysis produced a correlation coefficient of -0.78. What percent of the variation in price remains unexplained by this linear model?

A) 4.8%   B) 22%   C) 39.2%
D) 60.8%   E) 78%

11. A school district will use grade 1-5 students at one elementary school to study whether a new reading program can produce better reading comprehension scores. Half of the students will continue with the current program, and half will receive instruction with the new program. Which is the most appropriate design?

A) Completely randomized, randomly assigning students to reading programs.
B) Randomized blocks, blocked by reading program.
C) Randomized blocks, blocked by gender.
D) Randomized blocks, blocked by grade level.
E) Randomized blocks, blocked by both grade level and gender.

12. Researchers are conducting an experiment at a zoo to compare the effect of two food supplements on weight gain among meerkats. The zoo has a large population of meerkats that live in 8 groups, each with separate living areas. Researchers will randomly assigned four of the living areas to receive each food supplement. What are the experimental units in this study?

A) the two supplements
B) the individual meerkats
C) the 8 groups of meerkats
D) the weight gain of each meerkat
E) the mean weight gain for each of the 8 groups

13. One advantage of using a control group with a placebo in an experiment is that the experiment will be

A) better because of the larger number of treatment groups.
B) better able to account for variability arising from extraneous factors.
C) better able to avoid bias.
D) better able to establish which factor is cause and which is effect.
E) better able detect changes in the response variable.

14. Some pet shop owners will soon begin selling a new variety of turtle, but first they want to determine the best environment for raising them. They want to try three aquarium temperatures and two different mixtures of food. They will randomly assign the different combinations of temperature and food to 18 aquariums, each of which holds 4 turtles. Which of the following statements is correct?

A) This experiment is poorly designed because there is no control group.
B) This experiment is poorly designed because there is no replication.
C) The factor is temperature and the treatments are the 2 food mixtures.
D) The 5 treatments are the 2 food mixtures and the 3 temperatures.
E) The factors are food and temperature, and there are 6 treatments.

15. A consumer group doing marketing research on consumer preferences for soft drinks will conduct a taste test comparing cola C and cola P. Which of the following is not a good suggestion for an effective study?

A) Include people of different ages, genders, and other characteristics.
B) Randomly select which cola is served first.
C) Serve the colas in unmarked cups, rather than cans or bottles with a label.
D) Ask a specific question, such as "Did you prefer cola C?"
E) Use statistical analysis to check for a significant difference in preference.

16. In Las Vegas $1 slot machines average a 95% payout; in other words, the expected value of a $1 bet is $0.95. Given that the machines' outcomes are random, which of these is true?

A) A gambler who spends $100 will win $95 back.
B) If a gambler plays long enough, he'll get all but 5% of his money back.
C) There is a 95% chance that a gambler will lose money.
D) In the long run, the casinos' profits should be about 5% of what the gamblers bet.
E) A gambler who has lost many times in a row is more likely to win on the next bet.

17. In a large university, 56% of the undergraduate students are female, as are 42% of the graduate students. How many women would we expect there to be in a random sample of 50 undergraduate and 30 graduate students?

A) 18.8   B) 19.6   C) 37.8
D) 39.2   E) 40.6

18. What's the standard deviation of the number of women found in samples of college students like the one described in Question 17 above?

A) 2.49   B) 3.16   C) 4.43
D) 4.47   E) 6.21

19. The table shows the probability distribution for the number of points (X) a player can win by spinning a child's game spinner. What is the variance of X?

X	1	2	3	4	5
P(X)	0.4	0.2	0.2	0.1	0.1

A) 1.35   B) 1.81   C) 2.00
D) 2.30   E) 2.50

20. Among the seniors at a certain high school, 65% attended the prom and 45% went on the senior trip, but 25% of the seniors did not participate in either the prom or the trip. What's the probability that a senior who went on the trip attended the prom?

A) 0.20   B) 0.35   C) 0.54
D) 0.69   E) 0.78

21. The salmon filets a seafood shop from a supplier weigh an average of 5.6 ounces, but 12% of them weigh less than 4 ounces. The shop owner believes the weights are normally distributed. What's the standard deviation of the filet weights?

A) 0.73 oz.   B) 1.03 oz.   C) 1.17 oz.
D) 1.36 oz.   E) 3.40 oz.

22. The PSA is a screening test for prostate cancer often recommended for men over 50. Unfortunately, it's not very reliable. Further testing reveals that only 30% of the men whose PSA comes back positive actually do have prostate cancer. Suppose a lab processing several PSAs reports 8 positive results one day. What's the probability that no more than 2 of those men really have prostate cancer?

A) 0.093   B) 0.296   C) 0.379
D) 0.552   E) 0.774

23. A citrus grower offers gourmet gift boxes of 6 Ruby Red grapefruits. The grapefruits used in these boxes weigh an average of 18 ounces with a standard deviation of 1.2 ounces. Assuming the weights are approximately normally distributed, what is the probability that the average weight of the Ruby Reds in one of these boxes is between 17.5 and 18.5 ounces?

A) 0.323   B) 0.383   C) 0.693
D) 0.986   E) 0.997

24. Which of the following statements is true?

A) A P-value of 0.011 is weaker evidence against the null hypothesis than a P-value of 0.033.
B) If we reject the null hypothesis at $\alpha = 0.05$, we'd also reject it at $\alpha = 0.01$.
C) We can increase the power of a test by decreasing the significance level from 5% to 1%.
D) We can decrease the risk of a Type I error by increasing the significance level from 1% to 5%.
E) The farther the true parameter is from the value we hypothesized, the lower the risk of Type II error.

25. In June 2012 the Gallup organization conducted a poll using a random sample of 1004 adults, aged 18 and older, living in all 50 U.S. states and the District of Columbia. When asked, "On the whole, do you think immigration is a good thing or a bad thing for this country?" 66% responded "a good thing." The news report adds, "For results based on the total sample of national adults, one can say with 95% confidence that the maximum margin of sampling error is ± 4 percentage points."
Source: http://www.gallup.com/poll/155210/Americans-Positive-Immigration.aspx
Based on this report, we can conclude that:

A) The percentage of all adult Americans who think that immigration is a good thing for this country is between 62% and 70%.
B) There's a 5% chance that the proportion of all adults who think immigration is a good thing is not between 62% and 70%
C) In only 5% of all samples like this would the results differ by more than 4% from the proportion of all adults who think immigration is a good thing.
D) In 95% of all samples like this between 62% and 70% of the respondents would say immigration is a good thing.
E) Between 62% and 70% of Americans think immigration is a good thing for 95% of all immigrants.

26. How many babies must be randomly selected and tested in order to estimate the rate of jaundice among newborns to within $\pm 3\%$ with 92% confidence?

A) 251   B) 549   C) 752
D) 851   E) 1068

27. To a study the effectiveness of an interactive e-book, a researcher gave 20 students a pre-test, had them study from the e-book, and then gave them a post-test. The table below summarizes the data she collected.

	Pre-test	Post-test	Post-test – pre-test
Mean	78.9	81.4	2.5
Standard deviation	6.1	5.5	8.5

The researcher conducted a hypothesis test to see if using the e book would produce a significant increase in mean score. What P-value did she find?

A) 0.028   B) 0.087   C) 0.091
D) 0.094   E) 0.102

28. Maria has data from a random sample of 16 subjects and is constructing a 95% confidence interval for the population mean. Which value should she use for t^*?

A) 1.746   B) 1.753   C) 1.960
D) 2.120   E) 2.131

29. Could listening to rock music raise your blood pressure? To find out, researchers randomly divided 50 subjects

into two groups. For 10 minutes, people in one group listened to loud rock music and those in the other group listened to soft jazz. The researchers recorded each person's change in blood pressure, and plan to run a hypothesis test to see if there's a significant difference between the mean changes for the two groups. Which conditions must they check before proceeding?

I.   The subjects are a random sample of the population.
II. The two groups are independent.
III. The blood pressure changes for each group are approximately Normal.

A) III only   B) I and II only   C) I and III only
D) II and III only   E) I, II, and III

30. A supplier tells a large hospital stockroom that the facility should buy 30% small, 45% medium and 25% large surgical gloves. When the stockroom manager randomly surveys 100 workers about their size preferences, 23 request small, 57 medium, and 30 large gloves. The manager wonders this constitutes evidence that the supplier's recommendation won't meet the hospital's needs? What test should she run?

A) chi-square goodness of fit test
B) chi-square test of homogeneity
C) 1-sample *t*-test
D) 1-proportion *z*-test
E) 2-proportion *z*-test

31. An investigation of water plant growth compares the heights (in inches) of a species of plant growing in two different lakes. The null hypothesis is H$_0$: $\mu_1 = \mu_2$. Here are the summaries of the data collected:

	Mean	StdDev	n
Lake 1	8.6	3.8	25
Lake 2	7.6	2.3	25

What are the mean and standard error of the sampling distribution for the difference in sample means to be used by the hypothesis test?

A) 0 and 0.888   B) 0 and 1.5   C) 0 and 4.44
D) 1 and 0.888   E) 1 and 4.44

32. A study conducted in 2007 by The Baylor Department Sociology asked the following question: "Please indicate your level of agreement with the following statement about science: Humans evolved from other primates over millions of years." The researcher summarized the data in a table categorizing the responses by level of *Belief in Evolution* and the *Age* group of the respondent. The P-value for a chi-square test of independence based was 0.025. Which statement correctly interprets this P-value?

A) The probability that the variables *Age* and *Belief in Evolution* are independent is 0.025.
B) The probability that the variables *Age* and *Belief in Evolution* are independent is 0.975.

C) Given the data that were observed, the probability that the variables *Age* and *Belief in Evolution* are independent is 0.025.
D) If *Belief in Evolution* is independent of *Age,* the probability of seeing results at least this extreme is only 0.025.
E) If there is an association between *Belief in Evolution* and *Age,* the probability of seeing results at least this extreme is only 0.025.

33. The Pew Research Center asked a random sample of 1,047 adult (aged 18+) social media users and 623 teen (aged 12–17) social media users the following question: "How often do you witness online cruelty and meanness?" The table below summarizes the responses Pew obtained.

Witness cruelty and meanness	Teens (12–17)	Adults (18+)	Total
Frequently	75	73	148
Sometimes	181	188	369
Only once in a while	293	461	754
Never	68	304	372
Don't know	6	21	27
Total	623	1047	1670

To test for differences in the two age group distributions, what is the expected count for the cell that counts adults who reply "Sometimes"?

A) 167.0   B) 184.5   C) 188.0
D) 209.4   E) 231.34

34. In 2003 the U.S. Department of Education conducted a National Assessment of Adult Literacy by interviewing a random sample of 18,102 adults (aged 16+). Researchers assessed the respondents' literacy levels and also asked them how often they engaged in volunteer activities. The data are summarized in the table below.

		Literacy Level			
		Below Basic	Basic	Intermediate	Proficient
Volunteer activity	Once per week or more	263	818	1654	611
	Less than once per week	184	872	2068	782
	Never	2185	3762	4550	1051

Source: U.S. Department of Education, National Center for Education Statistics.

If we wish to test whether Volunteer Activity and Literacy Level are independent, how many degrees of freedom are there?

A) 4   B) 6   C) 11   D) 12   E) 18,801

35. Soda companies have begun marketing 10-calorie sodas. The companies believe men prefer the taste and name of

these drinks over traditional diet sodas. The table below summarizes preferences expressed by a random sample of 50 men and 50 women and also shows expected cell counts in parentheses.

	Diet	10-calorie
Men	12 (26.5)	38 (23.5)
Women	41 (26.5)	9 (23.5)

In calculating the χ^2 test statistic, what value is contributed by the cell for men who prefer 10-calorie drinks?

A) 0.62   B) 5.53   C) 8.95
D) 14.50   E) 210.25

36. Researchers are investigating the association between the temperature in degrees and sales sales of hot dogs (in hundreds) at an outdoor sports stadium for a random sample of days. They constructed a 95% confidence interval for the slope of the regression line for predicting *Sales* from *Temperature*, obtaining $(-0.24, 1.68)$. Which statement is correct?

A) There is not a significant association because the interval contains 0.
B) The association is probably positive, because most of the interval is above 0.
C) The association must be positive, because the interval contains 1.
D) The researchers cannot make any conclusion, because the margin of error is too large.
E) The researchers made an error, because the upper endpoint is greater than 1.

37. A company selling home furnaces claims that installing their new high efficiency furnace will "cut your heating bill in half." A consumer agency collects data for a random sample of homes and plots this year's heating costs with the new furnaces against previous costs with the old furnaces. Which are the most appropriate hypotheses to test the company's claim?

A) $H_0: \beta = 0, H_A: \beta \neq 0$
B) $H_0: \beta = 0, H_A: \beta > 0$
C) $H_0: \beta = 0, H_A: \beta < 0$
D) $H_0: \beta = 0, H_A: \beta < 0.5$
E) $H_0: \beta = 0.5, H_A: \beta \neq 0.5$

38. The regression analysis below models the relationship between the fuel efficiency (in miles per gallon) and horsepower (in 100s) for a random sample of 15 cars.

Dependent variable is: MPG
R squared = 82.6%   R squared (adjusted) = 81.3%
s = 2.435 with 15 − 2 = 13 degrees of freedom

Variable	Coefficient	s.e. of Coeff	t-ratio	prob
Constant	43.4518	2.057	21.1	≤ 0.0001
HP100	−7.0166	0.89	−7.86	≤ 0.0001

The highlighted value of 0.89 estimates the variability in

A) horsepower for this sample of cars.
B) fuel economy for this sample of cars.
C) slopes among this sample of cars.
D) slopes among all samples from this population of cars.
E) errors for predictions made by this model.

39. Which sample result would provide evidence that there is an association between two quantitative variables based on a hypothesis test for the true population slope β?

A) Any sample with $b_1 = 0$.
B) Any sample with $b_1 \neq 0$.
C) Any sample with $b_1 > 0$.
D) Any sample with b_1 significantly close to 0.
E) Any sample with b_1 significantly different from 0.

40. Before conducting inference for regression slope we need to check several assumptions by looking at the residuals. Which of these is important?

I.   The residual plot should not have observable patterns.
II. The residuals should have a fairly consistent vertical spread.
III. The distribution of residuals should be approximately normal.

A) III only   B) I and II only
C) I and III only   D) II and III only
E) I, II, and III

Free Response

1. Origami USA hosts a conference in New York City every summer. Classes are offered on a beginner, intermediate, and advanced levels of folding. Simple paper models use only a few folds, while the more complex models require more folds. The boxplot and statistics below summarize the number of folds required for each of the models taught in a random sample of 15 classes.

Variable	Folds
N	15
Mean	72
StDev	24.07429
Median	78
Range	88
Min	15
Max	103
Q1	61
Q3	90

a) Verify that minimum number of folds is an outlier. Show work and explain your result.
b) Do these data indicate that the conference is best suited for beginners or for more advanced folders? Use the graph and the summary statistics to justify your answer.

c) Would it be appropriate to use these data to create a confidence interval for the number of folds in all the models taught in this conference's classes? Check the condition(s) for inference describe any concern(s) you have about creating such an interval. (Do not find the confidence interval.)

2. A small town has a Farmer's Market each Thursday night in a downtown park. Local growers, as well as craft makers and food vendors, set up booths for three hours. The Chamber of Commerce believes that the Market helps other downtown businesses because it attracts more visitors to the area. However, the Chamber has recently received complaints from some of the downtown businesses that the Farmer's Market hurts their normal sales on Thursdays. The Chamber wishes to survey more merchants in the area to determine the impact of the Market on their businesses.

a) The Chamber of Commerce first considers inviting local merchants to fill out a poll on the Chamber's website. Describe a problem you see with this method.

b) The downtown business district encompasses stores along 20 blocks of street frontage. The Chamber can improve its design by interviewing merchants in of 3 blocks. Describe the procedure for this method.

c) The park where the Farmer's Market is located is in one corner of the downtown business district. The Chamber might also consider a sample of businesses stratified by distance from the Market. Explain an advantage such a stratified sample may have over the cluster sampling method.

d) Identify another way to stratify the downtown businesses, and explain why the Chamber should consider doing that.

3. Jane's Java serves both coffee and tea, each in a small or large size. History shows that 70% of the customers who order just one drink order coffee. Of those customers, 80% order a large; 60% of the customers who order tea order a large.

a) Compute the probability that a random customer who orders one drink will get a small tea.

b) Compute the probability that a random customer who orders one large drink gets coffee.

c) Jane's prices are $3.00 for large coffee, $2.50 for large tea, $2.00 for small coffee, and $1.50 for small tea. What is the expected price for a drink sold by Jane?

4. A study published in 2005 examined health and diet among a representative sample of Canadian women. Researchers reported that only 6.8% of women who were

vegetarians had high cholesterol, compared to 11.3% of non-vegetarians.

a) Explain two reasons why this study would not justify a newspaper headline proclaiming, "Want to Lower Your Cholesterol? Become a Vegetarian!" Include an alternative explanation for the reported difference.

b) If these results were based on data collected from 162 vegetarians and 177 non-vegetarians, do they provide statistical evidence that there's a difference in the rates of high cholesterol in the two populations?

5. Some people claim that students learn better from teachers that "look like themselves." Do public and private schools hire different racial compositions of teachers? The table below shows the distributions of race among 3894 people hired one year for to teach in a large city.

| | | Type of School | |
		Public	Private
Race	White	2829	423
	Black	239	20
	Hispanic/Latino	240	29
	Other	96	18
	Total	3404	490

Identify any evidence of differences in the distribution of race among teachers in public schools versus private schools.

6. Although the event may seem startling to their students, 119 statistics teachers participated in a fun run. Before the run, each person was asked to guess what his/her time would be (in seconds). After the run, the actual times were plotted against the teachers' guesses, and the assumptions for inference appeared to be reasonable. Here is the regression output:

Dependent Variable: actualtime
R-squared = 75.01%
s = 278.1940

Variable	Coefficient	Se(coeff)	t-ratio	P-Value
Intercept	156.3777	79.3023	1.9719	0.051
guess	0.887006	0.04734	18.739	<0.0001

a) Does this output provide evidence of an association between guessed time and actual time? Explain.

b) Assuming these 119 teachers represent a representative sample from a larger population who might participate in such a run, construct and interpret a 95% confidence interval for the slope of the regression line for that population.

c) Does the interval you constructed in part b) support your conclusion in part a)? Explain.

Appendix A: Selected Formulas

$Range = Max - Min$

$IQR = Q3 - Q1$

Outlier Rule-of-Thumb: $y < Q1 - 1.5 \times IQR$ or $y > Q3 + 1.5 \times IQR$

$$\bar{y} = \frac{\sum y}{n}$$

$$s = \sqrt{\frac{\sum(y - \bar{y})^2}{n - 1}}$$

$$z = \frac{y - \mu}{\sigma} \text{ (model based)}$$

$$z = \frac{y - \bar{y}}{s} \text{ (data based)}$$

$$r = \frac{\sum z_x z_y}{n - 1}$$

$$\hat{y} = b_0 + b_1 x \quad \text{where } b_1 = \frac{r s_y}{s_x} \text{ and } b_0 = \bar{y} - b_1 \bar{x}$$

$P(\mathbf{A}) = 1 - P(\mathbf{A}^C)$

$P(\mathbf{A} \cup \mathbf{B}) = P(\mathbf{A}) + P(\mathbf{B}) - P(\mathbf{A} \cap \mathbf{B})$

$P(\mathbf{A} \cap \mathbf{B}) = P(\mathbf{A}) \times P(\mathbf{B}|\mathbf{A})$

$$P(\mathbf{B}|\mathbf{A}) = \frac{P(\mathbf{A} \cap \mathbf{B})}{P(\mathbf{A})}$$

If $\mathbf{A}$ and $\mathbf{B}$ are independent, $P(\mathbf{B}|\mathbf{A}) = P(\mathbf{B})$

$E(X) = \mu = \sum x \cdot P(x)$	$Var(X) = \sigma^2 = \sum(x - \mu)^2 P(x)$
$E(X \pm c) = E(X) \pm c$	$Var(X \pm c) = Var(X)$
$E(aX) = aE(X)$	$Var(aX) = a^2 Var(X)$
$E(X \pm Y) = E(X) \pm E(Y)$	$Var(X \pm Y) = Var(X) + Var(Y),$ if X and Y are independent

Geometric: $P(x) = q^{x-1}p$ $\mu = \dfrac{1}{p}$ $^*\sigma = \sqrt{\dfrac{q}{p^2}}$

Binomial: $P(x) = \dbinom{n}{x} p^x q^{n-x}$ $\mu = np$ $\sigma = \sqrt{npq}$

$\hat{p} = \dfrac{x}{n}$ $\mu(\hat{p}) = p$ $SD(\hat{p}) = \sqrt{\dfrac{pq}{n}}$

Sampling distribution of $\bar{y}$:

(CLT) As n grows, the sampling distribution approaches the Normal model with

$$\mu(\bar{y}) = \mu_y \qquad SD(\bar{y}) = \frac{\sigma}{\sqrt{n}}$$

Inference:

Confidence interval for parameter $=$ **statistic $\pm$ critical value $\times$ SD(statistic)**

Test statistic $= \dfrac{Statistic - Parameter}{SD(statistic)}$

Parameter	Statistic	SD(statistic)	SE(statistic)
p	$\hat{p}$	$\sqrt{\dfrac{pq}{n}}$	$\sqrt{\dfrac{\hat{p}\hat{q}}{n}}$
$p_1 - p_2$	$\hat{p}_1 - \hat{p}_2$	$\sqrt{\dfrac{p_1 q_1}{n_1} + \dfrac{p_2 q_2}{n_2}}$	$\sqrt{\dfrac{\hat{p}_1 \hat{q}_1}{n_1} + \dfrac{\hat{p}_2 \hat{q}_2}{n_2}}$
μ	$\bar{y}$	$\dfrac{\sigma}{\sqrt{n}}$	$\dfrac{s}{\sqrt{n}}$
$\mu_1 - \mu_2$	$\bar{y}_1 - \bar{y}_2$	$\sqrt{\dfrac{\sigma_1^2}{n_1} + \dfrac{\sigma_2^2}{n_2}}$	$\sqrt{\dfrac{s_1^2}{n_1} + \dfrac{s_2^2}{n_2}}$
μ_d	$\bar{d}$	$\dfrac{\sigma_d}{\sqrt{n}}$	$\dfrac{s_d}{\sqrt{n}}$
σ_ε	$s_e = \sqrt{\dfrac{\Sigma(y - \hat{y})^2}{n - 2}}$		
β_1	b_1		$\dfrac{s_e}{s_x \sqrt{n - 1}}$
$*\mu_\nu$	$\hat{y}_\nu$		$\sqrt{SE^2(b_1) \cdot (x_\nu - \bar{x})^2 + \dfrac{s_e^2}{n}}$
$*y_\nu$	$\hat{y}_\nu$		$\sqrt{SE^2(b_1) \cdot (x_\nu - \bar{x})^2 + \dfrac{s_e^2}{n} + s_e^2}$

*Pooling: For testing difference between proportions: $\hat{p}_{pooled} = \dfrac{y_1 + y_2}{n_1 + n_2}$

For testing difference between means: $s_p = \sqrt{\dfrac{(n_1 - 1)s_1^2 + (n_2 - 1)s_2^2}{n_1 + n_2 - 2}}$

Substitute these pooled estimates in the respective SE formulas for both groups when assumptions and conditions are met.

Chi-square: $\chi^2 = \sum \dfrac{(obs - exp)^2}{exp}$

Appendix B: Guide to Statistical Software

Chapter 2: Displaying and Describing Categorical Data

DATA DESK

To make a bar chart or pie chart:
- Select the variable.
- In the **Plot** menu, choose **Bar Chart** or **Pie Chart**.

To make a frequency table:
- In the **Calc** menu choose **Frequency Table**.

Data Desk cannot make stacked bar charts.

COMMENTS

These commands treat data as categorical even if they are numerals. If you select a quantitative variable by mistake, you'll see an error message warning of too many categories. The **Replicate Y by X** command can generate variables from summary counts with values for each case, and thus suitable for these commands.

EXCEL

To make a bar chart:
- First make a pivot table (Excel's name for a frequency table). From the **Data** menu, choose **Pivot Table** and **Pivot Chart Report**.
- When you reach the Layout window, drag your variable to the row area and drag your variable again to the data area. This tells Excel to count the occurrences of each category. Once you have an Excel pivot table, you can construct bar charts and pie charts.
- Click inside the Pivot Table.

- Click the Pivot Table Chart Wizard button. Excel creates a bar chart.
- A longer path leads to a pie chart; see your Excel documentation.

COMMENTS

Excel uses the pivot table to specify the category names and find counts within each category. If you already have that information, you can proceed directly to the Chart Wizard.

JMP

JMP makes a bar chart and frequency table together:
- From the **Analyze** menu, choose **Distribution**.
- In the Distribution dialog, drag the name of the variable into the empty variable window beside the label "Y, Columns"; click **OK**.

To make a pie chart:
- Choose **Chart** from the **Graph** menu.
- In the Chart dialog, select the variable name from the Columns list.

- Click on the button labeled "Statistics," and select "N" from the drop-down menu.
- Click the "**Categories, X, Levels**" button to assign the same variable name to the x-axis.
- Under Options, click on the **second** button—labeled "**Bar Chart**"—and select "Pie" from the drop-down menu.

MINITAB

To make a bar chart:
- Choose **Bar Chart** from the **Graph** menu.
- Select "Counts of unique values" in the first menu, and select "Simple" for the type of graph. Click **OK**.

- In the Chart dialog, enter the name of the variable that you wish to display in the box labeled "Categorical variables."
- Click **OK**.

SPSS

To make a bar chart:
- Open the **Chart Builder** from the **Graphs** menu.
- Click the **Gallery** tab.
- Choose **Bar Chart** from the list of chart types.
- Drag the appropriate bar chart onto the canvas.
- Drag a categorical variable onto the x-axis drop zone.
- Click **OK**.

COMMENT

A similar path makes a pie chart by choosing **Pie chart** from the list of chart types.

STATCRUNCH

To make a bar chart or pie chart
1. Click on **Graph**.
2. Choose the type of plot » **With Data** or » **With Summary**.
3. Choose the variable name from the list of **Columns**; if using summaries, also choose the counts.

4. Choose **Frequency/Counts** or (usually) **Relative frequency/Percents**. Note that you may elect to group categories under a specified percentage as "Other."
5. Click on **Compute!**

TI-NSPIRE

To make a dot chart, bar chart, or pie chart from categorical data using a named list on a List & Spreadsheet page, press ▲ to get to the list name (type one in the top row if it has no name), then press ▲ once more so that the entire list is highlighted. Press (menu), ③ for Data, and ⑨ for Quick Graph. This will create a dot chart. To switch plot types, press (menu), ① for Plot Type, and ⑧ for Bar Chart or ⑨ for Pie Chart.

To create the plot on a full page, press (⌂), then [DATA/STAT] for Data and Statistics. Move the cursor to the axis and click on the text "Click to add variable." Use the arrows to select the variable and press [ENTER]. To switch plot type, press (menu), then press ① for Plot Type, then select the type of chart you would like.

To make a plot from a frequency table, with categories in one named list and frequencies in another, begin as instructed above using the categories list. Then press (menu), then ② for Plot Properties, then ⑨ for Add Y Summary List. Select the name of the list with the frequencies.

Chapter 3: Displaying and Summarizing Quantitative Data

DATA DESK

To make a histogram:
• Select the variable to display.
• In the **Plot** menu, choose **Histogram**.

To calculate summaries:
• In the **Calc** menu, open the summaries submenu.
  Options offer separate tables, a single unified table, and other formats.

EXCEL

In Excel, there is another way to find some of the standard summary statistics. For example, to compute the mean:
• Click on an empty cell.
• Go to the Formulas tab in the Ribbon. Click on the drop down arrow next to "AutoSum" and choose "**Average**."
• Enter the data range in the formula displayed in the empty box you selected earlier.
• Press **Enter**. This computes the mean for the values in that range.

To compute the standard deviation:
• Click on an empty cell.
• Go to the Formulas tab in the Ribbon and click the drop down arrow next to "AutoSum" and select "**More functions…**"
• In the dialog window that opens, select "**STDEV**" from the list of functions and click OK. A new dialog window opens. Enter a range of fields into the text fields and click **OK**.
Excel computes the standard deviation for the values in that range and places it in the specified cell of the spreadsheet.

JMP

To make a histogram and find summary statistics:
• Choose **Distribution** from the **Analyze** menu.
• In the **Distribution** dialog, drag the name of the variable that you wish to analyze into the empty window beside the label "**Y, Columns**."

• Click **OK**. JMP computes standard summary statistics along with displays of the variables.

MINITAB

To make a histogram:
• Choose **Histogram** from the **Graph** menu.
• Select "Simple" for the type of graph and click **OK**.
• Enter the name of the quantitative variable you wish to display in the box labeled "Graph variables." Click **OK**.

To calculate summary statistics:
• Choose **Basic statistics** from the **Stat** menu. From the **Basic Statistics** submenu, choose **Display Descriptive Statistics**.
• Assign variables from the variable list box to the Variables box. MINITAB makes a Descriptive Statistics table.

SPSS

To make a histogram in SPSS open the Chart Builder from the Graphs menu:
- Click the **Gallery** tab.
- Choose **Histogram** from the list of chart types.
- Drag the histogram onto the canvas.
- Drag a scale variable to the y-axis drop zone.
- Click **OK**.

To calculate summary statistics:
- Choose **Explore** from the **Descriptive Statistics** submenu of the **Analyze** menu. In the Explore dialog, assign one or more variables from the source list to the Dependent List and click the **OK** button.

STATCRUNCH

To make a histogram, dotplot, or stem-and-leaf plot:
- Click on **Graph**.
- Choose the type of plot.
- Choose the variable name from the list of **Columns**.
- (For a histogram) Choose **Frequency** or (usually) **Relative frequency**, and (if desired) set the axis scale by entering the **Start** value and **Bin width**.
- Click on **Compute!**

To calculate summaries:
- Click on **Stat**.
- Choose **Summary Stats » Columns**.
- Choose the variable name from the list of **Columns**.
- Click on **Compute!**

COMMENTS
- You may need to hold down the Ctrl or command key to choose more than one variable to summarize.
- Before calculating, click on additional summary statistics if so desired.

TI-NSPIRE

To plot a histogram using a named list, press press ▲ or use the cursor to click to get to the list name (type one in the top row if the list has no name), then press ▲ once more so that the entire list is highlighted. Press (menu), ③ for Data, and ⑨ for Quick Graph. This creates a dotplot. Then press (menu), ① for Plot Type, and ③ for Histogram.

To create the plot on a full page, press (⌂), and then select the fifth icon (the one that looks like a bar chart) for Data & Statistics. Move the cursor to "Click to add variable," and then press (✹) and select the list name. Then press (menu), ① for Plot Type, and ③ for Histogram.

Chapter 4: Understanding and Comparing Distributions

There are two ways to organize data when we want to compare groups. Each group can be in its own variable (or list, on a calculator). In this form, the experiment comparing cups would have four variables, one for each type of cup:

CUPPS	SIGG	Nissan	Starbucks
6	2	12	13
6	1.5	16	7
6	2	9	7
18.5	3	23	17.5
10	0	11	10
17.5	7	20.5	15.5
11	0.5	12.5	6
6.5	6	24.5	6

But there's another way to think about and organize the data. What is the variable of interest (the *What*) in this experiment? It's the number of degrees lost by the water in each cup. And the *Who* is each time she tested a cup. We could gather all of the temperature values into one variable and put the names of the cups in a second variable listing the individual results, one on each row. Now, the *Who* is clearer—it's an experimental run, one row of the table. Most statistics packages prefer data on groups organized in this way.

That's actually the way we've thought about the wind speed data in this chapter, treating wind speeds as one variable and the groups (whether seasons, months, or days) as a second variable.

Cup	Temperature Difference	Cup	Temperature Difference
CUPPS	6	SIGG	12
CUPPS	6	SIGG	16
CUPPS	6	SIGG	9
⋮			⋮
Nissan	2	Starbucks	13
Nissan	1.5	Starbucks	7
Nissan	2	Starbucks	7
⋮			⋮

DATA DESK

If the data are in separate variables:
- Select the variables and choose **Boxplot side by side** from the **Plot** menu. The boxes will appear in the order in which the variables were selected.

If the data are a single quantitative variable and a second variable holding group names:
- Select the quantitative variable as Y and the group variable as X.

- Then choose **Boxplot y by x** from the **Plot** menu. The boxes will appear in alphabetical order by group name.

Data Desk offers options for assessing whether any pair of medians differ.

EXCEL

Excel cannot make boxplots.

JMP

- Choose **Fit y by x**.
- Assign a continuous response variable to **Y, Response** and a nominal group variable holding the group names to **X, Factor**, and click **OK**. JMP will offer (among other things) dotplots of the data.

- Click the red triangle and, under **Display Options**, select Boxplots.

Note: if the variables are of the wrong type, the display options might not offer boxplots.

MINITAB

- Choose **Boxplot**...from the **Graph** menu.

If your data are in the form of one quantitative variable and one group variable:
- Choose **One Y** and **with Groups**.

If your data are in separate columns of the worksheet:
- Choose **Multiple Y's**.

SPSS

To make a boxplot in SPSS, open the **Chart Builder** from the Graphs menu:
- Click the **Gallery** tab.
- Choose **Boxplot** from the list of chart types.
- Drag a single or 2-D (side-by-side) boxplot onto the canvas.

- Drag a scale variable to the *y*-axis drop zone.
- To make side-by-side boxplots, drag a categorical variable to the *x*-axis drop zone.
- Click **OK**.

STATCRUNCH

To make a boxplot
1. Click on **Graph**.
2. Choose **Boxplot**.
3. Choose the variable name from the list of **Columns**.
4. Indicate that you want to **identify outliers**.
5. Click on **Compute!**

To make side-by-side boxplots
1. Click on **Graph**.
2. Choose **Boxplot**.
3. Choose the variable name from the list of **Columns**.
4. Choose the column that holds the categories to **Group by**.
5. Indicate that you want to **Plot groups for each column**.
6. Indicate that you want to **identify outliers**.
7. Click on **Compute!**

TI-NSPIRE

To compute summary statistics from a Calculator page using a named list, press (menu) (9) for Statistics, (1) for Stat Calculations, and (1) for One-Variable Statistics. Complete the dialog boxes. To do so from a Lists & Spreadsheet page, press (menu), (4) for Statistics, (1) for Stat Calculations, and (1) for One-Variable Statistics. Complete the dialog boxes.

To create a box plot using a named list, press ▲ several times or use the cursor or click to get to the list name (type one in the top row if the list has no name), then

press ▲ once more so that the entire list is highlighted. Press (menu), (3) for Data, and (9) for Quick Graph. Then press (menu), (1) for Plot Type, and (2) for Box Plot. To create the plot on a full page, press (🏠), then select the fifth icon (the one that looks like a bar chart) for Data & Statistics. Move the cursor to "Click to add variable," and then press (📷) and select the list name. Then press (menu), (1) for Plot Type, and (2) for Box Plot.

Chapter 5: The Standard Deviation as a Ruler and the Normal Model

DATA DESK

To make a "Normal Probability Plot" in Data Desk:
- Select the Variable.
- Choose **Normal Prob Plot** from the **Plot** menu.

COMMENTS

Data Desk places the ordered data values on the vertical axis and the Normal scores on the horizontal axis.

EXCEL

Excel offers a "Normal probability plot" as part of the Regression command in the Data Analysis extension, but (as of this writing) it is not a correct Normal probability plot and should not be used.

JMP

To make a "Normal Quantile Plot" in JMP:
- Make a histogram using **Distributions** from the **Analyze** menu.
- Click on the drop-down menu next to the variable name.
- Choose **Normal Quantile Plot** from the drop-down menu.
- JMP opens the plot next to the histogram.

COMMENTS

JMP places the ordered data on the vertical axis and the Normal scores on the horizontal axis. The vertical axis aligns with the histogram's axis, a useful feature.

MINITAB

To make a "Normal Probability Plot" in MINITAB:
- Choose **Probability Plot** from the **Graph** menu.
- Select "Single" for the type of plot. Click **OK**.
- Enter the name of the variable in the "Graph variables" box. Click **OK**.

COMMENTS

MINITAB places the ordered data on the horizontal axis and the Normal scores on the vertical axis.

SPSS

To make a Normal "P-P plot" in SPSS:
- Choose P-P from the **Graphs** menu.
- Select the variable to be displayed in the source list.
- Click the arrow button to move the variable into the target list.
- Click the **OK** button.

COMMENTS

SPSS places the ordered data on the horizontal axis and the Normal scores on the vertical axis. You may safely ignore the options in the P-P dialog.

STATCRUNCH

To make a Normal probability plot:
- Click on **Graph**.
- Choose **QQ Plot**.
- Choose the variable name from the list of **Columns**.
- Click on **Create Graph**.

To work with Normal percentiles:
- Click on **Stat**.
- Choose **Calculators » Normal**.
- Choose a lower tail ($\leq$) or upper tail ($\geq$) region.
- Enter the z-score cutoff, and then click on **Compute** to find the probability.

OR

Enter the desired probability, and then click on **Compute** to find the z-score cutoff.

TI-NSPIRE

To create a normal probability plot using a named list, press ▲ several times or use the cursor to click to get to the list name (type one in the top row if the list has no name), then press ▲ once more so that the entire list is highlighted. Press (menu), ③ for Data, and ⑨ for Quick Graph. Then press (menu), ① for Plot Type, and ④ for Normal Probability Plot.
To create the plot on a full page, press (⌂), and then select the fifth icon (the one that looks like a bar chart) for Data & Statistics. Move the cursor to "Click to add variable," and then press (✴) and select the list name. Then press (menu), ① for Plot Type, and ④ for Normal Probability Plot.

To compute the area under a normal curve from a Calculator page, press (menu), ⑤ for Probability (or ⑥ for Statistics), ⑤ for Distributions, and ② for Normal Cdf. Complete the dialog box.
To compute the value for a given percentile on a Calculator page, press (menu), ⑤ for Probability (or ⑥ for Statistics), ⑤ for Distributions, and ③ for Inverse Normal. Complete the dialog box.

Chapter 6: Scatterplots, Association, and Correlation

DATA DESK

To make a scatterplot of two variables:
- Click to select one variable as Y.
- Shift-click to select the other as X.
- Choose **Scatterplot** from the **Plot** menu.
- Then find the correlation by choosing **Correlation** from the scatterplot's HyperView menu.

Alternatively, select the two variables and choose **Pearson Product-Moment** from the **Correlations** submenu of the **Calc** menu.

COMMENTS

We prefer that you look at the scatterplot first and then find the correlation. But if you've found the correlation first, click on the correlation value to drop down a menu that offers to make the scatterplot.

EXCEL

To make a scatterplot in Excel:
- Select the columns of data to use in the scatterplot. You can select more than one column by holding down the control key while clicking.
- In the Insert tab, click on the **Scatter** button and select the **Scatter with only Markers** chart from the menu.

Unfortunately, the plot this creates is often statistically useless. To make the plot useful, we need to change the display:
- With the chart selected click on the Gridlines button in the Layout tab to cause the Chart Tools tab to appear.
- Within Primary Horizontal Gridlines, select None. This will remove the gridlines from the scatterplot.
- To change the axis scaling, click on the numbers of each axis of the chart, and click on the **Format Selection** button in the Layout tab.
- Select the Fixed option instead of the Auto option, and type a value more suited for the scatterplot. You can use the pop-up dialog window as a straightedge to approximate the appropriate values.

Excel automatically places the leftmost of the two columns you select on the x-axis, and the rightmost one on the y-axis. If that's not what you'd prefer for your plot, you'll want to switch them.
To switch the X- and Y-variables:
- Click the chart to access the **Chart Tools** tabs.
- Click on the **Select Data** button in the Design tab.
- In the pop-up window's Legend Entries box, click on **Edit**.
- Highlight and delete everything in the Series X Values line, and select new data from the spreadsheet. (Note that selecting the column would inadvertently select the title of the column, which would not work well here.)
- Do the same with the Series Y Values line.
- Press **OK**, then press **OK** again.

JMP

To make a scatterplot and compute correlation:
- Choose **Fit Y by X** from the **Analyze** menu.
- In the Fit Y by X dialog, drag the Y variable into the "**Y, Response**" box, and drag the X variable into the "**X, Factor**" box.
- Click the **OK** button.

Once JMP has made the scatterplot, click on the red triangle next to the plot title to reveal a menu of options:
- Select **Density Ellipse** and select 0.95. JMP draws an ellipse around the data and reveals the **Correlation** tab.
- Click the blue triangle next to Correlation to reveal a table containing the correlation coefficient.

MINITAB

To make a scatterplot:
- Choose **Scatterplot** from the **Graph** menu.
- Choose "Simple" for the type of graph. Click **OK**.
- Enter variable names for the Y variable and X variable into the table. Click **OK**.

To compute a correlation coefficient:
- Choose **Basic Statistics** from the **Stat** menu.
- From the Basic Statistics submenu, choose **Correlation**. Specify the names of at least two quantitative variables in the "Variables" box.
- Click **OK** to compute the correlation table.

SPSS

To make a scatterplot in SPSS:
- Open the Chart Builder from the **Graphs** menu. Then
- Click the **Gallery** tab.
- Choose Scatterplot from the list of chart types.
- Drag the scatterplot onto the canvas.
- Drag a scale variable you want as the response variable to the y-axis drop zone.
- Drag a scale variable you want as the factor or predictor to the x-axis drop zone.
- Click **OK**.

To compute a correlation coefficient:
- Choose **Correlate** from the **Analyze** menu.
- From the Correlate submenu, choose **Bivariate**.
- In the Bivariate Correlations dialog, use the arrow button to move variables between the source and target lists.

Make sure the **Pearson** option is selected in the Correlation Coefficients field.

STATCRUNCH

To make a scatterplot:
- Click on **Graph**.
- Choose **Scatter Plot**.
- Choose **X** and **Y** variable names from the list of **Columns**.
- Click on **Compute!**

To find a correlation:
- Click on **Stat**.
- Choose **Summary Stats » Correlation**.
- Choose two variable names from the list of **Columns**. (You may need to hold down the ctrl or command key to choose the second one.)
- Click on **Compute!**

TI-NSPIRE

To create a scatterplot using named lists, press ▲ several times or use the cursor to click to get to the list name (type one in the top row if the list has no name), then press ▲ once more so that the first list is highlighted. Then press (Shift) ▶ so that the second list is highlighted. press (menu), ③ for Data, and ⑨ for Quick Graph. (Note: the list on the left will be the explanatory variable in the plot.)
To create the plot on a full page, press ⑨, then select the fifth icon (the one that looks like a bar chart) for Data & Statistics. Move the cursor to "Click to add variable," and then press (⌘) and select the list name. Repeat for the other axis.

To find the correlation from a Calculator or Lists & Spreadsheets page, press (menu), ⑥ on a calculator page or ④ on a Lists & Spreadsheets page for Statistics, ① for Stat Calculations, and ④ for Linear Regression. Complete the Dialog boxes. Once the regression line has been calculated, the correlation (for the most recently calculated regression line) can be accessed by pressing VAR and selecting the variable stat.r.

Chapter 7: Linear Regression

DATA DESK

- Select the y-variable and the x-variable.
- In the **Plot** menu, choose **Scatterplot**.
- From the scatterplot HyperView menu, choose **Add Regression Line** to display the line.
- From the HyperView menu, choose **Regression** to compute the regression.
- To plot the residuals, click on the **HyperView** menu on the **Regression** output table.
- A menu drops down that offers scatterplots of residuals against predicted values (as well as other options).

COMMENTS

Alternatively, find the regression first with the **Regression** command in the **Calc** menu. Click on the x-variable's name to open a menu that offers the scatterplot.

EXCEL

- Click on a blank cell in the spreadsheet.
- Go to the **Formulas** tab in the Ribbon and click **More Functions → Statistical**.
- Choose the **CORREL** function from the drop-down menu of functions.
- In the dialog that pops up, enter the range of one of the variables in the space provided.
- Enter the range of the other variable in the space provided.
- Click **OK**.

COMMENTS

The correlation is computed in the selected cell. Correlations computed this way will update if any of the data values are changed.
Before you interpret a correlation coefficient, always make a scatterplot to check for nonlinearity and outliers. If the variables are not linearly related, the correlation coefficient cannot be interpreted.

JMP

- Choose **Fit Y by X** from the **Analyze** menu.
- Specify the *y*-variable in the Select Columns box and click the **y, Response** button.
- Specify the *x*-variable and click the **X, Factor** button.
- Click **OK** to make a scatterplot.

- In the scatterplot window, click on the red triangle beside the heading labeled "**Bivariate Fit …**" and choose **Fit Line**. JMP draws the least squares regression line on the scatterplot and displays the results of the regression in tables below the plot.

MINITAB

- Choose **Regression** from the **Stat** menu.
- From the Regression submenu, choose **Fitted Line Plot**.
- In the Fitted Line Plot dialog, click in the **Response Y** box, and assign the *y*-variable from the variable list.

- Click in the **Predictor X** box, and assign the *x*-variable from the Variable list. Make sure that the Type of Regression Model is set to Linear. Click the **OK** button.

SPSS

- Choose **Interactive** from the **Graphs** menu.
- From the interactive Graphs submenu, choose **Scatterplot**.
- In the Create Scatterplot dialog, drag the *y*-variable into the **y-axis target**, and the *x*-variable into the **x-axis target**.

- Click on the **Fit** tab.
- Choose **Regression** from the **Method** popup menu. Click the **OK** button.

STATCRUNCH

To compute a regression:
- Click on **Stat**.
- Choose **Regression » Simple Linear**.
- Choose X and Y variable names from the list of columns.
- Click on **Compute!** to see the regression analysis.
- Click on > to see the scatterplot.

COMMENTS

Remember to check the scatterplot to be sure a linear model is appropriate. Note that before you **Compute!** you can:
- enter an X-value for which you want to find the predicted Y-value;
- save all the fitted values;
- save the residuals;
- ask for a residuals plot.

TI-NSPIRE

To plot and find the equation of the regression line, first create a scatterplot. Using named lists, press ▲ several times so that the first list is highlighted. Then press [Shift] ▶ so that the second list is highlighted. Press (menu), ③ for Data, and ⑨ for Quick Graph. Then press (menu), ④ for Analyze, ⑥ for Regression, and ② for Show Linear. To find the equation of the regression line on a calculator page, press (menu), ⑥ for Statistics, ① for Stat Calculations, and ④ for Linear Regression. Complete the dialog boxes.

To see the plot on a full page, press (⌂), and then select the fifth icon (the one that looks like a bar chart) for Data & Statistics. Move the cursor to "Click to add variable," and then press (⊙) and select the list name. Repeat for the other axis. Then press (menu), ④ for Analyze, ⑥ for Regression, and ③ for Show Linear.

Chapter 8: Regression Wisdom

DATA DESK

- Click on the **HyperView** menu on the **Regression** output table. A menu drops down to offer scatterplots of residuals against predicted values, Normal probability plots of residuals, or just the ability to save the residuals and predicted values.
- Click on the name of a predictor in the regression table to be offered a scatterplot of the residuals against that predictor.

COMMENTS

If you change any of the variables in the regression analysis, Data Desk will offer to update the plots of residuals.

EXCEL

The Data Analysis add-in for Excel includes a Regression command. The dialog box it shows offers to make plots of residuals.

COMMENTS

Do not use the Normal probability plot offered in the regression dialog. It is not what it claims to be and is wrong.

JMP

- From the **Analyze** menu, choose **Fit Y by X**. Select **Fit Line**.
- Under Linear Fit, select **Plot Residuals**. You can also choose to **Save Residuals**.

- Subsequently, from the **Distribution menu**, choose **Normal quantile plot** or **histogram** for the residuals.

MINITAB

- From the **Stat** menu, choose **Regression**.
- From the **Regression** submenu, select Regression again.
- In the Regression dialog, enter the response variable name in the "Response" box and the predictor variable name in the "Predictor" box.
- To specify saved results, in the Regression dialog, click **Storage**.
- Check "Residuals" and "Fits." Click **OK**.

- To specify displays, in the Regression dialog, click **Graphs**.
- Under "Residual Plots," select "Individual plots" and check "Residuals versus fits."
- Click **OK**. Now back in the Regression dialog, click **OK**. Minitab computes the regression and the requested saved values and graphs.

SPSS

- From the **Analyze** menu, choose **Regression**.
- From the Regression submenu, choose **Linear**.
- After assigning variables to their roles in the regression, click the "**Plots ...**" button.

In the Plots dialog, you can specify a Normal probability plot of residuals and scatterplots of various versions of standardized residuals and predicted values.

COMMENTS

A plot of *ZRESID against *PRED will look most like the residual plots we've discussed. SPSS standardizes the residuals by dividing by their standard deviation. (There's no need to subtract their mean; it must be zero.) The standardization doesn't affect the scatterplot.

STATCRUNCH

To create a residuals plot:
- Click on **Stat**.
- Choose **Regression » Simple Linear** and choose X and Y.
- Scroll down in the Graphs box to indicate which type of residuals plot you want.
- Click on **Compute!**

COMMENTS

Note that before you **Compute!** you may choose to save the values of the residuals. Residuals becomes a new column, and you may use that variable to create a histogram or residuals plot.

TI-NSPIRE

To create a residual plot on a page that already has a scatterplot with regression line, (menu), (4) for Analyze, (7) for Residuals, then (2) for Show Residual Plot.
To create a residual plot on a page by itself, a regression line must have already been calculated. Press (⌂), then select the fifth icon (the one that looks like a bar chart)

for Data & Statistics. Click the cursor on "Click to add variable," and then [CURSOR] and select the list name. On the vertical axis, select the variable **stat.resid**.

Chapter 9: Re-expressing Data: Get It Straight!

DATA DESK

To re-express a variable in Data Desk, select the variable and Choose the function to re-express it from the **Manip > Transform** menu. Square root, log, reciprocal, and reciprocal root are immediately available. For others, make a derived variable and type the function. Data Desk makes a new derived variable that holds the re-expressed values. Any value changed in the original variable will immediately be re-expressed in the derived variable.

COMMENTS

Or choose **Manip > Transform > Dynamic > Box-Cox** to generate a continuously changeable variable and a slider that specifies the power. Set plots to **Automatic Update** in their HyperView menus and watch them change dynamically as you drag the slider.

EXCEL

To re-express a variable in Excel, use Excel's built-in functions as you would for any calculation. Changing a value in the original column will change the re-expressed value.

JMP

To re-express a variable in JMP, double-click to the right of the last column of data to create a new column. Name the new column and select it. Choose **Formula** from the **Cols** menu. In the Formula dialog, choose the transformation and variable that you wish to assign to the new column. Click the **OK** button. JMP places the re-expressed data in the new column.

COMMENTS

The log and square root re-expressions are found in the **Transcendental** menu of functions in the formula dialog.

MINITAB

To re-express a variable in MINITAB, choose **Calculator** from the **Calc** menu. In the Calculator dialog, specify a name for the new re-expressed variable. Use the

Functions List, the calculator buttons, and the **Variables list** box to build the expression. Click **OK.**

SPSS

To re-express a variable in SPSS, Choose **Compute** from the **Transform** menu. Enter a name in the Target Variable field. Use the calculator and Function List to build the

expression. Move a variable to be re-expressed from the source list to the Numeric Expression field. Click the **OK** button.

STATCRUNCH

To re-express a variable in StatCrunch, choose **Data » Compute Expression**. You can type an expression directly (e.g. log(var2)) or you can choose **Build** and click on the column(s) and function(s) of your choice.

Click on **Compute!** to place the re-expressed data in the new column.

TI-NSPIRE

To re-express data, create a new list and enter the formula in the cell in the second row. For example, if one column has a list named *time*, another list can be created

using the formula *log(time)*. Variable names can be accessed by pressing VAR, then ‹3› for Link To:.

Chapter 10: Understanding Randomness

DATA DESK

Generate random numbers in Data Desk with the **Generate Random Numbers ...** command in the **Manip** menu. A dialog guides you in specifying the number of variables to fill, the number of cases, and details about the values. For most simulations, generate random uniform values.

COMMENTS

Bernoulli Trials generate random values that are 0 or 1, with a specified chance of a 1.
Binomial Experiments automatically generate a specified number of Bernoulli trials and count the number of 1s.

EXCEL

The **RAND** function generates a random value between 0 and 1. You can multiply to scale it up to any range you like and use the INT function to turn the result into an integer.

COMMENTS

Published tests of Excel's random-number generation have declared it to be inadequate. However, for simple simulations, it should be OK. Don't trust it for important large simulations.

JMP

- In a new column, in the **Cols** menu choose **Column Info. . . .**
- In the dialog, click the **New Property** button, and choose **Formula** from the drop-down menu.

- Click the **Edit Formula** button, and in the **Functions(grouped)** window click on Random. **Random Integer (10)**, for example, will generate a random integer between 1 and 10.

MINITAB

- In the **Calc** menu, choose **Random Data**....
- In the Random Data submenu, choose **Uniform**....

A dialog guides you in specifying details of range and number of columns to generate.

SPSS

The **RV.UNIFORM(min, max)** function returns a random value that is equally likely between the min and max limits.

STATCRUNCH

To generate a list of random numbers:
- Click on **Data**.
- Choose **Simulate data » Uniform**.
  Enter the number of rows and columns of random numbers you want. (Often you'll specify the desired number of random values as **Rows** and just 1 **Column**.)
- Enter the interval of possible values for the random numbers ($a \leq x < b$).
- Click on **Compute!** The random numbers will appear in the data table.

COMMENTS

To get random *integers* from 0 to 99, set **a** $= 0$ and **b** $= 100$, and then simply ignore the decimal places in the numbers generated.

OR

- Click **Build**.
- Add Function **Floor**.
- Add Column.
- Click on **Okay**.
- Click on **Compute!**

TI-NSPIRE

To generate random integers from a Calculator page, press (menu), (5) for Probability, (4) for Random, and (2) for Integer. Then type the range for the random integers, such as randInt(1,6).
To create a list of random integers, type the length of the list as the third value, such as randInt(1,6,10).

To generate a list of random integers on a Lists & Spreadsheet page, click or arrow to the second row and type randInt(followed by the range and the length of the list, separated by commas. For example, to get a list of 10 integers from 1 to 6, type randInt(1,6,10).

Chapter 15: Random Variables

STATCRUNCH

To compute the mean and standard deviation for a discrete random variable, enter the values in one column and the probabilities in another. Then choose **Stat** >> **Calculators** >> **Custom**. Select the **Values** and the probabilities as the **Weights** and click **Compute!**

TI-NSPIRE

There are several ways to compute the mean and standard deviation for a discrete random variable. First, on a Lists & Spreadsheet page, enter the values in one named list and the probabilities in another. (To name the lists, type the variable names in the top row of the header of the column.) From a Calculator page, type mean(*variable name*, *frequency list name*) or stDevPop(*variable name*, *frequency*

list name). In a Lists & Spreadsheet page, you can do this within a cell by typing = first.
A second option is to press (menu), then (6) on a Calculator page or (4) on a Lists & Spreadsheets page for Statistics, (1) for Stat Calculations, (1) for One-Variable Statistics, then complete the dialog boxes.

Chapter 16: Probability Models

Most statistics packages offer functions that compute probabilities for various probability models. The only important differences among these functions are in what they are named and the order of their arguments. In these functions, pdf stands for "probability density function"—what we've been calling a probability model. The letters cdf stand for "cumulative distribution function," the technical term when we want to accumulate probabilities over a range of values. These technical terms show up in many of the function names.

DATA DESK

BinomDistr(*x, n, prob*) (pdf)
CumBinomDistr(*x, n, prob*) (cdf)

COMMENTS

These functions work in derived variables or in scratchpads.

EXCEL

Binomdist(*x, n, prob, cumulative*)

COMMENTS

Set cumulative = *true* for cdf, *false* for pdf.
Excel's function fails when *x* or *n* is large.

JMP

Binomial Probability (*prob, n, x*) (pdf)
Binomial Distribution (*prob, n, x*) (cdf)

MINITAB

- Choose **Probability Distributions** from the **Calc** menu.
- Choose **Binomial** from the Probability Distributions submenu.
- To calculate the probability of getting *x* successes in *n* trials, choose **Probability**.

- To calculate the probability of getting *x* or fewer successes among *n* trials, choose **Cumulative Probability**.
- For Geometric, choose **Geometric** from the Probability Distribution submenu.

SPSS

PDF.GEOM(*x, prob*)
CDF.GEOM(*x, prob*)

PDF.BINOM(*x, n, prob*)
CDF.BINOM(*x, n, prob*)

STATCRUNCH

To calculate binomial probabilities:
- Click on **Stat**.
- Choose **Calculators » Binomial**.
- Enter the parameters, **n** and **p**.

- Choose a specific outcome (=) or a lower tail ($\leq$ or $<$) or upper tail ($\geq$ or $>$) sum.
- Enter the number of successes **x**.
- Click on **Compute**.

TI-NSPIRE

To compute geometric and binomial probabilities, press (menu), ⑤ for Probability or ⑥ for Statistics, and ⑤ for Distributions. Select the menu item. Pdf is for

the probability distribution function; Cdf will display cumulative probabilities for a selected range. Complete the dialog box.

Chapter 18: Confidence Intervals for Proportions

DATA DESK

Data Desk does not offer built-in methods for inference with proportions.

COMMENTS

For summarized data, open a Scratchpad to compute the standard deviation and margin of error by typing the calculation. Then use **z-interval for individual μs.**

EXCEL

Inference methods for proportions are not part of the standard Excel tool set.

COMMENTS

For summarized data, type the calculation into any cell and evaluate it.

JMP

For a **categorical** variable that holds category labels, the **Distribution** platform includes tests and intervals for proportions. For summarized data, put the category names in one variable and the frequencies in an adjacent variable. Designate the frequency column to have the **role** of **frequency**. Then use the **Distribution** platform.

COMMENTS

JMP uses slightly different methods for proportion inferences than those discussed in this text. Your answers are likely to be slightly different, especially for small samples.

MINITAB

- Choose **Basic Statistics** from the **Stat** menu.
- Choose **1Proportion** from the Basic Statistics submenu.
- If the data are category names in a variable, assign the variable from the variable list box to the **Samples in columns** box. If you have summarized data, click the **Summarized Data** button and fill in the number of trials and the number of successes.
- Click the **Options** button and specify the remaining details.
- If you have a large sample, check **Use test and interval based on normal distribution**.
  Click the **OK** button.

COMMENTS

When working from a variable that names categories, MINITAB treats the last category as the "success" category. You can specify how the categories should be ordered.

SPSS

SPSS does not find confidence intervals for proportions.

STATCRUNCH

To create a confidence interval for a proportion using summaries:
- Click on **Stat**.
- Choose **Proportion Statistics » One sample » With Summary**.
- Enter the **Number of successes** (x) and **Number of observations** (n).
- Indicate **Confidence Interval for p** (Standard-Wald), and then enter the **Level** of confidence.
- Click on **Compute!**

To create a confidence interval for a proportion using data:
- Click on **Stat**.
- Choose **Proportion Statistics » One sample » With Data**.
- Choose the variable **Column** listing the **Outcomes**.
- Enter the outcome to be considered a **Success**.
- Indicate **Confidence Interval for p** (Standard-Wald), and then enter the **Level** of confidence.
- Click on **Compute!**

TI-NSPIRE

To compute a confidence interval for a population proportion from a Calculator page, press (menu), ⑥ for Statistics, ⑥ for Confidence Intervals, and ⑤ for 1-Prop z-interval. From a Lists & Spreadsheet page, press (menu), ④ for Statistics,

③ for Confidence intervals, and ⑤ for 1-Prop z interval. Complete the dialog box. Be sure to enter the number of successes, x, as a whole number, and the C level as a decimal, such as .99.

Chapter 19: Testing Hypotheses About Proportions

DATA DESK

Data Desk does not offer built-in methods for inference with proportions. The **Replicate Y by X** command in the **Manip** menu will "reconstruct" summarized count data so that you can display it.

COMMENTS

For summarized data, open a Scratchpad to compute the standard deviation and margin of error by typing the calculation. Then perform the test with the **z-test for individual μs** found in the Test command.

EXCEL

Inference methods for proportions are not part of the standard Excel tool set.

COMMENTS

For summarized data, type the calculation into any cell and evaluate it.

JMP

For a **categorical** variable that holds category labels, the **Distribution** platform includes tests and intervals of proportions. For summarized data:
- Put the category names in one variable and the frequencies in an adjacent variable.
- Designate the frequency column to have the **role of frequency**. Then use the **Distribution** platform.

COMMENTS

JMP uses slightly different methods for proportion inferences than those discussed in this text. Your answers are likely to be slightly different.

MINITAB

Choose **Basic Statistics** from the **Stat** menu:
- Choose **1Proportion** from the Basic Statistics submenu.
- If the data are category names in a variable, assign the variable from the variable list box to the **Samples in columns** box.
- If you have summarized **data**, click the **Summarized Data** button and fill in the number of trials and the number of successes.
- Click the **Options** button and specify the remaining details.

- If you have a large sample, check **Use test and interval based on Normal distribution**.
- Click the **OK** button.

COMMENTS

When working from a variable that names categories, Minitab treats the last category as the "success" category. You can specify how the categories should be ordered.

SPSS

SPSS does not offer hypothesis tests for proportions.

STATCRUNCH

To test a hypothesis for a proportion using summaries:
- Click on **Stat**.
- Choose **Proportion Statistics » One sample » With Summary**.
- Enter the **Number of successes** (x) and **Number of observations** (n).
- Indicate **Hypothesis Test for p**, then enter the hypothesized Null proportion, and choose the **Alternative** hypothesis.
- Click on **Compute!**

To test a hypothesis for a proportion using data:
- Click on **Stat**.
- Choose **Proportion Statistics » One sample » With Data**.
- Choose the variable **Column** listing the **Outcomes**.
- Enter the outcome to be considered a **Success**.
- Indicate **Hypothesis Test for p**, then enter the hypothesized Null proportion, and choose the **Alternative** hypothesis.
- Click on **Compute!**

TI-NSPIRE

To compute a hypothesis test for a population proportion from a Calculator page, press (menu), ⑥ for Statistics, ⑦ for Stat Tests, and ⑤ for 1-Prop *z*-test. From a Lists & Spreadsheet page, press (menu), ④ for Statistics, ④ for Stat Tests, and ⑤

for 1-Prop z Test. Complete the dialog box. From a Lists & Spreadsheet page can you select the option to draw the plot of the test statistic with the P-value shaded. Be sure to enter the number of successes, *x*, as a whole number.

Chapter 21: Comparing Two Proportions

DATA DESK

Data Desk does not offer built-in methods for inference with proportions. Use **Replicate Y by X** to construct data corresponding to given proportions and totals.

COMMENTS

For summarized data, open a Scratchpad to compute the standard deviations and margin of error by typing the calculation.

EXCEL

Inference methods for proportions are not part of the standard Excel tool set.

COMMENTS

For summarized data, type the calculation into any cell and evaluate it.

JMP

For a **categorical** variable that holds category labels, the **Distribution** platform includes tests and intervals of proportions:
- For summarized data, put the category names in one variable and the frequencies in an adjacent variable.
- Designate the frequency column to have the **role of frequency**. Then use the **Distribution platform**.

COMMENTS

JMP uses slightly different methods for proportion inferences than those discussed in this text. Your answers are likely to be slightly different.

MINITAB

To find a hypothesis test for a proportion:
- Choose **Basic Statistics** from the **Stat** menu.
- Choose **2Proportions** … from the Basic Statistics submenu. If the data are organized as category names in one column and case IDs in another, assign the variables from the variable list box to the **Samples in one column** box.
- If the data are organized as two separate columns of responses, click on **Samples in different columns**: and assign the variables from the variable list box. If you have summarized data, click the **Summarized Data** button and fill in the number of trials and the number of successes for each group.

- Click the **Options** button and specify the remaining details. Remember to click the **Use pooled estimate** of p **for test** box when testing the null hypothesis of no difference between proportions.
- Click the **OK** button.

COMMENTS

When working from a variable that names categories, MINITAB treats the last category as the "success" category. You can specify how the categories should be ordered.

SPSS

SPSS does not perform hypothesis tests for proportions.

STATCRUNCH

To do inference for the difference between two proportions using summaries:
- Click on **Stat**.
- Choose **Proportion Statistics » Two sample » With Summary**.
- Enter the **Number of successes** (x) and **Number of observations** (n) in each group.
- Indicate **Hypothesis Test**, then enter the hypothesized Null proportion difference (usually 0), and choose the Alternative hypothesis.

OR

Indicate **Confidence Interval**, and then enter the **Level** of confidence.
- Click on **Compute!**

To do inference for the difference between two proportions using data:
- Click on **Stat**.
- Choose **Proportion Statistics » Two sample » With Data**.
- For each group, choose the variable **Column** listing the **Outcomes**, and enter the outcome to be considered a **Success**.
- Indicate **Hypothesis Test**, then enter the hypothesized Null proportion difference (usually 0), and choose the Alternative hypothesis.

OR

Indicate **Confidence Interval**, and then enter the **Level** of confidence.
- Click on **Compute!**

TI-NSPIRE

To compute a confidence interval for the difference between two population proportions from a Calculator page, press (menu), ⟨6⟩ for Statistics, ⟨6⟩ for Confidence Intervals, and ⟨6⟩ for 2-Prop z-interval. From a Lists & Spreadsheet page, press (menu), ⟨4⟩ for Statistics, ⟨3⟩ for Confidence Intervals, and ⟨6⟩ for 2-Prop z Interval. Complete the dialog box. Be sure to enter each number of successes as a whole number, and the C level as a decimal, such as .99.

To compute a hypothesis test for the difference between two population proportions from a Calculate page, press (menu), ⟨6⟩ for Statistics, ⟨7⟩ for Stat Tests, and ⟨6⟩ for 2-Prop z-test. From a Lists & Spreadsheet page, press (menu), ⟨4⟩ for Statistics, ⟨4⟩ for Stat Tests, and ⟨6⟩ for 2-Prop z Test. Complete the dialog box. From a Lists & Spreadsheet page can you select the option to draw the plot of the test statistic with the P-value shaded. Be sure to enter each number of successes as a whole number.

Chapter 22: Inferences About Means

DATA DESK

- Select variables.
- From the **Calc** menu, choose **Estimate** for confidence intervals or **Test** for hypothesis tests.

- Select the interval or test from the drop-down menu and make other choices in the dialog.

EXCEL

Specify formulas. Find t* with the TINV(alpha, df) function.

COMMENTS

Not really automatic. There's no easy way to find P-values in Excel. For the examples in this chapter, substitute 0.05 for "alpha" in the TINV command.

JMP

- From the **Analyze** menu, select **Distribution**.
- For a confidence interval, scroll down to the "Moments" section to find the interval limits.
- For a hypothesis test, click the red triangle next to the variable's name and choose **Test Mean** from the menu.
- Then fill in the resulting dialog.

COMMENTS

"Moment" is a fancy statistical term for means, standard deviations, and other related statistics.

MINITAB

- From the **Stat** menu, choose the **Basic Statistics** submenu.
- From that menu, choose **1-sample t. . . .**
- Then fill in the dialog.

COMMENTS

The dialog offers a clear choice between confidence interval and test.

SPSS

- From the **Analyze** menu, choose the **Compare Means** submenu.
- From that, choose the **One-Sample t-test** command.

COMMENTS

The commands suggest neither a single mean nor an interval. But the results provide both a test and an interval.

STATCRUNCH

To do inference for a mean using summaries:
- Click on **Stat**.
- Choose **T Statistics » One sample » With Summary**.
- Enter the **Sample mean**, **Sample std dev**, and **Sample size**.
- Indicate **Hypothesis Test**, then enter the hypothesized Null mean, and choose the Alternative hypothesis.

OR

Indicate **Confidence Interval**, and then enter the **Level** of confidence.
- Click on **Compute!**

To do inference for a mean using data:
- Click on **Stat**.
- Choose **T Statistics » One sample » With Data**.
- Choose the variable **Column**.
- Indicate **Hypothesis Test**, then enter the hypothesized Null mean, and choose the Alternative hypothesis.

OR

Indicate **Confidence Interval**, and then enter the **Level** of confidence.
- Click on **Compute!**

TI-NSPIRE

To compute a confidence interval for a population mean from a Calculator page, press (menu), ⑥ for Statistics, ⑥ for Confidence Intervals, and ② for *t*-interval. From a Lists & Spreadsheet page, press (menu), ④ for Statistics, ③ for Confidence Intervals, and ② for t interval. Select between Data and Stats, (tab) to OK, and press (enter). Complete the dialog box. Be sure to enter the number of successes, *x*, as a whole number, and the C level as a decimal, such as .99.

To compute a hypothesis test for a population mean from a Calculator page, press (menu), ⑥ for Statistics, ⑦ for Stat Tests, and ② for *t*-test. From a Lists & Spreadsheet page, press (menu), ④ for Statistics, ④ for Stat Test, and ② for t Test. Select between Data and Stats, (tab) to OK, and (enter). Complete the dialog box. From a Lists & Spreadsheet page can you select the option to draw the plot of the test statistic with the P-value shaded.

Chapter 23: Comparing Means

There are two ways to organize data when we want to compare two independent groups. The data can be in two lists, as in the table at the start of this chapter. Each list can be thought of as a variable. In this method, the variables in the batteries example would be *Brand Name* and *Generic*. Graphing calculators usually prefer this form, and some computer programs can use it as well.

There's another way to think about the data. What is the response variable for the battery life experiment? It's the *Time* until the music stopped. But the values of this variable are in both columns, and actually there's an experiment factor here, too—namely, the *Brand* of the battery. So, we could put the data into two different columns, one with the *Times* in it and one with the *Brand*. Then the data would look as shown in the table to the right.

This way of organizing the data makes sense as well. Now the factor and the response variables are clearly visible. You'll have to see which method your program requires. Some packages even allow you to structure the data either way.

The commands to do inference for two independent groups on common statistics technology are not always found in obvious places. Here are some starting guidelines.

Time	Brand
194.0	Brand name
205.5	Brand name
199.2	Brand name
172.4	Brand name
184.0	Brand name
169.5	Brand name
190.7	Generic
203.5	Generic
203.5	Generic
206.5	Generic
222.5	Generic
209.4	Generic

DATA DESK

- Select variables.
- From the **Calc** menu, choose **Estimate** for confidence intervals or **Test** for hypothesis tests.
- Select the interval or test from the drop-down menu and make other choices in the dialog.

COMMENTS

Data Desk expects the two groups to be in separate variables.

EXCEL

- From the Data Tab, Analysis Group, choose **Data Analysis.**
- Alternatively (if the Data Analysis Tool Pack is not installed), in the Formulas Tab, choose More functions > Statistical > TTEST, and specify Type = 3 in the resulting dialog.
- Fill in the cell ranges for the two groups, the hypothesized difference, and the alpha level.

COMMENTS

Excel expects the two groups to be in separate cell ranges.
Notice that, contrary to Excel's wording, we do not need to assume that the variances are *not* equal; we simply choose not to assume that they *are* equal.

JMP

- From the **Analyze** menu, select **Fit y by x.**
- Select variables: a **Y, Response** variable that holds the data and an **X, Factor** variable that holds the group names. JMP will make a dotplot.
- Click the **red triangle** in the dotplot title, and choose **Unequal variances.** The *t*-test is at the bottom of the resulting table.
- Find the P-value from the Prob > F section of the table (they are the same).

COMMENTS

JMP expects data in one variable and category names in the other. Don't be misled: There is no need for the variances to be unequal to use two-sample *t* methods.

MINITAB

- From the **Stat** menu, choose the **Basic Statistics** submenu.
- From that menu, choose **2-sample t. . . .** Then fill in the dialog.

COMMENTS

The dialog offers a choice of data in two variables, or data in one variable and category names in the other.

SPSS

- From the **Analyze** menu, choose the **Compare Means** submenu.
- From that, choose the **Independent-Samples t-test** command. Specify the data variable and "group variable."
- Then type in the labels used in the group variable. SPSS offers both the two-sample and pooled-*t* results in the same table.

COMMENTS

SPSS expects the data in one variable and group names in the other. If there are more than two group names in the group variable, only the two that are named in the dialog box will be compared.

STATCRUNCH

To do inference for the difference between two means using summaries:
- Click on **Stat.**
- Choose **T Statistics » Two sample » With Summary.**
- Enter the **Sample mean, Standard deviation,** and sample **Size** for each group.
- De-select **Pool variances.**
- Indicate **Hypothesis Test,** then enter the hypothesized Null mean difference (usually 0), and choose the Alternative hypothesis.

OR

Indicate **Confidence Interval,** and then enter the **Level** of confidence.
- Click on **Compute!**

To do inference for the difference between two means using data:
- Click on **Stat.**
- Choose **T Statistics » Two sample » With Data.**
- Choose the variable Column for each group.
- De-select **Pool variances.**
- Indicate **Hypothesis Test,** then enter the hypothesized Null mean difference (usually 0), and choose the Alternative hypothesis.

OR

Indicate **Confidence Interval,** and then enter the **Level** of confidence.
- Click on **Compute!**

TI-NSPIRE

To compute a confidence interval for the difference between two population means from a Calculator page, press (menu), (6) for Statistics, (6) for Confidence Intervals, and (4) for 2-Sample *t*-interval. From a Lists & Spreadsheet page, press (menu), (4) for Statistics, (3) for Confidence Intervals, and (4) for 2-Sample t interval. Select between Data and Stats, (tab) to OK, and (enter). Complete the dialog box. Be sure to enter the C level as a decimal, such as .99.

To compute a hypothesis test for the difference between two population means from a Calculator page, press (menu), (6) for Statistics, (7) for Stat Tests, and (4) for 2-Sample *t*-test. From a Lists & Spreadsheet page, press (menu), (4) for Statistics, (4) for Stat Tests, and (4) for 2-Sample t Test. Select between Data and Stats, (tab) to OK, and (enter). Complete the dialog box. From a Lists & Spreadsheet page can you select the option to draw the plot of the test statistic with the P-value shaded.

Chapter 24: Paired Samples and Blocks

DATA DESK

- Select variables.
- From the **Calc** menu, choose **Estimate** for confidence intervals or **Test** for hypothesis tests.
- **Select** the interval or test from the drop-down menu, and make other choices in the dialog.

COMMENTS

Data Desk expects the two groups to be in separate variables and in the same "Relation"—that is, about the same cases.

EXCEL

- In Excel 2003 and earlier, select **Data Analysis** from the **Tools** menu.
- In Excel 2007, select **Data Analysis** from the **Analysis** Group on the **Data Tab**.
- From the **Data Analysis** menu, choose **t-test: paired two-sample for Means**. Fill in the cell ranges for the two groups, the hypothesized difference, and the alpha level.

COMMENTS

Excel expects the two groups to be in separate cell ranges.
Warning: Do not compute this test in Excel without checking for missing values. If there are any missing values (empty cells), Excel will usually give a wrong answer. Excel compacts each list, pushing values up to cover the missing cells, and then checks only that it has the same number of values in each list. The result is mismatched pairs and an entirely wrong analysis.

JMP

- From the **Analyze** menu, select **Matched Pairs**.
- Specify the columns holding the two groups in the **Y Paired Response** dialog.
- Click **OK**.

MINITAB

- From the **Stat** menu, choose the **Basic Statistics** submenu.
- From that menu, choose **Paired _t_** ...
- Then fill in the dialog.

COMMENTS

Minitab takes "First sample" minus "Second sample."

SPSS

- From the **Analyze** menu, choose the **Compare Means** submenu.
- From that, choose the **Paired-Samples _t_-test** command.
- Select pairs of variables to compare, and click the arrow to add them to the selection box.

COMMENTS

You can compare several pairs of variables at once. Options include the choice to exclude cases missing in any pair from all tests.

STATCRUNCH

To do inference for the mean of paired differences:
- Click on **Stat**.
- Choose **T Statistics » Paired**.
- Choose the **Column** for each variable.
- Check **Save differences** so you can look at a histogram to be sure the Nearly Normal condition is satisfied.

- Indicate **Hypothesis Test**, then enter the hypothesized Null mean difference (usually 0), and choose the Alternative hypothesis.

OR

  Indicate **Confidence Interval**, and then enter the **Level** of confidence.
- Click on **Compute!**

TI-NSPIRE

For inference on a matched pair design, compute a third list of differences such as _diff = time2-time1_. To do this, name the third list by typing the name of the variable in the top row of your new list. Then ENTER to get to the second row of the column. In that second row, type =, then press VAR and arrow to the name of the first variable,

press −, then VAR, then arrow down to the name of the second variable. Press ENTER. Then construct the confidence interval or conduct the hypothesis test in the same way as 1-sample procedures, using the list of differences.

Chapter 25: Comparing Counts

DATA DESK

- Select variables.
- From the **Calc** menu, choose **Contingency Table**.
- From the table's HyperView menu choose **Table Options**. (Or Choose **Calc** > **Calculation Options** > **Table Options**.)
- In the dialog, check the boxes for **Chi Square** and for **Standardized Residuals**. Data Desk will display the chi-square and its P-value below the table, and the standardized residuals within the table.

COMMENTS

Data Desk automatically treats variables selected for this command as categorical variables even if their elements are numerals. The **Compute** Counts command in the table's HyperView menu will make variables that hold the table contents (as selected in the Table Options dialog), including the standardized residuals.

EXCEL

Excel offers the function
CHITEST(actual_range, expected_range), which computes a chi-square value for homogeneity. Both ranges are of the form UpperLeftcell:LowerRightCell, specifying two rectangular tables that must hold counts (although Excel will not check for integer values). The two tables must be of the same size and shape.

COMMENTS

Excel's documentation claims this is a test for independence and labels the input ranges accordingly, but Excel offers no way to find expected counts, so the function is not particularly useful for testing independence. You can use this function only if you already know both tables of counts or are willing to program additional calculations.

JMP

- From the **Analyze** menu, select **Fit Y by X**.
- Choose one variable as the Y, response variable, and the other as the X, factor variable. Both selected variables must be Nominal or Ordinal.
- **JMP** will make a plot and a contingency table. Below the contingency table, **JMP** offers a **Tests** panel. In that panel, the Chi Square for independence is called a **Pearson ChiSquare**. The table also offers the P-value.
- Click on the **Contingency Table** title bar to drop down a menu that offers to include a **Deviation** and **Cell Chi square** in each cell of the table.

COMMENTS

JMP will choose a chi-square analysis for a **Fit Y by X** if both variables are nominal or ordinal (marked with an N or O), but not otherwise. Be sure the variables have the right type.

Deviations are the observed—expected differences in counts. Cell chi-squares are the squares of the standardized residuals. Refer to the deviations for the sign of the difference.

Look under **Distributions** in the **Analyze** menu to find a chi-square test for goodness-of-fit.

MINITAB

- From the **Stat** menu, choose the **Tables** submenu.
- From that menu, choose **Chi Square Test** . . . .
- In the dialog, identify the columns that make up the table. Minitab will display the table and print the chi-square value and its P-value.

COMMENTS

Alternatively, select the **Cross Tabulation** . . . command to see more options for the table, including expected counts and standardized residuals.

SPSS

- From the **Analyze** menu, choose the **Descriptive Statistics** submenu.
- From that submenu, choose **Crosstabs** . . . .
- In the **Crosstabs** dialog, assign the row and column variables from the variable list. Both variables must be categorical.
- Click the **Cells** button to specify that standardized residuals should be displayed.
- Click the **Statistics** button to specify a chi-square test.

COMMENTS

SPSS offers only variables that it knows to be categorical in the variable list for the Crosstabs dialog. If the variables you want are missing, check that they have the right type.

STATCRUNCH

To perform a Goodness-of-Fit test:
- Enter the observed counts in one column of a data table, and the expected counts in another.
- Click on **Stat**.
- Choose **Goodness-of-fit » Chi-Square test**.
- Choose the **Observed Column** and the **Expected Column**.
- Click on **Compute!**

COMMENTS

These Chi-square tests may also be performed using the actual data table instead of summary counts. See the StatCrunch **Help page** for details.

To perform a Test of Homogeneity or Independence:
- Create a table (without totals):
- Name the first column as one variable, enter the categories underneath.
- Name the adjacent columns as the categories of the other variable, entering the observed counts underneath.
- Click on **Stat**.
- Choose **Tables » Contingency » With Summary**.
- Choose the **Columns** holding counts.
- Choose the **Row labels column**.
- Enter the **Column label** name.
- Choose **Expected Count** (and, optionally, Contributions to Chi-Square).
- Click on **Compute!**

TI-NSPIRE

To conduct a χ^2 goodness of fit test, enter the observed and the expected values into two named lists. To name a list, type a variable name in the top row of the column. To conduct the test from a calculator page, press (menu), ⑥ for Statistics, ⑦ for Stat Tests, and ⑦ for χ^2 GOF. To conduct the test from the Lists & Spreadsheet page, press (menu), ④ for Statistics, ④ for Stat Tests, and ⑦ for χ^2 GOF. Complete the dialog box. From a Lists & Spreadsheet page can you select the option to draw the plot of the test statistic with the P-value shaded.

To conduct a χ^2 test of independence or homogeneity, first enter the data into a matrix. From a Calculator page, press (ctrl) (×⁺⁰) and select the 3 × 3 matrix icon. Enter the dimensions and (tab) to OK, and (enter). Then type the data into the matrix. Then press ▶ to exit the matrix, press (ctrl) (sto→) and a matrix name such as *obs* (for observed) to store the matrix. To complete the test from a Calculator page, press (⌂), ① for Calculator, (menu), ⑥ for Statistics, ⑦ for Stat Tests, and ⑧ for χ^2 2-way Test. To complete the test from the Lists & Spreadsheet page, press (menu), ④ for Statistics, ④ for Stat Tests, ⑧ for χ^2 2-way Test. Use the arrow keys to select the name of the observed matrix in the dialog box. The matrix of expected values will be called stat.expmatrix, and can be accessed by pressing [VAR] from a Calculator page.

Chapter 26: Inferences for Regression

DATA DESK

- Select *Y*- and *X*-variables.
- From the **Calc** menu, choose **Regression**.
- Data Desk displays the regression table.
- Select plots of residuals from the Regression table's HyperView menu.

COMMENTS

You can change the regression by dragging the icon of another variable over either the *Y*- or *X*-variable name in the table and dropping it there. The regression will recompute automatically.

EXCEL

- In Excel 2003 and earlier, select Data Analysis from the **Tools** menu. In Excel 2007, select Data Analysis from the **Analysis Group** on the Data Tab.
- Select Regression from the **Analysis Tools** list.
- Click the **OK** button.
- Enter the data range holding the Y-variable in the box labeled "Y-range."
- Enter the range of cells holding the X-variable in the box labeled "X-range."
- Select the **New Worksheet Ply** option.
- Select **Residuals** options. Click the **OK** button.

COMMENTS

The Y and X ranges do not need to be in the same rows of the spreadsheet, although they must cover the same number of cells. But it is a good idea to arrange your data in parallel columns as in a data table.

Although the dialog offers a Normal probability plot of the residuals, the data analysis add-in does not make a correct probability plot, so don't use this option. Excel calls the standard deviation of the residuals the "Standard Error." This is a common error. Don't be confused; it is not $SE(y)$, but rather s_e.

JMP

- From the **Analyze** menu, select **Fit Y by X**.
- Select variables: a Y, Response variable, and an X, Factor variable. Both must be continuous (quantitative).
- JMP makes a scatterplot.
- Click on the red triangle beside the heading labeled **Bivariate Fit** ... and choose **Fit Line**. JMP draws the least squares regression line on the scatterplot and displays the results of the regression in tables below the plot.
- The portion of the table labeled "Parameter Estimates" gives the coefficients and their standard errors, *t*-ratios, and P-values.

COMMENTS

JMP chooses a regression analysis when both variables are "Continuous." If you get a different analysis, check the variable types.

The Parameter table does not include the residual standard deviation s_e. You can find that as Root Mean Square Error in the Summary of Fit panel of the output.

MINITAB

- Choose **Regression** from the **Stat** menu.
- Choose **Regression** ... from the **Regression** submenu.
- In the Regression dialog, assign the Y-variable to the Response box and assign the X-variable to the Predictors box.
- Click the **Graphs** button.
- In the Regression-Graphs dialog, select **Standardized residuals**, and check **Normal plot of residuals** and **Residuals versus fits**.
- Click the **OK** button to return to the Regression dialog.
- Click the **OK** button to compute the regression.

COMMENTS

You can also start by choosing a Fitted Line plot from the **Regression** submenu to see the scatterplot first—usually good practice.

SPSS

- Choose **Regression** from the **Analyze** menu.
- Choose **Linear** from the **Regression** submenu.
- In the Linear Regression dialog that appears, select the Y-variable and move it to the dependent target. Then move the X-variable to the independent target.
- Click the **Plots** button.

- In the Linear Regression Plots dialog, choose to plot the *SRESIDs against the *ZPRED values.
- Click the **Continue** button to return to the Linear Regression dialog.
- Click the **OK** button to compute the regression.

STATCRUNCH

- Click on **Stat**.
- Choose **Regression » Simple Linear**.
- Choose X and Y variable names from the list of columns.
- Indicate that you want to see a residuals plot and a histogram of the residuals.
- Indicate **Hypothesis tests**, then enter the hypothesized null slope (usually 0) and choose the alternative hypothesis.

  OR

  Indicate **Confidence Interval**, then enter the **Level** of confidence.
- Click on **Compute!**
- Click on **Next** to see any plots you chose.

COMMENTS

Be sure to check the conditions for regression inference by looking at both the residuals plot and a histogram of the residuals.

TI-NSPIRE

To compute a confidence interval for a population slope, first enter the data into two named lists. To name a list, type a variable name in the top row of the column. To conduct the test from a calculator page, press (menu), (6) for Statistics, (6) for Confidence Intervals, and (7) for Linear Reg *t*-intervals. To construct the interval from the Lists & Spreadsheet page, press (menu), (4) for Statistics, (3) for Confidence Intervals, and (7) for Linear Reg t intervals. Select slope, (tab) to OK, and (enter). Complete the dialog box. Be sure to enter the C level as a decimal, such as .99.

To compute a hypothesis test for a population slope, first enter the data into two named lists. To name a list, type a variable name in the top row of the column. To conduct the test from a calculator page, press (menu), (6) for Statistics, (7) for Stat Tests, and (A) for Linear Reg *t*-test. To complete the test from the Lists & Spreadsheet page, press (menu), (4) for Statistics, (4) for Stat Tests, (A) for Linear Reg t test. Complete the dialog box.

Chapter 27: Analysis of Variance

DATA DESK

- Select the response variable as Y and the factor variable as X.
- From the **Calc** menu, choose **ANOVA**.
- Data Desk displays the ANOVA table.
- Select plots of residuals from the ANOVA table's HyperView menu.

COMMENTS

Data Desk expects data in "stacked" format. You can change the ANOVA by dragging the icon of another variable over either the Y or X variable name in the table and dropping it there. The analysis will recompute automatically.

EXCEL

- In Excel 2003 and earlier, select **Data Analysis** from the Tools menu.
- In Excel 2007, select **Data Analysis** from the Analysis Group on the Data Tab.
- Select **Anova Single Factor** from the list of analysis tools.
- Click the **OK** button.
- Enter the data range in the box provided.
- Check the **Labels in First Row** box, if applicable.
- Enter an alpha level for the F-test in the box provided.
- Click the **OK** button.

COMMENTS

The data range should include two or more columns of data to compare. Unlike all other statistics packages, Excel expects each column of the data to represent a different level of the factor. However, it offers no way to label these levels. The columns need not have the same number of data values, but the selected cells must make up a rectangle large enough to hold the column with the most data values.

JMP

- From the **Analyze** menu select **Fit Y by X**.
- Select variables: a quantitative Y, Response variable, and a categorical X, Factor variable.
- JMP opens the **Oneway** window.
- Click on the red triangle beside the heading, select **Display Options**, and choose **Boxplots**.

- From the same menu choose the **Means/ANOVA.t-test** command.
- JMP opens the oneway ANOVA output.

COMMENTS

JMP expects data in "stacked" format with one response and one factor variable.

MINITAB

- Choose **ANOVA** from the Stat menu.
- Choose **One-way** … from the **ANOVA** submenu.
- In the One-way Anova dialog, assign a quantitative Y variable to the Response box and assign a categorical X variable to the Factor box.
- Check the **Store Residuals** check box.
- Click the **Graphs** button.
- In the ANOVA-Graphs dialog, select **Standardized residuals**, and check **Normal plot of residuals** and **Residuals versus fits**.

- Click the **OK** button to return to the ANOVA dialog.
- Click the **OK** button to compute the ANOVA.

COMMENTS

If your data are in unstacked format, with separate columns for each treatment level, choose **One-way (unstacked)** from the **ANOVA** submenu.

SPSS

- Choose **Compare Means** from the **Analyze** menu.
- Choose **One-way ANOVA** from the **Compare Means** submenu.
- In the One-Way ANOVA dialog, select the Y-variable and move it to the dependent target. Then move the X-variable to the independent target.
- Click the **OK** button.

COMMENTS

SPSS expects data in stacked format. The **Contrasts** and **Post Hoc** buttons offer ways to test contrasts and perform multiple comparisons. See your SPSS manual for details.

STATCRUNCH

To compute an ANOVA:
- Click on **Stat**.
- Choose **ANOVA » One Way**.
- Choose the **Columns** for all groups. (After the first one, you may hold down the ctrl or command key to choose more.)

OR

Choose the **single column** containing the data **(Responses)** and the column containing the **Factors**.
- Click on **Compute!**

TI-NSPIRE

From a Calculator page, press (menu), ⟨6⟩ for Statistics, ⟨7⟩ for Stat Tests, and ⟨C⟩ for ANOVA. From a Lists & Spreadsheet page, press (menu), ⟨4⟩ for Statistics, ⟨4⟩ for Stat

Tests, and ⟨C⟩ for ANOVA. Select Data or Stats as an input method, then complete the dialog box.

Chapter 28: Multiple Regression

DATA DESK

- Select Y- and X-variable icons.
- From the **Calc** menu, choose **Regression**.
- Data Desk displays the regression table.
- Select plots of residuals from the Regression table's HyperView menu.

COMMENTS

You can change the regression by dragging the icon of another variable over either the Y- or an X-variable name in the table and dropping it there. You can add a predictor by dragging its icon into that part of the table. The regression will recompute automatically.

EXCEL

- In Excel 2003 and earlier, select **Data Analysis** from the **Tools** menu.
- In Excel 2007, select **Data Analysis** from the **Analysis Group** on the Data Tab.
- Select **Regression** from the **Analysis Tools** list.
- Click the **OK** button.
- Enter the data range holding the Y-variable in the box labeled "Y-range."
- Enter the range of cells holding the X-variables in the box labeled "X-range."
- Select the **New Worksheet** Ply option.
- Select **Residuals** options. Click the **OK** button.

COMMENTS

The Y and X ranges do not need to be in the same rows of the spreadsheet, although they must cover the same number of cells. But it is a good idea to arrange your data in parallel columns as in a data table. The X-variables must be in adjacent columns. No cells in the data range may hold nonnumeric values.

Although the dialog offers a Normal probability plot of the residuals, the data analysis add-in does not make a correct probability plot, so don't use this option.

JMP

- From the **Analyze** menu, select **Fit Model**.
- Specify the response, Y. Assign the predictors, X, in the **Construct Model Effects** dialog box.
- Click on **Run Model**.

COMMENTS

JMP chooses a regression analysis when the response variable is "Continuous." The predictors can be any combination of quantitative or categorical. If you get a different analysis, check the variable types.

MINITAB

- Choose **Regression** from the **Stat** menu.
- Choose **Regression** ... from the **Regression** submenu.
- In the Regression dialog, assign the Y-variable to the Response box and assign the X-variables to the Predictors box.
- Click the **Graphs** button.

- In the Regression-Graphs dialog, select **Standardized residuals**, and check **Normal plot of** residuals and **Residuals versus fits**.
- Click the **OK** button to return to the Regression **dialog**.
- Click the **OK** button to compute the regression.

SPSS

- Choose **Regression** from the **Analyze** menu.
- Choose **Linear** from the **Regression** submenu.
- When the Linear Regression dialog appears, select the Y-variable and move it to the dependent target. Then move the X-variables to the independent target.
- Click the **Plots** button.

- In the Linear Regression Plots dialog, choose to plot the *SRESIDs against the *ZPRED values.
- Click the **Continue** button to return to the **Linear** Regression dialog.
- Click the **OK** button to compute the regression.

STATCRUNCH

To compute a multiple regression:
- Click on **Stat**.
- Choose **Regression » Multiple Linear**.
- Choose the Y-variable name from the list of columns.
- Choose the X-variable names. (After the first one, you may need to hold down the ctrl or command key to choose more.)
- Click on **Compute!**

COMMENTS

Note that before you **Compute!** you may choose to save the residuals and/or fitted values in your data table.

TI-NSPIRE

To construct a multiple regression confidence interval from a Calculator page, press (menu), ⟨6⟩ for Statistics, ⟨6⟩ for Confidence Intervals, and ⟨8⟩ for Multiple Reg Intervals. From a Lists & Spreadsheet page, press (menu), ⟨4⟩ for Statistics, ⟨3⟩ for Confidence Intervals, and ⟨8⟩ for Multiple Reg Intervals. Enter the number of independent variables and complete the dialog box.

To construct a multiple regression test from a Calculator page, press (menu), ⟨6⟩ for Statistics, ⟨7⟩ for Stat Tests, and [B] for Multiple Reg Tests. From a Lists & Spreadsheet page, press (menu), ⟨4⟩ for Statistics, ⟨4⟩ for Stat Tests, and [B] for Multiple Reg Tests. Enter the number of independent variables and complete the dialog box.

Appendix C: Answers

Here are the "answers" to the exercises for the chapters and the unit reviews. Note that, these answers are only outlines of the complete solution. Your solution should follow the model of the Step-By-Step examples, where appropriate. You should explain the context, show your reasoning and calculations, and draw conclusions. For some problems, what you decide to include in an argument may differ somewhat from the answers here. But, of course, the numerical part of your answer should match the numbers in the answers shown.

Chapter 1

1. Categorical
3. Quantitative
5. Answers will vary.
7. *Who*—2500 cars; *What*—Distance from car to bicycle; *Population*—All cars passing bicyclists
9. *Who*—Coffee drinkers at a Newcastle University coffee station; *What*—Amount of money contributed; *Population*—All people in honor system payment situations
11. *Who*—24 blind patients; *What*—Response to embryonic stem cell treatment; *Population*—All patients with one of these two forms of blindness
13. *Who*—54 bears; *Cases*—Each bear is a case; *What*—Weight, neck size, length, and sex; *When*—Not specified; *Where*—Not specified; *Why*—To estimate weight from easier-to-measure variables; *How*—Researchers collected data on 54 bears they were able to catch; *Variable*—Weight; *Type*—Quantitative; *Units*—Not specified; *Variable*—Neck size; *Type*—Quantitative; *Units*—Not specified; *Variable*—Length; *Type*—Quantitative; *Units*—Not specified; *Variable*—Sex; *Type*—Categorical
15. *Who*—Arby's sandwiches; *Cases*—Each sandwich is a case; *What*—Type of meat, number of calories, and serving size; *When*—Not specified; *Where*—Arby's restaurants; *Why*—To assess nutritional value of sandwiches; *How*—Report by Arby's restaurants; *Variable*—Type of meat; *Type*—Categorical; *Variable*—Number of calories; *Type*—Quantitative; *Units*—Calories; *Variable*—Serving size; *Type*—Quantitative; *Units*—Ounces
17. *Who*—882 births; *Cases*—Each of the 882 births is a case; *What*—Mother's age, length of pregnancy, type of birth, level of prenatal care, birth weight of baby, sex of baby, and baby's health problems; *When*—1998–2000; *Where*—Large city hospital; *Why*—Researchers were investigating the impact of prenatal care on newborn health; *How*—Not specified exactly, but probably from hospital records; *Variable*—Mother's age; *Type*—Quantitative; *Units*—Not specified, (probably years); *Variable*—Length of pregnancy; *Type*—Quantitative; *Units*—Weeks; *Variable*—Birth weight of baby; *Type*—Quantitative; *Units*—Not specified, (probably pounds and ounces); *Variable*—Type of birth; *Type*—Categorical; *Variable*—Level of prenatal care; *Type*—Categorical; *Variable*—Sex; *Type*—Categorical; *Variable*—Baby's health problems; *Type*—Categorical
19. *Who*—Experiment subjects; *Cases*—Each subject is an individual; *What*—Treatment (herbal cold remedy or sugar solution) and cold severity; *When*—Not specified; *Where*—Not specified; *Why*—To test efficacy of herbal remedy on common cold; *How*—The scientists set up an experiment; *Variable*—Treatment; *Type*—Categorical; *Variable*—Cold severity rating; *Type*—Quantitative (perhaps ordinal categorical); *Units*—Scale from 0 to 5; *Concerns*—The severity of a cold seems subjective and difficult to quantify. Scientists may feel pressure to report negative findings of herbal product.

21. *Who*—Streams; *Cases*—Each stream is a case; *What*—Name of stream, substrate of the stream, acidity of the water, temperature, BCI; *When*—Not specified; *Where*—Upstate New York; *Why*—To study ecology of streams; *How*—Not specified; *Variable*—Stream name; *Type*—Identifier; *Variable*—Substrate; *Type*—Categorical; *Variable*—Acidity of water; *Type*—Quantitative; *Units*—pH; *Variable*—Temperature; *Type*—Quantitative; *Units*—Degrees Celsius; *Variable*—BCI; *Type*—Quantitative; *Units*—Not specified
23. *Who*—102 refrigerator models; *Cases*—Each of the 102 refrigerator models is a case.; *What*—Brand, cost, size, temperature performance, noise, ease of use, energy efficiency, estimated annual energy cost, overall rating, and exterior dimensions; *When*—2012; *Where*—United States; *Why*—To provide information to the readers of *Consumer Reports*; *How*—Not specified; *Variable*—Brand; *Type*—Categorical; *Variable*—Cost; *Type*—Quantitative; *Units*—Not specified (dollars); *Variable*—Size; *Type*—Quantitative; *Units*—Cubic feet; *Variable*—Temperature performance; *Type*—Categorical; *Variable*—Noise; *Type*—Categorical; *Variable*—Ease of Use; *Type*—Categorical; *Variable*—Energy efficiency; *Type*—Categorical; *Variable*—Estimated annual energy cost; *Type*—Quantitative; *Units*—Not specified (dollars); *Variable*—Overall rating; *Type*—Categorical; *Variable*—Exterior dimensions; *Type*—Quantitative; *Units*—Not specified (inches)
25. *Who*—Kentucky Derby races; *What*—Year, winner, jockey, trainer, owner, time; *When*—1875 to 2012; *Where*—Churchill Downs, Louisville, Kentucky; *Why*—Not specified (to see trends in horse racing?); *How*—Official statistics collected at race; *Variable*—Year; *Type*—Quantitative; *Units*—Year; *Variable*—Winner; *Type*—Identifier; *Variable*—Jockey; *Type*—Categorical; *Variable*—Trainer; *Type*—Categorical; *Variable*—Owner; *Type*—Categorical; *Variable*—Time; *Type*—Quantitative; *Units*—Minutes and seconds

Chapter 2

1. Answers will vary.
3. Answers will vary.
5. a) Yes; each is categorized in a single genre.
   b) Horror
7. a) Comedy
   b) It is easier to tell from the bar chart; slices of the pie chart are too close in size.
9. i. d, ii. a, iii. c, iv. b
11. 1755 students applied for admission to the magnet schools program. 53% were accepted, 17% were wait-listed, and the other 30% were turned away.
13. a) Yes. We can add because these categories do not overlap. (Each person is assigned only one cause of death.)
    b) $100 - (25.4 + 23.2 + 5.6 + 5.3 + 5.1) = 35.4\%$

c) Either a bar chart or pie chart with "other" added would be appropriate. A bar chart is shown.

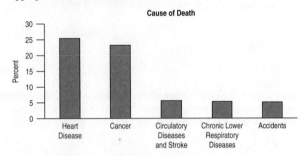

Cause of Death

15. a) The bar chart shows that grounding and collision are the most frequent causes of oil spills. Very few have unknown causes.
    b) A pie chart seems appropriate as well.

17. There's no title, the percentages total only 93%, and the three-dimensional display distorts the sizes of the regions.

19. In both the South and West, about 58% of the eighth-grade smokers preferred Marlboro. Newport was the next most popular brand, but was far more popular in the South than in the West, where Camel was cited nearly 3 times as often as in the South. Nearly twice as many smokers in the West as in the South indicated that they had no usual brand (12.9% to 6.7%).

21. a) The column totals are 100%.
    b) 19.5%   c) 19.2%
    d) i. 21.7%; ii. can't tell; iii. 0%; iv. can't tell

23. a) 82.5%   b) 12.9%   c) 11.1%   d) 13.4%   e) 85.7%

25. a) 73.9% 4-yr college, 13.4% 2-year college, 1.5% military, 5.2% employment, 6.0% other
    b) 77.2% 4-yr college, 10.5% 2-year college, 1.8% military, 5.3% employment, 5.3% other
    c) Many charts are possible. Here is a side-by-side bar chart.

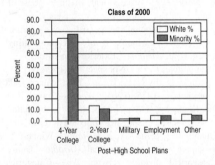

Class of 2000

d) The white and minority students' plans are very similar. The small differences should be interpreted with caution because the total number of minority students is small. There is little evidence of an association between race and plans.

27. a) 16.6%   b) 11.8%   c) 37.7%   d) 53.0%

29. 1755 students applied for admission to the magnet schools program: 53% were accepted, 17% were wait-listed, and the other 30% were turned away. While the overall acceptance rate was 53%, 93.8% of blacks and Hispanics were accepted, compared to only 37.7% of Asians and 35.5% of whites. Overall, 29.5% of applicants were black or Hispanic, but only 6% of those turned away were. Asians accounted for 16.6% of all applicants, but 25.4% of those turned away. Whites were 54% of the applicants and 68.5% of those who were turned away. It appears that the admissions decisions were not independent of the applicant's ethnicity.

31. a) 9.3%   b) 24.7%   c) 80.8%
    d) No, there appears to be no association between weather and ability to forecast weather. On days it rained, his forecast was correct 79.4% of the time. When there was no rain, his forecast was correct 81.0% of the time.

33. a) Low 20.0%, Normal 48.9%, High 31.0%
    b)

	Age		
Blood Pressure	**Under 30**	**30–49**	**Over 50**
Low	27.6%	20.7%	15.7%
Normal	49.0%	50.8%	47.2%
High	23.5%	28.5%	37.1%

c) **Blood Pressure by Age**

d) As age increases, the percent of adults with high blood pressure increases. By contrast, the percent of adults with low blood pressure decreases.

e) No, but it gives an indication that it might. We would need to follow the same individuals over time.

35. No, there's no evidence that Prozac is effective. The relapse rates were nearly identical: 28.6% among the people treated with Prozac, compared to 27.3% among those who took the placebo.

37. a) 4.8%   b) 49.7%.
    c) There are about 50% of each sex in each age group, but it ranges from 49% female in the youngest group to 54.5% in the second oldest age group. As the age increases, there generally a slight increase in the percentage of female drivers.

d) There is a slight association. As the age increases, there is a small increase in the percentage of female drivers.

39. a) 160 of 1300, or 12.3%
    b) Yes. Major surgery: 15.3% vs. minor surgery: 6.7%
    c) Large hospital: 13%; small hospital: 10%
    d) Large hospital: Major 15% vs. minor 5% Small hospital: Major 20% vs. minor 8%
    e) No. Smaller hospitals have a higher rate for both kinds of surgery, even though it's lower "overall."
    f) The small hospital has a larger percentage of minor surgeries (83.3%) than the large hospital (20%). Minor surgeries have a lower delay rate, so the small hospital looks better "overall."

41. a) 42.6%

b) A higher percentage of males than females were admitted: Males: 47.2% to females: 30.9%

c) Program 1: Males 61.9%, females 82.4%
Program 2: Males 62.9%, females 68.0%
Program 3: Males 33.7%, females 35.2%
Program 4: Males 5.9%, females 7.0%

d) The comparisons in c) show that males have a lower admittance rate in every program, even though the overall rate shows males with a higher rate of admittance. This is an example of Simpson's paradox.

Chapter 3

1. Answers will vary.

3. Answers will vary.

5. a) Unimodal (near 0) and skewed to the right. Many seniors will have 0 or 1 speeding tickets. Some may have several, and a few may have more than that.

b) Probably unimodal and slightly skewed to the right. It is easier to score 15 strokes over the mean than 15 strokes under the mean.

c) Probably unimodal and symmetric. Weights may be equally likely to be over or under the average.

d) Probably bimodal. Men's and women's distributions may have different modes. It may also be skewed to the right, since it is possible to have very long hair, but hair length can't be negative.

7. a) Bimodal. Looks like two groups. Modes are near 6% and 46%. No real outliers.

b) Looks like two groups of cereals, a low-sugar and a high-sugar group.

9. a) 78%

b) Skewed to the right with at least one high outlier. Most of the vineyards are less than 90 acres with a few high ones. The mode is between 0 and 30 acres.

11. a) Because the distribution is skewed to the right, we expect the mean to be larger.

b) Bimodal and skewed to the right. Center mode near 8 days. Another mode at 1 day (may represent patients who didn't survive). Most of the patients stay between 1 and 15 days. There are some extremely high values above 25 days.

c) The median and IQR, because the distribution is strongly skewed.

13. a) 45.5 points

b) 37 points and 55 points

c) No, the fences are at 10 points and 82 points.

d)

e) In the Super Bowl teams typically score a total of about 45 points, with half the games totaling between 37 and 55 points. In only one fourth of the games have the teams scored fewer than 37 points, and they once totaled 75.

15. a) The boxplots do not show clusters and gaps, nor locations of multiple modes.

b) Boxplots can give only general ideas about overall shape, and should not be used when more detail is needed.

17. a) Essentially symmetric, very slightly skewed to the right with two high outliers at 36 and 44. Most victims are between the ages of 16 and 24.

b) The slight increase between ages 22 and 24 is apparent in the histogram but not in the boxplot. It may be a second mode.

c) The median would be the most appropriate measure of center because of the slight skew and the extreme outliers.

d) The IQR would be the most appropriate measure of spread because of the slight skew and the extreme outliers.

19. a) The distribution is strongly skewed to the right, so use the median and IQR.

b) The IQR is 50, so the upper fence is the upper quartile +1.5 IQRs; that is, 78 + 75 = 153. There appear to be 3 to 5 parks that should be considered as outliers with more than 153 camp sites.

c)

d) The distribution is unimodal with a strong skew to the right. There are several outliers past the 1.5 × IQR upper fence of 153 camp sites. The median number of camp sites is 43.5 sites. The mean is 62.8 sites. The mean is larger than the median because it has been influenced by the strong skew and the outliers.

21. a) The standard deviation will be larger for set 2, since the values are more spread out. SD(set 1) = 2.2, SD(set 2) = 3.2.

b) The standard deviation will be larger for set 2, since 11 and 19 are farther from 15 than are 14 and 16. Other numbers are the same. SD(set 1) = 3.6, SD(set 2) = 4.5.

c) The standard deviation will be the same for both sets, since the values in the second data set are just the values in the first data set +80. The spread has not changed. SD(set 1) = 4.2, SD(set 2) = 4.2.

23. The mean and standard deviation because the distribution is unimodal and symmetric.

25. a) The mean is closest to $2.60 because that's the balancing point of the histogram.

b) The standard deviation is closest to $0.15 since that's a typical distance from the mean. There are no prices as far as $0.50 or $1.00 from the mean.

27. a) About 105 minutes

b) Yes, only 2 of these movies run that long.

c) They'd probably be about the same because the distribution is reasonably symmetric.

29. a) i. The middle 50% of movies ran between 97 and 115 minutes.

ii. On average, movie lengths varied from the mean run time by 17.3 minutes.

b) Because the distribution is reasonably symmetric, either measure of spread would be appropriate

31. The standard deviation would be much lower. (It actually becomes 4.6.) The IQR would be affected very little, if at all.

33. a) The median will be unaffected. The mean will be larger.

b) The range and standard deviation will increase; the IQR will be unaffected.

35. The publication is using the median; the watchdog group is using the mean, pulled higher by the several very successful movies in the long right tail.

37. a) Mean $525, median $450

b) 2 employees earn more than the mean.

c) The median, because of the outlier.

d) The IQR will be least sensitive to the outlier of $1200, so it would be the best to report.

39. a)

Stem	Leaf
33	1
33	5
34	1 2 3 3
34	6 6 7 8
35	
35	9
36	2 3
36	5 5 6

33|1 = $3.31/gallon

b) The distribution of gas prices is bimodal and skewed to the left (lower values), with peaks around $3.45, and $3.65. The lowest and highest prices were $3.31 and $3.66.

c) There is a rather large gap between the $3.48 and $3.59 prices.

41. a) Since these data are strongly skewed to the right, the median and IQR are the best statistics to report.

b) The mean will be larger than the median because the data are skewed to the right.

c) The median is 4 million. The IQR is 5 million (Q3 = 6 million, Q1 = 1 million).

d) The distribution of populations of the states and Washington, DC, is unimodal and skewed to the right. The median population is 4 million. Four states are outliers, one with a population of 34 million.

43. Reasonably symmetric, except for 2 low outliers, median at 36.

45. a)

Number of Hurricanes

b) Skewed to the right. Unimodal, mode near 2. No outliers.

47. a) This is not a histogram. The horizontal axis should split the number of home runs hit in each year into bins. The vertical axis should show the number of years in each bin.

b)

Runs

49. Skewed to the right, with one fairly symmetric cluster near 4.4, another small cluster at 5.6. Two stray values in middle seem not to belong to either group.

```
Stem | Leaf
  57 | 8
  56 | 27
  55 | 1
  54 |
  53 |
  52 | 9
  51 |
  50 | 8
  49 |
  48 | 2
  47 | 3
  46 | 034
  45 | 267
  44 | 015
  43 | 0199
  42 | 669
  41 | 22
```

41|2 = 4.12 pH

51. Histogram bins are too wide to be useful.

53. Neither appropriate nor useful. Zip codes are categorical data, not quantitative. But they do contain *some* information. The leading digit gives a rough East-to-West placement in the United States. So, we see that they have almost no customers in the Northeast, but a bar chart by leading digit would be more appropriate.

55. a) Median 284, IQR 10, Mean 282.98, SD 7.60

b) Because it's a bit skewed to the left, probably better to report Median and IQR.

c) Slightly skewed to the left. The center is around 284. The middle 50% of states scored between 278 and 288. Mississippi's was somewhat lower than other states' scores.

57. The histogram shows that the distribution of *Percent Change* is unimodal and skewed to the right. The states vary from a minimum

of 0.3% (Michigan) to 32.3% (Nevada) growth in the decade. The median was 7.0% and half of the states had growth between 3.5% and 11.8%. Not surprisingly, the top three states in terms of growth were all from the West: Nevada (32.3%), Arizona (28.6%), and Utah (24.7%).

% Change

Chapter 4

1. Answers will vary.

3. Answers will vary.

5. a) Prices appear to be both higher on average and more variable in Baltimore than in the other three cities. Prices in Chicago may be slightly higher than in Dallas and Denver, but the difference is very small.

b) There are outliers on the low end in Baltimore and Chicago and one high outlier in Dallas, but these do not affect the overall conclusions reached in part a).

7. The distributions are both unimodal and skewed to the right, with 2011 being more strongly skewed. Though there are potential outliers on the high end for both sets, none seem to be departures from the pattern. The median for 2011 is somewhat lower and the IQR is slightly larger than for 2007.

9. a) About 59%

b) Bimodal

c) Some cereals are very sugary; others are healthier low-sugar brands.

d) Yes

e) Although the ranges appear to be comparable for both groups (about 28%), the IQR is larger for the adult cereals, indicating that there's more variability in the sugar content of the middle 50% of adult cereals.

11. a)

Region

b) Growth rates in NE/MW states are tightly clustered near 5%. S/W states are more variable, and bimodal with modes near 14 and 22. The S/W states have an outlier as well. Around all the modes, the distributions are fairly symmetric.

13. a) They should be put on the same scale, from 0 to 20 days.

b) Lengths of men's stays appear to vary more than for women. Men have a mode at 1 day and then taper off from there. Women have a mode near 5 days, with a sharp drop afterward.

c) A possible reason is childbirth.

15. a) Both girls have a median score of about 17 points per game, but Scyrine is much more consistent. Her IQR is about 2 points, while Alexandra's is over 10.

b) If the coach wants a consistent performer, she should take Scyrine. She'll almost certainly deliver somewhere between 15 and 20 points.

But if she wants to take a chance and needs a "big game," she should take Alexandra. Alex scores over 24 points about a quarter of the time. (On the other hand, she scores under 11 points as often.)

17. Women appear to marry about 3 years younger than men, but the two distributions are very similar in shape and spread.

19. In general, fuel economy is higher in cars than in either SUVs or pickup trucks. The top 50% of cars get higher fuel economy than 75% of SUVs and all of the pickups. The distribution for pickups shows less spread.

21. Load factors are generally highest and least variable in the summer months (June–August). They are lower and more variable in the winter and spring.

23. The class A is 1, class B is 2, and class C is 3.

25. a) Probably slightly left skewed. The mean is slightly below the median, and the 25th percentile is farther from the median than the 75th percentile.

  b) No, all data are within the fences.

  c)

  d) The 48 universities graduate, on average, about 68% of freshmen "on time," with percents ranging from 43% to 87%. The middle 50% of these universities graduate between 59% and 75% of their freshmen in 4 years.

27. a) *Who:* Student volunteers

  What: Memory test

  Where, when: Not specified

  How: Students took memory test 2 hours after drinking caffeine-free, half-dose caffeine, or high-caffeine soda.

  Why: To see if caffeine makes you more alert and aids memory retention.

  b) Drink: categorical; test score: quantitative.

  c)

  d) The participants scored about the same with no caffeine and low caffeine. The medians for both were 21 points, with slightly more variation for the low-caffeine group. The high-caffeine group generally scored lower than the other two groups on all measures of the 5-number summary: min, lower quartile, median, upper quartile, and max.

29. a) About 36 mph

  b) Q_1 about 35 mph and Q_3 about 37 mph

  c) The range appears to be about 7 mph, from about 31 to 38 mph. The IQR is about 2 mph.

  d) We can't know exactly, but the boxplot may look something like this:

  e) The median winning speed has been about 36 mph, with a max of about 38 and a min of about 31 mph. Half have run between about 35 and 37 mph, for an IQR of 2 mph.

31. a) Boys

  b) Boys

  c) Girls

  d) The boys appeared to have more skew, as their scores were less symmetric between quartiles. The girls' quartiles are the same distance from the median, although the left tail stretches a bit farther to the left.

  e) Girls. Their median and upper quartiles are larger. The lower quartile is slightly lower, but close.

  f) $[14(4.2) + 11(4.6)]/25 = 4.38$

33.

There appears to be an outlier! This point should be investigated. We'll proceed by redoing the plots with the outlier omitted:

It appears that slow speed provides much greater accuracy. But the outlier should be investigated. It is possible that slow speed can induce an infrequent very large distance.

35. a)

  b) Mileage for U.S. models is typically lower, and less variable than for cars made elsewhere. The median for U.S. models is around 22 mpg, compared to 24 for the others. Both groups have some high outliers—most likely hybrid vehicles. (Other answers possible.)

37. a) Day 16 (but any estimate near 20 is okay).

  b) Day 65 (but anything around 60 is okay).

  c) Around day 50

39. a) Most of the data are found in the far left of this histogram. The distribution is very skewed to the right.

  b) Re-expressing the data by, for example, logs or square roots might help make the distribution more nearly symmetric.

41. a) The logarithm makes the histogram more symmetric. It is easy to see that the center is around 3.5 in log assets.

  b) That has a value of around 2,500 million dollars.

  c) That has a value of around 1,000 million dollars.

43. a) Fusion time and group.

b) Fusion time is quantitative (units = seconds). Group is categorical.

c) Both distributions are skewed to the right with high outliers. The boxplot indicates that visual information may reduce fusion time. The median for the Verbal/Visual group seems to be about the same as the lower quartile of the No/Verbal group.

Chapter 5

1. 1

3. McGwire's performance was more impressive from a comparison of z-scores (3.87 vs 3.16). But you'll have to decide if the steroid use makes it less impressive anyway.

5. Lowest score = 910. Mean = 1230. SD = 120. Q3 = 1350. Median = 1270. IQR = 240.

7. Your score was 2.2 standard deviations higher than the mean score in the class.

9. 65

11. In January, a high of 55 is not quite 2 standard deviations above the mean, whereas in July a high of 55 is more than 2 standard deviations lower than the mean. So it's less likely to happen in July.

13. The z-scores, which account for the difference in the distributions of the two tests, are 1.5 and 0 for Derrick and 0.5 and 2 for Julie. Derrick's total is 1.5, which is less than Julie's 2.5.

15. a) Megan   b) Anna

17. a) About 1.81 standard deviations below the mean.

b) 1000 ($z = -1.81$) is more unusual than 1250 ($z = 1.17$).

19. a) Mean = 1152 − 1000 = 152 pounds; SD is unchanged at 84 pounds.

b) Mean = 0.40(1152) = $460.80; SD = 0.40(84) = $33.60.

21. Min = 0.40(980) − 20 = $372;
median = 0.40(1140) − 20 = $436;
SD = 0.40(84) = $33.60; IQR = 0.40(102) = $40.80.

23. College professors can have between 0 and maybe 40 (or possibly 50) years' experience. A standard deviation of 1/2 year is impossible, because many professors would be 10 or 20 SDs away from the mean, whatever it is. An SD of 16 years would mean that 2 SDs on either side of the mean is plus or minus 32, for a range of 64 years. That's too high. So, the SD must be 6 years.

25. a)

b) 18.6 to 31.0 mpg   c) 16%

d) 13.5%   e) less than 12.4 mpg

27. Any weight more than 2 standard deviations below the mean, or less than 1152 − 2(84) = 984 pounds, is unusually low. We expect to see a steer below 1152 − 3(84) = 900 pounds only rarely.

29. a)

b) Between 52.5 and 102.5 points   c) 16%

d) 81.5%   e) 2.5%

31. Since the histogram is not unimodal and symmetric, it is not wise to have faith in numbers from the Normal model.

33. a) The distribution is not unimodal or symmetric as shown by the histogram. Also, 78% of the data lie within one standard deviation of the mean, but it would be about 68% if a normal model were appropriate.

b) The distribution would be unimodal and rather symmetric. The mean would increase and the standard deviation would decrease.

35. a) 16%   b) 13.56%

c) Because the Normal model doesn't fit perfectly.

d) Distribution is skewed to the right.

37. a) 2.5%

b) 2.5% of the 191 receivers, or 4.8 of them should gain more than 397.15 + 2 * 362.4 = 1122 yards. (In fact, 8 receivers exceeded this value.)

c) Data are strongly skewed to the right, not symmetric.

39. a) 12.2%   b) 71.6%   c) 23.3%

41. a) 1259.7 lb   b) 1081.3 lb   c) 1108 lb to 1196 lb

43. a) 1130.7 lb   b) 1347.4 lb   c) 113.3 lb

45. a)

b) 30.85%   c) 17.00%

d) 32 points   e) 212.9 points

47. a) 11.1%   b) (35.9, 40.5) inches   c) 40.5 inches

49. a) 5.3 grams   b) 6.4 grams

c) Younger because SD is smaller.

Part I Review

1. a)

b) Median 49 cents, IQR 7 cents.

c) The distribution is unimodal and left skewed. The center is near 50 cents; values extend from 42 cents to 53 cents.

3. a) If enough sopranos have a height of 65 inches, this can happen.

b) The distribution of heights for each voice part is roughly symmetric. The basses are generally slightly taller than the tenors. The sopranos and altos have about the same median height. Heights of basses and sopranos are more consistent than those of altos and tenors.

5. a) It means their heights are also more variable.

b) The z-score for women to qualify is 2.40, compared with 1.75 for men, so it is harder for women to qualify.

7. a) *Who*—People who live near State University; *What*—Age, attended college? Favorable opinion of State U?; *When*—Not stated; *Where*—Region around State U; *Why*—To report to the university's directors; *How*—Sampled and phoned 850 local residents.

b) Age—Quantitative (years); attended college?—categorical; favorable opinion?—categorical.

c) The fact that the respondents know they are being interviewed by the university's staff may influence answers.

9. a) These are categorical data, so mean and standard deviation are meaningless.

b) Not appropriate. Even if it fits well, the Normal model is meaningless for categorical data.

11. a)

b) The scores on Friday were higher by about 5 points on average. This is a drop of more than 10% off the average score and shows that students fared worse on Monday after preparing for the test on Friday. The spreads are about the same, but the scores on Monday are a bit skewed to the right.

c)

d) The changes (Friday–Monday) are unimodal and centered near 4 points, with a spread of about 5 (SD). They are fairly symmetric, but slightly skewed to the right. Only 3 students did better on Monday (had a negative difference).

13. a) Categorical
   b) Go fish. All you need to do is match the denomination. The denominations are not ordered. (Answers will vary.)
   c) Gin rummy. All cards are worth their value in points (face cards are 10 points). (Answers will vary.)

15. a) Annual mortality rate for males (quantitative) in deaths per 100,000 and water hardness (quantitative) in parts per million.
   b) Calcium is skewed right, possibly bimodal. There looks to be a mode down near 12 ppm that is the center of a fairly tight symmetric distribution and another mode near 62.5 ppm that is the center of a much more spread out, symmetric (almost uniform) distribution. Mortality, however, appears unimodal and symmetric with the mode near 1500 deaths per 100,000.

17. a) They are on different scales.
   b) January's values are lower and more spread out.
   c) Roughly symmetric but slightly skewed to the left. There are more low outliers than high ones. Center is around 40 degrees with an IQR of around 7.5 degrees.

19. a) Bimodal with modes near 2 and 4.5 minutes. Fairly symmetric around each mode.
   b) Because there are two modes, which probably correspond to two different groups of eruptions, an average might not make sense.
   c) The intervals between eruptions are longer for long eruptions. There is very little overlap. More than 75% of the short eruptions had intervals less than about an hour (62.5 minutes), while more than 75% of the long eruptions had intervals longer than about 75 minutes. Perhaps the interval could even be used to predict whether the next eruption will be long or short.

21. a)

The distribution is left skewed with a center of about 15. It has an outlier between 11 and 12.
   b) Even though the distribution is somewhat skewed, the mean and median are close. The mean is 15.0 and the SD is 1.25.
   c) Yes. 11.8 is already an outlier. 9.3 is more than 4.5 SDs below the mean. It is a very low outlier.

23. If we look only at the overall statistics, it appears that the follow-up group is insured at a much lower rate than those not traced (11.1% of the time compared with 16.6%). But most of the follow-up group were black, who have a lower rate of being insured. When broken down by race, the follow-up group actually has a higher rate of being insured for both blacks and whites. So the overall statistic is misleading and is attributable to the difference in race makeup of the two groups.

25. a)

   b) According to the model, reaction times are symmetric with center at 1.5 seconds. About 95% of all reaction times are between 1.14 and 1.86 seconds.
   c) 8.2%   d) 24.1%
   e) Quartiles are 1.38 and 1.62 seconds, so the IQR is 0.24 second.
   f) The slowest 1/3 of all drivers have reaction times of 1.58 seconds or more.

27. a)

   b) Mean 100.25, SD 25.54 pieces of mail.
   c) The distribution is somewhat symmetric and unimodal, but the center is rather flat, almost uniform.
   d) 64%. The Normal model seems to work reasonably well, since it predicts 68%.

29. a) Who—100 health food store customers; What—Have you taken a cold remedy?, and Effectiveness (scale 1 to 10); When—Not stated; Where—Not stated; Why—Promotion of herbal medicine; How—In-person interviews.
   b) Have you taken a cold remedy?—categorical. Effectiveness—categorical or ordinal.
   c) No. Customers are not necessarily representative, and the Council had an interest in promoting the herbal remedy.

31. a) 38 cars
   b) Possibly because the distribution is skewed to the right.
   c) Center—median is 148.5 cubic inches. Spread—IQR is 126 cubic inches.
   d) No. It's bigger than average, but smaller than more than 25% of cars. The upper quartile is at 231 cubic inches.
   e) No. 1.5 IQR is 189, and $105 - 189$ is negative, so there can't be any low outliers. $231 + 189 = 420$. There aren't any cars with engines bigger than this, since the maximum has to be at most 105 (the lower quartile) + 275 (the range) = 380.
   f) Because the distribution is skewed to the right, this is probably not a good approximation.
   g) Mean, median, range, quartiles, IQR, and SD all get multiplied by 16.4.

33. a) 44.0%
   b) If this were a random sample of all voters, yes.
   c) 38.4%   d) 0.87%
   e) 9.96%   f) 8.96%

35. a) Republican—3705, Democrat—3976, Neither—733; or
Republican—44.0%, Democrat—47.3%, Neither—8.7%.

b)

c) It appears that the older the voter, the less likely they are to lean Democratic and the more likely to lean Republican.

d) No. There is an association between age and affiliation. Younger voters tend to be more Democratic and less Republican.

37. a) 0.43 hour   b) 1.4 hours
c) 0.89 hour (or 53.4 minutes)
d) Survey results vary, and the mean and the SD may have changed.

Practice Exam Answers

I. 1. d   **3.** d   **5.** a   **7.** d   **9.** e

II. 1. a) Because the mean is so much smaller than the median, I expect the distribution is skewed to the left.

b) No outliers. Min > Q1 − 1.5IQR = 24 and Max < Q3 + 1.5IQR = 128

c)

d) The distribution of Dec–May high temperatures is roughly symmetric, while those for June–Nov are skewed to the left. June–Nov days were typically warmer, with a median high temperature about 22° greater than that for Dec–May. On over 75% of the June–Nov days the daily high was above the median for Dec–May. Variability in temperatures during the two periods was comparable, with nearly identical ranges (76° vs 70°) and IQRs indicating only slightly less consistency during Dec–May than June–Nov (33° vs 26°).

Chapter 6

1. a) Weight in ounces: explanatory; Weight in grams: response. (Could be other way around.) To predict the weight in grams based on ounces. Scatterplot: positive, straight, strong (perfectly linear relationship).

b) Ice cream cone sales: explanatory. Air-conditioner sales: response—although the other direction would work as well. To predict one from the other. Scatterplot: positive, straight, moderate.

c) Shoe size: explanatory; GPA: response. To try to predict GPA from shoe size. Scatterplot: no direction, no form, very weak.

d) Miles driven: explanatory; Gallons remaining: response. To predict the gallons remaining in the tank based on the miles driven since filling up. Scatterplot: negative, straight, moderate.

3. a) None   b) 3 and 4   c) 2, 3, and 4
d) 1 and 2   e) 3 and possibly 1

5. a)

b) Positive.   c) Linear.   d) Strong.   e) No.

7. There seems to be a very weak positive—or possibly no—relation between brain size and performance IQ.

9. a) True.

b) False. It will not change the correlation.

c) False. Correlation has no units.

11. Correlation does not demonstrate causation. The analyst's argument is that sales staff cause sales. However, the data may reflect the store hiring more people as sales increase, so any causation would run the other way.

13. a)

b) Unimodal, skewed to the right. The skew.

c) The positive, somewhat linear relation between batch number and broken pieces.

15. a) 0.006   b) 0.777   c) −0.923   d) −0.487

17. There may be an association, but not a correlation unless the variables are quantitative. There could be a correlation between average number of hours of TV watched per week per person and number of crimes committed per year. Even if there is a relationship, it doesn't mean one causes the other.

19. a) Yes. It shows a linear form and no outliers.

b) There is a strong, positive, linear association between drop and speed; the greater the coaster's initial drop, the higher the top speed.

21. The scatterplot is not linear; correlation is not appropriate.

23. The correlation may be near 0. We expect nighttime temperatures to be low in January, increase through spring and into the summer months, then decrease again in the fall and winter. The relationship is not linear.

25. The correlation coefficient won't change, because it's based on z-scores. The z-scores of the prediction errors are the same whether they are expressed in nautical miles or miles.

27. a) Assuming the relation is linear, a correlation of −0.772 shows a moderately strong relation in a negative direction.

b) Continent is a categorical variable. Correlation does not apply.

29. a) Actually, yes, taller children will tend to have higher reading scores, but this doesn't imply causation.

b) Older children are generally both taller and are better readers. Age is the lurking variable.

31. a) No. We don't know this from the correlation alone. There may be a nonlinear relationship or outliers.

b) No. We can't tell from the correlation what the form of the relationship is.

c) No. We don't know from the correlation coefficient.

d) Yes, the correlation doesn't depend on the units used to measure the variables.

33. This is categorical data even though it is represented by numbers. The correlation is meaningless.

35. a) The association is positive, moderately strong, and roughly straight, with several states whose HCI seems high for their median income and one state whose HCI appears low given its median income.

b) The correlation would still be 0.65.

c) The correlation wouldn't change.

d) DC would be a moderate outlier whose HCI is high for its median income. It would lower the correlation slightly.

e) No. We can only say that higher median incomes are associated with higher housing costs, but we don't know why. There may be other economic variables at work.

37. a)

b) Negative, linear, strong.   c) −0.909

d) There is a fairly strong linear relation in a negative direction between horsepower and highway gas mileage. Lower fuel efficiency is generally associated with higher horsepower.

39.

(Plot could have explanatory and predictor variables swapped.) Correlation is 0.199. There does not appear to be a relation between sodium and fat content in burgers, especially without the low-fat, low-sodium item. The correlation of 0.199 shows a weak relationship, even with the outlier included.

41. a) Yes, the scatterplot appears to be somewhat linear.

b) As the number of runs increases, the attendance also increases.

c) There is a positive association, but it does not *prove* that more fans will come if the number of runs increases. Association does not indicate causality.

43. There is a fairly strong, positive, linear relationship between the length of the track and the duration of the ride. In general, a longer track corresponds with a longer ride, with a correlation of 0.698.

45. a) We would expect that as one variable (say length of ride) increases, the rank will improve, which means it will decrease.

b) Duration has the strongest correlation ($r = -0.461$), but even that correlation is only moderate. And the scatterplot is not convincing. With the longest duration, the number-one ranked Bizarro seems to be pulling the correlation down. Much of the data appears to have a positive trend. There appear to be other factors that influence the rank of coaster more than any of the ones measured in this data set.

47. a)

The relation between position and distance is nonlinear, with a positive direction. There is very little scatter from the trend.

b) The relation is not linear.

c)

The relation between position number and log of distance appears to be roughly linear.

Chapter 7

1. 281 milligrams

3. The potassium content is actually lower than the model predicts for a cereal with that much fiber.

5. The model predicts that cereals will have approximately 27 more milligrams of potassium for every additional gram of fiber.

7. 81.5%

9. The true potassium contents of cereals vary from the predicted amounts with a standard deviation of 30.77 milligrams.

11. a) Model is appropriate.

b) Model is not appropriate. Relationship is nonlinear.

c) Model may not be appropriate. Spread is changing.

13. 300 pounds/foot. It's ridiculous to suggest an extra foot in length would add 3, 30, or 3000 pounds to a car's weight.

15. a) False. The line usually touches none of the points. We minimize the sum of the squared errors.

b) True.

c) False. It is the sum of the squares of all the residuals that is minimized.

17. a) $\widehat{Sales} = 8.10 + 0.913$ *Number of Sales People Working*.

b) It means that an additional 0.913 ($1000) or $913 of sales is expected with each additional sales person working.

c) It would mean that, on average, we expect sales of 8.10 ($1000) or $8100 with 0 sales people working. Doesn't really make sense in this context.

d) $24.55 ($1000) or $24,550. (24,540 if using the technology solution.)

e) 0.45 ($1000) or $450. ($460 with technology.)

f) Underestimated.

19. a) Thousands of dollars

b) 2.77 (the largest residual in magnitude)

c) 0.07 (the smallest residual in magnitude)

21. $R^2 = 93.2\%$ About 93% of the variance in *Sales* can be accounted for by the regression of *Sales* on *Number of Sales Workers*.

23. a) linearity assumption.

b) outlier condition.

c) equal spread condition.

25. a) *Price* (in thousands of dollars) is y and *Size* (in square feet) is x.

b) Slope is thousands of $ per square foot.

c) Positive. Larger homes should cost more.

27. A linear model on *Size* accounts for 71.4% of the variation in home *Price* for these data.

29. a) 0.845; + because larger homes cost more.

b) Price should be 0.845 SDs above the mean in price.

c) Price should be 1.690 SDs below the mean in price.

31. a) Predicted *Price* increases by about $0.061 × 1000, or $61.00, per additional sq ft.

b) 230.82 thousand, or $230,820.

c) $115,020; −$6000 is the residual.

33. a) R^2 does not tell whether the model is appropriate, but measures the strength of the linear relationship. High R^2 could also be due to an outlier.

b) Predictions based on a regression line are estimates of average values of y for a given x. The actual wingspan will vary around the prediction.

35. a) Probably not. Your score is better than about 97.5% of people, assuming scores follow the Normal model. Your next score is likely to be closer to the mean.

b) The friend should probably retake the test. His score is better than only about 16% of people. His score is likely to be closer to the mean.

37. a) Probably. The residuals show some initially low points, but there is no clear curvature.

b) The linear model on *Tar* content accounts for 92.4% of the variability in *Nicotine*.

39. a) $r = 0.961$

b) Nicotine should be 1.922 SDs below average.

c) Tar should be 0.961 SDs above average.

41. a) $\widehat{Nicotine} = 0.15403 + 0.065052\ Tar$

b) 0.414 mg

c) Predicted nicotine content increases by 0.065 mg of nicotine per additional milligram of tar.

d) We'd expect a cigarette with no tar to have 0.154 mg of nicotine.

e) 0.1094 mg

43. a) Yes. The relationship is straight enough, with a few outliers. The spread increases a bit for states with large median incomes, but we can still fit a regression line.

b) From summary statistics: $\widehat{HCI} = -156.50 + 0.0107\ MFI$; from original data: $\widehat{HCI} = -157.64 + 0.0107\ MFI$

c) From summary statistics: predicted HCI = 324.93; from original data: 324.87.

d) 223.09   e) $\widehat{z_{HCI}} = 0.65 z_{MFI}$ f) $\widehat{z_{MFI}} = 0.65 z_{HCI}$

45. a) $\widehat{Total} = 539.803 + 1.103 Age$

b) Yes. Both variables are quantitative; the plot is straight (although flat); there are no apparent outliers; the plot does not appear to change spread throughout the range of *Age*.

c) $559.65; $594.94   d) 0.14%

e) No. The plot is nearly flat. The model explains almost none of the variation in *Total Yearly Purchases*.

47. a) Moderately strong, fairly straight, and positive. Possibly some outliers (higher-than-expected math scores).

b) The student with 500 verbal and 800 math.

c) $r = 0.68$, indicating a positive, fairly strong linear relationship between math scores and verbal scores.

d) $\widehat{Math} = 209.6 + 0.675 \times Verbal$.

e) Every point of verbal score adds 0.675 points to the predicted average math score.

f) 547.1 points   g) 50.4 points

49. a) 0.685   b) $\widehat{Verbal} = 171.3 + 0.694 \times Math$.

c) The observed verbal score is higher than predicted from the math score

d) 518.5 points   e) 559.6 points

f) This regression equation cannot make predictions in the other direction.

51. a) The relationship is very weak. In fact, there may be no relationship at all between these variables.

b) The number of wildfires has been increasing by about 210 per year.

c) Yes, the intercept estimates the number of wildfires in 1985 as about 74,487.

d) The residuals are distributed around zero with a standard deviation of 11,920 fires. Compared to the observed values, most of which are between 60,000 and 90,000 fires, this residual standard deviation in our model's predictions is quite large.

e) Only 1.9% of the variation in the number of wildfires can be accounted for by the linear model on *Year*. This confirms the

impression from the scatterplot that there is very little association between these variables—that is, that there has been little change in the number of wildfires during this period.

53. a)

b) Negative, linear, strong.   c) Yes.   d) -0.944

e) *Age* accounts for 89.1% of the variation in *Advertised Price*.

f) Other factors contribute—options, condition, mileage, etc.

55. a) $\widehat{Price} = 17,767 - 862 \times Years$.

b) Every extra year of age decreases predicted value by $862.

c) The average new Corolla is predicted to cost $17,767.

d) $11,733

e) Negative; the actual price is less than expected.

f) -647

g) No; the extrapolation suggests the seller would have to pay the buyer to take the car.

57. a)

b) 92.3% of the variation in calories can be accounted for by the fat content.

c) $\widehat{Calories} = 211.0 + 11.06 \times Fat$.

d)

Residuals show no clear pattern, so the model seems appropriate.

e) Could say a fat-free burger still has 211.0 calories, but this is extrapolation (no data close to 0).

f) Every gram of fat adds 11.06 calories, on average.

g) 553.5 calories.

59. a) The regression was for predicting calories from fat, not the other way around.

b) $\widehat{Fat} = -15.0 + 0.083 \times Calories$. Predict 34.8 grams of fat.

61. a) $\widehat{\%\ Body\ Fat} = -27.4 + 0.25 \times Weight$.

b) Residuals look randomly scattered around 0, so conditions are satisfied.

c) *% Body Fat* increases, on average, by 0.25 percent per pound of *Weight*.

d) Reliable is relative. R^2 is 48.5%, but residuals have a standard deviation of 7%, so variation around the line is large.

e) 0.9 percent.

63. a) $\widehat{HighJump} = 2.681 - 0.00671 \times 800mTime$. High-jump height is lower, on average, by 0.00671 meters per additional second of 800-m race time.

b) 16.4%

c) Yes, the slope is negative. Faster runners tend to jump higher.

d) There is a slight tendency for less variation in high-jump height among the slower runners than among the faster ones.

e) Not especially. The residual standard deviation is 0.060 meters, which is not much smaller than the SD of all high jumps (0.066 meters). The model doesn't appear to do a very good job of predicting.

65. The sum of the squared vertical distances to any other line would be greater than 1790.

Chapter 8

1. a) The trend appears to be somewhat linear up to about 1940, but from 1940 to about 1970 the trend appears to be nonlinear. From 1975 or so to the present, the trend appears to be linear.

b) Relatively strong for certain periods.

c) No, as a whole the graph is clearly nonlinear. Within certain periods (ex: 1975 to the present) the correlation is high.

d) Overall, no. You could fit a linear model to the period from 1975 to 2003, but why? You don't need to interpolate, since every year is reported, and extrapolation seems dangerous.

3. a) The relationship is not straight.

b) It will be curved downward.

5. a) No. We need to see the scatterplot first to see if the conditions are satisfied, and models are always wrong.

b) No, the linear model might not fit the data everywhere.

7. a) Millions of dollars per minute of run time.

b) Costs for movies increase at approximately the same rate per minute.

c) On average dramas cost about $20 million less for the same runtime.

9. This observation was influential. After it was removed, the R^2 value and the slope of the regression line both changed by a large amount.

11. No; we cannot infer causation. In warm weather, more children go outside and play.

13. Individual student scores will vary greatly. The class averages will have much less variability and may disguise important trends.

15. a) The use of the Oakland airport has been growing at about 59,700 passengers/year, starting from about 282,000 in 1990.

b) 71% of the variation in passengers is accounted for by this model.

c) Errors in predictions based on this model have a standard deviation of 104,330 passengers.

d) No, that would extrapolate too far from the years we've observed.

e) The negative residual is September 2001. Air traffic was artificially low following the attacks on 9/11.

17. a) 1) High leverage, small residual.

2) No, not influential for the slope.

3) Correlation would decrease because outlier has large z_x and z_y, increasing correlation.

4) Slope wouldn't change much because the outlier is in line with other points.

b) 1) High leverage, probably small residual.

2) Yes, influential.

3) Correlation would weaken, increasing toward zero.

4) Slope would increase toward 0, since outlier makes it negative.

c) 1) Some leverage, large residual.

2) Yes, somewhat influential.

3) Correlation would increase, since scatter would decrease.

4) Slope would increase slightly.

d) 1) Little leverage, large residual.

2) No, not influential.

3) Correlation would become stronger and become more negative because scatter would decrease.

4) Slope would change very little.

19. 1) e   2) d   3) c   4) b   5) a

21. Perhaps high blood pressure causes high body fat, high body fat causes high blood pressure, or both could be caused by a lurking variable such as a genetic or lifestyle issue.

23. a) The graph shows that, on average, students progress at about one reading level per year. This graph shows averages for each grade. The linear trend has been enhanced by using averages.

b) Very close to 1.

c) The individual data points would show much more scatter, and the correlation would be lower.

d) A slope of 1 would indicate that for each 1-year grade level increase, the average reading level is increasing by 1 year.

25. a) *Cost* decreases by $2.13 per degree of average daily *Temp*. So warmer temperatures indicate lower costs.

b) For an avg. monthly temperature of 0°F, the cost is predicted to be $133.

c) Too high; the residuals (observed − predicted) around 32°F are negative, showing that the model overestimates the costs.

d) $111.70

e) About $105.70

f) No, the residuals show a definite curved pattern. The data are probably not linear.

g) No, there would be no difference. The relationship does not depend on the units.

27. a) 0.88

b) Interest rates during this period grew at about 0.25% per year, starting from an interest rate of about 0.64%.

c) Substituting 50 in the model yields a prediction of about 13%.

d) Not really. Extrapolating 20 years beyond the end of these data would be dangerous and unlikely to be accurate.

29. a) The two models fit comparably well, but they have very different slopes.

b) This model predicts the interest rate in 2000 to be 3.24%, much lower than the other model predicts.

c) We can trust the new predicted value because it is in the middle of the data used for the regression.

d) The best answer is "I can't predict that."

31. a) Stronger. Both slope and correlation would increase.

b) Restricting the study to nonhuman animals would justify it.

c) Moderately strong.

d) For every year increase in life expectancy, the gestation period increases by about 15.5 days, on average.

e) About 270.5 days.

33. a) Removing hippos would make the association stronger, since hippos are more of a departure from the pattern.

b) Increase.

c) No, there must be a good reason for removing data points.

d) Yes, removing it lowered the slope from 15.5 to 11.6 days per year.

35. a) Answers may vary. Using the data for 1955–2010 results in a scatterplot that is relatively linear with some curvature. You might use the data after 1955 only to predict 2015, but that would still call for extrapolation and would not be safe. The prediction is 27.19.

b) Not much, since the data are not truly linear and 2020 is 10 years from the last data point (extrapolating is risky).

c) No, that extrapolation of more than 50 years would be absurd. There's no reason to believe the trend from 1955 to 2005 will continue.

37.

a) There is a moderate, negative, linear relationship between birth rate and life expectancy. Paraguay is an outlier with a slightly higher than expected life expectancy.

b) $\widehat{lifexp} = 83.96 - 0.487$ *Birth Rate*

c) 53.3% of the variation in life expectancy is accounted for by the regression on birth rate.

d) Paraguay has an unusually large residual. Its higher than predicted life expectancy continues to stand out.

e) Yes. The residual plot is reasonably scattered.

f) R^2 increased to 65.4%. The new equation is

$$\widehat{lifexp} = 85.3 - 0.57 \; Birth\;Rate.$$

g) While there is an association, there is no reason to expect causality. Lurking variables may be involved.

39. a) The scatterplot is clearly nonlinear; however, the last few years—say, from 1970 on—do appear to be linear.

b) Using the data from 1970 to 2011 gives $r = 0.999$ and

$$\widehat{CPI} = -8967 + 4.57 \; Year. \text{ Predicted CPI in 2020} = 263$$

(an extrapolation of doubtful accuracy).

41. a) 1.57

b) When they were removed, R-squared increased, the slope increased and the prediction error decreased. Those points were definitely exerting influence on the regression line.

c) Yes. Using either equation we get a prediction of about 5.45, which is closer to its current condition.

Chapter 9

1. a) No re-expression needed.
   b) Re-express to straighten the relationship.
   c) Re-express to equalize spread.

3. a) There's an annual pattern in when people fly, so the residuals cycle up and down.
   b) No, this kind of pattern can't be helped by re-expression.

5. a) 16.44   b) 7.84   c) 0.36   d) 1.75   e) 27.59

7. a) Fairly linear, negative, moderately strong.
   b) The model predicts that gas mileage decreases an average 7.652 mpg for each thousand pounds of weight.
   c) No. Residuals show a curved pattern.

9. a) Residuals are more randomly spread around 0, with some low outliers.
   b) $\widehat{Fuel\;Consumption} = 0.625 + 1.178 \times Weight.$
   c) For each additional 1000 pounds of *Weight*, the model estimates an additional 1.178 gallons will be needed to drive 100 miles.
   d) 21.06 miles per gallon.

11. a) Although about 88% of the variation in GDP can be accounted for by this model, we should examine a scatterplot of the residuals to see if it's appropriate.
    b) No. The residuals show clear curvature.

13. No. The residual plot still has a very strong pattern.

15. a)

$$\widehat{Distance} = -65.9 + 5.98 \; Speed.$$

But residuals have a curved shape, so linear model is not appropriate

b)

$\sqrt{Distance}$ linearizes the plot.

c) $Predicted \; \sqrt{Distance} = 3.30 + 0.235 \times Speed$

d) 263.4 feet   e) 390.2 feet (an extrapolation)

f) Fairly confident, since $R^2 = 98.4\%$, and s is small

17. a) The relationship appears to have a bend. Given that inflation is exponential, this is not surprising.
    b) The residual plot has a curve. This relationship is not linear. It should be re-expressed before regression is performed.
    c) The residual plot now appears reasonably scattered. We can use this model, but since our model predicts $102.2 million for 2015, extrapolation is very likely!
    d) Not only did we extrapolate for 2015, it appears that salaries have flattened out and that exceeding A-Rod's $26 million may not occur for some time.

19. a)

Log(*Distance*) against position works pretty well.

$$\widehat{\log(Distance)} = 1.245 + 0.271 \times Position\;number.$$

b) Pluto's residual is not especially larger in the log scale. However, a model without Pluto predicts the 9th planet should be 5741 million miles. Pluto, at "only" 3707 million miles, doesn't fit very well, giving support to the argument that Pluto doesn't behave like a planet.

21. The predicted log(*Distance*) of Eris is 3.685, corresponding to a distance of 4841 million miles. That's short of the actual average distance of 6300 million miles.

23. a)

$$\widehat{\sqrt{Bdft}} = -4 + diam$$

The model is exact.

b) 36 board feet.   c) 1024 board feet.

25.

$$\widehat{\log Life} = 1.684 + 0.187077 \log Decade$$

27. The curvature in the residuals reveals that the linear model is inappropriate. The relationship cannot be made straight by the methods of this chapter.

29. a) *R*-squared is very high, but the residuals still have some pattern.
b) 86 years
c) No. The residuals have a pattern. Also, my friend might have other variables in her life that could change her life expectancy dramatically (for example, being a smoker, having longevity in her family history, and so on).

Part II Review

1. % over 50, 0.69.
% under 20, −0.71.
% Graduating on time, −0.51.
% Full-time Faculty, 0.09

3. a) There does not appear to be a linear relationship.
b) Nothing, there is no reason to believe that the results for the Finger Lakes region are representative of the vineyards of the world.
c) $\widehat{CasePrice} = 92.77 + 0.567 \times Years$.
d) Only 2.7 % of the variation in case price is accounted for by the ages of vineyards. Most of that is due to two outliers. We are better off using the mean price rather than this model.

5. a) $\widehat{TwinBirths} = -5202391 + 2659.9 \times Year$.
b) Each year, the number of twins born in a year is predicted to increase by approximately 2660.
c) 154,681.81 births. The scatterplot appears to be somewhat linear, and has been decreasing the two most recent years. There is no reason to believe that the increase will continue to be linear 5 years beyond the data, especially when the 2008 and 2009 trend is negative.
d) The residuals plot shows increasing variability and a pattern so the relation is not safe to use for predictions.

7. a) −0.520
b) Negative, not strong, somewhat linear, but with more variation as pH increases.
c) The BCI would also be about average.
d) The predicted *BCI* will be 1.56 SDs of *BCI* below the mean *BCI*.

9. a) $\widehat{Manatee\ Deaths} = -50.02 + 0.1392 \times Powerboat\ Registrations$ (in 1000s).
b) According to the model, for each increase of 10,000 motorboat registrations, the number of manatees killed increases by approximately 1.392 on average.
c) If there were 0 motorboat registrations, the number of manatee deaths would be −50.02. This is obviously a silly extrapolation.
d) The predicted number is 77.25 deaths. The actual number of deaths was 83. The residual is 83 − 77.25 = 5.75. The model underestimated the number of deaths by 5.75.
e) Negative residuals would suggest that the actual number of deaths was lower than the predicted number.
f) Over time, the number of motorboat registrations has increased and the number of manatee kills has increased. The trend may continue. Extrapolation is risky, however, because the government may enact legislation to protect the manatee.

11. a) −0.984   b) 96.9%   c) 32.95 mph   d) 1.66 mph
e) Slope will increase.
f) Correlation will weaken (become less negative).
g) Correlation is the same, regardless of units.

13. a) Weight (but unable to verify linearity).
b) As weight increases, mileage tends to decrease.
c) *Weight* accounts for 81.5% of the variation in *Fuel Efficiency*.

15. a) $\widehat{Horsepower} = 3.50 + 34.314 \times Weight$.
b) Thousands. For the equation to have predicted values between 60 and 160, the *X* values would have to be in thousands of pounds.
c) Yes. The residual plot does not show any pattern.
d) 115.0 horsepower.

17. a) The scatterplot shows a fairly strong linear relation in a positive direction. There seem to be two distinct clusters of data.
b) $\widehat{Interval} = 33.967 + 10.358 \times Duration$.
c) The time between eruptions increases by about 10.4 minutes per minute of *Duration* on average.

d) Since 77% of the variation in *Interval* is accounted for by *Duration* and the error standard deviation is 6.16 minutes, the prediction will be relatively accurate.
e) 75.4 minutes.
f) A residual is the observed value minus the predicted value. So the residual = 79 − 75.4 = 3.6 minutes, indicating that the model underestimated the interval in this case.

19. a) *r* = 0.888. Although *r* is high, you must look at the scatterplot and verify that the relation is linear in form.
b)

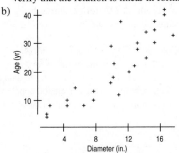

The association between diameter and age appears to be strong, somewhat linear, and positive.
c) $\widehat{Age} = -0.97 + 2.21 \times Diameter$.
d)

The residuals show a curved pattern (and two outliers).
e) The residuals for five of the seven largest trees (15 in. or larger) are positive, indicating that the predicted values underestimate the age.

21. Most houses have areas between 1000 and 5000 square feet. Increasing 1000 square feet would result in either 1000(.008) = 8 thousand dollars, 1000(.08) = 80 thousand dollars, 1000(.8) = 800 thousand dollars, or 1000(8) = 8000 thousand dollars. Only $80,000 is reasonable, so the slope must be 0.08.

23. a) The model predicts % smoking from year, not the other way around.
b) $\widehat{Year} = 2024.48 - 181.59 \times \%\ Smoking$.
c) The smallest % smoking given is 10, and an extrapolation to $x = 0$ is probably too far from the given data. The prediction is not very reliable in spite of the strong correlation.

25. The relation shows a negative direction, with lower temperatures generally associated with higher latitudes. The form is somewhat linear but perhaps with some slight curvature. Two cities are model outliers with high temperatures for their latitudes.

27. a) 71.9%
b) As latitude increases, the January temperature decreases.
c) $\overline{January\ Temperature} = 108.80 - 2.111 \times Latitude$.
d) As the latitude increases by 1 degree, the average January temperature drops by about 2.11 degrees, on average.
e) The *y*-intercept would indicate that the average January temperature is 108.8 when the latitude is 0 (at the equator). However, this is extrapolation and may not be meaningful.
f) 24.4 degrees.
g) The equation underestimates the average January temperature.

29. a) The scatterplot shows a strong, linear, positive association.
b) There is an association, but it is likely that training and technique have improved over time and affected both jump performances.
c) Neither; the change in units does not affect the correlation.
d) I would predict the winning long jump to be 0.913 SDs above the mean long jump.

31. a) No relation; the correlation would probably be close to 0.

b) The relation would have a positive direction and the correlation would be strong, assuming that students were studying French in each grade level. Otherwise, no correlation.

c) No relation; correlation close to 0.

d) The relation would have a positive direction and the correlation would be strong, since vocabulary and weight would increase with each grade level.

33. $\widehat{Calories} = 560.7 - 3.08 \times Time$.

Each minute extra at the table results in 3.08 fewer calories being consumed, on average. Perhaps the hungry children eat fast and eat more.

35. There seems to be a strong, positive, linear relationship with one high-leverage point (Northern Ireland) that makes the overall R^2 quite low. Without that point, the R^2 increases to 61.5%. Of course, these data are averaged across thousands of households, so the correlation appears to be higher than it would be for individuals. Any conclusions about individuals would be suspect.

37. a) 3.842   b) 501.187   c) 4

39. a) 30,818 pounds.

b) 1302 pounds.

c) 31,187.6 pounds.

d) I would be concerned about using this relation if we needed accuracy closer than 1000 pounds or so, as the residuals are more than ± 1000 pounds.

e) Negative residuals will be more of a problem, as the prediction would overestimate the weight of the truck; trucking companies might be inclined to take the ticket to court.

41. The original data are nonlinear, with a significant curvature. Using reciprocal square root of drain time gave a scatterplot that is nearly linear:

$$\widehat{1/\sqrt{Drain\ Time}} = 0.0024 + 0.219\ Diameter.$$

Practice Exam Answers

I.   **1.** D   **3.** B   **5.** C   **7.** D   **9.** E

    11. E   **13.** B   **15.** E

II. **1.** a) $\widehat{length} = 17.047 - 1.914\ time$

b) 92.3% of the variability of the length of the pencil can be explained by the number of hours it has been used.

c) The model suggests that new pencils are about 17.047 cm long, and then get about 1.914 cm shorter during each additional hour of use.

d) 6.597 cm

e) No. There are other variables that may affect this model, most importantly, the type of pencil and the way each person works. She may write slower, press harder, sharpen her pencil more often, etc.

3. a) Answers will vary. Two sample displays are shown here. Students should *not* stack bars or use any other display that has them adding the percentages, nor should they use pie charts.

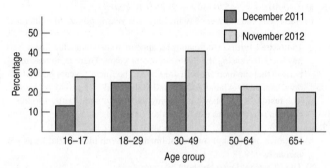

b) The percentage of book-readers who had read at least one e-book in the past year increased from 2011 to 2012 in every age category. In both years the 18–29 and 30–49 age groups had a larger proportion of e-book readers than the other age groups.

c) Students may look at the growth of e-book readership in terms of absolute differences in percentages, or in terms of percent difference.

	Age Group				
	16–17	**18–29**	**30–49**	**50–64**	**65+**
Absolute Difference	15	6	16	4	8
Percent Difference	115%	24%	64%	21%	67%

Either way, the growth in e-book readership among book readers is different for different age groups, so there is an association. For example, e-book reader growth in the 16–17 age group is far greater than in the over-50 age groups.

Chapter 10

1. a) Yes, who takes out the trash cannot be predicted before the flip of a coin.
   b) No, it is not random, since you will probably name a favorite team.
   c) Yes, your new roommate cannot be predicted before names are drawn.

3. A machine pops up numbered balls. If it is truly random, the outcome cannot be predicted and all possible outcomes would be equally likely. It is random only if the balls generate numbers in equal frequencies in the long run.

5. Use two-digit numbers 00–99; let 00–02 = defect, 03–99 = no defect

7. a) 45, 10   b) 17, 22

9. If the lottery is random, it doesn't matter which number you play; all are equally likely to win.

11. a) The outcomes are not equally likely; for example, tossing 5 heads does not have the same probability as tossing 0 or 9 heads, but the simulation assumes they are equally likely.
    b) The even-odd assignment assumes that the player is equally likely to score or miss the shot. In reality, the likelihood of making the shot depends on the player's skill.
    c) The likelihood for the first ace in the hand is not the same as for the second or third or fourth. But with this simulation, the likelihood is the same for each. (And it allows you to get 5 aces, which could get you in trouble in a real poker game!)

13. The conclusion should indicate that the simulation *suggests* that the average length of the line would be about 3.2 people. Future results might not match the simulated results exactly.

15. a) The component is one voter voting. An outcome is a vote for our candidate or not. Use two random digits, giving 00–54 a vote for your candidate and 55–99 for the underdog.
    b) A trial is 100 votes. Examine 100 two-digit random numbers, and count how many people voted for each candidate. Whoever gets the majority of votes wins that trial.
    c) The response variable is whether the underdog wins or not.

17. Answers will vary, but average answer will be about 51%.

19. Answers will vary, but average answer will be about 26%.

21. a) Answers will vary, but you should win about 10% of the time.
    b) You should win at the same rate with any number.

23. Answers will vary, but you should win about 10% of the time.

25. a) Sample response: Assign digit pairs 00–33 to represent a candidate passing on the first trial, 34–99 to represent failing the first time. Pairs 00–71 represent a pass on subsequent trials, and 72–99 represent a failure. In a random digit table, select pairs of digits, allowing repeated pairs, until a 'pass' is reached. Record the number of pairs needed to get a pass. Repeat many times.
    b) about 2.8 attempts.

27. Answers will vary, but average answer will be about 18%.

29. Do the simulation in two steps. First simulate the payoffs. Then count until $500 is reached. Answers will vary, but average should be near 10.2 customers.

31. Answers will vary, but average answer will be about 3 children.

33. Answers will vary, but average answer will be about 7.5 rolls.

35. No, it will happen about 40% of the time.

37. Answers will vary, but average answer will be about 37.5%.

39. Three women will be selected about 7.8% of the time.

Chapter 11

1. a) No. It would be nearly impossible to get exactly 500 males and 500 females from every country by random chance.
   b) A stratified sample, stratified by whether the respondent is male or female.

3. a) Voluntary response.
   b) We have no confidence at all in estimates from such studies.

5. a) The population of interest is all adults in the United States aged 18 and older.
   b) The sampling frame is U.S. adults with telephones.
   c) Some members of the population (e.g., many college students) don't have landline phones, which could create a bias.

7. a) Population—All U.S. adults.
   b) Parameter—Proportion who have used and benefited from alternative medicine.
   c) Sampling Frame—All Consumers Union subscribers.
   d) Sample—Those who responded.
   e) Method—The list of subscribers makes a convenience sample.
   f) Bias—Convenience Sample and Nonresponse. Subscribers are probably not representative of all U.S. adults, and those who respond may have strong feelings one way or another.

9. a) Population—Adults from that particular city.
   b) Parameter—Proportion who think drinking and driving is a serious problem.
   c) Sampling Frame—Patrons of this particular bar.
   d) Sample—Every 10th person leaving the bar.
   e) Method—Systematic sampling with a random start.
   f) Bias—Those interviewed had just left a bar. They may think drinking and driving is less of a problem than do other adults. Also, a particular bar would tend to cater to a certain demographic, which would not be representative of the population.

11. a) Simple random sample.
    b) No.
    c) They could stratify by distance from the dump site.

13. a) If the police used a list of registered automobiles, they would have to track down each of those owners, and unregistered vehicles would be unlikely to be on the list.
    b) They could use a systematic sample, checking every 20th vehicle that passes a checkpoint. They would randomly select a car out of the first 20, then check every 20th vehicle after that. They could also use a 20-sided die or a random number generator and roll for each car. If they roll a 1 the car gets checked.

15. Bias. Only people watching the news will respond, and their preference may differ from that of other voters. The sampling method may systematically produce samples that don't represent the population of interest.

17. a) Voluntary response. Only those who see the ad, have Internet access, *and* feel strongly enough will respond.
    b) Cluster sampling. One school may not be typical of all.
    c) Attempted census. Will have nonresponse bias.
    d) Stratified sampling with follow-up. Should be unbiased.

19. a) This is a multistage design, with a cluster sample at the first stage and a simple random sample for each cluster.
    b) If any of the three churches you pick at random is not representative of all churches, then you'll introduce more sampling error by the choice of that church.

21. a) This is a systematic sample.
    b) The sampling frame is patrons willing to wait for the roller coaster on that day at that time. It should be representative of the people in line, but not of all people at the amusement park.
    c) It is likely to be representative of those waiting for the roller coaster. Indeed, it may do quite well if those at the front of the line respond differently (after their long wait) than those at the back of the line.

23. a) Answers will definitely differ. Question 1 will probably get many "No" answers, while Question 2 will get many "Yes" answers. This is response bias.
    b) "Do you think standardized tests are appropriate for deciding whether a student should be promoted to the next grade?" (Other answers will vary.)

25. a) Biased toward yes because of "pollute." "Should companies be responsible for any costs of environmental cleanup?"
    b) Biased toward no because of "old enough to serve in the military." "Do you think the drinking age should be lowered from 21?"

27. a) Not everyone has an equal chance. Misses people with unlisted numbers, or without landline phones, or at work.
b) Generate random numbers and call at random times.
c) Under the original plan, those families in which one person stays home are more likely to be included. Under the second plan, many more are included. People without landline phones are still excluded.
d) It improves the chance of selected households being included.
e) This takes care of phone numbers. Time of day may be an issue. People without landline phones are still excluded.

29. a) Answers will vary.
b) Your own arm length. Parameter is your own arm length; population is all possible measurements of it.
c) Population is now the arm lengths of you and your friends. The average estimates the mean of these lengths.
d) Probably not. Friends are likely to be of the same age and not very diverse or representative of the larger population.

31. a) Assign numbers 001 to 120 to each order. Use random numbers to select 10 transactions to examine.
b) Sample proportionately within each type. (Do a stratified random sample.)

33. a) Select three cases at random; then select one jar randomly from each case.
b) Use random numbers to choose 3 cases from numbers 61 through 80; then use random numbers between 1 and 12 to select the jar from each case.
c) No. Multistage sampling.

35. a) Depends on the Yellow Page listings used. If from regular (line) listings, this is fair if all doctors are listed. If from ads, probably not, as those doctors may not be typical.
b) Not appropriate. This cluster sample will probably contain listings for only one or two business types.

37. a) Several terms are poorly defined. The survey needs to specify the meaning of "family" for this purpose and the meaning of "higher education." The term "seek" may also be poorly defined (for example, would applying to college but not being admitted qualify for seeking more education?).
b) i. Cluster sample.
ii. Stratified sample.
iii. Systematic sample.
c) This is not an SRS. Although each student may have an equal chance to be in the survey, groups of friends who choose to sit together will either all be in or out of the sample, so the selection is not independent.
d) i. This would suffer from voluntary response bias.
ii. This would be a convenience sample.
e) The proportion in the sample is a statistic. The proportion of all students is the parameter of interest. The statistic estimates that parameter, but is not likely to be exactly the same.

Chapter 12

1. a) No. There are no manipulated factors. Observational study.
b) There may be lurking variables that are associated with both parental income and performance on the SAT.

3. a) This is a retrospective observational study.
b) That's appropriate because MS is a relatively rare disease.
c) The subjects were U.S. military personnel, some of whom had developed MS.
d) The variables were the vitamin D blood levels and whether or not the subject developed MS.

5. a) This was a randomized, double-blind experiment.
b) Yes, such an experiment is the right way to determine whether maggots were as effective as surgery.
c) 100 men with wound on their lower limbs.
d) The treatments were sterile maggots and a traditional surgical procedure. The response was the percentage of dead tissue in the wounds.

7. a) Experiment.
b) 130 patients with eligible lacerations
c) Type of treatment, two levels.
d) 2 treatments.
e) Scarring, pain level, and speed of healing.
f) Completely randomized.
g) There is no discussion of blinding.
h) On these subjects, the two treatments worked equally well regarding scarring, but the adhesive was less painful and worked faster.

9. a) Observational study.
b) Prospective.
c) Men and women with moderately high blood pressure and normal blood pressure, unknown selection process.
d) Memory and reaction time.
e) As there is no random assignment, there is no way to know that high blood pressure *caused* subjects to do worse on memory and reaction-time tests. A lurking variable may also be the cause.

11. a) Experiment.
b) Postmenopausal women.
c) Alcohol—2 levels; blocking variable—estrogen supplements (2 levels).
d) 1 factor (alcohol) at 2 levels = 2 treatments.
e) Increase in estrogen levels.
f) Blocked.
g) Not blind unless beverages look, smell, and taste the same.
h) Indicates that alcohol consumption *for those taking estrogen supplements* may increase estrogen levels.

13. a) Observational study.
b) Retrospective.
c) Women in Finland, unknown selection process with data from church records.
d) Women's lifespans.
e) As there is no random assignment, there is no way to know that having sons or daughters shortens or lengthens the life span of mothers.

15. a) Observational study.
b) Prospective.
c) People with or without depression, unknown selection process.
d) Frequency of crying in response to sad situations.
e) There is no apparent difference in crying response (to sad movies) for depressed and nondepressed groups.

17. a) This is an experiment because the treatments were randomly assigned to the subjects.
b) The treatments were having at least one 15-minute neuromuscular warm-up exercise session per week and having no such warm-up session. The response variable was the number of ACL tears in each group.
c) Less variation in ACL tears would be expected among players of the same gender in the same sport, which would allow a difference between the treatments to be detected more easily.
d) The conclusions of the study cannot be generalized to males, females in other countries, nor to players of other sports.

19. a) The factors are type of pill (levels: standard pain reliever and placebo) and water (levels: ice water or no ice water).
b) The four treatments are standard pain reliever with ice water, standard pain reliever with no ice water, placebo with ice water, and placebo with no ice water. The response variable is the self-reported pain level.
c) Yes, the placebo is to prevent participants from knowing which pill they get. They cannot be blinded with respect to the ice water treatment.
d) There are several possibilities. Gender, age, frequency of migraines. But the reason given must be that there is reason to think that the variable is associated with the response variable of self-reported pain level.

21. They need to compare omega-3 results to something. Perhaps bipolarity is seasonal and would have improved during the experiment anyway.

23. a) Subjects' responses might be related to many other factors (diet, exercise, genetics, etc). Randomization should equalize the two groups with respect to unknown factors.

b) More subjects would minimize the impact of individual variability in the responses, but the experiment would become more costly and time consuming.

25. People who engage in regular exercise might differ from others with respect to bipolar disorder, and that additional variability could obscure the effectiveness of this treatment.

27. Answers may vary. Use a random-number generator to randomly select 24 numbers from 01 to 24 without replication. Assign the first 8 numbers to get no fertilizer, the second 8 numbers to get a half-dose, and the remaining numbers to get the full dose of fertilizer.

29. a) First, they are using athletes who have a vested interest in the success of the shoe by virtue of their sponsorship. They should choose other athletes. Second, they should randomize the order of the runs, not run all the races with their shoes second. They should blind the athletes by disguising the shoes if possible, so they don't know which is which. The timers shouldn't know which athletes are running with which shoes, either. Finally, they should replicate several times, since times will vary under both shoe conditions.

b) Because of the problems in (a), the results they obtain may favor their shoes. In addition, the results obtained for Olympic athletes may not be the same as for the general runner.

31. a) Allowing athletes to self-select treatments could confound the results. Other issues such as severity of injury, diet, age, etc., could also affect time to heal; randomization should equalize the treatment groups with respect to any such variables.

b) A control group could have revealed whether either exercise program was better (or worse) than just letting the injury heal.

c) Doctors who evaluated the athletes to approve their return to sports should not know which treatment the subject had.

d) It's hard to tell. The difference of 15 days seems large, but the standard deviations indicate that there was a great deal of variability in the times.

33. a) The differences among the Mozart and quiet groups were more than would have been expected from sampling variation.

b)

c) The Mozart group seems to have the smallest median difference and thus the *least* improvement, but there does not appear to be a significant difference.

d) No, if anything, there is less improvement, but the difference does not seem significant compared with the usual variation.

35. a) Observational, prospective study.

b) The supposed relation between health and wine consumption might be explained by the confounding variables of income and education.

c) No. While the variables have a relation, there is no causality indicated for the relation.

37. a) Arrange the 20 containers in 20 separate locations. Use a random-number generator to identify the 10 containers that should be filled with water.

b) Guessing, the dowser should be correct about 50% of the time. A record of 60% (12 out of 20) does not appear to be significantly different.

c) Answers may vary. You would need to see a high level of success—say, 90% to 100%, that is, 15 or more correct.

39. Randomly assign half the reading teachers in the district to use each method. Students should be randomly assigned to teachers as well. Make sure to block, by teacher if possible, and if not both by school and grade (or control grade by using only one grade). Construct an appropriate reading test to be used at the end of the year, and compare scores.

41. a) They mean that the difference is higher than they would expect from normal sampling variability.

b) An observational study.

c) No. Perhaps the differences are attributable to some confounding variable (e.g., people are more likely to engage in riskier behaviors on the weekend, routine procedures are rarely scheduled for weekends, etc.) rather than the day of admission.

d) Perhaps people have more serious accidents and traumas on weekends and are thus more likely to die as a result.

43. Answers may vary. This experiment has 1 factor (pesticide), at 3 levels (pesticide A, pesticide B, no pesticide), resulting in 3 treatments. The response variable is the number of beetle larvae found on each plant. Randomly select a third of the plots to be sprayed with pesticide A, a third with pesticide B, and a third with no pesticide (since the researcher also wants to know whether the pesticides even work at all). To control the experiment, the plots of land should be as similar as possible with regard to amount of sunlight, water, proximity to other plants, etc. If not, plots with similar characteristics should be blocked together. If possible, use some inert substance as a placebo pesticide on the control group, and do not tell the counters of the beetle larvae which plants have been treated with pesticides. After a given period of time, count the number of beetle larvae on each plant and compare the results.

45. Answers may vary. Find a group of volunteers. Each volunteer will be required to shut off the machine with his or her left hand and right hand. Randomly assign the left or right hand to be used first. Complete the first attempt for the whole group. Now repeat the experiment with the alternate hand. Check the differences in time for the left and right hands.

47. a) Jumping with or without a parachute.

b) Volunteer skydivers (the dimwitted ones).

c) A parachute that looks real but doesn't work.

d) A good parachute and a placebo parachute.

e) Whether parachutist survives the jump (or extent of injuries).

f) All should jump from the same altitude in similar weather conditions and land on similar surfaces.

g) Randomly assign people the parachutes.

h) The skydivers (and the people involved in distributing the parachute packs) shouldn't know who got a working chute. And the people evaluating the subjects after the jumps should not be told who had a real parachute either!

Part III Review

1. Observational prospective study. Indications of behavior differences can be seen in the two groups. May show a link between premature birth and behavior, but there may be lurking variables involved.

3. Experiment, matched by gender and weight, randomization within blocks of two pups of same gender and weight. Factor: type of diet. Treatments: low-calorie diet and allowing the dog to eat all it wants. Response variable: length of life. Can conclude that, on average, dogs with a lower-calorie diet live longer.

5. Completely randomized experiment, with the treatment being receiving folic acid or not (one factor, two levels). Treatments assigned randomly and the response variable is the number of/occurrence of additional precancerous growths. Neither blocking nor matching are mentioned, but in a study such as this one, it is likely that researchers and patients are blinded. Since treatments were randomized, it seems reasonable to generalize results to all people with precancerous polyps, though caution is warranted since these results contradict a previous study.

7. Sampling. Probably a simple random sample, although may be stratified by type of firework. Population is all fireworks produced each day.

Parameter is proportion of duds. Can determine if the day's production is ready for sale.

9. Observational retrospective study. Researcher can conclude that for anyone's lunch, even when packed with ice, food temperatures are rising to unsafe levels.

11. Experiment, with a control group being the genetically engineered mice who received no antidepressant and the treatment group being the mice who received the drug. The response variable is the amount of plaque in their brains after one dose and after four months. There is no mention of blinding or matching. Conclusions can be drawn to the general population of mice and we should assume treatments were randomized. To conclude the same for humans would be risky, but researchers might propose an experiment on humans based on this study.

13. Experiment. Factor is gene therapy. Hamsters were randomized to treatments. Treatments were gene therapy or not. Response variable is heart muscle condition. Can conclude that gene therapy is beneficial (at least in hamsters).

15. Sampling. Population is all oranges on the truck. Parameter is proportion of unsuitable oranges. Procedure is probably stratified random sampling with regions inside the truck being the strata. Can conclude whether or not to accept the truckload.

17. Observational retrospective study performed as a telephone-based randomized survey. Based on the excerpt, it seems reasonable to conclude that more education is associated with a higher Emotional Health Index score, but the insist on causality would be faulty reasoning.

19. Answers will vary. This is a simulation problem. Using a random digits table or software, call 0–4 a loss and 5–9 a win for the gambler on a game. Use blocks of 5 digits to simulate a week's pick.

21. Answers will vary.

23. a) Experiment. Actively manipulated candy giving, diners were randomly assigned treatments, control group was those with no candy, lots of dining parties.
   b) It depends on when the decision was made. If early in the meal, the server may give better treatment to those who will receive candy—biasing the results.
   c) A difference in response so large it cannot be attributed to natural sampling variability.

25. There will be voluntary response bias, and results will mimic those only of the visitors to sodahead.com and not the general U.S. population. The question is leading responders to answer "yes" though many might understand that the president's timing for his vacation had nothing to do with the events of the week.

27. a) Simulation results will vary. Average will be around 5.8 points.
   b) Simulation results will vary. Average will also be around 5.8 points.
   c) Answers will vary.

29. a) Yes.
   b) No. Residences without phones are excluded. Residences with more than one phone number had a higher chance.
   c) No. People who respond to the survey may be of age but not registered voters.
   d) No. Households who answered the phone may be more likely to have someone at home when the phone call was generated. These may not be representative of all households.

31. a) Does not prove it. There may be other confounding variables. Only way to prove this would be to do a controlled experiment.
   b) Alzheimer's usually shows up late in life. Perhaps smokers have died of other causes before Alzheimer's is evident.
   c) An experiment would be unethical. One could design a prospective study in which groups of smokers and nonsmokers are followed for many years and the incidence of Alzheimer's is tracked.

33.

Numerous subjects will be randomly assigned to see shows with violent, sexual, or neutral content. They will see the same commercials. After the show, they will be interviewed for their recall of brand names in the commercials.

35. a) May have been a simple random sample, but given the relative equality in age group, it may have been stratified.
   b) 38.2%
   c) We don't know. If data were collected from voting precincts that are primarily Democratic or primarily Republican, that would bias the results. Because the survey was commissioned by NBC News, we can assume the data collected are probably OK.
   d) Do party affiliations differ for different age groups?

37. The factor in the experiment will be type of bird control. I will have three treatments: scarecrow, netting, and no control. I will randomly assign several different areas in the vineyard to one of the treatments, taking care that there is sufficient separation that the possible effect of the scarecrow will not be confounded. At the end of the season, the response variable will be the proportion of bird-damaged grapes.

39. a) This was the control group. Subjects needed to have some sort of treatment to perceive they were getting help for their lower back pain.
   b) Even though patients were volunteers, their treatments were randomized. To generalize the results, we would need to assume these volunteers have the same characteristics of the general population of those with lower back pain (probably reasonable).
   c) It means that the success rates for the acupuncture groups were higher than could be attributed to chance when compared to the conventional treatment group. In other words, researchers concluded that both proper and "fake" acupuncture reduced back pain.

41. a) Use stratified sampling to select 2 first-class passengers and 12 from coach.
   b) Number passengers alphabetically, 01 = Bergman to 20 = Testut. Read in blocks of two, ignoring any numbers more than 20. This gives 65, 43, 67, 11 (selects Fontana), 27, 04 (selects Castillo).
   c) Number passengers alphabetically from 001 to 120. Use the random-number table to find three-digit numbers in this range until 12 different values have been selected.

43. Simulation results will vary. (Use integers 00 to 99 as a basis. Use integers 00 to 69 to represent a tee shot on the fairway. If on the fairway, use digits 00 to 79 to represent on the green. If off the fairway, use 00 to 39 to represent getting on the green. If not on the green, use digits 00 to 89 to represent landing on the green. For the first putt, use digits 00 to 19 to represent making the shot. For subsequent putts, use digits 00 to 89 to represent making the shot.)

Part III Practice Exam

MULTIPLE CHOICE

1. C	3. C	5. D	7. D	9. E
11. A	13. D	15. B	17. D	19. B

FREE RESPONSE

1. a)

Variable

b) The distribution of runs scored by Nationals players is strongly skewed high, while the Astros run production is fairly symmetrical with one high outlier at 80. The medians of the two distributions are fairly close (25 runs scored for the Nationals vs. 22 for the Astros). Run production for the Astros players is consistently low (an IQR of 25.5 runs) compared to the much larger spread for the Nats (IQR of 57 runs).

c) The Nationals' run production is very skewed to the right, which pulls the mean much higher than the median. Although the outlier of 80 runs increases the Astros' mean a bit, the rest of the data are very symmetrical, causing the mean to stay close to the median.

3. a) Assign each student a number label from 1 to 1200. Use a random number generator to select 120 different numbers.

b) Randomly select 30 students from each grade level.

c) Assign each homeroom a number label from 1 to 40. Use a random number generator to select 4 unique numbers. Survey all 30 students in each of those 4 homerooms.

d) Stratifying by grade would reduce the variation arising from differing computer usage among grade levels. Cluster sampling will make it easier to access students, but may not represent all the grade levels.

Chapter 13

1. a) $S = \{HH, HT, TH, TT\}$, equally likely.

b) $S = \{0, 1, 2, 3\}$, not equally likely.

c) $S = \{H, TH, TTH, TTT\}$, not equally likely.

d) $S = \{1, 2, 3, 4, 5, 6\}$, not equally likely.

3. In this context "truly random" should mean that every number is equally likely to occur.

5. There is no "Law of Averages." She would be wrong to think that they are "due" for a harsh winter.

7. There is no "Law of Averages." If at bats are independent, his chance for a hit does not change based on recent successes or failures.

9. a) There is some chance you would have to pay out much more than the $1500.

b) Many customers pay for insurance. The small risk for any one customer is spread among all.

11. a) 0.30   b) 0.80

13. a) 0.237   b) 0.763   c) 0.999

15. a) Legitimate.   b) Legitimate.

c) Not legitimate (sum more than 1).   d) Legitimate.

e) Not legitimate (can't have negatives or values more than 1).

17. A family may own both a computer and an HD TV. The events are not disjoint, so the Addition Rule does not apply.

19. When cars are traveling close together, their speeds are not independent, so the Multiplication Rule does not apply.

21. a) He has multiplied the two probabilities.

b) He assumes that being accepted at the colleges are independent events.

c) No. Colleges use similar criteria for acceptance, so the decisions are not independent.

23. a) 0.72   b) 0.89   c) 0.28

25. a) 0.5184   b) 0.0784   c) 0.4816

27. a) Repair needs for the two cars must be independent.

b) Maybe not. An owner may treat the two cars similarly, taking good (or poor) care of both. This may decrease (or increase) the likelihood that each needs to be repaired.

29. a) $511/1012 = 0.505$

b) $41/1012 + 41/1012 = 82/1012 = 0.081$

31. a) 0.071   b) 0.883

c) Responses are independent.

d) People were polled at random.

33. a) 0.5332   b) 0.9132

c) $(1 - 0.62) + 0.62(1 - 0.14)$ or $1 - (0.62)(0.14)$

35. a) 1) 0.30   2) 0.30   3) 0.90   4) 0.0

b) 1) 0.027   2) 0.128   3) 0.512   4) 0.271

37. a) Disjoint (can't be both red and orange).

b) Independent (unless you're drawing from a small bag).

c) No. Once you know that one of a pair of disjoint events has occurred, the other is impossible.

39. a) 0.0046   b) 0.125   c) 0.296   d) 0.421   e) 0.995

41. a) 0.027   b) 0.063   c) 0.973   d) 0.014

43. a) 0.024   b) 0.250   c) 0.543

45. 0.078

47. a) For any day with a valid three-digit date, the chance is 0.001, or 1 in 1000. For many dates in October through December, the probability is 0. (No three digits will make 10/15, for example.)

b) There are 65 days when the chance to match is 0. (Oct. 10–31, Nov. 10–30, and Dec. 10–31.) The chance for no matches on the remaining 300 days is 0.741

c) 0.259   d) 0.049

Chapter 14

1. 0.42   **3.** 0.675   **5.** 0.135

7. No $P(\mathbf{S}) = 0.323$ but $P(\mathbf{S}\,|\,\mathbf{FC}) = 0.625$. These are not the same.

9.

	United States	Not United States	Total
Log On Every Day	0.20	0.30	**0.50**
Do Not Log On Every Day	0.10	0.40	**0.50**
Total	**0.30**	**0.70**	**1.00**

11. a) 0.98   b) 0.02   c) 0.25

13. a) 0.31   b) 0.48   c) 0.31

15. a) 0.2025   b) 0.6965   c) 0.2404   d) 0.0402

17. a) 0.50   b) 1.00   c) 0.077   d) 0.333

19. a) 0.11   b) 0.27   c) 0.407   d) 0.344

21. a) 0.011   b) 0.222   c) 0.054   d) 0.337   e) 0.436

23. 0.21

25. a) 0.145   b) 0.118   c) 0.414   d) 0.217

27. a) 0.318   b) 0.955   c) 0.071   d) 0.009

29. a) 32%   b) 0.135

c) No, 7% of juniors have taken both.

d) No, the probability that a junior has taken a computer course is 0.23. The probability that a junior has taken a computer course *given* he or she has taken a Statistics course is 0.135.

31. a) 0.795
b) No, not quite. 79.5% of homes with phones have cell phones, but 83% of people have cell phones.
c) No. 58% of homes have both.

33. Yes, (Ace) = 4/52. (Ace|any suit) = 1/13.

35. a) 0.33
b) No. 9% of the chickens had both contaminants.
c) No. $P(C|S) = 0.64 \neq P(C)$. If a chicken is contaminated with salmonella, it's more likely also to have campylobacter.

37. No, only 32% of all men have high cholesterol, but 40.7% of those with high blood pressure do.

39. a) 72.8%
b) Probably. 71.5% of people with cell phones had landlines, and 72.8% of all people did.

41. No. Only 34% of men were Democrats, but over 39% of all voters were.

43. a) No, the probability that the luggage arrives on time depends on whether the flight is on time. The probability is 95% if the flight is on time and only 65% if not.
b) 0.695

45. 0.975

47. a) No, the probability of missing work for day-shift employees is 0.01. It is 0.02 for night-shift employees. The probability depends on whether they work day or night shift.
b) 1.4%

49. 57.1%

51. a) 0.20   b) 0.272   c) 0.353   d) 0.033

53. 0.563   **55.** Over 0.999

Chapter 15

1. a) 19   b) 4.2   **3.** 33 oranges   **5.** 4.58 oranges

7. a)

Amount won	$0	$5	$10	$30
P(Amount won)	$\frac{26}{52}$	$\frac{13}{52}$	$\frac{12}{52}$	$\frac{1}{52}$

b) $4.13   c) $4 or less (answers may vary)

9. a)

Children	1	2	3
P(Children)	0.5	0.25	0.25

b) 1.75 children   c) 0.875 boys

Boys	0	1	2	3
P(Boys)	0.5	0.25	0.125	0.125

11. $28,200   **13.** a) 7   b) 1.89   **15.** $5.44
17. 0.83   **19.** a) 1.7   b) 0.9   **21.** $\mu = 0.64, \sigma = 0.93$
23. a) $50   b) $100
25. a) No. The probability of winning the second depends on the outcome of the first.
b) 0.42   c) 0.08
d)

Games won	0	1	2
P(Games won)	0.42	0.50	0.08

e) $\mu = 0.66, \sigma = 0.62$

27. a)

Number good	0	1	2
P(Number good)	0.067	0.467	0.467

b) 1.40   c) 0.61
29. a) $\mu = 30, \sigma = 6$ b) $\mu = 26, \sigma = 5$
c) $\mu = 30, \sigma = 5.39$ d) $\mu = -10, \sigma = 5.39$
e) $\mu = 20, \sigma = 2.83$
31. a) $\mu = 240, \sigma = 12.80$ b) $\mu = 140, \sigma = 24$
c) $\mu = 720, \sigma = 34.18$ d) $\mu = 60, \sigma = 39.40$
e) $\mu = 600, \sigma = 22.63$

33. $1700, $141
35. a) 1.8   b) 0.87   c) Cartons are independent of each other.
37. a) 34.2 eggs   b) 0.87
c) The sum of the eggs is always 36, so the amount of variation in the good eggs is the same as the amount of variation in the bad eggs.
39. a) 1310   b) 200
41. a) $\mu = 13.6, \sigma = 2.55$
b) Assuming the hours are independent of each other.
c) A typical 8-hour day will have about 11 to 16 repair calls.
d) 19 or more repair calls would be a lot! That's more than two standard deviations above average.
43. a) $\mu = 23.4, \sigma = 2.97$
b) We assume each truck gets tickets independently.
45. a) There will be many gains of $150 with a few large losses.
b) $\mu = \$300, \sigma = \$8485.28$
c) $\mu = \$1,500,000, \sigma = \$600,000$
d) Yes. $0 is 2.5 SDs below the mean for 10,000 policies.
e) Losses are independent of each other. A major catastrophe with many policies in an area would violate the assumption.
47. a) 1 oz   b) 0.5 oz   c) 0.023   d) $\mu = 4$ oz, $\sigma = 0.5$ oz
e) 0.159   f) $\mu = 12.3$ oz, $\sigma = 0.54$ oz
49. a) 12.2 oz   b) 0.51 oz   c) 0.058
51. a) $\mu = 200.57$ sec, $\sigma = 0.46$ sec
b) No, $z = \dfrac{199.48 - 200.57}{0.461} = -2.36$. There is only 0.009 probability of swimming that fast or faster.
53. a) A = price of a pound of apples; P = price of a pound of potatoes; $Profit = 100A + 50P - 2$
b) $63.00   c) $20.62
d) Mean—no; SD—yes (independent sales prices).
55. a) $\mu = 1920, \sigma = 48.99; P(T > 2000) = 0.051$
b) $\mu = \$220, \sigma = 11.09$; No—$300 is more than 7 SDs above the mean.
c) $P(D - \frac{1}{2}C > 0) \approx 0.26$

Chapter 16

1. a) No. More than two outcomes are possible.
b) Yes, assuming the people are unrelated to each other.
c) No. The chance of a heart changes as cards are dealt so the trials are not independent.
d) No, 500 is more than 10% of 3000.
e) If packages in a case are independent of each other, yes.
3. a) Use single random digits. Let 0, 1 = LeBron. Count the number of random numbers until a 0 or 1 occurs.
b) Results will vary.
c) Results will vary.
d)

X	1	2	3	4	5	6	7	8	≥ 9
P(x)	0.2	0.16	0.128	0.102	0.082	0.066	0.052	0.042	0.168

5. a) Use single random digits. Let 0, 1 = LeBron. Examine random digits in groups of five, counting the number of 0's and 1's.
b) Results will vary.
c) Results will vary.
d)

X	0	1	2	3	4	5
P(x)	0.33	0.41	0.20	0.05	0.01	0.0

7. No. Departures from the same airport during a 2-hour interval may not be independent. All could be delayed by weather, for example.
9. a) 0.0819   b) 0.0064   c) 0.992
11. a) 5   b) 8 shots   c) 1.26 shots
13. 20 calls
15. a) 25   b) 0.185   c) 0.217   d) 0.693
17. a) 0.043   b) 0.999   c) 0.346
d) 0.188   e) 0.275   f) 0.913

19. a) 4.64   b) 1.396   c) 1.72 people
21. a) $\mu = 8.4, \sigma = 2.21$
   b) i. almost 1   ii. 0.965   iii. 0.136   iv. 0.306
23. $\mu = 20.28, \sigma = 4.22$
25. a) 0.118   b) 0.324   c) 0.744   d) 0.580
27. a) Success or failure: serve in or fault. 70% probability needs to stay constant and each serve independent. Counting the number of good serves out of 6 attempts.
   b) These assumptions are reasonable. (Although we might wonder if her probability of success changes under certain conditions. e.g., she becomes tired, nervous, more confident, etc.)
29. a) $\mu = 56, \sigma = 4.10$
   b) Yes, $np = 56 \geq 10$, $nq = 24 \geq 10$, serves are independent.
   c) In a match with 80 serves, approximately 68% of the time she will have between 51.9 and 60.1 good serves, approximately 95% of the time she will have between 47.8 and 64.2 good serves, and approximately 99.7% of the time she will have between 43.7 and 68.3 good serves.
   d) Normal, approx.: 0.014; Binomial, exact: 0.016
31. a) Assuming apples fall and become blemished independently of each other, Binom(300, 0.06) is appropriate. Since $np \geq 10$ and $nq \geq 10$, $N(18, 4.11)$ is also appropriate.
   b) Normal, approx.: 0.072; Binomial, exact: 0.085
   c) No, 50 is 7.8 SDs above the mean.
33. Normal, approx.: 0.053; Binomial, exact: 0.061
35. The mean number of sales should be 24 with SD 4.60. Ten sales is more than 3.0 SDs below the mean. He was probably misled.
37. a) 5   b) 0.066   c) 0.107   d) $\mu = 24, \sigma = 2.19$
   e) Normal, approx.: 0.819; Binomial, exact: 0.848
39. $\mu = 20, \sigma = 4$. I'd want *at least* 32 (3 SDs above the mean). (Answers will vary.)
41. Probably not. There's a more than 9% chance that he could hit 4 shots in a row, so he can expect this to happen nearly once in every 10 sets of 4 shots he takes. That does not seem unusual.
43. Yes. We'd expect him to make 22 shots, with a standard deviation of 3.15 shots. 32 shots is more than 3 standard deviations above the expected value, an unusually high rate of success.

Part IV Review

1. a) 0.34   b) 0.27   c) 0.069
   d) No, 2% of cars have both types of defects.
   e) Of all cars with cosmetic defects, 6.9% have functional defects. Overall, 7.0% of cars have functional defects. The probabilities here are estimates, so these are probably close enough to say the defects are independent.
3. a) C = Price to China; F = Price to France; Total = $3C + 5F$
   b) $\mu = \$5500, \sigma = \$672.68$ c) $\mu = \$500, \sigma = \$180.28$
   d) Means—no. Standard deviations—yes; ticket prices must be independent of each other for different countries, but all tickets to the same country are at the same price.
5. a) $\mu = -\$0.20, \sigma = \$1.89$ b) $\mu = -\$0.40, \sigma = \$2.67$
7. a) 0.999   b) 0.944   c) 0.993
9. a) 0.237   b) 0.015   c) 0.896
11. a) $\mu = 118.5, \sigma = 5.44$
   b) Yes, $np \geq 10$, $nq \geq 10$.
   c) Normal, approx: 0.059; binomial, exact: 0.073
13. a) 0.0173   b) 0.591
   c) Left: 960; right: 120; both: 120
   d) $\mu = 120, \sigma = 10.39$
   e) About 68% chance of between 110 and 130; about 95% between 99 and 141; about 99.7% between 89 and 151.
15. a) Men's heights are more variable than women's.
   b) Men (1.75 SD vs 2.4 SD for women)
   c) M = Man's height; W = Woman's height; M—W is how much taller the man is.
   d) 5.1"   e) 3.75"   f) 0.913

g) If independent, it should be about 91.3%. We are told 92%. This difference seems small and may be due to natural sampling variability.
17. a) The chance is 1.6×10^{-7}.   b) 0.952   c) 0.063
19. $-\$2080.00$
21. a) 0.717   b) 0.588
23. a) $\mu = 100, \sigma = 8$ b) $\mu = 1000, \sigma = 60$
   c) $\mu = 100, \sigma = 8.54$ d) $\mu = -50, \sigma = 10$
   e) $\mu = 100, \sigma = 11.31$
25. a) Many do both, so the two categories can total more than 100%.
   b) No. They can't be disjoint. If they were, the total would be 100% or less.
   c) No. Probabilities are different for boys and girls.
   d) 0.0524
27. a) 21 days
   b) 1649.73 som
   c) 3300 som extra. About 157-som "cushion" each day.
29. Perhaps. You'd expect 549.4 homeowners, with an SD of 13.46. 523 is 1.96 SDs below the mean; somewhat unusual.
31. a) 0.0176   b) 0.300   c) 0.26
33. a) 6   b) 15   c) 0.402
35. a) 38%   b) 41%   c) 19.6%
   d) 19.6% of classes that used calculators used computer assignments, while in classes that didn't use calculators, 22.4% used computer assignments. These are not close enough to think the choice is independent.
37. a) 1/11   b) 7/22   c) 5/11   d) 0   e) 19/66
39. a) $\mu = 18, \sigma = 3$
   b) 6
   c) No, σ is now 5.
   d) 10 or more
   e) What appears "surprising" in the short run becomes expected in a large number of trials.
41. a) 0.017   b) 0.824
43. a) 0.01   b) 0.0098   c) 0.366   d) First
   e) The chance of winning is 0.01 anywhere in line, so position does not matter.

MULTIPLE CHOICE

1. B	**3.** A	**5.** B	**7.** D	**9.** C
11. E	**13.** A	**15.** D	**17.** C	**19.** D
21. D	**23.** A	**25.** C		

FREE RESPONSE

1. a) 72.2%
   b) 43.3%
   c)

   d) No. Players listed as "probable" are more likely to play than those listed as "questionable," and players listed as "doubtful" are far less likely to play.
3. a) The patients are the units; the response variable is the pain rating.
   b) The treatments are the two dosage regimens.
   c) Number the patients 1–50. Generate a series of random numbers between 1 and 50. Ignoring repeats, assign the first 25 patients whose numbers are found to take 400 mg every 6 hours. Assign the remaining 25 patients to take 200 mg every 3 hours.

d) The control group would allow us to see how much improvement in pain levels might occur in 8 days even without the medication, making it clearer how much effect the medication may be having with either dosing regimen.

5. a) 0.512   b) $6.00   c) $6.633
d) Yes. When sold at full price, the expected profit on 100 cards is $600. If sold at a $500 discount, the expected profit would be only $100. The standard deviation of that profit on 100 cards is $66.33, making the possibility of a loss only about 1.5 standard deviations below the mean, not an unlikely outcome.

Chapter 17

1. All the histograms are centered near 0.05. As n gets larger, the histograms approach the Normal shape, and the variability in the sample proportions decreases.
3. a) Mean = 3.4, StDev = 1.517
b) 5, 4; 5,4; 5,3; 5,1; 4,4; 4,3; 4,1; 4,3; 4,1; 3,1
c) means: 4.5, 4.5, 4, 3, 4, 3.5, 2.5, 3.5, 2.5, 2
d) The mean is the same as that of the population. The standard deviation is smaller. The mean is an unbiased estimate of the population mean.
5. a) 0.536
b) 0.420. There is a difference of about 0.115.
c) 0.125
d) 0.117. There is a difference of about 0.008.
e) The normal model does not give a close approximation when np and nq are too small. (Even when they are large, it's still an approximation!)

7. a)

n	Observed mean	Theoretical mean	Observed st. dev.	Theoretical st. dev.
20	0.0497	0.05	0.0479	0.0487
50	0.0516	0.05	0.0309	0.0308
100	0.0497	0.05	0.0215	0.0218
200	0.0501	0.05	0.0152	0.0154

b) They are all quite close to what we expect from the theory.
c) The histogram is unimodal and symmetric for $n = 200$.
d) The success/failure condition says that np and nq should both be at least 10, which is not satisfied until $n = 200$ for $p = 0.05$. The theory predicted my choice.
9. a) No; highly skewed.
b) No, Normal does not give a good approximation.
11. a) Symmetric, because probability of heads and tails is equal.
b) 0.5   c) 0.125   d) $np = 8 < 10$
13. a) About 68% should have proportions between 0.4 and 0.6, about 95% between 0.3 and 0.7, and about 99.7% between 0.2 and 0.8.
b) $np = 12.5, nq = 12.5$; both are ≥ 10.
c)

0.3125 0.3750 0.4375 0.5000 0.5625 0.6250 0.6875
Proportion

$np = nq = 32$; both are ≥ 10.
d) Becomes narrower (less spread around 0.5).
15. This is a fairly unusual result: about 2.26 SDs below the mean. The probability of that is about 0.012. So, in a class of 100 this is certainly a reasonable possibility.

17. a)

0.547 0.598 0.649 0.700 0.751 0.802 0.853
Proportion Speeding

b) Both $np = 56$ and $nq = 24 \geq 10$. Drivers *may* be independent of each other, but if flow of traffic is very fast, they may not be. Or weather conditions may affect all drivers. In these cases they may get more or fewer speeders than they expect.
19. a) Assume that these children are typical of the population. They represent fewer than 10% of all children. We expect 20.4 nearsighted and 149.6 not; both are at least 10.
b)

0.045 0.070 0.095 0.120 0.145 0.170 0.195
Proportion

c) Probably between 12 and 29.
21. a) $E(\hat{p}) = 7\%, SD(\hat{p}) = 1.8\%$
b) Assume that clients pay independently of each other, that we have a random sample of all possible clients, and that these represent less than 10% of all possible clients. $np = 14$ and $nq = 186$ are both at least 10.
c) 0.048
23.

0.674 0.696 0.718 0.740 0.762 0.784 0.806
Retention Rate

These are not random samples, and not all colleges may be typical (representative). $np = 296, nq = 104$ are both at least 10.
25. Yes; if their students were typical, a retention rate of $522/603 = 86.6\%$ would be over 7 standard deviations above the expected rate of 74%.
27. 0.212. Reasonable that those polled are independent of each other and represent less than 10% of all potential voters. We assume the sample was selected at random. Success/Failure Condition met: $np = 208, nq = 192$. Both ≥ 10.
29. 0.0008 using $N(0.65, 0.048)$ model.
31. Answers will vary. Using $\mu + 3\sigma$ for "very sure," the restaurant should have 52 with-kids seats. Assumes customers at any time are independent of each other, a random sample, and represent less than 10% of all potential customers. $np = 36, nq = 84$, so Normal model is reasonable ($\mu = 0.30, \sigma = 0.042$).
33. a) Normal, center at μ, standard deviation $\sigma/\sqrt{n}$.
b) Standard deviation will be smaller. Center will remain the same. Shape will be closer to Normal.
35. a) The histogram is unimodal and slightly skewed to the right, centered at 36 inches with a standard deviation near 4 inches.
b) All the histograms are centered near 36 inches. As n gets larger, the histograms approach the Normal shape and the variability in the sample means decreases. The histograms are fairly Normal by the time the sample reaches size 5.

37. a)

n	Observed mean	Theoretical mean	Observed st. dev.	Theoretical st. dev.
2	36.314	36.33	2.855	2.842
5	36.314	36.33	1.805	1.797
10	36.341	36.33	1.276	1.271
20	36.339	36.33	0.895	0.899

b) They are all very close to what we would expect.

c) For samples as small as 5, the sampling distribution of sample means is unimodal and very symmetric.

d) The distribution of the original data is nearly unimodal and symmetric, so it doesn't take a very large sample size for the distribution of sample means to be approximately Normal.

39.

Mean GPA

Approximately Normal, $\mu = 3.4$, $\sigma = 0.07$. We assume that the students are randomly assigned to the seminars, less than 10% of all possible students, and that individual's GPAs are independent of one another.

41. a) As the CLT predicts, there is more variability in the smaller outlets.

b) If the lottery is random, all outlets are equally likely to sell winning tickets.

43. a) 21.1%   b) 276.8 days or more

c) $N(266, 2.07)$ d) 0.002

45. a) There are more premature births than very long pregnancies. Modern practice of medicine stops pregnancies at about 2 weeks past normal due date.

b) Parts (a) and (b)—yes—we can't use Normal model if it's very skewed. Part (c)—no—CLT guarantees a Normal model for this large sample size.

47. a) $\mu = \$2.00$, $\sigma = \$3.61$

b) $\mu = \$4.00$, $\sigma = \$5.10$

c) 0.190. Model is $N(80, 22.80)$.

49. a) $\mu = 2.825$, $\sigma = 1.331$

b) No. The score distribution in the sample should resemble that in the population, somewhat uniform for scores 1–4 and about half as many 5's.

c) Approximately $N\left(2.825, \dfrac{1.331}{\sqrt{40}}\right)$.

51. About 15%, based on $N(2.825, 0.1677)$.

53. a) $N(2.9, 0.045)$ b) 0.0131   c) 2.97 gm/mi

55. a) Can't use a Normal model to estimate probabilities. The distribution is skewed right—not Normal.

b) 4 is probably not a large enough sample to say the average follows the Normal model.

c) No. This is 3.16 SDs above the mean.

57. a) 0.0003. Model is $N(384, 34.15)$.

b) \$427.77 or more.

59. a) 0.734

b) 0.652. Model is $N(10, 12.81)$.

c) 0.193. Model is $N(120, 5.774)$.

d) 0.751. Model is $N(10, 7.394)$.

Chapter 18

1. She believes the true proportion is within 4% of her estimate, with some (probably 95%) degree of confidence.

3. a) Population—all cars; sample—those actually stopped at the checkpoint; p—proportion of all cars with safety problems; $\hat{p}$—proportion actually seen with safety problems (10.4%); if sample (a cluster sample) is representative, then the methods of this chapter will apply.

b) Population—general public; sample—those who logged onto the Web site; p—population proportion of those who favor prayer in school; $\hat{p}$—proportion of those who voted in the poll who favored prayer in school (81.1%); can't use methods of this chapter—sample is biased and nonrandom.

c) Population—parents at the school; sample—those who returned the questionnaire; p—proportion of all parents who favor uniforms; $\hat{p}$—proportion of respondents who favor uniforms (60%); should not use methods of this chapter, since not SRS (possible non-response bias).

d) Population—students at the college; sample—the 1632 students who entered that year; p—proportion of all students who will graduate on time; $\hat{p}$—proportion of that year's students who graduate on time (85.0%); can use methods of this chapter if that year's students (a cluster sample) are viewed as a representative sample of all possible students at the school.

5. a) Not correct. This implies certainty.

b) Not correct. Different samples will give different results. Many fewer than 95% will have 88% on-time orders.

c) Not correct. The interval is about the population proportion, not the sample proportion in different samples.

d) Not correct. In this sample, we *know* 88% arrived on time.

e) Not correct. The interval is about the parameter, not the days.

7. a) False   b) True   c) True   d) False

9. On the basis of this sample, we are 90% confident that the proportion of Japanese cars is between 29.9% and 47.0%.

11. a) (0.162, 0.280)

b) We're 95% confident that between 16.2% and 28% of all seafood packages sold in stores and restaurants in these three states are misidentified.

c) The size of the population is irrelevant. If Consumer Reports had a random sample, 95% of intervals generated by studies like this will capture the true fraud level.

13. a) 0.026

b) We're 90% confident that this poll's estimate is within $+/-2.6\%$ of the true proportion of people who are baseball fans.

c) Larger. To be more certain, we must be less precise.

d) 0.040   e) Less confidence.

15. a) (0.0465, 0.0491). The assumptions and conditions for constructing a confidence interval are satisfied.

b) The confidence interval gives the set of plausible values (with 95% confidence). Since 0.05 is outside the interval, that seems to be a bit too optimistic.

17. a) (12.7%, 18.6%)

b) We are 95% confident, based on this sample, that the proportion of all auto accidents thavt involve teenage drivers is between 12.7% and 18.6%.

c) About 95% of all random samples will produce confidence intervals that contain the true population proportion.

d) Contradicts. The interval is completely below 20%.

19. Probably nothing. Those who bothered to fill out the survey may be a biased sample.

21. a) Response bias (wording)   b) (55%, 61%)

c) Smaller—the sample size was larger.

23. a) (18.2%, 21.8%)

b) We are 98% confident, based on the sample, that between 18.2% and 21.8% of English children are deficient in vitamin D.

c) About 98% of all random samples will produce a confidence interval that contains the true proportion of children deficient in vitamin D.

25. a) Wider. The sample size is probably about one-fourth of the sample size for all adults, so we'd expect the confidence interval to be about twice as wide.

b) Smaller. The second poll used a slightly larger sample size.

27. a) (15.5%, 26.3%)   b) 612

c) Sample may not be random or representative. Deer that are legally hunted may not represent all sexes and ages.

29. a) 141   b) 318   c) 564

31. 1801   **33.** 384 total, using $p = 0.15$ **35.** 90%

Chapter 19

1. a) $H_0: p = 0.30$; $H_A: p < 0.30$

b) $H_0: p = 0.50$; $H_A: p \neq 0.50$

c) $H_0: p = 0.20$; $H_A: p > 0.20$

3. Statement d is correct.

5. No, we can say only that there is a 27% chance of seeing the observed effectiveness just from natural sampling variation. There is no *evidence* that the new formula is more effective, but we can't conclude that they are equally effective.

7. a) No. There's a 25% chance of losing twice in a row. That's not unusual.

b) 0.125   c) No, we expect that to happen 1 time in 8.

c) Maybe 5? The chance of 5 losses in a row is only 1 in 32, which seems unusual.

9. a) The new drug is not more effective than aspirin.

b) The new drug is more effective than aspirin.

c) There is not sufficient evidence to conclude that the new drug is better than aspirin.

d) There is evidence that the new drug is more effective than aspirin.

11. 1) Use p, not $\hat{p}$, in hypotheses.

2) The question was about failing to meet the goal, so H_A should be $p < 0.96$.

3) Did not check $0.04(200) = 8$. Since $nq < 10$, the Success/Failure Condition is violated. Didn't check 10% Condition.

4) $188/200 = 0.94$; $SD(\hat{p}) = \sqrt{\dfrac{(0.96)(0.04)}{200}} = 0.014$

5) z is incorrect; should be $z = \dfrac{0.94 - 0.96}{0.014} = -1.43$

6) $P = P(z < -1.43) = 0.076$

7) We lack evidence that the new instructions do not work.

13. a) $H_0{:}p = 0.30$; $H_A{:}p > 0.30$

b) Possibly an SRS; we don't know if the sample is less than 10% of his customers, but it could be viewed as less than 10% of all possible customers; $(0.3)(80) \geq 10$ and $(0.7)(80) \geq 10$. Wells are independent only if customers don't have farms on the same underground springs.

c) $z = 0.73$; P-value $= 0.232$

d) If his dowsing is no different from standard methods, there is more than a 23% chance of seeing results as good as those of the dowser's, or better, by natural sampling variation.

e) These data provide no evidence that the dowser's chance of finding water is any better than normal drilling.

15. a) $H_0{:}p_{2000} = 0.34$; $H_A{:}p_{2000} \neq 0.34$

b) Students were randomly sampled and should be independent. 34% and 66% of 8302 are greater than 10. 8302 students is less than 10% of the entire student population of the United States.

c) $z = -1.92$; P-value $= 0.055$

d) With such a large P-value, I fail to reject H_0. There is insufficient evidence to say there has been a change in the proportion of students who have no absences.

e) No. A difference this small, although statistically significant, is not meaningful. We might look at new data in a few years.

17. a) $H_0{:}p = 0.05$ vs. $H_A{:}p < 0.05$

b) We assume the whole mailing list has over 1,000,000 names. This is a random sample, and we expect 5000 successes and 95,000 failures.

c) $z = -3.178$; P-value $= 0.00074$, so we reject H_0; there is strong evidence that the donation rate would be below 5%.

19. a) $H_0{:}p = 0.66$, $H_A{:}p > 0.66$

b) The sample is representative. $240 < 10\%$ of all law school applicants. We expect $240(0.66) = 158.4$ to be admitted and $240(0.34) = 81.6$ not to be, both at least 10. $z = 0.63$; P-value $= 0.26$.

c) Because the P-value is high, there is not sufficient evidence to claim that LSATisfaction demonstrates improvement over the national average.

21. $H_0{:}p = 0.20$; $H_A{:}p > 0.20$. SRS (not clear from information provided); 22 is more than 10% of the population of 150; $(0.20)(22) < 10$. Do not proceed with a test.

23. $H_0{:}p = 0.03$; $p \neq 0.03$. $\hat{p} = 0.015$. One mother having twins will not affect another, so observations are independent; not an SRS; sample is less than 10% of all such births. However, the mothers at this hospital may not be representative of all teenagers; $(0.03)(469) = 14.07 \geq 10$; $(0.97)(469) \geq 10$. $z = -1.91$; P-value $= 0.0556$. With a P-value this low, cautiously reject H_0.

These data show some evidence that the rate of twins born to teenage girls at this hospital is less than the national rate of 3%. It is not clear whether this can be generalized to all teenagers.

25. $H_0{:}p = 0.25$; $H_A{:}p > 0.25$. SRS; sample is less than 10% of all potential subscribers; $(0.25)(500) \geq 10$; $(0.75)(500) \geq 10$. $z = 1.24$; P-value $= 0.1076$. The P-value is high, so do not reject H_0. These data do not show that more than 25% of current readers would subscribe; the company should not go ahead with the WebZine on the basis of these data.

27. $H_0{:}p = 0.40$; $H_A{:}p < 0.40$. Data are for all executives in this company and may not be able to be generalized to all companies; $(0.40)(43) \geq 10$; $(0.60)(43) \geq 10$. $z = -1.31$; P-value $= 0.0955$. Because the P-value is high, we fail to reject H_0. These data do not show that the proportion of women executives is less than the 40% of women in the company in general.

29. $H_0{:}p = 0.103$; $H_A{:}p > 0.103$. Assume this year is representative of recent and future years; $1782(0.103)$ and $1782(0.897)$ are both at least 10. $\hat{p} = 0.118$; $z = 2.06$; P-value $= 0.02$. Because the P-value is low, we reject H_0. These data provide evidence that the dropout rate has increased.

31. $H_0{:}p = 0.90$; $H_A{:}p < 0.90$. Assume these people are representative of all who lost luggage; $122(0.9)$, $122(0.1) \geq 10$. $\hat{p} = 0.844$; $z = -2.05$; P-value $= 0.0201$. Because the P-value is so low, we reject H_0. There is strong evidence that the actual rate at which passengers with lost luggage are reunited with it within 24 hours is less than the 90% claimed by the airline.

33. a) Yes; assuming this sample to be a typical group of people, $P = 0.0008$. This cancer rate is very unusual.

b) No, this group of people may be a typical for reasons that have nothing to do with the radiation.

Chapter 20

1. a) Let $p =$ probability of winning on the slot machine. $H_0{:}\ p = 0.01$ vs. $H_A{:}\ p \neq 0.01$

b) Let $p =$ proportion of patients cured by the new drug. $H_0{:}\ p = 0.3$ vs. $H_A{:}\ p \neq 0.3$

c) Let $p =$ proportion of clients now using the website. $H_0{:}\ p = 0.4$ vs. $H_A{:}\ p \neq 0.4$

3. a) False. A high P-value shows that the data are consistent with the null hypothesis, but provides no evidence for rejecting the null hypothesis.

b) False. It results in rejecting the null hypothesis, but does not prove that it is false.

c) False. A high P-value shows that the data are consistent with the null hypothesis but does not prove that the null hypothesis is true.

d) False. Whether a P-value provides enough evidence to reject the null hypothesis depends on the risk of a type I error that one is willing to assume (the α level).

5. a) $H_0{:}\ p = 0.40$ vs. $H_A{:}\ p \neq 0.40$. Two-sided.

b) $H_0{:}\ p = 0.42$ vs. $H_A{:}\ p > 0.42$. One-sided.

c) $H_0{:}\ p = 0.50$ vs. $H_A{:}\ p > 0.50$. One-sided.

7. a) Type I error. The actual value is not greater than 0.3 but they rejected the null hypothesis.

b) No error. The actual value is 0.50, which was not rejected.

c) Type II error. The null hypothesis was not rejected, but it was false. The true relief rate was greater than 0.25.

9. a) Two sided. Let p be the percentage of students who prefer Diet Pepsi. $H_0{:}\ p = 0.5$ vs. $H_A{:}\ p \neq 0.5$

b) One sided. Let p be the percentage of teenagers who prefer the new formulation. $H_0{:}\ p = 0.5$ vs. $H_A{:}\ p > 0.5$

c) One sided. Let p be the percentage of people who intend to vote for the override. $H_0{:}\ p = 2/3$ vs. $H_A{:}\ p > 2/3$.

d) Two sided. Let p be the percentage of days that the market goes up. $H_0{:}\ p = 0.5$ vs. $H_A{:}\ p \neq 0.5$

11. If there is no difference in effectiveness, the chance of seeing an observed difference this large or larger is 4.7% by natural sampling variation.

13. $\alpha = 0.10$: Yes. The P-value is less than 0.05, so it's less than 0.10. But to reject H_0 at $\alpha = 0.01$, the P-value must be below 0.01, which isn't necessarily the case.

15. a) There is only a 1.1% chance of seeing a sample proportion as low as 89.4% vaccinated by natural sampling variation if 90% have really been vaccinated.

b) We conclude that p is below 0.9, but a 95% confidence interval would suggest that the true proportion is between (0.889, 0.899). Most likely, a decrease from 90% to 89.9% would not be considered important. On the other hand, with 1,000,000 children a year vaccinated, even 0.1% represents about 1000 kids—so this may very well be important.

17. a) SRS so responses are independent, successes and failures both > 10; (0.486, 0.534).

b) Because 45% is not in the interval we have strong evidence that more than 45% of men identify themselves as the primary grocery shopper.

c) $\alpha = 0.01$; it's an upper tail test based on a 98% confidence interval.

19. a) (0.425, 0.475)

b) Since 50% is not in the confidence interval, we can reject the hypothesis that $p = 0.50$. It appears that his approval rating is lower.

21. a) The Success/Failure Condition is violated: only 5 pups had dysplasia.

b) We are 95% confident that between 5% and 26% of puppies will show signs of hip dysplasia at the age of 6 months.

23. a) Type II error

b) Type I error

c) By making it easier to get the loan, the bank has reduced the alpha level.

d) The risk of a Type I error is decreased and the risk of a Type II error is increased.

25. a) Power is the probability that the bank denies a loan that would not have been repaid.

b) Raise the cutoff score.

c) A larger number of trustworthy people would be denied credit, and the bank would miss the opportunity to collect interest on those loans.

27. a) The null is that the level of home ownership remains the same. The alternative is that it rises.

b) The city concludes that home ownership is on the rise, but in fact the tax breaks don't help.

c) The city abandons the tax breaks, but they were helping.

d) A Type I error causes the city to forego tax revenue, while a Type II error withdraws help from those who might have otherwise been able to buy a home.

e) The power of the test is the city's ability to detect an actual increase in home ownership.

29. a) It is decided that the shop is not meeting standards when it is.

b) The shop is certified as meeting standards when it is not.

c) Type I

d) Type II

31. a) The probability of detecting a shop that is not meeting standards.

b) 40 cars. Larger n.

c) 10%. More chance to reject H_0.

d) A lot. Larger differences are easier to detect.

33. a) One-tailed. The company wouldn't be sued if "too many" minorities were hired.

b) Deciding the company is discriminating when it is not.

c) Deciding the company is not discriminating when it is.

d) The probability of correctly detecting actual discrimination.

e) Increases power.

f) Lower, since n is smaller.

35. a) One-tailed. Software is supposed to decrease the dropout rate.

b) $H_0: p = 0.13$; $H_A: p < 0.13$

c) He buys the software when it doesn't help students.

d) He doesn't buy the software when it does help students.

e) The probability of correctly deciding the software is helpful.

37. a) $z = -3.21$, $p = 0.0007$. The change is statistically significant. A 95% confidence interval is (2.3%, 8.5%). This is clearly lower than

13%. If the cost of the software justifies it, the professor should consider buying the software.

b) The chance of observing 11 or fewer dropouts in a class of 203 is only 0.07% if the dropout rate is really 13%.

39. a) $H_A: p = 0.30$, where p is the probability of heads

b) Reject the null hypothesis if the coin comes up tails—otherwise fail to reject.

c) P(tails given the null hypothesis) $= 0.1 = \alpha$.

d) P(tails given the alternative hypothesis) $=$ power $= 0.70$

e) Spin the coin more than once and base the decision on the sample proportion of heads.

41. a) 0.0464   b) Type I   c) 37.6%

d) Increase the number of shots. Or keep the number of shots at 10, but increase alpha by declaring that 8, 9, or 10 will be deemed as having improved.

Chapter 21

1. 0.0161

3. a) We are 95% confident that, based on these data, the proportion of foreign-born Canadians is between 3.24% and 9.56% more than the proportion of foreign-born Americans.

b) If we were to take a large number of repeated samples of Canadians and Americans, we would expect 95% of the intervals to contain the true difference in the proportion of foreign-born citizens.

5. It's very unlikely that samples would show an observed difference this large if in fact there is no real difference in the proportions of boys and girls who access the internet with their smartphones.

7. The ads may be working. If there had been no real change in name recognition, there'd be only about a 3% chance the percentage of voters who heard of this candidate would be at least this much higher.

9. The responses are not from two independent groups, but are from the same individuals.

11. a) Stratified   b) 6% higher among males   c) 4%

d)

e) Yes; a poll result showing little difference is only 1–2 standard deviations below the expected outcome.

13. a) Yes. Random sample; less than 10% of the population; samples are independent; more than 10 successes and failures in each sample.

b) (0.055, 0.140)

c) We are 95% confident, based on these samples, that the proportion of American women age 65 and older who suffer from arthritis is between 5.5% and 14.0% more than the proportion of American men of the same age who suffer from arthritis.

d) Yes; the entire interval lies above 0.

15. a) 0.035   b) (0.356, 0.495)

c) We are 95% confident, based on these data, that the proportion of pets with a malignant lymphoma in homes where herbicides are used is between 35.6% and 49.5% higher than the proportion of pets with lymphoma in homes where no pesticides are used.

17. a) Experiment. Men were randomly assigned to have surgery or not.

b) (0.006, 0.080)

c) Since the entire interval lies above 0, there is evidence that surgery may be effective in preventing death from prostate cancer.

19. a) Yes, subjects were randomly divided into independent groups, and more than 10 successes and failures were observed in each group.

b) (4.7%, 8.9%)

c) Yes, we're 95% confident that the rate of infection is 5–9 percentage points lower. That's a meaningful reduction, considering the 20% infection rate among the unvaccinated kids.

21. a) $H_0: p_V - p_{NV} = 0$, $H_A: p_V - p_{NV} < 0$.
   b) Because 0 is not in the confidence interval, reject the null. There's evidence that the vaccine reduces the rate of ear infections.
   c) 2.5%   d) Type I
   e) Babies would be given ineffective vaccinations.
23. a) Prospective study
   b) $H_0: p_1 - p_2 = 0$; $H_A: p_1 - p_2 \neq 0$ where p_1 is the proportion of students whose parents disapproved of smoking who became smokers and p_2 is the proportion of students whose parents are lenient about smoking who became smokers.
   c) Yes. We assume the students are a representative sample; they are less than 10% of the population; samples are independent; at least 10 successes and failures in each sample.
   d) $z = -1.17$, P-value $= 0.2422$. These samples do not show evidence that parental attitudes influence teens' decisions to smoke.
   e) If there is no difference in the proportions, there is about a 24% chance of seeing the observed difference or larger by natural sampling variation.
   f) Type II
25. a) $(-0.065, 0.221)$
   b) We are 95% confident that the proportion of teens whose parents disapprove of smoking who will eventually smoke is between 22.1% less and 6.5% more than for teens with parents who are lenient about smoking.
   c) 95% of all random samples will produce intervals that contain the true difference in smoking rates between teens who have lenient vs. disapproving parents.
27. a) No; subjects weren't assigned to treatment groups. It's an observational study.
   b) $H_0: p_1 - p_2 = 0$; $H_A: p_1 - p_2 \neq 0$. $z = 3.56$, P-value $= 0.0004$. With a P-value this low, we reject H_0. There is a significant difference in the clinic's effectiveness. Younger mothers have a higher birth rate than older mothers. Note that the Success/Failure Condition is met based on the pooled estimate of p.
   c) We are 95% confident, based on these data, that the proportion of successful live births at the clinic is between 10.0% and 27.8% higher for mothers under 38 than in those 38 and older. However, the Success/Failure Condition is not met for the older women, since # Successes < 10. We should be cautious in trusting this confidence interval.
29. a) The polls are independent random samples satisfying the success/failure condition. $H_0: p_1 - p_2 = 0$; $H_A: p_1 - p_2 > 0$. $z = 1.18$, P-value $= 0.118$. With P-value this high, we fail to reject H_0. These data do not show evidence of a decrease in the voter support for the candidate.
   b) Type II
31. a) Assume these groups are independent and representative of all women; 94 and 20 are both at least 10. $H_0: p_1 - p_2 = 0$; $H_A: p_1 - p_2 \neq 0$. $z = -0.39$, P-value $= 0.6951$. With a P-value this high, we fail to reject H_0. There is no evidence of racial differences in the likelihood of multiple births, based on these data.
   b) Type II
33. a) We are 95% confident, that between 67.0% and 83.0% of patients with joint pain will find medication A effective.
   b) We are 95% confident, that between 51.9% and 70.3% of patients with joint pain will find medication B effective.
   c) Yes, they overlap. This might indicate no difference in the effectiveness of the medications. (Not a proper test.)
   d) We are 95% confident that the proportion of patients with joint pain who will find medication A effective is between 1.7% and 26.1% higher than the proportion who will find medication B effective.
   e) No. There is a difference in the effectiveness of the medications.
   f) To estimate the variability in the difference of proportions, we must add variances. The two one-sample intervals do not. The two-sample method is the correct approach.
35. Independent random samples; 417, 229, 78, and 76 are all at least 10. $H_0: p_{\text{urban}} - p_{\text{rural}} = 0$ vs. $H_A: p_{\text{urban}} - p_{\text{rural}} \neq 0$. Yes; $z = 3.19$. With a low P-value of 0.0013, reject the null hypothesis of no difference. There is strong evidence to suggest that the

percentages are different for the two groups: It appears that people from urban areas are more likely to agree with the statement than those from rural areas.
37. Yes; $z = 3.44$. With a low P-value of 0.0003, reject the null hypothesis of no difference. There's evidence of an increase in the proportion of parents checking the Web sites visited by their teens.

Part V Review

1. Call the omega-3 subjects Group 2. H_0: There is no difference in relapse rates, $p_1 - p_2 = 0$. H_A: The relapse rate in those who use omega-3 fatty acids is lower, $p_1 - p_2 > 0$.
3. a) 10.29
   b) Not really. The z-score is -1.11. Not any evidence to suggest that the proportion for Monday is low.
   c) Yes. The z-score is 2.26 with a P-value of 0.024 (two-sided).
   d) Some births are scheduled for the convenience of the doctor and/or the mother.
5. a) $H_0: p_1 = 0.40$; $H_A: p_1 < 0.40$
   b) Random sample; less than 10% of all California gas stations, $0.4(27) = 10.8$, $0.6(27) = 16.2$. Assumptions and conditions are met.
   c) $z = -1.49$, P-value $= 0.0677$
   d) With a P-value this high, we fail to reject H_0. These data do not provide evidence that the proportion of leaking gas tanks is less than 40% (or that the new program is effective in decreasing the proportion).
   e) Yes, Type II.
   f) Increase α, increase the sample size.
   g) Increasing α—increases power, lowers chance of Type II error, but increases chance of Type I error.
      Increasing sample size—increases power, costs more time and money.
7. a) The researcher believes that the true proportion of "A's" is within 10% of the estimated 54%, namely, between 44% and 64%.
   b) Small sample
   c) No, 63% is contained in the interval.
9. a) Pew uses a 95% confidence level. We can be 95% confident that the true proportion is within 2% of 13%—that is, that it is between 11% and 15%.
   b) The cell phone group would have the larger ME because its sample size is smaller.
   c) CI $= (78.5\%, 85.5\%)$
   d) The ME is 0.035, which is larger than the 2% ME in part a, largely because of the smaller sample size. It is larger than the ME in part b, mostly because 0.82 is smaller than 0.87 and proportions closer to 0.50 have larger MEs.
11. a) Bimodal!
   b) μ, the population mean. Sample size does not matter.
   c) $\sigma/\sqrt{n}$; sample size does matter.
   d) It becomes closer to a Normal model and narrower as the sample size increases.
13. a) For $\hat{p}$, $\mu = 0.80$, $\sigma = 0.028$
   b) Yes. $0.8(200) = 160$, $0.2(200) = 40$. Both ≥ 10. Assume shots are independent.
   c)

   d) 0.039
15. Let Group 1 be 1990 and Group 2 be 2000. H_0: There is no difference, $p_1 - p_2 = 0$. H_A: Early births have increased, $p_1 - p_2 < 0$. $z = -0.729$, P-value $= 0.2329$. Because the P-value is so high,

we do not reject H_0. These data do not show an increase in the incidence of early birth of twins.

17. a) Let 1 represent the magnesium sulfate group. H_0: There is no difference, $p_1 - p_2 = 0$. H_A: Treatment prevents deaths from eclampsia, $p_1 - p_2 < 0$.

b) Samples are random and independent; less than 10% of all pregnancies (or eclampsia cases); more than 10 successes and failures in each group.

c) 0.8008

d) There is insufficient evidence to conclude that magnesium sulfate is effective in preventing eclampsia deaths.

e) Type II f) Increase the sample size, increase α.

g) Increasing sample size: decreases variation in the sampling distribution, is costly. Increasing α: Increases likelihood of rejecting H_0, increases chance of Type I error.

19. a) It is not clear what the pollster asked or when the poll was conducted. Otherwise they did fine.

b) Stratified sampling. c) 4%

d) 95% e) Smaller sample size.

f) Wording and order of questions (response bias).

21. a) H_0: There is no difference, $p = 0.143$. H_A: The fatal accident rate is lower in girls, $p < 0.143$. $z = -1.67$, P-value $= 0.0479$. Because the P-value is low, we reject H_0. These data give some evidence that the fatal accident rate is lower for girls than for teens in general.

b) If the proportion is really 14.3%, we will see the observed proportion (11.3%) or lower 4.8% of the time by sampling variation.

23. a) One would expect many small fish, with a few large ones.

b) We don't know the exact distribution, but we know it's not Normal.

c) Probably not. With a skewed distribution, a sample size of five is not a large enough sample to say the sampling model for the mean is approximately Normal.

d) 0.961

25. a) Yes. $0.8(60) = 48, 0.2(60) = 12$. Both are ≥ 10.

b) 0.834

c) Higher. Bigger sample has smaller standard deviation for $\hat{p}$.

d) Answers will vary. For $n = 500$, the probability is 0.997.

27. a) Assume this is a representative group of less than 10% of patients; $335, 238 \geq 10$. 54.4 to 62.5%

b) Based on this study, with 95% confidence the proportion of Crohn's disease patients who will respond favorable to infliximab is between 54.4% and 62.5%.

c) 95% of all such random samples will produce confidence intervals that contain the true proportion of patients who respond favorably.

29. At least 423, assuming that p is near 50%.

31. a) Assume it's a representative sample, certainly less than 10% of all preemies and normal babies; more than 10 failures and successes in each group. 1.7% to 16.3% greater for normal-birth weight children.

b) Since 0 is not in the interval, there is evidence that preemies have a lower high school graduation rate than children of normal birth weight.

c) Type I, since we rejected the null hypothesis.

33. a) H_0: The computer is undamaged. H_A: The computer is damaged.

b) 20% of good PCs will be classified as damaged (bad), while all damaged PCs will be detected (good).

c) 3 or more. d) 20%

e) By switching to two or more as the rejection criterion, 7% of the good PCs will be misclassified, but only 10% of the bad ones will, increasing the power from 20% to 90%.

35. The null hypothesis is that Bush's disapproval proportion was 66%—the Nixon benchmark. The poll is a random sample of less than 10% of all voters; $1016(0.69)$ and $1016(0.31)$ are both at least 10. The one-tailed test has a z-value of 2.02, so the P-value is 0.0217. It looks like there is reasonable evidence to suggest that Bush's April 2008 ratings were worse than the Nixon benchmark low.

37. a) The company is interested only in confirming that the athlete is well known.

b) Type I: the company concludes that the athlete is well known, but that's not true. It offers an endorsement contract to someone who lacks name recognition. Type II: the company overlooks a well-known athlete, missing the opportunity to sign a potentially effective spokesperson.

c) Type I would be more likely, Type II less likely.

39. I am 95% confident that the proportion of U.S. adults who favor nuclear energy is between 7 and 19 percentage points higher than the proportion who would accept a nuclear plant near their area.

41. a) We are 95% confident that between 45.36% and 61.22% of patients with metastatic melanoma will have at least a partial response to vemurafenib based on this study.

b) 266 or 267 patients (using 0.5 or 0.53 as the proportion)

Practice Test

MULTIPLE CHOICE

1. D	3. C	5. D	7. E	9. B
11. D	13. E	15. B	17. E	19. B
21. B	23. A	25. C	27. C	29. E

FREE RESPONSE

1. a) Slightly skewed to the right. The mean is a bit larger than the median and the right tail is longer than the left tail.

b) Yes. Any point above $Q3 + 1.5IQR = 21 + 1.5(4) = 27$ is an outlier, and there is at least one value at 28. Any point below $Q1 - 1.5IQR = 17 - 1.5(4) = 11$ would be an outlier, but the minimum is 14 so there are no low outliers.

c) At least 25% of the cars get 17 mpg or less, so that's not "exceptionally gas-thirsty."

3. (*Answers may vary.*)

a) The wording of the question implies people are unhappy, which may influence employees to also say that they are unhappy, causing managers to overestimate employee dissatisfaction.

b) The fact that email responses would not be anonymous might lead employees to say they're happier than they really are, causing managers to underestimate employee dissatisfaction.

c) Those who chose not to reply may feel differently from those who did. If respondents tend to be (or at least claim to be) happier, this would lead managers to underestimate employee dissatisfaction.

5. a) $H_0: p = 0.0435$ $H_A: p > 0.0435$

We assume these 841 birds are representative of and fewer than 10% of all Chernobyl area barn swallows; $841(0.0435) = 36.6 \geq 10$ and $841(0.9565) = 804.4 \geq 10$. OK to do a 1-proportion z-test.

$$\hat{p} = \frac{112}{841} = 0.1332 \Rightarrow z = \frac{0.1332}{\sqrt{\frac{(0.0435)(0.9565)}{841}}} = 12.75; P = 0^+$$

Because the P-value is extremely small, we reject H_0. There is very strong evidence that the rate of albinism in barn swallows is abnormally high in the Chernobyl area.

b) Nothing; there may be other causes.

Chapter 22

1. a) SD's are 1 lb., 0.5 lbs, and 0.2 lbs respectively.

b) The distribution of pallets. The CLT tells us that the Normal model is approached in the limit regardless of the underlying distribution. As samples get larger the approximation gets better.

3. a) 1.74 b) 2.37 c) 0.0524 d) 0.0889

5. Shape becomes closer to Normal; center does not change; spread becomes narrower.

7. a) The confidence interval is for the population mean, not the individual cows in the study.

b) The confidence interval is not for individual cows.

c) We *know* the average gain in this study was 56 pounds!

d) The average weight gain of all cows does not vary. It's what we're trying to estimate.

e) No. There is not a 95% chance for another sample to have an average weight gain between 45 and 67 pounds. There is a 95% chance that another sample will have its average weight gain within two standard errors of the true mean.

9. a) No. A confidence interval is not about individuals in the population.

b) No. It's not about individuals in the sample, either.

c) No. We know the mean cost for students in the sample was $1196.

d) No. A confidence interval is not about other sample means.

e) Yes. A confidence interval estimates a population parameter.

11. a) Based on this sample, we can say, with 95% confidence, that the mean pulse rate of adults is between 70.9 and 74.5 beats per minute.

b) 1.8 beats per minute

c) Larger

13. a) Houses are independent; randomly sampled. Should check Nearly Normal by making a histogram. Sample size: 36 should be big enough

b) ($9,052.50, $10,067.50)

c) We are 95% confident that the interval ($9052.50, $10,067.50) contains the true mean loss in home value.

15. The assumptions and conditions for a *t*-interval are not met. The distribution is highly skewed to the right and there is a large outlier.

17. a) Yes. Randomly selected group; less than 10% of the population; the histogram is not unimodal and symmetric, but it is not highly skewed and there are no outliers, so with a sample size of 52, the CLT says $\bar{y}$ is approximately Normal.

b) (98.06, 98.51) degrees F

c) We are 98% confident, based on the data, that the average body temperature for an adult is between 98.06°F and 98.51°F.

d) 98% of all such random samples will produce intervals containing the true mean temperature.

e) These data suggest that the true normal temperature is somewhat less than 98.6°F.

19. a) Narrower. A smaller margin of error, so less confident.

b) Advantage: more chance of including the true value. Disadvantage: wider interval.

c) Narrower; due to the larger sample, the SE will be smaller.

d) About 252

21. a) (709.90, 802.54)

b) With 95% confidence, based on these data, the speed of light is between 299,709.9 and 299,802.5 km/sec.

c) Normal model for the distribution, independent measurements. These seem reasonable here, but it would be nice to see if the Nearly Normal Condition held for the data.

23. a) Given no time trend, the monthly on-time departure rates should be independent. Though not a random sample, these months should be representative. The histogram looks unimodal, but slightly left-skewed; not a concern with this large sample.

b) 80.22% < μ(OT Departure%) < 81.29%

c) We can be 90% confident that the interval from 80.22% to 81.29% holds the true mean monthly percentage of on-time flight departures.

25. $t = 2.2$ on 35df, $P = 0.034$. We reject the null hypothesis because 0.034 is small, and conclude that the loss of home values in this community does appear to be unusual.

27. The 95% confidence interval lies entirely above the 0.08 ppm limit, evidence that mirex contamination is too high and consistent with rejecting the null. We used an upper-tail test, so the P-value should therefore be smaller than $\frac{1}{2}(1 - 0.95) = 0.025$, and it was.

29. If in fact the mean cholesterol of pizza eaters does not indicate a health risk, then 7 of every 100 samples would have mean cholesterol levels as high (or higher) as observed in this sample.

31. a) Upper-tail. We want to show it will hold 500 pounds (or more) easily.

b) They will decide the stands are safe when they're not.

c) They will decide the stands are unsafe when they are in fact safe.

33. a) Decrease α. This means a smaller chance of declaring the stands safe if they are not.

b) The probability of correctly detecting that the stands are capable of holding more than 500 pounds.

c) Decrease the standard deviation—probably costly. Increase the sample size—takes more time for testing and is costly. Increase α—more Type I errors. Increase the "design load" to be well above 500 pounds—again, costly.

35. a) $H_0: \mu = 23.3$; $H_A: \mu > 23.3$

b) We have a random sample of the population. Population may not be normally distributed, as it would be easier to have a few much older men at their first marriage than some very young men. However, with a sample size of 40, $\bar{y}$ should be approximately Normal. We should check the histogram for severity of skewness and possible outliers.

c) $(\bar{y} - 23.3)/(s/\sqrt{40}) \sim t_{39}$ d) 0.1447

e) If the average age at first marriage is still 23.3 years, there is a 14.5% chance of getting a sample mean of 24.2 years or older simply from natural sampling variation.

f) We lack evidence that the average age at first marriage has increased from the mean of 23.3 years.

37. a) Probably a representative sample; the Nearly Normal Condition seems reasonable. (Show a Normal probability plot or histogram.) The histogram is nearly uniform, with no outliers or skewness.

b) $\bar{y} = 28.78, s = 0.40$ c) (28.36, 29.21) grams

d) Based on this sample, we are 95% confident the average weight of the content of Ruffles bags is between 28.36 and 29.21 grams.

e) The company is erring on the safe side, as it appears that, on average, it is putting in slightly more chips than stated.

39. a) Type I; he mistakenly rejected the null hypothesis that $p = 0.10$ (or worse).

b) Yes. These are a random sample of bags and the Nearly Normal Condition is met (Show a Normal probability plot or histogram.); $t = -2.51$ with 7 df for a one-sided P-value of 0.0203.

41. a) Random sample; the Nearly Normal Condition seems reasonable from a Normal probability plot. The histogram is roughly unimodal and symmetric with no outliers. (Show plot.)

b) (1187.9, 1288.4) chips

c) Based on this sample, the mean number of chips in an 18-ounce bag is between 1187.9 and 1288.4, with 95% confidence. The *mean* number of chips is clearly greater than 1000. However, if the claim is about individual bags, then it's not necessarily true. If the mean is 1188 and the SD deviation is near 94, then we'd expect about 2.5% of the bags to have fewer than 1000 chips, based on a Normal model. If in fact the mean is 1288, we'd expect about 0.1% to be below 1000.

43. 27 using z, 31 using t.

45. a) The Normal probability plot is relatively straight, with one outlier at 93.8 sec. Without the outlier, the conditions seem to be met. The histogram is roughly unimodal and symmetric with no other outliers. (Show your plot.)

b) $t = -2.63$, P-value $= 0.0160$. With the outlier included, we might conclude that the mean completion time for the maze is not 60 seconds; in fact, it is less.

c) $t = -4.46$, P-value $= 0.0003$. Because the P-value is so small, we reject H_0. Without the outlier, we see strong evidence that the average completion time for the maze is less than 60 seconds. The outlier here did not change the conclusion.

d) The maze does not meet the "one-minute average" requirement. Both tests rejected a null hypothesis of a mean of 60 seconds.

47. a) The study is well designed and the data show no outliers. We are 95% confident that the average wooden tip penetration is between 199 mm and 213.3 mm.

b) The experiment is well designed and the data show no outliers. We are 95% confident that the average stone tip penetration is between 212.3 mm and 236.8 mm.

49. a) We have a large sample size. We are 90% confident that the mean number of lawsuits will be between 4220 and 7249.

b) Since these are aggregate data taken over 8 years from 50 different states, the overall mean may not prove useful, especially to individual states. It is possible that this interval could be used on a federal level to see if the number of lawsuits is increasing.

51. a) $289.9 < \mu(\text{Drive Distance}) < 292.3$

b) These data are not a random sample of golfers. The top professionals are (unfortunately) not representative and were not selected at random. We might consider the 2011 data to represent the population of all professional golfers, past, present, and future.

c) The data are means for each golfer, so they are less variable than if we looked at all the separate drives.

Chapter 23

1. Yes. The high P-value means that we lack evidence of a difference, so 0 is a possible value for $\mu_{Meat} - \mu_{Beef}$.

3. a) Plausible values of $\mu_{Meat} - \mu_{Beef}$ are all negative, so the mean fat content is probably higher for beef hot dogs.

b) The difference is significant. c) 10%

5. a) False. The confidence interval is about means, not about individual hot dogs.

b) False. The confidence interval is about means, not about individual hot dogs.

c) True.

d) False. CI's based on other samples will also try to estimate the true difference in population means; there's no reason to expect other samples to conform to this result.

e) True.

7. a) 2.927   b) Larger

c) Based on this sample, we are 95% confident that students who learn Math using the CPMP method will score, on average, between 5.57 and 11.43 points better on a test solving applied Algebra problems with a calculator than students who learn by traditional methods.

d) Yes; 0 is not in the interval plausible differences are positive.

9. a) $H_0: \mu_C - \mu_T = 0$ vs. $H_A: \mu_C - \mu_T \neq 0$

b) Yes. Groups are independent, though we don't know if students were randomly assigned to the programs. Sample sizes are large, so CLT applies.

c) If the means for the two programs are really equal, there is less than a 1 in 10,000 chance of seeing a difference as large as or larger than the observed difference just from natural sampling variation.

d) On average, students who learn with the CPMP method do significantly worse on Algebra tests that do not allow them to use calculators than students who learn by traditional methods.

11. We must assume the samples were random or otherwise independent of each other. We also assume that the distributions are roughly normal, but the groups are large enough to proceed.

13. We are 95% confident that the mean purchase amount at Walmart is between $1.85 and $14.15 less than the mean purchase amount at Target.

15. a) (1.36, 4.64)

b) No; 5 minutes is beyond the high end of the interval.

17.

Random sample—questionable, but probably representative, independent samples, less than 10% of all cereals; boxplot shows no outliers—not exactly symmetric, but these are reasonable sample

sizes. Based on these samples, with 95% confidence, children's cereals average between 32.49% and 40.80% more sugar content than adult's cereals.

19. Random assignment; both distributions are unimodal and symmetric. $H_0: \mu_N - \mu_C = 0$ vs. $H_A: \mu_N - \mu_C > 0$; $t = 2.207$; P-value = 0.0168; $df = 37.3$. Because of the small P-value, we reject H_0. These data do suggest that new activities are better. The mean reading comprehension score for the group with new activities is significantly (at $\alpha = 0.05$) higher than the mean score for the control group.

21. a)

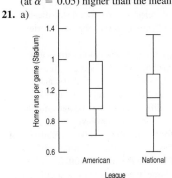

On average, American League stadiums see more home runs per game. Both distributions are reasonably symmetric with similar variability.

b) Based on these data, the average number of home runs hit per game in an American League stadium is between 0.9 and 1.16, with 95% confidence.

c) No. The boxplot indicates it isn't an outlier.

23. a) We want to work directly with the average difference. The two separate confidence intervals do not answer questions about the difference. The difference has a different standard deviation, found by adding variances.

b) $(-0.107, 0.212)$

c) Based on these data, with 95% confidence, American League stadiums average between 0.107 fewer home runs and 0.212 more home runs per game than National League stadiums.

d) No; 0 is in the interval.

25. These are not two independent samples. These are before and after scores for the same individuals (paired data).

27. a) These data do not provide evidence of a difference in ad recall between shows with sexual content and violent content.

b) Random assignment; large groups. $H_0: \mu_S - \mu_N = 0$ vs. $H_A: \mu_S - \mu_N \neq 0$. $t = -6.08$, df = 213.99, P-value = 5.5×10^{-9}. Because the P-value is low, we reject H_0. These data suggest that ad recall between shows with sexual and neutral content is different; those who saw shows with neutral content had higher average recall.

29. a) Groups are large and randomly assigned. $H_0: \mu_V - \mu_N = 0$ vs. $H_A: \mu_V - \mu_N \neq 0$. $t = -7.21$, df = 201.96, P-value = 1.1×10^{-11}. Because of the very small P-value, we reject H_0. There is a significant difference in mean ad recall between shows with violent content and neutral content; viewers of shows with neutral content remember more brand names, on average.

b) With 95% confidence, the average number of brand names remembered 24 hours later is between 1.45 and 2.41 higher for viewers of neutral content shows than for viewers of sexual content shows, based on these data ($df = 204.8$).

31. These were random samples, both less than 10% of properties sold. Prices of houses should be independent, and random sampling makes the two groups independent. The boxplots make the price distributions appear to be reasonably symmetric, and with the large sample sizes the few outliers don't affect the means much. Based on this sample, we're 95% confident that, in New York, having a waterfront is worth, on average, about $59,121 to 140,898 more in sale price.

33. a) $H_0: \mu_W = \mu_S$; $H_A: \mu_W < \mu_S$; assume these arrows are representative, samples are independent and reasonably symmetric.

$t = -3.2$, df $= 9.66$, p-value $= 0.005$. With a P-value < 0.05, we reject H_0. We found statistically significant evidence that the stone-tipped arrows penetrated deeper than the wooden-tipped arrows.

 b) We are 95% confident that the stone-tipped arrows penetrated 5.5 mm to 31.4 mm deeper. Depending on the animal being hunted, this fairly small difference may not be practically helpful.

35. $H_0: \mu_P = \mu_l$; $H_A: \mu_p > \mu_l$. The data was carefully collected from 2 independent groups. There are outliers in the lay individual's observations, but both data sets are symmetric. $t = 4.33$; df $= 11.6$; P-value $= 0.0005$, < 0.05 and we reject H_0. We found statistically significant evidence that the mean % collagen is higher in the priests' bones than the lay individuals.

37. $H_0: \mu_{big} - \mu_{small} = 0$ vs. $H_A: \mu_{big} - \mu_{small} \neq 0$; bowl size was assigned randomly; amount scooped by individuals and by the two groups should be independent. With 34.3 df, $t = 2.104$ and P-value $= 0.0428$. The low P-value leads us to reject the null hypothesis. There is evidence of a difference in the average amount of ice cream that people scoop when given a bigger bowl.

39. a) The 95% confidence interval for the difference is (0.61, 5.39). 0 is not in the interval, so scores in 1996 were significantly higher. (Or the t, with more than 7500 df, is 2.459 for a P-value of 0.0070.)

 b) Since both samples were very large, there shouldn't be a difference in how certain you are, assuming conditions are met.

41. Independent Groups Assumption: The runners are different women, so the groups are independent. The Randomization Condition is satisfied since the runners are selected at random for these heats.

Nearly Normal Condition: The boxplots show an outlier, but we will proceed and then redo the analysis with the outlier deleted. When we include the outlier, $t = 0.035$ with a two-sided P-value of 0.97. With the outlier deleted, $t = -1.14$, with P $= 0.2837$. Either P-value is so large that we fail to reject the null hypothesis of equal means and conclude that there is no evidence of a difference in the mean times for runners in unseeded heats.

43. With $t = -4.57$ ($df = 7.03$) and a very low P-value of 0.0013, we reject the null hypothesis of equal mean velocities. There is strong evidence that golf balls hit off Stinger tees will have a higher mean initial velocity.

45. a) We can be 95% confident that the interval 74.8 ± 178.05 minutes includes the true difference in mean crossing times between men and women. Because the interval includes zero, we cannot be confident that there is any difference at all ($df = 37.7$).

 b) Independence Assumption: There is no reason to believe that the swims are not independent or that the two groups are not independent of each other.

 Randomization Condition: The swimmers are not a random sample from any identifiable population, but they may be representative of swimmers who tackle challenges such as this.

 Nearly Normal Condition: the boxplots show no outliers. The histograms are unimodal; the histogram for men is somewhat skewed to the right. (Show your graphs.)

47. a) $H_0: \mu_R - \mu_N = 0$ vs. $H_A: \mu_R - \mu_N < 0$. $t = -1.36$, df $= 20.00$, P-value $= 0.0945$. Because the P-value is large, we fail to reject H_0. These data show no evidence of a difference in mean number of objects recalled between listening to rap or no music at all.

 b) Didn't conclude any difference.

Chapter 24

1. a) Paired.   b) Not paired.   c) Paired.

3. a) Randomly assign 50 hens to each of the two kinds of feed. Compare production at the end of the month.

 b) Give all 100 hens the new feed for 2 weeks and the old food for 2 weeks, randomly selecting which feed the hens get first. Analyze the differences in production for all 100 hens.

 c) Matched pairs. Because hens vary in egg production, the matched-pairs design will control for that.

5. a) Show the same people ads with and without sexual images, and record how many products they remember in each group. Randomly decide which ads a person sees first. Examine the differences for each person.

 b) Randomly divide volunteers into two groups. Show one group ads with sexual images and the other group ads without. Compare how many products each group remembers.

7. a) Matched pairs—same cities in different periods.

 b) There is a significant difference (P-value $= 0.0244$) in the labor force participation rate for women in these cities; women's participation seems to have increased between 1968 and 1972.

9. a) Use the paired t-test because we have pairs of Fridays in 5 different months. Data from adjacent Fridays within a month may be more similar than data from randomly chosen Fridays.

 b) We conclude that there is evidence (P-value 0.0212) that the mean number of cars found on the M25 motorway on Friday the 13th is less than on the previous Friday.

 c) We don't know if these Friday pairs were selected at random. If these are the Fridays with the largest differences, this will affect our conclusion. The Nearly Normal Condition appears to be met by the differences, but the sample size is small.

11. Adding variances requires that the variables be independent. These price quotes are for the same cars, so they are paired. Drivers quoted high insurance premiums by the local company will be likely to get a high rate from the online company, too.

13. a) The histogram—we care about differences in price.

 b) Insurance cost is based on risk, so drivers are likely to see similar quotes from each company, making the differences relatively smaller.

 c) The price quotes are paired; they were for a random sample of fewer than 10% of the agent's customers; the histogram of differences looks approximately Normal.

15. $H_0: \mu(Local - Online) = 0$ vs. $H_A: \mu(Local - Online) > 0$; with 9 df, $t = 0.83$. With a high P-value of 0.215, we don't reject the null hypothesis. These data don't provide evidence that online premiums are lower, on average.

17. a) No. The vehicles have no natural pairing.

 b) Possibly. The data are quantitative and paired by vehicle.

 c) The sample size is large, but there is at least one extreme outlier that should be investigated before applying these methods.

19. (7.17, 7.57). We are 95% confident that the interval from 7.17 mpg to 7.57 mpg captures the true improvement in highway gas mileage compared to city gas mileage.

21. The difference between fuel efficiency of cars and that of trucks can be large, but isn't relevant to the question asked about highway *vs.* city driving. Pairing places each vehicle in its own block to remove that variation from consideration.

23.

Data are paired for each city; cities are independent of each other; boxplot shows the temperature differences are reasonably symmetric, with no outliers. This is probably not a random sample, so we might be wary of inferring that this difference applies to all European cities. Based on these data, we are 90% confident that the average temperature in European cities in July is between 32.3°F and 41.3°F higher than in January.

25. Based on these data, we are 90% confident that boys, on average, can do between 1.6 and 13.0 more push-ups than girls (independent samples—not paired).

27. a) Paired sample test. Data are before/after for the same workers; workers randomly selected; assume fewer than 10% of all this company's workers; boxplot of differences shows them to be symmetric, with no outliers.

b) H_0: $\mu_D = 0$ vs. H_A: $\mu_D > 0$. $t = 3.60$, P-value $= 0.0029$. Because P < 0.01, reject H_0. These data show evidence that average job satisfaction has increased after implementation of the program.

c) Type I

29. H_0: $\mu_D = 0$ vs. H_A: $\mu_D \neq 0$. Data are paired by brand; brands are independent of each other; fewer than 10% of all yogurts (questionable); boxplot of differences shows an outlier (100) for Great Value:

With the outlier included, the mean difference (Strawberry – Vanilla) is 12.5 calories with a t-stat of 1.332, with 11 df, for a P-value of 0.2098. Deleting the outlier, the difference is even smaller, 4.55 calories with a t-stat of only 0.833 and a P-value of 0.4241. With P-values so large, we do not reject H_0. We conclude that the data do not provide evidence of a difference in mean calories.

31. a) Cars were probably not a simple random sample, but may be representative in terms of stopping distance; boxplot does not show outliers, but does indicate right skewness. A 95% confidence interval for the mean stopping distance on dry pavement is (131.8, 145.6) feet.

b) Data are paired by car; cars were probably not randomly chosen, but representative; boxplot shows an outlier (car 4) with a difference of 12. With deletion of that car, a Normal probability plot of the differences is relatively straight.

Retaining the outlier, we estimate with 95% confidence that the average braking distance is between 38.8 and 62.6 feet more on wet pavement

than on dry, based on this sample. (Without the outlier, the confidence interval is 47.2 to 62.8 feet.)

33. a) Paired Data Assumption: Data are paired by college. Randomization Condition: This was a random sample of public colleges and universities. 10% Condition: these are fewer than 10% of all public colleges and universities.

Normal Population Assumption: U.C. Irvine and New College of Florida seem to be outliers; we might consider removing them.

b) Having deleted the outliers, we are 90% confident, based on the remaining data, that nonresidents pay, on average, between $6377 and $9339 more than residents. If we retain the outliers, the interval is ($7144, $11,751).

c) Assertion is reasonable; without the outliers, $7000 is in the confidence interval. With the outliers, $7000 is actually lower than the values in the interval.

35. a) 60% is 30 strikes; H_0: $\mu = 30$ vs. H_A: $\mu > 30$. $t = 6.07$, P-value $= 3.92 \times 10^{-6}$. With a very small P-value, we reject H_0. There is very strong evidence that players can throw more than 60% strikes after training, based on this sample.

b) H_0: $\mu_D = 0$ vs. H_A: $\mu_D > 0$. $t = 0.135$, P-value $= 0.4472$. With such a high P-value, we do not reject H_0. These data provide no evidence that the program has improved pitching in these Little League players.

37. a) The data are clearly paired. Both races may have improved over time, but the pairwise differences are likely to be independent. We can only check the Nearly Normal Condition by using the computer files.

b) With 95% confidence we can say the mean time difference is between -15.81 minutes (women are faster) and $+8.67$ minutes (men are faster).

c) The interval contains 0, so we would not reject the hypothesis of no mean difference at $\alpha = 0.05$. We can't discern a difference between the female wheelchair times and the male running times.

Part VI Review

1. a) H_0: $\mu_{Jan} - \mu_{Jul} = 0$; H_A: $\mu_{Jan} - \mu_{Jul} \neq 0$. $t = -1.94$, df $= 43.68$, P-value $= 0.0590$. Since P < 0.10, reject the null. These data show a significant difference in mean age to crawl between January and July babies.

b) H_0: $\mu_{Apr} - \mu_{Oct} = 0$; H_A: $\mu_{Apr} - \mu_{Oct} \neq 0$. $t = -0.92$; df $= 59.40$; P-value $= 0.3610$. Since P > 0.10, do not reject the null; these data do not show a significant difference between April and October with regard to the mean age at which crawling begins.

c) These results are not consistent with the claim.

3. H_0: $p = 0.26$; H_A: p $\neq 0.26$. $z = 0.946$; P-value $= 0.3443$. Because the P-value is high, we do not reject H_0. These data do not show that the Denver-area rate is different from the national rate in the proportion of businesses with women owners.

5. Based on these data, we are 95% confident that the mean difference in aluminum oxide content is between -3.37 and 1.65. Since the interval contains 0, the means in aluminum oxide content of the pottery made at the two sites could reasonably be the same.

7. a) H_0: $p_{ALS} - p_{Other} = 0$; H_A: $p_{ALS} - p_{Other} > 0$. $z = 2.52$; P-value $= 0.0058$. With such a low P-value, we reject H_0. This is strong evidence that there is a higher proportion of varsity athletes among ALS patients than those with other disorders.

b) Observational retrospective study. To make the inference, one must assume the patients studied are representative.

9. $H_0: \mu = 7.41$; $H_A: \mu \neq 7.41$. $t = 2.18$; df = 111; P-value = 0.0313. With such a low P-value, we reject H_0. Assuming that Missouri babies fairly represent the United States, these data suggest that American babies are different from Australian babies in birth weight; it appears American babies are heavier, on average.

11. a) If there is no difference in the average fish sizes, the chance of seeing an observed difference this large just by natural sampling variation is less than 0.1%.

   b) If cost justified, feed them a natural diet. c) Type I

13. a) Assuming the conditions are met, from these data we are 95% confident that patients with cardiac disease average between 3.39 and 5.01 years older than those without cardiac disease.

   b) Older patients are at greater risk from a variety of other health issues, and perhaps more depressed.

15. a) Stratified sample survey.

   b) We are 95% confident that the proportion of boys who play computer games is between 7.0 and 17.0 percentage points higher than among girls.

   c) Yes. The entire interval lies above 0.

17. a) Bimodal.

   b) μ, the population mean. Sample size does not matter.

   c) $\sigma/\sqrt{n}$; sample size does matter.

   d) It becomes closer to a Normal model and less variable as the sample size increases.

19. Based on the data, we are 95% confident that the mean difference in words misunderstood is between −3.76 and 3.10. Because 0 is in the confidence interval, we would conclude that the two tapes could be equivalent.

21. a)

M − W Difference

The countries that appear to be outliers are Spain, Italy, and Portugal. They are all Mediterranean countries.

   b) $H_0: \mu_D = 0$; $H_A: \mu_D > 0$. $t = 5.56$; df = 10; P-value = 0.0001. With such a low P-value, we reject H_0. These data show that European men are more likely than women to read newspapers.

23. Based on the survey, we are 95% confident that the proportion of the American adults who would agree with the statement is between 56.7% and 63.1%

25. H_0: There is no difference in cancer rates, $p_1 - p_2 = 0$. H_A: The cancer rate in those who use the herb is higher, $p_1 - p_2 > 0$.

27. a) We are 95% confident that the mean difference in rainfall between Victorville and 29 Palms is (−57.6 cm, 43.2 cm).

   b) We are 95% confident that the mean difference in rainfall between Victorville and Mitchell's Cavern is (−37.2 cm, 66.0 cm).

   c) We are 95% confident that the mean difference in rainfall between 29 Palms and Mitchell's Cavern is (−15.2 cm, 58.8 cm).

   d) Zero is in the intervals. It appears that none of these regions are is statistically significantly different in average rainfall from the others.

29. Data are matched pairs (before and after for the same rooms); less than 10% of all rooms in a large hotel; uncertain how these rooms were selected (are they representative?). Histogram shows that differences are roughly unimodal and symmetric, with no outliers. A 95% confidence interval for the difference, before − after, is (0.58, 2.65) counts. Since the entire interval is above 0, these data suggest that the new air-conditioning system was effective in reducing average bacteria counts.

31. a) We are 95% confident that between 19.77% and 38.66% of children with bipolar symptoms will be helped with medication and psychotherapy, based on this study.

   b) 221 children

33. a) From this histogram, about 115 loaves or more. (Not Normal.) This assumes the last 100 days are typical.

b) Large sample size; CLT says $\bar{y}$ will be approximately Normal.

   c) From the data, we are 95% confident that on average the bakery will sell between 101.2 and 104.8 loaves of bread a day.

   d) 25

   e) Yes, 100 loaves per day is too low—the entire confidence interval is above that.

35. a) $H_0: p_{High} - p_{Low} = 0$; $H_A: p_{High} - p_{Low} \neq 0$. $z = -3.57$; P-value = 0.0004. Because the P-value is so low, we reject H_0. These data suggest the IRS risk is different in the two groups; it appears people who consume dairy products often have a lower risk, on average.

   b) Doesn't indicate causality; this is not an experiment.

37. Based on these data, we are 95% confident that seeded clouds will produce an average of between −4.76 and 559.56 more acre-feet of rain than unseeded clouds. Since the interval contains negative values, it may be that seeding is unproductive.

39. a) Randomizing order of the tasks helps avoid bias and memory effects. Randomizing the cards helps avoid bias as well.

   b) $H_0: \mu_D = 0$; $H_A: \mu_D \neq 0$

Boxplot of the differences looks symmetric with no outliers.

$t = -1.70$; P-value = 0.0999; do not reject H_0, because P > 0.05. The data do not provide evidence that the color or written word dominates.

41. a) Different samples give different means; this is a fairly small sample. The difference may be due to natural sampling variation.

   b) $H_0: \mu = 100$; $H_A: \mu < 100$

   c) Batteries selected are a SRS (representative); fewer than 10% of the company's batteries; lifetimes are approximately Normal.

   d) $t = -1.0$; P-value = 0.1666; do not reject H_0. This sample does not show that the average life of the batteries is significantly less than 100 hours.

   e) Type II.

43. a) $H_0: \mu_1 - \mu_2 = 0$; $H_A: \mu_1 - \mu_2 > 0$. $t = 0.90$; P-value = 0.1864. Since $P > 0.05$, we do not reject H_0. Weeklong study scores were not significantly higher.

   b) $H_0: p_1 - p_2 = 0$; $H_A: p_1 - p_2 \neq 0$. $z = -3.10$, P-value = 0.0019. With such a small P-value, we reject H_0. There is evidence of a difference in proportion for passing on Friday; it appears that cramming may be more effective.

   c) $H_0: \mu_D = 0$; $H_A: \mu_D > 0$. $t = 5.17$; P-value < 0.0001. These data show evidence that learning does not last for 3 days because mean score declined.

Practice Test

MULTIPLE CHOICE

1. B	**3.** A	**5.** B	**7.** C	**9.** E
11. B	**13.** D	**15.** E	**17.** E	**19.** A
21. C	**23.** D	**25.** C	**27.** A	**29.** E
31. B	**33.** E	**35.** D		

FREE RESPONSE

1. a) $\widehat{Low} = 108.798 - 2.111 Lat$

   b) The model suggest that average January low temperatures decrease about 2.111°F for every 1° increase in north latitude.

   c) Probably not. It suggests that the average January low temperature at the equator would be over 108°F.

   d) This city's average January low temperature is 4.2°F lower than the model predicts.

3. a) $(0.5)(0.2) + (0.5)(0.75) = 0.475$

b) $\dfrac{0.1}{0.475} \approx 0.21$

c) $100(0.525) + 1000(0.475) = \$527.50$

5. a) To see if the data suggest the two time populations are Normal.

b) Random allocation makes the groups independent; assume normality of population distributions. OK to use 2-sample t-interval: $(-19.99, 16.89)$. We can be 95% confident that the mean commuting time for the carbon frame bike is between 19.99 minutes shorter and 16.89 minutes longer than for the steel frame bike.

c) If the doctor ran this experiment repeatedly, we'd expect about 95% of the resulting intervals to capture the true difference in mean commuting times.

d) No; the interval offers the possibilities that the bike may be faster or slower.

Chapter 25

1. a) Chi-square test of independence. We have one sample and two variables. We want to see if the variable *Account Type* is independent of the variable *Trade Type*.

b) Other test. *Account Size* is quantitative, not counts.

c) Chi-square test of homogeneity. We want to see if the distribution of one variable, *Courses*, is the same for two groups (resident and nonresident students).

3. a) 10   b) Goodness-of-fit

c) H_0: The die is fair (all faces have $p = 1/6$).

H_A: The die is not fair.

d) Count data; rolls are random and independent; expected frequencies are all bigger than 5.

e) 5   f) $\chi^2 = 5.600$, P-value $= 0.3471$

g) Because the P-value is high, do not reject H_0. The data show no evidence that the die is unfair.

5. a) $(30, 30, 30, 30)$, 30 for each season.

b) 1.933   c) 3   d) 7.815

e) Do not reject the null hypothesis. There is not enough evidence to suggest that births are not distributed uniformly across the seasons.

7. a) Weights are quantitative, not counts.

b) Count the number of each kind of nut, assuming the company's percentages are based on counts rather than weights.

9. H_0: The police force represents the population (29.2% white, 28.2% black, etc.). H_A: The police force is not representative of the population. $\chi^2 = 16516.88$, df $= 4$, P-value $= 0.0000$. Because the P-value is so low, we reject H_0. These data show that the police force is not representative of the population. In particular, there are too many white officers in relationship to their membership in the community.

11. a) $\chi^2 = 5.671$, df $= 3$, P-value $= 0.1288$. With a P-value this high, we fail to reject H_0. Yes, these data are consistent with those predicted by genetic theory.

b) $\chi^2 = 11.342$, df $= 3$, P-value $= 0.0100$. Because of the low P-value, we reject H_0. These data provide evidence that the distribution is not as specified by genetic theory.

c) With small samples, many more data sets will be consistent with the null hypothesis. With larger samples, small discrepancies will show evidence against the null hypothesis.

13. a) $96/16 = 6$ b) Goodness-of-fit.

c) H_0: The number of large hurricanes remains constant over decades.

H_A: The number of large hurricanes has changed.

d) 15   e) P-value $= 0.63$

f) The very high P-value means these data offer no evidence that the numbers of large hurricanes has changed.

15. a) Independence

b) H_0: Breastfeeding success is independent of having an epidural.

H_A: There's an association between breastfeeding success and having an epidural.

17. a) 1   b) 159.34

c) Breastfeeding behavior should be independent for these babies. They are fewer than 10% of all babies; we assume they are representative. We have counts, and all the expected counts are at least 5.

19. a) 5.90   b) P-value < 0.005

c) The P-value is very low, so reject the null. There's evidence of an association between having an epidural and subsequent success in breastfeeding.

21. a) $\dfrac{(190 - 159.34)}{\sqrt{159.34}} = 2.43$

b) It appears that babies whose mothers had epidurals during childbirth are much less likely to be breastfeeding 6 months later.

23. These factors would not be mutually exclusive. There would be yes or no responses for every baby for each.

25. a)

	Did You Use the Internet Yesterday?	
	Yes	No
White	2506.38	895.62
Black	338.90	121.10
Hispanic/Other	445.73	159.27

b) 11.176   c) 2   d) 0.0037

e) Reject the null hypothesis; there is evidence that Race and Internet use are not independent.

27. a) 40.2%   b) 8.1%   c) 62.2%   d) 285.48

e) H_0: Survival was independent of status on the ship.

H_A: Survival depended on the status.

f) 3

g) We reject the null hypothesis. Survival depended on status. We can see that first-class passengers were more likely to survive than passengers of any other class.

29. First class passengers were most likely to survive, while 3rd-class passengers and crew were under-represented among the survivors.

31. a) Experiment—actively imposed treatments (different drinks)

b) Homogeneity

c) H_0: The rate of urinary tract infection is the same for all three groups.

H_A: The rate of urinary tract infection is different among the groups.

d) Count data; random assignment to treatments; all expected frequencies larger than 5.

e) 2   f) $\chi^2 = 7.776$, P-value $= 0.020$.

g) With a P-value this low, we reject H_0. These data provide reasonably strong evidence that there is a difference in urinary tract infection rates between cranberry juice drinkers, lactobacillus drinkers, and the control group.

h) The standardized residuals are

	Cranberry	Lactobacillus	Control
Infection	-1.87276	1.19176	0.68100
No Infection	1.24550	-0.79259	-0.45291

From the standardized residuals (and the sign of the residuals), it appears those who drank cranberry juice were less likely to develop urinary tract infections; those who drank lactobacillus were more likely to have infections.

33. a) Independence

b) H_0: *Political Affiliation* is independent of *Sex*. H_A: There is a relationship between *Political Affiliation* and *Sex*.

c) Counted data; probably a random sample, but can't extend results to other states; all expected frequencies greater than 5.

d) $\chi^2 = 4.851$, df $= 2$, P-value $= 0.0884$.

e) Because of the high P-value, we do not reject H_0. These data do not provide evidence of a relationship between *Political Affiliation* and *Sex*.

35. H_0: *Political Affiliation* is independent of *Region*. H_A: There is a relationship between *Political Affiliation* and *Region*. $\chi^2 = 13.849$, df $= 4$, P-value $= 0.0078$. With a P-value this low, we reject H_0. *Political Affiliation* and *Region* seem to be related. Examination of the residuals suggests that those in the West are more likely to be Democrat than Republican; those in the Northeast are more likely to be Republican than Democrat.

37. H_0: All digits 0–9 are equally likely to appear. H_A: Some numbers are more likely than others. Draws should be independent; these weeks should be representative of all draws, and are fewer than 10% of all possible Pick-3 lotteries. All expected values are at least 5. Goodness-of-Fit Test: df $= 9$, Chi-sq $= 6.46$, P-value $= 0.693$. These data fail to provide evidence that all the digits are not equally likely.

39. a) Homogeneity
b) H_0: The grade distribution is the same for both professors.
   H_A: The grade distributions are different.
c)

	Dr. Alpha	Dr. Beta
A	6.667	5.333
B	12.778	10.222
C	12.222	9.778
D	6.111	4.889
F	2.222	1.778

Three cells have expected frequencies less than 5.

41. a)

	Dr. Alpha	Dr. Beta
A	6.667	5.333
B	12.778	10.222
C	12.222	9.778
Below C	8.333	6.667

All expected frequencies are now larger than 5.
b) Decreased from 4 to 3.
c) $\chi^2 = 9.306$, P-value $= 0.0255$. Because the P-value is so low, we reject H_0. The grade distributions for the two professors seem to be different. Dr. Alpha gives fewer A's and more grades below C than Dr. Beta.

43. $\chi^2 = 14.058$, df $= 1$, P-value $= 0.0002$. With a P-value this low, we reject H_0. There is evidence of racial steering. Blacks are much less likely to rent in Section A than Section B.

45. a) $z = 3.74936$, $z^2 = 14.058$.
b) P-value $(z) = 0.0002$ (same as in Exercise 25).

47. $\chi^2 = 5.89$, df $= 3$, P $= 0.117$. Because the P-value is >0.05, these data do not provide evidence of an association between the mother's age group and the outcome of the pregnancy.

Chapter 26

1. The scatterplot shows a linear relationship with equal spread about a regression line throughout the range of *Acceptance Rates*. The residual plot has no structure, and there are not any striking outliers. The histogram of residuals is symmetric and bell-shaped. All conditions are satisfied to proceed with the regression analysis.

3. $s = 2.86$. Since the range of *Graduation Rates* is only about 17 percentage points, this number is fairly small and helps us to understand the amount of spread about the regression model.

5. The standard error for the slope tells us how much the slope of the regression equation would vary from sample to sample.

7. The administrators can conclude that there is a relationship between *Graduation Rates* and *Admission Rates* for these upper-echelon schools. It seems the lower the *Admission Rate*, the higher the *Graduation Rate*.

9. The administrators can conclude, with 95% confidence, that schools with *Admission Rates* that are lower by 1% will have, on average, *Graduation Rates* that are higher by between 0.203% to 0.299%.

11. a) $\widehat{24Error} = 132.3 - 2.01$ *Years since* 1970; according to the model, the error made in predicting a hurricane's path was about 132 nautical miles, on average, in 1970. It has been declining at a rate of about 2.01 nautical miles per year.
b) H_0: $\beta_1 = 0$; there has been no change in prediction accuracy. H_A: $\beta_1 \neq 0$; there has been a change in prediction accuracy.
c) With $t = -9.25$ and a P-value < 0.0001, I reject the null hypothesis and conclude that prediction accuracies have in fact been changing during this period.
d) 68.7% of the variation in hurricane prediction accuracy is accounted for by this linear model on time.

13. a) $\widehat{Budget} = -31.387 + 0.714$ *RunTime*. The model suggests that movies cost about \$714,000 per minute to make.
b) A negative starting value makes no sense, but the P-value of 0.07 indicates that we can't discern a difference between our estimated value and zero. The statement that a movie of zero length should cost \$0 makes sense.
c) Amounts by which movie costs differ from predictions made by this model vary, with a standard deviation of about \$33 million.
d) 0.154 \$m/min
e) If we constructed other models based on different samples of movies, we'd expect the slopes of the regression lines to vary, with a standard deviation of about \$154,000 per minute.

15. a) The scatterplot looks straight enough, the residuals look random and nearly normal, and the residuals don't display any clear change in variability.
b) I'm 95% confident that the cost of making longer movies increases at a rate of between 0.41 and 1.02 million dollars per additional minute.

17. a) H_0: $\beta_1 = 0$; there's no association between calories and sodium content in all-beef hot dogs. H_A: $\beta_1 \neq 0$: there is an association.
b) Based on the low P-value (0.0018), I reject the null. There is evidence of an association between the number of calories in all-beef hot dogs and their sodium contents.

19. a) Among all-beef hot dogs with the same number of calories, the sodium content varies, with a standard deviation of about 60 mg.
b) 0.561 mg/cal
c) If we tested many other samples of all-beef hot dogs, the slopes of the resulting regression lines would be expected to vary, with a standard deviation of about 0.56 mg of sodium per calorie.

21. I'm 95% confident that for every additional calorie, all-beef hot dogs have, on average, between 1.07 and 3.53 mg more sodium.

23. a) H_0: Difference in age at first marriage has not been changing, $\beta_1 = 0$. H_A: Difference in age at first marriage has been changing, $\beta_1 \neq 0$.
b) Residual plot looks randomly scattered, histogram is unimodal and a bit skewed, but shows no obvious skewness or outliers.
c) $t = -9.05$, P-value < 0.0001. With such a low P-value, we reject H_0. These data show evidence that difference in age at first marriage is decreasing.

25. Based on these data, we are 95% confident that the average difference in age at first marriage is decreasing at a rate between 0.022 and 0.036 years per year.

27. a) H_0: *Fuel Economy* and *Weight* are not (linearly) related, $\beta_1 = 0$. H_A: *Fuel Economy* changes with *Weight*, $\beta_1 \neq 0$. P-value < 0.0001, indicating strong evidence of an association.
b) Yes, the conditions seem satisfied. Histogram of residuals is unimodal and symmetric; residual plot looks OK, but some "thickening" of the plot with increasing values.
c) $t = -12.2$, P-value < 0.0001. These data show evidence that *Fuel Economy* decreases with the *Weight* of the car.

29. a) $(-9.57, -6.86)$ mpg per 1000 pounds.

b) Based on these data, we are 95% confident that *Fuel Efficiency* decreases between 6.86 and 9.57 miles per gallon, on average, for each additional 1000 pounds of *Weight*.

31. a) We are 95% confident that 2500-pound cars will average between 27.34 and 29.07 miles per gallon.

b) Based on the regression, a 3450-pound car will get between 15.44 and 25.36 miles per gallon, with 95% confidence.

33. a) Yes. $t = 2.73$, P-value $= 0.0079$. With a P-value so low, we reject H_0. There is a positive relationship between *Calories* and *Sodium* content.

b) No. $R^2 = 9\%$ and s appears to be large, although without seeing the data, it is a bit hard to tell.

35. Plot of *Calories* against *Fiber* does not look linear; the residuals plot also shows increasing variance as predicted values get large. The histogram of residuals is right skewed.

37. a) H_0: No (linear) relationship between *BCI* and *pH*, $\beta_1 = 0$.
H_A: There is a relationship, $\beta_1 \neq 0$.

b) $t = -7.73$ with 161 df; P-value < 0.0001

c) There seems to be a negative relationship; *BCI* decreases as *pH* increases at an average of 197.7 *BCI* units per increase of 1 *pH*.

39. a) H_0: No linear relationship between *Population* and *Ozone*, $\beta_1 = 0$. H_A: *Ozone* increases with *Population*, $\beta_1 > 0$. $t = 3.48$, P-value $= 0.0018$. With a P-value so low, we reject H_0. These data show evidence that *Ozone* increases with *Population*.

b) Yes, *Population* accounts for 84% of the variability in *Ozone* level, and s is just over 5 parts per million.

41. a) Based on this regression, each additional million residents corresponds to an increase in average ozone level of between 3.29 and 10.01 ppm, with 90% confidence.

b) The mean *Ozone* level for cities with 600,000 people is between 18.47 and 27.29 ppm, with 90% confidence.

43. a) 23 tablets.

b) Yes. The scatterplot is roughly linear with lots of scatter; plot of residuals vs. predicted values shows no overt patterns; Normal probability plot of residuals is reasonably straight.

c) H_0: No linear relationship between *Battery Life* and *Screen Brightness*, $\beta_1 = 0$. H_A: *Battery Life* decreases with brighter screens, $\beta_1 < 0$. $t = 2.85$; one-sided P-value $= 0.0048$. With a P-value this low, we reject H_0. These data provide strong evidence that brighter screens are associated with shorter battery lifetimes.

d) Not particularly. $R^2 = 27.9\%$ and $s = 1.913$ hours. Since the range of battery life is only about 9 hours, an s of 1.913 is quite large.

e) $\widehat{Hours} = 2.85 + 0.014\ Screen\ Brightness$

f) $(0.0055, 0.0225)$ hours per cd/m^2 units

g) *Battery life* increases, on average, between 0.0055 and 0.0225 *hours* per one cd/m^2 units, with 90% confidence.

45. a) H_0: No linear relationship between *Waist* size and *%Body Fat*, $\beta_1 = 0$. H_A: *%Body Fat* changes with *Waist* size, $\beta_1 \neq 0$. $t = 8.14$; P-value < 0.0001. There's evidence that *%Body Fat* seems to increase with *Waist* size.

b) With 95% confidence, mean *%Body Fat* for people with 40-inch waists is between 23.58 and 29.02, based on this regression.

47. a) The regression model is $\widehat{Midterm2} = 12.005 + 0.721\ Midterm1$

	Estimate	Std Error	t-ratio	P-value
Intercept	12.00543	15.9553	0.752442	0.454633
Slope	0.72099	0.183716	3.924477	0.000221

	RSquare	0.198982
	s	16.78107
	n	64

b) The scatterplot shows a weak, somewhat linear, positive relationship. There are several outlying points, but removing them only makes the relationship slightly stronger. There is no obvious pattern in the residual plot. The regression model appears appropriate. The small P-value for the slope shows that the slope is statistically distinguishable from 0 even though the R^2 value of 0.199 suggests that the overall relationship is weak.

c) No. The R^2 value is only 0.199 and the value of s of 16.8 points indicates that she would not be able to predict performance on *Midterm2* very accurately.

49. H_0: Slope of *Effectiveness* vs *Initial Ability* $= 0$; H_A: Slope $\neq 0$

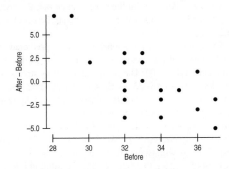

Scatterplot is straight enough. Regression conditions appear to be met. $t = -4.34$, df $= 18$, P-value $= 0.004$. With a P-value this small, we reject the null hypothesis. There is strong evidence that the effectiveness of the video depends on the player's initial ability. The negative slope observed that the method is more effective for those whose initial performance was poorest and less so for those whose initial performance was better. This looks like a case of regression to the mean. Those who were above average initially tended to be worse after training. Those who were below average initially tended to improve.

51. a) Scatterplot looks linear; no overt pattern in residuals; histogram of residuals roughly symmetric and unimodal.

b) H_0: No linear relationship between *Education* and *Mortality*, $\beta_1 = 0$. H_A: $\beta_1 \neq 0$. $t = -6.24$; P-value < 0.001. There is evidence that cities in which the mean education level is higher also tend to have a lower mortality rate.

c) No. Data are on cities, not individuals. Also, these are observational data. We cannot predict causal consequences from them.

d) $(-65.95, -33.89)$ deaths per 100,000 people.

e) *Mortality* decreases, on average, between 33.89 and 65.95 deaths per 100,000 for each extra year of average *Education*.

f) Based on the regression, the average *Mortality* for cities with an average of 12 years of *Education* will be between 874.239 and 914.196 deaths per 100,000 people.

Part VII Review

1. H_0: The proportions are as specified by the ratio 1:3:3:9; H_A: The proportions are not as stated. $\chi^2 = 5.01$; df $= 3$; P-value $= 0.1711$. Since $P > 0.05$, we fail to reject H_0. These data do not provide evidence to indicate that the proportions are other than 1:3:3:9.

3. a) H_0: *Mortality* and *calcium concentration* in water are not linearly related, $\beta_1 = 0$; H_A: They are linearly related, $\beta_1 \neq 0$.

b) $t = -6.73$; P-value < 0.0001. There is a significant negative relationship between calcium in drinking water and mortality.

c) $(-4.19, -2.27)$ deaths per 100,000 for each ppm calcium.

d) Based on the regression, we are 95% confident that mortality (deaths per 100,000) decreases, on average, between 2.27 and 4.19 for each part per million of calcium in drinking water.

5. 404 checks

7. H_0: *Income* and *Party* are independent. H_A: *Income* and *Party* are not independent. $\chi^2 = 17.19$; P-value $= 0.0018$. With such a small P-value, we reject H_0. These data show evidence that income level and party are not independent. Examination of components suggests Democrats are most likely to have low incomes; Independents are most likely to have middle incomes, and Republicans are most likely to have high incomes.

9. $H_0: p_L - p_R = 0$; $H_A: p_L - p_R \neq 0$. $z = 1.38$; P-value $= 0.1683$. Since P > 0.05, we do not reject H_0. These data do not provide evidence of a difference in musical abilities between right- and left-handed people.

11. a) $H_0: \mu_D = 0$; $H_A: \mu_D \neq 0$.
Boxplot of the differences indicates a strong outlier (1958). With the outlier kept in, the t-stat is 0, with a P-value of 1.00 (two sided). There is no evidence of a difference (on averag of actual and that predicted by Gallup. With the outlier taken out, the t-stat is still only -0.8525 with a P-value of 0.4106, so the conclusion is the same.
b) H_0: There is no (linear) relationship between predicted and actual number of Democratic seats won ($\beta_1 = 0$). H_A: There is a relationship ($\beta_1 \neq 0$). The relationship is very strong, with an R^2 of 97.7%. The t-stat is 22.56. Even with only 12 df, this is clearly significant (P-value < 0.0001). There is an outlying residual (1958), but without it, the regression is even stronger.

13. Conditions are met; df $= 4$; $\chi^2 = 0.69$; P-value $= 0.9526$. Since P > 0.05, we do not reject H_0. We do not have evidence that the way the hospital deals with twin pregnancies has changed.

15. a) Based on these data, the average annual rainfall in LA is between 11.65 and 17.39 inches, with 90% confidence.
b) About 46 years
c) No. The regression equation is $\widehat{Rain} = -51.684 + 0.033 \times Year$. $R^2 = 0.1\%$. For the slope, $t = 0.12$ with P-value $= 0.9029$.

17. a) 10.29
b) $\chi^2 = 6.56$; 6df; P-value $= 0.3639$. No evidence to suggest that births are not distributed equally across days of the week.
c) The standardized residuals for Monday and Tuesday are -1.02 and 2.09, respectively. Monday's value is not unusual at all. Tuesday's is borderline high, but we concluded that there is not evidence that births are not uniform. With 7 standardized residuals, it is not surprising that one is large.
d) Some births are scheduled for the convenience of the doctor and/or the mother.

19. a) Linear regression is meaningless—the data are categorical.
b) This is a two-way table that is appropriate. H_0: *Eye* and *Hair* color are independent. H_A: *Eye* and *Hair* color are not independent. However, four cells have expected counts less than 5, so the χ^2 analysis is not valid unless cells are merged. However, with a χ^2 value of 223.6 with 16 df and a P-value < 0.0001, the results are not likely to change if we merge appropriate eye colors.

21. a) $H_0: p_Y - p_O = 0$; $H_A: p_Y - p_O \neq 0$. $z = 3.56$; P-value $= 0.0004$. With such a small P-value, we reject H_0. We conclude there is evidence of a difference in effectiveness; it appears the methods are not as good for older women.
b) $\chi^2 = 12.70$; P-value $= 0.0004$. Same conclusion.
c) The P-values are the same; $z^2 = (3.563944)^2 = 12.70 = \chi^2$.

23. a) Positive direction, generally linear trend; moderate scatter.
b) H_0: There is no linear relationship between *Interval* and *Duration*. $\beta_1 = 0$. H_A: There is a linear relationship, $\beta_1 \neq 0$.
c) Yes; histogram is unimodal and roughly symmetric; residuals plot shows random scatter.
d) $t = 27.1$; P-value ≤ 0.001. With such a small P-value, we reject H_0. There is evidence of a positive linear relationship between duration and time to next eruption of Old Faithful.
e) The average time to next eruption after a 2-minute eruption is between 53.24 and 56.12 minutes, with 95% confidence.
f) Based on this regression, we will have to wait between 63.23 and 87.57 minutes after a 4-minute eruption, with 95% confidence.

25. a) $t = 1.42$, df $= 459.3$, P-value $= 0.1574$. Since P > 0.05, we do not reject H_0. There's no evidence the two groups differed in ability at the start of the study.

b) $t = 15.11$; P-value < 0.0001. The group taught using the accelerated Math program showed a significant improvement.
c) $t = 9.24$; P-value < 0.0001. The control group showed a significant improvement in test scores.
d) $t = 5.78$; P-value < 0.0001. The Accelerated Math group had significantly higher gains than the control group.

27. a) The regression—he wanted to know about association.
b) There is a moderate relationship between cottage cheese and ice cream sales; for every million pounds of cottage cheese, 1.19 million pounds of ice cream are sold, on average.
c) Testing if the mean difference is 0 (matched t-test). Regression won't answer this question.
d) The company sells more cottage cheese than ice cream, on average.
e) part (a)—linear relationship; residuals have a Normal distribution; residuals are independent with equal variation about the line. (c)—Observations are independent; differences are approximately Normal; less than 10% of all possible months' data.
f) About 71.32 million pounds.
g) (0.09, 2.29)
h) From this regression, every million pounds of cottage cheese sold is associated with an increase in ice cream sales of between 0.09 and 2.29 million pounds.

29. Based on these data, the average weight loss for the clinic is between 8.24 and 10.06 pounds, with 95% confidence. The clinic's claim is plausible.

31. $\chi^2 = 8.23$; P-value $= 0.0414$. There is evidence of an association between *cracker type* and *bloating*. Standardized residuals for the gum cracker are -1.32 and 1.58. Prospects for marketing this cracker are not good.

33. $\chi^2 = 40.715$; P-value < 0.0001. These data do indicate an association between education and family planning. More educated women have fewer unplanned pregnancies.

Practice Exam

MULTIPLE CHOICE

1. B	3. E	5. B	7. A	9. C
11. D	13. B	15. D	17. E	19. B
21. D	23. C	25. C	27. E	29. D
31. A	33. E	35. C	37. E	39. E

FREE RESPONSE

1. a) Q1 $- 1.5(IQR) = 61 - 1.5(90 - 61) = 17.5$; $15 < 17.5$
b) Advanced; 50% of the classes useat least 78 folds and fewer than 25% involve simpler models with 40-60 folds.
c) No; the sample size is small and there is an outlier.

3. a) 0.12   b) 0.757   c) $2.59

5. Chi-square test of homogeneity. H_0: The racial distribution of teachers hired for public schools is the same as that for private schools, H_A: The distributions are different. We have counts of teachers in each category. We assume the two groups are independent and that each teacher's race is independent of the others. Although these teachers are not a random sample, we will take these teachers to be representative of and less than 10% of other recent hirings. Expected counts (smallest is 14.35) are all at least 5. The conditions seem to be met, so we can use a Chi-Square model with $(4 - 1)(2 - 1) = 3$ degrees of freedom to conduct a Chi-Square test of homogeneity. $\chi^2 = 7.96$ with 3 df; P $= 0.047$. With P < 0.05 we reject H_0; there is evidence that the racial distribution of teachers hired by public schools is not the same as for those hired by private schools. It appears that Black teachers are underrepresented in private school hirings.

Appendix D: Photo and Text Acknowledgments

Photos

vii (top) Courtesy of David E. Bock, (middle) Courtesy of Paul F. Velleman, (bottom) Courtesy of Richard D. De Veaux **1** Shutterstock **3** David Cheskin/AP Images **4** Shutterstock **7** (top) Stefano Rellandini/Reuters, (bottom) Toria/Shutterstock **9** Shutterstock **14** Alamy **19** Alamy **25** peternile/Fotolia **27** Zoo Praha/AP Images **31** Alamy **43** Toru Hanai/Reuters **50** Olivier Juneau/Shutterstock **51** Alamy **56** Shutterstock **63** Newscom **69** Toru Hanai/Reuters **83** Getty Images **86** (top) Shutterstock, (bottom) Dreamstime **89** Erin Paul Donovan/Alamy **94** Ithaca Times **95** Getty Images **107** Myunggu Han/Newscom **108** Rickman/Alamy **112** Mark van Manen/Newscom **113** Robert Kneschke/Fotolia **117** Charlie Newham/Alamy **120** David Gallaher/Shutterstock **122** Alamy **123** Nayashkova Olga/Shutterstock **129** Myunggu Han/Newscom **150** Newscom **152** Alamy **157** Dreamstime **160** James Steidl/Fotolia **161** Fotolia **163** NASA **165** AP Images **166** Newscom **176** Kosoff/Shutterstock **179** Newscom **183** Alamy **185** Jeff Schmaltz/NASA **191** Mara Zemgaliete/Fotolia **196** (bottom) Kosoff/Shutterstock **209** Dick De Veaux **214** (top) Teresa Hoover Photography, (bottom) Newscom **215** Jim Cole/AP Images **216** Alamy **219** Jose/Fotolia **221** Dick De Veaux **232** Mark Phillips/Alamy **234** Ilker Canikligil/Shutterstock **238** Dmitry Deshevykh/Alamy **246** MarkPhillips/Alamy **267** Suzanne Tucker/Shutterstock **269** (top) Matthew Healey/Newscom, (middle) Newscom, (bottom) Reuters/Corbis **271** Jonathan D. Wilson/Shutterstock **272** Andrew Rich/iStockphoto **275** Suzanne Tucker/Shutterstock **280** Shutterstock **281** The New York Public Library/Art Resource, NY **282** Alamy **286** Monkey Business/Shutterstock **287** Justin Sullivan/Getty Images **288** (left) Jupiter Images, (right) James Woodson/Thinkstock **289** Sirena Designs/Fotolia **290** Jeff Greenberg/Alamy **297** Shutterstock **305** Alamy **306** iStockphoto **307** Alamy **308** (top) Andres Rodriguez/Fotolia, (bottom) Houghton Library **310** (top) Iurii Davydov/Shutterstock, (bottom) Getty Images **315** Galinka/Shutterstock **317** Orkhan Aslanov/Shutterstock **318** Shutterstock **319** Fotolia **320** iStockphoto **323** Alamy **343** Jessica Bethke/Shutterstock **345** Alamy **346** (top) Fotolia, (middle) iStockphoto, (bottom) Dreamstime LLC **347** Alamy **350** Chris Hepburn/iStockphoto **352** Carolyn Jenkins/Alamy **356** Jessica Bethke/Shutterstock **362** Mario Tama/Getty Images **363** (top) Shutterstock, (margin) U.S. Department of the Treasury **364** James Camp/Dreamstime **366** BST2012/Fotolia **369** Shutterstock **371** Randy Miramontez/Alamy **372** Dirk Ercken/Shutterstock **374** Fotolia **375** Dreamstime **376** Evgeny Murtola/Shutterstock **379** Alamy **383** Shutterstock **389** Getty Images **391** iStockphoto **393** (top) iStockphoto, (bottom) Shutterstock **396** Tijana/Fotolia **399** (top) Jeff Greenberg/Alamy, (bottom) Central Intelligence Agency **401** Rade Kovac/Shutterstock **403** Fotolia **406** Getty Images **413** (left) Newscom, (right) Isaac Brekken/AP Images **414** (left) Alamy **415** Tatiana Popova/iStockphoto **417** Shutterstock **420** Roman Okopny/iStockphoto **421** iStockphoto **423** (left) Pearson Education, (right) Abimelec Olan/iStockphoto **424** iStockphoto **425** Marilyn Nieves/iStockphoto **426** (top) Bryan Myhr/iStockphoto, (bottom) Henrik Jonsson/iStockphoto **428** Isaac Brekken/AP Images **438** Dick De Veaux **445** James Laurie/Shutterstock **447** North Wind Picture Archives/Alamy **451** Pavel L/Shutterstock **455** ZUMA Press, Inc./Alamy **457** Keith Brofsky/Photodisc/Thinkstock **460** Blend Images/SuperStock **463** James Laurie/Shutterstock **473** Ken Usami/Photodisc/Getty Images **476** Fotolia **477** Associated Press **481** Rudall30/Fotolia **483** Erlend Kvalsvik/iStockphoto **485** Tetra Images/Getty Images **487** Ken Usami/Photodisc/Getty Images **493** Tischenko Irina/Shutterstock **498** (top) Lisa F. Young/Shutterstock, (bottom) Lisa F. Young/Dreamstime **499** Lisa F. Young/Dreamstime **501** Teamdaddy/Shutterstock **505** Martin Richardson/Dorling Kindersley, Ltd. **509** Tischenko Irina/Shutterstock **516** Karl Weatherly/Photographer's Choice/Getty Images **518** AlcelVision/Fotolia **521** Paul Velleman **522** Custom Medical Stock Photo/Newscom **523** A. Barrington Brown/Science Source **525** Federal Highway Administration **529** Archmen/Fotolia **534** Karl Weatherly/Photographer's Choice/Getty Images **541** Monkey Business Images/Shutterstock **542** Marka/Alamy **544** Christopher Futcher/iStockphoto **545** (middle) CurvaBezier/iStockphoto, (bottom) Stockbyte/Thinkstock **548** (top) Pavel Ignatov/Fotolia, (bottom) Don Tremain/Getty Images **553** Pearson Education, Inc. **554** Jason Lugo/iStockphoto **555** Monkey Business Images/Shutterstock **574** Vava Vladimir Jovanovic/Shutterstock **575** Robert Crow/Shutterstock **577** International Statistical Institute **579** John Angerson/Alamy **581** A. Barrington Brown/Science Source **583** Scott Griessel/Fotolia **587** WavebreakmediaMicro/Fotolia **595** Vava Vladimir Jovanovic/

Shutterstock **605** Corbis **607** Juanmonino/iStockphoto **610** Jason Stitt/Fotolia **613** Jenifoto1/ Shutterstock **615** Aldomurillo/iStockphoto **618** (all) Pearson Education, Inc. **619** Getty Images **623** Corbis **634** Yuri Kadobnov/Getty Images **635** David Bock **638** Vladislav Gajic/Shutterstock **642** Darren Baker/Shutterstock **646** Yuri Kadobnov/Getty Images **655** Andy Crawford/Dorling Kindersley **672** Nomad Soul/Shutterstock **673** LindaYolanda/iStockphoto **674** Al Behrman/AP Images **675** (left) Malerapaso/iStockphoto, (right) MorePixels/iStockphoto **676** Rob/Fotolia **683** (top) Getty Images, (bottom) Pearson Education, Inc. **686** Peggy Greb/United States Department of Agriculture (USDA) **688** Simon Shepheard/Imagestate Media Partners Limited/Impact Photos/Alamy **689** Pearson Education, Inc. **696** Nomad Soul/Shutterstock **706** IvicaDrusany/Shutterstock **710** Getty Images **712** Dean Drobot/Shutterstock **717** Karl W./ Fotolia **718** Pearson Education, Inc. **721** Pearson Education, Inc. **728** IvicaDrusany/Shutter-stock **747** David Bock **27-1** Olivier/Fotolia **27-20** Pearson Education, Inc. **27-22** Tomas Skopal/ Shutterstock **27-25** Olivier/Fotolia **28-1** Losevsky Pavel/Shutterstock **28-13** Amlani/Dreamstime **28-19** Losevsky Pavel/Shutterstock

Text

10 (quote) "Pablo Picasso: A Composite Interview," by William Fifield in *The Paris Review* 32, Summer-Fall. **15** (maps) Copyright © 2012 M.E.J. Newman **38** (Exercise 20) MPAA **77** (Exercise 28) Data from http://www.pgatour.com/r/stats/info/xm.html?101 **115** (quote) Box, George E.P.; Norman R. Draper (1987). *Empirical Model-Building and Response Surfaces*, p. 424. **149** (Question 10) Data from http://ansonalex.com/infographics/facebook-user-statistics-2012-info-graphic/ **163** (quote) "New York State's Bridge Program in Brief," New York State Department of Transportation. **224** (Exercise 8) Data from http://www.cdc.gov/nchs **261-2** (Exercise 32) Data from "Trends in Twin Birth Outcomes and Prenatal Care Utilization in the United States, 1981-1997." **265** (Question 9) Chart by Douglas VanDerwerken from source data courtesy of TeamRankings.com **313** (Just Checking extract) Medicare & Medicaid Services **335** (Exercise 36) "The GOP," *The Weekly Standard* **475** (quote) John W. Tukey, *Annals of Mathematical Statistics* 33(1), 1962, p. 13. **504** (quote) From Marcello Truzzi, "On the Extraordinary: An Attempt at Clarification," The Zetetic Scholar, vol.1, no. 1, 1978, p. 11. **600** (Exercise 24) Data from http://www.transtats.bts.gov/HomeDrillChart.asp **631** (Exercises 39 and 40) NCES online **654** Exercise 33) College Board Online **703** (Exercise 36) Gallup World; (Exercise 37) Maryland State Lottery **731**(Exercise 12) United Nations Office on Drugs and Crime (UNDOC) **736** (Exercise 38) GISS Surface Temperature Analysis **746** (Exercise 12) Data from "Trends in Twin Birth Outcomes and Prenatal Care Utilization in the United States, 1981-1997."

Appendix E: Index

Appendix F: **Tables**

Row					TABLE OF RANDOM DIGITS					
1	96299	07196	98642	20639	23185	56282	69929	14125	38872	94168
2	71622	35940	81807	59225	18192	08710	80777	84395	69563	86280
3	03272	41230	81739	74797	70406	18564	69273	72532	78340	36699
4	46376	58596	14365	63685	56555	42974	72944	96463	63533	24152
5	47352	42853	42903	97504	56655	70355	88606	61406	38757	70657
6	20064	04266	74017	79319	70170	96572	08523	56025	89077	57678
7	73184	95907	05179	51002	83374	52297	07769	99792	78365	93487
8	72753	36216	07230	35793	71907	65571	66784	25548	91861	15725
9	03939	30763	06138	80062	02537	23561	93136	61260	77935	93159
10	75998	37203	07959	38264	78120	77525	86481	54986	33042	70648
11	94435	97441	90998	25104	49761	14967	70724	67030	53887	81293
12	04362	40989	69167	38894	00172	02999	97377	33305	60782	29810
13	89059	43528	10547	40115	82234	86902	04121	83889	76208	31076
14	87736	04666	75145	49175	76754	07884	92564	80793	22573	67902
15	76488	88899	15860	07370	13431	84041	69202	18912	83173	11983
16	36460	53772	66634	25045	79007	78518	73580	14191	50353	32064
17	13205	69237	21820	20952	16635	58867	97650	82983	64865	93298
18	51242	12215	90739	36812	00436	31609	80333	96606	30430	31803
19	67819	00354	91439	91073	49258	15992	41277	75111	67496	68430
20	09875	08990	27656	15871	23637	00952	97818	64234	50199	05715
21	18192	95308	72975	01191	29958	09275	89141	19558	50524	32041
22	02763	33701	66188	50226	35813	72951	11638	01876	93664	37001
23	13349	46328	01856	29935	80563	03742	49470	67749	08578	21956
24	69238	92878	80067	80807	45096	22936	64325	19265	37755	69794
25	92207	63527	59398	29818	24789	94309	88380	57000	50171	17891
26	66679	99100	37072	30593	29665	84286	44458	60180	81451	58273
27	31087	42430	60322	34765	15757	53300	97392	98035	05228	68970
28	84432	04916	52949	78533	31666	62350	20584	56367	19701	60584
29	72042	12287	21081	48426	44321	58765	41760	43304	13399	02043
30	94534	73559	82135	70260	87936	85162	11937	18263	54138	69564
31	63971	97198	40974	45301	60177	35604	21580	68107	25184	42810
32	11227	58474	17272	37619	69517	62964	67962	34510	12607	52255
33	28541	02029	08068	96656	17795	21484	57722	76511	27849	61738
34	11282	43632	49531	78981	81980	08530	08629	32279	29478	50228
35	42907	15137	21918	13248	39129	49559	94540	24070	88151	36782
36	47119	76651	21732	32364	58545	50277	57558	30390	18771	72703
37	11232	99884	05087	76839	65142	19994	91397	29350	83852	04905
38	64725	06719	86262	53356	57999	50193	79936	97230	52073	94467
39	77007	26962	55466	12521	48125	12280	54985	26239	76044	54398
40	18375	19310	59796	89832	59417	18553	17238	05474	33259	50595

Table Z

Areas under the
standard normal curve

	Second decimal place in z										
0.09	0.08	0.07	0.06	0.05	0.04	0.03	0.02	0.01	0.00	z
0.0001	0.0001	0.0001	0.0001	0.0001	0.0001	0.0001	0.0001	0.0001	0.0001	−3.8
0.0001	0.0001	0.0001	0.0001	0.0001	0.0001	0.0001	0.0001	0.0001	0.0001	−3.7
0.0001	0.0001	0.0001	0.0001	0.0001	0.0001	0.0001	0.0001	0.0002	0.0002	−3.6
0.0002	0.0002	0.0002	0.0002	0.0002	0.0002	0.0002	0.0002	0.0002	0.0002	−3.5
0.0002	0.0003	0.0003	0.0003	0.0003	0.0003	0.0003	0.0003	0.0003	0.0003	−3.4
0.0003	0.0004	0.0004	0.0004	0.0004	0.0004	0.0004	0.0005	0.0005	0.0005	−3.3
0.0005	0.0005	0.0005	0.0006	0.0006	0.0006	0.0006	0.0006	0.0007	0.0007	−3.2
0.0007	0.0007	0.0008	0.0008	0.0008	0.0008	0.0009	0.0009	0.0009	0.0010	−3.1
0.0010	0.0010	0.0011	0.0011	0.0011	0.0012	0.0012	0.0013	0.0013	0.0013	−3.0
0.0014	0.0014	0.0015	0.0015	0.0016	0.0016	0.0017	0.0018	0.0018	0.0019	−2.9
0.0019	0.0020	0.0021	0.0021	0.0022	0.0023	0.0023	0.0024	0.0025	0.0026	−2.8
0.0026	0.0027	0.0028	0.0029	0.0030	0.0031	0.0032	0.0033	0.0034	0.0035	−2.7
0.0036	0.0037	0.0038	0.0039	0.0040	0.0041	0.0043	0.0044	0.0045	0.0047	−2.6
0.0048	0.0049	0.0051	0.0052	0.0054	0.0055	0.0057	0.0059	0.0060	0.0062	−2.5
0.0064	0.0066	0.0068	0.0069	0.0071	0.0073	0.0075	0.0078	0.0080	0.0082	−2.4
0.0084	0.0087	0.0089	0.0091	0.0094	0.0096	0.0099	0.0102	0.0104	0.0107	−2.3
0.0110	0.0113	0.0116	0.0119	0.0122	0.0125	0.0129	0.0132	0.0136	0.0139	−2.2
0.0143	0.0146	0.0150	0.0154	0.0158	0.0162	0.0166	0.0170	0.0174	0.0179	−2.1
0.0183	0.0188	0.0192	0.0197	0.0202	0.0207	0.0212	0.0217	0.0222	0.0228	−2.0
0.0233	0.0239	0.0244	0.0250	0.0256	0.0262	0.0268	0.0274	0.0281	0.0287	−1.9
0.0294	0.0301	0.0307	0.0314	0.0322	0.0329	0.0336	0.0344	0.0351	0.0359	−1.8
0.0367	0.0375	0.0384	0.0392	0.0401	0.0409	0.0418	0.0427	0.0436	0.0446	−1.7
0.0455	0.0465	0.0475	0.0485	0.0495	0.0505	0.0516	0.0526	0.0537	0.0548	−1.6
0.0559	0.0571	0.0582	0.0594	0.0606	0.0618	0.0630	0.0643	0.0655	0.0668	−1.5
0.0681	0.0694	0.0708	0.0721	0.0735	0.0749	0.0764	0.0778	0.0793	0.0808	−1.4
0.0823	0.0838	0.0853	0.0869	0.0885	0.0901	0.0918	0.0934	0.0951	0.0968	−1.3
0.0985	0.1003	0.1020	0.1038	0.1056	0.1075	0.1093	0.1112	0.1131	0.1151	−1.2
0.1170	0.1190	0.1210	0.1230	0.1251	0.1271	0.1292	0.1314	0.1335	0.1357	−1.1
0.1379	0.1401	0.1423	0.1446	0.1469	0.1492	0.1515	0.1539	0.1562	0.1587	−1.0
0.1611	0.1635	0.1660	0.1685	0.1711	0.1736	0.1762	0.1788	0.1814	0.1841	−0.9
0.1867	0.1894	0.1922	0.1949	0.1977	0.2005	0.2033	0.2061	0.2090	0.2119	−0.8
0.2148	0.2177	0.2206	0.2236	0.2266	0.2296	0.2327	0.2358	0.2389	0.2420	−0.7
0.2451	0.2483	0.2514	0.2546	0.2578	0.2611	0.2643	0.2676	0.2709	0.2743	−0.6
0.2776	0.2810	0.2843	0.2877	0.2912	0.2946	0.2981	0.3015	0.3050	0.3085	−0.5
0.3121	0.3156	0.3192	0.3228	0.3264	0.3300	0.3336	0.3372	0.3409	0.3446	−0.4
0.3483	0.3520	0.3557	0.3594	0.3632	0.3669	0.3707	0.3745	0.3783	0.3821	−0.3
0.3859	0.3897	0.3936	0.3974	0.4013	0.4052	0.4090	0.4129	0.4168	0.4207	−0.2
0.4247	0.4286	0.4325	0.4364	0.4404	0.4443	0.4483	0.4522	0.4562	0.4602	−0.1
0.4641	0.4681	0.4721	0.4761	0.4801	0.4840	0.4880	0.4920	0.4960	0.5000	−0.0

For $z \leq -3.90$, the areas are 0.0000 to four decimal places.

Table Z (cont.)

Areas under the standard normal curve

z	0.00	0.01	0.02	0.03	0.04	0.05	0.06	0.07	0.08	0.09
				Second decimal place in z						
0.0	0.5000	0.5040	0.5080	0.5120	0.5160	0.5199	0.5239	0.5279	0.5319	0.5359
0.1	0.5398	0.5438	0.5478	0.5517	0.5557	0.5596	0.5636	0.5675	0.5714	0.5753
0.2	0.5793	0.5832	0.5871	0.5910	0.5948	0.5987	0.6026	0.6064	0.6103	0.6141
0.3	0.6179	0.6217	0.6255	0.6293	0.6331	0.6368	0.6406	0.6443	0.6480	0.6517
0.4	0.6554	0.6591	0.6628	0.6664	0.6700	0.6736	0.6772	0.6808	0.6844	0.6879
0.5	0.6915	0.6950	0.6985	0.7019	0.7054	0.7088	0.7123	0.7157	0.7190	0.7224
0.6	0.7257	0.7291	0.7324	0.7357	0.7389	0.7422	0.7454	0.7486	0.7517	0.7549
0.7	0.7580	0.7611	0.7642	0.7673	0.7704	0.7734	0.7764	0.7794	0.7823	0.7852
0.8	0.7881	0.7910	0.7939	0.7967	0.7995	0.8023	0.8051	0.8078	0.8106	0.8133
0.9	0.8159	0.8186	0.8212	0.8238	0.8264	0.8289	0.8315	0.8340	0.8365	0.8389
1.0	0.8413	0.8438	0.8461	0.8485	0.8508	0.8531	0.8554	0.8577	0.8599	0.8621
1.1	0.8643	0.8665	0.8686	0.8708	0.8729	0.8749	0.8770	0.8790	0.8810	0.8830
1.2	0.8849	0.8869	0.8888	0.8907	0.8925	0.8944	0.8962	0.8980	0.8997	0.9015
1.3	0.9032	0.9049	0.9066	0.9082	0.9099	0.9115	0.9131	0.9147	0.9162	0.9177
1.4	0.9192	0.9207	0.9222	0.9236	0.9251	0.9265	0.9279	0.9292	0.9306	0.9319
1.5	0.9332	0.9345	0.9357	0.9370	0.9382	0.9394	0.9406	0.9418	0.9429	0.9441
1.6	0.9452	0.9463	0.9474	0.9484	0.9495	0.9505	0.9515	0.9525	0.9535	0.9545
1.7	0.9554	0.9564	0.9573	0.9582	0.9591	0.9599	0.9608	0.9616	0.9625	0.9633
1.8	0.9641	0.9649	0.9656	0.9664	0.9671	0.9678	0.9686	0.9693	0.9699	0.9706
1.9	0.9713	0.9719	0.9726	0.9732	0.9738	0.9744	0.9750	0.9756	0.9761	0.9767
2.0	0.9772	0.9778	0.9783	0.9788	0.9793	0.9798	0.9803	0.9808	0.9812	0.9817
2.1	0.9821	0.9826	0.9830	0.9834	0.9838	0.9842	0.9846	0.9850	0.9854	0.9857
2.2	0.9861	0.9864	0.9868	0.9871	0.9875	0.9878	0.9881	0.9884	0.9887	0.9890
2.3	0.9893	0.9896	0.9898	0.9901	0.9904	0.9906	0.9909	0.9911	0.9913	0.9916
2.4	0.9918	0.9920	0.9922	0.9925	0.9927	0.9929	0.9931	0.9932	0.9934	0.9936
2.5	0.9938	0.9940	0.9941	0.9943	0.9945	0.9946	0.9948	0.9949	0.9951	0.9952
2.6	0.9953	0.9955	0.9956	0.9957	0.9959	0.9960	0.9961	0.9962	0.9963	0.9964
2.7	0.9965	0.9966	0.9967	0.9968	0.9969	0.9970	0.9971	0.9972	0.9973	0.9974
2.8	0.9974	0.9975	0.9976	0.9977	0.9977	0.9978	0.9979	0.9979	0.9980	0.9981
2.9	0.9981	0.9982	0.9982	0.9983	0.9984	0.9984	0.9985	0.9985	0.9986	0.9986
3.0	0.9987	0.9987	0.9987	0.9988	0.9988	0.9989	0.9989	0.9989	0.9990	0.9990
3.1	0.9990	0.9991	0.9991	0.9991	0.9992	0.9992	0.9992	0.9992	0.9993	0.9993
3.2	0.9993	0.9993	0.9994	0.9994	0.9994	0.9994	0.9994	0.9995	0.9995	0.9995
3.3	0.9995	0.9995	0.9995	0.9996	0.9996	0.9996	0.9996	0.9996	0.9996	0.9997
3.4	0.9997	0.9997	0.9997	0.9997	0.9997	0.9997	0.9997	0.9997	0.9997	0.9998
3.5	0.9998	0.9998	0.9998	0.9998	0.9998	0.9998	0.9998	0.9998	0.9998	0.9998
3.6	0.9998	0.9998	0.9999	0.9999	0.9999	0.9999	0.9999	0.9999	0.9999	0.9999
3.7	0.9999	0.9999	0.9999	0.9999	0.9999	0.9999	0.9999	0.9999	0.9999	0.9999
3.8	0.9999	0.9999	0.9999	0.9999	0.9999	0.9999	0.9999	0.9999	0.9999	0.9999

For $z \geq 3.90$, the areas are 1.0000 to four decimal places.

Two tail probability		0.20	0.10	0.05	0.02	0.01	
One tail probability		0.10	0.05	0.025	0.01	0.005	
Table T	df						df
Values of t_α	1	3.078	6.314	12.706	31.821	63.657	1
	2	1.886	2.920	4.303	6.965	9.925	2
	3	1.638	2.353	3.182	4.541	5.841	3
	4	1.533	2.132	2.776	3.747	4.604	4
	5	1.476	2.015	2.571	3.365	4.032	5
	6	1.440	1.943	2.447	3.143	3.707	6
	7	1.415	1.895	2.365	2.998	3.499	7
	8	1.397	1.860	2.306	2.896	3.355	8
	9	1.383	1.833	2.262	2.821	3.250	9
	10	1.372	1.812	2.228	2.764	3.169	10
	11	1.363	1.796	2.201	2.718	3.106	11
	12	1.356	1.782	2.179	2.681	3.055	12
	13	1.350	1.771	2.160	2.650	3.012	13
	14	1.345	1.761	2.145	2.624	2.977	14
	15	1.341	1.753	2.131	2.602	2.947	15
	16	1.337	1.746	2.120	2.583	2.921	16
	17	1.333	1.740	2.110	2.567	2.898	17
	18	1.330	1.734	2.101	2.552	2.878	18
	19	1.328	1.729	2.093	2.539	2.861	19
	20	1.325	1.725	2.086	2.528	2.845	20
	21	1.323	1.721	2.080	2.518	2.831	21
	22	1.321	1.717	2.074	2.508	2.819	22
	23	1.319	1.714	2.069	2.500	2.807	23
	24	1.318	1.711	2.064	2.492	2.797	24
	25	1.316	1.708	2.060	2.485	2.787	25
	26	1.315	1.706	2.056	2.479	2.779	26
	27	1.314	1.703	2.052	2.473	2.771	27
	28	1.313	1.701	2.048	2.467	2.763	28
	29	1.311	1.699	2.045	2.462	2.756	29
	30	1.310	1.697	2.042	2.457	2.750	30
	32	1.309	1.694	2.037	2.449	2.738	32
	35	1.306	1.690	2.030	2.438	2.725	35
	40	1.303	1.684	2.021	2.423	2.704	40
	45	1.301	1.679	2.014	2.412	2.690	45
	50	1.299	1.676	2.009	2.403	2.678	50
	60	1.296	1.671	2.000	2.390	2.660	60
	75	1.293	1.665	1.992	2.377	2.643	75
	100	1.290	1.660	1.984	2.364	2.626	100
	120	1.289	1.658	1.980	2.358	2.617	120
	140	1.288	1.656	1.977	2.353	2.611	140
	180	1.286	1.653	1.973	2.347	2.603	180
	250	1.285	1.651	1.969	2.341	2.596	250
	400	1.284	1.649	1.966	2.336	2.588	400
	1000	1.282	1.646	1.962	2.330	2.581	1000
	∞	1.282	1.645	1.960	2.326	2.576	∞
Confidence levels		80%	90%	95%	98%	99%	

Two tails diagram: $-t_{\alpha/2}$, 0, $t_{\alpha/2}$, with $\frac{\alpha}{2}$ in each tail.

One tail diagram: 0, t_α, with α in the tail.

Right tail probability		0.10	0.05	0.025	0.01	0.005
Table x	df					
Values of χ^2_α	1	2.706	3.841	5.024	6.635	7.879
	2	4.605	5.991	7.378	9.210	10.597
	3	6.251	7.815	9.348	11.345	12.838
	4	7.779	9.488	11.143	13.277	14.860
	5	9.236	11.070	12.833	15.086	16.750
	6	10.645	12.592	14.449	16.812	18.548
	7	12.017	14.067	16.013	18.475	20.278
	8	13.362	15.507	17.535	20.090	21.955
	9	14.684	16.919	19.023	21.666	23.589
	10	15.987	18.307	20.483	23.209	25.188
	11	17.275	19.675	21.920	24.725	26.757
	12	18.549	21.026	23.337	26.217	28.300
	13	19.812	22.362	24.736	27.688	29.819
	14	21.064	23.685	26.119	29.141	31.319
	15	22.307	24.996	27.488	30.578	32.801
	16	23.542	26.296	28.845	32.000	34.267
	17	24.769	27.587	30.191	33.409	35.718
	18	25.989	28.869	31.526	34.805	37.156
	19	27.204	30.143	32.852	36.191	38.582
	20	28.412	31.410	34.170	37.566	39.997
	21	29.615	32.671	35.479	38.932	41.401
	22	30.813	33.924	36.781	40.290	42.796
	23	32.007	35.172	38.076	41.638	44.181
	24	33.196	36.415	39.364	42.980	45.559
	25	34.382	37.653	40.647	44.314	46.928
	26	35.563	38.885	41.923	45.642	48.290
	27	36.741	40.113	43.195	46.963	49.645
	28	37.916	41.337	44.461	48.278	50.994
	29	39.087	42.557	45.722	59.588	52.336
	30	40.256	43.773	46.979	50.892	53.672
	40	51.805	55.759	59.342	63.691	66.767
	50	63.167	67.505	71.420	76.154	79.490
	60	74.397	79.082	83.298	88.381	91.955
	70	85.527	90.531	95.023	100.424	104.213
	80	96.578	101.879	106.628	112.328	116.320
	90	107.565	113.145	118.135	124.115	128.296
	100	118.499	124.343	129.563	135.811	140.177